Advanced Science

Chemistry

Chris Conoley & Phil Hills

The complete guide to Chemistry

William Collins's dream of knowledge for all began with the publication of his first book in 1819. A self-educated mill worker, he not only enriched millions of lives, but also founded a flourishing publishing house. Today, staying true to this spirit, Collins books are packed with inspiration, innovation and practical expertise. They place you at the centre of a world of possibility and give you exactly what you need to explore it.

Collins. Freedom to teach.

Published by Collins
An imprint of HarperCollinsPublishers
77-85 Fulham Palace Road
Hammersmith
London
W6 8JB

Browse the complete Collins catalogue at
www.collinseducation.com

10 9 8 7 6 5

ISBN 978-0-00-726747-7

International edition:

ISBN 978-0-00-726748-4

British Library Cataloguing in Publication Data. A Catalogue record for this publication is available from the British Library.

Commissioned by Penny Fowler
Project management by Laura Deacon
Edited by Patrick Roberts and Susan Watt
Proof read by Camilla Behrens and Rosie Parrish
Indexing by Jane Henley
Photo research by Alex Riley
Design by Newgen Imaging
Cover design by Angela English
Production by Arjen Jansen
Printed and bound in Hong Kong by Printing Express

Contents

1

Reactions, equations, energy and equilibria

1 REACTIONS, EQUATIONS, ENERGY AND EQUILIBRIA

Steel manufacture and the Humber Bridge

The Humber Bridge – one of the largest of its kind in the world

At its opening in 1981 the Humber Bridge was the largest single span suspension bridge in the world. It is now the fourth largest. It contains 27 500 tonnes of steel, all supplied by Corus (previously British Steel): 16 500 tonnes make up the deck section and 11 000 tonnes are in the cables that support it.

Each part of such a large construction needs a different type of steel with different properties to meet particular needs, which may include rigidity, flexibility, strength in tension and ease of being welded. The properties of steel depend on its composition. Steels contain a little carbon in the iron, and can include very small amounts of manganese, niobium, vanadium and titanium. Phosphorus and sulfur are impurities from iron ore that can make steel brittle. The number of times the steel sheet is rolled will also affect its final properties. It is the job of the metallurgist to obtain the balance of composition and treatment of the steel that suits a specific need.

For a gas or oil pipeline, a low carbon content and some niobium and vanadium in the steel make a clean, strong weld that will not crack or leak, while for reinforcing structures in buildings, more carbon gives steel greater rigidity and strength.

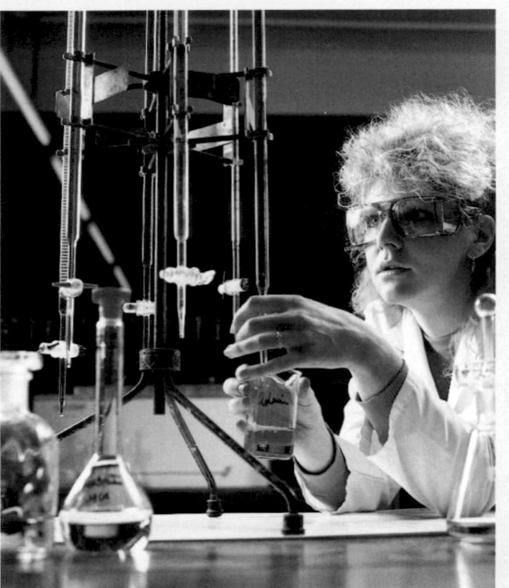

In the manufacture of steel, enormous amounts of water are used to cool the rolling mills. The water is recycled many times and can pick up contaminants, including cyanide, nitrite and chloride, from the steel. So chemists regularly check that the water eventually discharged meets environmental requirements

Laser welding is a recent development in working with sheet steel. In conventional 'oxyacetylene' welding, the parts to be joined are heated extensively to melt and join the surfaces. Then, when the steel cools and shrinks, there is inevitably distortion at the join where the steel does not cool evenly. With laser welding, the beam is focused to a small spot that moves along the joint. Only the surfaces of the sheets are heated, very intensely and for a very short time, so the process avoids distortion and saves a great deal of money.

INTRODUCTION

Nearly all the materials you touch, the polymers and dyes in clothes, the food on your plate, or the Paracetamol in the bathroom cabinet, have at some stage involved industrial chemists. This chapter will give you an insight into the world of the chemical industry. You will discover that organising the economics of a large manufacturing plant depends on understanding the chemistry of industrial processes.

Also, throughout this chapter you will find out more about chemical equations, the chemical reactions that they represent and how energy is transferred in reactions. You will also meet reversible reactions that can form dynamic equilibria.

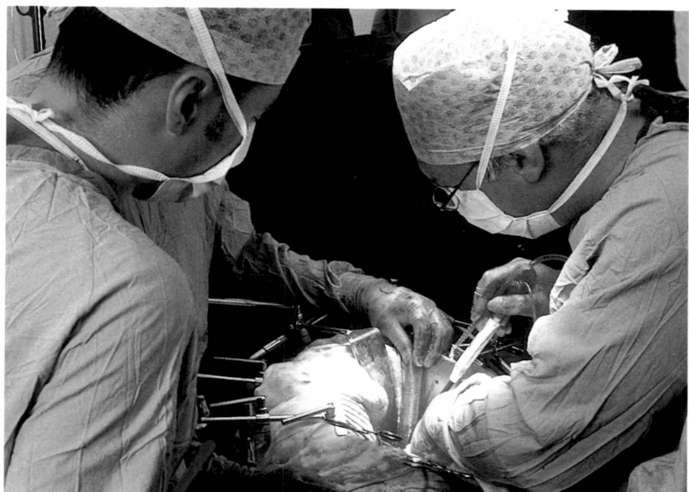

Fig 1.1 The artificial aorta that is fitted to a person's heart is made of a polyester fibre called Dacron

Fig 1.2 Oral rehydration powders contain sodium chloride, glucose, trisodium citrate and potassium chloride. As the treatment for victims of dehydration, this mixture dissolved in water has saved the lives of millions of people, especially children

1 CHEMICAL FORMULAE AND CHEMICAL EQUATIONS

Iron and steel have transformed the world around us in a way no other chemicals have. Think of their use in various means of travel – bridges, railways, cars, the Channel Tunnel and ships – in buildings such as skyscrapers, factories and warehouses, as well as in leisure products such as golf clubs and garden furniture.

Anyone fortunate enough to tour an iron- and steel-making plant comes away impressed by the large scale of it all. Yet the process is fairly simple and uses only four major raw materials: iron ore, coal (made into coke), limestone and air. The principles of iron making have not changed much since Roman times, so where does the chemist fit in today? Chemists have the best understanding of the manufacturing process, so they can, for example, calculate the exact amounts of raw materials required and advise on the best reaction conditions. Later in this chapter, you can read about the role of chemists in optimising the efficiency of processes, controlling the properties of steels, understanding reactions in equilibrium and avoiding industrial pollution.

In the blast furnace, coke, which is carbon, reacts with air to produce carbon monoxide. This can be represented by a word equation:

$$\text{carbon} + \text{oxygen} \rightarrow \text{carbon monoxide}$$

However, it is more convenient and much more informative to use formulae in an equation. A **molecular formula** tells you how many atoms of each element there are in a molecule of the compound. Oxygen has two atoms per molecule: O_2. But not all substances are made up of molecules. For example, sodium chloride is a giant lattice made up of sodium and chloride ions. For these it is more correct just to use the term **formula** or **formula unit**. So the equation could be written:

$$C + O_2 \rightarrow CO$$

Fig 1.3 The Queen Victoria blast furnace in Corus's Scunthorpe iron-making works

Fig 1.4 The base of the blast furnace: hot air at about 750 °C is blown into it

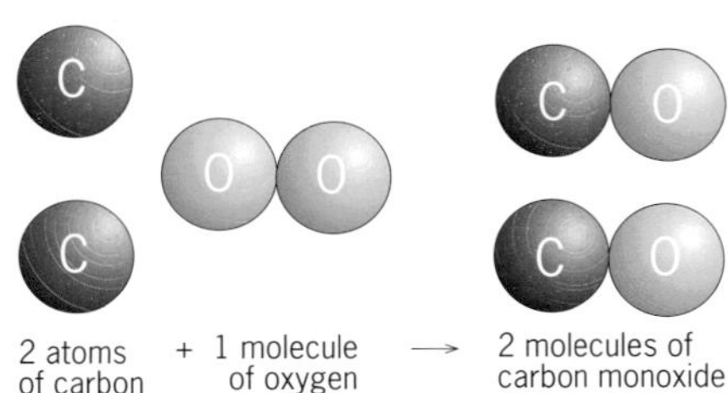

Fig 1.5 Combining 2 carbon atoms with 1 oxygen molecule

But this equation is unbalanced. There are more oxygen atoms on the left side of the equation than there are on the right, where one atom of oxygen has 'disappeared'. Atoms are not made or destroyed in chemical reactions, and so in any chemical equation the number of atoms of each element must be the same on both sides. This makes for a *balanced* equation.

It would be simple to balance the equation above by making CO into CO_2. But this changes the chemical carbon monoxide into carbon dioxide, a different gas with completely different properties. Instead, as shown in Fig 1.5 the balanced equation for the reaction is:

$$2C + O_2 \rightarrow 2CO$$

Chemical equations often include **state** symbols. These show the physical state of the reactants, usually at room temperature. So this equation may be written:

$$2C(s) + O_2(g) \rightarrow 2CO(g)$$

where (s) means the solid state and (g) means the gaseous state.

REMEMBER THIS

Two other state symbols are:
(l) which means the liquid state; and
(aq) which means the substance is dissolved in water to form an aqueous solution.

QUESTION 1

1 Write balanced equations for the following reactions that are all associated with iron and steel making:

a) Calcium carbonate ($CaCO_3$) decomposing to calcium oxide (CaO) and carbon dioxide (CO_2);

b) Manganese (Mn) reacting with oxygen (O_2) to form manganese(II) oxide (MnO);

c) Phosphorus (P) reacting with oxygen to form phosphorus(V) oxide (P_4O_{10}).

EXAMPLE

Q One type of iron ore is haematite. The chemical name for haematite is iron(III) oxide and its formula is Fe_2O_3. When haematite reacts with carbon monoxide in the blast furnace, carbon dioxide and iron are produced.

Work out the balanced equation for this reaction.

A Step 1. Write the word equation:

iron(III) oxide + carbon monoxide → iron + carbon dioxide

Step 2. Change the words into formulae:

$$Fe_2O_3 + CO \rightarrow Fe + CO_2$$

Step 3. Balance the equation by making the number of atoms of each element the same on both sides and add state symbols. (Remember that the formulae are fixed.)

$$Fe_2O_3(s) + 3CO(g) \rightarrow 2Fe(s) + 3CO_2(g)$$

As you get used to balancing equations, you will probably do these steps in your head.

2 RELATIVE ATOMIC MASS

The masses of different atoms can be compared using the **relative atomic mass scale**, also called the **A_r scale** (r stands for 'relative'). On this scale, the isotope carbon-12 is given an A_r of exactly 12. The isotope carbon-12 is the standard and the A_r values of all other atoms are measured relative to this standard. Most elements exist as two or more isotopes and the A_r is an average of the masses of all these different isotopes taking into account the percentages in which they occur naturally. This leads to the definition:

Relative atomic mass of an element is the average mass of the atom (taking into account all of its isotopes and their abundance) compared to 1/12 the mass of one atom of carbon-12.

Historically, the standard for the A_r scale was hydrogen. As the lightest element, it was given an A_r of 1. Carbon-12 is now used because it is easier to handle than gaseous hydrogen and is an abundant isotope. Table 1.1 shows some approximate relative atomic masses.

Notice in Table 1.1 that the relative atomic masses have no units. The relative atomic mass just shows *how many times heavier* one atom is compared to another. So one atom of silicon is 28 times heavier than one atom of hydrogen, and calcium atoms are twice as heavy as neon atoms.

Table 1.1 The approximate relative atomic masses of some elements

Element	Symbol	Relative atomic mass, A_r
hydrogen	H	1
carbon	C	12
oxygen	O	16
neon	Ne	20
silicon	Si	28
sulfur	S	32
calcium	Ca	40
iron	Fe	56
copper	Cu	64

? QUESTION 2

2 Work out how many times heavier the following are:
a) Fe atoms than H atoms;
b) S atoms than O atoms;
c) Fe atoms than O atoms.

You can find out more about isotopes in Chapter 2.

A full list of relative atomic masses is given in Appendix 2.

3 AMOUNTS AND THE MOLE

When chemists use the word *amount*, they are talking about the number of particles in a substance. The particles can be atoms, molecules, ions, electrons, etc. Amount may also mean other things, for instance the mass of a substance, which has units of grams or kilograms. To describe the amount of something such as eggs, you would probably use the unit dozens. When chemists talk about the amount of substance they use the unit **moles**.

The A_r of copper is 64, and so each copper atom is twice as heavy as a sulfur atom. Imagine that 64 grams of copper and 64 grams of sulfur (A_r of 32) are weighed out. The mass of each substance is the same, but to a chemist the amount of each substance is very different: 64 grams contains twice as many sulfur *atoms* as copper atoms: there is double the *amount* of sulfur atoms.

Using relative atomic masses, you can work out amounts of atoms or molecules in different masses of substances. Weigh out 32 grams of sulfur, and you have the same number of atoms as there are in 64 grams of copper. Also, weigh out the relative atomic mass in grams of different elements, and you are weighing the same *amount* of atoms each time. This amount is called the mole:

The relative atomic mass in grams of any element contains one mole of atoms.

The mole is a unit in the same way that the gram is a unit. The shortened form of the mole unit is **mol**. (Care: this is not short for molecule!) This is its definition:

One mole is the amount of substance that contains as many particles as there are atoms in exactly 12 g of carbon-12.

These are some examples:
The amount of atoms in 32 g sulfur is 1 mol.
In 32 g of copper there is 0.5 mol of copper atoms.
32 g of oxygen atoms is 2 mol of oxygen atoms.

To work out how many moles of atoms are in a particular mass of substance, use this equation:

$$\text{amount in moles} = \frac{\text{mass in grams}}{\text{mass of one mole (in grams)}}$$

REMEMBER THIS

The mass of one mole of a substance is the **molar mass**. Its unit is g mol^{-1}.

EXAMPLE

Q How many moles are there in 56 g silicon?

A Amount in moles = $\frac{\text{mass in grams}}{\text{mass of one mole (in grams)}}$

moles of silicon = $\frac{56\text{ g}}{28\text{ g}}$
= 2 moles

Q What is the mass of 0.25 mol of iron?

A Mass in grams = amount in moles × mass of 1 mole
mass of iron = 0.25 mol × 56 g
= 14 g

? QUESTION 3

3 **a)** Work out how many moles of atoms there are in:
56 g of iron; 20 g of calcium;
128 g of copper; 4 g of sulfur.
b) Give the mass of:
3.0 mol of calcium atoms;
0.3 mol of neon atoms.

One mole of substance is 6.02×10^{23} particles. This is a very large amount of atoms or molecules to try and imagine. It is so large that even all the sand grains around the whole coastline of the British Isles do not make a mole. The number of atoms or molecules in one mole is called the **Avogadro constant, L,** sometimes also given the symbol N_A.

4 RELATIVE MOLECULAR MASS

We saw on page 4 that carbon dioxide is made up of molecules each with the molecular formula CO_2. Fe_2O_3 is the formula unit of iron(III) oxide. It is not a molecular formula because this compound is not made up of separate molecules. Most equations that chemists deal with involve either molecules or formula units.

The **relative molecular mass, M_r,** is calculated using the relative atomic mass scale. Again, we use the standard carbon-12 as a comparison for the masses of molecules or formula units. Suppose we want to find the M_r of carbon dioxide (see Fig 1.6).

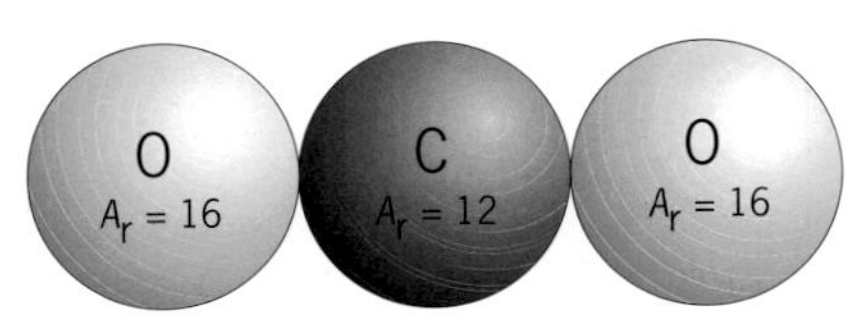

Fig 1.6 Working out the relative molecular mass of carbon dioxide

The M_r of CO_2 = A_r of C + ((A_r of O) × 2)
= 12 + (16 × 2) = 44

The M_r of Fe_2O_3 = ((A_r of Fe) × 2) + ((A_r of O) × 3)
= (56 × 2) + (16 × 3) = 160

In this case, the M_r refers to the formula Fe_2O_3 and is called the **relative formula mass**.

? QUESTION 4

4 Give the relative formula mass (M_r) of:
a) P_4O_{10} **b)** O_2 **c)** $CaSiO_3$

During your course, you will come across many formulae with brackets, for example, $Ca(OH)_2$. Fig 1.7 shows what this means.

Notice that the particles in $Ca(OH)_2$ are ions and have positive and negative charges. The charges on ions do not affect their A_r values.

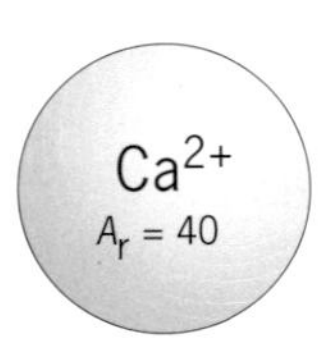

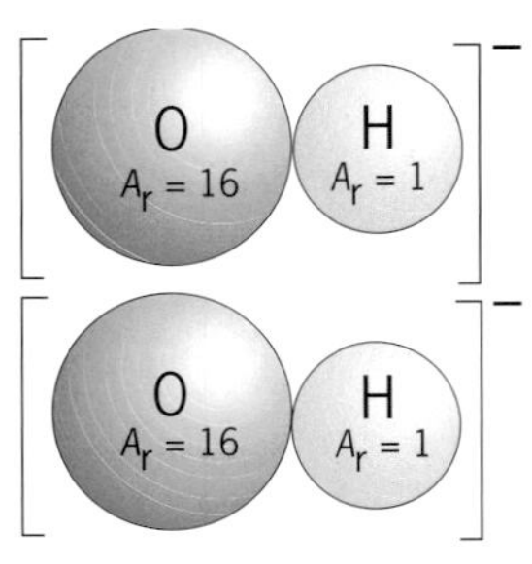

Fig 1.7 Diagram to explain the parts of the formula $Ca(OH)_2$

So:

M_r of $Ca(OH)_2$ = A_r of Ca + 2 × (A_r of O + A_r of H)
= 40 + 2 × (16 + 1)
= 40 + 2 × 17 = 74

? QUESTION 5

5 Give the M_r of: **a)** $Fe(OH)_2$
b) $Al_2(SO_4)_3$ **c)** $(CH_3CO)_2O$

5 USING EQUATIONS IN INDUSTRY

We have seen that the chemical formula CO means that 1 molecule contains 1 atom of carbon combined with 1 atom of oxygen. It also means that 1 mole of carbon atoms is combined with 1 mole of oxygen atoms. This is sometimes called the **stoichiometric ratio**. In the equation for the reaction of coke in the blast furnace:

$$2C \quad + \quad O_2 \quad \rightarrow \quad 2CO$$

This means:

2 atoms C + 1 molecule O_2 → 2 molecules CO

If instead of 2 atoms of carbon we had 2 million atoms, then:

2 million C atoms + 1 million O_2 molecules → 2 million CO molecules

We now use moles:

2 mol carbon atoms + 1 mol oxygen molecules → 2 mol carbon monoxide molecules

The carbon monoxide formed in the blast furnace **reduces** (removes oxygen from) the iron(III) oxide to iron:

$$Fe_2O_3(s) \quad + \quad 3CO(g) \quad \rightarrow \quad 2Fe(s) \quad + \quad 3CO_2(g)$$

So: 1 mol Fe_2O_3 + 3 mol CO → 2 mol Fe + 3 mol CO_2

REMEMBER THIS

Extraction of a metal from its ore always includes a reduction reaction. Many metal ores contain oxygen, but for those that don't, a suitable definition of reduction is to be found in Chapter 5.

Now that we know the amounts of the chemical substances in this equation, we can work out the masses of the substances involved using the A_r scale.

The mass of 1 mol of Fe_2O_3 is (56 × 2) g + (16 × 3) g = 160 g
The mass of 1 mol of CO is 12 g + 16 g = 28 g
The mass of 1 mol of Fe is 56 g = 56 g
The mass of 1 mol of CO_2 is 12 g + (16 × 2) g = 44 g

To work out the masses of substances involved in converting 160 g of iron(III) oxide to iron, we can use the following steps:

Step 1. Write the balanced equation:

$$Fe_2O_3(s) \quad + \quad 3CO(g) \quad \rightarrow \quad 2Fe(s) \quad + \quad 3CO_2(g)$$

Step 2. Convert the equation to amounts:

1 mol Fe_2O_3 + 3 mol CO → 2 mol Fe + 3 mol CO_2

Step 3. Work out the amount being used:

$$\text{Moles of } Fe_2O_3 \text{ in 160 g} = \frac{\text{mass in grams}}{\text{mass of one mole (in grams)}} = \frac{160}{160} = 1 \text{ mol}$$

Step 4. Scale the amounts in the equation:
In this case, only 1 mol Fe_2O_3 is used, so the amounts do not need to be scaled.

1 mol Fe_2O_3 + 3 mol CO → 2 mol Fe + 3 mol CO_2

Step 5. Convert amounts (moles) to masses:

160 g Fe_2O_3 + 3 × 28 g (= 84 g) CO →
2 × 56 g (= 112 g) Fe + 3 × 44 g (= 132 g) CO_2

So, from 160 g of Fe_2O_3, in theory 112 g of iron could be produced and 132 g of carbon dioxide given off. In the iron and steel industry, amounts this size are ridiculously small. A typical blast furnace can produce up to 10 000 tonnes a day of molten iron.

? QUESTION 6

6 Coke also reacts directly with iron ore in the hotter part of the blast furnace to give iron and carbon monoxide.

a) Work out the balanced equation for this reaction.

b) How many tonnes of carbon monoxide are produced in this part of the furnace for every tonne of iron produced?

For information about significant figures, refer to Appendix 1.

? QUESTION 7

7 Zinc metal can be produced from its ore (zinc oxide, ZnO) by heating it with carbon, in a furnace.

a) Write the balanced equation for this reaction.

b) Calculate which reactant is the limiting reactant if 8.1 tonnes of zinc oxide is reacted with 8.1 tonnes of carbon.

? QUESTION 8

8 What is the percentage yield of iron produced if 10.0 tonnes of iron(III) oxide yields 6.7 tonnes of iron?

Suppose we now want to find out how much iron can be produced from 10 tonnes of iron(III) oxide. We can still use the same method. The first step is identical, so we can start at Step 2.

Step 2. Convert the equation to amounts:

$$1 \text{ mol } Fe_2O_3 \text{ gives } 2 \text{ mol Fe}$$

Note: In this example we do not need to know the moles of CO or CO_2.

Step 3. Work out the amount being used (1 tonne = 1 000 000 g):

$$\text{Moles of } Fe_2O_3 \text{ in 10 tonnes} = \frac{\text{mass in grams}}{\text{mass of one mole (in grams)}}$$

$$= \frac{10\,000\,000 \text{ g}}{160 \text{ g}}$$

$$= 62\,500 \text{ mol}$$

Step 4. Scale the amounts in the equation:

$$1 \text{ mol } Fe_2O_3 \text{ produces } 2 \text{ mol Fe}$$
$$62\,500 \text{ mol } Fe_2O_3 \text{ produces } 125\,000 \text{ mol Fe}$$

Step 5. Convert amount (moles) of Fe to a mass:

$$125\,000 \text{ mol} \times 56 \text{ g} = 7\,000\,000 \text{ g}$$
$$= 7.0 \text{ tonnes of Fe (to two significant figs)}$$

6 YIELD AND PERCENTAGE YIELD

You should now be able to use balanced equations to calculate the masses of substances involved in reactions. We have found that 10 tonnes of iron(III) oxide can give 7.0 tonnes of iron. The mass of iron produced is known as the **yield**. But Corus could never achieve this yield in a real blast furnace. Therefore we say that the calculation from the balanced equation gives the **theoretical yield**, which is the maximum amount if all the reactants are converted into products. There are many reasons why the theoretical yield is not achieved: reactions may not be finished in the time available, or some of the product may be lost during the purification procedure.

In industry or the laboratory, if there is more than one reactant, the reactant that is in short supply is called the **limiting reactant**. This is the reactant that determines the maximum amount of product that can be made. The other reactants are said to be in excess. In the case of reducing iron(III) oxide in the blast furnace, the limiting reactant will be the iron(III) oxide because it is cheaper to have an excess of carbon monoxide.

The **actual yield** from a reaction can be found only by doing the reaction. From 10.0 tonnes of iron(III) oxide, it is going to be less than 7.00 tonnes of iron. Percentage yield is a convenient way of expressing how close the actual yield is to the theoretical yield.

$$\textbf{Percentage yield} = \frac{\text{actual yield}}{\text{theoretical yield}} \times 100\%$$

Industrial processes aim for 100 per cent yield. The closer they get, the less is the waste of raw material.

7 ATOM ECONOMY

Even with 100 per cent yield there may still be a huge wastage of resources through the production of unwanted by-products. This is where the term **atom economy** comes in. It was first put forward by Barry Trost in the United States in 1991, as part of a new way of thinking called **Green Chemistry**. He urged chemists to consider how many atoms from the starting materials actually ends up in useful products. If the atoms from the reactants do not end up in useful products, then they end up in waste products.

$$\textbf{Atom economy} = \frac{\text{mass of desired product}}{\text{total mass of reactants}} \times 100\%$$

Let's consider again the reduction of the iron ore, Fe_2O_3.

$$Fe_2O_3(s) + 3CO(g) \rightarrow 2Fe(s) + 3CO_2(g)$$

The desired product is iron and in the equation 2 moles are produced:
$2 \times 56\ g = 112\ g$
The masses of the reactants:

$$1\ mol\ Fe_2O_3\ (= 160\ g) + 3\ mol\ CO = 3 \times 28\ (= 84\ g)$$

Total masses of reactants = 244 g.

$$\textbf{Atom economy} = \frac{\text{mass of desired product}}{\text{total mass of reactants}} \times 100\% = \frac{112}{244} = 45.9\%$$

So almost 54 per cent of the starting materials are lost, in this case to the atmosphere where the carbon dioxide contributes to the greenhouse effect and global warming. However, if there were an economic use for carbon dioxide then the atom economy would be 100 per cent.

8 RATE OF REACTION

The aim of industrial chemistry is to get as close as possible to a 100 per cent yield. However, in producing any chemical economically, the yield from a reaction is just one of the factors involved. Another very important factor is the *rate* of reaction. It is no use having a high yield of iron in a process that is extremely slow.

As a simple definition:

The rate of a reaction is the amount of substance formed per unit of time.

The rate could be in moles per second. For industry, it is more convenient to give the rate as the mass of substance, such as kilograms or tonnes, formed per unit of time. Reactions can be very fast and uncontrolled, such as a gas explosion (Fig 1.8 overleaf). Reactions can also be very slow, as in the rusting of iron (Fig 1.9 overleaf).

An industrial chemist needs to control the rate of a reaction. This can either mean speeding it up or slowing it down. In industry, the rate often needs to be increased so that the amount of product formed in, say, a day is enough to make a profit.

You can read more about the principles of Green Chemistry at the end of this chapter.

If you are not sure how the mass of one mole of Fe_2O_3 and one mole of CO is calculated look back to page 7.

QUESTION 9

9 Another ore of iron has the formula Fe_3O_4. In a blast furnace, 10.0 tonnes of this ore makes 6.50 tonnes of iron.

a) Write down the balanced equation for the reaction of Fe_3O_4 with carbon monoxide to make iron and carbon dioxide.

b) How many tonnes of iron can be made from 10.0 tonnes of Fe_3O_4? Note: this will be the theoretical yield.

c) Calculate the percentage yield.

d) What is the atom economy for this reaction?

QUESTION 10

10 One ore of copper contains copper(I) sulfide. This is heated with air to obtain the copper.

$Cu_2S(s) + O_2(g) \rightarrow 2Cu(s) + SO_2(g)$

a) Calculate the atom economy for this reaction if copper is the only useful product.

b) What would the atom economy be if the sulfur dioxide produced was used to make sulfuric acid?

c) What is the environmental consequence of allowing sulfur dioxide to escape into the air?

You can find out more about rates of reaction in Chapters 8 and 26.

Fig 1.8 An explosion is a reaction with a very rapid rate

Fig 1.9 The owner of this fishing trawler knows that it will not be long before his boat starts leaking because of rust

FACTORS THAT AFFECT RATE OF REACTION

Chemists control the rate of a reaction by changing the conditions of the reaction. These are some of the conditions that affect reaction rates.

- **Temperature:** an increase in temperature makes a reaction go faster (except for those involving enzymes).
- **Pressure:** increasing pressure increases the rate of reactions that involve gases.
- **Concentration:** increasing the concentration normally increases the rate of reaction.
- **Surface area of reactant:** an increase in surface area increases the rate of reaction.
- **Catalyst:** the addition of a catalyst usually increases the rate of reaction.

Catalysts can be recovered chemically unchanged at the end of a reaction. You can find out more about catalysts on page 166 of Chapter 8 and pages 203–207 of Chapter 10.

Reactions involve the rearrangement of atoms when bonds are broken and others are made. This rearrangement seldom takes place spontaneously; the particles normally need to collide with each other. All the conditions listed above will change the collision rate between particles. And if the collision rate is increased then the rate of reaction will increase.

Changing temperature, pressure, concentration and surface area can control the rate of reaction. However, these factors can also reduce the yield of an industrial process, so there is often a compromise between rates of reaction and yield. Remember, a major concern is the economics of the process – that is, making product quickly enough to give the maximum profit. We look at this compromise in more detail in Section 14 of this chapter.

9 CHEMICAL REACTIONS AND ENERGY CHANGES

All chemical reactions involve a change of energy, meaning *the transfer of energy to or from chemicals* in the reaction. This is crucial not only to industry, but also to the reactions of life itself. In many chemical reactions, *energy is given out by the reactants as they form products*, causing the temperature of the surroundings to rise. Such reactions are known as **exothermic reactions**.

In the blast furnace (see Fig 1.13 overleaf), blasts of hot air at 750 °C are blown in at the base and start a reaction between coke and oxygen:

$$C + O_2 \rightarrow CO_2$$

This reaction is highly exothermic, raising the temperature at the base of the furnace to about 2000 °C. This is because the stored energy in carbon and oxygen is greater than the stored energy in carbon dioxide. Stored energy is known as **enthalpy**, symbol ***H***. It is not possible to measure enthalpy, but **enthalpy changes** can easily be found by measuring temperature changes during reactions at constant pressure. Enthalpy changes are given the symbols **Δ*H***; Δ is a Greek letter pronounced 'delta' and is used by chemists to mean 'change of'.

In the coke and oxygen reaction, the energy released is 394 kilojoules (kJ) for every mole of carbon that reacts. So we write:

$$C(s) + O_2(g) \rightarrow CO_2(g) \quad \Delta H = -394 \text{ kJ mol}^{-1}$$

Notice that Δ*H* has a negative sign. This is because as the reaction proceeds, *energy is lost*. This energy heats up the surroundings, in this case the contents of the blast furnace. We show the reaction and its enthalpy changes in an **energy level diagram,** also called an **enthalpy level diagram** (see Fig 1.10).

Not all reactions are exothermic. **Endothermic reactions** are the opposite of exothermic reactions. In endothermic reactions *energy is taken in by the reactants to form products*. The energy comes from the surroundings, which lose energy and cool down.

The carbon dioxide produced at the base of the blast furnace reacts with more coke to produce carbon monoxide. This is an endothermic reaction.

$$CO_2(g) + C(s) \rightarrow 2CO(g) \quad \Delta H = +173 \text{ kJmol}^{-1}$$

The plus sign shows that carbon monoxide *takes in energy when it is formed* (Fig 1.11). As this reaction occurs in the middle of the furnace (see Fig 1.13), this is one of the reasons why the blast furnace becomes cooler towards the top (see also Fig 1.13).

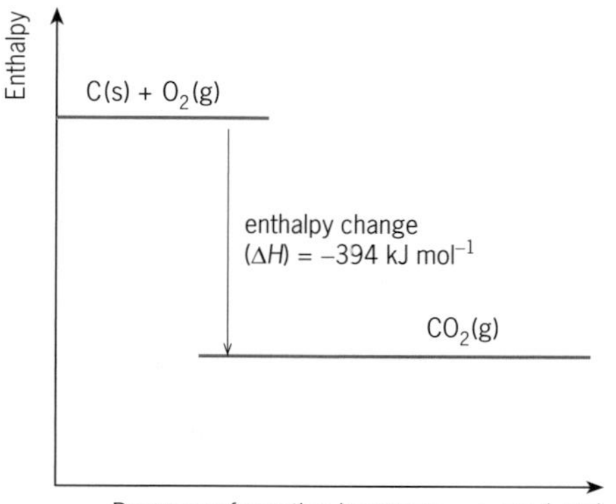

Fig 1.10 Enthalpy level diagram for the exothermic reaction $C(s) + O_2(g) \rightarrow CO_2(g)$

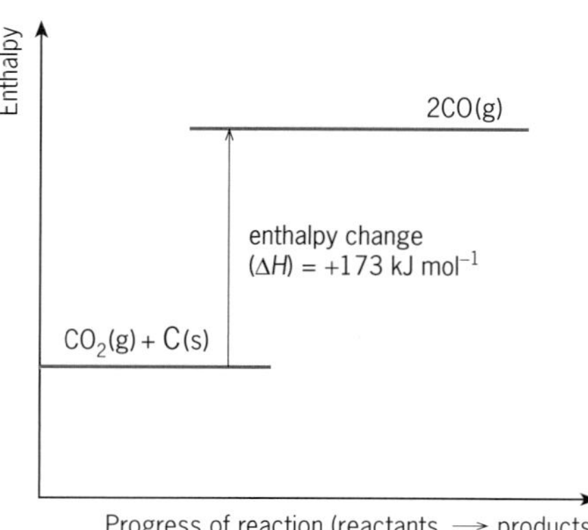

Fig 1.11 Enthalpy level diagram for the endothermic reaction $CO_2(g) + C(s) \rightarrow 2CO(g)$

REMEMBER THIS

State symbols should always be shown in chemical equations when dealing with energy from chemical reactions. Remember, (s) = solid, (l) = liquid and (g) = gas.

You can read more about enthalpy changes in Chapter 8.

REMEMBER THIS

This is an easy way to remember what exothermic and endothermic mean:

Energy **ex**its in **ex**othermic reactions. Look for the minus sign (Δ*H* is negative).

Energy **en**ters in **en**dothermic reactions. Look for the plus sign (Δ*H* is positive).

QUESTION 11

11 The following reactions are all associated with iron and steel manufacture. For each reaction, draw its energy level diagram and say whether the reaction is exothermic or endothermic.

a) $S(s) + O_2(g) \rightarrow SO_2(g)$
$\Delta H = -297 \text{ kJ mol}^{-1}$

b) $CaCO_3(s) \rightarrow CaO(s) + CO_2(g)$
$\Delta H = +178 \text{ kJ mol}^{-1}$

c) $Fe_2O_3(s) + 3CO(g) \rightarrow 2Fe(s) + 3CO_2(g)$
$\Delta H = -27 \text{ kJ mol}^{-1}$

REMEMBER THIS

An **ore** is a naturally occurring group of minerals from which a metal, or metals, can be extracted for a profit.

10 THE BLAST FURNACE

The blast furnace smelts the iron ore. Smelting means that the ore is melted, mixed with the other reactants and reduced to the metal. The reaction of iron ore with coke, limestone and hot air, produces the iron (Fig 1.12).

The iron-making process needs high-quality iron ore that contains at least 60 per cent iron. Most iron ores contain impurities such as sand (silicon(IV) oxide, SiO_2), sulfur compounds and phosphorus compounds so, if not pure enough, the ore has to be pre-refined to increase the percentage of iron.

The blast furnace lining has to last for many years to save the cost of replacing it and to minimise shutdown time for replacement work, since the loss of production may cost millions of pounds.

coke

iron ore

limestone

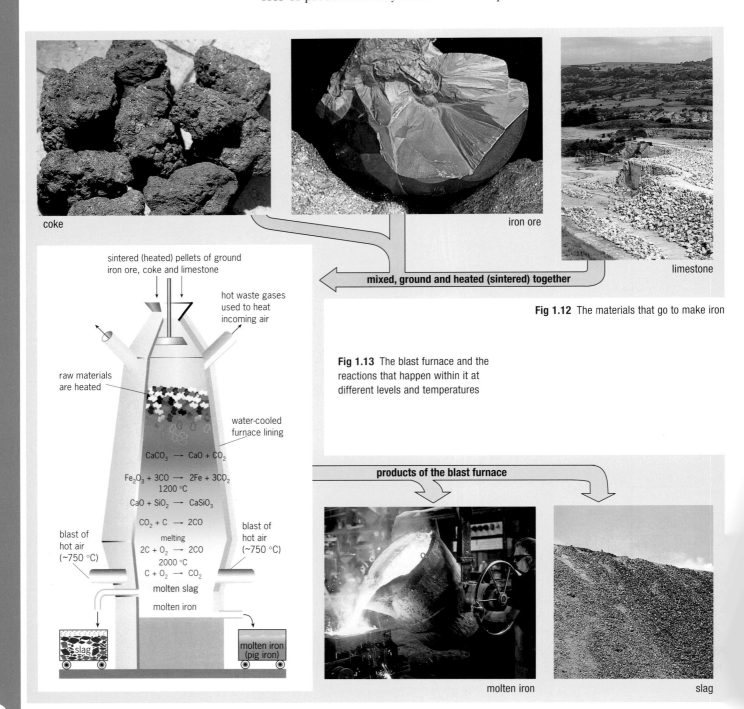

mixed, ground and heated (sintered) together

Fig 1.12 The materials that go to make iron

sintered (heated) pellets of ground iron ore, coke and limestone

hot waste gases used to heat incoming air

raw materials are heated

water-cooled furnace lining

$CaCO_3 \rightarrow CaO + CO_2$

$Fe_2O_3 + 3CO \rightarrow 2Fe + 3CO_2$

1200 °C

$CaO + SiO_2 \rightarrow CaSiO_3$

$CO_2 + C \rightarrow 2CO$

melting

$2C + O_2 \rightarrow 2CO$

2000 °C

$C + O_2 \rightarrow CO_2$

blast of hot air (~750 °C)

blast of hot air (~750 °C)

molten slag

molten iron

slag

molten iron (pig iron)

Fig 1.13 The blast furnace and the reactions that happen within it at different levels and temperatures

products of the blast furnace

molten iron

slag

CONDITIONS IN THE BLAST FURNACE

Some of the blast furnace reactions require high temperatures, so a great input of energy is needed. Fortunately, the reaction of carbon with oxygen to give carbon dioxide provides the energy, since it is an extremely exothermic reaction.

The blasts of hot air, which give the blast furnace its name, are enriched with oxygen. The concentration of oxygen, higher than in air, speeds up the reaction. The faster the rate of reaction, the more the energy produced by the exothermic reactions. In this way, the high temperatures are easier to maintain.

The carbon dioxide formed near the base travels upwards through the melt and solids, and is converted into carbon monoxide (see Fig 1.13). It is the carbon monoxide rather than the solid carbon that reduces most of the iron ore, since gases react faster than solids. The use of pelletised reactants ensures that carbon monoxide comes into very close contact with the iron(III) oxide. This helps to speed up the reaction because the surface area of the ore has been increased.

Finally, there are the reactions involving limestone. Limestone (calcium carbonate) decomposes into calcium oxide (lime) and carbon dioxide in the heat of the furnace:

$$CaCO_3(s) \rightarrow CaO(s) + CO_2(g)$$

The calcium oxide removes impurities such as sand (silicon(IV) oxide, SiO_2) as liquid slag. The reaction for sand is:

$$CaO(s) + SiO_2(s) \rightarrow CaSiO_3(s)$$

Slag and iron are both liquids at the high temperatures of the furnace. Conveniently, the less dense slag floats on top of the more dense iron, so they are tapped off separately at different levels.

The production of iron is a **continuous process**. In continuous processes, raw materials are constantly added and the products are continually removed. A continuous process is a very efficient and cost-saving way to produce materials in large quantities. By comparison, in **batch processes**, the reactor vessel must be closed down and reset to make another batch. This is expensive because there is 'dead time' when no product is being produced. However, a range of products can be made in the same vessel, although care is exercised to ensure no contamination occurs. For small quantities, the batch process is usually more cost effective.

? QUESTION 12

12 The temperature of the furnace may reach 2000 °C, especially near the base. Most materials melt at this temperature. What properties does the lining material of the blast furnace need to have?

? QUESTION 13

13 **a)** Coke, iron ore and limestone are mixed and ground together before they enter the blast furnace. Suggest why this is done.
b) The powder produced is heated (sintered) to make loosely packed pellets. What is the advantage of having pellets rather than powder in a blast furnace?

? QUESTION 14

14 Explain why companies prefer to use continuous processes instead of batch processes whenever possible.

DISPOSAL OF WASTE

The waste gases and slag could cause pollution problems, but ways have been found to minimise them. Much of the slag goes to build roads and make cement. Some of it is even used to insulate houses: Rockwool, used for non-flammable loft insulation, is produced by blowing air into the molten slag to make it light and fluffy.

11 MAKING STEEL FROM IRON

Iron from a blast furnace is impure, brittle and not very strong, so most iron produced is changed immediately into steel. Steel is the name given to countless **alloys** that contain iron, carbon and, usually, small amounts of other elements.

An alloy is a mixture of two or more elements, at least one being a metal. The elements are mixed when they are molten and allowed to cool down to form a uniform solid.

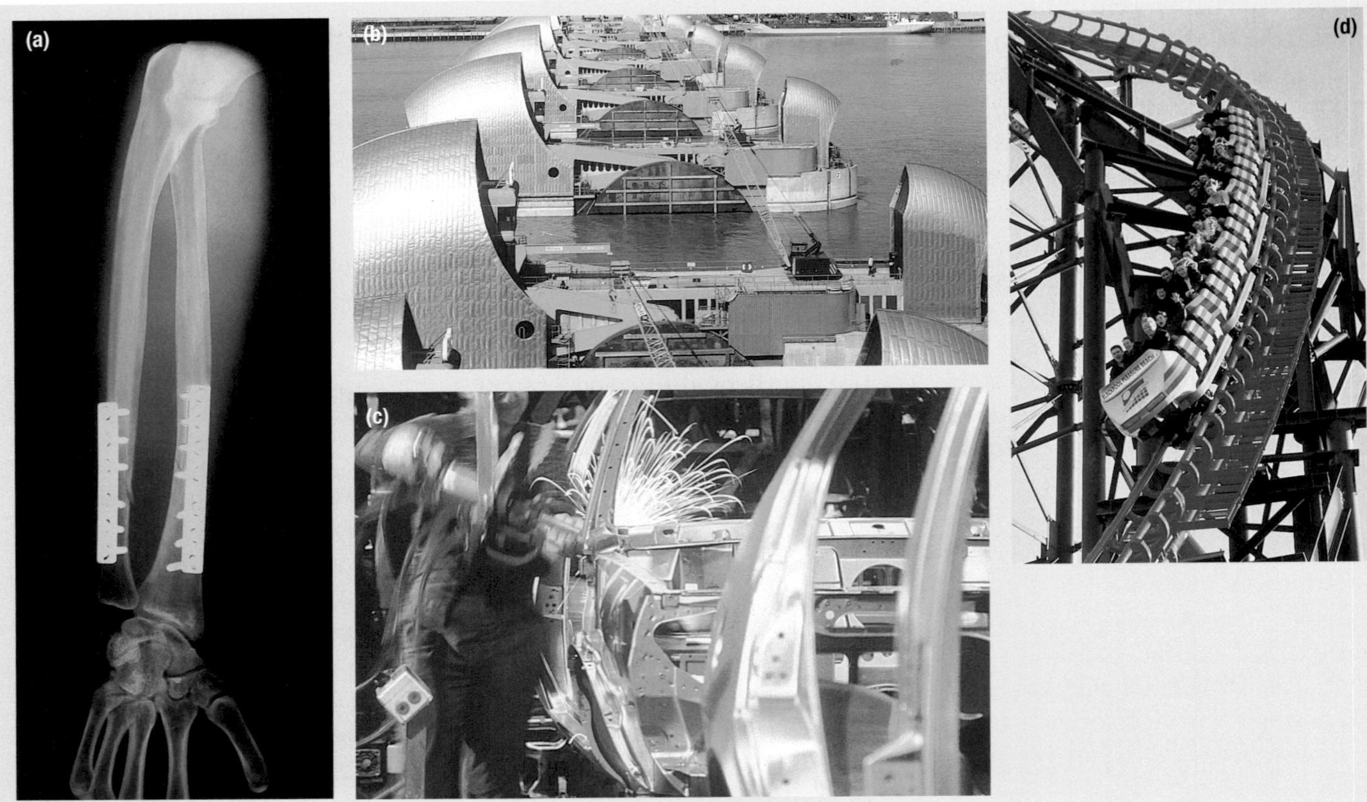

Fig 1.14 Different types of steel have properties that iron lacks, giving steel a very wide range of uses.
(a) Fractured bones are held together by permanent steel pins
(b) The Thames Barrier is mainly steel and concrete
(c) Well over half the components of a typical car are made of steel, including the body shell, clutch plates and engine
(d) The Fruit Drop at Blackpool is made of steel to meet high safety standards

Alloys have different properties from the elements that form them. It is the addition of different elements to iron that changes the properties of the alloy, and so makes steels such versatile materials.

CARBON STEELS

As the name suggests, carbon steels are alloys of mainly iron with some carbon. The sheet steel used in car bodies contains just 0.2 per cent of carbon, an amount that makes it easy to bend and shape.

As the carbon content of steel is increased, the steel becomes stronger and more rigid. Most of the steel used to construct a bridge, which needs some flexibility, contains between 0.3 and 0.6 per cent carbon. The steel used in drill bits has to be very hard and contains up to 1.5 per cent carbon. It is tempting to think that by increasing the carbon content the steel would carry on strengthening. Unfortunately, this is not the case, and at just 4 per cent carbon, steel becomes very brittle.

ALLOY STEELS

Alloy steels are steels that contain one or more other metals. These metals include manganese, tungsten, chromium and vanadium. Stainless steels are probably the best-known alloy steels, containing at least 12 per cent chromium. The chromium increases steel's rust resistance. A common stainless steel is called 18–8 and contains 18 per cent chromium and 8 per cent nickel, a composition you will find in some cutlery. Steels that contain tungsten are very hard-wearing. Adding molybdenum (with certain other elements) enables drill bits to retain their cutting edge, even when hot. Spacecraft use titanium steel because it can withstand the high temperatures of re-entry.

STEEL MAKING

The main process for converting iron into steel is the **basic oxygen process**. Impure iron from a blast furnace is known as pig iron and contains about 4 per cent carbon, together with other elements such as silicon, manganese and phosphorus. To convert the iron into steel, the carbon content is lowered and other elements are removed by reacting them with oxygen. An oxygen lance is lowered into the basic oxygen furnace (Fig 1.15) and oxygen is blown into the molten iron at twice the speed of sound.

The impurity elements form oxides. These are some of the reactions:

$$2C(s) + O_2(g) \rightarrow 2CO(g)$$
$$2Mn(s) + O_2(g) \rightarrow 2MnO(s)$$
$$4P(s) + 5O_2(g) \rightarrow P_4O_{10}(s)$$
$$Si(s) + O_2(g) \rightarrow SiO_2(s)$$

Being a gas, carbon monoxide bubbles out of the liquid mixture. Both SiO_2 and P_4O_{10} are acidic oxides and are removed by adding lime (calcium oxide), which is a basic oxide – hence the word 'basic' in the basic oxygen process. They react to form a slag that floats on top of the molten steel.

These chemical reactions with oxygen are highly exothermic and generate great heat, so the reactions inside the furnace keep the contents molten. In fact, the temperature has to be stopped from becoming too high. Scrap iron and steel are added to prevent the mixture from overheating, because as the mixture melts it takes in energy – melting is an endothermic process.

If the temperature rose unchecked, then the rate of the exothermic reactions would increase, causing yet more energy to be released and making the reaction faster still. Eventually, the process would be uncontrollable and the furnace lining would be damaged.

It is the job of industrial chemists known as metallurgists to ensure that the reaction conditions of the steel making process are optimised. They also establish and monitor the best mix of elements for a particular type of steel to match customer demand.

? QUESTION 15

15 Desulfurisation is a process that can be carried out on the impure molten iron from the blast furnace. Powdered magnesium is added to the molten iron and it reacts with the sulfur to form magnesium sulfide (MgS).

a) Write a balanced equation for this reaction.

b) How much magnesium would need to be added to remove 0.32 kg of sulfur?

c) In practice, much more magnesium would be required than the theoretical value you have calculated. Explain why.

d) Suggest why the magnesium is added in powdered form.

e) Liquid magnesium sulfide is less dense than the molten iron. Suggest how these two liquids may be separated.

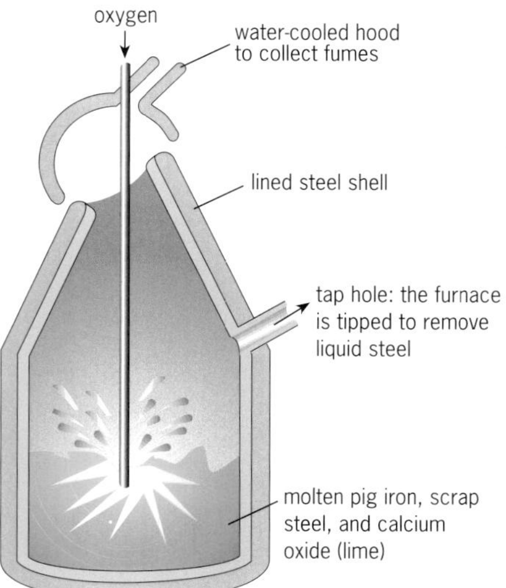

Fig 1.15 The arrangement for the basic oxygen process

Fig 1.16 The basic oxygen furnace for making steel being charged with molten iron

STEEL – MATCHING SUPPLY AND DEMAND

Steel manufacture is a business, and so the production of steels must be tied closely with long-term planning to sell it. An industrial plant the size of an iron and steel works cannot run economically if it has to start and stop manufacture to match short-term demands. A blast furnace works efficiently only if it is running all the time. Also, if furnace production is stopped, the furnace has to cool down and then be cleaned out. Starting it up again is a costly and difficult process that requires a lot of energy. But if the production of steel outstrips demand, the only economic solution may be to close down the whole steel works.

So to keep steel works operating, the staff who are in charge of sales constantly search for new markets.

SAVING ON ENERGY

We have seen that steel making has very expensive energy requirements. About 15 per cent of the cost of making steel is spent on the energy that the process consumes, so companies are continually on the look-out for ways to make savings.

Building the steel works next to the blast furnace allows molten iron to be used as soon as it is made. Since the iron does not cool much between processes, its energy is not wasted. Also, the energy from hot waste gases and slag can be transferred to other parts of the operation. The waste gases can even be burnt to release energy.

12 REACTIONS IN EQUILIBRIUM

The reactions we have so far discussed appear **irreversible** and for practical purposes they are. Let's take burning sulfur in oxygen as an example:

$$S(s) + O_2(g) \rightarrow SO_2(g)$$

The single arrow tells us that this reaction is irreversible and *goes to completion.*

REMEMBER THIS

The reaction of sulfur and oxygen to produce sulfur dioxide occurs in the blast furnace and although the sulfur is present in only small amounts, the sulfur dioxide produced enters the atmosphere and contributes to acid rain.

But many reactions can be reversed. A **reversible reaction** is one that can take place in either direction. Such a reaction may form an **equilibrium** if the circumstances allow. Equilibrium means a state of balance. Understanding chemical **equilibria** is fundamental, not only to theoretical chemistry, but to life itself.

The oxygen we breathe in forms an equilibrium with haemoglobin in red blood cells. In the lungs, oxygen reacts with haemoglobin to produce oxyhaemoglobin, which is transported to the body's tissues. Here, a different set of conditions reverses the reaction and releases the oxygen. We can represent the equilibrium as follows:

$$\text{haemoglobin} + \text{oxygen} \rightleftharpoons \text{oxyhaemoglobin}$$

In the lungs, which have high concentrations of oxygen, the position of the equilibrium is shifted to the side of the product, oxyhaemoglobin. The concentration of oxygen is much lower in the body tissue, so when the red blood cell reaches it the position of the equilibrium shifts in favour of the reactants and so oxygen is released. Understanding the changes in conditions that cause an equilibrium to shift is particularly important in industry, as we shall see in Section 12.

UNDERSTANDING DYNAMIC EQUILIBRIA

One of the first reversible reactions to be investigated was that between hydrogen and iodine more than a hundred years ago.

$$H_2(g) + I_2(g) \rightleftharpoons 2HI(g)$$

A German chemist called Max Ernst Bodenstein (1871–1942) sealed mixtures of hydrogen and iodine in glass bulbs and kept them at the same temperature in a thermostatically controlled bath. He repeated this procedure with glass bulbs of hydrogen iodide (HI). At certain time intervals he would stop the reactions in the bulbs by cooling them rapidly and analyse their contents. The results of his experiments are shown in Fig 1.17.

The vertical axis represents the amount of HI present in the reaction mixtures from the start of the reaction. The lower curve (drawn in blue) shows reaction mixtures of 0.5 moles of H_2 and 0.5 moles of I_2. If the reaction went to completion there would be 1 mole of HI in the flask and zero moles of H_2 and I_2. But this did not happen. Only 0.78 moles of HI were present. The red, upper curve shows that by starting with 1 mole of HI in the flask, at the same temperature, the same amount was left after 84 minutes (i.e. 0.78 mol). It also did not matter how much time elapsed after 84 minutes: the amounts of all three chemicals were identical. After 84 minutes the reaction had reached its **equilibrium position**.

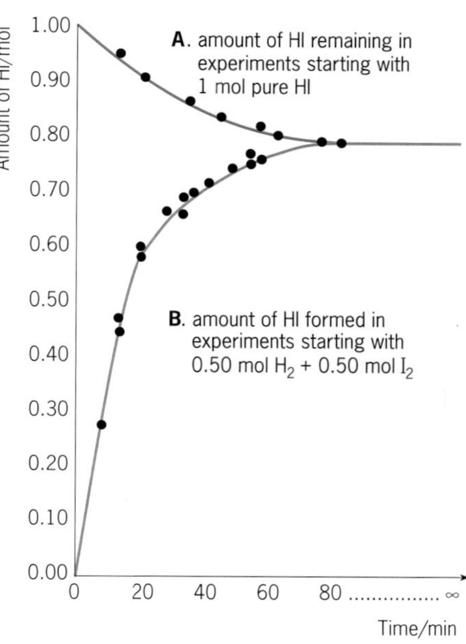

Fig 1.17 Experimental data produced in 1894 for the reaction: $H_2(g) + I_2(g) \rightleftharpoons 2HI(g)$ carried out at 448 °C

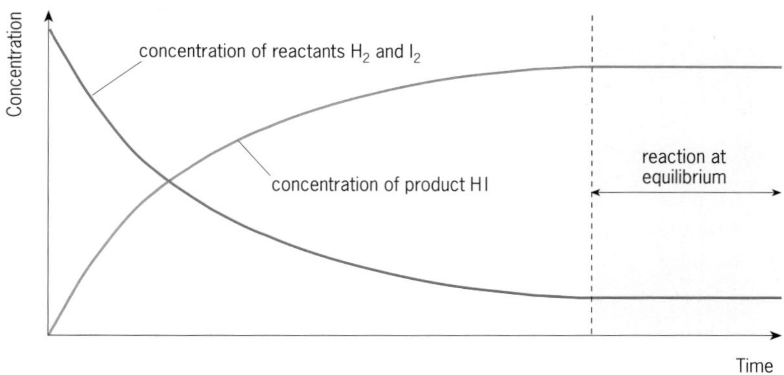

Fig 1.18 The changes in concentrations with time for the reaction: $H_2(g) + I_2(g) \rightleftharpoons 2HI(g)$ when the same concentrations of $H_2(g)$ and $I_2(g)$ are mixed

The concentrations of the gases in the flask do not change once equilibrium has been established (Fig 1.18). But does this mean that reactions stop when equilibrium is reached? The answer is no. This is where the term **dynamic** comes in. Although the concentrations of reactants and products remain constant, both the forward and reverse reactions are still proceeding. This means that, at a molecular level, H_2 molecules are colliding with I_2 molecules to form HI molecules:

$$H_2(g) + I_2(g) \rightarrow 2HI(g) \text{ (forward reaction)}$$

and HI molecules collide to give the reverse reaction:

$$2HI(g) \rightarrow H_2(g) + I_2(g) \text{ (reverse reaction).}$$

? QUESTION 16

16 **a)** In the experiment that Bodenstein performed, why does rapid cooling apparently stop the reaction?

b) If the starting mixture contained 1 mol H_2 and 1 mol I_2, how many moles of HI would be present if the reaction went to completion?

c) Predict the amount of HI that would actually be present in the mixture from **b)** when it reached equilibrium. Assume the same conditions as those shown in Fig 1.17.

? QUESTION 17

17 If you removed some HI from the equilibrium mixture, predict what would happen:

a) to the rate of the reverse reaction;

b) to the position of the equilibrium.

As the concentrations of reactants and products remain the same, the forward and reverse reactions must occur at the same rate.

A dynamic equilibrium is reached when the forward and reverse reactions occur at the same rate.

The forward and reverse reactions still continue so we call the equilibrium *dynamic*.

In Fig 1.19 you can see that, when the reaction between H_2 and I_2 starts, the rate of reaction is at its fastest. As the reaction continues, the rate of the forward reaction begins to slow. At the same time, the concentration of HI increases, which means that the rate of the reverse reaction increases. When the two rates are equal, a dynamic equilibrium is established.

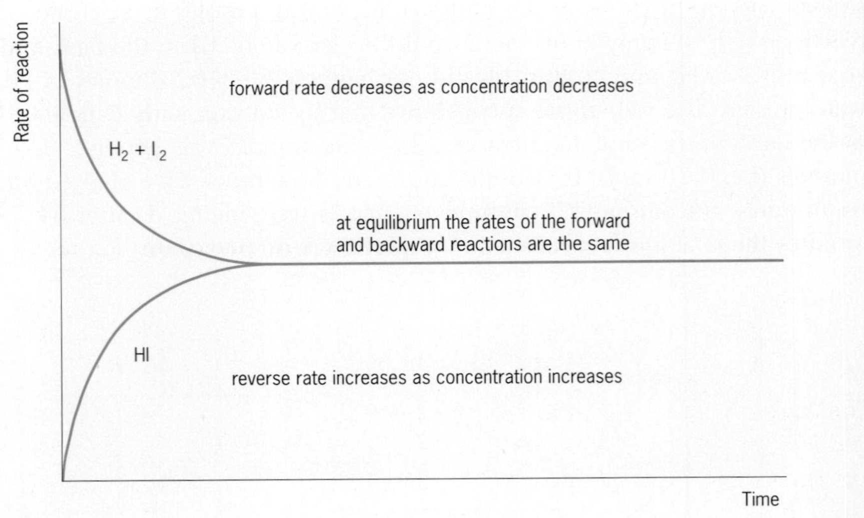

Fig 1.19 Graphs of rate of reaction with time for the forward and reverse reactions of: $H_2(g) + I_2(g) \rightleftharpoons 2HI(g)$

THE IMPORTANCE OF A CLOSED SYSTEM

The glass bulbs that Bodenstein used to do his experiments were sealed so that none of the gases could escape. This is an example of a **closed system**, in which no material is exchanged with the surroundings. Any loss of reactant or product affects their equilibrium concentrations, and from Section 8 (page 9) you may remember that rate of reaction depends on concentration. A lower concentration means the slowing down of a reaction so the equilibrium is destroyed.

The equilibrium mixture which Bodenstein investigated is an example of a **homogeneous equilibrium**, because the components are all in the same phase – the gaseous phase. If we consider a **heterogeneous equilibrium**, where the components are in different phases, the importance of a closed system may be easier to understand.

If we heat calcium carbonate in a closed vessel the following equilibrium is established:

$$CaCO_3(s) \rightleftharpoons CaO(s) + CO_2(g)$$

If we now let some of the carbon dioxide escape, the rate of the reverse reaction becomes slower, so the position of the equilibrium shifts towards the products. If we take the lid off completely, all the carbon dioxide escapes and in this **open system** an equilibrium cannot be established, so all of the calcium carbonate reacts to form products. The blast furnace is just such an open system, which is why all of the calcium carbonate reacts irreversibly.

13 THE EFFECTS OF CHANGING CONDITIONS ON EQUILIBRIUM SYSTEMS

The key to achieving the best yields from industrial processes that involve reversible reactions is to understand the effects of changing the conditions on the position of equilibrium. The position of equilibrium is the proportion of the concentration of products to the concentration of reactants in the equilibrium. If the products are present in high concentrations and there are hardly any reactants left, then we say that the position of the equilibrium lies to the right. If there is a high concentration of reactants still present and the concentration of products is very low, then we say the position of equilibrium lies to the left.

LE CHATELIER'S PRINCIPLE

In 1884, Henri Le Chatelier, a French chemist, formulated his famous principle:

The position of the equilibrium of a system changes to minimise the effect of any imposed change in conditions.

The conditions referred to in Le Chatelier's principle are temperature, pressure and concentration of a substance in an equilibrium mixture, and when a change in one or more of these is imposed this affects the position of equilibrium. Le Chatelier's principle applies to any reaction that is in dynamic equilibrium and allows chemists to manage and manipulate equilibrium reactions.

Fritz Haber discovered one of the most important industrial reactions in history in 1909. He succeeded in synthesising 100 g ammonia (NH_3) from nitrogen and hydrogen. The reaction was important because it allowed the manufacture of a nitrogen compound (ammonia) from nitrogen in the air. Before the **Haber process** was discovered, nitrogen was thought to be too unreactive to be used to manufacture a compound economically. The reaction on which the Haber process is based is:

$$N_2(g) + 3H_2(g) \rightleftharpoons 2NH_3(g)$$

and an understanding of Le Chatelier's principle explains how Haber succeeded in manufacturing a nitrogen compound when so many chemists had tried and failed previously.

? QUESTION 18

18 In which of the following is there a dynamic equilibrium?
a) An unopened can of coke;
b) An open can of coke;
c) A sealed bag containing a wet towel;
d) A wet towel hanging on a washing line.

You can find out more about how to calculate the position of an equilibrium in Chapter 18 and Chapter 28.

Change	Type or part of system	Effect on equilibrium
pressure increasing	if there are more moles of gaseous products than gaseous reactants	shift to left
	if there are more moles of gaseous reactants than gaseous products	shift to right
temperature increasing	if $\Delta H^{\ominus}_{\text{forward reaction}}$ is positive, i.e. the forward reaction is endothermic	shift to right
	if $\Delta H^{\ominus}_{\text{forward reaction}}$ is negative, i.e. the forward reaction is exothermic	shift to left
concentration increasing	of reactants	shift to right
	of products	shift to left
catalyst added		no effect

Fig 1.20 A summary of Le Chatelier's principle. $^{\ominus}$ in $\Delta H^{\ominus}$ means that the conditions are standard conditions – usually 100 kPa, 1 mol dm^{-3} and 298 K

HOW SCIENCE WORKS

Solving the nitrate crisis

Fig 1.21 Fritz Haber (1868–1934) in 1905

In the 21st century we will probably face an oil crisis. At the beginning of the 20th century, scientists were predicting that there would be a nitrate crisis. Already, valuable deposits of nitrates from bird droppings in Peru were exhausted. Although there were still vast supplies of sodium nitrate in the deserts of Chile, these supplies could not keep up with demand. So what was so important about nitrates and why were they in such high demand?

Nitrogen compounds are naturally present in the soil. If they were not, most plants would not be able to grow. Farmers have, for thousands of years, been replacing the nitrogen compounds taken out of the soil through harvesting crops by adding fertiliser – usually animal manure. However, a huge population increase in the nineteenth century in Europe and North America made these sources insufficient to support the more intensive agriculture required, so nitrate compounds were imported from wherever they could be found.

Nor was this the only use of nitrates. Explosives such as trinitrotoluene (TNT) and nitroglycerine required nitric acid, and this too needed nitrates as the raw material.

In 1909, Germany was preparing for war. Realising that British control of the seas would make it very difficult to import Chilean nitrate, German scientists were set the problem of producing nitrogen compounds from nitrogen in the air.

Nitrogen gas is so plentiful it might seem surprising that this route had not been used before to manufacture its compounds. But nitrogen gas is very unreactive and many chemists had tried, but none were successful in discovering a cost-effective process. In 1909, Fritz Haber, a German chemist, had succeeded in producing 100 g of ammonia (NH_3) from nitrogen and hydrogen. Between 1910 and 1913, Haber carried out more than 6000 experiments to find the most effective catalyst. In 1913 Carl Bosch, a chemical engineer employed by BASF, scaled up the process to produce the first industrial ammonia plant.

Today the rising population of the world relies on fertilisers produced using the Haber process. It was Haber's understanding of equilibria that made this process possible.

Fig 1.22 Fertilisers manufactured from ammonia support the lives of millions of people through the intensive growing of crops

> **? QUESTION 19**
>
> **19** What is the atom economy of the principal reaction of the Haber process?

TEMPERATURE CHANGES AND THE POSITION OF EQUILIBRIUM

When Haber started his research, several methods of making ammonia and nitrates had been invented. However, they were all costly in terms of energy and raw materials. Haber was interested in the most efficient way to make nitrogen compounds from nitrogen in the air. As you can read in the

How Science Works box (left), it was crucial to find a process that would do this, because nitrogen compounds are not only essential fertilisers, but also the necessary starting point for manufacturing explosives.

At 1000 °C, the position of equilibrium lies well towards the reactants. In fact, the reaction mixture contains only a tiny percentage of ammonia (0.0005 %–0.0012 %).

Using Le Chatelier's principle, we can predict the effect of an increase in temperature on the position of equilibrium:

$$N_2(g) + 3H_2(g) \rightleftharpoons 2NH_3(g) \quad \Delta H^\ominus = -92 \text{ kJ mol}^{-1}$$

According to Le Chatelier's principle, raising the temperature will cause the equilibrium to shift to absorb energy. This is because the change imposed on the system by increasing the temperature increases the energy of the system. This equilibrium is **exothermic** in the forward direction – the sign of ΔH tells you this. So raising the temperature shifts the equilibrium in the endothermic direction (i.e. to the left) so that energy is absorbed.

Another reaction that involves nitrogen occurs in the high temperatures of the car engine. As we have seen, under normal conditions nitrogen is very unreactive, but because the petrol engine can reach temperatures of 1000 °C nitrogen monoxide is one of the compounds produced. It is an atmospheric pollutant, which you can read more about on page 205, Chapter 10.

$$N_2(g) + O_2(g) \rightleftharpoons 2NO(g) \quad \Delta H^\ominus = +90 \text{ kJ mol}^{-1}$$

This reaction is endothermic so the forward reaction absorbs energy; thus, raising the temperature will cause the equilibrium to shift to the right. Even at the high temperatures in an engine the yield is less than 1 per cent, but this is still sufficient to cause pollution problems.

PRESSURE CHANGES AND THE POSITION OF EQUILIBRIUM

Haber expected that increasing the pressure would shift the position of the equilibrium to the right. As he wrote: "To begin with, it was clear that a change to the use of maximum pressure would be advantageous. It would improve the point of equilibrium and probably the rate of reaction as well. The compressor which we then possessed allowed gas to be compressed to 200 atmospheres."

He predicted this using Le Chatelier's principle. This time the *imposed change in condition* is an increase in pressure, so the equilibrium will shift to minimise this. In the equilibrium:

$$\begin{array}{lcll} N_2(g) & + & 3H_2(g) & \rightleftharpoons 2NH_3(g) \\ \text{1 mole} & + & \text{3 moles (= 4 moles)} & \rightleftharpoons \text{2 moles} \end{array}$$

There are fewer molecules of gas on the right-hand side – 2 moles instead of 4 moles. In a gas, pressure depends on the number of molecules, and therefore increasing the pressure causes the equilibrium to shift to the right.

By contrast, in this equilibrium:

$$\begin{array}{lcll} H_2(g) & + & I_2(g) & \rightleftharpoons 2HI(g) \\ \text{1 mole} & + & \text{1 mole (= 2 moles)} & \rightleftharpoons \text{2 moles} \end{array}$$

the number of molecules of gas is the same on both sides, so increasing the pressure has no effect on the equilibrium position. (Also, note that a change in pressure has no effect on equilibria that do not involve gases.)

REMEMBER THIS

$^\ominus$ in $\Delta H^\ominus$ means that the conditions are standard conditions at a particular temperature, usually 298 K. Standard conditions are 100 kPa and 1 mol dm^{-3} (see page 153).

? QUESTION 20

20 In the Haber, process, Le Chatelier's principle suggests that lowering the temperature should give a much higher yield. Why did Haber consider using high temperatures? (Hint: consider rates of reaction.)

? QUESTION 21

21 Predict the effect of increasing the temperature on the following equilibria. Explain your reasoning.

a) $2SO_2(g) + O_2(g) \rightleftharpoons 2SO_3(g) \quad \Delta H^\ominus = -98 \text{ kJ mol}^{-1}$

b) $4NH_3(g) + 5O_2(g) \rightleftharpoons 4NO(g) + 6H_2O(l) \quad \Delta H^\ominus = -908 \text{ kJ mol}^{-1}$

c) $CH_4 + H_2O(g) \rightleftharpoons CO(g) + 3H_2(g) \quad \Delta H^\ominus = +210 \text{ kJ mol}^{-1}$

? QUESTION 22

22 Use Le Chatelier's principle to predict the effect of increasing the pressure on each of these equilibria:

a) $CH_4(g) + H_2O(g) \rightleftharpoons CO(g) + 3H_2(g)$

b) $CO(g) + H_2O(g) \rightleftharpoons CO_2(g) + H_2(g)$

c) $2NO_2(g) \rightleftharpoons N_2O_4(g)$

CONCENTRATION CHANGES AND THE POSITION OF EQUILIBRIUM

Equilibria in the liquid phase are not affected by changes in pressure, because there is no significant change in volume, but they are affected by changes in concentration.

Esters are sweet-smelling organic compounds. They are mainly used as solvents, but some of them can also be used to flavour food to give, for example, the taste of raspberry or pear. In addition, they are added to plastics such as polyvinylchloride (PVC) to make them more flexible. One industrial reaction to produce an ester is:

$$CH_3COOH(l) + C_2H_5OH(l) \rightleftharpoons CH_3COOC_2H_5(l) + H_2O(l)$$

ethanoic acid + ethanol ⇌ ethyl ethanoate ester + water

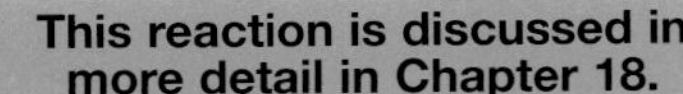
This reaction is discussed in more detail in Chapter 18.

According to Le Chatelier's principle, increasing the concentration of either of the reactants ethanol or ethanoic acid will shift the equilibrium to the right. This is because the imposed change in condition is an increase in concentration, so the equilibrium will shift to the product side to minimise this effect. In industry, because an increase in either reactant can be used to shift the equilibrium to form more ester, the cheaper reactant is used in excess.

Another way to manage the reversible reaction is to *remove* one of the products the moment it has formed. This prevents the reaction from achieving equilibrium, but it nevertheless will 'keep trying' to, as long as the product continues to be removed from the reaction mixture. In esterification, water is the easier product to remove.

THE EFFECT OF ADDING A CATALYST ON THE POSITION OF EQUILIBRIUM

A catalyst does not affect the position of equilibrium, but it does increase the rate of both the forward and reverse reactions, decreasing the time taken to reach equilibrium. Haber carried out more than 6000 experiments just to find the most effective catalyst, because he needed to speed up the reaction between nitrogen and hydrogen to obtain ammonia more quickly.

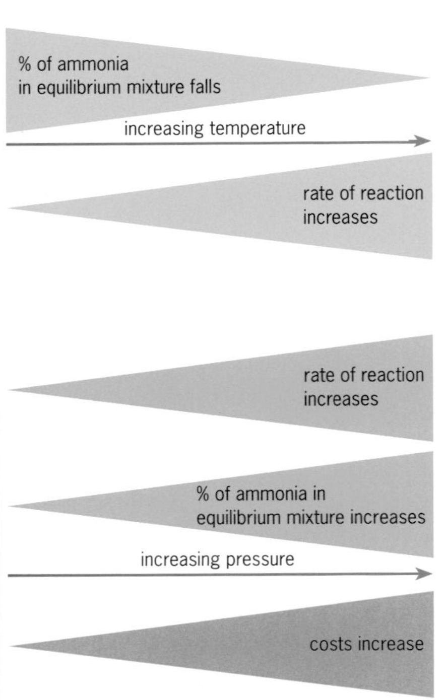

Fig 1.23 The effects of changing conditions on the Haber process

14 OBTAINING AN ECONOMIC YIELD IN THE HABER PROCESS

Le Chatelier's principle suggests that low-temperature and high-pressure conditions would maximise ammonia production (see Fig 1.23):

$$N_2(g) + 3H_2(g) \rightleftharpoons 2NH_3(g) \quad \Delta H^{\circ} = -92\ \text{kJ mol}^{-1}$$

However, the chemical industry does not use these conditions, for economic reasons. In choosing the right temperature, manufacturers must consider:

- the yield, that is, the percentage of ammonia in the equilibrium mixture;
- the rate of reaction.

The rate of reaction is related to temperature. If the temperature is too low the reaction proceeds too slowly. Therefore, manufacturers use a combination of a moderate temperature and a catalyst to ensure that the rate of ammonia production is high.

High pressure gives a fast rate of reaction for a gas-phase reaction, but using higher pressures drastically increases the cost of building and running the plant because the pipes that carry the gaseous mixtures need to withstand high pressures, and much more energy is used to maintain these pressures.

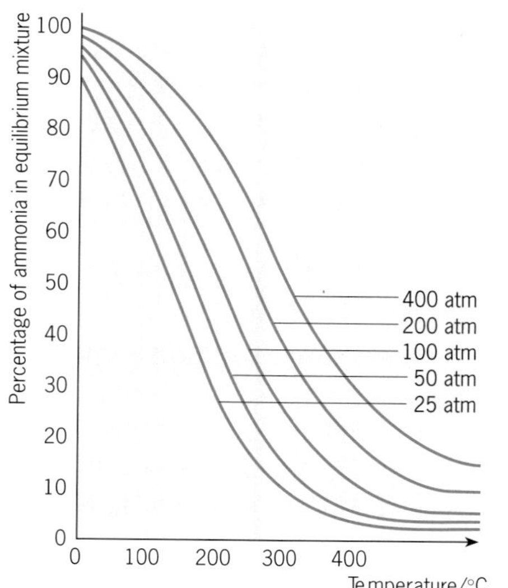

Fig 1.24 The effect of temperature on the percentage of ammonia in the equilibrium mixture

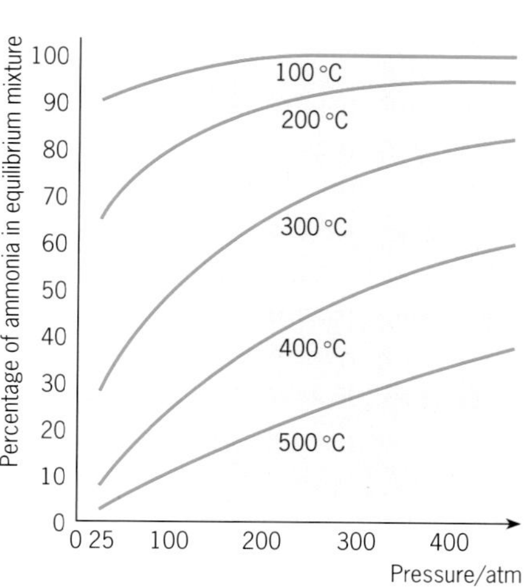

Fig 1.25 The percentage of ammonia in the equilibrium mixture as pressure changes

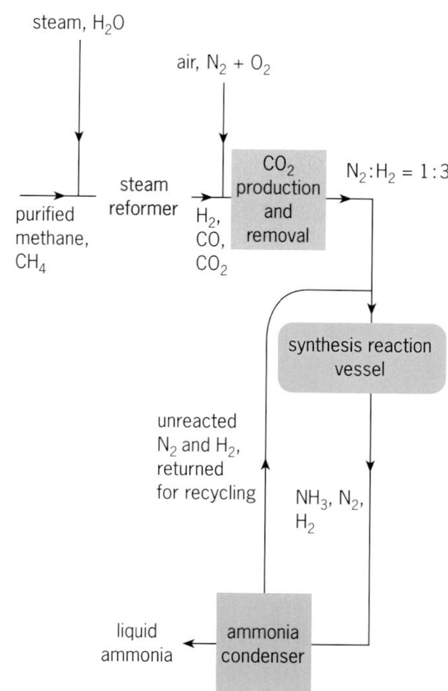

Fig 1.26 Flow diagram of the ammonia manufacturing process

Most manufacturers use:

- a temperature of between 350 and 550 °C. This is a compromise temperature at which the rate of reaction is fast enough to give a reasonable yield in a short time.
- a pressure of 100 to 200 atmospheres. Higher pressures are very costly in terms of running a plant and also more expensive to construct, as thicker-walled pipes are required.
- a finely divided iron catalyst, which speeds up both the forward and reverse reactions so that the equilibrium is attained faster.

A flow diagram for the complete process of ammonia manufacture is shown in Fig 1.26. Using the conditions outlined, the percentage yield of ammonia is about 20–30 per cent in a single run. The unreacted nitrogen and hydrogen are recycled, which eventually gives yields of up to 98 per cent.

Recently, alternative catalysts have been developed that operate at lower temperatures and pressures, giving considerable savings in energy and so making the process 'greener'.

? QUESTION 23

23 In the steam-reforming stage shown in Fig 1.26, hydrogen is made by reacting methane and steam at 850 °C and 30 atmospheres pressure.

$$CH_4(g) + H_2O(g) \rightleftharpoons CO(g) + 3H_2(g)$$

$\Delta H^{\ominus} = +210 \text{ kJ mol}^{-1}$

a) Use Le Chatelier's principle to explain why such a high temperature is used.

b) What is the effect on the position of equilibrium of using an increased pressure?

c) Why is a higher pressure of 30 atmospheres used? (Hint: think about the rate of the reaction.)

? QUESTION 24

24 In the synthesis reaction vessel in Fig 1.26 ammonia is formed.

a) Explain why the iron catalyst is finely divided, rather than in large lumps.

b) What is the effect of reducing the concentration of ammonia in the equilibrium mixture by removing it in the ammonia condenser?

15 ECONOMICS, SUSTAINABLE DEVELOPMENT AND THE UK CHEMICAL INDUSTRY

The chemical industry as a whole is the largest export earner in the UK, bringing in a surplus of £8 billion to the economy. So it is not surprising that some of Britain's largest companies are in the chemical industry. They contribute towards our quality of life by supplying the products we buy and the raw materials used to make them, and by providing jobs and earning income for the nation.

The chemical principles of energy changes, yield, atom economy and rate of reactions have been described for making iron and steel and for the manufacture of ammonia. For any industrially produced chemical, the same factors influence manufacturing costs. Other economic decisions that have to be made include which raw materials to start from, and which chemical reactions to use. These are some of the economic principles applied:

- **Raw materials** These must be as cheap as possible.
- **Location of the plant** This should be as near as possible to its raw materials. It should have good rail and road links. Iron and steel plants are often close to ports for easy export and import. It should also be near an appropriately skilled workforce.
- **Type of process** Research chemists are constantly seeking processes that minimise costs.
- **Energy costs** The enormous amount of energy that iron and steel making requires accounts for a significant part of the total cost of a finished product. In all chemical production processes, energy transfer is carefully controlled to optimise energy use.
- **Safety** Before any new plant is built or any existing plant modified, the safety of the workers on the site and of the people who live in the area is given top priority. All potential risks are assessed and minimised in the design, from the possibility of a pipe leaking to the threat of an explosion.
- **Pollution control** A growing public awareness of the problems of the emission of gases and other pollutants means that industry must take steps to reduce pollution. However, this often may increase process costs.

REMEMBER THIS

One of the principles of Green Chemistry is to invest in new processes that produce fewer pollutants when manufacturing chemicals.

The costs of a process include both fixed and variable costs. The fixed costs are incurred whether 1 tonne or 1 million tonnes are manufactured. These include the capital costs of setting up the plant, the rent, council taxes and payment of any loans, and depreciation in value of the plant. Variable costs change with output. If there is no production, then these will not be incurred. Examples of variable costs include the costs of raw materials, fuel, labour and transport.

DO THE BENEFITS OF A CHEMICAL PLANT OUTWEIGH THE DISADVANTAGES?

This is a question to which there is no right answer. We are ready to enjoy the tremendous benefits that manufactured chemicals have brought us – the polymers in our clothing, the paint on our buildings, the materials that make these buildings, the fertilisers that help us to grow more food for more people, the medicinal drugs that combat so many potentially fatal diseases – the list goes on.

The chemical industry is part of our society and we have come to need many of its products. However, for every chemical process there are costs as well as benefits. But any cost/benefit analysis depends on the viewpoint of the person doing the analysis.

We can easily recognise the costs of damage to the environment through chemical pollution. There is no doubt that nowadays we are more aware than we used to be of pollution and its consequences and that chemists have improved the detection methods. We have seen that pressure to take action has led to more effective laws to control emissions of toxic and other waste products.

Now that we understand much more about the consequences of pollution, a main concern in designing new chemical plants is the possible effect on the environment. Also, as a society, we need to decide which chemicals are essential to maintain a healthy and comfortable lifestyle, and which are not.

GREEN CHEMISTRY AND SUSTAINABLE DEVELOPMENT

A response to growing public anxiety about the environmental damage caused by the manufacture of chemicals has led chemists to embrace Green Chemistry. The ideas of Green Chemistry started in the 1990s and they bring together a range of techniques and technologies that try to make chemical manufacture more environmentally friendly and more sustainable.

Sustainable development involves balancing the need for economic development, decent standards of living and respect for the environment, so that resources are available for future generations. This includes re-using and recycling materials, rather than making new products from the Earth's natural resources. The WorldWide Fund for Nature (WWF) estimates that if everyone on the planet lived as we do in the UK, then we would need the resources from three Earths rather than the one planet that we have. This is why one of the ideas of Green Chemistry – that of atom economy – is so important, because it is leading chemists to design new processes that use more of the raw materials to make useful products, rather than producing waste that must be disposed of in the environment.

Returning to iron and steel making, we said at the outset of this chapter that no other process had so transformed the world around us. Yet mining the iron ore can devastate large areas of land with its quarries and large spoil heaps. Likewise, slag from the blast furnace and steel making used to be piled high. Now it is used to make cement; so not only is the Humber Bridge made of British steel, the concrete in it is also made of British steel slag! This increases the atom economy of iron manufacture.

Chemists are at the forefront of dealing with pollution, both in its detection and in the development of control procedures. The waste gases from the blast furnace are harmful, so it is chemists who have developed ways to reduce these emissions. Carbon monoxide is a waste gas from iron and steel manufacture and is toxic to all vertebrates. Any trace of sulfur, particularly in iron sulfide ores, leads to the production of sulfur dioxide, which causes acid rain (see more about atmospheric pollutants on pages 207–209).

The chemical industry still suffers from a negative public image, despite the essential nature of many of its products. But a world without manufactured chemicals would now be unimaginable.

Recycling 1 tonne of iron, say from food cans:

- **saves 1.5 tonnes of iron ore**
- **saves 0.5 tonnes of coke**
- **saves 1.28 tonnes of solid waste**
- **reduces air emissions by 86%**
- **reduces water pollution by 76%**
- **reduces energy usage by 75%.**
- **reduces energy usage by 75%.**

SUMMARY

After studying this chapter, you should know and understand the following.

- A **molecular formula** tells you how many atoms of each element there are in a molecule of the compound.
- The term **formula** applies to ionic compounds as well as to molecules. It is the number of atoms of the different elements that make up the smallest complete unit of a substance.
- In the **relative atomic mass** (A_r), scale, the average mass of the atom of an element (taking into account all of its isotopes and their abundances) is compared to 1/12 the mass of one atom of carbon-12.
- The **relative molecular mass** (M_r) is calculated using the relative atomic mass scale using the same standard, carbon-12. The A_r values of each atom in the molecule are added together to give the M_r. When dealing with ionic compounds we use the term **relative formula mass**.
- The **amount** is the number of particles (atoms, molecules, ions, electrons, etc.) in a substance. The unit of amount is the mole (mol).
- The **mole** is defined as the amount of substance that contains as many particles as there are atoms in exactly 12 g of carbon-12.
- The relative atomic mass in grams of any element contains one mole of atoms.

$$\text{Amount in moles} = \frac{\text{mass in grams}}{\text{mass of one mole (in grams)}}$$

- The mass of one mole is also known as the **molar mass**.
- The **Avogadro constant** is the number of atoms or molecules in one mole which is 6.02×10^{23}.
- The **theoretical yield** is the maximum amount obtainable if all the reactants are converted into products. This is calculated from the balanced equation.
- $$\text{Percentage yield} = \frac{\text{actual yield}}{\text{theoretical yield}} \times 100\%$$
- **Limiting reactant** is the reactant that determines the maximum amount of product that can be made. The other reactants are said to be in excess.
- **Atom economy** is a measure of how many of the atoms in the reactants actually end up in desirable products.

$$\text{Atom economy} = \frac{\text{mass of desired product}}{\text{total mass of reactants}} \times 100\%$$

- The rate of reaction is affected by temperature, pressure, concentration, surface area of reactants, and catalysts.
- The **enthalpy change**, ΔH, is a measure of the transfer of energy into or out of a reacting system at constant pressure.
- **Exothermic reactions** give out energy and cause the temperature of the surroundings to rise. ΔH is negative.
- **Endothermic reactions** take in energy and cause the temperature of the surroundings to fall. ΔH is positive.
- Iron ore, coke, limestone and air are the raw materials used in a blast furnace. The iron ore is reduced to iron by carbon monoxide. A high carbon content makes the iron very brittle.
- Steels are alloys of iron with carbon and other elements such as manganese and vanadium. The percentage of the elements alloyed with iron gives a particular steel its unique properties.
- A **reversible reaction** is one that can take place in either direction.
- A **dynamic equilibrium** is reached when the forward and reverse reactions occur at the same rate. This will only occur in a closed system.
- Le Chatelier's principle states that the position of the equilibrium of a system changes to minimise the effect of any imposed change in conditions.
- The Haber process is used to manufacture ammonia. Applying Le Chatelier's principle gives industrial chemists an understanding of the effect of changing temperature and pressure on the equilibrium position.

Practice questions and a How Science Works assignment for this chapter are available at www.collinseducation.co.uk/CAS

2

The nucleus and radioactivity

2 THE NUCLEUS AND RADIOACTIVITY

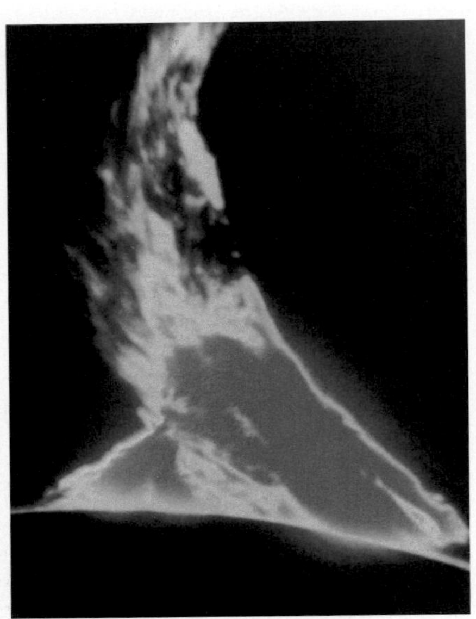

Jet of glowing gas. Evidence of nuclear fusion at the Sun's core is given by jets of glowing gas on the surface of the Sun

Birth of the elements

The eight most abundant elements of the Earth's crust are oxygen, silicon, aluminium, iron, calcium, sodium, potassium and magnesium. Hydrogen is the ninth, at less than 1 per cent. Yet 92 per cent of all atoms in the Universe are hydrogen, 7 per cent are helium, and all the other elements comprise only 1 per cent. So the Earth is very rich in its range of elements, much richer than some other parts of the Universe, and we may ask: why are the elements so unevenly distributed?

The reason lies in the way the Universe, and later the Earth, first began. It is generally thought that the Universe started with tremendous fury in the Big Bang, just less than 14 thousand million years ago, when a super-hot, super-dense 'soup' of sub-atomic particles exploded with unimaginable force. It was then that hydrogen atoms were first created. Some of these fused to give helium, and so a second element was born.

After about a thousand million years, stars began to form. Gaseous hydrogen with a little helium came together in local areas, contracting and becoming denser. The temperature rose to levels at which other light elements could form, and even more heat was produced. Stars continue to be born in this way from the dispersed material of the Universe.

The heavier elements appeared only in the relatively few very massive stars. Those with a much larger mass than the Sun became unstable after a few million years and exploded, propelling atoms out into the cosmos, to become the raw material of a second generation of stars with the heavier elements already formed.

It is thought that the Solar System was formed from such recycled material about 4500 million years ago. The Sun itself is a gigantic fusion reactor, its heat and light helping to sustain life on Earth. The Earth contains some 90 light and heavy elements from the remains of earlier stars. In its intensely hot core, it retains some of the heat from when it was formed. More heat is produced by radioactive elements, which decay and release energy. In fact, as well as the heat of the core and from the Sun, this decay process is significant in maintaining the Earth's present-day heat balance.

A HISTORY OF IDEAS ABOUT THE ELEMENTS

In 580 BC, the Ancient Greek philosopher Thales suggested that water was the fundamental 'element' from which all matter in the Universe was composed. Other Ancient Greek thinkers added earth, air and fire, proposing that these four elements made up the world. 200 years later Aristotle added a fifth, 'aether', making up the heavens. This idea persisted for almost 2000 years until Robert Boyle published a book called *The Sceptical Chymist* in 1661.

Boyle's book was a turning point in chemistry. It put forward the first modern concept of an **element** as something that cannot be changed into

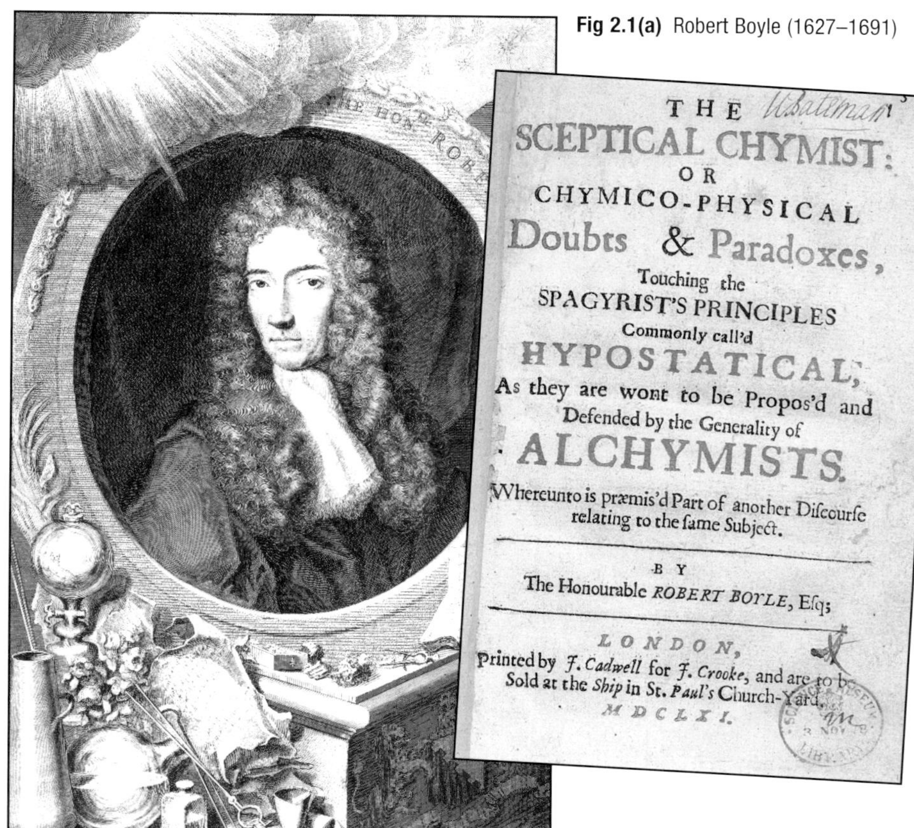
THE
SCEPTICAL CHYMIST:
OR
CHYMICO-PHYSICAL
Doubts & Paradoxes,
Touching the
SPAGYRIST'S PRINCIPLES
Commonly call'd
HYPOSTATICAL;
As they are wont to be Propos'd and
Defended by the Generality of
ALCHYMISTS.
Whereunto is præmis'd Part of another Discourse
relating to the same Subject.

BY
The Honourable ROBERT BOYLE, Esq;

LONDON,
Printed by J. Cadwell for J. Crooke, and are to be
Sold at the Ship in St. Paul's Church-Yard.
MDCLXI.

Fig 2.1(a) Robert Boyle (1627–1691)

Fig 2.1(b) The title page of Boyle's book, *The Sceptical Chymist*. In it he first introduced the idea of the 'chemist'. He also proposed the idea that elements cannot be changed into anything simpler and that they could be discovered by experimentation

Fig 2.2 John Dalton (1766–1844)

anything simpler. The book also marked the beginning of chemistry as an experimental science, since Boyle was urging chemists to carry out practical investigations rather than merely observe, think and make deductions, as the Greeks had done.

His advice was heeded. In 1803, the English chemist, John Dalton, summed up 150 years of progress through experiment with his **atomic theory**. He used the Greek word 'atomos', which means 'cannot be cut', to describe the particles that make up elements. The idea that all matter was made up of tiny, indivisible particles had been proposed in 450 BC, again by an Ancient Greek, Democritus. But now, Dalton could support it with experimental results. These are the main points of his theory:

- Atoms are indivisible and indestructible.
- All atoms of the same element have identical mass and identical chemical properties.
- When atoms react, they join together to form 'compound atoms' (now called molecules).

Today, we still hold Dalton's view of an element as a substance composed of only one type of atom. So the element iron is made up of only iron atoms, which are different from copper atoms, which make up the element copper.

We now know that an atom can be subdivided into smaller parts called **sub-atomic particles**, discovered after Dalton's time. But the atom is still the smallest particle with all the *chemical* properties of an element, and Dalton's atomic theory provided the foundation on which other scientists have built.

Fig 2.3 The symbols Dalton used to represent the elements known at the time

1 BUILDING MODELS OF THE ATOM

A **model** is the scientist's attempt to explain observations. Dalton's model of the atom suggested that atoms could not be sub-divided into smaller particles. This model was accepted until the end of the 19th century, when new discoveries and observations meant the model had to be changed.

THE DISCOVERY OF THE ELECTRON

In 1897, J. J. Thomson was working at the Cavendish Laboratories in Cambridge. He was experimenting with a ray that was emitted from the **cathode** (the conducting material connected to the negative pole of a battery). This ray, called the **cathode ray**, had been discovered in the 1870s when an electric current was passed through a gas at low pressure (Fig. 2.4).

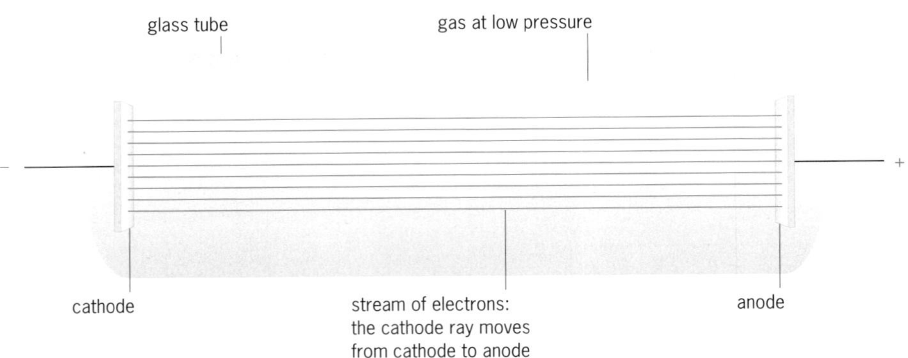

Fig 2.4 A simplified diagram of a cathode ray tube. The high voltage between the cathode and anode strips electrons from the atoms of the gas

REMEMBER THIS

The cathode ray is used to produce a picture on older televisions, which is why the back of the set is so large. Flat screens do not use cathode rays.

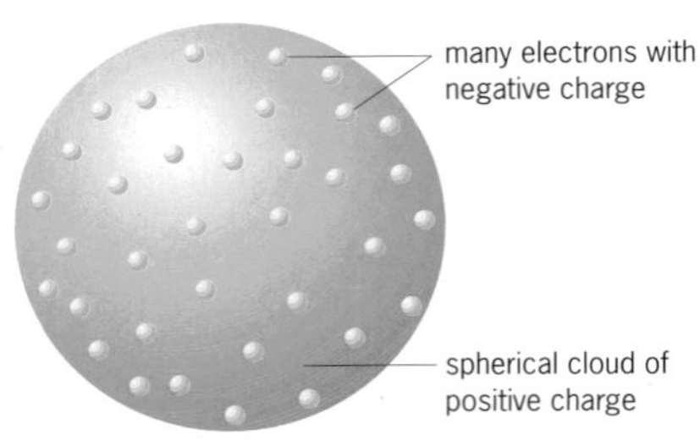

Fig 2.5 Thomson's plum pudding model of the atom. Negative and positive charges cancel out

Thomson found evidence that cathode rays were made up of particles, which he called **electrons**. He calculated the mass of an electron to be 2000 times less than the mass of a hydrogen atom and, as they were attracted to the positive **anode**, he knew they must be negatively charged. (In 1906, this work gained him a Nobel Prize.) Since these particles must be being produced from the atoms inside the apparatus, he realised he had discovered a **sub-atomic particle**. This meant that a new model of the atom was required to explain the observation.

As no other sub-atomic particles had been discovered, Thomson's model of the atom was made up of negatively charged electrons. It was known that atoms had no net charge and were electrically neutral, so the electrons were thought by Thomson to be embedded in a sphere of positive charge, which had no mass. To account for the mass of atoms, Thomson assumed that they must all have thousands of electrons. His model was known as the 'plum pudding' model because the electrons were pictured like raisins in a pudding.

QUESTION 1

1 In Thomson's model of the atom, how many electrons would be needed to account for the mass of the hydrogen atom?

RUTHERFORD'S MODEL OF THE ATOM

Ernest Rutherford, a New Zealander working under Thomson at Cambridge, began investigating a newly discovered phenomenon, **radioactivity**. Over several years of work, he found that the radioactivity, which a French scientist called Henri Becquerel had first observed (see the How Science Works box on the facing page), was composed of three main types of radiation. Rutherford named the first two **alpha** (α) and **beta** (β) radiation. When the third was eventually discovered, it was called **gamma** (γ), see Fig 2.6.

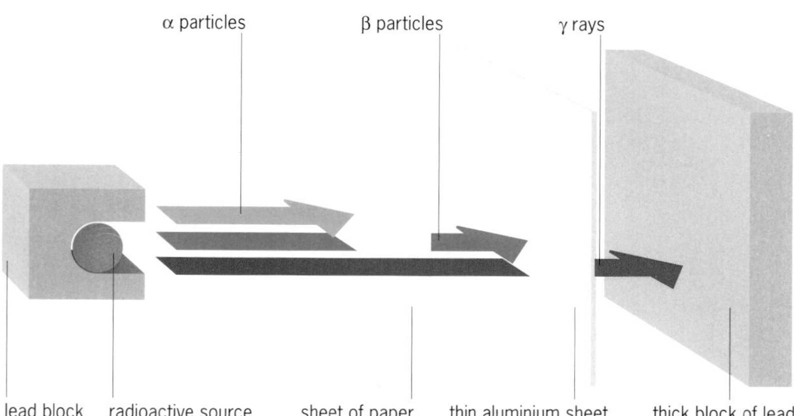

Fig 2.6 The different penetrating properties of the three types of radiations. A lead block has to be about 10 cm thick to halt gamma rays

HOW SCIENCE WORKS

The discovery of radioactivity

While Thomson was experimenting on cathode rays, another momentous discovery was being made. The Frenchman Henri Becquerel was investigating crystals that phosphoresce – that is, chemicals that glow in the dark after being exposed to sunlight. Becquerel assumed that sunlight was causing the crystals to give out X-rays, which had just been discovered (see page 35). X-rays were not understood then, which is why they were called 'X'. But they were known to make a dark, blurred image, called 'fogging', on photographic film, even when it was sealed in opaque (light-proof) black paper.

Becquerel decided to test his assumption. He put crystals of a uranium salt on top of a photographic film, which was sealed in black paper that kept out all light. He let the Sun shine on the crystals and, as he expected, when the film was developed, it had become fogged.

On 26 February 1896, clouds shut out the Sun, so Becquerel just put the crystals, unexposed to sunlight, in a drawer on top of some sealed photographic film. Three days later, he decided to develop the film anyway, just to check that there was no image. To his amazement, he found a clear area of darkening on the film below the crystals (Fig 2.7). He concluded that some unknown kind of radiation that is not dependent on the Sun and that could pass through the black paper must have come from the crystals. It took the work of others to discover more about this radiation.

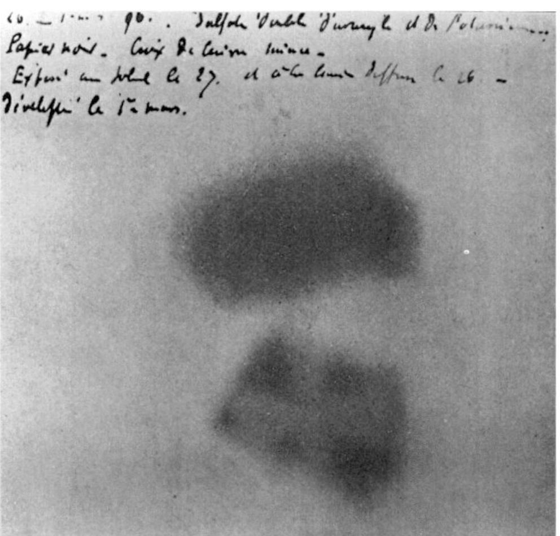

Fig 2.7 Becquerel's photographic film on which he had put crystals unexposed to sunlight: the first recorded evidence of radioactivity

With her husband Pierre, Marie Curie began to study this phenomenon. She investigated other uranium compounds and found that they, too, gave off this new radiation. She called it radioactivity. (Radioactivity is the spontaneous break-up of atomic nuclei, giving out rays called radiation.)

The Curies observed that pitchblende, an ore known to contain uranium, was much more radioactive than they expected from the uranium in the ore alone. The Curies later discovered two new radioactive elements, radium and polonium. It was radium that gave pitchblende its high radioactivity. The radiation of radium had so much energy that it burned the skin, and while uranium fogged film after a few hours, radium did so instantly. Marie Curie also discovered another radioactive element, thorium.

Fig 2.8 Marie Curie (1867–1934) was the first person to receive two Nobel prizes. The first was for physics (1903), which she shared with her husband and Becquerel for work on radioactivity. The second was for chemistry (1911) for her discovery of radium and polonium (named after her homeland, Poland). At the time of her discoveries it was not realised that the radiation to which she was exposed would cause the leukaemia that was to end her life

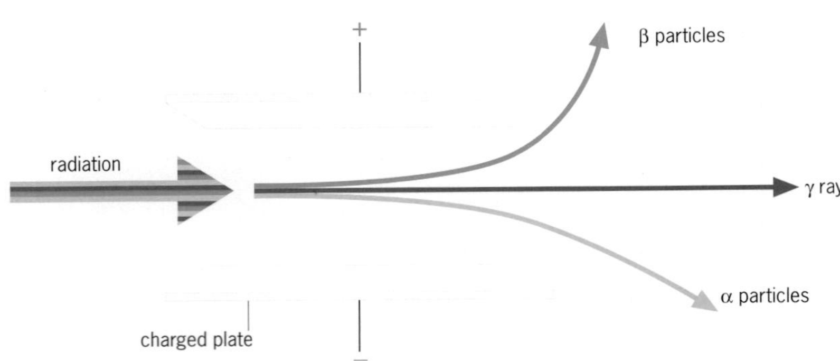

Fig 2.9 Deflection of the three types of radiations by charged plates

> **? QUESTION 2**
>
> 2 Fig 2.9 shows how the three different types of radiation are affected by an electric field.
> **a)** What do you deduce about the charge on the different types of radiation?
> **b)** Why do you think the beta particles are deflected so much more than the alpha particles?

Rutherford showed that alpha radiation was composed of particles that were four times heavier than hydrogen. In fact, they are the same as positively charged helium atoms. (We refer to them as helium nuclei.) The beta particles were soon found to be very high-energy electrons, which of course are negatively charged. Gamma rays were identified later as electromagnetic rays similar to X-rays (see page 41), but of much higher energy. Being electromagnetic rays, gamma rays were found to have no particle properties, and therefore no mass. You can find out more about radioactivity in the How Science Works box on page 31.

RUTHERFORD'S SCATTERING EXPERIMENT

Rutherford moved to Manchester University to set up his own research group, which included Hans Geiger and Ernest Marsden. In their investigations into the structure of the atom, they fired alpha particles at gold foil, using the equipment shown in Fig 2.10(a). From Thomson's model of the atom, it was expected that a few alpha particles would be deflected slightly from a straight route when they passed through foils. Even the thinnest foils were about 2000 atoms thick, and some of the alpha particles were sure to collide with electrons. Marsden was asked to see if any alpha particles were scattered through large angles. Rutherford said, 'I may tell you that I did not believe they would be, since we knew that the alpha particle was a very fast, massive particle, with a great deal of energy.'

But then, a few days later, Geiger announced that some alpha particles had been recorded as bouncing back from the gold foil. Rutherford recalled, 'It was quite the most incredible event that has happened to me in my life. It was almost as incredible as if you fired a fifteen inch [artillery] shell at a piece of tissue paper, and it came back and hit you.' The results of the scattering experiment are a good example of how science works. To check the reliability of the results, the experiment was repeated many times.

It was now clear to Rutherford that Thomson's model of the atom could not be correct. If it were, then the alpha particles should have shot through the foil with hardly any deflection. But one in 8000 had apparently been halted in its tracks and deflected back. Most particles passed through without any deflection, and Rutherford concluded that they were travelling through empty space. The positively charged alpha particles deflected through larger angles must have encountered (and been repelled by) another positive charge of considerable mass, which he called the **nucleus**.

> **? QUESTION 3**
>
> 3 A tennis ball has a radius of 3.1 cm. Assume it represents a nucleus. What is the radius of the atom on this scale to the nearest kilometre?

From the angles of deflection and other data, Rutherford calculated that the radius of the gold nucleus was 10^{-14} metre. (We now know that the radius is closer to 10^{-15} metre.) He calculated, too, that the radius of the atom was approximately 10^{-10} metre. This means that the nucleus was 100 000 times smaller in diameter than the atom – an astonishing difference. If a full stop on this page represented the nucleus, then the outer limit of the atom would be about 25 metres away. No wonder that only a very few alpha particles came back.

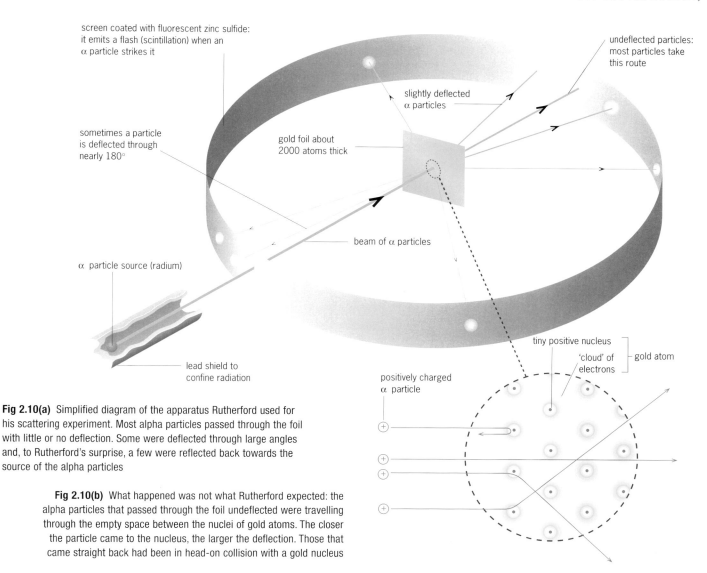

screen coated with fluorescent zinc sulfide: it emits a flash (scintillation) when an α particle strikes it

undeflected particles: most particles take this route

slightly deflected α particles

sometimes a particle is deflected through nearly 180°

gold foil about 2000 atoms thick

beam of α particles

α particle source (radium)

lead shield to confine radiation

tiny positive nucleus

'cloud' of electrons

gold atom

positively charged α particle

Fig 2.10(a) Simplified diagram of the apparatus Rutherford used for his scattering experiment. Most alpha particles passed through the foil with little or no deflection. Some were deflected through large angles and, to Rutherford's surprise, a few were reflected back towards the source of the alpha particles

Fig 2.10(b) What happened was not what Rutherford expected: the alpha particles that passed through the foil undeflected were travelling through the empty space between the nuclei of gold atoms. The closer the particle came to the nucleus, the larger the deflection. Those that came straight back had been in head-on collision with a gold nucleus

Rutherford proposes his model

Rutherford now proposed his model of the atom – a minute, central nucleus containing all the positive charge of the atom and almost all of the mass, surrounded by empty space in which electrons orbited the nucleus, rather like planets round the Sun.

Fig 2.11(a) Ernest Rutherford (1871–1937) in his laboratory

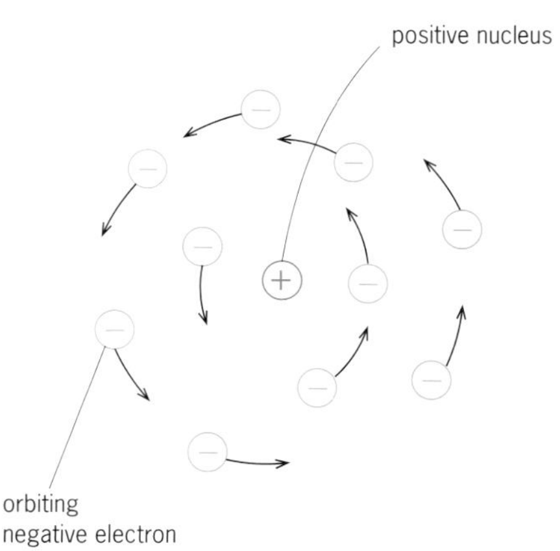

Fig 2.11(b) Rutherford's model of the atom proposed after the scattering experiment shown in Fig 2.10(a)

This model of the atom raised some difficult questions. In particular, since opposite charges attract, why did the electrons not fall into the nucleus? The Danish scientist, Niels Bohr, came up with the answer to this question, and you can read more about this in Chapter 3. The idea of the minute, central, positive nucleus is fundamental to all models of the atom accepted today.

PROTONS AND NEUTRONS

The proton was the next sub-atomic particle to be discovered, and again the alpha particle was the tool used in research. When alpha particles were fired through hydrogen gas, positive particles emerged which were about the same mass as the hydrogen atom. In 1919, Rutherford said that these same particles could be knocked out of other atoms, suggesting that they must be present in the other atoms. He gave the name **protons** to these positive particles. The model of a nucleus made up of protons was now taking shape.

A year later, Rutherford suggested that protons were not the only type of particle in the nucleus, and that there must be particles of equal mass but neutral charge in the nucleus as well. In suggesting this, Rutherford was almost a lone voice in the scientific community. Then, in 1932, James Chadwick, another member of his research team, was able to show that this neutral particle did exist.

Chadwick fired alpha particles at beryllium atoms. Instruments that detected charged particles were unable to detect any radiation, as in Fig 2.12(a). Yet when paraffin wax was put between the beryllium and the detector, a shower of protons was detected, as in Fig 2.12(b). Rutherford guessed that the protons were coming from the paraffin because some sort of radiation was hitting it. He said this radiation was like the effect of an invisible man who cannot be seen directly, but who is known to be there because he collides with other people in a crowd.

The invisible particle in the radiation of Chadwick's experiment – with no charge and with its mass equal to that of the proton – was named the **neutron**. In this way, protons and neutrons became known to scientists.

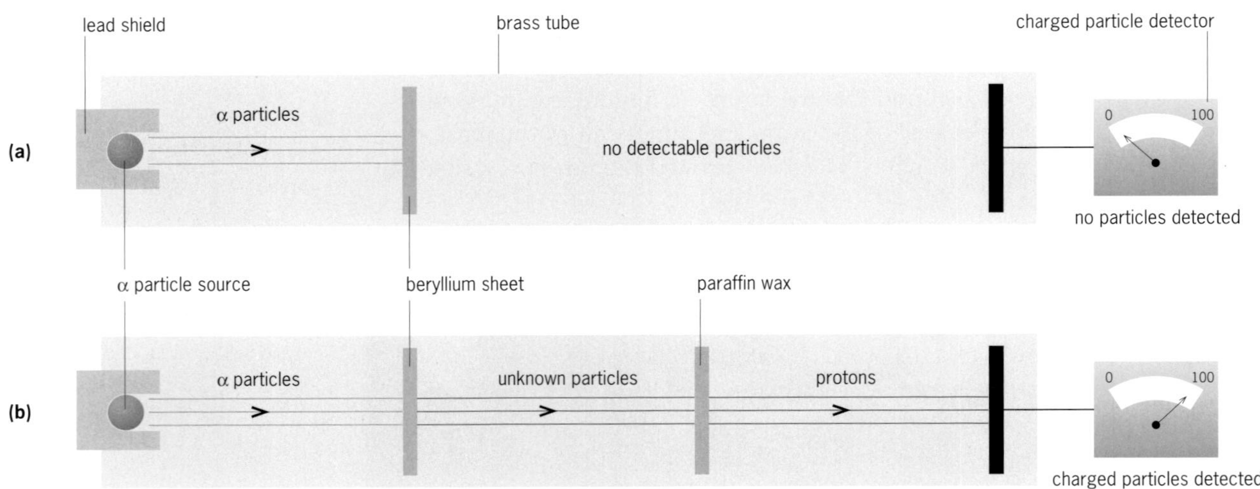

Fig 2.12 A simplified diagram of Chadwick's apparatus. The unknown particles were neutrons, which knocked protons from the paraffin wax

2 DESCRIBING THE ATOM

ATOMIC NUMBER (PROTON NUMBER)

Five years before Rutherford announced the discovery of the proton, Henry Moseley, another of his research team, had been working with X-rays. These were discovered in 1895 by Wilhelm Röntgen in Germany. Röntgen had been trying to make substances fluoresce by bombarding them with cathode rays (streams of high-energy electrons). Fluorescing substances absorb energy of one wavelength, usually outside the visible spectrum, and give out energy at another wavelength that is visible. (For more information about the spectrum, refer to Chapter 3.) Some substances emitted what Röntgen called an X-ray. The X-ray could penetrate some matter, and would blacken photographic film, as Röntgen was able to show by making an image of his wife's hand.

Moseley bombarded different metal targets with cathode rays (Fig 2.14), and measured the frequency of the X-rays that emerged from the anode. Frequency increased with the increase in mass of the metal atom.

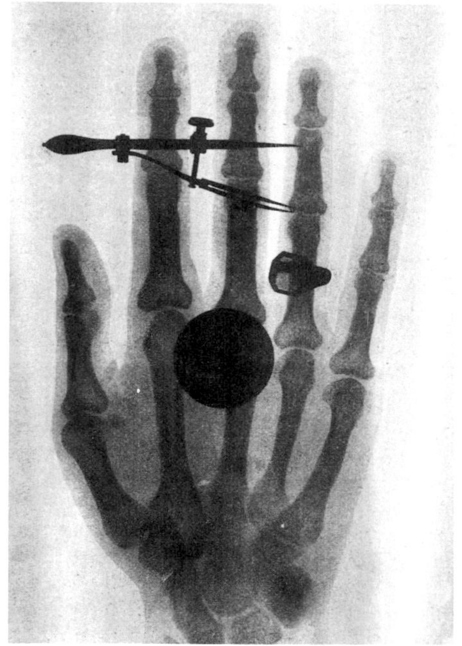

Fig 2.13 The first X-ray image of a human, taken by Röntgen of his wife's hand, showing her wedding ring, a coin and a compass

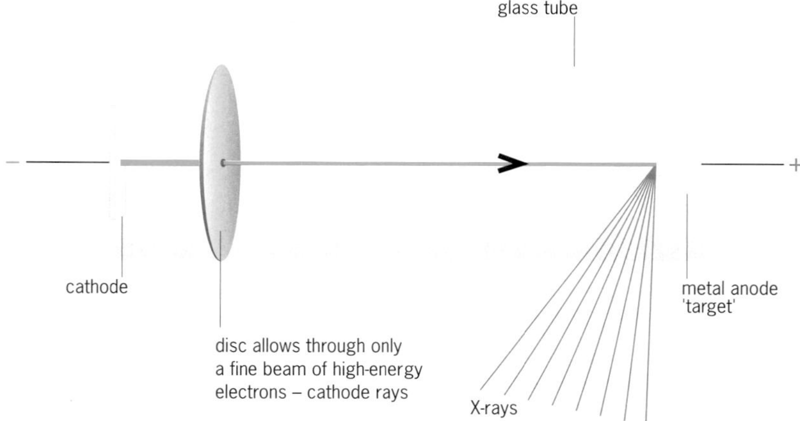

Fig 2.14 Moseley's experiment, in which he bombarded different metal targets with a beam of high-energy electrons – cathode rays – and measured the frequency of X-rays produced

From Moseley's results, Rutherford calculated the positive charge on the nucleus, which he called the **atomic number**:

The atomic number is the number of protons in the atom.

It has the symbol **Z**, and is also known as the **proton number**.

The atomic number tells us the following:

- What the element is: only atoms of the same element have the same atomic number.
- The element's numbered position in the Periodic Table: the Periodic Table is arranged in atomic number order (see Chapter 5).
- The number of electrons in a neutral atom: the charge on an electron is equal, but opposite, to the charge on a proton.

Information we now know about the three fundamental particles in an atom is summarised in Table 2.1.

? QUESTION 4

4 Sodium is the 11th element in the Periodic Table.

a) Work out its atomic number, the number of protons in its nucleus and the number of electrons that orbit outside the nucleus.

b) How will your answer to **a)** differ if you now consider a sodium ion, Na^+?

Table 2.1 Some properties of the three main sub-atomic particles

Particle	Symbol	Relative mass	Relative charge	Location
proton	$^{1}_{1}p$	1	+1	in nucleus
neutron	$^{1}_{0}n$	1	0	in nucleus
electron	$^{0}_{-1}e$	$\frac{1}{1850}$ or 0.0005[a]	−1	orbiting nucleus

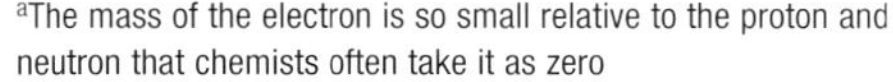

[a]The mass of the electron is so small relative to the proton and neutron that chemists often take it as zero

MASS NUMBERS (NUCLEON NUMBERS) AND ISOTOPES

The mass of the atom is almost entirely the mass of the protons and neutrons in its nucleus. Even the 92 electrons in an atom of uranium have hardly any effect on its mass. Since each proton and neutron contributes equally to an atom's mass, we call the total of both in an atom the **mass number**.

The mass number is the total number of protons and neutrons in the atom.

It has the symbol **A**.

As protons and neutrons are constituents of the nucleus, they are referred to as **nucleons**. So A is also known as the **nucleon number**. (Note that the mass number is not the *mass* of a nucleus.)

$$\text{Mass number} = \text{number of protons} + \text{number of neutrons}$$
$$A = Z + N$$

Chemists summarise information about an atom in this way:

$${}^{A}_{Z}\text{X}$$

mass number — A; atomic number — Z; X — element symbol

The atomic number Z is always the same for a particular element, and it indicates the position of the element in the Periodic Table. It is often left out of symbol notation, since the element symbol itself identifies the atom.

EXAMPLE

Q An atom of phosphorus has 16 neutrons. Write the full symbol notation for this atom.

A First you need to find out the atomic number of phosphorus. The atomic number, Z, for phosphorus is 15, and its symbol is P.

$A = Z + N$, so $A = 15 + 16 = 31$

The full symbol notation is ${}^{31}_{15}\text{P}$.

It specifies a particular atomic number and mass number, and hence represents an atom with a particular number of protons and a particular number of neutrons.

Q How many protons, neutrons and electrons does the following atom have?

Fluorine, ${}^{19}_{9}\text{F}$

A $Z = 9$, so the number of protons = 9. The atom is neutral, so the number of electrons = 9.

$A = 19$, so $Z + N = 19$.

Since $Z = 9$, $N = 10$.

The number of neutrons (N) = 10.

Now try question 5 in the margin.

? QUESTION 5

5 Work out the name, mass number, atomic number, and number of protons, electrons and neutrons in these atoms:

a) ${}^{56}_{26}\text{Fe}$

b) ${}^{200}_{80}\text{Hg}$

The **mass spectrometer** made it possible to determine the masses of atoms very accurately (see Chapter 11), and it became clear that not all atoms of the same element had the same mass. This meant that there had to be a change to Dalton's atomic theory, which said that all atoms of an element had an identical mass.

The accurate masses gave rise to the idea of the **isotope**. In Greek, isotope means 'same place', but it is more helpful to think of it as meaning 'alternative'. Some elements, such as fluorine, have only one nuclear arrangement so there are no isotopes of fluorine. The element chlorine, on the other hand, has two isotopes because there are two possible nuclear arrangements, each with a different mass.

Table 2.2 shows that isotopes are atoms of the same element and so have the same atomic number. The only difference between the isotopes is their mass number, which indicates that the nucleus of one isotope has more neutrons than that of another. Thus:

Isotopes are atoms with the same atomic number, but different mass numbers.

Isotopes of the same element have identical chemical properties because they have the same number of electrons. It is the number of electrons that determines the chemical properties of an element. However, their physical properties differ slightly: heavier isotopes are denser and have higher boiling points.

Table 2.2 The two naturally occurring isotopes of chlorine

Isotope	$^{35}_{17}Cl$	$^{37}_{17}Cl$
mass number, *A*	35	37
atomic number, *Z*	17	17
number of protons	17	17
number of neutrons, *N*	18	20
number of electrons	17	17

RELATIVE ISOTOPIC MASS

Chapter 1 explained that the relative atomic mass, A_r, compares the masses of atoms to the mass of one atom of carbon-12, an isotope of carbon whose relative atomic mass is taken as exactly 12. Its symbol is $^{12}_{6}C$, or ^{12}C, because, as we have seen, once the element is known, the atomic number is fixed. In nature, most elements are composed of isotopes.

The relative mass of an isotope (its relative isotopic mass) is the mass of one atom of that isotope compared to $\frac{1}{12}$ the mass of one atom of carbon-12.

Some elements do not have any isotopes so all their atoms have the same number of neutrons. For these elements the mass number is very nearly the same as the relative atomic mass. For example, naturally occurring fluorine has just one type of atom, ^{19}F, and the A_r of fluorine is 18.998 40. But for all except the most accurate work, 19 is used.

QUESTION 6

6 Most of the carbon in nature is the isotope $^{12}_{6}C$. But two other isotopes exist, $^{13}_{6}C$ and $^{14}_{6}C$. $^{14}_{6}C$ is radioactive and is used to estimate the age of very old objects. Draw and complete a table similar to Table 2.2 for the isotopes of carbon.

RELATIVE ATOMIC MASS

The two isotopes of chlorine are ^{35}Cl and ^{37}Cl. If these two isotopes had the same abundance, then the A_r of chlorine would be 36. But ^{35}Cl has a 75 per cent abundance and ^{37}Cl has a 25 per cent abundance.

This means that for every 100 atoms, 75 have a relative isotopic mass of 35, and 25 have a relative isotopic mass of 37.

The A_r is calculated as follows.

$$A_r \text{ chlorine} = \frac{(75 \times 35)}{100} + \frac{(25 \times 37)}{100}$$

$$= \frac{3550}{100}$$

$$= 35.5$$

This leads to the statement:

The relative atomic mass of an element is the average mass of the atom (taking into account all of its isotopes and their abundance) compared to $\frac{1}{12}$ the mass of one atom of carbon-12.

REMEMBER THIS

Carbon-12 is another way of representing the isotope ^{12}C.

QUESTION 7

7 Study Table 2.3 and work out the A_r of bromine and magnesium.

Table 2.3 Isotopes of bromine and magnesium

Element	Isotope	Abundance
bromine	^{79}Br	50%
	^{81}Br	50%
magnesium	^{24}Mg	78.6%
	^{25}Mg	10.1%
	^{26}Mg	11.3%

STRETCH AND CHALLENGE

Nuclear fusion

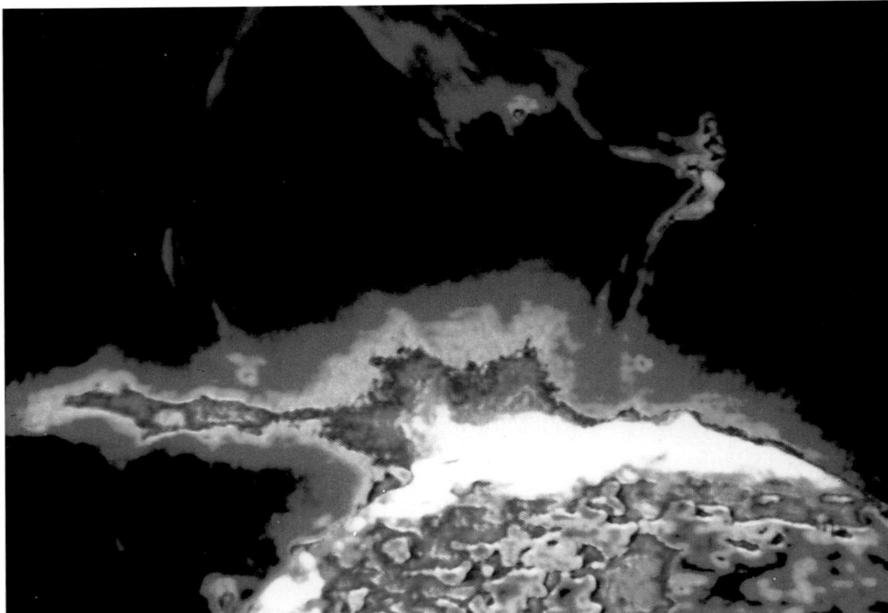

Fig 2.15 The Sun can be thought of as an immense fusion reactor that provides energy which reaches all parts of the Solar System

The Sun's energy comes from the nucleus of atoms, or rather from countless billions of nuclei. When atoms could be weighed very accurately, their mass was found to be less than the sum of the masses of their sub-atomic particles. The missing mass is called the **mass defect**. Einstein explained this in perhaps the most famous equation of all time,

$$E = mc^2$$

E is the energy in joules, m is the missing mass in kilograms and c is the speed of light in metres per second (3×10^8 m s^{-1}). The missing mass is released as energy when nucleons (protons and neutrons) fuse to form a nucleus. This energy is called the **binding energy**. The speed of light is an extremely large number, which is squared, so just a small mass produces a very large amount of energy.

Let's consider deuterium, which is an isotope of hydrogen ($^2_1$H). It is made up of one proton and one neutron.

The 'missing' mass of its nucleus is 3.965×10^{-30} kg.

Energy released when the proton and neutron fuse = mc^2,

$= 0.003\,965 \times 10^{-27} \times (3.00 \times 10^8)^2$
$= 3.57 \times 10^{-13}$ J (to 3 sig.fig.)

This result is a very small amount of energy, but it is only for one nucleus. Consider a mole (6.02×10^{23}) of protons and neutrons which fuse; the energy released is 2.15×10^{11} J (215 million kJ). Burning one mole of carbon in oxygen gives a mere 394 kJ. The energy released in chemical reactions is tiny compared to the awesome amount released from nuclear reactions.

NUCLEAR FUSION AS AN ENERGY SOURCE?

It is the dream of scientists to provide a cheap source of power that has harmless by-products, by fusing nuclei and harnessing the vast amounts of energy released. The reaction that is most promising involves deuterium ($^2_1$H) and another isotope of hydrogen, tritium ($^3_1$H). Deuterium is abundant on the Earth, with three in every 20 000 hydrogen atoms in water being deuterium atoms. Tritium is relatively easily manufactured from lithium, also an abundant element.

$$^2_1\text{H} + {}^3_1\text{H} \rightarrow {}^4_2\text{He} + {}^1_0\text{n} + \text{energy}$$

The energy released in this fusion reaction is 1700 million kJ per mole.

In the reaction, the atoms are stripped of their electrons to form a state of matter called a plasma, and heated to temperatures in excess of 100 million °C.

There have been a number of experimental reactors, called tokamaks, where the hot gas is confined in a torus-shaped vessel using a magnetic field. Now the latest experimental fusion reactor, Iter (International Thermonuclear Experimental Reactor), is to be built in France and will be operational by 2017. The hope is that Iter will produce more energy than it consumes. If it does, Iter may be the forerunner of many fusion reactors.

Fig 2.16 An experimental tokamak fusion reactor and (below) a deuterium plasma inside it. For a fusion reactor to work, it needs to keep a plasma at a high enough temperature and density for long enough for nuclei to fuse

SCIENCE IN CONTEXT

Elements formed in stars

The nuclear fusion reaction in Question 8 (below) is just one of many reactions of large stars in which the nuclei of light elements fuse to make heavier elements. All these reactions give out energy. But when iron nuclei fuse, they *take in* energy rather than release it, and this fusion marks the end for a star. Iron builds up right at its centre. When fusing, iron nuclei take up too much of the star's energy, the core collapses with incredible speed and force, and its density increases hundreds of millions of times.

Then, the collapsing matter rebounds. In a stupendous explosion, the outer layers of the dying star are blown off into the cosmos and, for a short time, we see the star as a supernova, shining as brightly as an entire galaxy. Look at Fig 2.17 to see Supernova 1987A as its matter was blasted into space, to become part of a new generation of stars.

Our own Sun is a second-generation star containing recycled remnants from past supernovae, and the many elements found on Earth come from giant stars that exploded as supernovae billions of years ago.

Fig 2.17 On 23 February 1987, a brilliant star-like object about 170 000 light years away was first seen from the Earth. A supergiant star had collapsed in on itself and then exploded. It became known as Supernova 1987A. The photo shows the Tarantula nebula, an enormous cloud of ionised gas, and below it, the exploding star

3 DESCRIBING NUCLEAR REACTIONS

REPRESENTING SUB-ATOMIC PARTICLES

We have used the standard notation for mass number, symbol *A*, and atomic number, symbol *Z*, to represent atoms of elements and their isotopes. Soon we will be looking at nuclear reactions using this notation. These reactions involve atoms, and some also include sub-atomic particles. So that we can balance nuclear equations, we use a system of notation for the sub-atomic particles that matches the one we use for atoms, shown in Table 2.4.

The diagram above right explains what the numbers mean for the sub-atomic particles in Table 2.4.

We know that protons and neutrons determine the mass numbers and atomic numbers of atoms. The information above shows us that the numbers given to the sub-atomic particles are connected with these mass numbers and atomic numbers, and we will find this is useful in writing nuclear equations.

$^{1}_{1}p$: upper 1 — contributes to the mass number of an element; lower 1 — contributes to the atomic number of an element

$^{1}_{0}n$: upper 1 — contributes to the mass number of an element; lower 0 — does not contribute to an element's atomic number

$^{0}_{-1}e$: upper 0 — the negligible mass of an electron: nearly zero; lower −1 — the charge on an electron

Table 2.4 Properties of sub-atomic particles

Particle	Symbol	Relative mass	Relative charge	Location
proton	$^{1}_{1}p$	1	+1	in nucleus
neutron	$^{1}_{0}n$	1	0	in nucleus
electron	$^{0}_{-1}e$	0.000 5	−1	orbiting nucleus

BALANCING NUCLEAR EQUATIONS

The following is an example of a **nuclear equation**:

$$^{1}_{1}H + ^{2}_{1}H \rightarrow ^{3}_{2}He + \text{energy}$$

hydrogen + deuterium → helium isotope

Notice that the mass numbers and the atomic numbers balance. In any nuclear equation, atomic numbers and mass numbers must balance in this way. The major fusion reaction of the Sun is thought to be between nuclei of deuterium and nuclei of another hydrogen isotope, tritium, to give helium and a neutron.

$$^{2}_{1}H + ^{3}_{1}H \rightarrow ^{4}_{2}He + ^{1}_{0}n + \text{energy}$$

? QUESTION 8

8 Our Sun is a relatively small star. In stars that are about 30 times larger than the Sun, much higher temperatures allow helium to fuse to form other elements. Complete this nuclear equation and work out the identity of element X.

$^{4}_{?}He + ^{4}_{?}He \rightarrow ^{?}_{?}X + \text{energy}$

4 THREE RADIATIONS

Radioactivity is the spontaneous breakdown (disintegration) of a nucleus and the emission of radiation. 'Spontaneous' means something that happens without anything seeming to cause it.

Some naturally occurring isotopes are unstable and will break down. As a result, they release at least one of the three principal types of radiation mentioned earlier in the chapter – alpha (α), beta (β), and gamma (γ) radiation. All three can knock electrons off atoms they collide with, and so they are called **ionising radiations**.

Table 2.5 Some properties of the three main types of radiation (See also Fig 2.6, page 31)

Type	Nature	Speed	Charge	Relative mass	Distance travelled in air	Stopped by
α	helium nucleus, He^{2+} or $^{4}_{2}He$, i.e. 2 protons + 2 neutrons	10% light speed	+2	4	a few centimetres	paper, skin or clothing
β	high-speed electron	90% light speed	−1	0.000 5	a few metres	thin aluminium sheet
γ	electromagnetic wave of high energy	light speed	0	0	a few kilometres	10 cm sheet of lead or several metres of concrete

NUCLEAR EQUATIONS AND RADIOACTIVE DECAY

Alpha decay

Any atom with an atomic number greater than 82 will be radioactive, and most of these elements decay by emitting alpha radiation.

Americium, with an atomic number of 95, is the radioactive isotope used in smoke detectors. The nuclear equation for its **alpha decay** is:

$$\underset{\text{americium}}{^{241}_{95}\text{Am}} \rightarrow \underset{\text{neptunium}}{^{237}_{93}\text{Np}} + \underset{\text{alpha particle}}{^{4}_{2}\text{He}}$$

Study this equation carefully. We have already seen that mass numbers and atomic numbers need to balance on each side of the equation. There are two other points to note:

1 In alpha decay equations, we do not write He^{2+}. This is because the alpha particle comes from the nucleus of a larger atom – it is not a helium atom which has lost electrons. However, alpha particles do eventually gain electrons and become helium atoms.

2 In nuclear equations, we always write the atomic number, even though we know the name of the element.

Now look closely at what happens to americium. It becomes a new element because its atomic number decreases by 2. The event of one element changing into another in this way is called a **transmutation**.

Beta decay

Some isotopes of lighter elements have too many neutrons and are unstable. Such isotopes decay by emitting a beta particle (a high-speed electron). When a nucleus releases a beta particle, a neutron changes to a proton, and this makes the atom of the new element more stable.

? QUESTION 9

9 Ernest Rutherford and Frederick Soddy were the first people ever to observe a transmutation. They discovered that radium (^{226}Ra) decayed by releasing an alpha particle to give an element never before known. Use the Periodic Table in Appendix 2 to help you decide what this element is called, and write a nuclear equation to describe this decay.

SCIENCE IN CONTEXT

Smoke detectors use radioactivity

Smoke can be the first sign that a house is on fire, and detecting smoke before the fire has a chance to take hold can be a real life saver.

One type of smoke detector makes use of the radioactivity of americium-241 to detect smoke. A small sample of americium-241 in the chamber of the detector gives off alpha radiation, which keeps the air in the chamber permanently ionised. This allows a small current to keep flowing between two electrodes, as shown in Fig 2.18.

Smoke entering the sensing chamber interferes with the ionised particles and allows them to recombine with electrons. This reduces the current, and when the current falls, the smoke alarm is set off.

A smoke detector, well sited inside a house, will often respond to a level of smoke concentration that the people in it are not aware of. Smoke detectors save hundreds of lives every year, and also protect people's homes.

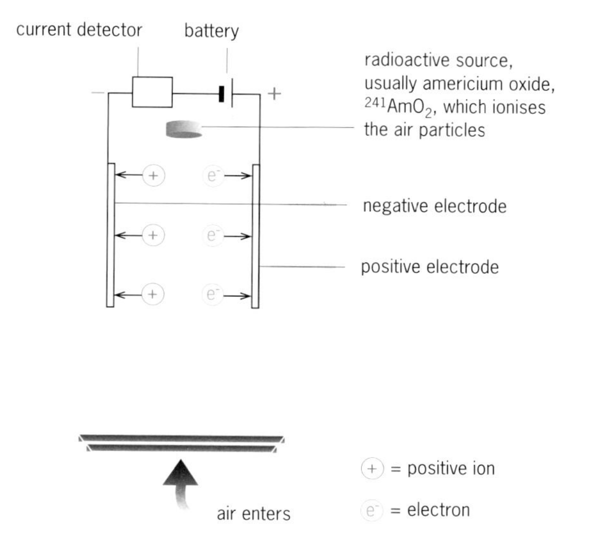

Fig 2.18 The smoke-sensing chamber of a smoke detector

QUESTION 10

10 Read the Science in Context box on smoke detectors.

a) Why is alpha radiation used in smoke detectors rather than beta or gamma radiation? (Hint: look at Table 2.5 on the opposite page.)

b) The useful life of a smoke detector is about 10 years. Why should it not be thrown away with household waste?

The next nuclear equation is the equation for **beta decay**. Note that in this equation, the electron is given values of zero and –1. Look back to page 39 to remind you of this numbering.

$$^{1}_{0}n \rightarrow\ ^{1}_{1}p +\ ^{0}_{-1}e \quad (^{0}_{-1}e \text{ is a beta particle})$$

The nuclear reaction of carbon-14 is a good example of beta decay, and beta particles are detected in radiocarbon dating:

$$^{14}_{6}C \rightarrow\ ^{14}_{7}N +\ ^{0}_{-1}e$$

Since a neutron changes into a proton, the atomic number of the new element is increased by one, but the mass number stays the same.

Gamma rays

Gamma rays arise when the nucleus has excess energy – when it is in an excited state. The nucleus is usually excited after a nuclear decay, for instance after releasing alpha or beta particles. It loses its excess energy by emitting gamma rays, which are very high-energy electromagnetic radiation. Gamma radiation on its own does not result in the formation of a new element.

QUESTIONS 11–12

11 Strontium-90 used to be produced in nuclear explosions. It decays by beta radiation. Write the nuclear equation for this decay, and state what new element has formed.

12 15 million potassium-40 atoms decay inside each person's body every hour. How many grams is this? (Hint: how many atoms in a mole?)

5 HALF-LIFE AND THE RATE OF RADIOACTIVE DECAY

Radioactive isotopes never stop decaying and, unlike chemical reactions, their rate of decay is not affected by pressure and temperature. The rate of decay of a radioisotope is measured as the **half-life, $t_{1/2}$**, which is the time it takes for half of the isotope to decay, and is unique to every isotope (see Table 2.6). The most stable are those with the longest half-lives, and the least stable have the shortest half-lives. For this reason, many of the very unstable isotopes which can be artificially synthesised do not exist in nature.

Iodine-131 is a very unstable isotope. Half of it will decay in 8.1 days ($t_{1/2}$ = 8.1 days). It is useful in medical diagnosis, to detect liver and brain tumours and to investigate the human thyroid gland.

Cobalt-60 is a much more stable isotope. It takes more than five years for cobalt to decay to half its original mass. Its beta and gamma rays are used to sterilise medical equipment and to attack cancer cells in the human body.

At the other end of the scale, uranium-238 was formed before the Solar System existed, and takes 4.5×10^9 years for half of it to decay.

When atomic weapons were tested in the 1950s and 1960s, the isotope strontium-90 was released into the Earth's atmosphere for the first time, and was deposited throughout the world. It entered the human food chain through water supplies and, in particular, through milk from cows grazing on contaminated pastures. Since strontium is chemically very similar to calcium, it becomes incorporated into people's teeth and bones, and is linked to an increased risk of developing leukaemia and bone cancer.

Strontium-90 has a half-life of 28 years, so starting with 20 mg, 28 years later half will have decayed, leaving 10 mg of strontium-90. In a further

? QUESTION 13

13 **a)** Why is it important that radio-isotopes for use in the body have a short half-life?

b) Xenon-133 is a gas which is used to form a picture of the ventilation pathways of the lungs. It has a half-life of 5.3 days. How much of a 100 g sample of xenon-133 will remain after 53 days?

Table 2.6 The half-life of some radioactive isotopes

Isotope	Nuclear equation	Half-life ($t_{1/2}$)
polonium-212	$^{212}_{84}Po \rightarrow ^{208}_{82}Np + ^{4}_{2}He$	3×10^{-7} seconds
sodium-24	$^{24}_{11}Na \rightarrow ^{24}_{12}Mg + ^{0}_{-1}e$	15.0 hours
iodine-131	$^{131}_{53}I \rightarrow ^{131}_{54}Xe + ^{0}_{-1}e$	8.1 days
cobalt-60	$^{60}_{27}Co \rightarrow ^{60}_{28}Ni + ^{0}_{-1}e$	5.3 years
uranium-238	$^{238}_{92}U \rightarrow ^{234}_{90}Th + ^{4}_{2}He$	4.5×10^9 years

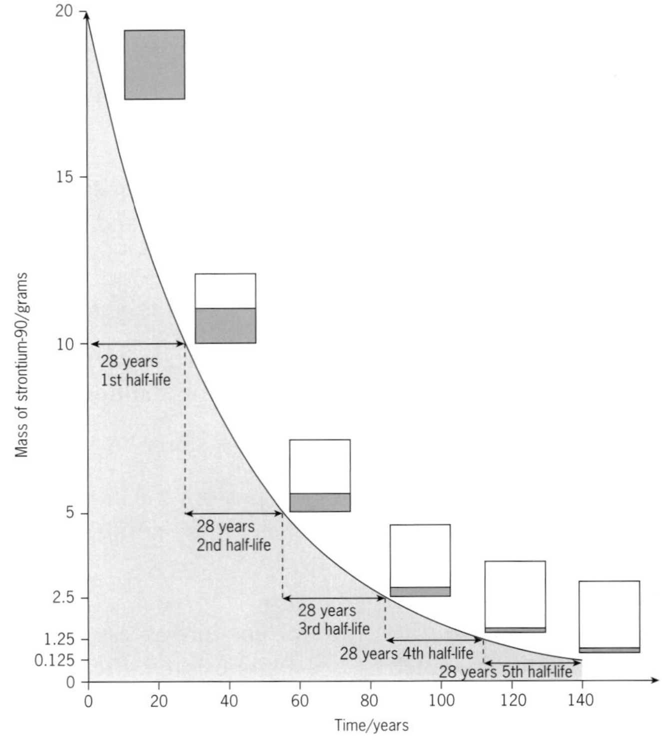

Fig 2.19 Graph of the decay of 20 mg of $^{90}_{38}Sr$, a beta particle emitter

28 years there will be only 5 mg. After 84 years, which is three half-lives, only 2.5 mg will remain, and so on. No matter how much strontium-90 there is to begin with, half disintegrates every 28 years. Fig 2.19 is a graph of this decay, known as **exponential decay**, and its shape is typical of all radioactive isotopes.

EXAMPLE

Q In experiments, cobalt-60 has been used to irradiate strawberries to prevent them developing mould. (At the time of publication, irradiation of food is not permitted in the UK.) As Table 2.6 shows, the half-life of cobalt-60 is 5.3 years. If a food company were to start with 2.00 g of the isotope, how much would be left after 21.2 years?

A Find out how many half-lives 21.2 years is:

$$21.2 \div 5.3 = 4 \text{ half-lives}$$

2 g: $t_{1/2}$ = 5.3 years → **1 g:** $t_{1/2}$ = 5.3 years → **0.5 g:** $t_{1/2}$ = 5.3 years → **0.25 g:** $t_{1/2}$ = 5.3 years → **0.125 g**

You can see that the starting amount of the sample has been reduced four times by a half:

$$2 \times \frac{1}{2} \times \frac{1}{2} \times \frac{1}{2} \times \frac{1}{2} = 2 \times \left(\frac{1}{2}\right)^4 = 0.125 \text{ g}$$

Q All the isotopes of technetium are radioactive. They do not occur naturally on Earth and all are artificially made. Technetium-99 has a half-life of 6 hours. It is used in medicine to assess the damage to heart muscles after a heart attack. A sample has an initial count rate of 3000 counts per minute (c.p.m.). How long will it take for the count rate to fall to 94 c.p.m?

A **3000:** $t_{1/2}$ = 6 h → **1500:** $t_{1/2}$ = 6 h → **750:** $t_{1/2}$ = 6 h → **375:** $t_{1/2}$ = 6 h → **187.5:** $t_{1/2}$ = 6 h → **93.75** ≈ **94 c.p.m.**

As shown, it takes 5 half-lives for activity to reduce to 94 c.p.m. Therefore, 6 × 5 = 30 hours.

When the count rate does not correspond to an exact number of half-lives, a graph can be drawn, similar to the curve of Fig 2.19, and the time corresponding to the count rate can be read off from it.

6 DATING AND OTHER USES OF RADIOACTIVITY

It is because radioactive isotopes have fixed half-lives, which temperature or pressure cannot alter, that we can use half-lives to date objects, and even date the Earth itself. We also use radioisotopes elsewhere, particularly in medicine.

Carbon-14 dating

The best-known radioactive dating technique uses carbon-14, which has a half-life of 5730 years. The Earth is continually bombarded by particles from the cosmos. These 'cosmic rays' crash into atoms in the upper atmosphere and cause them to release high-energy neutrons. They, in turn, smash into the nuclei of nitrogen-14 atoms in the upper atmosphere, and produce carbon-14 which is radioactive.

$${}^{14}_{7}\text{N} + {}^{1}_{0}\text{n} \rightarrow {}^{14}_{6}\text{C} + {}^{1}_{1}\text{H}$$

Carbon-14 reacts with oxygen to form $^{14}CO_2$, and this is used by plants in photosynthesis to make carbohydrates (see Fig 2.20). Through food chains, the isotope becomes incorporated into all living things, including ourselves. As soon as the carbon-14 is produced it starts to decay by emitting beta particles:

$${}^{14}_{6}\text{C} \rightarrow {}^{14}_{7}\text{N} + {}^{0}_{-1}\text{e}$$

The amount of carbon-14 in the atmosphere stays roughly constant because, over time, the rate of decay of carbon-14, and its rate of production by cosmic rays, have become balanced. It is important to realise that the amount of carbon-14 produced each year is very small indeed – only about 7.5 kg in total – and that the ratio of carbon-14 to carbon-12 all around us and inside us is also very small. There are 12 million million carbon-12 atoms for every one carbon-14 atom. This ratio of carbon-12 to carbon-14 in any organism stays constant while it is alive, because any carbon-14 that decays is replaced in the turnover of food and nutrients from its environment. When the organism dies, it no longer takes in carbon-14, so the amount of carbon-14 starts to decrease.

Imagine that archaeologists find a wooden carving they think is thousands of years old. They find that its carbon-12 to carbon-14 ratio is 24 million million to 1 – that is, the carving has half the amount of carbon-14 that is found in wood cut down today. The carving is therefore one half-life old, or about 5730 years.

QUESTION 14

14 The amount of carbon-14 becomes too small to measure accurately when it falls below 1 per cent of the amount originally present when the organism was alive. How many half-lives are involved in the amount of carbon-14 falling to below 1 per cent? About how many years will this go back in time? (Hint: the carbon-14 present when the organism was alive is 100 per cent.)

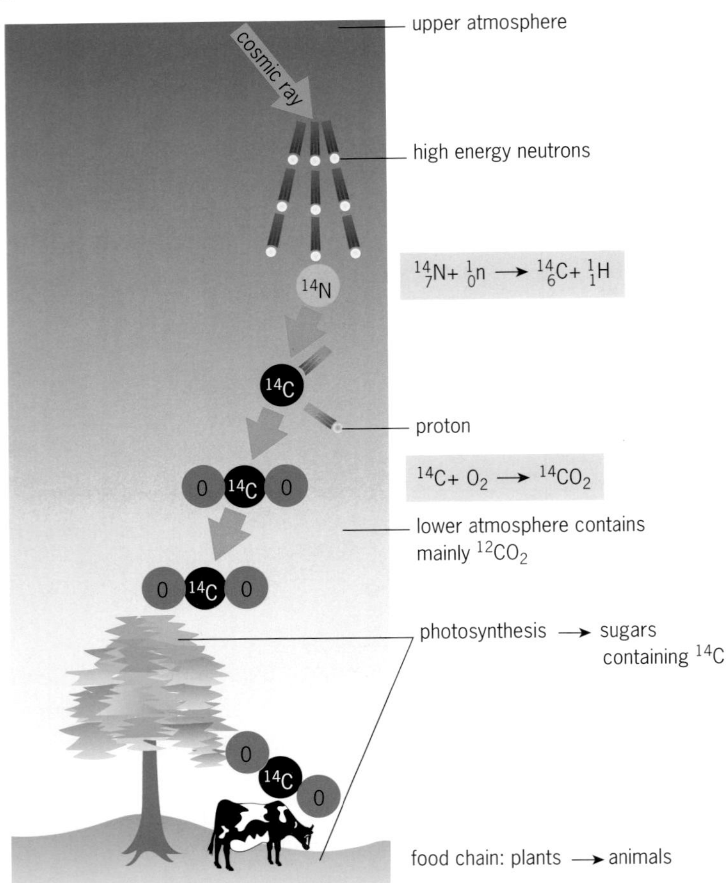

Fig 2.20 The path taken by carbon-14 as it enters the food chain

Fig 2.21 To radiocarbon date a bone, it is cut into pieces and the collagen is extracted chemically. It is this material that is analysed for its carbon-14 content in a counting chamber that operates like a Geiger counter. Depending on age, the count rate may vary from several disintegrations per hour to a few per day. For very old material, counting may last several months

HOW SCIENCE WORKS

Dating the Turin shroud

This has been one of the most famous uses of radiocarbon dating. The Shroud of Turin is a linen cloth, over 4 metres long, bearing two images of a man – one of the front and the other of the back of someone who appears to have been crucified. For centuries many people have believed that the cloth wrapped the body of Jesus after his death. According to the first reliable records, the cloth appeared in France in 1350, and was later taken to the Italian city of Turin.

In recent years, chemical tests were done on small samples of the cloth. They revealed that the image was not painted on to the cloth by any known method, and that the blood stains are definitely human.

In 1988, laboratories in Oxford, Zurich and Arizona were each given pieces 2 centimetres square to date independently. The linen was found to have come from flax grown somewhere between 1260 and 1390. For the cloth to have been the burial shroud of Jesus, the flax would have had to have grown some time before his death. But some people still claim that a burst of energy during the resurrection could have made the flax seem younger than it is. Many questions remain unanswered, including the way that the image was formed.

Fig 2.22 The face on the Shroud of Turin which is thought by many people to have been the burial shroud of Jesus

SCIENCE IN CONTEXT

Radioactivity – safe or unsafe?

Radioactive materials have proved themselves useful to society, especially in medicine. Yet we often hear that exposure to radiation is dangerous. Is there a contradiction?

We have looked at three types of radiation – alpha, beta and gamma – which come from the breakdown of unstable nuclei. These are called ionising radiation because when they pass through molecules they can knock off electrons and form ions. This disrupts molecules. In living matter, many of the molecules are complex and contain weak bonds. DNA, the carrier of genetic information, is an example. So exposure to ionising radiation can alter or destroy important biomolecules, and can result in illness and even death.

Whether ionising radiation causes damage depends on two properties: the penetrating power of the radiation and its ability to ionise (and so disrupt) molecules. Alpha particles are relatively massive, so can be very damaging, but their penetrating power is very low, so they do not progress far in living tissue. Beta particles penetrate further, but they are lighter and so cause less ionisation of molecules. Gamma radiation is the most penetrating of the three, but because it has no mass or charge, it is the least ionising.

In using radioisotopes, particularly in medicine, we balance the risk of the damage that a form of radiation may do to biomolecules with the benefits that using it can bring. We are exposed to natural radiation all the time and our own molecules contain radioisotopes, but we also have mechanisms to repair radiation damage. We cannot eliminate exposure altogether, but we can minimise and control it.

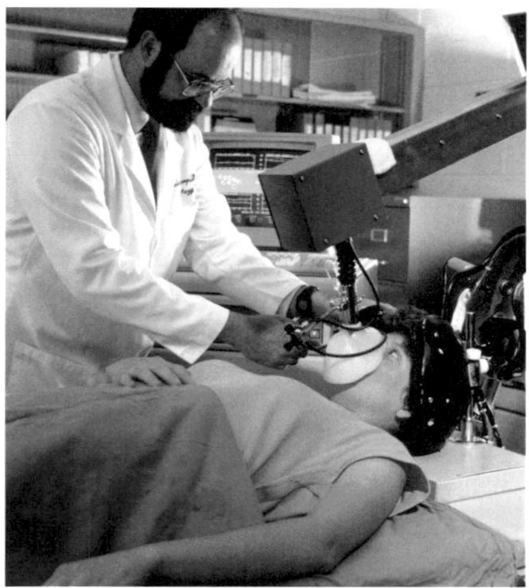

Fig 2.23 The patient is breathing a gas containing xenon-133 which becomes concentrated in the blood vessels of the brain. The gamma rays emitted enable an image of the blood flow in the brain to be seen

SCIENCE IN CONTEXT

Radioactivity made useful

Radioactive materials are being used in an increasing number of ways.

Treating cancer

Cancerous cells are abnormal cells that divide at a rapid rate, producing tumours that invade surrounding tissues. Some types of radiation can cause a cancer. At the same time, cancerous cells are easily destroyed by carefully controlled doses of radiation, which do not affect other surrounding cells. Over a period of weeks, a tumour in a patient can be made inactive by exposure to doses of gamma rays. The radioisotope generally used for this treatment is cobalt-60.

Tracers for diagnosis and treatment

A radioisotope has the same chemical properties as any other atom of the same element, and molecules of a substance 'labelled' with a radioisotope can therefore be traced because their radiation can be detected. For example, iron-59 can be introduced into haemoglobin to follow the production of red blood cells in bone marrow. Iodine-131 as a label in sodium iodide ($Na^{131}I$) is used to investigate the activity of the thyroid gland and to diagnose and treat diseases of this gland. Labelled iodine also helps to detect liver and brain tumours. Sodium-24 as sodium chloride ($^{24}NaCl$) solution is put into the blood to follow blood flow and locate clots and obstructions in vessels. Tracers must have a half-life that is long enough to allow detection but short enough not to cause undue damage to body cells.

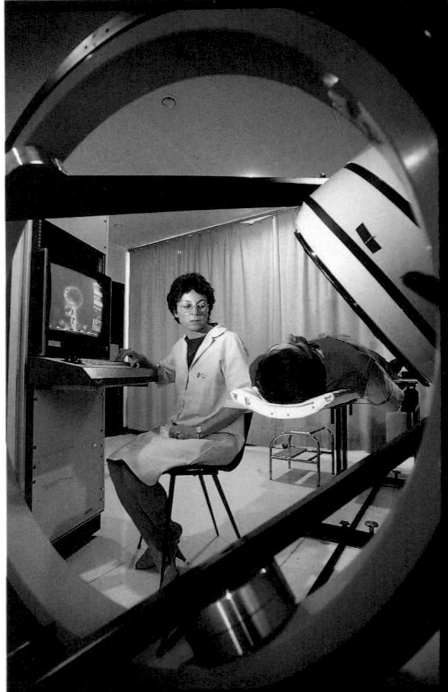

Fig 2.24 The head of a patient with bone cancer is being scanned. A short-lived radiotracer that concentrates more strongly in cancerous bone than in normal bone is used to diagnose whether or not the cancer is spreading

Tracers in agriculture

Labelled sodium iodide in boreholes indicates the route taken by water underground. Barium sulfate, which is insoluble in water, can be labelled with barium-140, and the labelled compound $^{140}BaSO_4$ is used to monitor how silt is deposited in rivers. Phosphorus-32 in phosphate compounds helps to trace the uptake of fertilisers by plants.

SUMMARY

Having studied this chapter, you should know and understand the following.

- An atom has a minute central nucleus of protons and neutrons which is surrounded by electrons.
- A proton has a relative mass 1, and a relative charge of +1.
- A neutron has a relative mass of 1, and no relative charge.
- An electron has a relative mass of 5.5×10^{-4}, and a relative charge of –1.
- Atomic (proton) number = number of protons in the atom.
- Mass (nucleon) number = number of protons and neutrons in the atom.
- Isotopes have the same atomic number, but different mass numbers. They have identical chemical properties, but their physical properties are slightly different.
- The relative atomic mass of an element is the average mass of the atom (taking into account all the naturally occurring isotopes and their abundance) compared to $\frac{1}{12}$ the mass of one atom of carbon-12.
- Radioactivity is the spontaneous disintegration of the nucleus of an atom. Three kinds of radiation emitted from the nucleus are: alpha particles (helium nuclei), beta particles (electrons) and gamma rays (electromagnetic radiation).
- The half-life of a radioactive isotope is the time it takes for half of the isotope to decay. The smaller the half-life, the more unstable the nucleus.
- Radioisotopes can be used as 'tracers' in the body, and to date objects.

Practice questions and a How Science Works assignment for this chapter are available at www.collinseducation.co.uk/CAS

3

A close look at electrons

3 A CLOSE LOOK AT ELECTRONS

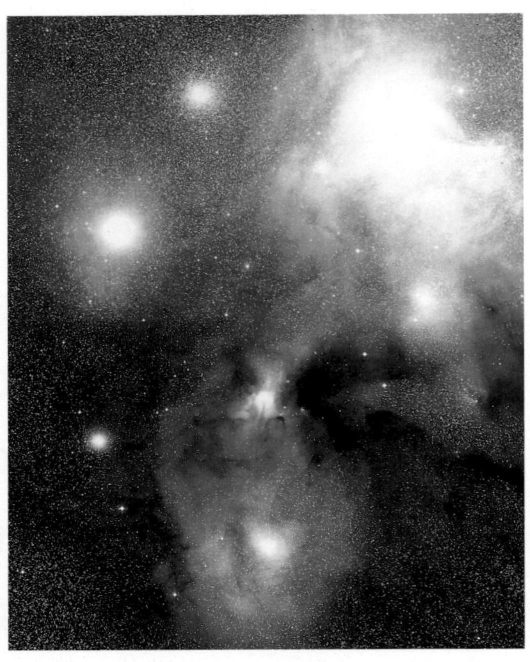

Interstellar clouds. These contain some very interesting molecules, which are identified by their spectra. Some may represent the forerunners of the molecules of life itself

Molecules in interstellar space

The giant gas clouds in interstellar space – the space between stars – contain vast amounts of very thinly dispersed atoms, ions and molecules. We know what this material consists of because each type of atom, ion or molecule absorbs and emits energy of particular wavelengths across the electromagnetic spectrum, depending on the arrangement of its electrons. From Earth, we can detect and record the patterns of these absorbed and emitted wavelengths: each type of particle has its own pattern, which reveals its presence, rather like a fingerprint. Hydrogen, for example, emits radio waves at a wavelength of 21 cm.

Chemists are very interested in the molecules of interstellar space. So far, they have identified over 200. Some are also common on Earth, such as hydrogen chloride, carbon monoxide, water and ethanol. Others are unusual, and several were discovered in space before they were even identified in the laboratory. An example is a three-member carbon ring, C_3H_2, an interstellar molecule widespread in our Galaxy and common in others, but too unstable to be made on Earth.

So how do these molecules form? The temperatures and pressures of interstellar space are just too low for atoms to meet, collide and join together as they do on Earth. However dust particles in space provide a surface on which reactions are catalysed. Also gases could condense on the grains, be bombarded by cosmic radiation and form more complex molecules. Some scientists think that these dust particles may have fallen onto young planets, seeding them with the chemicals for life. There is recent evidence that glycine (NH_2CH_2COOH), an amino acid that is one of the building blocks of our proteins, exists in interstellar space.

INTRODUCTION

The glorious colours of fireworks, the discovery of helium in the Sun before it was discovered on Earth (page 56), our knowledge of the atoms and molecules in distant stars – all these are connected with electrons. To a chemist, the most interesting part of an atom is its electrons, because it is the electrons, not the protons or neutrons, that account for every chemical reaction.

We cannot observe electrons directly, but in this chapter you will find out how our knowledge about them has grown and developed. The fact that we can know things without direct observations is a key point in understanding the nature of scientific knowledge.

Scientists in the 19th century investigated the light that materials absorb and emit, and went on to discover new elements. In the 20th century, our understanding of the nature of light led scientists to release the energy of lasers (see page 57).

Fig 3.1 The spectacular colours of fireworks are caused by the electrons in the atom releasing energy at different wavelengths in the visible spectrum

1 ALL THE COLOURS OF THE RAINBOW

In 1666, Sir Isaac Newton first recorded the fact that, when visible light was passed through a prism, it was split to produce a **continuous spectrum** of rainbow colours that contained all the wavelengths of light.

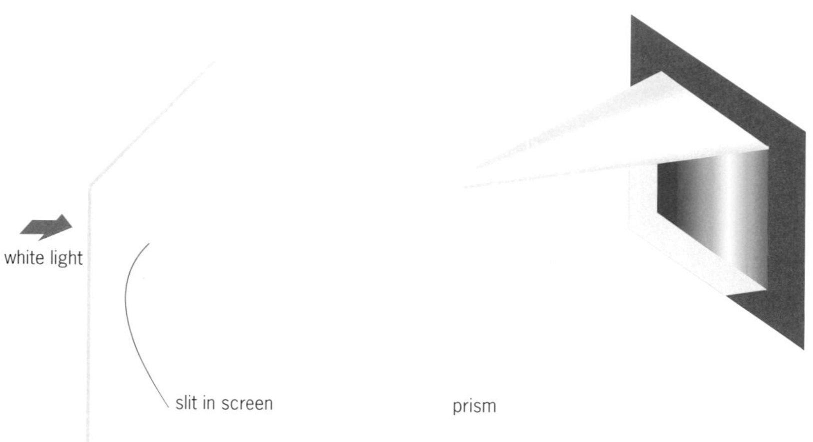

Fig 3.2(a) Light from a white light source passes through a slit and then through a prism. It is split into its different wavelengths, producing a continuous spectrum – a rainbow effect

Fig 3.2(b) Isaac Newton, who used a prism to split visible light and produce a continuous spectrum

Robert Bunsen is best known for inventing the Bunsen burner. But, with Wilhelm Kirchhoff, he invented an even more important instrument for the progress of chemistry – the spectroscope. When put in a flame, compounds that contain sodium were known to colour the flame yellow; potassium compounds coloured it lilac. The spectroscope took these observations and Newton's visible spectrum a step further.

In the spectroscope, light passes through a narrow slit and a prism, and forms a spectrum. When Bunsen and Kirchhoff viewed the sodium flame with this light, they saw bright lines (images of the slit) in the yellow part of the spectrum and some less bright lines in other parts. They viewed light from compounds of other metals in the same way, and soon realised that each element had its own characteristic *fingerprint* of lines in different parts of the spectrum.

With this technique, they discovered the elements rubidium and caesium while examining the spectrum of a lithium ore. They found strong red and blue lines (Fig 3.4) which could not be accounted for by the elements known to be in the ore. Rubidium is taken from the Latin for 'red' and caesium is named after the Latin for 'blue'.

Fig 3.3 Robert Bunsen (centre) and Wilhelm Kirchhoff (left)

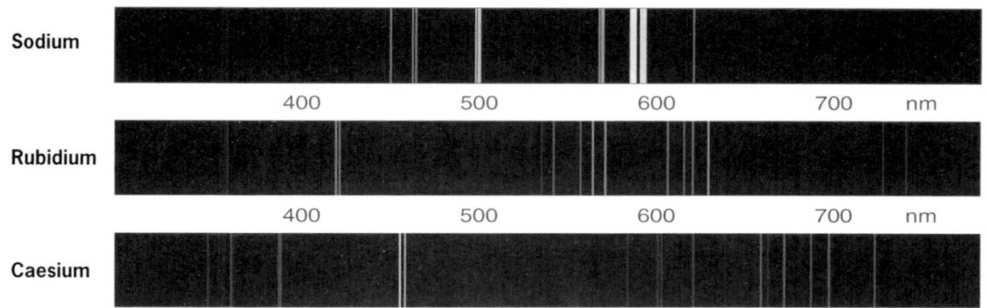

Fig 3.4 The line emission spectra of sodium, and of rubidium and caesium, both discovered by Bunsen and Kirchhoff

As elements were being discovered and identified by their spectral fingerprints, astronomers were soon pointing spectroscopes at the stars to look for characteristic lines of elements and finding that the cosmos was made up of the same elements as the Earth.

REMEMBER THIS

The line spectra we met on the previous page are examples of **discontinuous spectra** because there are discrete lines that do not merge into one another.

2 THE ELECTROMAGNETIC SPECTRUM

The 'light' our eyes detect is just a small part of a very much wider **electromagnetic spectrum**. This spectrum is made up of all the types of **electromagnetic radiation**, and includes X-rays used by dentists, the microwave radiation used to heat food in microwave ovens and radio waves that bring us radio and television signals. Fig 3.5 shows the **continuous spectrum** of electromagnetic radiation, called continuous because all the wavelengths are represented. We shall now look more closely at electromagnetic radiation to help us understand electrons and how they are arranged in atoms.

As its name suggests, electromagnetic radiation is made up of two components: an electrical and a magnetic one. Electromagnetic radiation is one of the ways energy is transmitted through space. It is why we feel the warmth of the Sun on Earth, and why we suffer sunburn if we have too much ultraviolet radiation.

Fig 3.5 The electromagnetic spectrum. Notice that visible light – the radiation our eyes can detect – is only a very small part of the spectrum

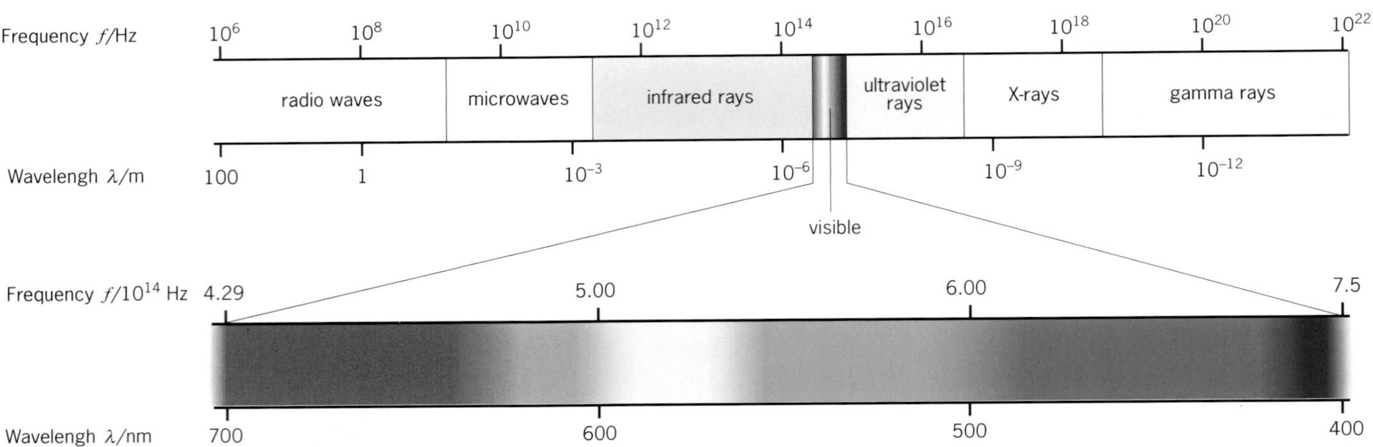

3 LIGHT AS WAVES

Electromagnetic radiation is usually thought of as **waves** and, like any wave, it can be described by its **wavelength**, **frequency** and **speed**. Wavelength is measured in metres (m), and its symbol is the Greek letter **λ** (lambda). Frequency, ***f***, is measured in hertz (Hz), which means cycles per second, and which can also be written as s^{-1}. The speed of electromagnetic waves is the speed of light, with the symbol **c**, which is measured in metres per second ($m\,s^{-1}$).

Fig 3.6 shows what these terms mean, and that when the frequency is higher, the wavelength is smaller. The terms are related in a simple equation:

$$\mathbf{c} = \lambda \times f$$
$$(m\,s^{-1}) \quad (m) \quad (s^{-1})$$

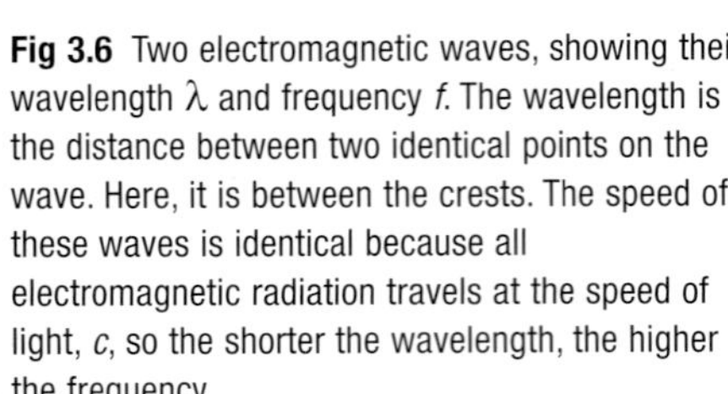

Fig 3.6 Two electromagnetic waves, showing their wavelength λ and frequency *f*. The wavelength is the distance between two identical points on the wave. Here, it is between the crests. The speed of these waves is identical because all electromagnetic radiation travels at the speed of light, *c*, so the shorter the wavelength, the higher the frequency

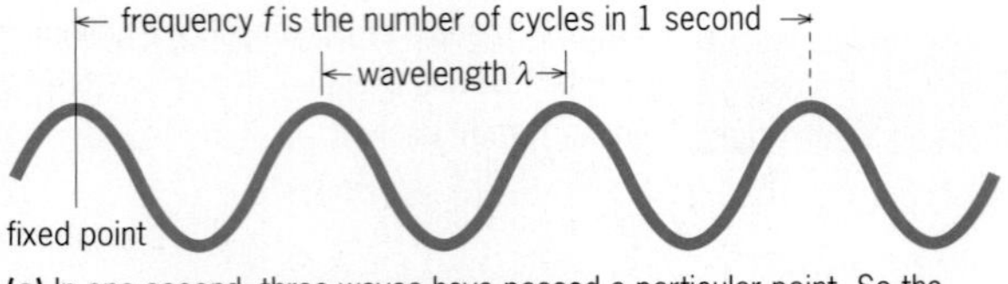

(a) In one second, three waves have passed a particular point. So the frequency is three cycles per second, or 3 hertz (this is far lower than the frequency of actual electromagnetic radiation)

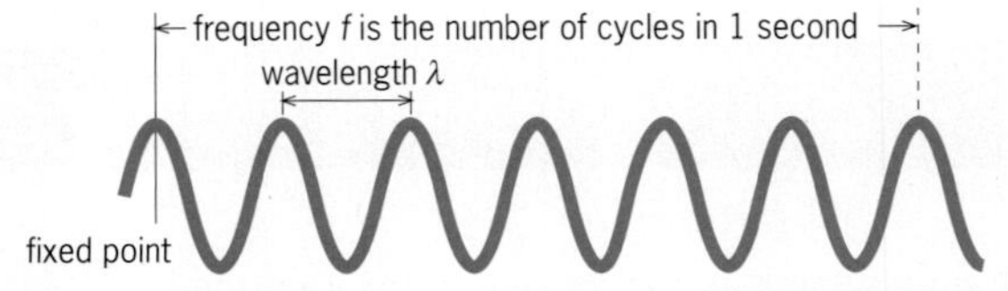

(b) In one second, six cycles have passed the same fixed point. So the frequency is 6 cycles per second, or 6 hertz (again, far lower than frequencies found in the electromagnetic spectrum)

EXAMPLE

Q The average wavelength of a red traffic light is 700 nm. What is the frequency of this radiation?

A First, you need to convert the average wavelength to metres: 700 nm is 700×10^{-9} m, and the speed of light c is 3.00×10^{8} m s^{-1}.

Inserting these values into the equation:

$$3.00 \times 10^{8} = \lambda \times 700 \times 10^{-9}$$

Rearranging: $$\lambda = \frac{3.00 \times 10^{8}}{700 \times 10^{-9}} = 4.29 \times 10^{14}$$

So the frequency of the red light is 4.29×10^{14} Hz.

QUESTION 1

Barium nitrate is added to fireworks to colour them. When heated, this compound emits light of frequencies around 5.45×10^{14} Hz. Calculate the wavelength, and use Fig 3.5 to work out what colour the fireworks will be.

4 LIGHT AS PARTICLES

Up to the beginning of the 20th century, scientists were agreed on a model of electromagnetic energy in which it was radiated and absorbed by matter in the form of waves. The model did not explain all the observations, notably why heated elements emitted radiation as discontinuous spectra with separate lines.

In 1900, Max Planck proposed that particles such as atoms and molecules absorb and emit energy in discrete (separate) amounts or packets called '**quanta**'. Planck's equation is:

$$E = hf$$

where E is the energy in joules (J), h is Planck's constant, with a value of 6.626×10^{-34} J s, and f is the frequency in hertz (Hz or s^{-1}).

A **quantum of energy** is a precise packet of energy. These packets can have different energy values, depending on their source, but you cannot have parts of packets, only whole ones.

THE PHOTOELECTRIC EFFECT

In 1905, Albert Einstein used Max Planck's equation to explain another phenomenon that baffled scientists, the **photoelectric effect**. This is the release of electrons by some metals when light is shone on them. It was found that the light had to be above a minimum frequency (and so a minimum energy) before electrons were emitted. It did not matter how intense (bright) the electromagnetic radiation was below this *threshold frequency*; only with a high enough frequency were electrons released. Then, the more intense the light, the greater the number of electrons released. According to the old wave-only model, the electrons would absorb even low-energy radiation and eventually release it. Einstein developed Planck's idea of quanta, saying that some of the properties of electromagnetic radiation could be explained only if light were thought of as consisting of *particles*, which we now call **photons**. He also said that each photon is associated with a particular quantum (amount) of energy, linked to its frequency by Planck's equation: $E = hf$.

Fig 3.7 Albert Einstein. By the age of 26, while an assistant in a Swiss patent office, he had written three papers in his spare time that were to change 20th-century science

QUESTION 2

The microwave radiation used in ovens has a frequency of 1.0×10^{10} Hz.

a) What is the wavelength of this radiation?

b) What is the energy of one photon?

c) What is the energy of one mole of these photons?

QUESTION 3

Exposure to both gamma radiation and ultraviolet light can cause cancer. Use Fig 3.5 and Planck's equation to suggest why gamma radiation is much more harmful than ultraviolet light.

The photoelectric effect could now be explained. Negative electrons are held in the atom by electrostatic forces of attraction to the positive protons in the nucleus. When a photon collides with an electron, the photon gives up its energy to that electron. If the energy is high enough, the electron breaks free of the atom. The more intense the radiation, the greater the number of photons, so more electrons are released from materials that the radiation reaches.

Fig 3.8 In digital cameras, photons enter the lens and strike a grid of pixels, where they release electrons due to the photoelectric effect. The charge from the electrons provides the digital values for each pixel, and depends on the amount of light

THE WAVE–PARTICLE NATURE OF LIGHT

In 1905, Einstein wrote three papers. One was on his famous theory of special relativity, another an explanation of Brownian motion. His third, an explanation of the photoelectric effect, gained him a Nobel Prize in 1921, three years after his great friend Max Planck received one. All three papers had a profound effect on 20th-century science and Einstein was seen by many as one of the two greatest scientists that ever lived, the other being Isaac Newton.

We have noted some observations suggesting that light is made up of particles and others suggesting it is made up of waves. So we now think of light as having a *dual* nature, of both particles and waves. This **wave–particle duality** is not just confined to light; matter, too, can sometimes behave as if it were made up of waves, as you will see on page 57.

5 THE ATOMIC EMISSION SPECTRUM OF HYDROGEN

When hydrogen is placed in a discharge tube at low pressure and with a high voltage between the two plates, some of the bonds in the hydrogen molecules (H_2) are broken to give separate hydrogen atoms. When the radiation emitted from the discharge tube is passed through a spectroscope, a series of characteristic lines shows up in the visible spectrum.

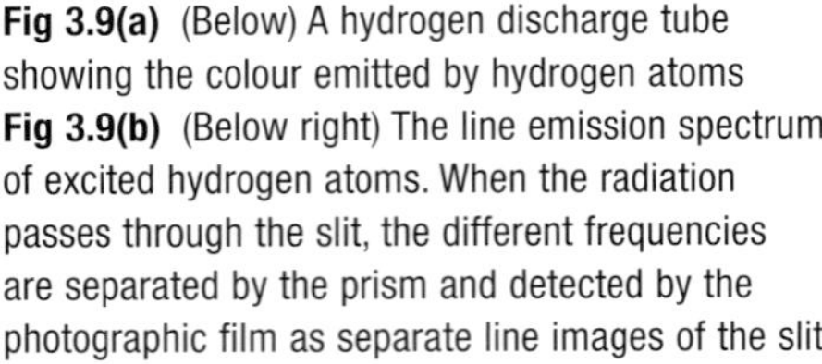

Fig 3.9(a) (Below) A hydrogen discharge tube showing the colour emitted by hydrogen atoms
Fig 3.9(b) (Below right) The line emission spectrum of excited hydrogen atoms. When the radiation passes through the slit, the different frequencies are separated by the prism and detected by the photographic film as separate line images of the slit

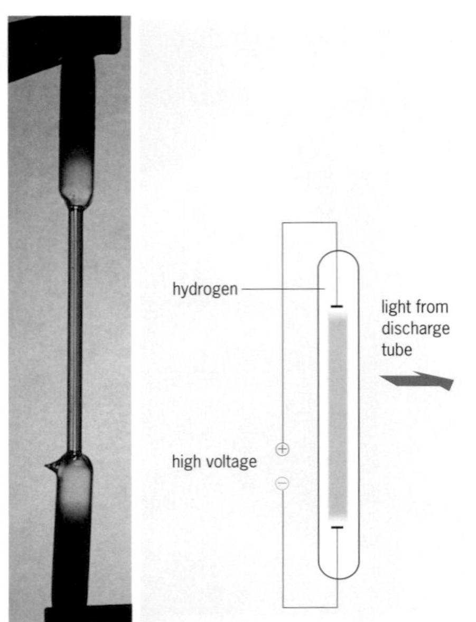

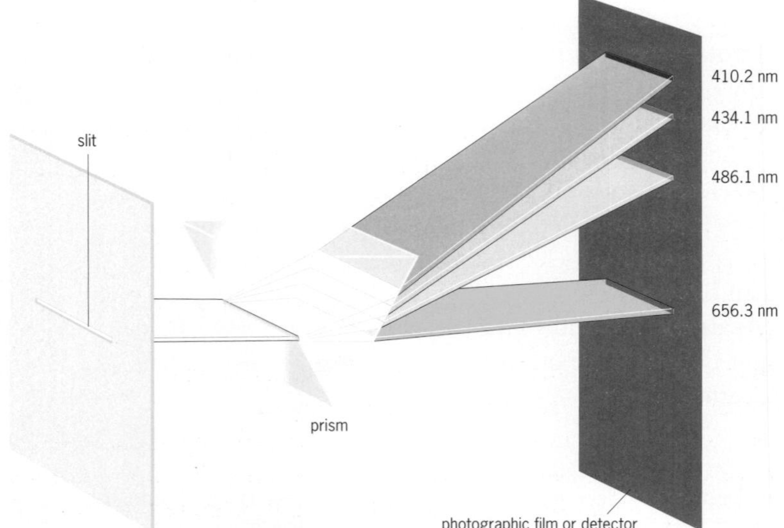

THE BALMER SERIES

You read earlier about Bunsen and Kirchhoff's work in identifying elements by their characteristic spectral lines. These spectra are called **line emission spectra** for the following reason. When an element's atoms are given energy, they absorb it and then emit radiation as discrete (separate) lines, at specific frequencies (and hence energies) for that element. For example, hydrogen, which has just one electron, has several prominent lines in the visible part of its emission spectrum (Fig 3.10, page 54). These are known as the **Balmer series**, after a Swiss music teacher who worked out a mathematical relationship between the lines.

Line emission spectra are used in atomic emission spectroscopy to determine the concentration of metal ions in blood serum and other biological fluids (see page 558).

OTHER SERIES

Other series of lines for hydrogen are found in different parts of the electromagnetic spectrum. The **Lyman series** is found in the ultraviolet section and the **Paschen series** in the infrared. Both these series are named after their discoverers.

Like the photoelectric effect, the different series of lines (characteristic for an element) was another 19th-century mystery that could not be explained by the theories of that time. An explanation had to wait for the ideas of Planck and Einstein to be developed.

ABSORPTION SPECTRA

Electrons absorb energy at specific frequencies, producing a series of black lines on a coloured background. This is called an absorbtion spectrum (see Fig 3.14 on page 56). The background is the visible electromagnetic spectrum, while the black lines are frequencies that are missing because electrons have absorbed these energies. The black lines occur in the same place as the coloured lines in the line emission spectrum, showing that electrons absorb radiation at specific frequencies and then release the energy at the same frequency.

6 NIELS BOHR'S MODEL OF THE HYDROGEN ATOM

In Chapter 2 we were left with the Rutherford model of the atom. However, there are problems with this model. Rutherford proposed that electrons orbited the nucleus, rather like planets round the Sun. Planetary motion was well understood by this time – the Sun's gravity pulls planets towards it, while their acceleration, caused by being in a circular orbit, creates a balancing force. (For an object to travel in a circle, it is constantly changing direction, so it has to accelerate constantly.)

Negative electrons in circular motion are attracted to the positive protons in the nucleus by electrostatic forces. If they orbited the nucleus like planets, their acceleration would keep them from falling into the nucleus. But electrons are charged particles, and *accelerating* charged particles were known to emit electromagnetic radiation and lose energy. If Rutherford's model was correct, instead of a few separate lines, a continuous spectrum should have been observed, with the atom emitting light all the time and the electron losing its energy and falling into the nucleus, causing the hydrogen atom to collapse. Clearly, this does not happen!

Another model of the atom was needed to explain the observations. Niels Bohr used the quantum ideas developed by Planck and Einstein to propose his model for the hydrogen atom. Bohr's model still had the hydrogen electron orbiting the nucleus. But the orbits, or **energy levels** that the hydrogen

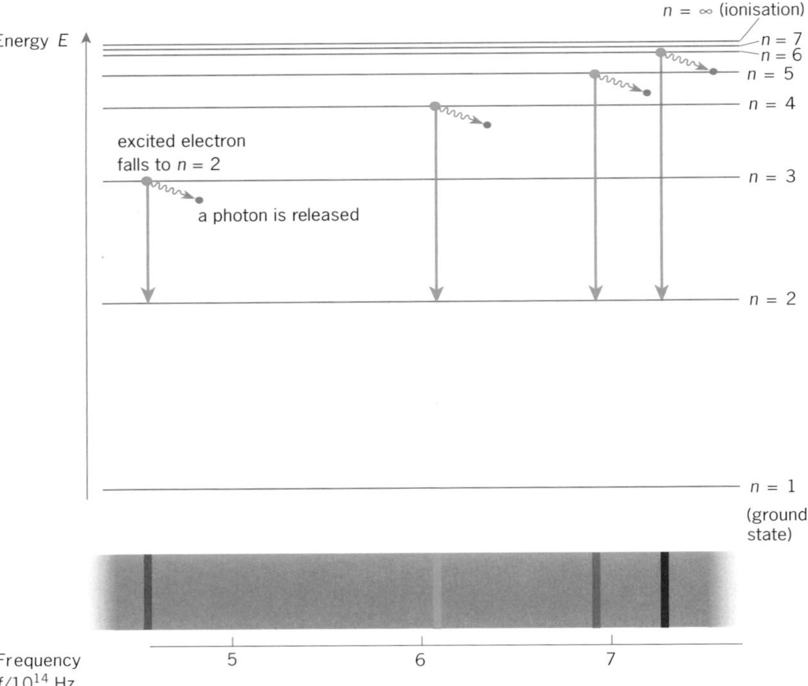

Fig 3.10 The Balmer series for the hydrogen atom, and its line emission spectrum. The lines in the visible spectrum are produced when the excited electron falls back to the $n = 2$ energy level. The higher the energy level from which it falls, the higher the frequency of the photon emitted. The lines become closer at higher values of n, and merge at $n = \infty$

Fig 3.11(a) Niels Bohr (1885–1962). For his work on the structure of atoms, Bohr received a Nobel Prize in 1922

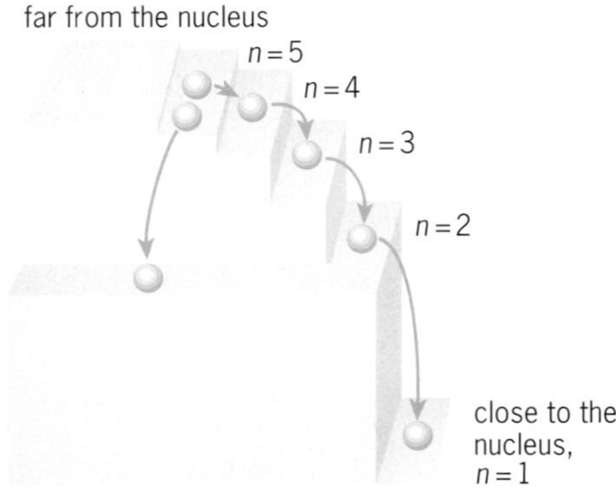

Fig 3.11(b) A staircase model for the energy levels of a hydrogen atom. Note the ball can fall down one level at a time, or it can fall more than one level

QUESTION 4

4 Look back to Fig 3.9(b). It shows the lines moving closer together towards one end of the spectrum for hydrogen. Are they closer together at the higher energy end of the spectrum or the lower energy end?

electron could occupy, were *quantised* – that is, they had fixed energy values. With this model, hydrogen's line emission spectrum could be explained.

In Bohr's model, the electron normally occupies the lowest possible energy level, called the **ground state**. This is the energy level of the electron when it is not excited (see Fig 3.10). Raising the electron to an excited state (giving it energy), say by an electric discharge, causes it to move up to a higher energy level by absorbing a quantum of energy (Fig 3.10). When it returns to the lower, ground state energy level, it releases this quantum of energy as a photon of light of a specific frequency, so giving a line in the emission spectrum.

Imagine that the hydrogen electron is like a ball on a staircase, as in Fig 3.11(b). The ball can rest on any step, but it cannot stop in between. This is the case with the electron. The ball needs energy to go up the step, and when it falls back down it releases this energy. The lines in an emission spectrum are closer together at one end because the higher energy levels that the electron can occupy are also closer together. This happens as the electron moves away from the nucleus, as shown in Fig 3.11(b).

When the electron is closest to the nucleus, it is at its lowest energy level. Moving the electron away from the nucleus requires energy, and the further away it is moved, the more energy it requires. So the more energy an electron receives, the higher it can rise through the energy levels, and the more energy it will release as it falls back down again.

Each energy level is given a number, called the **principal quantum number, *n***. The term 'principal quantum number' is still used to describe the main energy levels of electrons in an atom. When the hydrogen's electron is in the $n = 1$ level, it is not excited, so this is the ground state.

The Balmer series of lines is for the energy transitions when the excited hydrogen electron falls back from higher energy levels to $n = 2$, see Fig 3.10. We see the lines because they are in the visible spectrum. The higher the energy level the electron falls from, the higher the frequency of the emitted photon. Note that the energy levels eventually merge at $n = \infty$ (infinity). This is when the atom has become ionised and lost its electron, which has escaped from the nucleus's attraction.

SCIENCE IN CONTEXT

Using the light of electron transitions

Neon is used everywhere in advertising signs. An electric current is passed through the gas at low pressure. The fast-moving electrons of the electric current excite electrons in the neon atoms into higher energy levels and, when they return to lower energy levels, orange-red light is emitted. The colour can be varied by adding other atoms such as argon or mercury, or by colouring the glass tube the neon is in.

Street lights usually contain sodium or mercury and they work on the same principle as neon lights. When excited mercury atoms return to their ground state, the radiation they emit has frequencies in the ultraviolet, yellow, green and blue parts of the spectrum.

Fig 3.12 Excited electrons in atoms of neon produce the colours for this sign in Tokyo, Japan

Sodium lights have replaced mercury ones across the UK road network because the radiation emitted by excited electrons in sodium atoms when they return to their lower levels is centred upon yellow. This has longer wavelengths than the light from mercury and is not as readily scattered by fog, so it can illuminate further. Also, sodium atoms require less energy to excite their electrons, and sodium is not as toxic as mercury.

Fluorescent lights and low-energy light bulbs in the home or office contain low-pressure mercury vapour. The inside of the lighting tube is coated with a phosphor, which absorbs the energy of ultraviolet light when its electrons are excited. On returning to the ground state, these produce many frequencies of light in the visible range, which combine to give white light.

The advantage of fluorescent lights and low energy light bulbs over filament lamps is that they use less energy, since nearly all the energy is radiated in the visible spectrum. They feel cool to touch, while ordinary light bulbs with tungsten filaments waste energy as they become very hot. However, they do contain up to 50 mg of mercury, which is an environmental hazard when they are thrown away.

The Lyman series of lines is found in the ultraviolet part of the spectrum. It is caused by the excited hydrogen electron returning from higher levels to the $n = 1$ ground-state energy level. As this is the lowest level nearest to the nucleus, far more energy is released when an electron excited to a particular energy level returns to $n = 1$ than when it returns to $n = 2$, and so the lines show up in the more energetic ultraviolet part of the spectrum.

Each line of hydrogen's emission spectrum represents one electron transition, a movement from a higher to a lower level. As large numbers of hydrogen atoms are involved, all the possible transitions are represented, which gives the full spectrum of lines.

Fig 3.13 The Lyman series for a hydrogen atom. In this series, the transitions of the electron are to $n = 1$

REMEMBER THIS

A substance fluoresces when it takes in light of one wavelength, usually outside the visible spectrum, and gives out light of another, often in the visible spectrum. This is described on page 35.

IONISATION ENERGY (IONISATION ENTHALPY)

From the Lyman series we can work out the energy needed to remove an electron completely from a hydrogen atom. This is the ionisation energy for hydrogen. It is the energy required to take the electron from the ground state, at $n = 1$, to where the energy levels converge at $n = \infty$, when the electron is free of the attraction of the nucleus. We shall look more closely at ionisation energy on pages 58–59.

QUESTION 5

The Paschen series is one of the series of spectral lines for hydrogen, this time in the infrared part of the spectrum and for movement of the electron to the $n = 3$ level.

a) Draw a diagram similar to the Lyman series diagram in Fig 3.13 to show the first three electron transitions of the Paschen series.

b) Which transition in this series will result in the lowest-energy line in the infrared spectrum?

SUMMARY OF THE BOHR MODEL OF THE ATOM

- Electrons exist only in certain permitted energy levels and in these levels they do not emit energy.
- Electrons move to higher energy levels by absorbing quanta of energy. They return to lower energy levels by emitting these quanta as photons of light, which show up as lines in different parts of the electromagnetic spectrum.

The Bohr model of the atom successfully explained the lines in the emission spectrum of the hydrogen atom. It worked for hydrogen, the simplest atom with just one electron, but it did not predict accurately the spectral lines of atoms with several electrons.

The Bohr model is important because it used the idea of *quantised* energy levels to explain atomic structure and provided a foundation on which others could build. The Nobel Prize went to Bohr in 1922, one year after Einstein had received the prize for his explanation of the photoelectric effect.

QUESTION 6

Read the box on the absorption spectrum of the Sun below. In the spectrum, what is the colour of the helium line that Janssen observed?

SCIENCE IN CONTEXT

DISCOVERING SODIUM

Almost 50 years before Kirchhoff and Bunsen used emission spectra to identify elements (see page 49), the German scientist Josef von Fraunhofer was experimenting with light. When he passed sunlight through a very fine slit and a high-quality prism, some wavelengths were missing from the sunlight's absorption spectrum. They showed up on his spectrum as black lines in the otherwise continuous spectrum. These lines represented energies that were being absorbed by some unknown material.

Fraunhofer recorded the wavelengths of several hundred black lines, which are now named after him. Two were close together in the yellow part of the spectrum at almost 600 nm. At exactly the same wavelengths, Bunsen and Kirchhoff found two characteristic yellow lines in the emission spectrum of sodium (look back to Fig 3.4, on page 49). They concluded that sodium must be in the Sun's atmosphere. Using this hypothesis, they soon discovered other elements in the Sun.

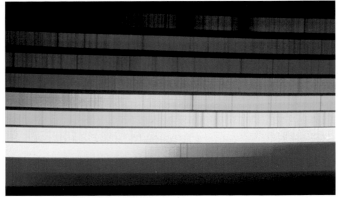

Fig 3.14 Fraunhofer found over 600 dark lines in the visible spectrum of the Sun. These represent wavelengths missing from a continuous spectrum of light from the Sun. Some of the lines have been matched with the elements responsible for them

To explain the Fraunhofer lines more fully, we need to look at the Bohr model of the atom again. When excited, electrons fall from higher energy levels to lower levels: they *emit* energy to give a line *emission* spectrum. In reverse, they reach the higher (excited) energy levels by *absorbing* quanta of energy (photons).

So, in an *absorption* spectrum, the quanta of energy the electrons absorb in going to these higher energy levels will be represented as black lines for the wavelengths missing from the spectrum.

For Fraunhofer's spectrum for sunlight, the light has passed from the interior of the Sun through the Sun's atmosphere. Assuming that there is sodium vapour in the Sun's atmosphere and that its electrons are being excited to higher energy levels, then the corresponding points in the spectrum for the light reaching Earth will be missing – hence the dark lines for sodium.

DISCOVERING HELIUM

In 1868, the French astronomer Pierre Jules Janssen took his spectroscope to India to view a total eclipse of the Sun. In the spectrum of light from the Sun's corona, he saw a bright line very close to the two sodium lines. This line could not be accounted for by the line spectra of any known element. In the same year the English scientist, Norman Lockyer, also observed this line and suggested that it was caused by a new element. It was given the name helium, after the Greek word *helios* for the Sun.

At the time, Lockyer was ridiculed for his suggestion that he had identified a new element. Yet this discovery later helped helium to be discovered on Earth. In 1895, a quarter of a century after helium was found on the Sun, it was isolated by the Scottish chemist William Ramsey. He found the gas trapped in a uranium ore, where it had been formed as a product of radioactive decay.

SCIENCE IN CONTEXT

Lasers – amplifying the energy from electron transitions

The letters in the word 'laser' stand for 'light amplification by stimulated emission of radiation'. Lasers, invented in 1960, have revolutionised medicine, technology and science. The ruby laser was the first, and it works by using a burst of ultraviolet light to excite electrons in atoms of a ruby rod. In the same instant, a few electrons fall back to lower energy levels and produce photons. These photons are bounced back and forth by mirrors at the ends of the rod and stimulate other excited electrons to fall back to lower levels, emitting a simultaneous burst of photons of the same frequency. This laser pulse is very intense, and provided one of the mirrors is partially transparent, it can pass through it. The ruby laser produces a wavelength in the red part of the spectrum. Other lasers use other materials and produce wavelengths from infrared to X-ray.

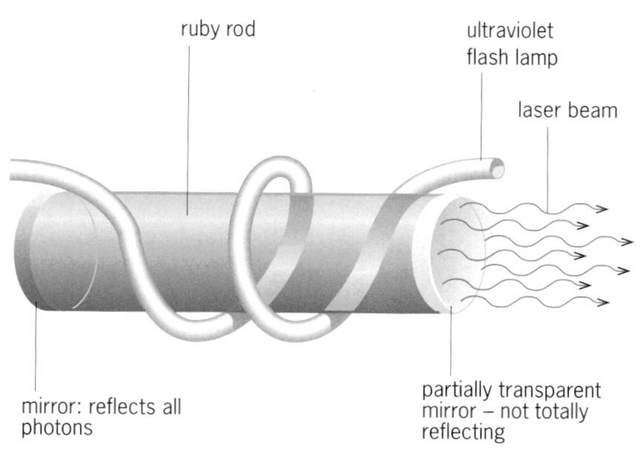

Fig 3.15 A ruby laser emitting laser light

STRETCH AND CHALLENGE

Electrons as waves producing images of smaller and smaller objects

In 1923, Louis de Broglie proposed that, just as light could behave as a particle or a wave, so could matter. He suggested that electrons, regarded as particles, also had wave-like properties. Four years later, a beam of electrons was diffracted (bent and scattered) by a metal crystal to form a pattern. This diffraction could only be explained by assuming that the electrons were behaving as waves.

The electron microscope uses the wave-like behaviour of electrons. It allows us to see images of very small objects in much greater detail than we can see with ordinary microscopes, which rely on visible light. Light is visible to us between wavelengths of about 400 to 700 nm. To form an image of an object, the object cannot be smaller than half the shortest wavelength, that is, 200 nm (or a 20 thousandth of a millimetre).

Electrons travelling at high speed have very short wavelengths, and they allow images of objects measuring 2×10^{-3} nm to be made and magnified. For this reason, the electron microscope is a very helpful tool in biological and chemical research.

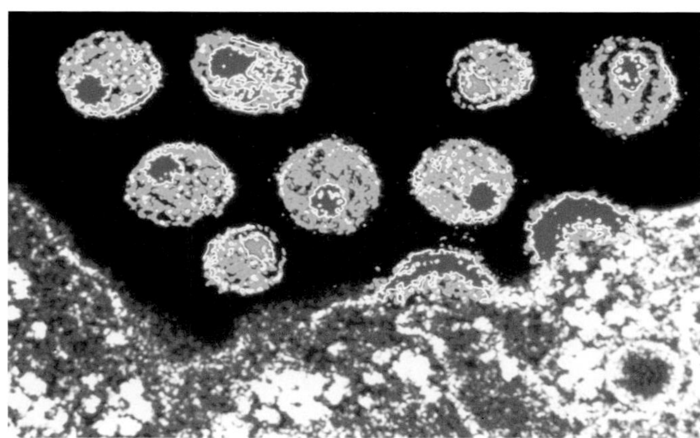

Fig 3.16(a) (left) An electron micrograph of AIDS viruses budding off from the surface of a human T-lymphocyte, a white blood cell. Such images of organisms that cause disease help scientists to seek a way to combat infection

In 1986, the scanning tunnelling microscope (STM) took electron microscopy one stage further. With the aid of a computer to process the data and enhance the image, scientists used the charge on the electrons around atoms to make images of separate atoms and molecules. A newer technique involves probing the surface of a substance with a fine conducting tip. The instrument that does this is the scanning probe microscope (SPM). SPMs have enabled the nanotechnology revolution to take off (see page 145).

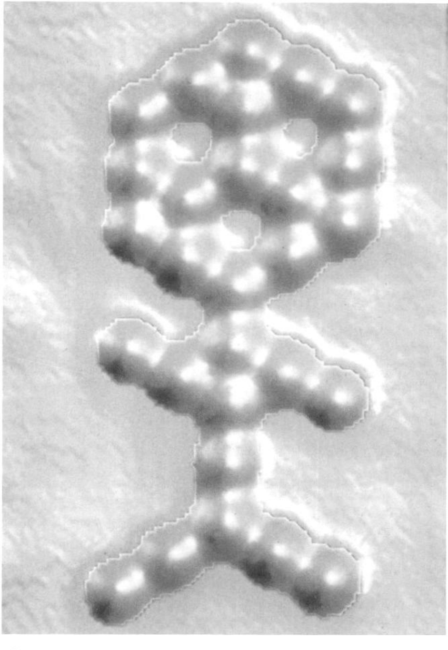

Fig 3.16(b) (above) A scanning tunnelling microscope was used to produce this piece of 'atomic art'. To make this molecular person, 28 carbon monoxide (CO) molecules were positioned on a piece of platinum by using a charged probe wire

7 A MODERN MODEL OF HOW ELECTRONS ARE ARRANGED

Once it was realised that electrons could behave like waves, a new model of the atom was possible. This model is based on some very complicated mathematics to describe the wave properties of electrons. Erwin Schrödinger takes the credit for devising an equation that describes the energy levels for electrons in hydrogen and other atoms. (The mathematics is beyond the scope of this book.) His model is known as the **quantum mechanical model**.

ELECTRON SHELLS

Electron shells correspond to the energy levels that Bohr first identified in his model of the atom. The first shell has a **principal quantum number** $n = 1$, the second shell's principal quantum number is $n = 2$, and so on, as in Table 3.1. The first shell is closest to the nucleus and has the lowest energy. As the principal quantum number increases, so does its energy.

Each shell can hold a maximum number of electrons. You will already have met shells in a previous chemistry course, and you may remember that the arrangement of electrons is known as its **electron configuration**, also referred to as **electronic structure**.

REMEMBER THIS

The energy level of an electron is given a principal quantum number, and this number is also given to the shell it is in.

Table 3.1 Principal quantum numbers for electrons, and the maximum number of electrons each shell can contain

Principal quantum number, *n*	Shell	Max. no. of electrons in shell
1	first	2
2	second	8
3	third	18
4	fourth	32

QUESTION 7

7 If you think like a mathematician, you may be able to work out a simple formula containing the principal quantum number, *n*, which can be used to predict the maximum number of electrons in the *n*th shell. Hence, calculate the number of electrons in the fifth shell. (Hint: the same formula can be used to predict the maximum number in the first four shells.)

QUESTION 8

8 Make drawings like that in Fig 3.17 of the following atoms:

a) aluminium ($Z = 13$)
b) carbon ($Z = 6$)
c) chlorine ($Z = 17$)
(Hint: refer also to Table 3.1.)

EXAMPLE

Q Magnesium has an atomic (proton) number (Z) of 12. Work out the arrangement of electrons in the shells of a magnesium atom.

A The atomic number tells you there are 12 electrons. (The atomic number is the number of protons in an atom, and as the atom has no net charge, it has the same number of electrons. If you need to remind yourself about this, look back to page 35.)

The shells of lowest energy are filled first, so:

- shell 1 will take 2 electrons, which is all it can hold;
- shell 2 will take the next 8 electrons, the maximum it can hold;
- shell 3 will take the remaining two.

So the electron configuration of magnesium is 2,8,2. You can use a *dot-and-cross* diagram to represent this, as shown in Fig 3.17.

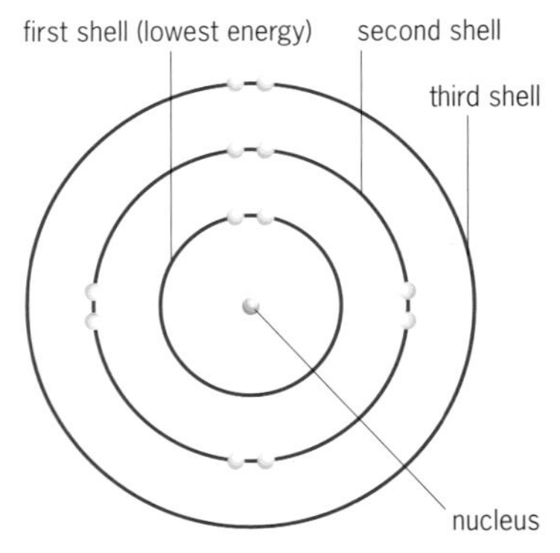

Fig 3.17 The arrangement of the electrons in the shells of a magnesium atom. This is known as a dot-and-cross diagram because the electrons are represented by dots (as here), or crosses, or both. The first shell is at the lowest energy. The second and third shells are at increasingly higher energies. It requires little additional energy to remove electrons from the third shell

EVIDENCE FOR SHELLS FROM SUCCESSIVE IONISATION ENERGIES

Earlier in this chapter, we met the term ionisation energy, also referred to as ionisation enthalpy, which is the energy that an electron must be given to remove it from an atom (or ion). Now let's look at this term more closely. In order to measure ionisation energy, the atoms have to be separated, and this means they must be in the gaseous state. For convenience, the value for ionisation energies is given for one mole of atoms (or ions):

The energy required to remove one mole of electrons from one mole of gaseous atoms to form one mole of ions with a single positive charge is called the first ionisation energy.

The first ionisation energy may be referred to as standard molar ionisation energy.

So the first ionisation energy is the energy required to do this:

$$X(g) \rightarrow X^+(g) + e^- \quad \Delta H = \text{first ionisation energy in kJ mol}^{-1}$$

where X is any element, (g) tells you that it is gaseous, and e^- is the symbol for an electron.

The **second ionisation energy** is the energy required to remove a second mole of electrons:

$$X^+(g) \rightarrow X^{2+}(g) + e^-$$

Let's look again at magnesium with its electron configuration of 2,8,2.

First ionisation energy: $Mg(g) \rightarrow Mg^+(g) + e^-$

$\Delta H = +738\text{ kJ mol}^{-1}$

Second ionisation energy: $Mg^+(g) \rightarrow Mg^{2+}(g) + e^-$

$\Delta H = +1451\text{ kJ mol}^{-1}$

Third ionisation energy: $Mg^{2+}(g) \rightarrow Mg^{3+}(g) + e^-$

$\Delta H = +7733\text{ kJ mol}^{-1}$

The ionisation energy increases as each successive electron is removed. This is the reason: the positive nuclear charge stays the same, because there are still 12 protons in the nucleus, and each time an electron is removed, the remaining ones are attracted more strongly by the nucleus. Notice that there is a large jump in the energy required to remove the third electron. This is because we are breaking into the second shell, which is closer to the nucleus. This shell is also less **shielded** by inner electron shells from the positive charge of the nucleus, because there is only one full shell between it and the nucleus.

From these values, you can see that there is a large range in the ionisation energies of magnesium. It would be difficult to choose a scale on which to plot them. But if we convert the ionisation energies into logarithms to the base 10 ($\log_{10}$), we condense the scale and make it more manageable, as in Fig 3.18. Don't worry if you are not familiar with logarithms. Using them here is just a way of making the pattern of the graph easier to see.

QUESTION 9

a) Write an equation for the 11th ionisation energy of magnesium (omitting the energy value).
b) The energy required to remove the 11th electron is almost 170 000 kJ mol^{-1}, but the tenth ionisation energy is 'only' 35 500 kJ mol^{-1}. Explain this huge jump in energy to remove the 11th electron.

QUESTION 10

Sketch a graph like Fig 3.18 for $\log_{10}$ of the successive ionisation energies of sodium, and explain its shape. (Hint: sodium has an electron configuration of 2,8,1.)

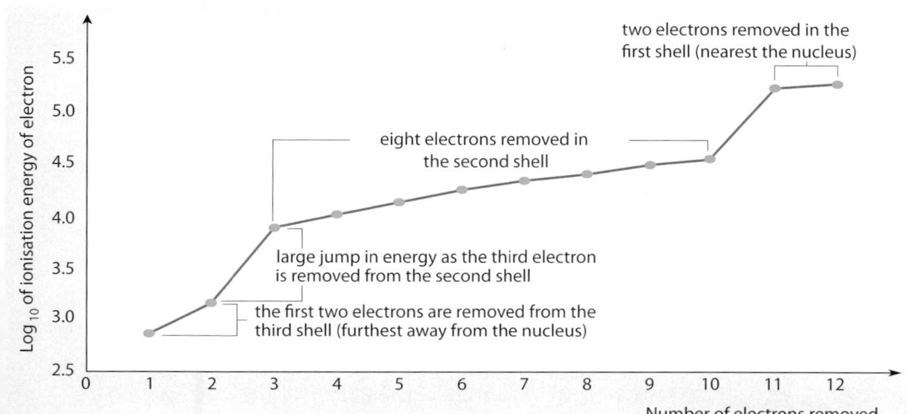

Fig 3.18 Graph of the successive ionisation energies (as logarithms to base 10) of the magnesium atom against the number of electrons removed. The values for electrons closest to the nucleus are on the right

ARRANGING ELECTRONS IN SUBSHELLS

Let us now return to the atomic emission spectra of elements. When looked at in finer detail, the lines on these spectra are seen to be divided into more lines. Each finer line represents the energy level of a **subshell**. So electrons arranged in shells are also subdivided into subshells, or **sub-levels**. The fine lines result from electron transitions (movement of electrons) between the subshells. The subshells are known by letters:

- **s** subshell contains 2 electrons
- **p** subshell contains 6 electrons
- **d** subshell contains 10 electrons
- **f** subshell contains 14 electrons

The letters were given in the early 20th century and refer to spectral lines. Some of the lines were **s**harp, hence **s**, some were more spread out or **d**iffuse, and some of the lines were brighter and called **p**rincipal lines. (Note: in this book we do not go into detail about the **f** subshell.)

We can now see in Table 3.2 how the shells are split into subshells.

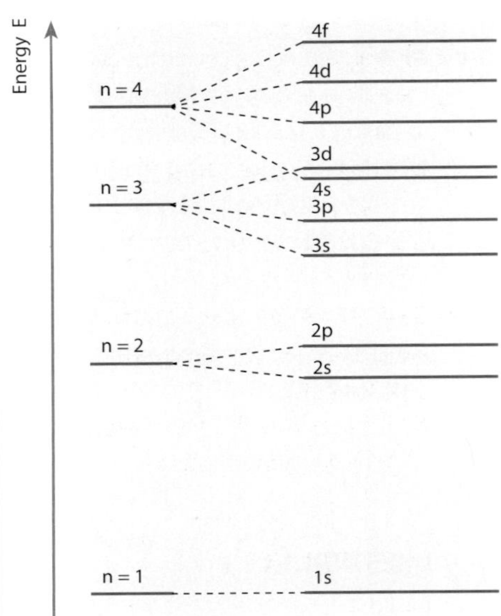

Fig 3.19 The energies of the various subshells in an atom with many electrons

Table 3.2 The system for arranging electrons in subshells

Principal quantum number, *n*	Shell number	Subshells	Maximum number of electrons	
1	1	1s	2	Total = 2
2	2	2s	2	Total = 8
		2p	6	
3	3	3s	2	Total = 18
		3p	6	
		3d	10	
4	4	4s	2	Total = 32
		4p	6	
		4d	10	
		4f	14	

Notice that the subshell (sub-level) takes the number of the principal quantum number or shell.

The energies of the subshells are shown in Fig 3.19. Notice that the 3d subshell has a higher energy than the 4s subshell. This has important consequences for the chemistry of the transition elements, which you can read more about in Chapter 24.

QUESTION 11

11 **a)** What is meant by the first ionisation energy of an element?
b) Explain why the first ionisation energy of neon is higher than that of lithium.
c) Explain why the first ionisation energy decreases going down Group 2. How does this provide evidence for electron shells (energy levels)?

EVIDENCE FOR SUBSHELLS FROM FIRST IONISATION ENERGIES

The evidence for subshells also comes from ionisation energies. If we plot the first ionisation energy of different elements against their atomic numbers, a regular pattern emerges (Fig 3.20). The repeating pattern of a property is called **periodicity**. It forms the basis of the Periodic Table, and you can find out more about it in Chapter 5.

Several features of the graph in Fig 3.20 point to the existence of subshells. Let's look at the eight elements lithium to neon, which have electrons in shell 2. We would expect an increase in the first ionisation energy as the atomic number increases. This is because the number of protons in the nucleus increases, which increases the nuclear charge. If the electrons were all in the same energy level, we would expect the graph to be a straight line, showing a steady increase in the first ionisation energy as the number of protons in the nucleus increases.

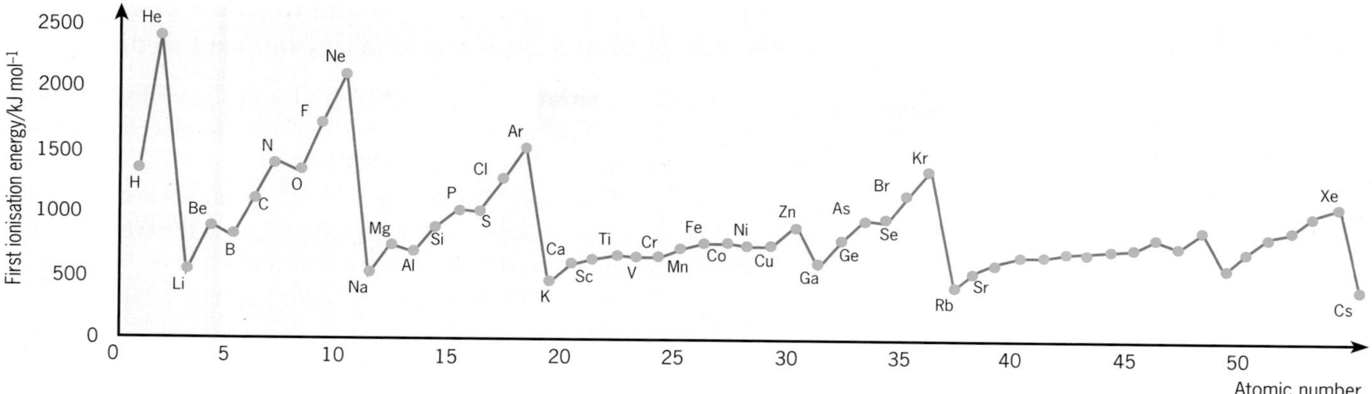

Fig 3.20 Graph showing the periodicity of first ionisation energies as they vary with atomic number (proton number)

The 2,3,3 pattern of first ionisation energies tells us that the electrons removed from this second electron shell are not arranged such that they are all in the same energy level.

From reading this section on subshells, you may be able to explain the 2,3,3 pattern. Refer to Fig 3.20 as you read on.

The first part of the 2,3,3 pattern is caused by electrons being taken from the 2s subshell. There is an increase in first ionisation energy from Li to Be because Be has an extra proton attracting the outer electrons. Then, instead of a further increase from Be to B, there is a decrease in first ionisation energy as the electron in B is taken from the 2p subshell: this electron is at a higher energy level than the 2s subshell and further from the nucleus.

In moving from B to N, there is the expected increase in first ionisation energy, but another dip occurs at O. After N, the electrons in the 2p subshell start to pair up. (This pairing of electrons in subshells is covered in the next section.) There is more electrostatic repulsion between paired electrons, which means that the fourth electron in the 2p subshell is easier to remove, explaining the dip at O.

You can see the same pattern repeated for elements starting with Na when the next shell is being filled. After Mg, there is a drop at Al, and another drop at S. The 3p subshell is complete at Ar.

After this, there is an interruption to the 2,3,3 pattern. There is the expected increase from K to Ca as electrons are taken from the same 4s subshell. Then we have electrons removed from the slightly higher energy 3d subshell (refer back to Fig 3.19) before the 3,3 pattern returns, beginning with Ga, as the 4p subshell electrons are removed.

QUESTION 12

Read about the 2,3,3 pattern for first ionisation energies of the elements lithium to neon and look at Fig 3.20. Then answer the following questions:

a) Why is there an increase in first ionisation energy from sodium to magnesium?

b) Explain why there is a drop at i) aluminium and ii) sulfur.

QUESTION 13

The first ionisation energy of beryllium (Be) is 900 kJ mol^{-1}, while that of boron (B) is 801 kJ mol^{-1}. Explain this, using information about subshells in the text.

8 ATOMIC ORBITALS

Another great scientist of the 20th century was Werner Heisenberg, whose work has clarified our current model of the way electrons are arranged in atoms. In 1927, he said that you could determine either the speed of an electron or its position, but not both at the same time. This is now known as the **Heisenberg uncertainty principle**. While it applies to any particle, it becomes important only when the particle is very tiny. Heisenberg's mathematics has enabled electron arrangements to be worked out in even more detail.

We have seen that lines appear in the emission spectra of excited atoms, representing the wavelengths (energies) of photons emitted by excited electrons returning to lower energy levels. When this light is passed through a magnetic field, even more lines show up, and these are evidence of what we can now call **atomic orbitals**.

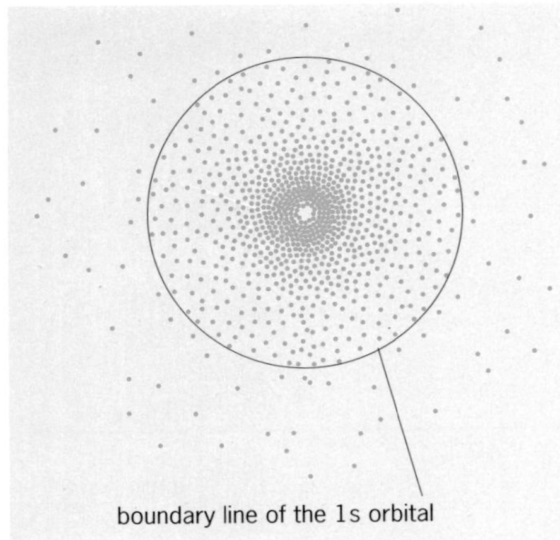

Fig 3.21(a) Electron density plot showing a section through the 1s orbital. If we could map the position of an electron as a dot at regular intervals, then over a period of time 90 per cent of the dots could be enclosed by drawing a boundary that is spherical. This drawing represents a slice through the sphere. Notice that the electron charge is concentrated close to the nucleus and tails off further out

An atomic orbital is a region around the atom in which there is a high probability of finding an electron at any moment in time.

Advanced mathematics can be used to plot the probable position of an electron at one instant, followed by further plots at other instants until an **electron density plot** or **electron density map** is produced (Fig 3.21(a)).

The existence of electrons in orbitals helps the chemistry of atoms and molecules to fall into place, and so is tremendously useful. In the s subshell there is only one orbital, the s orbital. If we draw around the region of the atom where the electron spends most of its time, then we find the s orbital is spherical (Fig 3.21(a)).

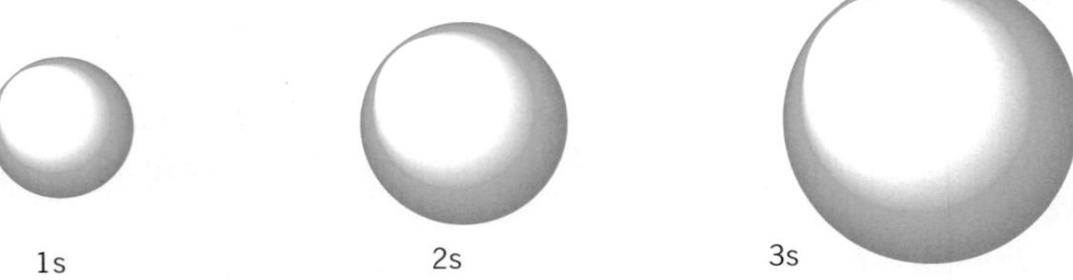

Fig 3.21(b) The boundary surface of 1s, 2s and 3s orbitals. In each case, the imaginary 'surface' is the limit of the space in which the 1s, 2s and 3s electrons are likely to be found. Also in each case, the orbital is centred on the nucleus, and the boundary represents a 90 per cent probability of finding an electron within it

There are three p orbitals in the p subshell. They are dumb-bell shaped and arranged at right angles to each other, with the nucleus in the centre of each dumb-bell.

The d subshell contains five d orbitals. We shall look at the shapes of d orbitals in Chapter 24. There are seven f orbitals in the f subshell.

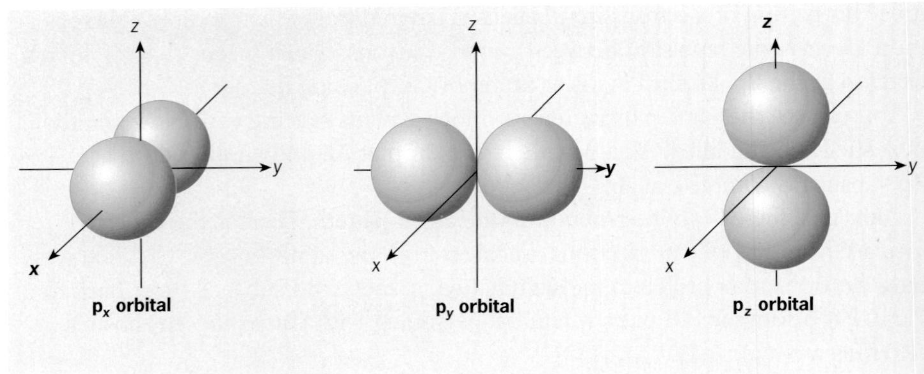

Fig 3.22 The three p orbitals found in the p subshell

SPINNING ELECTRONS

Each atomic orbital can hold a maximum of two electrons. As the electrons move in their orbitals, they are also spinning on their own axes. If there are two electrons, one spins clockwise, while the other spins anticlockwise. This keeps the repulsion of the electrons to a minimum. Electron spin is often represented by box diagrams, in which each box is an atomic orbital and arrows represent the electrons spinning in opposite directions, as in Fig 3.23.

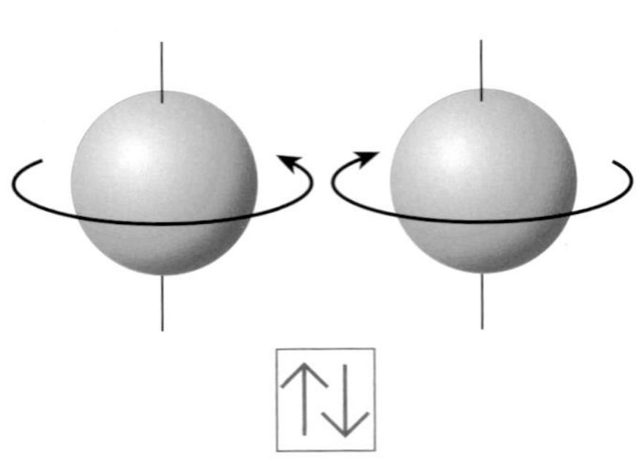

Fig 3.23 Two electrons in the same atomic orbital spin in opposite directions, which is represented by a box diagram with arrows showing the opposite spin of each electron. The box is the atomic orbital

ELECTRON ADDRESSES

So we are now in a position to give *addresses* to electrons. In the same way that a letter can be addressed with the country, the town, the street and the house number so that it arrives at only one unique destination, so we can give a unique 'address' to represent the state of an electron in an atom. The idea of a unique state for every electron is known as the **Pauli exclusion principle**.

According to this principle, an electron is precisely defined by:

- its shell – the main energy level;
- its subshell – the sub-energy level division of the shell;
- its atomic orbital – each subshell possesses at least one atomic orbital, and all orbitals in a subshell possess the same energy;
- and its direction of spin – a maximum of two electrons of opposite spin can occupy one atomic orbital: orbitals in the same subshell are singly filled first, with electrons of parallel spin.

Fig 3.24 shows the electron configuration of the magnesium atom in its ground state. Remember, this is the lowest energy state of the atom, so the electrons occupy orbitals in subshells of the lowest possible energy.

Notice that:

- the s subshells contain one s atomic orbital;
- the p subshells contain three p orbitals;
- each orbital contains a maximum of two electrons with opposite spin.

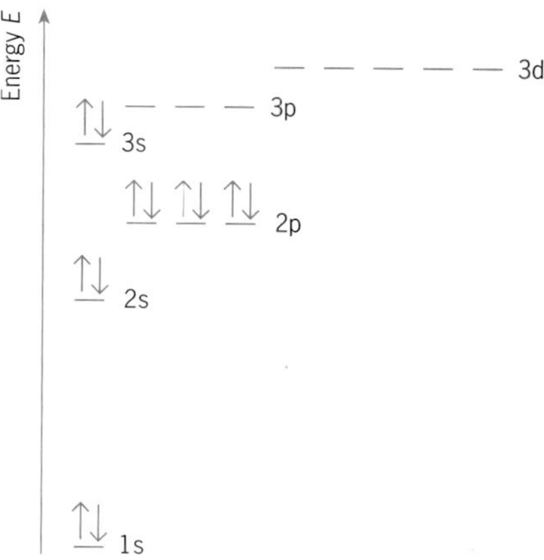

Fig 3.24 The arrangement of electrons in the atomic orbitals of a magnesium atom. The atom is in its ground state, so no electrons have been excited into higher energy orbitals

9 THE ELECTRON CONFIGURATIONS OF THE FIRST 36 ELEMENTS

Let us now look at how electrons fill orbitals in an atom. The orbitals of lowest energy are occupied first, and this is known as the **aufbau principle**. Aufbau is a German word for building up. So when you are building up the electron configuration of an element, remember that each electron goes into the lowest energy orbital available.

Hydrogen (*Z* = 1)

The lowest energy orbital is in the first shell that has a principal quantum number $n = 1$ and is an s orbital, so it is written 1s.

1s

[↑] electron configuration $1s^1$

Helium (*Z* = 2)

1s

[↑↓] electron configuration $1s^2$

Note: when you say $1s^2$, it is 'one s two', not 'one s squared'!

Lithium (*Z* = 3)

1s 2s

[↑↓] [↑] electron configuration $1s^2\,2s^1$

The lithium atom has three electrons, and the aufbau principle tells us that the third electron electron will occupy the next orbital of lowest energy, which is in the $n = 2$ shell, so it will start with 2 and be an s orbital, and therefore it is 2s.

Beryllium (*Z* = 4)

1s 2s

[↑↓] [↑↓] electron configuration $1s^2\,2s^2$

QUESTION 14

Draw the electron configuration of the Mg^{2+} ion in a similar way to Fig 3.24.

REMEMBER THIS

Another name for shell is energy level, and for subshell is sub-level.

Next, we come to filling the p subshell. Notice that each orbital is occupied singly at first. This is because two electrons in the same orbital exert a repulsion that raises the energy of a doubly filled orbital above that of a singly filled one. Notice that the spins are all in the same direction. The repulsion that results helps to keep the electrons apart.

Boron (*Z* = 5)

1s	2s	2p		
↑↓	↑↓	↑		

electron configuration $1s^2 2s^2 2p^1$

Carbon (*Z* = 6)

1s	2s	2p		
↑↓	↑↓	↑	↑	

electron configuration $1s^2 2s^2 2p^2$

Nitrogen (*Z* = 7)

1s	2s	2p		
↑↓	↑↓	↑	↑	↑

electron configuration $1s^2 2s^2 2p^3$

Oxygen (*Z* = 8)

1s	2s	2p		
↑↓	↑↓	↑↓	↑	↑

electron configuration $1s^2 2s^2 2p^4$

Fluorine (*Z* = 9)

1s	2s	2p		
↑↓	↑↓	↑↓	↑↓	↑

electron configuration $1s^2 2s^2 2p^5$

Neon (*Z* = 10)

1s	2s	2p		
↑↓	↑↓	↑↓	↑↓	↑↓

electron configuration $1s^2 2s^2 2p^6$

EXAMPLE

Q What is the electron configuration of sulfur (Z = 16)?

A The atomic number is 16, so there are 16 electrons. We already know that the arrangement in the first two shells is:

$$1s^2 2s^2 2p^6$$

This takes ten electrons, leaving six electrons to place. Remember to start with the lowest energy orbital available, the 3s. This orbital takes two electrons:

$$1s^2 2s^2 2p^6 3s^2$$

leaving another four to make the electron configuration for sulfur:

$$1s^2 2s^2 2p^6 3s^2 3p^4$$

The next eight elements follow the same pattern, filling up the s and p orbitals of the third shell.

Potassium and calcium come next. The 3d subshell is available, but remember (Fig 3.19, page 60) that the 4s subshell has a lower energy, so it is filled first.

? QUESTION 15

15 **a)** Write down the electron configuration of sodium (Z = 11), chlorine (Z = 17) and aluminium (Z = 13).

b) Write down the electron configuration of Na^+, Cl^- and Al^{3+}.

Potassium (*Z* = 19)

1s	2s	2p			3s	3p			3d					4s
↑↓	↑↓	↑↓	↑↓	↑↓	↑↓	↑↓	↑↓	↑↓						↑

electron configuration $1s^2 2s^2 2p^6 3s^2 3p^6 4s^1$

Notice that we do not write $3d^0$ in the electron configuration. We leave it out. Similarly, we can draw the box diagram for calcium like this:

Calcium (*Z* = 20)

1s	2s	2p			3s	3p			4s
↑↓	↑↓	↑↓	↑↓	↑↓	↑↓	↑↓	↑↓	↑↓	↑↓

electron configuration $1s^2 2s^2 2p^6 3s^2 3p^6 4s^2$

FILLING THE d ORBITALS (*Z* = 21–30)

Table 3.3 Electron configurations of titanium (*Z* = 21) to zinc (*Z* = 30)

Z	Element	Electron configuration	Electron spin diagram: 1s	2s	2p	3s	3p	3d	4s
21	Sc	$1s^2 2s^2 2p^6 3s^2 3p^6 3d^1 4s^2$	↑↓	↑↓	↑↓ ↑↓ ↑↓	↑↓	↑↓ ↑↓ ↑↓	↑ □ □ □ □	↑↓
22	Ti	$1s^2 2s^2 2p^6 3s^2 3p^2 3d^2 4s^2$	↑↓	↑↓	↑↓ ↑↓ ↑↓	↑↓	↑↓ ↑↓ ↑↓	↑ ↑ □ □ □	↑↓
23	V	$1s^2 2s^2 2p^6 3s^2 3p^2 3d^3 4s^2$	↑↓	↑↓	↑↓ ↑↓ ↑↓	↑↓	↑↓ ↑↓ ↑↓	↑ ↑ ↑ □ □	↑↓
24	Cr	$1s^2 2s^2 2p^6 3s^2 3p^6 3d^5 4s^1$	↑↓	↑↓	↑↓ ↑↓ ↑↓	↑↓	↑↓ ↑↓ ↑↓	↑ ↑ ↑ ↑ ↑	↑
25	Mn	$1s^2 2s^2 2p^6 3s^2 3p^6 3d^5 4s^2$	↑↓	↑↓	↑↓ ↑↓ ↑↓	↑↓	↑↓ ↑↓ ↑↓	↑ ↑ ↑ ↑ ↑	↑↓
26	Fe	$1s^2 2s^2 2p^6 3s^2 3p^6 3d^6 4s^2$	↑↓	↑↓	↑↓ ↑↓ ↑↓	↑↓	↑↓ ↑↓ ↑↓	↑↓ ↑ ↑ ↑ ↑	↑↓
27	Co	$1s^2 2s^2 2p^6 3s^2 3p^6 3d^7 4s^2$	↑↓	↑↓	↑↓ ↑↓ ↑↓	↑↓	↑↓ ↑↓ ↑↓	↑↓ ↑↓ ↑ ↑ ↑	↑↓
28	Ni	$1s^2 2s^2 2p^6 3s^2 3p^6 3d^8 4s^2$	↑↓	↑↓	↑↓ ↑↓ ↑↓	↑↓	↑↓ ↑↓ ↑↓	↑↓ ↑↓ ↑↓ ↑ ↑	↑↓
29	Cu	$1s^2 2s^2 2p^6 3s^2 3p^6 3d^{10} 4s^1$	↑↓	↑↓	↑↓ ↑↓ ↑↓	↑↓	↑↓ ↑↓ ↑↓	↑↓ ↑↓ ↑↓ ↑↓ ↑↓	↑
30	Zn	$1s^2 2s^2 2p^6 3s^2 3p^6 3d^{10} 4s^2$	↑↓	↑↓	↑↓ ↑↓ ↑↓	↑↓	↑↓ ↑↓ ↑↓	↑↓ ↑↓ ↑↓ ↑↓ ↑↓	↑↓

Table 3.3 shows the electron configurations of the next ten elements. The 4s subshell has already been filled at calcium, so now the 3d subshell, which is at a slightly higher energy level, can begin to fill. Electrons spinning in the same direction first occupy the 3d orbitals singly, since this helps to minimise electron repulsion.

Notice that the chromium atom and the copper atom have only one electron in the 4s atomic orbital. The reason for this is explained in Chapter 24.

GALLIUM (*Z* = 31) TO KRYPTON (*Z* = 36)

From gallium to krypton the 4p orbitals are filled because they have the next highest energy level (see Fig 3.19).

Gallium (*Z* = 31)

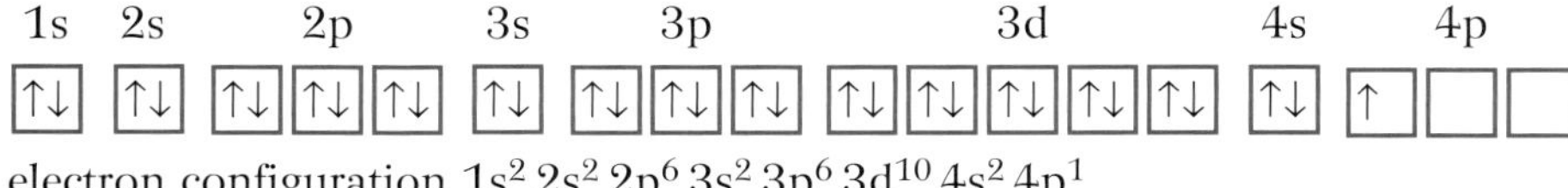

electron configuration $1s^2\ 2s^2\ 2p^6\ 3s^2\ 3p^6\ 3d^{10}\ 4s^2\ 4p^1$

Krypton (*Z* = 36)

1s	2s	2p	3s	3p	3d	4s	4p
↑↓	↑↓	↑↓ ↑↓ ↑↓	↑↓	↑↓ ↑↓ ↑↓	↑↓ ↑↓ ↑↓ ↑↓ ↑↓	↑↓	↑↓ ↑↓ ↑↓

electron configuration $1s^2\ 2s^2 2p^6\ 3s^2\ 3p^6\ 3d^{10}\ 4s^2\ 4p^6$

QUESTIONS 16–17

16 Write down the electron configurations of arsenic (*Z* = 33), selenium (*Z* = 34) and bromine (*Z* = 35).

17 Period 5 begins with rubidium, atomic number 37, and continues to xenon, atomic number 54. Work out the electron configurations of the following elements. The Periodic Table in Appendix 2 will give you the necessary atomic numbers.

a) rubidium
b) yttrium
c) cadmium
d) indium
e) iodine
f) xenon

10 ELECTRON CONFIGURATIONS AND THE PERIODIC TABLE

When we know the electron configurations of the elements, we can organise and explain the chemical properties of the elements, because it is the electrons that determine these properties. Long before electrons were even known to exist, Mendeleev grouped together elements with similar properties in his Periodic Table. You can read much more about this in Chapter 5. However, from your previous course you will know about the Periodic Table, so let's see how electron configurations fit in.

Group	1	2											3	4	5	6	7	0
Period 1						H $1s^1$												He $1s^2$
2	Li $2s^1$	Be $2s^2$											B $2p^1$	C $2p^2$	N $2p^3$	O $2p^4$	F $2p^5$	Ne $2p^6$
3	Na $3s^1$	Mg $3s^2$											Al $3p^1$	Si $3p^2$	P $3p^3$	S $3p^4$	C1 $3p^5$	Ar $3p^6$
4	K $4s^1$	Ca $4s^2$	Sc $3d^14s^2$	Ti $3d^24s^2$	V $3d^34s^2$	Cr $3d^54s^1$	Mn $3d^54s^2$	Fe $3d^64s^2$	Co $3d^74s^2$	Ni $3d^84s^2$	Cu $3d^{10}4s^1$	Zn $3d^{10}4s^2$	Ga $4p^1$	Ge $4p^2$	As $4p^3$	Se $4p^4$	Br $4p^5$	Kr $4p^6$

Fig 3.25 Electron configuration for the outermost subshell of the first 36 elements

Fig 3.25 shows the electron configuration for the outermost subshell of the first 36 elements. In Period 4, containing the transition elements, electron configurations show the two outermost subshells, because it is the d subshell that is being filled. Notice that the elements in each of the eight main groups end with the same number of electrons in their outer subshell. So Group 7 always has a p^5 subshell and Group 2 always has a filled s^2 orbital. The Periodic Table is often divided into blocks, as in Fig 3.26, according to the subshell being filled.

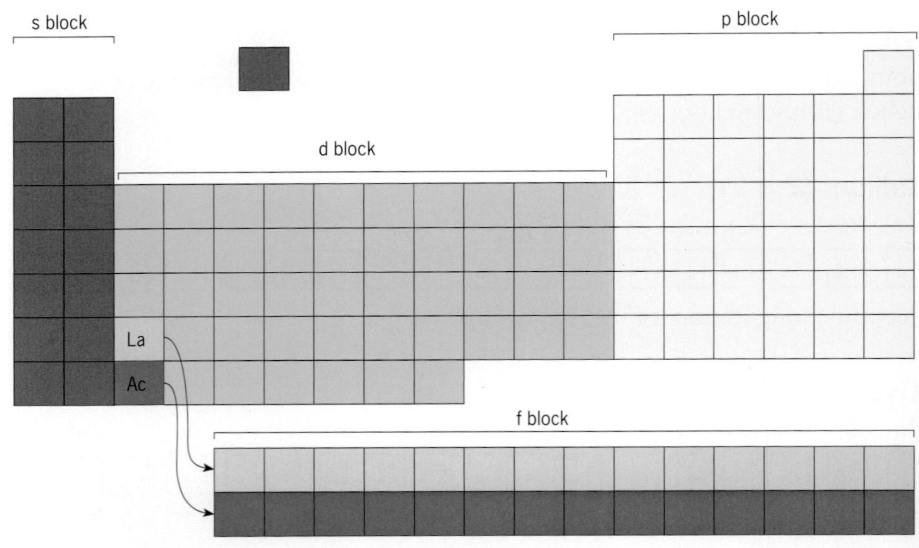

Fig 3.26 The Periodic Table, showing the different blocks labelled according to which outer orbital is being filled

So when we know how electrons are arranged in atoms, we have a better understanding of how the Periodic Table is arranged. It groups elements together with similar properties, and we can now see that it is the electron configuration that determines this similarity in properties.

? QUESTION 18

18 Predict the electron configuration of the outermost subshell of:

a) polonium;
b) radium;
c) astatine;
d) radon.

(Hint: you will need to look at the Periodic Table in Appendix 2.)

SUMMARY

When you have studied this chapter, you should be able to understand the following ideas.

- Evidence for the arrangement of electrons in energy levels comes from the line emission spectra of atoms.
- Energy levels are quantised, which means that each level has a fixed amount of energy.
- An electron can move from a lower to a higher energy level by absorbing a quantum (packet) of energy. It releases the same quantum of energy when it falls back, as a photon of light of a particular frequency (which gives a line in the emission spectrum).
- The arrangement of electrons in an atom is called its electron configuration.
- Energy levels and sub-levels are also known as shells and subshells.
- Shells are numbered 1, 2, 3 and so on. The shell number is also known as the principal quantum number.
- The higher the shell number, the higher is its energy and the further the shell is from the nucleus.
- Successive ionisation energies of atoms provide evidence of shells (energy levels).
- Within each shell there are subshells (sub-levels) s, p, d and f, of different energies. Shell 1 contains only an s subshell. Shell 2 contains two subshells, s and p, while shell 3 contains three subshells s, p and d.
- A plot of first ionisation energy against atomic number provides evidence of subshells. In each subshell the electrons are arranged in atomic orbitals.
- Atomic orbitals are regions in the atom where there is a high probability of finding the electron.
- There is one s atomic orbital in the s subshell, three p orbitals in the p subshell, five d orbitals in the d subshell and seven f orbitals in the f subshell.
- Each atomic orbital can accommodate two electrons of opposite spins.
- The Periodic Table is arranged in blocks, s, p, d and f, according to the subshell that is being filled.

Practice questions and a How Science Works assignment for this chapter are available at www.collinseducation.co.uk/CAS

4

Chemical bonding and the shapes of molecules and ions

4 CHEMICAL BONDING AND THE SHAPES OF MOLECULES AND IONS

SCIENCE IN CONTEXT

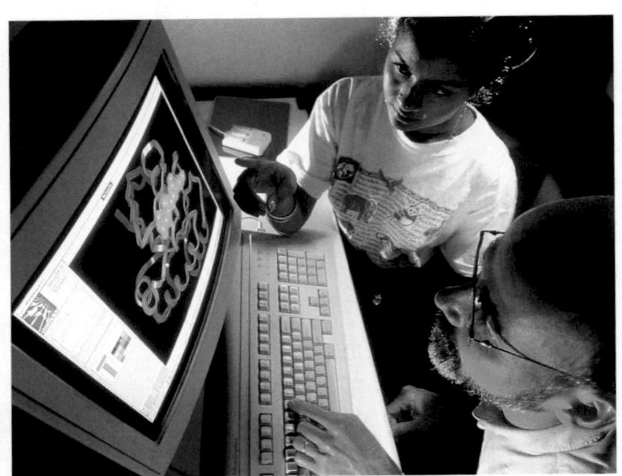

Computers aid drug design. The geometry of biological molecules is studied on screen, and this helps to design drugs that are likely to fit receptor sites and alter biological activity

Designing drug molecules

Most of the drugs used to treat illnesses today have been discovered after a process of search and trial. Many thousands of compounds were screened for any biological activity before a promising compound was found. Now, this time-consuming and enormously expensive process is being dramatically speeded up by computer-aided drug design, based on knowing the three-dimensional shape of chemicals and on how a molecule's shape affects its biological activity.

Drugs work on complex chemical receptor sites around the body, rather like keys fitting into locks. Researchers first use X-ray crystallography to work out the shape and structure of the receptor site they are interested in. Once this is known, they can start to design a drug that fits into the site and so may turn out to be useful. It is here that computers help, since molecular shapes are easily visualised using computer graphics.

During on-screen design, different groups of atoms are made to probe the surface of the site, and gradually the shape of the active part of a drug, the pharmacophore, can be built up. Investigators then see if the designed compound is known or if it can be synthesised. Databases of known compounds are searched for compounds that have the right pharmacophore.

While computers will not replace laboratory chemistry, they greatly reduce the cost and time of finding suitable compounds to investigate further.

1 NOBLE GASES AND STABILITY

Until 1962, chemists thought that the elements in Group 0 were inert (totally unreactive and so unable to form compounds). The group was even called 'the inert gases'. However, in 1962, the British chemist Neil Bartlett synthesised the first Group 0 compound, xenon hexafluoroplatinate. This caused a sensation, and chemists rushed to make more. Three years later there were textbooks devoted to noble gas chemistry. Only helium and neon deserve the title 'inert' as they still seem not to form any compounds, so Group 0 now bears the name of 'noble'.

The noble gases were discovered more than 50 years before chemists made compounds of them. Why did it take so long? It is largely because chemists learned about elements through their reactions, and because the noble gas elements were believed to be so unreactive chemists did not really attempt to make any compounds with them.

OUTER SHELL STABILITY OF NOBLE GASES

In 1916, American chemists Gilbert Lewis and Irvin Langmuir separately realised that the outer shells of all the noble gases (except helium) contained eight electrons, and they suggested this electron configuration made the noble gases unreactive. They put forward the idea that when atoms of other elements formed compounds they gained, lost or shared electrons to make a noble gas outer shell of eight – or two for the lighter elements close to helium.

The ionisation energy of the noble gases also confirms this stability (see Fig 3.20, page 61); as you go across a period, the first ionisation energy rises to a maximum at Group 0. As the configuration is so stable a great deal of energy is needed to remove an electron.

QUESTION 1

What is the outer shell electron configuration (using s and p orbital notation) of the noble gases? (Hint: look back at Chapter 3, page 58. Electron configuration means the number of electrons and their arrangement.)

2 IONIC BONDING

A **chemical bond** is an electrostatic force of attraction between two atoms or ions. **Ions** are formed when atoms lose or gain electrons and become charged.

An ionic bond is the electrostatic attraction that forms between oppositely charged ions.

Sometimes ionic bonds are called **electrovalent bonds**.

A common ionic compound is sodium chloride, an industrially important raw material. The dot-and-cross diagram in Fig 4.1(a) shows how the atoms of sodium and chlorine achieve noble gas configurations. Usually, only the outer electrons are shown, as the inner shells stay the same.

The sodium atom loses one electron, leaving a noble gas core of 2,8. However, because it still has 11 protons it now has a +1 positive charge. It has formed a positive ion: positive ions are called **cations**.

The chlorine atom gains one electron to make the noble gas configuration of 2,8,8: it has one more electron than protons in the nucleus, giving it a charge of –1. Negative ions are called **anions**. Opposite charges attract by electrostatic attraction and this holds the sodium ions and chloride ions together.

Sodium ions and chloride ions are roughly spherical, and the negative charge is distributed evenly all over the spheres. We know this from X-ray diffraction patterns (page 78). X-rays are diffracted by electrons, which allows us to build up **electron density maps**. These show convincingly that the ions are roughly spherical (see Fig 4.1(b)). Each sodium ion attracts several chloride ions and vice versa, so the ionic bonding is not just between one sodium ion and one chloride ion. The ions form an ordered three-dimensional structure known as a **lattice**, shown in Fig 4.2(a). The formula unit (see Chapter 1, pages 3–4) of this lattice is NaCl. As the ionic bonds in the lattice are strong, it takes a lot of energy to separate the ions, and so sodium chloride has a high melting point.

Na • + ×Cl (7 ×) → Na^+ [•Cl (7 ×)]$^-$

2,8,1   2,8,7   2,8   2,8,8

Fig 4.1(a) Dot-and-cross diagram for the outer electrons of sodium chloride. The transfer of an outer shell electron from sodium to chlorine results in two oppositely charged ions. Both ions have a noble gas configuration. Dot-and-cross diagrams are sometimes called Lewis structures in honour of Gilbert Lewis

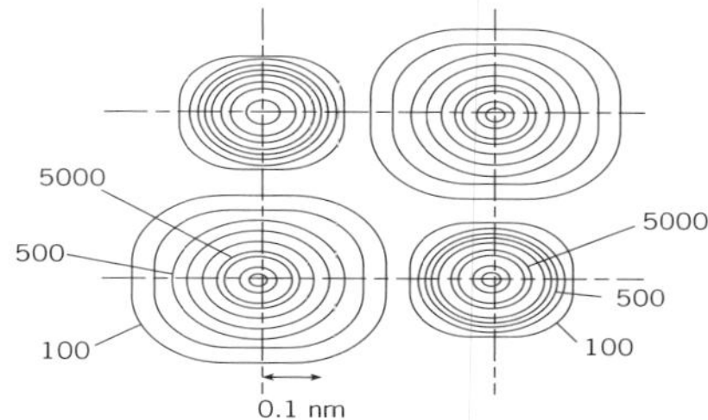

Fig 4.1(b) Electron density map for sodium chloride. The lines join together regions of the same electron density, in a similar way to contours of a map joining together regions of the same height. Electron densities are measured in electrons per cubic nanometre

QUESTION 2

When sodium and chlorine form ions, which noble gas configuration does each have?

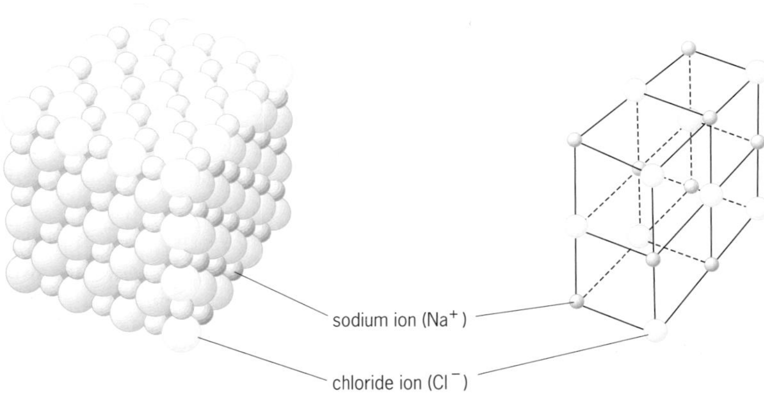

Fig 4.2(a) Part of a sodium chloride lattice. Each ion is surrounded by six oppositely charged ions. (Find out more about ionic lattices and their properties in Chapters 21 and 22)

Fig 4.2(b) Crystals of sodium chloride

REMEMBER THIS

Although we draw dots and crosses, one electron is really the same as any other, and you may well see diagrams with all dots. We use dots and crosses purely for the convenience of showing which atom they come from. Note also that the dots and crosses do not represent the positions of electrons in an atom (see Chapter 3, page 62).

EXAMPLE

Q Work out the dot-and-cross diagrams for calcium fluoride and aluminium oxide.

Calcium fluoride

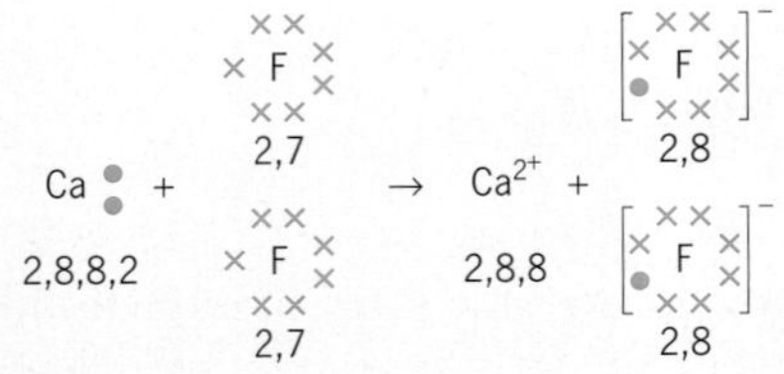

Fig 4.3

Calcium must lose two electrons to achieve a noble gas configuration, and fluorine needs only one. So there are two fluoride ions for each calcium ion, which gives a formula of CaF_2.

Aluminium oxide

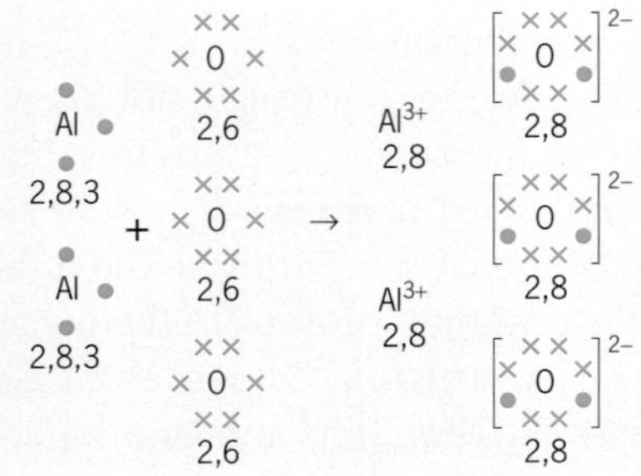

Fig 4.4

From Fig 4.4, the formula is Al_2O_3.

QUESTION 3

3 Work out dot-and-cross diagrams for:

a) potassium fluoride;
b) magnesium oxide;
c) calcium chloride;
d) sodium oxide.

State the formula of each compound.

QUESTION 4

4 **a)** Fluorine, chlorine, bromine and iodine are all in Group 7. How many electrons are there in the outer shell of each?

b) How many electrons are in the outer shell of:

i) Group 1 elements;
ii) Group 2 elements?

QUESTION 5

5 Sodium ions and chloride ions are isoelectronic with which noble gases?

IONIC BONDING AND ATOMIC ORBITALS

So far we have seen the formation of ions in terms of dot-and-cross diagrams. This is a good model to explain how ions are formed, but how does it fit in with the atomic orbital model of the atom on pages 61–63? Let's look at some examples.

In sodium chloride:

$$\begin{array}{ccccccc} \text{Na} & + & \text{Cl} & \rightarrow & \text{Na}^+ & + & \text{Cl}^- \\ 1s^2 2s^2 2p^6 3s^1 & & 1s^2 2s^2 2p^6 3s^2 3p^5 & \rightarrow & 1s^2 2s^2 2p^6 & & 1s^2 2s^2 2p^6 3s^2 3p^6 \end{array}$$

1 electron moves

We can write electron configurations using a simplified notation. We represent the inner shells (sometimes known as the noble gas core) by the appropriate noble gas:

$$[\text{Ne}]3s^1 + [\text{Ne}]3s^2 3p^5 \rightarrow [\text{He}]2s^2 2p^6 + [\text{Ne}]3s^2 3p^6$$

Both ions have attained noble gas configurations on the right: they are **isoelectronic** with the noble gases, meaning that they have the same electron configuration.

The 3s electron is removed from the sodium atom because the 3s orbital is the highest occupied energy orbital in the outer shell, and so the electron needs the least amount of energy to transfer it. Electrons in the inner shells are not removed during a normal chemical reaction: their orbitals are at a much lower energy because they are part of a stable full shell closer to the nucleus, so their removal would require very high energies.

For calcium fluoride:

Ca	+	F	→	Ca^{2+}	+	F^-
$1s^22s^22p^63s^23p^64s^2$		$1s^22s^22p^5$		$1s^22s^22p^63s^23p^6$		$1s^22s^22p^6$
$[Ar]4s^2$		$[He]2s^22p^5$		$[Ne]3s^23p^6$		$[He]2s^22p^6$
		F				F^-
		$1s^22s^22p^5$				$1s^22s^22p^6$
		$[He]2s^22p^5$				$[He]2s^22p^6$

THE ENERGY CONSIDERATIONS OF IONIC BONDING

Notice that ionic bonds are formed between metal atoms, which lose electrons, and non-metal atoms, which gain electrons. Look back at Fig 3.20, page 61, which shows that metals have low first ionisation energies. So the energy required to remove electrons from metals to attain a noble gas structure is relatively low. Non-metals, particularly oxygen and those in Group 7, have a strong affinity (attraction) for electrons. **Electron affinity** can be measured and is the energy change when *gaseous atoms* attract electrons. Fluorine has a very large electron affinity:

$$F(g) + e^- \rightarrow F^-(g) \quad \Delta H = -328 \text{ kJ mol}^{-1}$$

To be strictly correct, this is the **first electron affinity** of fluorine; the definition of first electron affinity is:

The first electron affinity is the energy change when one mole of gaseous atoms accepts one mole of electrons to form one mole of singly charged anions.

Look at the first ionisation energy of sodium:

$$Na(g) \rightarrow Na^+(g) + e^- \quad \Delta H = +496 \text{ kJ mol}^{-1}$$

We can see that the energy released by fluorine gaining an electron is not enough to remove an electron from sodium. So why do sodium and fluoride ions form the compound sodium fluoride? It is because of the very large amount of energy released when solid sodium fluoride forms from these gaseous ions:

$$Na^+(g) + F^-(g) \rightarrow Na^+F^-(s) \quad \Delta H = -918 \text{ kJ mol}^{-1}$$

The energy released when one mole of solid sodium fluoride is formed from its gaseous ions is called the lattice energy.

It is this highly exothermic **lattice energy** also called **lattice enthalpy**, caused by the strong electrostatic attraction of oppositely charged ions, which makes sodium fluoride much more stable than the elements sodium and fluorine. This shows that, while ions being formed need to achieve full noble gas shells, we have to look at all the energy changes to see the full picture.

? QUESTION 6

6 Write out the full and simplified electron configurations (using s, p, d notation) to show how ionic bonding arises in:

a) potassium fluoride;
b) magnesium chloride;
c) aluminium oxide.

? QUESTION 7

7 What does the minus sign tell you about the first electron affinity of fluorine?

REMEMBER THIS

Energy is released when gaseous atoms each attract an electron. Since the atoms lose energy, the value for the first electron affinity is negative. However, the value of the **second electron affinity** of oxygen is endothermic. This is because energy is required to overcome the repulsion of the negative charge on the O^- ion as it attracts a second electron.

$O^-(g) + e^- \rightarrow O^{2-}(g)$

$\Delta H = +798 \text{ kJ mol}^{-1}$

You can read more about lattice energy, electron affinity and first ionisation energy in the formation of ionic compounds in Chapter 22.

3 COVALENT BONDING AND THE FORMATION OF MOLECULES

Many of the compounds we come across in everyday life are covalent compounds. The wood or plastic of a table top, the clothes we wear, nearly all the food we eat and most of the body's chemicals have atoms that are joined together by **covalent bonds**.

So what is a covalent bond? The answer again lies in the stability of noble gas electron configurations: with the exception of helium, there are eight electrons in the outer shell of a noble gas. For elements in Groups 4 to 7 in the Periodic Table, the loss of four or more electrons to achieve a noble gas configuration requires a great deal of energy, which could not be paid back by the formation of a stable lattice. So ions with a charge of 4+ or more do not occur under the usual conditions of a chemical reaction. Instead, the atoms of these elements can achieve noble gas configurations by *sharing* electrons.

Let us look at the example of methane, CH_4 (Fig 4.5), found in natural gas. Its molecule contains one carbon atom and four hydrogen atoms.

```
  •                      H
                         •×
• C •   +   4 ×H   →   H ×C •H
  •                      •×
                         ×•
                         H
```

Fig 4.5 Dot-and-cross diagram for methane

By sharing outer shell electrons, hydrogen has the noble gas configuration of helium (i.e. 2) and carbon has the neon configuration (2,8).

Now we can answer the question: what is a covalent bond?

A covalent bond is a pair of electrons shared between two atoms.

The force that holds the carbon and hydrogen atoms together in this bond is the electrostatic attraction between their positive nuclei and the shared pair of negative electrons.

We represent the shared pair of electrons as a line, as shown for the C–H bond in Fig 4.6 and the methane molecule in Fig 4.7.

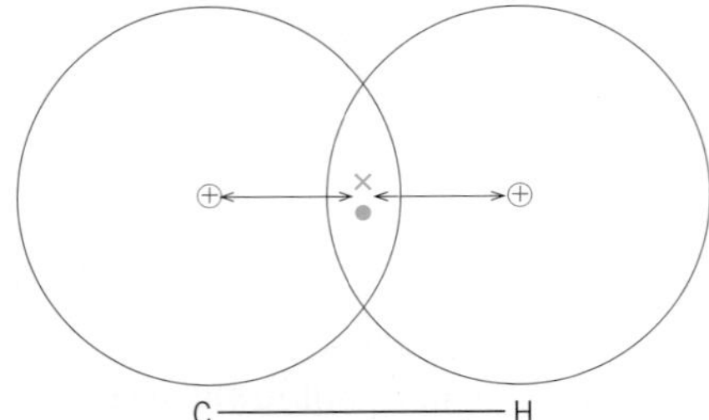

Fig 4.6 The C–H bond: electrostatic forces between the shared pair of electrons and the nuclei hold the carbon and hydrogen atoms together in the covalent bond

```
    H
    |
H — C — H
    |
    H
```

Fig 4.7 The methane molecule

The bonding in a fluorine molecule (Fig 4.8(a)) is also covalent. Fluorine is in Group 7, so has 7 electrons in its outer shell. The evidence for electron sharing can be seen in the electron density map in Fig 4.8(b), showing that there is high electron density between the nuclei.

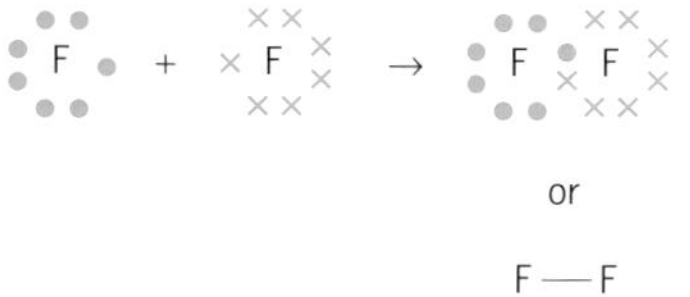

Fig 4.8(a) The fluorine molecule

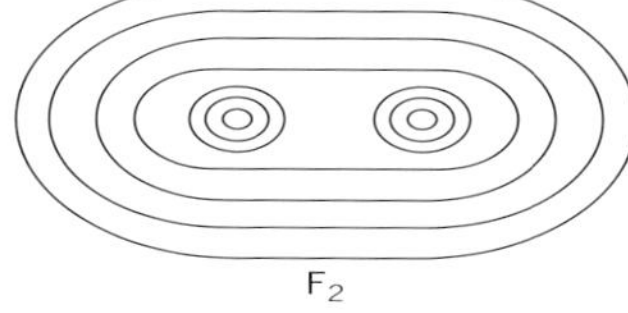

Fig 4.8(b) Electron density map of the fluorine molecule. The lines show regions of the same electron density. Unlike the sodium chloride electron density map, there is electron density shared between the two nuclei

REMEMBER THIS

The representation of a pair of electrons by a line is sometimes extended to all the electron pairs in a molecule. For example:
$\overline{\underline{|F}}-\overline{\underline{F|}}$

Each fluorine atom has achieved an **octet** (8 electrons) through sharing a pair of electrons. In this molecule there is one **bonding pair** of electrons and three **non-bonding pairs** or **lone pairs** on each atom. While lone pairs play no part in the bond, they are important in determining the shape of a molecule. We look at this more closely on page 79.

QUESTION 8

Draw electron dot-and-cross diagrams for the following molecules:

a) ammonia, NH_3
b) hydrogen, H_2
c) chlorine, Cl_2

EXAMPLE

Q Use dot-and-cross diagrams to show the covalent bonding in water, H_2O.

A The oxygen atom has 6 electrons (it is in Group 6). Draw out the separate atoms first, and then combine them in the molecule.

2H• + O → H O H

Fig 4.9

MULTIPLE BONDS

Sometimes sharing one pair of electrons is not enough to make up a noble gas configuration. Carbon dioxide is a very common molecule held together by **double bonds** (Fig 4.10).

O + C + O ⟶ O C O

O=C=O

Fig 4.10 The carbon dioxide molecule

Fig 4.11 Bonds in carbon dioxide

Carbon shares two pairs of electrons with each oxygen, so all three atoms achieve the noble gas octet. Count the electrons round each atom to be convinced of this. Double bonds are shown by two lines, as in Fig 4.11.

Nitrogen gas, N_2, forms 79 per cent of the Earth's atmosphere. It is a very unreactive molecule because its very strong **triple bond** needs to be broken before it can react.

N + N ⟶ N N   N≡N

Fig 4.12 The nitrogen molecule

EXAMPLE

Q Methanal, HCHO, reacts with phenol to make the glue that holds plywood together. Draw out the dot-and-cross diagram for methanal and hence show the covalent bonds.

A **Step 1.** Draw dot-and-cross diagrams for the separate atoms of the molecule, Fig 4.13.

Fig 4.13

Step 2. Arrange the atoms in the order they are in the molecule (if you know this) and put in shared pairs of electrons for those atoms that can only form a single bond – in this case, the hydrogen atoms, Fig 4.14.

Fig 4.14

Step 3. Now make up noble gas configurations for the other atoms using one, two or three shared pairs, Fig 4.15.

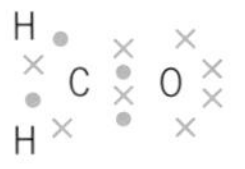

Fig 4.15

Now you can draw the bonds for each of the shared pairs, Fig 4.16.

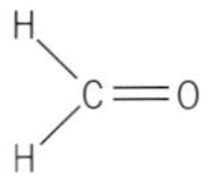

Fig 4.16

? QUESTION 9

9 Draw dot-and-cross diagrams for the following molecules and draw separate diagrams to show the covalent bonds.

a) oxygen
b) ethene, C_2H_4
c) hydrogen cyanide, HCN

4 COORDINATE (DATIVE COVALENT) BONDS

When atoms bond together, the electron pair is usually made up of one electron from each atom. But often, both the electrons in a covalent bond come from just one of the atoms. This bond is called a **dative covalent bond**, or a **coordinate bond**.

The poisonous gas carbon monoxide is a good example; see Fig 4.17(a). As in Fig 4.17(b), an arrow is sometimes used to show which atom has donated the pair of electrons. But since the bond, once formed, is exactly the same as any other covalent bond, we often draw it as in Fig 4.17(c).

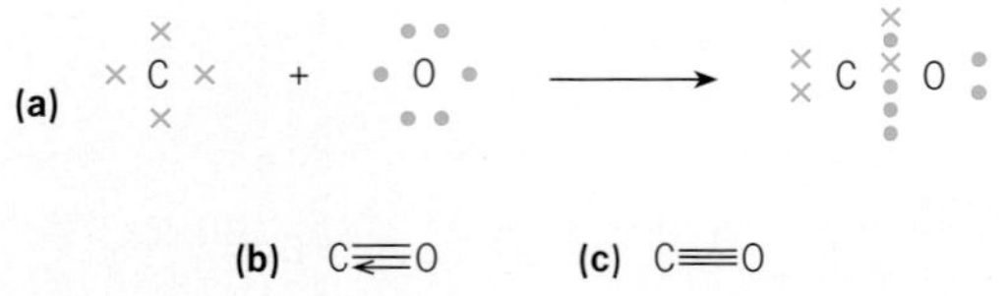

(a)

(b) C≡O **(c)** C≡O

Fig 4.17(a)–(c) The carbon monoxide molecule

Another example is the ammonium ion, formed when an ammonia molecule bonds with a proton (H^+). Note that when a hydrogen atom loses its electron, a nucleus that consists of one proton is left.

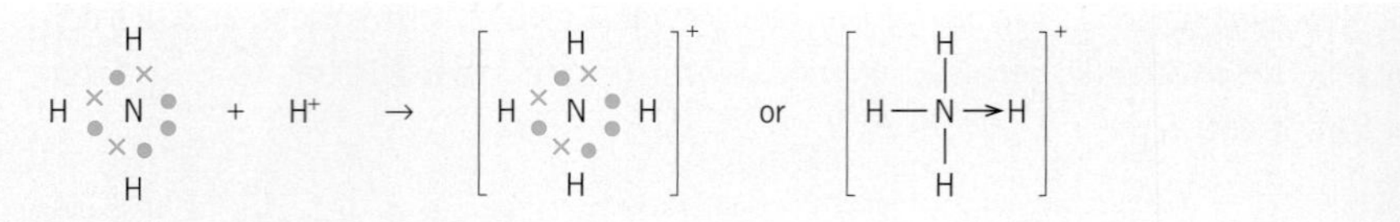

Fig 4.18 Formation of the ammonium ion

The whole species has a charge of +1, because there is still one more proton than there are electrons.

? QUESTION 10

10 Draw the dot-and-cross diagram for the oxonium ion, H_3O^+. (Hint: this is made up of water and a proton.)

5 EXCEPTIONS TO THE OCTET RULE

We work out how a molecule is bonded using noble gas configurations. Since the atoms of the noble gases (apart from helium) contain eight outer shell electrons, this is sometimes called the **octet rule**. However, some atoms can expand their octets.

MORE ELECTRONS THAN AN OCTET

At the start of this chapter we saw that the first noble gas compound was made in 1962, more than 50 years after noble gases were discovered. Chemists had not tried earlier since they were convinced that an atom with eight outer shell electrons would be inert (unreactive). One of the first compounds to be synthesised was xenon tetrafluoride. Fig 4.19(b) shows the dot-and-cross diagram for this molecule.

Before chemists found that noble gases could expand their octets, many other compounds were known that broke the octet rule, for example phosphorus(V) chloride, PCl_5 (Fig 4.20), and sulfur dioxide (Fig 4.21).

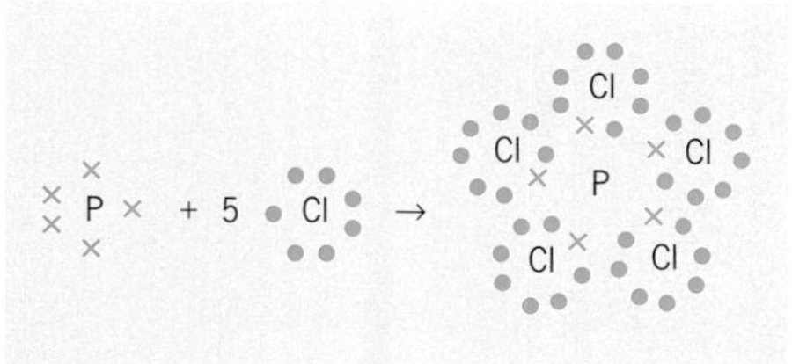

Fig 4.20 The phosphorus(V) chloride molecule

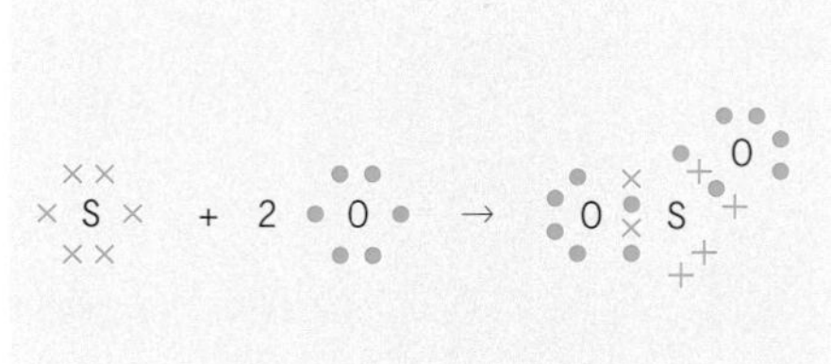

Fig 4.21 The sulfur dioxide molecule

FEWER ELECTRONS THAN AN OCTET

Some compounds have less than an octet of electrons. Aluminium chloride as a gas exists as $AlCl_3$, as in Fig 4.22(a). (Aluminium chloride has another surprise: it has covalent bonds between metal and non-metal atoms. We shall return to this on page 86 when we consider polar covalent bonds.)

Though there are only six electrons round the aluminium, we can produce an imaginary dot-and-cross diagram with a stable octet, Fig 4.22(b). But this does not agree with the experimental evidence obtained using X-rays, which shows three single covalent bonds around the aluminium. When aluminium chloride gas is cooled, the formula of the molecule is found to be Al_2Cl_6, as in Fig 4.22(c). There are two dative covalent bonds, shown with arrows: you can see that every atom now has its noble gas octet.

Many compounds of boron are called electron deficient because boron has just six electrons round its atom, as in boron trifluoride, BF_3, shown in Fig 4.23.

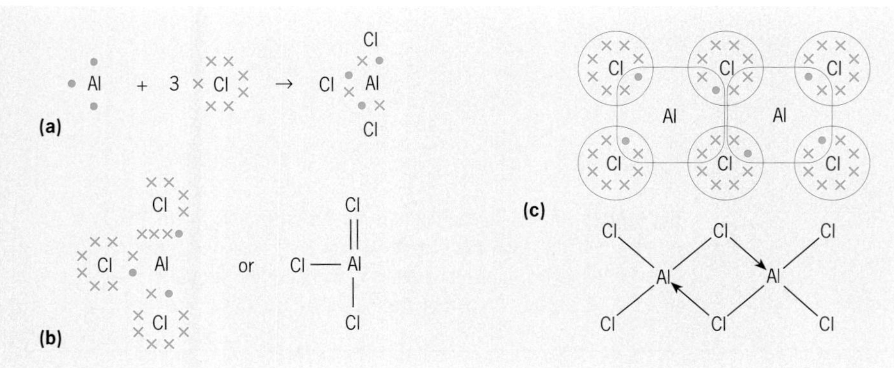

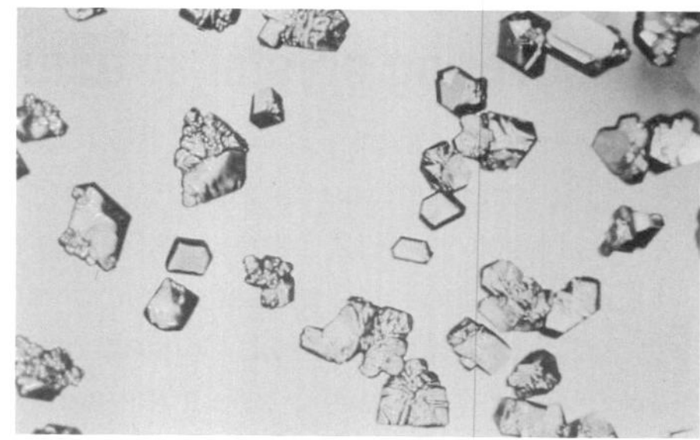

Fig 4.19(a) Xenon tetrafluoride crystals. This was one of the first noble gas compounds to be made

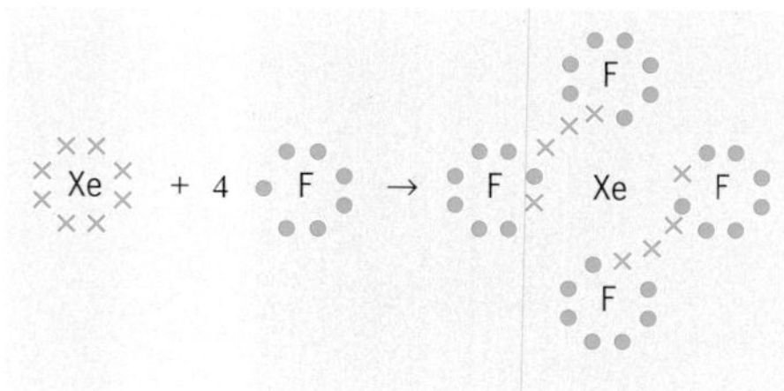

Fig 4.19(b) The xenon tetrafluoride molecule

QUESTION 11

11 **a)** Draw a dot-and-cross diagram for sulfur(VI) fluoride, SF_6.

b) Draw, using lines, the bonds in the sulfur dioxide molecule. You should also show the lone pair on the sulfur atom.

QUESTION 12

12 BF_3 reacts with NH_3 to form the molecule BF_3NH_3. In this molecule, boron has eight electrons in its outer shell. Draw the dot-and-cross diagram for this molecule and a diagram of the molecule showing the bonds as lines.

Fig 4.22(a)–(c) The aluminium chloride gas molecule (a) as a gas with six electrons round the aluminium atom, (b) with an octet round Al – *this form does not exist*, (c) in the cooled gas

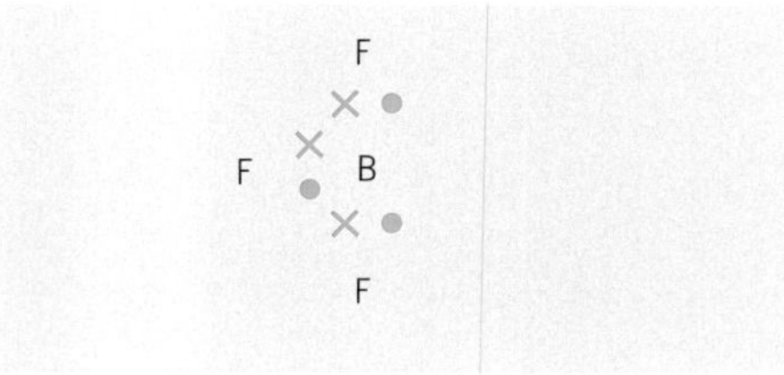

Fig 4.23 The boron trifluoride molecule

Discovering crystal structures using X-ray crystallography

X-ray crystallography is one of the most effective techniques used to establish the structure of compounds. It has its origins back in 1912, when the German scientist Max von Laue first suggested that atoms in crystals might diffract (bend and scatter) X-rays. Crystals diffract X-rays because the wavelength of X-rays is about the same as the distance between the nuclei in a crystal. As the X-rays scatter, some of the waves constructively interfere – they follow parallel wave paths, see Fig 4.24(a) – and this leads to spots on a photographic film, as shown in the diffraction pattern of Fig 4.24(b).

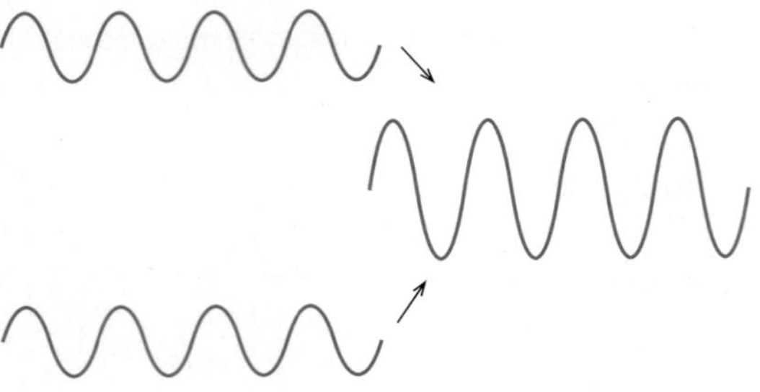

Fig 4.24(a) X-rays that follow parallel paths have a crest and trough at the same point as each other. They are described as in phase and constructively interfere to reinforce each other

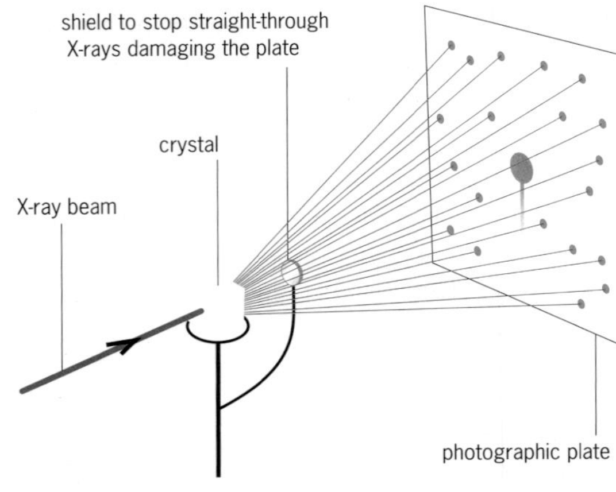

Fig 4.24(b) Arrangement for obtaining the X-ray diffraction pattern of a crystal

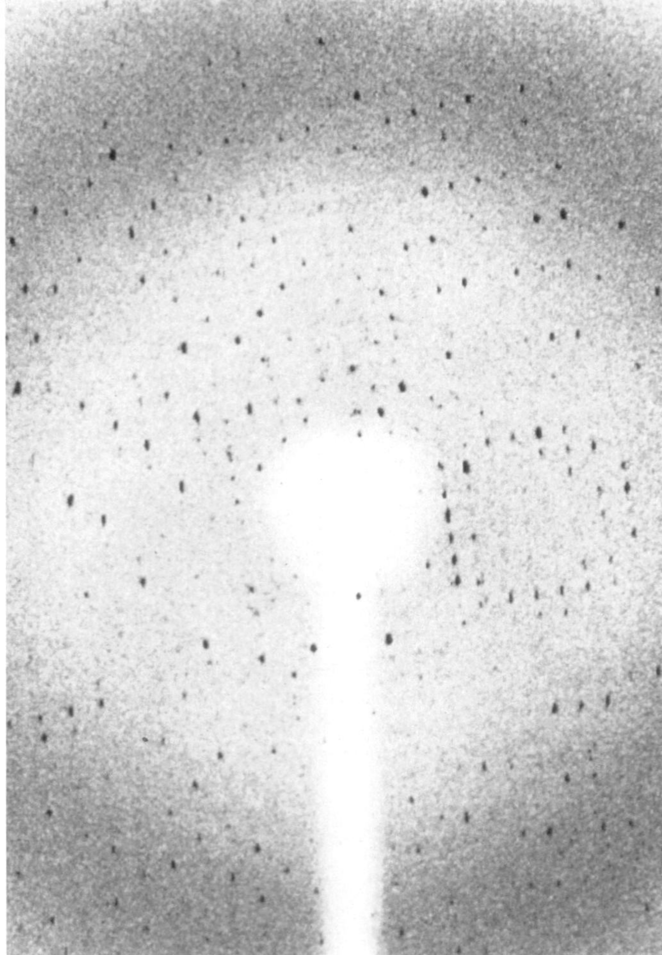

The pattern of spots must now be matched to the positions of atoms or ions. The English scientists William and Lawrence Bragg, a father-and-son team, worked out the mathematical calculations to match patterns to three-dimensional positions, and received the Nobel Prize for Physics in 1915.

Nowadays, the X-ray diffraction pattern is usually detected electronically, and computers carry out the calculations.

X-ray crystallography is a very important tool for the chemist. It has been used recently to confirm the structure of the newest form of carbon, buckminsterfullerene (see pages 136, 145), and has determined the structure of well over 1000 proteins.

Fig 4.24(c) X-ray diffraction pattern for the protein lysozyme. This is an enzyme, found in body fluids (such as tears and saliva), that destroys bacteria. The X-ray image helped researchers to work out the structure of lysozyme (see Fig 19.33) and the part of the molecule that acts against bacteria

6 THE SHAPES OF MOLECULES AND IONS

We now have a way of working out the *bonds* in a molecule; and the model we have used will also tell us about a molecule's *shape*, which is often important in determining the properties of the molecule. For example, the taste of substances is thought to depend on the shape of its molecules and on how they fit into taste receptor molecules on the tongue. In the same way, the shape of drug molecules can determine just how effective they are in treating illness (see the Science in Context box on page 70).

SCIENCE IN CONTEXT

Dorothy Hodgkin, pioneer in X-ray crystallography

At the age of 11, Dorothy Hodgkin started at secondary school in Beccles, Suffolk. The school allowed pupils to follow physics and chemistry for one period each week. However, because she was female she could not do physics as it clashed with domestic science. Fortunately, this was the only time that Hodgkin's gender seems to have affected her scientific career.

At Oxford University she joined a newly formed X-ray crystallography group, and by 1945 she had used the technique to work out the structure of penicillin. She pioneered the use of the computer to help interpret the X-ray images, and by 1956, after eight years of work, she had unravelled the complex structure of vitamin B_{12}, a molecule of more than 90 atoms. Once its structure was known, chemists could synthesise the vitamin and use it to combat pernicious anaemia. For this work, she became the only British female scientist so far to receive the Nobel Prize, which she was awarded in 1964 for chemistry.

Fig 4.25 Dorothy Mary Crowfoot Hodgkin (1910–1994)

By 1969 she had achieved a lifetime ambition: 25 years earlier she had been given a tiny sample of insulin, and she was able to work out the structure of this complex molecule, which contains more than 800 atoms.

ELECTRON-PAIR REPULSION AND SHAPES OF MOLECULES AND IONS

Dot-and-cross diagrams are used to represent two types of electron pairs. A pair of electrons involved in bonding is called a **bonding pair**, while a pair of electrons not involved in bonding is known as a **non-bonding pair** or **lone pair**. We now look at a model of molecular shape. In this model, pairs of electrons repel each other and move as far apart from each other as possible.

Beryllium chloride (Fig 4.26(a)) has two bonding pairs and no lone pairs around its central atom – the only pairs we are interested in here since the three lone pairs round each chlorine atom do not affect the molecule's shape. As Fig 4.26(b) shows, the position of minimum repulsion has a bond angle of 180°. We describe the molecular shape as **linear** because the atoms lie in a straight line.

In this model of molecular shape, we group together the bonding pairs of electrons in double and triple bonds, so that one group is the equivalent of one electron pair. Carbon dioxide, with two double bonds (Fig 4.27(b)), then has a linear shape because it has two groups of electrons.

(a) Cl — Be — Cl

(b) Cl — Be — Cl (180°)

Fig 4.26(a), (b) Beryllium chloride is a linear molecule

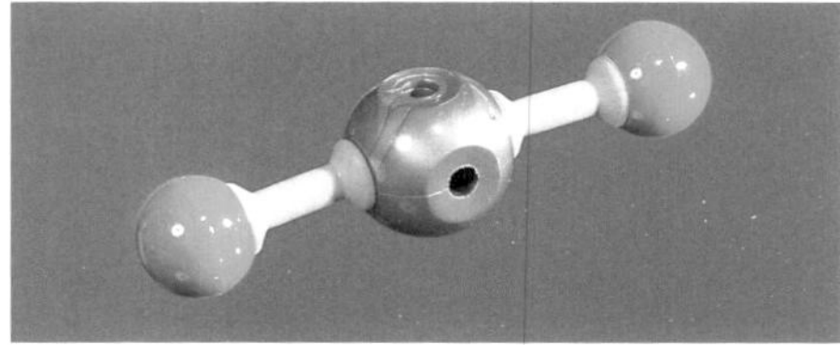

Fig 4.26(c) Ball and stick model of beryllium chloride

(a)

(b) O═C═O

Fig 4.27(a), (b) Carbon dioxide is a linear molecule

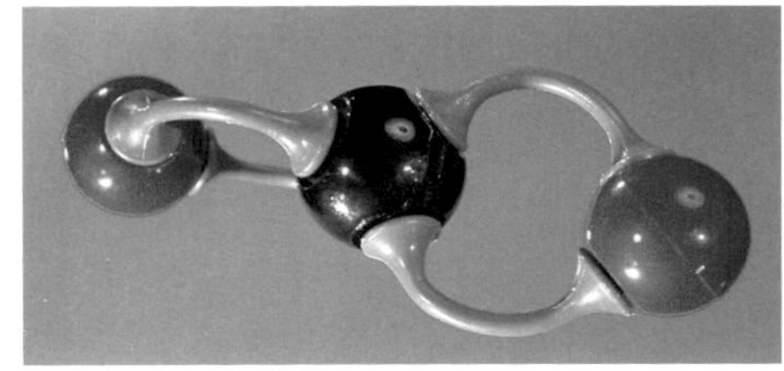

Fig 4.27(c) Ball-and-stick model of carbon dioxide

QUESTION 13

Explain why hydrogen cyanide, HCN, is a linear molecule. You may have already drawn the dot-and-cross diagram in answer to question 9(c).

Fig 4.28(a) Ball-and-stick model of boron trichloride

Boron trichloride has three bonding pairs and no lone pairs round the central boron (Fig 4.28(b)).

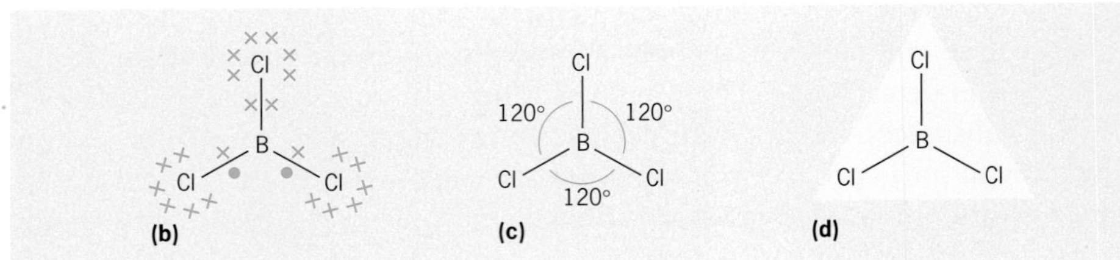

Cl
B
Cl
Cl
(b)
120°
120°
120°
(c)
(d)

Fig 4.28(b)–(d) Boron trichloride is a trigonal planar molecule

QUESTION 14

14 The carbonate ion CO_3^{2-} can be represented by the dot-and-cross diagram in Fig 4.29.

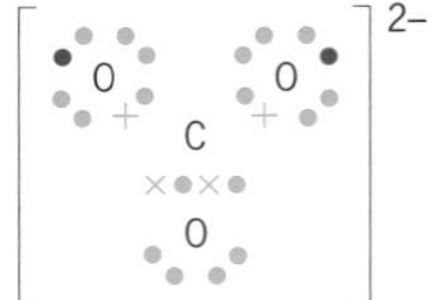

O
O
C
O
2–

Fig 4.29
The electrons shown in red are the two extra electrons, which give the carbonate ion its 2– charge. Explain what shape you would expect this ion to have.

The furthest apart these bonding pairs can get is 120°, which gives a **trigonal planar** shape. Trigonal tells us that the atoms are where the points of a triangle would be, and planar means that the molecule would lie flat on a plane.

Ethene, C_2H_4, is used in industry to make a wide range of important chemicals including polythene and antifreeze. The double bond is one group of electrons and counts as one bonding pair (Fig 4.30), so there are the equivalent of three bonding pairs of electrons around each carbon atom. This gives a trigonal planar shape round each carbon atom.

(a) (b)

Fig 4.30(a), (b) Ethene is a trigonal planar molecule

Fig 4.30(c) Ball-and-stick model of ethene

REMEMBER THIS

Sometimes the model of electron pairs repelling each other is called the Valence Shell Electron Pair Repulsion (VSEPR) theory, the valence shell being the outermost shell.

Methane (Fig 4.31(a), (c)) has four bonding pairs round the carbon. This molecule does not lie in one plane, but has a three-dimensional structure in which all the bond angles are 109.5°. The shape is tetrahedral because the hydrogen atoms are at the points of a tetrahedron. All carbon atoms with four bonding pairs have a tetrahedral structure.

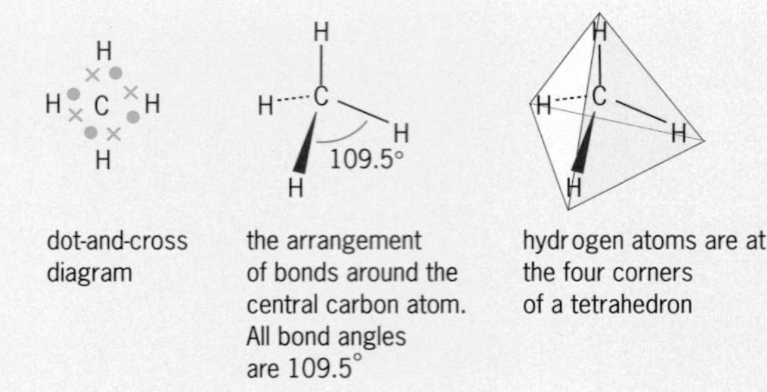

dot-and-cross diagram
the arrangement of bonds around the central carbon atom. All bond angles are 109.5°
hydrogen atoms are at the four corners of a tetrahedron
109.5°

Fig 4.31(a) Three ways to represent the structure of methane, a tetrahedral molecule

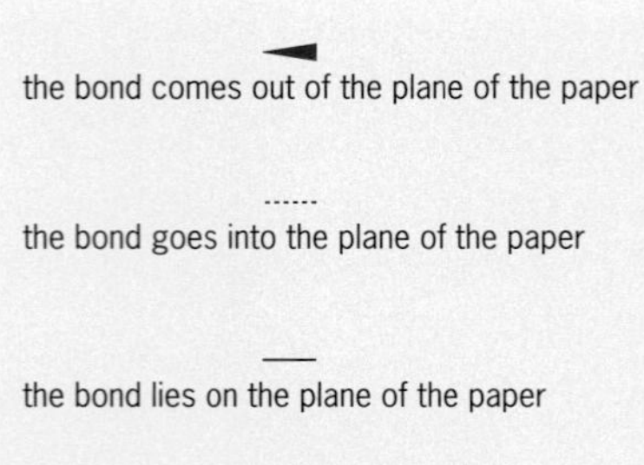

the bond comes out of the plane of the paper
the bond goes into the plane of the paper
the bond lies on the plane of the paper

Fig 4.31(b) Ways to represent the directions of bonds

Fig 4.31(c) Ball-and-stick model of methane

Fig 4.32(a) The ethane molecule: the hydrogen atoms are arranged tetrahedrally around the carbon atoms. The carbon atom can rotate about its single bond, so the hydrogen atoms at one end could be in any position relative to those at the other end. But they tend to remain in the position shown in (b), in which the repulsion of the bonding pairs of electrons is the least, and which is the most stable arrangement. However, the energy difference between this and other positions is very small, which results in free rotation about the carbon–carbon single bond at room temperature

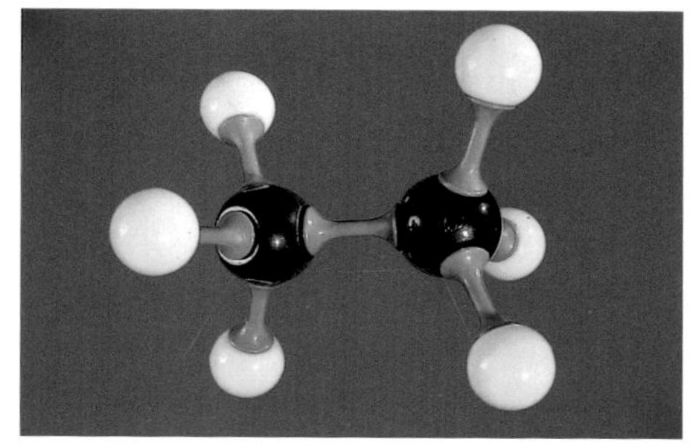

Fig 4.32(b) Ball-and-stick model of ethane

Phosphorus(V) fluoride, PF_5 (Fig 4.33), has five bonding pairs and no lone pairs, so the shape that puts the bonding pairs furthest apart is **trigonal bipyramidal** (two triangular pyramids base to base).

Sulfur hexafluoride, SF_6 (Fig 4.34), has an **octahedral shape** in which all bond angles are 90° because it has six bonding pairs and no lone pairs. This gas has some remarkable properties (see Science in Context box, page 82).

QUESTION 15

15 Draw the shape of the butane molecule, C_4H_{10}. (Hint: it is a chain of four carbon atoms: look at the shape of the ethane molecule (Fig 4.32) to help you.)

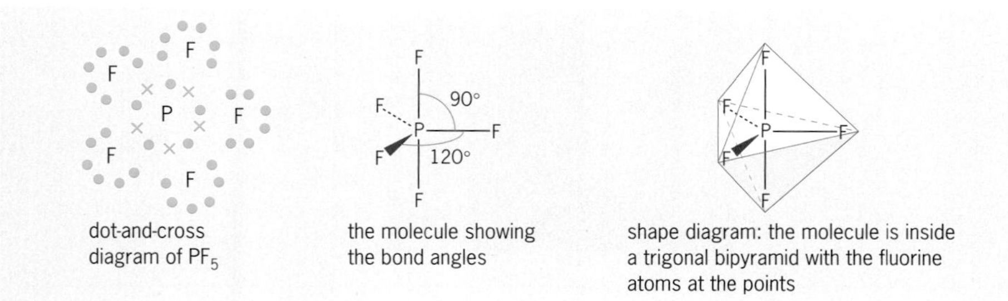

dot-and-cross diagram of PF_5
90°
120°
the molecule showing the bond angles
shape diagram: the molecule is inside a trigonal bipyramid with the fluorine atoms at the points

Fig 4.33(a) Three ways to represent the phosphorus(V) fluoride molecule

Fig 4.33(b) Ball-and-stick model of PF_5

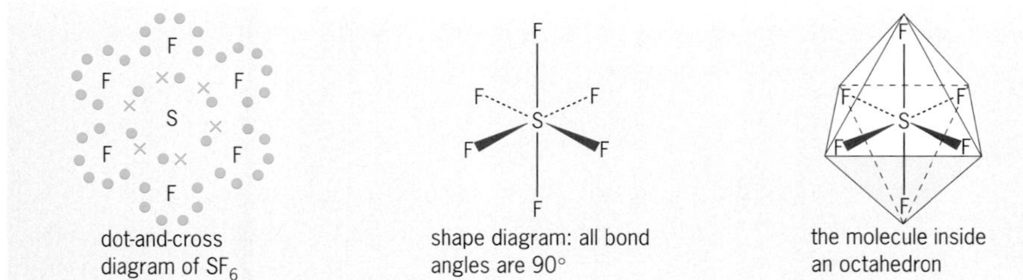

dot-and-cross diagram of SF_6
shape diagram: all bond angles are 90°
the molecule inside an octahedron

Fig 4.34(a) Three ways to represent the sulfur hexafluoride molecule

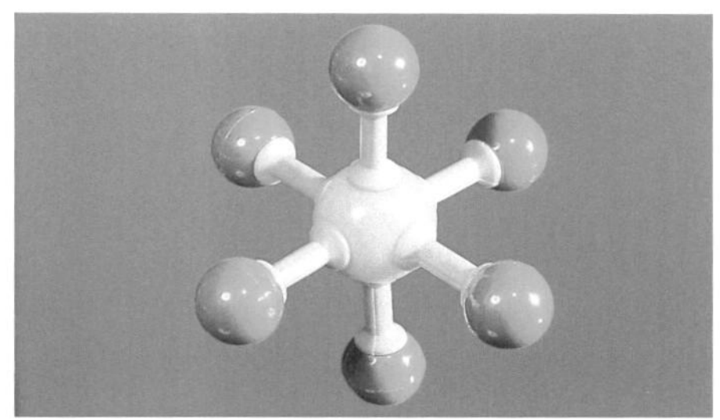

Fig 4.34(b) Ball-and-stick model of SF_6

The shapes of complex ions

Many metal ions form **complex ions**, often more simply known as **complexes**. The central metal ion is surrounded by species – atoms, ions or molecules. These species donate lone pairs of electrons to form dative covalent (coordinate) bonds and are called **ligands**. It is common for six ligands to be bonded to a central metal cation. Since there are six bonding pairs of electrons these also have an octahedral shape. One example is the hydrated magnesium ion, $[Mg(H_2O)_6]^{2+}$ (Fig 4.35). You can read more about complexes and their shapes in Chapter 24.

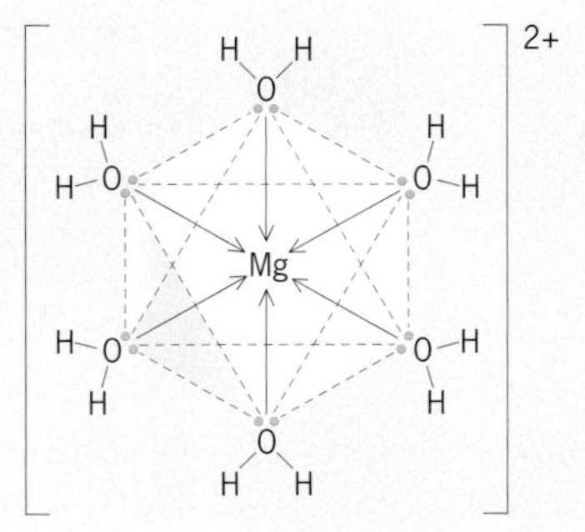

Fig. 4.35 The Mg^{2+} ion accepts lone pairs of electrons from the water molecules to form six dative covalent bonds. The six bonding pairs repel each other to form an octahedral shape

QUESTION 16

16 Draw the dot-and-cross diagram for the ion AlF_4^-. Predict its shape. (Hint: although this is a negative ion, the extra electron it has is used to form an Al–F bond.)

Sulfur hexafluoride gas – the most potent greenhouse gas known

Sulfur hexafluoride, SF_6, does not occur naturally; it is manufactured. It is non-toxic, odourless, colourless, very unreactive and, at five times denser than air, one of the densest gases known. Since the 1970s it has been used extensively in equipment to transmit and distribute electricity. It is an extremely good insulator and so is used in high-voltage transformers and circuit breakers. It is likely that an electricity substation near you has this gas inside it. It is also very good at preventing arcing, a potential fire hazard when high-voltage circuits are switched on and off. About 80 per cent of its use worldwide is in electricity transmission.

In this country it is also used as an inert gas in the production of magnesium alloys. It provides a blanket, preventing oxygen in the air from oxidising molten magnesium.

Apart from these major uses, it is used to trace air flow in the ventilation systems of buildings. In 2007 it was even used in the London Underground to find out how toxic gases from a terrorist attack would travel through tube stations. It is also used during lung radiography. Manufacturers of trainers used it in gas sacs in the soles of their trainers as a cushion. It is also a highly effective sound insulator in double glazing.

If this remarkable substance seems too good to be true, that's because it is. There are real concerns over its growing use. Molecule for molecule, it is the most potent greenhouse gas ever evaluated, being over 23 000 times more warming than carbon dioxide. Its very stability means that it has an atmospheric lifetime of 3200 years. It is one of the gases that countries signed up to the Kyoto Protocol are pledged to reduce; see page 202, Chapter 10. Although its concentration in the atmosphere is very small, it is increasing year on year and measures are being put in place to reduce and restrict its use.

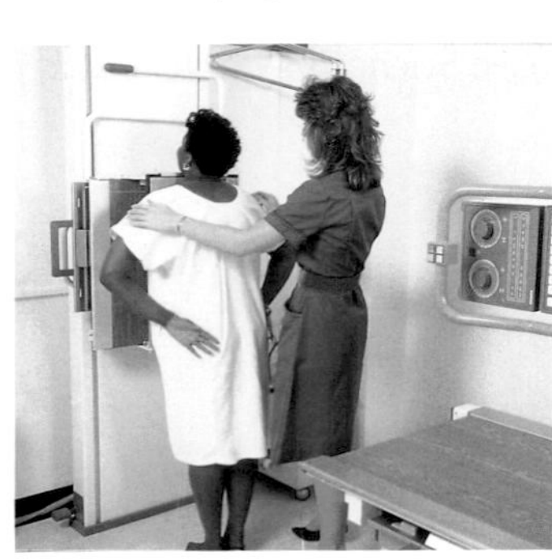

Fig. 4.36 Sulfur hexafluoride is sometimes used in lung radiography. It is breathed in by the person having the radiograph

QUESTION 17

17 Using the electron-pair repulsion theory, predict the shape of $[Be(H_2O)_4]^{2+}$ and draw it.

REPULSION BY LONE PAIRS

So far, we have looked at compounds with no lone (non-bonding) pairs of electrons around the central atom. When we name the shape of a molecule, we are looking at the position of the atoms only. We do not include the lone pairs, yet they are critical in determining that shape.

Ammonia

Ammonia has three bonding pairs of electrons and one lone pair (Fig 4.37) around the central atom. According to the electron-pair repulsion theory, we could expect the shape to be based on a tetrahedron. However, if we look at the arrangement of the atoms, the actual shape of the ammonia molecule is best described as **pyramidal**.

Although ammonia's shape is based on a tetrahedron, the bond angles are not the expected 109.5°. The three bonds have been squeezed closer together because the lone electron pair exerts a **greater repulsion** than the bonding pairs.

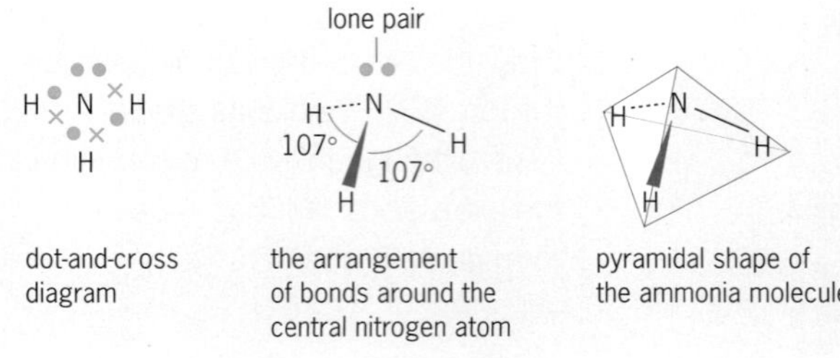

Fig 4.37(a) Three ways to represent the ammonia molecule

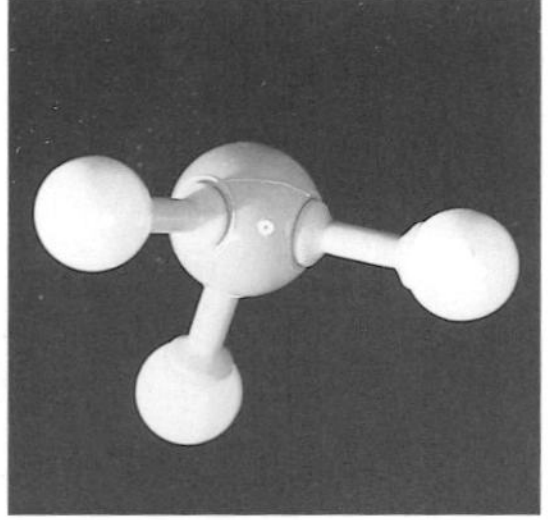

Fig 4.37(b) Ball-and-stick model of ammonia

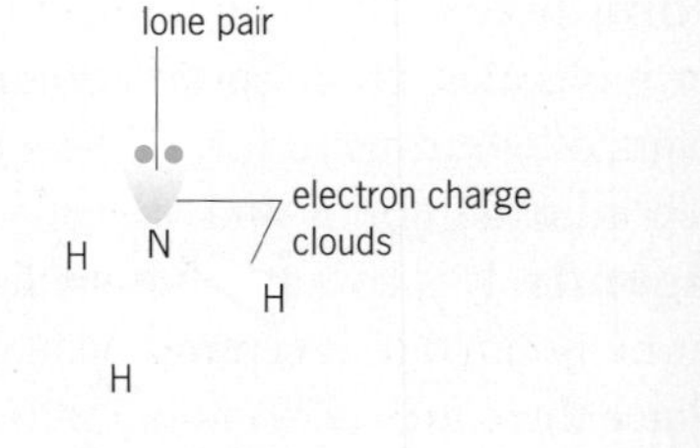

Fig 4.37(c) The ammonia molecule showing the bonding pairs and the lone pair as electron charge clouds

Think of the pairs of electrons as electron clouds and the reason for the greater repulsion becomes clear. The bonding-pair clouds are not as spread out as the lone-pair cloud, because the bonding electrons are held between the nuclei of two atoms and attracted to both, so the bonding-pair clouds are relatively thin. Only one nucleus holds the lone-pair cloud, so its electrons are pulled closer to the nucleus, it spreads out more and is therefore fatter than the bonding-pair clouds, as in Fig 4.37(c). This is why lone pairs have greater repulsion.

Water

In the ammonia molecule there are two types of repulsion:

- bonding pair–bonding pair
- lone pair–bonding pair

In the water molecule (Fig 4.38) there is an extra repulsion:

- lone pair–lone pair

We can put these three types of repulsion in order of increasing repulsion:

- bonding pair–bonding pair repulsion
- lone pair–bonding pair repulsion
- lone pair–lone pair repulsion

(increasing repulsion ↓)

You can see that because of this, the two lone pairs in water exert the greatest repulsion and narrow its bond angle in comparison to that in ammonia. The shape of the water molecule is called **non-linear**.

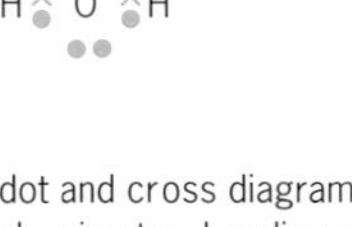

dot and cross diagram showing two bonding pairs and two lone pairs

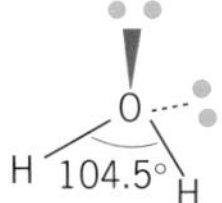

the oxygen atom and hydrogen atoms lie in the plane of the paper and give the molecule its characteristic V-shape

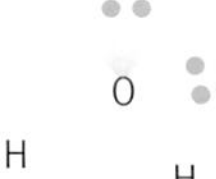

lone pairs and bonding pairs are shown as electron charge clouds

Fig 4.38(a) Three ways to represent the water molecule

QUESTION 18

Draw the shape of:

a) CCl_4

b) NH_4^+ (see page 76 if you are unsure of the dot-and-cross diagram)

c) PCl_3

In each case, state the number of bonding pairs and lone pairs and use this information to estimate the bond angles.

d) Predict the shape and bond angles of SO_2 (see Fig. 4.21 page 77 for the dot-and-cross diagram). (Hint: remember that double bonds are treated as if they have the same repulsion as a single bond pair.)

Fig 4.38(b) Ball-and-stick model of water

7 COVALENT BONDING AND THE OVERLAP OF ATOMIC ORBITALS

The simple dot-and-cross model agrees with much of the experimental evidence. But how does our atomic orbital model of atoms in Chapter 3 treat covalent bonding? A covalent bond is formed by the overlap of two atomic orbitals (one from each atom). Each orbital must be occupied by a single electron, so that when they overlap, the bond formed contains two electrons.

σ AND π BONDS

In hydrogen chloride, a covalent bond forms when an s orbital overlaps with a p orbital, as in Fig 4.39. Note that the s orbital is spherical and the p orbital has two lobes. This kind of covalent bond is known as a **π bond (sigma bond)**. A σ bond is formed when two atomic orbitals overlap at one point: there is always a σ bond between two atoms if they are covalently bonded. A σ bond gives the greatest possible electron density between the two nuclei.

In double and triple bonds there is a sideways overlap between two p orbitals, and a **π bond (pi bond)** forms (Fig 4.40). Ethene has a carbon–carbon double bond made up of a σ bond and a π bond. The π bond is what makes ethene such

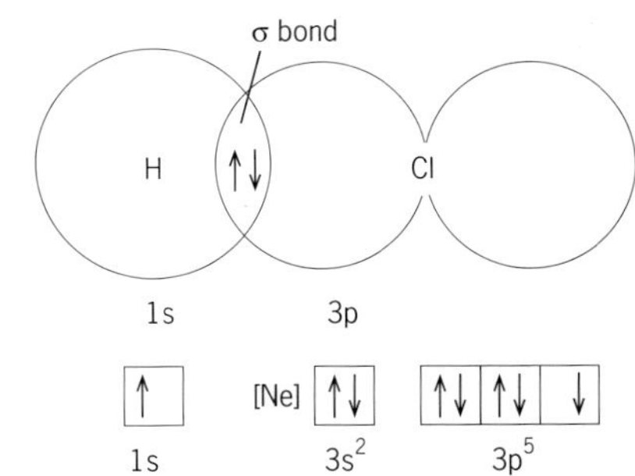

Fig 4.39 Hydrogen chloride: a σ bond forms by the overlap of two singly filled atomic orbitals

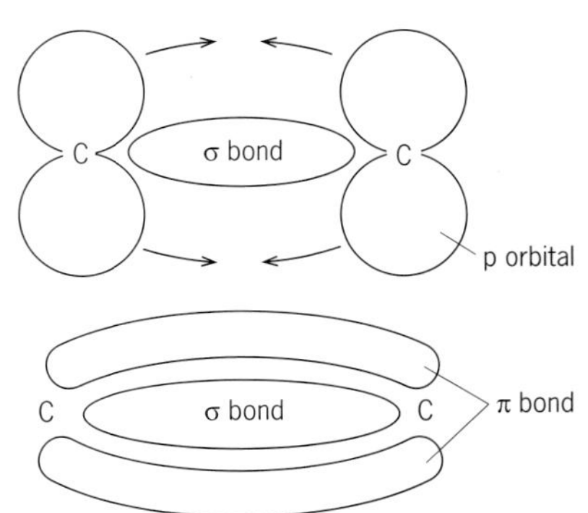

Fig 4.40 A carbon–carbon double bond is made up of one σ bond and one π bond. The π bond forms by the sideways overlap of two p orbitals. This arrangement is found in molecules such as ethene, C_2H_4. (The other atoms that would be joined to the carbon atoms are not shown.)

a versatile industrial chemical. Being above and below the plane of the nuclei, the π bond is more loosely held by the carbon nuclei than the σ bond and therefore allows ethene to be very reactive. To find out more about ethene, see Chapter 14.

STRETCH AND CHALLENGE

A closer look at the orbitals in carbon

Let us look at just how carbon forms its covalent bonds. It has the electron configuration $[He]2s^22p^2$, with two singly occupied p orbitals. So on the atomic orbital overlap model it should form only two bonds. But we know that carbon forms four bonds. To do this it promotes a 2s electron to a 2p orbital and, together, these become four sp^3 orbitals, as shown below. These mixed orbitals all have the same energy and are sometimes called **hybrid** orbitals.

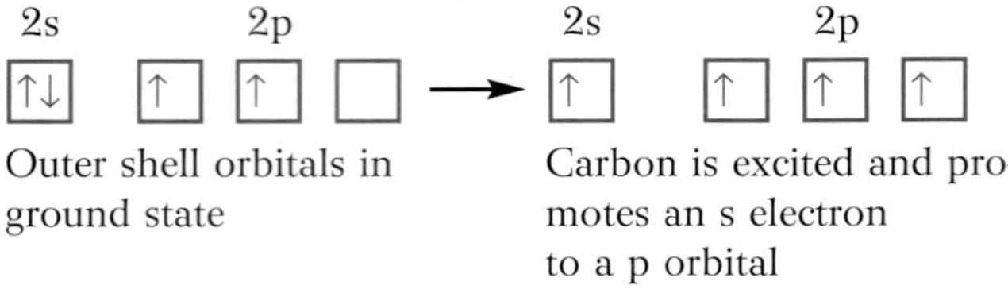

In singly bonded carbon compounds such as methane, for example, carbon's one 2s orbital and three 2p orbitals together form four sp^3 orbitals.

sp^3

↑ ↑ ↑ ↑

Then, as in Fig 4.41, the σ bonds in methane form when the carbon sp^3 orbitals overlap with hydrogen's 1s orbital. The sp^3 orbitals repel each other equally, to give a tetrahedral shape to the methane molecule.

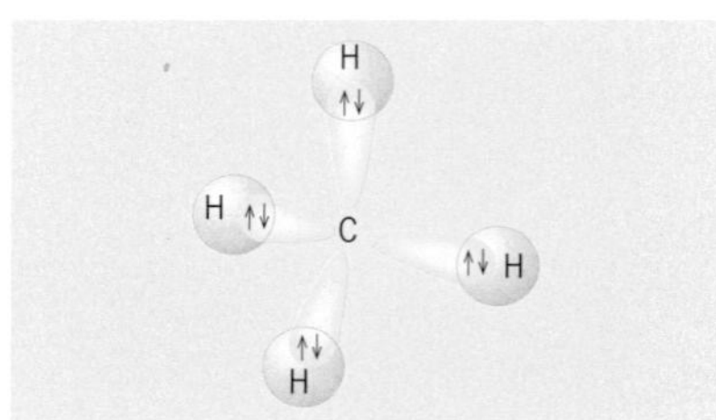

Fig 4.41 In methane, each of the four σ bonds is formed when a hydrogen 1s orbital overlaps with a carbon sp^3 orbital

In ethene, there are three σ bonds and a π bond around each carbon, as shown in Fig 4.42. In this case, after the 2s electron is excited to the 2p orbital, three sp^2 orbitals form, leaving a single 2p orbital.

π bond
σ bonds
σ bond

Fig 4.42 The bonding in ethene

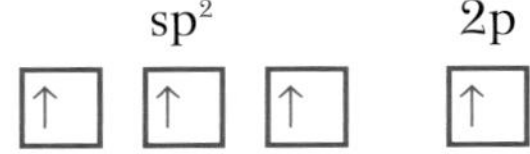

The sp^2 orbitals repel each other to give the trigonal planar shape shown in Fig 4.30. The p orbitals overlap to give a π bond, as in Fig 4.43.

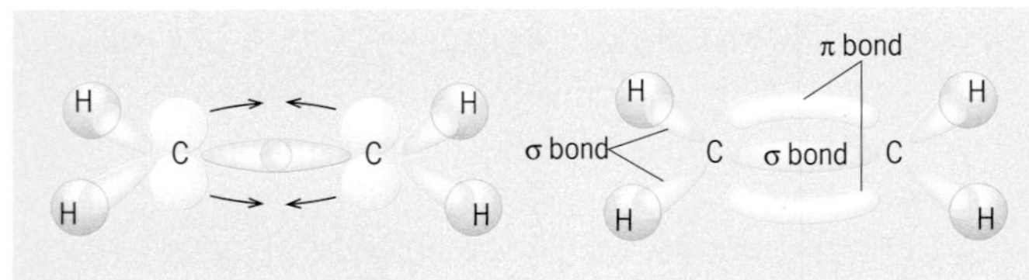

Fig 4.43 The sp^2 orbitals in ethene are all in one plane, while the p orbitals overlap to form the π bond

Ethyne is another hydrocarbon molecule. Each carbon atom has two σ bonds and two π bonds around it, one to the hydrogen atom and one to the other carbon atom (Fig 4.44(a)).

Fig 4.44(a) The bonding in ethyne

sp 2p

↑ ↑ ↑ ↑

Fig 4.44(b) The s electron is excited to give two sp orbitals and two p orbitals

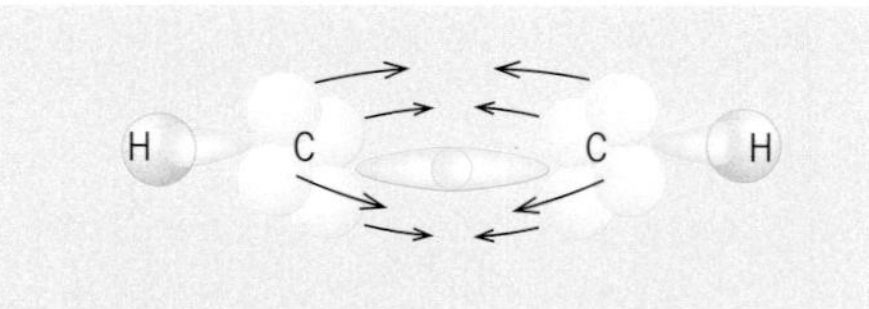

Fig 4.44(c) The sp orbitals repel each other to give a linear shape, and the p orbitals overlap to give two π bonds

? QUESTION 19

19 Explain why sp^2 and sp orbitals are given these names.

8 POLAR COVALENT BONDS AND ELECTRONEGATIVITY

We know that a covalent bond is a shared pair of electrons, but is the pair of electrons shared equally? If we look at the hydrogen molecule, H_2, the electrons are shared equally between the two atoms.

The force of attraction between the two nuclei and the electrons in the covalent bond holds the hydrogen molecule together: being the same, both atoms exert the same force, so the electrons are shared equally.

However, in hydrogen chloride the electrons are not shared equally. The chlorine atom attracts the shared pair more than the hydrogen atom. This means that chlorine has a slight excess of negative charge, which is not balanced out by the protons in its nucleus. We say that the chlorine has a partial negative charge, which is shown by the symbol δ–. Hydrogen has a δ+ charge, because the electrons in the covalent bond are nearer the chlorine atom, which gives hydrogen a partial positive charge. This bond is called a **polar covalent bond**, or just a **polar bond**.

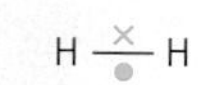

Fig 4.45 In a hydrogen molecule the electrons in a pair are equally shared

To remind yourself about the electrostatic forces of attraction that are present in a covalent bond, see Fig 4.6, page 74.

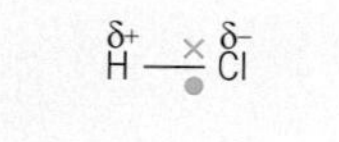

δ+ δ–
H — Cl

Fig 4.46 Electrons in this covalent bond spend more time nearer the chlorine atom, giving it a slightly negative charge and the hydrogen a slightly positive charge

In any covalent bond between two different atoms the sharing of electrons is likely to be unequal, because different atoms have different powers to attract bonding pairs of electrons.

The power of an atom in a molecule to attract the bonding electrons to itself is called its electronegativity.

The most electronegative element is fluorine. It is given a value of 4.0 on a scale devised by one of the 20th century's most famous scientists, Linus Pauling. Next comes oxygen at 3.5 and the third most electronegative elements are chlorine and nitrogen, both having a value of 3.0.

SCIENCE IN CONTEXT

Pauling and his work on chemical bonds

In a life that spanned much of the 20th century, the American Linus Pauling was one of the most influential chemists. He worked briefly with Niels Bohr (see Chapter 3) and in 1954 he received the Nobel Prize for Chemistry for his work on chemical bonds. His book *The Nature of the Chemical Bond* has deeply influenced scientists. It explains how atoms combine and helped predict the way compounds react. He put forward the ideas of hybrid orbitals and the partial ionic character of covalent bonds.

Shortly after the Second World War, helped by his wife, he spearheaded a campaign for nuclear disarmament. This led in 1963 to a limited test ban treaty. For this he received the Nobel Peace Prize in 1963 and became the first person (and to date, the only person) to receive two Nobel prizes awarded solely.

Fig 4.47 Linus Carl Pauling, 1901–1995

In the Periodic Table (Fig 4.49 overleaf), we see that electronegativity increases up a group and across a period.

Fig 4.48 shows the difference in electronegativity between hydrogen and fluorine in a hydrogen fluoride molecule. Fluorine is the more electronegative element, so it attracts electrons more and has a partial negative charge, δ–.

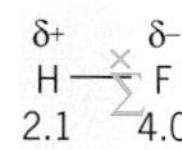

Fig 4.48 The large difference in electronegativity values causes hydrogen fluoride to be very polar

QUESTION 20

20 Use Fig 4.49 to decide whether the following bonds are polar. Where appropriate show the negative and positive poles using δ– and δ+. Rank the polarity of these bonds putting the most polar first:

C–Cl, P–H, H–F, O–H, F–F, N–H, C–I

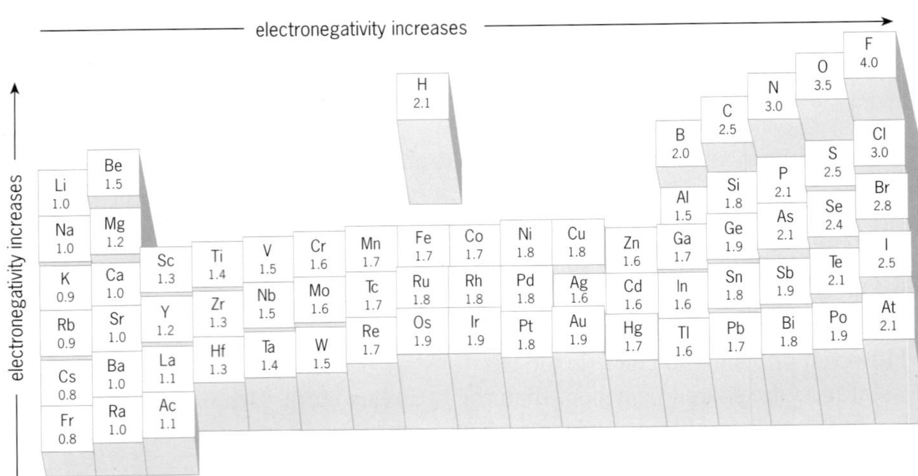

Fig 4.49 The Periodic Table showing electronegativity values

In the hydrogen fluoride molecule the difference between the electronegativities is 1.9. The larger the difference between the electronegativities, the more polar is the covalent bond, and a value of 1.9 means that the hydrogen fluoride molecule is very polar.

REMEMBER THIS

Even though bonds within them may be polar, the molecules themselves may be non-polar if the bonds are symmetrically arranged such that their polarities cancel each other out.

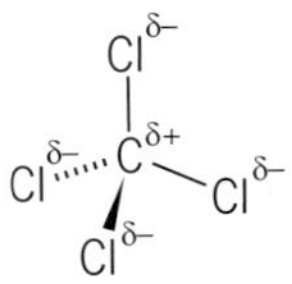

Cl^δ– Cl^δ– C^δ+ Cl^δ– Cl^δ–

Fig 4.50 The tetrachloromethane bond is non-polar because the four polar bonds are symmetrically arranged and cancel each other out

Core charge

Why do different atoms have different powers to attract electrons in a covalent bond? To answer this question we must look at the **core charge** of atoms and at their size.

The core charge is the negative charge of the inner electron shells plus the positive charge due to the protons in the nucleus.

On page 72 we saw that the inner shells of an atom can be called the **noble gas core**. As seen in Fig 4.51, chlorine has a noble gas core of 10 electrons in its two inner shells. There are 17 positive protons in the nucleus, so the outer bonding electrons experience a core charge of 7+. Fluorine also has a core charge of 7+, because it has two electrons in its inner shell and nine protons in the nucleus. It is more electronegative than chlorine because fluorine is a smaller atom, so the outer bonding electrons are closer to its nucleus. With a core charge of 7+ and being a small atom, fluorine is the strongest of all the atoms in attracting electrons to itself.

A carbon–chlorine bond is polar because the shared pair is attracted more to the larger core charge. However, a bond with even greater polarity is the carbon–fluorine bond, because the fluorine atom is so small (see Fig 4.51).

To summarise, the two factors that help to determine electronegativity are the core charge of an atom and its size. Small atoms with high core charges are the most electronegative atoms.

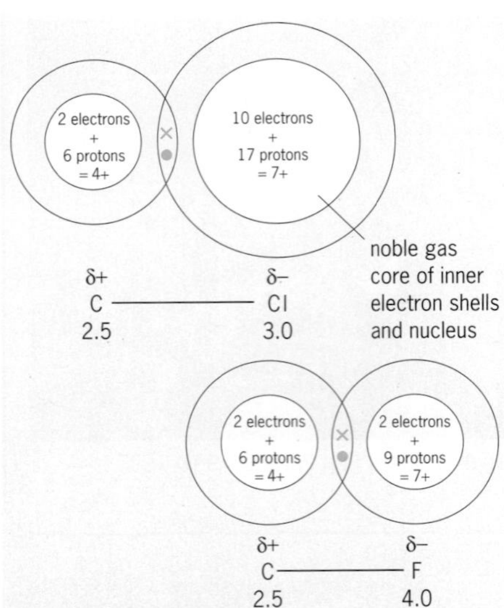

Fig 4.51 Inner shell electrons and protons in carbon–chlorine and carbon–fluorine

THE IONIC CHARACTER OF MANY COVALENT BONDS

The more polar a covalent bond, the greater the electron shift towards the more electronegative element. In some compounds this shift is so great that, instead of covalent bonds, ions are formed.

In fact, very few covalent bonds share electrons exactly equally. Most have some ionic character. The greater the difference in electronegativity between the two atoms in a bond, the more ionic is its character. When the electronegativity difference is more than about 2.0, the bond is more ionic than covalent, and we call it ionic. Thus there are pure covalent bonds and pure ionic bonds, but most compounds lie between these two extremes.

THE COVALENT CHARACTER OF IONIC BONDS

In the same way that covalent bonds have some ionic character, so all ionic bonds have some degree of covalent character. We think of ions as spheres of negative charge that surround the nucleus, but negative ions (anions) can be **polarised** by positive ions (cations). Negative ions have more electrons than protons, so the electron cloud is held more loosely than in an atom. A positive ion may cause some distortion of this electron cloud and this is called **polarising** the negative ion.

To work out how much covalent character there is likely to be in an ionic bond, first we need to know the sort of cation that has the most polarising power over an anion, and the sort of anion that is most easily polarised.

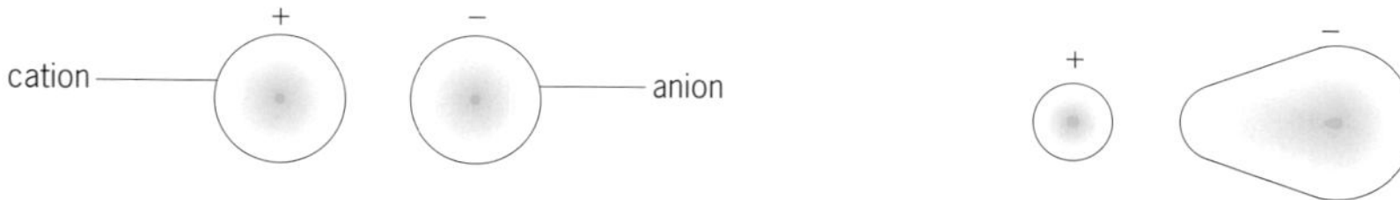

Fig 4.52(a) There is no polarisation of the anion by the cation, so there is no covalent character

Fig 4.52(b) The cation is polarising the anion, so there is some covalent character in the ionic bond

A cation has high polarising power if:

- it has a **high positive charge** (to attract the electrons in the anion)
- it is a **small cation** (the nucleus has more attraction for the electrons in the anion).

These cations have a **high charge density.**

An anion is easily polarised if:

- it has a **high negative charge** (outer electrons are more loosely held by the nucleus)
- it is a **large anion** (the nucleus is further away from the outer electrons, so has less of an attraction for them).

These ideas were first proposed by the Polish chemist Kasimir Fajans in 1923 and are often known as Fajans' rules. You can read more about the ionic and covalent character of compounds in Chapter 21.

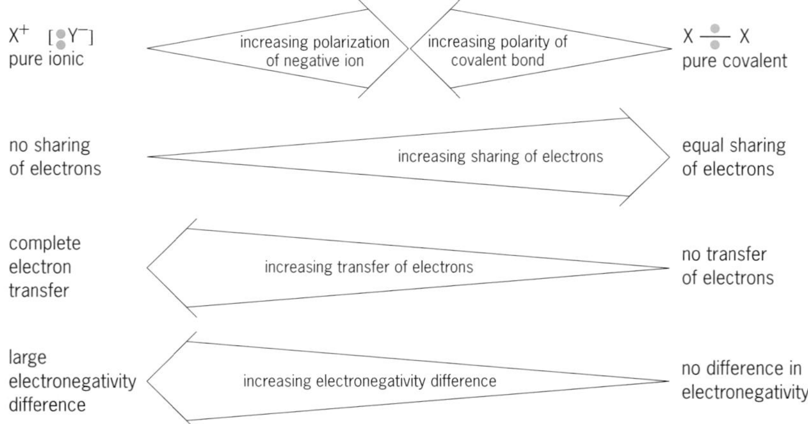

Fig 4.53 Ionic and covalent are two extremes of bonding while in-between a gradual transition takes place

QUESTION 21

Work out the core charge of oxygen and sulfur in Group 6 of the Periodic Table. What will the core charge of the other elements in this group be? Explain why the electronegativity of these atoms increases on going up the group.

REMEMBER THIS

Ionic radius and the charge on an ion can be treated as a single property of the ion, called **charge density**. You can read more about this in Chapter 22.

QUESTION 22

Which of the following compounds show the most covalent character? Sodium chloride, magnesium chloride or aluminium chloride?

EXAMPLE

Q Explain why the covalent character of the chlorides formed with Group 1 elements of the Periodic Table increases on going up the group.

A All Group 1 ions have a 1+ charge. On going down the group, an extra electron shell is added, so the cations are larger, which means they become less polarising.

The anion is the chloride ion in each compound. It is going to be most readily polarised by the smallest positive ion, which in this case is lithium, and least distorted by the largest ion, which is the francium ion. Thus, covalent character increases on going up Group 1.

SUMMARY

When you have finished studying this chapter, you should understand and be able to use the following ideas.

- Usually when atoms bond together in a compound they lose, gain or share electrons to make stable noble gas configurations of outer shell electrons.
- Ionic bonding involves the complete transfer of electrons between atoms to create oppositely charged ions, which are held together by strong electrostatic forces of attraction. Lattice energy (lattice enthalpy) is the energy released when one mole of lattice is formed from its gaseous ions.
- A covalent bond is a shared pair of electrons between two atoms. It is an electrostatic attraction between the pair of negative electrons and the positively charged nuclei. Double and triple bonds can also form between atoms. When the pair of electrons in a covalent bond comes from one atom it is called a dative covalent (coordinate) bond. Sometimes atoms in a covalent compound have more or less than the usual noble gas octet.
- The shape of molecules and ions can be determined by counting up the number of electron pairs around the central atom. Bonding pairs exert the least repulsion on each other; lone pair–bonding pair repulsion is greater; while two lone pairs have the most repulsion.
- A single covalent bond is formed by the overlap of two atomic orbitals to form a σ bond. π bonds occur in double or triple bonds (together with σ bonds). π bonds are formed by the sideways overlap of p orbitals.
- Ionic and covalent bonds are two extreme forms of bonding. Most bonds are neither purely ionic nor purely covalent. Pure covalent bonds are formed between atoms with the same electronegativities. Polar covalent bonds arise when there is a difference in electronegativities between the atoms in the bond and the pair of electrons is not equally shared. Electronegativity is the power of an atom within a molecule to attract bonding pairs of electrons to itself.
- All ionic compounds have some degree of covalent character. Small highly charged cations are best at distorting the electron cloud around anions. Large highly charged anions are the most easily distorted.

Practice questions and a How Science Works assignment for this chapter are available at www.collinseducation.co.uk/CAS

5

The Periodic Table, redox and oxidation number

5 THE PERIODIC TABLE, REDOX AND OXIDATION NUMBER

SCIENCE IN CONTEXT

Modern day alchemy – the discovery of 'super-heavy' elements

Seeking elements. Research is carried out at CERN, the European centre for particle physics, into the nature of the elements and their subatomic particles

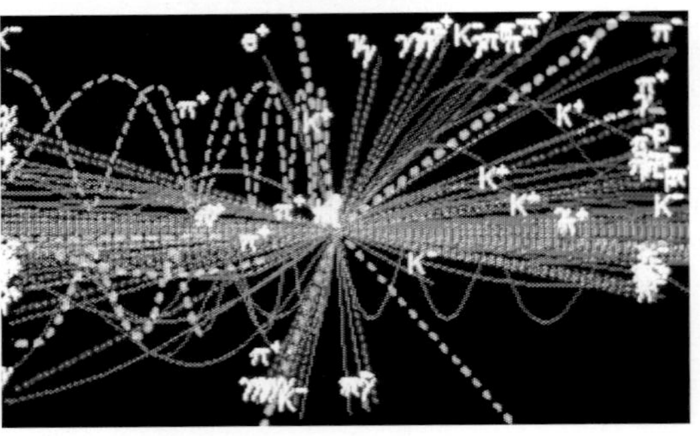

Ever since Mendeleev proposed his theory of the periodic classification of elements, scientists have set themselves the task of discovering yet more elements. During the past 30 years or so, the search has been on to discover elements that are not normally found on Earth. Scientists have bombarded atoms with high-energy particles in an effort to achieve the alchemist's dream of transmutation, the changing of one element into another.

In 1999 scientists in California claimed to have made the 'super-heavy element' atomic number 118. They predicted from its position in the Periodic Table that the element would be the seventh noble gas. Three years later these scientists retracted their discovery, because they were unable to repeat their findings. However between 2002 and 2005, collaborative research involving Russian and American scientists continued, and eventually sufficient evidence was collected to announce the discovery of element number 118. The scientists' results indicated that they had made just a few atoms of this new element, which then spontaneously broke down into simpler elements.

The element was given the symbol Uuo and the name ununoctium. It may be many years before scientists can agree to give it a two letter atomic symbol and a proper name, because there has to be international agreement about the names of elements.

Fig 5.1 Dmitri Mendeleev – father of the Periodic Table. By bringing together all the facts known then about the elements, Mendeleev made a system of classification that led to many predictions and the discovery of new elements

INTRODUCTION

For thousands of years, people have been curious about the nature of materials in the world around them, and have had theories on the simplest forms of matter, or 'elements', that make up these materials. However, only in the past 200 years, since modern experimental work began, have chemists been sure that they have identified **elements** – substances that could not chemically be changed to anything simpler. Only then could the chemists start to devise useful theories on how elements are related to each other, theories that would help scientists progress in their knowledge and understanding of materials.

The greatest breakthrough in this recent endeavour was presented to the world of chemistry on 17 February 1869, when the Russian chemist, Dmitri Mendeleev, published his work on the properties of elements. Assembling the observations and discoveries of earlier workers, he had assembled the symbols of the chemical elements in the order of their atomic masses. Mendeleev's particular arrangement became the first modern Periodic Table, and was to prove invaluable to chemists. It greatly accelerated the discovery of new elements and understanding of their properties, and enabled scientists to predict the full range of elements that were eventually found or artificially made.

This chapter describes how the Periodic Table developed. We explore its importance, both in explaining chemical trends in the elements, long before the structure of the atom was understood, and in predicting the discovery of unknown elements.

1 EARLY IDEAS ABOUT THE ELEMENTS

More than 2000 years ago the Ancient Greeks pictured all the material in the world as being composed of four 'elements' – earth, water, air and fire. The idea may seem far-fetched now, but the Ancient Greeks were thinkers and not experimenters. It is interesting, however, that their model reflected the three states of matter – solid, liquid and gas – and fire, which has the capacity to transform materials.

QUESTION 1

Suggest why metals such as gold, silver, copper and mercury were the first elements discovered.

Fig 5.2 A caricature from about 1800 showing alchemists trying to discover the Philosopher's Stone, a process they believed would transform metals into gold

Fig 5.3 Fire, charcoal and copper ore were used in the process of smelting copper shown in this engraving from 1574

Civilisations, some of them older than the Ancient Greeks, were able to produce metallic elements such as copper and mercury by mixing their ores with charcoal and heating them in a simple furnace. Other elements, including gold and silver, were found naturally in their elemental state. The drive to produce metals and their alloys arose not because people were interested in chemistry for its own sake, but because of the need to make tools and weapons that were hard-wearing and shatterproof.

The early Arab civilisation was noted for its practical interest in chemicals, and has given us words currently used in chemistry such as alchemy and alkali. In the Middle Ages, particularly, alchemists in Europe tried hard to find a way to make gold from lead and other metals, not knowing that this was chemically impossible.

The pace of discovery of the elements is shown in Fig 5.4. Notice that where the graph is nearly vertical, several elements were discovered at the same time. Typically, this happened when a new theory or idea was proposed, or new apparatus was developed.

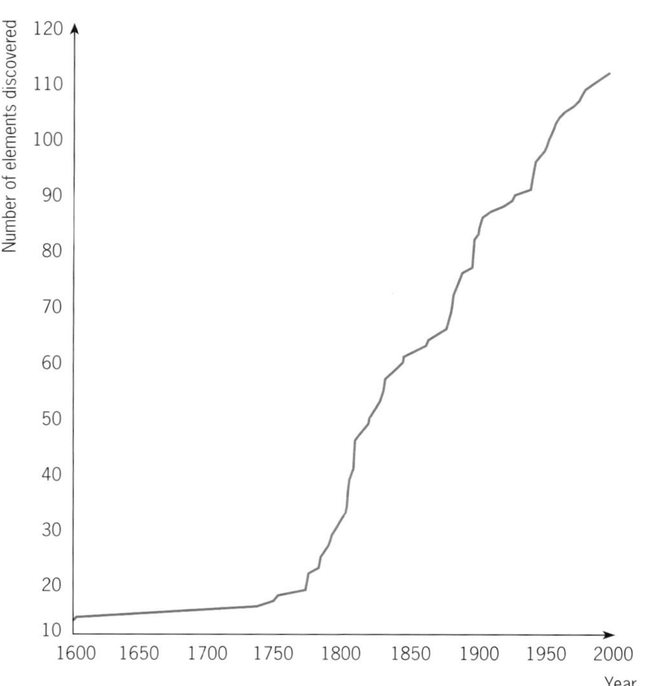

Fig 5.4 Graph to show the number of elements known from 1600 to the present. Notice how the graph steepens as the rate of new discoveries increases, for example at the time of Mendeleev's work

? QUESTION 2

2 What development happened in 1896 which allowed several elements to be discovered? (Hint: look back to page 31.)

For example, towards the end of the 18th century, glassware designed for gas preparation led to the discovery of oxygen, chlorine and hydrogen, and air was shown to be a mixture of gases, not a single substance. Later, the new technique of electrolysis enabled Humphry Davy to discover another group of elements – sodium, potassium, magnesium, barium and strontium.

DALTON'S ATOMIC THEORY

In Dalton's atomic theory (see Chapter 2), he proposed that an atom was the smallest part of an element and could not be split, and that atoms of a particular element had a characteristic mass. We now call this mass the element's **relative atomic mass**, but in Dalton's time it was known as the atomic weight.

For a definition of relative atomic mass, see Chapter 2, p.37.

Through the early 19th century, chemists continued their search for new elements, gradually improving their techniques, but working in an unsystematic way, as they lacked a basis for predicting the existence of unknown elements.

DÖBEREINER AND HIS TRIADS

Johann Döbereiner was among the scientists who were looking for ways to classify the elements. In 1829, he reported on patterns he had found among the elements known at the time. He noticed that elements with similar properties could be grouped into threes, or triads, and that there was a mathematical pattern in the values for their relative atomic masses. One of his triads was lithium, sodium and potassium, elements that Davy had discovered using electrolysis.

? QUESTION 3

3 Look at the following triads. The relative atomic mass is written underneath each element.

lithium	sodium	potassium
7	23	39
sulfur	selenium	tellurium
32	79	128
chlorine	bromine	iodine
35.5	80	127
calcium	strontium	barium
40	88	137

What is the mathematical connection between the atomic masses in these triads? Suggest an explanation for this connection.

NEWLANDS' OCTAVES

Döbereiner had begun to bring order to the elements, showing patterns that existed in their relative atomic masses. Then, in 1864, John Newlands arranged the elements in the order of their relative atomic mass. He, too, found a pattern, noticing that elements with similar chemical properties were eight positions away from each other, rather like the notes in a musical octave. (At this time, the noble gases, which would have added another position, had not been discovered.)

Look at Table 5.1. If you start with lithium as '1' and count eight elements, you come to sodium. The two elements are very similar physically and chemically. Count another eight elements and you reach potassium, also very similar to lithium and sodium.

Table 5.1 Newlands' octaves

hydrogen	lithium	beryllium	boron	carbon	nitrogen	oxygen
fluorine	sodium	magnesium	aluminium	silicon	phosphorus	sulfur
chlorine	potassium	calcium	chromium	titanium	manganese	iron

Though the rule of octaves works for some elements, there are exceptions. Look at sulfur and iron, which are, again, eight elements apart and yet have very different properties. It needed another system of grouping to make sense of these differences, but Newlands was too preoccupied with the idea of an octave to devise an explanation.

2 MENDELEEV AND THE DEVELOPMENT OF THE PERIODIC TABLE

The Russian chemist Mendeleev was fascinated by the elements. He wrote down each known element and its properties on a separate card, and began to put them in a logical order (see Table 5.2). He ignored hydrogen, for it did not seem to fit anywhere, and started with lithium. Guided by atomic weights, he put beryllium, boron, carbon, nitrogen, oxygen and fluorine in a column. The next element was sodium. He put it next to lithium as both were reactive soft metals. Then the elements magnesium, aluminium, silicon, phosphorus, sulfur and chlorine duly took their place.

Mendeleev continued his list with potassium and calcium. The obvious place for the next card, for titanium, seemed to be below calcium. Then Mendeleev made a brilliant decision. He recognised that a gap had to be left between calcium and titanium for an element that was yet to be discovered. The chemical properties of titanium were more like those of carbon and silicon than of boron and aluminium, and so Mendeleev placed titanium next to silicon.

Mendeleev concluded, 'the elements, if arranged according to their atomic weights [now known as relative atomic masses], exhibit an evident periodicity of properties.' Mendeleev's original Periodic Table is shown in Fig 5.5.

Notice that Mendeleev first placed the elements in columns. Soon afterwards, he decided on the arrangement of the Periodic Table that is familiar to us. Each element had its own number and fixed position in the table, and Mendeleev left several gaps for elements that were then unknown, but which he believed existed.

Mendeleev's table made a considerable impact among chemists of his time, because it was so useful to them. It revealed clear patterns and trends in the properties of elements, both known and unknown. For the first time, chemists could guess at the total number of elements that might exist and, in noting the gaps, they were spurred on to make discoveries of new elements.

Mendeleev used his table to predict the properties of scandium and gallium, and also germanium, which he first named 'eka-silicon'. Within 20 years, all three elements had been discovered. Table 5.3 shows the properties he expected germanium to have, and how close his predictions proved to be. Altogether, Mendeleev predicted properties for ten unknown elements, and was later proved correct in eight cases.

Table 5.2 Mendeleev's early lists of elements

lithium	sodium	potassium
beryllium	magnesium	calcium
boron	aluminium	?
carbon	silicon	titanium
nitrogen	phosphorus	vanadium
oxygen	sulfur	
fluorine	chlorine	

			Ti=50	Zr=90	?=180.
			V=51	Nb=94	Ta=182.
			Cr=52	Mo=96	W=186.
			Mn=55	Rh=104,4	Pt=197,4
			Fe=56	Ru=104,4	Ir=198.
			Ni=Co=59	Pl=106,6	Os=199.
H=1			Cu=63,4	Ag=108	Hg=200.
	Be=9,4	Mg=24	Zn=65,2	Cd=112	
	B=11	Al=27,4	?=68	Ur=116	Au=197?
	C=12	Si=28	?=70	Sn=118	
	N=14	P=31	As=75	Sb=122	Bi=210
	O=16	S=32	Se=79,4	Te=128?	
	F=19	Cl=35,5	Br=80	I=127	
Li=7	Na=23	K=39	Rb=85,4	Cs=133	Tl=204
		Ca=40	Sr=87,6	Ba=137	Pb=207.
		?=45	Ce=92		
		?Er=56	La=94		
		?Yt=60	Di=95		
		?In=75,6	Th=118?		

Fig 5.5 Mendeleev's Periodic Table, published in his book *Principles of Chemistry* in 1869. In it he first set out his periodic law, which earned him international fame. The question marks on their own before the equal signs are for unknown elements. Mendeleev published his version of the Periodic Table with horizontal periods and vertical groups in 1870

Table 5.3 Properties of 'eka-silicon', germanium and tin

	Silicon	Predicted properties for eka-silicon	Actual values for germanium (eka-silicon)	Tin
Atomic mass	28	72	72.59	118
Density/g cm^{-3}	2.3	5.5	5.3	7.3
Appearance	grey non-metal	grey metal	grey metal	white metal
Formula of oxide	SiO_2	EkO_2	GeO_2	SnO_2
Formula of chloride	$SiCl_4$	$EkCl_4$	$GeCl_4$	$SnCl_4$
Reaction with acid	none	very slow	slow with concentrated acid	slow

Group / Series	0	I	II	III	IV	V	VI	VII	VIII
1		Hydrogen H 1.008	—	—	—	—	—	—	
2	Helium He 4.0	Lithium Li 7.03	Beryllium Be 9.1	Boron B 11.0	Carbon C 12.0	Nitrogen N 14.04	Oxygen O 16.00	Fluorine F 19.0	
3	Neon Ne 19.9	Sodium Na 23.05	Magnesium Mg 24.3	Aluminium Al 27.0	Silicon Si 28.4	Phosphorus P 31.0	Sulfur S 32.06	Chlorine Cl 35.45	
4	Argon Ar 38	Potassium K 39.1	Calcium Ca 40.1	Scandium Sc 44.1	Titanium Ti 48.1	Vanadium V 51.4	Chromium Cr 52.1	Manganese Mn 55.0	Iron Fe 55.9; Cobalt Co 59; Nickel N 59 (Cu)
5		Copper Cu 63.6	Zinc Zn 65.4	Gallium Ga 70.0	Germanium Ge 72.3	Arsenic As 75	Selenium Se 79	Bromine Br 79.95	
6	Krypton Kr 81.8	Rubidium Rb 85.4	Strontium Sr 87.6	Yttrium Y 89.0	Zirconium Zr 90.6	Niobium Nb 94.0	Molybdenum Mo 96.0	—	Ruthenium Ru 101.7; Rhodium Rh 103.0; Palladium Pd 106.5 (Ag)
7		Silver Ag 107.9	Cadmium Cd 112.4	Indium In 114.0	Tin Sn 119.0	Antimony Sb 120.0	Tellurium Te 127	Iodine I 127	
8	Xenon Xe 128	Caesium Cs 132.9	Barium Ba 137.4	Lanthanum La 139	Cerium Ce 140	—	—	—	— — —
9	—	—	—	—	—	—	—	—	
10	—	—	—	Ytterbium Yb 173	—	Tantalum Ta 183	Tungsten W 184	—	Osmium Os 191; Iridium Ir 193; Platinum Pt 194.9 (Au)
11		Gold Au 197.2	Mercury Hg 200.0	Thallium Tl 204.1	Lead Pb 206.9	Bismuth Bi 208	—	—	
12	—	—	Radium Rd 224	—	Thorium Th 232	—	Uranium U 239		

Fig 5.6 By 1905, Mendeleev's Periodic Table had a few new elements, but still looked similar to his earlier version

Mendeleev positioned elements on the basis of their properties. Where elements did not fit in order of their atomic weights (relative atomic masses), he assumed that these had been determined incorrectly. So, for example, he placed cobalt before nickel, though cobalt has a greater mass.

ATOMIC (PROTON) NUMBER

We have seen that, on the basis of their properties, a few elements in the Periodic Table had to be placed out of sequence in terms of their relative atomic masses. We now know that relative atomic mass depends on the overall structure of the nucleus, and that it is the order of *atomic number* of elements that determines the trends in their properties.

The atomic number (or proton number) of an element is the number of protons found in one atom of that element.

QUESTION 4

4 Look at a modern Periodic Table (opposite page).

a) Cobalt and nickel are not in the order of their atomic masses. Explain why this is so.

b) Write down the names of two more pairs of elements that are not arranged in order of relative atomic mass in the Periodic Table.

REMEMBER THIS

In older versions of the Periodic Table, the group numbers are roman numerals – Group 7 is Group VII, for example.

REMEMBER THIS

In some modern versions of the Periodic Table, Groups 1, 2, 3, 4, 5, 6 and 7 in Fig 5.7 are referred to as 1A, 2A, 3A, 4A, 5A, 6A and 7A.

PERIODS AND GROUPS IN THE PERIODIC TABLE

The table Mendeleev devised in 1905 (Fig 5.6) has much in common with a modern version of the Periodic Table (see Fig 5.7). *Horizontal rows* of elements represent **periods**, numbered 1 to 7. Note how the seven modern periods compare to Mendeleev's series, and how they differ.

The *vertical columns* 1 to 7 and 0 in Fig 5.7 represent the **groups** of elements. Elements in a group have very similar chemical properties. Notice in Fig 5.6 that each group has two columns. These correspond to a group in the s and p block and another group taken from the d-block elements (Fig 5.8, see later). We just refer to Group 1 as being all the elements in the vertical column headed by lithium, and similarly, Group 7 is all the elements headed by fluorine.

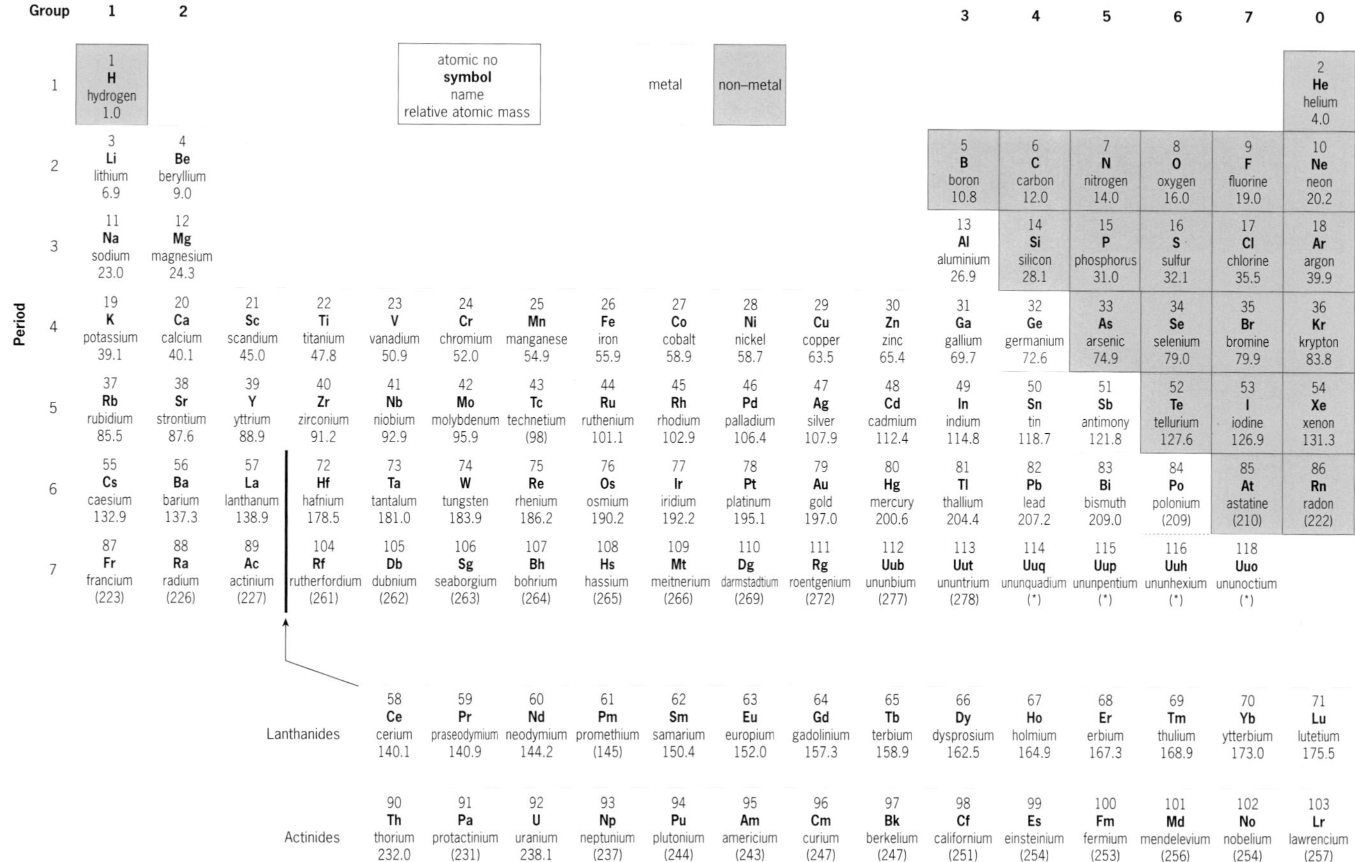

Fig 5.7 A modern version of the Periodic Table, with 109 elements (relative atomic mass in brackets = mass number of most stable isotope)

ELECTRON CONFIGURATION

As you may have read in Chapter 3, the electron configuration of an atom describes the arrangement of its electrons. Electrons occupy a series of shells round the nucleus, numbered 1, 2, 3, 4, and so on. Each shell has particular subshells labelled s, p, d, f, etc. Each subshell has its own energy level. Within a shell the energy levels of the subshells increase in the order s, p, d, f, etc. The chart sums up this information:

Shell	Subshells			
1	1s			
2	2s,	2p		
3	3s,	3p,	3d	
4	4s,	4p,	4d,	4f

⟶ increasing energy level of subshell

A maximum number of electrons can occupy each subshell:

Subshell	s	p	d	f
Maximum number of electrons	2	6	10	14

In each shell, electrons fill the subshells in the same order. Usually, a subshell has to be fully occupied before further electrons start to fill the next level. We indicate the number of electrons in a subshell by a raised number and describe the electrons in an atom as:

$1s^2\ 2s^2\ 2p^6\ 3s^2\ 3p^6\ 3d^{10}\ 4s^2\ 4p^6\ 4d^{10}$, etc.

The chemical properties of an element depend on its electron configuration.

QUESTION 5

Which element is found in Period 4 and Group 6?

QUESTION 6

The names for elements 104 to 111 have only recently been decided. Suggest reasons for the choice of each name.

REMEMBER THIS

Some modern period tables have introduced groups 1 to 18, where groups 1 and 2 are in the s-block, groups 3 to 12 are groups in the d-block and 13 to 18 are groups in the p-block. This is of limited use because the group number will not correspond to the number of electrons in the outer shell of the atom.

? QUESTION 7

7 Write down the electron configurations for each of the following elements: nitrogen, chlorine, calcium and titanium.

Go back to page 63 if you wish to find out more about electron configurations.

? QUESTION 8

8 For each of these electron configurations, state the group to which the element belongs:

a) $1s^2\ 2s^2\ 2p^6\ 3s^2\ 3p^6\ 3d^{10}\ 4s^2\ 4p^5$
b) $1s^2\ 2s^2\ 2p^6\ 3s^2\ 3p^6\ 3d^{10}\ 4s^2\ 4p^1$
c) $1s^2\ 2s^2\ 2p^6\ 3s^2\ 3p^6\ 3d^{10}\ 4s^2\ 4p^6\ 4d^{10}\ 5s^2\ 5p^2$

? QUESTION 9

9 In which block of the Periodic Table would you place the elements with the following electron configurations?

a) $1s^2\ 2s^2\ 2p^6\ 3s^2\ 3p^3$
b) $1s^2\ 2s^2\ 2p^6\ 3s^2\ 3p^6\ 3d^{10}\ 4s^2\ 4p^5$
c) $1s^2\ 2s^2\ 2p^6\ 3s^2\ 3p^6\ 3d^{10}\ 4s^2\ 4p^6\ 4d^6\ 5s^2$
d) $1s^2\ 2s^2\ 2p^6\ 3s^2\ 3p^6\ 3d^{10}\ 4s^2\ 4p^6\ 4d^{10}\ 5s^2$

Since elements are arranged in the Periodic Table in order of their atomic number and chemical properties, it is no surprise to see that elements are also ordered according to their electron configurations. Another strength of Mendeleev's Periodic Table is that it reflected electron configurations before electrons were even discovered.

We can see that electron configurations of the outer shells of elements in the Periodic Table show regular trends or **periodicity**. (We look again at periodicity in Section 3.) The elements of any particular group have the same number of electrons in the same type of outer shell. For example, the Group 6 elements oxygen, sulfur, selenium, tellurium and polonium all have six electrons in their outer shell, namely $ns^2\ np^4$, where n is the period number.

If an element is in Period 4, then its outer electrons occupy either 4s or 4p orbitals; if an element is in Period 6, then its outer electrons occupy either 6s or 6p orbitals.

'BLOCKS' IN THE PERIODIC TABLE

Refer to Fig 5.8 as you read this section. The **s block** contains the elements with an electron configuration that ends with electrons in an s subshell, meaning that the outermost electrons are in s subshells. A typical s-block element is strontium, with an electron configuration of $1s^2\ 2s^2\ 2p^6\ 3s^2\ 3p^6\ 3d^{10}\ 4s^2\ 4p^6\ 5s^2$, so its outer electrons are in an s subshell.

The **p block** contains elements that have an electron configuration that ends with electrons in a p subshell, so the outermost electrons are in a p subshell. A typical p-block element is chlorine, with an electron configuration of $1s^2\ 2s^2\ 2p^6\ 3s^2\ 3p^5$, so the outer electrons are in a p subshell.

The **d block** contains elements in Periods 4, 5 and 6. They are often called the transition elements or transition metals. Most of the d-block elements have an incomplete set of d electrons in their second-to-outermost subshell, and either one or two electrons in the outermost s subshell. A typical element of the d block is iron, with the electron configuration $1s^2\ 2s^2\ 2p^6\ 3s^2\ 3p^6\ 3d^6\ 4s^2$, where $3d^6$ is an incomplete subshell of shell 3.

Ten electrons can fill a d subshell, and the d block is also 10 elements wide. In Period 4, for example, scandium on the far left has one electron in the d subshell, and on the right zinc has 10. Electrons are added one at a time to the 3d subshell as you go across the d-block part of the Period.

The f block contains elements with an incomplete set of electrons in an f subshell. These elements are often called the lanthanides and actinides, named after the elements lanthanum and actinium just before them in Periods 6 and 7.

s block

	1	2
1s	H	
2s	Li	Be
3s	Na	Mg
4s	K	Ca
5s	Rb	Sr
6s	Cs	Ba
7s	Fr	Ra

d block

3d	Sc	Ti	V	Cr	Mn	Fe	Co	Ni	Cu	Zn
4d	Y	Zr	Nb	Mo	Tc	Ru	Rh	Pd	Ag	Cd
5d	La	Hf	Ta	W	Re	Os	Ir	Pt	Au	Hg
6d	Ac									

(4f, see below)
(5f, see below)

p block

	3	4	5	6	7	0
						He
2p	B	C	N	O	F	Ne
3p	Al	Si	P	S	Cl	Ar
4p	Ga	Ge	As	Se	Br	Kr
5p	In	Sn	Sb	Te	I	Xe
6p	Ti	Pb	Bi	Po	At	Rn

f block

Lanthanides 4f	Ce	Pr	Nd	Pm	Sm	Eu	Gd	Tb	Dy	Ho	Er	Tm	Yb	lu
Actinides 5f	Th	Pa	U	Np	Pu	Am	Cm	Bk	Cf	Es	Fm	Md	No	Lr

Fig 5.8 The blocks of the Periodic Table

3 PATTERNS IN THE PERIODIC TABLE

Examining the Periodic Table, we have seen that the properties of elements are often repeated at regular intervals. This characteristic is called **periodicity**. Now we have seen that the fundamental structure of atoms, represented by their electron configurations, also shows similar patterns in periodicity.

ATOMIC RADII

Look at Fig 5.9, which shows the atomic radii of the elements.

1 **H** 30																	2 **He**
3 **Li** 152	4 **Be** 111											5 **B** 88	6 **C** 77	7 **N** 70	8 **O** 66	9 **F** 64	10 **Ne**
11 **Na** 186	12 **Mg** 160											13 **Al** 143	14 **Si** 117	15 **P** 110	16 **S** 104	17 **Cl** 99	18 **Ar**
19 **K** 231	20 **Ca** 197	21 **Sc** 160	22 **Ti** 146	23 **V** 131	24 **Cr** 125	25 **Mn** 129	26 **Fe** 126	27 **Co** 126	28 **Ni** 124	29 **Cu** 128	30 **Zn** 133	31 **Ga** 122	32 **Ge** 122	33 **As** 121	34 **Se** 117	35 **Br** 114	36 **Kr**
37 **Rb** 244	38 **Sr** 215	39 **Y** 180	40 **Zr** 157	41 **Nb** 143	42 **Mo** 136	43 **Tc** 136	44 **Ru** 133	45 **Rh** 134	46 **Pd** 138	47 **Ag** 144	48 **Cd** 149	49 **In** 168	50 **Sn** 140	51 **Sb** 141	52 **Te** 137	53 **I** 133	54 **Xe**
55 **Cs** 262	56 **Ba** 217	57 **La** 188	72 **Hf** 157	73 **Ta** 143	74 **W** 137	75 **Re** 137	76 **Os** 134	77 **Ir** 135	78 **Pt** 138	79 **Au** 144	80 **Hg** 155	81 **Tl** 171	82 **Pb** 175	83 **Bi** 146	84 **Po** 140	85 **At** 140	86 **Rn**
(87) **Fr** 270	88 **Ra** 220	89 **Ac** 200															

Fig 5.9 Atomic (covalent) radii of the elements measured in picometres (10^{-12} m)

The atomic radius of an isolated atom cannot be measured as it is not possible to know where exactly its outer boundary is. There are several definitions for the atomic radius of an atom, but we shall use the following one.

The atomic radius of an element X is half of the distance between the nuclei of two atoms joined by a single covalent bond X–X.

This is also referred to as the **covalent radius** of an element. The force between the negative electrons and the positive nucleus is the force that holds the electrons in the atom. The larger the force, the stronger is the attraction between the electrons and the nucleus and so the smaller the covalent radius.

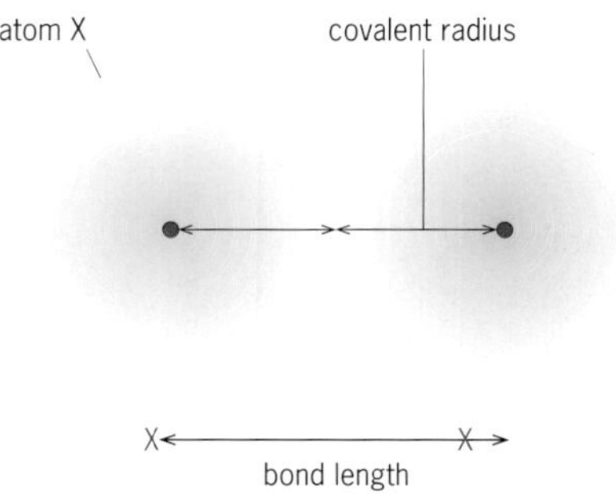

Fig 5.10 The bond length X–X in X_2 is twice the covalent radius of X

Change in atomic radius within a group

Look at any group in the Periodic Table. Going down the group, the number of (positively charged) protons per atom increases, and hence the nuclear charge increases. At the same time, the number of electrons per atom increases. You might expect that, with this increase in nuclear charge, the electrons would be held more tightly, to make the atomic radius smaller. However, as the atomic number increases, and there are more occupied electron shells, the inner electron shells shield the outermost electrons from the full positive charge, and so the outermost electrons move further away from the nucleus. Sometimes we refer to an 'effective' nuclear charge, which takes this shielding into consideration.

QUESTION 10

Describe the trend in atomic radius

a) down a group;

b) across a period.

? QUESTION 11

11 You can estimate the bond length of a covalent bond by adding together the two atomic radii of the atoms concerned. For example, the bond length of H–Cl can be estimated by adding the atomic radius of chlorine to that of hydrogen.

a) Use the information in the Periodic Table in Fig 5.9 to calculate the following bond lengths: i) the O–H bond in a water molecule, ii) the bond length of the N–H bond in ammonia, NH_3.

b) Explain why you cannot use the data in the Periodic Table to estimate the bond length of the carbon–carbon double bond in ethene, C_2H_4.

c) The Periodic Table does not have a value for the atomic radius of helium. Why not?

? QUESTION 12

12 Study the information below which is for elements in Group 5.

Nitrogen: a.r. 70; n.c. +7; e.c. $1s^2\ 2s^2\ 2p^3$

Phosphorus: a.r. 110; n.c. +15; e.c. $1s^2\ 2s^2\ 2p^6\ 3s^2\ 3p^3$

Arsenic: a.r. 121; n.c. +33; e.c. $1s^2\ 2s^2\ 2p^6\ 3s^2\ 3p^6\ 3d^{10}\ 4s^2\ 4p^3$

Antimony: a.r.141; n.c. +51 e.c. $1s^2\ 2s^2\ 2p^6\ 3s^2\ 3p^6\ 3d^{10}\ 4s^2\ 4p^6\ 4d^{10}\ 5s^2\ 5p^3$

a.r. = atomic radius in picometres (1×10^{-12} m);
n.c. = nuclear charge from protons;
e.c. = electron configuration.

Use this information to explain the change in atomic radius in Group 5.

You can read more about X-ray crystallography on page 78.

To summarise, as the atomic number of an element in a group increases the atomic radius increases because:

- the number of shielding shells increases
- the force of attraction of the nucleus for the outermost electrons weakens.

This increase in the atomic radius occurs despite the increase in the positive charge of the nucleus; essentially the effective nuclear charge is reduced.

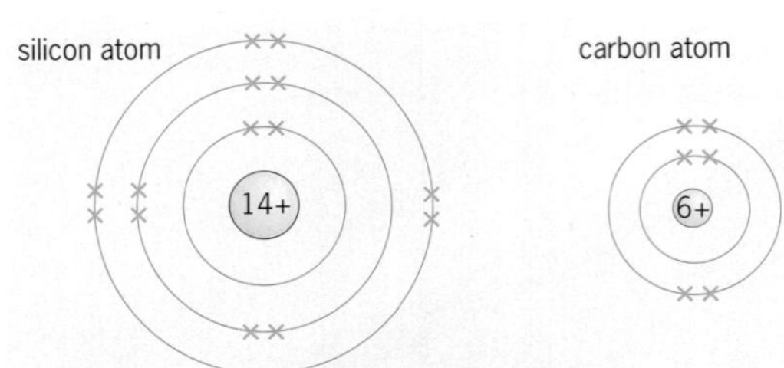

Fig 5.11 A silicon atom has a larger atomic radius than a carbon atom, because although it has more protons it also has an extra shell of shielding inner electrons

Change in atomic radius across a period

Look at one of the short periods and notice that, as you go across the Periodic Table, the atomic radius tends to become smaller as the atomic number increases. The nuclear charge is effectively shielded only by the electrons in the inner shells, and not by the electrons in the outer shell. Going across the period, the nuclear charge increases, but the same outer electron shell is filling up, so there is no increase in shielding by inner electron shells. This means that as the atomic number increases the electrons in the outer shell are more strongly attracted towards the nucleus, and so the atomic radius decreases.

To summarise: across a period the atomic radius decreases because:

- the number of shielding shells remains constant
- the positive charge of the nucleus increases
- the force of attraction between the nucleus and the outermost electrons strengthens.

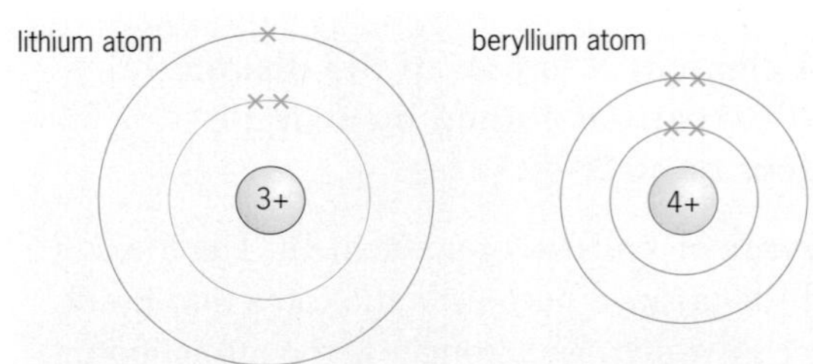

Fig 5.12 A beryllium atom has a smaller atomic radius than lithium because it has an increased nuclear charge, but no extra shielding inner electron shells

IONIC RADII

The ionic radii of ions have been determined experimentally from X-ray crystallography studies on ionic crystals.

A sodium ion, Na^+, is formed when a sodium atom loses one electron. Its ionic radius is 95 picometres, which is smaller than the atomic radius

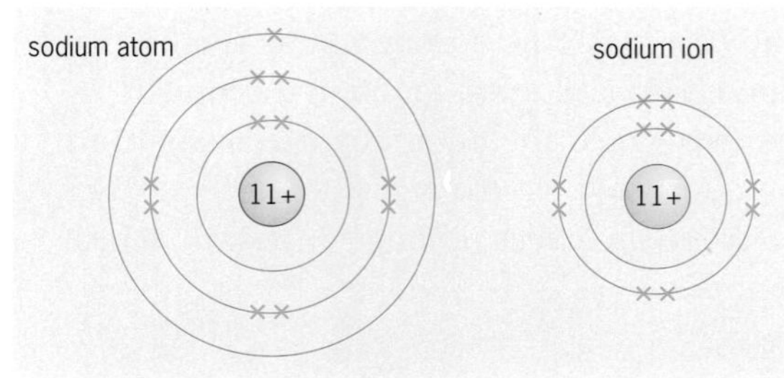

Fig 5.13 A sodium ion has a smaller radius than a sodium atom as the ion has one more proton than electrons and so the excess positive charge pulls the outer electrons closer to the nucleus. In addition, a sodium ion has one less shielding shell of electrons

of sodium. This is because there is one less occupied shell and the ion has one proton more than its number of electrons, so the outer electrons are held with a greater attraction and are drawn closer towards the nucleus.

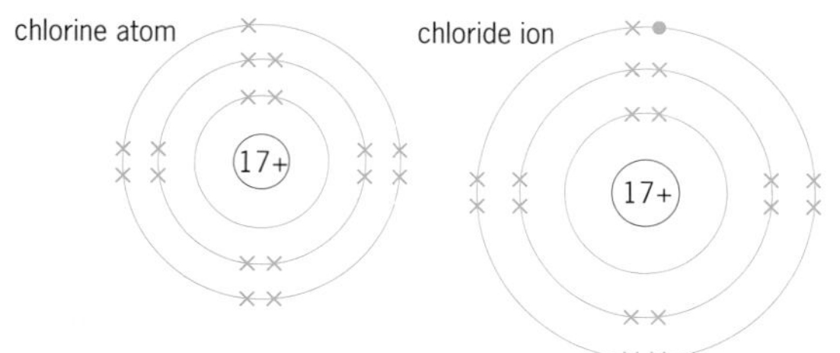

Fig 5.14 A chloride ion has a larger radius than a chlorine atom as the ion has one more electron than protons and so the outer electrons experience less electrostatic attraction from the nucleus

A chloride ion, Cl^-, is formed when a chlorine atom gains one electron. A chloride ion has a larger radius than a chlorine atom since it has one electron more than its number of protons, so the outer electrons are held with less attraction.

FIRST IONISATION ENERGY

The Periodic Table features trends in the energy required to remove electrons from an atom. In Chapter 3 we defined the first ionisation energy as the energy required for one mole of gaseous atoms to lose one mole of electrons, to form one mole of gaseous ions with a single positive charge.

The process can be represented by the equation below, in which the electron is designated by the symbol e^-. It is important to include the state symbols for this process because ionisation energy only refers to gaseous atoms becoming gaseous ions.

$X(g) \rightarrow X^+(g) + e^-$ ΔH_{IE} = positive (first ionisation energy measured in kJ mol^{-1})

The first ionisation energy is necessarily endothermic, since electrons are firmly attracted electrostatically to the positive nucleus. The greater the attraction the larger the first ionisation energy.

Look at Fig 5.15, which shows the first ionisation energies for the first 54 elements of the Periodic Table. Many of these first ionisation energies were determined using atomic emission spectroscopy.

QUESTION 13

a) A sodium cation and a magnesium cation have the same number of electrons. Explain why the ionic radius of Mg^{2+} is much smaller than that of Na^+.

b) Look at the table below, which shows the radius of some ions and atoms.

Species	P^{3-}	S^{2-}	Cl^-	Ar	K^+	Ca^{2+}	Sc^{3+}
Radius/pm	212	184	181	154	133	99	81

i) Explain why a sulfide ion has a larger radius than a sulfur atom.
ii) The species in the table are isoelectronic, meaning that they all have the same electron configuration. Explain the trend shown by the radii of these species.

You can read more about atomic emission spectroscopy in Chapter 27.

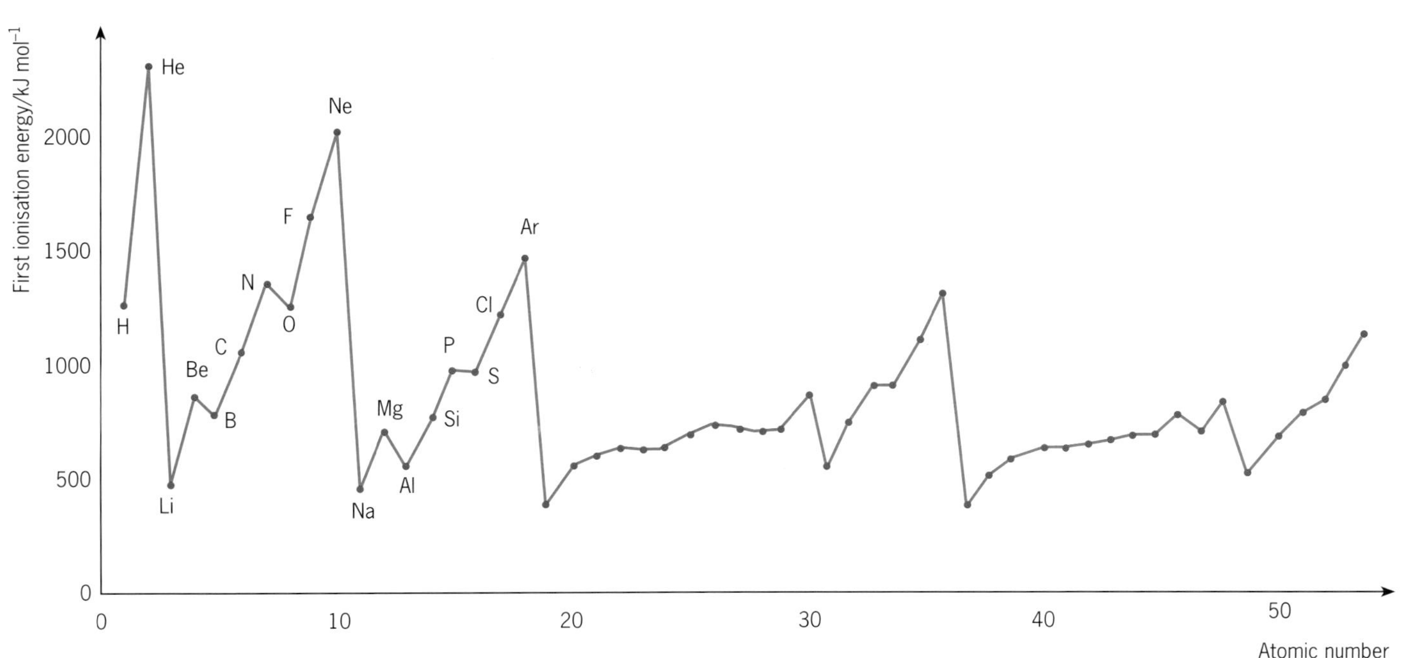

Fig 5.15 Graph of first ionisation energy against atomic number

You can read much more about the reasons for the variation of the first ionisation energies of elements on page 59 in Chapter 3 and in Chapter 22.

The line joining the values for the first ionisation energies of the elements shows a regular up-and-down pattern with peaks and troughs. It is difficult to remove an electron from the gaseous atom of the elements at the peaks, because there is a strong attraction between the outer electrons and the positive nucleus. The electron configurations of these elements are often referred to as being stable. Much less energy is needed to remove electrons from elements that appear in a trough, because there is a much weaker attraction between the outer electrons and the positive nucleus.

QUESTION 14

14 Explain why the first ionisation energy of sodium is much lower than the first ionisation energy of chlorine.

QUESTION 15

15 Why does helium have a higher first ionisation energy than any other element?

QUESTION 16

16 **a)** Explain in terms of electron configuration why an element in Group 3 has a lower first ionisation energy than the elements in Groups 2 or 4 of the same period.
b) Explain in terms of electron configuration why an element in Group 6 has a lower first ionisation energy than the elements in Groups 5 or 7 of the same period.

QUESTION 17

17 Look at the first ionisation energies for the elements scandium to zinc in the first series of the transition block in Fig 5.15. The element with the highest first ionisation energy is zinc. Account for this in terms of its electron configuration.

Change in first ionisation energy in a group

The first ionisation energies decrease down a group. This is because as the atomic number increases:

- the number of shielding shells increase
- the force of attraction of the nucleus for the outermost electrons weakens
- the atomic radius increases
- it is easier to remove the outermost electron from the atom (less energy needed).

Essentially, as the atomic number increases the effective nuclear charge decreases.

Change in first ionisation energy across a period

The general trend of the first ionisation energy is that it increases across a period so that all the elements of Group 0, the noble gases, are at peaks. The reason for this trend is that as the atomic number increases:

- the number of shielding shells remains constant
- the positive charge of the nucleus increases
- the force of attraction between the nucleus and the outermost electrons strengthens
- the atomic radius decreases
- it is more difficult to move the outermost electron from the atom (more energy is needed).

As a result, the first ionisation energy of the noble gases (Group 0) in a period will always be the highest. The electron configurations of the noble gases are very stable. They are shown in Table 5.4.

Table 5.4 Electron configuration of the noble gases

Atom	Atomic number	Electron configuration
He	2	$1s^2$
Ne	10	$1s^2\ 2s^2\ 2p^6$
Ar	18	$1s^2\ 2s^2\ 2p^6\ 3s^2\ 3p^6$
Kr	36	$1s^2\ 2s^2\ 2p^6\ 3s^2\ 3p^6\ 3d^{10}\ 4s^2\ 4p^6$
Xe	54	$1s^2\ 2s^2\ 2p^6\ 3s^2\ 3p^6\ 3d^{10}\ 4s^2\ 4p^6\ 4d^{10}\ 5s^2\ 5p^6$

Periodicity in ionisation energy

We have already noted that the first ionisation energy of the noble gases is always the highest within a period. This is not the only pattern that emerges from the graph of first ionisation energy against atomic number (Fig 5.15). Elements in Groups 1, 3 and 6 also appear in troughs; that is, their first ionisation energy is less than those of the elements with an atomic number plus or minus one. This pattern is repeated in every period. These troughs are associated with particular electron configurations. For example, elements in Group 1 have only one electron in their outer shell and so have the maximum number of inner shielding electrons compared to outer shell electrons for the period.

You can read more about first ionisation energy and electron configuration on pages 59 and 63 in Chapter 3.

4 METALS, NON-METALS AND METALLOIDS

The 110 or more known elements can be divided into three main groups: the metals, the non-metals and the metalloids or semi-metals. Metals and non-metals often have opposite chemical and physical properties.

METALS

Metals make up over 80 per cent of all the elements and occupy the centre and left-hand side of the Periodic Table. They have many physical properties in common. A metal is generally a good conductor of heat and electricity and it often has a lustre – a shine. It is likely to be malleable, that is, it can be shaped by hammering, and ductile, meaning that it can be drawn into a wire. Most metals have relatively high melting and boiling points.

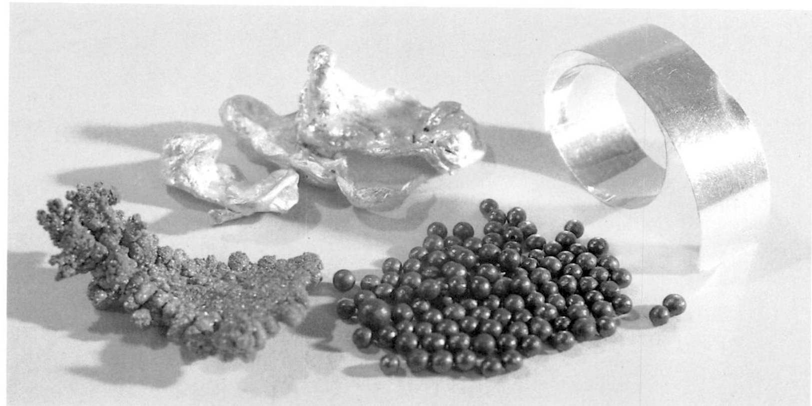

Fig 5.16 Metals. Clockwise from top left: zinc, silver foil, lead shot, copper crystals

Fig 5.17 Non-metals. Clockwise from top left: sulfur, bromine, phosphorus, carbon, iodine

Chemical properties of metals

The most striking property of a metal atom is its ability to lose electrons. Commonly, metal atoms lose one, two or three electrons to form cations. They will *lose* electrons so as to obtain a stable electron configuration like that found in one of the noble gases. (You will have met this idea in Chapter 3.)

The relationship between the structure and bonding of elements and their physical properties is explored in more detail in Chapter 7.

Table 5.5 The main properties of metals and non-metals

Metals
- readily form cations by loss of electrons
- normally have 1, 2 or 3 electrons in their outer shell of electrons
- normally form basic or amphoteric oxides
- react with non-metals to form ionic compounds
- do not react with other metals
- are normally reducing agents

Non-metals
- readily form anions by gain of electrons
- normally have 4, 5, 6, 7 or 8 electrons in their outer shell of electrons
- normally form acidic or neutral oxides
- react with other non-metals to form molecular compounds
- react with metals to form ionic compounds
- are normally oxidising agents

NON-METALS

Non-metals occupy the right-hand side of the Periodic Table. Non-metal solids are normally poor conductors of heat and electricity, rarely have a lustre and are normally brittle and non-ductile. Many non-metals are gases at room temperature and pressure, although other non-metals have very high melting and boiling points.

Chemical properties of non-metals

Non-metals, unlike metals, tend to *gain* electrons to become anions or to make covalent bonds. It is clear, then, that metals will tend to react with non-metals, and a metal atom 'gives away' electrons to a non-metal atom. So an *ionic* substance is formed when a non-metal reacts with a metal. When a non-metal reacts with another non-metal a covalent substance is formed, since it is impossible for one non-metal atom to gain electrons and another to lose them to make ions.

METALLOIDS

If an element cannot be classified as a metal or a non-metal, it is called a metalloid or semi-metal and it will possess some of the properties of metals

Fig 5.18 The redox reaction between aluminium powder and iodine powder is highly exothermic. The purple vapour in the figure is iodine vaporised during the reaction

Fig 5.19 Sodium burning in chlorine

and some of the properties of non-metals. A good example of a metalloid is silicon, which is shiny, but brittle and a poor conductor of heat and electricity.

Since a metalloid element has some of the properties of a metal and some of a non-metal, it is quite likely to be able to gain electrons or to lose electrons.

5 REDOX REACTIONS

We have described how metal atoms lose electrons when they react and that non-metal atoms gain electrons. So when a metal reacts with a non-metal, electrons are transferred from metal atoms to non-metal atoms. A reaction that involves a transfer of electrons is called a **redox reaction**.

REACTION BETWEEN SODIUM AND CHLORINE

Consider the reaction of sodium with chlorine. Sodium is a typical metal, and chlorine is a typical non-metal. When they react, one electron is transferred from a sodium atom to a chlorine atom to form a sodium ion and a chloride ion. This transfer of electrons results in both the sodium ion and the chloride ion obtaining a stable electron configuration like that of a noble gas.

The reaction may be written as follows:

$$Na \rightarrow Na^+ + e^-$$
$$\tfrac{1}{2}Cl_2 + e^- \rightarrow Cl^-$$

Taken together, these equations become:

$$Na + \tfrac{1}{2}Cl_2 + e^- \rightarrow Na^+ + Cl^- + e^-$$

You can see that the electron can be cancelled out on both sides of the equation, so that this reaction can be written as:

$$Na + \tfrac{1}{2}Cl_2 \rightarrow Na^+ + Cl^-$$

The equation becomes more familiar if the product is given its normal formula which ignores the charges on the ions:

$$Na + \tfrac{1}{2}Cl_2 \rightarrow NaCl$$

The final equation hides the fact that during the reaction an electron is transferred from the sodium to the chlorine and that this is a redox reaction.

In a redox reaction, one particle loses electrons and another particle gains them.

OXIDATION AND REDUCTION

Redox is a shorthand word for reduction–oxidation.

In the process of **oxidation**, a particle loses electrons – it is an electron donor. In **reduction**, a particle gains electrons – it is an electron acceptor. A particle here can be an atom or a molecule or an ion.

Looking at the reaction between sodium and chlorine, you can see that sodium is oxidised and chlorine is reduced. Redox is best remembered by using OIL RIG:

Oxidation Is Loss and Reduction Is Gain of electrons.

It is impossible to have a redox reaction in which only reduction or only oxidation takes place. Where there are two reactants, when an atom, molecule or ion loses electrons, it follows that the other reactant must gain electrons.

Since the main property of metal atoms is to lose electrons, and since 'oxidation is loss', **metals are always oxidised** by other substances when they react. The most reactive of the metals have atoms that lose electrons easily and they are oxidised easily. Conversely, the noble metals, such as gold, have atoms that lose electrons with difficulty and are not oxidised easily.

Since non-metal atoms gain electrons they are normally reduced when they react with other substances. A reactive non-metal, such as fluorine, is reduced very easily, but a non-reactive non-metal is reduced with difficulty.

QUESTION 18

The most reactive metal in Group 1 is francium, Fr. Suggest why this might be so, in terms of the ease of loss of electrons from a francium atom.

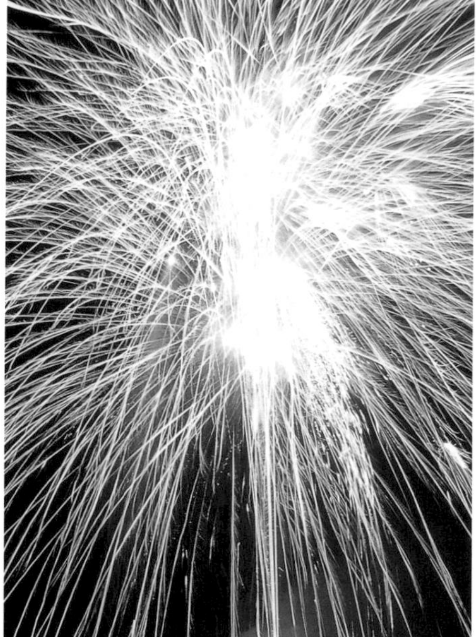

Fig 5.20 In an oxidation reaction, magnesium gives fireworks and flares a brilliant white colour

Fig 5.21 In the thermite process, a very vigorous redox reaction, aluminium reduces iron oxide to iron

QUESTION 19

For each of the following processes decide whether it is oxidation, reduction or neither oxidation nor reduction:

a) $Na \rightarrow Na^{+} + e^{-}$
b) $Ca \rightarrow Ca^{2+} + 2e^{-}$
c) $Cl_2 + 2e^{-} \rightarrow 2Cl^{-}$
d) $H_2O(l) \rightarrow H_2O(g)$
e) $Fe^{2+} \rightarrow Fe^{3+} + e^{-}$
f) $2I^{-} \rightarrow I_2 + 2e^{-}$
g) $O_2 + 4e^{-} \rightarrow 2O^{2-}$

HOW SCIENCE WORKS

Oxidation and reduction in organic chemistry

The term oxidation was originally used exclusively to describe the reaction of a substance with oxygen to form an oxide. This definition has been extended to take account of recent knowledge about the redistribution of electrons during a reaction. Nevertheless, in organic chemistry, it is still useful to consider oxidation as the addition of oxygen to an atom or molecule or the removal of hydrogen from an atom or molecule.

In an equation or reaction scheme, oxidation is often shown by the symbol [O]. For example, the oxidation of ethanol, CH_3CH_2OH, to make ethanal, CH_3CHO, can be represented as:

$$C_2H_5OH + [O] \rightarrow CH_3CHO + H_2O$$

In organic chemistry, it is often easier to consider reduction as the addition of hydrogen to an atom or molecule, or the removal of oxygen. In an equation, reduction is often shown by the symbol [H]. For example, the reduction of nitrobenzene, $C_6H_5NO_2$, to form phenylamine, $C_6H_5NH_2$, can be represented as:

$$C_6H_5NO_2 + 6[H] \rightarrow C_6H_5NH_2 + 2H_2O$$

Oxidising agent or oxidant

The **oxidising agent** or **oxidant** is the species (atom, molecule or ion) that gains electrons. That is, it is the substance that does the oxidising. To be an oxidising agent, a species must be able to accept electrons easily. Many non-metals have electron configurations that are only one, two or three electrons short of the electron configurations of a noble gas. This means that the atoms of these non-metals can gain electrons to make a stable octet of electrons. Consequently, such non-metals are oxidising agents.

Fig 5.22 Magnesium burning in chlorine

Reducing agent or reductant

The **reducing agent** or **reductant** is the species (atom, molecule or ion) that loses electrons. That is, it is the substance that does the reducing (and is itself oxidised). To be a good reducing agent, a species must be able to lose electrons easily. Many metals have electron configurations with an outer shell that contains only one, two or three electrons. This means that the atoms of these metals can easily lose electrons to make a stable octet of electrons. Consequently, such metals are reducing agents.

HALF EQUATIONS

It is often convenient to write the equation for a redox reaction as two half equations, one for the oxidation part and the other for the reduction part. A half equation is a balanced equation for an oxidation or a reduction that shows the atom, ion or molecule gaining or losing electrons. A half equation always includes electrons, e^-.

Consider the reaction of magnesium with chlorine to form magnesium chloride. This is the reaction of a metal with a non-metal, so it is a redox reaction. Magnesium is the reducing agent and chlorine the oxidising agent. First, think of the oxidation process. It involves the loss of electrons. Magnesium atoms can lose electrons to form magnesium ions. Since each atom has two electrons in its outer shell, it needs to lose two electrons to achieve a stable octet. The half equation for this process is:

$$Mg \rightarrow Mg^{2+} + 2e^-$$

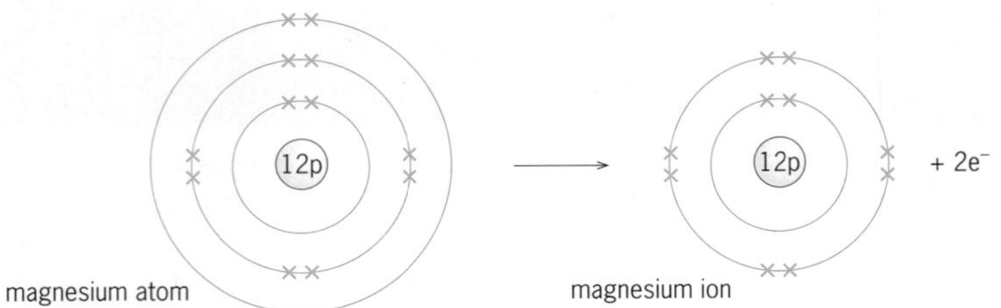

Fig 5.23 Electron configurations to show the half equation $Mg \rightarrow Mg^{2+} + 2e^-$

Fig 5.23 represents this half equation with the full electron configurations of a magnesium atom and a magnesium ion.

Now consider the reduction process. This must involve chlorine because, being a non-metal, it can gain electrons. A chlorine atom has seven electrons in its outer shell, so it needs to gain one electron to achieve a stable octet and form a chloride ion. Cl is not written in the half equation, since chlorine exists as molecules, Cl_2, at room temperature. The half equation therefore is:

$$Cl_2 + 2e^- \rightarrow 2Cl^-$$

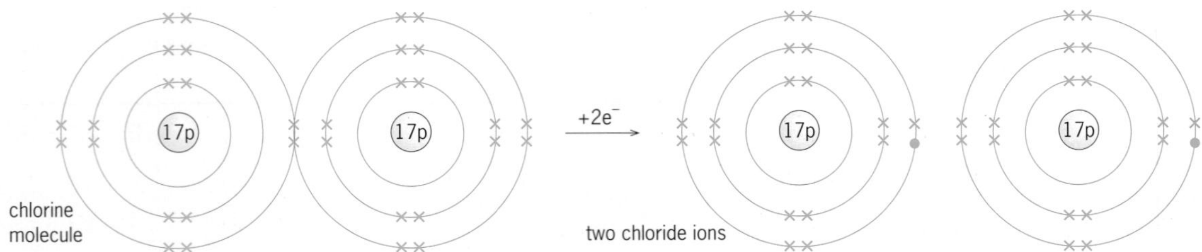

Fig 5.24 Electron configurations to show the half equation $Cl_2 + 2e^- \rightarrow 2Cl^-$

Notice that this is a balanced equation. The atoms balance and the charges balance (each side of the equation has two negative charges).

Do not try to balance the symbol e^-, just its charge. Remember, e^- is an electron, not an atomic symbol. Only atoms and charges are balanced in an equation.

There is a second point to note: the number of electrons in this half equation is the same as in the magnesium equation. That is, it states exactly how the redox reaction is taking place. Two electrons are being transferred from a magnesium atom to a chlorine molecule. When the two half equations are added together, the electrons cancel out.

The two half equations added together give:

$$Mg + Cl_2 + 2e^- \rightarrow Mg^{2+} + 2Cl^- + 2e^-$$

So, the full equation is:

$$Mg + Cl_2 \rightarrow Mg^{2+} + 2Cl^-$$

Even the full equation is often simplified still further to show the two ions formed as the formula unit $MgCl_2$:

$$Mg + Cl_2 \rightarrow MgCl_2$$

Balancing redox reactions using half equations

It is fundamental to a redox reaction that the number of electrons lost by one species always equals the number of electrons gained by another. Knowing this can help you to construct balanced equations for redox reactions.

You will have noticed that in two of the half equations in question 20, the number of electrons lost is not equal to those gained. This apparent imbalance can be very useful in constructing a balanced equation. For example, the half equations associated with the reaction of magnesium with nitrogen to form magnesium nitride are:

Oxidation $Mg \rightarrow Mg^{2+} + 2e^-$
Reduction $N_2 + 6e^- \rightarrow 2N^{3-}$

The oxidation half equation involves two electrons, but the reduction half equation involves six. It is important to realise that the half equations do not each need to involve the same number of electrons. They need only show the oxidation and the reduction parts, each as a self-contained process. Since the electrons gained by the nitrogen molecule have to be lost from the magnesium, this would suggest that three magnesium atoms are involved, which together would lose the six electrons required by nitrogen. So, the half equations could be written as:

Oxidation $3Mg \rightarrow 3Mg^{2+} + 6e^-$
Reduction $N_2 + 6e^- \rightarrow 2N^{3-}$

Now each half equation has the same number of electrons. Therefore, when the two half equations are added together the electrons are the same on both sides (and can be cancelled out):

$$3Mg + N_2 + 6e^- \rightarrow 3Mg^{2+} + 2N^{3-} + 6e^-$$

So the full equation is written as:

$$3Mg + N_2 \rightarrow Mg_3N_2$$

with the formula unit being used for magnesium nitride instead of the separate ions.

The equations for these types of redox reaction are quite easy to work out without this analysis of the half equations. But for more complicated redox reactions, this method of balancing equations is invaluable.

QUESTION 20

Write down the two half equations for each of the following redox reactions:

a) Calcium reacting with chlorine to form calcium chloride.
b) Magnesium reacting with fluorine to form magnesium fluoride.
c) Magnesium reacting with oxygen to form magnesium oxide.
d) Sodium reacting with nitrogen to form sodium nitride.

QUESTION 21

a) Write down the half equations for the reaction of aluminium with oxygen. Use the half equations to write the full equation for this redox reaction.

b) Use the following half equations to write the full equation for the following redox reactions:

i) reaction between zinc and iron(III) ion:

$$Zn \rightarrow Zn^{2+} + 2e^-$$
$$Fe^{3+} + e^- \rightarrow Fe^{2+}$$

ii) reaction of iron(II) ion with acidified manganate(VII) ion:

$$Fe^{2+} \rightarrow Fe^{3+} + e^-$$
$$MnO_4^- + 8H^+ + 5e^- \rightarrow Mn^{2+} + 4H_2O$$

iii) reaction of iron(II) ion with acidified dichromate(VI) ion:

$$Fe^{2+} \rightarrow Fe^{3+} + e^-$$
$$Cr_2O_7^{2-} + 14H^+ + 6e^- \rightarrow 2Cr^{3+} + 7H_2O$$

REMEMBER THIS

The terms 'oxidation number' and 'oxidation state' are interchangeable. (Both terms are used in this book.)

6 OXIDATION NUMBER

So far, we have looked only at the reactions of metals and non-metals to form ionic compounds. In these, it is easy to see the transfer of electrons in terms of electrons being gained and lost (which is precisely what happens during the formation of an ionic bond). However, non-metals can react with one another, and these too are redox reactions. One of the non-metals is oxidised and the other is reduced. The compound formed is covalent, involving the sharing of electrons. So, how can we tell which element is oxidised, since there is no actual loss of electrons to produce ions? We use a concept called **oxidation number**.

OXIDATION NUMBER IN IONIC SUBSTANCES

For bonding to take place, there must be a redistribution of electrons to create attraction between the atoms involved. The driving force for this redistribution is the formation of a stable set of outer electrons, often an octet. Each atom in a substance is assigned an oxidation number. This is the number of outer electrons of the atom that are involved in forming a stable set of electrons. The oxidation number is given a positive or a negative sign to indicate whether electrons have been added or removed from the atom to achieve the stable set.

Consider sodium chloride with the formula NaCl. It is composed of the sodium ion and the chloride ion. To form the stable octet of the sodium ion, a sodium atom **loses one electron**. So, in sodium chloride, sodium has an oxidation number of +1.

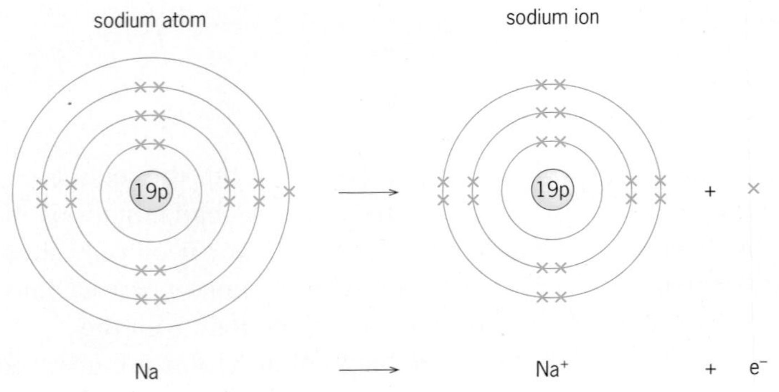

Fig 5.25 The formation of the sodium ion

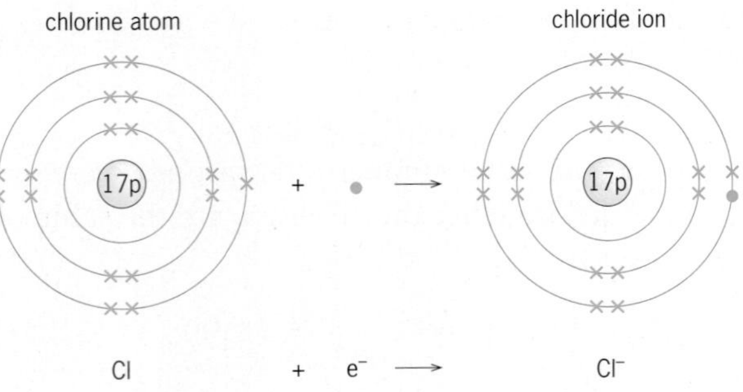

Fig 5.26 The formation of the chloride ion

To form the stable octet of the chloride ion, a chlorine atom has to **gain one electron**. So, in sodium chloride, chlorine has an oxidation number of –1. Note that the oxidation number corresponds to the charge on the ion. This is not a coincidence and makes it easy to calculate the oxidation number in simple ions:

oxidation number =
(no. of outer electrons in the atom before bonding) –
(no. of outer electrons in the atom or ion after bonding)

It is assumed for this equation that a positive ion, such as Mg^{2+}, has 0 electrons in its outer shell.

The oxidation number is the charge on the ion.

EXAMPLE

Q What is the oxidation number of magnesium and oxygen in magnesium oxide?

A The formula of magnesium oxide is MgO. It is an ionic compound composed of a magnesium ion Mg^{2+} and an oxide ion O^{2-}.

So, the oxidation number of magnesium is +2, and the oxidation number of oxygen is –2.

A magnesium atom has two outer electrons and a magnesium ion has no outer electrons. Therefore, the equation for the oxidation number also gives +2 for magnesium.

QUESTION 22

What is the oxidation number of each of the atoms in the following ionic compounds?

a) Aluminium oxide, Al_2O_3, containing the ions Al^{3+} and O^{2-}

b) Calcium chloride, $CaCl_2$, containing the ions Ca^{2+} and Cl^-

c) Magnesium nitride, Mg_3N_2 containing the ions Mg^{2+} and N^{3-}

d) Copper(II) chloride, $CuCl_2$

e) Potassium sulfide, K_2S

f) Barium fluoride

OXIDATION NUMBER IN COVALENT COMPOUNDS

The simple equation for oxidation number does not work with compounds bonded covalently, because the electrons are shared rather than transferred. So the equation must be modified.

Given the connection between electronegativity and polar covalent bonds, first the more electronegative atom in the bond is identified. Once this is done, all the electrons being shared are assumed to be in the outer shell of this atom. Now it is possible to use the equation to calculate the oxidation number of the element. Another way is to pretend that the polar covalent bond is actually ionic and then work out the charge on the ions formed.

Take, for example, covalent hydrogen fluoride (Fig 5.27). Since fluorine is the more electronegative atom, it will form a polar covalent bond in which the fluorine end of the molecule is negative. To work out the oxidation number of the fluorine atom, assume that all the electrons being shared in the covalent bond are in the outer shell of the fluorine atom, and proceed as follows:

Number of electrons in outer shell of fluorine	= 7
Number of electrons in outer shell of fluorine after bonding (fluorine is the more electronegative atom)	= 8
Therefore: oxidation number of fluorine	= 7 – 8 = –1

Now consider the hydrogen atom. In this case, the hydrogen atom starts with one electron in its outer shell, but after bonding it has none, since it has a lower electronegativity than fluorine. Therefore, the hydrogen atom has an oxidation number of 1 – 0 = +1.

Remember that this is a theoretical exercise in that the bond between the fluorine atom and the hydrogen atom is always *covalent*.

The more electronegative atom in a bond has a negative oxidation number that corresponds to the number of electrons it is contributing to the covalent bonds it forms. The less electronegative atom in a bond has a positive oxidation number that corresponds to the number of electrons it has contributed to the covalent bonds it forms.

You can read about electronegativity on page 85 and polar covalent bonds on page 74.

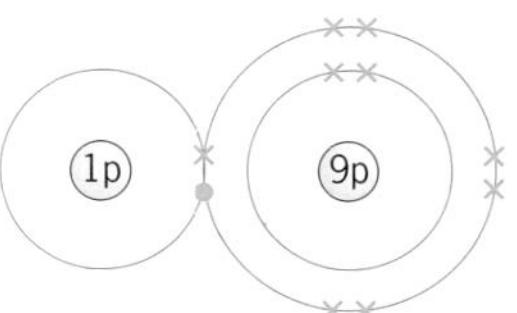

Fig 5.27 Dot-and-cross diagram for hydrogen fluoride

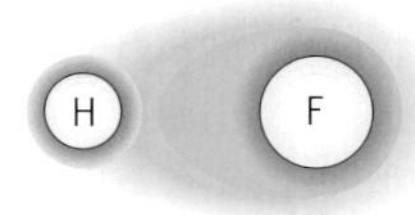

Fig 5.28 Distortion of the electron cloud towards fluorine in hydrogen fluoride

QUESTION 23

a) Draw the dot-and-cross diagram for tetrachloromethane (carbon tetrachloride).

b) Which atom is more electronegative, carbon or chlorine?

c) Work out the oxidation number for each atom.

EXAMPLE

Q What are the oxidation numbers of each atom in sulfuric acid?

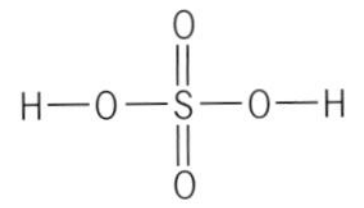

Fig 5.29 The displayed formula of sulfuric acid

A The most electronegative atom is oxygen. Each oxygen atom has two covalent bonds round it, so it has an oxidation number of −2.

The least electronegative atom is hydrogen. Each hydrogen atom has just one covalent bond connecting it to another atom, so it has an oxidation number of +1.

This leaves the oxidation number of sulfur to be found, as follows:

Sum of the known oxidation numbers

$$= 4 \times (-2) + 2 \times (+1) = -6$$

(Note that the oxidation number of every atom is used.)

Given the sum of all the oxidation numbers of the atoms in a compound is zero, it follows that:

oxidation number of sulfur

$$= 0 - (-6) = +6$$

The displayed formula confirms that the sulfur atom is indeed surrounded by six covalent bonds.

Oxidation numbers from displayed formulae

Another way to work out the oxidation number of an atom in a compound is to draw the displayed formula. First, decide which is the most electronegative atom. This atom will have a negative oxidation number. Then, count the number of covalent bonds around the atom to find the oxidation number. Do the same for the least electronegative atom. This atom will have a positive oxidation number. To work out the oxidation number of any other atom, an additional rule is needed, namely:

The sum of the oxidation numbers of each atom in the formula of a compound is zero.

REMEMBER THIS

When an atom can have several oxidation states in different compounds, the oxidation state is normally included in the names of the compounds. In $KClO_3$, for instance, chlorine normally has an oxidation number of −1, so it is important to state that chlorine's oxidation number here is +5, hence the name potassium chlorate(V).

Oxidation numbers and elements

So far, the discussion of oxidation numbers has concentrated on atoms in compounds. But what about the oxidation number of an atom in an element? Consider oxygen, which exists as a diatomic molecule, O_2. The bonding is covalent, but since both atoms have the same electronegativity, the molecule is non-polar. This means it is impossible to apply any of the rules used so far. In fact, the oxidation number of each atom in this case is taken as zero.

Oxidation numbers in compounds

Table 5.6 shows some simple rules about oxidation numbers that should help you to calculate them in compounds. Oxidation numbers should be whole numbers, so why does the sulfur in question 25(f) have an oxidation number of +2.5? The reason is that the compound has at least two different sulfur atoms with different oxidation numbers. So, what is worked out is an *average* oxidation number over all four sulfur atoms.

Oxidation numbers of atoms in ions

Many ions have both ionic and covalent bonds present, but this poses little difficulty in determining the oxidation number of each element. What has to be remembered is that:

The sum of the oxidation numbers of all the atoms is the same as the charge on the ion.

QUESTION 24

24 **a)** What is the oxidation number of each atom in nitric acid (Fig 5.30)?

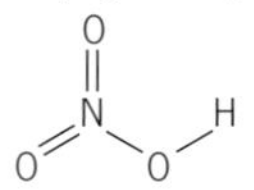

Fig 5.30 The displayed formula of nitric acid

b) i) What is the oxidation number of each atom in chloric(V) acid (Fig 5.31)?

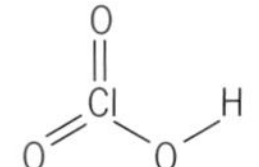

Fig 5.31 The displayed formula of chloric(V) acid

ii) What is the significance of the '(V)' in chloric(V) acid?

Table 5.6 Rules for oxidation numbers

1 The oxidation number of an atom in an element is always 0.
2 The oxidation number of a simple ion is taken as its charge.
3 The oxidation number of fluorine in a compound is always –1.
4 The oxidation number of a Group 1 element in a compound is always +1.
5 The oxidation number of a Group 2 element in a compound is always +2.
6 The oxidation number of oxygen in a compound is nearly always –2.
7 The oxidation number of a Group 7 element in a compound is often –1.
8 The oxidation number of hydrogen in a compound is +1, unless the hydrogen atom is bonded to a metal ion, in which case it has an oxidation number of –1.
9 The oxidation number of a metal in a compound is always positive.
10 The sum of the oxidation numbers of all the atoms in the formula of a compound is always zero.
11 The sum of the oxidation numbers of all the atoms in an ion is always the charge on the ion.

EXAMPLE

Q What are the oxidation numbers of potassium, chlorine and oxygen in potassium chlorate(V), $KClO_3$?

A Being in Group 1, potassium must have an oxidation number of +1. Since oxygen almost always has an oxidation number of –2, it is assumed to be –2 in this case. This leaves the chlorine oxidation number to be found. To work this out, first add up all the known oxidation numbers and then see what the chlorine oxidation number must be to end up with zero as the total:

Sum of the known oxidation numbers

$$= (+1) + 3 \times (-2) = -5$$

Sum of all the oxidation numbers

$$= -5 + \text{oxidation number of chlorine} = 0$$

Therefore, the oxidation number of the chlorine must be +5.

This explains why $KClO_3$ is called potassium chlorate (V). The '(V)' refers to the oxidation state of the chlorine in the compound.

QUESTION 25

a) What is the oxidation number of each of the elements in potassium sulfate, K_2SO_4?
b) What is the oxidation number of each of the elements in potassium sulfite, K_2SO_3?
c) What is the oxidation number of oxygen in F_2O?
d) What is the oxidation number of oxygen in H_2O_2?
e) What is the oxidation number of sulfur in $Na_2S_2O_3$?
f) What is the oxidation number of sulfur in $Na_2S_4O_6$?

REMEMBER THIS

The fertiliser KNO_3 is called potassium nitrate(V) because the oxidation state of the nitrogen is +5. In contrast, KNO_2 is called potassium nitrate(III) because the oxidation state of the nitrogen is +3.

EXAMPLE

Q What is the oxidation number of chromium in the dichromate ion $Cr_2O_7^{2-}$?

A The oxidation number of oxygen is almost always –2.

The sum of all the known oxidation numbers (due to oxygen) is:

$$7 \times (-2) = -14$$

The sum overall of the oxidation numbers must be the charge on the ion, –2.

So, the sum of the oxidation numbers of the two chromium atoms

$$= -2 - (-14) = +12$$

This means that each chromium atom has an oxidation number of +6, which explains why the dichromate ion is more accurately referred to as the dichromate(VI) ion.

The displayed formula method can also be used to work out the oxidation numbers of atoms in ions. The charge on an atom also contributes to the oxidation number.

QUESTION 26

a) What is the oxidation number of iron in the ferrate ion FeO_4^{2-}?
b) Suggest the systematic name for this ion.
c) What is the oxidation state of manganese in MnO_4^- and MnO_4^{2-}?
d) What is the oxidation number of the following?
i) copper in $CuCl_4^{2-}$
ii) nitrogen in NH_4^+
iii) sulfur in $S_2O_3^{2-}$

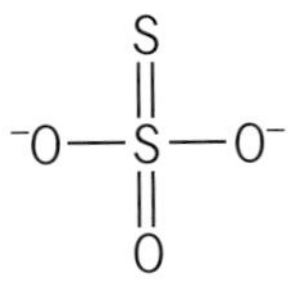

Fig 5.32 The displayed formula of the thiosulfate ion

The thiosulfate ion

The thiosulfate ion has the formula $S_2O_3^{2-}$. The oxidation number of each oxygen atom is –2, which makes +2 the oxidation number of each sulfur atom.

The displayed formula of the thiosulfate ion (Fig 5.32) shows that the two sulfur atoms are not identical. One is bonded only to a sulfur atom, while the other is bonded to three oxygen atoms as well as a sulfur atom. Using the displayed formula to determine the oxidation numbers shows this difference between the sulfur atoms. One sulfur atom has an oxidation state of +4. This is the atom attached to the oxygen atoms. The other sulfur atom has an oxidation number of 0, because covalent bonds between atoms of the same element are not counted.

HOW SCIENCE WORKS

Naming inorganic compounds

As the number of inorganic compounds discovered and synthesised increased all the time, it became clear that the naming of them had to be standardised. Traditional names, such as ammonia, are of no use if they do not convey information about the structures, and help chemists to remember them: 'ammonia' offers no clue to the fact that the compound contains nitrogen and hydrogen.

Chemists recognised that the oxidation number should be stated in the name of a compound if there could be any doubt as to its value. For example, the oxidation number of fluorine in any compound is always –1, so there is no reason to specify it. However, in different compounds chromium can have oxidation numbers that range from +2 to +6. So, it is helpful to specify the oxidation numbers in their names. Hence, CrO, Cr_2O_3 and CrO_3 are distinguished as follows: chromium(II) oxide, chromium(III) oxide and chromium(VI) oxide.

Under the rules introduced for naming compounds, the name of a simple inorganic chemical starts with the name of the least electronegative element (often a metal), followed by the name of the most electronegative element or the name of the anion involved. So, K_2CrO_4 is named potassium chromate(VI). The '(VI)' refers to the oxidation number of chromium in the anion.

QUESTION 27

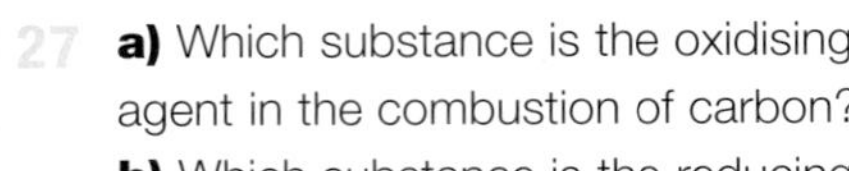

27 **a)** Which substance is the oxidising agent in the combustion of carbon?
b) Which substance is the reducing agent in the combustion of carbon?

QUESTION 28

28 Sodium reacts with hydrogen to give a substance called sodium hydride, NaH. Explain in terms of electron transfer why this is a redox reaction.

REDOX REACTIONS AND OXIDATION NUMBER

The oxidation number is a most useful way of explaining whether a reaction is an example of a redox reaction. During oxidation, the oxidation number of an element becomes more positive; during reduction, it becomes more negative. Do not confuse this with what happens to the oxidising agent and to the reducing agent. The oxidising agent contains an element for which the oxidation number becomes more negative during the reaction. The reducing agent contains an element for which the oxidation number becomes more positive during the reaction.

During the combustion of coal at a power station, carbon reacts to form carbon dioxide. This is a redox reaction between two non-metals:

$$C(s) + O_2(g) \rightarrow CO_2(g)$$

During the reaction, the oxidation number of carbon changes from 0 in the element to +4 in carbon dioxide. This means that carbon is oxidised. At the same time, the oxidation number of oxygen has changed from 0 to –2, so oxygen has been reduced.

Consider next the reaction of two non-metals that does not involve oxygen, in which it is not so easy to spot the oxidising agent. The hydrogen chloride is made by reacting hydrogen and chlorine together:

$$H_2(g) + Cl_2(g) \rightarrow 2HCl(g)$$

During the reaction, the oxidation number of the hydrogen atom changes from 0 to +1 (oxidation), while the oxidation number of the chlorine atom changes from 0 to –1 (reduction). That is, chlorine is the oxidising agent. The non-metal, which has the higher electronegativity, will be the oxidising agent in a reaction between two non-metals. (Electronegativity values are given on page 86.)

Disproportionation

In question 29b) iv) you should find something unusual when you look at the change of oxidation number of chlorine. The numbers should go up and down. The oxidation number of chlorine in the chlorine molecule is 0, while in sodium chloride it is –1 and in sodium chlorate(I), NaOCl, it is +1. This means that chlorine is both the oxidising and the reducing agent at the same time. A reaction in which a substance can be both oxidised and reduced at the same time is called a disproportionation reaction.

Another disproportionation reaction takes place when hydrogen peroxide, H_2O_2, decomposes. Look at Fig 5.33, which shows the equation for the decomposition of hydrogen peroxide.

The oxidation numbers are written under each atom. Where there are two atoms, then the oxidation number is written twice to reinforce the idea that each atom has its own oxidation number. The arrows link the atoms in which the oxidation number changes. You can see that H_2O_2 has been both oxidised and reduced in the reaction, so that disproportionation has happened.

This diagrammatic analysis is a very useful way of explaining what is happening in a redox reaction.

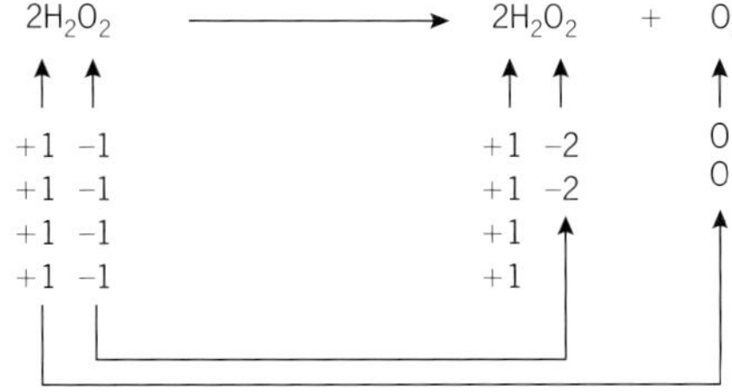

Fig 5.33 The decomposition of hydrogen peroxide is an example of disproportionation. Hydrogen peroxide is oxidised to oxygen and reduced to water.

7 PREDICTIONS USING THE PERIODIC TABLE

One of the uses of the Periodic Table is that it enables you to make informed guesses about the chemical properties of an element. The Periodic Table allows you to:

- predict whether an element is a metal or non-metal;
- predict the charge of any ion that may be formed;
- predict the relative reactivity of the element compared to other elements in its group;
- predict whether the element is a reducing or oxidising agent.

Take as an example the element radium. Radium is highly radioactive, so it would be impossible to study its chemistry in a school laboratory. However, with the aid of the Periodic Table we can make some sensible predictions about this element. Firstly, it is in Group 2. Its atomic number is 88 and its electron configuration is:

$$1s^2\ 2s^2\ 2p^6\ 3s^2\ 3p^6\ 3d^{10}\ 4s^2\ 4p^6\ 4d^{10}\ 4f^{14}\ 5s^2\ 5p^6\ 5d^{10}\ 6s^2\ 6p^6\ 7s^2$$

QUESTION 29

a) Phosphorus(III) chloride reacts with excess chlorine to form phosphorus(V) chloride:

$$PCl_3(l) + Cl_2(g) \rightarrow PCl_5(s)$$

i) What are the changes in the oxidation number for each element?
ii) Which substance is the oxidising agent?

b) For each of the following reactions, use the change in oxidation numbers to deduce the oxidising and reducing agents.
i) $2Cu^{2+} + 4I^- \rightarrow 2CuI + I_2$
ii) $MnO_2 + 4HCl \rightarrow MnCl_2 + Cl_2 + 2H_2O$
iii) $2Cu + 4HCl + O_2 \rightarrow CuCl_2 + 2H_2O$
iv) $2NaOH + Cl_2 \rightarrow NaOCl + NaCl$

QUESTION 30

Work out the oxidation numbers of every chlorine atom or ion in the following equation. Use your answers to explain why the following reaction is an example of disproportionation.

$$6NaOH + 3Cl_2 \rightarrow 5NaCl + NaClO_3 + 3H_2O$$

QUESTION 31

Use the Periodic Table to make some predictions about selenium, Se.

(It would have been easier to state that all its inner shells were full and that the outer electrons were $7s^2$) This means that, to obtain a stable electron configuration, a radium atom must lose two electrons. In a reaction, the radium atom will always form a radium ion:

$$Ra \rightarrow Ra^{2+} + 2e^-$$

This property of forming an ion establishes that radium is a metal, which reacts with non-metals to form ionic compounds. Now let us make some predictions about radium.

It would be reasonable to suggest that radium could react with chlorine to give $RaCl_2$, and that its oxide would have the formula RaO. The effective nuclear charge for radium is quite small compared to that of other elements in Group 2, because the large number of electrons in its six inner shells shield the outer electrons from the nuclear charge. So we can reasonably guess that radium will be the most reactive metal in Group 2. As a metal, radium will probably be a good conductor of heat and electricity and is likely to be ductile and malleable.

When element number 118 was discovered, scientists predicted that it would have properties very similar to that of radon, since it was directly underneath radon in the Periodic Table. Unfortunately, not enough of the element has been made to test these predictions.

SUMMARY

As a result of studying this chapter you should understand that:

- In the Periodic Table, elements are arranged in order of increasing atomic number.
- Columns of elements in the Periodic Table are in the same group.
- Rows of elements in the Periodic Table are called periods.
- Elements in a group have atoms with the same number of electrons in their outer shell.
- Elements in the same period have atoms with the same number of electron shells.
- The first ionisation energy decreases and the covalent radius increases down a group, because of the increased shielding effect by the inner electron shells.
- The first ionisation energy tends to increase and the covalent radius decreases across a period, because of the increasing nuclear charge.
- A positive ion is smaller than the atom from which it is formed and a negative ion is larger than the atom from which it is formed.
- Properties of elements, such as ionisation energy and atomic (covalent) radius, show a periodic variation (periodicity) with increasing atomic number.
- The Periodic Table can be used to predict the properties of elements.
- Reduction is the gain of electrons by a species. Reduction can be recognised by a decrease in the oxidation number of an atom.
- Oxidation is the loss of electrons from a species. Oxidation can be recognised by an increase in the oxidation number of an atom.
- Metal atoms lose electrons and are oxidised during a reaction. Metals are reducing agents.
- Non-metal atoms gain electrons and are reduced during a reaction between a metal and a non-metal. Non-metals are often oxidising agents.
- During a redox reaction there is a transfer of electrons.
- During a disproportionation reaction the same substance is simultaneously oxidised and reduced.

Practice questions and a How Science Works assignment for this chapter are available at www.collinseducation.co.uk/CAS

6

Gases, liquids, solids and solutions

6 GASES, LIQUIDS, SOLIDS AND SOLUTIONS

The chemistry of airbags

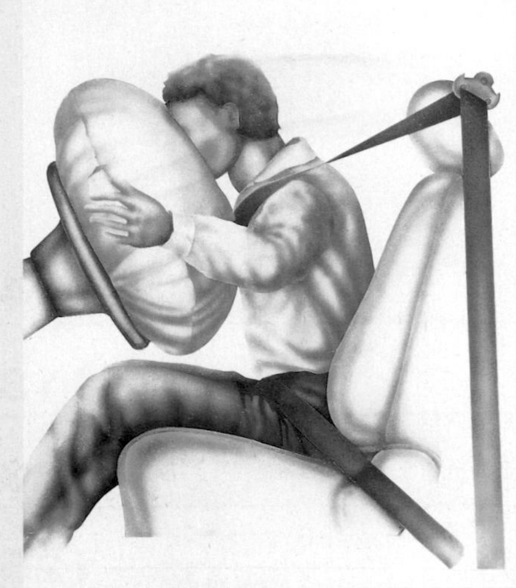

Airbags are a familiar addition to car safety these days. But a crash happens very fast, so how does the airbag inflate in time to protect the driver from injury? Airbags are usually inflated by a very fast gas-producing chemical reaction to blow up the bag.

First, the rapid deceleration of the car triggers an electric current, and the energy from this current explodes a detonator cap. The explosion makes the solid substance sodium azide (NaN_3) decompose and release nitrogen gas extremely rapidly. The nitrogen inflates the airbag, cushioning the driver and reducing the likelihood of injury.

This is the equation for the decomposition of sodium azide:

$$2NaN_3(s) \rightarrow 2Na(s) + 3N_2(g)$$

Sodium azide is a solid, so its particles are arranged very close together. The nitrogen it produces is a gas with widely spaced particles. For a relatively small amount of solid, the volume of gas that inflates the airbag is very large. At the same time, because the particles are now widely spaced, they can be compressed and so cushion the driver. The whole process takes about 40 milliseconds. Then, the specially designed bag deflates rapidly, so that the driver has vision and movement after the crash. This also applies where airbags are fitted for other passengers.

Airbags work, therefore, because of the different arrangements of particles in the different states of matter, an application of chemistry that regularly saves lives.

? QUESTION 1

1 When you read the Science in Context box above about inflating air bags you will notice that sodium metal is produced. It would be dangerous to have this reactive metal inside the car, so in practice the sodium azide is mixed with potassium nitrate. This removes the reactive sodium metal as soon as it is formed. The potassium nitrate converts the sodium into sodium oxide (Na_2O), also producing potassium oxide (K_2O) and nitrogen. Write a balanced equation, including state symbols, for this reaction.

1 GASES

The Greeks thought that air was one of four elements that made up the Earth (see the start of Chapter 2). Later, alchemists produced what they called 'airs' or 'vapours' in their efforts to change common metals, such as iron and lead, into gold. It was a Dutch scientist, Jan van Helmont, who realised that not all these 'airs' were the same, though they would fill any container they were put into. The Greeks had a name for the substance that they thought the gods had changed into the four elements of Earth – chaos. In 1624, van Helmont used this word for 'airs'. When pronounced in Dutch the word became 'gas'. It is now recognised as a **state of matter**, together with the other two states, solid and liquid.

VOLUMES OF GASES

Towards the end of the 18th century, chemists were experimenting with reactions that involved gases. The French chemist Joseph Gay-Lussac studied a large number of these reactions and realised that when gases reacted their volumes were in a simple whole-number ratio, provided he measured the volumes at the same temperature and pressure. If the product was a gas this volume too was in a simple ratio to the reactants.

When Gay-Lussac reacted 5 dm^3 of hydrogen gas with 5 dm^3 of chlorine gas, he produced 10 dm^3 of hydrogen chloride (Fig 6.1). The simple whole-number ratio for this reaction is:

1 volume of hydrogen : 1 volume of chlorine : 2 volumes of hydrogen chloride

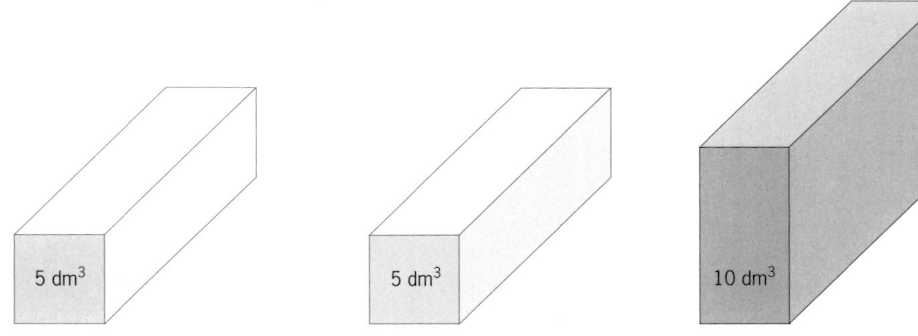

Fig 6.1 1 volume of hydrogen + 1 volume of chlorine $\longrightarrow$ 2 volumes of hydrogen chloride

In 1808, Gay-Lussac summed up his experimental observations in his law of combining volumes:

When gases react, they do so in volumes that bear a simple whole-number ratio to each other and to the volumes of any gaseous products, provided the temperature and pressure are the same when the volumes are measured.

REMEMBER THIS

A decimetre is 10 cm. So a cubic decimetre (dm^3) is $10 \times 10 \times 10 = 1000$ cm^3. It is the same as a litre.

Fig 6.2 Joseph Gay-Lussac (1778–1850) is most famous for his experiments with gases. He also made the highest balloon flight of his day. At over 7 kilometres, the record stood for a very long time

AVOGADRO'S LAW

In 1811, an Italian named Amedeo Avogadro put forward a theory to explain the experimental findings of Gay-Lussac:

Equal volumes of different gases contain equal numbers of molecules at the same temperature and pressure.

It took about 50 years of confusion before Avogadro's theory was accepted. One of the stumbling blocks was John Dalton, who in 1803 had produced his atomic theory (see page 29). He thought that all gaseous elements were made up of separate atoms. He did not understand, for instance, that hydrogen and chlorine were made up of H_2 molecules and Cl_2 molecules. If Dalton had been right, and hydrogen and chlorine gas had been made up of atoms, then Avogadro's theory would have suggested that these atoms were split.

QUESTION 2

2 What is the simple whole-number ratio of gas volumes for the reaction of 5 dm^3 of hydrogen with 2.5 dm^3 of oxygen producing 5 dm^3 of steam?

Fig 6.3 Amedeo Avogadro (1776–1856), lawyer and later a professor of physics. It was not until four years after his death that another distinguished Italian, Stanislao Cannizzaro, rediscovered Avogadro's hypothesis and used it to explain Gay-Lussac's observations

Let's try to follow Dalton's argument to see if it is logical:

1 volume of hydrogen + 1 volume of chlorine → 2 volumes of hydrogen chloride

Since, according to Avogadro's theory, each volume must contain the same number of particles if they are measured at the same temperature and pressure, then:

1 million atoms of hydrogen + 1 million atoms of chlorine → 2 million molecules of hydrogen chloride

So: 1 atom H + 1 atom Cl → 2 molecules HCl

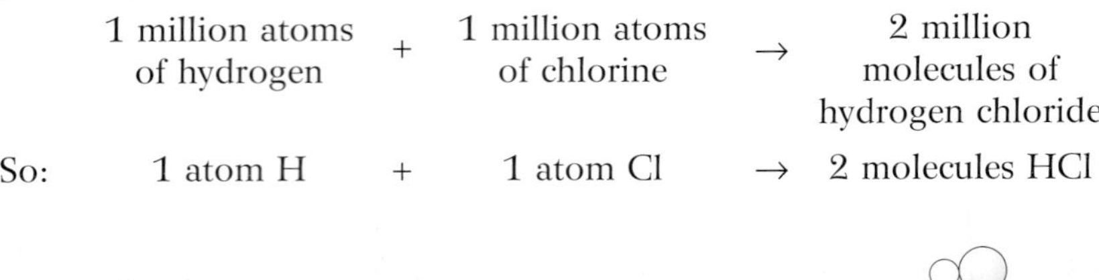

Fig 6.4

This is wrong. If it were correct, there would have to be two hydrogen atoms created from one. This confusion held up progress in chemistry for half a century.

However, if we accept that hydrogen and chlorine are made up of diatomic molecules, then the observations of Gay-Lussac are explained easily by Avogadro:

1 million molecules H_2 + 1 million molecules Cl_2 → 2 million molecules HCl

So: 1 molecule H_2 + 1 molecule Cl_2 → 2 molecules HCl

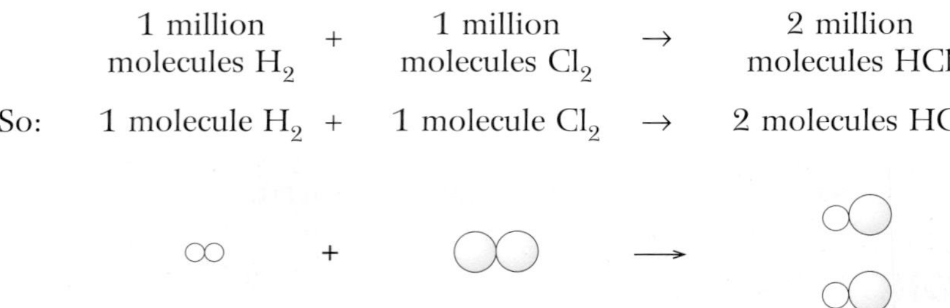

Fig 6.5

Avogadro's theory is also called Avogadro's law, because gases at low pressures obey it.

MOLAR GAS VOLUME

We first introduced the **mole** in Chapter 1. It is the unit chemists use for the **amount** of a substance and is **6.02 × 10²³ particles**. It follows from Avogadro's law that the volume of one mole of any gas – **the molar gas volume V_m** – must be the same under identical conditions of temperature and pressure, because we are dealing with the same number of particles.

At 273 K (0 °C) and a pressure of 101 kilopascals, kPa (1 atmosphere), 1 mole of gas occupies 22.4 dm^3. These conditions of temperature and pressure are called **standard temperature and pressure (s.t.p.)**. So:

The molar gas volume at s.t.p. is 22.4 dm^3.

At 298 K and 101 kPa (sometimes called room temperature and pressure), the molar gas volume is 24 dm^3.

? QUESTION 3

3 Dalton thought that the formula of water was HO. It was known that two volumes of hydrogen reacted with one volume of oxygen to give two volumes of steam. Explain how this observation suggests that the formula of water is H_2O.

REMEMBER THIS

One mole is the amount of substance that contains as many particles as there are atoms in exactly 12 g of carbon-12. The number of particles in a mole is 6.02×10^{23} and is called the **Avogadro constant, L**. You can find out more about the mole on page 5.

? QUESTION 4

4 What is 298 K in degrees Celsius (°C)?

EXAMPLE

Q What is the volume at s.t.p. of:
a) 1 mol hydrogen, **b)** 2 mol hydrogen chloride, **c)** 0.5 mol oxygen?

Remember: mol is the shortened form of the mole unit.

A a) Since one mole of any gas occupies 22.4 dm^3 at s.t.p., the volume occupied by hydrogen must be 22.4 dm^3.

b) The amount of gas is 2 moles, so the volume of hydrogen chloride is double the molar volume at s.t.p., which is 44.8 dm^3 (22.4×2).

c) Oxygen will occupy half the molar gas volume, since there is 0.5 mole present.

$$22.4 \times 0.5 = 11.2\ dm^3$$

QUESTION 5

5 The volume of an oxygen cylinder is 6 dm^3. How many moles of gas are left in the cylinder when it is empty under room conditions (298 K and 101 kPa)? (Hint: when the cylinder is 'empty', the oxygen gas left in it is at the pressure and temperature of the surrounding room.)

CALCULATIONS FROM CHEMICAL EQUATIONS

An understanding of the laws of Gay-Lussac and Avogadro is very useful to chemists. It allows them to use equations to work out the volumes of gases required for a particular reaction, or to predict the volumes of gases that are produced by a given reaction.

EXAMPLE

Q A coal-fired power station uses fuel that contains 0.5 per cent sulphur by mass. What volume of sulfur dioxide at 298 K and 101 kPa is released into the atmosphere by burning 100 tonnes of coal?

A Step 1. Write the balanced equation for the reaction:

$$S(s) + O_2(g) \rightarrow SO_2(g)$$

(Remember: (s) = solid, (l) = liquid, (g) = gas.)

Step 2. Convert the equation to amounts:

$$1 \text{ mol S gives } 1 \text{ mol } SO_2$$

Note: In this example we do not need to know the number of moles of O_2.

Step 3. Work out the amount being used (1 tonne is 1 000 000 g):

100 tonnes of coal contains 0.5 per cent S.

So the mass of S = 0.5 tonnes = 500 000 g.

$$\text{Moles of S in 0.5 tonnes} = \frac{\text{mass (in grams)}}{\text{mass of one mole (in grams)}}$$

$$= \frac{500\,000 \text{ g}}{32 \text{ g}}$$

$$= 15\ 625 \text{ mol (ignoring sig. figs)}$$

Step 4. Scale the amounts in the equation:

$$1 \text{ mol S produces } 1 \text{ mol } SO_2$$
$$15\,625 \text{ mol S produces } 15\,625 \text{ mol } SO_2$$

Step 5. Calculate the volume of gas:

1 mole of gas occupies 24 dm^3 at 298 K and 101 kPa.

$$\text{Volume of } SO_2 = 15\,625 \times 24 = 375\,000\ dm^3$$
$$= 380\,000\ dm^3 \text{ (to two sig. figs)}$$

Q What volume of oxygen is required to burn 600 cm^3 of natural gas under room conditions? Assume that natural gas is pure methane.

A Step 1. Write the balanced equation for the reaction:

$$CH_4(g) + 2O_2(g) \rightarrow CO_2(g) + 2H_2O(l)$$

Step 2. Convert the equation to volumes:

Since: 1 mol CH_4 reacts with 2 mol O_2
1 volume CH_4 reacts with 2 volumes O_2

Step 3. Use the simple whole number ratio of volumes to work out the actual volume used:

$$600\ cm^3\ CH_4 \text{ reacts with } 2 \times 600\ cm^3\ O_2$$

So the volume of oxygen used is 1200 cm^3.

QUESTION 6

6 Petrol is mainly a mixture of hydrocarbons, of which octane, C_8H_{18}, is typical. Calculate the volume of carbon dioxide produced by burning 1.0 kg of octane at room conditions.

QUESTION 7

7 Explosives work by producing large numbers of gaseous molecules very quickly. Nitroglycerine detonates according to this equation:

$$4C_3H_5(NO_3)_3(l) \rightarrow 12CO_2(g) + 10H_2O(g) + 6N_2(g) + O_2(g)$$

A sample of nitroglycerine explodes to produce 1 dm^3 of oxygen. What will be the total volume of gas produced during the explosion, assuming that pressure and temperature remain constant?

2 THE GAS LAWS

Gas is in some ways the simplest of the three states of matter. Through doing countless experiments, scientists have investigated the physical properties of different gases. They have found that, with minor variations, all gases behave in a similar way under room conditions and that there are definite relationships between pressure, volume, temperature and amount. These relationships have been defined in what we call the **gas laws**.

THE RELATIONSHIP BETWEEN PRESSURE AND VOLUME: BOYLE'S LAW

If you have read the beginning of Chapter 2 (page 29) you will know that Robert Boyle had a deep influence on modern chemistry in urging chemists to do experiments to find elements. He also conducted his own experiments on gases.

In 1662 he carried out a series of experiments using a glass tube 5 metres high, shaped like a letter J and closed at the short end. He trapped an amount of air in the short end of the J tube by pouring mercury into the other end. He found that the more mercury he added, the smaller the volume of air became. He had thus discovered that the volume of air depended on the pressure of the mercury upon it, a finding he summarised in what we now know as **Boyle's law**:

The volume of a fixed amount of gas is inversely proportional to its pressure at constant temperature.

We can express this mathematically as:

$$p \propto \frac{1}{V} \quad \text{or} \quad pV = \text{a constant}$$

? QUESTION 8

8 What temperature and pressure are considered to be room conditions? (See page 116.)

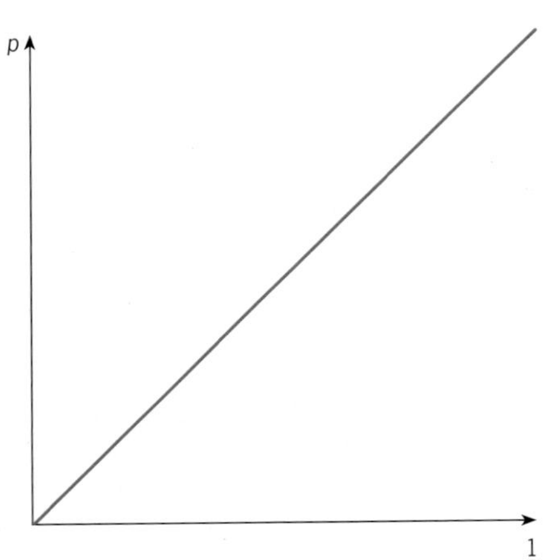
p
1/V

Fig 6.6 Graph showing the effect of pressure on the volume of a fixed amount of gas at constant temperature

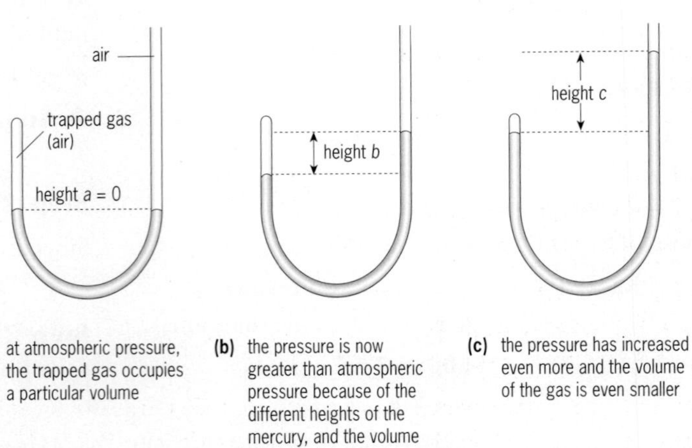
air
trapped gas (air)
height a = 0
height b
height c
(a) at atmospheric pressure, the trapped gas occupies a particular volume
(b) the pressure is now greater than atmospheric pressure because of the different heights of the mercury, and the volume of trapped air has decreased
(c) the pressure has increased even more and the volume of the gas is even smaller

Fig 6.7 Robert Boyle's J tube experiment. Boyle was the first to carry out quantitative experiments on gases. He measured the volume of gas and the difference in heights of the mercury columns to work out the pressure. Note that the temperature is kept constant throughout the experiment

We can show this graphically by plotting pressure against 1/volume as in Fig 6.6.

So if we double the pressure on a gas the volume will halve, and if the volume increases four times then the pressure on the gas must have decreased to a quarter, provided the temperature stays constant.

? QUESTION 9

9 A balloon containing 1 dm^3 of helium gas at 100 kPa is allowed to float up 6 kilometres into the air. At this height, the pressure is 50 kPa. What is the volume of the balloon?

SCIENCE IN CONTEXT

Breathing and Boyle's law

When we are at rest we breathe in and out about twelve times a minute. As we breathe in, the volume of our chest cavity expands, increasing the volume of our lungs. This causes the pressure of air inside the lungs to decrease. Since the pressure of the air outside our bodies is now greater than that inside, air flows in. The reverse happens when we breathe out: the chest volume reduces, which increases the pressure of air in the lungs. The air is now at a higher pressure than that outside, so air flows out.

Fig 6.8

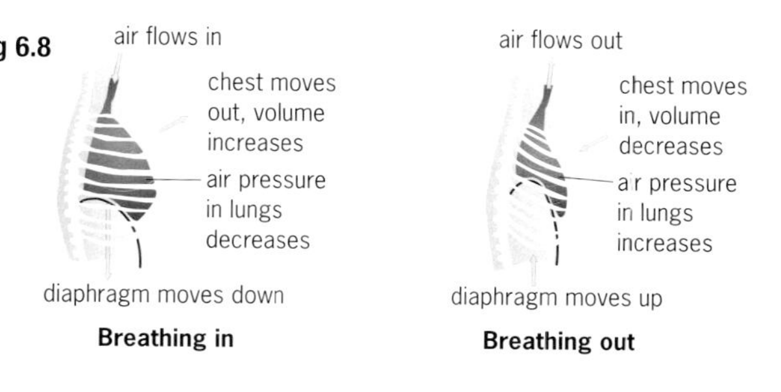
air flows in
chest moves out, volume increases
air pressure in lungs decreases
diaphragm moves down
Breathing in
air flows out
chest moves in, volume decreases
air pressure in lungs increases
diaphragm moves up
Breathing out

Pressure p multiplied by volume V is always equal to the same constant for a fixed amount of gas at the same temperature. So we can write:

$$\boldsymbol{p_1V_1} = \mathrm{k} = \boldsymbol{p_2V_2} \quad \text{or} \quad \boldsymbol{p_1V_1} = \boldsymbol{p_2V_2}$$

where $\boldsymbol{p_1}$ and $\boldsymbol{V_1}$ are one set of conditions, and $\boldsymbol{p_2}$ and $\boldsymbol{V_2}$ are another set.

We can use this equation, which comes from Boyle's law, to work out the effect of changing the pressure on a particular volume of a gas. It can also be used to find the effect on the pressure exerted by a gas of changing the volume of that gas.

QUESTION 10

10 A 5 cm^3 bubble of gas rises up from the ocean floor, where it is at a pressure of 2000 kPa. The pressure just below the surface is 100 kPa. What is the volume of the bubble when it reaches this point?

EXAMPLE

Q A 400 cm^3 mixture of petrol vapour and air is taken into the cylinder of a car engine at 200 °C and a pressure of 100 kPa. The piston compresses this gaseous mixture to 50 cm^3. What is the pressure of the compressed gas if the temperature does not change?

A

Initial conditions	**Final conditions**
$p_1 = 100$ kPa	$p_2 = ?$
$V_1 = 400\ cm^3$	$V_2 = 50\ cm^3$

Since $p_1V_1 = p_2V_2$:

$100 \times 400 = p_2 \times 50$

$p_2 = 800$ kPa

THE RELATIONSHIP BETWEEN VOLUME AND TEMPERATURE: CHARLES' LAW

During the 18th century, several experimenters, including Boyle, noticed that temperature had an effect on the volume of a gas. In about 1800, two French scientists carried out experiments on the relationship between volume and temperature, as a spin-off from their ballooning activities. Jacques Charles was the first and Gay-Lussac, working independently, was the second.

Both workers discovered that, when a fixed amount of gas was kept at constant pressure, the volume varied in proportion to the temperature. Gay-Lussac also noticed that if the volume of gas at 0 °C was taken, then for every 1 °C drop in temperature the volume decreased by 1/273 under the same conditions of pressure and amount. This suggested that at –273 °C there would be a zero volume of gas.

It was Lord Kelvin, a British scientist, who realised the significance of this 50 years later. He called –273 °C the **absolute zero** temperature, below

Fig 6.9 During 1783, hot-air balloon flights became possible thanks to the enterprising Montgolfier brothers. In the same year, Jacques Charles filled a balloon with hydrogen, as seen here, for the first ascent in a hydrogen balloon and the second-ever manned balloon flight

? QUESTION 11

11 What are the values of the following temperatures on the absolute temperature scale:

a) 25 °C

b) 100 °C

c) –50 °C?

(Hint: temperature in K = 273 + temperature in °C.)

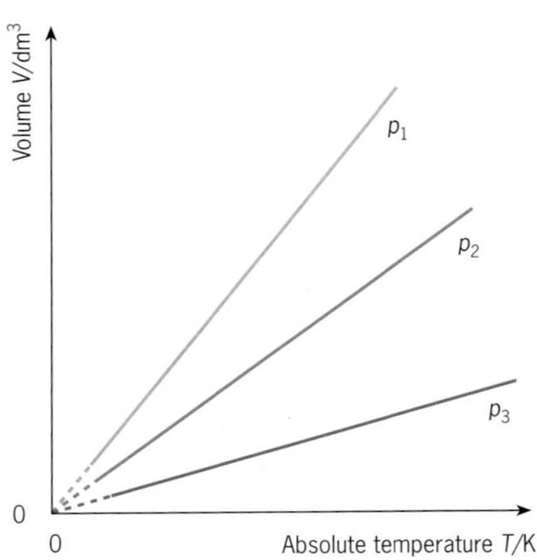

Fig 6.10 Three graphs plotted at three different pressures (p_1, p_2, p_3) demonstrate Charles' law. All gases condense before absolute zero and the dashed portion of the graph shows the extrapolation, assuming that gases do not condense

? QUESTION 12

12 A sample of carbon dioxide in a syringe occupies 50.0 cm^3 at 20.0 °C. What will the volume of CO_2 be at 100.0 °C if the pressure is constant?

which it was impossible to go. This gave rise to a new temperature scale – the absolute temperature scale. Absolute zero has a value of 0 K (zero kelvin), and a 1 K rise in temperature is the same as a 1 °C rise in temperature. This means that at 0 °C the temperature on the Kelvin scale is 273 K.

The relationship between volume and temperature is called **Charles' law**:

The volume of a fixed amount of gas is directly proportional to its absolute temperature at constant pressure.

This can be expressed mathematically:

$$\boldsymbol{V \propto T} \text{ or } \boldsymbol{V = \mathrm{k}T}$$

where k is a constant and T is absolute temperature. Note that k is not the same as the constant in Boyle's law.

For a fixed amount of gas at the same pressure, the same constant will apply, so we can write:

$$\frac{V_1}{T_1} = \mathrm{k} = \frac{V_2}{T_2} \text{ so } \frac{V_1}{T_1} = \frac{V_2}{T_2}$$

EXAMPLE

Q A balloon is filled with 1250 cm^3 of helium gas at 25.0 °C. Overnight the temperature cools to 10.0 °C. What is the new volume of the balloon, assuming the pressure is constant?

A The temperatures must first be converted into absolute temperatures. To do this, 273 is added to the temperature in Celsius so:

25 °C = 273 + 25 = 298 K, and 10 °C = 273 + 10 = 283 K

Initial conditions	**Final conditions**
V_1 = 1250 cm^3	V_2 = ? cm^3
T_1 = 298 K	T_2 = 283 K

Since:

$$\frac{V_1}{T_1} = \frac{V_2}{T_2}$$

$$\frac{1250}{298} = \frac{V_2}{283}$$

$$V_2 = \frac{1250 \times 283}{298}$$

V_2 = 1190 cm^3 (to three significant figures)

THE IDEAL GAS EQUATION

Another way of expressing Avogadro's law on page 115 is:

The volume of a gas is directly proportional to its amount at constant temperature and pressure.

This can be expressed mathematically as:

$$\boldsymbol{V \propto n}$$

where n = amount in moles.

So the volume of a gas is related to three other properties: amount, temperature and pressure.

Avogadro's law	$V \propto n$	(constant p and T)
Boyle's law	$V \propto 1/p$	(constant n and T)
Charles' law	$V \propto T$	(constant n and p)

Combining these:

$$V \propto \frac{nT}{p} \quad \text{or} \quad V = \frac{RnT}{p}$$

where **R** is a constant called the **gas constant**. Rearranging this equation gives the **ideal gas equation**:

$$pV = nRT$$

When p is measured in pascals (Pa), V in cubic metres (m^3), n in moles (mol) and T in kelvins (K), **R** is **8.31 J K^{-1} mol^{-1}**. These units are internationally agreed and are called **SI units**, after the French words *Système Internationale*.

Gases that obey the ideal gas equation exactly are called **ideal gases**. In reality, no gas is an ideal gas, but the variations from ideal behaviour are quite small over wide ranges of temperature and pressure. This enables us to use the ideal gas equation as a very useful tool to relate the four properties of volume, temperature, pressure and amount.

REMEMBER THIS

Ideal gases obey the ideal gas equation exactly under all conditions.

EXAMPLE

Q At 90.0 °C how many moles of nitrogen are present in a flask of volume 750.0 cm^3 at 100 kPa pressure?

A First convert the units into SI units.

The volume in cm^3 must be converted into m^3. There are 100^3 cm^3 in 1 m^3 (i.e. 10^6 cm^3), so:

$$V = 750.0 \times 10^{-6} = 7.50 \times 10^{-4}\ m^3$$

$$p = 100 \times 10^3\ Pa$$

$$T = 273 + 90.0 = 363\ K$$

Then substitute into the ideal gas equation:

$$pV = nRT \quad \text{so } \boldsymbol{n} = \frac{pV}{RT}$$

$$n = \frac{100 \times 10^3 \times 7.50 \times 10^{-4}}{8.31 \times 363}$$

$$n = 0.0249\ mol$$

? QUESTION 13

13 What is the molar gas volume at s.t.p. using SI units?

? QUESTION 14

14 **a)** What is the volume in cm^3 of 2 mol fluorine gas at 27 °C and 100 kPa?

b) The balloon used by Jacques Charles for his historic flight contained 1300 moles of hydrogen. What was its volume at 17 °C and 100 kPa?

USING THE IDEAL GAS EQUATION TO CALCULATE RELATIVE MOLECULAR MASS

Nowadays, mass spectrometers are used to measure relative molecular mass M_r accurately. However, when this instrument is not available the M_r of gases and volatile liquids can be calculated using the ideal gas equation:

$$pV = \frac{mRT}{M_r}$$

One way of doing this is to use a gas syringe (Fig 6.11). If the substance is a gas, it can be passed into a syringe of known mass and the syringe reweighed to give the mass of gas. The volume of gas is read from the syringe and the temperature and pressure are measured. Then these values are simply inserted into the ideal gas equation.

REMEMBER THIS

Amount in moles:

$$n = \frac{\text{mass in grams}}{\text{mass of one mole (in grams)}}$$

and M_r is numerically equal to the mass of one mole: see pages 6 and 7.

Also see Chapter 11, page 215, for information about the mass spectrometer.

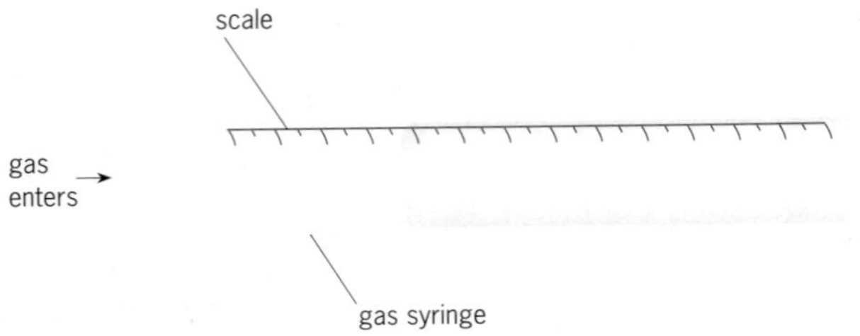

Fig 6.11 Using a gas syringe to measure the M_r of a gas

EXAMPLE

Q The gas propane contains hydrogen and carbon only. When 100.0 cm^3 of it is passed into a syringe, the mass is found to be 0.178 g. The pressure and temperature are 298 K and 100 kPa, respectively. Calculate the M_r.

A The measurements are converted into SI units, with the exception of the mass, which is left as grams. (The SI unit for mass is the kilogram.)

$$p = 100 \times 10^3 \text{ kPa}, V = 100 \times 10^{-6} \text{ m}^3, m = 0.178 \text{ g},$$

$$M_r = ?, \text{R} = 8.31 \text{ J K}^{-1}\text{mol}^{-1} \text{ and } T = 298 \text{ K}.$$

Insert these values into the ideal gas equation:

$$pV = \frac{m\text{R}T}{M_r}$$

$$100 \times 10^3 \times 100 \times 10^{-6} = \frac{0.178}{M_r} \times 8.31 \times 298$$

Rearranging: $$M_r = \frac{0.178 \times 8.31 \times 298}{100 \times 10^3 \times 100 \times 10^{-6}}$$

$$M_r = 44.1$$

? QUESTION 15

15 A gaseous compound containing sulfur and fluorine is used to trace leaks in air conditioning systems. If 100.0 cm^3 has a mass of 0.586 g at 27.0 °C and 100.0 kPa pressure, what is the M_r of the compound?

? QUESTION 16

16 Often a small volume of air is in the gas syringe. Why is it important to know the exact volume of this air?

The same method is used to determine the M_r of a volatile liquid, only this time the end of the syringe is sealed with a self-sealing rubber cap through which a known mass of the liquid under investigation is injected. Since the liquid is volatile it has a low boiling point, and if the syringe is heated to above the boiling point the volume of the gas produced can be measured on the syringe. In practice, there is usually a small volume of air in the syringe. So long as this volume is known at the temperature of the experiment, it can be subtracted from the volume of the vaporised sample.

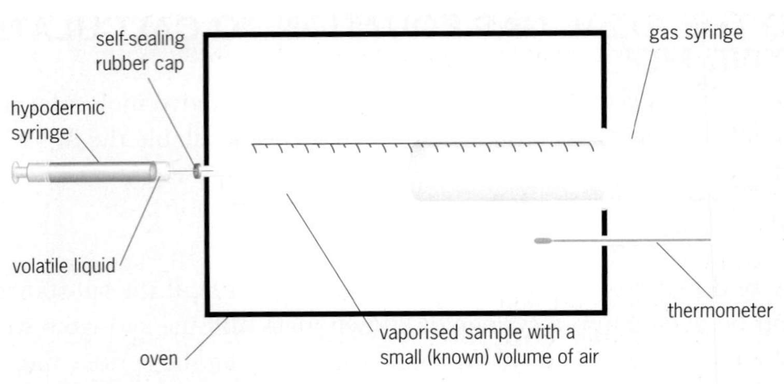

Fig 6.12 The apparatus used to measure the M_r of a volatile liquid

EXAMPLE

Q The following results were obtained from a sample of volatile liquid injected into a syringe:

Mass of liquid sample	0.180 g
Volume of air	10 cm^3
Volume of vapour + air	100 cm^3
Temperature	80 °C
Pressure	100 600 Pa

Calculate the M_r of the sample.

A Volume of vapour = 100 cm^3 – 10 cm^3 = 90 cm^3.

Once again, the measurements are converted into SI units, with the exception of the mass which is left as grams.

$p = 100\,600$ Pa, $V = 90 \times 10^{-6}$ m^3, $m = 0.180$ g,

$M_r = ?$, R = 8.31 J K^{-1} mol^{-1} and $T = 273 + 80 = 353$ K.

Insert these values into the ideal gas equation:

$$pV = \frac{mRT}{M_r}$$

$$100\,600 \times 90 \times 10^{-6} = \frac{0.180}{M_r} \times 8.31 \times 353$$

Rearranging: $$M_r = \frac{0.180 \times 8.31 \times 353}{100\,600 \times 90 \times 10^{-6}}$$

$$M_r = 58$$

? QUESTION 17

17 In a gas syringe, 0.121 g of an alcohol is vaporised to produce 80 cm^3 of gas at 370 K at 101 kPa pressure. Calculate the M_r of the alcohol.

3 A KINETIC–MOLECULAR MODEL FOR HOW GASES BEHAVE

So far, we have considered experimental observations of the behaviour of gases that relate pressure, volume, temperature and amount through the ideal gas equation. The concise statements of these observations we have called **laws**, because they are true for all gases under a range of conditions.

However, scientists soon began to produce **models** to explain gas behaviour. The simple model that we accept today for the behaviour of ideal gases is called the **kinetic–molecular model** or the **kinetic theory**, after the Greek work *kinein*, to move. Several notable scientists developed the model in the nineteenth century, in particular Ludwig Boltzmann, an Austrian, and James Maxwell, a Scot.

In the kinetic–molecular model, ideal gases are assumed to be made up of particles (atoms, molecules or ions) that:

- are very widely separated;
- have negligible (zero) volume;
- exert no force of attraction on each other;
- move continuously and randomly;
- have perfectly **elastic collisions** with each other and the container walls (this means that there is no net gain or loss of energy in collisions);
- have an average kinetic energy that is directly proportional to the absolute temperature of the sample.

REMEMBER THIS

Kinetic energy is the energy due to the motion of the particles. Not all molecules move at the same speed, or velocity, at any particular temperature above 0 K, so molecules possess a range of kinetic energies at a particular temperature.

For example, in any sample of gas, some molecules will be moving very fast, while others will be moving fairly slowly. This range of speeds leads to a range of energies. You can read more about this in Chapter 8 (page 172) and Chapter 26, in which we discuss the Maxwell–Boltzmann distribution of molecular energies.

Now that we have a model we can see how well it fits the experimental observations – the true test of any model.

Gases can be compressed – the kinetic–molecular model explains this because the particles are widely spaced.

Boyle's law – the collisions of particles with the container walls cause pressure. Two factors determine the pressure: the number of collisions in a certain time on a specific area, and the force of these collisions. If we keep the temperature and amount of gas particles constant and decrease the volume of the container that the gas is in, more particles will collide with a specific area of container wall, so the pressure will increase.

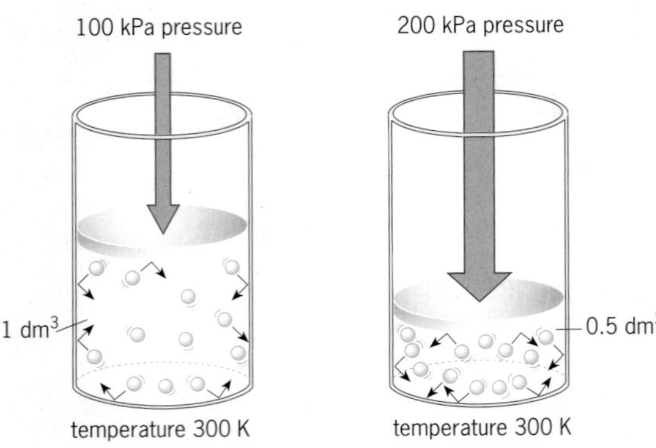

Fig 6.13 Boyle's law explained by the kinetic theory. There is a fixed amount of gas at constant temperature. If the pressure is doubled, the volume is decreased by half

Charles' law – according to the kinetic–molecular model, the average kinetic energy of the particles is proportional to the absolute temperature. As the temperature of a gas increases, so does the average kinetic energy of its particles. This means that the particles will collide with the container walls more often and with greater force. So doubling the absolute temperature doubles the average kinetic energy of particles in a gas, which leads to a doubling of volume provided the pressure stays constant.

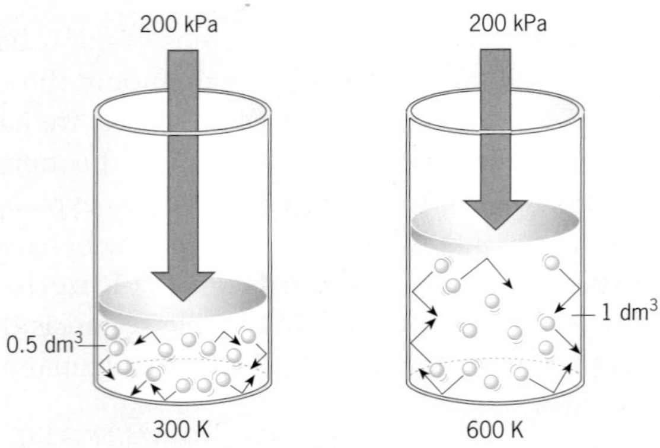

Fig 6.14 Charles' law explained by the kinetic theory. The amount of gas and pressure remain the same, but as the temperature doubles the average kinetic energy of the molecules also doubles to cause a doubling of the volume

? QUESTION 18

18 **a)** Use the kinetic–molecular model to explain why the pressure of a gas decreases when the volume increases, provided the amount and temperature stay constant.

b) How many particles of gas are there in the container in Fig 6.13? (Hint: you will need to use the ideal gas equation.)

c) On page 120 one way of stating Avogadro's law is given as: The volume of a gas is directly proportional to its amount at constant temperature and pressure. Explain this, using the kinetic theory.

REAL GASES

So far, we have looked at ideal gases that obey the ideal gas equation exactly under any conditions. In the calculations, we have assumed that gases behave ideally, and at room conditions this assumption is a good one. However, the kinetic–molecular model assumes that particles in a gas have negligible (zero) volume. This is reasonable at room temperature and pressure, when the particles are very widely spaced. But as the particles in a gas squeeze closer together, their volume becomes significant. The kinetic–molecular model also assumes that there are no attractive forces between molecules. This, again, is reasonable provided that the particles are far apart, but, as they draw closer together, forces of attraction are increasingly important. The higher the temperature, the faster the particles move, which also tends to make the forces of attraction between particles negligible.

REMEMBER THIS

If there were no attractive forces between particles, gases would never liquefy and liquids would never solidify.

Thus the conditions that allow a gas to approach ideal gas behaviour are high temperatures and low pressures.

Conditions in which real gases do not show ideal behaviour are those that bring the particles close together.

Fig 6.15(a) At low pressures the particles are far apart, so their volume is negligible compared with the volume of the gas

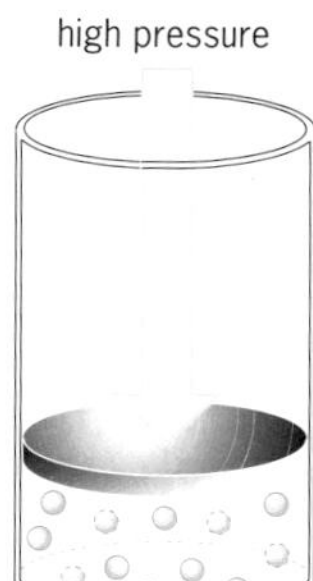

Fig 6.15(b) At high pressures the gas particles take up a larger proportion of the volume, so the volume of the particles becomes significant. Also, because the particles are closer together, attractive forces can operate

At low temperatures the speed of the particles is slower, so forces of attraction between them can operate. At the boiling point of a substance the forces of attraction are great enough to allow the gas to condense to a liquid. So the closer the temperature of a gas is to its boiling point, the more it will deviate from ideal behaviour. This can be seen for nitrogen gas from the graphs in Fig 6.16. If nitrogen were an ideal gas it would obey the ideal gas equation, $pV = n\mathrm{R}T$. So if we had one mole of nitrogen:

$$\frac{pV}{\mathrm{R}T} = 1$$

We can plot $pV/\mathrm{R}T$ against pressure, and at any pressure the value should be 1. However, at a low temperature of 203 K the graph shows wide deviation from this value, while at higher temperatures the deviation from ideal behaviour is much less.

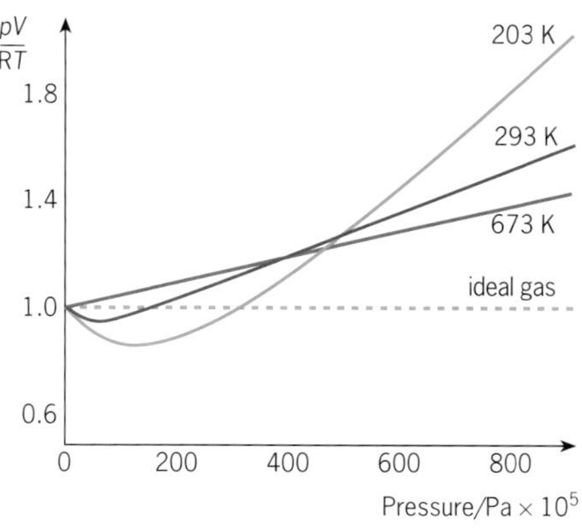

Fig 6.16 Graphs of $pV/\mathrm{R}T$ against pressure for nitrogen gas at three different temperatures. Notice that as the temperature increases, the gas begins to approach ideal behaviour

4 HOW PARTICLES ARE ARRANGED IN LIQUIDS AND SOLIDS

The kinetic–molecular model can be extended to include the liquid and solid states of matter. The particles in a liquid are much closer together than they are in a gas, and because of this they exert considerable attractive forces on each other. The particles are also free to move, but are much slower than in the gas.

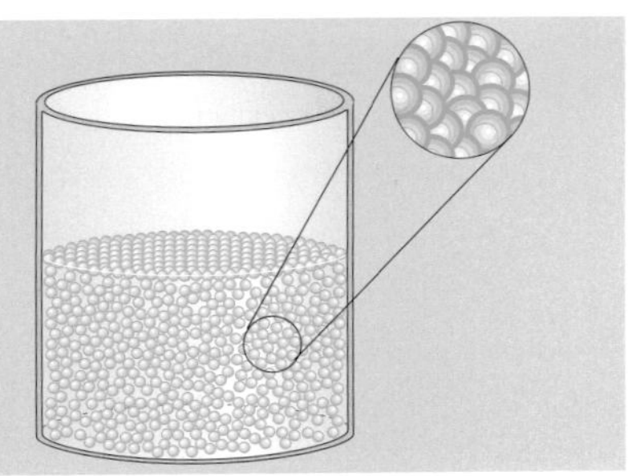

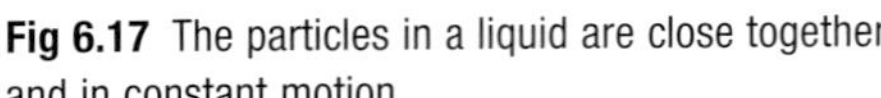

Fig 6.17 The particles in a liquid are close together and in constant motion

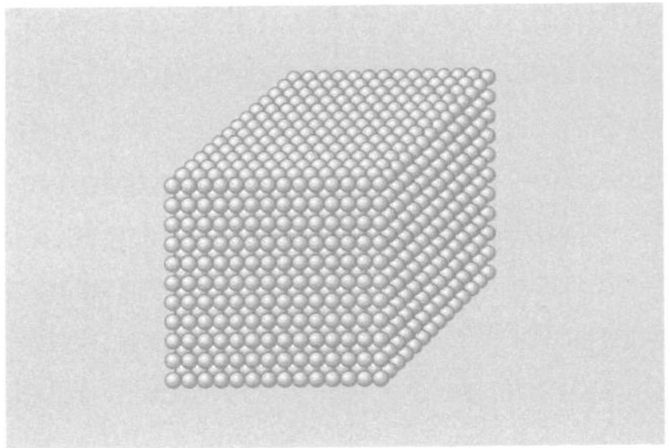

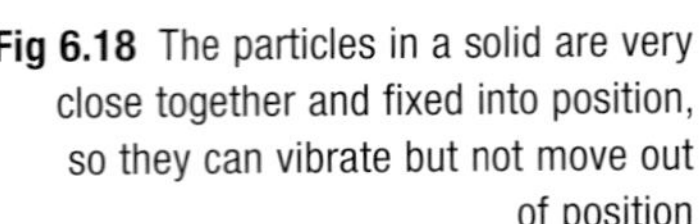

Fig 6.18 The particles in a solid are very close together and fixed into position, so they can vibrate but not move out of position

REMEMBER THIS

X-ray diffraction is used to investigate crystalline lattices – the regular arrangement of particles. You can read more about this on page 78.

In the solid state, particles are even closer than liquids, so while it is possible to compress liquids slightly, solids are almost incompressible. The particles are also fixed into position, held by attractive forces, vibrating but not free to move out of position. This three-dimensional arrangement of particles gives solids a definite shape and is called a lattice. Most solids are crystalline, which means that the particles are arranged in an orderly way. You can read more about some of the different attractive forces in crystalline lattices in Chapters 7 and 21.

However, not all solids are crystalline. Glass may have a fixed shape and so we call it a solid, but the arrangement of its particles has the disorder associated with a liquid.

QUESTION 19

19 Explain the following in terms of the kinetic–molecular model:

a) 1300 dm^3 of steam condenses to only 1 dm^3 of water.

b) Gases are easily compressed, but liquids are only slightly compressible.

c) The density of a liquid is much higher than that of a gas.

d) A liquid takes the shape of the container it is in.

e) Solids do not flow, whereas gases and liquids do.

THE KINETIC–MOLECULAR MODEL USED TO EXPLAIN CHANGES IN STATE

It takes energy to change a substance from a solid to a liquid and then to a gas, as shown in Fig 6.19. As energy is transferred to a solid, the particles vibrate more until they break out of their fixed lattice positions. This is when the solid melts. There is no change in temperature during this time, because all the energy transferred to the substance goes into overcoming the forces that hold the particles in the lattice. The energy needed to change one mole of substance from a solid to a liquid at the melting point is called the **enthalpy change of fusion (ΔH_m)**. When all the solid has melted, the temperature begins to rise and the particles have more energy and move faster.

At the boiling point, the liquid begins to change into a gas and the temperature again stays constant, since all the energy taken in by the liquid is used to overcome the forces that hold the particles close together in the liquid state. The **enthalpy change of vaporisation (ΔH_b)** is the energy that must be supplied to change one mole of liquid into a gas at a particular temperature. When all the liquid has boiled to form a gas, the temperature rises again as the particles move faster. The process of changing state is a physical process because no chemical bonds are broken.

QUESTION 20

20 In Fig 6.19, notice that the flat portion of the heating curve is much less at the melting point (T_m) than at the boiling point (T_b). What does this tell you about the relative magnitudes of ΔH_m and ΔH_b?

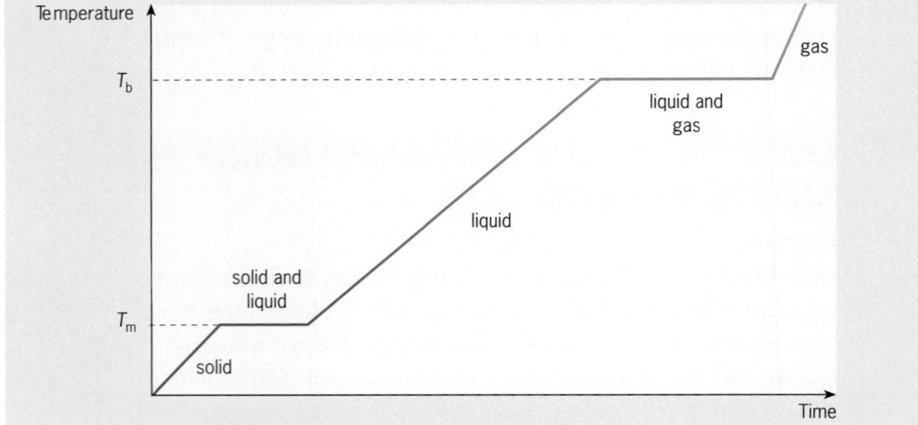

Fig 6.19 A plot of temperature against time as energy is supplied at a constant rate to a substance that starts in the solid state

SCIENCE IN CONTEXT

Liquid crystals: a new state of matter

You often come across liquid crystal displays (LCDs): the numbers displayed on a digital watch use liquid crystals, you may have seen liquid crystal thermometers that change colour depending on the temperature, and laptop computer screens also use liquid crystals. LCDs are commonplace now, but they have only been around for about 40 years.

Fig 6.20(a) A laptop computer with a liquid crystal display

Fig 6.20(b) A digital watch showing its LCD. This was an early use of liquid crystals

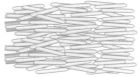

Fig 6.20(c) The nematic arrangement of molecules in a liquid crystal state. Much of the order of the crystalline state has disappeared, but the molecules are still, on average, parallel to each other

When crystals melt, the ordered crystalline arrangement of their particles breaks down into the disordered liquid state. However, some crystals melt to give particles in an ordered state. This effect was first noticed over 100 years ago by an Austrian botanist, Friedrich Reinitzer. He was heating cholesteryl benzoate to try to determine its molecular structure, and noticed that at 146 °C a cloudy liquid was produced. The liquid went completely clear at 178 °C. This process was reversible on cooling. Reinitzer had discovered a new state of matter that was intermediate between the liquid and solid states.

The liquid crystal state was an interesting curiosity, but thought to be of no practical use since there were no stable compounds that showed the liquid crystal state at room temperature. In the late 1960s, the search was on to replace the heavy cathode-ray tubes in aircraft. (Older televisions use a cathode-ray tube.) There was renewed interest in trying to produce chemicals that showed liquid crystal properties at room temperature. In 1972 at Hull University, the world's first stable, room-temperature liquid crystal was made.

Liquid crystals nearly always contain thin, rigid organic molecules. There are three different arrangements of molecules in the liquid crystal state. The one that is used in digital watches and computer display screens is called *nematic*. In a digital watch or calculator display, a mirror at the back of the display reflects light from the surroundings. An electric field alters the orientation of the liquid crystal molecules in a thin film; this causes the display to go dark. The computer display of a laptop is lit from the back of the display by fluorescent tubes and the liquid crystals are arranged in pixels (small parts of the image) with red, green or blue filters above them to give a coloured image. By altering the electric field in each pixel the orientation of the liquid crystal molecules is changed and a coloured image is generated.

Another arrangement of liquid crystals gives different colours at different temperatures and these are used in thermometers.

Liquid crystals are now very big business and a lot of research is under way to produce new molecules that show this interesting state of matter.

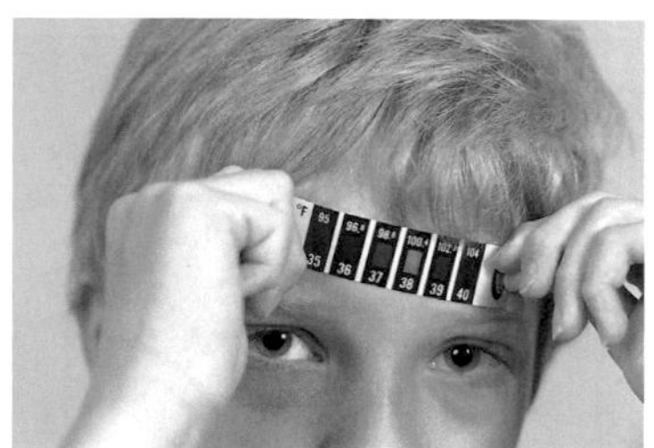

Fig 6.20(d) Using a liquid crystal thermometer

Fig 6.20(e) This camcorder uses a colour LCD screen

VAPOUR PRESSURE

Although a liquid changes into a gas at its boiling point, we know that a liquid can change into a gas at temperatures below the boiling point. If a glass of water is left on a windowsill for a week, some of it changes into a gas and the volume of water decreases. The process is **evaporation**. At any temperature, the particles in a liquid have a range of kinetic energies. Some particles have enough energy to break free of the forces that hold them in the liquid state and

REMEMBER THIS

The vapour pressure is independent of the volume of the closed vessel, the volume of the liquid and the volume of the vapour.

QUESTION 21

21 Look at the graphs in Fig 6.21.
a) Which liquid has the higher vapour pressure at any particular temperature?
b) What does the answer to part **a)** suggest about the relative strengths of forces between the molecules in ethanol and water?

they evaporate. Since it is the particles with the most kinetic energy that evaporate, the particles with lower kinetic energies are left in the liquid, so the average kinetic energy of the liquid drops and the temperature of the liquid falls. On a windy day in winter, weather forecasters talk about the wind chill factor. This is because air moving over the skin causes water to evaporate from the skin faster, so the apparent temperature we feel is much colder than the air temperature.

If the liquid is in a closed vessel, then the rate at which the particles escape from the liquid surface equals the rate at which the particles rejoin the liquid. When the rate of evaporation is the same as the rate of condensation, we say that the liquid is in **equilibrium** with its vapour (or gas). This is a **dynamic equilibrium** (see Chapter 1 page 17.). The pressure caused by the vapour above the liquid at equilibrium is called its **vapour pressure**. As the temperature increases, so does the average kinetic energy of the particles in the liquid. This means that more particles have sufficient energy to escape into the gas and so the vapour pressure increases. When the vapour pressure of a liquid is the same as the external pressure above its surface, then the liquid boils (see Fig 6.21). For example, water boils when its vapour pressure equals atmospheric pressure. This is why water boils at 70 °C on the top of Mount Everest, where the atmospheric pressure is very much less than it is at sea level.

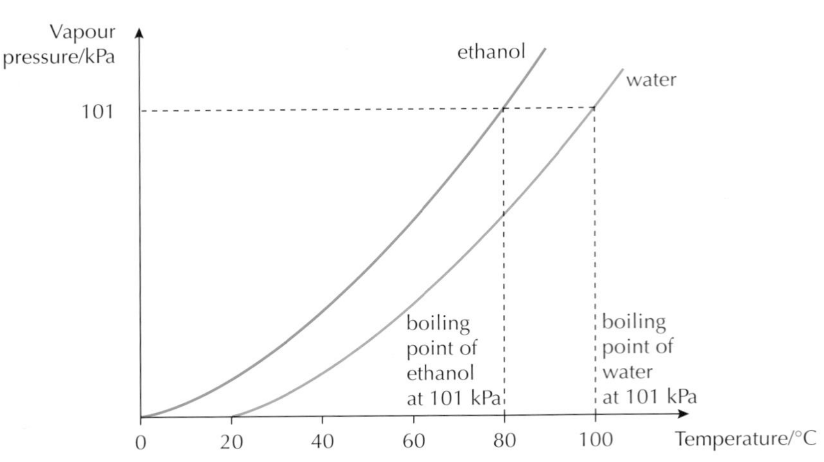

Fig 6.21 Graphs showing the relationship between vapour pressure and temperature for two liquids, ethanol and water. These liquids boil when the vapour pressure is equal to the external atmospheric pressure. At sea level this is 101 kPa

STRETCH AND CHALLENGE

Sublimation and freeze-dried coffee

When a solid changes straight to a gas without passing through the liquid state, the change of state is known as **sublimation**. The reason that some solids can do this is that they possess relatively high vapour pressures. Iodine and carbon dioxide are two examples of solids with high vapour pressures, so at atmospheric pressure their solid forms sublime.

Fig 6.22 A flask containing solid iodine in equilibrium with its purple vapour

While we expect ice to melt rather than sublime, it can change straight to water vapour: you may have noticed that frozen puddles gradually shrink in freezing weather. This phenomenon is used in the production of instant coffee. The coffee is brewed and then frozen in a container from which air is removed by a vacuum pump. Lowering the pressure causes the ice in the frozen brewed coffee to sublime. When nearly all the ice has been removed, the coffee is said to be freeze-dried and is ready for packaging. This method of removing water leaves the flavour molecules intact, and so freeze-dried instant coffee has a much better flavour than coffee that is dried by slow heating.

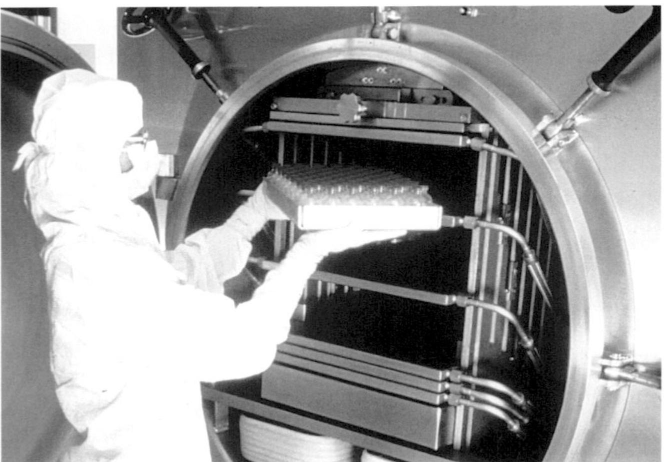

Fig 6.23 In the freeze-dry process, the sample is put in a chamber and frozen very fast. Then the chamber is evacuated and the temperature raised slightly, when any ice turns to vapour

THE UNUSUAL BEHAVIOUR OF WATER – HYDROGEN BONDING

Water is a remarkable liquid. Indeed, if you compare its boiling point to those of the hydrides of other Group 6 elements you could be forgiven for thinking that it should not really be a liquid at all at room temperature (see Fig 6.24).

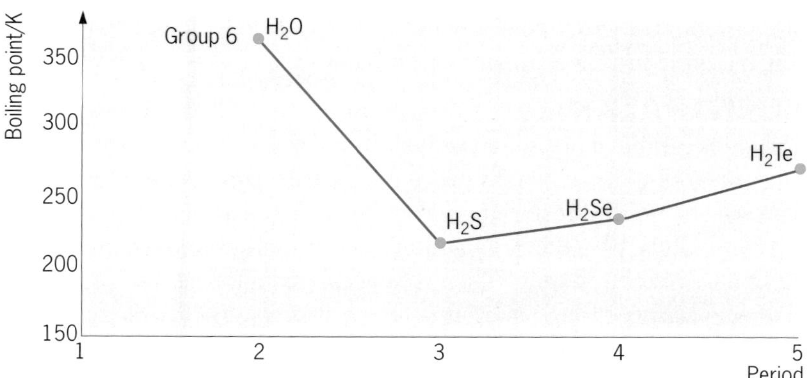

Fig 6.24 Boiling points of the Group 6 hydrides

Water's surprisingly high boiling point and melting point results from a special intermolecular bonding known as **hydrogen bonding**. Hydrogen bonds form when hydrogen is covalently bonded to the very electronegative elements oxygen, fluorine and nitrogen.

In water, the hydrogen–oxygen covalent bond is very polar. Hydrogen has only one electron, so when the oxygen atom in water pulls the bonded pair of electrons towards itself, it leaves the hydrogen atom's nucleus of a single proton more exposed. The resulting δ+ on the hydrogen atom is attracted to the lone pair (δ–) of the oxygen atom to form a very strong intermolecular attraction (Fig 6.25).

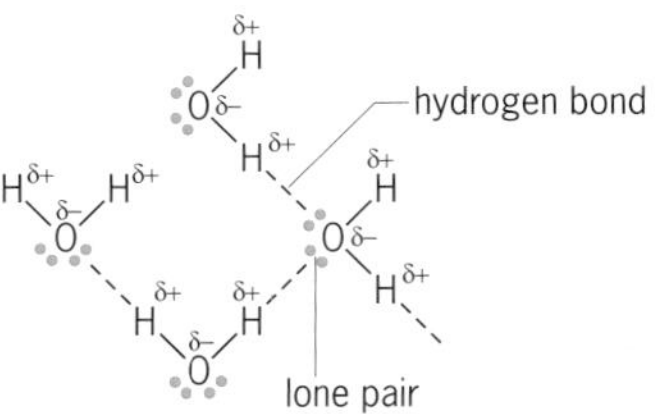

Fig 6.25 Hydrogen bonding in water

Hydrogen bonds are very strong compared to other intermolecular forces. This means that for water to boil these strong hydrogen bonds have to be broken before the molecules can escape into the vapour, which leads to the relatively high boiling point of 100 °C at sea level (101 kPa).

Hydrogen bonding also explains why water expands on freezing. As water cools, the molecules come closer together and so the density increases until, at 4 °C, the density is at a maximum. Below this temperature, the water molecules start to move apart because of the increasing number of hydrogen bonds they form. When water freezes, the water molecules are held apart by hydrogen bonds in a three-dimensional lattice. Thus the ice is less dense than water, which explains why ice is one of the very few solids to float on its liquid.

? QUESTION 22

22 If water followed the trend of the other Group 6 hydrides, what would you predict its boiling point to be in:

a) Kelvin;

b) degrees Celsius?

REMEMBER THIS

Electronegativity is the power of an atom within a molecule to attract electrons to itself. You can read more about this and about polar bonds in Chapter 4, pages 85–87.

For information about hydrogen bonding in alcohols, see Chapter 13, page 261; and for information about hydrogen bonding in proteins and how this makes your hair curl, see Chapter 19, page 384.

? QUESTION 23

23 Hydrogen bonding is also present in ammonia and hydrogen fluoride. Draw a diagram similar to Fig 6.25 to show this hydrogen bonding.

? QUESTION 24

24 Ice is less dense than liquid water, so ice floats. Most solids sink in their liquids. Imagine what would happen if ice sank in the oceans. For one thing, it would probably never melt, as it would be away from the warming of the Sun's rays. Describe the effects you think this would have on our weather systems and on life on Earth.

5 AQUEOUS SOLUTIONS AND CONCENTRATION

Aqueous solutions have water as the solvent. When we write equations and we want to show that a reactant is dissolved in water, we write the state symbol (aq). So a dilute solution of sulfuric acid is written: H_2SO_4(aq).

Water covers most of the surface of our planet, and many reactions take place in aqueous solution. Nearly all the chemical reactions of our bodies occur in the aqueous environment of our cells.

REMEMBER THIS

One mole is the amount of substance that contains as many particles as there are atoms in exactly 12 g of carbon-12. The number of particles in a mole is 6.02×10^{23} and is called the **Avogadro constant, L**. You can find out more about the mole on page 5 of Chapter 1. Now work through the Examples.

CONCENTRATION

Concentration can be expressed in terms of the mass of a substance in a certain volume of solution. The units often used are **grams per cubic decimetre** or **g dm^{-3}**. So if 40 grams of sodium hydroxide are present in one cubic decimetre the concentration is 40 g dm^{-3}. Chemists, however, are usually interested in the *amount* of a substance present. When we considered gases at the beginning of this chapter, it was not long before we came across the mole. It is the same with solutions. We need to know how many moles of solute are present in a particular volume of solution, because this tells us the number of particles that are there.

The concentration of a solute in a solution is measured in **moles per cubic decimetre – mol dm^{-3}**.

EXAMPLE

Q Work out how many moles of solute are dissolved in these solutions:

a) 250 cm^3 of 1.0 mol dm^{-3} NaOH(aq);

b) 20 cm^3 0.50 mol dm^{-3} HCl(aq).

A a) 1 dm^3 = 1000 cm^3

Number of moles NaOH in 1000 cm^3 = 1 mol

So number of moles NaOH in 250 cm^3 $= 1 \times \frac{250}{1000}$

$= 0.25$ mol

b) Number of moles of HCl in 1000 cm^3 = 0.50 mol

So number of moles HCl in 20 cm^3 $= 0.50 \times \frac{20}{1000}$

$= 0.010$ mol

Q What is the concentration (in mol dm^{-3}) of the following aqueous solutions?

a) 0.25 mol HCl dissolved in 50 cm^3;

b) 5.85 g of NaCl dissolved in 250 cm^3.

A a) There are 0.25 mol HCl in 50 cm^3

So in 1000 cm^3, the number of moles $= 0.25 \times \frac{1000}{50}$

$= 5.0$ mol

Concentration of HCl in solution = 5.0 mol dm^{-3}

b) Moles of NaCl in 5.85 g

$$\frac{\text{mass in grams}}{\text{mass of one mole (in grams)}} = \frac{5.85\text{g}}{58.5\text{g}} = 0.100 \text{ mol}$$

There are 0.100 mol NaCl in 250 cm^3

So in 1000 cm^3, the number of moles $= 0.100 \times \frac{1000}{250}$

$= 0.400$ mol

Concentration of NaCl solution = 0.400 mol dm^{-3}

QUESTION 25

25 **a)** Calculate how many moles of solute are present in these solutions:

i) 100 cm^3 of 2 mol dm^{-3} H_2SO_4(aq)

ii) 25.0 cm^3 of 5.00 mol dm^{-3} HNO_3(aq)

iii) 5 dm^3 of 0.2 mol dm^{-3} KCl(aq)

iv) 10 cm^3 of 0.5 mol dm^{-3} Na_2CO_3(aq)

v) 50 cm^3 of 0.050 mol dm^{-3} $NaNO_3$(aq)

b) Calculate the concentration of the following aqueous solutions:

i) 0.40 mol $CaCl_2$ dissolved in 250 cm^3

ii) 16 g $CuSO_4$ dissolved in 100 cm^3 (A_r: Cu = 64, S = 32, O = 16)

PREPARING SOLUTIONS OF KNOWN CONCENTRATION

Chemists often need to make up very accurately a solution of known concentration. To do this a volumetric flask is used. On its long neck is etched a line that indicates a precise volume (Fig 6.26(a)). The required amount of substance to be dissolved to make up the solution is weighed accurately on a balance. It is then dissolved in a small volume of distilled water in a beaker (Fig.6.26(b)) and the solution is poured into the volumetric flask (Fig.6.26(c)). Distilled water is then used to rinse any remaining solution from the beaker into the flask. More distilled water is added until the meniscus in the neck of the flask just lies on the etched line (Fig 6.26(d)).

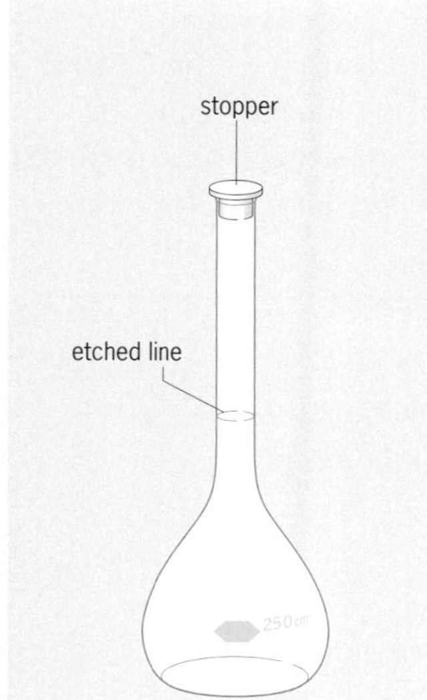

Fig 6.26(a) A 250 cm^3 volumetric flask

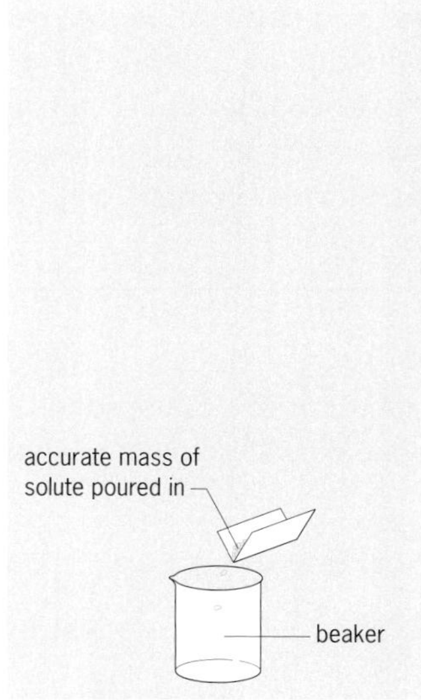

Fig 6.26(b) A known mass of solute is poured into a beaker followed by a small volume of distilled water to dissolve it

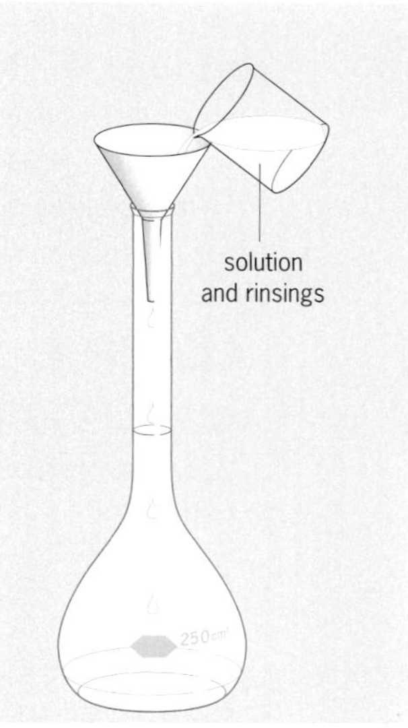

Fig 6.26(c) The solution is poured in and the beaker rinsed with more distilled water. The rinsings are also poured into the flask to ensure all the dissolved solute has been transferred

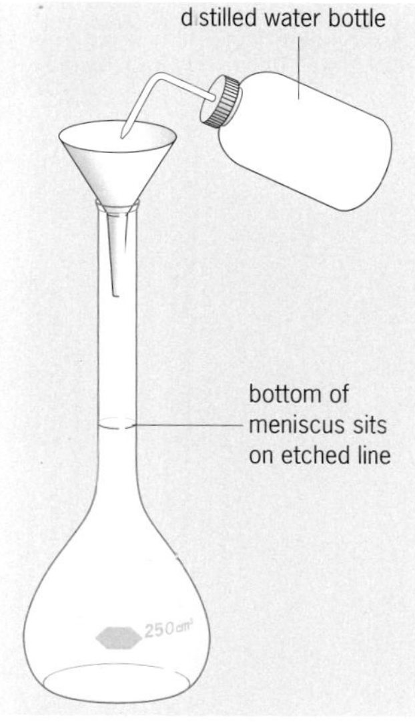

Fig 6.26(d) The solution is topped up with distilled water until the bottom of the meniscus just sits on the etched line

EXAMPLE

Q What mass of sodium hydroxide is required to make up 250.0 cm^3 of a 0.50 mol dm^{-3} sodium hydroxide solution? [A_r: Na = 23, O = 16, H = 1].

A Step 1. Work out the amount (moles) required:

$$1\ dm^3 = 1000\ cm^3$$

Number of moles NaOH required in 1000 cm^3 = 0.50 mol

So number of moles NaOH required in 250 cm^3 $= 0.50 \times \frac{250}{1000}$

$= 0.125$ mol

Step 2. Convert amount (moles) of NaOH into a mass:

Mass in grams = amount in moles × mass of 1 mole

Mass of 1 mol of NaOH = 23 + 16 + 1 g = 40 g

So mass of sodium hydroxide = 0.125 mol × 40 g

= 5.0 g

Therefore 5.0 g must be dissolved in 250 cm^3 to make a 0.5 mol dm^{-3} solution.

REMEMBER THIS

The concentration of a solution is expressed per dm^3 of solution and not per dm^3 of solvent. If one mole of solid is added to 1 dm^3 of water the volume of the solution will probably be slightly greater than 1 dm^3, so the concentration will not be exactly 1 mol dm^{-3}.

? QUESTION 26

26 What mass of solute is required to make up the following solutions?

a) 500 cm^3 of 2.0 mol dm^{-3} $CuSO_4$(aq)

b) 2.0 dm^3 of 0.125 mol dm^{-3} Na_2CO_3(aq)

c) 250 cm^3 of 0.25 mol dm^{-3} KOH(aq)

REMEMBER THIS

The choice of indicator depends on the type of acid and base used. (You can read more about this in Chapter 17.)

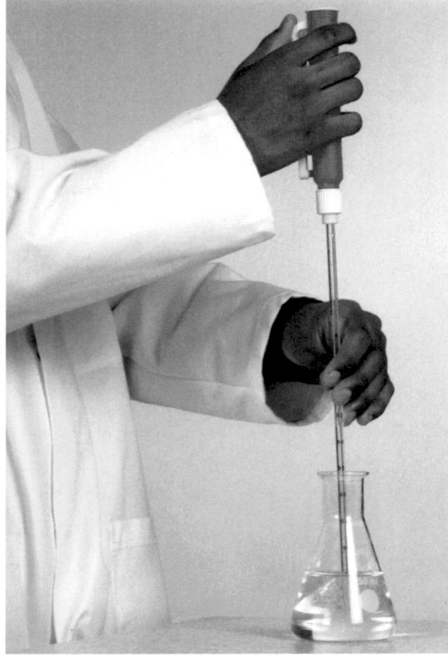

Fig 6.27(a) A pipette is used to measure an accurate volume of base solution (alkali)

Fig 6.27(b) The base solution (alkali) has been coloured using a few drops of indicator. The acid solution is added from a burette until the alkali is completely neutralised. This is the end point and is shown by the indicator changing colour. At least two titrations need to be done to confirm the exact volume of acid solution required

6 TITRATIONS

A **titration** is an experimental technique that uses the known volumes of two solutions to determine the amounts of substances that are reacting very accurately. This technique is often referred to as **volumetric analysis**.

In a titration one of the solutions is of known concentration. If we know the balanced equation for a reaction, then we know how many moles of each reactant will react. Using this information we can calculate the concentration of the other solution.

Acids are **neutralised** by bases. If we wish to discover the concentration of a dilute hydrochloric acid solution we can perform an **acid–base titration**. Sodium hydroxide of known concentration is used as the base to neutralise exactly the hydrochloric acid. The reaction is followed using an **acid–base indicator**, which changes colour at the **neutralisation point** or **end point** or **equivalence point**. Fig 6.27 shows how an acid–base titration may be performed.

EXAMPLE

Q What is the concentration of a dilute solution of hydrochloric acid if 26.1 cm^3 is exactly neutralised by 25.0 cm^3 of 0.100 mol dm^{-3} sodium hydroxide?

A Step 1. Write the balanced equation:

$$NaOH(aq) + HCl(aq) \rightarrow NaCl(aq) + H_2O(aq)$$

Step 2. Convert the equation into amounts:

$$1 \text{ mol NaOH} + 1 \text{ mol HCl} \rightarrow 1 \text{ mol NaCl} + 1 \text{ mol } H_2O$$

Step 3. Work out the amount of sodium hydroxide being used:

Number of moles NaOH in 1000 cm^3 = 0.100 mol

So number of moles NaOH in 25.0 $cm^3 = 0.100 \times \frac{25.0}{1000} = 0.002\,50$ mol

Step 4. Scale the amounts in the equation:

1 mol NaOH reacts with 1 mol HCl

0.002 50 mol NaOH reacts with 0.00250 mol HCl

So there are 0.00250 mol HCl in 26.1 cm^3 solution.

Step 5. Scale the amount in the known volume to 1 dm^3:

0.002 50 mol HCl in 26.1 cm^3 solution

In 1 dm^3 (1000 cm^3) moles HCl $= 0.002\,50 \times \frac{1000}{26.1}$

Therefore concentration of HCl = 0.095 8 mol dm^{-3}.

Q Clover leaves contain ethanedioic acid, $(COOH)_2$. 1.01 g of clover leaves are crushed and the ethanedioic acid is extracted using a small volume of distilled water. This solution is exactly neutralised by 17.2 cm^3 of 0.200 mol NaOH. What is the mass of ethanedioic acid in this sample of leaves?

A Step 1. Write the balanced equation:

$$(COOH)_2(aq) + 2NaOH(aq) \rightarrow (COO^-Na^+)_2(aq) + 2H_2O(l)$$

EXAMPLE (Cont.)

Step 2. Convert the equation into amounts:

1 mol $(COOH)_2$ + 2 mol NaOH → 1 mol $(COO^-Na^+)_2(aq)$ + 2 mol $H_2O(l)$

Step 3. Work out the amount of sodium hydroxide being used:

Number of moles NaOH in 1000 cm^3 = 0.200 mol

So number of moles NaOH in 17.2 cm^3 = $0.200 \times \frac{17.2}{1000}$ = 0.003 44 mol

Step 4. Scale the amounts in the equation:

1 mol $(COOH)_2$ reacts with 2 mol NaOH

0.001 72 mol $(COOH)_2$ reacts with 0.003 44 mol NaOH

So there are 0.001 72 mol $(COOH)_2$ in the clover leaf solution.

Step 5. Convert the amount (moles) into a mass:

Mass in grams = amount in moles × mass of 1 mole

Mass of 1 mol of $(COOH)_2$ = (12 + (16 × 2) + 1 g) × 2 = 90 g

So the mass of ethanedioic acid in 1.01 g clover leaves = 0.001 72 mol × 90 g

= 0.155 g $(COOH)_2$

? QUESTION 27

27 Vinegar contains ethanoic acid, CH_3COOH. What is the concentration in mol dm^{-3} and g dm^{-3} if 25.0 cm^3 of 1.00 mol dm^{-3} sodium hydroxide is neutralised by 26.3 cm^3 vinegar? The balanced equation for this reaction is:

$CH_3COOH(aq) + NaOH(aq) \rightarrow CH_3COO^-Na^+(aq) + H_2O(l)$

Table 6.1

Substance	Entropy (*S*)/J K^{-1} mol^{-1}
He(g)	126.0
Ne(g)	146.2
$H_2O(l)$	69.9
$H_2O(g)$	188.7
$C_{(graphite)}$	5.7
$C_{(diamond)}$	2.4

7 ENTROPY

Entropy is a measure of the disorder of a system. Think of your bedroom – when it is tidy everything has a place and there is a high degree of order. We could say that the entropy of your room is low. However, it is all too easy for the entropy to rise as things are moved out of place and it becomes untidy.

When particles are in a solid they cannot move out of position, they can only vibrate. In the solid state there is usually a high degree of order and a limited number of ways to arrange the particles, so the entropy of a solid is usually low. The liquid state tends to have a higher entropy than the solid state, because the particles can move about, so many more arrangements are possible. A gas has widely spaced particles that move very rapidly in all directions. All gases have high entropies.

The entropies of different substances can be determined for a particular temperature. Entropy is given the symbol ***S*** and is measured in joules per kelvin per mole (J K^{-1} mol^{-1}). Table 6.1 gives the entropy values for some substances at 298 K and 101 kPa.

You can see from Table 6.1 that water vapour has a much higher entropy than water liquid. This is to be expected, since water molecules are more spread out in the vapour and more arrangements are possible. Diamond has a lower entropy than graphite, which means that diamond has a more ordered structure. Neon and helium are both gases and so have relatively high entropies. However, neon has heavier atoms. This tends to bring its energy levels closer together, which in fact provides more possibilities of arranging quanta of energy.

Entropy, then, is not just about the number of possible ways of arranging particles. It is also about how the energies of the particles are arranged in the energy levels, an aspect of entropy that we shall continue in Chapter 8.

? QUESTION 28

28 Predict whether each of the following leads to an increase or decrease in the entropy of the substances involved:

a) Ice melting
b) Water vapour condensing
c) Cooling oxygen from 60°C to 20°C
d) Dissolving sugar in water
e) Subliming solid iodine
f) Freezing liquid bromine

REMEMBER THIS

At zero Kelvin, perfect crystals have zero entropy. This is the third law of thermodynamics. Perfect crystals have only one arrangement of particles – they are perfectly ordered in all directions. At 0 K, molecular motion virtually stops, so there is only one arrangement of particles and energy, hence zero entropy.

SUMMARY

After studying this chapter, you should know and understand the following points and be able to use the equations.

- Equal volumes of gases contain equal numbers of molecules at the same temperature and pressure (Avogadro's law).
- 1 mole of any gas at standard temperature and pressure occupies 22.4 dm^3. Under room conditions the molar gas volume is 24 dm^3.
- Boyle's law relates the pressure and volume of a fixed amount of gas and is expressed mathematically as: $p \propto 1/V$, where p = pressure and V = volume.
- Charles' law relates volume and temperature of a gas and is expressed mathematically as: $V \propto T$, where T = absolute temperature in kelvin.
- The ideal gas equation relates pressure, volume and temperature for a fixed amount of gas: $pV = n\text{R}T$.
- The kinetic–molecular model explains the behaviour of ideal gases by making a number of assumptions, such as zero intermolecular forces and that the volume of particles is negligible.
- Real gases approach ideal behaviour when at a high temperature or low pressure.
- The kinetic–molecular model can be extended to include solids and liquids.
- Vapour pressure is the pressure above a liquid when it is in equilibrium with its vapour. Vapour pressure increases with temperature.
- Hydrogen bonding is a strong intermolecular force between hydrogen bonded to an F, N or O atom and the lone pair of another F, N or O atom. It accounts for the anomalous properties of water.
- The concentration of a solute in a solution is measured in mol dm^{-3} or g dm^{-3}.
- Titrations are a valuable way to analyse amounts in solutions.
- Entropy is a measure of the disorder of a system. Solids usually have low entropies, whereas gases have high entropies.

Practice questions and a How Science Works assignment for this chapter are available at www.collinseducation.co.uk/CAS

7

Structure and bonding of the elements

7 STRUCTURE AND BONDING OF THE ELEMENTS

Fullerenes

The geodesic dome. Designed by Buckminster Fuller for Expo 67

The United States pavilion at the 1967 World Fair Expo in Montreal was a giant geodesic dome, a revolutionary structure designed by the pioneering American engineer and architect Robert Buckminster Fuller.

Little did he know that 18 years later his design was to be repeated at atomic level in the Nobel Prize-winning discovery of forms of carbon that made up the class of structures named after him – the buckminsterfullerenes. These are net-like spheres (and tubes), each of many atoms, possibly opening up new areas of organic chemistry.

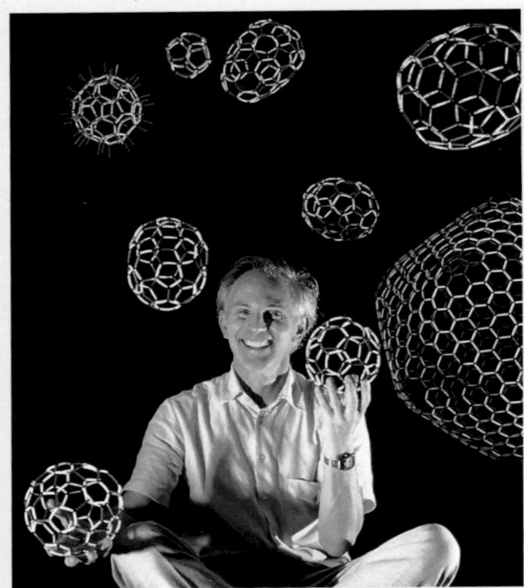

Fullerenes. Sir Harry Kroto holding buckminsterfullerene, C_{60}, in his left hand, with other fullerenes he discovered

Until Harry Kroto and his team at Sussex University discovered the chemical class of fullerenes in 1985, chemists believed that carbon had only two crystalline forms: graphite and diamond. Now there were three. Scientists were astonished, because – so they thought – carbon had been exhaustively researched over many decades. Equally astonishing was the discovery that fullerenes are present in soot.

Fullerene model. A football is a model of a fullerene

Like Buckminster Fuller's dome, the design of which is held in place by natural forces, the structure of a crystalline form such as a fullerene is constrained by the forces that hold its component particles in place. It is this definite internal arrangement that determines the physical properties of all elements in the crystalline state.

1 STRUCTURE AND THE PERIODIC TABLE

On pages 94–96, you can read how elements in the same group of the Periodic Table have similar chemical properties because they have a similar arrangement of their outer electrons. For example, all the elements in Group 7 have an outer electron configuration that ends in p^5. The periodicity of properties, such as ionisation energy, is explained in terms of the regular recurring pattern of electron configurations from one period to another.

This chapter takes certain physical properties of elements and relates them to the structure and bonding of elements in the solid state. We look particularly at the electrical conductivity, melting points and boiling points of the elements in the second and third periods of the Periodic Table.

2 KINETIC–MOLECULAR MODEL FOR SOLIDS

In any discussion of solid-state structures we need to look at the kinetic–molecular model of matter as it applies to solids.

Fig 7.1 shows a typical arrangement of the particles in a solid. These are their common characteristics:

- The particles may be ions, atoms or molecules.
- They are closely packed in an ordered pattern, occupying relatively fixed positions.
- This ordered arrangement of particles is called a **lattice**.
- The particles in a solid are fixed in position, but are able to vibrate.
- The particles in a solid are attracted to one another.

We look at the nature of some of the attractive forces later in the chapter.

For more information on the kinetic–molecular model of matter, see page 123.

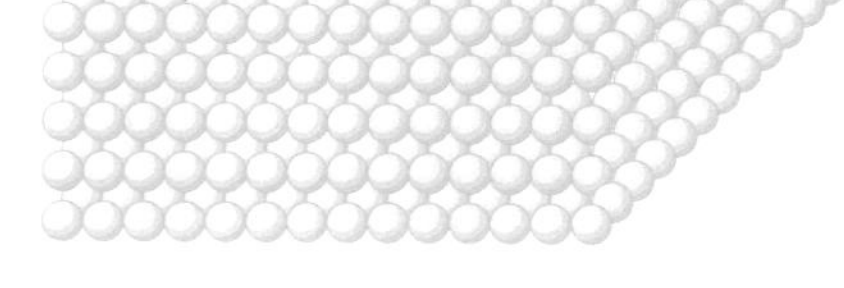

Fig 7.1 The arrangement of particles in a solid

TYPES OF SOLID

When a liquid substance is cooled sufficiently, it freezes and forms a solid. Normally, the particles of the substance take up ordered positions and form a crystalline solid. Even elements that are gases at room temperature and pressure solidify when the temperature is low enough. For helium, the temperature has to be lowered to within 3 degrees of absolute zero before it becomes a solid.

Crystalline solids

In a **crystalline** solid, the particles are arranged in a definite repeating pattern throughout the solid. Such a solid is said to be homogeneous. This three-dimensional repeating pattern of particles in crystals is usually referred to as a **crystal lattice**.

Many elements form crystalline solids and have a crystal lattice that can be described geometrically. An example is diamond, one of the allotropes of carbon.

A crystalline solid has a definite melting point because all the attractive forces that hold the particles in the lattice have the same strength. So the same amount of energy is required to overcome the forces between every pair of neighbouring particles.

Amorphous solids

Some elements do not always form crystalline structures. They form **amorphous** solids. Carbon black or soot is an example. In an amorphous solid, the particles do not take up a definite regular pattern during freezing. This is usually the case when the freezing process is so fast that the particles have no time to become arranged in an orderly way.

QUESTION 1

1 Describe the arrangement of particles in a liquid and in a gas.

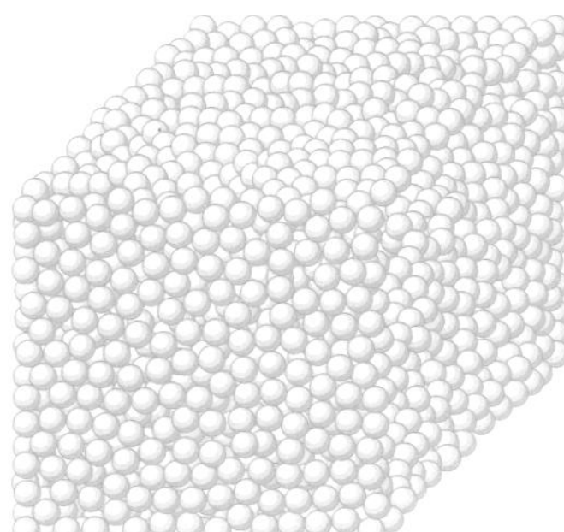

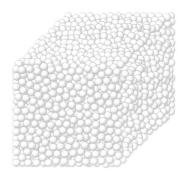

Fig 7.2 The arrangement of particles in an amorphous solid. It is very much like the arrangement in a liquid, but particles are free to move in a liquid and not in a solid

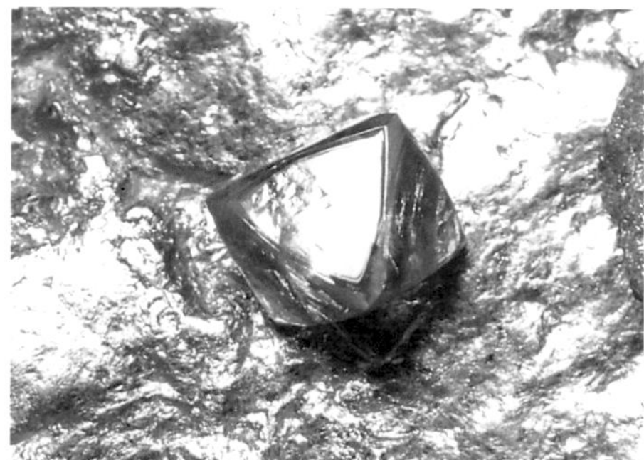

Fig 7.3 Diamond has a highly ordered crystal lattice, which gives it a geometric shape. Diamond is seen here on graphite, another structured form of carbon. (Soot is an amorphous form of carbon)

Metallic glasses are amorphous solids, formed when liquid metal cools at a rate of a million degrees per second. The particles occupy completely random positions within the amorphous solid and so do not form a crystal lattice. Nevertheless, the positions of the particles are relatively fixed.

Usually, amorphous solids don't have a definite melting point, but soften gradually on heating. This is because the forces that hold their particles together have different strengths, and so different amounts of energy are needed to break them.

An element with an amorphous form always has at least one crystalline structure that is more stable than the amorphous form. Hence, an amorphous solid will slowly change into a crystalline solid. In this change of form, the particles are rearranged from a disordered state into an ordered pattern. However, if the forces between the particles in the amorphous form of a solid are strong, it could take a very long time to change from its amorphous form to its crystalline form.

3 PATTERNS ACROSS THE PERIODIC TABLE

You can read about the crystals of compounds in Chapter 21.

We have already described the periodicity in properties, such as first ionisation and atomic radii, that depend on the electron configuration of the elements (see Chapter 5). The periodicity shown by the physical properties of elements, such as boiling point or electrical conductivity, depends on the structure and bonding rather than the electronic configuration.

MELTING AND BOILING POINTS

Figs 7.4 and 7.5 show how the melting points and boiling points of the elements vary with atomic number (proton number). Although we can see a pattern – the melting point tends to rise to a maximum around Group 4 and then falls to the noble gases – it cannot really be described as a periodic function. In fact, if the melting point graph extended to elements with higher atomic numbers, we would see even less of a pattern.

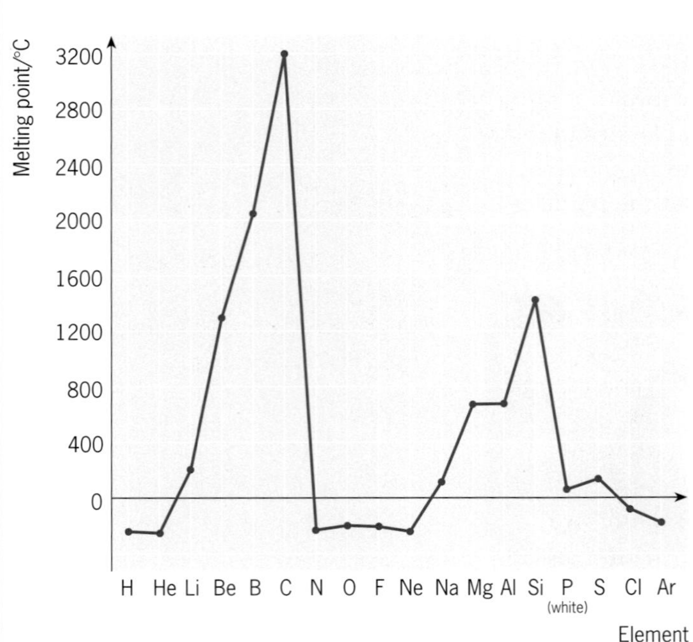

Fig 7.4 Melting points of elements in Periods 2 and 3

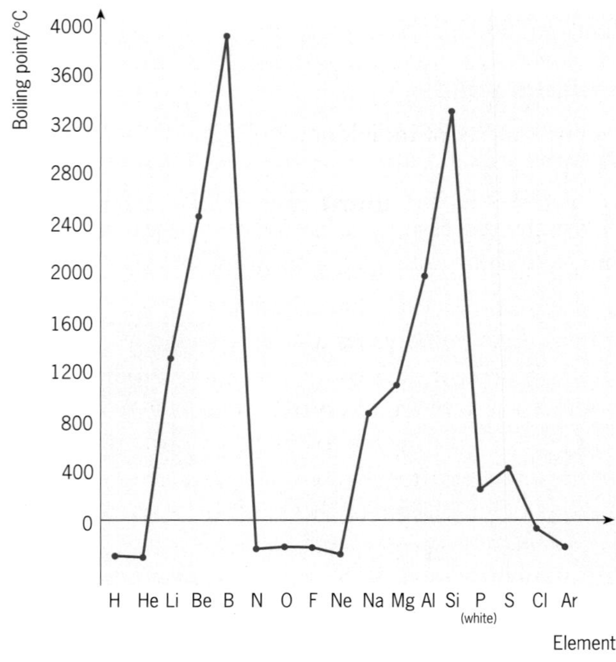

Fig 7.5 Boiling points of elements in Periods 2 and 3. Carbon sublimes above 4000 °C

Put simply, it is the strength of the attraction between the particles in an element's crystal lattice that determines the value of the melting point, rather than the electron configuration of the element. The greater the force between the particles, the higher the melting point. Much the same can be said of the boiling point, but in this case it is the magnitude of the attraction between the particles in the liquid phase.

Figs 7.4 and 7.5 show that, in terms of the melting and boiling points, there are broadly three types of element:

- metals that have reasonably high melting points;
- non-metals that have very high melting points (such as carbon and silicon);
- non-metals that have very low melting points.

The difference in the melting points of elements depends on their crystal structure and the strength of the bonding between particles (atoms, molecules or ions) of the elements.

ELECTRICAL CONDUCTIVITY

Metals are good electrical conductors and non-metals are poor electrical conductors. For electrical conduction to occur, there must be charged particles that are free to move. In the case of an element, the charged particles are its electrons. Provided electrons in an element are free to move when subjected to a potential difference (i.e. when a voltage is applied), the element is a good electrical conductor. There is a complication: the electrical conductivity of elements can change under varying conditions of temperature and pressure, but this is dealt with on page 142.

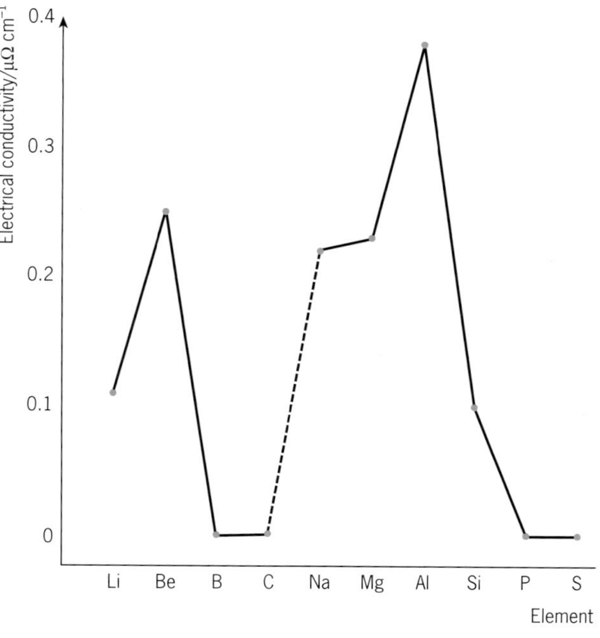

Fig 7.6 The electrical conductivity of non-gaseous elements in Periods 2 and 3 at 298 °C

METALLIC AND NON-METALLIC ELEMENTS

We can understand the differences in the melting points, boiling points and electrical conductivities of the elements by looking at the types of particle which make up the crystal lattices.

A metal crystallises into a giant structure of closely packed metal *ions*. No molecules are present. By contrast, a typical non-metal is very different: its structure normally contains *covalent bonds*, so it consists of individual molecules. Hence, the arrangement of particles in a non-metal is described as a **molecular lattice**.

The molecules may be **giant**, and we can consider the whole crystal as being one molecule, as in diamond and graphite. Alternatively, the lattice may be made up of repeating units of **simple** molecules, as in the case of solid iodine, which has a lattice that contains a repeating pattern of iodine molecules, I_2. The noble gases are monatomic, so the crystal lattice of a solid noble gas is based on a repeating pattern of atoms.

Across the whole range of crystalline elements, there is a wide variation in the strengths of the attractive forces between the particles that form their different crystal structures. Evidence for this is the wide variation in the melting and boiling points of these elements shown in Figs 7.4 and 7.5. Solids with small molecules, such as hydrogen, nitrogen and oxygen, have low melting points because the forces between them in the crystalline state are very weak. Graphite and diamond have very strong forces of attraction – covalent bonds – between their atoms, which gives them extremely high melting points.

? QUESTION 2

2 Describe the changes in arrangement, motion and the strength of attraction between particles when a solid melts into a liquid.

4 METALLIC BONDING

Look at a lump of any pure metallic element. It is not obvious that it is crystalline: it does not have the regular faces that a crystal of diamond has. Nevertheless, a powerful microscope reveals its crystalline structure.

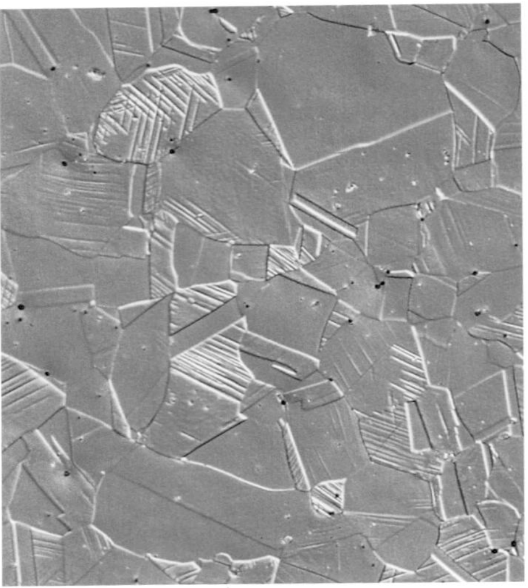

Fig 7.7 The surface of stainless steel showing its crystalline nature

Fig 7.8 Metal crystals growing on zinc in lead nitrate solution

? QUESTION 3

3 Write down the ionic equation for the displacement reaction between:
a) zinc and aqueous silver ions;
b) zinc and aqueous lead(II) ions.

We can see crystals forming in the displacement reaction between zinc and aqueous silver nitrate or lead(II) nitrate:

$$Zn(s) + 2Ag(NO_3)_2(aq) \rightarrow Zn(NO_3)_2(aq) + 2Ag(s)$$

$$Zn(s) + Pb(NO_3)_2(aq) \rightarrow Zn(NO_3)_2(aq) + Pb(s)$$

CLOSE PACKING

The existence of crystals in a piece of metal means that the particles in the metal must have a regular arrangement. These particles are, in fact, positive ions and not metal atoms. All of these ions are the same size, so they can close pack in a manner similar to the most space-saving way of packing spheres in a layer: each sphere is surrounded by six others (Fig 7.9).

In some metals, close packing is achieved through a repeating pattern in which each positive ion is in contact with six others in the same layer. Such a crystal will

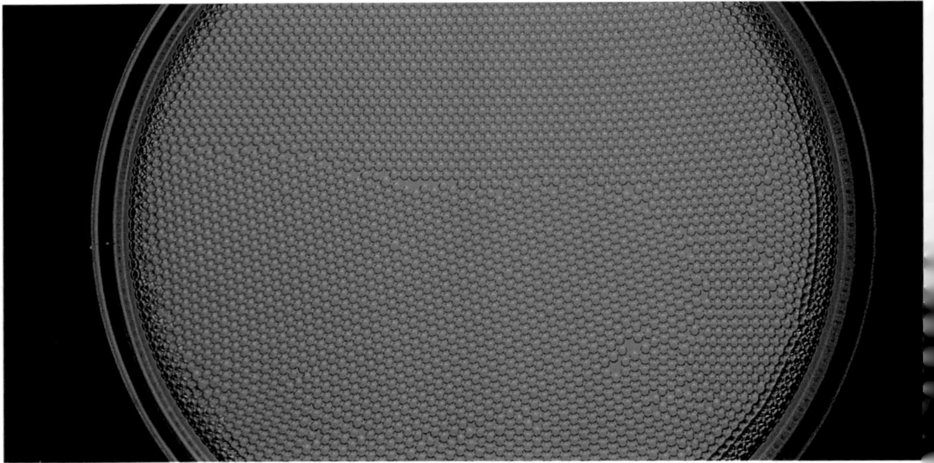

Fig 7.9 Bubbles packed together in a close-packed array

There are three different particle arrangements for metals. Each one is an example of close packing

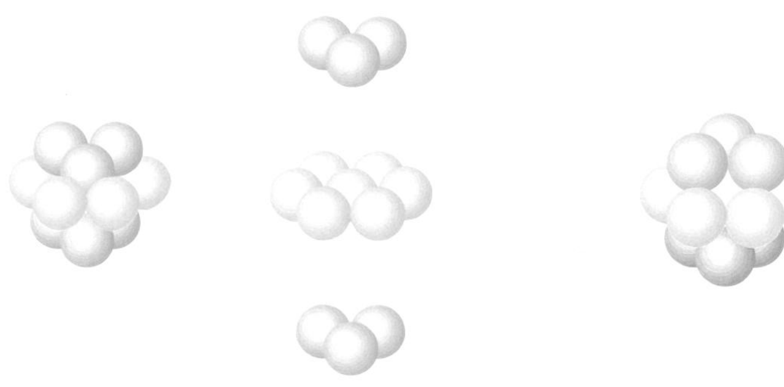

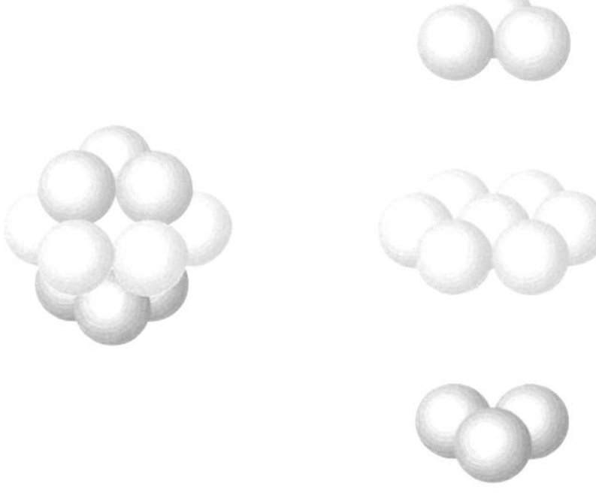

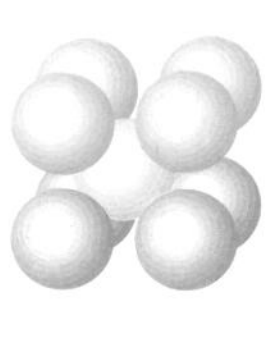

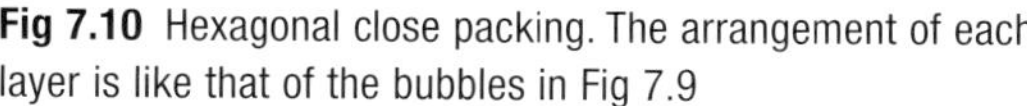

Fig 7.10 Hexagonal close packing. The arrangement of each layer is like that of the bubbles in Fig 7.9

Fig 7.11 Cubic close packing. The arrangement is also like that of the bubbles in Fig 7.9

Fig 7.12 Body-centred cubic close packing

have not just one but many layers of positive ions. Each additional layer is arranged to maximise the number of ions that can be packed.

However, there is more than one way of close-packing positive ions in a three-dimensional metallic crystal lattice, as Figs 7.10 to 7.12 show.

The **coordination number** is the number of nearest neighbours that a particle has when it is in a crystal lattice. A **unit cell** is the simplest pattern of particles for which repetition in three dimensions produces a crystal lattice.

Since metal ions are closely packed, it is not surprising that most metals have high densities.

? QUESTION 4

4 Give the coordination number (number of nearest neighbours) of a metal ion in:

a) hexagonal close packing;

b) cubic close packing;

c) body centred cubic packing.

REMEMBER THIS

Density depends on ionic radius, atomic mass and the type of close packing.

POSITIVE IONS IN A SEA OF DELOCALISED ELECTRONS

Since they all have the same charge, why don't the metal ions repel one another? Clearly, they don't, otherwise metals would not be strong and mostly hard. The answer lies in the electrons that have been removed to form the positive ions. These electrons are free to move throughout the metal crystal. The electrons are said to be **delocalised**. The delocalised electrons are often referred to as a 'sea of electrons', because they occupy the whole space between the closely packed metal ions (Fig 7.13). The positive metal ions are not repelled from one another because each is electrostatically attracted towards the sea of delocalised electrons.

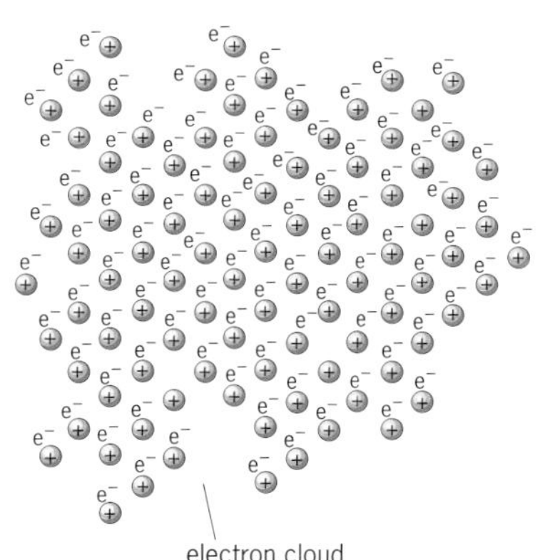

Fig 7.13 The sea of electrons model

You can read about the presence of delocalised electrons in a benzene molecule on page 298.

REMEMBER THIS

Metallic bonding is often described as closely packed metal ions in a sea of electrons.

Electrical conductivity of metals

All metals are good electrical conductors because of the presence of delocalised electrons throughout the metal lattice. It is only the electrons in the outer shell that can be delocalised. The inner electrons are still localised around the nucleus of the metal atom. This means that, for example, 1 mole of sodium has 1 mole of delocalised electrons, since it forms Na^+, whereas 1 mole of magnesium has 2 moles of delocalised electrons, since it forms Mg^{2+}.

When an electrical potential is placed across a strip of metal, electrons move from the negative to the positive area of the metal, and as a result a current flows. Normally, there is a resistance to the flow of electrons in a metal. This results in some of the electrical energy being lost as heat.

? QUESTION 5

5 The melting point of magnesium, which has two outer electrons, is much higher than that of sodium, which has just one outer electron. Why do you think this is so?

SCIENCE IN CONTEXT

Superconductors

Superconductors are materials that conduct electricity with little or no resistance. So when a current flows through a superconductor there should be little or no heating effect. The superconducting materials presently available work only at low temperatures, but each year new materials are discovered that raise the operating temperatures of superconductors. However, we are still some time away from a superconductor that will work at room temperature.

Fig 7.14 Levitating magnets, an example of superconductivity. This is a disc of yttrium–barium–copper oxide floating above a nitrogen-cooled cylinder of superconducting ceramic

Superconductors offer the promise of loss-free power transmission, super-fast electronic circuits, powerful electromagnets and ultra-sensitive magnetic detectors. Figure 7.15 shows a magnetic levitation ultra-high speed train now in commercial use in China. This uses a superconductor to make an extremely powerful electromagnet.

Fig 7.15 The Yamanashi Maglev high-speed train on its test track. Maglev stands for magnetic levitation. The high-speed train (now in service in Pudong, Shanghai) 'floats' on powerful superconducting magnets, virtually eliminating friction between train and track. The train can run at speeds of up to 500 km per hour

? QUESTION 6

6 Which of the following metals is likely to have the greatest electrical conductivity: aluminium, magnesium or sodium? Explain your answer.

REMEMBER THIS

In transition elements, d electrons can also be delocalised throughout the metal lattice. (See Chapter 24.)

Electrons and energy levels are covered in Chapter 3.

Other physical properties of metals

All the other typical physical properties of metals can be explained by reference to the model of bonding in which the positive ions are immersed in a sea of delocalised electrons.

- A metal is strong because when its structure is deformed by applied stress and the positive ions move, the delocalised electrons move as well, thus maintaining their attraction to the positive ions.
- Kinetic energy is transferred easily from one delocalised electron to another, which explains why metals are such good thermal conductors.
- The presence of delocalised electrons also explains why metals are shiny and have a lustre. These electrons are easily promoted to higher energy levels, from which they fall to lower levels, emitting the light that makes the metal shine.
- Metals have fairly high melting and boiling points because the attraction between the delocalised electrons and the positive ions is strong. The greater the number of delocalised electrons the stronger this attraction, so magnesium has a higher melting point than sodium.

5 GIANT MOLECULES

Some elements have extremely high melting points. This is because the particles – atoms – in the crystal lattice are strongly attracted to one another. In a giant molecule the atoms are held together by strong covalent bonds. The whole crystal of a giant molecule should be considered as one molecule. For such a structure to melt, each atom must be able to move freely. For this to happen, every covalent bond must be broken. Covalent bonds are very strong, so need a large amount of energy to break. This means the melting point, and consequently the boiling point, is very high. Carbon and silicon, which have the highest melting points in their respective periods, exist as giant molecules. Elements that have a giant molecular structure are also said to have a macromolecular or giant covalent structure.

ALLOTROPY OF CARBON

Diamond and graphite are different crystalline forms of carbon that both have a giant molecular structure. These crystalline forms are called **allotropes**. An element is said to have allotropes (show allotropy) when it exists in the same state in more than one structural form. Oxygen, for example, can exist in two forms in the gaseous state: dioxygen (O_2) and ozone (O_3). The allotropes of an element have the same or similar chemical properties, but their physical properties are different, because each allotrope has its own crystal structure. In the case of diamond and graphite, the physical properties are markedly different. For example, graphite is a good conductor of electricity, but diamond is an extremely poor conductor. These differences in properties are associated with the different internal crystal structures.

> **? QUESTION 7**
>
> 7 Diamond and graphite both burn in oxygen to give a gaseous product. What is the name of this gaseous product?

Diamond

Diamond is the hardest substance known and is the least compressible. It is a better conductor of heat at room temperature than any other material, and when completely pure it is transparent. These extreme properties make diamond technologically very useful. Its hardness makes it useful as an industrial abrasive and as a cutting tool in industry and surgery, and because of its excellent heat conductivity it is used as a heat sink to cool electronic components rapidly. Certain impurities implanted in diamond can make it a semiconductor. These applications are a direct result of the internal structure of a diamond crystal (Fig 7.17), coupled with the fact that the carbon atoms in diamond are more closely packed than the atoms in any other material.

The atoms in a crystal of diamond are bonded into one giant molecule. Each carbon atom is covalently bonded to four other carbon atoms, and so each carbon atom has four nearest neighbours. That is, its coordination number is 4. When just four bonding pairs of electrons surround an atom, they are arranged tetrahedrally (see page 80). This is the case with diamond. All outer electrons of the carbon atoms are involved in the formation of covalent bonds. There is no possibility of delocalised or mobile electrons. As a consequence, diamond cannot conduct electricity.

Fig 7.16 A diamond-tipped oil drill part

Fig 7.17 The internal structure of diamond

Fig 7.18 A unit cell for diamond

> **✔ REMEMBER THIS**
>
> In diamond each carbon atom is sp^3 hybridised, whereas in graphite each carbon atom is sp^2 hybridised. (You can read about sp^3 and sp^2 hybridisation on page 84.)

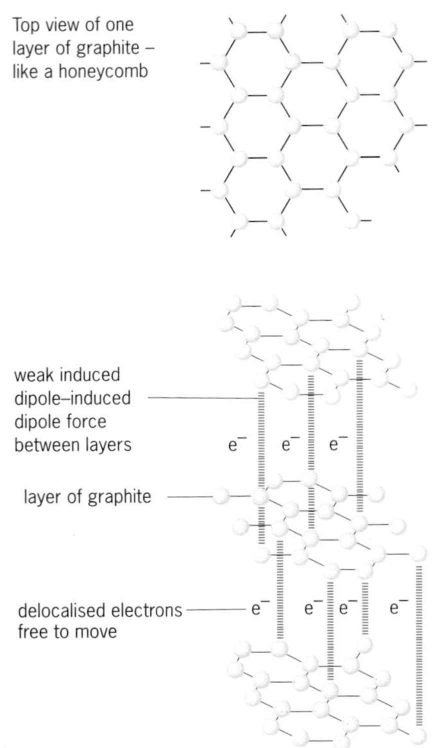

Fig 7.19 The internal structure of graphite

Graphite

Figs 7.19 and 7.20 show that the arrangement of carbon atoms in graphite is considerably different from that in diamond. In graphite, each carbon atom is covalently bonded to three other carbon atoms. All the bond angles are 120°, which is what would be expected if only three bonding electron pairs surrounded the carbon atom.

However, a check on the number of electrons reveals that there is one electron per carbon atom left over. These electrons are delocalised and explain the electrical conductivity of graphite. Although graphite is composed of giant molecules, its atoms are arranged in layers that can slide past one another. Weak van der Waals (induced dipole–induced dipole) forces hold the layers together.

The internal structure of graphite accounts for its:

- extremely high melting point, since many very strong covalent bonds have to be broken to allow the carbon atoms to move freely;
- electrical conductivity, since the delocalised electrons are free to move when subjected to a potential difference;
- brittleness, since it is easy to split one layer from another as only the weak van der Waals forces need to be broken.

Graphite is used to make electrodes (particularly for high-temperature work, since they will not melt), electrical contacts and brushes for electrical motors. It is also used in lubricants and in pencil leads. Composite materials that contain graphite are used to make racquets, fishing rods and golf clubs.

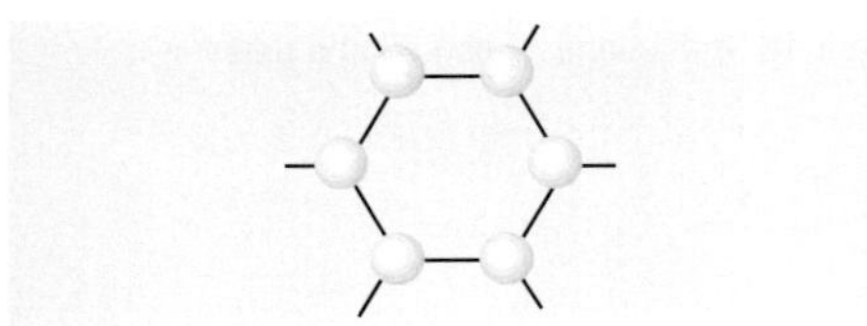

Fig 7.20 A unit cell for graphite

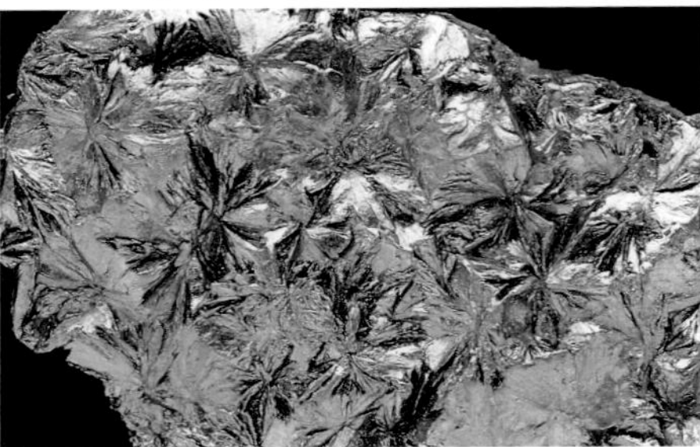

Fig 7.21 When graphite is ground and compressed, it forms the 'lead' in pencils

STRETCH AND CHALLENGE

Stability of graphite and diamond

Given its structure, it is surprising that graphite is the most stable form of carbon. The logical conclusion of this would be that diamond should change into graphite at room temperature. Fortunately, this does not happen, because to change the structure of the diamond lattice, all of its covalent bonds would have to be broken. This would need an enormous amount of energy, which is not available at room temperature. Chemists refer to this property as having a huge **activation energy**.

In the same way, it is difficult to convert graphite into diamond, because it has a very large activation energy. To convert graphite into diamond requires an extremely high temperature and pressure. In fact, a pressure of approximately 60 000 atmospheres and a temperature of 1500 °C are needed before diamond becomes more stable than graphite.

The conversion of graphite to diamond is endothermic:

$$C_{(graphite)} \rightarrow C_{(diamond)} \quad \Delta H = +1.9\ \text{kJmol}^{-1}$$

Natural diamonds are produced by this combined action of heat and pressure deep under the Earth's surface. Artificially, graphite and carbon-containing compounds, such as coal, or even peanut butter, are processed at about 100 000 atmospheres and 2000 °C to make industrial diamonds. Despite the commercial importance of industrial diamonds, only about 100 tonnes per year are needed world-wide. Most of these industrial diamonds are used in abrasive coatings and in cutting tools. The quality of industrial diamonds is steadily improving and soon they may compete with natural diamonds to be used in jewellery.

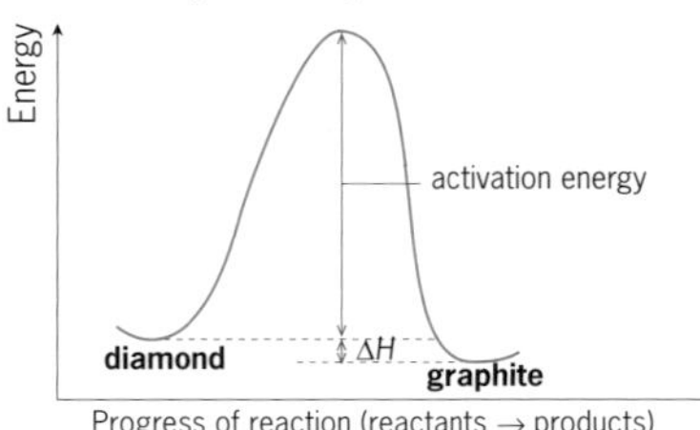

Fig 7.22 Reaction profile for the conversion of diamond to graphite. The profile is not drawn to scale because of the magnitude of the activation energy. To convert diamond to graphite requires extremely high temperatures to provide sufficient energy to overcome the activation energy

Fullerenes

Since the mid-1980s, a new class of allotropes of carbon has been identified. They were given the name 'fullerenes' (see the Science in Context box on page 136). The most widely studied fullerene is C_{60}, which is a self-contained molecule that exists in both crystalline and amorphous forms. An interesting aspect of the C_{60} molecule, apart from its sheer size, is its shape, which resembles a football with hexagonal and pentagonal panels. So, originally called buckminsterfullerene, C_{60} is now almost always known as 'buckyball'. Since the discovery of C_{60}, many other arrangements have been identified: for example, C_{70}, C_{72} and C_{84}.

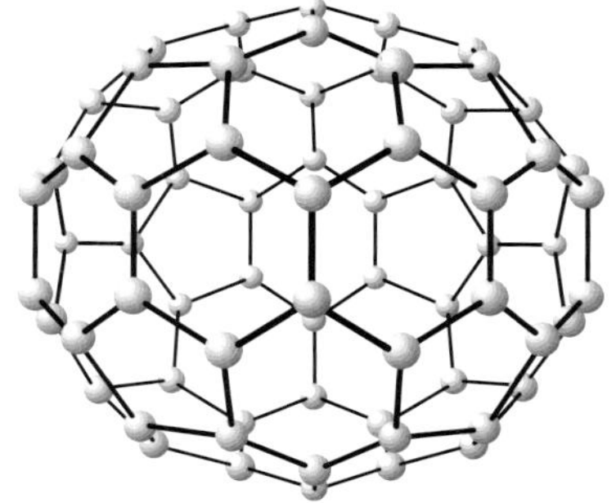

Fig 7.23 Fullerene C_{72}

> **QUESTION 8**
>
> 8 **a)** Which feature of the structure of graphite makes it useful as a lubricant?
> **b)** Explain why graphite is a suitable material with which to make high-temperature crucibles.
> **c)** Which allotrope of carbon would you expect to have the higher melting point – graphite or diamond? Explain your answer.

SILICON

Silicon is in Group 4 of the Periodic Table. From this we can reasonably predict that the silicon atom forms four covalent bonds. Silicon forms a giant molecular structure in which every silicon atom is surrounded by four covalent bonds, similar to the structure of diamond.

Since silicon and diamond have the same internal structures, their physical properties should be similar – and they are. Silicon has a very high melting point, it is very hard and in its pure state it is a very poor conductor of electricity.

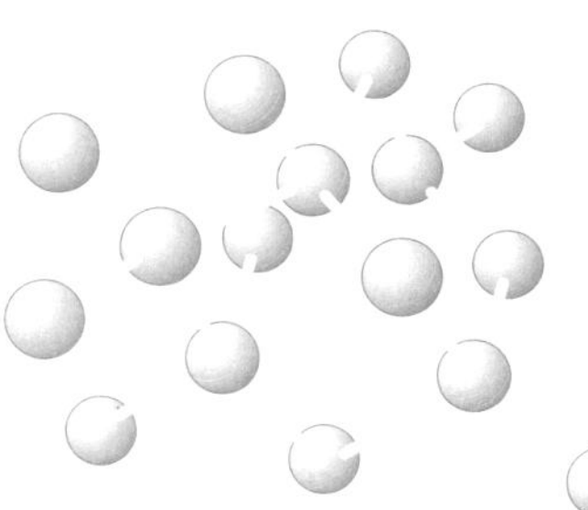

Fig 7.24 The internal structure of silicon. Note the similarity to diamond, the only difference being the longer covalent bond

> **QUESTION 9**
>
> 9 **a)** Explain why silicon has a very high melting point.
> **b)** Silicon carbide, SiC, is used as an abrasive in grinding wheels. It has a high melting point and does not conduct electricity:
> i) What type of structure do you think silicon carbide has?
> ii) Draw the unit cell for silicon carbide.

HOW SCIENCE WORKS

Molecular nanotechnology

The three-dimensional structural control of materials, processes and devices at the atomic scale is called nanotechnology. It will lead to a fundamental breakthrough in the way materials, devices and systems are understood, designed and manufactured.

Following the discovery of buckminsterfullerene in 1986 there has been rapid development in a range of similar molecules called nanotubes. Carbon nanotubes are a new form of carbon, first identified in 1991 by Sumio Iijima, a Japanese scientist. The honeycomb-shaped walls of nanotubes may consist of either multiple layers of carbon atoms or just a single layer.

Scientists consider carbon nanotubes as one of the building blocks of the 21st-century nanotechnological revolution. This is because they possess many remarkable properties. For example, it is estimated that they are 100 times stronger than steel at only one-sixth of the mass, they conduct electricity better than copper and transmit heat better than diamond.

Possible applications for nanotubes include flat-screen LCD screens, hydrogen storage, drug delivery, catalysts, memory-storage devices and many more yet to be discovered.

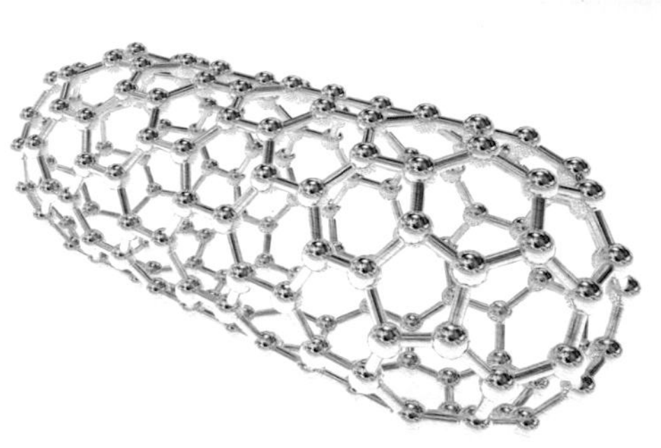

Fig 7.25 Carbon nanotubes

STRETCH AND CHALLENGE

Semiconductors

The biggest use for silicon is in semiconductor components for electronic circuits. You may wonder how it is that a non-metallic element with a diamond-like structure can be made to conduct electricity. **Semiconductors** are materials that, to put it simply, have been altered to allow electrons to pass through them, but not as easily as through metals. We know that metals conduct because they have a sea of delocalised electrons and that an insulator such as diamond does not conduct because all its electrons are locked into covalent bonds.

In silicon, it is possible to excite electrons into vacant higher energy orbitals. When this happens, these electrons become free to move. But at room temperature there is not sufficient energy to promote more than a small fraction of electrons and therefore the electrical conductivity stays exceedingly low.

A way to improve the electrical conductivity of silicon is to provide extra electrons that are easier to excite. This improvement is made by adding trace amounts of an element such as arsenic, a process called **doping**. The added element is called a **dopant**. Doping is a powerful way to increase the electrical conductivity of certain materials in the solid state.

Boron is also used to dope silicon, but the mechanism is not the same as for arsenic. With boron, the number of electrons is reduced, which causes the formation of electron vacancies, known as **holes**. This encourages electrons to move to fill the holes, and thereby generates new holes. These holes are filled by other electrons moving in, and so on. There is thus a flow of charge through the silicon and so it conducts electricity.

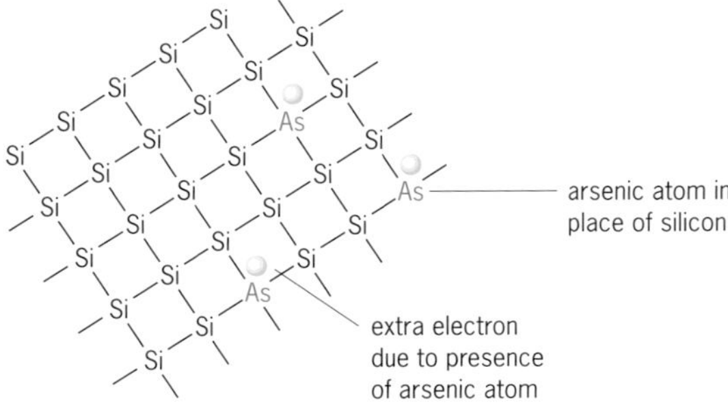

Fig 7.26 n-type silicon

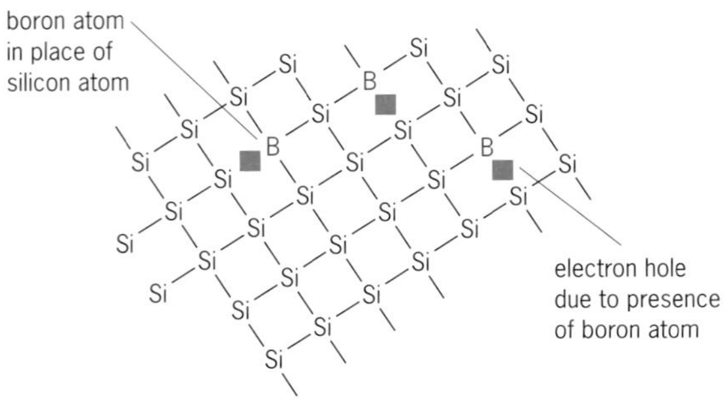

Fig 7.27 p-type silicon

SCIENCE IN CONTEXT

The solar cell

The Earth receives more energy in sunlight in two days than is stored in all the known energy resources. So, if just a fraction of this energy could be harnessed it would provide a major alternative supply of electricity. Many solar energy plants already exist around the world, but many more are needed if they are to become a real alternative to the other energy sources.

Solar cells are used to convert radiant energy into electrical energy. One form of solar cell consists of two joined layers of silicon. One layer is doped with arsenic or phosphorus and is called n-type silicon (arsenic and phosphorus each have five electrons in their outer shells). This layer has *extra* electrons available for conduction, generated by the small number of dopant atoms. The other layer is doped with boron and is called p-type silicon. This layer has a *shortage* of electrons (a surplus of holes) because boron has only three outer electrons. The cell is connected into an external circuit.

Electrons flow from the n-type to the p-type layer. Eventually, equilibrium is established with a potential difference between the two layers. When sunlight falls on the cell, the equilibrium is disturbed and electrons move from the p-type to the n-type layer. The electrons return to the p-type layer via the external circuit, and thereby generate an electric current.

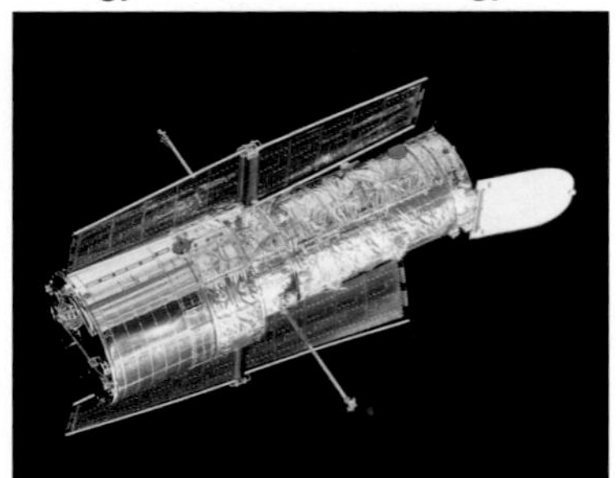

Fig 7.28 Solar cells on the Hubble Telescope

Fig 7.29 Solar cells on a roof

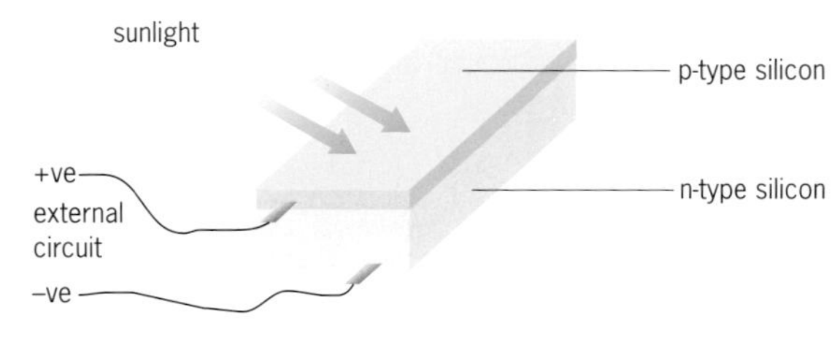

Fig 7.30 A solar cell

6 SIMPLE-MOLECULE LATTICES

Some non-metals, such as solid nitrogen, oxygen, chlorine and iodine, form crystal lattices that consist of the simple molecules (N_2, O_2, Cl_2, I_2) held in place by attractive forces between the molecules. These forces are known as **intermolecular forces** because they are forces between molecules rather than within a molecule. The intermolecular forces are weak between simple molecules, and therefore the non-metals mentioned have relatively low melting points and boiling points. (The intermolecular forces are often referred to as **van der Waals** forces.) These weak forces explain why many non-metals are gases at room temperature and atmospheric pressure. It is easy to separate the molecules, but difficult to atomise them, because the covalent bonds within each molecule (the intramolecular bonds) are so much stronger.

Elements with a simple molecular structure do not conduct electricity, since they do not have electrons or ions that are free to move. All the electrons are localised.

DIATOMIC MOLECULES

As mentioned before, many gases, including chlorine, oxygen, nitrogen and hydrogen, form diatomic molecules. They cannot have a permanent dipole as both atoms must have equal electronegativity. So it is difficult to imagine forces of attraction between these neutral molecules.

An induced dipole–induced dipole attraction is set up between molecules. A temporary dipole is set up in a *molecule* when the electrons in the covalent bond that links the atoms move more to one end of the molecule than to the other. It is important to realise that the dipole is temporary and the electrons may also move more to the other end of the molecule. As a result of the temporary dipole, one end of the molecule carries a very small positive charge and the other end a very small negative charge. This small charge separation is not permanent and so in another instant in time the charges can switch ends. Look at Fig 7.31, which shows the way an intermolecular force may be set up in a hydrogen molecule. Once a temporary dipole is set up within a molecule, it induces a similar dipole in neighbouring molecules and a very weak intermolecular force is set up.

The intermolecular force is known as an **induced dipole–induced dipole interaction**. When the temperature is sufficiently low, this force may be strong enough to hold the hydrogen molecules in a solid crystalline lattice.

The melting point of hydrogen is very low, since the two electrons in a hydrogen molecule are firmly attracted to the nuclei of their atoms. The larger the molecule, the stronger the intermolecular force, because there are more electrons and the outer electrons are under a weaker attractive force from the nuclei. So it is easier to produce a temporary dipole and the induced dipole–induced dipole attraction becomes stronger. In Group 7 this is illustrated by the change in physical state at room temperature and pressure. Fluorine, F_2, a small molecule with few electrons, has weak induced dipole–induced dipole attractions, whereas bromine, Br_2, a liquid, has more electrons, a larger molecule and stronger induced dipole–induced dipole attractions. In the case of iodine, I_2, the strength of the intermolecular force is sufficient to make iodine a solid at room temperature.

It is important to understand that in these types of simple molecular crystal lattice it is attraction *between molecules* that holds the lattice together. The atoms within the molecule are covalently bonded to one another, but the strength of this covalent bonding does not determine the magnitude of the melting point.

? QUESTION 10

10 The electrical conductivity of silicon increases as the temperature increases. Explain this property in terms of the ease with which electrons can be excited.

? QUESTION 11

11 **a)** What is the number of electrons in the outer shell of a silicon atom and of an arsenic atom?
b) Explain how arsenic can provide extra electrons when added in trace amounts to silicon.

? QUESTION 12

12 **a)** Write down the electron configuration for boron.
b) Explain how adding trace amounts of boron to silicon can produce electron holes.

A temporary dipole can also be known as an instantaneous dipole.

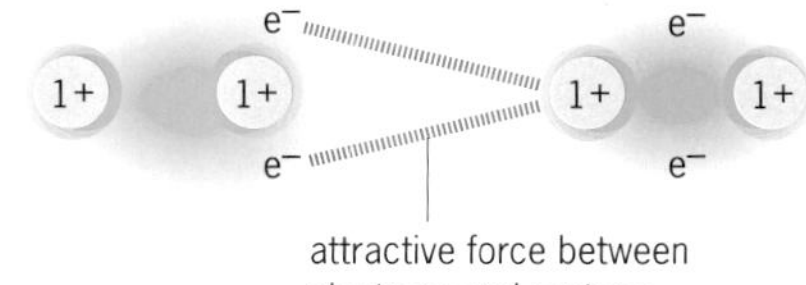

Once an instantaneous dipole forms, attractive forces distort the electron cloud on a neighbouring molecule, producing a temporary induced dipole.

The distortion (greatly exaggerated in the figure) results in only extremely weak forces of attraction.

Fig 7.31 The formation of induced dipoles in a hydrogen molecule

? QUESTION 13

13 Which halogen is likely to have the highest melting point? Explain your answer.

? QUESTION 14

14 Suggest why phosphorus and sulfur are solids at room temperature and atmospheric pressure, but nitrogen and oxygen are gases.

PHOSPHORUS AND SULFUR

Phosphorus and sulfur both form simple molecular lattices, but a discussion of the structures is complicated since both elements exhibit allotropy. Suffice to say that phosphorus forms a lattice in which P_4 molecules are arranged in a fixed pattern, while sulfur has a molecule with the formula S_8. The P_4 and S_8 molecules are held in position in their respective lattices by weak induced dipole–induced dipole interactions.

Phosphorus and sulfur are both solids at room temperature and atmospheric pressure because the induced dipole–induced dipole interactions are much stronger than those in the diatomic molecules of nitrogen and oxygen, since there are more electrons in molecules of phosphorus and sulfur. Sulfur has a higher melting point than phosphorus because it has larger molecules with more electrons, so the intermolecular force is stonger.

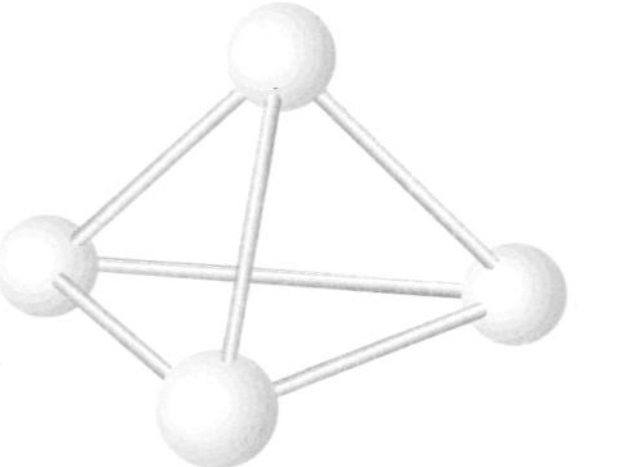

Fig 7.32 The P_4 molecule is a regular tetrahedron

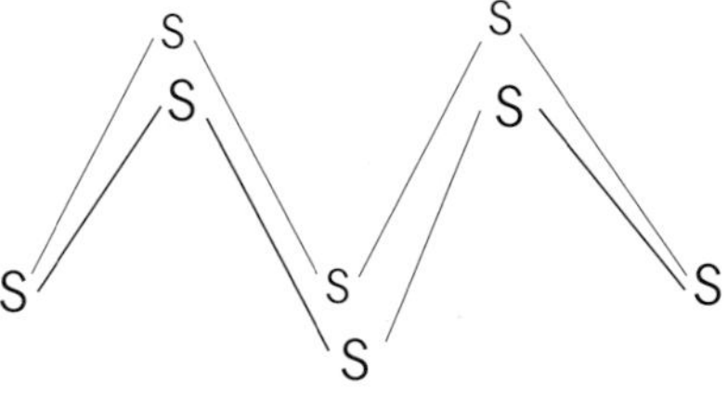

Fig 7.33 The puckered ring of S_8. All S–S lengths are identical

THE NOBLE GASES

The noble gases exist as monatomic molecules; that is, just as atoms. All the noble gases have low melting points. Helium has the lowest melting point of any element, and therefore it must have the weakest intermolecular forces between its atoms.

The electrons in an atom are in constant motion and do not occupy set positions or set orbits (see page 58). Therefore, it is possible for both of the electrons in a helium atom to arrive simultaneously on the same side of the atom, as in Fig 7.34. This side of the helium atom thus becomes very slightly negatively charged, while the other side becomes very slightly positively charged.

The electron cloud around the helium nucleus becomes distorted (it is no longer spherical), and it causes a similar distortion in neighbouring helium atoms. In other words, a temporary or instantaneous dipole is induced in neighbouring atoms.

A very weak attraction then exists between the slightly positive side of one helium atom and the slightly negative side of a neighbouring helium atom. This is a very weak intermolecular force because it relies on an asymmetric distribution of electrons, which has a low probability since both electrons are close to the nucleus, so are firmly attracted to the nucleus. At the same time, they repel each other.

Once this force between helium atoms in the solid phase has been broken, the attraction between helium atoms in the liquid phase is negligible, and so helium boils only 4 degrees higher than the temperature at which it melts.

The larger the noble gas atom (hence the greater the number of electrons), the greater the likelihood of this asymmetric distribution of electrons. This is because the outer electrons are further away from the nucleus and are shielded from the nuclear charge. This means that the outer electrons can occupy a larger region and so undergo a smaller electron–electron repulsion, even when asymmetrically distributed. Therefore, the melting and boiling points of the noble gases increase with increasing atomic (proton) number.

e⁻ 2+ e⁻ 2+ e⁻ e⁻

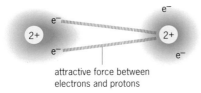

attractive force between electrons and protons

An instantaneous dipole forms on the left atom, causing an attractive force on the nucleus of the right atom, producing a temporary induced dipole.

The distortion (greatly exaggerated in the figure) results in only extremely weak forces of attraction.

Fig 7.34 The formation of induced dipoles in helium

? QUESTION 15

15 Which of the noble gases has the highest melting point?

7 A SUMMARY OF THE STRUCTURE AND BONDING OF ELEMENTS IN PERIOD 3

The physical properties of elements in Periods 2 and 3 vary considerably. Tables 7.1 and 7.2 summarise their physical properties.

Table 7.1 Summary of the physical properties of Period 2 elements

	Lithium	Beryllium	Boron	Carbon	Nitrogen	Oxygen	Fluorine	Neon
Bonding	metallic	metallic	covalent	covalent	covalent	covalent	covalent	–
Structure	giant	giant	giant	giant	simple	simple	simple	simple
Melting point	low	high	very high	very high	very low	very low	very low	very low
Electrical conductivity of solid	high	high	low	low*	low	low	low	low

*high for graphite but low for diamond

Table 7.2 Summary of the physical properties of Period 3 elements

	Sodium	Magnesium	Aluminium	Silicon	Phosphorus	Sulfur	Chlorine	Argon
Bonding	metallic	metallic	metallic	covalent	covalent	covalent	covalent	–
Structure	giant	giant	giant	giant	simple	simple	simple	simple
Melting point	low	high	high	very high	low	low	very low	very low
Electrical conductivity of solid	high	high	high	low	very low	very low	very low	very low

SUMMARY

After studying this chapter, you should know the following.

- A crystalline solid has a regular arrangement of particles called a lattice that can be produced by the repetition in three dimensions of a unit cell. Crystalline solids have definite melting points.
- An amorphous solid has a closely packed structure that is disordered, and therefore does not have a unit cell. Amorphous solids melt over a range of temperatures, and many soften on heating.
- The structure and bonding within a crystal lattice determine the physical properties of an element.
- The melting point of a solid is determined by the strength of the force of attraction between particles in a lattice.
- A metal consists of closely packed positive ions in a sea of delocalised electrons. It has a giant structure.
- Metals conduct electricity because of the movement of the delocalised electrons in the presence of a potential difference.
- Metals are typically hard, strong, shiny and good thermal conductors. Many have high melting points. These properties can be explained by the structure and bonding.
- Non-metals form molecular lattices.
- Non-metals that form a molecular lattice have a regular arrangement of simple molecules held together by weak intermolecular forces, called van der Waals forces (induced dipole–induced dipole interactions). Such non-metals have low melting points and low boiling points and do not conduct electricity.
- The strength of the induced dipole–induced dipole interaction in a diatomic element depends on the molecular size (indicated by the number of electrons in the molecule).
- A non-metal that forms a giant molecular or macromolecular lattice has a regular arrangement of atoms held together by strong covalent bonds. The whole crystal is assumed to be the molecule. Such non-metals have high melting points.
- Non-metals are generally poor electrical conductors because they have no free electrons.
- Different forms of the same element in the same state are known as allotropes. Allotropes have different physical properties, but similar chemical properties.
- Diamond, graphite and fullerene (C_{60}) are three allotropes of carbon. Carbon and diamond have a giant molecular structure. Graphite has mobile delocalised electrons, so it conducts electricity. Diamond does not conduct electricity.

Practice questions and a How Science Works assignment for this chapter are available at www.collinseducation.co.uk/CAS

8

Fuels, energy changes and reaction rates

8 FUELS, ENERGY CHANGES AND REACTION RATES

Hydrogen fuel

Hydrogen explosion Hydrogen is dangerous to transport, as the *Hindenburg* disaster of 1937 showed. The *Hindenburg* was a hydrogen-filled airship that blew up, killing 35 people

World supplies of oil are likely to run out before the end of the 21st century, but by then other energy sources should have been developed. A possibility is hydrogen, derived from water. Hydrogen has several advantages: when it burns it produces no harmful products, it gives out more energy per gram than natural gas or petrol, and existing car engines need only minor modifications to use it as a fuel.

The reason we don't start splitting water now to obtain hydrogen is that the process requires energy. In the UK, this would have to come from burning fossil fuels, which would cancel out any environmental advantages. Alternatively, more efficient solar cells are under development using nanoparticle coatings to capture the energy from sunlight and produce an electric current. However, the cells are not yet efficient enough to make hydrogen economically. The possibility of producing hydrogen from the photosynthetic reactions of genetically modified algae is also being investigated, but so far the algae produce too little hydrogen to make this project commercially viable.

Hydrogen fuel Hydrogen has become a fuel in space exploration: the US Space Shuttle piggy-backs a lift on giant tanks that contain 1.74 million litres of liquid hydrogen and 0.65 million litres of liquid oxygen. This fuel supplies the shuttle's main engines

Another problem is how to carry hydrogen around, since to stay liquid and occupy a convenient volume it would have to be stored below −253 °C. One promising area of research is the use of metal hydrides (solid compounds of a metal and hydrogen): the hydrogen could be produced from them as needed by heating.

The BMW Hydrogen 7 car can run on petrol or hydrogen. The hydrogen is stored as a liquid in a thermally insulated tank

WHAT IS A FUEL?

A fuel is a substance that releases energy that can do work. Most fuels release this energy during combustion reactions. So what properties make an ideal fuel? Petrol is a popular fuel, but is it ideal?

For its mass, petrol produces a lot of energy, which makes the expense of transporting and storing it worthwhile. It does not produce solid waste that needs to be dumped. But an ideal fuel produces no harmful by-products when it burns, and petrol does. Even if the by-products were just carbon dioxide and water, we now know that carbon dioxide is a major greenhouse gas and its continued output at today's levels is leading to global warming.

? QUESTION 1

1 Magnesium produces energy on burning. Why don't we class magnesium as a fuel?

You can read more about greenhouse gases and some of the other harmful by-products of using petrol as a fuel on pages 199–209.

1 HOW A FUEL RELEASES ITS ENERGY: EXOTHERMIC REACTIONS AND ENTHALPY CHANGES

When natural gas burns it releases its energy to the surroundings, which could be the water in a central heating system. This is an **exothermic**

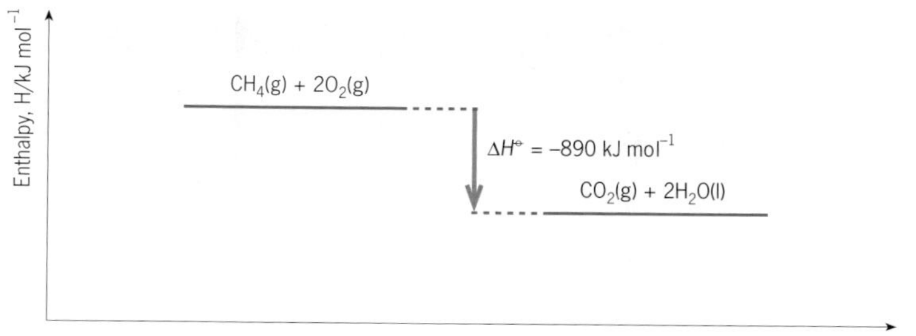

Fig 8.1 Enthalpy level diagram for methane burning

reaction between methane (in natural gas) and oxygen.

$$CH_4(g) + 2O_2(g) \rightarrow CO_2(g) + 2H_2O(l)\ \Delta H^{\ominus}_{298} = -890\ \text{kJ mol}^{-1}$$

The stored energy of the reactants, methane and oxygen, is higher than the stored energy of the products of the reaction, carbon dioxide and water. The difference in energy is released to the surroundings when methane and oxygen react. We can see this in Fig 8.1.

STANDARD CONDITIONS

The symbol $^{\ominus}$ tells us that the conditions for this enthalpy change are standard conditions at a particular temperature. Standard conditions are 100 kPa (1 atmosphere) pressure and, for solutions, a concentration of 1 mol dm^{-3}. The particular temperature for these standard conditions is usually 298 K (25 °C). A small 298 following $\Delta H^{\ominus}$ (pronounced 'delta H standard') shows that the standard temperature is used. The substances in the chemical reaction (the reactants and products) are known as the **system**, and all substances in the system are in their **standard states**. The standard state is the most stable physical state at 1 atmosphere pressure and the particular temperature, usually 298 K (25 °C). So the standard state of methane is a gas and the standard state of water is a liquid.

Standard enthalpy change of a reaction is the enthalpy change under standard conditions at 298 K and 100 kPa, with all substances in their standard states.

This refers to the mole quantities given in the balanced equation for the reaction.

The **surroundings** are everything that is not the system. It could be the reaction vessel, the air around the reaction vessel, the liquid in a thermometer and anything else that is not a reactant or product. But the surroundings must be associated thermally with the system, which means that heat can transfer between system and surroundings.

2 STANDARD ENTHALPY CHANGE OF COMBUSTION

Methane burning in oxygen is called a **combustion reaction**. The enthalpy change is an enthalpy change of reaction, but if we burn one mole of methane completely in oxygen it can also be referred to as an **enthalpy change of combustion**.

Standard enthalpy change of combustion $\Delta H^{\ominus}_{c,298}$ is the enthalpy change when one mole of a substance burns completely in oxygen under standard conditions (298 K and 100 kPa) with all reactants and products in their standard state.

REMEMBER THIS

Enthalpy, H, is the stored energy in a compound. Enthalpy changes are transfers of energy into or out of the system at constant pressure. (We first met these terms on page 11.)

ΔH (pronounced 'delta H') is the **enthalpy change** of the reaction. It is measured under conditions of constant pressure. In exothermic reactions, ΔH is negative. In endothermic reactions, ΔH is positive.

kJ mol^{-1} means the energy released per mole of equation. This means the mole quantities given in the equation release so many kilojoules.

REMEMBER THIS

State symbols should always be shown in chemical equations when dealing with energy from chemical reactions. The enthalpy change responsible for gaseous water (steam) being produced is not the same as the enthalpy change responsible for liquid water being produced.

State symbols are:
(s) = solid
(l) = liquid
(g) = gas
(aq) = aqueous

REMEMBER THIS

You will sometimes see the standard enthalpy change of reaction written as the **standard molar enthalpy change of reaction**. Similarly, the standard enthalpy change of combustion is some times called the **standard molar enthalpy change of combustion**.

Some definitions of standard enthalpy change use a standard pressure of 101.3 kPa.

? QUESTION 2

2 Why do we need to include the state symbols in the equation in the Example below? (Hint: the answer is on page 153.)

See page 198 to find out more about the composition of petrol.

? QUESTION 3

3 Draw energy level diagrams for the standard enthalpy changes of combustion of hydrogen and propane.

$\Delta H^{\ominus}_{c}$ (H_2) = –286 kJ mol^{-1}
$\Delta H^{\ominus}_{c}$ (C_3H_8) = –2219 kJ mol^{-1}

So the standard enthalpy change of combustion of methane is written:

$$\Delta H^{\ominus}_{c,298}\ (CH_4) = -890\ \text{kJ mol}^{-1}$$

and because the standard temperature is 298 K, we often write:

$$\Delta H^{\ominus}_{c} = -890\ \text{kJ mol}^{-1}$$

In the definition for standard enthalpy change of combustion we say *burns completely* to show clearly that only carbon dioxide and water are formed. With not enough oxygen, there may be incomplete combustion, which would produce carbon monoxide or even carbon (soot). The enthalpy changes for both these reactions will affect any measurement of enthalpy change of combustion.

In practice, when methane burns in oxygen, the temperature is much higher than 25 °C, so adjustments must be made to calculate the value under standard conditions.

EXAMPLE

Q Explain fully what is meant by:

$$\Delta H^{\ominus}_{c,298}\ (H_2) = -286\ \text{kJ mol}^{-1}$$

A This is the standard enthalpy change of combustion of hydrogen. It is the enthalpy change when one mole of hydrogen is completely burnt in oxygen under standard conditions of 100 kPa pressure and 298 K, with reactants and products in their standard states at this temperature.

$$H_2(g) + \tfrac{1}{2}O_2(g) \rightarrow H_2O(l)\quad \Delta H^{\ominus}_{c} = -286\ \text{kJ mol}^{-1}$$

Notice that this includes the state symbols. Also, $\tfrac{1}{2}O_2$ does not mean half a molecule of oxygen, but half a mole.

We could have written:

$$2H_2(g) + O_2(g) \rightarrow 2H_2O(l)\quad \Delta H^{\ominus}_{r} = -572\ \text{kJ mol}^{-1}$$

The standard enthalpy change of this reaction, $\Delta H^{\ominus}_{r}$ (small r for reaction; or just $\Delta H^{\ominus}$) is twice that for the standard enthalpy change of combustion of hydrogen. This is because ΔH_r applies to the mole quantities in the equation – 2 moles of hydrogen – whereas the standard enthalpy change of combustion is for burning *one* mole of hydrogen.

ALTERNATIVE FUELS TO PETROL

Table 8.1 gives the standard enthalpy changes of combustion of some fuel alternatives to petrol. There is no standard enthalpy change for petrol because it is a complex mixture of about 100 compounds, mainly hydrocarbons, of which most are alkanes (see page 192).

The standard enthalpy change of combustion of octane, one of the constituents of petrol, is –5470 kJ mol^{-1}, which is a good indication of why petrol is used in cars.

Table 8.1 The standard enthalpy changes of combustion of fuel alternatives to petrol

Fuel	Main constituent	Formula and standard state	$\Delta H^{\ominus}_{c}$ of main constituent (kJ mol^{-1})
hydrogen	hydrogen	$H_2(g)$	–286
compressed natural gas (CNG)	90% methane	$CH_4(g)$	–890
liquid petroleum gas (LPG)	95% propane	$C_3H_8(g)$	–2219
methanol	methanol	$CH_3OH(l)$	–726
ethanol	ethanol	$C_2H_5OH(l)$	–1367

SCIENCE IN CONTEXT

Biofuels

Biofuels are fuels that are made from the products of living things, usually plants. Biodiesel and bioethanol are two biofuels whose production is increasing rapidly, as countries seek alternatives to fossil fuels to power vehicles and reduce carbon dioxide emissions. By 2010 the British government requires that 5 per cent of all fuel sold at the pump must be biofuels. The United States is also intending to dramatically increase its use of biofuels in vehicle transport.

Sugar cane and wheat can be fermented to produce ethanol, which is blended with petrol or used on its own. There is enough fuel in a field of wheat the size of a football pitch to run a family car for 12 000 miles. One source of biodiesel is oilseed rape.

Growing crops for fuel is the subject of heated debate. Some people say biofuels are the answer to increasing levels of carbon dioxide in the atmosphere: the crops remove carbon dioxide from the air to make sugars and oils, and the gas returns to the atmosphere when fuels made from the crops are burnt, so there is no net carbon dioxide increase. This means that biofuels could be called **carbon neutral.** But other analysts point out that this ignores the fossil fuels that must be burnt to provide the fuel for planting making the fertiliser, and for harvesting, transporting and processing the crop. In this case biofuels are definitely not carbon neutral.

Fig 8.2 The oil from this field of oilseed rape can be converted into biodiesel fuel

Fig 8.3 Brazil pioneered the use of ethanol as a vehicle fuel. It is made from the fermented and distilled juice of sugar cane

There are also concerns that using agricultural land to grow fuel crops will lead to less food crops and therefore food shortages in some parts of the world.

3 MEASURING ENTHALPY CHANGES

The process of measuring energy changes is known as **calorimetry**, after the old unit for heat, a calorie. Any vessel that is used to measure enthalpy changes is known as a **calorimeter**.

ENTHALPY CHANGE OF COMBUSTION

See Table 8.1 for some enthalpy changes of combustion. Being combustion reactions, they are all exothermic. We can find these values by measuring the increase in temperature of a certain mass of water to which energy from the combustion reaction is transferred, making its temperature rise. One gram of water requires 4.2 joules of energy to raise its temperature by 1°C; this is water's **specific heat capacity**.

Specific heat capacity (c) is the energy required to raise the temperature of 1 gram of substance by 1 K. It is measured in J g^{-1} K^{-1}.

So the specific heat capacity of water is 4.2 J g^{-1} K^{-1} (joules per gram per kelvin).

We can use a simple equation to work out the energy transferred to the water:

Energy transferred, q =	m	×	c	×	ΔT
	mass of water		specific heat capacity of water		temperature change
	g		J g^{-1} K^{-1}		K

REMEMBER THIS

A temperature rise of 1 K is the same as a temperature rise of 1 °C.

REMEMBER THIS

The specific heat capacity of water is unusually high. It takes energy to break the hydrogen bonds between clusters of water molecules. This means that the energy does not go towards increasing the kinetic energy of molecules and hence the temperature of the water. You can read about hydrogen bonding of water molecules on page 129 of Chapter 6.

QUESTION 4

4 How many kilojoules of energy will be required to raise the temperature of 1 kg of water by **a)** 1°C, **b)** 20°C?

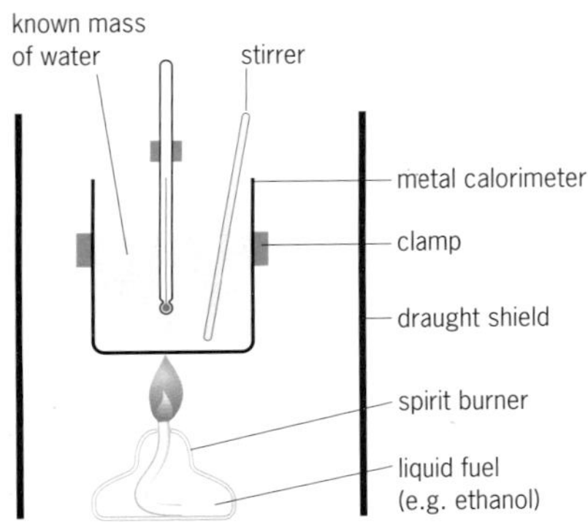

Fig 8.4 Measuring the enthalpy change of combustion of a liquid fuel

SIMPLE DETERMINATION OF THE ENTHALPY CHANGE OF COMBUSTION OF A LIQUID FUEL

A simple experimental set-up to measure the enthalpy change of combustion of ethanol is shown in Fig 8.4. The experiment is carried out as follows:

- **Step 1.** Weigh the spirit burner with ethanol in it at the beginning of the experiment and take the water temperature.
- **Step 2.** Allow the spirit burner to heat up a known mass of water by about 20 °C. Stop heating and take the temperature of the water. This gives the temperature rise in Kelvin.
- **Step 3.** Weigh the spirit burner again. This gives the mass of the ethanol used.

The next Example explains how this experiment may be used to calculate the enthalpy change of combustion for ethanol.

? QUESTION 5

5 Read the method for measuring enthalpy change of combustion and examine Fig 8.4.
a) Why is the calorimeter made of metal and not glass?
b) Why does a temperature rise in °C give you the temperature rise in Kelvin?
c) The draught shield is used to reduce loss of energy to the surroundings. What other things could be done to minimise energy losses in this experiment?

? QUESTION 6

6 The experiment in the Example gives an inaccurate result. Compare the value of the enthalpy change of combustion of ethanol in the Example with the value given in Table 8.1, on page 154. What aspects of the experiment lead to such a low calculated value?

EXAMPLE

Q 0.40 g ethanol raises the temperature of 100.00 g water in the metal calorimeter by 21.0 °C. Calculate the enthalpy change of combustion of ethanol.

A **Step 1.** Work out the energy transferred to the water:

Energy transferred to the water, $q = mc\Delta T$

$= 100.00 \times 4.2 \times 21.0 = 8820\text{ J} = 8.82\text{ kJ}$ (ignoring significant figures)

Step 2. Calculate the amount of fuel used in moles:

Mass of 1 mole of ethanol (C_2H_5OH)

$= (12 \times 2) + (1 \times 5) + 16 + 1 = 46\text{ g}$

$$\text{Moles of ethanol in } 0.40\text{ g} = \frac{\text{mass in grams}}{\text{mass of one mole (in grams)}}$$

$$= \frac{0.40}{46} = 0.0087\text{ mol (correct to 2 sig. figs)}$$

Step 3. Work out the enthalpy change of combustion, that is, the energy transferred when 1 mole of ethanol burns:

0.0087 mol ethanol releases 8.82 kJ

$$1\text{ mole of ethanol releases } 8.82 \times \frac{1}{0.0087} = 1014\text{ kJ}$$

So the enthalpy change of combustion of ethanol = -1000 kJ mol^{-1} (correct to 2 sig. figs)

STRETCH AND CHALLENGE

The bomb calorimeter

The bomb calorimeter (Fig 8.6) measures enthalpy changes of combustion. The weighed sample is inside a stainless steel container – the bomb – filled with oxygen under pressure, and the fuel is ignited electrically.

The principle is the same as for the simplified experiment on page 156. Energy is transferred from the combusted fuel to the surrounding water, and the temperature rise is measured. However, this apparatus gives a much more accurate value than the simple version, provided the readings are taken quickly, because energy losses to the surroundings are reduced to almost zero. In our simplified experiment on page 156, another potential error arises when the spirit burner is weighed, as some of the ethanol may evaporate between weighings.

Enthalpy change is transfer of energy into or out of the system *at constant pressure*. As the bomb has a constant *volume*, it does not measure enthalpy change exactly. However, because the difference is small, we need not be concerned with the mathematical correction used to adjust the value to constant pressure.

The bomb calorimeter (Cont.)

THE ENERGY VALUES OF FOOD

The bomb calorimeter is used to measure the enthalpy changes of combustion of different foods. Although our bodies don't burn foods in quite the same way as we burn natural gas in a cooker, the outcome is the same. Oxygen is still required and the energy we obtain from compounds in foodstuffs is the same as if they were burnt in a bomb calorimeter. The crucial difference is that when we burn fuels in cooking or in the combustion engine of a car, most of the released energy is wasted and goes to heat up the surroundings.

When we 'burn' glucose in our bodies, there is not just one high-temperature reaction, but several small steps with small transfers of energy, each step catalysed by complex molecules called enzymes. The human machine is nowhere near 100 per cent efficient, and we do lose energy to our surroundings (hence, our body warmth), but much of the energy of the step-wise reactions is either used to maintain electrical and chemical body functions or stored, and so does not emerge as heat. Of course, when we store energy it is usually as fat, so taking in more energy than we require will make us obese.

QUESTION 7

7 Look back to page 156 and the enthalpy of combustion experiment.

a) i) If some of the ethanol evaporates between weighings and so is not burnt, would this lead to a higher or a lower value for the calculated enthalpy change of combustion?

ii) How could the weighings be carried out in the bomb calorimeter experiment to avoid any losses through evaporation?

b) i) Study Fig 8.6, which shows the bomb calorimeter, and explain the features of the apparatus that minimise energy loss to the surroundings.

ii) What is the purpose of the stirrer?

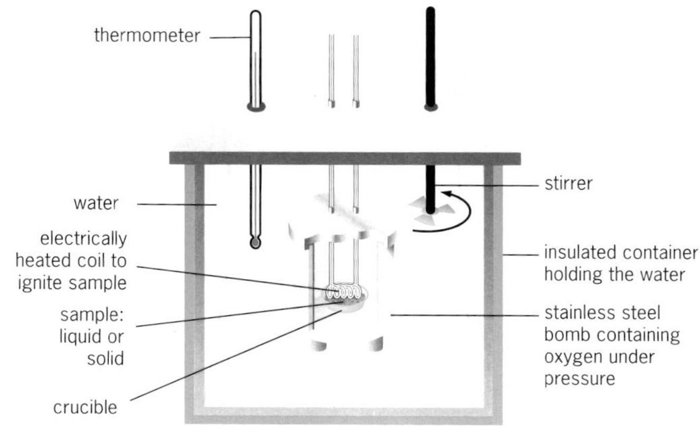

Fig 8.6 The bomb calorimeter. The sample is ignited electrically and as it burns the energy is transferred to the water with minimal loss. The thermometer measures the rise in temperature

ENTHALPY CHANGE OF NEUTRALISATION

Many reactions take place in aqueous solution and the enthalpy changes of these reactions can be measured using an expanded polystyrene cup fitted with a lid (Fig 8.5). Expanded polystyrene is a very good insulator and also absorbs little energy from the reaction itself.

When an acid reacts completely with an alkali the reaction is called a **neutralisation**. The reaction between aqueous sodium hydroxide and aqueous hydrochloric acid can be represented by the equation:

$$HCl(aq) + NaOH(aq) \rightarrow NaCl(aq) + H_2O(l)$$

In this reaction all the reactants are present as ions, so:

$$H^+(aq) + Cl^-(aq) + Na^+(aq) + OH^-(aq) \rightarrow Na^+(aq) + Cl^-(aq) + H_2O(l)$$

You will notice that the ions reacting are only $H^+(aq)$ and $OH^-(aq)$, while $Na^+(aq)$ and $Cl^-(aq)$ take no part in the reaction and are called **spectator ions**. So the neutralisation reaction is more simply written as

$$H^+(aq) + OH^-(aq) \rightarrow H_2O(l)$$

So standard enthalpy change of neutralisation is defined as:

The enthalpy change when an acid and a base react to form one mole of water under standard conditions (298 K, 100 kPa).

The experiment to measure this enthalpy change is carried out as follows:

- **Step 1.** Measure out accurately a known volume and amount of aqueous acid and take the temperature of the solution.

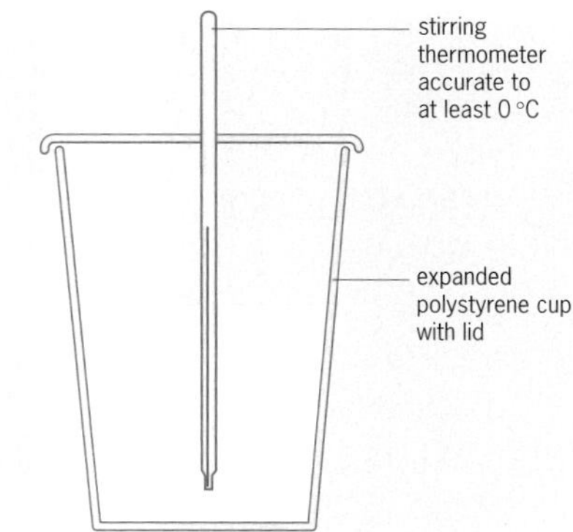

Fig 8.5 A simple calorimeter to measure the enthalpy change of neutralisation

REMEMBER THIS

The burning of fuels and the oxidation of carbohydrates, such as glucose, in respiration, are important exothermic processes.

- **Step 2.** Mix the acid from Step 1 with a known volume and amount of alkali in a polystyrene cup and stir the mixture thoroughly.
- **Step 3.** Calculate the temperature rise. This is usually done by taking the temperature of the alkali in the polystyrene cup each minute for four minutes. On the fifth minute the acid is added, the mixture stirred and then the temperature measured again every minute from the sixth minute until the 12th minute. A graph is drawn (Fig 8.7 overleaf) and the temperature rise as the acid is added to the alkali is extrapolated.

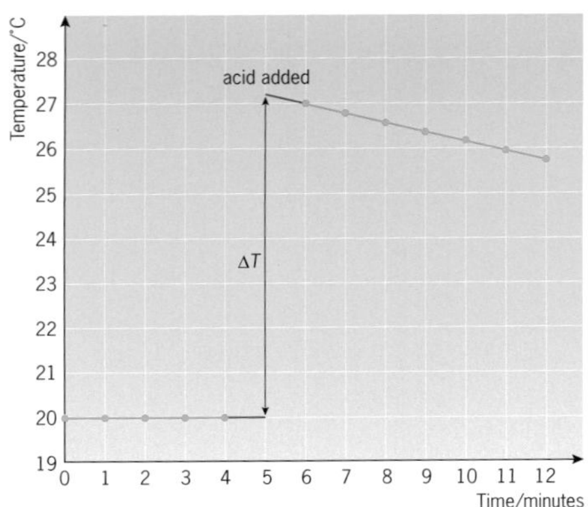

Fig 8.7 Graph of temperature against time plotted in an experiment to measure the enthalpy change of a neutralisation reaction

EXAMPLE

Q An experiment to measure the enthalpy change of neutralisation is carried out. When 50.0 cm^3 HCl(aq) is added to 50.0 cm^3 NaOH(aq), both of 1.0 mol dm^{-3} concentration, the temperature rise is 6.0 °C. Calculate the enthalpy change of neutralisation.

A **Step 1.** Work out the energy transferred to the water. The acid and alkali are very dilute so the volume of water is assumed to be:

$$50.0 \text{ cm}^3 \text{ HCl(aq)} + 50.0 \text{ cm}^3 \text{ NaOH(aq)} = 100.0 \text{ cm}^3$$

The mass of 100 cm^3 water = 100 g (1.0 cm^3 water has a mass of 1.0 g)

Energy transferred to the water, q = $mc\Delta T$

$= 100.0 \times 4.2 \times 6.0 = 2520$ J = 2.52 kJ (ignoring significant figures)

Step 2. Calculate the amount of acid and alkali used in moles:

Number of moles NaOH in 1000 cm^3 = 1.0 mol

So number of moles NaOH (and also HCl) in 50.0 cm^3 = $1.0 \times \dfrac{50.0}{1000}$

= 0.050 mol

To remind yourself how to do this calculation see page 130.

Step 3. Work out the enthalpy change of neutralisation, that is, the energy transferred when 1 mol of water is formed from 1.0 mol NaOH and 1.0 mol HCl.

When 0.050 mol of water is formed the energy released = 2.52 kJ

1 mole of water being formed releases $2.52 \times \dfrac{1}{0.050} = 50.4$ kJ

So enthalpy change of neutralisation = –50 kJ mol^{-1} (correct to 2 sig. figs)

The data book value $\Delta H^{\circ}_{\text{neut}} = -57.2$ kJ mol^{-1}, so heat energy was lost to the surroundings in this experiment.

REMEMBER THIS

Neutralisation is an exothermic reaction.

? QUESTION 8

8 Calculate a value for the enthalpy change of neutralisation when 25.0 cm^3 of 1.0 mol dm^{-3} H_2SO_4(aq) is neutralised by 50.0 cm^3 NaOH(aq). The temperature rise on mixing is 8.2 °C. (Hint: remember 1 mol H_2SO_4 produces 2 mol H^+(aq).)

REMEMBER THIS

When measuring the enthalpy change of a reation that takes place in solution we make two asssumptions:

i) The specific heat capacity of the solution is the same as water (4.2.J g^{-1}K^{-1})

ii) The density of the solutions is the same as that of water, i.e. 1cm^3 of solutions weight 1g.

REMEMBER THIS

The combustion of fuels is an oxidation reaction. Combustion reactions are exothermic.

4 RELEASING ENERGY FROM FUEL MOLECULES

We know that fuel molecules need oxygen to combust and so release energy. Where does this energy come from? To answer this question we need to look at the energies of the bonds that hold atoms together, and what happens when these bonds are broken and made.

BREAKING BONDS: BOND ENTHALPY

Bond enthalpy or **bond energy** is the energy required to break a bond between two atoms in a gaseous molecule. However, because the energy required to break one bond is so small, we define bond enthalpy as the energy required to break *one mole* of bonds. Thus:

Bond enthalpy, *E*, is the energy required to break one mole of bonds of the same type in gaseous molecules under standard conditions (298 K, 100 kPa).

Consider hydrogen, which has an H—H bond:

$$\text{H—H(g)} \rightarrow \text{H(g)} + \text{H(g)} \quad \Delta H^\ominus = +436 \text{ kJ mol}^{-1}$$

Bond enthalpy, *E*, is often written as:

$$E(\text{H—H}) = +436 \text{ kJ mol}^{-1}$$

Notice that **bond breaking is an endothermic process**. This is because we are *putting in* energy to overcome the force of electrostatic attraction in the bond.

The bond energy in oxygen molecules is for breaking a *double* bond:

$$\text{O=O(g)} \rightarrow \text{O(g)} + \text{O(g)} \quad \Delta H^\ominus = +498 \text{ kJ mol}^{-1}$$

So $E(\text{O=O}) = +498 \text{ kJ mol}^{-1}$. This is larger than for the H–H bond, because double bonds are stronger than single bonds and so require more energy to be broken.

Up to now, we have considered bond energies only in *diatomic* molecules (molecules that contain two atoms). There are two O—H bonds in the triatomic water molecule. The energy required to break the first O—H bond is not the same as the energy required to break the second:

$$\text{H—O—H(g)} \rightarrow \text{H(g)} + \text{O—H(g)} \quad \Delta H^\ominus = +502 \text{ kJ mol}^{-1}$$

$$\text{O—H(g)} \rightarrow \text{O(g)} + \text{H(g)} \quad \Delta H^\ominus = +427 \text{ kJ mol}^{-1}$$

The first bond is in a molecular environment in which two O—H bonds exist, and more energy is required to break this bond. The second O—H bond is in a changed environment, and clearly this has an effect on its bond energy. A total of 929 kJ mol^{-1} is required to break both the O—H bonds in water, so we say the *average* or mean bond enthalpy for the O—H bond in this molecule is +464 kJ mol^{-1}.

We also find differences in the bond enthalpy of the O—H bond in other molecules because of the different molecular environments of the bond. For example, the bond enthalpy of the O—H bond in methanol is +437 kJ mol^{-1}. Usually, the differences in bond enthalpies are not very great and we average them to give a good approximation that we can use.

Table 8.2 Average bond enthalpies of some common bonds

Bond	E(X–Y)/ kJ mol^{-1}
C—C	+347
C=C	+612
C≡C	+838
C—H	+413
C—O	+358
C=O (in CO_2)	+805
H—H	+436
O—H	+464
O=O	+498

MAKING BONDS

So far, we have only considered *breaking* bonds, which requires energy, but we still have not answered

REMEMBER THIS

A covalent bond is the electrostatic attraction between the shared pair of electrons and the positive nuclei. The higher the bond enthalpy, the closer together the two nuclei are. The distance between the nuclei is called the **bond length.** Bonds with a high bond enthalpy are short and strong.

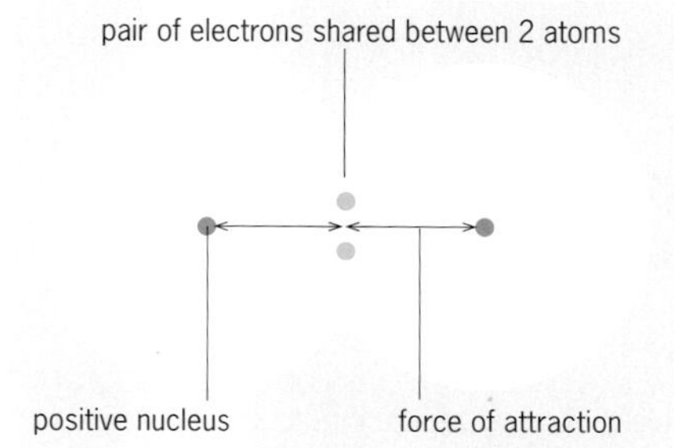

Fig 8.8 The force of attraction between the shared pair of electrons and the positive nuclei keep these two hydrogen atoms together. Breaking this covalent bond will require energy

? QUESTION 9

9 The nitrogen molecule is very unreactive. Why is this so? (Hint: if you are not sure, look at Fig 4.12, page 75).

? QUESTION 10

10 **a)** Two bond enthalpies in Table 8.2 are not average bond enthalpies. Identify these and explain your reasoning.

b) Use Table 8.2 to calculate the amount of energy to atomise one mole of gaseous methanol (convert the molecule into its separate, gaseous atoms).

$CH_3OH(g) \rightarrow C(g) + 4H(g) + O(g)$

c) Methanol is normally a liquid at 298 K. Explain why it must be a gas in this calculation. (Hint: what other energy is involved if methanol is not gaseous?)

? QUESTION 11

11 **a)** Compare the value given in the text for the enthalpy change for the combustion of hydrogen in oxygen to the enthalpy change of combustion of hydrogen in the Example on page 154. Explain the difference.

b) For endothermic reactions, what differences would you expect to see in the energy level diagram?

REMEMBER THIS

The enthalpy profile diagram (energy level diagram) may also be called a reaction pathway diagram.

? QUESTION 12

12 Methanol is a fuel used in motor racing. Use bond energy values in Table 8.2, page 159, to calculate the energy released when one mole of methanol is burnt.

$$CH_3OH(g) + 1\tfrac{1}{2}O_2(g) \rightarrow CO_2(g) + 2H_2O(g)$$

the question: where does the energy come from that is released when fuels burn in oxygen?

If it takes energy to break bonds, the same amount of energy must be released when bonds are formed. This reflects the **first law of thermodynamics** (also called the **law of conservation of energy**), thus:

Energy can neither be created nor destroyed.

This certainly applies to chemical reactions. Take hydrogen:

$$H\text{–}H(g) \rightarrow H(g) + H(g) \quad \Delta H^{\ominus} = +436 \text{ kJ mol}^{-1}$$

$$H(g) + H(g) \rightarrow H\text{–}H(g) \quad \Delta H^{\ominus} = -436 \text{ kJ mol}^{-1}$$

Note that **bond making is exothermic.**

Energy changes in bond breaking and making

We are now in a position to look at energy changes from bond breaking and bond making. For example, when hydrogen burns in oxygen:

$$H\text{–}H(g) \quad + \tfrac{1}{2}O\text{=}O(g) \rightarrow \qquad H\text{–}O\text{–}H(g)$$

$$E(H\text{–}H) \quad + \tfrac{1}{2}E(O\text{=}O) \rightarrow \quad 2 \times \; -E(O\text{–}H)$$

$$+436 \qquad + \tfrac{1}{2}(498) \qquad\qquad 2 \times (-464)$$

The enthalpy change for this reaction is:

$$\Delta H^{\ominus} = +436 + 249 + (-928) = -243 \text{ kJ mol}^{-1}$$

So:

The energy given out when fuels burn is the difference between the energy required to break bonds and the energy released when bonds are made.

We can show this on an enthalpy profile diagram (energy level diagram) such as Fig 8.9.

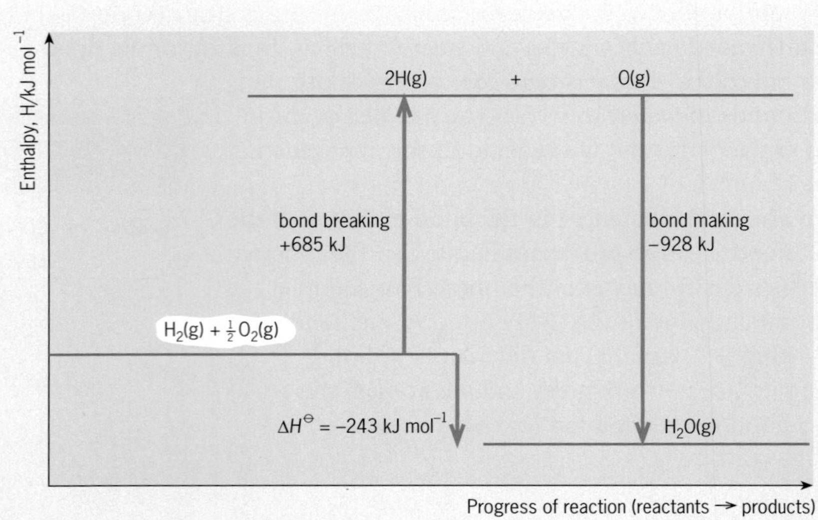

Fig 8.9 The enthalpy profile diagram (energy level diagram) for hydrogen burning in oxygen, showing the energy for breaking and making bonds and the enthalpy change of combustion

USING BOND ENERGIES TO ESTIMATE ENTHALPY CHANGES OF REACTION

Average bond energies can be used to estimate the enthalpy changes for different reactions.

EXAMPLE

Q Calculate the energy released when 1 mole of methane burns completely in oxygen using average bond energy data in Table 8.2 on page 159.

A

Step 1. Write down the equation for the reaction:

$$CH_4(g) + 2O_2(g) \rightarrow CO_2(g) + 2H_2O(g)$$

Step 2. Draw out full structural formulae to show all the atoms and bonds:

```
  H
  |
H–C–H + 2O=O → O=C=O + 2H–O–H
  |
  H
```

Step 3. Write down which bond energies you will need to look up, and decide whether bonds are broken (endothermic) or formed (exothermic). Use the data in Table 8.2.

Bond breaking	Bond making
$4E(C–H) + 2E(O=O)$	$2 \times -E(C=O) + [4 \times -E(O–H)]$
$4 \times +413 + 2 \times +498$	$2 \times -805 + 4 \times -464$
+2648	–3466

$$\Delta H^{\ominus} = +2648 + (-3466)$$
$$= -818 \text{ kJ mol}^{-1}$$

The standard enthalpy change of combustion of methane is –890 kJ mol^{-1}. This is different from the value obtained in the Example above because the standard state of water is liquid at 298 K. *When dealing with bond energies, all the molecules must be in the gaseous state*, and energy is released when gaseous water condenses.

The standard enthalpy change of combustion of methanol is –726 kJ mol^{-1}. This is less than the value for methane, so why is there a difference? Fuels release their energy when they form bonds with oxygen. The two compounds have the same number of carbon and hydrogen atoms, but one O–H bond is already formed in the methanol molecule.

```
  H          H
  |          |
H–C–H      H–C–O–H
  |          |
  H          H
methane    methanol
```

When methane reacts, all four O–H bonds must be made, which gives a more exothermic reaction. Fuels that contain oxygen are sometimes called oxygenates; they usually have a lower enthalpy change of combustion than fuels with the same number of carbon atoms, but without oxygen.

? QUESTION 13

13 Read the Science in Context box on page 162.

a) Why is water not used to put out petrol fires?

b) Look at Table 8.3 (page 162) and the information about petrol just under it. Think of another disadvantage of using methanol as a fuel.

Fig 8.10 Refuelling a racing car that runs on methanol

REMEMBER THIS

Fuels that are oxygenates are partially oxidised (have oxygen added to them).

SCIENCE IN CONTEXT

Methanol as a fuel

Methanol powers the engines of Formula One motor racing (see Question 12 and Fig 8.10). These are some of the advantages that make it an ideal fuel for high performance engines:

- It has a high octane number, meaning that it does not *knock* (prematurely explode) when the gases are compressed by a piston in an engine (see page 197).
- Although flammable, it is less volatile than petrol, and if a fire starts it is easier to put out because methanol is miscible (mixes) with water.
- Methanol is an oxygenate, and so less oxygen is required to fully combust it in a car engine. (In contrast, in cars using petrol, carbon monoxide is a major pollutant, formed because not enough oxygen is taken in for full combustion.)
- It is fairly cheap to produce. It can be made from natural gas or coal, or from methane through the decomposition of rotting waste (making it a biofuel).
- It can be blended with petrol (as is done in several countries).

Too good to be true? Well, there are problems using methanol.

- It is a highly toxic alcohol, so the vapour is a hazard when refuelling.
- It produces methanal as a pollutant from the exhaust, an aldehyde (see page 316) that is carcinogenic (causes cancer).
- It absorbs water from the air and is very corrosive in existing petrol tanks.

Will methanol become more popular as a fuel? A recent study suggests that by 2050 it could provide half of the global liquid fuel requirements, given advances in technology to overcome these problems.

? QUESTION 14

14 **a)** Write out the equation for the standard enthalpy change of combustion of octane.
b) Work out which bonds are formed, and how many of each one.

REMEMBER THIS

$$\text{Energy density} = \Delta H_c^\ominus \times \frac{1000}{\text{mass of 1 mole}}$$

? QUESTION 15

15 **a)** What is the energy density of propane, which makes up 95 per cent of liquid petroleum gas (LPG)? You will need information from Table 8.1 on page 154 to answer this question.
b) i) What is the volume of 1 kg of hydrogen at 25 °C and 100 kPa? 1 mole of any gas at this temperature occupies 24 dm^3 (see page 116).
ii) 1 litre of petrol has a mass of 740 g. How many litres of petrol are there in 1 kg?

BREAKING AND MAKING MORE BONDS

Octane (molecular formula C_8H_{18}), one of the chemicals in petrol, has 18 C–H bonds and seven C–C bonds. Methane has just four C–H bonds. While this means that you need more energy to break the bonds in a mole of octane, far more energy is released when new bonds are formed with oxygen.

The standard enthalpy change of combustion of octane is –5470 kJ mol^{-1}.

octane H–C–C–C–C–C–C–C–C–H (each carbon also bonded to two H atoms)

Energy density

Most means of transport have to carry their own fuel around with them – think of petrol in the tank of a car. So we want to obtain the maximum energy we can from any mass of fuel. With transport fuels we need to know the amount of energy released by 1 kg of fuel. We call this the **energy density** of the fuel. To calculate the energy density we need to know the standard enthalpy change of combustion ($\Delta H_c^\ominus$) and the mass of one mole of the fuel.

Table 8.3 The energy density values of four fuels

Fuel	Formula	$\Delta H_c^\ominus$/kJ mol^{-1}	Mass of 1 mole/g	Energy density/kJ kg^{-1}
hydrogen	$H_2(g)$	–286	2	143 000
methane	$CH_4(g)$	–890	16	27 800
methanol	$CH_3OH(l)$	–726	32	22 700

Petrol has an energy density of approximately 46 000 kJ kg^{-1}, which makes it a very concentrated energy source. It is a mixture of many hydrocarbons and you can read more about it in Chapter 10.

SCIENCE IN CONTEXT

Metal hydrides: the key to on-board hydrogen storage?

One major disadvantage of hydrogen as a fuel is how to store it. Liquefying it means cooling it to –253 °C, which costs four times more than making an equivalent amount of petrol. It also takes energy to do this. Then it has to be kept cold. The expense is worth it for the Space Shuttle, page 152, but it is no surprise that other storage methods are being sought for more general uses of hydrogen as a fuel.

One possible method is to store hydrogen as a solid. This doesn't mean freezing it, which takes too much energy, but combining it as a metal hydride. Magnesium hydride, MgH_2, is 7.7 per cent hydrogen by mass.

A litre of magnesium hydride contains almost as much hydrogen as a litre of liquefied hydrogen, though it is a lot heavier. The hydrogen is released by this reaction:

$$MgH_2(s) + H_2O(l) \rightarrow Mg(OH)_2(s) + H_2(g)$$

In magnesium hydride, magnesium is ionically bonded to hydrogen in a small whole-number ratio. But another type of metal hydride – an interstitial hydride – can soak up hydrogen rather like a sponge. The metal is bathed in hydrogen. At the metal surface, the hydrogen molecule splits into its atoms and the atoms occupy holes in the metal lattice. Very large quantities of hydrogen can be absorbed and released on heating. There remains the problem of the weight of the hydride, but chemists are actively searching for new alloy hydrides of lower weight.

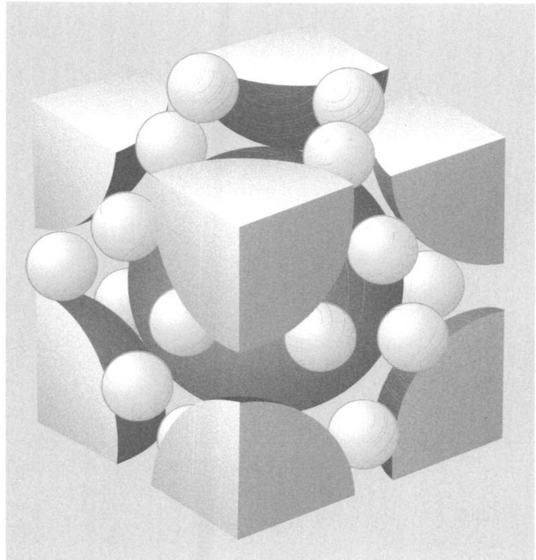

Fig 8.11 An interstitial metal hydride (metal = niobium). The smaller hydrogen atoms are trapped between the larger metal atoms in the lattice

Fig 8.12 The hydrogen in this prototype bus was stored as an interstitial metal compound. Its petrol engine needed only slight modification. The bus ran in Augusta, Georgia, USA in 1997–98.

From Table 8.3, hydrogen would appear to have excellent prospects as a fuel of the future. But there is still the problem of how to store it on board a vehicle. If you try Question 15(b), you will see why.

5 ACTIVATION ENERGY, ENERGY PROFILES AND THE COLLISION THEORY

HOW A REACTION GETS STARTED

As we have seen, octane, one of the components of petrol, has an enthalpy change of combustion of $-5470\ kJ\ mol^{-1}$. So why doesn't it burst into flames as soon as it is exposed to oxygen in the air? The reason is that it takes energy to break bonds. This must happen before oxygen molecules and octane molecules can react together.

To break all the bonds in oxygen and octane vapour would require $+7437\ kJ\ mol^{-1}$ of energy – a very large amount. So, in fact, it appears difficult for octane to start reacting at all. Yet we know that a match thrown on to petrol causes it to react spectacularly. Clearly, not *all* the bonds need to be broken before other bonds start forming, and once bonds start to form, energy is released to break other bonds. This keeps the reaction going. (Look back at page 159 and calculate the energy required to break all the bonds in octane and oxygen.)

The minimum energy required for a reaction to start is called the **activation energy, E_a**. When this energy is supplied to molecules in the system, the bonds begin to stretch and break. Reactions usually occur because the molecules collide with enough energy to make this happen. Sometimes there is sufficient energy in the system for a reaction to occur at room temperature, but for petrol in an engine, energy needs to be supplied to reach the activation energy level. This energy comes from a spark.

Chemists draw **energy profiles**, or **enthalpy profiles**, for reactions, such as those in Fig 8.13. They show how the energy changes as the reactions proceed.

REMEMBER THIS

In an enthalpy profile diagram the vertical axis (*y* axis) is usually labelled enthalpy rather than energy.

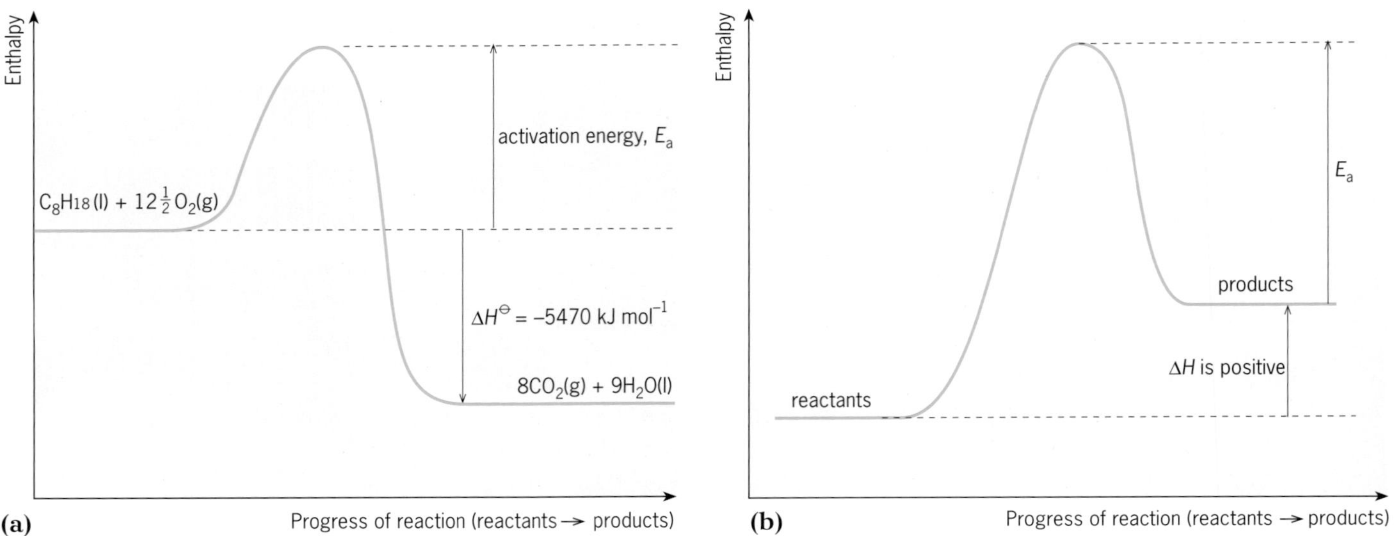

Fig 8.13 Energy profile for **(a)** the exothermic reaction of octane with oxygen, and **(b)** an endothermic reaction. Notice that activation energy E_a is always endothermic

? QUESTION 16

16 Draw an energy profile diagram for:

a) methanol burning in oxygen (see Table 8.1, page 154);

b) nitrogen reacting with oxygen to give nitrogen monoxide:

$$\frac{1}{2}N_2(g) + \frac{1}{2}O_2(g) \rightarrow NO(g) \quad \Delta H^{\ominus} = +90 \text{ kJ mol}^{-1}$$

This occurs in the high temperatures of car engines and leads to pollution (see page 205).

c) Why is the reaction in **b)** not an enthalpy change of combustion?

STRETCH AND CHALLENGE

Why do endothermic reactions take place?

When a spark ignites petrol, there is a **spontaneous reaction** with oxygen, which produces carbon dioxide and water. It does not matter that energy was needed to start this reaction – once started, it is spontaneous (takes place of its own accord).

Let's consider the spontaneous combustion of octane, which is one of the compounds in petrol:

$$2C_8H_{18}(g) + 25O_2(g) \rightarrow 16CO_2(g) + 18H_2O(g)$$

We know that this is an exothermic reaction and that energy is released to the surroundings because the energies of octane and oxygen are higher than the energies of carbon dioxide and water (see Fig 8.13). So we could say that for a reaction to occur it must be exothermic.

However, this does not explain spontaneous reactions that are endothermic, such as the reaction between nitrogen and oxygen to produce nitrogen monoxide:

$$N_2(g) + O_2(g) \rightarrow 2NO(g) \quad \Delta H = +180 \text{ kJ mol}^{-1}$$

Why do endothermic reactions take place? (Cont.)

As this reaction is a major cause of pollution from vehicle exhausts, we know that it happens spontaneously in the high temperature of a petrol engine. To understand why spontaneous reactions occur we must seek another explanation that involves entropy.

ENTROPY CHANGES AND SPONTANEOUS REACTIONS

We first met entropy on page 133. Entropy is a measure of the disorder of a system. The entropy of a gas is higher than the entropy of a solid, because there are more ways of arranging molecules in a gas, and so there is more disorder.

For a reaction to occur spontaneously there must be an overall increase in entropy.

This is the **second law of thermodynamics**. Thermodynamics is the study of energy transfer. We have already encountered the **first law of thermodynamics** (see page 160):

Energy can neither be created nor destroyed in physical and chemical processes.

One of the most important endothermic reactions for life on this planet is photosynthesis. Although it is endothermic, there must be an increase in entropy or it would not be able to occur at all. You can read more about entropy in Chapter 25.

COLLISION THEORY

The collision theory is a model that chemists use to explain how reactions occur. When two molecules react, bonds are broken or made, and there is usually a rearrangement of the atoms. According to collision theory, for this to happen the molecules first have to collide. The collision model was developed from the kinetic–molecular model of gases discussed on pages 123 to 125.

In 1 cm^3 of gas at atmospheric pressure and room temperature, it has been calculated that there are about 10^{27} collisions between two molecules every second. If all these collisions were to lead to reactions, then all reactions that involve gases would be over in microseconds. For example, nitrogen and oxygen molecules, which under normal conditions coexist without reacting, would long ago have reacted to fill the atmosphere with polluting nitrogen oxides. This means that only a certain number of collisions result in reactions. These are called **effective collisions** and occur when molecules collide with the minimum energy needed to start a reaction, the **activation energy**.

REMEMBER THIS
Activation energy is sometimes referred to as **activation enthalpy.**

ACTIVATION ENERGY AND STABILITY

The higher the activation energy for a reaction, the less likely it is to occur. Even though octane looks very unstable compared with carbon dioxide and water, the high energy of activation acts as an **energy barrier** to the reaction (see Fig 8.13(a)). In petrol, this prevents the reaction from occurring when the tank is filled at the garage, and is the reason why there are 'No smoking' signs!

Reaction kinetics is the study of *rates* of reactions and we look at this study in more detail at the end of this chapter and in Chapter 26. We say petrol is **kinetically stable**, even though it is **energetically unstable**.

REMEMBER THIS
Octane is energetically unstable because it is at a higher energy level than carbon dioxide and water. Sometimes we use the term **thermodynamically unstable** instead of energetically unstable.

REMEMBER THIS

Catalysts are not put into the overall chemical equation for reactions because they are chemically unchanged at the end of the reaction.

LOWERING THE ENERGY BARRIER: CATALYSIS

In your previous chemistry course you probably learnt that **catalysts** usually increase the rate of chemical reactions. We also mention this in Chapter 1. Catalysts play a crucial role in the lives of everybody. Protein catalysts called enzymes dramatically increase the rates of thousands of different chemical reactions that take place in our bodies every second. And catalysts are involved in the manufacture of most of the chemicals on which people rely. As a definition:

Catalysts are substances that alter the rate of a chemical reaction but can be recovered unchanged at the end of the reaction.

But they do become involved temporarily in the reaction, because:

Catalysts work by providing an alternative reaction pathway with a lower energy of activation.

While catalysts are not used in the combustion of petrol, they are used in catalytic converters to speed up reactions that control emissions from the exhausts of cars (see pages 203 to 205).

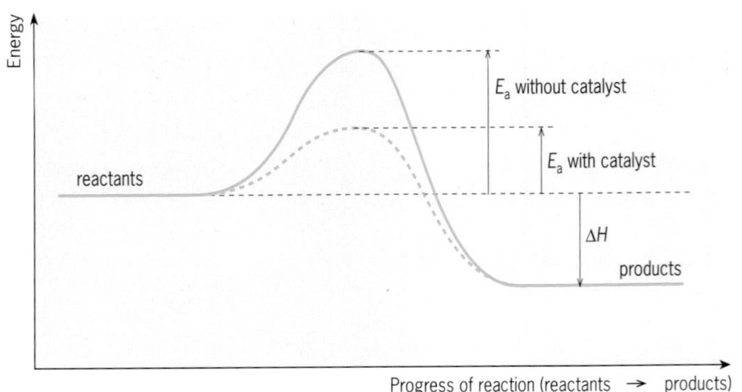

Fig 8.14a) Energy profile (enthalpy profile) of a reaction with a catalyst and without a catalyst. Notice that the overall enthalpy change is the same, whichever route is followed. Otherwise, it would break the law of conservation of energy (first law of thermodynamics)

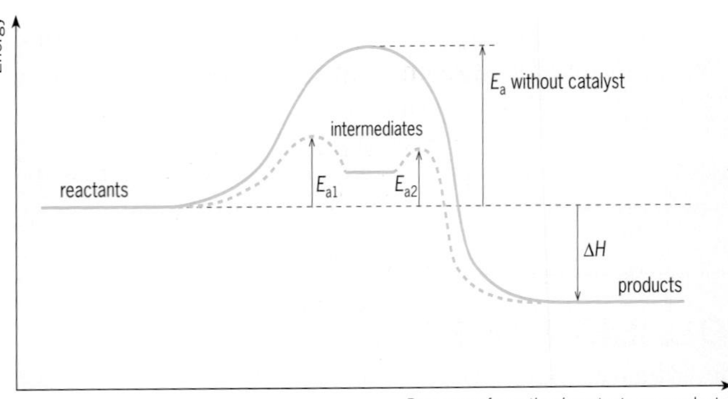

Fig 8.14b) Energy profile (enthalpy profile) of a reaction with and without a catalyst. Here, the formation of intermediates E_{a1} and E_{a2} are the relevant activation energies of the catalysed reaction pathway.

Catalysts often work by forming intermediates. In this case the energy profile diagram becomes a little more complicated (see Fig 8.14(b)). The intermediates are of a higher energy than the reactants or products but the overall reaction pathway is still of lower energy.

Sometimes catalysts slow down the rate of a reaction. When they do, they are called negative catalysts or **inhibitors**. These do not work by raising the activation energy, but instead usually react with molecules produced in the intermediate stages of a reaction and remove them from the reaction pathway.

REMEMBER THIS

ΔH°_{f} can be used instead of $\Delta H^{\circ}_{f,298}$, because we assume that the temperature is 298 K unless another temperature is quoted.

Standard molar enthalpy change of formation is another name for standard enthalpy change of formation. You will find standard enthalpy changes of formation for compounds in almost any chemistry data book.

6 ANOTHER ENTHALPY CHANGE: THE ENTHALPY CHANGE OF FORMATION

Like the enthalpy change of combustion on page 155, the **enthalpy change of formation** is just one more enthalpy change of reaction. But this time, instead of burning one mole of compound completely in oxygen, we *form one mole of compound from its elements.*

The standard enthalpy change of formation $\Delta H^{\circ}_{f,298}$ is the enthalpy change when one mole of a compound is formed from its elements under standard conditions, that is, 298 K and 100 kPa pressure.

The standard enthalpy change of formation of water:

$$\Delta H_f^{\circ} (H_2O(l)) = -286 \text{ kJ mol}^{-1}$$

applies to this reaction:

$$H_2(g) + \tfrac{1}{2}O_2(g) \rightarrow H_2O(l)$$

Notice that the value above is also the standard enthalpy change of combustion of hydrogen.

The equation for the standard enthalpy change of formation of methanol is:

$$C(s) + 2H_2(g) + \tfrac{1}{2}O_2(g) \rightarrow CH_3OH(l) \quad \Delta H_f^{\circ} = -239 \text{ kJ mol}^{-1}$$

WHY STANDARD ENTHALPY CHANGES OF FORMATION ARE IMPORTANT

You can see the value of knowing enthalpy changes of combustion – because, for example, they allow us to work out the energy we can obtain from a fuel. But why bother with the standard enthalpy changes of formation?

The enthalpy of a substance is its energy content. We cannot measure this and give it an absolute value. It is all the energy associated with the particles in the substance, which includes the nuclei, the electrons and the movements of whole molecules.

We can only measure enthalpy *changes*, and to do this we must have a reference point. The one that scientists have chosen is the **standard enthalpy change of formation**. It applies to any compound, while for elements it is *zero*. It is a very important piece of information about a compound, and once you know the value for each of the different substances in a reaction, you can calculate the enthalpy change for that whole reaction.

However, before we consider this, we must look at how we obtain values for the standard enthalpy change of formation.

7 MEASURING THE STANDARD ENTHALPY CHANGE OF FORMATION

Some standard enthalpy changes of formation can be measured experimentally. For example, we can measure ΔH_f° of carbon dioxide by using a bomb calorimeter and combusting graphite. The ΔH_f° of magnesium oxide can be measured in the same way.

When a compound is easily synthesised from its elements under normal conditions, we can usually measure ΔH_f° directly. However, many standard enthalpy changes of formation cannot be measured directly, which is where we need an indirect approach using **energy cycles**.

THE INDIRECT METHOD FOR CALCULATING ENTHALPY CHANGES OF FORMATION USING ENERGY CYCLES

This method relies on **Hess's law**, which states:

If a change can be brought about by more than one route, then the overall enthalpy change for each route must be the same provided that the starting and finishing conditions are the same for each route.

This is conservation of energy again – energy cannot be created or destroyed in chemical reactions, so the energy changes for a reaction must be the same, whether it takes place in one step or in a whole series of steps. We can show this by constructing an energy cycle like that shown in Fig 8.15.

? QUESTION 17

17 Write equations for:

a) $\Delta H_f^{\circ} (CO_2(g)) = -393 \text{ kJ mol}^{-1}$

b) $\Delta H_f^{\circ} (CH_4(g)) = -75 \text{ kJ mol}^{-1}$

c) When these compounds are formed from their elements, in which direction is energy transferred – to the compound or to its surroundings? Are they more or less stable than the elements that make them up?

REMEMBER THIS

Carbon exists in two forms, **diamond** and **graphite**. The most stable of these forms is graphite, which is the standard state used for standard enthalpy changes of formation. Sometimes you will see $C_{graphite}$ written in the equation.

To find out about a third form of carbon, buckminsterfullerene, read Chapter 7.

? QUESTION 18

18 To measure $\Delta H_f^{\circ} (CO_2(g))$, graphite is burnt in the bomb calorimeter rather than diamond. Both are forms of carbon. Why is graphite chosen?

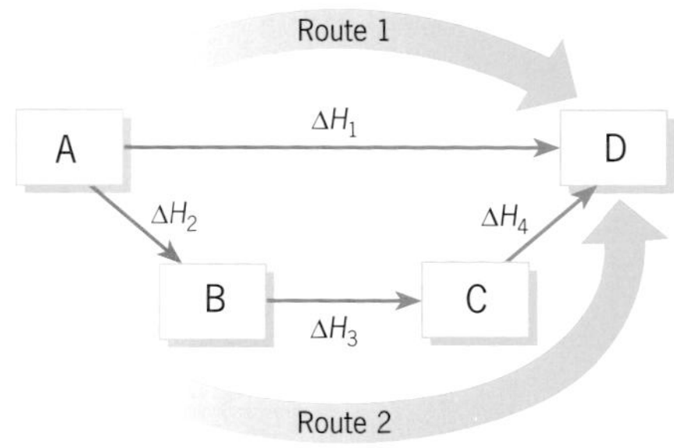

Fig 8.15 An energy cycle illustrating Hess's law. It does not matter whether the reaction takes place by the one-step route or goes through several intermediate steps. The overall enthalpy change for route 2 must be the same as that for route 1, so:

$$\Delta H_1 = \Delta H_2 + \Delta H_3 + \Delta H_4$$

See Chapter 22 for information about another energy cycle, the Born-Haber cycle, used for ionic compounds.

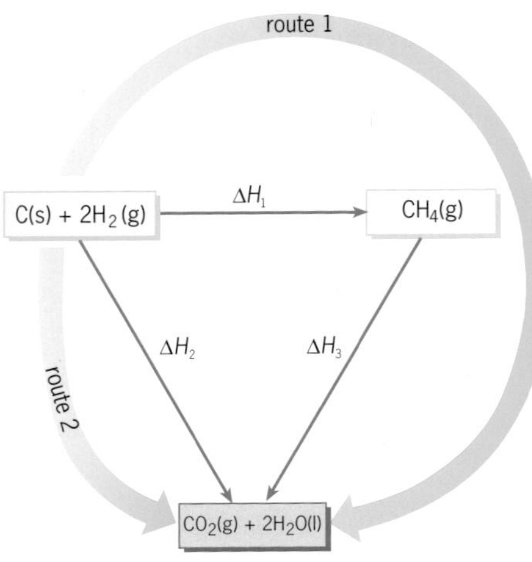

Fig 8.16 A Hess's law energy cycle being used to calculate $\Delta H^\ominus_f(CH_4(g))$ indirectly

Methane cannot be prepared directly from its elements, but we can use Hess's law to calculate its standard enthalpy change of formation. By burning carbon, hydrogen and methane, we can measure their standard enthalpy changes of combustion directly. We can therefore devise an energy cycle with two routes in it, as shown in Fig 8.16. By Hess's law:

the overall enthalpy change for route 1 = the enthalpy change for route 2.

So: $$\Delta H_1 + \Delta H_3 = \Delta H_2$$

ΔH_1 is what we are trying to find, in this case the standard enthalpy change of formation of methane, $\Delta H^\ominus_f(CH_4(g))$.
For route 1:

$$\Delta H_2 \text{ is } \Delta H^\ominus_c \text{ for carbon (graphite)} = -393 \text{ kJ mol}^{-1}$$

$$\textbf{plus } 2 \times \Delta H^\ominus_c \text{ for hydrogen} = 2 \times -286 \text{ kJ mol}^{-1}$$

(Remember: $\Delta H^\ominus_c$ for hydrogen is for burning one mole of hydrogen. In this energy cycle there are two moles of hydrogen.)

ΔH_3 = standard enthalpy change of combustion of methane, $\Delta H^\ominus_c(CH_4(g))$ = –890 kJ mol^{-1}.

So: $$\Delta H_1 = \Delta H_2 - \Delta H_3 = -393 + 2(-286) - (-890)$$

$$\Delta H^\ominus_f(CH_4(g)) = -75 \text{ kJ mol}^{-1}$$

EXAMPLE

Q Calculate the standard enthalpy change of formation of methanol, given that:

$$\Delta H^\ominus_c(CH_3OH(l)) = -726 \text{ kJ mol}^{-1}$$
$$\Delta H^\ominus_c(C(s)) = -393 \text{ kJ mol}^{-1}$$
$$\Delta H^\ominus_c(H_2(g)) = -286 \text{ kJ mol}^{-1}$$

A **Step 1.** Write out the enthalpy change you have been asked to calculate – in this case $\Delta H^\ominus_f(CH_3OH(l))$:

$$C(s) + 2H_2(g) + \tfrac{1}{2}O_2(g) \rightarrow CH_3OH(l)$$

Step 2. Construct an energy cycle with two alternative routes. It helps if you always put the enthalpy change you want to calculate along the top of the cycle and call it ΔH_1.

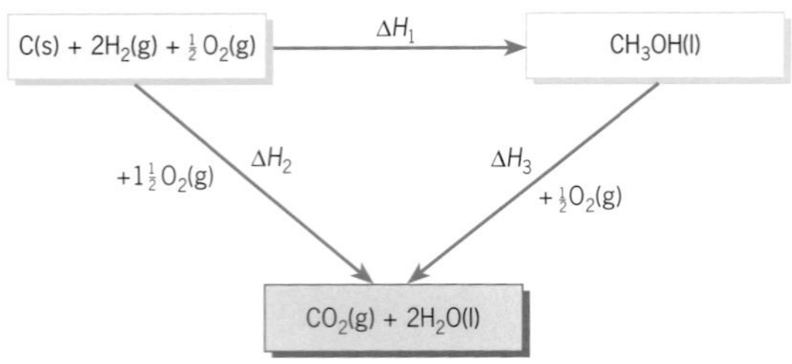

Fig 8.17 A Hess's law energy cycle to calculate $\Delta H^\ominus_f(CH_3OH(l))$

Step 3. Write out the enthalpy changes for the two routes:

$$\Delta H_1 + \Delta H_3 = \Delta H_2$$

Step 4. Decide what enthalpy changes are represented by ΔH_1, ΔH_2 and ΔH_3. It is a good idea to do this on the energy cycle diagram first.

$$\Delta H_1 = \Delta H^\ominus_f(CH_3OH(l))$$

that is, the enthalpy change we wish to calculate.

$$\Delta H_2 = \Delta H^\ominus_c(C(s)) + [2 \times \Delta H^\ominus_c(H_2(g))]$$
$$\Delta H_3 = \Delta H^\ominus_c(CH_3OH(l))$$

Step 5. Rearrange the equation in Step 3 and insert the values of the enthalpy changes that are known:

$$\Delta H_1 = \Delta H_2 - \Delta H_3$$
$$\Delta H^\ominus_f(CH_3OH(l)) = \Delta H^\ominus_c(C(s)) + [2 \times \Delta H^\ominus_c\ \Delta H_2(g))] - \Delta H^\ominus_c(CH_3OH(l))$$

$$\Delta H^\ominus_f(CH_3OH(l)) = -393 + 2(-286) - (-726) = -239 \text{kJmol}^{-1}$$

Thus the standard enthalpy change of formation of methanol is –239 kJ mol^{-1}.

? QUESTION 19

19 Work out the standard enthalpy change of formation of:

a) propane (see Table 8.1 and the Example above for the values you need);

b) carbon monoxide $\Delta H^\ominus_c(CO(g)) = -283$ kJ mol^{-1}).

c) Why is it impossible to obtain $\Delta H^\ominus_f(CO(g))$ directly?

8 USING STANDARD ENTHALPY CHANGES OF FORMATION TO CALCULATE ENTHALPY CHANGES OF REACTION

Knowing standard enthalpy changes of formation means we can calculate enthalpy changes of reaction. To do this, we again apply Hess's law.

EXAMPLE

Q The Space Shuttle Orbiter uses methylhydrazine as fuel, which is oxidised by dinitrogen tetroxide to provide the energy for propulsion. Calculate the standard enthalpy change $\Delta H_r^\ominus$ for this reaction:

$$4CH_3NHNH_2(l) + 5N_2O_4(l) \rightarrow 4CO_2(g) + 12H_2O(l) + 9N_2(g)$$

These are the standard enthalpy changes of formation:

$$\Delta H_f^\ominus\ (CH_3NHNH_2(l)) = +54\ \text{kJ mol}^{-1}$$

$$\Delta H_f^\ominus\ (N_2O_4(l)) = -20\ \text{kJ mol}^{-1}$$

$$\Delta H_f^\ominus\ (CO_2(g)) = -393\ \text{kJ mol}^{-1}$$

$$\Delta H_f^\ominus\ (H_2O(l)) = -286\ \text{kJ mol}^{-1}$$

Fig 8.18 The Space Shuttle Orbiter is powered by methylhydrazine and dinitrogen tetroxide

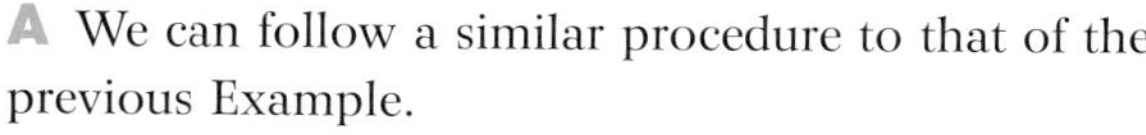

A We can follow a similar procedure to that of the previous Example.

Step 1. Write out the enthalpy change you have been asked to calculate. This has already been given in the question, so this step is not required.

Step 2. Construct an energy cycle with two alternative routes. Remember that it helps if you always put the enthalpy change you want to calculate along the top and call it ΔH_1.

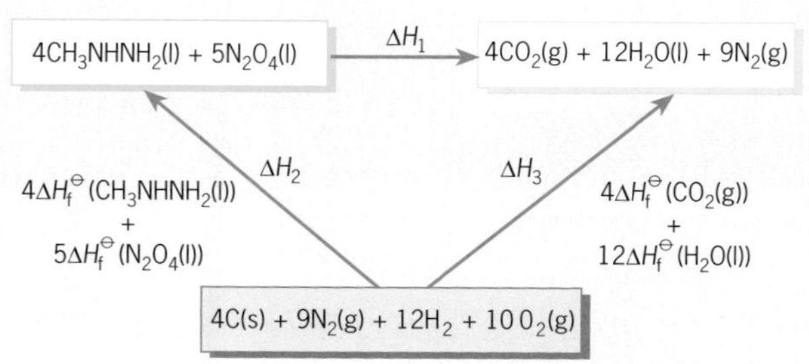

Fig 8.19 A Hess's law energy cycle to calculate $\Delta H_r^\ominus$

Step 3. Write out the enthalpy changes for the two routes:

$$\Delta H_2 + \Delta H_1 = \Delta H_3$$

Notice that this is different from the previous Example.

Step 4. Decide what enthalpy changes are represented by ΔH_1, ΔH_2 and ΔH_3. Remember that it is a good idea to do this on the energy cycle diagram first.

$\Delta H_1 = \Delta H_r^\ominus$, that is, the enthalpy change we wish to calculate.

$\Delta H_2 = 4\Delta H_f^\ominus\ (CH_3NHNH_2(l)) + 5\Delta H_f^\ominus\ (N_2O_4(l))$

$\Delta H_3 = 4\Delta H_f^\ominus\ (CO_2(g)) + 12\Delta H_f^\ominus\ (H_2O(l))$

Step 5. Rearrange the equation in Step 3 and insert the values of the enthalpy changes that are known:

$$\Delta H_1 = \Delta H_3 - \Delta H_2$$

$$\Delta H_r^\ominus = 4\Delta H_f^\ominus\ (CO_2(g)) + 12\Delta H_f^\ominus\ (H_2O(l)) - [4\Delta H_f^\ominus\ (CH_3NHNH_2(l)) + 5\Delta H_f^\ominus\ (N_2O_4(l))]$$

$$\Delta H_r^\ominus = 4(-393) + 12(-286) - [4(+54) + 5(-20)]$$

$$= -5004 - [+116] = -5120\ \text{kJ mol}^{-1}$$

So:

$$4CH_3NHNH_2(l) + 5N_2O_4(l) \rightarrow 4CO_2(g) + 12H_2O(l) + 9N_2(g)$$
$$\Delta H_r^\ominus = -5120\ \text{kJ mol}^{-1}$$

The standard state of dinitrogen tetroxide is a gas at 298 K. However, in this case we have used the standard enthalpy change of formation of the liquid dinitrogen tetroxide, because this is how it is carried in the Orbiter. This is the standard enthalpy change of formation in its non-standard state.

? QUESTION 20

20 **a)** Work through the Example on page 169. Why don't we need to look up the standard enthalpy change of formation of nitrogen?

b) Methanol can be produced from coal and the final stage of this process is:

$$CO(g) + 2H_2(g) \rightarrow CH_3OH(l)$$

Calculate the standard enthalpy change for this reaction using the following information:

$\Delta H_f^{\ominus}$ (CO(g)) = −110 kJ mol^{-1}

$\Delta H_f^{\ominus}$ (CH_3OH(l)) = −239 kJ mol^{-1}

REMEMBER THIS

The study of rates of reaction is called reaction kinetics (from the Greek word for 'to move'). You can read more about reaction kinetics in Chapter 26.

The rate of a reaction can be simply defined as the amount of substance formed per unit of time (e.g. moles per second).

Look closely at Step 5 in the Example on page 169 and you will notice that:

$$\boldsymbol{\Delta H_r^{\ominus} = \Sigma\Delta H_f^{\ominus}\ \text{(products)} - \Sigma\Delta H_f^{\ominus}\ \text{(reactants)}}$$

The symbol Σ means 'the sum of', and remember that the standard enthalpy changes of formation must be multiplied by the number of moles given in the equation.

The equation above in bold can be used to calculate the enthalpy change for any chemical reaction, which makes it very useful. You do not need to draw an energy cycle, but as you can see, it is still an application of Hess's law.

9 RATES OF REACTION AND THE COLLISION THEORY

In Chapter 1, pages 9–10, we considered the importance of being able to control reaction rates in industry. The rate at which a reaction is going to occur and how it can be slowed down (or accelerated) are crucial information. For example, unwanted chemical reactions that lead to food deterioration are slowed down by putting food in a refrigerator or freezer; desired chemical reactions in the cooking of food are accelerated by the energy from an oven. Billions of chemical reactions are happening all around us all the time – from those that occur inside our bodies to those involved in rusting and rotting, and in the life processes of plants and animals. Understanding how reaction rates can be accelerated or slowed down is extremely important and the collision theory introduced on page 165 of this chapter will help us to do this.

Fig 8.20 The gas explosion that caused this fire took only a fraction of a second

Fig 8.21 Divers find that wrecks do not alter much over decades. The rusting of the *Titanic*, which sank in the North Atlantic in 1912, is such a slow reaction that it has been possible to salvage many iron artefacts from the ship more than 90 years later

In the collision theory, for a reaction to occur between two particles (atoms, molecules or ions) they must collide with a certain minimum energy called the **activation energy, E_a**. Fig 8.22 is an energy profile diagram for an exothermic reaction, showing the activation energy. The higher this activation energy is, the fewer the number of molecules that will have sufficient energy to react at any particular temperature, so the rate of reaction will be slow. At the very top of the energy barrier, bonds are forming and being broken, and the molecules are in a **transition state** between reactants and products (Fig 8.23). The idea of a transition state comes from another model of reaction kinetics called the **transition state theory**. This model looks at

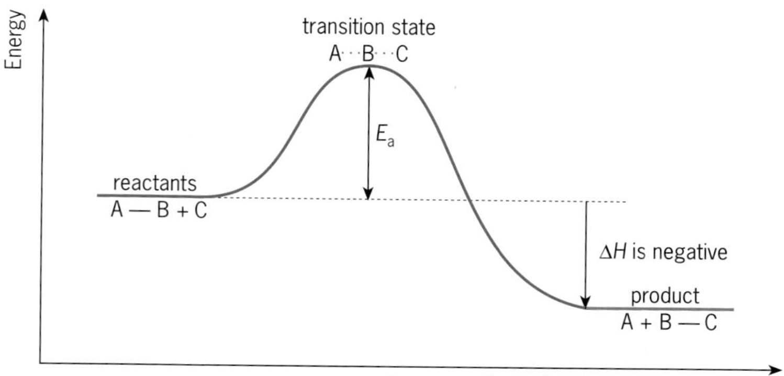

Fig 8.22 An energy profile diagram for an exothermic reaction

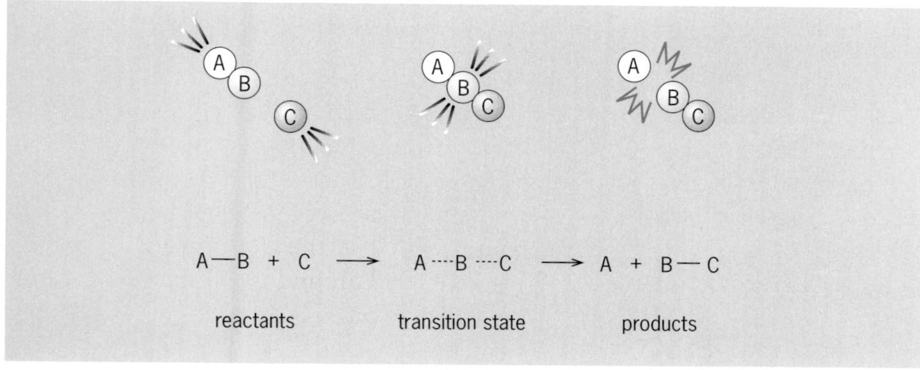

Fig 8.23 An effective collision between two particles has the required activation energy. The transition state shows bonds breaking and forming as a result of the collision

> **REMEMBER THIS**
> Particles must not only collide with the required activation energy; they must also collide with the correct orientation for a reaction to occur.

the collision in detail. While collision theory is a good model for simple gaseous reactions, the transition state theory is applicable more generally. However, all subsequent discussions in this chapter are based on the collision theory.

EXPLAINING THE EFFECT OF CONCENTRATION AND PRESSURE ON REACTION RATE

Increasing the concentrations of certain reacting species accelerates the reactions in which they are involved. Can collision theory explain this observation?

To react, particles must first collide. So, increasing the number of particles in a given volume must increase the number of collisions and therefore, it would be reasonable to assume, increase the number of effective collisions (Fig 8.24).

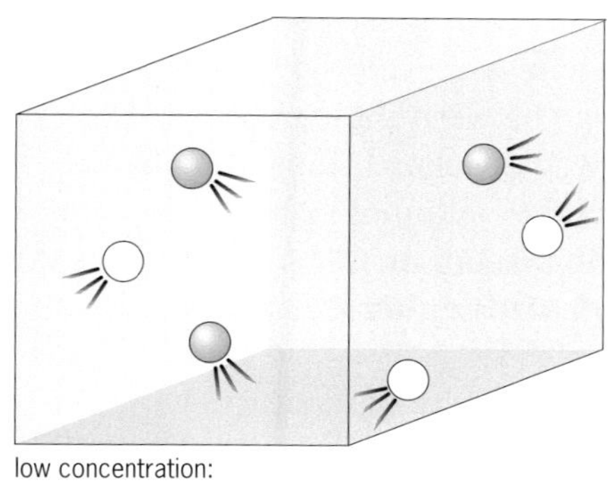

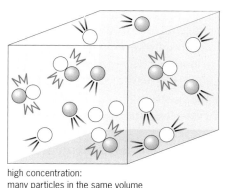

low concentration:
few particles in a given volume

high concentration:
many particles in the same volume

Fig 8.24 Effect of increasing the concentration of reacting species

Collision theory also explains why increasing the pressure under which gaseous reactions take place increases their rate of reaction. This is illustrated in Fig 8.25. Note that, at a given temperature, increasing the pressure in gases increases the concentration of molecules.

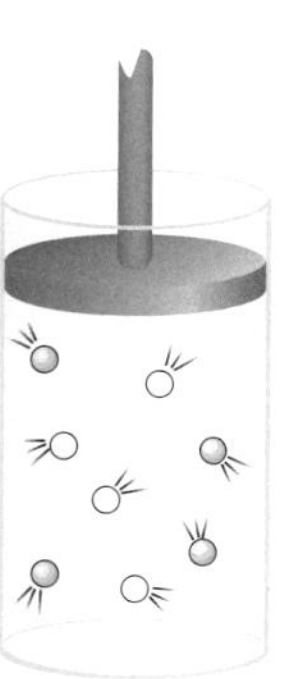

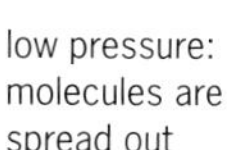

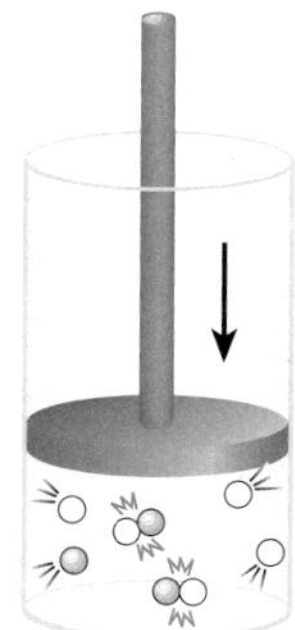

low pressure:
molecules are spread out

high pressure:
volume has been decreased (at constant temperature), forcing the molecules closer together

Fig 8.25 The effect of increasing the pressure of reacting gases, held at a constant temperature

EXPLAINING THE EFFECT OF INCREASING THE SURFACE AREA OF SOLID REACTANTS ON THE RATE OF REACTION

If solid particles are large, they have a small surface area compared to the amount of reactant molecules they contain, and only the reactant molecules at the surface can take part in collisions with other molecules. If a solid particle is ground into a fine powder, then many more molecules are available for effective collisions. Miners have long known about the hazards of dust explosions, but dust explosions are not confined to underground mines. Fig 8.26 shows the effect of a dust explosion in a grain silo. Milling flour also has the same risks.

Fig 8.26 In 1977, in New Orleans, an explosion of dust killed 37 people. A spark ignited very fine dust in grain silos and the terrifyingly fast rate of combustion with the oxygen in the air caused the explosion

10 ENERGY DISTRIBUTION AMONG MOLECULES IN A GAS

Molecules have energy due to their motion. It is called **kinetic energy**. At any instant in time, some molecules have very high energies relative to the rest, because they are moving very fast, while other molecules have very low energies relative to the rest, because they are moving very slowly. Most molecules, however, have energies in a range between these extremes. The distribution of energies among molecules was calculated statistically by the Scottish physicist James Clerk Maxwell in 1859, and more generally applied by the Austrian physicist Ludwig Eduard Boltzmann in 1871. The result is the **Maxwell–Boltzmann distribution of molecular energies**, shown in Fig 8.27.

The Maxwell–Boltzmann distribution explains why, at a particular temperature, only a certain number of collisions are effective ones. Fig 8.28 shows the effect of increasing temperature. At the higher temperatures, a greater proportion of molecules has higher energies and so a greater proportion possesses energies equal to, or above, the activation energy, giving more effective collisions.

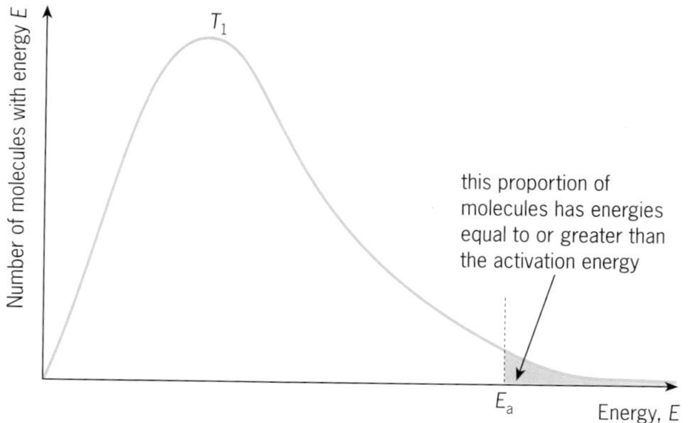

Fig 8.27 Maxwell–Boltzmann distribution curve at a particular temperature

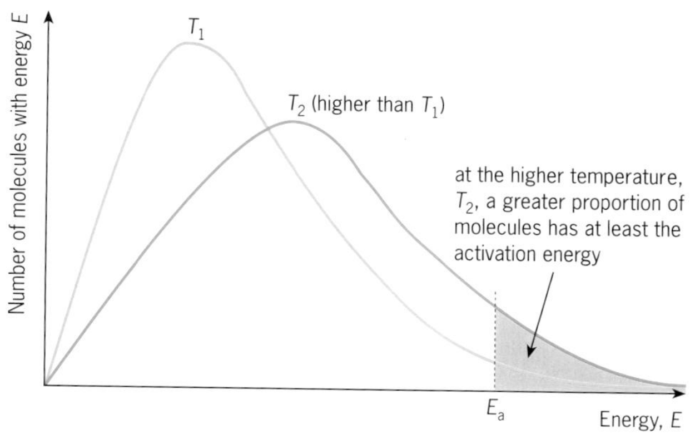

Fig 8.28 Maxwell–Boltzmann distribution curve at temperatures T_1 and T_2

THE MAXWELL–BOLTZMANN DISTRIBUTION EXPLAINS THE EFFECT OF TEMPERATURE ON REACTION RATE

As Fig 8.28 shows, increasing the temperature increases the proportion of molecules that has the minimum activation energy needed for them to react when they collide. This means that the rate of reaction increases with temperature. In fact, only a small temperature increase leads to a large increase in rate.

For every 10 °C rise, the rate of most reactions approximately doubles.

REMEMBER THIS

The Maxwell–Boltzmann distribution curve is sometimes referred to as the **Boltzmann distribution**

THE MAXWELL–BOLTZMANN DISTRIBUTION EXPLAINS THE EFFECT OF A CATALYST ON REACTION RATE

By having its energy of activation lowered, a reaction occurs at a faster rate; this is what a catalyst does. A catalyst provides an alternative route for a reaction to take place. To explain this in terms of the collision theory, consider again the Maxwell–Boltzmann distribution of molecular energies. If the activation energy is lowered, a greater proportion of molecules has sufficient energy to make effective collisions. This is illustrated in Fig 8.29.

For more information about catalysts, see pages 203–207

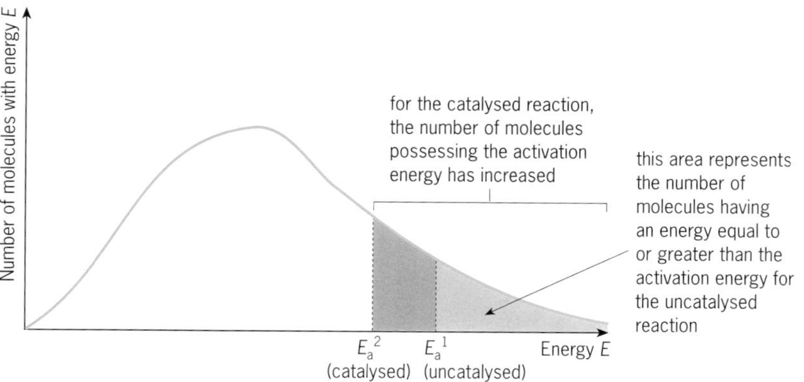

Fig 8.29 Maxwell–Boltzmann distribution of molecular energies, showing the effect of adding a catalyst. Note that, by lowering the activation energy, there is a large increase in the number of molecules that possess at least the activation energy

SUMMARY

When you have studied this chapter, you should be able to understand the following ideas:

- The enthalpy (*H*) of a compound is the energy stored in it.
- Enthalpy changes (ΔH) are transfers of energy into or out of a system at constant pressure.
- The standard enthalpy change of reaction ($\Delta H^{\ominus}_{r,298}$) is the enthalpy change when the mole quantities expressed in a balanced equation react under standard conditions (298 K, 100 kPa) with all substances in their standard states.
- The standard enthalpy change of combustion ($\Delta H^{\ominus}_{c,298}$) and the standard enthalpy change of formation ($\Delta H^{\ominus}_{f,298}$) are both standard enthalpy changes of reaction.
- Some enthalpy changes can be measured directly, using a bomb calorimeter.
- Other enthalpy changes are measured indirectly by using Hess's law.
- Average bond enthalpies can give a good estimate of the enthalpy change of a reaction.
- Activation enthalpy (energy) is the minimum energy required for a reaction to occur.
- High activation energies make reactants kinetically stable, even if they are energetically unstable.
- Catalysts lower the activation energy by providing an alternative reaction pathway.
- Collision theory is a model that explains how reacting molecules must collide with sufficient energy, the activation energy, for a reaction to occur.
- The Maxwell–Boltzmann distribution of energies in a sample of gaseous molecules explains why increasing temperature or the use of a catalyst speeds up reactions.

Practice questions and a How Science Works assignment for this chapter are available at www.collinseducation.co.uk/CAS

9

Oil and the petrochemical industry

9 OIL AND THE PETROCHEMICAL INDUSTRY

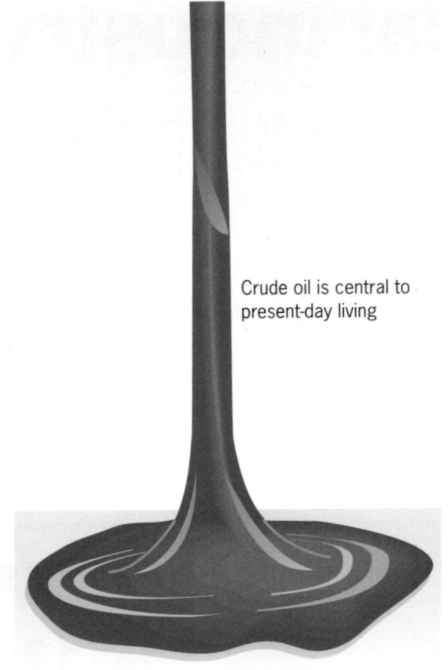
Crude oil is central to present-day living

The Oil Age?

The Stone Age, the Iron Age and the Bronze Age are periods of history during which humans began to exploit particular substances and so made massive improvements to their lives over a short period of time. In our own time, it is oil that has had the greatest impact on our lives.

From 1859, when the first oil well was sunk by Edwin Drake in the United States, societies have relied increasingly on the production of oil, especially for fuel: 90 per cent of oil is burnt, very much of it used to provide energy for cheap transport. However, if the present time were to become known as the Age of the Car, this would ignore what we now do with the remaining 10 per cent of oil. This is transformed through chemical reactions into medicines, paints, insecticides, dyes, detergents, plastics and other chemicals on which we now depend.

Oil is suitably named when called 'black gold', to indicate how precious it is. Maybe future historians will look back and call the age we live in the Oil Age, but it will probably have lasted a brief 200 years from the drilling of the first well to the exhaustion of supplies. In fact, we may well see the end of the Oil Age in our own lifetimes.

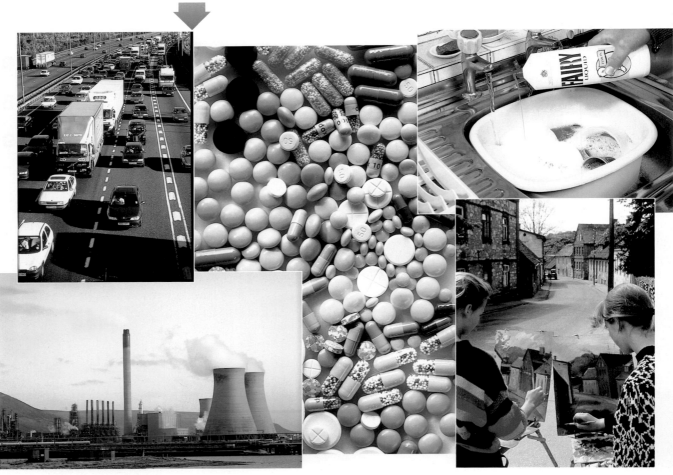

1 CRUDE OIL

The oil that comes out of the ground is known as crude oil, or petroleum (from the Latin *petra* for rock and *oleum* for oil). It is a thick, dark, smelly liquid. Two thousand years ago, the Chinese occasionally came across it when they dug for brine; then, they used it as a fuel. Inhabitants of the Middle East found it lying on the ground, and because the **volatile** (easily evaporated) components had evaporated, a sticky tar called pitch remained. This they used to waterproof their wooden boats. The ancient Mexicans even used it as chewing gum!

Crude oil is a very complex mixture of hundreds of different compounds. Most are **hydrocarbons**, compounds of carbon and hydrogen only. There are also compounds that contain nitrogen, oxygen and sulfur and even some molecules that have metals in them.

Since crude oil is a mixture, its composition varies. Oil from some wells may contain up to 98 per cent hydrocarbons, while oil from others may have as little as 50 per cent. Hydrocarbons are the most sought-after component, used for energy and for making thousands of compounds, from plastics to drugs, that are essential to modern society. Hydrocarbon composition also varies; there is more about this on page 184.

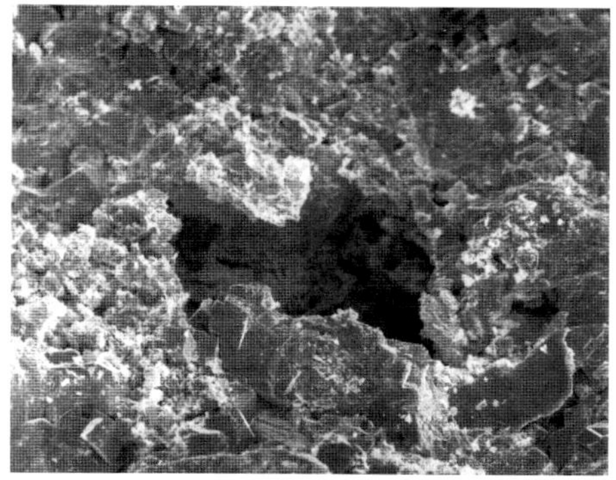

Fig 9.1 A microscope picture of sandstone grains. The spaces in between the grains are known as pores. In an oil reservoir, these pores contain crude oil and water

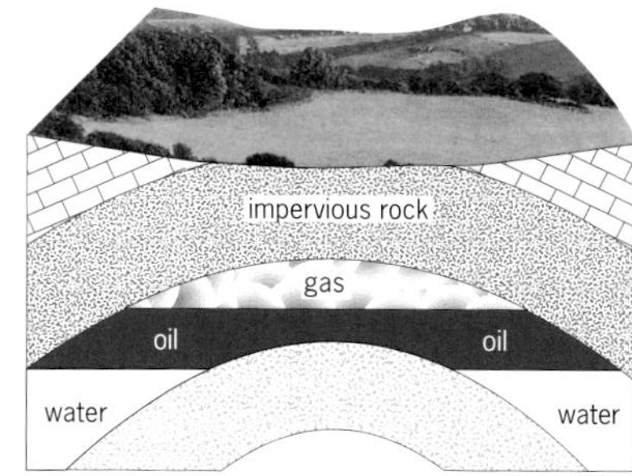

Fig 9.2 Crude oil trapped by a layer of impervious rock to form an oil reservoir

HOW OIL WAS FORMED

The process that formed the oil we find today began up to 400 million years ago, as tiny marine animals and plants died, sank to the bottom of the oceans and became trapped in mud. No oxygen was present in this mud, and bacteria decomposed them anaerobically. An **anaerobic reaction** is one that occurs without oxygen. The decay products stayed in the mud and were acted on by pressure, radioactivity and high temperatures, until finally the mixture we know as oil was created. This took millions of years. As evidence that crude oil came from living organisms, some of the hydrocarbon molecules resemble those found in rubber, cholesterol and vitamin A.

Under pressure, the mud eventually turned into rock, squeezing out the oil into porous rock above. In some areas, the oil seeped to the surface; but fortunately in others it became trapped in the porous rock, which became a reservoir because there was an impervious layer of rock above it.

QUESTION 1

Natural gas is associated with oil deposits. What is the principal component of natural gas?

SCIENCE IN CONTEXT

Recovering oil from the ground

There are widely differing estimates of how much oil is left on Earth, and of how much can actually be extracted. So estimates of how long oil supplies will last vary even more widely.

When a new oil reservoir is discovered and wells are sunk, the oil often flows quite freely to the surface, squeezed out by the pressure of rock above it. This is called **primary recovery**. It recovers only 10 to 30 per cent of the oil in the reservoir. **Secondary recovery** methods may involve flooding the reservoir with steam. The energy from the steam makes heavier components of the oil less viscous and drives the oil to the well. Even so, much of the oil still stays in the ground. Whether secondary methods are employed depends on economics. When the price of oil is low, it may cost more to retrieve the extra oil from the ground than it is worth. **Tertiary methods** are being developed for the 21st century, and these require new chemistry and new techniques. Detergents allow oil and water to mix, and may be used to help to extract some of the deposits of heavier oil. The use of polymer solutions and alkalis may also help wrest the oil from the pores of the rock. There are even plans to use carbon dioxide from a Norwegian power station to force oil to the surface and at the same time bury the CO_2. If these methods succeed, thousands of millions more barrels of oil could be extracted. As the price of oil rises this tertiary extraction becomes cost-effective.

Fig 9.3 Drilling on the Brent Charlie oil rig in the North Sea

2 WHY ARE THERE SO MANY CARBON COMPOUNDS?

It was once thought that the chemicals produced in living organisms were different from those found in non-living materials. They were thought to have an associated 'life force' and were therefore known as **organic chemicals**. The rocks and minerals of the Earth, and the chemicals made from them, were called **inorganic chemicals**. But in 1828, a German chemist, Fredrich Wöhler, managed to produce an organic compound, urea (found in urine), from an inorganic compound, ammonium cyanate, just by heating it in his laboratory:

$$\underset{\text{ammonium cyanate}}{NH_4OCN} \xrightarrow{\text{heat}} \underset{\text{urea}}{NH_2CONH_2}$$

This simple experiment helped to dispel the myth of a life force, but we still refer to the chemistry of carbon compounds as **organic chemistry**, and that of all other compounds as inorganic chemistry. During the 19th century, many organic chemicals were synthesised (made) and this has continued at an increasing rate, particularly during the past few decades. There are now over eight million recorded organic chemicals, and most of these do not occur in nature. Some are completely new compounds, while others are variations of those found in nature, designed to enhance the properties of particular groups of atoms. The search for new compounds and new ways to synthesise known compounds is a very exciting aspect of chemistry. (You can read about how computers are helping to design drug molecules on page 70.)

REMEMBER THIS

Metal carbonates, carbon dioxide and carbon monoxide are considered to be inorganic chemicals.

Synthesis is the production of one compound from two or more other substances.

See page 77 to remind yourself about atoms that can expand their octets.

3 THE SPECIAL NATURE OF CARBON

There are about 100 000 known inorganic substances, a tiny number compared with the vast array of carbon compounds. To explain this, we need to consider just what is so special about carbon.

CARBON CANNOT EXPAND ITS OCTET

Carbon forms covalent bonds. It is in Group 4, so it has four electrons in its outer shell, which means that it can form four covalent bonds. In common with all the elements in Period 2, it cannot expand its outer shell to hold more than eight electrons. This means that its compounds are resistant to chemical attack. We know that methane, CH_4, in natural gas does not react with water at normal temperatures and pressures. Silicon is also in Group 4, but it can expand its octet. As a result, the silane SiH_4 reacts vigorously with water. The **mechanism** for this reaction involves water molecules each donating a lone pair of electrons to the silicon atom.

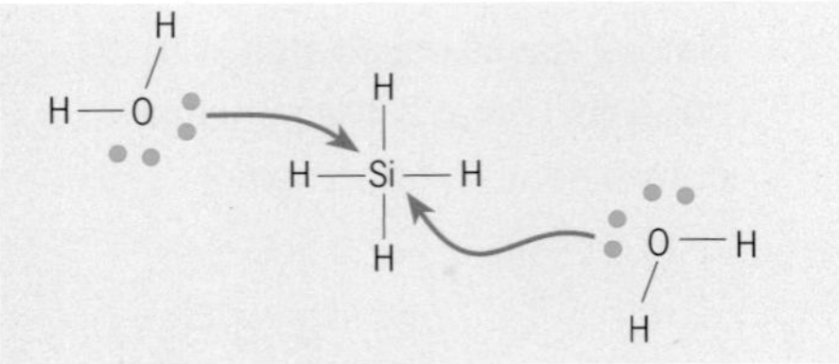

Fig 9.4 SiH_4 reacting with water molecules. Notice that when the water molecules form their covalent bonds with this silane, there are 12 outer-shell electrons around the silicon atom

CARBON CAN FORM STRONG BONDS WITH ITSELF

The ability to form bonds between atoms of the same element in a compound is called **catenation**. While this is not unique to carbon, the C–C bond energy is very high, as you can see from Table 9.1. This means that carbon can form chains of carbon atoms. On page 164 we described octane, a component of petrol, which has a chain of eight carbon atoms:

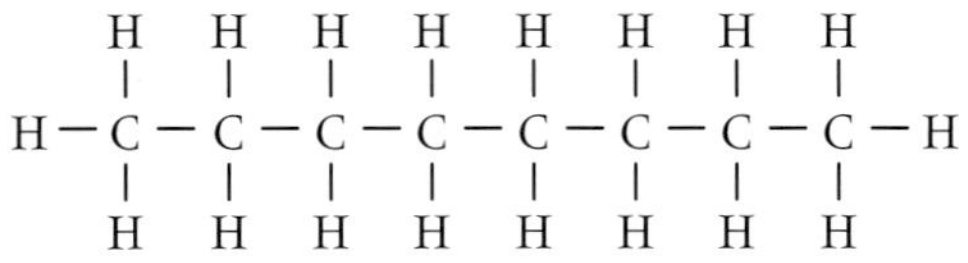

```
    H   H   H   H   H   H   H   H
    |   |   |   |   |   |   |   |
H − C − C − C − C − C − C − C − C − H
    |   |   |   |   |   |   |   |
    H   H   H   H   H   H   H   H
```

QUESTIONS 2–4

2 Why doesn't carbon form ionic bonds?

3 Draw the dot-and-cross diagrams for methane and water. You can check your diagrams by looking back to pages 74 and 75.

4 When water donates a lone pair of electrons to silicon, what is the name of the bond formed? See page 76 if you are not sure.

However, in poly(ethene), commonly called polythene, there are thousands of carbon atoms in a chain:

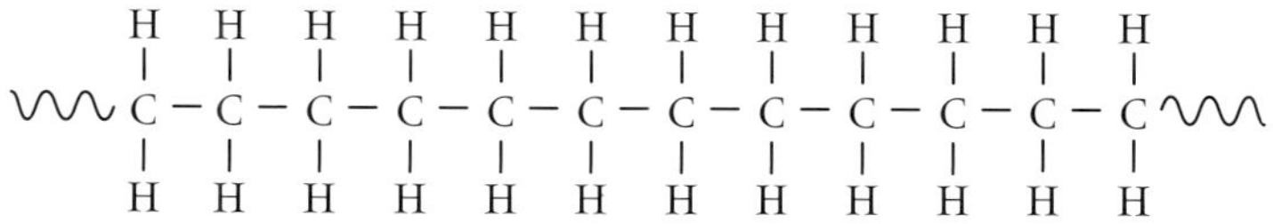

```
     H   H   H   H   H   H   H   H   H   H   H   H
     |   |   |   |   |   |   |   |   |   |   |   |
~~~ C - C - C - C - C - C - C - C - C - C - C - C ~~~
     |   |   |   |   |   |   |   |   |   |   |   |
     H   H   H   H   H   H   H   H   H   H   H   H
```

Both octane and poly(ethene) are energetically unstable in air under normal conditions, because their products, carbon dioxide and water, are at a much lower energy level. The reaction with oxygen is **spontaneous** at room temperature. This means that once started the reaction takes place of its own accord, but poly(ethene) bags and petrol don't suddenly catch fire when exposed to air. This is because of the large activation energy, which makes them kinetically stable (see page 165). This activation energy is the energy required to break the C–C and C–H bonds.

Although silicon atoms form short chains with each other, they are not stable under normal conditions. Take, for example, the silane:

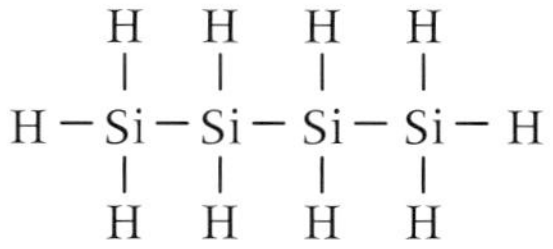

```
     H    H    H    H
     |    |    |    |
H - Si - Si - Si - Si - H
     |    |    |    |
     H    H    H    H
```

This compound is unstable and decomposes to silicon and hydrogen at room temperature if no oxygen is present. In air, it bursts into flame for two reasons. First, the reaction with oxygen has a low energy of activation; second, the Si–O bond is very strong. This means that silicon dioxide, which is sand, is energetically very stable. The stability of the Si–O bond is the reason why so much of the Earth's rocks and soils are made up of silicon dioxide and silicates.

No other element forms chains with itself like carbon does. But its uniqueness does not stop at chains – it can form rings as well, such as the example in Fig 9.5.

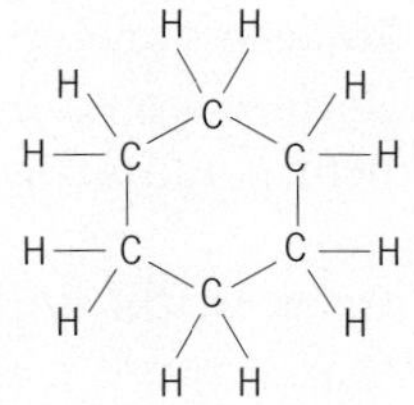

Fig 9.5 Cyclohexane: an example of a ring structure

Table 9.1 Some average bond energies (enthalpies) between atoms of elements near carbon in the Periodic Table. Also included are bond energies of C—H, Si—H and Si—O

Bond	Bond energy/kJ mol^{-1}
C—C	347
Si—Si	226
N—N	158
P—P	198
C—H	413
Si—H	318
Si—O	466

REMEMBER THIS

Bond enthalpy or **bond energy** is the energy required to break a bond between two atoms in a gaseous molecule, leaving two separate atoms. See page 159 for more information. The higher the bond enthalpy, the shorter the bond length and the higher the bond strength.

You can find out more about spontaneous reaction, activation energy and kinetic stability by looking back at Chapter 8 and also by reading Chapter 26. Poly(ethene) is discussed in detail on pages 288 and 395. Information about the structure of silicon dioxide is in Chapter 21.

Carbon can form double and triple bonds

The strength of the C–H and C–C bonds means that compounds such as octane are fairly unreactive with other chemicals, the reaction with oxygen being a notable exception. However, the inclusion of double and triple carbon–carbon bonds changes all this. Ethene, for example, is the backbone of the organic chemical industry, and it is the π-bond that makes it so reactive (see Fig 9.6).

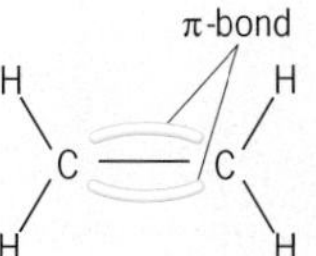

Fig 9.6 A molecule of ethene showing the π-bond

See page 80 to remind yourself about the bonding in ethene. There is much more about its chemistry in Chapter 14.

4 REPRESENTING HYDROCARBON MOLECULES

MOLECULAR AND STRUCTURAL FORMULAE

We first met molecular formulae on page 3. Structural formulae show more information about the arrangement of atoms in a molecule. The molecular formula of methane is CH_4 and that of propane is C_3H_8. These can be drawn out to show how the atoms are bonded together.

These representations of molecules are often referred to as **displayed formulae**, **graphical formulae** or even **full structural formulae**. We shall use the term *displayed formula* when we draw out all the atoms and

REMEMBER THIS

A molecular formula tells you how many atoms of each element there are in a molecule of the compound (see page 3).

QUESTION 5

5 What is the molecular formula of octane? You can see its structural formula on page 178.

REMEMBER THIS
In this book we use structural formula to mean shortened structural formula.

See page 80 in Chapter 4 to remind yourself about why carbon forms a tetrahedral arrangement.

QUESTION 6

6 Draw the displayed formulae (showing all atoms and bonds) of ethane, CH_3CH_3 and pentane, $CH_3CH_2CH_2CH_2CH_3$.

See page 80 to remind yourself about the shape of ethene.

QUESTION 7

7 **a)** Work out which of the following compounds will be unsaturated from the molecular formulae:

C_4H_{10}, C_7H_{14} and C_9H_{18}

(Hint: remember that carbon always has four bonds.)

b) Draw the displayed formulae of the following compounds:

$CH_3CH_2CH{=}CH_2$, $CH_3CH{=}CHCH_3$

What do you notice about the molecular formulae of these two compounds?

bonds. Structural formulae can also be written in as shortened form. For example, the shortened (or condensed) structural formula of propane can be shown as:

$$CH_3{-}CH_2{-}CH_3 \text{ or even } CH_3CH_2CH_3$$

None of these accurately represents the three-dimensional shape of the molecule. The ball-and-stick models are shown in Fig 9.7.

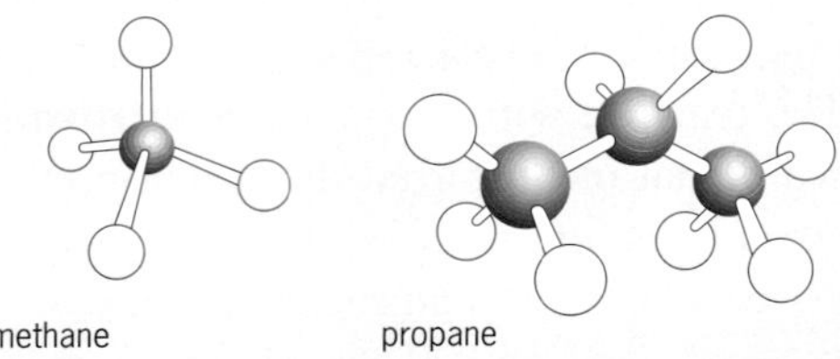

Fig 9.7 Ball-and-stick models of methane and propane

Notice the tetrahedral arrangement of the hydrogen atoms around the carbon atoms. This can be represented as in Fig 9.8.

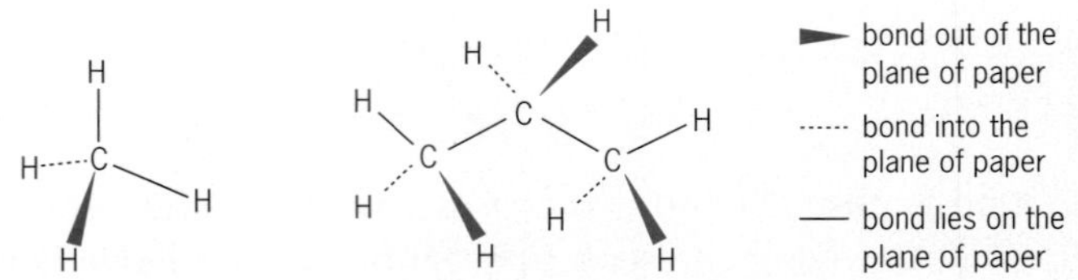

Fig 9.8 Shape diagram of methane and propane

To keep drawing three-dimensional representations of the shapes of molecules would be very inconvenient, which is why we use displayed or structural formulae instead.

Compounds with carbon–carbon double bonds

Ethene and propene have carbon–carbon double bonds in their compounds. They are known as **unsaturated hydrocarbons**, because the double bond can allow them to bond with more atoms. Any hydrocarbon with double or triple carbon–carbon bonds in it is called **unsaturated**. Hydrocarbons with no double or triple bonds are called **saturated hydrocarbons**.

The structural formulae of ethene and propene are:

$$CH_2{=}CH_2 \quad \text{and} \quad CH_3CH{=}CH_2$$

Their displayed formulae are shown in Fig 9.9. Notice that we show the bonds around a double-bonded carbon at about 120°.

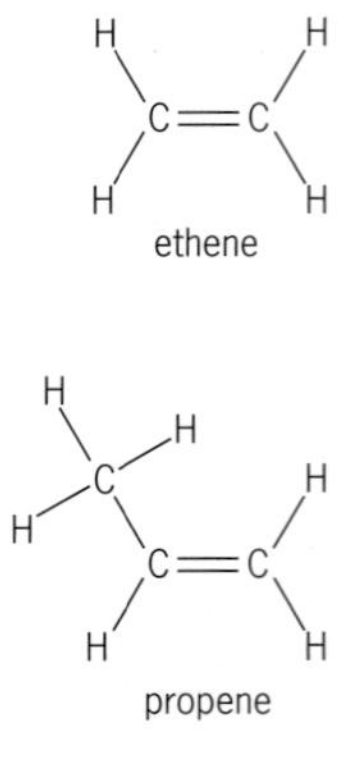

Fig 9.9 The displayed formulae of ethene and propene

Ring compounds

As we noted earlier in the chapter, carbon can form rings as well as chains, which adds enormously to the variety of possible organic compounds. Three examples are given in Fig 9.10.

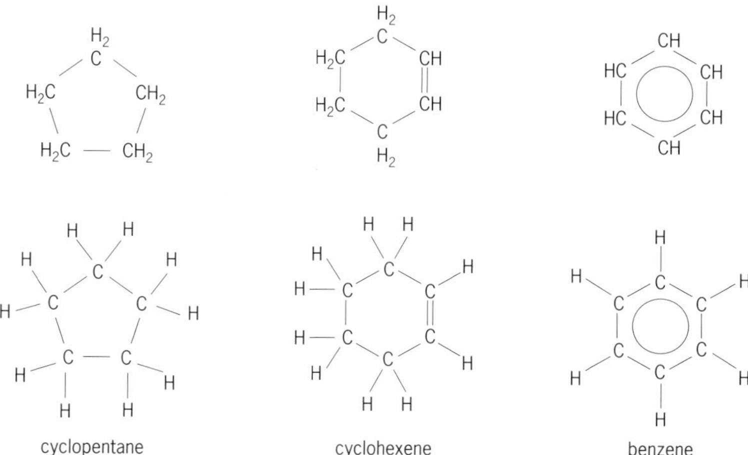

Fig 9.10 The structural and displayed formulae of cyclopentane, cyclohexene and benzene

Benzene has a ring inside its carbon skeleton, and from the above formula it does not seem to have four bonds around each carbon. The ring represents a **delocalised π-cloud** of electrons, which gives benzene a different set of properties from other ring compounds. **Delocalised electrons** are electrons that do not belong to any one carbon atom.

Skeletal formulae

Formulae that show only how the carbon atoms are bonded are called **skeletal formulae**. These depict the carbon skeleton of the molecule without any hydrogen atoms. However, when there are other atoms (such as oxygen or nitrogen) in the molecule, these are shown. Hexane is a hydrocarbon found in crude oil and used in petrol. Its skeletal formula, shown in Fig 9.11, is the typical zig-zag shape of a hydrocarbon chain.

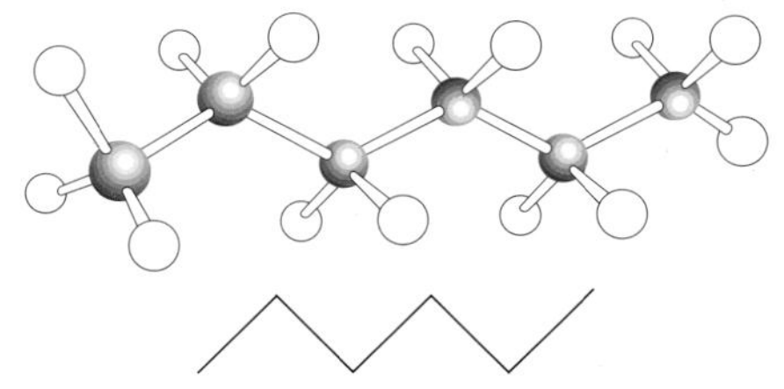

Fig 9.11 Ball-and-stick model of hexane and its skeletal formula

Oct-3-ene has the structural formula:

$$CH_3CH_2CH_2CH_2CH{=}CHCH_2CH_3$$

Its skeletal formula is shown in Fig 9.12.

The skeletal formulae you will encounter most are those of ring compounds. Fig 9.13 illustrates the skeletal formulae of the compounds in Fig 9.10.

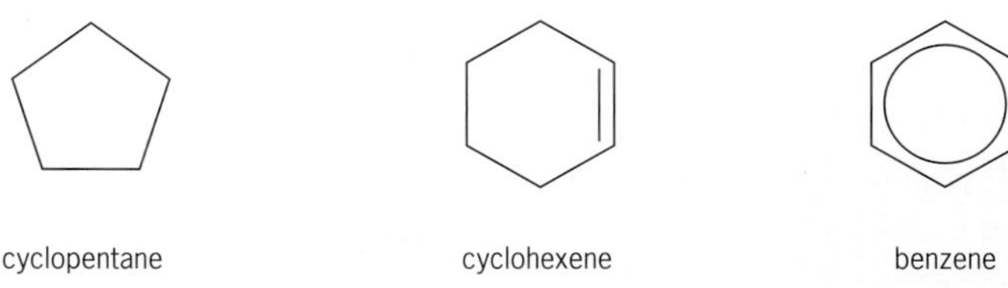

Fig 9.13 The skeletal formulae of cyclopentane, cyclohexene and benzene

QUESTION 8

a) Draw the displayed formula of cyclobutane, C_4H_8.
b) Compare this molecular formula with the molecular formulae in part **b)** of Question 7. What do you notice?

See Chapter 15 for more information on benzene and its structure.

QUESTION 9

Draw the skeletal formulae of:
a) octane;
b) $CH_3CH_2CH{=}CH_2$;
c) $CH_3CH{=}CHCH_3$.
Note: there are two ways of drawing the last skeletal formula. This is because there is no rotation about a double bond (see Chapter 14).

Fig 9.12 The skeletal formula of oct-3-ene

QUESTION 10

Draw the structural formulae and displayed formulae of the compounds below:

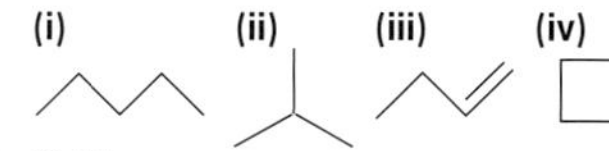

Fig 9.14

5 ISOMERISM

If you have done the questions in the previous section, you may have noticed that sometimes compounds with the same molecular formula can have different arrangements of their atoms. This is called **isomerism.**

Molecules with the same molecular formula but with different arrangements of their atoms are called **isomers.**

There are two types of isomerism: **structural isomerism** and **stereoisomerism.** These are summarised in Fig 9.15.

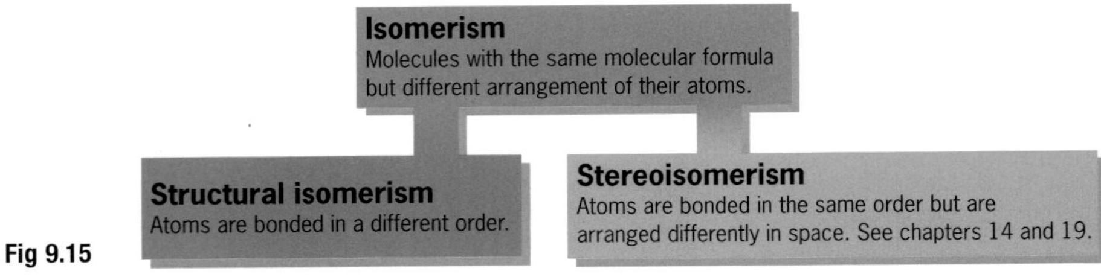

Isomerism
Molecules with the same molecular formula but different arrangement of their atoms.
Structural isomerism
Atoms are bonded in a different order.
Stereoisomerism
Atoms are bonded in the same order but are arranged differently in space. See chapters 14 and 19.

Fig 9.15

We shall deal in this chapter with structural isomerism. From Fig 9.15 you can see that **structural isomers have the same molecular formula, but different structures.** Their structural formulae are different because the atoms are bonded in a different order.

Consider propane, found in liquefied petroleum gas fuel. It has the molecular formula C_3H_8, but only *one* arrangement is possible for its molecule. Even though the carbon atoms can rotate about a single bond, there is still only one structure, which is shown in Fig 9.16.

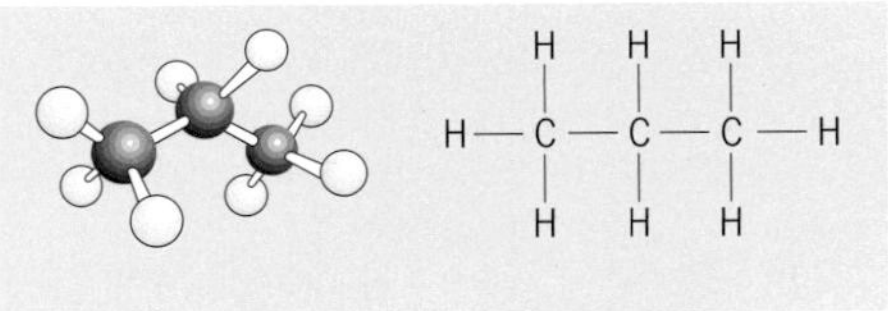

Fig 9.16 Ball-and-stick model of propane and its displayed formula

The displayed formula used to show the rotation of the carbon atom in Fig 9.17 looks exaggerated, so it is tempting to think that it is not the same structure. The carbon atom chain has been shaded so that you see immediately that the three carbon atoms are still in one unbranched chain.

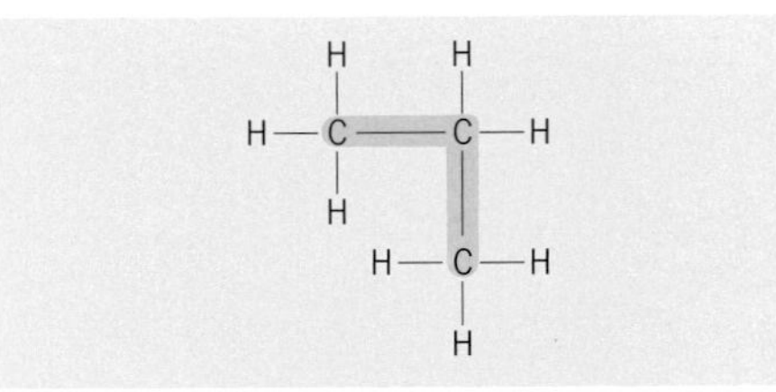

Fig 9.17 The tinted area shows that the carbon atoms are still part of the same unbranched chain

Now we consider the molecular formula C_4H_{10}. It is possible to have the two different structures shown in Fig 9.18. These two isomers have different properties, even though they have the same molecular formula. For example, butane boils at a higher temperature than methylpropane.

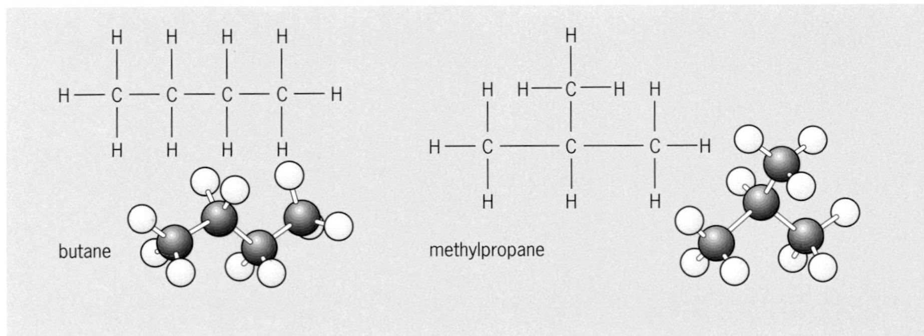

Fig 9.18 Ball-and-stick models of butane and methylpropane and their structural formulae

As the number of carbons in the molecular formula goes up, the number of isomers increases dramatically. For the molecular formula $C_{10}H_{22}$, there are 75 isomers, but when the formula is $C_{30}H_{62}$, the number of possible isomers is more than 400 million! However, hardly any of these have been synthesised and few occur in nature, but these numbers do illustrate how carbon forms so many different compounds.

? QUESTION 11

11 Explain why the structures below are not additional isomers of butane:

a)

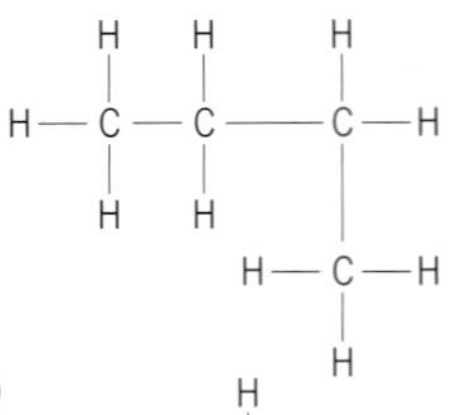

b)

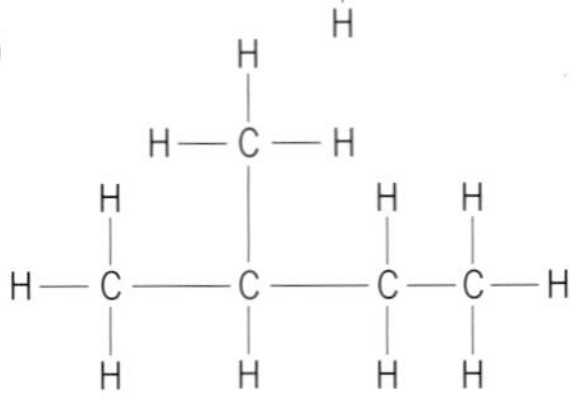

EXAMPLE

Q Draw the displayed formulae (full structural formulae) of the possible isomers of C_5H_{12}.

A You are unlikely to be asked to find the number of isomers of hydrocarbons with molecular formulae of more than seven carbon atoms. The only way you can approach this problem is by making attempts and then improving them, but try to be systematic.

Always start with the longest hydrocarbon chain you can make from the given formula, as in Fig 9.19.

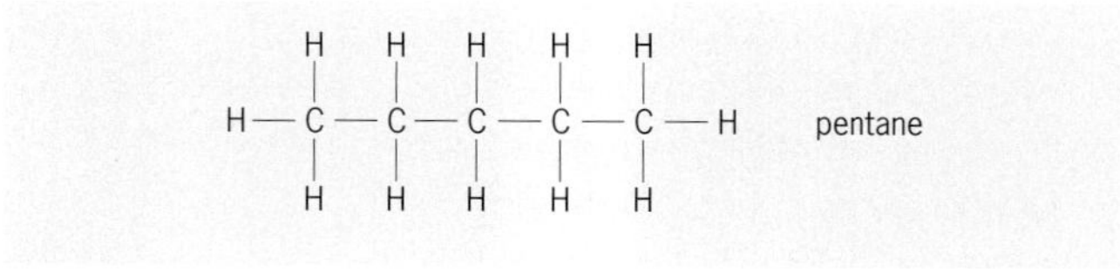

pentane

Fig 9.19

Now look for possible branches. But remember, the carbon atoms can rotate about a single bond. In this case, there are only two more possible structures that have branches from the longest chain, as in Fig 9.20.

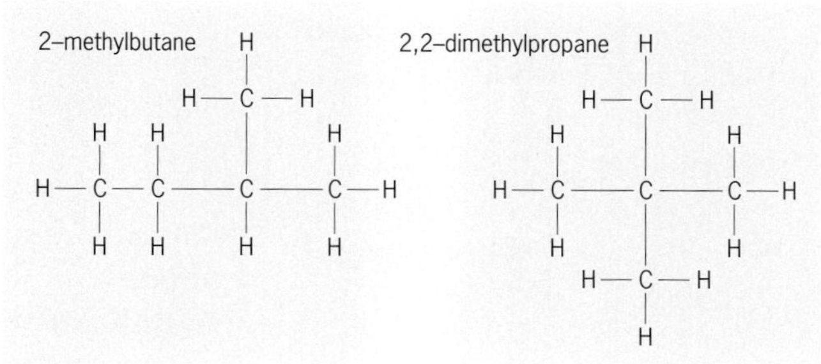

2–methylbutane

2,2–dimethylpropane

Fig 9.20

A common mistake is to draw a third structure, as shown in Fig 9.21. However, this is actually the structure of 2-methylbutane, and all that has happened is that the molecule has been turned round.

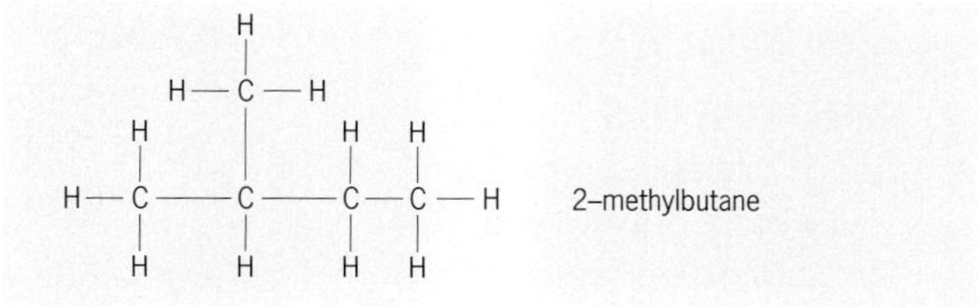

2–methylbutane

Fig 9.21

We have not yet explained how we name the different hydrocarbons. This is easier than it looks. We deal with naming some of these on pages 192–194.

6 MORE STRUCTURAL ISOMERS

So far, we have only looked at saturated hydrocarbon chains, but structural isomers occur whenever there are different ways of bonding the atoms together – for example, when the position of a double bond changes. An example of this is shown in Fig 9.22.

Fig 9.22 Structural isomers

Earlier in this chapter, we said that carbon could bond to atoms other than hydrogen. When this happens, the properties of the molecule change. When we substitute an atom, or a group of atoms, into a hydrocarbon molecule, we call the atom or group a **functional group**.

QUESTION 12

12 Draw the structural formulae of isomers with the molecular formula C_6H_{14}.
(Hint: if you draw more than five, they will not all be different isomers.)

QUESTION 13

13 **a)** What is the molecular formula of the two isomers in Fig 9.18?
b) Why is the structure below not another isomer?

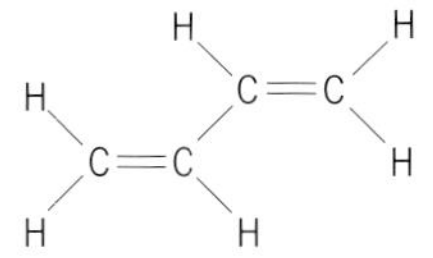

There is more about functional groups on page 190.

A functional group largely determines the set of properties a molecule possesses. The C=C double bond in a molecule is a functional group, because it gives a characteristic set of properties to the molecule. Cl is a halogen functional group and the molecular formula C_3H_7Cl has two isomers, which are shown in Fig 9.23.

1-chloropropane	H H H │ │ │ H—C—C—C—Cl │ │ │ H H H
2-chloropropane	H H H │ │ │ H—C—C—C—H │ │ │ H Cl H

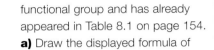

Fig 9.23 The two isomers of C_3H_7Cl

QUESTION 14

14 There are two possible isomers with the formula C_2H_6O. One is an ether with a functional group C—O—C. The other isomer has a different functional group and has already appeared in Table 8.1 on page 154.
a) Draw the displayed formula of both isomers. You can check one of the displayed formulae by looking at page 264.
b) Identify the functional group of the isomer of ether. You can check your answer by looking at Table 9.4 on page 190.

7 THE HYDROCARBONS IN CRUDE OIL

Crude oil is a mixture, and because of this its composition varies. Crude oil is composed mainly of hydrocarbons, and we have already seen that there can be several hundreds of these because of carbon's ability to form stable chains, branched chains and rings. No two deposits of oil contain exactly the same hydrocarbons in the same proportions. However, we can divide the hydrocarbons found in oil into three major classes of compound: **alkanes**, **cycloalkanes** and **arenes** (also known as aromatic hydrocarbons). We have met representatives of all three classes earlier in this chapter.

ALKANES

Alkanes are saturated hydrocarbons. They can either be branched chains or straight chains. Table 9.2 summarises some of the alkanes featured in this chapter.

REMEMBER THIS

Saturated hydrocarbons have no double bonds or triple bonds. The carbon atoms are bonded to the maximum possible number of hydrogen atoms.

Table 9.2 Some of the alkanes discussed in this chapter

Alkane	Molecular formula	Structural formula	Displayed formula
methane	CH_4	CH_4	H │ H—C—H │ H
ethane	C_2H_6	CH_3CH_3	H H │ │ H—C—C—H │ │ H H
propane	C_3H_8	$CH_3CH_2CH_3$	H H H │ │ │ H—C—C—C—H │ │ │ H H H
butane	C_4H_{10}	$CH_3CH_2CH_2CH_3$	H H H H │ │ │ │ H—C—C—C—C—H │ │ │ │ H H H H
methylpropane	C_4H_{10}	CH_3CHCH_3 │ CH_3	H H H │ │ │ H—C——C——C—H │ │ │ H H—C—H H │ H

QUESTION 15

15 What is the name given to molecules, such as butane and methylpropane, that have the same molecular formula?

Look at the molecular formulae of the alkanes in Table 9.2 and you should spot a mathematical relationship. For every carbon atom in each formula, there are twice that number of hydrogen atoms, plus another two. This can be represented by the **general formula C_nH_{2n+2}**, in which n is the number of carbon atoms. Let's try out this general formula with methane:

$n = 1$ because there is one carbon atom.

So the molecular formula for methane is:

$$C_1H_{(2 \times 1)+2} = CH_4$$

QUESTION 16

What is the molecular formula of alkanes with nine carbon atoms?

CYCLOALKANES

Most of the cycloalkanes found in crude oil are based on five- or six-membered carbon rings, as shown in Fig 9.24.

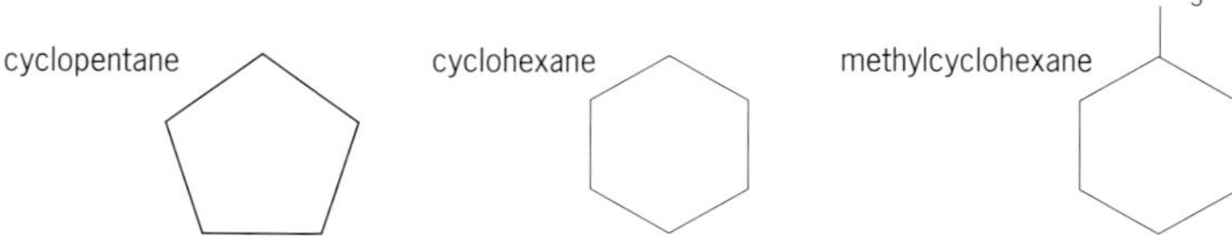

Fig 9.24 Three examples of cycloalkanes

See Chapter 15 for more information on arenes.

ARENES

Arenes or **aromatic hydrocarbons** contain at least one benzene ring (see Fig 9.25). The name 'aromatic' originally came from the characteristic smell of some naturally occurring compounds that contain a benzene ring.

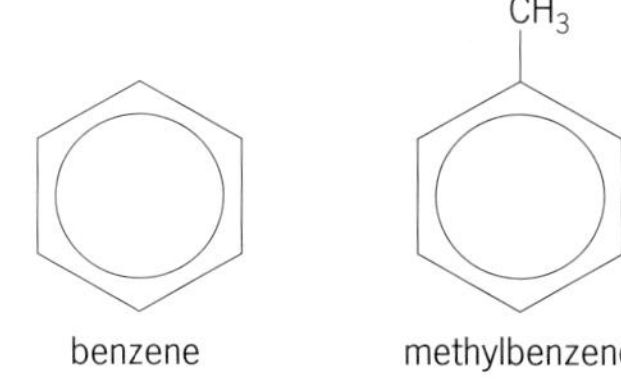

Fig 9.25 Two examples of arenes

SCIENCE IN CONTEXT

Peak oil

It may seem obvious, but at some stage the world production of oil has to reach a peak. Dr M. King Hubbert, a US scientist, first put this idea forward in 1956. The graph he produced is called Hubbert's Curve (see Fig 9.26). He used his curve to accurately predict the decline in US oil production in the 1970s. Now, many scientists believe we may reach a peak in oil production by 2010, after which oil production will start to decline. In 2007 the world produced almost 90 million barrels of oil a day. Each barrel contains 159 litres. This means that every second our world consumed 166 000 litres of oil. Our thirst for oil is increasing as the economies of China and India develop at a rapid rate and older industrialised nations, such as the UK and USA, fail to reduce their demand. But not all scientists believe we will reach Hubbert's peak. As the price of oil spirals ever upwards there will be new incentives to find new oil fields and produce more oil from existing ones. However, if oil production does peak as predicted, then the world faces a much more uncertain future.

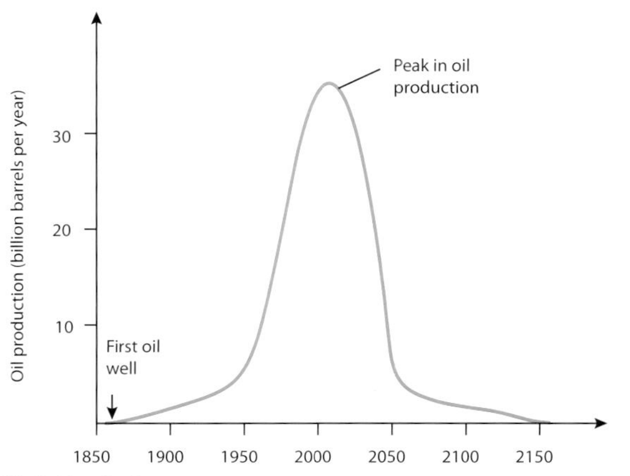

Fig 9.26 Hubbert's Curve

8 MAKING CRUDE OIL USEFUL

Crude oil as it comes out of the ground is of little practical use. To make it useful, it must first be separated into its components, which is done by **fractional distillation**.

FRACTIONAL DISTILLATION

Fractional distillation is the separation of a mixture of compounds by their different boiling points. When crude oil is separated there are five major fractions, as shown in Table 9.3.

Table 9.3 The major fractions of crude oil and their uses

Fraction	Boiling point range/°C	Number of carbon atoms	Percentage of crude oil	Uses
refinery gas	less than 20	C_1 to C_4	1 to 2	fuel and as a feedstock for petrochemicals
gasoline/ naphtha	20–120	C_5 to C_{12}	15 to 30	petrol for transport and as a feedstock for petrochemicals; (the part of the fraction so used is called naphtha)
kerosene	120–220	C_{10} to C_{16}	10 to 15	fuel for jets, paraffin for heating
diesel oil	220–350	C_{15} to C_{25}	15 to 25	fuel for transport, power plants and heating
residue	more than 350	$>C_{25}$	40 to 50	oil-fired power stations, polishing waxes, lubricating oils, bitumen on roads

When crude oil reaches a refinery it is heated to about 400 °C and much of it vaporises. As vapour and liquid, it is fed into a fractionating column, which is about 60 metres high (see Fig 9.27). The liquid collects at the bottom of the column and is called the residue. The vapour passes on up the column and as it cools it turns back to liquid, collecting in trays at various heights. The smaller the vapour molecules, the further the vapour can travel up the column before it condenses. This is because smaller molecules form liquids with lower boiling points (see also pages 194–196). At the top of the column, gases that contain very small molecules are collected. These hydrocarbons boil below 20 °C.

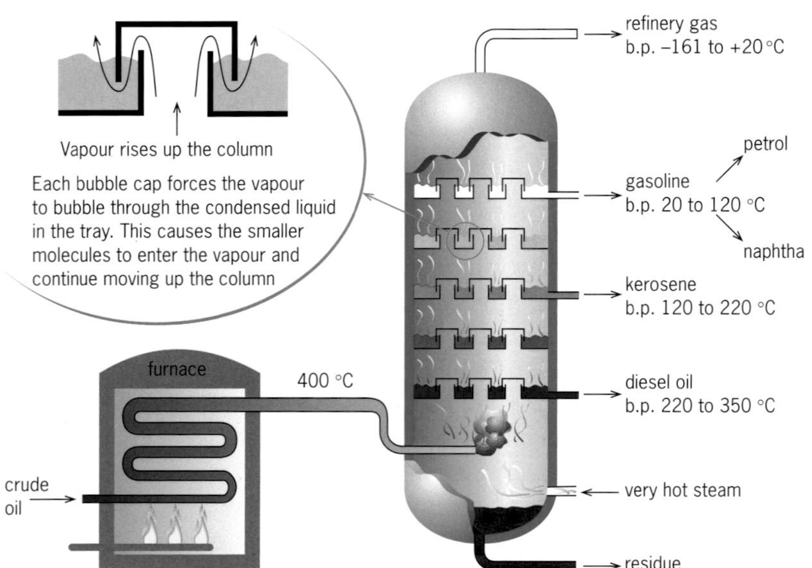

Fig 9.27 The primary fractional distillation of crude oil takes place in a fractionating column. Here the column is cut away to show some of the trays. Each tray contains many bubble caps, although only three are shown

This is the first distillation – the **primary distillation** – and the fractions coming from the column are further separated by a variety of processes. Before a fraction can be used, the sulfur must be removed (this forms a valuable by-product).

Fig 9.28 A fractionating column in a refinery

MEETING THE DEMAND: USING ALL THE FRACTIONS

The gasoline fraction is the main source of petrol for car and other vehicle engines, but the amount of gasoline fraction produced is never enough to meet the demand for petrol. Typically, about 40 per cent of the output from a distillation column may be required as petrol. Compare this percentage with that given in Table 9.3 for the gasoline fraction. So, other fractions have to be altered by chemical processes to produce the extra petrol needed.

CRACKING LARGER MOLECULES

Petrol for use in cars and other vehicles requires alkanes in the range from C_5 to C_{10}. To meet demand, longer alkanes from other fractions are shortened by being split. This process is known as **cracking**. There are two basic processes for splitting alkane chains:

- using high temperatures and pressures, which is called **thermal cracking**;
- using catalysts, which is called **catalytic cracking** (or 'cat cracking' for short), where the pressure is slight but the temperature is still high.

Whichever process is employed, sufficient energy must be supplied to split the very strong C–C and C–H bonds (look back at Table 9.1, page 179). The advantage of thermal cracking is that molecules in the **residue** can be cracked. Catalytic cracking only works on the **distillate** (the liquid that has been distilled) such as that from the diesel oil fraction. However, catalytic cracking tends to produce more branched-chain alkanes, which are better for use in petrol as they promote efficient combustion.

Take, for example, the molecule of dodecane, $C_{12}H_{26}$:

$$CH_3CH_2CH_2CH_2CH_2CH_2CH_2CH_2CH_2CH_2CH_2CH_3 \xrightarrow{\text{zeolite catalyst}}$$

$$CH_3\underset{\displaystyle CH_3}{\overset{}{C}}HCH_2CH_2\underset{\displaystyle CH_3}{\overset{}{C}}HCH_3 + CH_3CH_2CH{=}CH_2$$

(with CH_3 branches on the second and fifth carbons: $CH_3CH(CH_3)CH_2CH_2CH(CH_3)CH_3$)

Notice that in this case the molecule has been split and a branched-chain alkane produced. Branched chains are useful in petrol blending because they have higher octane numbers (see page 198). Of course, the molecule can split in almost any way, so a huge variety of hydrocarbons, both straight chains and branched, are produced. These are separated by fractional distillation.

The composition of petrol is discussed in greater detail in Chapter 10.

QUESTION 17

a) In which fraction will dodecane be found? (Hint: see Table 9.3.)

b) Write equations to show the cracking reactions of dodecane to form:

i) ethene and a straight-chain alkane;

ii) propene and a branched-chain alkane.

If you cannot remember the formulae for ethene and propene, look at page 180.

Bond breaking in thermal and catalytic cracking

The main difference between thermal cracking and catalytic cracking is the way in which covalent bonds in the carbon chains are broken. A covalent bond consists of a pair of electrons that is shared between two atoms (see page 74). The splitting or **fission** of the covalent bond in thermal cracking leaves each species produced with one electron of this pair (Fig 9.29). This type of bond splitting is called **homolytic fission**.

When a species has an unpaired electron it is called a **free radical** (or sometimes just **radical**). The unpaired electron makes the free radical very reactive and in the reaction mixture it will react again to produce either an alkane or an alkene.

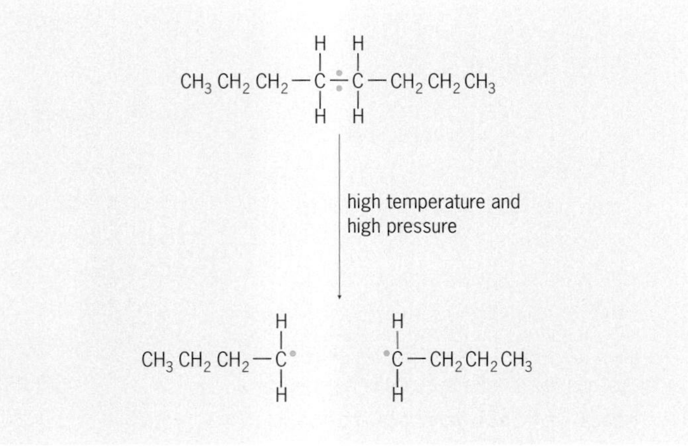

Fig 9.29 Homolytic fission in thermal cracking produces two free radicals

You can read more about homolytic fission in chapters 10 and 12.

Thermal cracking requires high temperatures (between 400 and 900 °C) and high pressures (up to 7000 kPa), so it is expensive. On completion, it produces a high percentage of alkenes, which are used to produced most of the organic chemicals on which our modern world relies.

Catalytic cracking involves a different type of bond splitting known as **heterolytic fission**. In this case the shared pair of electrons both go to one of the species produced when the covalent bond is broken. A positive and a negative ion are produced.

QUESTION 18

18 The cracking of alkanes always produces at least one alkene molecule. Why is this?

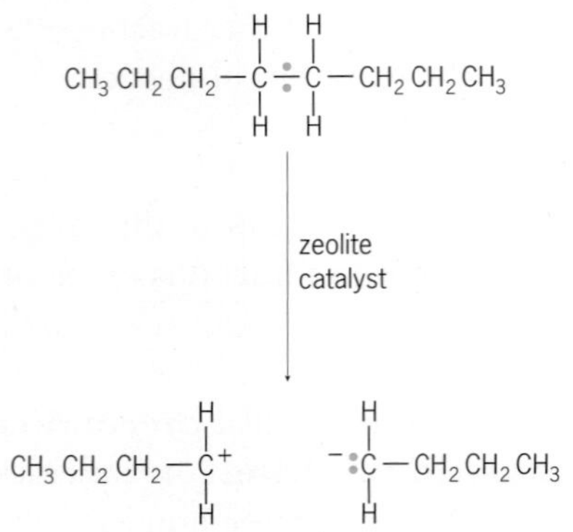

$CH_3CH_2CH_2-CH_2 \div CH_2-CH_2CH_2CH_3$

zeolite catalyst

$CH_3CH_2CH_2-CH_2^+ \quad ^-{:}CH_2-CH_2CH_2CH_3$

Fig 9.30 Heterolytic fission produces a positive and negative ion

The positive ion in Fig 9.30 is called a **carbocation**. This reacts with other alkane molecules to produce a high percentage of branched-chain alkanes, which are ideal for use in petrol (see Chapter 10 page 198).

Catalytic cracking is carried out at about 450 °C with a slight pressure and is the major method of producing petrol and also aromatic hydrocarbons.

QUESTION 19

19 **a)** Name the molecules shown in Fig 9.31 and write balanced equations for the two reforming reactions.
b) What valuable by-product is produced in the catalytic-reforming reaction?

REFORMING AND ISOMERISATION

Catalytic **reforming** is used to modify molecules to suit demand. Straight-chain alkanes can be converted into cycloalkanes and arenes. Different catalysts are used, depending on the product required. The catalysts used are often bimetallic catalysts such as platinum–rhodium. An example is given in Fig 9.31. Notice that during reforming, whatever happens to the molecule, the total number of carbon atoms remains the same.

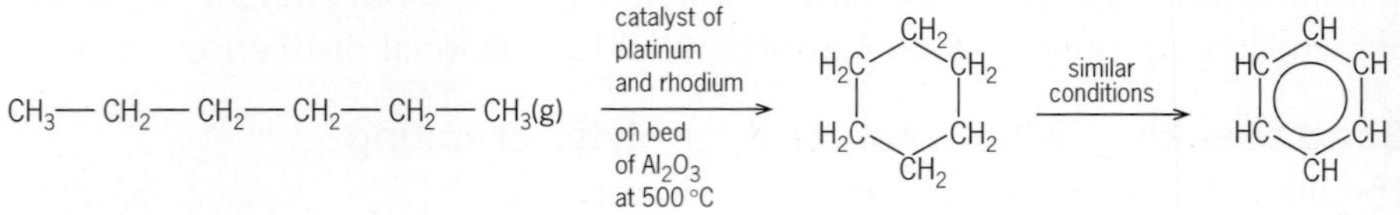

$CH_3-CH_2-CH_2-CH_2-CH_2-CH_3(g)$ → (catalyst of platinum and rhodium on bed of Al_2O_3 at 500 °C) cyclohexane (H_2C, CH_2 ring) → (similar conditions) benzene (HC, CH ring)

Fig 9.31 Catalytic reforming to produce cycloalkanes and aromatic compounds

QUESTION 20

20 **a)** Why is the process that produces branched-chain alkanes from straight-chain ones called isomerisation?
(Hint: look back to page 182.)
b) Read the Science in Context box on zeolites. Explain how zeolites can be used to separate branched-chain alkanes from unbranched ones.

We have already mentioned the need to produce branched-chain alkanes to increase the octane number of petrol. This can be done by taking a straight-chain alkane, heating it to break one of the C–C bonds and allowing the molecule to reform, which often produces a branched chain. This process is called **isomerisation**.

$$CH_3CH_2CH_2CH_2CH_3 \xrightarrow{\text{platinum catalyst; heat}} CH_3CH_2CH(CH_3)CH_3$$

Zeolites

Zeolite means 'boiling stone' (from the Greek *zeein* for boiling and *lithos* for stone). The name was coined by a Swedish scientist, Baron Cronstedt, in 1756, when he discovered a mineral that released steam and appeared to boil when heated. This name now applies to about 40 naturally occurring minerals that display the same property when heated. Many more zeolites are made synthetically. The steam comes from water trapped in tiny pores and channels in the zeolite. These channels are the key to understanding how this remarkable substance is such an effective catalyst. It took some 200 years from Cronstedt's discovery until synthetic zeolites were first used commercially in 1959 to crack hydrocarbons.

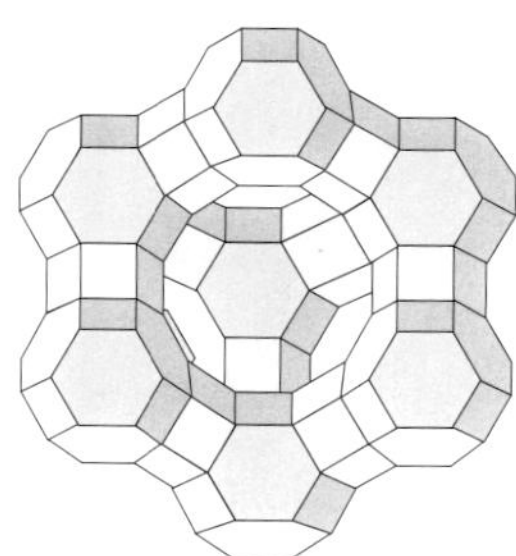

Fig 9.32 Zeolite Y is used in cracking hydrocarbon molecules

Zeolites are a family of aluminium silicates. Their silicon, oxygen and aluminium atoms provide a regular network of pores and interconnecting channels, rather like a sponge (see Fig 9.32). The size of the pores is critical to the use of zeolites as catalysts for the host chemicals. Pore sizes (in picometres, 10^{-12} m) range from 300 pm to 1000 pm, which is why the pores trap water molecules (which have an effective diameter of about 270 pm). If the pore size is large enough, the zeolite can accommodate hydrocarbon molecules, which are cracked inside the structure.

Synthetic rather than natural zeolites are now used for catalytic cracking, and there is much research into ways to alter the sizes of pores and channels by inserting different atoms into the structure. Zeolites have saved the oil industry billions of pounds because they are so efficient and produce more commercially desirable products than the clay catalysts previously used.

The uses for zeolites don't stop with cracking. In New Zealand, which has vast reserves of natural gas but hardly any oil, a synthetic zeolite, ZSM-5, converts methanol into petrol. Methane is first converted into methanol, then ZSM-5 removes water from the molecules to leave hydrocarbon chains trapped inside the zeolite. These are then driven out of the zeolite as petrol.

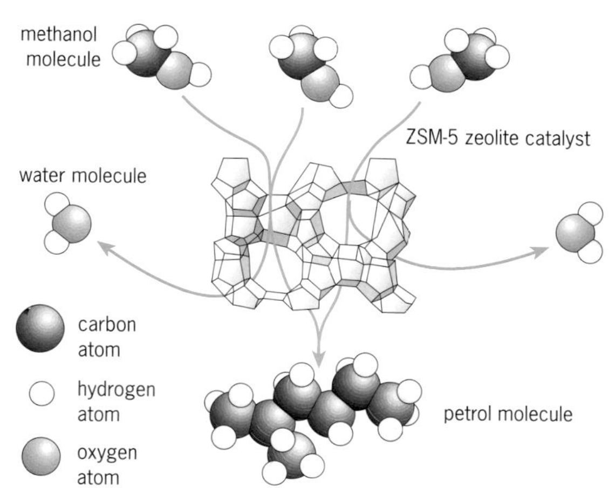

Fig 9.33 Zeolite ZSM-5 converts methanol produced from natural gas into petrol

Zeolites in washing powder

Almost certainly you have zeolites in your house. Look at a packet of washing powder and you will probably find more than a quarter of the powder is made from zeolites. The function of zeolites here is to remove calcium ions, which cause water hardness, and replace them with sodium ions, which soften the water. About three-quarters of the world's production of zeolites ends up in detergents.

Zeolites also function as molecular sieves. For example, detergents need to be biodegradable. The largest part of the detergent molecule is a hydrocarbon chain. If this chain is branched, microorganisms at the sewage works cannot break down the detergent.

At the refinery, unbranched hydrocarbons for detergent manufacture are separated from those with branched chains by passing them into the channels of a zeolite. The branched chains are too large to enter the zeolite, so just the unbranched hydrocarbons pass through it.

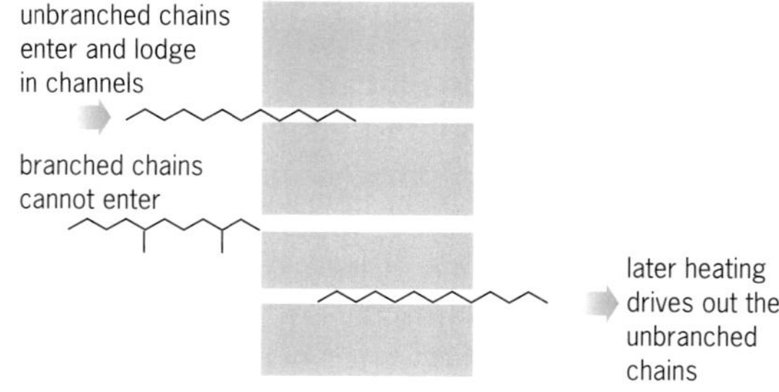

Fig 9.34 Zeolite acting as a molecular sieve. Unbranched chains can enter channels, whereas branched chains cannot. The zeolite is removed and the unbranched chains are driven from the zeolite by heating

There is speculation that zeolites may have provided the organising influence that produced the biologically active molecules that form the basis of life on Earth.

9 PETROCHEMICALS

More than 90 per cent of crude oil is burnt; the rest forms the basis of the petrochemical industry, which produces almost all of our organic chemicals. We have already said that alkanes are rather unreactive. However, the cracking process produces small-chain alkenes, of which ethene is the most important. Ethene provides a synthetic route to many organic chemicals. The reactivity of its double bond provides a way to insert many different functional groups and thereby allow the full potential of hydrocarbons to be realised.

FUNCTIONAL GROUPS

A functional group is an atom or group of atoms that determines the properties of a molecule. It is usually the functional group that provides the reaction site of the molecule. Table 9.4 summarises some of the functional groups you will meet in subsequent chapters.

Table 9.4

Functional group	Name of class of compound	Typical example	Use of example
—C—X X = F, Cl, Br, I	halogenoalkanes	chloroethane, C_2H_5Cl	starting material for the production of time-release capsules for medicines
—O—H	alcohols	ethanol, C_2H_5OH	industrial solvent
—C(=O)—H	aldehydes	ethanal, CH_3CHO	making synthetic rubber
>C=O	ketones	propanone, CH_3COCH_3	industrial and domestic solvents, e.g. nail polish remover
—C(=O)—OH	carboxylic acids	ethanoic acid, CH_3COOH	production of pharmaceuticals and in vinegar
—N(H)—H	amines	ethylamine, $C_2H_5NH_2$	important intermediate for many industrial compounds

SUMMARY

After studying this chapter, you should know that:

- Carbon is unique in its formation of stable chains produces and rings, millions of different chemicals.
- Crude oil is a hydrocarbon mixture of alkanes, cycloalkanes and arenes (aromatic hydrocarbons).
- Components of crude oil can be separated by fractional distillation in a fractionating column. The five major fractions separated in a fractionating column are refinery gases, gasoline (includes petrol and naphtha), kerosene, diesel oil and residue.
- Larger hydrocarbon molecules are broken down (cracked) into smaller ones for petrol and to produce alkenes, such as ethene, for the petrochemical industry.
- Thermal cracking produces a high percentage of alkenes via a free radical mechanism. It takes place at high temperature and pressure.
- Catalytic cracking produces branched-chain alkanes suitable for motor fuel and aromatic hydrocarbons. It takes place via a carbocation mechanism. It still uses a high temperature, but only slight pressure.
- Reforming is a way to produce cycloalkanes and arenes from straight-chain alkanes.
- Isomerisation converts straight-chain alkanes into branched-chain isomers.
- Crude oil is a finite resource. Its use as a feedstock for most organic chemicals has changed the world, yet most of it is burnt as fuel.
- Saturated hydrocarbons have no double or triple carbon–carbon bonds in their molecules.
- Unsaturated hydrocarbons possess double or triple carbon–carbon bonds.
- Isomerism occurs when molecules have the same molecular formula, but different ways of arranging their atoms. There are two types of isomerism: structural isomerism and stereoisomerism.
- Structural isomers have the same molecular formula, but different structures because the atoms are bonded in a different order.
- Functional groups give a molecule characteristic properties.

Practice questions and a How Science Works assignment for this chapter are available at www.collinseducation.co.uk/CAS

10

Alkanes, the car engine and atmospheric pollution

10 ALKANES, THE CAR ENGINE AND ATMOSPHERIC POLLUTION

HOW SCIENCE WORKS

Smog in 2001 Milan enveloped in mist and smog

Winter smogs and summer smogs

People have become too reliant on the car. In the UK, for example, nearly a quarter of all car trips are less than 2 miles in length. Of all the journeys we make of between 1 and 2 miles, by any means, 60 per cent are done by car and only 6 per cent by bus; yet 50 years ago bus and coach travel were far more important than travel by car.

As a consequence of car travel we now face two types of major air pollution event: winter smogs and summer smogs. Winter smogs happen on cold, calm days when a 'lid' of cold air, less than 100 metres from the ground, traps pollutants such as nitrogen oxides, carbon monoxide and sulfur oxides. This happens in our major towns and cities and during these episodes more deaths than usual are recorded.

In the summer, on very hot, sunny days, chemical reactions in the atmosphere lead to the production of low-level ozone. The first two weeks in August 2003 saw the hottest temperatures ever recorded and more than 600 premature deaths were thought to have occurred in the UK as a direct result of increased ozone levels. A major feature of summer smogs is that they may start in urban centres, but recirculating air over Europe, or even from Africa, can sweep in low-level ozone from thousands of miles away.

The amount of carbon dioxide in the atmosphere is increasing and the burning of petrol in cars is a major contributor to this increase. Chemists and engineers can do only so much to reduce car emissions. So people must somehow be weaned from their addiction to the car if the present trend towards serious global warming and climate change, with its alarming consequences, is to be stopped.

REMEMBER THIS

Remember that **saturated hydrocarbons** have no double or triple bonds. The carbon atoms are bonded to the maximum possible number of hydrogen atoms (see page 178). The term refers to a time when chemists added hydrogen to various organic compounds and those that would not take up extra hydrogen were called saturated.

QUESTION 1

1 Look at Table 10.1. The next two straight-chain alkanes in this series have 11 and 12 carbon atoms per molecule. What are their molecular formulae and structural formulae?

1 ALKANES

Petrol and diesel, the most common fuels for road transportation, are both mixtures that consist mainly of alkane hydrocarbons. Chapter 8 explains where the energy comes from when these fuels are burnt in air, and shows that petrol and diesel are both concentrated energy sources. Chapter 9 shows that both fuels are the products of the fractional distillation of crude oil. In this chapter, we look more closely at these two fuels and the properties of the alkanes that they contain. We also look at the environmental consequences of using petrol and diesel.

NAMING ALKANES

Alkanes are saturated hydrocarbons. Around each carbon atom are four σ bonds (sigma bonds) to hydrogen atoms or other carbon atoms. In Chapter 9, page 185, we found that alkanes could be represented by the general formula $\mathbf{C}_n\mathbf{H}_{2n+2}$. Table 10.1 lists the first six straight-chain alkanes.

Past four, the number of carbon atoms in a molecule is given by a prefix derived from either the Latin or Greek word for the number. For example, in pentane, *pent* tells you that there are five carbon atoms, and *ane* tells you that the compound is saturated. (The suffix 'ane' was coined by August Hofmann, an eminent German chemist.)

Table 10.1 The molecular formulas, structural formulae and names of the first ten straight-chain alkanes

Molecular formula	Structural formula	Name
CH_4	CH_4	methane
C_2H_6	CH_3CH_3	ethane
C_3H_8	$CH_3CH_2CH_3$	propane
C_4H_{10}	$CH_3CH_2CH_2CH_3$	butane
C_5H_{12}	$CH_3CH_2CH_2CH_2CH_3$	pentane
C_6H_{14}	$CH_3CH_2CH_2CH_2CH_2CH_3$	hexane
C_7H_{16}	$CH_3CH_2CH_2CH_2CH_2CH_2CH_3$	heptane
C_8H_{18}	$CH_3CH_2CH_2CH_2CH_2CH_2CH_2CH_3$	octane
C_9H_{20}	$CH_3CH_2CH_2CH_2CH_2CH_2CH_2CH_2CH_3$	nonane
$C_{10}H_{22}$	$CH_3CH_2CH_2CH_2CH_2CH_2CH_2CH_2CH_2CH_3$	decane

Try this with nonane, Fig 10.1.

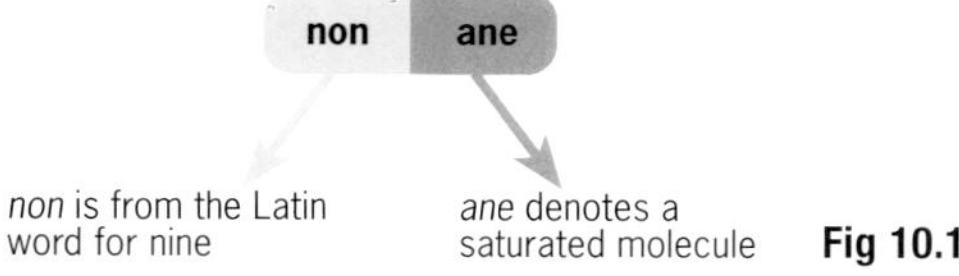
non
ane
non is from the Latin word for nine
ane denotes a saturated molecule

Fig 10.1

So the molecular formula is C_9H_{20}. Another word you will see for the system of naming compounds is **nomenclature**.

Naming straight-chain alkanes is thus straightforward, but what about branched chains? Look at Table 10.2, which shows some **alkyl groups**. Alkyl groups are alkanes with one hydrogen removed so that they can bond with other atoms.

Table 10.2 The names and structural formulae of some alkyl groups

Name	Structural formula
methyl	CH_3—
ethyl	CH_3CH_2—
propyl	$CH_3CH_2CH_2$—
butyl	$CH_3CH_2CH_2CH_2$—

The name of this branched-chain alkane is methylpropane:

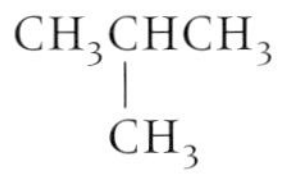

$$\begin{array}{l} CH_3\underset{\displaystyle CH_3}{\underset{|}{C}H}CH_3 \end{array}$$

The longer carbon chain has three carbons (hence propane) and a methyl side group. Sometimes, the molecule is called 2-methylpropane to show that the methyl group is attached to the second carbon in the propane chain. But in the case of this molecule, the 2- is usually left out, as this is the only place the methyl group can go.

Fig 10.2 shows the structural formula of 2-methylbutane. Note that the carbon atoms in the longer chain are numbered so that the position of attachment of the alkyl group is the lowest number possible, 2.

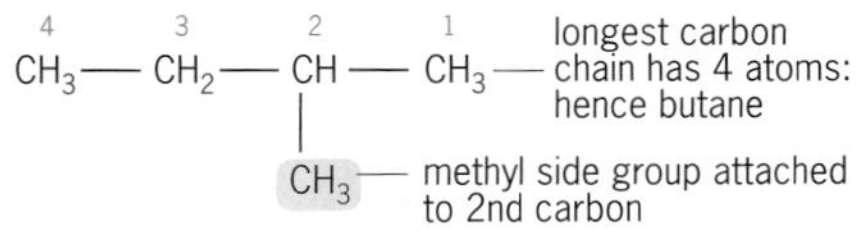

$$\overset{4}{CH_3}—\overset{3}{CH_2}—\underset{\displaystyle CH_3}{\underset{|}{\overset{2}{CH}}}—\overset{1}{CH_3}$$

longest carbon chain has 4 atoms: hence butane

methyl side group attached to 2nd carbon

Fig 10.2 2-methylbutane

REMEMBER THIS

You will need to remember at least the prefixes of the first eight alkanes.

QUESTION 2

a) Draw the structural formula and the displayed formula of nonane. (Hint: look back at page 179 if you need to remind yourself about displayed formulae.)

b) What is the name of the straight-chain alkane with a molecular formula $C_{10}H_{22}$? Look at Table 10.3 on page 194 to check your answer.

REMEMBER THIS

Sometimes it is convenient to refer to an alkyl group as R in a structural formula.

QUESTIONS 3–5

a) What is the name of the alkyl group with five carbon atoms?

b) Work out the general formula for the alkyl group.

Draw the straight-chain isomer of methylpropane. (Hint: if you are not sure how to do this look back at pages 181 and 182.)

a) Why is this molecule not a structural isomer of 2-methylbutane?

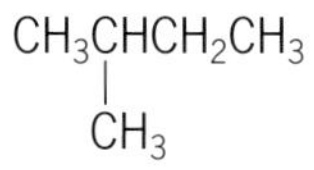

$$CH_3\underset{\displaystyle CH_3}{\underset{|}{C}H}CH_2CH_3$$

To check your answer, see page 183.

b) Sometimes this molecule is just called methylbutane. Why?

QUESTION 6

Draw the structural formulae of 3-ethyl-2,2-dimethylhexane and 4-propylheptane.

Note that in the first compound, ethyl comes before methyl: by convention, side groups are put in alphabetical order. See the Example box on page 194.

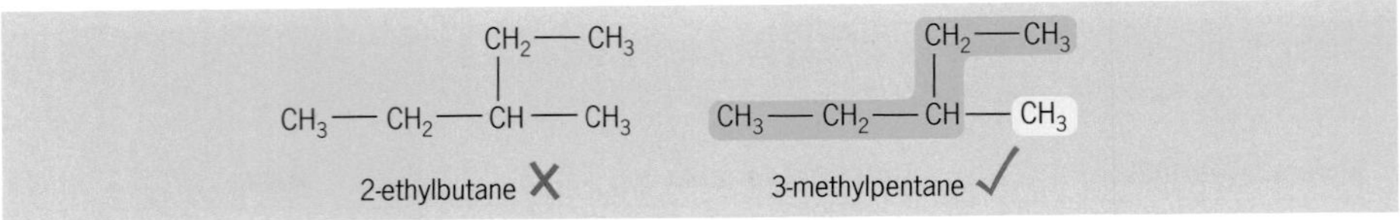

CH_3—CH_2—CH—CH_3 (with CH_2—CH_3 on the CH) — 2-ethylbutane ✗

CH_3—CH_2—CH—CH_3 (with CH_2—CH_3 on the CH) — 3-methylpentane ✓

Fig 10.3

The molecule in Fig 10.3 is **not** called 2-ethylbutane, because the longest continuous carbon chain has five carbon atoms, and hence its name is 3-methylpentane.

EXAMPLE

Q The compound shown in Fig 10.4 used to be called iso-octane. It is important in working out the octane numbers of fuels (see page 198). What is its systematic name?

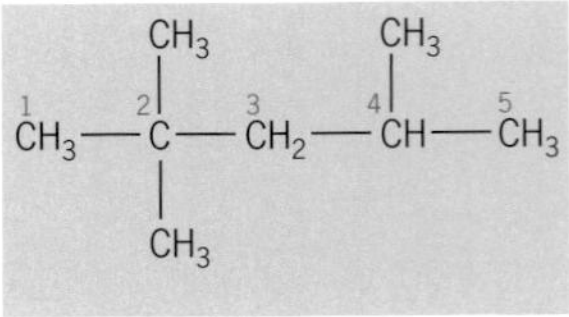

CH_3(1)—C(2)(CH_3)(CH_3)—CH_2(3)—CH(4)(CH_3)—CH_3(5)

Fig 10.4

A Step 1. Look for the longest carbon chain (called the **parent alkane**) and name it. In this case, the chain has five carbon atoms, so it will be called pentane.

Step 2. Name every side group. The methyl group is the only side group, but there are three of them. We must ensure the name includes all the side groups, so we use *tri*, and hence trimethylpentane. If there were two groups, we would use *di*, and if four, *tetra*.

Step 3. Indicate the position of all of the side groups on the longest carbon chain by numbering the carbon atoms. Remember to use the lowest numbers possible. When there are two or more different alkyl groups, they are placed in alphabetical order irrespective of their numbered position on the chain. (See the compound named in Question 6.)

We have numbered the carbon atoms on the molecule so that two of the methyl groups have the lowest possible number, 2. You will notice that the numbers in a name are separated by commas, that the numbers and letters are separated by hyphens and that every side group has a number.

So the **systematic name** of iso-octane is:

2,2,4-trimethylpentane

The system we have used to name this compound is recommended by the International Union of Pure and Applied Chemistry (IUPAC) and is accepted by chemists throughout the world. However, if you enter the chemical industry, you may still hear non-systematic names such as iso-octane.

? QUESTION 7

7 Look at Table 10.3. As the number of carbon atoms in each molecule increases, what is the increase in the number of hydrogen atoms?

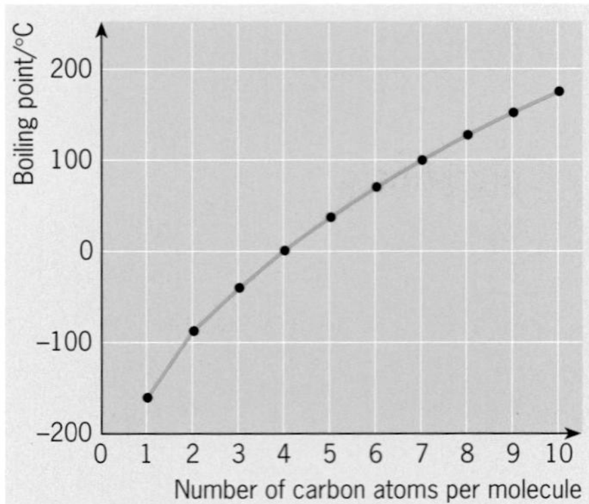

Fig 10.5 The graph shows the gradual change of boiling points of straight chain alkanes as the number of carbon atoms per molecule increases

ALKANES FORM A HOMOLOGOUS SERIES

We have already seen that alkanes can be represented by a general formula, C_nH_{2n+2}. This makes them a **homologous series**.

The compounds in a homologous series have similar chemical properties, and each subsequent member differs from the previous one by a CH_2 group. They can be represented by a general formula.

As you go through this book you will encounter other homologous series. Chemists find it useful to classify organic compounds into homologous series, because of the similar chemical properties of the members. This means they can be studied as a group, rather than studying each individual compound.

Table 10.3 Three physical properties of the first ten straight chain alkanes. Note the gradual change in these physical properties

Alkane	Structural formula	Melting point/°C	Boiling point/°C	Density/g cm^{-3}
methane	CH_4	−182	−164	0.466
ethane	CH_3CH_3	−183	−89	0.572
propane	$CH_3CH_2CH_3$	−190	−42	0.585
butane	$CH_3(CH_2)_2CH_3$	−138	−1	0.601
pentane	$CH_3(CH_2)_3CH_3$	−130	36	0.626
Hexane	$CH_3(CH_2)_4CH_3$	−95	69	0.660
heptane	$CH_3(CH_2)_5CH_3$	−91	98	0.684
octane	$CH_3(CH_2)_6CH_3$	−57	126	0.703
nonane	$CH_3(CH_2)_7CH_3$	−54	151	0.718
decane	$CH_3(CH_2)_8CH_3$	−30	174	0.730

Physical properties of alkanes

The term **physical properties** refers to those properties of a substance that can be measured or observed without changing the composition or identity of the substance. Physical properties include melting point, boiling point, colour and density. In a homologous series the physical properties gradually change as the number of CH_2 groups increases.

The steady change in the boiling points of the alkanes is used to separate them when crude oil is passed into a fractionating column.

The effect of branched chains on the boiling point

The occurrence of branching on a carbon chain reduces the boiling point. The more branching there is in an alkane, the lower will be its boiling point compared with its straight-chain structural isomer. Pentane and its isomers (Fig 10.6) clearly demonstrate this.

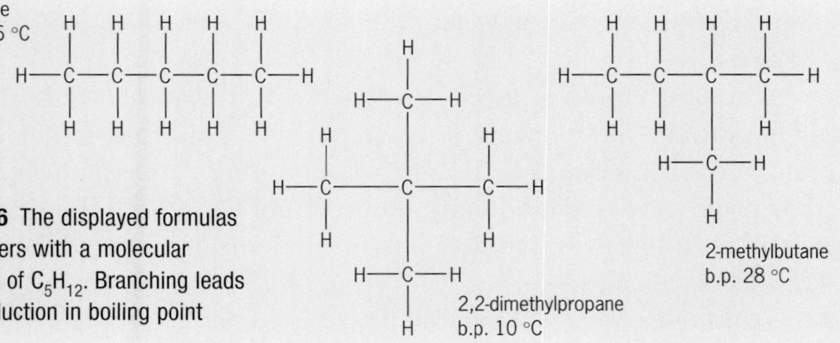

pentane
b.p. 36 °C

2,2-dimethylpropane
b.p. 10 °C

2-methylbutane
b.p. 28 °C

Fig 10.6 The displayed formulas of isomers with a molecular formula of C_5H_{12}. Branching leads to a reduction in boiling point

Branching also affects the **volatility** of an alkane. Volatility describes the tendency of a liquid to evaporate. Branched chain alkanes are very important in blending petrol because of their volatility, as you can see on page 198.

2 INTERMOLECULAR FORCES IN ALKANES: INDUCED DIPOLE–INDUCED DIPOLE

We have seen two trends in the boiling points of alkanes. Straight-chain alkanes show an increase in boiling point as the number of carbon atoms increases, and branching reduces the boiling point when the number of carbon atoms remains the same.

The particles in a liquid are held close together by forces. The stronger the forces are, the higher is the boiling point of the liquid. The forces that hold alkane molecules in a liquid are weak **intermolecular forces**, which is why alkanes have relatively low boiling points.

Fig 4.46, page 85, shows that a molecule of hydrogen chloride is **polar** because the centres of negative and positive charge do not coincide. In a polar molecule there is a **permanent dipole**: one part of the molecule always has a slightly positive charge and another part always has a slightly negative charge. Highly polar molecules can often attract each other quite strongly. Alkanes are sometimes referred to as **non-polar** molecules because they have no permanent dipole. So how do intermolecular forces arise and cause the molecules to attract each other? And how is the increase in carbon atoms related to the trend of increasing boiling point?

The electrons in atoms and molecules are in constant motion. Although alkanes have no permanent dipole, the electron charge cloud around their molecules is not always evenly distributed. At any one instant, there may be a **temporary dipole** when the centres of positive and negative charge do not

For the separation of the components of crude oil, see page 186.

QUESTION 8

a) In Table 10.3, the density measurements for each alkane were taken at 298 K, except for those that are gases at this temperature. For these compounds, measurements were taken at just below their boiling points. Which density measurements were not taken at 298 K? (See page 120 if you are not sure how to convert Kelvin to degrees Celsius.)

b) i) Dodecane has 12 carbon atoms in its molecule. Predict its boiling point and density by drawing a graph similar to that in Fig 10.5 (opposite).
ii) By drawing a graph of melting point against the number of carbon atoms in a molecule, predict how many carbon atoms the first alkane to be a solid at 298 K will have.

To remind yourself about particles in solids, liquids and gases, see pages 123 to 126.

REMEMBER THIS

Intermolecular means between molecules. Remember that δ+ and δ– show the positive and negative poles in a molecule. The symbol δ tells you that it is a partial charge.

QUESTION 9

Why do you think highly polar molecules can often attract each other quite strongly? To check your answer, look at page 238.

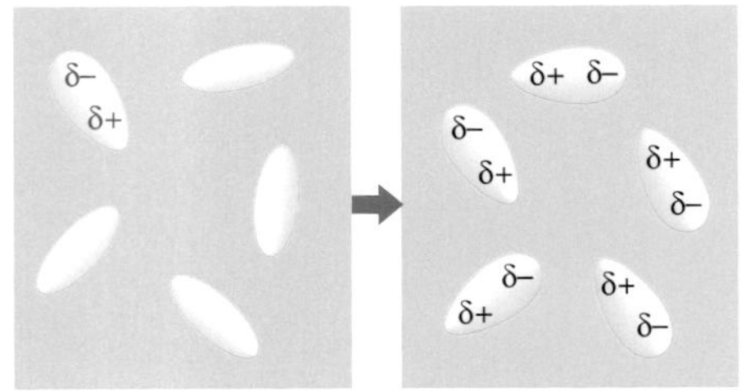

Fig 10.7 Left: An instantaneous dipole forms due to the uneven distribution of the electron cloud **Right:** This induces dipoles in neighbouring molecules, which then attract each other

REMEMBER THIS

Because the first dipole that is formed at any one instant is called an instantaneous dipole or a temporary dipole, induced dipole–induced dipole forces are sometimes referred to as **instantaneous dipole–induced dipole forces** or **temporary dipole–induced dipole forces.** You may also see these intermolecular forces refered to as **London forces**

QUESTION 10

10 Explain why the boiling points of the noble gases increase as you descend Group 0.

Read more about induced dipole–induced dipole forces in noble gases and other elements on page 148.

QUESTION 11

11 **a)** Explain why the volatility of petrol in Spain is different from that of petrol in the UK.

b) Increasing the volatility of petrol can be done by using short straight-chain alkanes. What other alkanes can be used to increase volatility?

coincide. This **instantaneous dipole** induces dipoles in neighbouring molecules. A positive charge at one end of a molecule induces a negative charge in a molecule close to it.

These opposite charges can now attract. The force between them is known as an **induced dipole–induced dipole force**, or a **van der Waals force** after the Dutch physicist Johannes van der Waals (1837–1923). In the next instant, the movement of electrons in the molecules leads to a collapse of these temporary dipoles and elsewhere in the substance other instantaneous dipoles form. Averaged over time, alkane molecules do not have a permanent dipole, but billions of instantaneous dipoles are continuously occurring, providing a weak intermolecular force.

The more electrons there are in a molecule, the stronger are the induced dipole–induced dipole forces, because there can be a greater distortion of the charge cloud. As the number of carbon atoms increases in the straight-chain alkanes, so do the number of electrons, which in turn increases the induced dipole–induced dipole forces between the molecules, and so leads to increased boiling points.

Although induced dipole–induced dipole forces are relatively weak, they are extremely important. They explain, for example, why noble gases such as helium can be condensed at 4 K. If there were no forces between helium atoms, they could never form a liquid. Induced dipole–induced dipole forces are present in all substances, but they are most important in those that are non-polar, such as alkanes.

We can use the same model of induced dipole–induced dipole forces to explain why branching in alkanes reduces the boiling point. Look back at Fig 10.6. You can see that pentane is a straight chain, which allows the molecules to line up beside each other, thereby giving a greater surface area over which these weak forces can act. Branching reduces this surface area and so the intermolecular forces are correspondingly weaker, which leads to lower boiling points.

Getting the volatility right

For petrol to burn in an engine, it must vaporise and be mixed with air. So a crucial property of petrol is high volatility. On extremely cold winter mornings, the petrol must be volatile enough to vaporise and mix with air. If it is not, the engine won't start. However, on hot summer days, if the petrol is too volatile, it may vaporise in the fuel feed-pipe and so cause a vapour lock that prevents the petrol mixing with air and reaching the engine. Again, the engine won't start. Another drawback of high volatility is that if the petrol vaporises too readily, too much of it enters the engine and there won't be enough oxygen for it to burn efficiently.

In the UK, the composition or blend of petrol is changed four times a year, according to the season. So, during the winter, petrol contains more of the alkanes with low boiling points, such as butane and pentane, while in the summer, the proportion of volatile components is reduced and more compounds of low volatility are incorporated.

3 REACTIONS OF THE ALKANES

REMEMBER THIS

Electronegativity is the power of an atom in a molecule to attract electrons to itself. See page 85 (Chapter 4).

Alkanes are quite unreactive because of the nature of the C–C and C–H bonds. First, both are very strong bonds that are difficult to break. Second, both these bonds are non-polar as the electronegativities of carbon and hydrogen are similar. The non-polar nature of the molecules makes them unreactive to polar reagents and ions, and they do not react with acids, bases, metals or oxidising agents such as potassium manganate(VII). It is their very unreactivity that makes them so useful as lubricants and plastics.

The few reactions that alkanes can undergo are of fundamental importance. Two of these reactions – cracking and reforming – are used to produce petrol with desirable properties (see pages 187 to 188). Another involves halogens (for more about the reaction of halogens with alkanes, see pages 240 to 245). Cracking also produces alkenes, which are the bedrock of the petrochemical industry. For more about the petrochemical industry, see pages 190 and 274.

4 COMBUSTION OF ALKANES

We know that a spark produces a spontaneous reaction between petrol vapour and oxygen. The spark provides the activation energy for this reaction. Alkanes burn in oxygen, releasing a great deal of energy, which is one of the reasons why they are such popular fuels. They all produce carbon dioxide and water when there is enough oxygen for them to burn completely.

Butane is used to increase the volatility of petrol. Its reaction with oxygen is:

$$C_4H_{10}(g) + 6\tfrac{1}{2}O_2(g) \rightarrow 4CO_2(g) + 5H_2O(l) \quad \Delta H^{\ominus}_c = -2877 \text{ kJ mol}^{-1}$$

Heptane is another component of petrol. Its reaction with oxygen is:

$$C_7H_{16}(g) + 11O_2(g) \rightarrow 7CO_2(g) + 8H_2O(l) \quad \Delta H^{\ominus}_c = -4817 \text{ kJ mol}^{-1}$$

If there is insufficient oxygen, alkanes are not completely oxidised. They produce carbon monoxide, which is very poisonous, and even carbon (soot). This is a real problem when burning petrol in vehicle engines.

QUESTION 12

a) What is the name given to reactions that release energy to the surroundings?

b) What are the meanings of the terms spontaneous reaction and activation energy?

c) Alkanes and oxygen need a spark before they react. So chemists say that they are kinetically stable but energetically unstable. Explain what they mean by this statement.

d) What are the desirable properties of a fuel?

The answers can be found in Chapter 8.

To check what is meant by $\Delta H^{\ominus}_c$ see page 153 Chapter 8.

SCIENCE IN CONTEXT

The petrol engine

The petrol engine is classed as an internal combustion engine because the fuel burns inside the engine and the products of this combustion drive the engine directly. Fig 10.8 illustrates the four-stroke cycle on which vehicle engines work.

The drive to make petrol engines more fuel-efficient centres upon how much the petrol–air mixture can be compressed. In engines of the 1920s, each piston compressed the mixture to a quarter of its original volume. But in today's engines the petrol–air mixture in each cylinder is compressed to one-tenth of its original volume, thereby extracting much more energy from the combustion of the hydrocarbons.

Fig 10.8 How a four-stroke petrol engine works

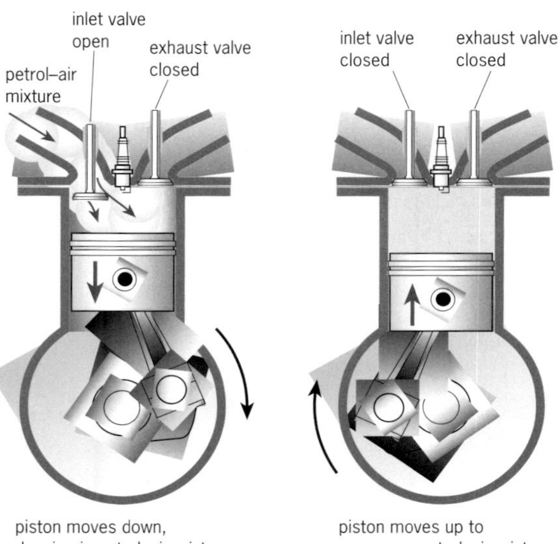

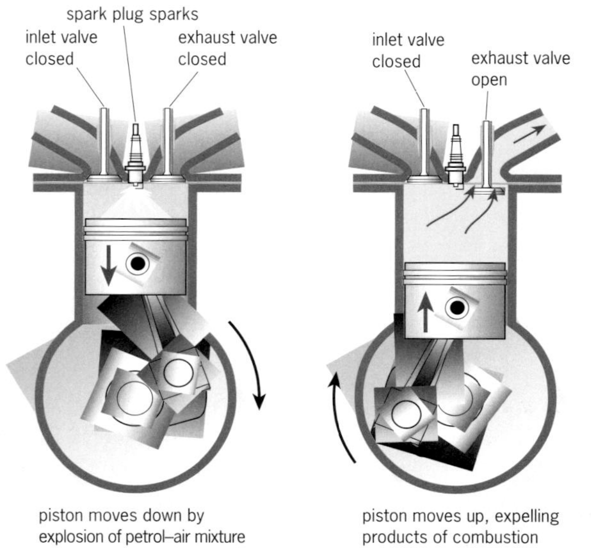

REMEMBER THIS

2,2,4-trimethylpentane used to be known as iso-octane, which is why the scale is called octane number or octane rating.

OCTANE NUMBER OF A PETROL BLEND

As the petrol–air mixture is compressed in a cylinder (see Fig 10.8), it becomes hot and sometimes ignites without the aid of a spark from the spark plug **(auto-ignition)**. When this happens, the fuel does not burn smoothly, but explodes in different parts of the cylinder, which generates a rapid knocking noise. **Knocking** can damage the cylinder and piston head and, because the petrol–air mixture does not ignite at the right time, it leads to a loss of power and inefficient use of fuel.

It is the job of the petrol blender to produce a knock-free fuel. So blenders use a measure of a blended petrol's resistance to knocking, called octane number. Branched-chain alkanes burn more smoothly in an engine than straight-chain alkanes. 2,2,4-Trimethylpentane has a low tendency to auto-ignite when compressed and is given an octane number of 100. The unbranched-chain alkane heptane knocks readily, even under mild compression. It is given an octane number of 0. Mixtures of these two alkanes are used to assign octane numbers to petrol blends. So if, under test conditions, a particular petrol blend knocks at the same compression as a mixture of 90 per cent 2,2,4-trimethylpentane and 10 per cent heptane, the octane number of that petrol blend is 90.

ADDING LEAD COMPOUNDS TO PETROL

The advantage of high-octane fuels is that they can be highly compressed, which gives more power per piston stroke and a more efficient use of fuel. In the 1920s, it was discovered that adding small amounts of a certain lead compound to petrol significantly increased the octane number, a practice that was adopted by all the major petrol suppliers.

HOW SCIENCE WORKS

Blending unleaded petrol

If lead is not used to improve the octane number, an alternative way must be found. One solution is to dissolve into the blended petrol small straight-chain alkanes such as butane. The shorter the chain, the less likely is auto-ignition of the mixture. As the alkane chain becomes longer, so the tendency to auto-ignite increases and the octane number decreases. However, too many short-chain alkanes with low boiling points increase the volatility of the petrol too much. This could lead to vapour lock (page 196), and it increases evaporation of hydrocarbons from the fuel tank and engine – yet another source of atmospheric pollution.

Using branched-chain alkanes is another way to raise the octane number of petrol. The more branched the chain, the higher its octane number. Branched-chain alkanes are produced at the refinery by cracking and reforming reactions; more about this on pages 187–188.

A third way is to add aromatic hydrocarbons. Some 40 per cent of the petrol used in the UK contains such additives. One of them is benzene, a known carcinogen (cancer-causing agent), which can make up to 5 per cent by volume of a petrol. Benzene is now being linked to childhood leukaemia, and is thought to be the major cause in some cases. The level of benzene inside a car can be up to ten times higher than outside.

Fig 10.9 This child is at exhaust fume level

The octane number can also be improved by adding **oxygenates**. We first came across oxygenates on page 161. Oxygenates are partly oxidised fuels that already contain oxygen in their molecules. Methanol (CH_3OH) has an octane number of 114 and ethanol (C_2H_5OH) is 111, so they burn smoothly under high compression. Also, because they have an oxygen atom already in the molecule, when they are combusted they require less oxygen in the petrol–air mixture. So petrols that contain oxygenates produce less carbon monoxide when burnt.

However, the use of lead in petrol to raise the octane number is now banned in most countries for two reasons, both of them connected to the pollution of the environment by lead discharged through exhausts. The first is that lead is a poison that accumulates in the bodies of humans and other animals. Evidence suggests that when children breathe in airborne lead, their IQ level is liable to be lowered, and for adults, their ability to concentrate is reduced. Since the advent of unleaded petrol in the UK in 1986, the concentration of airborne lead in the environment has been reduced by 80 per cent. The second reason why leaded petrol is being phased out is that it inhibits the action of catalysts in catalytic converters, which are essential to reduce pollution from vehicle exhausts.

5 POLLUTION FROM VEHICLES AND CATALYTIC CONVERTERS

At the outset of this chapter, we pointed to the problem of vehicle emissions. Now we are going to look in more detail at the chemistry behind the pollution associated with the burning of hydrocarbon fuels, such as petrol in vehicle engines.

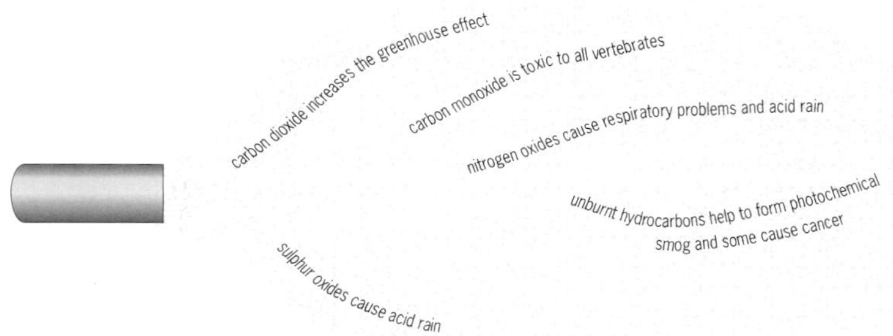
carbon dioxide increases the greenhouse effect
carbon monoxide is toxic to all vertebrates
nitrogen oxides cause respiratory problems and acid rain
unburnt hydrocarbons help to form photochemical smog and some cause cancer
sulphur oxides cause acid rain

Fig 10.10 Uncontrolled exhaust emissions cause increasing concern

QUESTION 13

13 Methanol and ethanol are members of the homologous series the alcohols.

a) What is the general formula of this homologous series?

You can check your answer by looking at page 261.

b) How would you expect the physical and chemical properties to change in this homologous series?

CARBON DIOXIDE AND THE GREENHOUSE EFFECT

When a hydrocarbon completely combusts, it produces just two products: carbon dioxide and water. Both of these are referred to as greenhouse gases, but what are greenhouse gases and how did they get their name?

QUESTION 14

14 Decane is one of the components of petrol. Write the equation for its complete combustion in air.

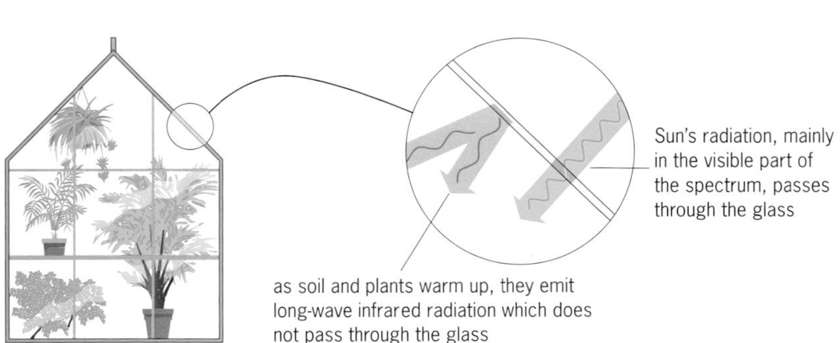
Sun's radiation, mainly in the visible part of the spectrum, passes through the glass
as soil and plants warm up, they emit long-wave infrared radiation which does not pass through the glass

Fig 10.11 Visible light from the Sun enters the greenhouse through the glass. Plants and soil absorb this energy and emit mainly long-wave infrared radiation. The long-wave infrared radiation cannot pass out through the glass, so the greenhouse warms up inside

QUESTION 15

15 **a)** Which radiation transfers more energy: infrared or visible light?
b) Explain why the Earth does not radiate light in the visible region of the electromagnetic spectrum.
(Hint: if you are not sure, look back at pages 50 to 51.)

REMEMBER THIS

A quantum is an indivisible unit or packet of radiation energy (also called a photon). A molecule can exist only in certain fixed vibrational energy levels. When it absorbs a quantum of energy, it becomes excited and jumps to a higher energy level. In this level, it vibrates more. This is similar to electrons absorbing quanta of energy and moving to higher electron energy levels. See page 54 and Chapter 27.

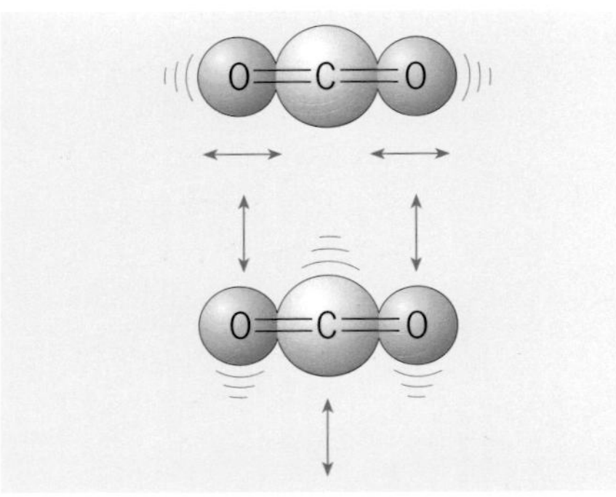
O=C=O
O=C=O

Fig 10.13 C=0 bonds in carbon dioxide absorb infrared radiation, increasing their vibrations

The atmosphere of the Earth acts like glass in a greenhouse. For a start, the atmosphere reflects 30 per cent of the Sun's energy back into space. The remaining 70 per cent, which is mainly visible spectrum light, passes through the atmosphere to strike the Earth. Some of this energy is used by plants for photosynthesis, and some is used to evaporate water from the oceans, lakes and vegetation, but most of it warms the surface of the planet. The warm Earth's surface in turn emits its own radiation, and this is where certain gases in the atmosphere act like greenhouse glass. The energy radiated by the Earth's surface is in the long-wave infrared (IR) region of the spectrum. The greenhouse gases absorb this IR radiation, and their molecules become excited.

Three of the most important greenhouse gases are water, carbon dioxide and methane. Quanta of infrared radiation are absorbed by O–H, C=0 and C–H bonds, which increase their vibration and thus promote them to higher vibrational energy levels. When they fall back to a lower vibrational energy level, they radiate these quanta of energy in all directions. Some of this energy is transferred through collision of air particles increasing their kinetic energy and warming the atmosphere, but the rest is transmitted back to the Earth's surface. This is the greenhouse effect.

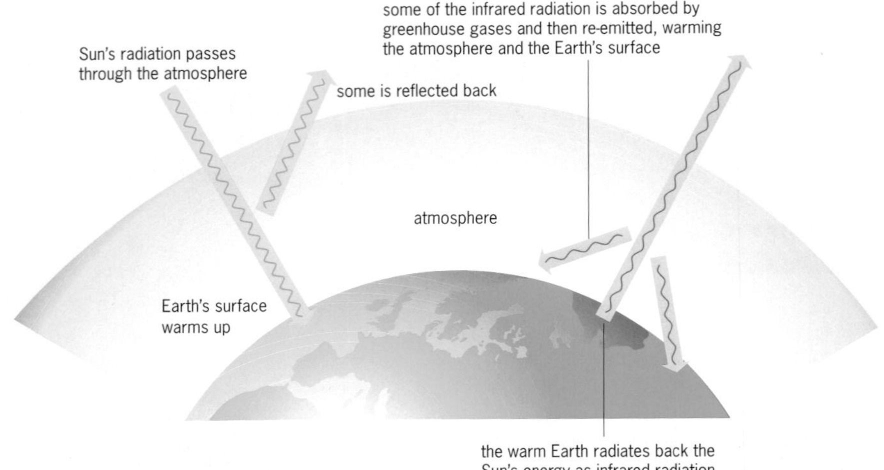
some of the infrared radiation is absorbed by greenhouse gases and then re-emitted, warming the atmosphere and the Earth's surface
Sun's radiation passes through the atmosphere
some is reflected back
atmosphere
Earth's surface warms up
the warm Earth radiates back the Sun's energy as infrared radiation

Fig 10.12 The Sun's visible radiation passes through the atmosphere to strike the Earth. The Earth warms up and emits infrared radiation. Some of this is trapped by the greenhouse gases, as shown

Why we need the greenhouse effect in moderation

If carbon dioxide and water were not present in the atmosphere, the Earth would be many degrees cooler. We only have to look at our two closest

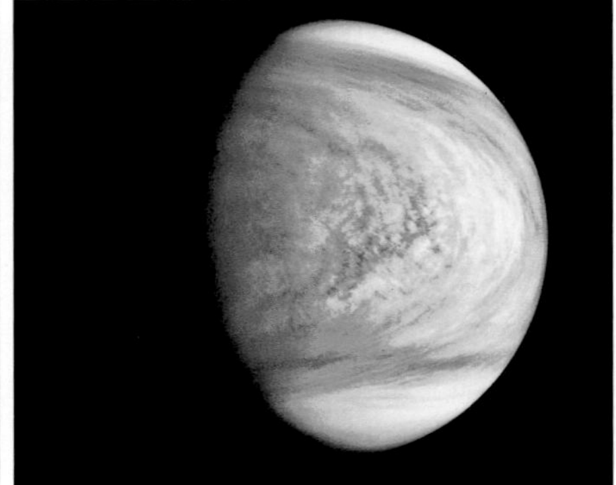

Fig 10.14 The extremely dense atmosphere seen here around Venus is mostly carbon dioxide, which keeps its surface roasting hot

Fig 10.15 Earth enjoys moderate temperatures because greenhouse gases are present in its atmosphere

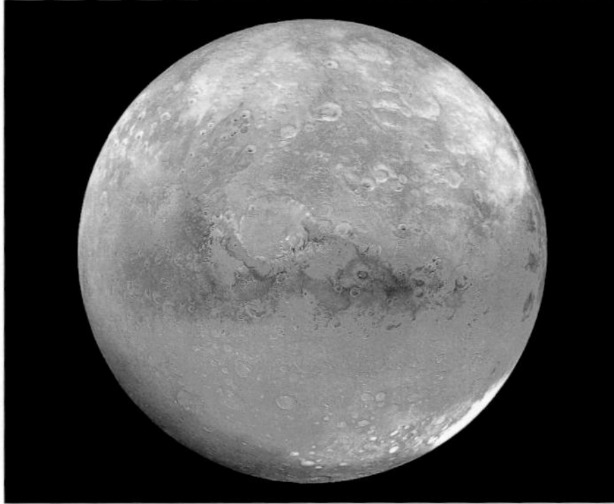

Fig 10.16 Mars has little atmosphere to prevent infrared radiation from escaping. The temperature of its surface is never greater than 40 °C and can plunge to –80 °C

planetary neighbours to see the effect of greenhouse gases. Venus has an extremely dense atmosphere, 90 times denser at the suface than that of Earth. It is 96 per cent carbon dioxide and keeps the temperature on Venus to a scorching 450 °C. The atmosphere of Mars has about the same proportion of carbon dioxide as that on Venus, but is only 1 per cent as dense as the Earth's atmosphere. Mars is further from the Sun than Venus so you would expect Mars to be cooler, but it has greater temperature fluctuations and cools during its night to –80 °C because there are not enough carbon dioxide molecules to trap the IR radiation.

The greenhouse effect and climate change

In 2007 there were about 2850 gigatonnes of carbon dioxide in the Earth's atmosphere, but its concentration in air is small at 0.0383 per cent, or 383 parts per million (ppm) by volume. We know from analysing polar ice cores that the atmospheric concentration of carbon dioxide has varied between 180 ppm and 300 ppm during the last 650 000 years, and at the beginning of the Industrial Revolution, 250 years ago, there were about 280 ppm of carbon dioxide. Most of the increase in carbon dioxide concentration we see today has happened in the past 50 years. This is clearly shown by Fig 10.17, which is one of the most important graphs in history, produced from results of the Mauna Loa Observatory in Hawaii.

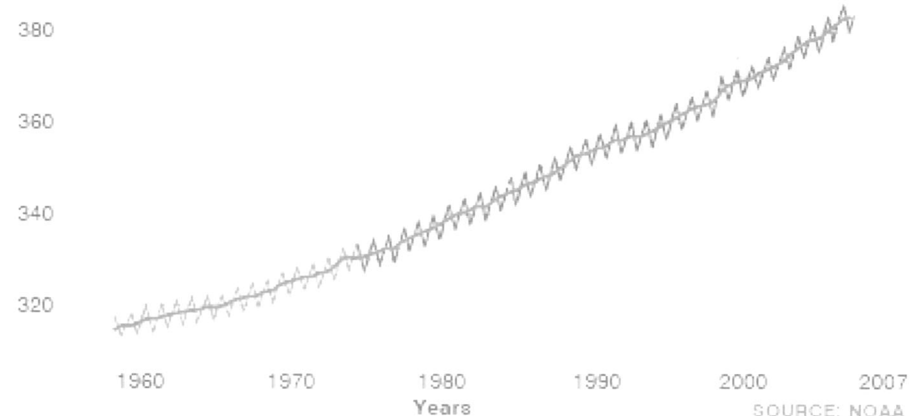

Fig 10.17 The readings taken by the Mauna Loa Observatory, Hawaii, show clearly the trend of increasing carbon dioxide concentration in the atmosphere

At present, the rate of increase in carbon dioxide is about 2 ppm per year, which means that about 14 Gt is added to the atmosphere. However, more than 28 Gt of carbon dioxide is produced each year, much of it from burning fossil fuels for electricity generation or transport. So, where does half the carbon dioxide go? Some is removed by plants during photosynthesis, and a lot more dissolves in the oceans, lakes and waterways, but there is evidence that oceans are absorbing less CO_2 than they did a few years ago.

Recognising that global climate change could result from increasing concentrations of greenhouse gases in the atmosphere, the World Meteorological Society and the United Nations established the Intergovernmental Panel on Climate Change (IPCC) in 1988. This organisation assesses all the evidence from thousands of scientists researching into climate change, in order to come to conclusions. Its latest report was produced in 2007. On the eve of its publication, the Paris authorities switched off the lights around the city to acknowledge the seriousness of the

QUESTION 16

The Earth is in an orbit between those of Mars and Venus, and it has a similar origin. When the atmospheres of Mars and Venus have such a high percentage of CO_2, suggest why the Earth has as little as 0.0383 per cent?

QUESTION 17

Each year there is a seasonal variation in carbon dioxide concentration.

a) What causes the troughs and during which season of the year does a trough occur?

b) What causes the peaks and in what season? (Hint: think about what happens to deciduous trees in the winter.)

REMEMBER THIS

Something that removes carbon in the form of carbon dioxide is known as a **sink**. So plants and open waters are sinks for carbon. Crude oil is another sink for carbon, and so are the bodies of animals. The amount of time carbon spends in a sink is known as the **residence time**.

REMEMBER THIS

A gigatonne (Gt) is 10^9 tonnes.

REMEMBER THIS

Carbon capture and storage is the name given to the removal of waste carbon dioxide. It can be liquefied and injected deep into the oceans. Carbon dioxide can also be stored in geological formations. This is happening now in some of the Norwegian oil fields that have come to the end of their production. It can also the reacted with metal oxides to form stable carbonate minerals.

? QUESTION 18

18 **a)** What carbon dioxide emission targets is the UK currently working towards?

b) There are more than 400 million cars in the world today. Assume that petrol consists only of octane, which has a density of 0.7 g cm^{-3}:

i) What is the mass of 1 dm^3 of petrol?

ii) Write the equation for the complete combustion of octane.

iii) What mass of CO_2 is produced by burning 1 dm^3 of petrol?

iv) How many tonnes of carbon dioxide are produced daily if 400 million cars use 1 dm^3 of fuel a day?

(Hint: you may need to look at pages 5 and 6 to remind yourself of work on moles and equations.)

report's findings. The report says that it is 90 per cent certain that human activities are causing global warming and climate change. This was important, because many scientists in the past have argued that global warming is part of a natural cycle and not the result of human activity. For its work the IPCC was awarded the Nobel Prize, together with former US Vice President Al Gore, who has also done much to publicise the worrying implications of climate change.

The Kyoto Protocol was the first international attempt at curbing greenhouse gas emissions. It was hammered out in 1997 at the city in Japan in that bears its name. It committed industrialised nations to legally binding targets to limit the emissions of six greenhouse gases (most important being carbon dioxide) to 5.2 per cent below their 1990 levels by the period 2008–2012. In 2005 the Kyoto Protocol came into force when sufficient countries had signed up to account for 55 per cent of the 1990 carbon dioxide emissions. The number of countries that have now signed is 141, but significantly the USA and Australia refuse to support the agreement. The USA alone previously accounted for almost one-quarter of all carbon dioxide emissions. This country was the biggest emitter of carbon dioxide until it was overtaken by China, which was building two coal-fired power stations each week in 2007. However, it would be unfair for developed countries to point the finger at China. The average Chinese person emits 3.5 tonnes of carbon dioxide per year, compared with the average Briton who emits 10 tonnes or the average US citizen who emits 20 tonnes.

The situation is now considered very urgent and many scientists believe that political action is falling a long way short of what is needed to avert lasting damage to Earth's climate.

Table 10.4 Global warming potentials of some atmospheric gases. (Data taken from the Intergovernmental Panel on Climate Change: 1992 Supplement)

Gas	Global warming potential
CO_2	1
CH_4	11
N_2O	270
CFC12	7100

Other greenhouse gases

Carbon dioxide is not the most potent greenhouse gas. If its concentration in the Earth's atmosphere were to double, the effect would probably be an increase in temperature of 1.5–4.5 °C. However, the concentration in the atmosphere of hydrocarbons such as methane, chlorofluorocarbons (CFCs), dinitrogen oxide (N_2O) and ozone (O_3) are also increasing because of human activity, and these strongly absorb IR radiation. Clearly, the build-up of all greenhouse gases needs careful monitoring.

SCIENCE IN CONTEXT

Methane galore

We know from analysing the air in polar-ice cores that the amount of atmospheric methane, a potent greenhouse gas, is increasing. Methane-producing bacteria decompose carbohydrates, such as glucose and cellulose, into methane and carbon dioxide, which are discharged into the air. From glucose:

$$C_6H_{12}O_6 \rightarrow 3CO_2(g) + 3CH_4(g)$$

The bacterial process is anaerobic: it does not require oxygen, so it happens in bodies of stagnant water. For example, paddy fields produce large amounts of methane because the water and mud that cover the rotting vegetation provide the right conditions for anaerobic bacteria to work. Cattle, too, produce enormous amounts of methane (each cow discharges about 500 dm^3 a day in belches from partly digested food in its gut). And when methane reacts in the atmosphere, it produces mostly ozone, another greenhouse gas.

Fig 10.18 The growing of rice contributes significantly to atmospheric methane

Modelling the global climate

If we want to work out what the effects on our climate will be of increasing concentrations of carbon dioxide in the atmosphere, we must turn to computer models. These models use complex mathematical equations that are based on known physical laws to make predictions about weather and climate.

The problem is that a model is only as good as the information that scientists put into it. The climate is very complex, involving complex interactions. For example, as the Earth warms up, more water vapour will evaporate from the oceans. Water is a potent greenhouse gas, so you might expect it to warm the Earth. However, increasing water vapour leads to increased cloud cover, which prevents the Sun's radiation from reaching the Earth's surface by day but at night traps some of the infrared radiation, keeping air near the surface warmer. Even the height at which the clouds form has an impact on our climate. The planet's icecaps, too, act as giant reflectors, reflecting the Sun's radiation back into space; if these diminish then more radiation will strike the Earth's surface, producing in turn more infrared radiation.

Another aspect that climate modellers must consider is the effect of aerosols. These are minute atmospheric particles such as sulfates and soot particles that are produced naturally from forest fires, as well as by humans from fossil-fuel power stations and other industrial activities. (Sulfate particles arise mainly from the burning of sulfur in fuels.) These particles reflect sunlight back into space.

At present the oceans, plants and soils absorb half of the carbon dioxide produced by humans. Latest climate-model predictions expect this to decrease, so as you can see very sophisticated models of how our climate may change as a result of the greenhouse effect are required.

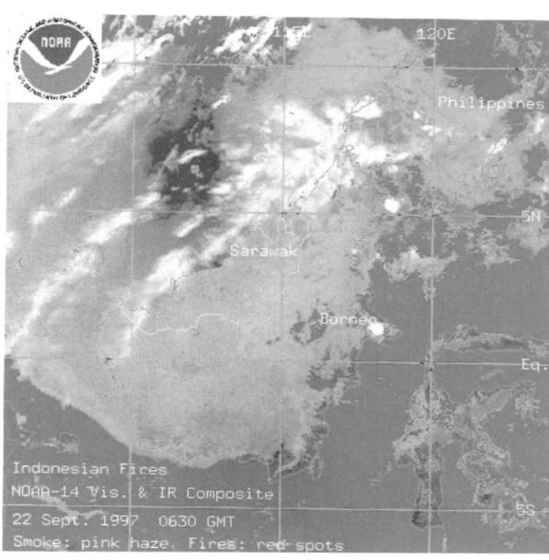

Fig 10.19 Forests burning in Indonesia in 1997 set up dust clouds that covered the whole region. In this satellite image, the dust is pink and the fires are red.

REMEMBER THIS

Even a 1 °C rise in global temperature would produce unpredictable and possibly devastating climate changes.

CARBON MONOXIDE EMISSIONS AND CATALYTIC CONVERTERS

In the UK, transport is responsible for 90 per cent of all emissions of carbon monoxide. When petrol does not have enough oxygen for complete combustion, carbon monoxide is produced. It is toxic to all vertebrates (and some invertebrates) and reacts quickly with haemoglobin in red blood cells to produce carboxyhaemoglobin. Once carbon monoxide has reacted with haemoglobin molecules, they do not carry oxygen. Mild symptoms of carbon monoxide poisoning include headaches, dizziness and tiredness. Severe symptoms are fainting and a possibly fatal coma.

The World Health Organization guidelines for carbon monoxide are often exceeded in urban areas all over the world, with health consequences. For example, in New York when carbon monoxide concentrations exceed recommended levels, the incidence of heart attacks increases.

Catalysts can be used to reduce exhaust emissions. Platinum, for example, catalyses the oxidation of carbon monoxide:

Pt catalyst

$$2CO(g) + O_2(g) \longrightarrow 2CO_2(g)$$

A catalytic converter is fitted to a vehicle exhaust system to speed up this reaction (see Fig 10.20). It is sited near the engine so that it heats up quickly, because platinum does not start catalysing until its temperature is 240 °C.

REMEMBER THIS

Catalysts are substances that speed up the rate of a chemical reaction and can be recovered chemically unchanged at the end of the reaction. A catalyst works by providing an alternative reaction pathway that has a lower activation energy. See page 166 and Chapter 26 for more information.

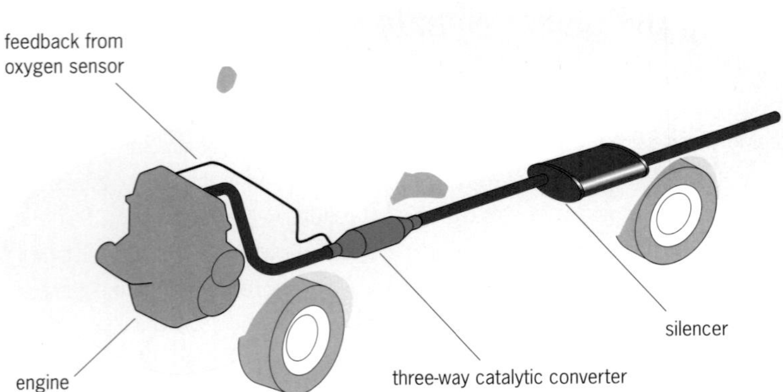

Fig 10.20 The three-way catalytic converter is sited near the engine so that it warms up quickly. An oxygen sensor monitors the oxygen flowing through the exhaust system, and this feeds back to control the fuel–air mixture that enters the engine. Too little oxygen flowing over the catalyst slows down oxidation of carbon monoxide and unburnt hydrocarbons

Fig 10.21 The catalytic converter unit

This temperature is reduced to about 150 °C when platinum is mixed with rhodium. The mixture of platinum and rhodium is known as a three-way catalyst, because, in addition to catalysing the oxidation of carbon monoxide, it catalyses two other reactions that involve emission pollutants (see below). Only 1–2 g of each element is used, but because they are coated onto a honeycomb filter of aluminium oxide, the surface area of the catalyst is equivalent to that of two football pitches. The use of **inert** supports to increase the surface area of catalysts reduces the amount of catalyst required and maximises the surface area available for reaction.

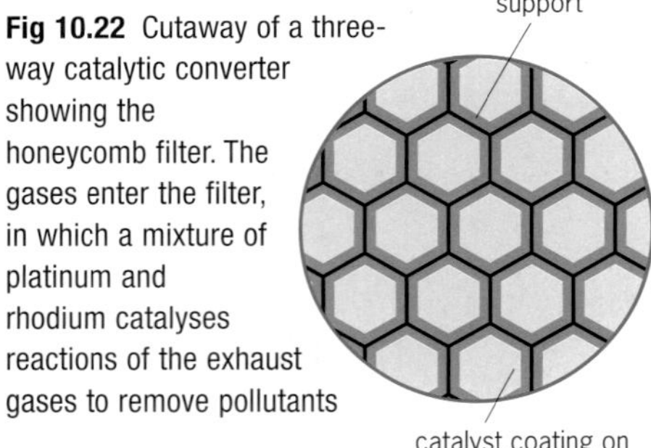

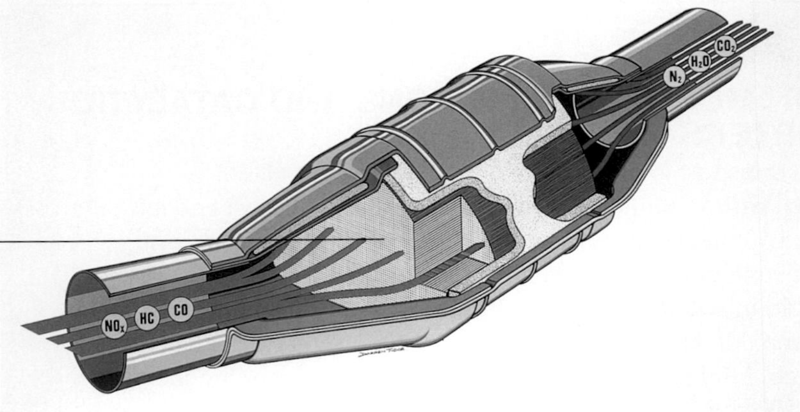

Fig 10.22 Cutaway of a three-way catalytic converter showing the honeycomb filter. The gases enter the filter, in which a mixture of platinum and rhodium catalyses reactions of the exhaust gases to remove pollutants

UNBURNT HYDROCARBONS C_xH_y AND CATALYTIC CONVERTERS

When there is too little oxygen in a petrol engine, unburnt hydrocarbons are present in the exhaust emissions, together with carbon monoxide. Also, volatile hydrocarbons, such as butane, evaporate from the petrol tank on a warm day when a vehicle is stationary. There is increasing concern about **volatile organic compounds (VOCs)**, especially those in the environment. In the UK, almost 40 per cent are from fuel evaporation and exhaust emissions. Long-term exposure to hydrocarbon emissions can impair lung function, while even-short-term exposure can irritate the lung lining. Also, some VOCs, such as benzene, are known carcinogens.

? QUESTION 19

19 **a)** Why is a very large surface area of catalyst required in a catalytic converter?

b) Why is the catalyst applied to its support so that an irregular surface is provided rather than a smooth one?

The platinum in a converter catalyses the oxidation of C_xH_y in the exhaust emissions. This is the second way in which pollutants are removed in a three-way catalytic converter:

$$C_xH_y \xrightarrow{O_2,\ \text{Pt catalyst}} CO_2 + H_2O$$

The catalyst needs to warm up before it becomes effective, so it is during cold starts that most emissions of exhaust hydrocarbons occur. Enough oxygen is also required through the exhaust to oxidise C_xH_y and carbon monoxide. For this reason, an oxygen sensor is fitted just before the catalytic converter to feed back information about oxygen concentration to the vehicle's fuel injection system (see Fig 10.20).

QUESTIONS 20–21

On a car journey, when are exhaust emissions of carbon monoxide and C_xH_y likely to be greatest?

a) Write a balanced equation for the oxidation of benzene (C_6H_6).

b) What environmental problem is worsened by the oxidation of carbon monoxide and C_xH_y?

SCIENCE IN CONTEXT

Minimising hydrocarbon evaporation

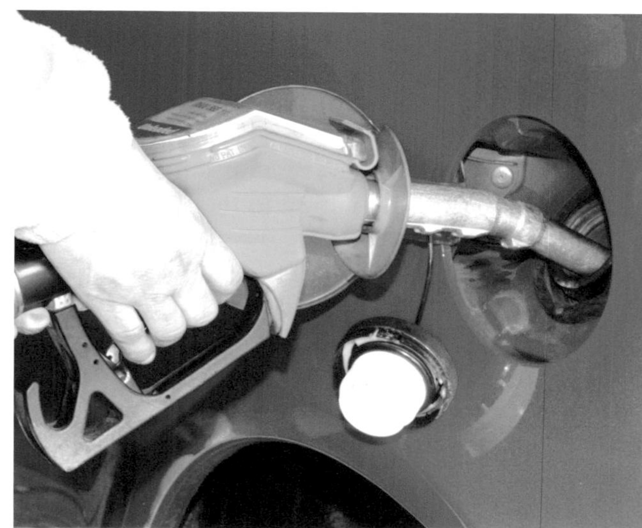

Evaporation of hydrocarbons can occur at any time from the blending of petrol at the refinery to the refuelling of a vehicle at a petrol pump. The UK is now well on the way to introducing a closed system for the loading and transport of petrol, so that all hydrocarbon vapours are trapped and recycled before they escape. Hydrocarbon vapours still escape at the petrol pump, almost always because of a poor seal between the tank and the nozzle.

Fig 10.23 The flap round this petrol hose nozzle is essential in preventing volatile hydrocarbons from escaping into the atmosphere

OXIDES OF NITROGEN AND CATALYTIC CONVERTERS

Air contains mostly nitrogen. Under normal conditions, nitrogen is very unreactive. However, a petrol engine reaches temperatures of 1000 °C, which supplies enough energy to split the very strong triple bond in nitrogen. Nitrogen then reacts with oxygen to form nitrogen oxides (NO_x) – mainly nitrogen monoxide (NO).

NO can be further oxidised in the air to give nitrogen dioxide (NO_2). While NO is colourless, NO_2 is brown; when the atmospheric conditions allow, this can build up as a brown haze in large cities. NO_2 contributes to **acid rain**, reacting with water to form nitrous and nitric acids:

$$2NO_2(g) + H_2O(l) \rightarrow HNO_2(aq) + HNO_3(aq)$$

NO_2 also catalyses the oxidation of sulfur dioxide in the atmosphere (see page 207) and causes respiratory diseases such as bronchitis. The How Science Works box at the start of this chapter mentions the build-up of nitrogen oxides in winter smogs. Nitrogen dioxide, in particular, often far exceeds World Health Organization guidelines and is linked to more deaths occuring than are usual.

REMEMBER THIS

The chemical reactions that lead to the production of acid rain are complicated and beyond the scope of this book. This account is a much-simplified one.

QUESTION 22

Write a balanced equation for the formation of NO.

QUESTION 23

23 Dinitrogen oxide (N_2O) is also formed in vehicle engines. This is non-toxic to humans and is used as an anaesthetic; one of its names is laughing gas. Seven per cent of N_2O in the atmosphere comes from vehicle exhausts. What major environmental problem is linked to a build-up of N_2O? To check your answer, look at Table 10.4, page 202.

REMEMBER THIS

Adsorb is the word used when reactants are weakly bonded to a surface. Do not confuse this with **absorb**, used for substances that enter the material like a sponge soaking up water.

The third way in which a three-way catalytic converter works is to reduce NO_x back to nitrogen and oxygen, this time using a rhodium (Rh) catalyst:

$$2NO_x(g) \xrightarrow{\text{Rh catalyst}} N_2(g) + xO_2(g)$$

Too much oxygen passing from the engine into the catalytic converter reduces the efficiency of this reaction. So, achieving the correct fuel–air mixture is critical to the efficient function of a catalytic converter, hence the need for an oxygen sensor in the exhaust system (Fig 10.20).

6 HETEROGENEOUS CATALYSIS AND CATALYST POISONING

The catalysts in converters are always solids, and the reactants are always gases. So the catalysts are said to be **heterogeneous catalysts**, because they are in a different physical state to the reactants. The reactants are **adsorbed** onto the catalyst surface, which means they weakly bond to it. This holds the reactant molecules close together and also allows their covalent bonds to weaken. This provides the alternative route of lower activation energy (discussed on page 166 and Chapter 26). New bonds form and the product molecules are **desorbed**.

The strength of the weak bonds formed at the catalyst surface with the reactant molecules and product molecules is critical to the efficient function of

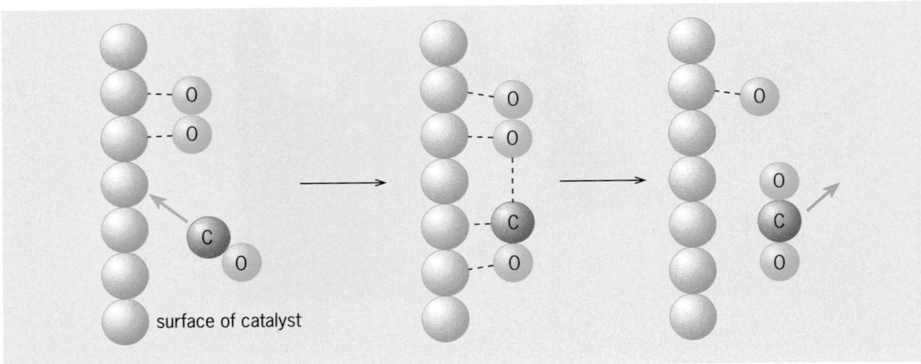

Fig 10.24 The catalytic oxidation of carbon monoxide. CO molecules and O_2 molecules form weak bonds on the catalyst surface. The adsorbed molecules are held close together and their internal covalent bonds weaken, which allows a reaction with lower activation energy to occur. The molecules are then desorbed from the catalyst surface

REMEMBER THIS

The reduction of NO to N_2O takes place by adsorption of NO onto the catalyst surface and desorption of N_2 from the catalyst surface .

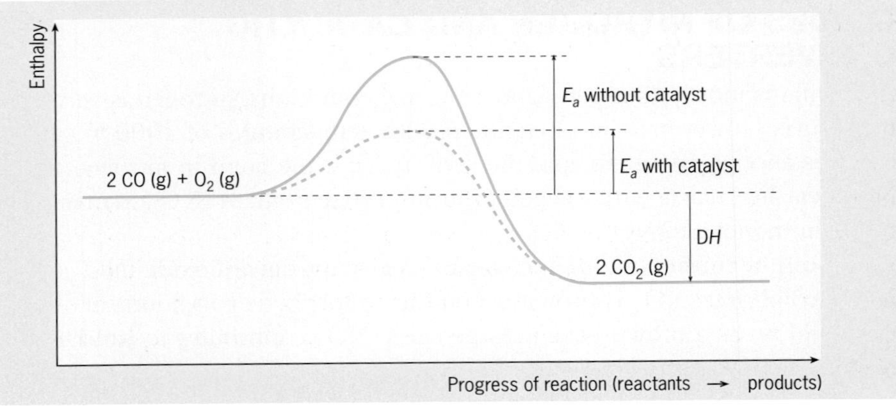

Fig 10.25 An energy (enthalpy) profile diagram showing the effect of using a platinum catalyst on the oxidation of carbon monoxide

the catalyst. If the bonds formed are too weak the reactant molecules will not be held in place, and if the bonds are too strong the product molecules will not be able to leave. This is why platinum and rhodium are good catalysts to use in catalytic converters. Tungsten, on the other hand, forms bonds with reactants that are too strong, while those formed with silver are too weak.

Just two tanks of leaded petrol are enough to render a catalytic converter useless. This is called **catalyst poisoning**. Lead poisons the catalyst because it is adsorbed more strongly than the reactant molecules and so it blocks the **active sites** at which the reactants bond. (It was for this reason that the US government made it illegal to use leaded petrol in cars fitted with catalytic converters, and required that the opening to the petrol tank in these cars be made too small to take a leaded-petrol nozzle.)

REMEMBER THIS
To find out about other heterogeneous catalysts see Chapter 1 page 20 (iron in the Haber process) and Chapter 14, pages 285 (nickel in the hydrogenation of unsaturated fats).

7 SULFUR OXIDES, ACID RAIN AND HOMOGENEOUS CATALYSIS

Most emissions of sulfur oxides, which are toxic, are from the burning of sulfur in fossil fuels. Sulfur dioxide is produced in large quantities by the reaction:

$$S \text{ (in fuel)} + O_2(g) \rightarrow SO_2(g)$$

and can be further oxidised into sulfur trioxide in the atmosphere:

$$2SO_2(g) + O_2(g) \rightarrow 2SO_3$$

Nitrogen dioxide can act as a catalyst in this reaction, which is another reason why nitrogen dioxide emissions should be reduced where possible. The equations for this catalysis are given in Question 24. As the physical state of the nitrogen dioxide is the same as that of the reactants, it is called a **homogeneous catalyst**.

QUESTION 24

24 Explain why NO_2 acts as a catalyst in these two reactions:

$$SO_2(g) + NO_2(g) \rightarrow SO_3(g) + NO(g)$$
$$NO(g) + \tfrac{1}{2}O_2(g) \rightarrow NO_2(g)$$

Dilute solutions of sulfuric acid in rainwater are the main cause of acid rain. (NO_x in the atmosphere also contributes to acid rain by forming nitrous and nitric acid solutions – see page 205.)

REMEMBER THIS
H_2SO_3 is sulfurous acid. It is a weak acid and only exists in dilute solution.

Acid rain has had serious and far-reaching consequences, killing trees in forests and lowering the pH in lakes so that fish die. Burning petrol and diesel in vehicle engines is not the main contributor to sulfur dioxide in the atmosphere; it accounts for only 2 per cent of emissions in the UK. By far the worst culprit is the burning of coal in power stations (see page 439).

Sulfur dioxide also poisons the catalyst in catalytic converters by forming strong bonds with the active sites at the surface. Fortunately, mixing the catalyst with aluminium oxide alleviates this because the sulfur dioxide is held in preference by the oxide. The normal conditions of engine running are oxidising, but during acceleration the conditions in the converter allow the sulfur dioxide to be reduced to hydrogen sulfide, which is then expelled and so the catalyst is not permanently poisoned.

Fig 10.26 These statues on the outside of Exeter Cathedral show the eroding effects of acid rain

8 PHOTOCHEMICAL SMOG

The word 'smog' was first used to describe the combination of smoke, fog and sulfur dioxide that used to build up in London and cause hundreds of extra deaths. The Clean Air Act of 1956 made this type of smog a thing of the past in the UK. However, photochemical smog caused by exhaust emissions from vehicles is a problem in many urban areas. This is also called summer smog; Los Angeles and Beijing are two infamous examples. The atmosphere around a large city forms a vast mixing bowl for chemical reactions, and sorting out just what causes photochemical smog has not been easy. Even now, it is not fully understood.

LIGHT AND THE FORMATION OF SECONDARY POLLUTANTS

Any pollutant in an exhaust emission is called a **primary pollutant**. A **secondary pollutant** is formed in air as a result of the chemical reactions

REMEMBER THIS
NO_2 is really a secondary pollutant, as NO is the principal gas emitted from the exhaust. It is oxidised to NO_2 in the atmosphere.

REMEMBER THIS

'Hetero' comes from the Greek word for 'different'. 'Homo' is from the Greek for 'same'.)

Cracking of long-chain hydrocarbon molecules involves heterolytic or homolytic fission, depending on the conditions (see pages 187–188).

The halogenation of alkanes is an example of homolytic fission (see page 241).

REMEMBER THIS

$E = hf$

where E = energy

h = Planck's constant and

f = frequency

of a primary pollutant. Sunlight supplies the energy to initiate the reactions that form the secondary pollutants ozone (O_3) and organic nitrates, such as PAN. Reactions in which the energy is supplied by light are called **photochemical reactions**.

NO_2 absorbs photons of ultraviolet light that supply the energy (*hf*) to split one of the covalent bonds holding N and O together:

$$NO_2(g) \xrightarrow{hf} NO(g) + O(g)$$

The splitting of the covalent bond is sometimes called **bond fission**. There are two ways in which a covalent bond can split: heterolytic fission and homolytic fission. In **heterolytic fission** (Fig 10.27), both electrons from the bond go to one atom. The atom that gains an electron becomes negatively charged, while the atom that loses an electron becomes positively charged. In **homolytic fission** (Fig 10.28), when the bond breaks the bonding pair of electrons are equally shared, so that each atom in the bond gains one electron. The atoms are not charged because the number of protons is balanced by the number of electrons.

$$X{:}Y \longrightarrow X^+ + {:}Y^-$$

Fig 10.27 Heterolytic fission

$$X{:}Y \longrightarrow X\bullet + \bullet Y$$

Fig 10.28 Homolytic fission

Free radicals result from homolytic fission. Free radicals are species with an unpaired electron, which often makes them highly reactive. So, in our example, X and Y are both free radicals. The unpaired electron on a free radical can be shown by a raised dot, X•.

Take the homolytic fission of NO_2 by light, Fig 10.29.

$$O{=}N{:}O \xrightarrow{hf} \bullet N{=}O + \bullet O \bullet$$

Fig 10.29 Homolytic fission of NO_2 by light

Both NO and O are free radicals. In fact, atomic oxygen has two unpaired electrons and is called a **diradical**.

Oxygen atoms are very reactive; one of their reactions involves the production of ozone (O_3):

$$O(g) + O_2(g) \rightarrow O_3(g)$$

The presence of ozone in the stratosphere is essential to prevent too much ultraviolet light penetrating to the lower atmosphere (troposphere). However, its build-up close to the Earth's surface is dangerous as it causes respiratory problems, and in high concentrations it produces coughing and nausea. It is also involved in a series of other reactions with unburnt hydrocarbons, which produce yet more ozone and a group of very unpleasant molecules called organic nitrates (Fig 10.30).

? QUESTION 25

25 Use s,p,d notation to write the full electron configuration of an oxygen atom.

If you are not sure how to do this, look back at pages 63–64.

Why do you think it has two unpaired electrons?

For more information about ozone in the stratosphere, see page 254.

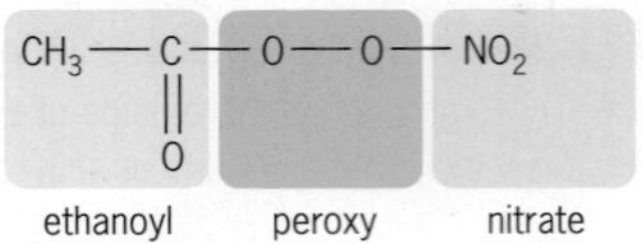

Fig 10.30 The organic nitrate peroxyethanoyl nitrate (often called PAN after its older name peroxyacetyl nitrate). PAN causes breathing difficulties and makes eyes water. It also damages plants

SOLUTIONS TO VEHICLE POLLUTION

Particularly through the development of catalytic converters, chemists have reduced exhaust emissions of some pollutants. Catalysts need to heat up before they can start to operate, so emissions on short journeys or in cold weather are particularly high. Because of the tremendous problems with vehicle pollution in Los Angeles, Californian legislation ruled that by 2003 10 per cent of all cars sold must be zero-emission vehicles (ZEV). However motor manufacturers, supported by the US government, successfully challenged this legislation. Even though it is still the subject of fierce argument, California's attempts to reduce air pollution from vehicles is having an effect, as other countries have been influenced to introduce new regulations on vehicle emissions, forcing motor manufacturers to respond.

People need to change their attitude to the car. Unless personal car use is reduced, we shall have to continue to accept growing deterioration of the environment.

REMEMBER THIS

In practice, ZEV really means an electric car. But is this truly zero-emission when fossil fuels are burned to provide the electricity to charge up the batteries?

Fig 10.31 The cycle path and tramway in Amsterdam, which encourage people to leave their cars at home

Fig 10.32 The Supertram is one way in which Sheffield is making public transport more attractive

SUMMARY

After studying this chapter, you should know the following:

- Alkanes are named systematically using IUPAC conventions.
- A homologous series is a group of compounds that have similar chemical properties, and each subsequent member differs from the previous member by a CH_2 group. They can be represented by a general formula.
- Trends in melting point, boiling point and volatility can be explained by weak intermolecular forces, called induced dipole–induced dipole forces or van der Waals forces.
- As the chain length of alkanes increases, so does the number of electrons, which increases the strength of the induced dipole–induced dipole forces.
- Branching reduces the surface area over which induced dipole–induced dipole forces can act.
- Alkanes are non-polar molecules, so do not react with polar reagents.
- Strong C–C and C–H bonds make alkanes relatively unreactive.
- Complete combustion of alkanes produces CO_2 and H_2O.
- Branched-chain alkanes combust smoothly under pressure in a petrol engine and have high octane numbers.
- CO_2 in the atmosphere is an important greenhouse gas. Increasing its concentration is almost certainly leading to global warming and climate change.
- Exhaust emissions include CO, unburnt hydrocarbons and NO_x. Three-way catalytic converters reduce the concentrations of these pollutants, but need to heat up before they can start operating.
- Heterogeneous catalysts are in a different physical state to the reactants.
- Homogeneous catalysts are in the same physical state as the reactants.
- SO_2 and SO_3, produced when sulfur-containing fossil fuels are burnt, are the main cause of acid rain. They are also toxic.
- Free radicals are produced by the homolytic fission of covalent bonds, for which light can provide the energy.
- Ozone is a secondary pollutant.
- Photochemical smog is produced by the interaction of sunlight with vehicle exhaust emissions.

Practice questions and a How Science Works assignment for this chapter are available at www.collinseducation.co.uk/CAS

11

Analysing, identifying and separating substances

11 ANALYSING, IDENTIFYING AND SEPARATING SUBSTANCES

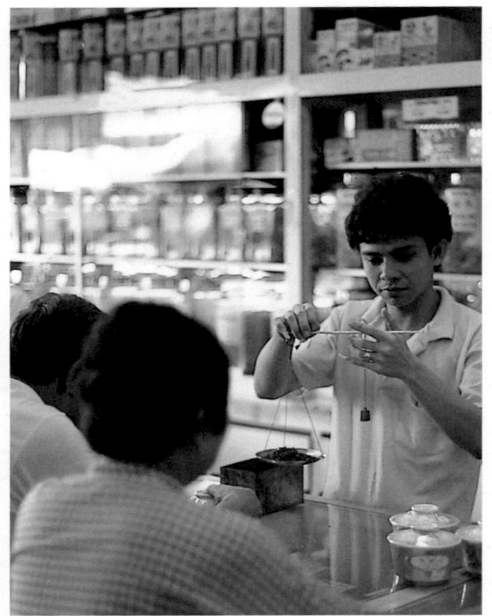
Chinese chemist Herbal medicine has always been the norm in China and is gaining greater acceptance in Europe and America

Medicines from plants

The first recorded use of plants for medicinal purposes was found on Egyptian papyrus dated about 1500 BC. All over the world, from tribal medicine in Africa to folk remedies in Britain, many thousands of plants have been used to cure all sorts of ailments. Often, this knowledge has been handed down by word of mouth, and there is now increasing concern that much of it will be lost for ever as cultures change, tribes disperse and modern medicines of the industrialised world dominate.

The bark of the cinchona tree was used for centuries by the people of Peru to treat malaria, and we now know that it contains the anti-malarial drug quinine. The Chinese use herbal medicines extensively: one of their herbal cures for asthma has been shown to contain ephedrine, which is used in modern medicine to enlarge the air passages of the lungs. Even aspirin has its origins in willow bark.

Interest in tribal and folk remedies has been reawakened since the 1960s. Chemists investigate thousands of plants each year to see if they contain biologically active chemicals that might be developed into medicinal drugs. There is real hope that newly documented remedies will lead to drugs that treat cancer, heart disease and HIV/AIDS.

INTRODUCTION

Many organic chemicals come from crude oil, many come from plants and animals and others are made synthetically. Chemists use separation techniques to isolate these substances, followed by analytical techniques to find out their chemical structure. The methods chemists use to identify the formulae of these organic compounds and determine their structures are the subject of this chapter.

SCIENCE IN CONTEXT

Pharmaceutical libraries and stage synthesis

Although you may think of a library as a collection of books, to a pharmaceutical company it is a huge collection of compounds that may be useful as drugs. Only a few years ago most of the compounds in these libraries were naturally occurring compounds made by plants, animals or microorganisms. Chemists would make these complex compounds one at a time, and each would then be screened for any pharmaceutical activity. This was a slow and expensive process. Typically, a medicinal chemist might synthesise a hundred compounds in a year.

In the past few years there has been a revolution in building these libraries. Thousands of likely drug molecules can be built up in stages in a matter of hours using computerised syringes. Basic building blocks of chemicals are reacted at each stage; they combine and the resulting molecules become larger and larger.

Building up molecules in stages

Suppose the reactant molecules that a chemist wishes to combine are called A, B and C.

At stage 1, the A, B and C molecules are placed in separate reaction vessels and shaken with polymer beads, often polystyrene. These beads are tiny and one gram will contain approximately one million. The A, B, or C molecules covalently bond to the polymer beads. This makes it easier to wash and separate the growing molecules at the end of each stage. The reactant molecules, attached to their polymer beads, are then split equally into three reaction vessels, so that each one contains a third of each reactant.

At stage 2, more of A, B or C is added to each of the three reaction vessels giving nine possible product molecules, all attached to polymer beads:

A-A, A-B, A-C, B-A, B-B, B-C, C-A, C-B, C-C

The polymer beads in each of the three reaction vessels are washed to remove any unreacted reagent. They are again split into three equal portions and mixed together in another three reaction vessels.

Pharmaceutical libraries and stage synthesis (Cont.)

At stage 3, A, B or C is again added to give 27 different product molecules. Each is still attached to its polystyrene bead. The beads are washed and the mix and split is repeated a fourth time.

At stage 4, A, B or C is added. There are now 81 different product molecules.

Large-scale screening

By the sixth stage there are 46 656 product molecules. It would be no use producing this number of molecules if it took years to screen each one. In Chapter 4, page 70, we described drugs as fitting into the receptor sites of molecules. These molecules are usually enzymes (see page 384). To screen the thousands of molecules produced from combinatorial chemistry rapidly for potential drug activity, the molecules are split from their polymer beads and reacted with enzymes. Only those that affect enzymes by fitting into the receptor sites are developed further.

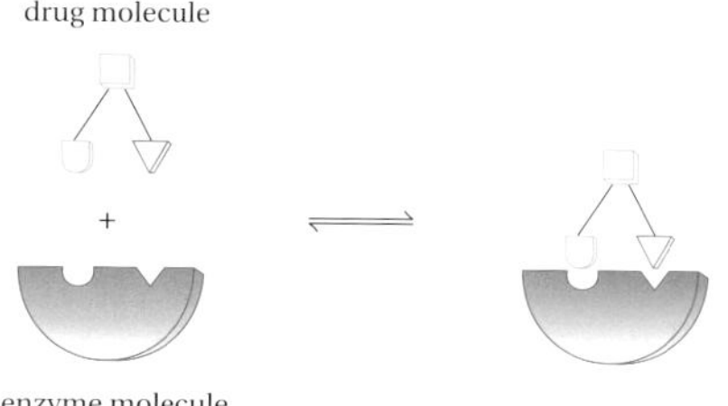

Fig 11.1 For a molecule to have drug activity it must fit into a receptor site

1 WORKING OUT THE FORMULA OF A COMPOUND

We are now going to see how you can work out the formulae of completely unknown compounds, such as those that might be present in a medicinal herb.

PERCENTAGE COMPOSITIONS AND EMPIRICAL FORMULAE

One very useful piece of information about any compound is its **percentage composition**. This means the percentage by mass of each element in the compound.

If you have just discovered what you think is a new compound, finding the percentage composition is one of the first investigations you would carry out. You can either decompose a known mass of the compound into its constituent elements, or burn it in oxygen and weigh the products formed, such as carbon dioxide and water (called **combustion analysis**). The masses of any other elements present in the compound are found by other means.

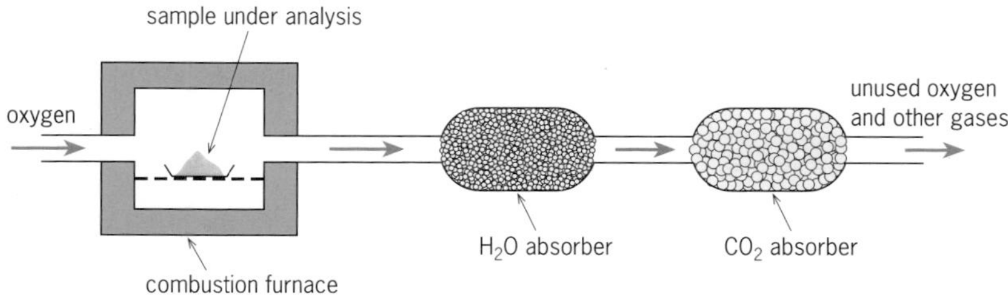

Fig 11.3 Combustion analysis for carbon and hydrogen. A few milligrams of the compound being analysed are completely combusted. The carbon and hydrogen in the compound form CO_2 and H_2O, respectively. The increased masses of the two absorbent materials are measured to give the masses of CO_2 and H_2O produced. Given that the mass of the original sample is known, the percentage composition can be calculated

Once the percentage composition is known, the **empirical formula** of a compound can be calculated.

The empirical formula is the simplest whole-number ratio of the number of atoms of each element in a compound.

For example, benzene has a molecular formula of C_6H_6, but its empirical formula is CH.

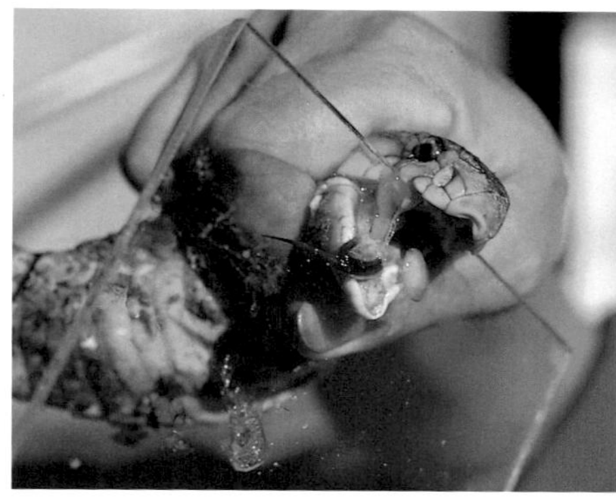

Fig 11.2 The venom of some snakes immobilises their prey by dramatically reducing blood pressure. This discovery has already led to greater understanding of how to treat high blood pressure, a potentially fatal condition

QUESTION 1

Why is combustion analysis done mostly on organic compounds? (Hint: think about the products formed.)

REMEMBER THIS

Remember that the molecular formula tells you the number of atoms of each element in a molecule of a compound. See page 3.

QUESTION 2

What are the empirical formulae of ethane C_2H_6 and ethanoic acid CH_3COOH?

EXAMPLE

Q Combustion analysis shows that the compound that gives cinnamon its characteristic aroma has a percentage composition of 81.8 per cent carbon, 6.1 percent hydrogen and 12.1 per cent oxygen. Calculate its empirical formula. (A_r: C = 12.0, H = 1.0, O = 16.0)

A The simplest way to deal with this problem is to imagine that you have 100.0 g of the compound so that the masses of the elements are easy to work out as shown below. Remember that the amount in moles is the mass in grams divided by the mass of one mole in grams.

	Carbon	**Hydrogen**	**Oxygen**
Mass in grams:	81.8	6.1	12.1
Amount in moles:	$\frac{81.8}{12.0} = 6.82$	$\frac{6.1}{1.0} = 6.1$	$\frac{12.1}{16.0} = 0.756$
Simplest ratio: (divide by the smallest number)	$\frac{6.82}{0.756} = 9.02$	$\frac{6.1}{0.756} = 8.1$	$\frac{0.756}{0.756} = 1.00$
Simplest whole-number ratio:	9	8	1

So its empirical formula is C_9H_8O

Another way of calculating the empirical formula from combustion analysis data is to use the actual masses of the different compounds produced, as shown in the next Example.

REMEMBER THIS

The **relative atomic mass** of an element is the average mass of the atom (taking into account all of its isotopes and their abundance) compared to 1/12 the mass of one atom of carbon-12. See page 37.

Isotopes are atoms of the same element with the same atomic number, but different mass numbers. See page 37. **Relative isotopic mass** is also defined on page 37.

? QUESTION 3

3 **a)** A hydrocarbon is found to contain 84.5 per cent carbon. What is its empirical formula?
b) An amino acid contains 0.601 g carbon, 0.799 g oxygen, 0.125 g hydrogen and 0.349 g nitrogen. Calculate its empirical formula.

? QUESTION 4

4 0.200 g of a compound found in vinegar on complete combustion gave 0.293 g CO_2 and 0.120 g H_2O. The compound contained C, H and O only. Work out its empirical formula.

EXAMPLE

Q 0.100 g of a sugar, known to contain only carbon, hydrogen and oxygen, is completely combusted to give 0.147 g CO_2 and 0.0600 g H_2O. Calculate its empirical formula.

A First, we need to calculate the masses of C, H and O in the compound.

There are 12.0 g of C in 44.0 g CO_2

(Remember: 1 mol CO_2 = 12.0 + (16.0 × 2) g)

Therefore, the mass of C in 0.147 g = $0.147 \times \frac{12.0}{44.0} = 0.0401$ g

There are 2.0 g H in 18.0 g H_2O (1 mol H_2O = (1.0 × 2) + 16.0 g)

Therefore, the mass of H in 0.0600 g = $0.0600 \times \frac{2.0}{18.0} = 0.0067$ g

Since we now know the masses of C and H in 0.100 g of sugar, the rest of the mass must result from O:

$$0.0401 + 0.0067 = 0.0468 \text{ g}$$

Therefore, mass of O in 0.100 g = 0.100 – 0.0468 = 0.053 g

We can proceed now as we did in the previous Example:

	Carbon	**Hydrogen**	**Oxygen**
Mass in grams	0.0401	0.0067	0.053
Amount in moles	$\frac{0.0401}{12.0} = 0.00334$	$\frac{0.0067}{1.0} = 0.0067$	$\frac{0.053}{16.0} = 0.0033$
Simplest ratio (divide by the smallest number)	$\frac{0.00334}{0.0033} = 1.0$	$\frac{0.0067}{0.0033} = 2.0$	$\frac{0.0033}{0.0033} = 1.0$
Simplest whole-number ratio:	1	2	1

So its empirical formula is CH_2O

FINDING THE MOLECULAR FORMULA

The empirical formula tells you the simplest ratio of different atoms in a compound. It does not tell you the actual number of atoms of each element in a molecule. To find the molecular formula, you need to know the **relative molecular mass, M_r**, of a compound, which is:

$$M_r = \frac{\text{mass of 1 molecule of the compound}}{\frac{1}{12}\text{ mass of one atom of carbon-12}}$$

The molecular formula may be the same as the empirical formula, or it may be a multiple of it. In Question 4, the empirical formula of the substance found in the vinegar is CH_2O. So we have:

$$\text{Relative mass of } CH_2O = 12 + (2 \times 1) + 16 = 30$$

However, M_r (CH_2O) = 60, so because 30 × 2 = 60, the empirical formula must be multiplied by 2 to find the molecular formula:

$$(CH_2O) \times 2 = C_2H_4O_2$$

In the second Example (left), the empirical formula of the sugar is also CH_2O, but in this case, M_r = 180. This is 30 × 6, so the molecular formula is:

$$(CH_2O) \times 6 = C_6H_{12}O_6$$

So, when we analyse a compound, we need to know its relative molecular mass. This is where the mass spectrometer comes in.

? QUESTION 5

5 What are the molecular formulae of the following compounds?

a) An arene hydrocarbon with empirical formula CH and M_r = 78.

b) An acid in ant sting with empirical formula CH_2O_2 and M_r = 46.

c) Caffeine with empirical formula $C_4H_5N_2O$ and M_r = 194.

2 MASS SPECTROMETRY

Mass spectrometry is the most accurate method of determining relative atomic and molecular masses, but it has many other applications. The mass spectrometer is used by geologists to date rocks, by anaesthetists to analyse compounds in a patient's breath, by pharmaceutical companies to determine the structure of novel compounds and by the oil industry to work out where samples of crude oil originated. A mass spectrometer has even been taken to Mars to analyse rocks and dust on the planet's surface and gases in its atmosphere. Clearly, the potential of mass spectrometry is enormous.

The mass spectrometer was developed in 1919 by the eminent English physicist Francis Aston, from apparatus used by J.J. Thomson (the discoverer of the electron, see page 30). In 1922, Aston received a Nobel prize for his work in developing the mass spectrometer.

Aston used his mass spectrometer to show that neon gas was composed of **isotopes**. Neon atoms were ionised, separated, and then made to hit a photographic plate. Two lines were produced, one much darker than the other, corresponding to relative isotopic masses of 20 and 22. By measuring the relative darkness of the two lines, Aston found that neon-20 atoms were ten times more abundant than neon-22.

From this information, he worked out the average atomic mass of neon to be 20.2. As mass spectrometers became more accurate, a third isotope of neon, neon-21, was discovered. It did not show up in Aston's instrument because only 0.26 per cent of naturally occurring neon is ^{21}Ne (only 26 atoms in 10 000).

HOW A MASS SPECTROMETER WORKS

All mass spectrometers have three basic operations:

- producing gaseous ions from a sample;
- separating the ions according to their mass (and charge);
- detecting the ions.

QUESTION 6

6 Why is a very low pressure maintained inside the mass spectrometer?

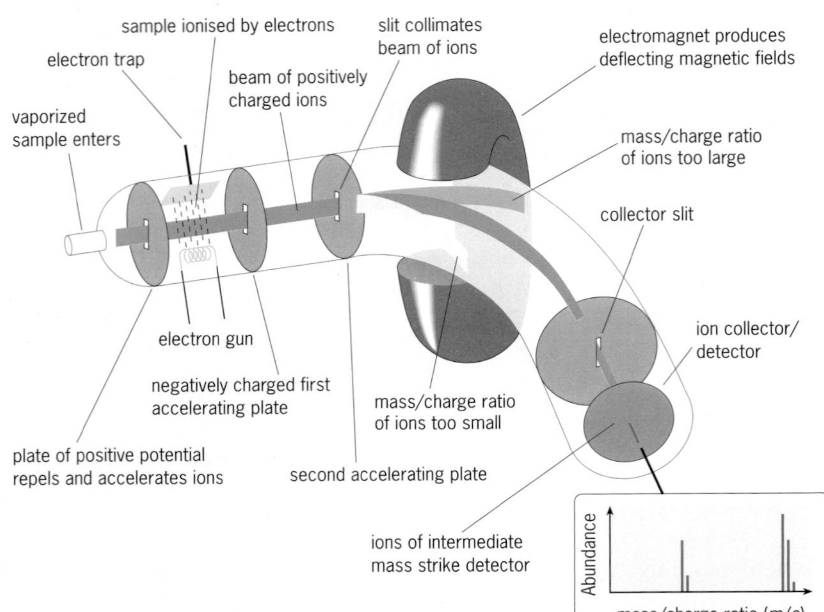

Fig 11.5 A simplified diagram of the workings of a mass spectrometer

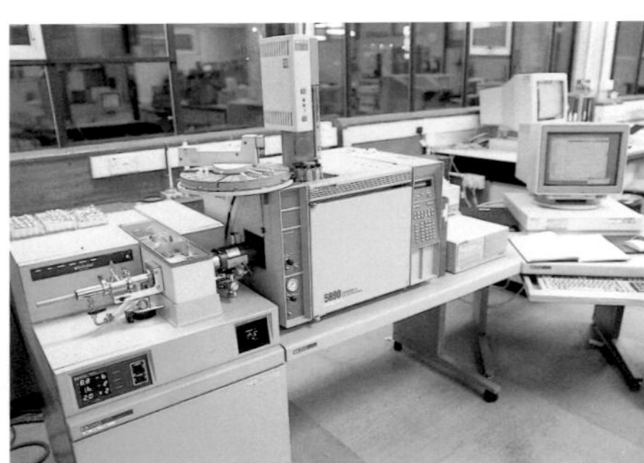

Fig 11.4 A modern mass spectrometer with digital readout and screen monitor. The sample is injected on the left of the apparatus

The type of mass spectrometer that dominated much of the 20th century was based on Ashton's apparatus and it is still in use today (see Fig 11.5). A vaporised sample is injected into the instrument. The exceptional sensitivity of the instrument means that only nanograms (10^{-9} g) of a sample are required. This is then ionised to form positive ions by bombarding the atoms or molecules with high-energy electrons from an electron gun. The bombarding electrons knock electrons from the atoms or molecules in the sample, creating positively charged ions.

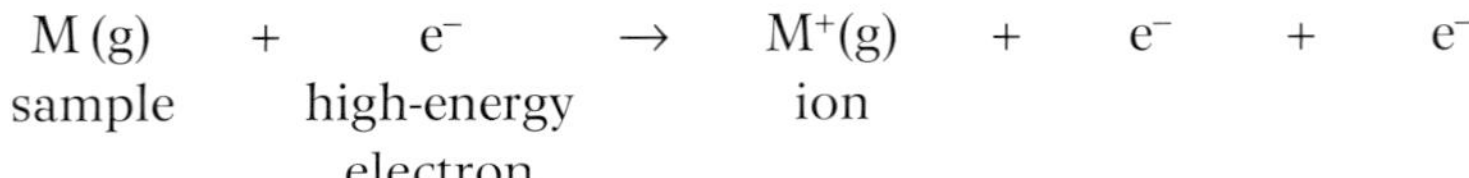

M (g)	+	e^-	→	$M^+(g)$	+	e^-	+	e^-
sample		high-energy electron		ion				

The ions are separated according to their mass and charge, first by accelerating them in an electric field of several thousand volts between two negatively charged plates. Next, they enter a strong magnetic field, which deflects them into a series of separate circular paths according to their **mass/charge ratio**. Positive ions with higher mass/charge ratios are deflected less than those with lower ratios. So by varying precisely the strength of the magnetic field, ions of a particular mass/charge ratio can be focused on the detector in sequential order to build up a spectrum.

On the mass spectra shown in this book, the x-axis is labelled mass/charge ratio (m/e). Although 2+ ions do occur (when two electrons are removed), 1+ ions are far more abundant. For these, the charge e = +1, which makes the mass/charge ratios of the peaks equal to the relative masses of the ions. This is why you will sometimes see the x-axis labelled simply: mass.

REMEMBER THIS

As mass spectometers have become smallar they have been used to analyse the frozen hydrocarbon surface of Titan, the largest moon of the planet Saturn, and to search for evidence of life on Mars.

REMEMBER THIS

Mass spectrometers are used to determine the ratio of carbon-12 to carbon-14 in dead organic matter. This is the basis of carbon-14 dating. See page 43, Chapter 2.

SCIENCE IN CONTEXT

Time-of-flight mass spectrometer

It was during the 1950s that time-of-flight mass spectrometers (TOF-MS) were first developed. The principle is simpler than the single beam mass spectrometer on the previous page. To separate ions according to their masses and charges, instead of deflecting ions in a known magnetic field, the ions are simply all accelerated together by a pulsed electric field. A voltage is applied across the accelerator plates, which creates a burst of ions all with the same kinetic energy. These are released at the same time into a region of the mass spectrometer where there are no fields, so they are allowed to drift.

Since kinetic energy = ½ mv^2 (m = mass of ion and v = velocity (speed) of ion), the speed of each ion from the accelerating electric field to the detector depends on its mass. It also depends on its charge, since the greater the charge on an ion the more it is accelerated. This means that the time taken to reach the detector will be different for ions of different mass/charge ratios. As with the single beam mass spectrometer, most of the ions produced are singly charged, so the peaks of the mass spectrum show the relative masses.

Although TOF-MS was developed more than 50 years ago, it is more recently that accurate timing of the ions in the drift region was developed, since ions complete the path to the detector in times of less than a nanosecond. This type of mass spectrometer can now be made much smaller than the single-beam instrument and there are portable versions in many research labs around the world.

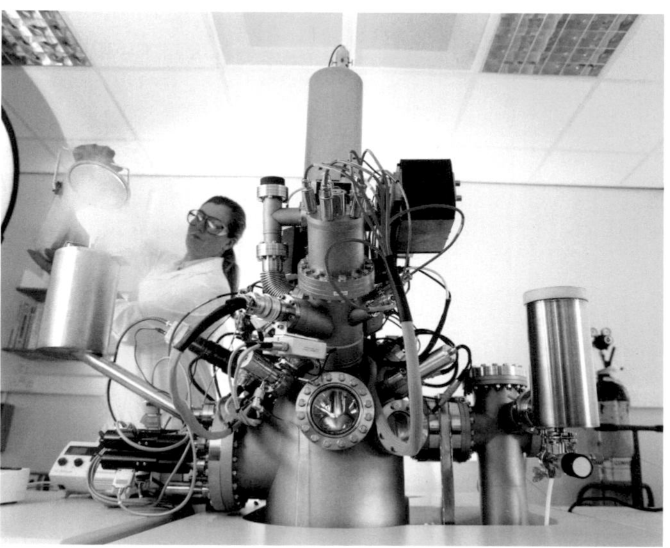

Fig 11.7 The TOF-MS is now smaller and cheaper, meaning that many research laboratories all over the world can have one

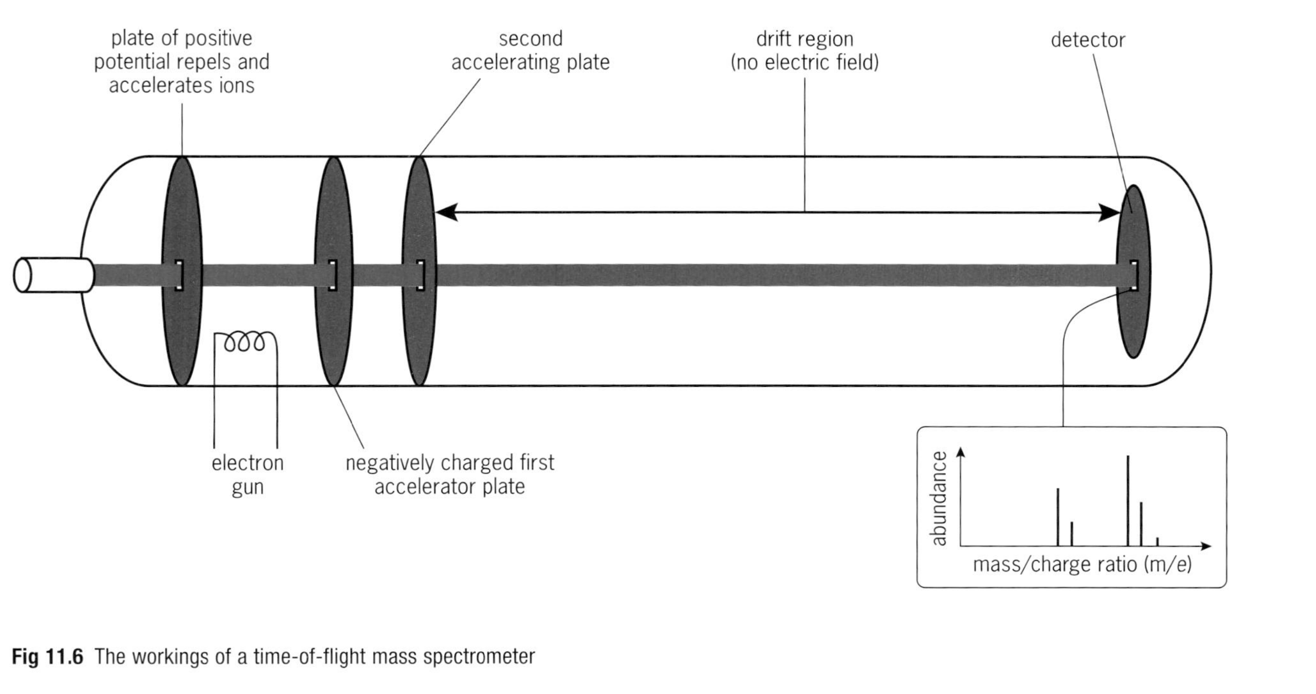

Fig 11.6 The workings of a time-of-flight mass spectrometer

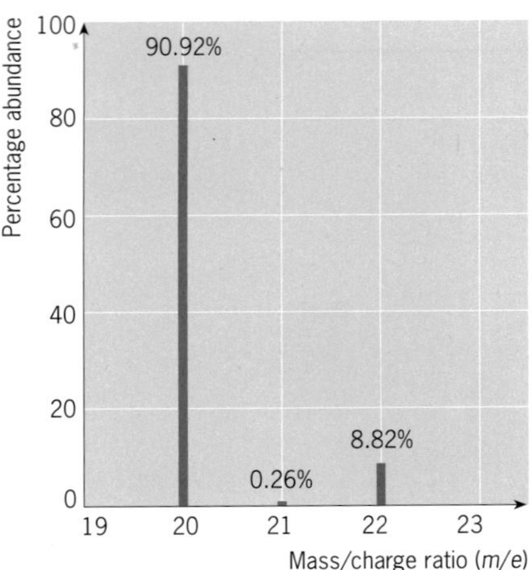

Fig 11.8 The mass spectrum of naturally occurring neon, showing the percentage abundance of each isotope. It is easy to see why Aston's crude instrument failed to reveal the third isotope

REMEMBER THIS

The molecular ion is sometimes called the **parent ion**.

QUESTION 7

7 Calculate germanium's A_r from its mass spectrum, shown in Fig 11.9.

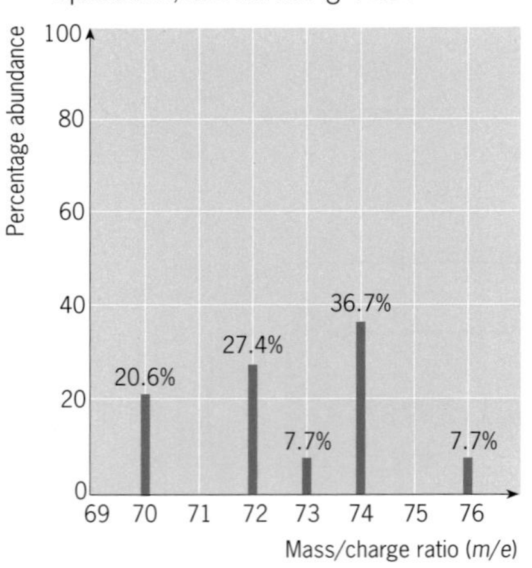

Fig 11.9 The mass spectrum of germanium

QUESTION 8

8 An acidic substance found in the sting of an ant is analysed using high-resolution mass spectrometry, which gives M_r = 46.0054. A database of compounds is searched and the following molecules are revealed to have values of M_r in this region: C_2H_6O, CH_2O_2 and NO_2. Using accurate values of A_r, work out which molecule is present in an ant sting.

FINDING RELATIVE ATOMIC MASSES

Relative atomic masses can be worked out very accurately using a mass spectrometer. (This was one of the first uses for the instrument.) Take neon, for example, whose mass spectrum is shown in Fig 11.8.

To calculate its A_r, we follow the same procedure as that on page 37:

$$A_r(\text{neon}) = \frac{(90.92 \times 20) + (0.26 \times 21) + (8.82 \times 22)}{100} = \frac{2018}{100}$$
$$= 20.18$$

Germanium has five naturally occurring isotopes (Fig 11.9) and its relative atomic mass can be calculated in the same way (see Question 7).

FINDING RELATIVE MOLECULAR MASSES AND MOLECULAR FORMULAE

The mass spectrometer is more commonly used to find relative molecular masses. When a compound such as butane, C_4H_{10}, is analysed, its mass spectrum shows a whole series of peaks (Fig 11.10). The peak with the highest mass is the **molecular ion, M⁺**. It is the molecule with one electron removed.

A simple calculation of M_r tells you to expect the molecular ion to have its peak at 58, but Fig 11.10 also shows a peak at 59. This peak results from those molecules of butane that contain an atom of carbon-13 (^{13}C). Known as the **M+1 peak**, it is always smaller than the molecular-ion peak because there are far fewer naturally occurring carbon-13 atoms. However, as shown later, the M+1 peak can help you to work out the number of carbon atoms in the molecule.

The other peaks in the mass spectrum are caused by butane molecular ions that have broken into fragments. This fragmentation is important in working out the identity of molecules.

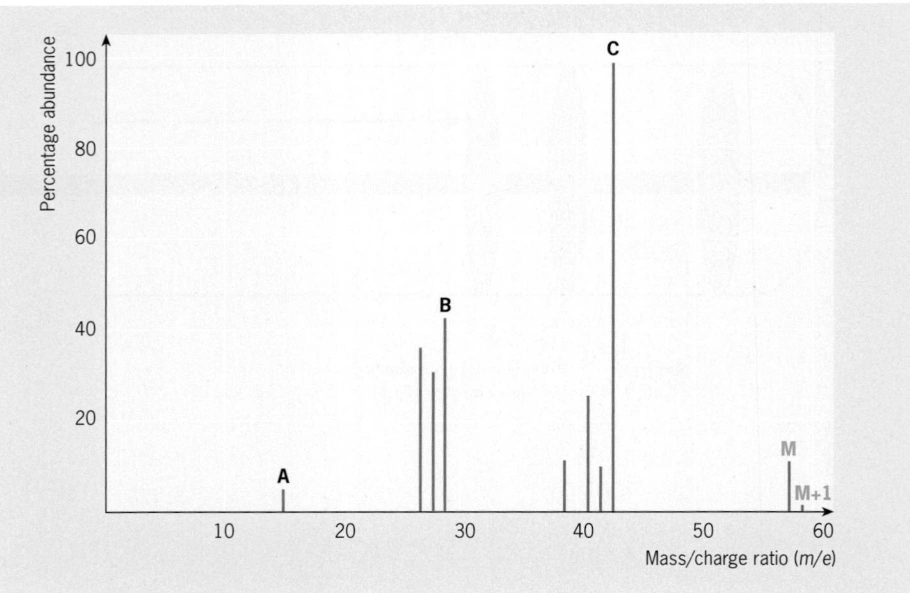

Fig 11.10 The mass spectrum of butane

High-resolution mass spectrometry

The mass spectrum in Fig 11.10 is from a spectrometer set for low resolution. At higher resolutions, it may be possible to deduce the molecular formula.

We know that butane has an approximate M_r of 58, but C_4H_{10} is not the only possible molecular formula for this value of M_r: all the molecules shown in Table 11.1 have an approximate M_r of 58. However, very accurate values for M_r can be obtained using **high-resolution mass spectrometry**. So it is possible to determine which compound is present if accurate values for relative isotopic masses are used.

Table 11.1 Relative isotopic masses: 12C = 12.0000; 1H = 1.0078; 14N = 14.0031; 16O = 15.9949

Formulae	Accurate M_r (*m/e* values for M^+ ion)
C_4H_{10}	58.0780
C_3H_6O	58.0417
$C_2H_2O_2$	58.0054
$C_2H_6N_2$	58.0530

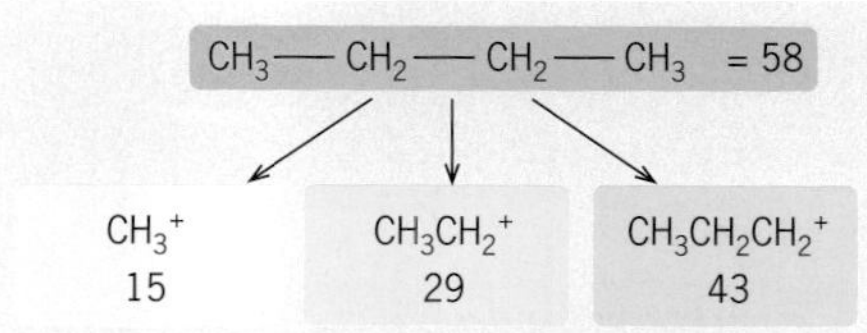

Fig 11.11 Fragmentation of butane

USING FRAGMENTATION TO DETERMINE STRUCTURE

The mass spectrum of butane in Fig 11.10 has several peaks because of the way that the butane molecular ion fragments in the mass spectrometer. These fragments can be used to identify parts of the molecule and, in many cases, build up a picture of the structure of the whole molecule. This is illustrated in Fig 11.11, which we can use to identify some of the peaks in Fig 11.10.

A is from CH_3^+, **B** from $C_2H_5^+$ and **C** from $C_3H_7^+$. The other peaks in the spectrum result from the loss of hydrogen atoms from these fragments. Typical ion fragments produced in the mass spectrometer are shown in Table 11.2.

Table 11.2 The masses of some typical positive ion fragments

Ion	Mass/charge ratio (*m/e*)	Ion	Mass/charge ratio (*m/e*)
CH_3^+	15	CH_3CO^+	43
CO^+	28	$C_3H_7^+$	43
$C_2H_4^+$	28	$C_2H_5O^+$	45
CHO^+	29	$C_4H_9^+$	57
$C_2H_5^+$	29	$C_6H_5^+$	77

Apart from deciding the identity of the peaks, the other important thing to look for is the differences in masses between the peaks. A difference of 15 almost certainly shows the loss of a methyl fragment, and so it is highly likely that the original molecule contains a CH_3 group. A different of 77 indicate the loss of a phenyl group ($C_6H_5^+$).

When a molecular ion fragments, the chemistry becomes rather complicated, and so we discuss only a small part of it here. In Fig 11.10, the molecular-ion peak (M) is 58. The loss of a methyl fragment produces the peak at 43. The butane molecular ion has fragmented to give $C_3H_7^+$ and a free radical $CH_3^•$, which is neutral and so is not affected by the electric and magnetic fields of the mass spectrometer.

$$C_4H_{10}^+ \rightarrow C_3H_7^+ + CH_3^•$$

There are other characteristic differences in the masses between peaks that give clues to which other groups have been lost as free radicals. For example, a loss of 17 suggests that an OH group may have been removed, and a loss of 29 suggests a C_2H_5 group.

Let's now look at the spectrum of ethanol, CH_3CH_2OH (Fig 11.12).

QUESTIONS 9–11

9 **a)** Why are the *m/e* values of the ions in Table 11.2 also their masses?

b) Which fragment is responsible for the peak at *m/e* = 28 of the butane mass spectrum?

The most abundant ion gives the strongest signal at the detector. This is called the **base peak** and is given a relative abundance of 100 per cent. All other abundances are percentages of the base peak.

c) What is the mass of the base peak in butane?

10 Free radicals are highly reactive species with unpaired electrons.

a) Why is $C_4H_{10}^{+•}$ sometimes written $C_4H_{10}^+$?

b) Why is the particle $CH_3^•$ not detected by the mass spectrometer?

11 What groups may have been removed when there is a mass difference of

a) 28; **b)** 45; **c)** 77?

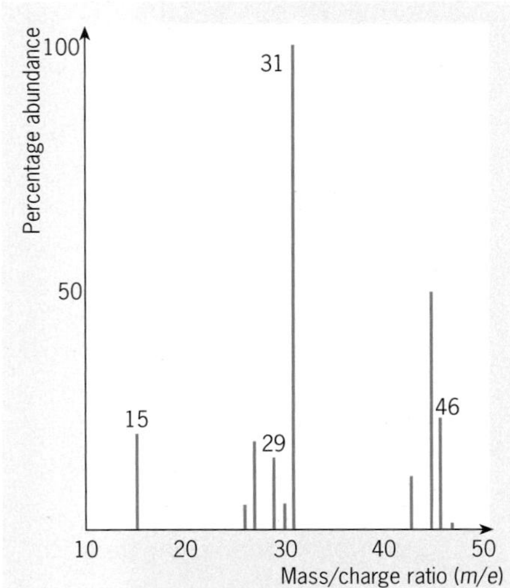

Fig 11.12 The mass spectrum of ethanol

The base peak is given a relative abundance of 100 percent. See Question 9.

The molecular-ion peak is 46, which corresponds to $CH_3CH_2OH^+$. The base peak is at 31. The difference between 46 and 31 is 15, so it is probable that a CH_3 group has been lost. Therefore, the ion responsible for the base peak is CH_2OH^+. The peak at 29 is from $CH_3CH_2^+$, and that at 15 is CH_3^+.

Notice that there is an M+1 peak at 47 because of the isotope carbon-13. In the next section, we see how this peak can be used to determine the number of carbon atoms present in the molecule.

QUESTION 12

12 The compound responsible for the sting of an ant has the mass spectrum shown in Fig 11.13. It has a molecular formula CH_2O_2.

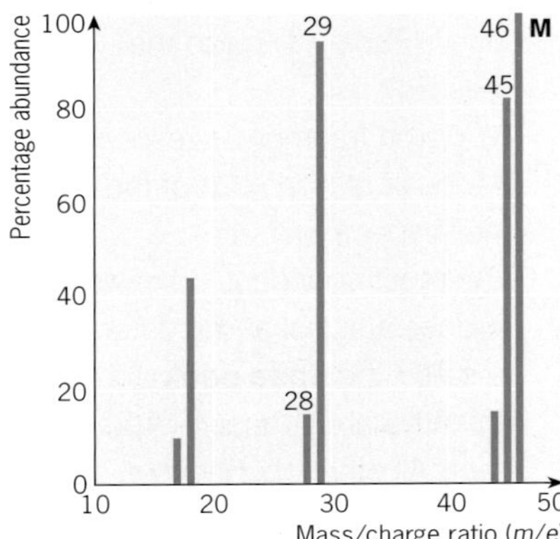

Fig 11.13

a) What group of atoms could be lost from the molecular ion to form a peak at *m/e* 29?
b) Which ions are probably responsible for the peaks at *m/e* = 28, 29 and 45?
c) Suggest a displayed formula for this compound.

ISOTOPE PEAKS

The presence of isotopes in a compound will give characteristic peaks. We met one of these peaks when we examined butane and ethanol (see Figs 11.10 and 11.12). The M+1 peak resulted from the isotope carbon-13. Carbon-13 is present naturally as 1.1 per cent of all carbon atoms, information that allows us to work out the number of carbon atoms in a molecule. If there were one carbon atom in a molecule, the M+1 peak height would be 1.1 per cent of the molecular-ion peak, because 1.1 per cent of molecules would contain carbon-13. So, in a molecule that contains five carbon atoms, about 5.5 per cent will be carbon-13. That is, the M+1 peak height will be about 5.5 per cent of the M peak height.

[M+2] and [M+4] peaks

Chlorine and bromine atoms also give characteristic peaks because of their isotopic compositions. In the case of chlorine, 75 per cent is ^{35}Cl and 25 per cent is ^{37}Cl. This means that when a molecule contains a chlorine atom, there are two peaks at M and **M+2** in the ratio of 75:25, or 3:1 (see Fig 11.14). Although we refer to these peaks as M and M+2, they both are molecular ions. Similarly M+2 and M+4 peaks arise if there are two chlorine atoms or two bromine atoms in a molecule. (See Question 13(d).)

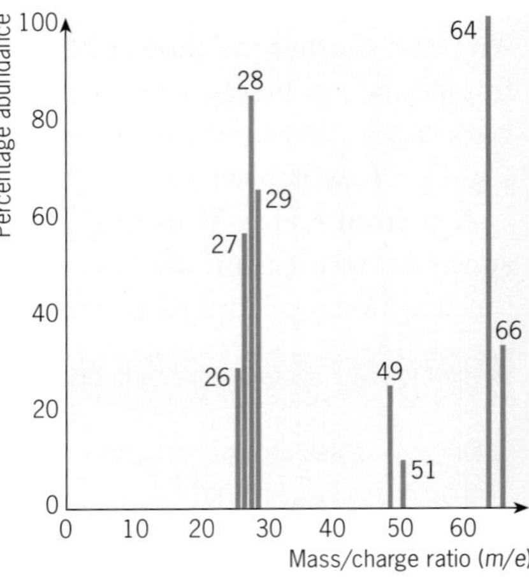

Fig 11.14 The mass spectrum of chloroethane

QUESTION 13

13 **a)** A component of petrol gives an M+1 peak that is 11.1 per cent of the M peak. How many carbon atoms does the molecule contain?
b) In Fig 11.14, the peaks at *m/e* = 49 and 51 are in the ratio 3:1.
i) Which ions are responsible for these peaks?
ii) Both peaks at 64 and 66 are from molecular ions. Identify these ions.
iii) Which group has been lost from the M peak to give a peak at 49?
c) Bromine has two isotopes, ^{79}Br and ^{81}Br, which are present in equal amounts, that is, 50:50. Sketch the mass spectrum you would expect for bromoethane.
d) When analysed, chlorine gas gives peaks at *m/e* = 74, 72 and 70.
i) Explain which ions are responsible for these peaks.
ii) Why is the peak at *m/e* = 74 the smallest?

REMEMBER THIS

The more stable an X^+ species produced in high -resolution spectron, the higher the peak. This is noticable with carbocations and RCO^+ irons.

3 INFRARED SPECTROSCOPY

Atoms in molecules vibrate about their coralent bonds. These vibrations have definite (quantised) energy levels, in the same way that electrons have quantised energy levels in atoms (see pages 51–52). To increase the vibration in a molecule from one level to the next, a definite amount of energy must be absorbed. These amounts of energy correspond to radiation in the infrared part of the electromagnetic spectrum.

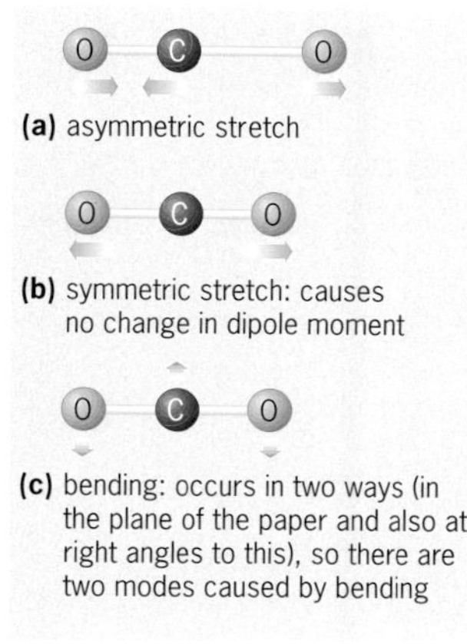

(a) asymmetric stretch
(b) symmetric stretch: causes no change in dipole moment
(c) bending: occurs in two ways (in the plane of the paper and also at right angles to this), so there are two modes caused by bending

Fig 11.15 Ways in which the carbon dioxide molecule vibrates

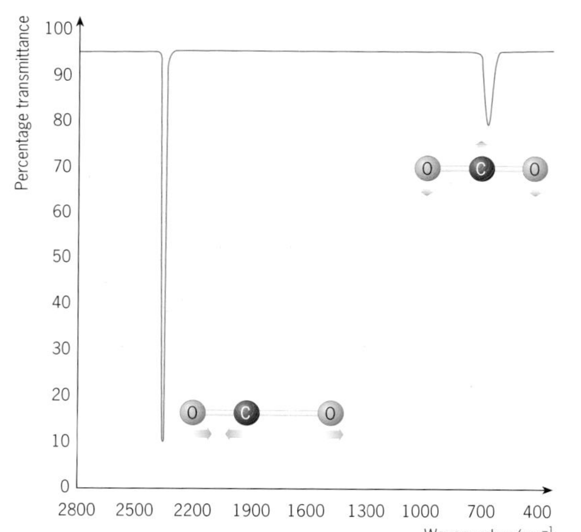

Fig 11.16 The infrared spectrum of carbon dioxide, showing the two modes of vibration that absorb energy in this region. Note both types of bending occur at the same energy levels and produce just a single peak. When two energy levels coincide they are said to be degenerate

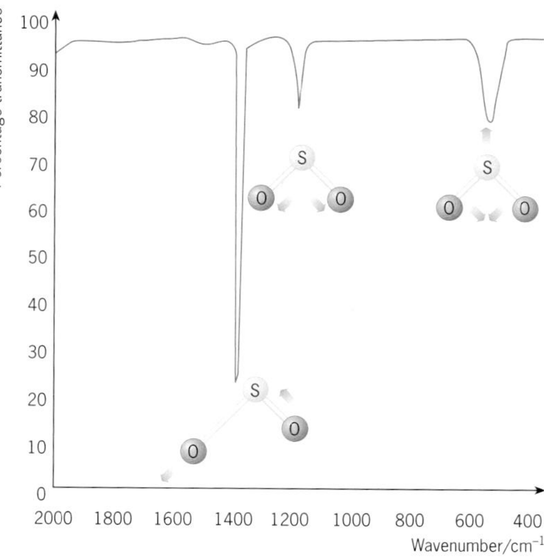

Fig 11.17 The infrared spectrum of sulfur dioxide, showing the vibrations responsible for the peaks

For infrared radiation to be absorbed, the vibrations must cause a change in the dipole moment of the molecule. Therefore, the symmetrical bonds in N_2 and O_2 do not absorb infrared radiation. This is why N_2 and O_2 are not greenhouse gases – they cannot absorb the infrared radiation emitted by the Earth. However, we know from Chapter 10 that carbon dioxide is an important greenhouse gas because it can absorb infrared radiation, due to the ways in which it can vibrate.

Fig 11.15 shows four vibrational modes of carbon dioxide. There is no change in the dipole moment in Fig 11.15(b), and so there is no absorption of infrared radiation in this mode. However, the infrared spectrum of carbon dioxide (Fig 11.16) shows that the modes shown in (a) and (c) absorb in this region.

Notice that in Fig 11.16 the *x*-axis represents **wavenumber**. This is the reciprocal of wavelength ($1/\lambda$) and is usually measured in cm^{-1}. Using it makes the numbers more manageable, since the wavelength of the infrared region lies between 2.5×10^{-5} m and 2.5×10^{-6} m (wavenumber 400 cm^{-1} to 4000 cm^{-1}).

In the case of sulfur dioxide, all three vibrational modes lead to infrared absorption, as shown in Fig 11.17. This is because the V-shaped SO_2 molecule has a different shape from the linear CO_2 molecule, so all three modes lead to a change in the dipole moment. Notice that the absorptions do not occur at the same wavenumbers. This is because the energy required to excite vibration depends on the strength of the bond. As the C=O bond is stronger than the S=O bond, the corresponding vibrations occur at a higher wavenumber in CO_2.

As molecules become more complex, so do their vibrational modes. You may meet terms such as scissoring or rocking. Do not worry about these. When you have to interpret an infrared spectrum, look for one or two peaks that are characteristic of particular bonds.

QUESTION 14

14 **a)** Why is there no absorption peak in the infrared spectrum of hydrogen molecules?

b) Why is there only one absorption peak in the infrared spectrum of HCl?

c) The absorption peaks of HCl and HI are 2886 cm^{-1} and 2230 cm^{-1}, respectively. What does this tell you about the bond energies of H–Cl and H–I?

REMEMBER THIS

A **dipole** occurs in a molecule when one part of the molecule has a slightly positive charge and another part has a slightly negative charge. Dipole moment measures the size of the charges and the distance that separates them. HCl is a polar molecule whose vibrations cause fluctuations in the dipole moment as the distance that separates the charges varies, so it absorbs infrared radiation. See pages 85 and 86.

REMEMBER THIS

The infrared spectrum in Fig 11.16 appears to be upside down. But notice that the *y*-axis represents transmittance, so the sample is at maximum transmittance (allowing the passage of all infrared radiation) when it is not absorbing radiation.

REMEMBER THIS

Equation for the velocity of light:

$$c = \lambda f$$

where c = velocity of light,
λ = wavelength, f = frequency.
So, $1/\lambda$ is directly proportional to f. Since energy $E = hf$, the higher the wavenumber (and the higher the frequency), the higher the energy (h = Planck's constant). (See Chapter 3.)

STRETCH AND CHALLENGE

The infrared spectrometer

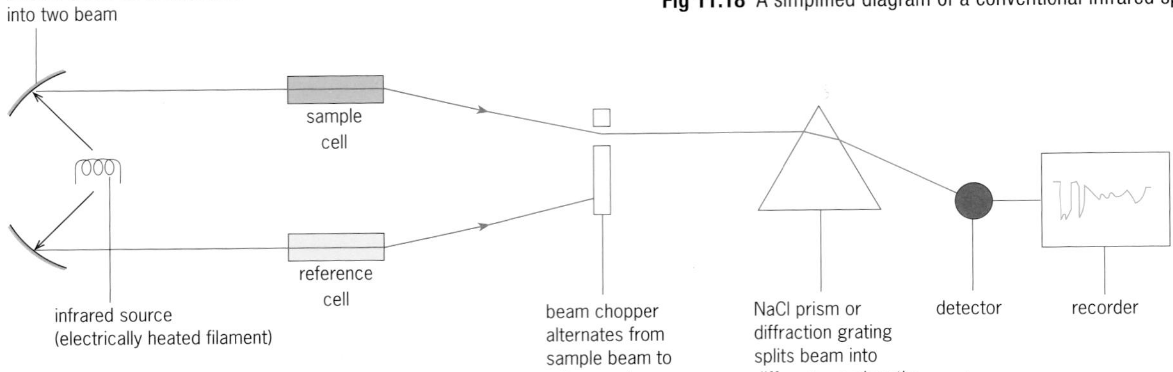

Fig 11.18 A simplified diagram of a conventional infrared spectrometer

The spectrometer shown in Fig 11.18 is a double-beamed instrument, in which one infrared beam passes through the sample and the other passes through the reference cell. The reference cell may contain air, or the solvent of a solid being analysed as a solution. The purpose of the reference is to eliminate absorptions caused by CO_2 and water vapour in the air or by the solvent. The difference in the intensities of the two beams is measured by the detector at each wavenumber and fed to the recorder, which produces a spectrum. When the sample does not absorb, there is no peak at that wavenumber; but when it does absorb radiation, the difference in intensities of the two beams produces a peak at that wavenumber.

Notice that when a prism is fitted instead of a diffraction grating, it is made from sodium chloride rather than glass. This is because sodium chloride is transparent to most infrared frequencies, whereas glass absorbs it strongly.

A more sophisticated instrument is the Fourier transform infrared spectrometer, in which a single beam that contains all the required infrared frequencies is passed through the sample. The emerging beam is fed into a computer, which produces an infrared spectrum by a mathematical technique known as Fourier transformation. This instrument has enabled us to understand how photochemical smog develops by analysing samples taken from a 1.6 km long path of air over Los Angeles and determining concentrations of pollutants down to parts per billion (per 10^9).

PREPARING SAMPLES FOR ANALYSIS

Gases are passed into the sample cell (about 10 cm long), which has sodium chloride windows at each end. Liquids are smeared into a thin film between two potassium bromide or sodium chloride discs. (Sodium chloride is not transparent to all infrared, so potassium bromide, which does not absorb in the infrared spectrum, can be used instead.)

Solids are made either into a solution, or into a disc or a **mull**. To make a disc, the solid is finely ground with potassium bromide and pressed into a mould under very high pressure. The advantage of a disc is that it produces the spectrum of the pure compound. A mull is made by grinding up the sample with a drop of nujol, a long-chain hydrocarbon. The mull is then smeared between discs, as for a liquid sample. The disadvantage of using a mull is that nujol absorbs infrared radiation, which shows up on the spectrum.

Fig 11.19 Preparing a mull

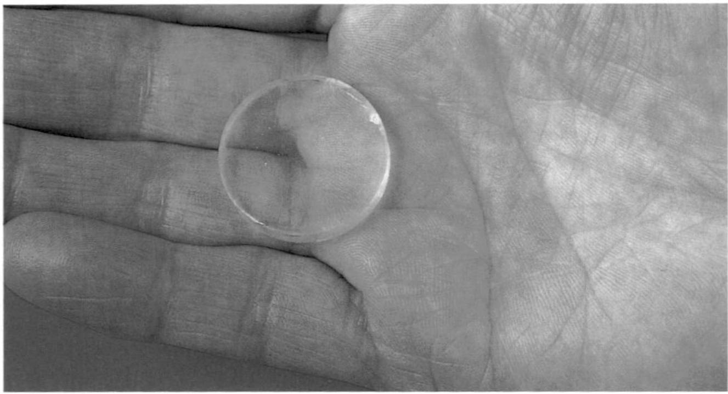

Fig 11.20 A potassium bromide disc

INTERPRETING INFRARED SPECTRA

The spectrum of each and every compound acts like a unique fingerprint and thus makes it possible to identify any compound, often through a computer database. When a totally new compound is discovered, perhaps isolated from a medicinal herb, it will not match any spectrum in the database. But we can still infer a great deal about the compound from the peaks in its spectrum. For example, the functional group C=O absorbs in the range 1680–1750 cm^{-1}, so a peak in this region is a strong indication that C=O is present.

Table 11.3 contains some examples of bonds and their characteristic infrared absorption bands. Notice that some bonds give strong absorptions while others are weaker. The explanation lies with bond polarity. When they vibrate, very polar bonds, such as C=O and O–H, are subject to a greater change in their dipole moments than are non-polar bonds, such as C–H. So polar bonds absorb more energy, and thereby give a stronger peak.

REMEMBER THIS

The fingerprint of a complex molecule arises through the vibrations of the whole molecule; this can be used to check batches of drugs for purity. The identifying peaks of the fingerprint are usually below 1600 cm^{-1}.

REMEMBER THIS

Arenes are aromatic compounds and have a benzene ring. For more information on arenes, see Chapter 15.

For more information on hydrogen bonding, see page 261.

Table 11.3 Bonds and their characteristic infrared absorption bands. Intensity classification: s = strong, m = medium

Bond	Location	Wavenumber range/cm^{-1}	Intensity
O—H	alcohols and phenols, free (not hydrogen bonded)	3580–3670	s
N—H	primary amines	3350–3500	m
O—H	alcohols and phenols, hydrogen bonded	3230–3550	s (broad)
C—H	alkanes, alkenes, arenes	2840–3030	m–s
O—H	carboxylic acids, hydrogen bonded	2500–3300	m (broad)
C≡N	nitriles	2200–2280	m
C=O	aldehydes, ketones, carboxylic acids, esters	1680–1750	s
C=C	alkenes	1610–1680	m
C—O	alcohols, ethers, esters	1000–1300	s
C—Cl		700–800	s

Let's now look at the infrared spectra in Figs 11.21, 11.22 and 11.23. Notice that we have not attempted to identify all the peaks in these three spectra. The more atoms and bonds there are in a molecule, the more complex its spectrum becomes.

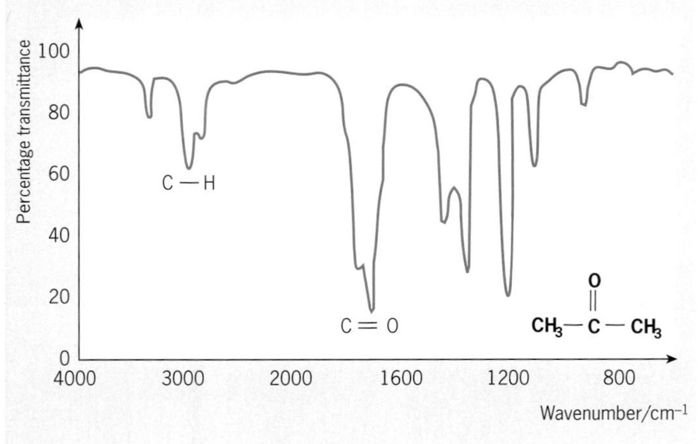

Fig 11.21 The infrared spectrum of propanone

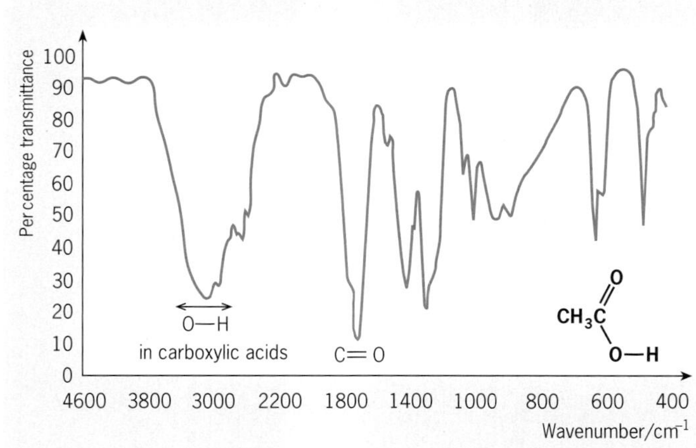

Fig 11.22 The infrared spectrum of ethanoic acid

QUESTION 15

15 Fig 11.23 shows the infrared spectrum of butan-1-ol, $CH_3CH_2CH_2CH_2OH$. Using Table 11.3, identify the peaks marked A and B.

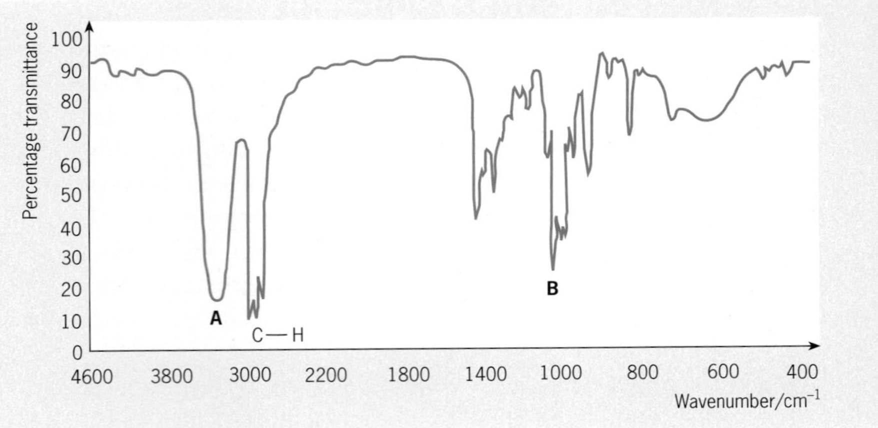

Fig 11.23 The infrared spectrum of butan-1-ol

Infrared spectroscopy is rarely used on its own to identify unknown compounds. However, coupled with data from mass spectrometry, gas–liquid chromatography or combustion analysis, it gives important information about the presence of functional groups.

HOW SCIENCE WORKS

Using infrared spectroscopy

At the police station

All alcoholic drinks slow down the body's response times. So drinking and driving can be a fatal combination and anyone driving over the legal limit of 35 micrograms per 100 cm^3 of breath is likely to be prosecuted.

The roadside breathalyser gives the first indication that a driver may be over the limit. When a breathalyser test proves positive, the next step is to take the driver to the police station for a more accurate analysis, which is now often carried out using infrared absorption.

Fig 11.24 The Intoximeter 3000 shown here has been widely adopted by the British police to analyse breath. It checks the C—H absorption at about 3000 cm^{-1}

At the roadside and garage

Infrared spectroscopy is very quick and devices that measure pollution from individual cars can be placed at the roadside or used in the annual MOT test to monitor exhaust emissions. In these cases, the C≡O bond in carbon monoxide, the bonds in nitrogen oxides and the C—H bonds in unburnt hydrocarbons all give characteristic absorptions (Fig 11.25).

In the atmosphere

Crucial evidence for the rise of carbon dioxide levels in the atmosphere has been provided by the Mauna Loa observatory in

Fig 11.25 Cars fitted with catalytic converters should emit no more than 0.3% CO and 200 ppm (parts per million) unburnt hydrocarbons

Hawaii (Fig 11.26). The graph of the results from 1958 to the present day is one of the most famous in history and is cited by scientists from around the world when they urge for a reduction in the present output of carbon dioxide emissions. (See page 221.)

Fig 11.26 The rise in atmospheric CO_2 levels is monitored here at the Mauna Loa Observatory in Hawaii. (See page 201.)

To follow the progress of a reaction

If a reaction involves one functional group changing into another, then the progress of an industrial reaction can be followed using infrared spectroscopy.

QUESTION 16

16 Read the How Science Works box on page 224 before answering these questions. Look at the infrared spectrum for ethanol in Fig 11.27.

a) Identify the bond responsible for the broad peak at about 3400 cm^{-1}.

b) Breath contains water vapour. Why can't the peak at 3400 cm^{-1} be used to check for alcohol concentration?

c) Diabetics may have propanone on their breath. Look at Fig 11.21 and suggest why this could be a problem for the Intoximeter 3000 (see Fig 11.24).

d) The maximum output of unburnt hydrocarbons from car exhausts is given in parts per million (ppm). What is the maximum *percentage* of C_xH_y permissible?

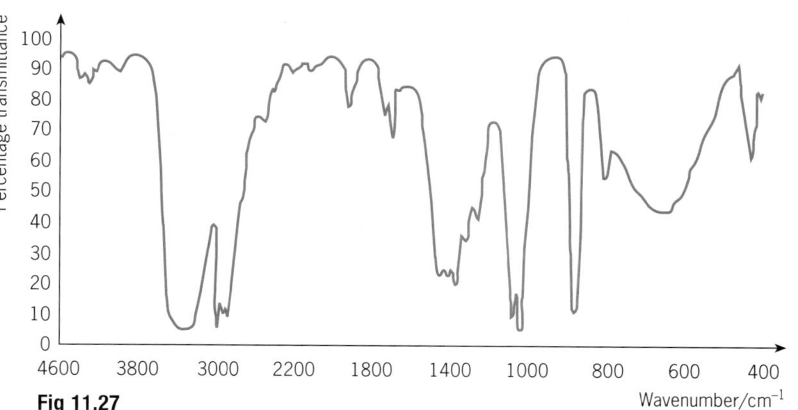

Fig 11.27

4 NUCLEAR MAGNETIC RESONANCE SPECTROSCOPY

In **nuclear magnetic resonance (NMR) spectroscopy**, organic chemists have a very powerful technique for determining the detailed structures of compounds. Like infrared spectroscopy, it uses absorption – but in NMR it is the absorption of radio waves by certain nuclei when they are in a strong magnetic field. When a new organic compound is discovered, its NMR spectrum is often the first to be looked at because it gives detailed information about the hydrogen nuclei (protons) in the molecule. It can also be used to investigate other nuclei, such as ^{13}C.

A hydrogen nucleus acts like a tiny bar magnet because it spins, generating a minute magnetic field. When a strong external magnetic field is applied to a compound that contains hydrogen atoms, many of their spinning nuclei (protons) line up so that their magnetic fields are in the same direction as that of the external field (like a group of compass needles, each one pointing north). However, some of the nuclei directly oppose the external magnetic field. (Imagine a compass needle pointing south!) This requires more energy, so these nuclei are at a higher energy level than the rest.

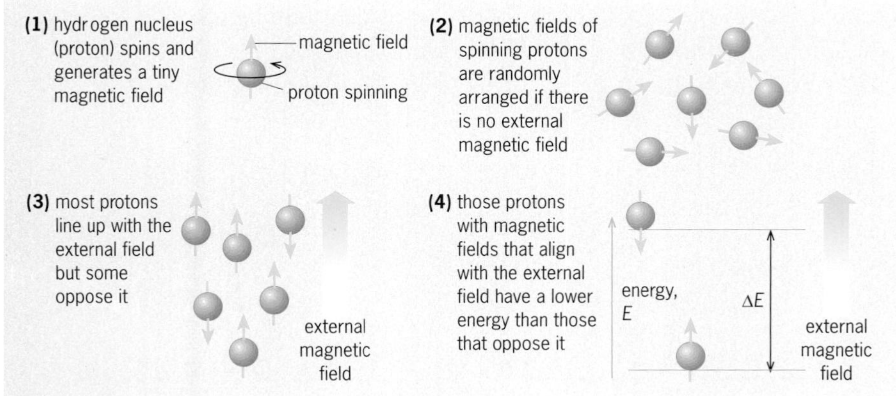

Fig 11.28 How hydrogen nuclei behave when an external magnetic field is applied to their molecular environment

REMEMBER THIS

Spectroscopy such as infrared and NMR involves the interaction of materials with energy from regions of the electromagnetic spectrum.

REMEMBER THIS

Nuclear magnetic resonance spectroscopy is also called magnetic resonance imaging (MRI), especially when used in medicine.

REMEMBER THIS

Remember that hydrogen ^1_1H has a nucleus that consists of only one proton.

QUESTION 17

17 **a)** As well as ^{13}C and ^1H, the following nuclei also exhibit the property known as spin and generate a magnetic field: ^{19}F, ^{31}P, ^{15}N. Look at the mass numbers and identify a pattern.

b) Why do you think a ^{13}C NMR gives a simpler spectrum than ^1H NMR?

QUESTION 18

18 An NMR spectrum is often referred to as a PMR spectrum when the nucleus involved is hydrogen. What does PMR mean?

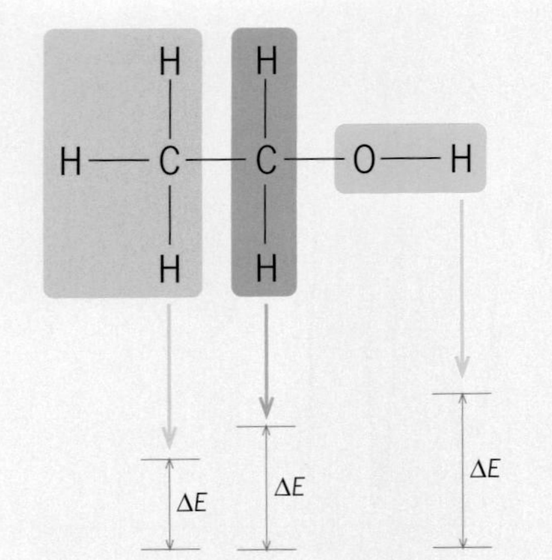

Fig 11.29 The hydrogen nuclei (protons) in ethanol are in three different chemical environments. Each gives different proton NMR energy levels

REMEMBER THIS

Notice that δ (chemical shift) is in ppm (parts per million). This relates to the difference in frequency and applied fields, but a full explanation is beyond the scope of this book.

If the compound is now irradiated with a pulse of radio waves of a particular frequency, the lower-energy nuclei absorb these waves and thereby flip to the higher level. This causes a peak in the NMR spectrum.

Both the energy needed to promote a hydrogen nucleus (proton) to a higher energy level and the strength of the applied magnetic field experienced by the nucleus depend on the chemical environment of the nucleus. So, hydrogen atoms in different environments in a molecule absorb radio waves in slightly different parts of the spectrum. Take, for example, the ethanol molecule (CH_3CH_2OH) shown in Fig 11.29.

The hydrogen nuclei are in three different chemical environments. This means that the energy gap between the higher and lower energy levels is different for each of these environments. So, when a pulse of radio waves is applied, more of the lower energy nuclei jump to the higher energy level as they absorb radio waves of a particular frequency. In the case of ethanol, the three peaks in the spectrum correspond to the three different frequencies absorbed by the detector.

A compound called tetramethylsilane, $Si(CH_3)_4$, or TMS for short, has all 12 protons (hydrogen nuclei) in the same chemical environment. It therefore gives a very sharp signal. This is used as the standard against which all signals caused by other proton chemical environments are measured, because its protons absorb at a frequency well away from the frequencies of most interest to chemists. TMS is given a value of zero, and the difference between the protons in TMS and the protons in other chemical environments is known as the **chemical shift**, symbol δ.

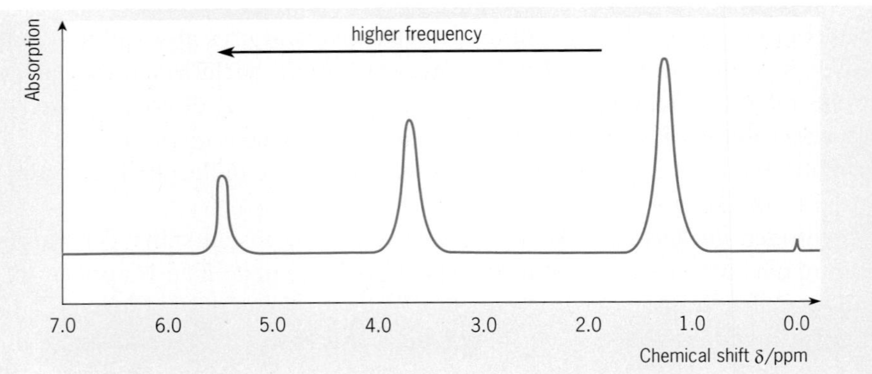

Fig 11.30 The low-resolution NMR spectrum of ethanol

QUESTION 19

19 Give the number of peaks the following molecules show in their NMR spectra:

a) benzene (C_6H_6)

b) propanone (CH_3COCH_3)

c) methanol (CH_3OH)

The NMR spectrum of ethanol is shown in Fig 11.30. The area under each peak is proportional to the number of protons in the particular environment. Many NMR spectrometers give these areas as an 'integration trace' superimposed on the NMR spectrum. By measuring the height of each step on the integration trace, the ratio of the protons in each environment can be worked out. This is shown in Fig 11.31, with the ratio of protons.

REMEMBER THIS

Chemical shifts give an indication of the types of proton to be found in a molecule. Some typical values are shown in Table 11.4. Chemists use these values when they are trying to identify molecules and work out structures.

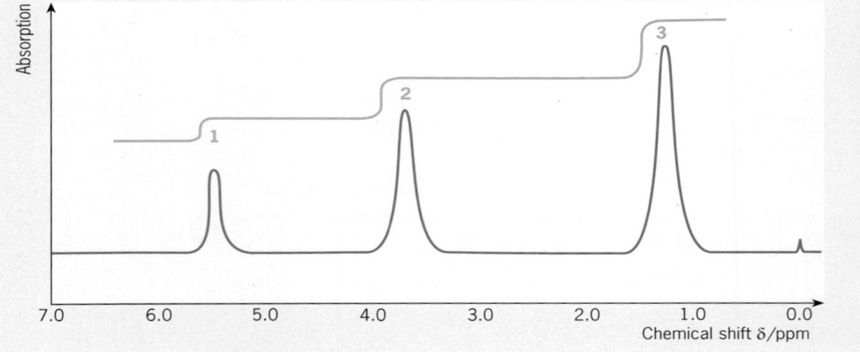

Fig 11.31 The low-resolution NMR spectrum of ethanol, showing the integration trace and the ratio of protons

Table 11.4 Some typical chemical shift values for different types of proton
***These values vary according to the solvent, the nature of R and concentration.**

Type of proton	Chemical shift (ppm)	Type of proton	Chemical shift (ppm)
$R—CH_3$	0.9	$R—CH_2—Hal$	3.2–3.7
$R—CH_2—R$	1.3		
R_3CH	2.0	$R—O—CH_3$	3.8
		$R—O—H$	4.5*
$CH_3—C(=R)—OR$	2.0	$RHC=\mathbf{CH_2}$	4.9
		$R\mathbf{H}C=CH_2$	5.9
$R—C(=O)—CH_3$	2.1	$C_6H_5—OH$	7*
$C_6H_5—CH_3$	2.3	$C_6H_5—H$	7.3
		$R—C(=O)—H$	9.7*
$C_6H_5—CH_2—R$	2.3—2.7		
$R—C=C—H$	2.6	$R—C(=O)—O—H$	11.5*

Using Table 11.4, let's work through another NMR spectrum and determine the structure of the compound.

EXAMPLE

Q A hydrocarbon has a molecular formula C_8H_{10}. Work out its displayed formula from the low-resolution NMR spectrum in Fig 11.32.

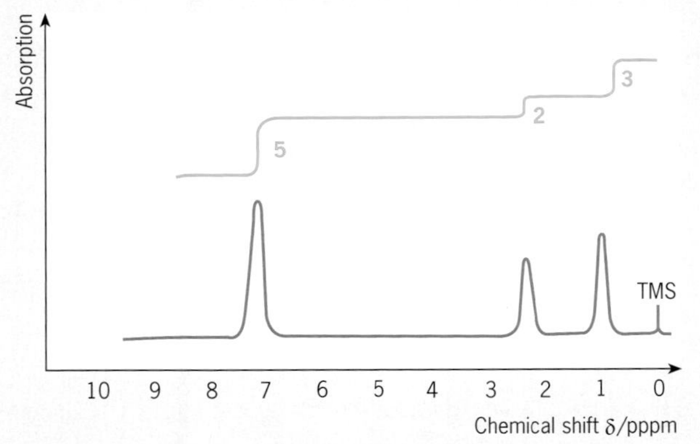

Fig 11.32

A As there are three peaks, there must be three types of proton (three different chemical environments).

The peak at 7.2 is probably due to a benzene ring. The integration trace gives a ratio of 5:2:3. This adds up to ten, which in this case is the actual number of hydrogen atoms in the molecule. So there are five hydrogen atoms on the benzene ring. In other words, it is **monosubstituted**.

(Monosubstituted means that only one hydrogen in the benzene ring has been substituted by another atom or group.)

Since there are two more peaks, the peak at 2.4 corresponds to $C_6H_5—CH_2—R$, not $C_6H_5—CH_3$. From the integration trace, we can see that it does indeed have two hydrogen atoms.

The integration trace shows that three hydrogen atoms are responsible for the peak at 1.0. Even though the value of the peak does not correspond precisely to the value in Table 11.4, it is certainly CH_3.

These data give the structural formula in Fig 11.33. The compound is ethylbenzene.

Fig 11.33

QUESTION 20

20 Using Table 11.4, draw the NMR spectrum you might expect for ethanal (Fig 11.34).

Fig 11.34

HOW SCIENCE WORKS

Using the NMR spectrometer

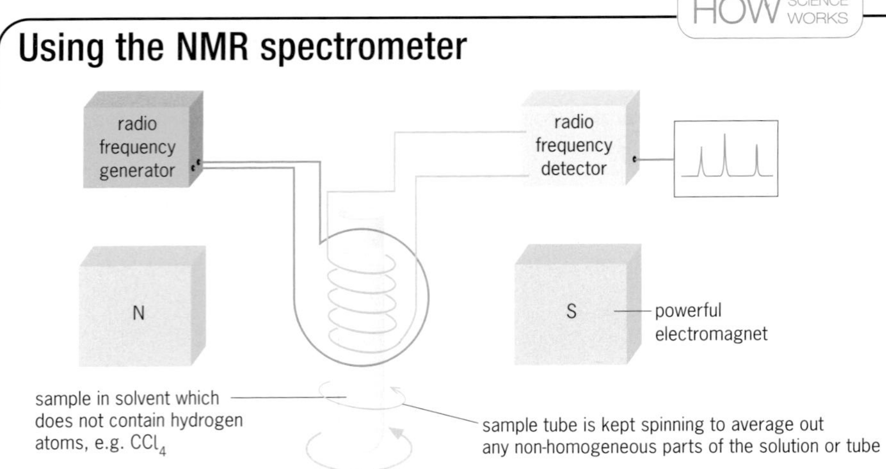

Fig 11.35 A schematic diagram of an NMR spectrometer

The sample is dissolved in a solvent that does not contain hydrogen atoms, such as CCl_4, or a solvent in which hydrogen has been replaced by its isotope deuterium. These solvents are called **deuterated solvents**, an example of which is $CDCl_3$. The solution is then placed in a strong, magnetic field and a pulse of radio waves passed through it, exciting hydrogen nuclei into their higher spin energy levels. The absorption of radio waves at particular frequencies is detected and recorded.

Either the radio frequency or the magnetic field can be varied to give characteristic peaks.

REMEMBER THIS

Deuterated solvents contain deuterium, D, instead of hydrogen (e.g. CD_3COCD_3, which is propanone, CH_3COCH_3, with the hydrogen atoms exchanged for deuterium). The deuterium nucleus contains a proton and a neutron and does not respond to NMR at the same radio frequencies.

HIGH-RESOLUTION NMR SPECTROSCOPY

The NMR spectra of ethanol and ethylbenzene (Figs 11.30 and 11.32, respectively) are at low resolution. High-resolution instruments give more peaks, as shown in the high-resolution NMR spectrum of ethylbenzene in Fig 11.36. These extra peaks yield more information about the structure of molecules.

When different types of proton are next to each other in a molecule, they exert an effect on their respective magnetic fields, known as **spin–spin splitting** or **spin–spin coupling**. Look closely at the displayed formula of ethylbenzene in Fig 11.33. The CH_3 protons are next to the CH_2 protons, so they have a direct effect on each other. However, the protons in the benzene ring are not next to any other type of proton.

The CH_3 protons experience three different magnetic fields because of the way the CH_2 protons spin, while the CH_2 protons experience four different magnetic fields because of the way the CH_3 protons spin, as shown in Fig 11.37.

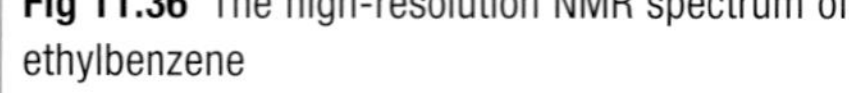

Fig 11.36 The high-resolution NMR spectrum of ethylbenzene

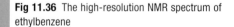

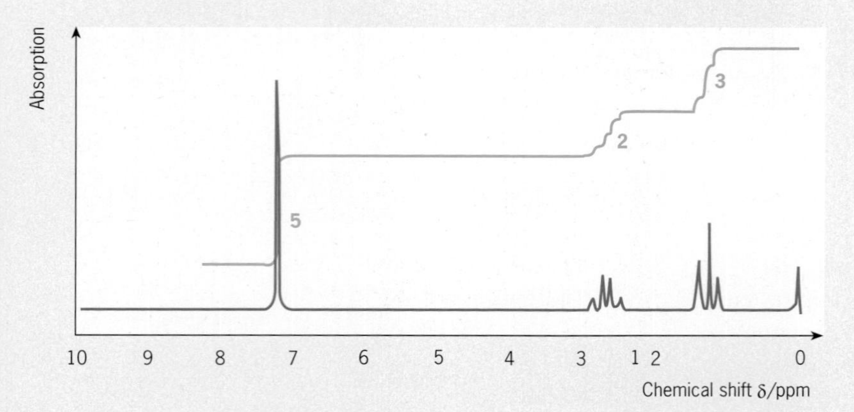

So, with high-resolution NMR, we can work out the number of nearest-neighbour hydrogen atoms there are in different groups in a molecule. This is made easy to remember by the ***n* + 1 rule:**

If a proton has *n* protons as nearest neighbours then its absorption peak will be split into *n* + 1.

This helps us to gain greater insight into the structure of unknown molecules.

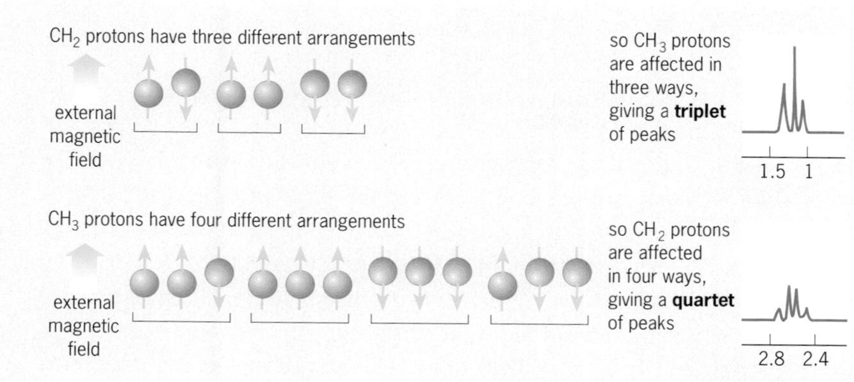

Fig 11.37 Spin–spin splitting in ethylbenzene

QUESTION 21

21 There are two types of proton in CH_3CHO. How many arrangements are there of the magnetic field for the CHO proton? How many peaks will there be in the CH_3 part of the spectrum?

Using D_2O to identify labile protons

Certain functional groups, such as OH and NH_2, rapidly exchange protons with neighbouring molecules. These protons are said to be **labile**. They usually appear in high-resolution NMR spectra as single peaks, because they do not interact with neighbouring protons.

To identify functional groups with labile protons, heavy water (deuterium oxide, D_2O) is used. In the case of ethanol, there is an exchange that produces CH_3CH_2OD, and the peak corresponding to the proton in the OH group disappears.

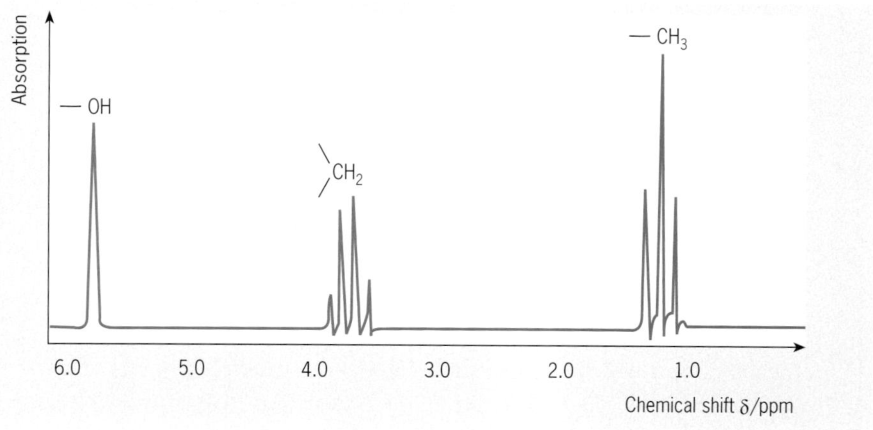

Fig 11.38 The high-resolution NMR spectrum of ethanol

HOW SCIENCE WORKS

Imaging the human body with NMR

Nuclear magnetic resonance is becoming increasingly important in medicine, where it is more usually known as magnetic resonance imaging. The most common nucleus in the body is that of hydrogen, and whole-body scanners investigate tissues by producing images based on the fact that protons in different tissues have different **relaxation times**. When protons are excited and flipped into higher energy levels, they return to the lower energy stage, emitting energy. The time they take to do this is called the relaxation time. This is detected and the data transformed by computer into an image. With NMR, doctors can see an image of the soft tissues in the body, which X-rays cannot distinguish. An added advantage is that, unlike X-rays, no side-effects are known, and therefore scans can be safely taken of a patient at regular intervals.

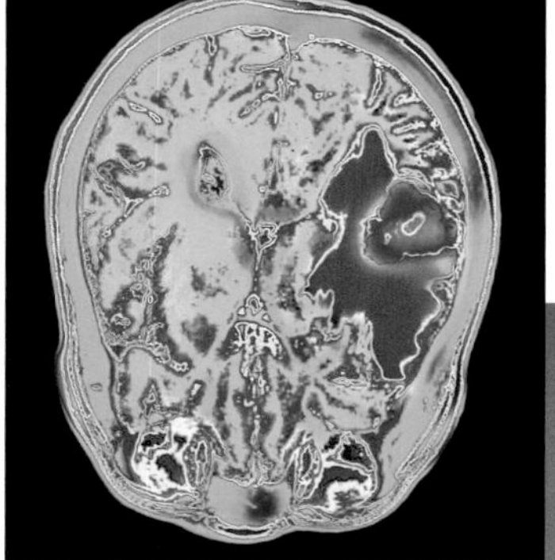

Fig 11.39 (left) A brain tumour is shown by magnetic resonance imaging as a magenta coloured area surrounded by damaged brain tissue in scarlet. Protons in tumours have longer relaxation times than protons in normal tissue, and this provides a way of locating them

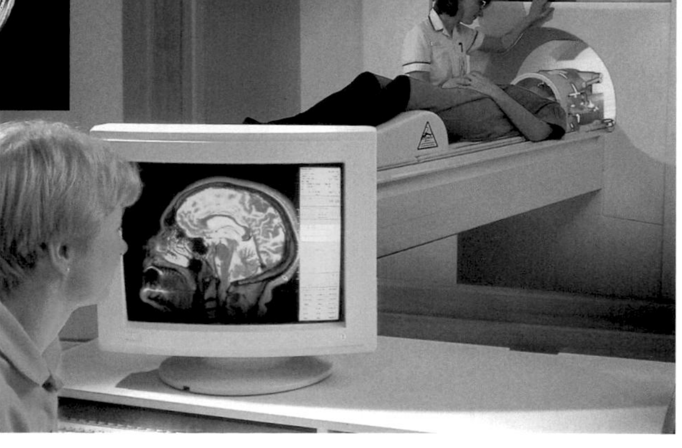

Fig 11.40 (right) Scanning a patient by NMR. His body is lying along the central axis of a superconducting electromagnet

5 CHROMATOGRAPHY

Chromatography is a method of separating and identifying the components of a mixture. (The name is derived from two Greek words, *chroma* for colour and *graphe* for writing.) The technique was discovered and developed by a Russian botanist, Mikhail Tswett, who first reported it in 1903. He was using powdered calcium carbonate packed into a tube to separate coloured plant pigments called chlorophylls (see How Science Works, page 231). Before this, chemists mainly used crystallisation techniques to separate and purify substances. Today, chromatography in one form or another is used in nearly all chemical laboratories, not just to separate and purify, but also to analyse substances, most of which are colourless.

All forms of chromatography operate on the same principle. The mixture is introduced into two different phases, of which one remains stationary while the other flows over it. The phases are known as the **stationary phase** and the **mobile phase**, respectively. The components of the mixture distribute themselves differently between the two phases, according to their affinity for each phase.

There are two main types of chromatography: **partition chromatography** and **adsorption chromatography**.

REMEMBER THIS

A **phase** is a homogeneous part of a system that is physically distinct. It is separated from the other parts by a boundary surface. Petrol is a mixture, but it is in one phase, as is a solution of sodium chloride in water. But petrol and water form two physically distinct phases because they do not mix.

REMEMBER THIS

Adsorption of reactants by heterogeneous catalysts is a mechanism by which reactions are accelerated. You can read about this on page 206.

REMEMBER THIS

A chromatogram is the record of a separation of components achieved through chromatography.

Partition chromatography

In partition chromatography, the stationary phase is a non-volatile liquid film held on an inert solid surface. The mobile phase is a liquid or gas. The components to be separated distribute themselves between the two phases according to how soluble they are in each.

Adsorption chromatography

In adsorption chromatography, the stationary phase is a solid and the mobile phase is a liquid or gas. The components to be separated are **adsorbed** on (bonded to) the solid surface of the stationary phase. Those that are only weakly adsorbed travel faster in the mobile phase than those that are strongly adsorbed.

PAPER CHROMATOGRAPHY

Chromatography using paper is the method you are probably most familiar with. It is an example of partition chromatography (Fig 11.41). The stationary phase is water adsorbed on the cellulose fibres of the paper. The mobile phase is another liquid solvent.

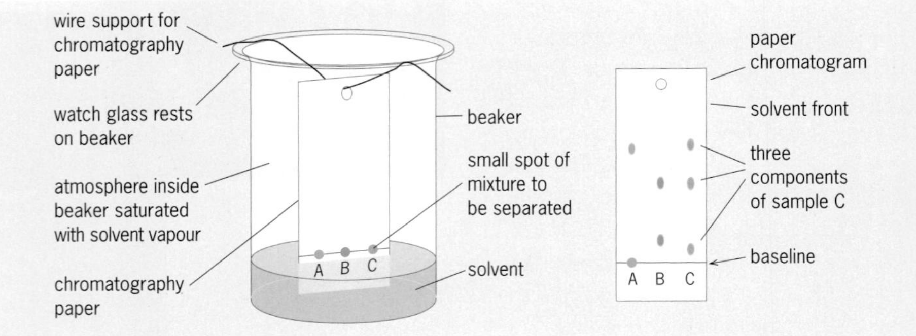

Fig 11.41 The principle of paper chromatography. The watch glass prevents the solvent evaporating

? QUESTION 22

22 Compound Y is very soluble in the mobile phase and almost insoluble in the stationary phase. Compound X has greater solubility in the stationary phase than in the mobile phase. Which will travel further during chromatography?

The sample is spotted along a **base-line** drawn in pencil on the paper (Fig 11.41). When the spots are dry, the paper is placed in the solvent, which rises up the paper, taking the components of the sample with it. As they travel up the paper, the components separate according to their different solubilities in the mobile and stationary phases. When the **solvent front** (the leading edge of the solvent) is almost at the top of the paper, the **chromatogram** is taken out and dried (Fig 11.41). If the components are colourless, the chromatogram needs to be **developed** to make them visible. Sometimes, just heating the chromatogram does this. If not, some other method, such as exposure to ultraviolet light, is used.

Components can be identified by the distances they have travelled up the paper compared with the distance travelled by the solvent front. This is known as the **retention factor** or **retardation factor** R_f:

$$R_f = \frac{\text{distance moved by component from base-line}}{\text{distance moved by solvent front from base-line}}$$

To obtain the separated components as samples, their spots are cut out and they are dissolved off in solvents.

QUESTION 23

23 Look at Fig 11.41.

a) Why is the base-line drawn in pencil?

b) Why is the air inside the beaker saturated with solvent vapour before the chromatography paper is put in?

c) Calculate the R_f values for the two components in sample **B**. (Hint: measure to the centre of the spot.)

d) Explain why one of the components in sample **A** has not risen off the base-line.

HOW SCIENCE WORKS

Column chromatography

This was the first way in which chromatography was carried out, and it is still the simplest. When Tswett was separating plant pigments, he used a column of calcium carbonate powder as the stationary phase, and a hydrocarbon liquid mixture as the mobile phase. He placed the sample on top of the column, where it made contact with the hydrocarbon mixture, which dissolved the pigments in the sample.

Tswett's arrangement is still used today. The column is kept topped up with fresh solvent, which washes the pigment down through the inert solid stationary phase (usually aluminium oxide powder or silica gel). The components that are adsorbed most strongly by the stationary phase take the longest time to flow through the column.

When the components are coloured, they can be identified by eye. But if they are colourless, other techniques are used: for example, some components fluoresce in ultraviolet light. Once separation is finished, the solvent is removed by evaporating it off.

Fig 11.42 Modern column chromatography in operation

Column chromatography in laboratories is usually used to separate minute quantities of mixtures, while in industry it is used in large-scale separation processes that require columns several metres high.

High-performance liquid chromatography (HPLC)

The performance of column chromatography can be improved by using very fine powder as the stationary phase. In this case, gravity alone is not enough to force the mobile liquid phase through the column, and therefore high pressure must be applied. For this reason, this technique is also called high-pressure liquid chromotagraphy, but the abbreviation is the same: HPLC. This is a very efficient process that requires short columns of between 10 and 30 cm in length. Many of the components separated by HPLC absorb ultraviolet light, so it is generally used to identify them.

HPLC has many applications, particularly for the separation and identification of non-volatile substances. Foodstuffs, for example can be checked for additives and contaminants. The detection of steroids in athletes' body fluids is another example. HPLC can also be linked to a mass spectrometer to identify compounds. This is called HPLC–MS. Fint of all HPLC separates compounds, than the mass spectrometer analyses them. This combination of instruments has been used to identify compounds on the surface of Mars.

? QUESTION 24

24 Is column chromatography adsorption or partition chromatography?

Two-way chromatography

Sometimes the components of a mixture are not separated completely because they have similar solubilities in a particular solvent. To overcome this problem the paper is rotated 90° and a different solvent is used to ensure a separation. This process is called **two-way chromatography**.

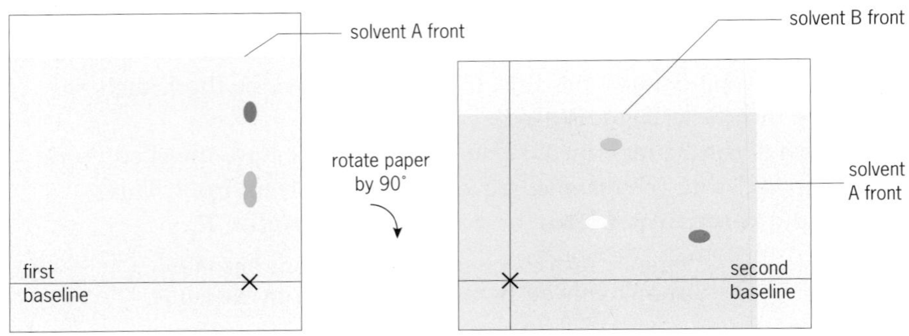

Fig 11.43 An incomplete separation is achieved with solvent A so the chromatogram is turned on its side and placed in a different solvent, B. The component that appears green is now separated into blue and yellow components

GAS–LIQUID CHROMATOGRAPHY (GLC) OR (GC)

An unreactive gas, such as nitrogen or helium, is used as the **carrier gas** (the mobile phase) in GLC. The stationary phase consists of a liquid coating adsorbed on the particles of a finely powdered inert solid. The powder fills a coiled narrow-bore tube called a column. The temperature of the column is controlled by an oven. The arrangement is shown in Fig 11.44. Typically, the tube that forms the column can be 5 to 10 m long, with a bore of 2 to 10 mm.

? QUESTION 25

25 **a)** Why does the liquid used in GLC need to have a high boiling point?
b) Explain why GLC is an example of partition chromatography.

✔ REMEMBER THIS

In GLC, a chromatogram is the record of the analysis displayed on a screen. In paper chromatography, it is the components themselves separated on the paper.

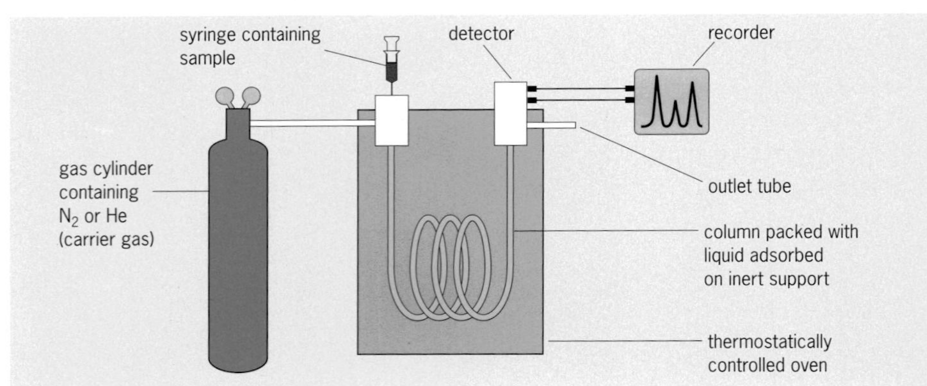

Fig 11.44 The main features of a gas–liquid chromatography apparatus

The vaporised sample is injected into the carrier gas as it streams into the column at an even rate. As the components of the sample are carried through the column, those that are more soluble in the stationary phase move at a slower rate. Volatile components are carried more quickly than non-volatile ones, which therefore have a greater opportunity to dissolve in the liquid stationary phase. So the volatile components emerge first from the column.

The time each component remains in the column is known as its **retention time**. This depends on such factors as the flow rate of the carrier gas, the temperature of the oven, and the length and diameter of the column. Each retention time is characteristic of a particular component, which allows it to be separated and identified. The area under each peak displayed on the recorder is the measure of the amount of that component present. The relative amounts of components in a mixture can therefore be calculated, an operation now done by computer. This leads to the percentage composition of the components of a mixture.

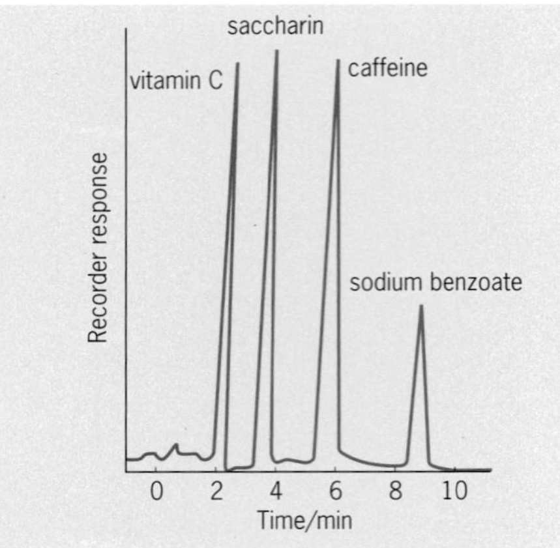

Fig 11.45 A gas–liquid chromatogram of additives in a soft drink

GLC is very sensitive and is able to detect and measure minute amounts of substances. It has therefore become a standard method for detecting banned

SCIENCE IN CONTEXT

Revealing the secrets of old masters

When you next go round an art gallery and look at the work of a great artist who perhaps lived four or five centuries ago, give some thought to its chemistry. Paint pigments provide the colour, but they do not stick to the canvas or wooden panel on their own. They need to be bound together and to the supporting surface. For some 5000 years before the Renaissance, artists mainly used egg yolk (called tempera) as the binding medium. Then, in the 1400s, Italian and Flemish artists started to switch to oils that dried, such as linseed, poppy or walnut oil. This is where GLC comes in.

Fig 11.46 *The Virgin and Child Embracing* by Sassoferratto (1609–1685), during restoration of the painting

When a gallery wants to restore a painting, it is essential to know what binding medium the artist used. In the oils used, long chain fatty acids are present as esters (see page 357) and these have characteristic retention times in a gas chromatogram. By performing GLC, the different oils used can be identified and then the same oils are mixed with pigments for restoring the picture.

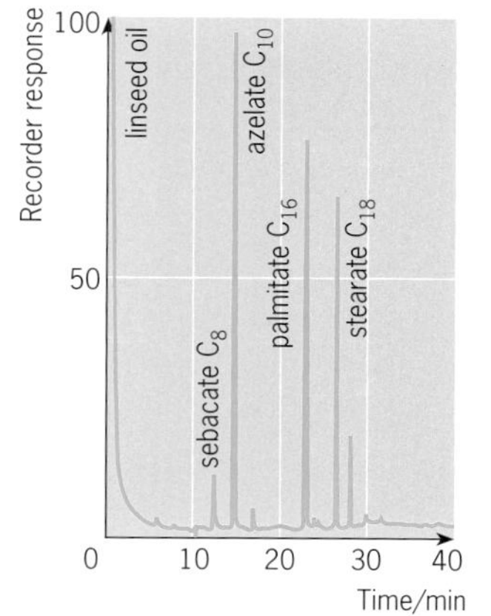

Fig 11.47 Chromatograms of known oils can be compared with the chromatogram of a tiny sample of paint removed from the picture to be restored, to determine which drying oil the artist used. Shown here is the chromatograph of a paint sample from a picture painted in about the year 1500

substances in sport, and for determining the amount of alcohol in a motorist's blood when it is close to the legal limit. When connected to a mass spectrometer, it is used to identify substances in foods and medicinal plants.

THIN-LAYER CHROMATOGRAPHY (TLC)

TLC is similar to paper chromatography, except that the stationary phase is a fine, inert powder, such as silica gel, which is made into a paste and spread in a thin, even layer over a plastic or glass plate. The layer of paste is then baked on to the plate. Spots of the mixture are applied near the bottom of the plate. As the solvent rises by capillary action up through the spaces between the inert powder particles, the components of the mixture are carried up and separated. When the solvent is almost at the top of the plate, the plate is taken out and dried. R_f values can be obtained in the same way as for paper chromatography.

TLC has three advantages over paper chromatography: it is quicker, the results are more easily reproduced and the separations are more efficient. The technique is used in industry to follow the course of a reaction by studying which components are present. The separated components can be retrieved by scraping off the spots on the stationary phase and dissolving them in a suitable solvent. TLC is also widely used in forensic science to identify explosive residues and to detect cannabis.

QUESTION 26

26 A database of retention times is compiled for each GLC machine using known compounds. Why is it essential to keep the conditions constant after forming this database?

REMEMBER THIS

GLC analysis of blood or urine samples provides a more accurate result than infrared testing of breath (see How Science Works on page 224) if the breath-alcohol concentration is between 40 and 50 ∝ g per 100 cm^3.

REMEMBER THIS

Gas chromatography has its limitations:

- similar compounds often have similar returntion times;
- unknown components have not been through the apparatus, so there are no retention times that can be used for reference.

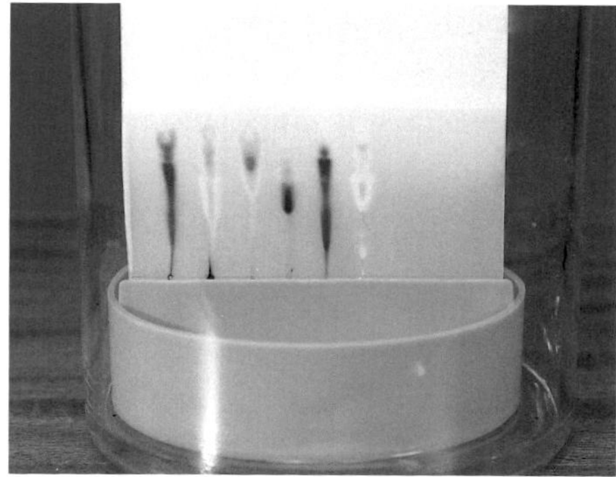

Fig 11.48 Thin-layer chromatography of felt-tip pen inks

QUESTION 27

27 Is TLC adsorption or partition chromatography?

6 COMBINING TECHNIQUES

REMEMBER THIS

Further analytical tools are electrophoresis (Chapter 19), ultraviolet/visible absorption spectroscopy (Chapter 27) and atomic emission spectroscopy (Chapter 27).

As you have seen, the chemist has many techniques to call upon when determining the structure of an unknown molecule, and they are hardly ever used in isolation.

Gas–liquid chromatography is an excellent way to separate large numbers of components, whereas mass spectrometry is not the instrument to use if you want to identify substances in complex mixtures because the spectra are too complicated. By combining the two techniques, the separation is first carried out by gas–liquid chromatography, and then the separated components are analysed by the mass spectrometer. This is known as GC-MS. GC denotes gas chromotography, and MS mass spectrometry. When the mass spectrum of a compound is obtained it is compared to a database of mass spectra using a computer. This often gives a positive identification of the component. The use of GC–MS is important in forensics, environmental analysis, airport security and space probes.

In the same way, infrared spectrometry gives good evidence of the presence of functional groups, but its information needs to be pooled with that from other techniques to gain the complete picture.

Combustion analysis gives information about the percentage composition of elements in a compound by burning a sample in oxygen. The gases produced, such as CO_2, H_2O and SO_2, can be fed directly into gas–liquid chromatography apparatus, where they are separated and analysed, and their amounts determined.

Another example is the combination of NMR, which determines the different environments of hydrogen nuclei, with mass spectroscopy, which gives fragmentation patterns and an accurate M_r. The use of a combination of analytical techniques makes the determination of structural formulas more likely.

SUMMARY

After studying this chapter, you should know and understand the following:

- Percentage composition by mass of different elements in a compound may be obtained by combustion analysis.
- Empirical formulas can be determined from percentage composition data.
- A mass spectrometer separates streams of positive ions according to their masses and charges.
- Mass spectra of elements can be used to work out accurate values of A_r.
- When molecules are ionized in the mass spectrometer, they often break up into fragments that can give information about the structure of a molecule.
- High-resolution mass spectrometry gives highly accurate values of M_r, which may lead to the molecular formula of a compound.
- Infrared spectroscopy can be used to identify functional groups that absorb infrared radiation because of their vibrations.
- Nuclear magnetic resonance (NMR) spectroscopy gives information on the location and number of hydrogen atoms in a molecule, because spinning protons create a magnetic field.
- Chemical shift in an NMR spectrum occurs because protons in different chemical environments experience slight variations in the externally applied magnetic field.
- Chromatography involves the distribution of the components of a mixture between a stationary phase and a mobile phase, and provides a way of separating compounds.
- There are several types of chromatography. They include paper chromatography, thin-layer chromatography (TLC), column chromatography and gas–liquid chromatography (GLC).

Practice questions and a How Science Works assignment for this chapter are available at www.collinseducation.co.uk/CAS

12

Halogenoalkanes

12 HALOGENOALKANES

A polyfluorohydrocarbon may well prove to be the right compound for a synthetic blood substitute. It would enable emergency medical workers at accident and injury sites to give life-saving transfusions immediately

Fluorohydrocarbons

After a serious accident, such as a plane crash in a remote area, blood transfusions may have to be carried out under difficult conditions, but there may not be time to identify and match a victim's blood group beforehand. Ideally, the rescue team needs a synthetic blood substitute that will suit any victim and do the vital job of carrying large amounts of oxygen efficiently around the body.

The blood substitute must also be inert and non-toxic. It needs to be chemically stable so that it has a long shelf-life. Some polyfluorohydrocarbons have these properties, and so the search is now on for the polyfluorocarbon compound that is the best synthetic blood substitute.

For over 50 years, it was because of these properties – of being inert, non-toxic and chemically stable – that the related chemicals chlorofluorocarbons (CFCs) were thought to be the best choice for industrial and domestic applications as refrigerants, electrical insulators and aerosol propellants.

Then, in the 1980s, the same properties were found to make CFCs an environmental hazard: scientists confirmed that CFCs released into the air break down the ozone in the stratosphere. They warned that, even if CFCs were banned, because they are so stable it would be a long time before the ozone could build up to its original level. Since then, CFCs are being replaced in their previous applications by alternative compounds.

1 HALOGENO-HYDROCARBONS

A **halogeno-hydrocarbon** (also called a halogenated hydrocarbon) is a compound that contains carbon, hydrogen and at least one of the halogens. This means that the molecule contains at least one polar covalent bond, namely, the carbon–halogen bond.

A **halogenoalkane** or haloalkane is an alkane in which one or more of the hydrogen atoms have been substituted by halogen atoms.

See pages 85–86 for information on polar covalent bonds.

REMEMBER THIS

Halogenoalkanes are sometimes also called haloalkanes.

USES OF HALOGENO-HYDROCARBONS

Halogenated hydrocarbons are synthesised as useful chemicals in their own right, as well as being intermediates in the synthesis of other chemicals.

As chemicals in their own right

Chloroalkanes have been used as solvents and anaesthetics for at least a century, but they are poisonous and damage vital organs in the body, including the liver. Synthetic methods were then developed that successfully introduced several different halogen atoms into a carbon skeleton. This changed the situation immediately. The new chlorofluorocarbons were non-toxic, low-boiling liquids or gases at room temperature and chemically inert. They seemed just the answer for many applications, for example as the fluids for air conditioning systems, refrigerators, aerosols and blowing plastic foams.

Fig 12.1 The sales of aerosol cans that contain CFCs have been phased out since the damage that CFCs do to the atmosphere became understood

Later, other halogen atoms were introduced and the halons were developed for fire extinguishers. Unfortunately, as the How Science Works box (opposite) points out, the very properties that make halogeno hydrocarbons so useful have caused a global environmental problem – ozone depletion in the upper atmosphere.

Some chlorinated hydrocarbons have been synthesised as insecticides and pesticides. But again, their chemical stability leads to environmental problems.

You can read more about pesticides and insecticides in Chapter 23.

See page 192 to find out how to name the carbon skeleton.

```
H         Cl
 \       /
  C ═══ C
 /       \
H         H
```

Fig 12.2 The displayed formula for vinyl chloride shows it has the hydrocarbon skeleton of ethene, so its systematic name is chloroethene

As synthetic intermediates

The manufacture of many pharmaceutical products and polymers requires halogenated hydrocarbons as intermediates in their organic synthesis. A chlorine or bromine atom in an organic molecule often increases its reactivity and allows a change of functional groups or an extension of the carbon skeleton. The polymer poly(chloroethene), or polyvinyl chloride (PVC), is manufactured from chloroethene.

NAMING HALOGENO-HYDROCARBONS

Although many halogeno-hydrocarbons have traditional names, such as chloroform ($CHCl_3$) or vinyl chloride (CH_2CHCl), it is important that you can systematically name halogeno-hydrocarbons.

The names of all organic compounds are based on the carbon skeleton and the functional groups present. In halogeno-hydrocarbons the functional group is the carbon–halogen bond. The presence of a halogen is indicated by the use of a prefix, namely *fluoro*, *chloro*, *bromo* or *iodo*. The number and location of the halogen atoms must also be specified. The number of each halogen atom is specified by the use of *mono*, *di*, *tri*, *tetra*, *penta*, *hexa*, and so on, before the prefix. For example, the trichloromethane molecule contains three chlorine atoms, and the dibromo-dichloroethane molecule contains two chlorine atoms and two bromine atoms.

Isomers are substances that have the same molecular formula, but their atoms are arranged differently. They may have different structural or displayed formulae, or different arrangements about a double bond. This means that, for example, dibromo-dichloroethane is not sufficient to describe fully one particular compound. Fig 12.3 shows the four different structural isomers that could each be called dichloro-dibromoethane.

This means that the *position* of the halogen atom must also be specified in the name. We use the system of numbering each carbon atom which we started to use on page 192. In Fig 12.3, structure **A** is called 2,2-dibromo-1,1-dichloroethane, and structure **B** is called 1,2-dibromo-1,1-dichloroethane. Notice that the order of the halogens in the name is given alphabetically.

```
A                  B                   C                  D
    Cl  Br             Cl  H              Cl  Cl              Cl  Br
    |   |              |   |              |   |               |   |
 H—C——C—H          Br—C——C—H          H—C——C—H           H—C——C—Cl
    |   |              |   |              |   |               |   |
    Cl  Br             Cl  Br             Br  Br              H   Br
```

Fig 12.3 The displayed formulae for all the isomers of dichloro-dibromoethane

? QUESTIONS 1–2

1 How many halogen atoms are there in one molecule of 1,2,2,3-tetrachloro-5-iodoheptane?

2 **a)** What is the systematic name for compounds with structures labelled **C** and **D** in Fig 12.3?

b) Draw the displayed formulae for the following halogeno-hydrocarbons:

i) 2,2-dichloropropane;

ii) 1,2,3,4-tetrachlorocyclohexane;

iii) 1,2-dichloro-3,3,4-triiodoheptane;

iv) 1,2-difluoroethene;

v) tetrafluoroethene.

✔ REMEMBER THIS

Sometimes, all the hydrogen atoms in an alkane are replaced by halogen atoms. If the halogen atoms are all identical, the prefix 'per' is often used. So that, for example, C_2F_6 is often referred to as perfluoroethane rather than hexafluoroethane.

See page 182 for more information on isomerism.

? QUESTION 3

3 **a)** What is the name of C_4F_{10}?

b) What is the name of $CH_3CH_2CHClCH_3$?

c) What is the name of C_6Cl_{14}?

2 PHYSICAL PROPERTIES OF HALOGENOALKANES

The uses of halogenoalkanes generally depend on their physical properties. One notable property is their low boiling point, which accounts for many of their applications (see the Science in Context box opposite).

The boiling point of a substance depends on the nature and strength of the *inter*molecular forces in the liquid.

INTERMOLECULAR FORCES IN HALOGENOALKANES

Halogenoalkanes are all covalently bonded and form simple molecules. The presence of simple molecules in the liquid phase makes halogenoalkanes solvents that are electrically non-conducting.

Halogen atoms tend to withdraw electrons from a carbon–halogen bond. In a monohalogenoalkane, the halogen atom will be slightly negative (δ–) and the carbon atom it is bonded to will be slightly positive (δ+). We say that the molecule has a **permanent dipole** and that it possesses a **dipole moment**, a measurable degree of polarity.

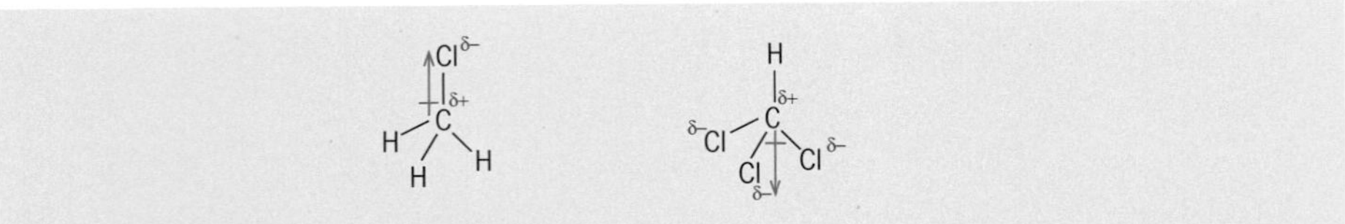

chloromethane

trichloromethane

Fig 12.4 The high electronegativity of chlorine gives both chloromethane and trichloromethane a dipole moment. The dipole moment is shown in blue

Chloromethane and trichloromethane are both molecules with a dipole moment, so they have permanent dipoles. The positive part of one molecule can attract the negative part of another molecule. This intermolecular force is referred to as a **permanent dipole–permanent dipole interaction**.

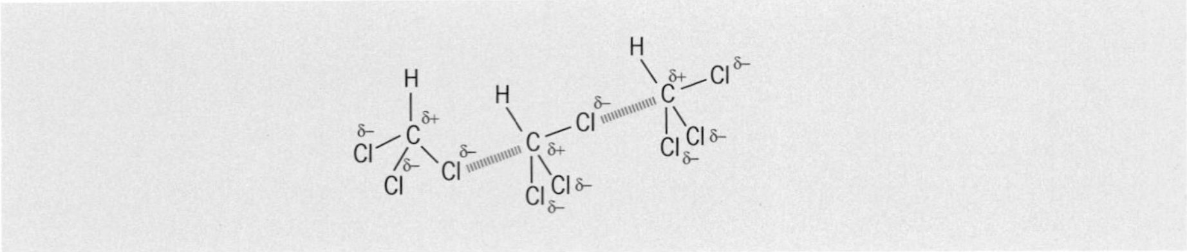

Fig 12.5 The permanent dipole causes a weak electrostatic attraction between the negative chlorine atom and the positive carbon atom in a neighbouring trichloromethane molecule

See page 195 for information on induced dipole–induced dipole forces.

When a large alkyl group is present as well, part of the intermolecular attraction consists of induced dipole–induced dipole van der Waals attractions.

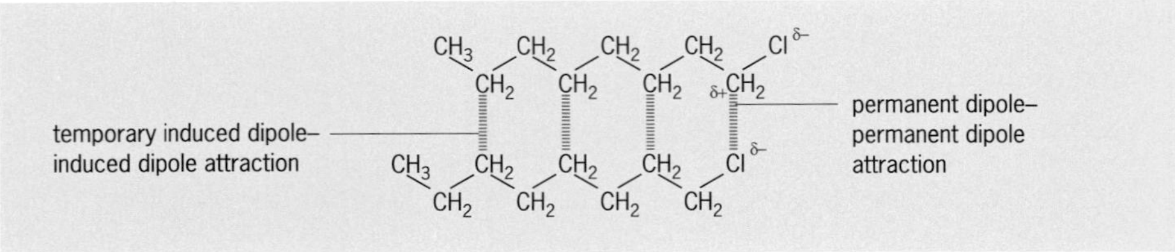

temporary induced dipole–induced dipole attraction

permanent dipole–permanent dipole attraction

Fig 12.6 The intermolecular forces in 1-chlorooctane result from both induced dipole–induced dipole and permanent dipole–permanent dipole interactions

SCIENCE IN CONTEXT

Chlorofluorocarbons and halons

The most well-known and infamous of the halogenoalkanes are the chlorofluorocarbons, known as CFCs. As their name suggests, these are halogenoalkanes in which every hydrogen atom has been replaced by either a chlorine or a fluorine atom. Table 12.1 shows some CFCs.

Table 12.1 Some typical CFCs and their previous uses

Molecular formula	Name	CFC designation	Uses
$CFCl_3$	trichloro-fluoromethane	CFC11	blowing foam plastics, refrigeration, air conditioning
CF_2Cl_2	dichloro-difluoromethane	CFC12	blowing foam plastics, refrigeration, air conditioning, aerosol propellant, sterilisation and food freezing
$C_2F_3Cl_3$	trichloro-trifluoroethane	CFC113	solvent
$C_2F_4Cl_2$	dichloro-tetrafluoroethane	CFC114	blowing foam plastics, refrigeration, air conditioning
C_2F_5Cl	chloro-pentafluoroethane	CFC115	refrigeration, air conditioning

Fig 12.7 In an old fridge, the CFC refrigerant circulates through pipes at the back. When it is disposed of, the CFCs must be extracted carefully to avoid the gases escaping into the atmosphere

Fig 12.8 Halon fire extinguishers, once used on fires near electrical equipment, are now replaced by carbon dioxide gas extinguishers as seen here

When a hydrogen atom is included in a CFC, the compound is called HCFC (hydrogen-chlorofluorocarbon).

Another variety of fully halogenated alkanes is called the halons. These contain bromine atoms as well as fluorine and chlorine atoms. Halons are used extensively in fire extinguishers because they are chemically unreactive and non-toxic. Being much denser than air, they effectively blanket a fire, which keeps out the air and so extinguishes the fire.

Typical halons include trifluoro-bromomethane (Halon 1301), and bromo-chloro-difluoromethane (Halon 1211). Unfortunately, these halons also cause the same environmental problems as CFCs, and research is under way to find substitutes.

BOILING POINTS OF CFCs

Boiling occurs when intermolecular forces are overcome, so the molecules can separate from each other and spread out. Thus, the stronger the intermolecular attraction, the higher the boiling point. The type of intermolecular forces in CFCs are similar to those in other halogenoalkanes, but the dipole moment of the molecules is likely to be much smaller since the individual bond polarities tend to cancel out. This means that the attraction due to permanent dipole–permanent dipole interactions is liable to be quite weak. As a result, even though the relative molecular masses of CFCs are quite large compared with, say, the monohalogenoalkanes, they will have lower boiling points.

? QUESTION 4

4 **a)** Suggest the molecular formula of: i) CFC13, ii) CFC112.

b) Suggest the name for CFC112.

? QUESTION 5

5 Why is it important that halons are more dense than air?

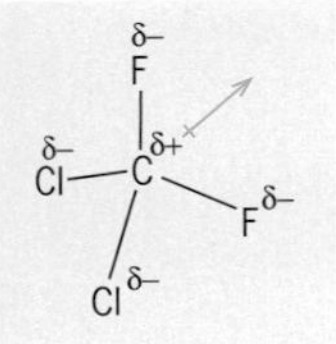

Fig 12.9 The bond polarities nearly cancel out one another, so that dichloro-difluoromethane has a small dipole moment

Table 12.2 Boiling points of some halogenated alkanes

Compound	Boiling point/°C
1-fluorobutane	(see question **6(a)** below)
1-chlorobutane	77
1-bromobutane	100
1-iodobutane	130

? QUESTIONS 6–7

6 **a)** Predict the boiling point of 1-fluorobutane.
b) What does the difference in boiling points of the compounds in Table 12.2 suggest about the strength of intermolecular attraction in 1-chlorobutane and 1-iodobutane?
c) The boiling point of 2-chloro-2-methylpropane is 51 °C, and that of 1-chlorobutane is 77 °C, even though they have the same molecular mass. Explain why there is this difference. (Hint: read page 196 for some ideas.)

7 Draw in the bond polarities for trichloro-fluoromethane, and hence draw in the dipole moment of the molecule.

REMEMBER THIS

The electron shells are shown here as circles, but in Chapter 4 they are not. Either way is acceptable.

? QUESTION 8

8 Draw a dot-and-cross diagram of each of the following and use this to decide which particle is a free radical: Cl^-, Cl^+, I, BF_3, CH_3.

As CFCs have weak intermolecular attraction, they are volatile liquids and vaporise easily at low temperatures. Many CFCs have two properties that make them efficient refrigerants: they absorb heat energy readily when vaporising, and they can be easily compressed back into liquids, releasing this energy. Provided a heat exchanger efficiently removes the heat released during liquefaction, net cooling of the refrigerant CFC will occur. If the intermolecular forces were larger, it would be difficult to vaporise the CFC and so little heat would be absorbed.

3 PREPARATION OF HALOGENOALKANES BY FREE RADICAL SUBSTITUTION

Even before the wide-scale use of halons and CFCs, the preparation of halogenoalkanes had become an important synthetic reaction because halogenoalkanes are extremely useful in making other chemicals.

The main methods of preparation of halogenoalkanes are described in Fig 12.10. One route involves a direct substitution of a hydrogen atom by a halogen atom. It is called **free radical substitution**.

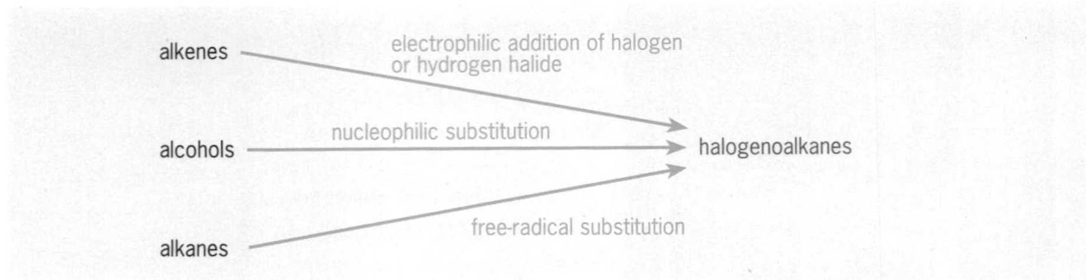

Fig 12.10 A summary of the preparation routes for halogenoalkanes. Electrophilic addition is described in Chapter 14, and the conversion of alcohols into halogenoalkanes is described later in this chapter and in Chapter 13

FREE RADICALS

Alkanes are very unreactive compounds, with the exception of their reaction with oxygen. Their lack of polar covalent bonds is the reason for this behaviour. Nucleophiles are particles that seek out electron-deficient centres and donate pairs of electrons to form a covalent bond, while electrophiles are particles that seek out electron-rich centres and accept pairs of electrons to form a covalent bond. It is difficult for alkanes to react with nucleophiles or electrophiles, because alkanes are non-polar molecules.

Alkanes require a different type of particle with which they can react, that is, one which is highly reactive and does not need to seek out a polar covalent bond before it can react. The particle in question is called a **free radical** (see also page 208). This is an atom or group of atoms that possesses at least one unpaired electron in its outer shell.

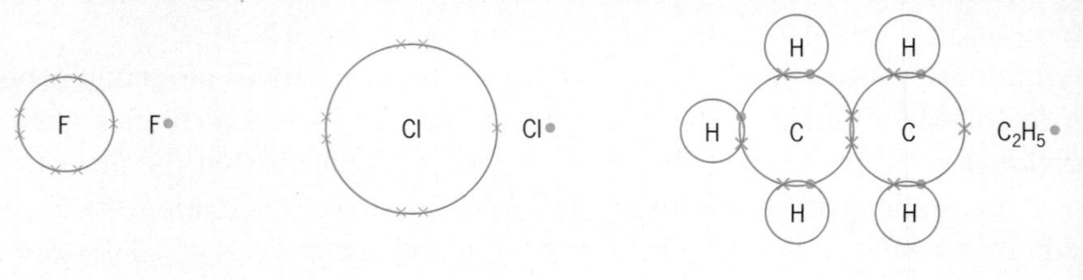

Fig 12.11 Dot-and-cross diagrams of three important free radicals

Most free radicals are extremely reactive because they pair up their unpaired electron with an electron removed from a covalent bond. Alternatively, two free radicals can react with each other, pairing up the unpaired electrons to form a covalent bond.

A free radical is usually indicated by a dot on the right-hand side of the formula of the particle. For example, a chlorine atom Cl is a free radical, which is made clear by writing Cl•.

HOMOLYTIC FISSION

As they are so reactive, free radicals cannot be stored in reagent bottles. They must be made in the reaction vessel, ready to react immediately. To make a free radical, a covalent bond is broken in such a way that the electrons in the shared pair of the covalent bond become two single electrons, one electron per atom. This is known as **homolytic fission** of a covalent bond.

Fig 12.12 is a representation of this process, here to make a chlorine free radical from a chlorine molecule. A curly half-headed arrow shows the movement of one electron.

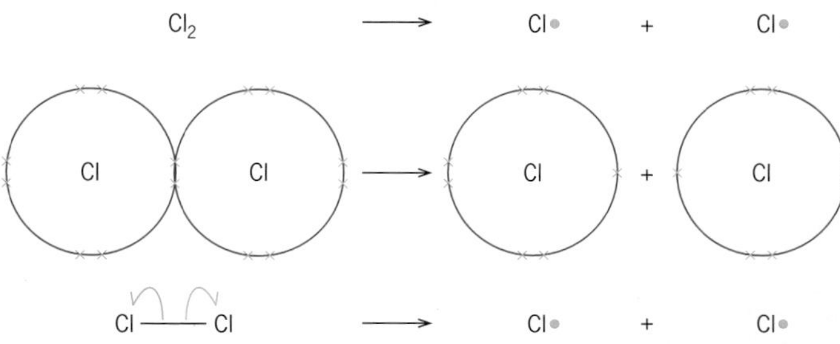

Fig 12.12 Chlorine free radicals are formed by the homolytic fission of a single bond in a chlorine molecule

The homolytic fission of a covalent bond requires energy (it is an endothermic process). For example, the production of the chlorine free radical requires half of the bond dissociation energy of a chlorine molecule:

$$\tfrac{1}{2}Cl_2(g) \rightarrow Cl\bullet(g)$$

FREE-RADICAL SUBSTITUTION OF METHANE

Methane is a typical alkane, which reacts with chlorine in the presence of ultraviolet light to give a large variety of products.

Fig 12.13 gives you an idea of the variety of **substitution** products that result from the free-radical chlorination of methane. Substitution is the process by which a hydrogen atom is swapped for a chlorine atom.

But there are also other methane products whose origin is less obvious. The only way to explain this large collection of products is to examine the mechanism of the reaction.

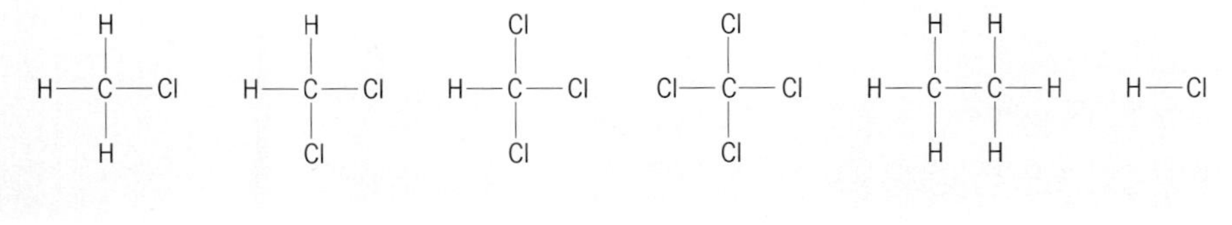

Fig 12.13 The displayed formulae of six products of the free-radical chlorination of methane

? QUESTION 9

9 **a)** What is the energy required to make 1 mole of fluorine free radicals from fluorine molecules under standard conditions? (Refer to data tables.)

b) Predict whether bromine, chlorine or fluorine will be the most reactive towards ethane.

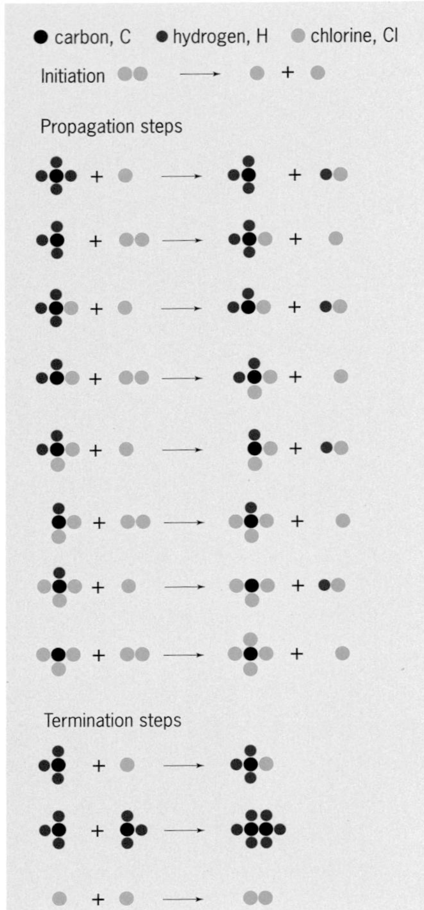

Fig 12.14 The complete mechanism of the free-radical substitution of methane

REACTION MECHANISMS

The **mechanism** of a reaction details the smaller steps in the process that results in the formation of the end products from the starting materials. In particular, a mechanism seeks to explain which bonds are broken and formed, and when and how they are broken and formed.

Free-radical substitution is split into three, clearly defined steps:

- **Initiation** This is the formation of the free radical.
- **Propagation** This involves reactions that maintain the presence of the free radicals.
- **Termination** This involves reactions that remove the free radicals.

Initiation

We have already described the formation of a chlorine free radical. Ultraviolet radiation is used to supply the energy to break the chlorine molecule into free radicals as follows:

$$Cl_2(g) \xrightarrow{\text{ultraviolet light}} 2Cl\bullet(g)$$

This equation tends to mislead, since it suggests that all the chlorine molecules are changed into free radicals. In fact, only an extremely small percentage of chlorine molecules are dissociated to form free radicals, and most remain as chlorine molecules.

Propagation

Reactions take place when the appropriate particles collide. Since free radicals are so reactive, they almost invariably react with any particle they collide with. In the chlorine free-radical substitution of methane, once the very first free radical is produced, it is most likely to collide with either a chlorine molecule or a methane molecule. The collision with the chlorine molecule does not lead to any change.

$$Cl\bullet(g) + Cl_2(g) \rightarrow Cl_2(g) + Cl\bullet(g)$$

When the highly reactive free radical collides with a methane molecule, it gains an electron from one of the covalent bonds in the molecule. In this case it is the C–H bond. This gives:

$$CH_4(g) + Cl\bullet(g) \rightarrow CH_3\bullet(g) + HCl(g)$$

This reaction produces a new free radical, $CH_3\bullet$, which is the methyl free radical. The methyl free radical is also very reactive so as soon as it collides with a particle, it reacts. Which particle it collides with is just a question of chance or probability. The most likely will be the particles that are in the greatest concentration. Certainly, at the start of the reaction, these will be either methane or chlorine molecules:

$$CH_3\bullet(g) + Cl_2(g) \rightarrow CH_3Cl(g) + Cl\bullet(g)$$

Notice that the two steps just described regenerate the chlorine free radical, which explains why they are called propagation steps.

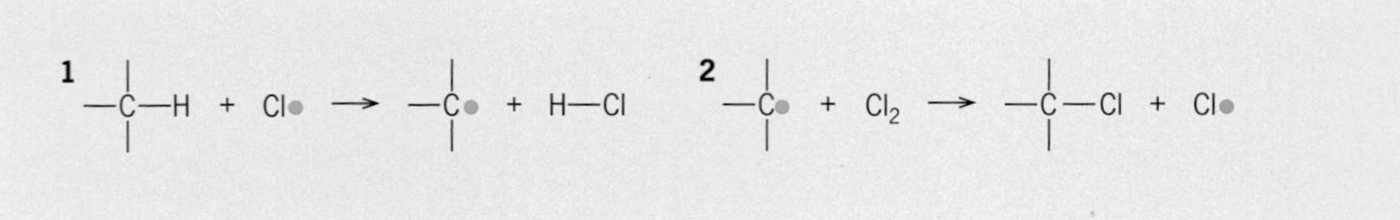

1 $-\overset{|}{\underset{|}{C}}-H + Cl\bullet \rightarrow -\overset{|}{\underset{|}{C}}\bullet + H-Cl$

2 $-\overset{|}{\underset{|}{C}}\bullet + Cl_2 \rightarrow -\overset{|}{\underset{|}{C}}-Cl + Cl\bullet$

Fig 12.15 The propagation steps in the free-radical substitution of a C–H bond. Notice that the chlorine free radical is regenerated at the end of the second step

? QUESTION 10

10 Explain why the collision of a methyl free radical with a methane molecule does not lead to a reaction.

Fig 12.15 generalises these two steps, explaining the process as a characteristic reaction of a C–H bond associated with a carbon atom with four single bonds.

The product chloromethane still has three hydrogen atoms each bonded to carbon atoms. So, when a chlorine free radical collides with a chloromethane molecule, a reaction will take place:

$$CH_3Cl(g) + Cl\bullet(g) \rightarrow CH_2Cl\bullet(g) + HCl(g)$$

$$CH_2Cl\bullet(g) + Cl_2(g) \rightarrow CH_2Cl_2(g) + Cl\bullet(g)$$

Provided there is sufficient chlorine present, this reaction can continue until all the hydrogen atoms have been substituted. The composition of the reaction mixture varies with the time allowed for the reaction and the mole ratio of methane and chlorine.

Termination

The propagation steps cannot go on forever, because eventually chance takes a hand and a free radical collides with another free radical. Although this is not very likely, it is not impossible – particularly when there are many millions of collisions per second.

When two free radicals collide, they react and form a covalent bond. As any two free radicals may collide, note that all of the following are possible termination steps.

$$Cl\bullet(g) + Cl\bullet(g) \rightarrow Cl_2(g)$$

$$Cl\bullet(g) + CH_2Cl\bullet(g) \rightarrow CH_2Cl_2(g)$$

$$CH_3\bullet(g) + CH_3\bullet(g) \rightarrow C_2H_6(g)$$

Also note that one of the products is ethane, a molecule that does not even contain chlorine.

FLUORINATION OF ALKANES

Fluorine is a much more reactive halogen than chlorine, and so it should react much more readily with alkanes. The bond dissociation energy for a fluorine molecule is 158 kJ mol^{-1}, considerably smaller than that for a chlorine molecule at 244 kJ mol^{-1}. The reason for this low bond energy is the repulsion experienced by each atom because of the very small bond length of 254 pm in a fluorine molecule. This makes the initiation step for fluorine much easier than that for chlorine.

The reaction of fluorine with alkanes does not need the presence of ultraviolet light and it proceeds much faster than the reaction of chlorine. Consequently, it is difficult to produce monofluoro derivatives of alkanes. The reaction is often explosive and produces the perfluoro derivative (all H atoms are substituted by F atoms). The reaction of fluorine with alkanes is slowed down by diluting the fluorine with nitrogen.

? QUESTIONS 11–13

11 **a)** How could you ensure that you obtain a large proportion of chloromethane during the free-radical substitution of methane with chlorine?

b) How could you ensure that you obtain a large proportion of tetrachloromethane during the free-radical substitution of methane with chlorine?

12 Write down two more possible termination steps during the free-radical chlorination of methane.

13 Ethane reacts with chlorine in the presence of ultraviolet light to form a mixture of products, including chloroethane, dichloroethane, trichloroethane, tetrachloroethane, hydrogen chloride and ethane. Account for the formation of each of these products by writing down the detailed mechanism for the reaction.

? QUESTION 14

14 Write down the mechanism of the reaction of methane with excess fluorine by a free-radical substitution.

SCIENCE IN CONTEXT

Uses of organofluorines

Organofluorine compounds occur rarely in nature, but because of the wide range of their applications – from inert solvents in the electronics industry to highly active pharmaceuticals – their development and production are increasingly important.

The three-dimensional shape of an organofluorine molecule determines its biological activity. Substitution of a fluorine atom for a hydrogen atom rarely changes the shape since the change in bond length is small, being from 120 pm (120×10^{-12} m) in a C—H bond to 147 pm in a C—F bond. The C—F bond is also very strong and resistant to reaction with water (hydrolysis), which is why pharmaceuticals that contain this bond are liable to remain active in the aqueous medium of a living cell for a long time.

Perfluoro compounds are very unreactive, which has led to yet further applications for which being chemically inert is vital. Perfluorodecalin, for example, dissolves oxygen very efficiently and has been successfully transfused into human volunteers as a blood substitute.

Other perfluorohydrocarbons are being developed to provide a fluid that will preserve donor organs before

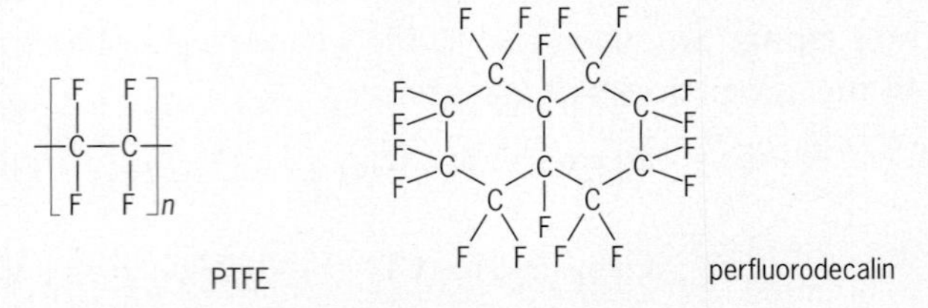

Fig 12.16 The displayed structures of PTFE and perfluorodecalin

transplantation. There are still problems with these perfluoro compounds, since to be of use they must remain as an emulsion in water. Unfortunately, they tend to separate during storage.

Polytetrafluoroethene (PTFE) is a perfluorinated polymer that is used to make non-stick surfaces, such as on saucepans and frying pans.

Coatings are being developed from polymers that include fluoro-substituted alkanes, since these side chains make the surface much easier to clean by their ability to shrug off stains and dirt. Clothes have been developed with such a coating, and it may be possible to make a surface that could remain graffiti free.

Fig 12.17 Will organofluorines provide surfaces for building materials that will make them easy to wipe free of graffiti?

BROMINATION OF ALKANES

Bromine is less reactive than chlorine, and the free-radical reaction with alkanes is much slower than with chlorine. This is surprising since the bond energy for a bromine molecule is only 151 kJ mol^{-1}. The reason for this is related to the strength of the carbon–halogen bond that is formed, since a carbon–chlorine bond is much stronger than a carbon–bromine bond, and hence more stable.

When a few drops of liquid bromine are added to hexane, a liquid alkane, and the mixture is left in direct sunlight, there is a very slow reaction. Eventually, the orange colour of the bromine disappears, leaving a mixture of all the structural isomers of monobromoalkane as the main products:

$$C_6H_{14}(l) + Br_2(l) \rightarrow C_6H_{13}Br(l) + HBr(g)$$

FREE-RADICAL SUBSTITUTION AS A SYNTHETIC REACTION

Although free-radical substitution is a very successful way of introducing halogen atoms into an alkane molecule, it suffers from a major drawback: often a large variety of products is formed.

The aim of synthesis is to make a compound in high yield and purity, using few reactions and involving cheap starting materials. The large number of products formed during free-radical substitution makes it a poor tool in organic synthesis.

The only way to make the reaction useful is to minimise the number of possible products, so that a relatively high yield of the required product is obtained. In practice, this means using reactants that have very few hydrogen atoms to be substituted.

? QUESTION 15

15 Draw the displayed formulae of all the structural isomers of monobromohexane.

STRETCH AND CHALLENGE

Free-radical chlorination in carboxylic acids and methylbenzene

SUBSTITUTION IN METHYLBENZENE

Look at the reaction in Fig 12.18. Methylbenzene has three C–H bonds not attached to the benzene ring; these hydrogen atoms are chemically identical and behave in a similar way to the hydrogen atoms in methane. Therefore, it is quite easy to synthesise chloromethylbenzene by passing chlorine into boiling methylbenzene in the presence of ultraviolet light. By keeping the available chlorine concentration low, the monosubstituted product chloromethylbenzene acid is the major product.

? QUESTION 16

16 How might you modify the reaction conditions to make trichloromethylbenzene by the chlorination of methylbenzene? Write an equation for this reaction.

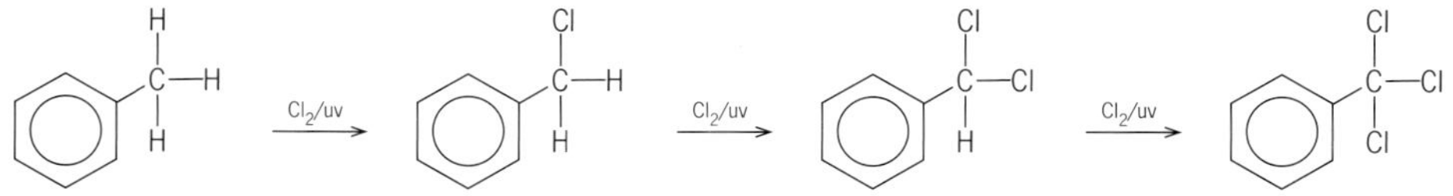

$$C_6H_5CH_3 \xrightarrow{Cl_2/uv} C_6H_5CH_2Cl \xrightarrow{Cl_2/uv} C_6H_5CHCl_2 \xrightarrow{Cl_2/uv} C_6H_5CCl_3$$

Fig 12.18 The reaction of methylbenzene with chlorine

SYNTHESIS OF THE AMINO ACID GLYCINE

Ethanoic acid also contains three hydrogen atoms that are similar to those found in methane and so ethanoic acid can be chlorinated to give chloroethanoic acid:

$$CH_3COOH(l) + Cl_2(g) \rightarrow CH_2ClCOOH(l) + HCl(g)$$

Amino acids are the building blocks of proteins. Glycine (aminoethanoic acid) is the simplest amino acid. It has a two-carbon skeleton, so ethanoic acid is an ideal candidate for its synthesis.

Fig 12.19 shows a synthetic route that could be used to make glycine, employing a free-radical chlorination reaction.

$$CH_3COOH \xrightarrow{Cl_2/uv} ClCH_2COOH \xrightarrow[(ii)\ H^+(aq)]{(i)\ NH_3} NH_2{-}CH_2{-}COOH$$

ethanoic acid — free-radical substitution — chloroethanoic acid — nucleophilic substitution — glycine

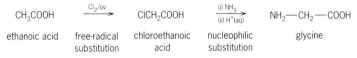

Fig 12.19 The synthesis of the amino acid glycine from ethanoic acid

You can read much more about the mechanism of the halogenation of alcohols on pages 265–266.

4 PREPARATION OF HALOGENOALKANES BY THE HALOGENATION OF ALCOHOLS

Monohalogenoalkanes are best prepared by replacing the hydroxyl group in an alcohol with a halogen atom. A variety of reagents can be used for this conversion and some of these are shown in Table 12.3. Note that R stands for an alkyl group such as methyl, ethyl or propyl.

REMEMBER THIS

'Reflux' means to heat a liquid in a flask with a condenser attached. Any vapour that evaporates cools in the condenser and drips back into the flask.

Table 12.3 Reagents and conditions that can be used to make halogenoalkanes from alcohols. Concentrated hydrochloric acid and concentrated hydrobromic acid are often generated *in situ* by the reaction of the corresponding sodium halide with concentrated sulphuric acid:

$$NaBr + H_2SO_4 \rightarrow NaHSO_4 + HBr$$

Halogenoalkane	Reagent	Conditions	Equation
chloroalkane	concentrated hydrochloric acid	heating under reflux	$ROH + HCl \rightarrow RCl + H_2O$
chloroalkane	phosphorus(V) chloride	heating under reflux	$ROH + PCl_5 \rightarrow RCl + HCl + POCl_3$
bromoalkane	concentrated hydrobromic acid	heating under reflux	$ROH + HBr \rightarrow RBr + H_2O$
bromoalkane	phosphorus(III) bromide	heating under reflux	$3ROH + PBr_3 \rightarrow 3RBr + H_3PO_3$
iodoalkane	red phosphorus and iodine	heating under reflux	$3ROH + PI_3 \rightarrow 3RI + H_3PO_3$

5 REACTIONS OF HALOGENOALKANES

? QUESTION 17

17 **a)** What is the major product when 3-methylhexan-2-ol is heated with red phosphorus and iodine?
b) Which reagents do you need to make i) iodoethane, ii) 3-bromohexane?

Halogenoalkanes are saturated compounds, which means that they have no double or triple carbon–carbon bonds in their molecules. This limits the reactions available, since saturated compounds cannot undergo addition reactions. Halogenoalkanes undergo two major types of reaction, **substitution** and **elimination**. During substitution, the halogen atom is swapped for one atom or a group of atoms. The elimination reaction involves the loss of a molecule of hydrogen halide to produce an alkene.

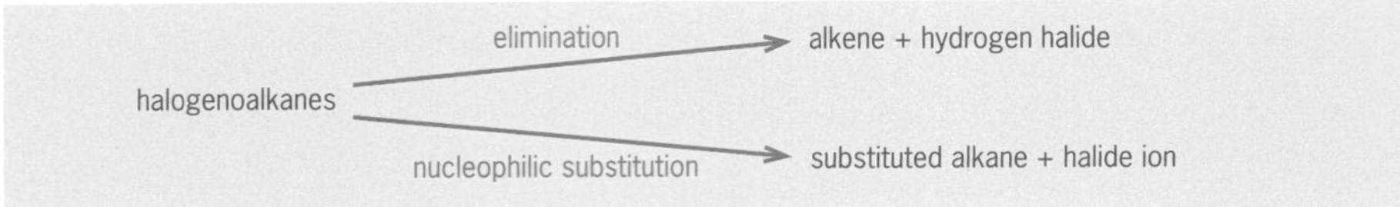

Fig 12.20 The two different reactions of halogenoalkanes

SUBSTITUTION REACTIONS

Polar covalent bond

The carbon–halogen bond is a polar covalent bond. The carbon atom is electron deficient, because the pair of electrons within the covalent bond is drawn closer to the highly electronegative halogen atom. A polar covalent bond, such as a carbon–chlorine bond, is open to attack by a species (an ion or a molecule) that seeks out electron-deficient centres.

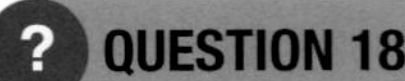

18 State which one of the following compounds has the most electron-deficient carbon atom: bromomethane, chloromethane, fluoromethane or iodomethane.

Fig 12.21 The electron cloud in carbon–halogen bonds

Nucleophiles

A **nucleophile** is a chemical species (ion or molecule) that donates an electron *pair* in order to make a covalent bond with an electron-deficient centre. A nucleophile has a complete outer shell of electrons and at least one lone pair. It is this lone pair of electrons that is donated to form the covalent bond.

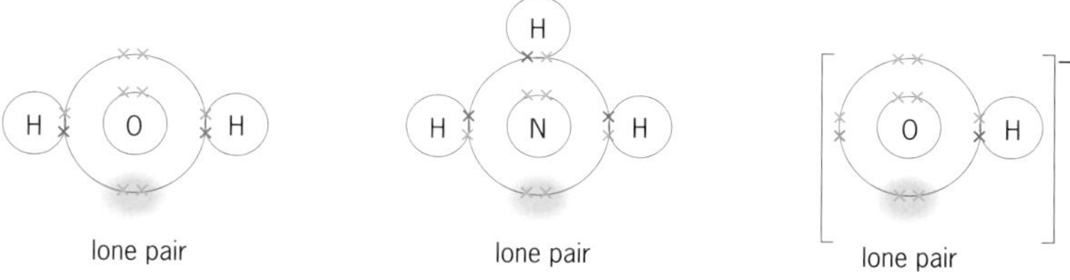

Fig 12.22 Dot-and-cross diagrams of nucleophiles H_2O, NH_3, and OH^-

When we describe a species as a good nucleophile, the term 'good' refers to the ability of the species to donate an electron pair to form a covalent bond. The halide ions, for example, do not very easily donate the lone pair of electrons in their outer shells, and are therefore classed as poor nucleophiles.

Nucleophilic substitution

The electron-deficient, δ+, carbon atom of the polar carbon–halogen bond is attacked by a nucleophile, which donates a pair of electrons to make a covalent bond. Since a carbon atom is normally surrounded by no more than four covalent bonds (four bonding pairs of electrons), this donation must result in the breaking of another bond.

Fig 12.23 shows that when a nucleophile reacts with a halogenoalkane, a substitution reaction takes place. This reaction is called **nucleophilic substitution**, since the species that starts the substitution is a nucleophile.

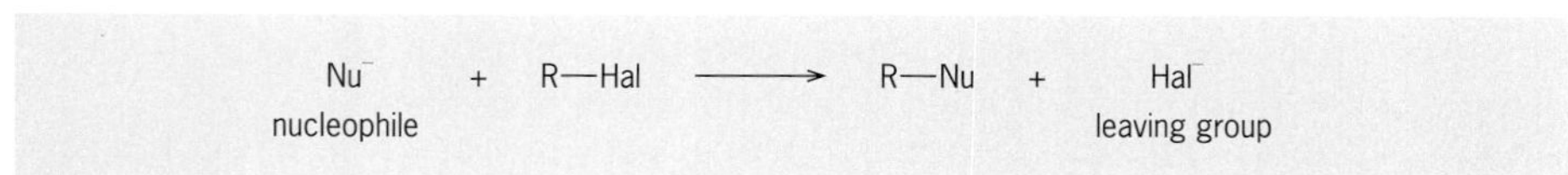

Fig 12.23 Nucleophilic substitution of a halogenoalkane by a nucleophile

During the reaction, a carbon–nucleophile bond is made and a carbon–halogen bond is broken. The halogen is sometimes called the **leaving group**, because it is the species that is lost. In this case, the leaving group must be able to gain the pair of bonding electrons. This is not a problem for a halogen atom, since it forms a halide ion, which easily supports a negative charge.

During nucleophilic substitution, **heterolytic fission** of a covalent bond occurs. In heterolytic fission, one of the two atoms joined by the covalent bond that breaks receives both electrons from the bond, so one atom then carries a positive charge and the other a negative charge.

Energy considerations during nucleophilic substitution

It is possible to estimate the enthalpy change of reaction by considering the bond energies of the carbon–nucleophile bond and carbon–halogen bond. Although this ignores the energy changes due to the formation of the nucleophile and the enthalpy changes due to interactions with the solvent, it is nevertheless a good way to make some predictions.

? QUESTION 19

19 Draw a dot-and-cross diagram for each of the following and use it to decide whether the particle is a nucleophile:

a) NH_2^-

b) CN^-

c) BCl_3

On page 159, we describe bond breaking as an endothermic process, which requires the transfer of energy *from* the surroundings, and bond formation as an exothermic process, which involves energy transfer *to* the surroundings.

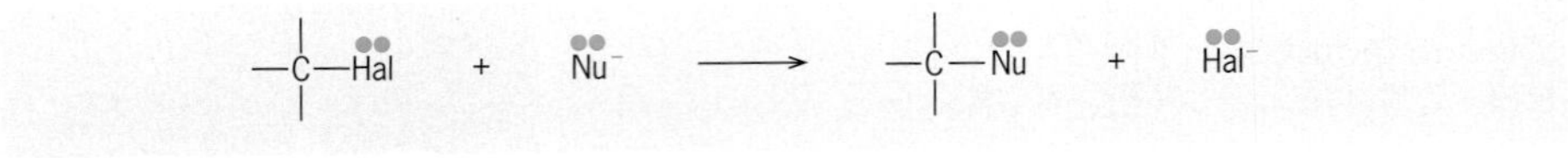

$$-\overset{|}{\underset{|}{C}}-\ddot{Hal} + \ddot{Nu}^- \longrightarrow -\overset{|}{\underset{|}{C}}-\ddot{Nu} + \ddot{Hal}^-$$

Fig 12.24 Bond breaking and bond forming for enthalpy calculations

An estimate for the enthalpy change (ΔH) for the nucleophilic substitution can be calculated using the equation:

ΔH = (bond energy of carbon–halogen) – (bond energy of carbon–nucleophile)

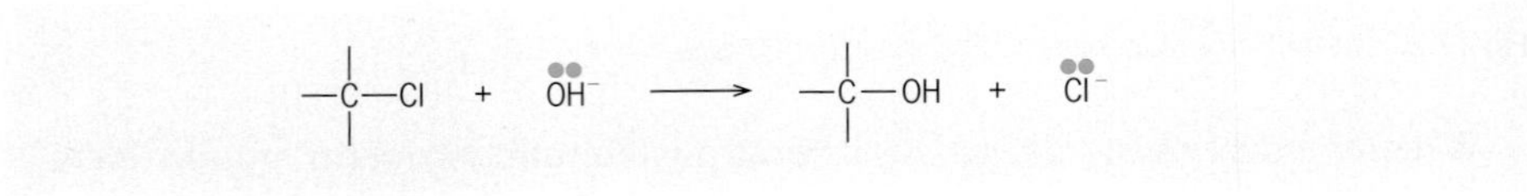

$$-\overset{|}{\underset{|}{C}}-Cl + \ddot{O}H^- \longrightarrow -\overset{|}{\underset{|}{C}}-OH + \ddot{Cl}^-$$

Fig 12.25 During the reaction of a chloroalkane and an aqueous hydroxide ion, a carbon–chlorine bond is broken and a carbon–oxygen bond is made. This means that the enthalpy change of reaction can be estimated as –20 kJ mol^{-1} (from bond enthalpy values). This value should be a good estimate of the enthalpy change for the reaction between any chloroalkane and a hydroxide ion

A useful guide to the feasibility of a reaction can be obtained by looking at the enthalpy change of reaction. An exothermic reaction is often more feasible than an endothermic reaction that has a positive ΔH. In the same way, a reaction that is highly exothermic is often more feasible than one that is less exothermic. (This discussion ignores entropy considerations, which are difficult to predict in the example we are looking at.)

This analysis of the energy changes during reactions leads to some simple ideas. Consider the reaction of hydroxide ion with fluoroalkanes, chloroalkanes, bromoalkanes and iodoalkanes. The same bond is made each time, namely the carbon–oxygen bond. But the bond that is broken, the carbon–halogen bond, becomes weaker and weaker, and so the overall enthalpy change is more negative. So, nucleophilic substitution is easier with iodoalkanes than with fluoroalkanes. The strength of the bond to be broken in the reaction, the carbon–halogen bond, plays an important part in determining whether nucleophilic substitution will take place.

This also means the nucleophile must make a strong bond with the carbon atom rather than a weak bond, which releases little energy into the surroundings. Since the carbon–halogen bonds are quite weak, the halide ions do not act as nucleophiles in this reaction.

You can read more about the feasibility of reactions on page 164 and Chapter 25.

? QUESTION 20

20 Estimate the enthalpy change of reaction for:

a) reaction between hydroxide ion and fluoroethane;

b) reaction between hydroxide ion and bromoethane.

You will need to look up the appropriate bond energies.

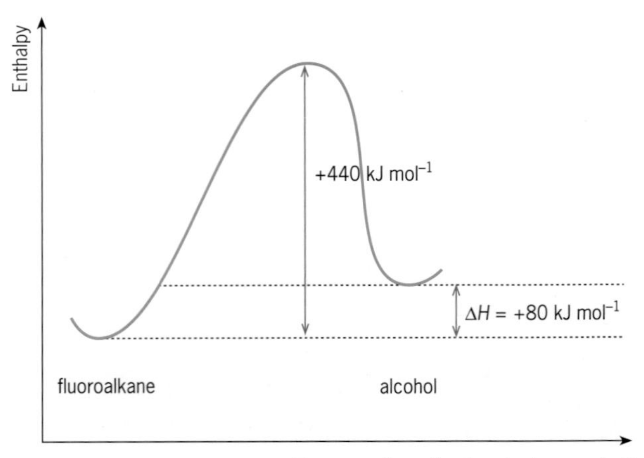

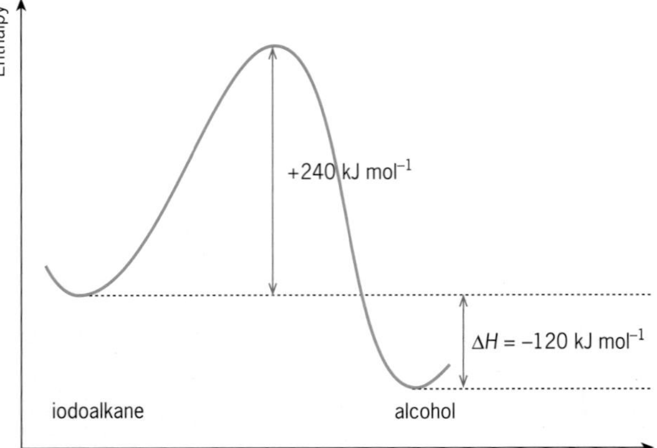

Fig 12.26 Energy profile diagrams for the reaction between a fluoroalkane or an iodoalkane with a hydroxide ion

Mechanism of nucleophilic substitution

So far, we have only discussed which bonds have been broken and which ones have been formed. The next question to ask is: what is the sequence of bond making or bond breaking? One mechanism assumes that the two processes take place together, so that as one bond is forming the other bond is breaking. This mechanism is called S_N2, since it involves two particles colliding in the slowest step of the mechanism. This mechanism is shown in Fig 12.27. The curly arrow shows the movement of an electron pair. Primary halogenoalkanes, RCH_2X (where R is an alkyl group and X is Cl, Br and I), react in this way.

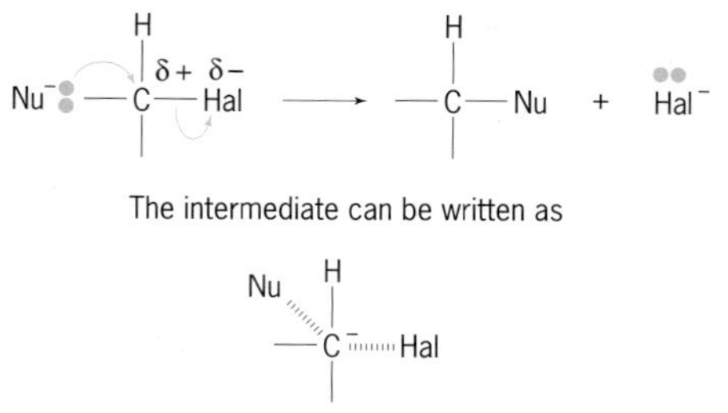

REMEMBER THIS

S_N2 stands for substitution, nucleophilic and second order. Second-order reactions have two particles that collide in the slowest step of the mechanism. You can read more about orders of reaction and mechanisms in Chapter 26.

REMEMBER THIS

A primary halogenoalkane has one alkyl or aryl group attached to the carbon of the carbon–halogen bond, a secondary halogenoalkane has two alkyl or aryl groups, and a tertiary halogenoalkane has three alkyl or aryl groups.

Fig 12.27 This mechanism is known as S_N2, since two species, the nucleophile and the halogenoalkane, are involved in the only step. Primary halogenoalkanes and many secondary halogenoalkanes show this mechanism during nucleophilic substitution

The nucleophile donates a pair of electrons to the electron-deficient carbon and at the same time the pair of electrons in the carbon–halogen bond move towards the electronegative halogen atom, eventually breaking the covalent bond and forming a halide ion. The structure of the intermediate shows that, at some time during the mechanism, both the nucleophile and the halogen atom are attached to the carbon atom.

HOW SCIENCE WORKS

S_N1 nucleophilic substitution

Primary chloro-, bromo- and iodoalkanes react with nucleophiles by the mechanism known as S_N2. However, tertiary chloro-, bromo- and iodoalkanes react with nucleophiles by a different mechanism known as S_N1.

S_N1 involves two steps rather than one and the slowest step involves only one particle – namely, the tertiary halogenoalkane. This step does not involve a collision between particles; instead the carbon–halogen bond is broken heterolytically to form a carbonium ion (often called a carbocation) and the halide ion. Once the positive carbonium ion is formed, it will react in a much faster step with the nucleophile.

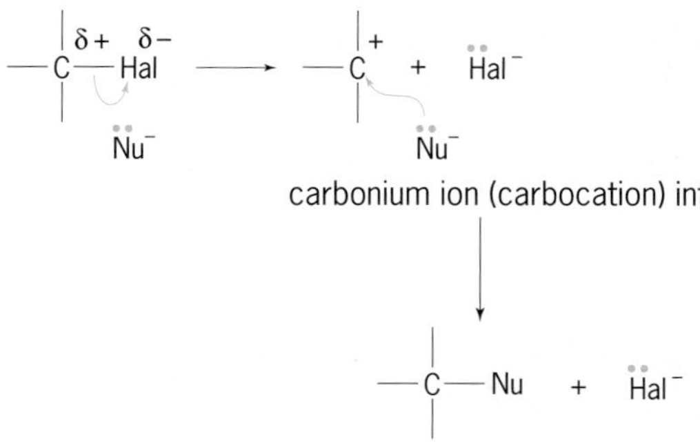

δ+ δ− C—Hal → C⁺ + Hal⁻

Nu⁻ Nu⁻

carbonium ion (carbocation) intermediate

↓

C—Nu + Hal⁻

Fig 12.28 The S_N1 mechanism

It is the stability of the carbonium ion that can be formed by the halogenoalkane that determines whether the nucleophilic substitution takes place by the S_N1 or the S_N2 mechanism. Tertiary carbonium ions (those with a positive carbon atom surrounded by three alkyl groups) are much more stable than secondary or primary carbonium ions. Nucleophilic substitution of secondary halogenoalkanes often involves both mechanisms.

NUCLEOPHILIC SUBSTITUTION REACTIONS OF HALOGENOALKANES

Hydroxide ions, ammonia, amines and cyanide ions are all good nucleophiles and will take part in the nucleophilic substitution reaction with chloroalkanes, bromoalkanes and iodoalkanes. As already described, the carbon–iodine bond is the weakest carbon–halogen bond of the three, so iodoalkanes readily take part in nucleophilic substitution. The nucleophilic substitution reactions of chloroalkanes are much slower.

REMEMBER THIS

A strong bond has a large bond enthalpy and a weak bond has a small bond enthalphy.

Hydrolysis

Halogenoalkanes react extremely slowly, if at all, with water. However, as we have already seen, they undergo a nucleophilic substitution reaction with aqueous hydroxide ions. (The hydroxide ion is a better mucleophite than a water molecule because it can more easily donate its lone pair to an electron-deficient centre.) This reaction is known as hydrolysis and involves boiling aqueous sodium hydroxide, for example, with the halogenoalkane. A hydroxyl group is substituted for the halogen atom and so an alcohol is produced. So, 1-bromobutane would be hydrolysed to form butan-1-ol:

$$CH_3CH_2CH_2CH_2Br + NaOH \rightarrow CH_3CH_2CH_2CH_2OH + NaBr$$

In these reactions, the hydroxide ion is behaving as a nucleophile.

Fluoroalkanes are not hydrolysed since the carbon–fluorine bond is too strong. The ease with which halogenoalkanes can be hydrolysed increases as the atomic number of the halogen increases and the carbon–halogen bond becomes weaker and longer.

Although the C–Cl bond is much more polar than the C–I bond, because chlorine is more electronegative than iodine, experiments have shown iodoalkanes are hydrolysed much more rapidly than chloroalkanes. This is because the C–I bond is much weaker than the C–Cl bond.

QUESTION 21

21 What is the product of the reaction of 2-iodopentane with aqueous potassium hydroxide?

Analysis of halogenoalkanes

Aqueous silver nitrate can be used to detect the presence of aqueous halide ions, but it will not detect the halogen in a halogenoalkane since the halogen is present in a covalent bond. To detect the halogen, the halogenoalkane must first be hydrolysed to produce aqueous halide ions.

The halogenoalkane is heated with aqueous sodium hydroxide. During hydrolysis, aqueous halide ions are produced. The hydrolysis mixture is acidified with dilute nitric acid, then aqueous silver nitrate is added. The halide ion present in the mixture is precipitated as the silver halide:

$$Ag^{+}(aq) + X^{-}(aq) \rightarrow AgX(s), \text{ where } X = Cl, Br \text{ or } I$$

The colours of the precipitate are given in Table 12.4 and shown in Fig 12.29.

You can read more about the reaction of aqueous silver nitrate with aqueous halide ions on page 465.

Table 12.4 Test for halide ion in the hydrolysis mixture

Halide ion in hydrolysis mixture	Add excess nitric acid followed by aqueous silver nitrate
chloride	white precipitate
bromide	pale cream precipitate
iodide	pale yellow precipitate

Fig 12.29 The resultant precipitates in the halide ion test

Reaction with ammonia

Ammonia is another nucleophile (Fig 12.22), since the lone pair on nitrogen can be donated to an electron-deficient centre to make a covalent bond. Ammonia reacts with halogenoalkanes to form an amine. For example, 1-bromobutane will produce butylamine:

$$CH_3CH_2CH_2CH_2Br + NH_3 \rightarrow CH_3CH_2CH_2CH_2NH_2 + HBr$$

There are two serious drawbacks with this reaction: ammonia reacts with one of the products (HBr), and the amine produced is also a nucleophile that can further react with the halogenoalkane.

In Section 2 of this chapter, we met a similar problem: a reaction that gives a poor yield and many products is of little use as a synthetic method.

Luckily, the reaction conditions can be manipulated to produce good yields. If excess concentrated ammonia is used, the acidic hydrogen halide product reacts with the excess ammonia to produce an ammonium salt. So, in our example, ammonium bromide would be produced:

$$HBr + NH_3 \rightarrow NH_4Br$$

The excess ammonia also means that there is always a vast excess of ammonia nucleophile compared with the product. This ensures the maximum yield of amine. The best reaction conditions involve heating the alcoholic halogenoalkane with excess ammonia in a sealed tube.

SCIENCE IN CONTEXT

Preparation of amphetamine

The drug amphetamine is a stimulant that gives a sense of well-being and speeds up the metabolism, so that someone who takes it before doing a sport would feel less fatigued. It is prepared using a halogeno compound as an intermediate. Fig 12.30 shows the synthetic route employed.

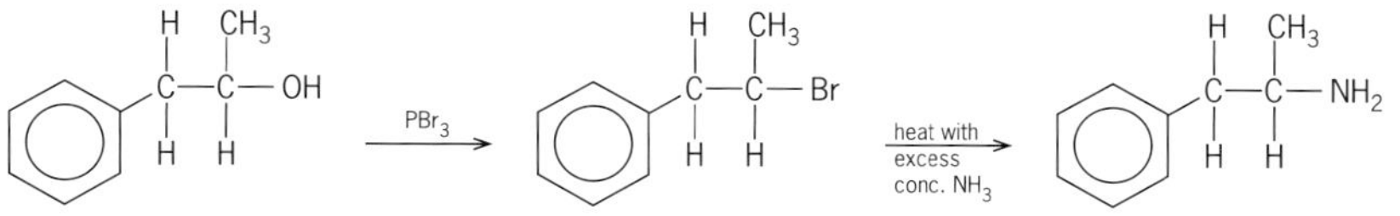

PBr$_3$

heat with excess conc. NH$_3$

Fig 12.30 The synthesis of amphetamine

An alcohol, 2-phenylpropan-2-ol, is halogenated to give 2-bromo-2-phenylpropane, which then allows a nucleophilic substitution to take place using excess ammonia. The bromo compound is a synthetic intermediate that allows the conversion of an alcohol into an amine while keeping the same carbon skeleton.

Reaction with amines

Amines are nucleophiles because of the lone pair on the nitrogen atom, so they can react with halogenoalkanes to give secondary, tertiary or quaternary ammonium salts.

You can read more about amines in Chapter 28.

? QUESTION 22

22 Draw the displayed formulae of the products of the following reactions:

a) excess ammonia with 2-chlorobutane;

b) excess ammonia with 3-iodopentane;

c) ethylamine with excess 2-iodopropane.

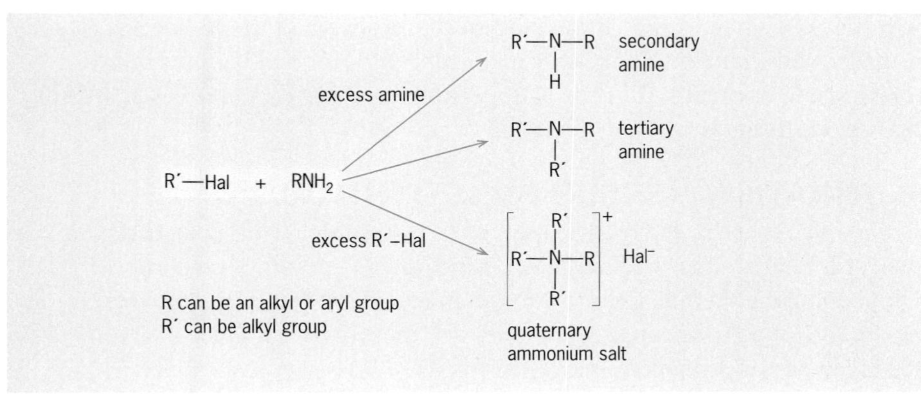

Fig 12.31 The reactions of amines with halogenoalkanes

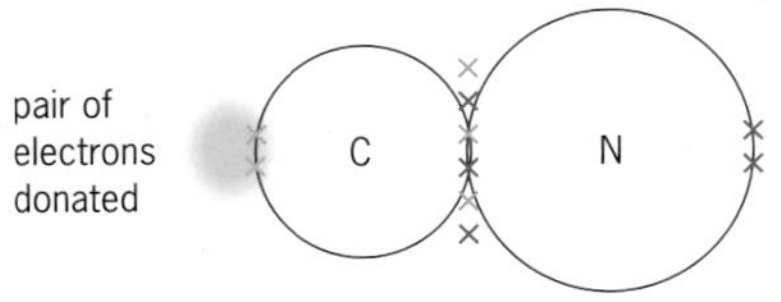

Fig 12.32 The dot-and-cross diagram for the cyanide ion

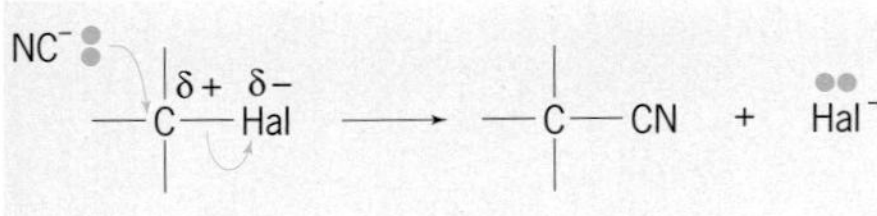

Fig 12.33 The mechanism of the reaction of the cyanide ion with a halogenoalkane

? QUESTION 23

23 Draw the displayed formula for the organic product of the reaction between 3-iodo-2-methylhexane with potassium cyanide in ethanolic solution.

Reaction with the cyanide ion

Whenever possible, the starting material chosen for a synthesis already has the correct carbon skeleton. The synthesis of amphetamine in Fig 12.30 illustrates this point, where the nine-carbon skeleton is in place in the starting material. There are times, however, when the chosen starting material does not have the correct carbon skeleton and part of the synthesis is to construct the correct skeleton. An important synthetic reaction is the *extension* of the carbon skeleton by one carbon.

One convenient way involves the nucleophilic substitution of a halogen atom with a cyanide ion. Fig 12.32 shows the dot-and-cross diagram for the cyanide ion. The negative charge resides on the carbon atom, and it is the non-bonding pair on this carbon that is donated to an electron-deficient carbon.

A nucleophilic substitution reaction with the cyanide ion leads to the formation of a carbon–carbon bond rather than a carbon–nitrogen bond. The mechanism of this is clearly shown in Fig 12.33. The net effect is to increase the carbon skeleton by one carbon. The resulting functional group is called a **nitrile**.

The normal reaction conditions are potassium cyanide or sodium cyanide in ethanol as a solvent. An organic solvent is chosen so that the halogenoalkane dissolves and thus allows more intimate contact with the ionic sodium cyanide. 1-bromobutane reacts with potassium cyanide in ethanol to form pentanenitrile, so changing a four-carbon skeleton into a five-carbon skeleton.

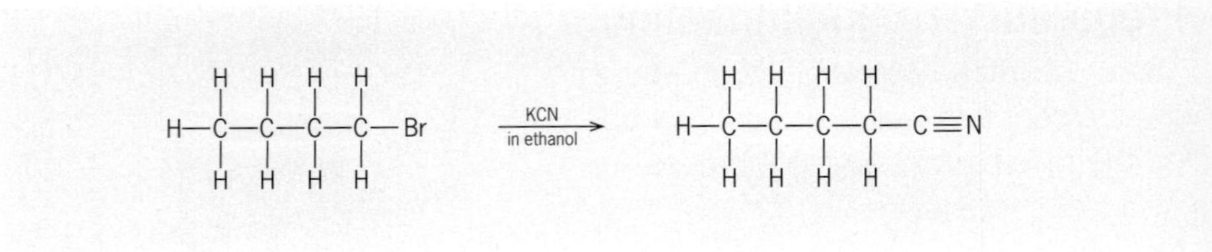

Fig 12.34 Changing a four-carbon skeleton into a five-carbon skeleton

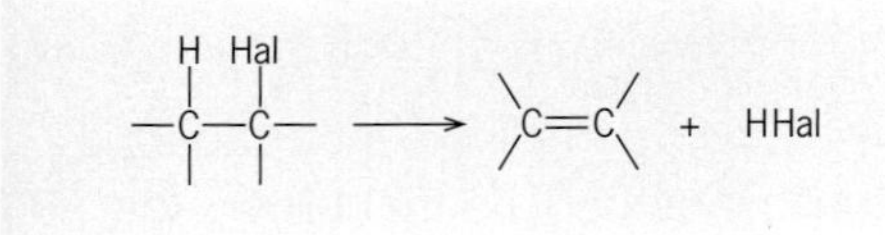

Fig 12.35 A general example of elimination from a halogenoalkane

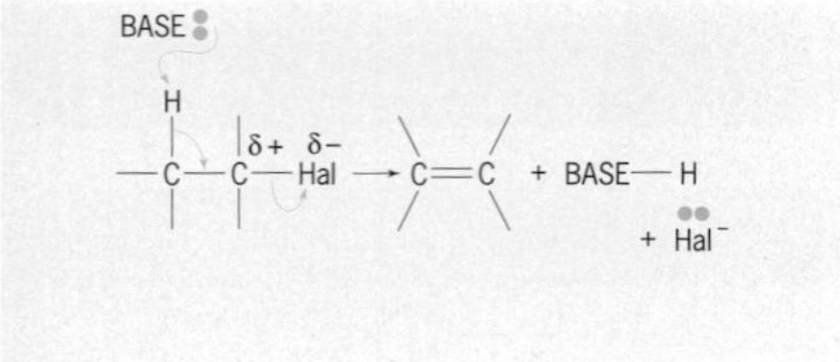

Fig 12.36 The mechanism for the elimination of a hydrogen halide from a halogenoalkane

ELIMINATION REACTIONS

When halogenoalkanes are heated strongly in the absence of air, an elimination reaction takes place with the formation of an alkene and the hydrogen halide. This reaction is quite impractical to use for manufacture, so the elimination reaction is encouraged by the use of a strong base. Strong bases are proton acceptors and will accept the proton lost during elimination. This, in turn, helps the heterolytic fission of the carbon–halogen bond to give a halide ion. The hydroxide ion is a strong base in ethanol solution. A suitable reagent for the elimination reaction in halogenoalkanes is potassium hydroxide in ethanol (ethanolic KOH).

Fig 12.36 shows a possible mechanism for the elimination, indicating the help the base gives in the elimination of the proton and the halide ion. Just as in nucleophilic substitution, the actual order of bond making and bond breaking can vary with different halogenoalkanes and different bases. (A full discussion of this is beyond the scope of this book.)

In heterolytic fission, a covalent bond is broken so that both electrons in the bond move to one of the atoms **(see page 282).**

ELIMINATION VERSUS SUBSTITUTION

Perhaps you may have already thought of this problem: the hydroxide ion is a nucleophile as well as a strong base, which means that in a reaction with a halogenoalkane there is always the possibility of substitution *and* elimination products. It would be more accurate to say that there will always be a mixture of both types of product.

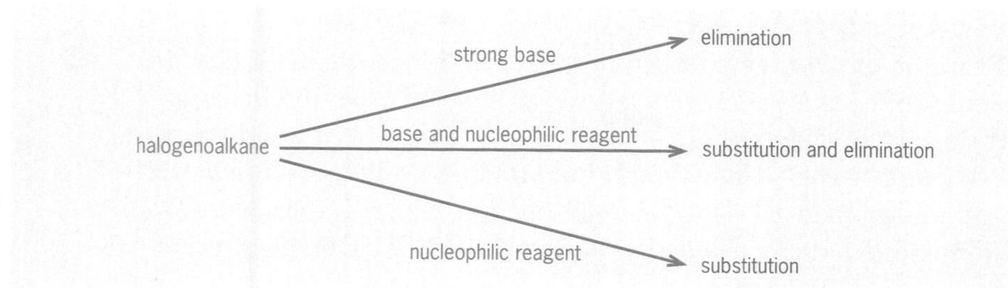

Fig 12.37 A summary of elimination and substitution reactions

The proportion of elimination and substitution products can be varied by changing both the temperature and the solvent used for the reaction. Table 12.5 summarises how the proportions can be changed.

Table 12.5 A summary of changing conditions for competing reactions

	Elimination	Substitution
Nucleophile	poor nucleophile	good nucleophile
Basicity	strong base	weak base
Solvent	ethanol	water
Temperature	high	low

? QUESTION 24

24 **a)** 1-iodopropane reacts with hot aqueous sodium hydroxide to give both elimination and substitution products. Draw the displayed formulae of both sets of products, and explain with the aid of equations how they arise.

b) 1-iodopentane is refluxed under heating with ethanolic sodium hydroxide. Predict the major product of the reaction.

See Chapter 28 for information on bases.

There is a further complication: there can often be more than one elimination product because there may be several different hydrogen atoms that can be lost.

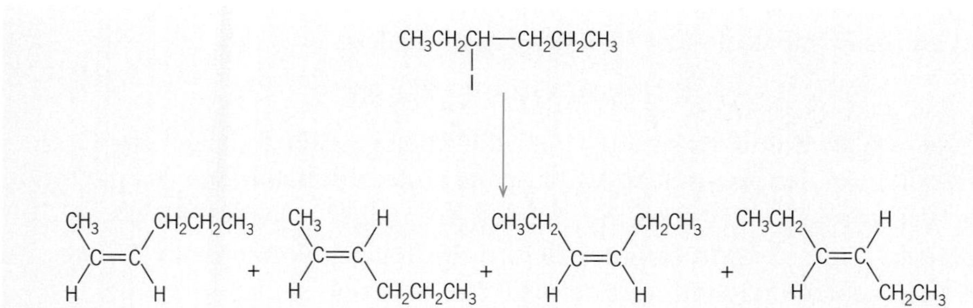

Fig 12.38 Elimination products of 3-iodohexane

6 REACTIONS OF CHLOROFLUOROCARBONS (CFCs)

The most remarkable thing about CFCs is their complete lack of reactivity towards reagents that react with monohalogenoalkanes. CFCs do not undergo nucleophilic substitution or elimination reactions. This lack of reactivity can be explained by the inability of carbon to expand its octet and by the strength of the C–F bond.

Once CFCs are released into the environment there are no natural ways in which the compounds can be broken down. The normal decomposition processes that involve hydrolysis with water, reaction with oxygen, or bacterial or microbial decay do not take place. This naturally leads to a build-up of CFCs in the environment, and there is a risk that they will increase within the cells of living organisms. In the stratosphere, the CFCs can react, but only because of the presence of ultraviolet light.

You can read more about the lack of reactivity of tetrachloromethane in Chapter 21.

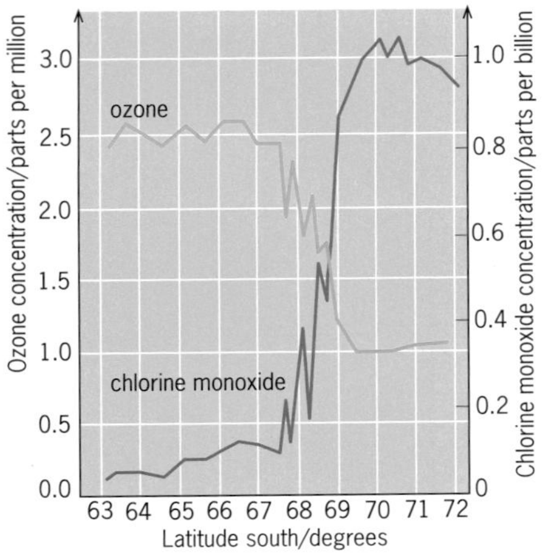

Fig 12.39 Chlorine monoxide concentration and ozone concentration at different latitudes

FREE-RADICAL REACTIONS OF CFCs

CFCs contain two types of carbon–halogen bonds: the C–F and the C–Cl bonds. The C–F bond is very strong, as shown by its bond energy of 439 kJ mol^{-1}, whereas the C–Cl bond is weaker with a bond energy of 330 kJ mol^{-1}. This means that in the presence of ultraviolet light a C–Cl bond, rather than a C–F bond, will undergo homolytic fission to form a chlorine free radical. The equation shows the homolytic fission of a CFC:

$$CF_3Cl(g) \rightarrow CF_3\bullet(g) + Cl\bullet(g)$$

OZONE DEPLETION

We have seen that the chlorine free radical is very reactive and will react with many other particles. In the stratosphere, it collides with ozone molecules to form chlorine monoxide and oxygen:

$$Cl\bullet(g) + O_3(g) \rightarrow ClO\bullet(g) + O_2$$

Chlorine monoxide is a free radical itself and this reaction is one of the propagation steps. The concentrations of ozone and chlorine monoxide in the stratosphere seem to confirm this reaction. The graphs in Fig 12.39 show that where there is a high chlorine monoxide concentration, there is a low ozone concentration, and vice versa.

Chlorine monoxide reacts with oxygen atoms present in the stratosphere to regenerate the chlorine free radical:

$$ClO\bullet(g) + O(g) \rightarrow Cl\bullet(g) + O_2(g)$$

The regeneration of the chlorine free radical completes the propagation step. The chlorine free radical can then react with more ozone.

The net result of these two reactions is the conversion of an oxygen atom and an ozone molecule into two oxygen molecules:

$$O(g) + O_3(g) \rightarrow 2O_2(g)$$

It is estimated that, on average, one chlorine free radical (hence one CFC molecule) can destroy over 100 000 ozone molecules before the chain reaction stops. It stops when the chlorine free radical collides with other molecules, such as hydrocarbons, to form hydrogen chloride that can leave the stratosphere dissolved in water. The rate of ozone depletion by the presence of chlorine free radicals now exceeds its rate of formation in parts of the stratosphere, especially over the Antarctic. The result is a 'hole' in the ozone layer that could have damaging consequences for humans and other life forms on Earth (see Stretch and Challenge box, page 255).

With the growing evidence of increasing ozone depletion being linked to CFC concentrations in the upper atmosphere, scientists convinced many governments to ban the use of CFC in new products. The hope is that without extra CFCs reaching the upper atmosphere, eventually the ozone holes over the poles will become smaller and smaller. This could take decades, because of the stability of CFCs.

? QUESTIONS 25–26

25 Draw the dot-and-cross diagram for chlorine monoxide to show it has one unpaired electron.

26 State a function of the chlorine free radical in the process described in the text.

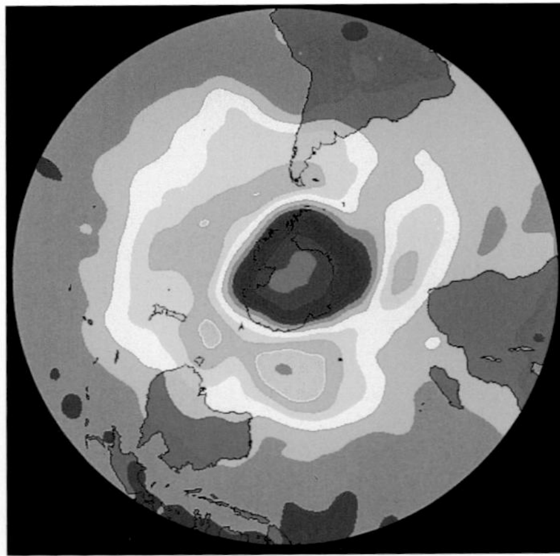

Fig 12.40 The 'ozone hole' (purple and grey) over the Antarctic. The consequences include damage by excess ultraviolet radiation to skin and the DNA of plant and animal cells, which could lead to mutations

STRETCH AND CHALLENGE

The nature of ozone

There are two forms of oxygen, O_2 and O_3. They are called allotropes. Ozone, O_3, is thermodynamically unstable with respect to oxygen, O_2:

$$2O_3(g) \rightarrow 3O_2(g)$$

Ozone is pale blue as a gas and deep blue as a liquid. It has a sharp, irritating odour and produces headaches in humans. Even at one part per million in air, it is poisonous. But the presence of ozone in the stratosphere (a layer 12 to 50 km above the Earth's surface) acts as a protective barrier that prevents much of the high energy UV-B radiation from the Sun reaching the Earth and destroying organisms. It is a strange paradox that we are worried both about an increase in ozone concentration in the lower atmosphere and about a decrease in its concentration in the stratosphere.

Fig 12.41 Photochemical smog containing ozone over Los Angeles, USA. (You can read more about photochemical smogs on page 227)

Ozone is present in the lower atmosphere, formed there by the reaction between nitrogen oxides and volatile organic compounds in the presence of sunlight. It is produced along with photochemical smogs. In the past 100 years, the levels of ozone concentration have doubled over much of Europe and North America.

The effects of ozone in the lower atmosphere include an estimated 5 to 10 per cent loss in potential crop yield. It damages textiles and makes their colours fade faster. Rubber (such as car tyres) deteriorates faster, and human lungs function less well, as the airways are irritated, which causes coughing.

Ozone in the stratosphere is produced by the reaction of oxygen atoms and oxygen molecules:

$$O(g) + O_2(g) \rightarrow O_3(g)$$

Stratospheric ozone absorbs UV-B radiation by undergoing homolytic fission to form an oxygen atom and an oxygen molecule. If the ozone layer were absent or depleted, much more UV-B would reach the Earth's surface and severely damage plant, marine and human organisms. For example, the incidence of skin cancers and cataracts would increase, crop yields would decrease and fish stocks would plummet.

SUMMARY

After studying this chapter, you should know that:

- Bromoalkanes, chloroalkanes and iodoalkanes can be prepared by the reaction of alcohols with concentrated hydrobromic acid, concentrated hydrochloric acid and concentrated hydroiodic acid, generated *in situ* by using the sodium halides and concentrated sulfuric acid, respectively, or by the reaction of the appropriate phosphorus halides with an alcohol.
- Alkanes react with chlorine, fluorine and bromine by free-radical substitution to give a mixture of halogeno and polyhalogenoalkanes.
- Free-radical substitution involves three steps: initiation by ultraviolet light, propagation and termination.
- Fluoroalkanes are very unreactive because of the strength of the C–F bond.
- Bromoalkanes, chloroalkanes and iodoalkanes react by nucleophilic substitution with cyanide ion to form nitriles, with ammonia to give amines and with aqueous hydroxide ion to give alcohols.
- Bromoalkanes, chloroalkanes and iodoalkanes can eliminate hydrogen halides to form alkenes when treated with hot ethanolic alkali.
- CFCs and halogenoalkanes have weak permanent dipole–permanent dipole interactions in the liquid state, so often have low boiling points.
- CFCs are non-toxic and inert, and have a low boiling point.
- CFCs are responsible for ozone depletion in the upper atmosphere because they provide chlorine free radicals.

Practice questions and a How Science Works assignment for this chapter are available at www.collinseducation.co.uk/CAS

13 Alcohols

13 ALCOHOLS

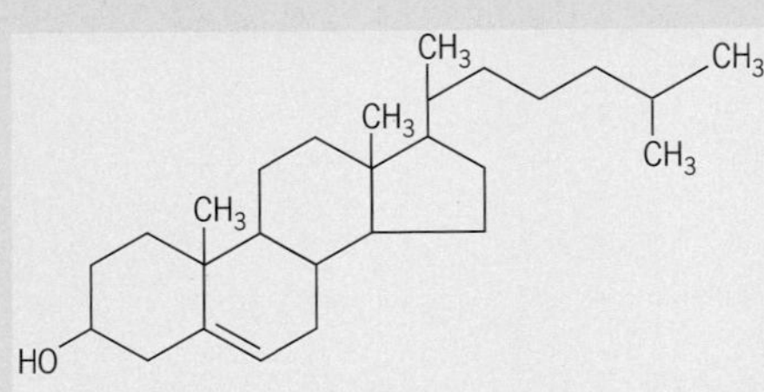

Cholesterol The structure is complicated, with four rings and two functional groups

An artery of the heart (coronary artery) The wall is orange. The build-up of cholesterol is yellow. Blood is confined to the blue area

Cholesterol and health

Cholesterol is a member of the class of chemicals called steroids, which are both naturally occurring and synthetic. The most abundant steroid is cholesterol, which (as the 'ol' at the end of its name suggests) contains an alcohol group. It is found in practically all animal tissues, and forms an important part of cell membranes. Cholesterol is particularly abundant in the brain and spinal cord. It is estimated that the total amount of cholesterol in a 75 kg person is about 250 g.

High levels of cholesterol in the blood are associated with two medical conditions, gallstones and atherosclerosis (hardening and thickening of the arteries when cholesterol from the blood builds up on their inside walls). Blood cholesterol levels can now be measured easily, and people with a high level are advised to reduce their intake of cholesterol-rich foods and saturated fats.

Cholesterol was first isolated in 1770, but it took almost another two centuries before the full structure was established. As the diagram shows, cholesterol has a complicated structure that contains four rings and two functional groups.

1 THE HYDROXYL GROUP

Phenols are compounds that contain a hydroxyl group directly attached to a benzene ring. You can read more about phenols on page 310 in Chapter 15.

Alcohols are organic compounds that contain carbon, hydrogen and oxygen. They contain the hydroxyl functional group, OH, which is an oxygen atom bonded to a hydrogen atom. The hydroxyl group is bonded directly to an alkyl or cycloalkyl group (see Fig 13.1).

There is sometimes confusion in the use of the name alcohol, because it is the name of a class of compounds as well as the name of a particular compound. Alcohol is the popular term for ethanol. But in this book, you can safely assume that when we refer to alcohols we mean the class of compounds – not the single compound.

R—O—H

R = alkyl group
e.g. CH_3CH_2–
or
cycloalkyl group
e.g. [cyclohexyl ring]

Fig 13.1 The generalised structure of alcohols

REMEMBER THIS

Make sure that you can recognise an alcohol when it is part of a very complicated molecule. Look for the hydroxyl group attached to a carbon atom.

ALCOHOLS IN NATURE

Many naturally occurring products are alcohols, including sugars and other carbohydrates, some fragrances, vitamins, pheromones, amino acids and steroids.

Serine is an amino acid used by organisms in the construction of proteins. The structure of serine is shown in Fig 13.2.

HO—CH_2—C(H)(NH_2)—COOH

Fig 13.2 The structure of serine, an amino acid that is also an alcohol

A lack of the alcohol vitamin A in a person's diet causes night blindness and dry skin. And without sufficient vitamin C – another compound that contains alcohol functional groups – a person would suffer from scurvy. The structures of vitamins A and C are shown in Fig 13.3.

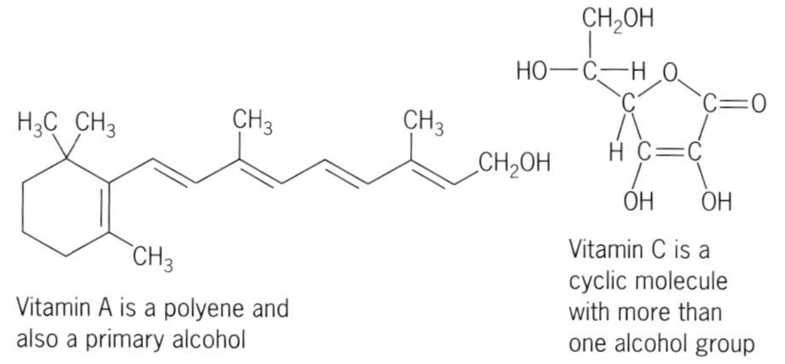

Vitamin A is a polyene and also a primary alcohol

Vitamin C is a cyclic molecule with more than one alcohol group

Fig 13.3 The structures of vitamin A and vitamin C

Fig 13.4 Citrus fruit, such as the oranges shown here, are rich in vitamin C, which contains the hydroxyl functional group

The fragrance of scented flowers is the result of a complex mixture of compounds, many of which are alcohols. The structures of three of these are shown in Fig 13.5.

Propane-1,2,3-triol is used to make fats and oils that the human body uses as energy stores and insulation, and also in the construction of membranes that surround living cells.

Polysaccharides contain many hydroxyl groups, as shown by the structure of the sugar maltose in Fig 13.6.

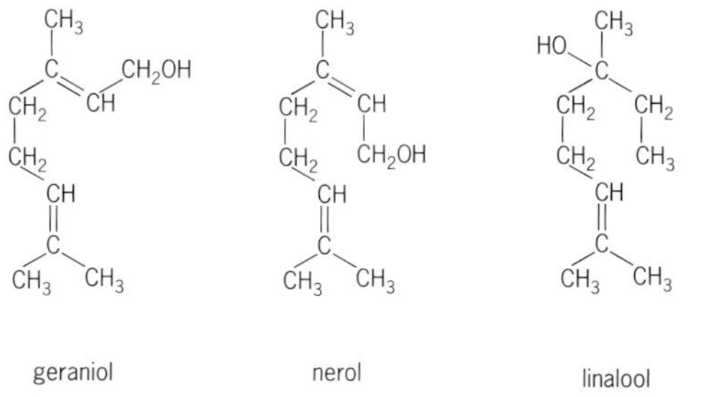

geraniol

nerol

linalool

Fig 13.5 The structures of three of the many compounds responsible for the fragrance of flowers. Geraniol, for example, is a constituent of rose fragrance

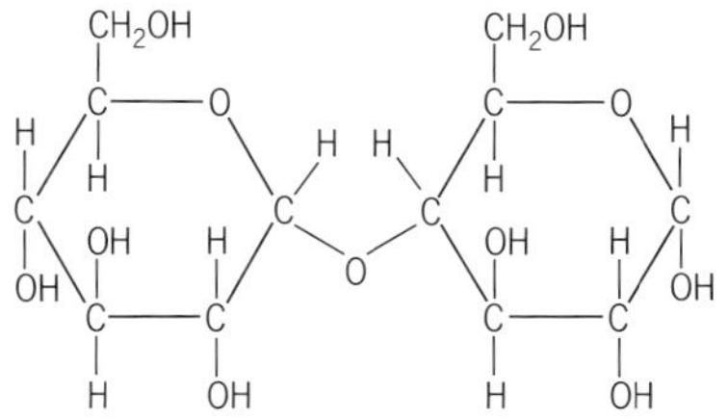

Fig 13.6 The structure of maltose, a polyhydric alcohol that contains eight hydroxyl groups

ALCOHOLS IN USE

Alcohols with low relative molecular mass, such as methanol, ethanol, the propanols and the butanols, are liquids and are used extensively as solvents, both for chemical reactions and in the production of paints and cosmetics. Many alcohols have the ability to dissolve both polar and non-polar compounds. This property is explained later (see page 261) in terms of the different types of intermolecular attraction between alcohol molecules and other molecules.

Alcohols are also very useful intermediates in the synthesis of compounds, since they have a functional group that can undergo different types of reaction.

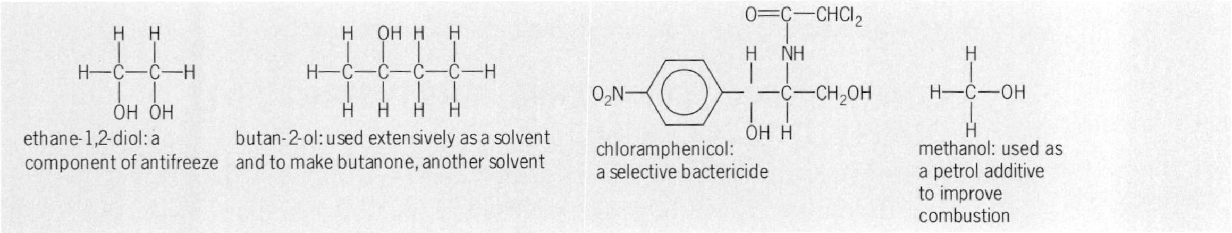

ethane-1,2-diol: a component of antifreeze

butan-2-ol: used extensively as a solvent and to make butanone, another solvent

chloramphenicol: a selective bactericide

methanol: used as a petrol additive to improve combustion

Fig 13.7 The displayed formulae of four typical alcohols with their main uses

NAMING ALCOHOLS

The name of an alcohol is derived from the name of its carbon chain, using the suffix *ol*. The position number for the hydroxyl group must also be specified in the name. The rule is that the carbon atom attached to the hydroxyl group is given the lowest number possible. When there is more than one functional group present, the position numbers are determined by the most important group for naming purposes. (For example, the position of the hydroxyl group is more important than the position of a double bond or of a halogen atom, but less important than the position of a carbonyl group.) The presence of two or more hydroxyl groups is designated by *di*, *tri*, and so on, placed immediately before the suffix *ol*.

? QUESTION 1

1 Draw the displayed formula for:
a) 2-methylpropan-1-ol;
b) 2-chloroethan-1-ol;
c) 3-hydroxypropanoic acid.

propan-1-ol
The OH group is on the number 1 carbon.

2-methylpropan-2-ol
The OH group is on the number 2 carbon, since you start at the end of a carbon chain.

3-chloropropan-1-ol
The OH group is more important than the Cl for naming purposes, so it is on the number 1 carbon.

3-methylpentan-3-ol
You must choose the longest carbon chain containing the OH group as the skeleton.

E-but-2-en-1-ol
The OH group is more important than the C=C double bond for naming purposes, so it is on the number 1 carbon atom.

4-hydroxybutanoic acid
The COOH group is more important than the OH for naming purposes. Notice the *hydroxy* prefix is used instead of the suffix *ol*.

Fig 13.8 Six typical alcohols with their carbon position-numbers

2 TYPES OF ALCOHOL

There are two ways of classifying alcohols. One is based on the number of carbon atoms attached to the carbon to which the hydroxyl group, OH, is bonded. The other is based on the number of hydroxyl groups present per molecule.

PRIMARY, SECONDARY AND TERTIARY ALCOHOLS

A primary alcohol has one carbon attached to the carbon to which the OH group is bonded, a secondary alcohol has two carbon atoms attached and a tertiary alcohol has three carbon atoms attached. Although all alcohols have some reactions in common, their other reactions are different and depend on whether they are primary, secondary or tertiary alcohols.

For convenience, methanol is classified as a primary alcohol, even though it does not have a carbon atom bonded to the carbon atom to which the OH group is attached.

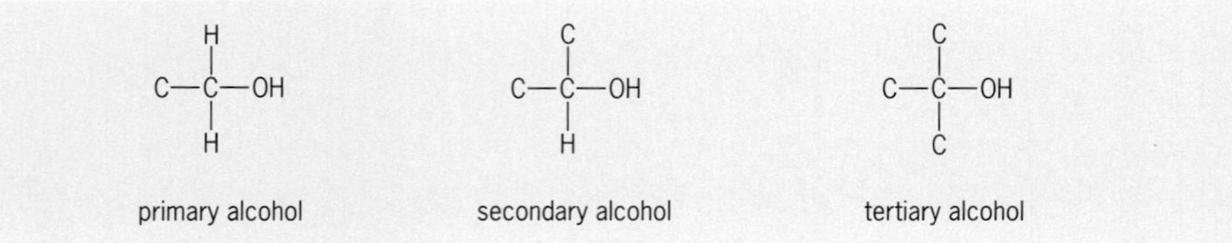

Fig 13.9 The basic structures of primary, secondary and tertiary alcohols

MONOHYDRIC, DIHYDRIC, TRIHYDRIC AND POLYHYDRIC ALCOHOLS

A monohydric alcohol has one hydroxyl group, a dihydric alcohol has two, a trihydric alcohol has three and a polyhydric alcohol has many hydroxyl groups per molecule.

Ethane-1,2-diol is a dihydric alcohol used in engine antifreeze. Glucose is a polyhydric alcohol with many hydroxyl groups per molecule.

? QUESTION 2

2 **a)** Draw the displayed formula for each of the following alcohols and state whether they are primary, secondary or tertiary:
i) propan-1-ol;
ii) 2-methylpentan-3-ol;
iii) propane-1,2,3-triol;
iv) cyclohexanol.
b) Look at the displayed formulae for geraniol, nerol and linalool in Fig 13.5. Classify each as a primary, secondary or tertiary alcohol.

? QUESTION 3

3 **a)** What type of alcohol is propane-1,3-diol?
b) Draw the displayed formula of a dihydric alcohol whose molecular formula is $C_4H_{10}O_2$.
c) Look back to the structures in Figs 13.5 and 13.6. Classify each compound as a monohydric, dihydric or polyhydric alcohol.

3 ALCOHOLS AS A HOMOLOGOUS SERIES

Methanol, ethanol, propan-1-ol, butan-1-ol and pentan-1-ol are alcohols that are members of a homologous series (see page 194). They are all primary alcohols with similar chemical properties, and can be represented by the general formula $C_nH_{2n+1}OH$.

As in all homologous series, the physical properties of each member show observable trends as the number of carbon atoms per molecule increases. Table 13.1 summarises this.

Table 13.1 Physical properties of the homologous series of primary alcohols

Alcohol	Formula	Melting point/°C	Boiling point/°C
methanol	CH_3OH	−98	65
ethanol	CH_3CH_2OH	−117	78.5
propan-1-ol	$CH_3CH_2CH_2OH$	−127	(see Question 4)
butan-1-ol	$CH_3CH_2CH_2CH_2OH$	−90	116
decan-1-ol	$CH_3(CH_2)_8CH_2OH$	6	228

BOILING POINTS

Table 13.1 shows that the boiling points of the primary alcohols $C_nH_{2n+1}OH$ increase with the carbon chain length, n. For an alcohol to boil, its molecules must overcome the two types of intermolecular force that keep them together. One is the weak induced dipole–induced dipole force that operates between all molecules as a result of the temporary dipoles within the molecules. This force becomes stronger as the molecule becomes larger and contains more electrons. The induced dipole–induced dipole force in alcohols is the attraction between the alkyl part of one molecule and the alkyl part of another molecule.

The other intermolecular force is the much stronger **hydrogen bond**. The hydroxyl group has a polar covalent bond in which the oxygen atom is slightly negative ($\delta-$) and the hydrogen atom is slightly positive ($\delta+$). This is because the oxygen is much more electronegative than hydrogen. This results in the molecule having a permanent dipole. The negative oxygen atom attracts the positive hydrogen atom of the hydroxyl group of another molecule, forming the hydrogen bond.

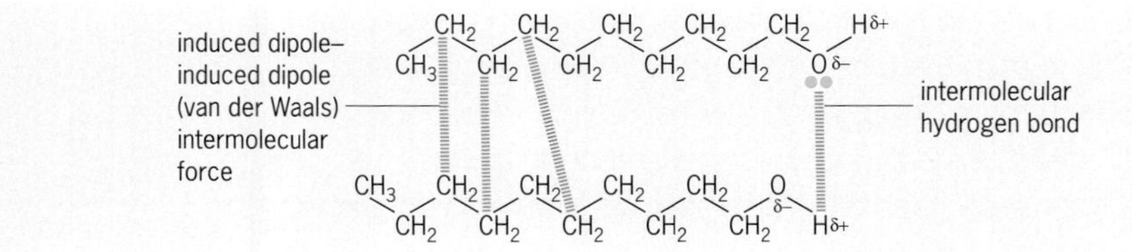

Fig 13.11 Hydrogen bonding and induced dipole–induced dipole forces in decan-1-ol. The intermolecular forces of attraction in decan-1-ol are induced dipole–induced dipole forces between alkyl groups and hydrogen bonds between hydroxyl groups. With a long-chain alkyl group, the strength of the induced dipole–induced dipole force is nearly the same as that in the corresponding alkane

SOLUBILITY IN WATER

The solubility of the primary alcohols $C_nH_{2n+1}OH$ decreases as the number of carbon atoms per molecule increases. Methanol and ethanol mix completely with water in all proportions, whereas nonan-1-ol, in which $n = 9$, is virtually insoluble in water. For a molecule to dissolve in water, it must be able to interact with water molecules. Water is a polar solvent, so the two hydroxyl groups in a water molecule can form intermolecular hydrogen bonds with methanol or ethanol molecules.

REMEMBER THIS

Methylated spirits (meths) is used as a solvent. It consists mostly of ethanol, not methanol as the name suggests.

For more information on induced dipole–induced dipole forces, see pages 147 and 195.

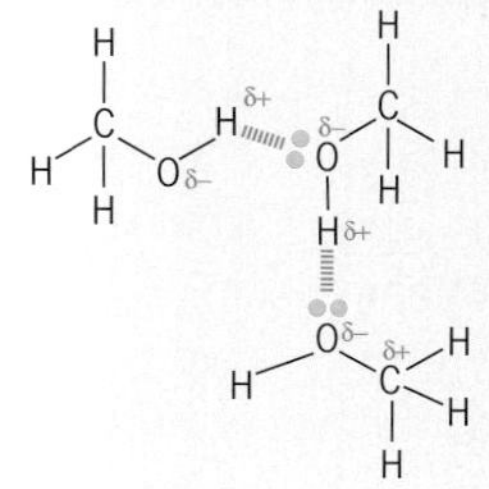

Fig 13.10 Hydrogen bonding in methanol

for information on hydrogen bonding in water, see page 129.

? QUESTIONS 4–5

4 Using data from Table 13.1, estimate the boiling point of propan-1-ol.

5 **a)** Explain why the boiling point of butan-1-ol is higher than that of methanol.

b) Explain why the boiling points of decane and decan-1-ol are quite similar, but those of methane and methanol are very different.

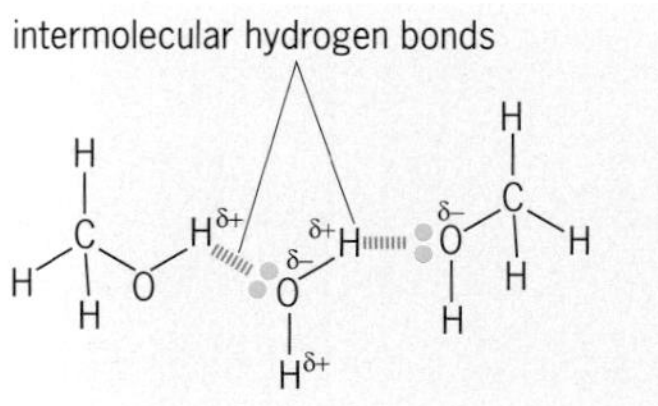

Fig 13.12 Intermolecular hydrogen bonds between methanol and water

There is more about dissolving in Chapters 21 and 22.

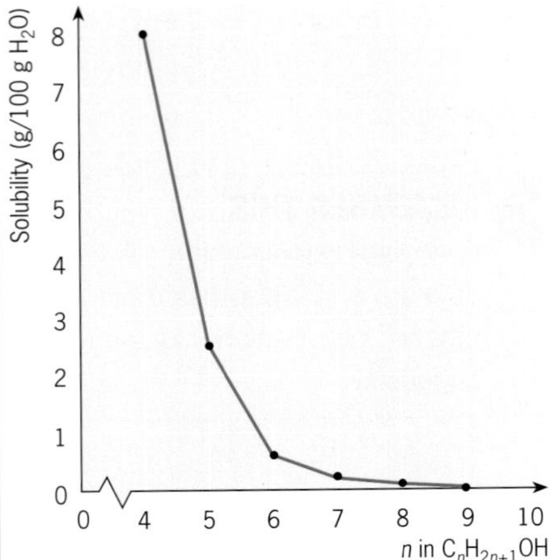

Fig 13.13 Variation of solubility of the homologous series of primary alcohols in water

? QUESTIONS 6–7

6 Is methanol more soluble in an alkane solvent, such as nonane, than in water? What is the case with nonan-1-ol?
Explain your answers, using ideas about intermolecular forces.

7 Describe the structural differences between the ring form of glucose in Fig 13.15 and the chain form in Fig 13.14.

In the case of nonan-1-ol, its hydroxyl groups form hydrogen bonds with the water molecules, but its long carbon chains form induced dipole–induced dipole interactions only with other nonan-1-ol molecules. Therefore, the predominant intermolecular attractions are between nonan-1-ol's molecules rather than between nonan-1-ol and water molecules.

SUGARS

Sugars are a group of naturally occurring polyhydric alcohols that normally have a carbon chain or ring of six or five carbon atoms. They also contain one ketone group or one aldehyde group (see Chapter 16).

glucose   fructose

Fig 13.14 Structures of two simple sugars

Fig 13.15 Intermolecular forces between a glucose molecule and water molecules. (For clarity, an interaction is not shown at every hydroxyl group)

Glucose exists in both a chain form (Fig 13.14) and a ring form (Fig 13.15). Both forms have a molecule with five hydroxyl groups. It readily dissolves in water because its hydroxyl groups can form hydrogen bonds with water molecules. It is this solubility that enables us to taste sugar in food or drink. There is more information on sugars and other carbohydrates on page 324.

SCIENCE IN CONTEXT

Artificial sweeteners

Diabetics do not produce sufficient insulin to change the excess glucose in their blood into glycogen, a storage carbohydrate. So they must carefully regulate their intake of glucose to stop the blood sugar level from getting too high. But like other people, diabetics like sweet-tasting foods. The solution is to use artificial sweeteners, such as sorbitol, which is used to sweeten diabetic chocolate and other confectionery.

Fig 13.16 The structure of sorbitol

The 3-D shape of the sorbitol molecule is very like that of the glucose molecule. The receptors in the tongue respond to sorbitol in the same way that they respond to glucose, and so sorbitol tastes sweet. However, despite sorbitol's similar taste to glucose, it is not metabolised by the body, and a diet too rich in sorbitol can cause diarrhoea.

Sorbitol is also used to sweeten dry wines. Adding extra sugar would sweeten the wine, but would also promote unwanted additional fermentation with the production of carbon dioxide, and the pressure could burst the bottle. Since sorbitol is not metabolised by yeast, further fermentation is not a problem.

The market for artificial sweeteners goes beyond helping diabetics and wine-making. Many low-calorie products have now replaced sugar as an ingredient. However, many artificial sweeteners chemically decompose with heating, so they cannot be used to sweeten products that are cooked, such as cake mixtures.

Artificial sweeteners (Cont.)

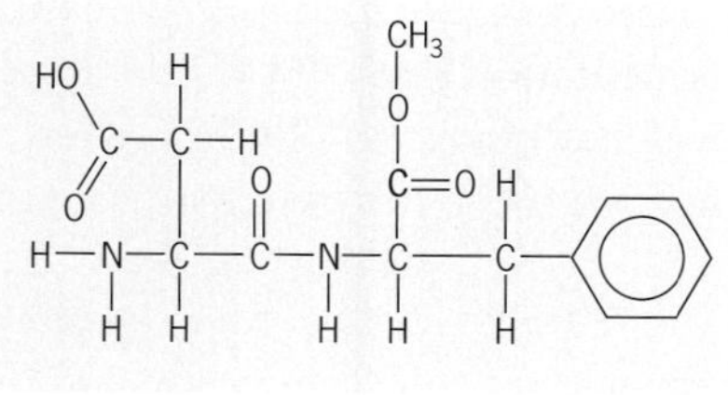

Fig 13.17 The structure of aspartame. Notice that it is completely different from the structure of glucose in Figs 13.14 and 13.15

Most artificial sweeteners are much sweeter than sugar, so a much smaller mass of the substance will be needed. Saccharin, for example, is about 300 times sweeter than sugar, though it has a bitter aftertaste. Another sweetener is aspartame, which is 200 times sweeter than sugar, but some people experience unpleasant effects from it, such as headaches and nausea.

Molecules of artificial sweeteners have a different shape from those of the simple sugars, so chemists are not yet sure which molecular shapes are responsible for the sweetness. Recent research has attempted to make sugar-like molecules by substituting some OH groups with chlorine atoms. But there must be extensive testing before such products can be put into foods, in order to ensure that they are completely free of side effects.

QUESTION 8

8 **a)** Describe the structural differences between sorbitol and glucose.

b) Why does sorbitol dissolve in water?

4 PREPARATION OF ALCOHOLS

There are five ways to make alcohols, as summarised in Fig 13.18. It follows that alcohols feature in a large number of reactions, which makes them good synthetic intermediates.

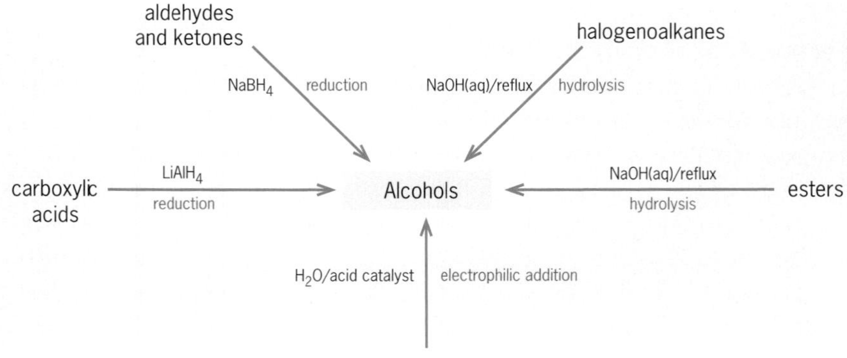

Fig 13.18 Preparative routes to alcohols

Further information on the processes given in Fig 13.18 can be found in this book, as follows: Hydrolysis of halogenoalkanes, page 250; Reduction of aldehydes and ketones, page 322; Reduction of carboxylic acids, page 347; Hydrolysis of esters, page 357.

MANUFACTURE OF ETHANOL

Ethanol is a compound in great demand as a solvent, a fuel and a synthetic intermediate in the production of other chemicals. It is, of course, present in alcoholic drinks, for which it is produced by the fermentation of glucose contained in grapes, for example.

Fermentation of sugars

Ethanol is one of the oldest manufactured chemicals, going back to the time when someone is thought to have first accidentally discovered the fermentation process – possibly 5000 years ago.

During fermentation (anaerobic respiration), aqueous glucose is converted into carbon dioxide and aqueous ethanol. The reaction is catalysed by the enzymes present in yeast:

$$C_6H_{12}O_6(aq) \rightarrow 2CH_3CH_2OH(aq) + 2CO_2(g)$$

You can read more about enzymes and denaturing on page 385.

Fig 13.19 Yeast floating on the liquid of a brewing fermentation vat

There is information on the use of fermentation to produce an alternative fuel on page 155.

Further information on the addition reactions of ethene is given in Chapter 14.

? QUESTION 9

9 **a)** Write down the equation for the addition reaction between concentrated sulfuric acid and ethene.
b) Write down the equation for the hydrolysis of ethyl hydrogensulfate to give ethanol.
c) What is the name of the alcohol produced when cyclopentene is hydrated?

The reaction is carried out between temperatures of 25 °C and 45 °C; any warmer and the enzymes in yeast are denatured and no longer function as a catalyst. It is also important that oxygen is not present, because if it is ethanol will be oxidised to produce ethanoic acid. Fermentation does not produce pure ethanol, but instead a dilute solution of ethanol is formed. To obtain higher concentrations of ethanol, the aqueous ethanol has to be fractionally distilled.

Hydration of ethene

We have described the manufacture of alkenes including ethene from the cracking of larger hydrocarbon molecules in Chapter 8. Large quantities of ethene are used to manufacture ethanol in a reaction known as hydration. This is achieved by reacting ethene and steam together at 300 °C and 6000 kPa (60 atmospheres) pressure in the presence of a phosphoric(V) acid catalyst:

$$C_2H_4(g) + H_2O(g) \rightarrow CH_3CH_2OH(g)$$

Ethene can also be hydrated in the laboratory, but different conditions must be used. Ethene is bubbled through concentrated sulfuric acid and undergoes an addition reaction to form ethyl hydrogensulfate, which is hydrolysed to give ethanol:

$$C_2H_4 + H_2SO_4 \longrightarrow CH_3CH_2OSO_3H$$

$$CH_3CH_2OSO_3H + H_2O \xrightarrow{60\,°C} H_2SO_4 + CH_3CH_2OH$$

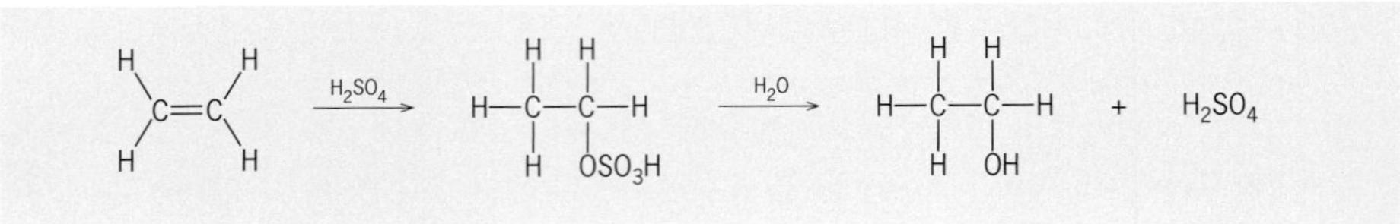

$$\mathrm{H_2C{=}CH_2} \xrightarrow{H_2SO_4} \mathrm{H{-}CH_2{-}CH_2(OSO_3H)} \xrightarrow{H_2O} \mathrm{H{-}CH_2{-}CH_2(OH)} + H_2SO_4$$

Fig 13.20 Flow diagram showing hydration of ethene

Fermentation or hydration?

In the manufacture of ethanol by fermentation and by hydration, each method has its advantages and disadvantages. Ethanol made by fermentation is a renewable resource, since the sugar required for the process comes from plants that can be grown again in the following season to make more sugar. However, the hydration process uses ethene that is made by cracking of long-chain hydrocarbons obtained from crude oil, which is a non-renewable resource. Fermentation is a batch process, so the quality of the product cannot be controlled as easily as in the continuous process of hydration. The batch process of fermentation allows a chemical company to match supply with demand fairly easily, since extra batches can be included if there is a need for extra ethanol.

Hydration of ethene produces pure ethanol, whereas fermentation produces a dilute solution of ethanol, which must be purified by distillation.

STRETCH AND CHALLENGE

The manufacture of methanol

Methanol is used in industry and in the laboratory as a solvent and reagent; also as car antifreeze and as an engine fuel. Though an alcohol, methanol is poisonous and can cause blindness if inhaled or drunk.

Until about 1925, methanol was made by heating wood in the absence of air, and collecting and condensing the volatile substances given off. This is why methanol is sometimes called wood spirit. It is now manufactured by the reaction of carbon monoxide or carbon dioxide with hydrogen. Both reactions are carried out at about 400 °C and under a pressure of some 1500 kPa, using heterogeneous catalysts (for more about heterogeneous catalysts, see Chapter 26):

$$CO(g) + 2H_2(g) \rightarrow CH_3OH(g)$$

STRETCH AND CHALLENGE

Alternative routes to hydrocarbon fuels

The catalysed conversion of either carbon monoxide or carbon dioxide into methanol provides a route (at least in theory) for converting the products of combustion of a fuel into a practical liquid fuel. The methanol itself can be further converted into other useful products.

These synthetic routes provide a way for alkanes and alkenes to be synthesised without the need of a raw material such as crude oil. Perhaps more research will be needed in this direction if people are going to continue to rely on hydrocarbons as an energy source and as a raw material for the petrochemical industry.

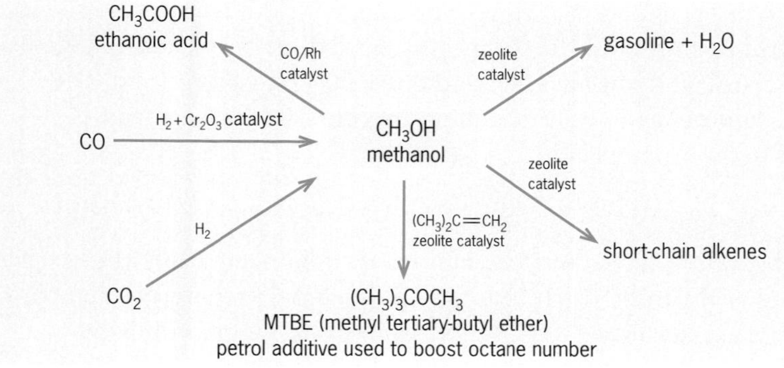

Fig 13.21 The catalytic conversion of methanol

? QUESTION 10

10 **a)** Write down the equation for the reaction of carbon dioxide with hydrogen to make methanol.
b) Suggest possible sources for the carbon dioxide and hydrogen needed to synthesise methanol.
c) Discuss whether gasoline made by the route shown in Fig 13.21 is renewable or non-renewable.

5 REACTIONS OF ALCOHOLS

THE NATURE OF THE HYDROXYL FUNCTIONAL GROUP

The hydroxyl functional group in an alcohol contains one covalent bond and is attached to a carbon atom by another covalent bond. This means that the functional group can undergo two types of reaction. The first involves the breaking of the carbon–oxygen (C–O) single bond, and the second involves the breaking of the oxygen–hydrogen (O–H) single bond.

The C–O bond can be broken either by a substitution reaction or by an elimination reaction.

When the O–H bond is broken, an alcohol behaves as an acid. This is because the breaking of the bond results in the transfer of a hydrogen ion. Since the oxygen atom of the hydroxyl group has two lone pairs of electrons, an alcohol can also behave as a nucleophile, reacting with a molecule that has an electron-deficient centre. Finally, an alcohol can take part in redox reactions.

The large range of reactions available to this functional group helps to explain why alcohols are very good synthetic intermediates.

The competition between substitution and elimination in halogenoalkanes is discussed on pages 252–253.

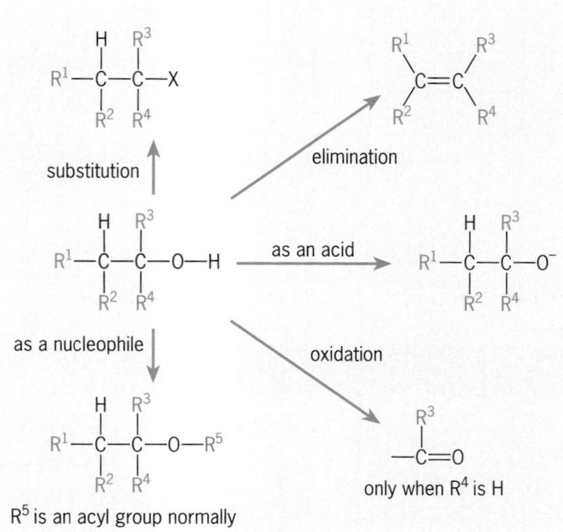

Fig 13.22 General summary of the reactions of alcohols. R^1, R^2, R^3 and R^4 are either alkyl groups or hydrogen atoms

SUBSTITUTION REACTIONS: HALOGENATION OF ALCOHOLS

The conversion of an alcohol into a halogenoalkane is a substitution reaction that involves the swapping of a halogen atom for a hydroxyl group. However, a halide ion is not a good enough nucleophile to substitute a hydroxyl group directly. The strength of the C–O bond is a great barrier to the substitution reaction.

The role of bond energy in nucleophilic substitution is covered on pages 247–248.

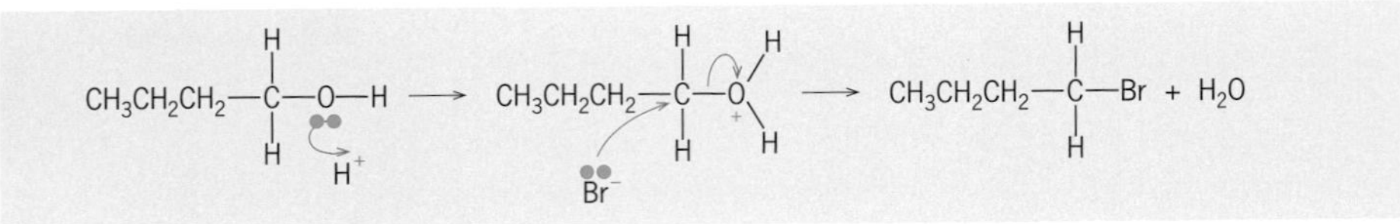

$$CH_3CH_2CH_2{-}CH_2{-}O{-}H \xrightarrow{H^+} CH_3CH_2CH_2{-}CH_2{-}O^+H_2 \xrightarrow{Br^-} CH_3CH_2CH_2{-}CH_2{-}Br + H_2O$$

Fig 13.23 The mechanism of the substitution of an alcohol using concentrated hydrobromic acid

The only way that the substitution reaction can succeed is if the C–O bond is substantially weakened. An acidic catalyst can be used to do this.

A lone pair of electrons on the oxygen atom is donated to a proton in an acid–base type interaction. The intermediate formed has a very weak C–O bond, since the oxygen has developed a positive charge. Nucleophilic substitution by a bromide ion is now very easy, with the water molecule as the leaving group. Normally, the reaction is carried out by refluxing together a mixture of alcohol, sodium bromide and concentrated sulfuric acid. The concentrated sulfuric acid and the sodium bromide together generate concentrated hydrobromic acid:

$$H_2SO_4 + NaBr \rightarrow HBr + NaHSO_4$$

For the substitution of the hydroxyl group by a halogen atom, the corresponding acid can be used. Iodo compounds are prepared by refluxing the alcohol with concentrated hydroiodic acid, bromo compounds by refluxing with concentrated hydrobromic acid and a trace of concentrated sulfuric acid as a catalyst, and chloro compounds by refluxing with concentrated hydrochloric acid using zinc chloride as a catalyst.

There are alternative ways to achieve this substitution. All of them involve weakening the C–O bond by a reaction with a phosphorus atom. Typically, phosphorus(V) chloride, phosphorus(III) bromide, or a mixture of phosphorus and iodine (essentially phosphorus(III) iodide) is refluxed with the alcohol to obtain the corresponding halogenoalkane. Representing an alcohol by the formula ROH, where R is an alkyl group, the equations for these three reactions are:

$$PCl_5 + ROH \rightarrow RCl + POCl_3 + HCl$$

$$PBr_3 + 3ROH \rightarrow 3RBr + H_3PO_3$$

$$PI_3 + 3ROH \rightarrow 3RI + H_3PO_3$$

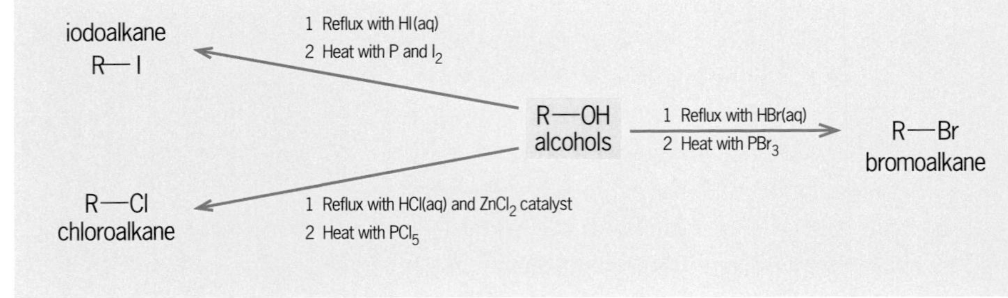

Fig 13.24 Summary of the preparation of halogenoalkanes from alcohols

Phosphorus(V) chloride for working out structures

Phosphorus(V) chloride reacts with hydroxyl groups in alcohols and carboxylic acids to produce fumes of hydrogen chloride, which are easily detected. So this reaction is used to analyse compounds to identify hydroxyl groups and establish whether they are alcohols or carboxylic acids.

? QUESTION 11

11 Predict the products of the following reactions:

a) pentan-2-ol refluxed with concentrated hydrobromic acid and a trace of concentrated sulfuric acid;

b) cyclohexanol refluxed with concentrated hydroiodic acid;

c) propan-1-ol refluxed with concentrated hydrochloric acid and zinc chloride catalyst.

? QUESTION 12

12 How would you prepare:

a) iodoethane;

b) 2-chloropentane;

c) 2-iodo-3-methylhexane?

? QUESTION 13

13 A compound has the formula C_2H_6O. It does not give fumes of hydrogen chloride when reacted with phosphorus(V) chloride. Suggest a structural formula for the compound.

ELIMINATION REACTIONS: DEHYDRATION TO FORM ALKENES

As we have already described, alcohols can react by substitution and elimination. The elimination reaction is more frequently referred to as dehydration, since it involves the loss of a molecule of water as well as the formation of an alkene (Fig 13.25). In dehydration, an alcohol is heated with a suitable catalysts such as concentrated sulfuric acid or concentrated phosphoric acid. Alternatively, the hot alcohol vapour is passed over a heated aluminium oxide catalyst.

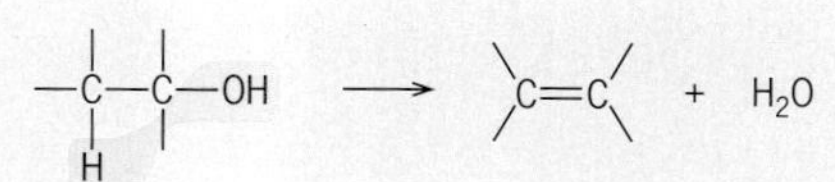

Fig 13.25 The dehydration of an alcohol to form an alkene

There is more on the dehydration of alcohols on page 280.

Laboratory dehydration

In the laboratory, it is convenient to take an alcohol and reflux it with a dehydrating agent. Concentrated sulfuric acid can be used, but concentrated phosphoric(V) acid is usually preferred, because it minimises the number of side products formed. Fig 13.26 shows the dehydration of cyclohexanol to give cyclohexene.

With some alcohols, several alkenes may be produced, but normally the most-substituted alkene is the major product.

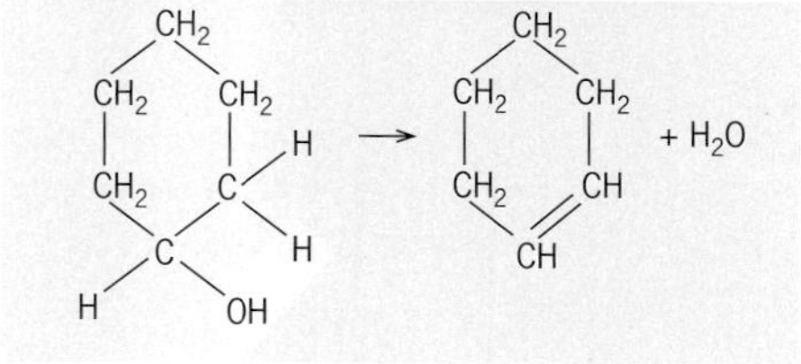

Fig 13.26 The dehydration of cyclohexanol to give cyclohexene

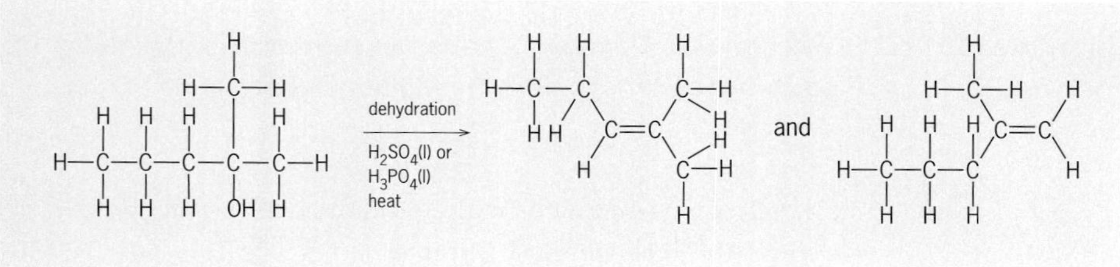

Fig 13.27 The dehydration products of 2-methylpentan-2-ol

QUESTION 14

14 **a)** Draw the dehydration product of cyclohexanol.

b) Draw all the alkenes that can be made by the dehydration of butan-2-ol.

c) Which of these alkenes is likely to be formed in the greatest amount?

6 OXIDATION OF ALCOHOLS

As a working definition in organic chemistry, *oxidation* can be considered as either the gain of an oxygen atom by a substance or the loss of two hydrogen atoms from it.

Since reduction is the chemical opposite of oxidation, *reduction* can be considered as either the gain of two hydrogen atoms or the loss of one oxygen atom.

These atoms are always denoted by [O] and 2[H]. The square brackets signify that we are not referring to atomic hydrogen or oxygen. The substance that supplies the oxygen, [O], and/or removes the hydrogen, 2[H], is called the **oxidising agent** or oxidant. The **reducing agent** or reductant removes the oxygen, [O], and/or supplies the hydrogen, 2[H].

To read about redox reactions in more detail, look at Chapter 5.

OXIDATION OF PRIMARY ALCOHOLS INTO ALDEHYDES

Consider butan-1-ol (Fig 13.28). The removal of two hydrogen atoms produces a new functional group – the aldehyde group, CHO.

In the oxidation of a primary alcohol, the reagent is acidified aqueous potassium dichromate(VI), a mixture of aqueous potassium dichromate(VI) and dilute sulfuric acid.

The conditions must be chosen to prevent any subsequent oxidation of the aldehyde to form a carboxylic acid. Either the reaction is carried out at room temperature and the oxidising agent is added to the alcohol so that the oxidising agent is never in excess, or the reagents are heated and the aldehyde product is allowed to distil out from the reaction vessel before it can be oxidised further.

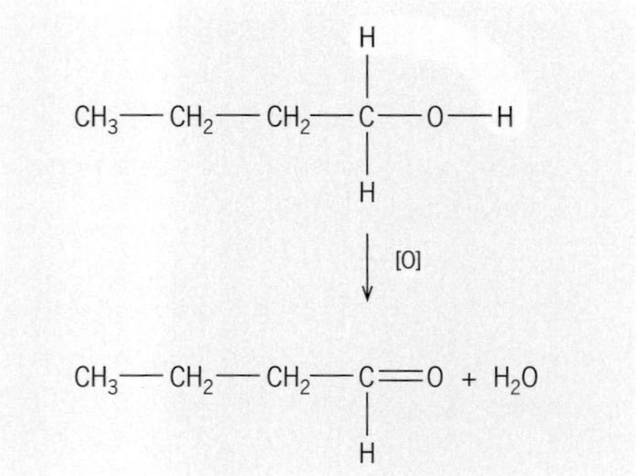

Fig 13.28 The formation of butanal from butan-1-ol

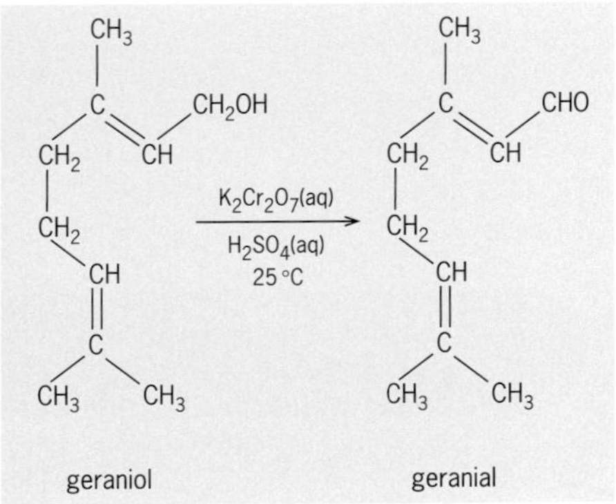

CH_3 CH_2OH CH_2 CH CH_2 CH C CH_3 CH_3 — $K_2Cr_2O_7(aq)$, $H_2SO_4(aq)$, 25 °C → CH_3 CHO CH_2 CH CH_2 CH C CH_3 CH_3

geraniol geranial

Fig 13.29 The oxidation of geraniol to geranial completely changes the odour of the compound. Geraniol has a rose scent and geranial that of a citrus fruit

During the oxidation, orange dichromate(VI) ions are reduced to form blue–green chromium(III) ions. The equation can be written as:

$$RCH_2OH + [O] \rightarrow RCHO + H_2O$$

In this equation, the oxidising agent is not given in full, but is represented by the oxygen [O] that it supplies to remove the two hydrogen atoms. This is a much simpler expression of the reaction than the full stoichiometric or ionic equation for the reaction.

The oxidation of ethanol by acidified potassium dichromate(VI) to give ethanal can be written as:

$$CH_3CH_2OH + [O] \rightarrow CH_3CHO + H_2O$$

STRETCH AND CHALLENGE

Stoichiometric and half equations

The stoichiometric equation for the oxidation of a primary alcohol is very complicated and tends to hide the processes taking place. For example, in oxidation by potassium dichromate(VI):

$$3RCH_2OH + K_2Cr_2O_7 + 4H_2SO_4 \rightarrow 3RCHO + K_2SO_4 + Cr_2(SO_4)_3 + 7H_2O$$

Even the slightly simpler ionic equation is still complicated:

$$3RCH_2OH + Cr_2O_7^{2-} + 8H^+ \rightarrow 3RCHO + 2Cr^{3+} + 7H_2O$$

It is easier to look at the oxidation half equation for the reaction (see Chapter 21). During the reaction, the alcohol loses two electrons, as in the half equation:

$$RCH_2OH \rightarrow RCHO + 2H^+ + 2e^-$$

The electrons supplied by the oxidation of the alcohol reduce the dichromate(VI) ion.

$$Cr_2O_7^{2-} + 14H^+ + 6e^- \rightarrow 2Cr^{3+} + 7H_2O$$

? QUESTIONS 15–18

15 Write an equation to show that pentan-1-ol is oxidised to the aldehyde pentanal.

16 Draw the structure of the products obtained when the two primary alcohols in Fig 13.5 are oxidised by acidified potassium dichromate(VI) at room temperature.

17 Draw the displayed formula of the product of the reaction of acidified potassium dichromate(VI) with 2-methylhexan-1-ol, at room temperature.

18 **a)** Write an equation to show the oxidation of butan-1-ol by refluxing with acidified potassium dichromate(VI).

b) Draw the displayed formula of the product of the reaction of 2-methyl-hexan-1-ol with hot acidified potassium dichromate(VI).

OXIDATION OF PRIMARY ALCOHOLS INTO CARBOXYLIC ACIDS

By changing the conditions of oxidation, it is possible to remove two hydrogen atoms and gain one oxygen atom to produce a carboxylic acid. The oxidation is represented by:

$$RCH_2OH + 2[O] \rightarrow RCO_2H + H_2O$$

The alcohol is refluxed with acidified potassium dichromate(VI). The acid used is dilute sufuric acid.

Ethanol is oxidised to give ethanoic acid by refluxing with acidified potassium dichromate(VI). The reaction can be represented as:

$$CH_3CH_2OH + 2[O] \rightarrow CH_3COOH + H_2O$$

The half equation for the oxidation of ethanol into the carboxylic acid involves the loss of four electrons from ethanol. This corresponds to the gain of [O] and the loss of 2[H]:

$$H_2O + CH_3CH_2OH \rightarrow CH_3COOH + 4H^+ + 4e^-$$

Ethanol can sometimes be oxidised by atmospheric oxygen. This can happen during fermentation, giving the ethanol a vinegary taste:

$$CH_3CH_2OH + O_2 \rightarrow CH_3COOH + H_2O$$

Fig 13.30 The ethanoic acid in wine vinegars is obtained by the oxidation of ethanol in air

OXIDATION OF SECONDARY ALCOHOLS INTO KETONES

Secondary alcohols can be oxidised to give ketones rather than aldehydes. The alcohol is refluxed with the acidified potassium dichromate(VI). Sulfuric acid is used to acidify the reagent.

The reaction involves the removal of two hydrogen atoms. So, for example, propan-2-ol is oxidised to give propanone as follows:

$$3CH_3CHOHCH_3 + Cr_2O_7^{2-} + 8H^+ \rightarrow 3CH_3COCH_3 + 2Cr^{3+} + 7H_2O$$

Notice that we have used the ionic formulae and left out the potassium ions, which remain unchanged during the reaction. The half equation for the oxidation of a secondary alcohol involves two electrons and is similar to the half equation for a primary alcohol being oxidised to give an aldehyde. Another, simpler way to represent the oxidation of propan-2-ol is:

$$CH_3CHOHCH_3 + [O] \rightarrow CH_3COCH_3 + H_2O$$

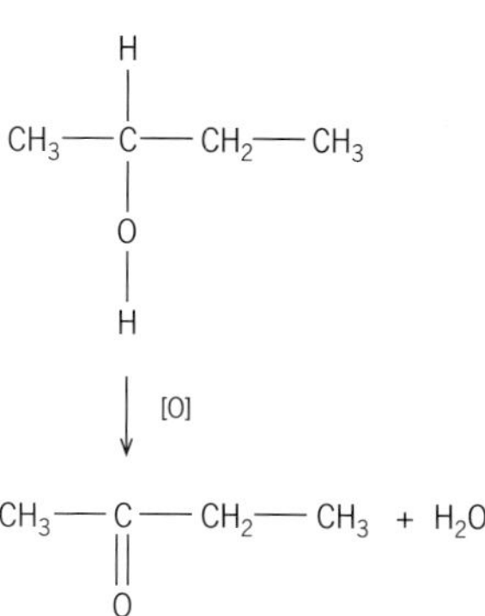

Fig 13.31 Equation describing the oxidation of butan-2-ol

HOW SCIENCE WORKS

Cutting costs to make propanone

Propanone is an important industrial and domestic solvent. It is manufactured from propan-2-ol, but acidified potassium dichromate(VI) is not used because the reagents are expensive and the special conditions required are difficult to maintain in a large industrial plant. Instead, propan-2-ol vapour is passed over a copper catalyst at 300 °C, which removes hydrogen molecules in a reaction called dehydrogenation (Fig 13.32).

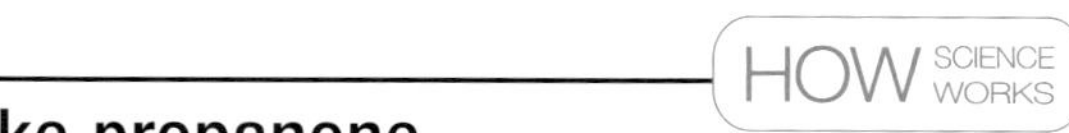

Fig 13.32 The dehydrogenation of propan-2-ol

OXIDATION OF TERTIARY ALCOHOLS

It is impossible to oxidise a tertiary alcohol without breaking a carbon–carbon bond. Acidified potassium dichromate(VI) has no effect on tertiary alcohols. (Look back to Fig 13.9 for the structure of a tertiary alcohol.)

DISTINGUISHING PRIMARY, SECONDARY AND TERTIARY ALCOHOLS

Since the three classes of alcohols give different reactions with acidified potassium dichromate(VI) they can be distinguished from one another by the products and colour changes in these reactions, as Table 13.2 shows.

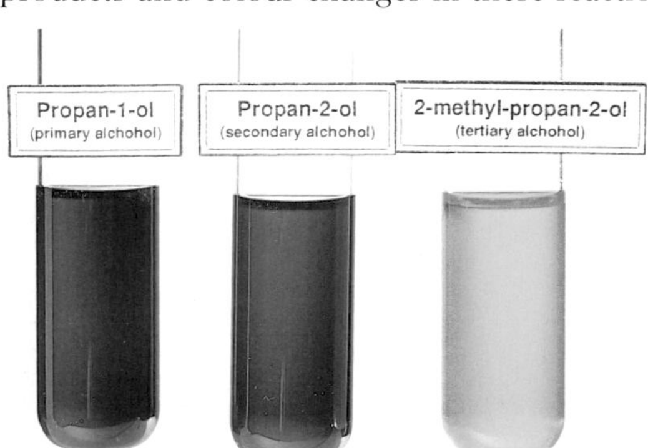

Fig 13.33 During the oxidation of primary and secondary alcohols using acidified potassium dichromate(VI), the colour changes from **orange** to **green,** but **tertiary alcohols do not change**

Table 13.2 Distinguishing alcohols

Alcohol	Acidified potassium dichromate(VI) at room temperature		Refluxing under heating with acidified potassium dichromate(VI)	
	Product	Observation	Product	Observation
primary	aldehyde	orange to green	carboxylic acid	orange to green
secondary	ketone	orange to green	ketone	orange to green
tertiary	no reaction	stays orange	no reaction	stays orange

? QUESTIONS 19–20

19 **a)** What is the name of the organic product when cyclohexanol reacts with acidified potassium dichromate(VI)?
b) Write an equation for the oxidation of butan-2-ol by acidified dichromate(VI) ions.
(Hint: use the given equation and substitute butan-2-ol for propan-2-ol.)

20 Write down a possible product from the reaction of vitamin C (Fig 13.3) with acidified potassium dichromate(VI).

? QUESTION 21

21 Name the product, if any, when each one of the following is mixed with acidified potassium dichromate(VI) at room temperature:
a) butan-1-ol; **b)** butan-2-ol;
c) 2-methylpropan-2-ol.

Tests to distinguish between aldehydes and ketones are covered on pages 320–324.

? QUESTION 22

22 **a)** Why is it important to develop carbon-neutral biofuels?
b) Methanol can also be used as a fuel.
i) Write an equation to show the complete combustion of methanol.
ii) Is methanol a carbon-neutral fuel? Explain your answer in detail.

? QUESTION 23

23 A straight-chain alcohol with the formula $C_5H_{11}OH$ reacts with acidified potassium dichromate(VI) to give a ketone and, in a separate test, also gives a yellow precipitate with alkaline aqueous iodine. Suggest an identity for the alcohol that is consistent with this data.

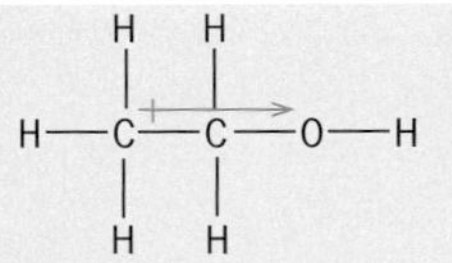

$pK_a = 15.5$

The methyl group is electron releasing compared with a hydrogen atom, so it strengthens the O–H bond

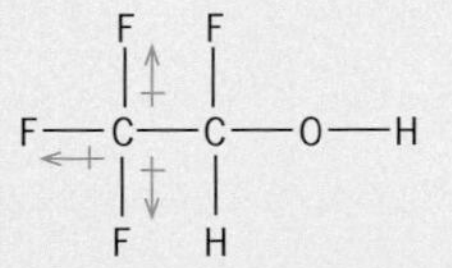

$pK_a = 12.4$

Fluorine atoms are highly electronegative and withdraw electrons from the O–H bond, which weakens it

Fig 13.34 Structures and pK_a values for two alcohols: the larger the value of pK_a the weaker the acid. (Read about pK_a on page 332.)

? QUESTION 24

24 Explain why 2-chloroethanol is a stronger acid than ethanol.
(Hint: what is the effect of the very electronegative chlorine atom on the strength of the O—H bond?)

Read more about the strengths of acids and weak acids in Chapter 17.

? QUESTION 25

25 Write down the equation for the reaction of methanol with sodium. What is the name of the alkoxide produced?

COMBUSTION OF ALCOHOLS

Simple alcohols such as methanol or ethanol will combust completely to form carbon dioxide and water. The presence of the oxygen atom in the molecule of methanol or ethanol means that less oxygen is needed for complete combustion than with the corresponding alkanes.

$$C_2H_5OH + 3O_2 \rightarrow 2CO_2 + 3H_2O$$

Alcohols with many carbon atoms per molecule are much more likely to burn incompletely in air producing a dirty flame and lots of soot.

Ethanol is a possible biofuel because it can be made by fermentation of sugars from plants. The combustion products, carbon dioxide and water, can be incorporated into plant carbohydrate by photosynthesis. It is therefore possible for ethanol to be a carbon-neutral biofuel, having no net effect on carbon dioxide concentrations in the air.

TRI-IODOMETHANE TEST FOR CH_3CHOH GROUP

Alkaline aqueous iodine reacts with ethanol and secondary alcohols containing the CH_3CHOH group to give a yellow precipitate of tri-iodomethane. This reaction is used as a chemical test in analysis and identification (see also page 325).

7 ALCOHOLS AS ACIDS

RELATIVE ACIDITY OF ALCOHOLS

An acid is a proton donor. An alcohol can behave as an acid because it can donate a proton if the O–H bond is broken. So, the weaker the O–H bond is, the stronger is the acidity of the alcohol. The strength of the O–H bond can be changed by changing the atoms attached to it. For example, electron-releasing groups strengthen the O–H bond and thereby make the alcohol a weaker acid.

Alkyl groups are electron releasing compared with the hydrogen atom, so it follows that the order of increasing acid strength in alcohols is: tertiary, secondary, primary (strongest). Water, which also contains the O–H bond, is a stronger acid than primary alcohols.

REACTION OF SODIUM WITH ALCOHOLS

Sodium reacts violently with water to form hydrogen and aqueous sodium hydroxide. In the reaction, an O–H bond is broken:

$$2H_2O(l) + 2Na(s) \rightarrow 2NaOH(aq) + H_2(g)$$

Ethanol reacts similarly with sodium, to produce hydrogen and a compound called sodium ethoxide. In the reaction, the sodium fizzes and produces a colourless solution. The reaction is slower than that of sodium with water, because the O–H bond in ethanol is harder to break than the O–H bond in water:

$$2C_2H_5OH + 2Na \rightarrow 2C_2H_5O^-Na^+ + H_2$$

Notice that the two equations show the same pattern.

Sodium reacts with other alcohols, to produce hydrogen and what is known as an alkoxide. The name of the alkoxide is derived from the alkyl chain of the alcohol:

$$2ROH + 2Na \rightarrow 2RO^-Na^+ + H_2$$

The alcohol in this example acts as a very weak acid, since it loses a proton. RO^-Na^+ is known as a sodium alkoxide.

8 ALCOHOLS AS NUCLEOPHILES

The oxygen atom of the OH group can donate a pair of electrons to make a covalent bond, which allows the group to act as a nucleophile.

Alcohols are poor nucleophiles, but they can react with electron-deficient centres provided the deficiency is sufficiently high. Typically, alcohols behave as nucleophiles in their reactions with carboxylic acids or their derivatives. Probably the most important reaction is that with a carboxylic acid to form an **ester**. Esters are sweet-smelling substances used in perfumes and in food flavourings.

REMEMBER THIS

Heterolytic fission involves breaking a covalent bond so that one atom receives both of the shared pair of electrons from the covalent bond.

The reactions of alcohols with carboxylic acids and their derivatives are described in more detail in Chapter 18.

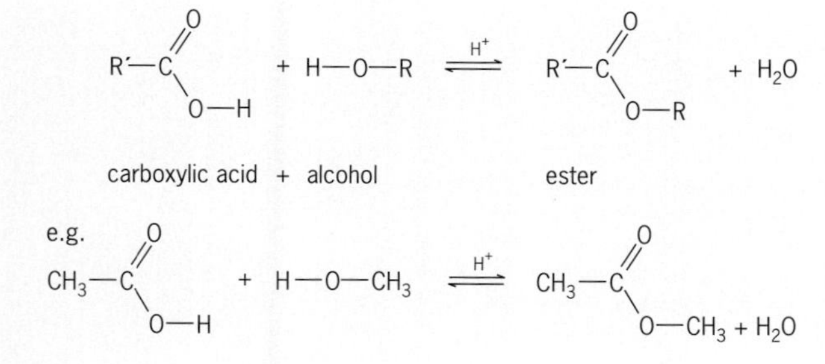

$$R'COOH + H{-}O{-}R \xrightleftharpoons{H^+} R'COOR + H_2O$$

carboxylic acid + alcohol → ester

e.g.

$$CH_3COOH + H{-}O{-}CH_3 \xrightleftharpoons{H^+} CH_3COOCH_3 + H_2O$$

Fig 13.35 The esterification reaction

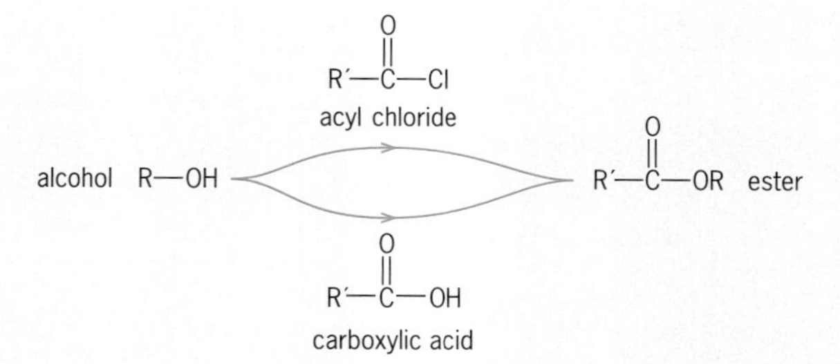

Fig 13.36 Reaction network showing an alcohol behaving as a nucleophile

SUMMARY

After studying this chapter, you should know:

- Alcohols contain the hydroxyl, OH, functional group attached directly to a carbon atom.
- Short-chain alcohols are soluble in water, but long-chain alcohols are insoluble.
- Primary alcohols can be oxidised by acidified potassium dichromate(VI) to give aldehydes. The aldehyde is often distilled from the reaction mixture as it is being made to prevent further oxidation.
- Primary alcohols can be oxidised by hot acidified potassium dichromate(VI) to give carboxylic acids.
- Secondary alcohols can be oxidised to give ketones, but tertiary alcohols cannot be oxidised.
- Alcohols completely combust to form carbon dioxide and water.
- Alcohols can be converted by substitution into halogenoalkanes, using phosphorus halides and the appropriate hydrohalic acid, HBr(aq), HCl(aq) or HI(aq).
- Alcohols can be dehydrated to give alkenes.
- Alcohols can behave as nucleophiles and react with carboxylic acids and acid chlorides to form esters.
- Many alcohols are used as solvents.
- Ethanol is manufactured by the anaerobic fermentation of aqueous glucose or by the hydration of ethene in the presence of an acid catalyst.
- Alcohols react with sodium to give sodium alkoxides.

Practice questions and a How Science Works assignment for this chapter are available at www.collinseducation.co.uk/CAS

14

Alkenes

14 ALKENES

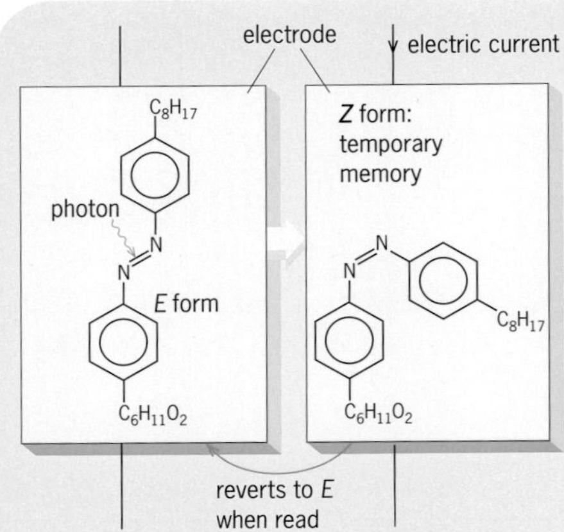

Memory molecules A memory molecule changes shape to its *Z* form when it absorbs a photon of ultraviolet light. This form could correspond to the binary digit 1 or 'on'. The *E* form of the molecule would then correspond to the binary digit 0 or 'off'

Memory molecules

Chemists are beginning to discover molecules that can be made to act like the memory of a computer, which stores binary information by switching between two states – 'on' or 'off'. The molecular shape of certain molecules that contain double bonds alters when they undergo a reversible photochemical change, and this is the 'switch' that chemists are exploiting. During the photochemical change, the molecule keeps its structure but changes its 3D shape. The two forms are called *E-Z* isomers.

When the molecule changes from an *E*-isomer into a *Z*-isomer, the distinct change in shape offers a way of storing information. For example, the *E* form could be designated 'off' (corresponding to binary digit 0) and the *Z* form designated 'on' (corresponding to binary digit 1). Given that the change easily reverses, information storage, retrieval and erasure become feasible. The hope is that, by depositing such memory molecules on an electrode, a hundred million bits of information could be stored on an area the size of a fingernail.

1 THE PETROCHEMICAL INDUSTRY

Fig 14.1 Introducing alternative energy sources, such as this wind farm, may reduce the demand for crude oil. But alternative energy sources will have to be better developed and more freely available before they supply a sizeable fraction of the energy we consume

The fractional distillation of crude oil is described on page 186.

Cracking is described on pages 187–188.

The petrochemical industry produces very many organic substances from crude oil and natural gas. About 90 per cent of the world's output of basic organic chemicals is now derived from these two fossil fuels. Petrochemicals include substances as diverse as solvents, detergents, pesticides, insecticides, polymers, plastics, drugs, paints and textiles. Without doubt, many improvements in the quality of life since the early 1900s are a direct consequence of the development of the petrochemical industry.

Dmitri Mendeleev, who originated the Periodic Table, wrote that crude oil was too valuable to be burnt as a fuel but should be used as a source for organic chemicals. That was in 1872, and many people would argue that it is still the case.

CRACKING AND THE PRODUCTION OF ALKENES

In an earlier chapter we described crude oil as a complex mixture of organic compounds. Even after the fractional distillation of crude oil, the fractions obtained are still complex mixtures (although the components of a fraction do have similar boiling points). Most of the hydrocarbons present in crude-oil fractions are alkanes, which normally are unreactive. However, alkanes can undergo the process called **cracking** to produce a variety of organic molecules that can be used to synthesise other, more generally useful, organic compounds. Cracking converts saturated hydrocarbons into unsaturated hydrocarbons. In particular, a group of versatile and important unsaturated hydrocarbons, called **alkenes**, are made and are much more reactive than alkanes.

IMPORTANCE OF ALKENES

Why are alkenes so important to the petrochemical industry? The main reason is that the carbon–carbon double bond of an alkene molecule makes it much more reactive than an alkane molecule. So, compounds with different carbon chains or with different functional groups can readily be synthesised from alkenes.

The chemistry of the alkene functional group is described in this chapter, which illustrates the wide range of substances that can be made from quite simple alkenes.

Another reason for their importance is that alkenes are readily available in large amounts from the cracking of crude oil.

SCIENCE IN CONTEXT

Alkenes and alkynes as fuels

Surplus alkenes can be used both as fuels and as starting materials for petrochemical processes. Being hydrocarbons, alkenes can burn, and so transfer energy to their surroundings.

Similarly, the alkyne ethyne (C_2H_2), which has a triple bond, is mixed with oxygen and burnt to produce an extremely high-temperature flame, which is used to cut iron and steel. The equipment is commonly known as the oxyacetylene torch (acetylene is the older name for ethyne).

$$C_2H_2(g) + 2\tfrac{1}{2}\ O_2(g) \rightarrow 2CO_2(g) + H_2O(g)$$

Fig 14.2 The temperature of an oxyacetylene flame is high enough to cut easily through steel

HOW SCIENCE WORKS

Natural gas as a source of petrochemicals

Natural gas is an abundant resource. Even if no new gas fields are discovered, present known sources should last until about 2050. So, could natural gas be used as a source of petrochemicals as well as a fuel?

Unfortunately, natural gas is mainly methane, which is a molecule that does not contain any carbon–carbon double bonds. However, much research is in progress to find ways of using methane to make alkenes to provide an alternative supply of raw materials for the petrochemical industry. For example, in a process known as an oxidative coupling, methane is reacted with oxygen at high temperatures in the presence of an oxide catalyst. In one such reaction, methane is converted into ethene – a precursor of a large number of petrochemicals:

$$2CH_4 + O_2 \rightarrow C_2H_4 + 2H_2O$$

The coupling reaction is believed to have a free-radical mechanism. As a result, the selectivity is low, so many other products are formed. Oxidative coupling has to be closely controlled since, with the wrong proportions of methane and oxygen, an explosive combustion reaction could occur.

QUESTION 1

1 When decane, $C_{10}H_{22}$, is cracked many products are formed.

a) In one cracking reaction decane forms propene, C_3H_6. Suggest the molecular formula of the other hydrocarbon produced in this reaction.

b) In another cracking reaction decane forms ethene, C_2H_4. Write a balanced equation for this cracking reaction.

QUESTION 2

2 Write down the equations for the complete combustion of ethene (C_2H_4) and of propene (C_3H_6).

QUESTION 3

3 **a)** Write down the equation for the complete combustion of methane in air.

b) Suggest how the risk of explosion during the oxidative coupling reaction can be reduced.

2 ALKENES AND ISOMERISM

Alkenes are a homologous series of unsaturated hydrocarbons that have a molecule with one carbon–carbon double bond. This means that their carbon chain has two fewer hydrogen atoms than the same carbon chain in alkanes, and has the general formula C_nH_{2n}. As already explained, it is the presence of the double bond that gives alkenes much greater reactivity than alkanes.

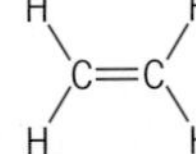

Fig 14.3 The displayed formula of ethene

NAMING ALKENES

The presence of the carbon–carbon double bond (C=C) is identified by the suffix *ene*. So, for example, ethene (Fig 14.3) has a two-carbon chain with one C=C bond, and propene has a three-carbon chain with one C=C bond. Notice that the displayed formula of ethene has been drawn with angles of about 120° between the bonds, which is their actual value. It is not essential to draw these bond angles accurately, but it is good practice, and will help you to remember them.

The name of an alkene must also specify the position of the C=C bond and the number of carbon atoms in the longest chain that contains the C=C bond. The use of *eth*, *prop*, *but*, and so on, to indicate the length of the carbon chain is described on page 193. We therefore proceed as follows:

- First, locate the longest chain that contains the double bond.
- Then, number the carbon atoms in the chain from the end that gives the lowest possible position number to the C=C bond.

Fig 14.4 shows the displayed formula of an alkene. The longest carbon chain that includes the C=C bond is shaded. It contains six carbon atoms, which means that *hex* is used to indicate the chain length. The carbon atoms must be numbered so that the C=C double bond has the lowest number possible. We say it is between carbon numbers 2 and 3 (not 4 and 5), and so the name of the compound will end with hex-2-ene. Notice that only the number of the first carbon atom of the double bond is included in this part of the name. Finally, the position of the methyl side chain must be specified in the name. In this example, the methyl group is attached to carbon 4. So the full name must be 4-methylhex-2-ene.

> **? QUESTION 4**
>
> 4 Using the number of electron pairs around a carbon atom, explain why the bond angles in ethene are about 120°. (Hint: if you have difficulty doing this, read page 80.)

Naming a carbon skeleton and numbering its carbon atoms are covered on pages 192–195.

Fig 14.4 The displayed formula of 4-methylhex-2-ene

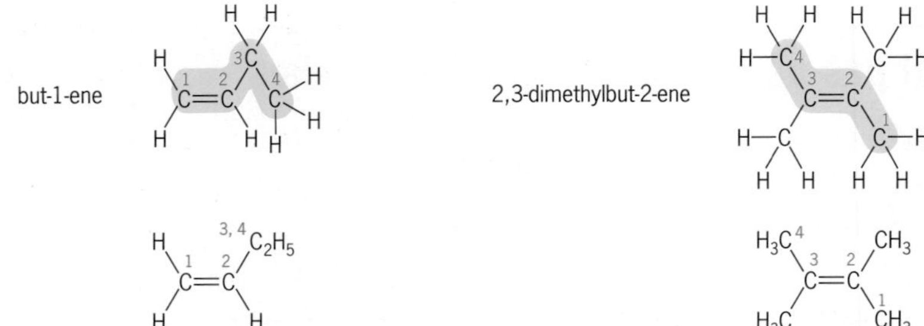

Fig 14.5 The displayed formulae of but-1-ene and 2,3-dimethylbut-2-ene. The lower formulae are easier to write and give practically as much information as the upper ones. The coloured patches show the longest carbon chain that contains the C=C bond

> **? QUESTION 5**
>
> 5 **a)** Draw the displayed formula for i) pent-1-ene, ii) cyclopentene and iii) penta-1,3-diene.
>
> **b)** Name each of the compounds in Figure 14.6.
>
> i)
>
> ii)
>
> iii)
>
> **Fig 14.6**

STEREOISOMERISM

The C=C bond is stronger than the C–C bond, but it is locked in position and cannot rotate in the way that the C–C bond does. This gives rise to **stereoisomerism** in compounds with double bonds. Compounds that are stereoisomers of one another have the same make-up of atoms, and their atoms are bonded in the same order (same structural formula), but the arrangements of the atoms in space are different.

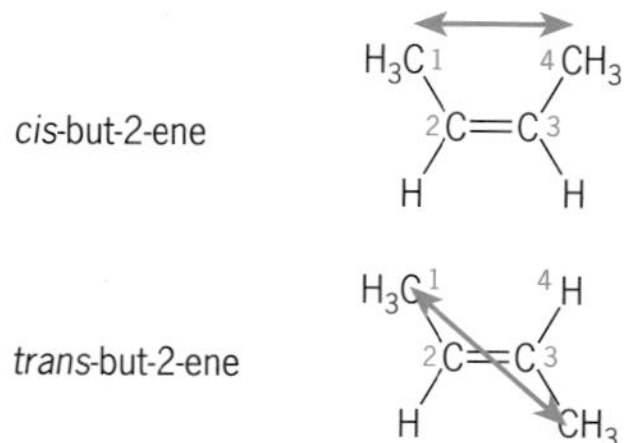

Fig 14.7 The displayed formulae of geometric isomers of but-2-ene. The distance between carbon 1 and carbon 4 is different in these two forms, because the shape of the *cis* molecule is different from that of the *trans* molecule

Fig 14.7 shows two compounds that have the same name, but-2-ene. But are they really the same? Each molecule has its own molecular shape, defined by the bond length and the bond angles within the molecule. Simple geometry shows that the length from carbon 1 to carbon 4 is not the same in both compounds. So they cannot be identical – they are isomers.

These two isomers have different physical properties. For example, the melting point of one form, *cis*-but-2-ene, is –139 °C, but that of the other form, *trans*-but-2-ene, is –106 °C. However, the chemical properties are very similar as both molecules contain the same type and number of bonds.

This type of stereoisomerisation is known as *cis–trans* or *E–Z* isomerisation. The prefix *cis* (Latin) means 'on the same side', and so the *cis* isomer has the two methyl groups on the same side of the double bond. The prefix *trans* (also Latin) means 'on the other side', and so the *trans* isomer has the methyl groups on opposite sides of the double bond. The *trans* isomer has the greater distance between carbon 1 and carbon 4.

Figure 14.8 summarises the different types of isomerism.

? QUESTION 6

6 **a)** Draw the displayed formula for i) *cis*-pent-2-ene and ii) *trans*-hex-3-ene.

b) What is the full name of each of the compounds in Figure 14.9?

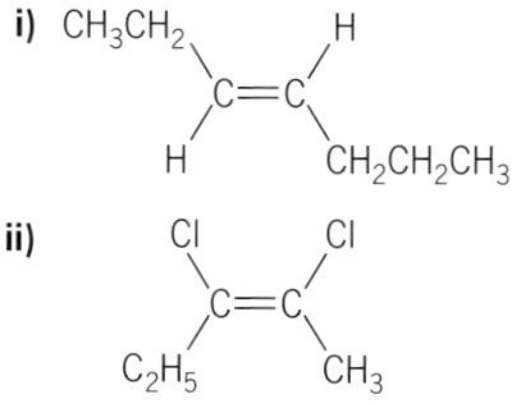

i) CH_3CH_2(H)C=C(H)$CH_2CH_2CH_3$ — CH_3CH_2 and H on the left carbon (top and bottom), H and $CH_2CH_2CH_3$ on the right carbon (top and bottom)

ii) Cl(C_2H_5)C=C(Cl)CH_3 — Cl and C_2H_5 on the left carbon (top and bottom), Cl and CH_3 on the right carbon (top and bottom)

Fig 14.9

c) State which of the following molecules can have *cis* and *trans* isomers: prop-1-ene, cyclopentene, hex-1-ene, hex-2-ene, hex-3-ene and 2,3-dimethylbut-2-ene.

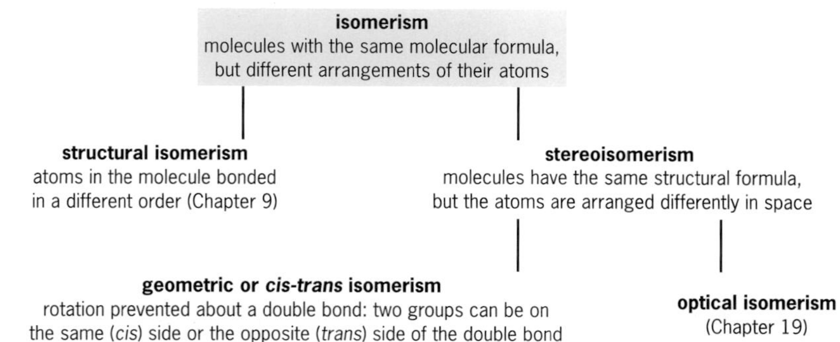

Fig 14.8 Different types of isomerism

Nature of the double bond

Cis–trans isomers exist because the C=C bond is not free to rotate. This is unlike the behaviour of the C–C carbon bond, which has free rotation about the bond, and therefore (see Fig 14.11) the positions of the two methyl groups in butane are not fixed. In but-2-ene, the carbon–carbon double bond hinders rotation, and so the two methyl groups are locked in their positions.

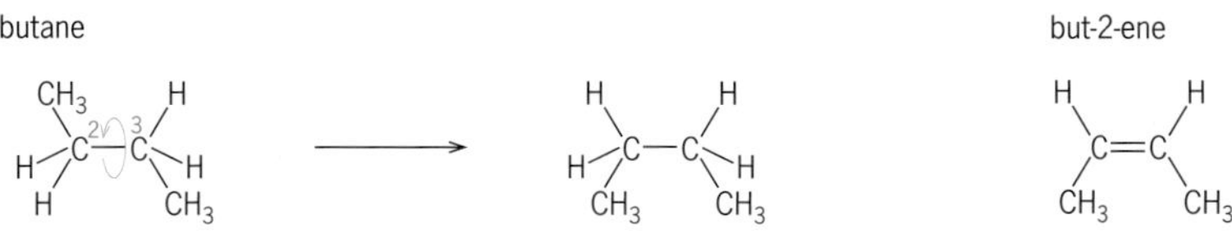

Fig 14.11 Butane has free rotation at the C—C bond, while in but-2-ene the C=C bond prevents rotation. When the single C—C bond between the carbon 2 and carbon 3 atoms rotates, the methyl groups also rotate. Although the conformations look like *cis* and *trans*, rotation of the C—C bond changes one conformation into the other, and back again. The double bond cannot rotate, so the *cis* form cannot become the *trans* form

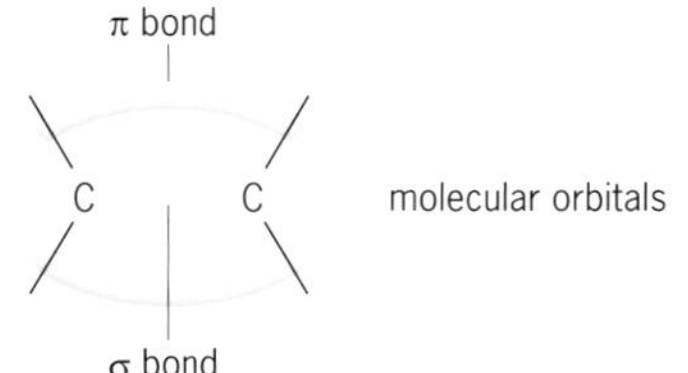

Fig 14.10 The orbital model of the double bond showing the two types of bond

Hybrid atomic orbitals and the nature of the double bond are also covered on pages 84 and 180.

The inability of the double bond to rotate is explained by the nature of the bond itself. Any double bond is composed of two different types of bond, the σ bond and the π bond (Fig 14.10).

The σ bond is formed by the overlap of two sp^2 hybrid atomic orbitals, one on each carbon atom. The bond lies mainly in the region directly *between* the two nuclei. The π bond is formed by the side-to-side overlap of two 2p atomic orbitals, one on each carbon atom. As a result, the π bond lies *above and below* the σ bond. The two carbon atoms are thus bonded together in two different ways, with the π bond preventing rotation about the carbon–carbon bond.

Note that the overlap of the atomic orbitals is less for the π bond, so the energy needed to break the π bond is less than that needed to break the σ bond.

QUESTION 7

7 Use the bond energies given in Table 8.2 page 159, to estimate the strength of the σ bond and that of the π bond in the carbon–carbon double bond.

HOW SCIENCE WORKS

E–Z isomerisation

Cis isomers have groups on the same side of the double bond and *trans* isomers on opposite sides. Look at Figure 14.12; in this case it is not possible to tell which isomer is *cis* and which one is *trans*.

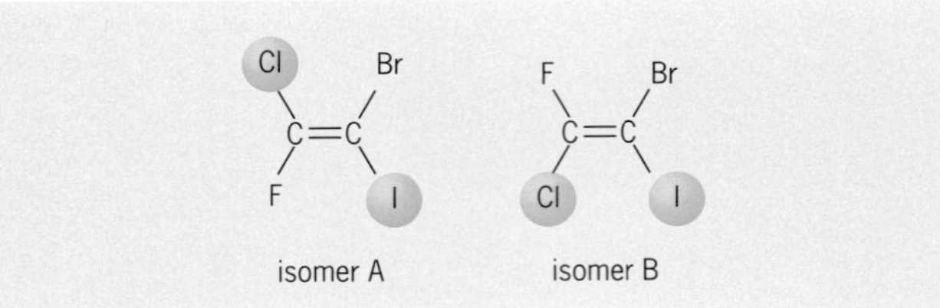

Fig 14.12 A and B are two isomers of 2-bromo-1-chloro-1-fluoro-2-iodo-ethene. It is not possible to use cis and trans to distinguish them

Chemists working on behalf of IUPAC have introduced a system to classify geometric isomers where it is impossible to tell whether they are *cis* or *trans*. It is called the *E–Z* system. The *Z* isomer often, but not necessarily, corresponds to the *cis* isomer and the *E* to the *trans* isomer.

Z comes from the German word *zusammen*, meaning 'together', and *E*, from the German word *entgegen*, meaning 'opposite'.

Look at the geometric isomers of but-2-ene in Fig 14.7.

Cis-but-2-ene is more correctly called *Z*-but-2-ene and *trans*-but-2-ene is more correctly called *E*-but-2-ene.

The general strategy of the *E–Z* system is to analyse the two groups at each end of the double bond. Each different group has been given a priority by IUPAC, as shown in Figure 14.13. The rules for assigning *E* or *Z* are as follows:

- Rank the two groups attached to the carbon atom at one end of the double bond.
- Rank the two groups attached to the carbon atom at the other end of the double bond.
- Look at the two higher-priority groups.
- If the two higher-priority groups are on the same side of the double bond, it is the *Z* isomer.
- If the two higher-priority groups are on the opposite sides of the double bond, then it is the *E* isomer.

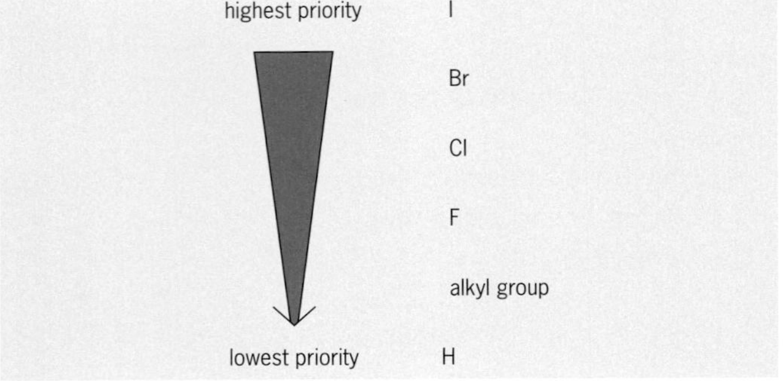

Fig 14.13 Simple ranking system for groups in *E-Z* isomerisms. You will not normally be expected to remember the priority rules but you should be able to apply them

Look back at Figure 14.12. The higher-priority groups are shaded green. They are I and Cl. It is now possible to identify the *E* and *Z* isomers. Isomer A is the *E* isomer and isomer B is the *Z* isomer.

Now consider 2-chloro-but-2-ene. Figure 14.14 shows the two geometric isomers.

The higher-priority groups are shaded in green. They are the Cl and the CH_3 group. Isomer C is *E*-2-chloro-but-2-ene and isomer D is *Z*-2-chloro-but-2-ene. Notice that using the *cis–trans* notation isomer C is *cis* and isomer D is *trans*.

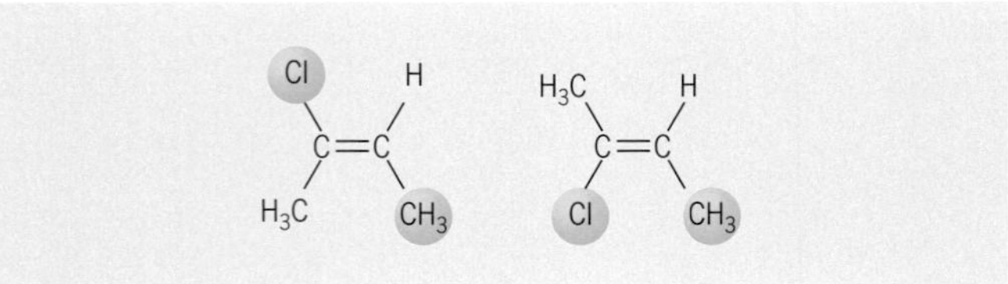

Fig 14.14 The geometric isomers of 2-chloro-but-2-ene

QUESTION 8

8 Read the How Science Works box opposite, then:

a) draw the displayed formula for *E*-hept-3-ene;

b) draw the displayed formula of *Z*-3-chloro-hex-2-ene.

c) What is the full name of each of the compounds in Fig 14.15:

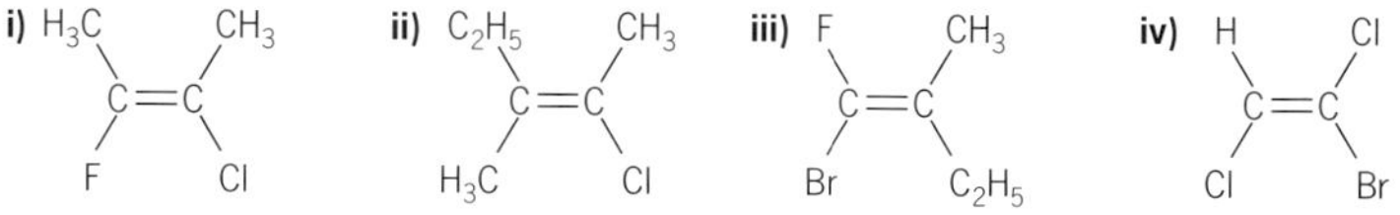

i) H_3C, CH_3 / C=C / F, Cl

ii) C_2H_5, CH_3 / C=C / H_3C, Cl

iii) F, CH_3 / C=C / Br, C_2H_5

iv) H, Cl / C=C / Cl, Br

Fig 14.15

STRETCH AND CHALLENGE

Interchange of *cis* and *trans* isomers and the photochemistry of vision

The only way to change a *cis* isomer into a *trans* isomer is first to break the π bond, and thereby allow free rotation about the σ bond, and then to re-form the π bond. This can be done by a **photochemical reaction** using ultraviolet light.

When ultraviolet light of the correct frequency is absorbed by the molecule, the π bond is broken, to form a **diradical**. A diradical is a particle that contains two atoms each with an unpaired electron in its outer shell.

During the time that the diradical exists, there is free rotation about the σ bond. As free radicals are highly reactive, the diradical very quickly re-forms the double bond to pair up the electrons. The result is a mixture of the *cis* and the *trans* isomer.

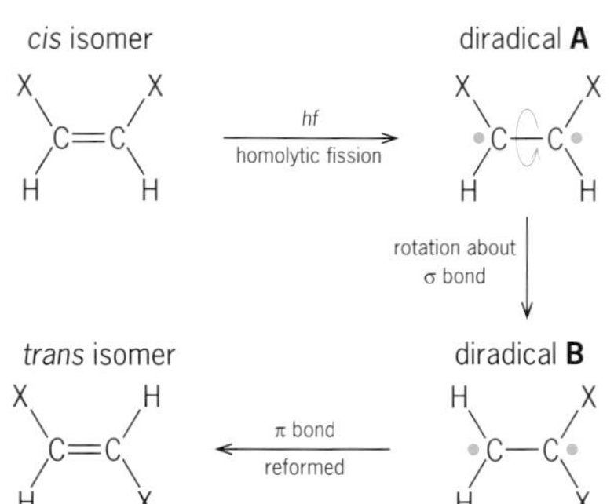

Fig 14.16 Photochemical *cis*–*trans* transformations. The π bond can re-form in either structure A or B, to produce both *cis* and *trans* isomers

The property of *cis*–*trans* isomerism is not limited to compounds with carbon–carbon double bonds. Compounds with carbon–nitrogen and nitrogen–nitrogen double bonds also form geometric isomers. For example, Fig 14.17 shows a candidate for the kind of memory molecule mentioned in the Science in Context box on page 274. The completely different shapes of the two isomers are obvious.

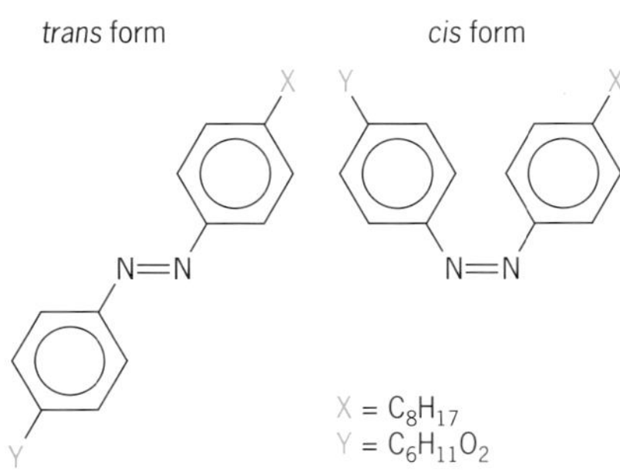

Fig 14.17 A possible memory molecule. The distance between X and Y is much shorter in the *cis* isomer than in the *trans* isomer. The *trans* isomer can be changed into the *cis* isomer by a photochemical reaction

PHOTOCHEMISTRY OF VISION

Light is detected on the inner layer of the eye known as the retina. The retina consists of two types of cell that can respond to light: rods and cones. Rods respond to differences in intensity of light, but not to colour. Cones respond to colour. However, both types of cell function in a similar way.

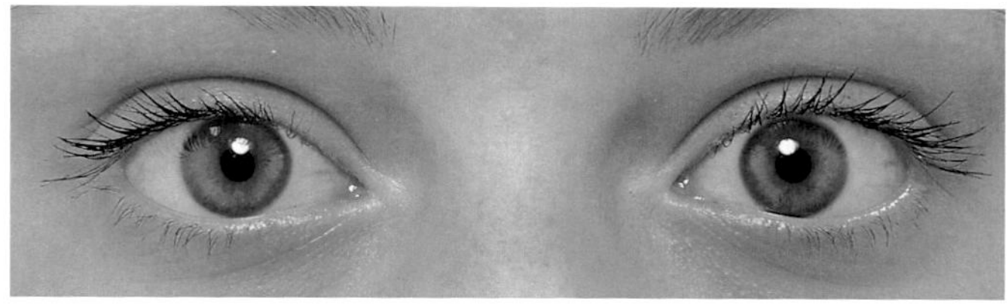

Fig 14.18 Detection of light by the retina is a direct result of the photochemical transformation of a *cis* isomer into a *trans* isomer

Light is detected because retinal, a molecule inside the cells of the retina, undergoes a photochemical reaction. When a photon of light reaches a *cis*-retinal molecule, it is changed into the *trans*-retinal form, as in Fig 14.19, and so its shape changes.

Retinal interacts with a protein responsible for producing the nerve impulses sent along the optic nerve to the brain; when the form changes from *cis* to *trans*, a nerve impulse is generated. It is not yet fully understood how the change of shape within the protein molecule produces a nerve impulse, but the energy from the light is in some way transferred, and an electrical signal results.

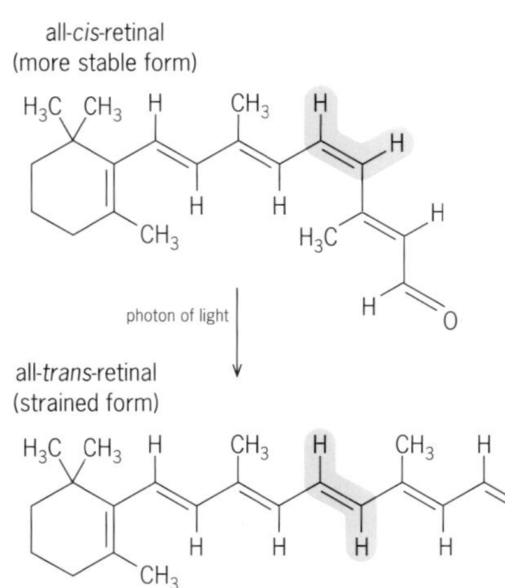

Fig 14.19 *Cis*- and *trans*-retinal. The two isomers have completely different overall shapes. The coloured patches highlight the *cis* and *trans* double bonds at which the difference arises

PHYSICAL PROPERTIES OF ALKENES

The alkenes form a homologous series. The first three members are gases, but all the higher members are liquids or solids. The increase in the boiling point of compounds as the number of carbon atoms per molecule increases is typical for all homologous series, not just the alkenes.

Table 14.1 Melting and boiling points of four alkenes

Alkene	Structural formula	Melting point/°C	Boiling point/°C
ethene	$CH_2{=}CH_2$	–169	–105
propene	$CH_3CH{=}CH_2$	–185	–48
but-1-ene	$CH_3CH_2CH{=}CH_2$	–185	–6
pent-1-ene	$CH_3CH_2CH_2CH{=}CH_3$	–165	30

? QUESTION 9

9 **a)** Explain why the boiling points of the alkenes increase as the number of carbon atoms per molecule increases. (Hint: read page 195 about the boiling points of the alkanes.)

b) Predict the boiling point of hex-1-ene.

3 PREPARATION OF ALKENES

CRACKING ALKANES

Saturated hydrocarbons can be cracked to give alkenes (see page 187). Though cracking can be carried out in the laboratory, this method is unsuitable because the mixture it produces is difficult to separate into individual alkenes.

Fig 14.20 Cracking liquid paraffin in the laboratory

HOW SCIENCE WORKS

Free-radical mechanism of cracking

Thermal cracking involves heating long-chain alkanes at a very high temperature so that the molecules break down to give hydrocarbons with shorter carbon chains. Since the C—C bond in alkanes is non-polar, the breaking of this bond is believed to involve **homolytic fission** with the formation of alkyl free radicals. Homolytic fission is normally initiated by ultraviolet light, but a very high temperature provides sufficient thermal energy for the process.

Cracking lacks selectivity, since its progress is determined by which C—C bond is first broken, and then by the type of collision in which the free radical is involved. Typically, the cracking of a hydrocarbon such as decane leads to a collection of short-chain alkanes, alkenes and hydrogen. (Free radicals and homolytic fission are described on pages 240 and 241; cracking is described on pages 187–188.)

There is more about the dehydration of alcohols on page 267.

DEHYDRATION OF ALCOHOLS

A suitable way to prepare alkenes is to dehydrate an alcohol. This reaction is an example of elimination and involves the loss of a molecule of water (Fig 14.21). Normally, an acidic catalyst, such as concentrated phosphoric(V) acid or concentrated sulfuric acid, is used to aid elimination of the water (Fig 14.22). Industrially, the dehydration is usually catalysed by passing the alcohol vapour over heated aluminium oxide.

Sometimes, the dehydration of an alcohol leads to the formation of two or more alkenes, which can be either structural or geometric isomers (Fig 14.23).

$$CH_3-CH_2-\overset{H}{\underset{H}{C}}-\overset{H}{\underset{H}{C}}-O-H \xrightarrow{-H_2O} \underset{CH_3-CH_2}{\overset{H}{}}\!\!>C{=}C<\!\!\underset{H}{\overset{H}{}}$$

Fig 14.21 Butan-1-ol loses a water molecule in an elimination reaction to form but-1-ene

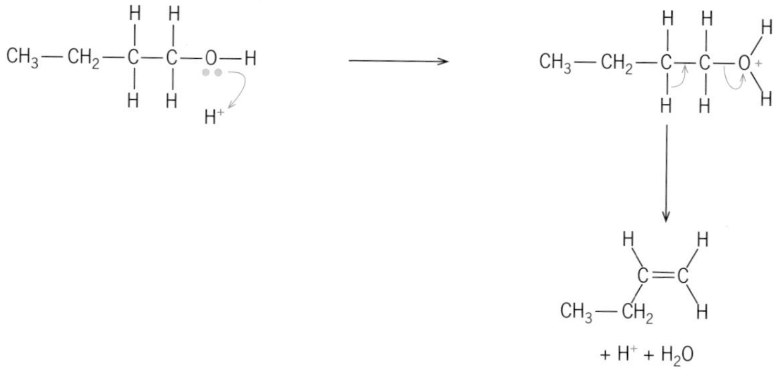

Fig 14.22 Dehydration of an alcohol (butan-1-ol) is catalysed by an acid. The reaction starts with protonation of the alcohol. After elimination of the water molecule, the proton is regenerated

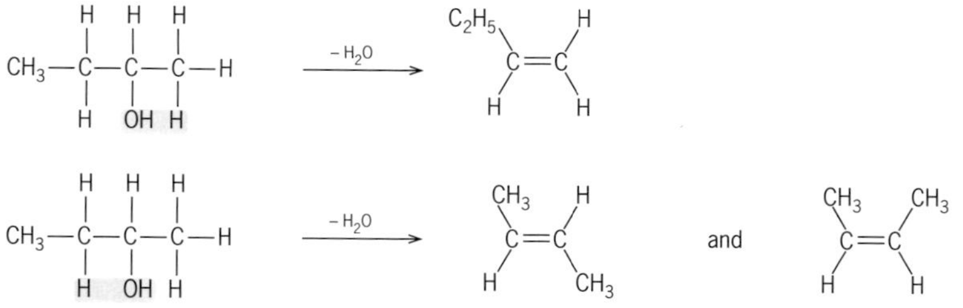

Fig 14.23 Dehydration of butan-2-ol. Three isomeric butenes can be made by the removal of different hydrogen atoms

Normally, the most stable alkene is produced in a greater proportion than the other isomers. This is the alkene with most alkyl groups attached to the carbon atoms that form the double bond (the most substituted alkene).

ELIMINATION OF HYDROGEN HALIDES FROM HALOGENOALKANES

When a halogenoalkane is heated with a strong base, such as ethanolic sodium hydroxide, the corresponding alkene is produced. Again, the reaction may give more than one alkene and may be complicated by substitution products as well (Fig 14.24).

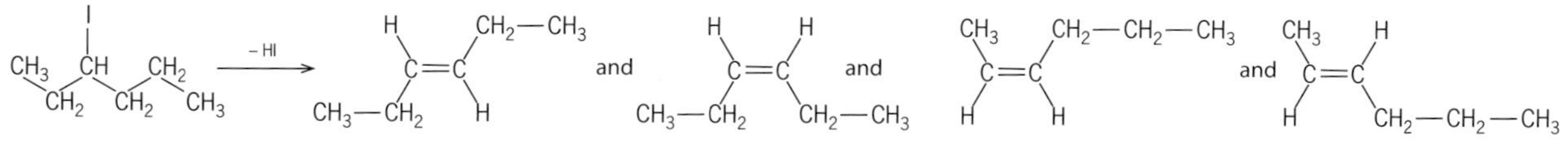

Fig 14.24 Elimination of hydrogen iodide from 3-iodohexane

QUESTION 10

10 **a)** State which alcohol you would need to make i) ethene, ii) propene and iii) cycloheptene.
b) i) Write down all the possible isomers that can be formed by the dehydration of pentan-2-ol.
ii) Which of these is likely to be produced in the greatest amount?

QUESTION 11

11 Draw the structures of all of the alkenes that can be produced when 2-iodohexane is heated with ethanolic sodium hydroxide.

For more information about the elimination of hydrogen halides from halogenoalkanes, see page 252.

4 REACTIONS OF ALKENES

The double bond in an alkene is non-polar, but it is electron rich because of the π bond. When the π bond is broken, the carbon atoms remain joined together by the σ bond.

ELECTROPHILES

An **electrophile** is a particle that can accept a pair of electrons to make a covalent bond. Often, an electrophile has only six electrons in its outer shell,

QUESTION 12

12 Explain why the π bond can be broken more easily than the σ bond.

so that by accepting two others it achieves a stable octet. A typical electrophile is Cl^+, with only six electrons in its outer shell. Metal ions are not electrophiles because, although they are positive, they cannot gain a pair of electrons to make a covalent bond. In fact, a metal ion is often positively charged to preserve a stable octet of electrons.

Electrophiles are generated by the **heterolytic fission** of a covalent bond. In heterolytic fission, the bond is broken as both bonding electrons migrate to the atom that is more electronegative, leaving the less electronegative atom positive. Fig 14.25 shows the heterolytic fission of a covalent bond. Again, we follow the convention of using a curly arrow to show the migration of a pair of electrons.

? QUESTION 13

13 State which of the following particles are electrophiles:
Br^+, CH_3, CH_3^+, H_2O and Na^+.

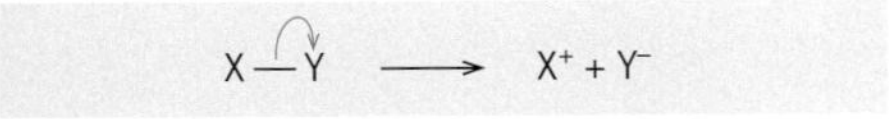

$X{-}Y \longrightarrow X^+ + Y^-$

Fig 14.25 Heterolytic fission of a covalent bond. Atom Y is normally more electronegative than atom X. It therefore accepts the bonding pair of electrons during heterolytic fission

ELECTROPHILIC ADDITION

Almost all double bonds will undergo a reaction known as **addition**. During addition, two substances react together to give one product and the original double bond becomes a single bond, as shown in Fig 14.26. In the case of the C=C bond, an electrophile is often needed to break the bond. Since the initial step in the mechanism involves an electrophile, it is called **electrophilic addition**.

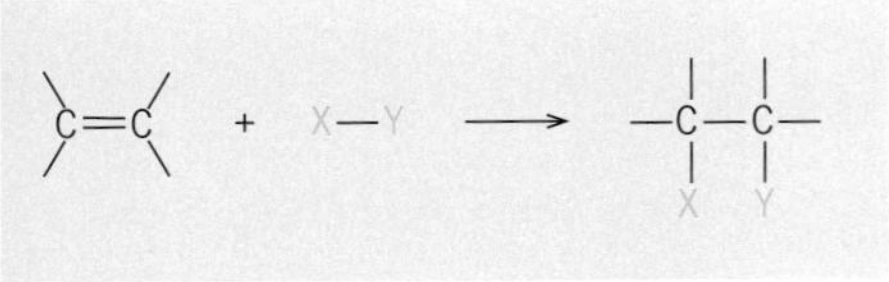

Fig 14.26 A generalised addition reaction. During addition, two reactants form one product and the double bond becomes a single bond

Mechanism of electrophilic addition

The first stage in electrophilic addition is the formation of the electrophile. (We will return later to discuss this stage in more detail, because it is not always as simple as is shown in Fig 14.27.) The electrophile 'attacks' the double bond. At the same time, a pair of electrons migrates from the double bond towards the electrophile. As seen in Fig 14.27, the effect of this is to break the carbon–carbon double bond and to make a single bond to the electrophile. In the final stage, the nucleophile reacts with the carbocation that has been formed, to complete the addition.

The net effect is to add atoms at the double bond to form a saturated compound. It is this ability to react by addition that accounts for the difference in reactivity between an alkane and an alkene.

The absence of reactivity in alkanes is covered on page 196.

REMEMBER THIS

A carbocation is also known as a carbonium ion. It is a positive ion that has a carbon atom attached to only three other atoms or groups of atoms, R_3C^+.

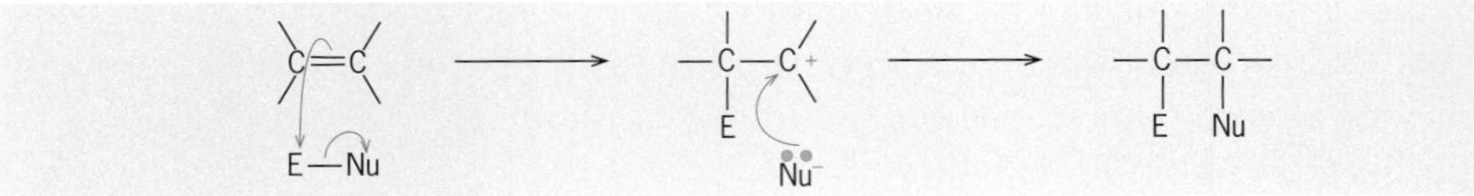

Fig 14.27 A generalised mechanism for electrophilic addition to an alkene. E = electrophile, Nu = nucleophile

ADDITION OF HYDROGEN HALIDES

Hydrogen bromide reacts with alkenes to produce bromoalkanes. The other hydrogen halides react with alkenes in a similar way. Hydrogen halides act as electrophiles, because the hydrogen end of the molecule is electron deficient as a result of the difference in electronegativity between the hydrogen and halogen atoms. The hydrogen end of the molecule can accept an electron pair, while simultaneously the hydrogen–halogen bond is broken, as shown in Fig 14.27. The reaction with hydrogen iodide is particularly favoured because of the relatively low bond energy of the H–I bond.

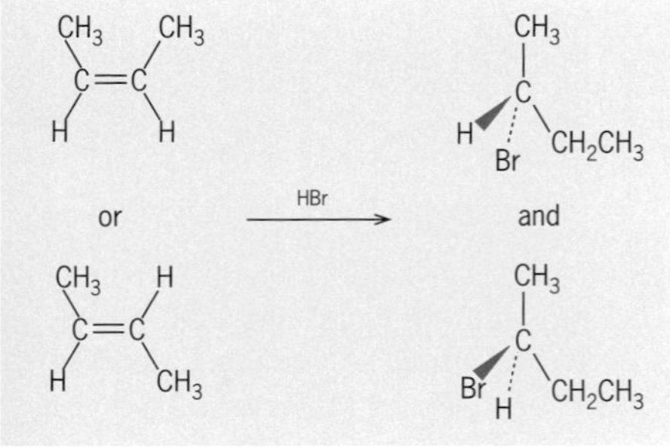

Ethene reacts with hydrogen chloride to give chloroethane. Likewise, both *cis*- and *trans*-but-2-ene yield 2-bromobutane. It does not matter what the starting isomer is, because after the addition the saturated product has acquired a single bond that can freely rotate, with loss of the *cis–trans* isomerism. In fact, as Fig 14.28 shows, two products are formed, which in this case are **optical isomers**.

The production of many petrochemicals starts with this type of electrophilic addition to give a halogenoalkane. The introduction of the carbon–halogen bond offers further possibilities for organic synthesis.

Fig 14.28 The addition of hydrogen bromide to either *cis*- or *trans*-but-2-ene gives the two optical isomers of 2-bromobutane

Optical isomerism is another form of steroisomerism: see page 277.

The reactions of halogenoalkanes are covered in detail in Chapter 12.

STRETCH AND CHALLENGE

Addition of hydrogen halides to unsymmetric alkenes: Markovnikov's rule

An unsymmetric alkene, such as propene, is an alkene that has different groups attached to the carbon atoms of the C=C bond.

The electrophilic addition of a hydrogen halide to an unsymmetric alkene can give more than one addition product, according to which end of the double bond the electrophile attacks. For example, the electrophilic addition of hydrogen bromide to propene can yield two products, 1-bromopropane and 2-bromopropane, depending on which end of the double bond is attacked by the electrophile (essentially a proton). As Fig 14.30 shows, the two products arise from different carbocations, which have different stabilities.

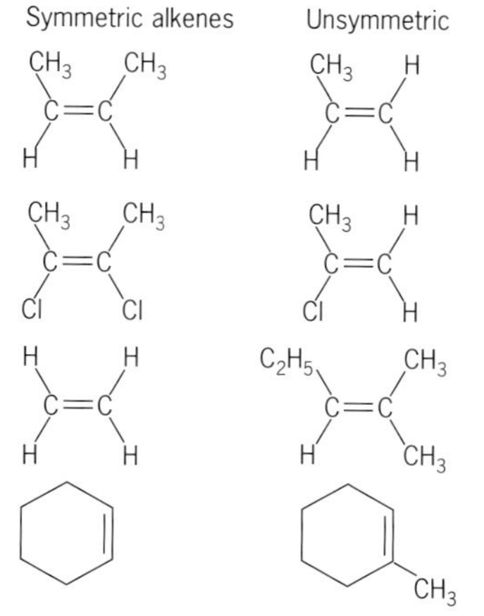

Symmetric alkenes Unsymmetric

Fig 14.29 The displayed formulas of four symmetric and four unsymmetric alkenes

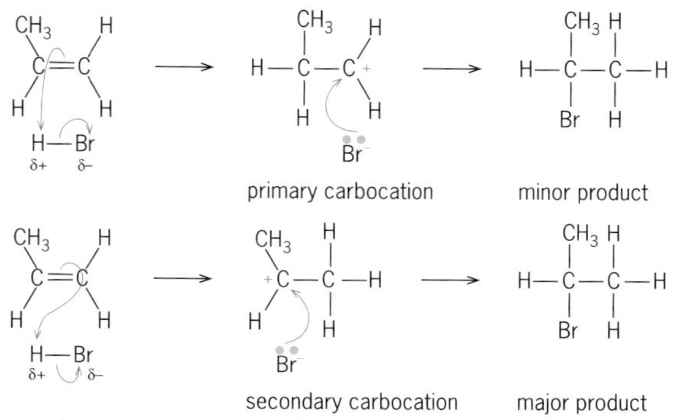

primary carbocation minor product

secondary carbocation major product

Fig 14.30 The mechanism of the electrophilic addition of hydrogen bromide to propene

primary carbocation: one alkyl or aryl group attached

secondary carbocation: two alkyl or aryl groups attached

tertiary carbocation: three alkyl or aryl groups attached

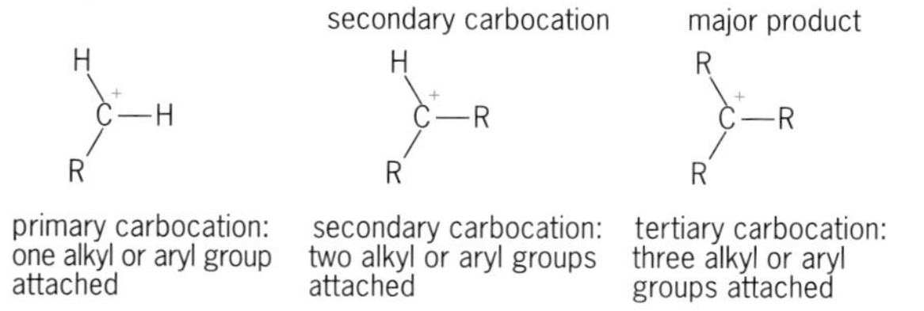

Fig 14.31 The three types of carbocation

A **primary carbocation** (Fig 14.31) is a species in which a positive carbon atom is attached to just one other carbon atom (of the alkyl or aryl R group). It is much less stable than a **secondary carbocation**, which is much less stable than a **tertiary carbocation**. The source of this stability is the alkyl or aryl group, which tends to donate electron density to the positive carbon atom and thereby stabilises the positive charge. It follows that two alkyl (or aryl) groups give more stability than one, and three give more stability than two. This action of pushing the electron density is called an **inductive effect**. The major product is formed from the most stable carbocation. Therefore, in the example of the addition of hydrogen bromide to propene, the major product is 2-bromopropane.

Markovnikov's rule predicts the way that hydrogen halides are added to unsymmetric alkenes:

An electrophile adds to an unsymmetric alkene so that the most stable carbocation is formed as an intermediate.

The term 'electrophile' is used to express the general concept, although the electrophile is often an electron-deficient hydrogen atom or a proton.

QUESTIONS 14–16

14 Hydrogen bromide adds to ethene to form bromoethane. Describe the mechanism of this reaction. (Hint: the hydrogen atom in HBr is electron deficient.)

15 Which of the carbocations in Fig 14.32 is the most stable?

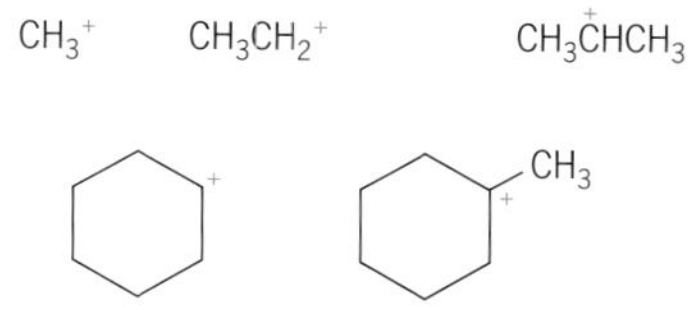

CH_3^+ $CH_3CH_2^+$ $CH_3\overset{+}{C}HCH_3$

Fig 14.32 Five different carbocations

16 Predict the major product in the reaction of hydrogen bromide with

a) *cis*-but-2-ene;
b) cyclohexene;
c) pent-1-ene;
d) *E*-hex-z-ene;
e) *Z*-1, 2-dichloroethene.

? QUESTION 17

17 Why is it difficult to make primary alcohols by the hydration of an alkene? (Hint: consider the acid-catalysed addition of water to propene.)

HYDRATION OF ALKENES

The hydration of an alkene is the electrophilic addition of water to give an alcohol. This reaction needs an acidic catalyst, since water itself is not a good electrophile. In the laboratory, sulfuric acid is used as the catalyst. With reactive double bonds, dilute acid is sufficient, but with alkenes, such as propene and ethene, concentrated acid is required.

Fig 14.33 details the use of concentrated sulfuric acid as a catalyst. The proton transferred by the acid is needed to start the reaction as it facilitates the breaking of the double bond. At the end of the reaction, the acid is regenerated.

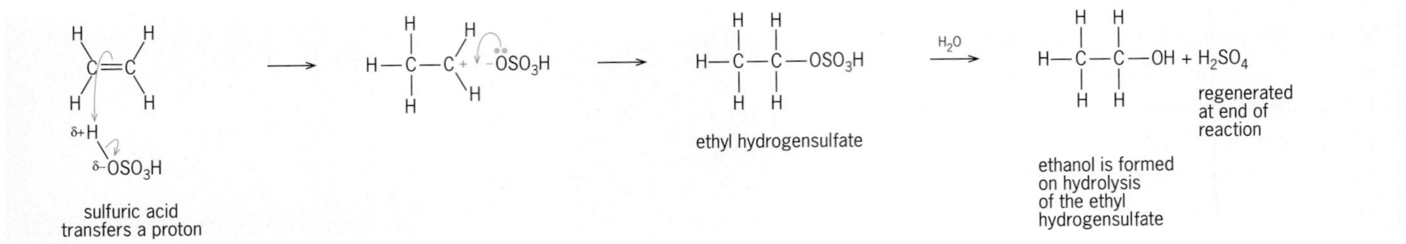

sulfuric acid transfers a proton

ethyl hydrogensulfate

regenerated at end of reaction

ethanol is formed on hydrolysis of the ethyl hydrogensulfate

Fig 14.33 The mechanism of the electrophilic addition of water, as shown by the hydration of ethene using concentrated sulfuric acid and subsequent hydrolysis of the intermediate

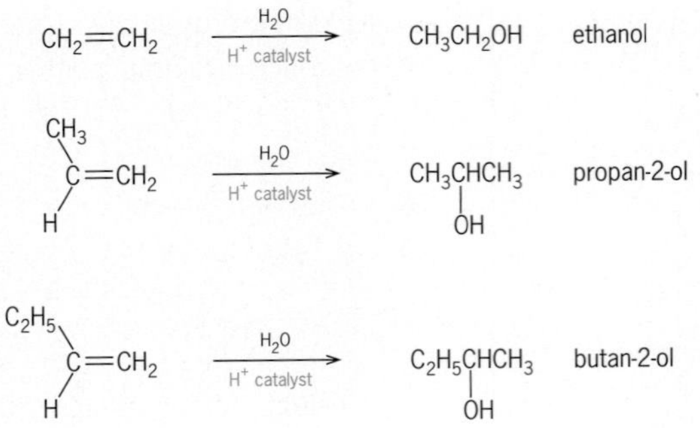

Fig 14.34 The hydration of ethene, propene and butene. Because of Markovnikov's rule, the hydration of propene and but-1-ene does not produce a large proportion of the primary alcohols propan-1-ol and butan-1-ol. Instead, the secondary alcohols propan-2-ol and butan-2-ol are formed

In the petrochemical industry, alkenes are hydrated to give several important solvents. Hydration is achieved either by the catalytic addition of water as steam to the alkene at 7000 kPa (70 atmospheres) pressure at 300 °C, using H_3PO_4 as a catalyst, or by using the concentrated sulfuric acid route just described. Ethene is converted into ethanol, propene into propan-2-ol, and but-1-ene into butan-2-ol (Fig 14.34). All three alcohols are extremely useful industrial solvents.

REACTION WITH HALOGENS

Halogens can add to double bonds to give dihalogenoalkanes. For example, bromine reacts with propene to give 1,2-dibromopropane:

$$CH_3CH{=}CH_2 + Br_2 \rightarrow CH_3CHBrCH_2Br$$

Mechanism of the reaction

Compared with a hydrogen halide, it is much more difficult to visualise a bromine *molecule* as an electrophile, since it is not a polar molecule and both of its atoms have a stable octet of electrons. Nevertheless, a bromine molecule is able to accept an electron pair, provided that, at the same time, the bromine–bromine single bond is broken by *heterolytic* fission.

As the bromine molecule approaches the double bond, the pair of electrons in the Br—Br bond moves to one of the bromine atoms (Fig 14.35). Why should this happen? After all, there is no positive or negative end to the bromine molecule, as there is in the addition of hydrogen bromide. The reason is that a temporary dipole is created in the bromine molecule as it approaches the π electrons of the double bond. It is as though the electrons in

? QUESTION 18

18 Draw the displayed formulae of the products of the reactions between:

a) bromine and ethene;

b) chlorine and cyclohexene;

c) chlorine and ethene.

? QUESTION 19

19 Chlorine will react with ethene to give 1,2-dichloroethane. The mechanism is identical to the bromination of ethene. Write down the mechanism for this reaction.

the bromine–bromine single bond are repelled by the π electrons. This temporary dipole produces an electron deficiency at the end of the molecule nearer the double bond, and thereby allows the electrophilic attack. The intermediate carbocation then reacts with a bromide ion to complete the electrophilic addition.

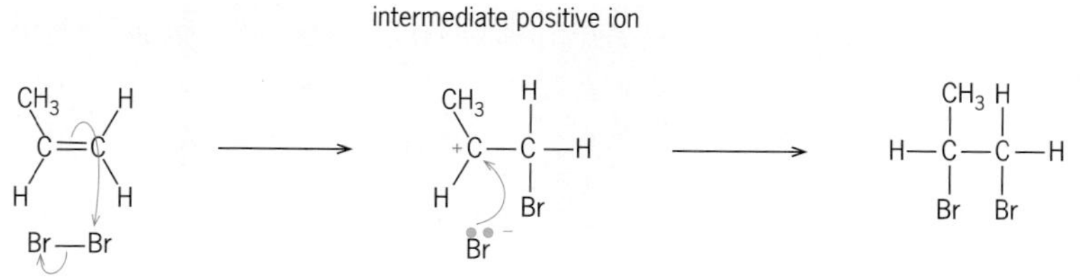

Fig 14.35 The mechanism of the electrophilic addition of bromine to propene. The bromine molecule has a temporary dipole that can be shown by including partial charges: $Br^{\delta+}$--$Br^{\delta-}$

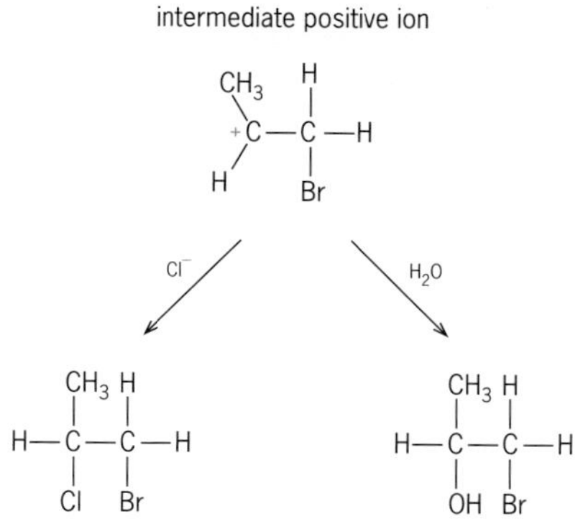

However, it is possible to intercept the carbocation with other nucleophilic reagents (Fig 14.36). So, if a chloride ion is also present in the reaction mixture, it can react as a nucleophile; and if the reaction is carried out in the presence of water, a water molecule can react with the carbocation.

This means that the choice of solvent used for the bromination is important. If it were aqueous bromine, for example, it would be impossible to stop the interception of the carbocation by a water molecule acting as a nucleophile.

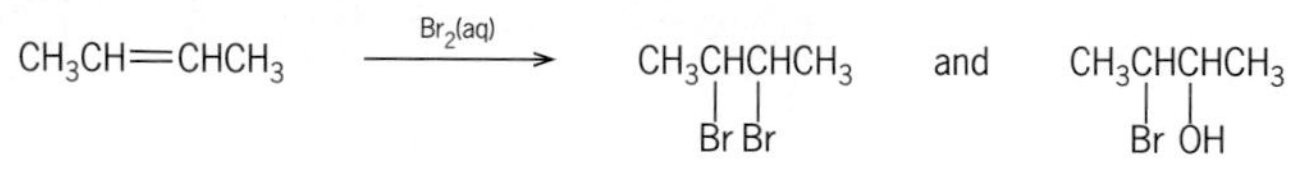

Fig 14.36 The intermediate positive ion (carbocation) can react with other nucleophiles, such as a chloride ion or water, if they are present

$$CH_3CH{=}CHCH_3 \xrightarrow{Br_2(aq)} CH_3CH(Br)CH(Br)CH_3 \text{ and } CH_3CH(Br)CH(OH)CH_3$$

Fig 14.37 Aqueous bromine reacts with but-2-ene to give two products

TEST FOR UNSATURATION

Aqueous bromine may be used to test for the presence of a C=C bond, since the reagent changes colour during electrophilic addition. Aqueous bromine is orange, but after reaction with an alkene it forms colourless products. Alternatively, bromine in an inert solvent, such as tetrachloromethane or hexane, can be used. In both of these solvents the same colour change is observed: orange to colourless.

? QUESTION 20

20 Explain the formation of two products during the reaction between aqueous bromine and but-2-ene. (Hint: read about the interception of a carbocation by other nucleophiles.)

HYDROGENATION

Alkenes can also undergo other sorts of addition reactions, such as catalytic hydrogenation. In this case, hydrogen is added to the C=C bond to make a saturated compound (Fig 14.38). The reaction mechanism is not electrophilic addition, but often involves **heterogeneous catalysis** (catalyst and reactant in different phases) using a nickel catalyst and high pressure (Fig 14.40).

Hydrogenation is not an important reaction in the petrochemical industry because it leads to the formation of saturated (unreactive) compounds, and petrochemical feedstocks are often already saturated. As a synthetic reaction, it is used when the raw material is derived from a natural source. For example, margarines are produced by the hydrogenation of polyunsaturated compounds derived from plant oils (Fig 14.39).

For more about heterogeneous catalysis, see page 206.

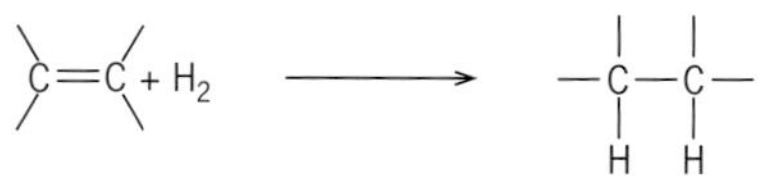

Fig 14.38 Hydrogenation converts an alkene to an alkane

Fig 14.39 The unsaturated compounds in sunflower oil have some of their double bonds hydrogenated to change the oil into a solid to make margarine

HOW SCIENCE WORKS

The mechanism of hydrogenation

Chemists develop models to describe the way that chemical reactions take place. These models are based on experimental observations involving rates of reaction. Figure 14.40 shows the mechanism (model) for the heterogeneous catalysis of the reaction between ethene and hydrogen.

1 Ethene molecule approaches nickel surface

2 Ethene molecule forms an intermediate with nickel surface

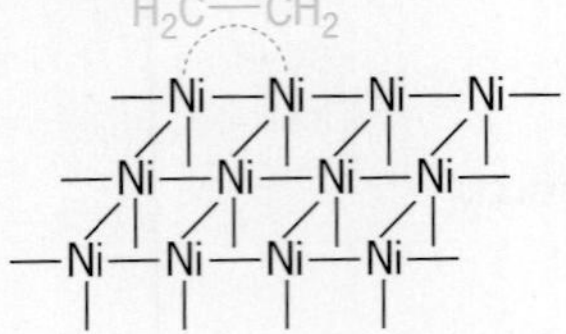

3 Hydrogen molecule approaches nickel surface

4 Hydrogen molecule forms an intermediate with nickel surface adjacent to ethene

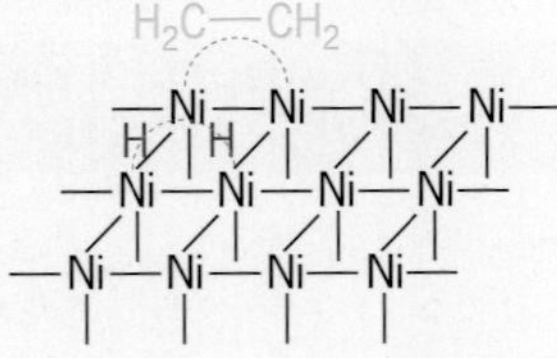

5 Hydrogen and ethene form a 'common intermediate' and addition occurs

6 Ethane molecule leaves nickel surface

Fig 14.40 Heterogeneous catalysis in the hydrogenation of ethene

QUESTION 21

21 A sample of a natural product, 0.0100 moles, with molecular formula $C_{22}H_{26}$ needs 0.0502 moles of hydrogen for complete hydrogenation. How many C=C bonds are present in one molecule of the natural product?

Working out structures

Hydrogenation is used to determine the number of C=C bonds in a compound. One mole of C=C bond requires one mole of hydrogen to fully hydrogenate it. Therefore the number of C=C bonds in the compound can be estimated by comparing the number of moles of hydrogen that have reacted with a fixed amount of compound.

SCIENCE IN CONTEXT

Unsaturated and saturated fatty acids

Between 25 and 50 per cent of the normal energy content of the diet from a person living in the United Kingdom comes from fats and oils. This percentage is beginning to decline as people become more aware of the dangers of eating too much fatty food and are able to recognise fatty foods from improved food labelling.

There is no structural difference between a fat and an oil; they are both members of a group of organic compounds called esters. Fats are solid at room temperature and oils are liquids.

Fats are the result of the reaction between a carboxylic acid and an alcohol called propane-1,2,3-triol. The carboxylic acids are still referred to as fatty acids. Fatty acids are often known by their older, non-systematic names; these names were often derived from their sources. Stearic acid, for instance, was derived from the Greek word *stear* for tallow or fat.

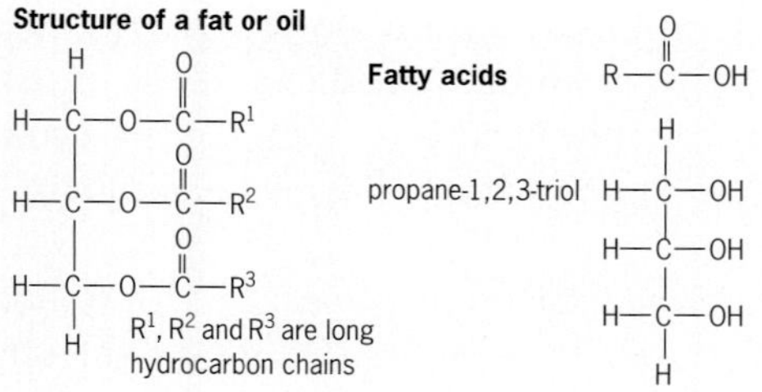

Structure of a fat or oil

Fatty acids

propane-1,2,3-triol

R^1, R^2 and R^3 are long hydrocarbon chains

Fig 14.41 Structures of fats and oils and of the alcohol and acids that react to make them

Some of the fatty acids in Table14.2 are unsaturated, because they have a molecule with at least one carbon–carbon double bond; some are polyunsaturated, with more than one carbon–carbon double bond per molecule; and others are saturated. Fats made from saturated fatty acids are called saturated fats and those from unsaturated fatty acids are called unsaturated fats. Medical research has recognised that unsaturated and, in particular, polyunsaturated fats are much better for your health than saturated ones. The presence of large quantities of saturated fats in a diet is believed to be part of the cause of atherosclerosis, a thickening of the arterial wall.

Vegetable oils tend to have a greater proportion of unsaturated fatty acids than animal fats and so it is believed that a diet having more vegetable oils may be better for you. There are some exceptions, however, since coconut oil contains around 90 per cent saturated fats.

Margarine and low-fat spreads are made from vegetable oils. These products contain vegetable oils that have been solidified. Vegetable oils are solidified by hydrogenating some of the carbon–carbon double bonds present – essentially by making the vegetable fats more saturated. The hydrogenation reaction uses a high pressure of hydrogen in the presence of a nickel catalyst.

Table 14.2 The composition of eight fats and oils

Fat or oil	Saturated fatty acids		Unsaturated (one C=C double bond)		Poly-unsaturated
	C_{16}	C_{18}	C_{16}	C_{18}	C_{18}
tallow (cow)	24–32%	14–32%	1–3%	35–48%	2–4%
lard (pork)	28–30%	12–18%	1–3%	41–48%	6–7%
butter	23–26%	10–13%	5%	30–40%	4–5%
whale	11–18%	2–4%	13–18%	33–38%	
olive	5–15%	1–4%	0–1%	69–84%	4–12%
coconut	4–10%	1–5%		2–10%	0–2%
soya bean	6–10%	2–6%		21–29%	54–67%
linseed	4–7%	2–5%		9–38%	28–91%

Fig 14.42 Whale meat contains a very low percentage of saturated fatty acids but it is ethically incorrect to eat endangered species

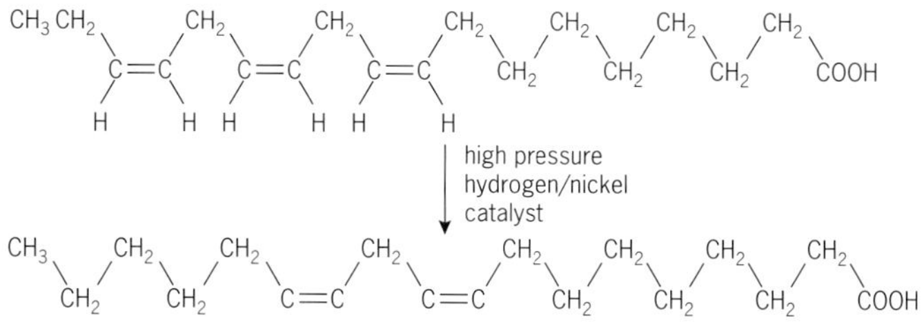

high pressure hydrogen/nickel catalyst

Fig 14.43 The partial hydrogenation of linolenic acid still gives an unsaturated fatty acid. Sometimes during the reaction one of the carbon-carbon double bonds in the product changes from *cis* to *trans* (*Z* to *E* orientation)

Natural fatty acids, such as linolenic acid, are *Z* or *cis* isomers. During hydrogenation some of the carbon–carbon double bonds react with hydrogen, but others remain intact. Unfortunately the double bonds remaining are sometimes changed from the *Z* form to the *E* form (*cis* to *trans*). These *E*-form fatty acids are not required by the body and there is some evidence that they may also contribute towards certain heart diseases. It is possible that the promotion of polyunsaturated fats as healthier foods may have inadvertently introduced another health risk from *E* or *trans* fatty acids.

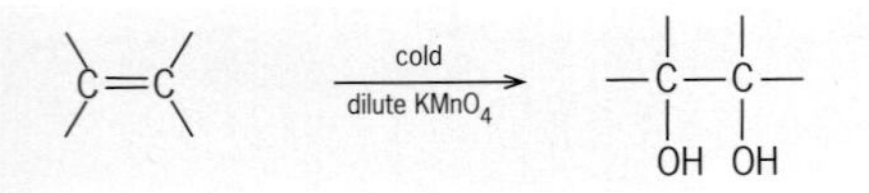

Fig 14.44 The oxidative addition of alkenes to give diols. When the alkene is ethene, the product is ethane-1,2-diol

OXIDATIVE ADDITION

Alkenes react with cold acidified dilute potassium manganate(VII) to give a diol product (with two OH groups). The reaction involves both oxidation and addition to the double bond, so it is sometimes called an oxidative addition. During the reaction, the purple colour of the manganate(VII) ion disappears. When ethene is bubbled into cold acidified dilute potassium manganate(VII), ethane-1,2-diol (a component of antifreeze) is produced:

$$CH_2{=}CH_2 + H_2O + [O] \rightarrow CH_2OHCH_2OH$$

Notice that the oxidation is represented in the equation in its simplified form using [O] (see page 268).

This reaction is not used to prepare diols, since an excess of potassium manganate(VII) will oxidise the diols further.

Alkaline aqueous potassium manganate(VII) will also produce a diol with alkenes. This time, a brown precipitate of MnO_2 is also produced.

? QUESTION 22

22 State the main organic product when each of the following is treated with cold dilute potassium manganate(VII):

a) prop-1-ene;
b) but-2-ene;
c) cyclopentene.

ADDITION POLYMERISATION

Alkene molecules can be joined to give molecules with very long chains, known as polymers. The constituent alkene molecules of polymers are called monomers. In polymerisation, thousands of monomers are combined to form a single polymer chain. An alkene forms an addition polymer when a double bond is converted into a single bond and two extra single bonds are created. The resulting polymer consists of repeating units called monomer units.

For example, ethene is converted into poly(ethene), or polythene, by the action of a catalyst, high pressure and a moderate temperature (Fig 14.45). Table 14.3 lists five addition polymers and their main uses.

There is more information on addition polymers and polymerisation in Chapter 20.

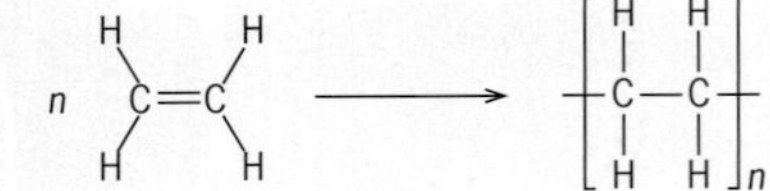

Fig 14.45 Poly(ethene) is formed by the repeated catalysed addition of ethene monomers to give a saturated polymer

Table 14.3 Five common addition polymers and their uses

Alkene monomer		Polymer	Structure of polymer (monomer unit)	Uses
ethene	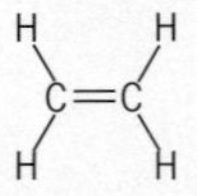 $H_2C{=}CH_2$	poly(ethene)	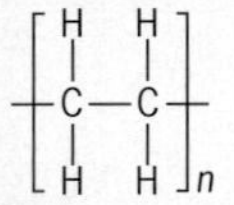 $[-CH_2-CH_2-]_n$	bags, insulation for wires, squeezy bottles
propene	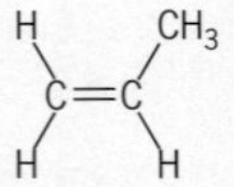 $H_2C{=}CH(CH_3)$	poly(propene)	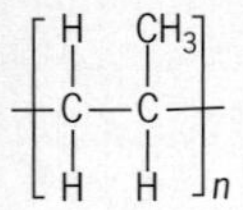 $[-CH_2-CH(CH_3)-]_n$	bottles, plastic plates, clothing, carpets, crates, ropes and twine
phenylethene	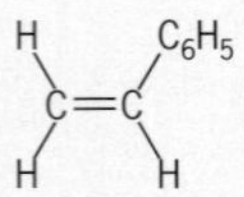 $H_2C{=}CH(C_6H_5)$	poly(phenylethene)	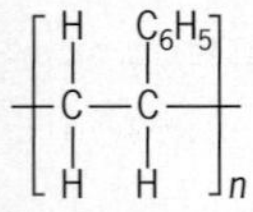 $[-CH_2-CH(C_6H_5)-]_n$	insulation, food containers, model kits, flowerpots, housewares
chloroethene	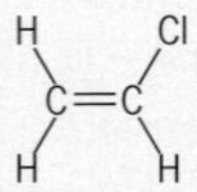 $H_2C{=}CHCl$	poly(chloroethene)	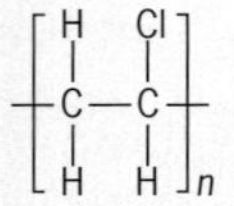 $[-CH_2-CHCl-]_n$	synthetic leather, water pipes, floor covering, guttering, window frames, curtain rails, wall cladding
methyl-2-methylpropenoate	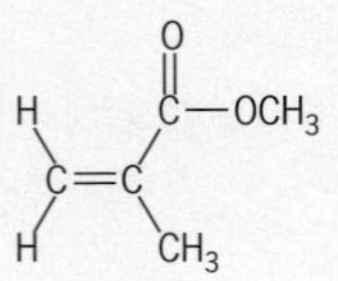 $H_2C{=}C(CH_3)COOCH_3$	poly(methyl-2-methylpropenoate)	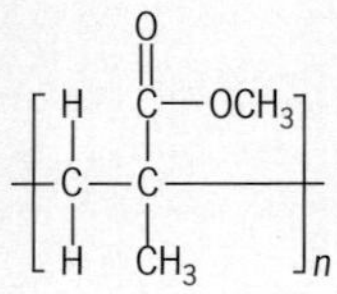 $[-CH_2-C(CH_3)(COOCH_3)-]_n$	light fittings, car rear lights, record player lids, tap tops, simple lenses

STRETCH AND CHALLENGE

Manufacture of poly(chloroethene) from ethene

Poly(chloroethene) is an important polymer, as seen from those uses listed in Table 14.3. It is manufactured from chlorine and ethene in a series of simple chemical reactions. First, ethene is reacted with chlorine to form 1,2-dichloroethane in a typical electrophilic addition:

$$Cl_2 + CH_2{=}CH_2 \rightarrow CH_2ClCH_2Cl$$

Next, the 1,2-dichloroethane from this reaction is heated very strongly to eliminate hydrogen chloride to form the monomer chloroethene. This involves a typical reaction of halogenoalkanes:

$$CH_2ClCH_2Cl \rightarrow CH_2{=}CHCl + HCl$$

Hydrogen chloride is a hazardous chemical that must not be allowed to reach the atmosphere. However, it is also a source of valuable chlorine. Therefore, the hydrogen chloride generated in the second reaction is recycled to make more 1,2-dichloroethane:

$$CH_2{=}CH_2 + \tfrac{1}{2}O_2 + 2HCl \xrightarrow[\text{catalyst}]{Cu^{2+}} CH_2ClCH_2Cl + H_2O$$

Chloroethene from the second reaction is a known carcinogen, and so it too is not allowed to escape from the reaction vessel into the atmosphere. Instead, this monomer is kept in a liquid state by increasing the pressure; it is then polymerised in the presence of a free-radical initiator to form poly(chloroethene):

$$nCH_2{=}CHCl \rightarrow \left[CH_2{-}CHCl \right]_n$$

The polymerisation is highly exothermic and has a free-radical chain mechanism. To avoid explosions the rate of reaction is carefully controlled by adding 'scavenger' chemicals to remove any free radicals.

Fig 14.46 Poly(methyl-2-methylpropenoate) is a transparent polymer used for car lights

Fig 14.47 Houses with unplasticised polyvinyl chloride (uPVC) wall fittings, window frames and doors

5 ALKENES AS PETROCHEMICAL STARTING MATERIALS

With their double bonds, alkenes can take part in many useful synthetic reactions. Almost all involve electrophilic additions and the introduction of functional groups to the carbon chain. It is these functional groups that allow further synthetic reactions and provide a route to a large number of organic substances.

ETHENE, PROPENE AND BUTENE

Fig 14.49 features six substances that can be synthesised directly from ethene. Most of the reactions involve electrophilic addition to the double bond and demonstrate why alkenes are of much more use than alkanes in the manufacture of petrochemicals. Alkanes have such a limited range of reactions that ethane could only be converted easily into one of the substances in Fig 14.48, chloroethane.

Propene and butene (Figs 14.50, 14.51) are two more very useful alkenes that can be converted into a variety of useful substances.

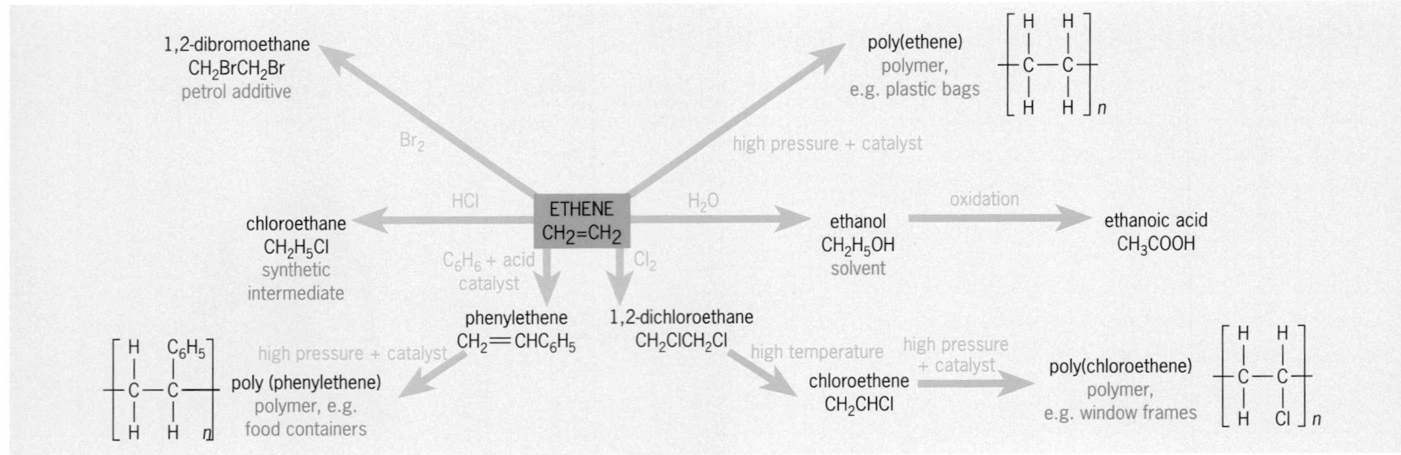

Fig 14.48 Addition and polymerisation reactions of ethene lead to a variety of useful petrochemicals

QUESTION 23

23 **a)** Identify the reactions in Fig 14.50 that are examples of electrophilic addition.

b) Identify the reactions in Fig 14.48 that are examples of polymerisation.

Fig 14.49 Poly(propene) is used to make marine rope

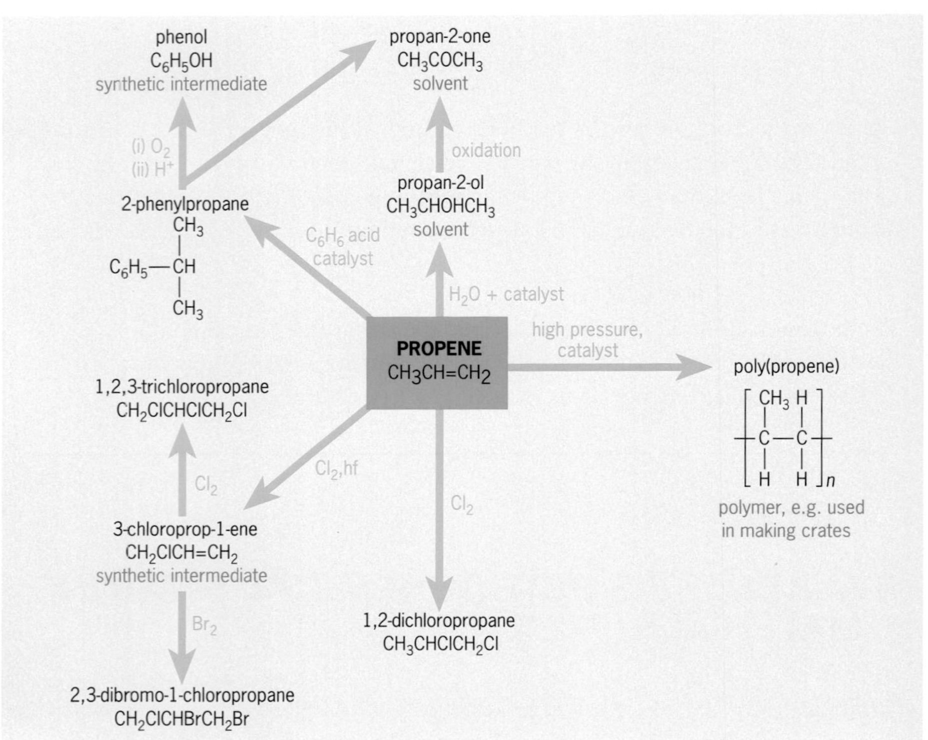

Fig 14.50 Propene is converted into many useful petrochemicals and synthetic intermediates

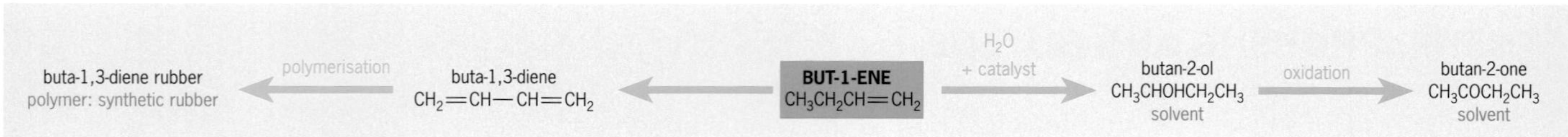

Fig 14.51 But-1-ene is converted into solvents and synthetic rubber

SUMMARY

After studying this chapter, you should know the following:

- Alkenes are more reactive than alkanes because of the presence of an electron-rich π bond.
- Alkenes are manufactured from long-chain alkanes by thermal or catalytic cracking.
- Alkenes are a major source of petrochemicals, such as polymers and plastics, solvents, detergents and pharmaceuticals.
- Alkenes burn in excess air to give carbon dioxide and water. Gaseous alkenes are sometimes used as fuels.
- Alkenes react by electrophilic addition with halogens to give dihalogenoalkanes, hydrogen to give alkanes, hydrogen halides to give halogenoalkanes, and water (catalysed by acid) to give alcohols.
- The addition of an electrophile to an unsymmetric alkene forms the most stable carbocation.
- Alkenes can be oxidised by cold dilute potassium manganate(VII) to give diols.
- Alkenes are monomers that form addition polymers.
- The double bond involves the overlap of two different types of atomic orbital to form a σ bond and a π bond.
- Unsymmetric alkenes show *E–Z* or geometric (*cis–trans*) isomerism because of the absence of free rotation about the carbon–carbon double bond.

Practice questions and a How Science Works assignment for this chapter are available at www.collinseducation.co.uk/CAS

15

Aromatic compounds and arenes

15 AROMATIC COMPOUNDS AND ARENES

SCIENCE IN CONTEXT

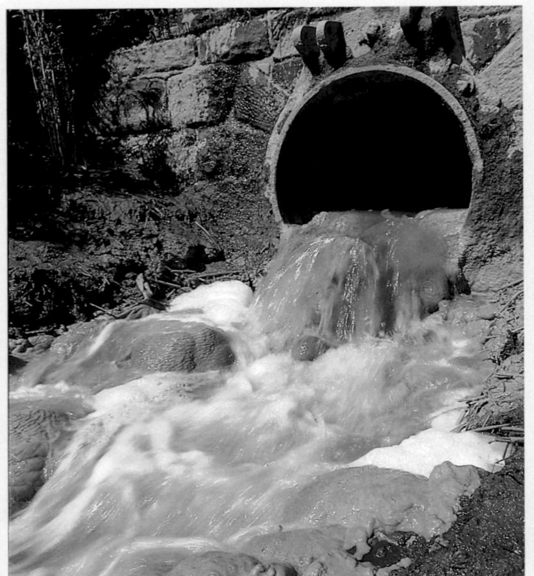

Pollution The use of envirocats will greatly reduce the pollution of waterways by effluents

Envirocats

Benzene is one of the most industrially important organic molecules. It features in the manufacture of petrochemicals, which are solvents, detergents, insecticides, dyes and polymers. The benzene molecule is normally unreactive, so to make these compounds requires special conditions such as high pressures and high temperatures. Almost all the reactions of benzene require catalysts, many of them hazardous environmental pollutants that can pollute water supplies and poison the organisms in them.

So there is an urgent need to develop environmentally friendly catalysts – *envirocats* – that can be separated easily by filtration from mixtures after they have reacted. Envirocats are based upon clay materials, which are made acidic by adding metal salts. Since they can be filtered from any aqueous effluent, envirocats can be prevented from harming water supplies and aquatic life. In addition, since envirocats are solid, their controlled disposal is easy. They have the economic benefit of conventional catalysts, too – envirocats can be used again and again, but eventually they lose their catalytic activity.

1 AROMATIC COMPOUNDS AND THEIR USES

In 1825, a young scientist at the Royal Institution of London was asked to remove the oily residue that collected in the gas cylinders of gas lamps and analyse it. He found that the residue contained a previously unknown hydrocarbon, for which the molecular formula was later shown to be C_6H_6. The scientist was Michael Faraday, one of the greatest scientists of the 19th century, though his name is seldom linked with organic chemistry.

The molecule of every aromatic compound contains at least one benzene ring. The name 'aromatic', meaning 'with an odour or smell', was given to these compounds because the first few to be discovered and used do, indeed, have odours. However, now we know that this was just a coincidence. Many aromatic compounds have no odour, whereas others have a particularly nasty smell. Also, some fragrant compounds do not contain a benzene ring (see page 259). Aromatic compounds can contain other functional groups as well as alkyl groups.

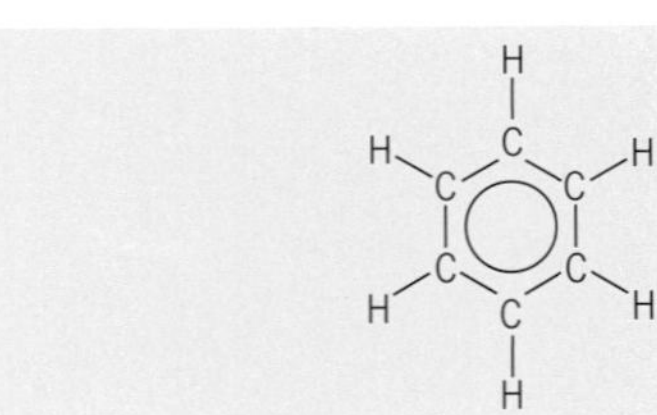

Fig 15.1 The displayed formula of benzene

The benzene ring is based upon a hexagonal arrangement of six carbon atoms and six delocalised π electrons (see page 181), which give the ring unique stability. Fig 15.1 shows benzene itself. The hydrogen atoms in the benzene molecule can be replaced by alkyl, aryl or other functional groups, as Fig 15.2 shows. Note the way in which the benzene ring is represented in compounds: it is a skeletal formula. The atomic symbols are not included, so remember that the benzene ring consists of six hydrogen atoms and six carbon atoms. The significance of the circle drawn inside the hexagon is explained on page 298.

? QUESTION 1

1 Draw the structure of each of these aromatic compounds:

a) C_7H_8
b) C_6H_5Cl
c) C_8H_{10}

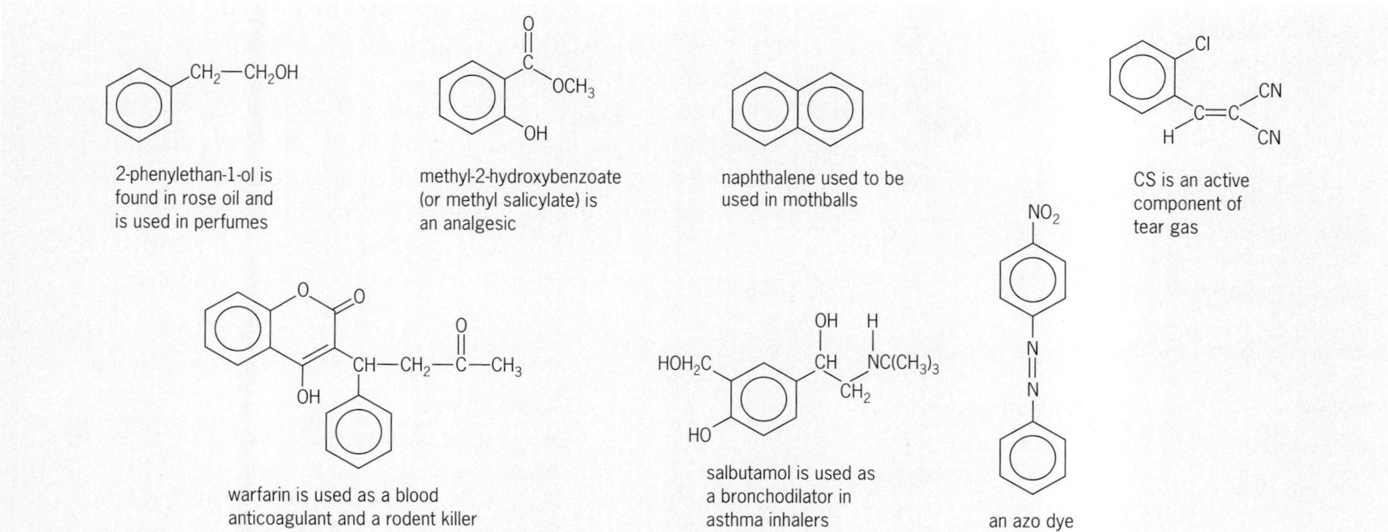

Fig 15.2 Seven aromatic compounds and their uses. Note that each compound has at least one benzene ring

SOME USES OF AROMATIC COMPOUNDS

Fig 15.2 features compounds that have one thing in common: the benzene ring. They are therefore all called aromatic compounds. All but two (2-phenylethan-1-ol and naphthalene) have to be synthesised industrially. The starting point for some of these compounds is benzene, C_6H_6, so the benzene part of their carbon skeleton is in place. Chemical changes are then needed to substitute one or more of the hydrogen atoms with one or more of the alkyl or aryl groups.

Fig 15.3 Salbutamol, an aromatic compound that contains alcohol and amine functional groups, is the world's most widely used bronchodilator: a bronchodilator relaxes the airways of people with breathing problems

2 STRUCTURE OF BENZENE

Benzene is one of the class of hydrocarbons known as arenes. Arenes are unsaturated hydrocarbons, characterised by great stability and non-reactivity, which distinguishes them from alkenes. All arenes contain a delocalised π system of electrons in their benzene rings.

> **REMEMBER THIS**
> **Arenes** are strictly hydrocarbons.
> **Aromatic** compounds can contain other functional groups.

KEKULÉ AND THE FIRST STRUCTURE OF BENZENE

Following the discovery of benzene, chemists had the challenge to try and determine its structure. In the early 19th century, benzene's molecular formula, C_6H_6, did not fit the way that chemists thought carbon atoms bonded in organic molecules.

It was 40 years later that August Kekulé proposed the ring structure of benzene, the first time a chemist realised that carbon atoms could form rings as well as chains. His brilliantly original idea was attributed to a dream he had about a snake biting its tail. Kekulé drew the structure of benzene as cyclohexa-1,3,5-triene (Fig 15.5).

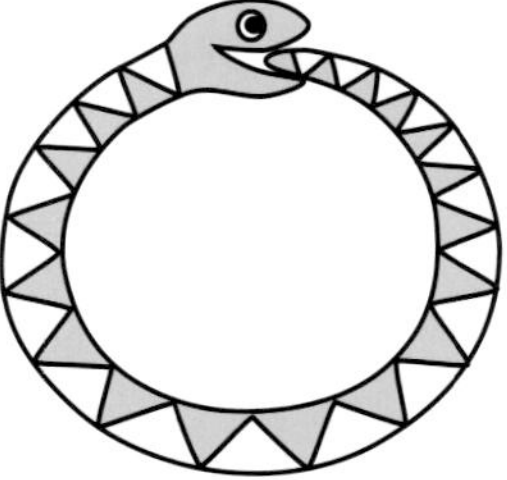

Fig 15.4 A snake biting its tail featured in Kekulé's inspirational dream

Kekulé's structure suggests that benzene has three single bonds and three double bonds, so that it should react in the same way as cyclohexene and be able to undergo electrophilic addition. It further suggests that benzene should have long carbon–carbon bonds, C–C, and short carbon–carbon bonds, C=C.

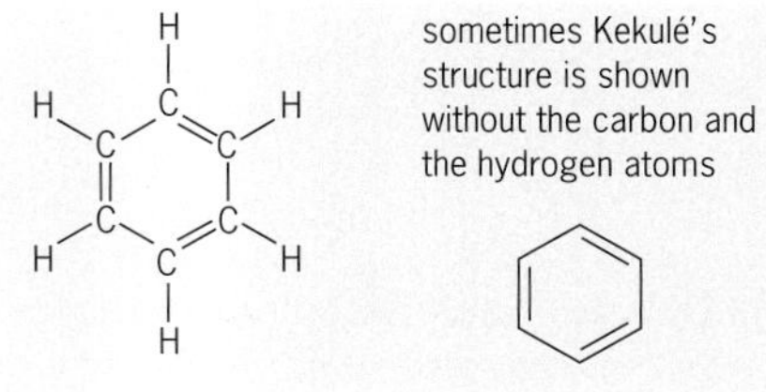

sometimes Kekulé's structure is shown without the carbon and the hydrogen atoms

Fig 15.5 Kekulé's structure of benzene (cyclohexa-1,3,5-triene)

Faults in Kekulé's model

Structural investigations using X-ray crystallography and infrared spectroscopy show that the benzene molecule is a regular hexagon of carbon atoms, with

REMEMBER THIS

The abbreviation pm stands for picometre, a unit of length equal to 10^{-12} m.

For information about x-ray crystallography, see page 78, and for information about infrared spectroscopy, read pages 220–224.

the lengths of all six carbon–carbon bonds equal to 140 pm. This value lies between the average bond length for a single bond, 154 pm, and the average bond length for a double bond, 134 pm.

A further complication is that benzene does not behave like an alkene. It reacts by electrophilic *substitution* rather than by electrophilic addition (see page 282). Table 15.1 shows some of the differences between cyclohexene, an alkene with a six-carbon ring, and benzene. The reactions of the two compounds are very different. This suggests that benzene cannot be an alkene and therefore cannot have Kekulé's structure of cyclohexa-1,3,5-triene.

Table 15.1 Comparison of the reactions of benzene and cyclohexene

Reagent	Cyclohexene (alkene)		Benzene (arene)	
bromine	electrophilic addition to give 1,2-dibromocyclohexane	CH_2, CH_2, CH—Br, CH_2, CH—Br, CH_2 (ring)	electrophilic substitution to give bromobenzene if catalyst used	Br
chlorine	electrophilic addition to give 1,2-dichlorocyclohexane	CH_2, CH_2, CH—Cl, CH_2, CH—Cl, CH_2 (ring)	electrophilic substitution to give chlorobenzene if catalyst used	Cl
hydrogen	addition to give cyclohexane		addition to give cyclohexane	
cold aqueous acidified potassium permanganate	oxidative addition to give cyclohexane-1,2-diol	CH_2, CH_2, CH—OH, CH_2, CH—OH, CH_2 (ring)	no reaction	
hydrogen bromide	electrophilic addition to give bromocyclohexane	CH_2, CH_2, CH_2, CH_2, CH—Br, CH_2 (ring)	no reaction	
oxygen (combustion)	carbon dioxide and water		carbon dioxide and water	
iodomethane	no reaction		electrophilic substitution to give methylbenzene if catalyst used	CH_3

Enthalpy change of hydrogenation

Unsaturated compounds, such as alkenes, react with hydrogen under pressure in the presence of a nickel catalyst to form alkanes (Figs 15.6 and 15.7). This reaction is known as **hydrogenation** (see page 285).

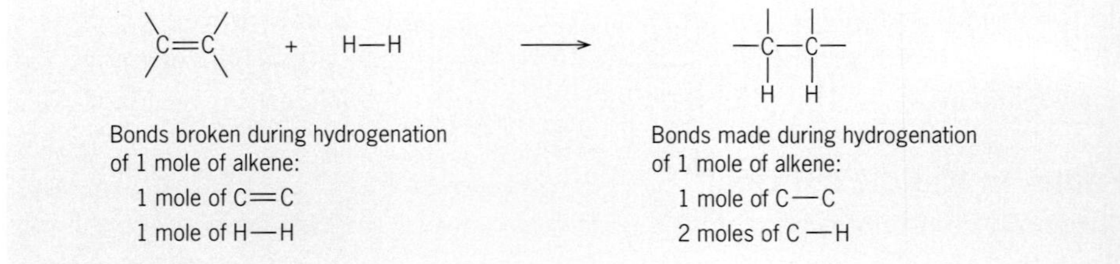

C=C + H—H → —C—C— H H
Bonds broken during hydrogenation of 1 mole of alkene:
1 mole of C=C
1 mole of H—H
Bonds made during hydrogenation of 1 mole of alkene:
1 mole of C—C
2 moles of C—H

Fig 15.6 Bond breaking and bond making during hydrogenation. Whatever groups are attached to the double bond, the bonds made and broken during hydrogenation are the same. When a molecule has a C=C bond, 1 mole of molecules will involve the breaking of 1 mole of C=C and 1 mole of H—H to make 1 mole of C—C and 2 moles of C—H

The **enthalpy change of hydrogenation** is defined as the enthalpy change that occurs when one mole of an alkene reacts with hydrogen to form an alkane. Studies of the hydrogenation of different alkenes show that for a molecule with one C=C bond, the enthalpy change of hydrogenation is about -120 kJ mol^{-1}. For a molecule with two double bonds, the enthalpy change of hydrogenation is about -240 kJ mol^{-1}.

Kekulé's structure for benzene has three C=C bonds, so we might expect the enthalpy change of hydrogenation for benzene to be about -360 kJ mol^{-1}. However, experimental data shows that its value is only -208 kJ mol^{-1}, which is a considerably smaller enthalpy change than the theoretical value. The only conclusions that can be drawn from the experimental data are that the benzene ring does not contain three C=C bonds, and that benzene is more stable than cyclohexa-1,3,5-triene.

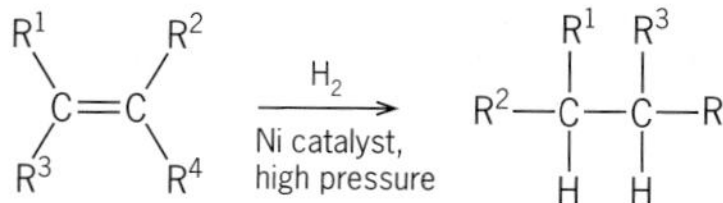

Fig 15.7 Hydrogenation of alkenes to form saturated compounds called alkanes

QUESTION 2

a) Write down equations to show the hydrogenation of:

i) cyclohexene;

ii) cyclohexa-1,3-diene;

iii) cyclohexa-1,3,5-triene (Kekulé's structure).

b) Predict the enthalpy change of hydrogenation for:

i) cyclohexene;

ii) cyclohexa-1,3-diene;

iii) cyclohexa-1,3,5-triene.

Fig 15.8 Benzene molecules as seen by a scanning tunnelling electron microscope, which shows the electron distributions in the molecules

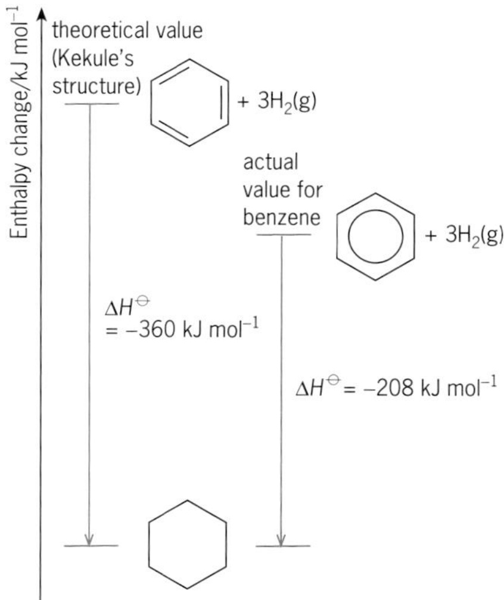

Fig 15.9 Energy level diagram for the enthalpy change of hydrogenation for benzene

MODERN THEORIES OF THE BENZENE MOLECULE

The value of the enthalpy change of hydrogenation of benzene is conclusive evidence that the benzene ring does not have the C=C bonds that Kekulé predicted. In addition, the enthalpy change transfers much less energy to the surroundings than his model predicts, which indicates that the bonding of the carbon atoms in the benzene ring is much stronger than Kekulé's model predicted, and so benzene is more stable.

HOW SCIENCE WORKS

Resonance structure model of the benzene molecule

Kekulé's model of benzene is inaccurate in the ways we have seen, and does not match the known physical dimensions of the benzene molecule. Fig 15.10 shows the structure of a benzene molecule as a hybrid of two Kekulé structures. The structure of benzene is called a **resonance hybrid**. Despite the name 'resonance hybrid', this model does not imply that the hybrid structure is continually changing from one to the other, but rather that it is a cross between the two structures, and possesses some of the characteristics of each. This means that each carbon–carbon bond has some single-bond and some d ouble-bond character. For this model it is often better to draw the structure of the benzene ring as shown in Fig 15.11. The circle represents a bond that contains six π electrons that are delocalised around the ring. This model has now been superseded by the orbital overlap model.

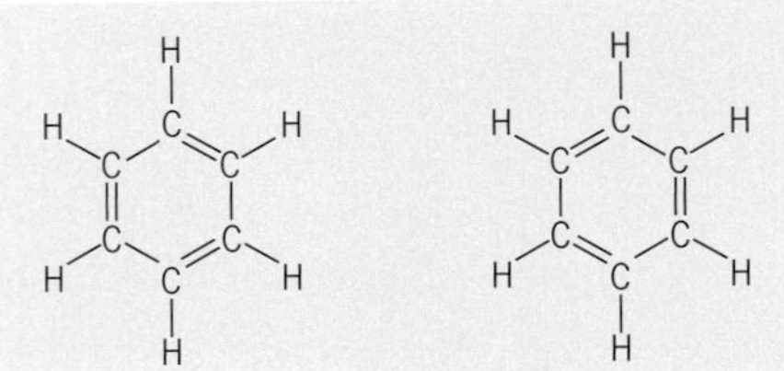

Fig 15.10 Two Kekulé structures of benzene that contribute to the resonance hybrid model of benzene

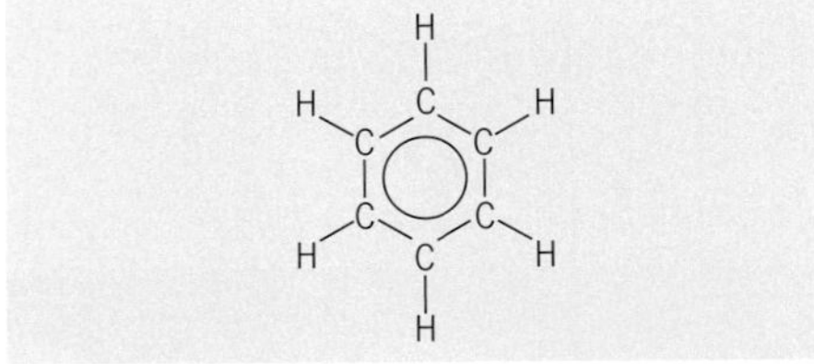

Fig 15.11 The displayed formula of benzene, in which the six delocalised π electrons are represented by a circle

Orbital overlap is covered on page 83.

Orbital overlap model

Covalent bonds are formed when atomic orbitals overlap (Fig 15.12). The C–H bond is a σ bond formed by the overlap of a hydrogen 1s atomic orbital with an atomic orbital on carbon. The C–C bond is a σ bond formed by the overlap of two atomic orbitals, one from each carbon atom.

In benzene, there is a third type of bond, which is formed by the side-to-side overlap of six 2p atomic orbitals on carbon. This forms an orbital that spreads, rather like a ring doughnut, above and below the plane of the carbon atoms that possess the six electrons. These electrons are not found between any particular atoms and so are known as delocalised π electrons.

(a)

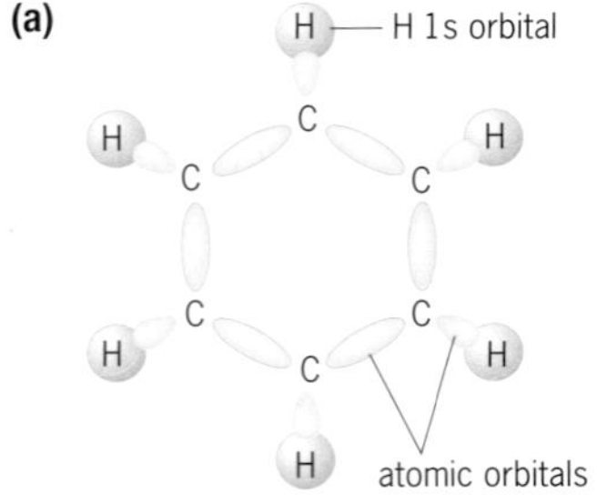

C–H σ bonds are made by the overlap of a hydrogen atomic orbital with a carbon atomic orbital. C–C σ bonds are made by the overlap of two carbon atomic orbitals

(b)

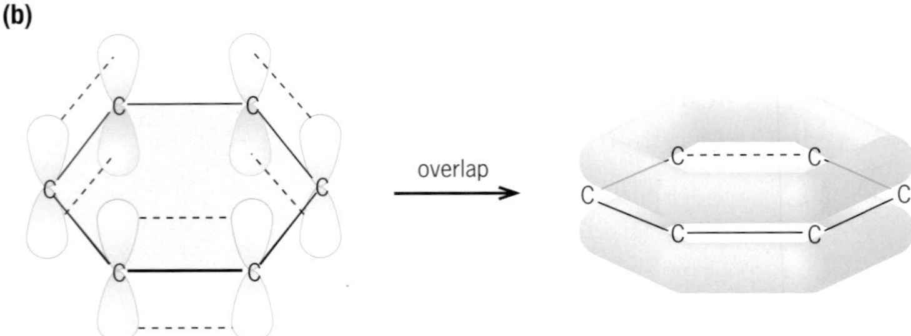

π bonds are formed by side-by-side overlap of all six 2p atomic orbitals

Fig 15.12 Molecular orbital description of benzene: (a) σ bonds; (b) π bonds

Stabilisation energy

Earlier in the chapter, we saw that the enthalpy change of hydrogenation of benzene is about 150 kJ mol^{-1} less than expected from the Kekulé model. This suggests that benzene is about 150 kJ mol^{-1} more stable than cyclohexa-1,3,5-triene (Kekulé's structure). This energy difference is sometimes referred to as the stabilisation energy of benzene.

Fig 15.13 In combustion, benzene burns with a smoky flame. It transfers less energy to its surroundings than would be expected from the Kekulé structure. This is because of the stability given by the delocalised electrons in the benzene ring

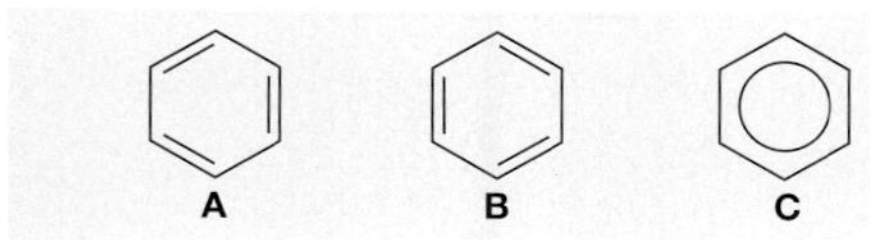

Fig 15.14 Three representations of the benzene ring. These are all skeletal formulae

The idea of stabilisation energy means that to break the benzene ring you need *more* energy than is predicted from Kekulé's model.

Structural formula for benzene

Fig 15.14 shows three ways to represent the displayed formula for benzene. Note that neither the hydrogen atoms nor the carbon atoms are shown. It is a skeletal formula. Every time you see one of these skeletal formulae of the benzene ring, remember that there are six hydrogen atoms attached to the ring. The circle inside the third hexagon represents the delocalised electrons. Each representation in Fig 15.14 has its advantages and disadvantages. Formulae A and B show three discrete double bonds (which do not exist in benzene), and formula C shows that some electrons are delocalised, but not how many. Nowadays, formula C is the most accepted way to represent the structure of benzene.

QUESTION 3

3 **a)** Write down the balanced equation for the complete combustion of benzene in air.
b) Would you expect the enthalpy change of combustion predicted from the bond energies of Kekulé's model to be the same as the experimentally measured value? Explain your answer.
(Hint: you may need to read page 160 about the enthalpy changes during bond breaking and bond making.)

REMEMBER THIS

In a displayed or structural formula, a shared pair of electrons is shown by a straight line between atoms.

QUESTION 4

4 What is the molecular formula for:
a) naphthalene;
b) pyrene?

3 NAMING AROMATIC COMPOUNDS

Arenes is the group name we give to benzene and the hydrocarbons that are just benzene rings joined together (fused). Arenes have similar chemical properties because their benzene rings have delocalised electrons. Fig 15.15 shows three arenes and their traditional names.

We have already described aromatic compounds in general as those that contain at least one benzene ring. Aromatic compounds are not necessarily hydrocarbons, but can contain a wide variety of other functional groups.

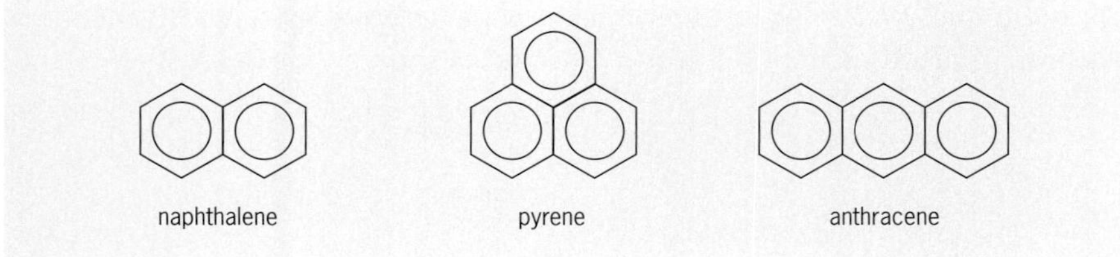

Fig 15.15 Three arenes: they all have fused benzene rings

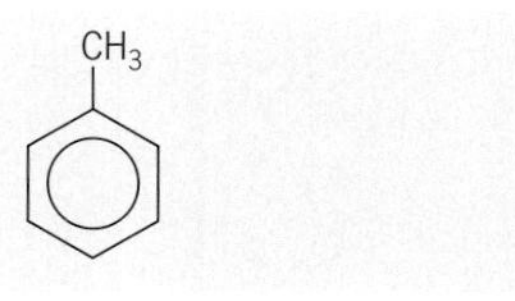

Fig 15.16 Methylbenzene

QUESTIONS 5–6

5 **a)** Draw the three structural isomers of dimethylbenzene.
b) Explain why there is not a compound called 1,5-dimethylbenzene.
6 Draw the structures of:
a) ethylbenzene;
b) 1,2-diethylbenzene;
c) 1,3-diethylbenzene.

SUBSTITUTED ARENES

To name aromatic compounds, we use the name of the parent arene together with a prefix or suffix to indicate the groups of atoms attached to the benzene ring. So, for example, the prefix 'methyl' in the name methylbenzene indicates that the methyl group has replaced one of the hydrogen atoms in the benzene ring (Fig 15.16). When more than one hydrogen atom is

QUESTION 7

7 **a)** Draw the structures of:

i) 1-chloro-3-iodobenzene;

ii) 1-chloro-4-iodobenzene.

b) A student names a compound 4,5-dichlorobenzene. Explain why this name must be incorrect.

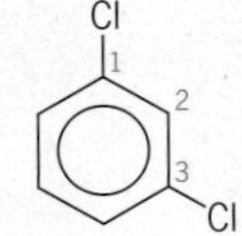

Fig 15.17 Naming 1,3-dichlorobenzene Note: this is 1,3-dichlorobenzene and not 1,5-dichlorobenzene, to keep the position-numbers as low as possible. The structure could equally well have been drawn as its mirror image, in which case the numbering would have been anticlockwise

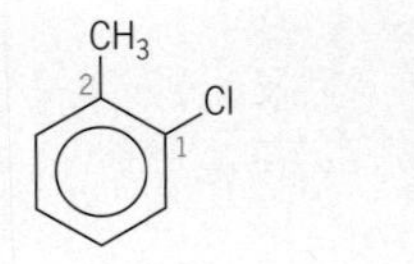

Fig 15.18 1-chloro-2-methylbenzene

replaced by methyl groups, as in dimethylbenzene, we specify the positions of the methyl groups by numbering the carbon atoms. So there are three structural isomers that could be called dimethylbenzene: 1,2-dimethylbenzene, 1,3-dimethylbenzene and 1,4-dimethylbenzene.

1,3-dichlorobenzene has a benzene ring with two chlorine atoms in place of hydrogen atoms. The first chlorine is attached to the carbon we number as carbon 1, and the second to carbon 3. A benzene ring is numbered clockwise or anticlockwise, depending on which direction gives the lower position-numbers. Fig 15.17 shows an example of the procedure.

As Fig 15.18 shows, 1-chloro-2-methylbenzene has two different groups attached to the benzene ring. Notice that we number the carbon attached to the chloro group as carbon 1, rather than the carbon attached to the alkyl group. (Halogens are named before alkyl groups.)

C_6H_5-

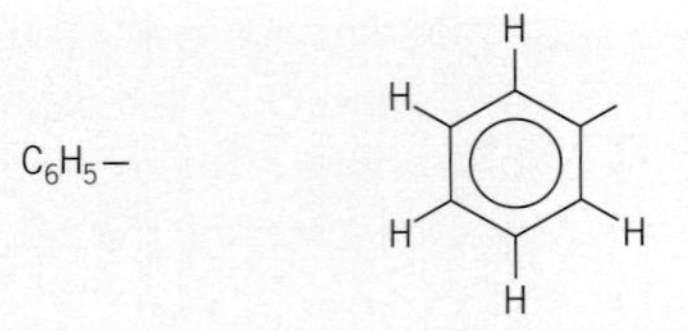

two representations of the phenyl group

Fig 15.19 The phenyl group

QUESTION 8

8 **a)** Write down the displayed formula for:

i) 2-phenylethan-1-ol;

ii) phenylethene.

b) What is the name of each of the four compounds in Fig 15.21?

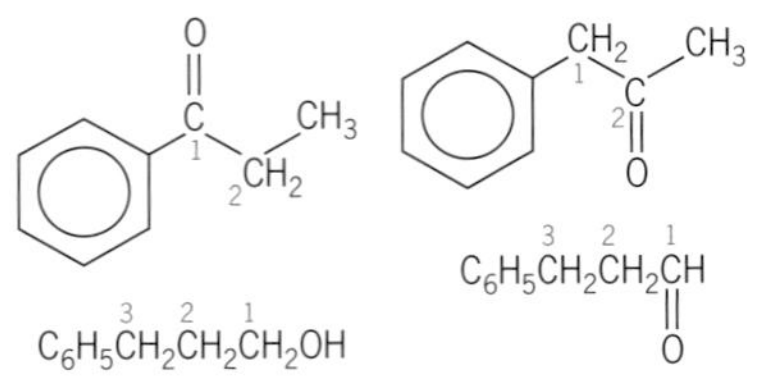

$C_6H_5CH_2CH_2CH_2OH$

$C_6H_5CH_2CH_2CHO$

Fig 15.21

Phenyl group

The phenyl group is C_6H_5 (Fig 15.19). It is found attached to carbon chains and to other functional groups. The phenyl group is an example of an **aryl** group. An aryl group is the aromatic equivalent of an alkyl group (see page 193) and always contains at least one benzene ring. In phenylamine, for example, the phenyl group is attached to the amine functional group. Phenylamine has the formula $C_6H_5NH_2$.

phenylethanone or $C_6H_5COCH_3$

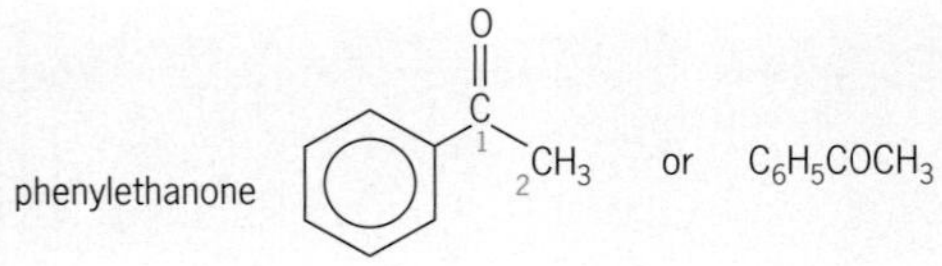

Fig 15.20 Phenylethanone. The position number of the phenyl group is not given, because the phenyl group has to be on carbon 1 for the compound to be a ketone

Compounds with some substituents (groups) on the benzene ring retain their traditional names. We refer to phenol instead of hydroxybenzene. In phenol, the carbon attached to the OH group is called carbon 1 (Fig 15.22).

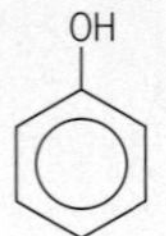

Fig 15.22 Phenol

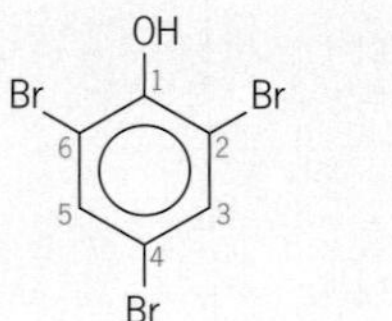

Fig 15.23 2,4,6-Tribromophenol

Benzoic acid and benzaldehyde

Benzoic acid and benzaldehyde (Fig 15.24) are two more aromatic compounds that have retained their older, pre-systematic names. Note carefully that both of these compounds contain seven carbon atoms rather than six, because the functional groups contain a carbon atom.

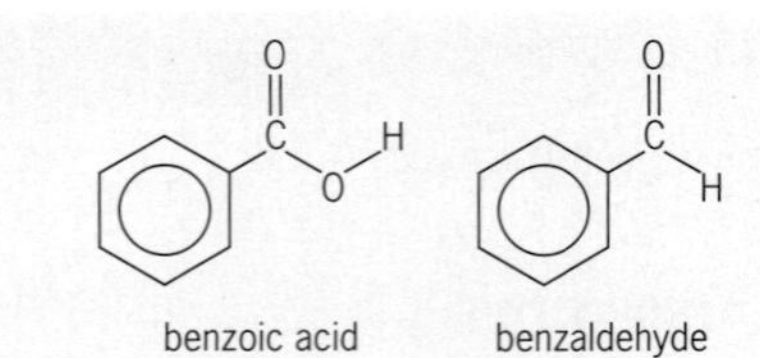

Fig 15.24 Benzoic acid and benzaldehyde

QUESTION 9

9 Draw the structures of:

a) 2,4,6-trichlorophenol;

b) 2,4,6-trimethylphenol.

SCIENCE IN CONTEXT

Arenes in exhaust fumes

Petrol contains aromatic hydrocarbons, which help to improve its octane rating. (See an account of octane rating on page 198.)

In an internal combustion engine, complete combustion of the fuel (petrol or diesel) rarely occurs. Some carbon monoxide, soot and smoke is always produced. Analysis of the soot by gas–liquid chromatography and spectroscopy shows that it contains the arenes benzene, naphthalene, phenanthrene, pyrene, coronene and ovalene (Fig 15.25). The presence of benzene is of particular concern, since it is a known carcinogen.

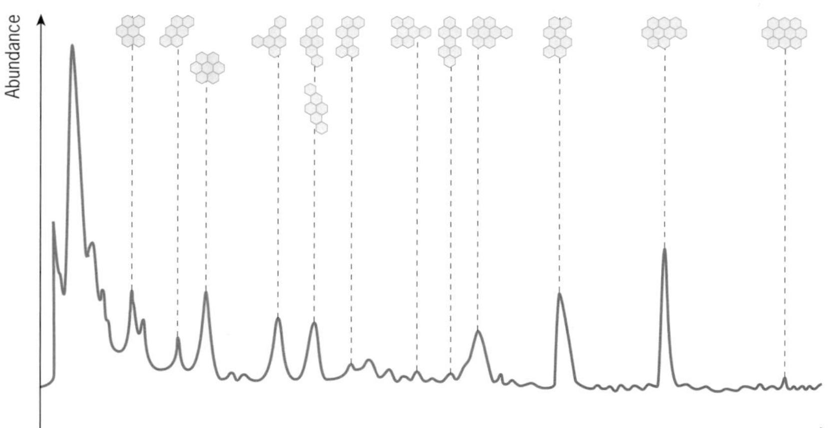

Fig 15.25 Gas–liquid chromatography and spectroscopic analysis of the volatile components of the soot from diesel-engine exhaust reveal many different arenes (structurally represented here as hexagons)

Fig 15.26 There is much concern about the amount of arenes emitted in vehicle exhaust fumes, since many arenes are known carcinogens

You can read about gas–liquid chromatography on page 232.

QUESTION 10

Deduce the molecular formula for each of the arenes shown in Fig 15.25.

4 DEMAND FOR AROMATIC COMPOUNDS

The synthesis of aromatic compounds often starts with one of the arenes. Aromatic compounds are needed for new technologies and new synthetic methods, and so the demand for arenes continues to expand. Until the 1940s, the main source of commercially produced benzene was coal, but today the principal source is crude oil.

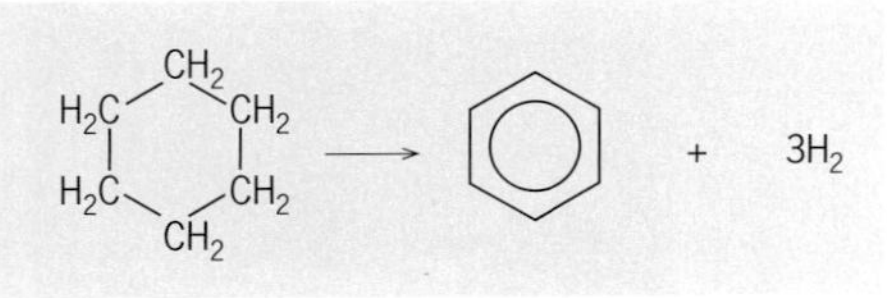

Fig 15.27 Reforming of cyclohexane to give benzene. Note that the carbon skeleton (the six-membered ring) remains intact

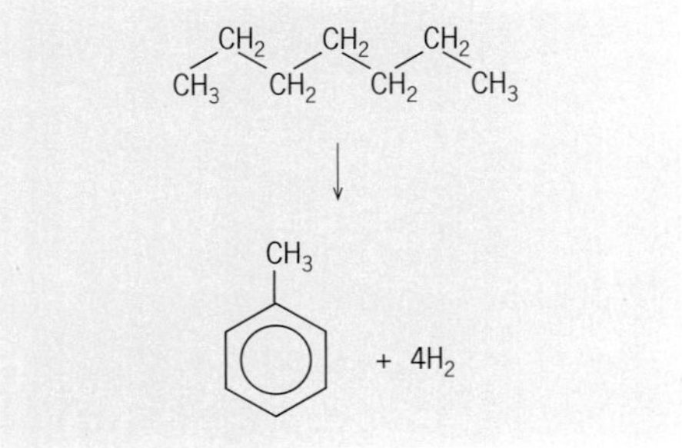

Fig 15.28 Reforming of heptane

Cracking and reforming are covered on pages 187–188.

? QUESTION 11

11 **a)** Into which aromatic hydrocarbon can 1,2-dimethylcyclopentane be reformed?

b) Write a balanced equation for this reaction.

REMEMBER THIS

An electrophile is a species that accepts a lone pair of electrons and forms a covalent bond.

AROMATIC COMPOUNDS FROM CRUDE OIL

Benzene and other aromatic compounds form only a small proportion of crude oil, and fractional distillation does not produce sufficient benzene or other aromatic compounds to match demand. So, as in the case of alkenes, cracking and reforming are processes used to increase the supply of arenes.

Catalytic reforming

In catalytic reforming, C_6 to C_{10} hydrocarbon vapours are heated in the presence of hydrogen at high pressure and high temperature, typically 900 °C, using a transition-element catalyst.

During the reforming process, the number of carbon atoms per molecule remains the same, but a change occurs in the carbon skeleton and/or the number of hydrogen atoms per molecule. For example, cyclohexane can be reformed to give benzene (Fig 15.27), and heptane can be reformed to give methylbenzene (Fig 15.28). In the second example, note that not only are hydrogen atoms lost, but also the carbon skeleton is rearranged.

Manufacture of benzene

Naphtha is a light fraction of crude oil that contains liquid alkanes of low relative molecular mass (see page 186). To make benzene, naphtha vapour is passed over a catalyst, such as platinum or molybdenum(VI) oxide, on aluminium oxide at 500 °C and 10–20 atmospheres of pressure. Compounds such as hexane found in naphtha are converted into ring compounds such as cyclohexane, which is then dehydrogenated to make benzene.

$$CH_3CH_2CH_2CH_2CH_2CH_3 \rightarrow C_6H_{12} \rightarrow C_6H_6$$

5 ELECTROPHILIC SUBSTITUTION REACTIONS OF BENZENE

In almost all synthetic schemes that start with an aromatic hydrocarbon (such as benzene), the substitution of one or more of the hydrogen atoms occurs. This substitution is normally electrophilic, and so only those groups that can behave as electrophiles are able to replace hydrogen atoms attached to the benzene ring.

MECHANISM OF ELECTROPHILIC SUBSTITUTION

The presence of the delocalised electrons means that benzene is electron rich. Therefore, an electrophile that is able to accept an electron pair donated by the ring is the most likely type of reagent to react with benzene. This is very similar to the first step in the electrophilic addition of an alkene (see page 282). The resulting positive species is much more stable than a normal carbocation, since it is possible to delocalise the positive charge around the ring (Fig 15.29). A proton is then eliminated from the positive ion to restore the stability of the benzene ring. We saw the unique stability of the benzene ring when looking at the bonding in benzene. This stability is also apparent in the mechanism of electrophilic substitution.

intermediate has positive charge delocalised over five carbon atoms

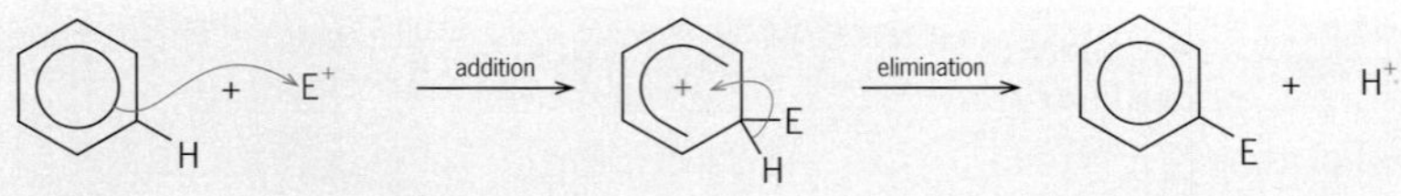

Fig 15.29 The mechanism of electrophilic substitution of benzene

Electrophilic substitution clearly needs an electrophile, but what is not obvious is that the electrophile must be an exceptionally good electrophile. This is because it has to react with a molecule made extremely stable by the presence of the π system of delocalised electrons. Benzene has a delocalised π system of electrons over all six atoms of the ring. But Fig 15.29 shows that in the positive intermediate, the π system involves only five of the carbon atoms. In practice, this means that the electrophile normally needs to be a fully positive species that is formed as a result of the heterolytic fission of a covalent bond (Fig 15.31). An induced dipole within a molecule is not normally enough to achieve electrophilic attack in benzene.

It is not easy to form a sufficiently powerful electrophile to attack a benzene molecule, so a catalyst is used in the reaction mixture to generate the electrophile. Typical electrophiles that can be generated in this way include Br^+, Cl^+, R^+, RCO^+ and NO_2^+. (R is an aryl or alkyl group.) You will notice that the charge carried by the atoms involved is not often what you would have expected by considering their electronic arrangements, which offers an explanation for why catalysts are needed to generate the electrophiles.

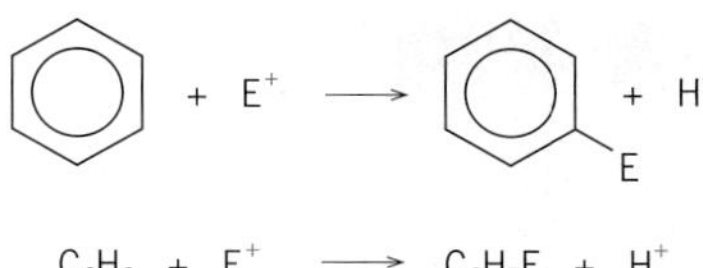

$$C_6H_6 + E^+ \longrightarrow C_6H_5E + H^+$$

Fig 15.30 The overall reaction of electrophilic substitution

$$E{-}X \longrightarrow E^+ + :X^-$$

Fig 15.31 The electrophile is formed by the heterolytic fission of a covalent bond, so that both electrons in the bond go to atom X

QUESTIONS 12–13

12 **a)** Write down the equation for the reaction of Cl^+ with benzene.

b) Draw the mechanism for the reaction of benzene with NO_2^+.

13 What is the electric charge you would expect to be carried by a chlorine atom when it forms an ion?

HALOGENATION OF BENZENE

Chlorine and bromine can both be converted into electrophiles that will attack a benzene ring. The resulting substitution reaction makes chlorobenzene or bromobenzene.

$$C_6H_6 + X_2 \rightarrow C_6H_5Cl + HX \text{ (where X is Br or Cl)}$$

Notice that the equation is written with molecular formulae rather than structural or displayed formulae. This is acceptable provided that there is only one structure possible.

Chlorination of benzene

When chlorine is bubbled through a mixture of refluxing benzene in the presence of a catalyst (such as aluminium chloride or iron(III) chloride), electrophilic substitution takes place to produce chlorobenzene:

$$C_6H_6(l) + Cl_2(g) \xrightarrow{FeCl_3,\ reflux} C_6H_5Cl(l) + HCl(g)$$

For more information on anhydrous aluminium chloride, see Chapter 21.

We already know the mechanism of this reaction: Cl^+ is represented by E^+ in Fig 15.29. The major question is: how does aluminium chloride help to generate the electrophile?

We might expect aluminium chloride to be ionic, since it is a metal compound. However, in the anhydrous state it has a high degree of covalent character, so it is better when drawing a dot-and-cross diagram to show the electrons as shared (Fig 15.32). This structure has the molecular formula $AlCl_3$, though there is strong evidence to suggest that the formula is actually Al_2Cl_6.

The dot-and-cross diagram of Fig 15.32 shows that $AlCl_3$ is an electron-deficient molecule, since the outer shell of the aluminium atom has only six electrons.

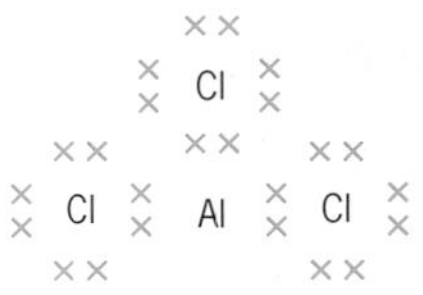

Fig 15.32 Dot-and-cross diagram for aluminium chloride

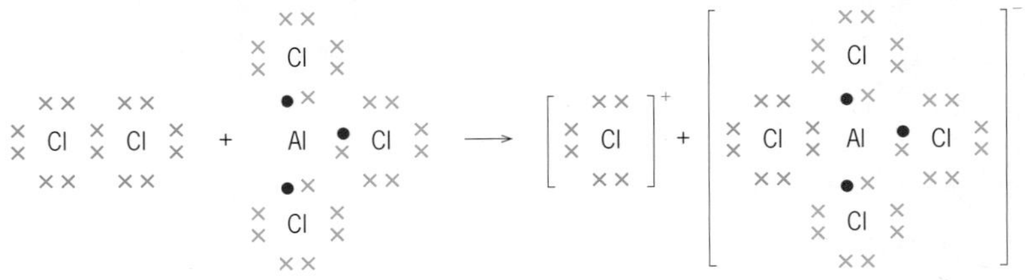

Fig 15.33 Dot-and-cross diagrams to show the formation of Cl^+. The electron deficiency of the aluminium chloride is transferred to the chlorine electrophile Cl^+. Note that the aluminium atom has eight electrons in its outer shell after the reaction, while the Cl^+ ion has six

QUESTION 14

14 Write down the equation for the reaction between a proton (hydrogen ion) and a tetrachloroaluminate ion.

Anhydrous aluminium chloride transfers its electron deficiency to a chlorine atom, to form the electrophile Cl^+ (Fig 15.33) and a tetrachloroaluminate ion, $AlCl_4^-$. The catalyst is often referred to as a halogen carrier. Once the electrophile is generated, the reaction proceeds to give the substitution product. The proton eliminated by the benzene ring at the end of the reaction can then react with the $AlCl_4^-$ ion to give hydrogen chloride.

Anhydrous iron(III) chloride acts as a catalyst in the same way. In fact, iron filings could be used as a catalyst, because iron(III) chloride would be formed in the reaction vessel by the reaction between iron and chlorine.

QUESTION 15

15 Explain, using dot-and-cross diagrams, how the electrophile Br^+ can be produced from bromine and anhydrous aluminium chloride.

Bromination of benzene

When a solution of bromine in benzene is refluxed with aluminium chloride, electrophilic substitution takes place to form bromobenzene:

$$C_6H_6 + Br_2 \rightarrow C_6H_5Br + HBr$$

Notice that the catalyst is not included in the overall chemical equation, even though it is intimately involved in the production of the electrophile. We do not have to use aluminium bromide as the catalyst, because aluminium chloride will work instead, in which case the bromo-trichloroaluminate ion, $AlBrCl_3^-$, is formed instead of $AlBr_4^-$.

ALKYLATION OF BENZENE

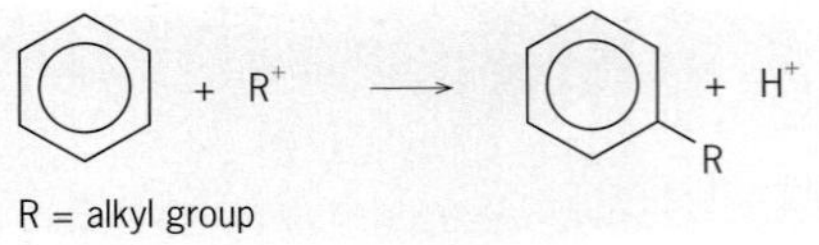

R = alkyl group

Fig 15.34 The alkylation of benzene: the Friedel–Crafts reaction

This reaction is known as the Friedel–Crafts reaction. In it, a hydrogen atom is substituted by an alkyl group (Fig 15.34). The importance of this reaction is that a carbon–carbon bond is made. This time the problem is to make a carbocation that can act as an electrophile.

There are two main ways in which the electrophile (the carbocation) can be generated. One involves the heterogeneous fission of a halogenoalkane, and the other the electrophilic addition of a proton to an alkene (Fig 15.35).

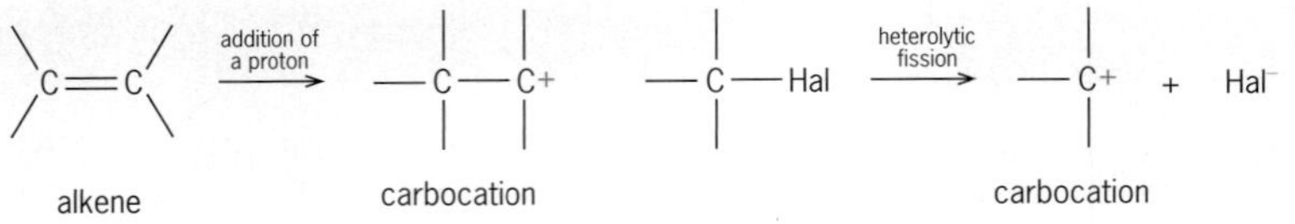

addition of a proton
heterolytic fission
alkene
carbocation
carbocation

Fig 15.35 Generation of carbocations by two routes

Carbocations from halogenoalkanes

Typically, benzene, a halogenoalkane and anhydrous aluminium chloride are refluxed together to give an alkylbenzene. For example, iodomethane gives methylbenzene:

$$C_6H_6 + CH_3I \xrightarrow{AlCl_3,\ reflux} C_6H_5CH_3 + HI$$

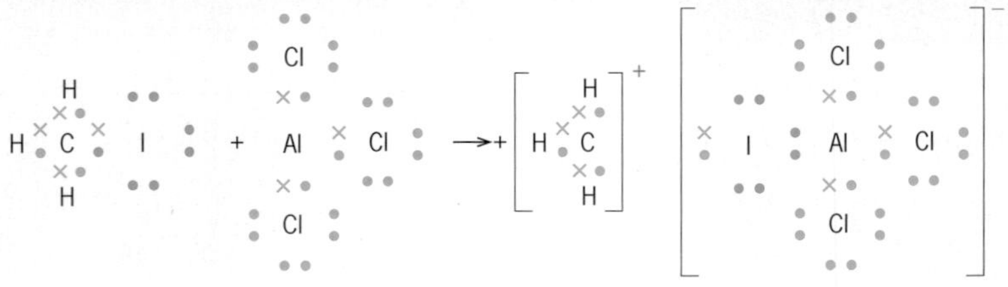

Fig 15.36 Dot-and-cross diagram to show the formation of a methyl carbocation from iodomethane

QUESTION 16

16 Draw the structure of the main organic product formed during the reaction between iodomethane and benzene in the presence of aluminium chloride.

Let's look at the formation of the carbocation that permits this reaction. Fig 15.36 shows that it is similar to the way in which the electrophile Cl^+ is formed (see Fig 15.33). Again, the aluminium chloride can transfer its electron deficiency, this time to an alkyl group, to form the carbocation.

Carbocations from alkenes

Electrophilic addition to alkenes is covered on pages 282–285.

Alkenes typically react by electrophilic addition (Fig 15.37). When the electrophile is a proton, a carbocation is formed as an intermediate. When this is formed in the presence of benzene, the carbocation can react with benzene, rather than completing the addition reaction.

Many different types of acid catalyst can be used in this reaction, including concentrated sulfuric acid, concentrated phosphoric(V) acid and aluminium chloride.

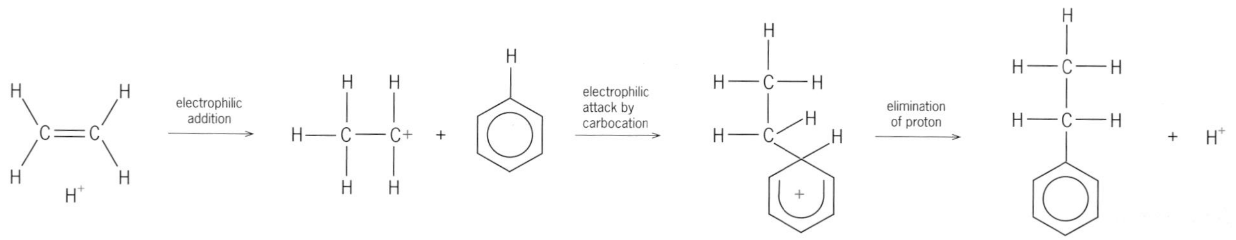

$$CH_2{=}CH_2 + H^+ \xrightarrow{\text{electrophilic addition}} CH_3{-}CH_2^+ \;+\; C_6H_6 \xrightarrow{\text{electrophilic attack by carbocation}} [C_6H_6CH_2CH_3]^+ \xrightarrow{\text{elimination of proton}} C_6H_5CH_2CH_3 + H^+$$

Fig 15.37 The mechanism of the acid-catalysed reaction of an alkene (in this case ethene) with benzene. Note that the proton H^+ reacts to generate the carbocation, but is later regenerated; hence it is a catalyst

STRETCH AND CHALLENGE

Manufacture of poly(phenylethene)

Poly(phenylethene), common name polystyrene, is familiar to us as the material of hot drinks cups, ceiling tiles and packaging for fragile objects. It is manufactured from two readily available hydrocarbons, ethene and benzene (Fig 15.38).

Fig 15.38 The manufacture of poly(phenylethene)

$$CH_2{=}CH_2 + C_6H_6 \xrightarrow[HCl(aq)]{AlCl_3(s)} C_6H_5CH_2{-}CH_3 \xrightarrow[650\ ^\circ C]{Fe_2O_3\ \text{catalyst}} C_6H_5CH{=}CH_2 \xrightarrow[\text{peroxide initiator}]{\text{free-radical}} [{-}CH(C_6H_5){-}CH_2{-}]_n$$

Ethene reacts with benzene in the presence of aluminium chloride, with hydrochloric acid as an acid catalyst, to form ethylbenzene. This is an example of electrophilic substitution. The purpose of the acid catalyst is to generate an ethyl carbocation that acts as the electrophile.

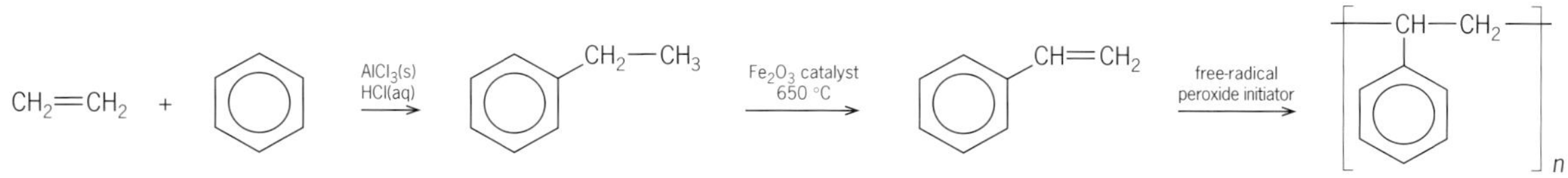

The ethylbenzene from the first reaction provides the correct carbon skeleton for the next stage. This involves a dehydrogenation reaction, in which two hydrogen atoms are lost to form phenylethene. Finally, the phenylethene is polymerised normally under the influence of a free-radical initiator.

Fig 15.39 Mobile telephones are made from poly(phenylethene)

Fig 15.40 Expanded poly(phenylethene), also known as polystyrene, is used for packaging material

ACYLATION OF BENZENE

Acylation is the substitution of a hydrogen atom by an acyl group, RCO (Fig 15.41).

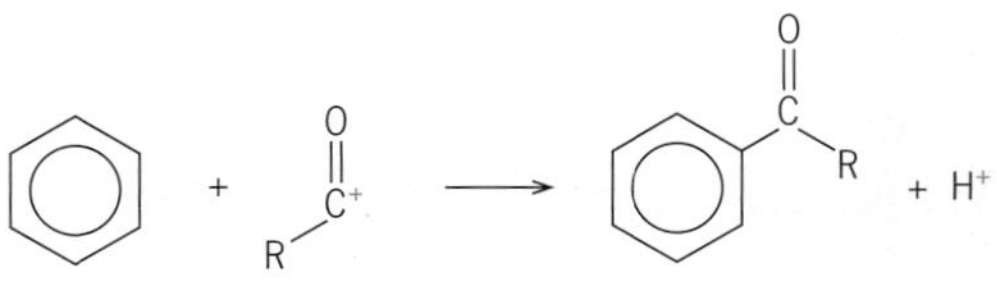

$$C_6H_6 + RC^+{=}O \longrightarrow C_6H_5COR + H^+$$

Fig 15.41 Acylation of the benzene ring. A hydrogen atom is substituted by the acyl group, RCO

? QUESTION 17

17 **a)** Write down the structure of the product formed in the reaction between propanoyl bromide, CH_3CH_2COBr and benzene.
b) Draw the structure of the reagent used to make 1-phenylbutan-1-one from benzene.

This reaction requires the formation of an acylium ion, RCO^+, to act as an electrophile. The acylium ion is produced by the heterolytic fission of a carbon–chlorine bond in an acyl chloride catalysed by anhydrous aluminium chloride (Fig 15.42). For example, ethanoyl chloride, aluminium chloride and benzene react together to give phenylethanone (Fig 15.43).

$$R{-}C({=}O){-}Cl + AlCl_3 \longrightarrow R{-}C({=}O)^+ + [AlCl_4]^-$$

acylium ion

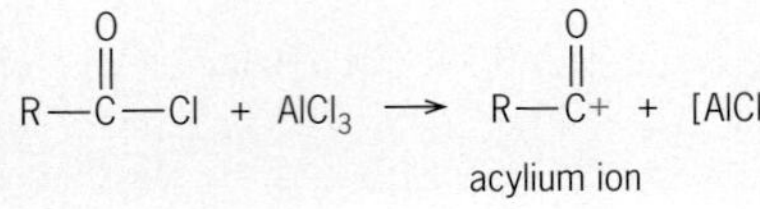

Fig 15.42 Generation of an acylium ion

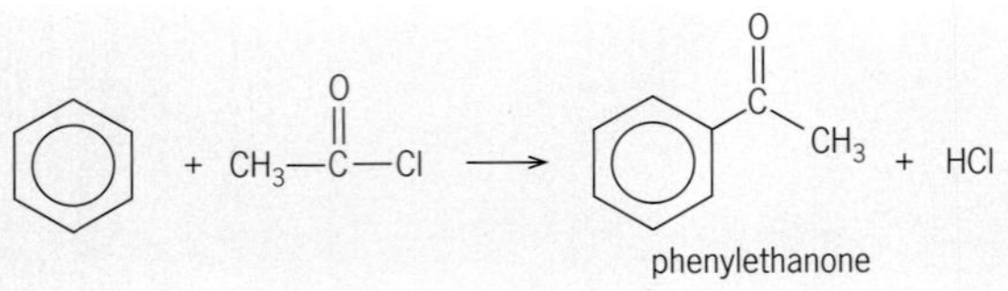

$$C_6H_6 + CH_3{-}C({=}O){-}Cl \longrightarrow C_6H_5{-}C({=}O){-}CH_3 + HCl$$

phenylethanone

Fig 15.43 Reaction of ethanoyl chloride with benzene

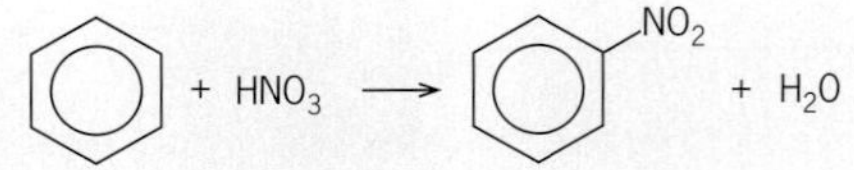

Fig 15.44 Nitration of benzene

NITRATION OF BENZENE

It is possible to introduce the nitro group into the benzene ring (Fig 15.44) by means of a nitrating mixture – in this case, concentrated nitric acid and concentrated sulfuric acid. Nitrobenzene is formed when benzene is heated at less than 60 °C with a nitrating mixture. Keeping the temperature below 60 °C stops the formation of di- or trinitrobenzene.

The electrophile in the reaction (Fig 15.45) is the nitryl or nitronium ion, NO_2^+, which is generated by a reaction between the two acids:

$$2H_2SO_4 + HNO_3 \rightarrow NO_2^+ + 2HSO_4^- + H_3O^+$$

This equation represents the overall process that starts with an acid–base reaction using two acids. The concentrated sulfuric acid donates a proton to the nitric acid, and then the protonated nitric acid loses a molecule of water:

$$H_2SO_4 + HNO_3 \rightarrow HSO_4^- + H_2NO_3^+$$

$$H_2NO_3^+ \rightarrow H_2O + NO_2^+$$

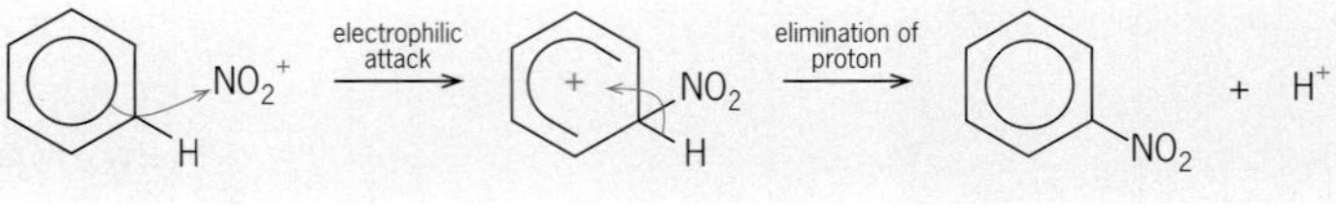

$$C_6H_6 + NO_2^+ \xrightarrow{\text{electrophilic attack}} [C_6H_6NO_2]^+ \xrightarrow{\text{elimination of proton}} C_6H_5NO_2 + H^+$$

Fig 15.45 Mechanism of the nitration of benzene. The nitryl ion, NO_2^+ is the electrophile and accepts an electron pair from the delocalised π system of the benzene ring. The positively charged intermediate is stabilised because it has a delocalised π system, but only over five carbon atoms. In this way, the positive charge is spread around the five atoms. Finally, a proton is eliminated from the intermediate to form a benzene ring. The proton reacts with HSO_4^- to regenerate the sulfuric acid

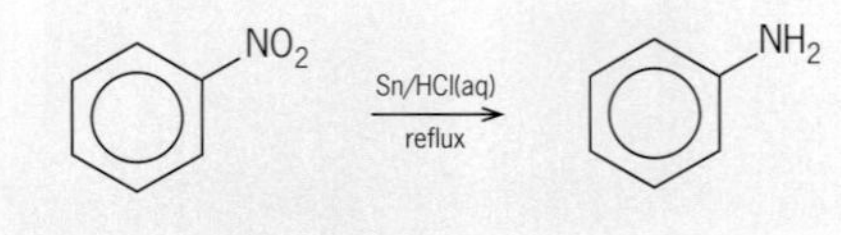

Fig 15.46 The reduction of nitrobenzene to form phenylamine. This provides a route to a reactive intermediate in the manufacture of dyes (see Chapter 27)

The nitration of benzene allows other nitrogen-containing functional groups to be introduced, since, for example, it is easy to reduce the nitro group to an amine, to give phenylamine. It is otherwise impossible to introduce the NH_2 group into a benzene ring: NH_2 is impossible to generate.

SULFONATION OF BENZENE

Sulfonation involves the reaction of concentrated sulfuric acid on its own with benzene (Fig 15.47). It is believed that sulfur trioxide, SO_3, is the electrophile. In the reaction, a hydrogen atom in the benzene ring is replaced by the sulfonic acid functional group to produce benzenesulfonic acid. Aromatic sulfonic acids are important components of detergents.

In the presence of cold, dilute, aqueous sodium hydroxide, benzene sulfonic acid is converted into the salt sodium sulfonate:

$$C_6H_5SO_3H + NaOH \rightarrow C_6H_5SO_3^-Na^+ + H_2O$$

However, benzenesulfonic acid is hydrolysed by boiling with aqueous sodium hydroxide to form sodium phenoxide (Fig 15.48). In this way, the OH group can be introduced into a benzene ring, even though it is impossible to generate OH^+ (the electrophile that would be necessary to produce phenol directly from benzene). So, in two easy steps, phenols can be produced from benzene or substituted benzenes.

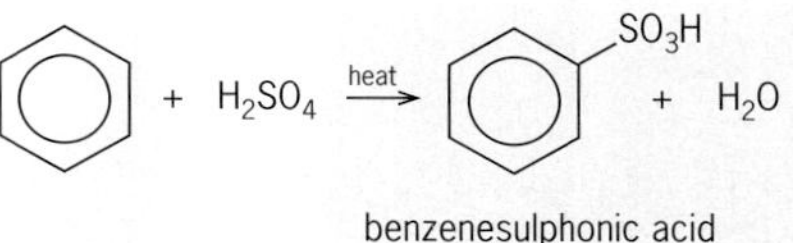

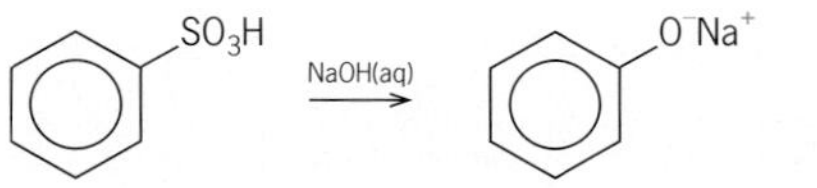

Fig 15.47 The sulfonation of benzene

SO_3H $\xrightarrow{NaOH(aq)}$ O^-Na^+

Fig 15.48 The hydrolysis of benzenesulfonic acid to form sodium phenoxide

? QUESTION 18

18 Write down the equation to show the hydrolysis of benzenesulfonic acid to form sodium phenoxide.

6 ADDITION REACTIONS OF BENZENE

Although it does not contain carbon–carbon double bonds, benzene is clearly an unsaturated compound, and under certain conditions will undergo an addition reaction to give a cyclohexane ring. The reaction is difficult, because the exceptional stability of the benzene ring has to be overcome.

HYDROGENATION OF BENZENE

Benzene is converted into cyclohexane when it is reacted with hydrogen under pressure in the presence of a nickel catalyst:

$$C_6H_6 + 3H_2 \rightarrow C_6H_{12}$$

It is impossible to stop the reaction to give either cyclohexene or cyclohexadiene, which is further evidence that benzene has a delocalised system of electrons rather than individual C=C bonds.

? QUESTION 19

19 What is the molecular formula of the hydrocarbon obtained by the hydrogenation of naphthalene?

ADDITION OF CHLORINE TO BENZENE

In the presence of chlorine and a catalyst, such as aluminium chloride or iron(III) chloride, benzene undergoes electrophilic substitution to form chlorobenzene. If the conditions are changed, and chlorine is bubbled through refluxing benzene in the presence of ultraviolet light without a catalyst, an addition reaction occurs with chlorine to give stereoisomers of 1,2,3,4,5,6-hexachlorobenzene.

? QUESTION 20

20 Write down the equation to show the addition of chlorine to benzene.

7 ELECTROPHILIC SUBSTITUTION REACTIONS OF SUBSTITUTED BENZENES

Electrophilic substitutions are characteristic of all compound with a benzene ring, not just arenes and other aromatic hydrocarbons. This raises the question: what happens to a molecule that has two functional groups – for example, in phenol, the benzene ring and the hydroxyl group? The answer is simple: the compound, in this case phenol, undergoes two sets of reactions, one involving the benzene ring and the other the hydroxyl group.

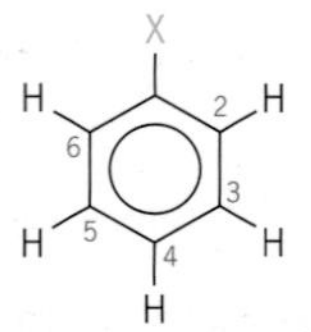

Fig 15.49 In a substituted benzene, there are three sets of hydrogen atoms: on carbon atoms 2 and 6, on carbon atoms 3 and 5, and on carbon atom 4

Consider a substituted benzene C_6H_5X, in which X is an atom or a group of atoms such as a methyl group. As the compound contains the benzene ring, it will undergo electrophilic substitution. It may also undergo another set of reactions because of the presence of X.

Limiting the discussion to reactions of the benzene ring in the substituted benzene, C_6H_5X, two questions arise. The first is: will the reaction be faster or slower than that with benzene? The second is: which hydrogen atom will be substituted? When the reaction is faster, X is said to **activate** the benzene ring. When the reaction is slower, X is said to **deactivate** the benzene ring.

Now look at Fig 15.49, which shows a substituted benzene. Three sets of hydrogen atoms can be substituted. Not all of the hydrogen atoms shown will be substituted. The group X *directs the substitution* so that only certain hydrogen atoms are substituted. The effect of the substituent X on electrophilic substitution is summarised in Table 15.2.

Table 15.2 Summary of the effects of substituents on electrophilic substitution of the benzene ring

Substituent	Reactivity compared with benzene	Direction of substitution
CH_3 and other alkyl groups	activating	2, 4 and/or 6
Cl, Br, and I	deactivating	2, 4 or 6
NO_2	deactivating	3 or 5
OH and OCH_3	highly activating	2, 4 and 6

ELECTROPHILIC REACTIONS OF PHENOL

Phenol consists of the benzene ring and the attached hydroxyl group. What effect does the hydroxyl group have on electrophilic substitution? The answer is easy to find out experimentally: aqueous bromine reacts with a solution of phenol in aqueous sodium hydroxide at room temperature to give 2,4,6-tribromophenol, which is a white precipitate. The reaction proceeds without a catalyst – the substitution does not require a catalyst to generate the electrophile. Also, more than one hydrogen atom is substituted.

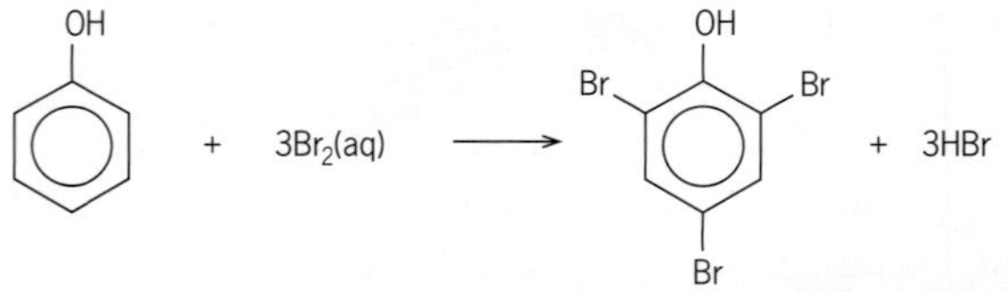

Fig 15.50 Reaction of aqueous bromine with phenol

Therefore the hydroxyl group is *highly activating*, since the reaction takes place much faster and more easily than with benzene. The hydroxyl group also directs the substitution to positions 2, 4 and 6 in the benzene ring (Fig 15.51).

The reason why phenol reacts much faster than benzene is that the lone pair of electrons on the oxygen atom becomes part of the delocalised π system. The π system is therefore extended over *seven* atoms rather than six (six carbon atoms and one oxygen atom). This has the effect of *increasing the electron density* in the π system, thereby making the benzene ring much more susceptible to electrophilic attack.

Phenol is very susceptible to electrophilic attack and is a very useful synthetic intermediate, as shown in Fig 15.52.

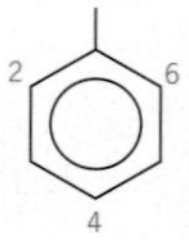

Fig 15.51 The positions to which the hydroxyl group in phenol directs substitution

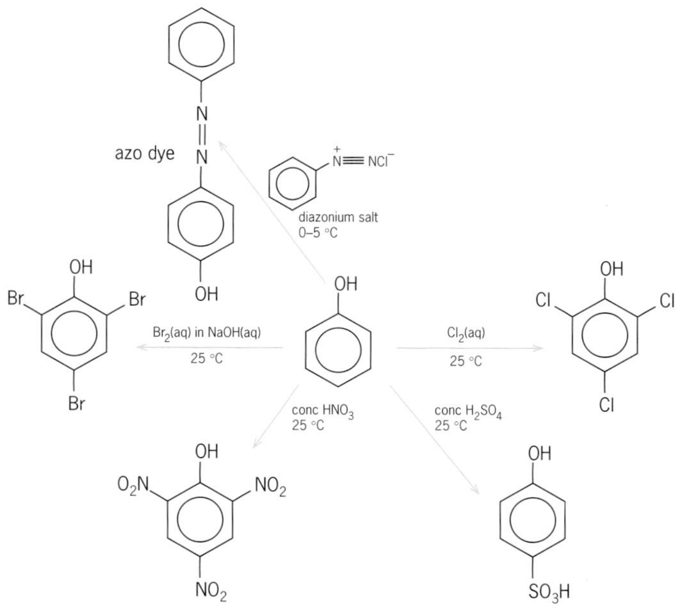

Fig 15.52 Five important electrophilic substitution reactions of phenol. Note that because phenol has an activated benzene ring, the conditions are less extreme than those required for substitution in benzene

ELECTROPHILIC SUBSTITUTION REACTIONS OF PHENYLAMINE

Phenylamine ($C_6H_5NH_2$) behaves towards electrophilic reagents in the same way as phenol, because the NH_2 functional group is very similar to the OH group in terms of its effect on the π system of the benzene ring. Both the nitrogen atom in the NH_2 group and the oxygen atom in the OH group have lone pairs.

The π system is again extended over seven atoms, this time to include the nitrogen atom of the amine group. The net effect is to increase the electron density of the π system, so making electrophilic substitution much easier than in benzene. Phenylamine reacts with aqueous chlorine to give 2,4,6-trichlorophenylamine, and with aqueous bromine to give the tribromo derivative.

ELECTROPHILIC SUBSTITUTION REACTIONS OF METHYLBENZENE

The traditional name for methylbenzene is toluene. As Table 15.2 indicates, the methyl substituent activates the benzene ring and directs the electrophilic substitution to positions 2, 4 and 6. In practice, the electrophilic substitution of methylbenzene always leads to a mixture of products.

Nitration of methylbenzene

The explosive compound trinitrotoluene (TNT), or more correctly 2,4,6-trinitromethylbenzene, is made by the nitration of methylbenzene (Fig 15.53). The reaction has three stages: first, one nitro group is substituted for a hydrogen atom, then a second group and finally a third group.

The mononitro compound is produced by heating concentrated nitric acid and concentrated sulfuric acid at 60 °C. Above this temperature, dinitro compounds are produced, and the trinitro compound is only produced above 120 °C.

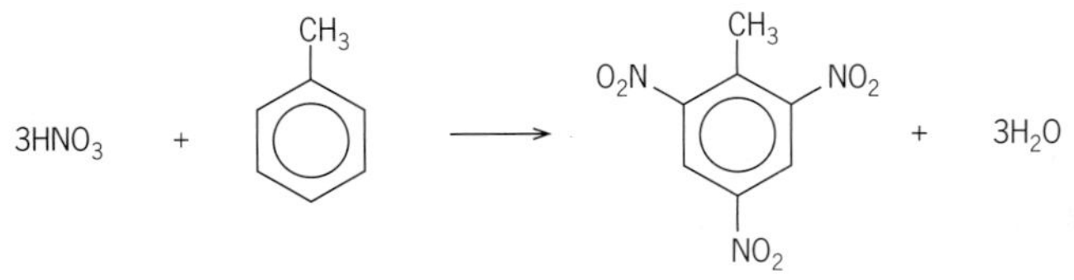

Fig 15.53 Making TNT by nitrating methylbenzene

QUESTION 21

a) Write equations to show the reaction of phenylamine with:
i) aqueous chlorine;
ii) aqueous bromine.
b) Use the information in Table 15.2 to predict the product of the nitration of:
i) chlorobenzene;
ii) 1,3-dichlorobenzene;
iii) nitrobenzene.

Fig 15.54 TNT used in a limestone quarry

QUESTION 22

a) Which mononitromethylbenzenes are formed when methylbenzene is nitrated? Draw their structures.
b) Which dinitromethylbenzenes are formed when methylbenzene is nitrated? Draw their structures.

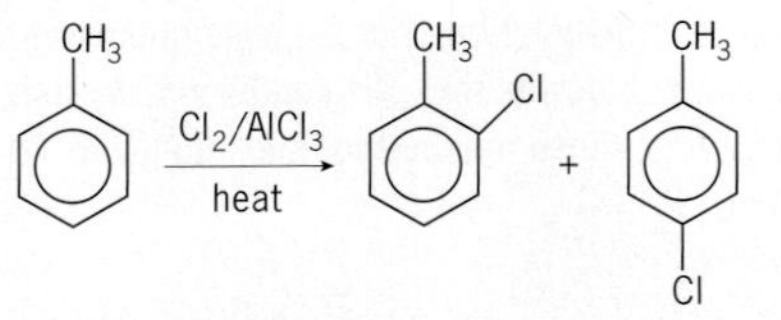

CH_3 $\xrightarrow[\text{heat}]{Cl_2/AlCl_3}$ CH_3, Cl + CH_3, Cl

Fig 15.55 The chlorination of methylbenzene in the presence of a halogen carrier

Chlorination of methylbenzene

When chlorine is bubbled through hot methylbenzene in the presence of aluminium chloride or iron(III) chloride, there is electrophilic substitution of one of the hydrogen atoms bonded to the carbon atoms in the benzene ring, with the formation of a mixture of 1-chloro-2-methylbenzene and 1-chloro-4-methylbenzene.

STRETCH AND CHALLENGE

Direction of electrophilic substitution in substituted benzenes

Table 15.2 shows that substituted benzenes direct the position of electrophilic substitution either to carbon atoms 2, 4, and 6 or to carbon atoms 3 and 5. To explain these substitution patterns, one model used is based on the resonance hybrid model of the structure of benzene, discussed on page 298.

Substituents on the benzene ring with lone pairs such as –OH, $-OCH_3$ and –Cl always give a 2, 4, 6 substitution pattern. Figure 15.56 shows the movement of electron pairs that causes this substitution pattern. Notice that the movement of electron pairs, starting from the lone pair, always results in the possible donation of an electron pair at carbon atoms 2, 4 and 6. When the lone pair is donated easily into the benzene ring as in the case of phenol then this effect will also activate the benzene ring.

Groups such as $-NO_2$ and $-COCH_3$ have groups that can accept an electron pair and in this case the substitution pattern is always at carbon atoms 3 and 5. Figure 15.56 also shows how this can be explained by the movement of electron pairs. Notice that this time it is because electron density is preferentially removed from carbon atoms 2, 4 and 6 that substitution at these positions is made less likely.

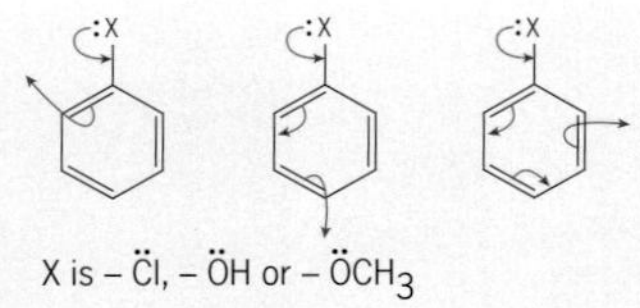

:X :X :X

X is – C̈l, – ÖH or – ÖCH_3

Fig 15.56(a) The movement of electron pairs shown in red shows that electron pairs can only be donated towards an electrophile from carbon atoms 2, 4 or 6.

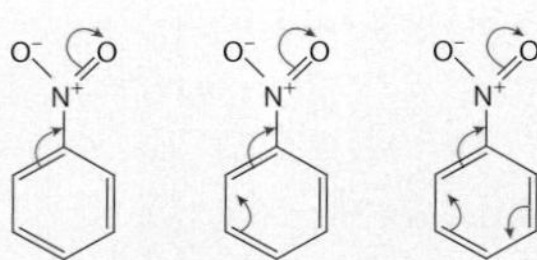

Fig 15.56(b) With nitrobenzene the movement of electron pairs is away from the benzene ring, and in particular away from carbon atoms 2, 4 or 6.

8 OTHER REACTIONS OF SUBSTITUTED BENZENES

We have already described the effect of substituents on the typical reaction of a compound with a benzene ring, electrophilic substitution. The benzene ring can also modify the reactions of functional groups attached to the benzene ring.

PHENOLS

Phenol is the name of the single compound C_6H_5OH, whereas the term phenols is reserved for a group of compounds that contain a hydroxyl group attached to a benzene ring. We described the electrophilic substitution reactions of phenol earlier on in this chapter, reactions that occur because of the presence of the benzene ring. Phenols also undergo reactions because of the presence of the hydroxyl group. Although alcohols also contain a hydroxyl group, the reactions of phenols and alcohols are quite different. The presence of the benzene ring modifies the reactivity of the hydroxyl group.

Phenols as acids

Phenols are classified as weak acids because they can act as proton donors when in aqueous solution. In water, the proton is donated to a water molecule:

$$C_6H_5OH(aq) \rightleftharpoons C_6H_5O^-(aq) + H^+(aq)$$

This dissociation lies very much on the left-hand side, but phenols are much more acidic than alcohols. The reason for this is the presence of the benzene ring. We have already discussed on page 308 that the lone pair on oxygen is donated into the benzene ring to produce an extended system. This not only increases the electron density of the system of the benzene ring, but also reduces the electron density between the oxygen and hydrogen atoms in the hydroxyl group. The O–H bond is thus weakened compared to the O–H bond in an alcohol and is easier to break.

You can also consider the stability of the anion produced when a proton is donated by a phenol molecule. The phenoxide ion, $C_6H_5O^-$, is much more stable than an alkoxide ion, such as $C_2H_5O^-$, because the negative charge is delocalised around the benzene ring. As a result phenols show some reactions that are typical of weak acids. Phenol reacts with the reactive metal sodium and with the alkali aqueous sodium hydroxide, but it is not sufficiently acidic to produce carbon dioxide from carbonates.

> **? QUESTION 23**
>
> 23 Explain, using curly arrows, why the electrophilic substitution occurs on carbon 3 in phenylethanone, $C_6H_5COCH_3$.

Reaction with aqueous sodium hydroxide

Phenol reacts with aqueous sodium hydroxide to produce sodium phenoxide, an ionic compound. In this reaction phenol transfers a proton to a hydroxide ion to form water:

$$C_6H_5OH(s) + OH^-(aq) \rightarrow C_6H_5O^-(aq) + H_2O(l)$$

As the state symbols in the equation show, phenol appears to dissolve in aqueous sodium hydroxide, but this must be seen as a chemical reaction rather than the physical process of dissolving. Alcohols do not show this reaction.

Reaction of sodium with phenol

Phenol reacts with sodium to produce sodium phenoxide. As in the case of alcohols, the reaction involves the heterolytic fission of the O–H bond. The reaction must be carried out in an inert solvent to ensure that the sodium and the phenol, both solids, do make contact.

$$2C_6H_5OH + 2Na \rightarrow 2C_6H_5O^-Na^+ + H_2$$

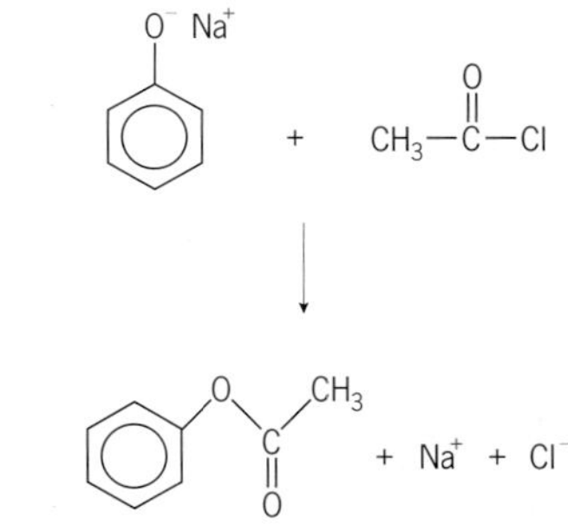

Fig 15.57 The reaction between ethanoyl chloride and sodium phenoxide

Reaction of phenols as nucleophiles

Alcohols can form esters when they react with carboxylic acids or acyl chlorides. Phenols do not react with carboxylic acids to make esters, but will react with acid chlorides to make esters.

Normally, sodium phenoxide is used (aqueous sodium hydroxide and phenol) as the reagent, because the phenoxide ion is a much better nucleophile than a neutral phenol molecule. The sodium hydroxide present also neutralises the hydrogen chloride formed as a by-product. With ethanoyl chloride, the ester phenyl ethanoate is formed (Fig 15.57):

$$C_6H_5O^-Na^+(aq) + CH_3COCl(l) \rightarrow CH_3COOC_6H_5(s) + NaCl(aq)$$

> **? QUESTION 24**
>
> 24 Draw the structure of the organic product of the following reactions:
>
> **a)** sodium phenoxide and propanoyl chloride;
>
> **b)** sodium phenoxide and benzoyl chloride.

Testing for phenols

Phenols will react with aqueous iron(III) chloride to form a violet colour.

STRETCH AND CHALLENGE

The phenols in tea

Tea is almost certainly the most popular hot drink in the world. In the United Kingdom alone, each person drinks on average three cups every day. There are reasons why this simple infusion of dry plant leaves is so popular. It has a characteristic taste and attractive colour derived from the compounds in the leaves.

Every cup of tea contains about 40 mg of caffeine, which is a stimulant to the central nervous system. There are also amino acids, carbohydrates and mineral ions. Complex phenols that are water soluble give the tea the astringent taste and refreshing effect we find appealing. As tea leaves are processed, some compounds are oxidised, which gives the brown colour characteristic of tea.

Fig 15.58 shows one of the phenolic components (epigallocatechin gallate) of tea. You can see that it is a much more complicated molecule than the compound phenol itself, with several benzene rings incorporated in the structure.

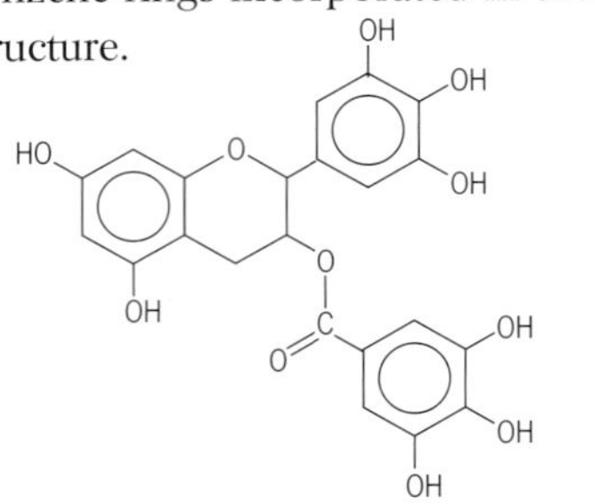

Fig 15.58 The structure of one of the phenolic constituents of tea

Fig 15.59 Tea leaves contain water-soluble phenols that contribute to the taste and colour of a cup of tea

STRETCH AND CHALLENGE

Relative acidities of substituted phenols

Phenols are weak acids and will react with aqueous sodium hydroxide to give phenoxide ions. Substituents on the benzene ring will affect the acidity of a phenol, so 4-methylphenol is a slightly weaker acid than phenol itself. In a similar way 4-chlorophenol is a stronger acid than phenol. The lone pair in the –OH group interacts with the π system of the benzene ring so that increases or decreases in the electron density within the π system have the same effect on the –OH group.

In Figure 15.60 the group X can be either electron-releasing or electron-accepting. If X is $-CH_3$, then this group is electron-releasing and tends to push electron density into the ring and hence towards the –OH group. This will strengthen the O–H bond, making it more difficult to break to form a proton and so making the 4-methylphenol a weaker acid. Another way of explaining the same thing is that the phenoxide ion becomes less stable because of the electron-releasing effect of the methyl group.

In Figure 15.60, if X is a chlorine atom then a chlorine atom will withdraw electrons from the ring. This will weaken the O–H bond in the phenol and stabilise the phenoxide ion, making 4-chlorophenol a stronger acid than phenol itself.

Figure 15.61 shows 4-nitrophenol; in this we can show the movement of electron pairs that will stabilise the phenoxide ion because it increases the π system across all the atoms. As a result, 4-nitrophenol is a much stronger acid than 4-chlorophenol.

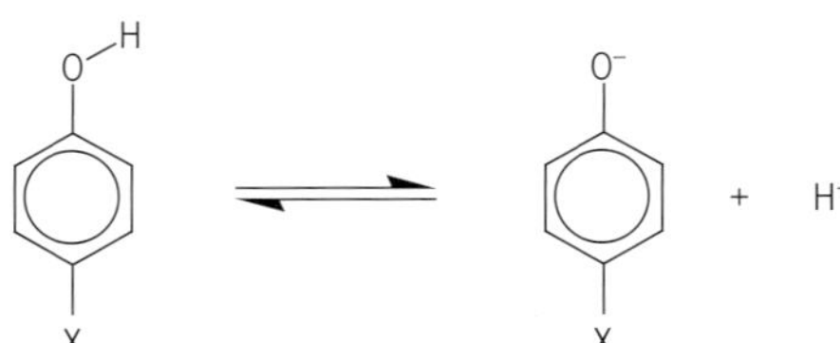

Fig 15.60 The ionisation of substituted phenols

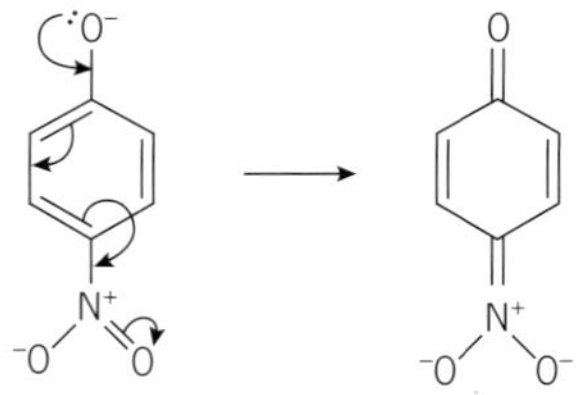

Fig 15.61 4-Nitrophenol forms a phenoxide ion that can be stabilised by forming an extended π-system over all the atoms

QUESTION 25

25 Rank the following phenols in order of increasing acid strength. Explain your answers.

a) 4-bromophenol, 4-chlorophenol and 4-fluorophenol.

b) 4-methylphenol, 2,4-dimethylphenol and 2,4,6-trimethylphenol.

c) 2-nitrophenol, 3-nitrophenol and 2,4-dinitrophenol.

METHYLBENZENE

Methylbenzene contains a methyl group attached to the benzene ring. We should expect that the methyl group would react in the same way as the methyl group in other hydrocarbons, such as ethane or methane. This is true for some reactions, but in others the proximity of the benzene ring gives the methyl group added reactivity.

Side-chain oxidation

Methylbenzene is oxidised by acidified potassium manganate(VII) to form benzoic acid:

$$C_6H_5CH_3 + 3[O] \rightarrow C_6H_5COOH + H_2O$$

This is called side-chain oxidation, since it is the side chain rather than the ring that is oxidised.

Side-chain oxidation occurs with any alkyl group, but always forms benzoic acid, regardless of the chain length of the alkyl group (Fig 15.62). When the benzene ring contains more than one alkyl group, each one is oxidised in the same way. This reaction can be used to determine the substitution pattern within an aromatic hydrocarbon. For example, if the side-chain oxidation of a hydrocarbon of molecular formula C_8H_{10} gives benzoic acid, then the hydrocarbon must be ethylbenzene, whereas if benzene-1,2-dicarboxylic acid is formed then the hydrocarbon would be 1,2-dimethylbenzene.

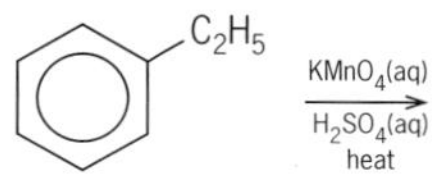

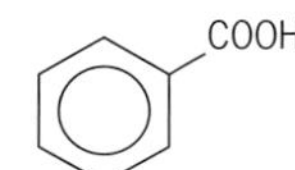

Fig 15.62 Side-chain oxidation of ethylbenzene to form benzoic acid

? QUESTION 26

a) The side-chain oxidation of an aromatic hydrocarbon C_8H_{10} gives benzene-1,4-dicarboxylic acid. What is the name of the hydrocarbon?

b) Draw the structural formulae of the products of the side-chain oxidation of i) butylbenzene, ii) octylbenzene and iii) 1,2,4-trimethylbenzene.

Free-radical chlorination

In Chapter 12, we described the free-radical substitution reaction of methane with chlorine. Methylbenzene reacts in a similar way with chlorine in the presence of ultraviolet light. Only the hydrogen atoms in the methyl group are substituted by chlorine atoms; those hydrogen atoms attached to the benzene ring are unaffected.

$$C_6H_5CH_3(l) + Cl_2(g) \xrightarrow{hf} C_6H_5CH_2Cl(l) + HCl(g)$$

$$C_6H_5CH_2Cl(l) + Cl_2(g) \xrightarrow{hf} C_6H_5CHCl_2(l) + HCl(g)$$

$$C_6H_5CHCl_2(l) + Cl_2(g) \xrightarrow{hf} C_6H_5CCl_3(l) + HCl(g)$$

? QUESTION 27

a) Draw the structures of the products of the reaction between chlorine and methylbenzene in the presence of aluminium chloride.

b) Draw the structures of the products of the reaction between chlorine and methylbenzene in the presence of ultraviolet light.

9 BENZENE AND THE PETROCHEMICAL INDUSTRY

Benzene is an important source of many petrochemicals. The electrophilic substitution reaction followed by other synthetic transformations allows a variety of different substituted benzenes to be produced. Some of the important reactions of benzene are shown in Fig 15.63.

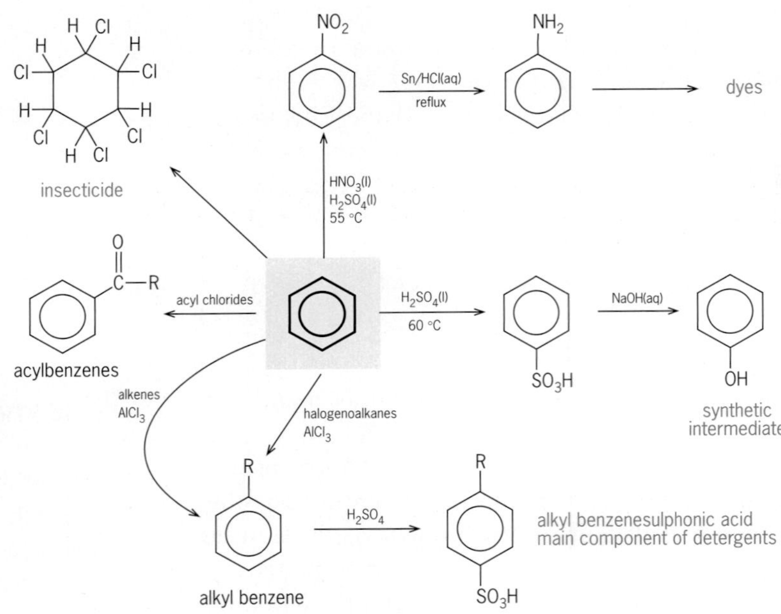

Fig 15.63 A summary of the reactions of benzene

SUMMARY

After studying this chapter, you should know the following:

- The benzene ring is a regular hexagon of carbon atoms joined by six σ bonds. Six π electrons are delocalised in an orbital above and below the ring.
- Benzene is made by cracking the naphtha fraction and by reforming C_6 alkanes or cycloalkanes.
- Arenes react by electrophilic substitution rather than electrophilic addition: an electrophile is substituted for a hydrogen atom, which is lost as a proton.
- An arene is alkylated by reaction with a halogenoalkane in the presence of aluminium chloride, and is acylated by reaction with an acyl chloride or acyl bromide in the presence of aluminium chloride.
- An arene is alkylated by reaction with alkenes in the presence of an acidic catalyst.
- An arene is chlorinated or brominated by the reaction between chlorine or bromine and the arene in the presence of a halogen carrier, such as iron(III) chloride or aluminium chloride.
- A nitrating mixture consists of concentrated sulfuric acid and concentrated nitric acid, and produces NO_2^+, which can react with arenes to form nitroarenes.
- Phenol and phenylamine have highly activated benzene rings, and with electrophiles form trisubstituted substitution products at positions 2, 4 and 6.
- Methylbenzene reacts faster than benzene towards electrophilic reagents and forms substitution products at positions 2, 4 or 6.
- Phenol is a weak acid and reacts with sodium and sodium hydroxide to give sodium phenoxide.
- Phenol is a nucleophile and reacts with acid chlorides to give esters.

Practice questions and a How Science Works assignment for this chapter are available at www.collinseducation.co.uk/CAS

16
Aldehydes and ketones

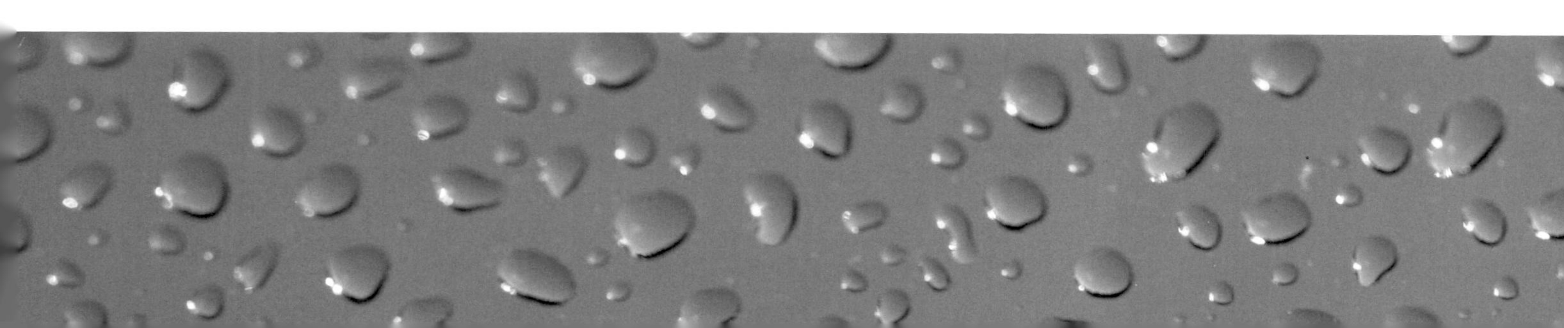

16 ALDEHYDES AND KETONES

SCIENCE IN CONTEXT

Sex hormones and the pill

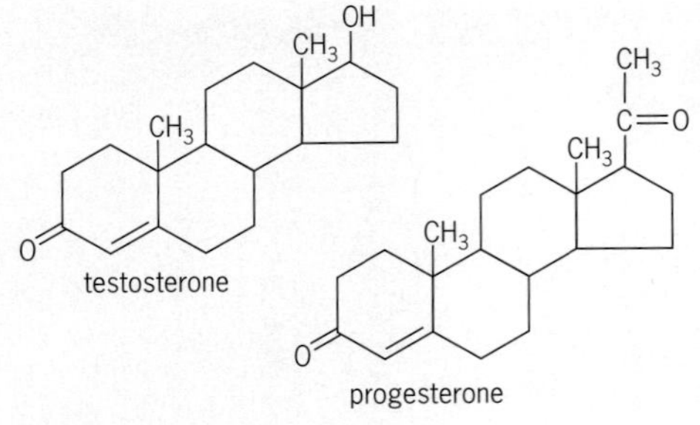

Steroid structure All steroids possess the four-ring structure. You can read about cholesterol, another steroid, on page 258.

The most publicised hormones must be progesterone and testosterone. The reason for their fame is sex.

Progesterone is the female sex hormone, which prepares the uterus for pregnancy after an egg has been fertilised, and which prevents the ovaries from releasing any more eggs. Testosterone is the male sex hormone responsible for sexual development and drive in males, and for muscle growth. Although they have very different functions, these hormones are structurally very similar (they are steroids with a carbonyl functional group), and it takes only a simple chemical reaction in the laboratory to convert progesterone into testosterone.

Once their structures became known in the 1940s, chemists set about synthesising them, and soon synthetic sex hormones featured in a number of medical applications. However, few people would have predicted the social revolution that was to follow when synthetic progestins, a group of chemicals that mimic the action of progesterone, were developed. The synthetic progestin norethynodrel was the basis of the first oral contraceptive, known simply as 'the pill', introduced in 1960. There is perhaps no more striking example of the influence of chemistry on our lives.

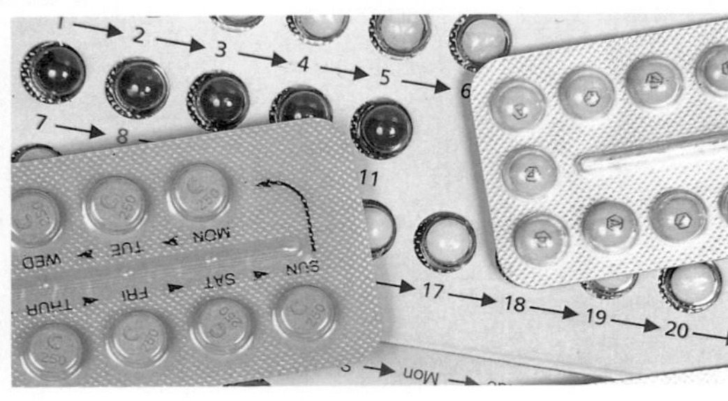

Oral contraceptives These pills liberated women from the fear of unwanted pregnancies and thereby revolutionised sexual behaviour. However, the occurrence of these chemicals in recycled drinking water have been partly blamed for reducing the sperm count of males

1 RECOGNISING AND NAMING ALDEHYDES AND KETONES

Fig 16.1 The carbonyl functional group

Aldehydes and ketones contain perhaps the most important functional group in organic chemistry – the **carbonyl group**, C=O (Fig 16.1). Hence, they are known as **carbonyl compounds**. The carbonyl functional group is found in many important biological molecules, from insect pheromones to human sex hormones. It occurs in the molecules in our eyes that are responsible for vision, and gives lemons their characteristic flavour. It is also involved in the manufacture of many important industrial chemicals, from plastics to solvents.

In aldehydes (Fig 16.2), the carbon atom of the carbonyl group (the carbonyl carbon) is bonded to at least one hydrogen atom, while in ketones (Fig 16.3) it is bonded to two carbon atoms, each from either an alkyl or an aryl group.

> **REMEMBER THIS**
>
> A **functional group** is an atom or group of atoms that gives a molecule a characteristic set of properties (see page 190).
>
> A **pheromone** is a chemical that communicates information between members of the same species.
>
> An **aryl group** contains a benzene ring.

Fig 16.2 An aldehyde. R can be hydrogen, an alkyl group or an aryl group

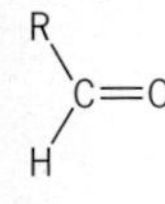

Fig 16.3 A ketone. The R′ group may be different from the R group. R cannot be hydrogen, otherwise the compound is an aldehyde

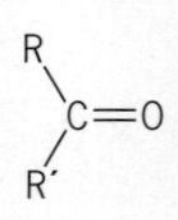

Four naturally occurring carbonyl compounds are shown in Fig 16.4.

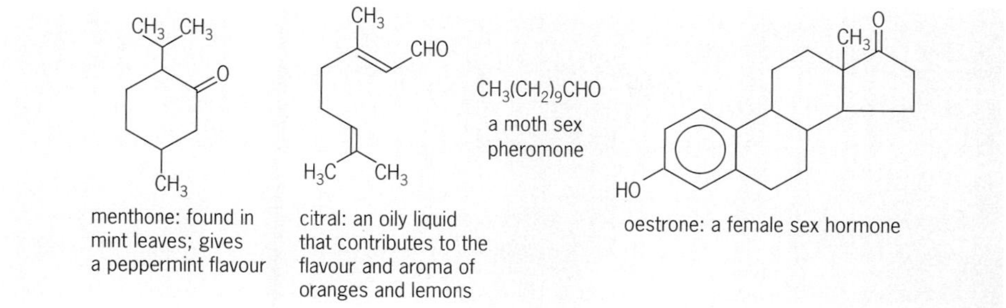

menthone: found in mint leaves; gives a peppermint flavour

citral: an oily liquid that contributes to the flavour and aroma of oranges and lemons

$CH_3(CH_2)_9CHO$ a moth sex pheromone

oestrone: a female sex hormone

Fig 16.4 Four naturally occurring carbonyl compounds

NAMING ALDEHYDES AND KETONES

The systematic naming of aldehydes and ketones is simple. The names of aldehydes end in *al*, and those of ketones end in *one*. The number of carbon atoms in the chain (including the carbon of the functional group) provides the rest of the name, which is based on the alkane name with the end *e* removed.

Eight of the more important aldehydes and ketones are listed in Table 16.1.

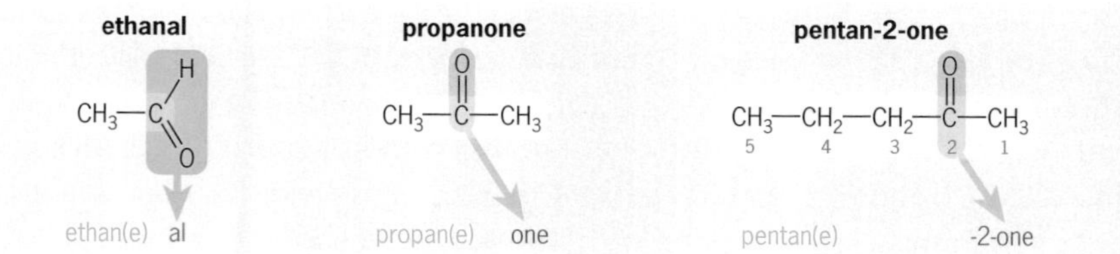

Fig 16.5 Deriving the names of aldehydes and ketones

SCIENCE IN CONTEXT

Methanal

Methanal, previously known as formaldehyde, is still used by many industrial chemists. In fact, it is the most-used industrial aldehyde. Methanal is produced by the air oxidation of methanol, using an iron or silver catalyst:

$$CH_3OH(l) + \tfrac{1}{2}O_2(g) \xrightarrow{\text{Fe or Ag catalyst, 500 °C}} HCHO(g) + H_2O(l)$$

Methanal is used to make plastics such as Bakelite (one of the first plastics) and other phenolic resins (for which it is reacted with phenol), urea–formaldehyde resins (for which it is reacted with urea), melamine resins and Formica.

Your kitchen almost certainly contains surfaces and equipment that are products of reactions with methanal – kitchen work tops and pan handles are just two examples. So does your bathroom, because methanal is used as a preservative in some shampoos and bath foams.

A 40 per cent aqueous solution of methanal – better known as formalin – is used to preserve biological specimens (Fig 16.6) and as a disinfectant and fungicide.

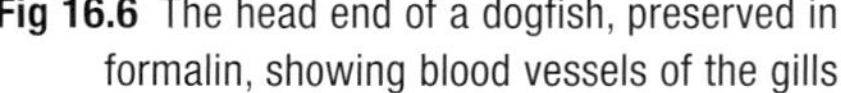

Fig 16.6 The head end of a dogfish, preserved in formalin, showing blood vessels of the gills

QUESTION 1

a) Which of the carbonyl compounds in Fig 16.4 are aldehydes and which are ketones?

b) Oestrone contains another functional group. What is its name?

Table 16.1 Some important aldehydes and ketones

Structural formula	Systematic name
$HCHO$	methanal
CH_3CHO	ethanal
CH_3CH_2CHO	propanal
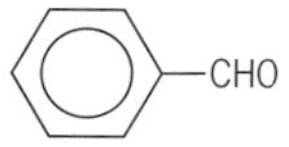 C_6H_5–CHO	benzaldehyde
CH_3COCH_3	propanone
$CH_3CH_2COCH_3$	butanone
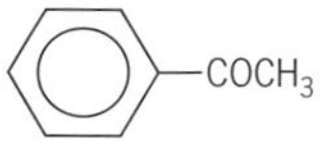 C_6H_5–$COCH_3$	phenylethanone
$CH_3CH_2CH_2COCH_3$	pentan-2-one

REMEMBER THIS

The carbon atom of the functional group counts towards the number in the carbon chain. The naming of alkanes is covered on page 192.

QUESTION 2

a) Work out the molecular formulae of propanone and propanal.

i) What do you notice?

ii) What is the name given to compounds with this feature?

b) The unpleasant odour of rancid butter is caused by butanal. Write down its structural formula.

c) Write down the structural formula of pentan-3-one.

REMEMBER THIS

Non-systematic or traditional names are still common in organic chemistry. Thus, you may see methanal referred to as formaldehyde, ethanal as acetaldehyde, and propanone as acetone.

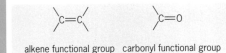

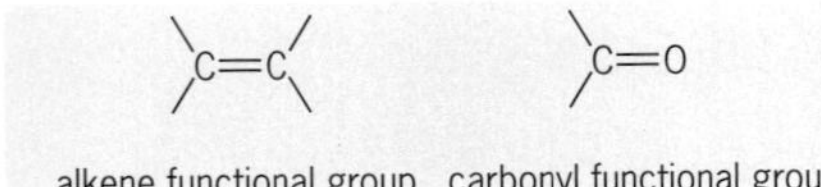

Fig 16.7 Alkene and carbonyl functional groups

2 THE CARBONYL GROUP AND NUCLEOPHILIC ADDITION

As the carbonyl functional group has a double bond, it undergoes *addition* reactions – just like the double bond of the functional group in alkenes (see page 282).

Alkenes undergo **electrophilic addition** reactions because the loose electron cloud of the π bond (see page 83) is attractive to electrophiles. There is a π bond in the carbonyl group as well, but it is between two atoms of different electronegativities, C and O. So the density of the electron cloud is greater at the more electronegative oxygen end, thereby making the bond polar.

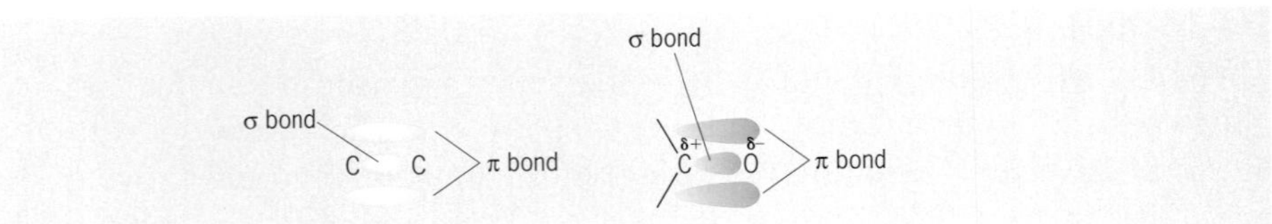

Fig 16.8 The difference between the π bonds in the alkene and carbonyl functional groups

> **REMEMBER THIS**
>
> An **electrophile** is a species that can accept a lone pair of electrons to form a covalent bond. A π bond is formed by the sideways overlap of two p orbitals.
>
> 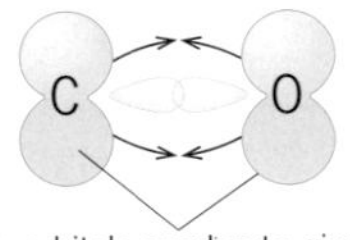
>
>
> **Fig 16.9** The sideways overlap of two orbitals

Since C=O is polar, the electron-deficient carbonyl carbon is susceptible to attack by nucleophiles. We have already come across nucleophilic reactions with the polar carbon–halogen bond of halogenoalkanes (page 247), and with the polar C–O bond of alcohols (pages 265–266). In both these cases, substitution reactions occur. However, the carbonyl group has a *double* bond, and therefore as the nucleophile attacks and forms a covalent bond with the carbonyl carbon, the π bond splits and forms another single covalent bond, which results in an addition reaction.

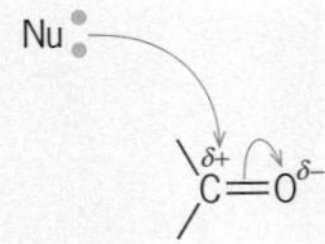

Fig 16.10 Nucleophile Nu: attacks the carbonyl carbon and the π bond breaks

> **? QUESTION 3**
>
> 3 Apart from the aromatic compounds, all the other aldehydes and ketones in Table 16.1 are very soluble in water.
>
> **a)** What type of intermolecular bonding explains this solubility? (Hint: if you are not sure of your answer, look back at page 261.)
>
> **b)** Draw a diagram to show the intermolecular bonding between an ethanal molecule and a water molecule.
>
> **c)** Why do you think aromatic carbonyl compounds are not very soluble in water?
>
> **d)** Aldehydes and ketones have lower melting and boiling points than alcohols of similar molecular mass. Explain why.

> 
>
> **REMEMBER THIS**
>
> A **nucleophile** has a lone pair of electrons with which it can form a covalent bond. It is attracted to a centre of positive charge.

NUCLEOPHILIC ADDITION BY HYDROGEN CYANIDE

The reaction of hydrogen cyanide (HCN) with the carbonyl group is a very important *synthesis* reaction, because it adds another carbon atom to the molecule.

Fig 16.11 shows two nucleophilic addition reactions. The reaction with propanone is used in the production of poly(methyl-2-methylpropenoate), better known by its trade name of Perspex. The first reaction in the synthesis of Perspex is the production of a hydroxynitrile (cyanohydrin).

The nucleophilic addition reaction of hydrogen cyanide was one of the first reaction mechanisms ever to be investigated; this was by a British chemist, Arthur Lapworth, in 1903. A **reaction mechanism** shows the steps by which a reaction takes place. The nucleophile is the cyanide ion (CN^-), rather

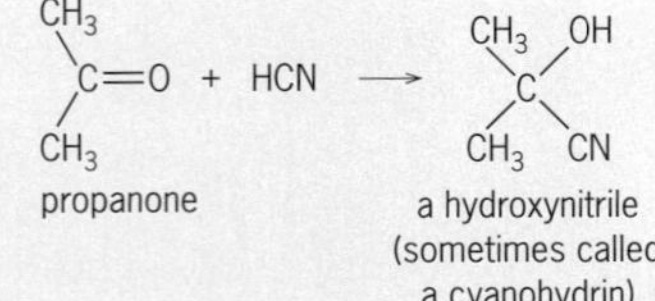

CH_3CHO (ethanal) + HCN → $CH_3CH(OH)CN$ (another hydroxynitrile)

Fig 16.11 Two nucleophilic addition reactions

> **? QUESTION 4**
>
> 4 Write the equation for the reaction of benzaldehyde with HCN.

than the HCN molecule (Fig 16.12).

HCN is a poor nucleophile, and if CN^- ions are not present, the reaction is very slow. Either the addition of potassium cyanide (KCN) can provide the CN^- ions or they can be generated from HCN by adding an alkali:

$$HCN + OH^- \rightarrow H_2O + CN^-$$

A proton then bonds to the oxygen atom. Often this proton is transferred from HCN or H_2O (Fig 16.13).

The net result is the addition of HCN to the molecule. Notice that in this case the proton was transferred from HCN, and a CN^- ion is regenerated.

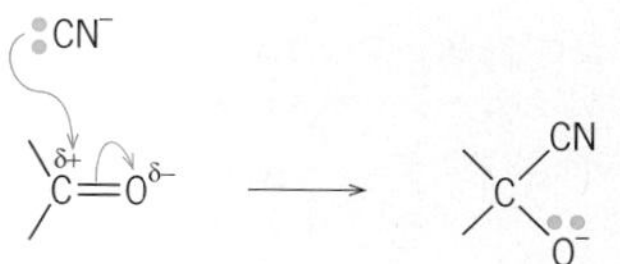

Fig 16.12 The CN^- nucleophile attacks the ketone group

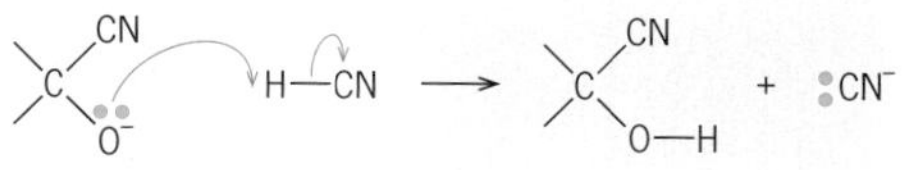

Fig 16.13 Proton bonds to the oxygen

Hydroxynitrile: the millipede's defence

To deter predators, if not finish them off for good, one species of millipede uses the deadly gas HCN – hydrogen cyanide – yet manages not to harm itself. This is because inside the millipede are separate stores of a hydroxynitrile derivative of benzaldehyde, which is harmless, and an enzyme that catalyses the compound's reaction to form benzaldehyde and HCN; the two are not mixed until they leave the animal's body. (Note this is the reverse of the reaction in Question 4.)

When the millipede is attacked, it discharges both compounds. The enzyme becomes mixed with the hydroxynitrile, and the hapless attacker is enveloped in hydrogen cyanide.

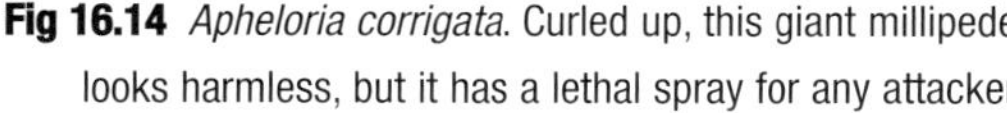

Fig 16.14 *Apheloria corrigata*. Curled up, this giant millipede looks harmless, but it has a lethal spray for any attacker

REMEMBER THIS

Hydrogen cyanide, HCN, is highly toxic and a concentration of 270 ppm of the gas causes death within ten minutes. KCN is also extremely toxic.

NUCLEOPHILIC ADDITION OF H^-

The hydride ion (H^-) comes from, $NaBH_4$ (sodium tetrahydridoborate(III)). The mechanism (Fig 16.15) is similar to that of the addition of hydrogen cyanide, only in this case the nucleophile is H^-. Once the $NaBH_4$ has reacted with the carbonyl compound, water is added to provide the protons.

Notice that an alcohol functional group is produced. This reaction is also called a *reduction*. The equation for the reduction of aldehydes in general is written:

$$RCHO + 2[H] \rightarrow RCH_2OH$$

H is put in brackets to signify that *it comes from a reducing agent*, which in this case is $NaBH_4$ (this is a simplified way to balance the equation and does not represent atomic hydrogen). This reaction is discussed again on page 322, where we look at the reduction of aldehydes and ketones.

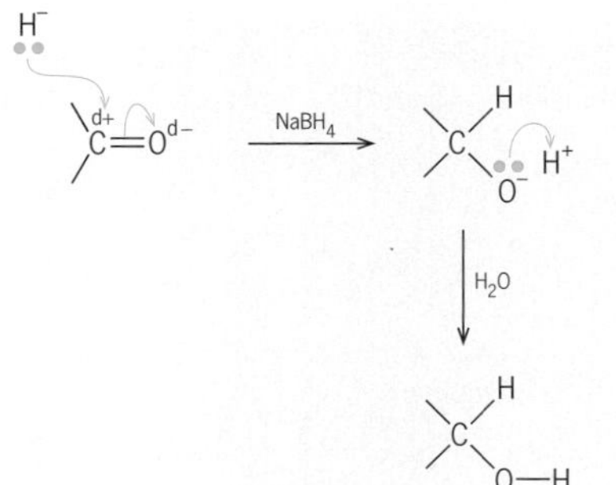

Fig 16.15 Nucleophilic addition takes place in the first step, and water then transfers the proton to make an alcohol functional group

QUESTION 5

5 Ethanal and propanone both react with $NaBH_4$ to produce alcohols.

a) Identify the alcohols formed, and draw their structural formulas.

b) Which of the alcohols is a primary alcohol and which is a secondary alcohol?

c) Write a balanced equation for the reaction of propanone with $NaBH_4$. You should use [H] in this equation. (Hint: if you need to be reminded about these alcohols, look at page 260.)

REMEMBER THIS

$LiAlH_4$ (lithium tetrahydridoaluminate(III)) is a more powerful reducing agent than $NaBH_4$. It also reacts with water, which is why it is dissolved in dry ethoxyethane. However, $NaBH_4$ is a more convenient reagent as it can be used in aqueous or alcoholic solution and the reaction is less vigorous.

Fig 16.16 2,4-Dinitrophenylhydrazone is a brightly coloured precipitate and its formation is a test for carbonyl groups, as shown in this test tube

3 REACTION WITH 2,4-DINITROPHENYLHYDRAZINE

2, 4-dinitrophenylhydrazine (also called Brady's reagent) reacts with carbonyl groups to give orange-coloured precipitates, called 2, 4-dinitrophenylhydrazones. This reaction is used to test for the presence of carbonyl group in unknown compounds.

The coloured precipitates from the reaction can be purified by recrystallisation to give products that have very precise melting points, which can therefore be used to identify a particular aldehyde or ketone.

The *ethanal derivative* ethanal 2,4-dinitrophenylhydrazone (Fig 16.17) melts at exactly 168 °C, so its melting point can be used to identify ethanal, a method that used to be important before different types of spectrometers became widely available for identifying substances (see Chapter 11). Before spectrometers, the best way to recognise unknown carbonyl compounds was to produce from the liquid a solid derivative with a sharp melting point.The melting point is also a good way to check the purity of a sample, since any impurity lowers the melting point.

QUESTION 6

6 Write balanced equations for the reaction of 2,4-dinitrophenylhydrazine with phenylethanone and benzaldehyde. (Hint: see Table 16.1 for the formulae of these carbonyl compounds.)

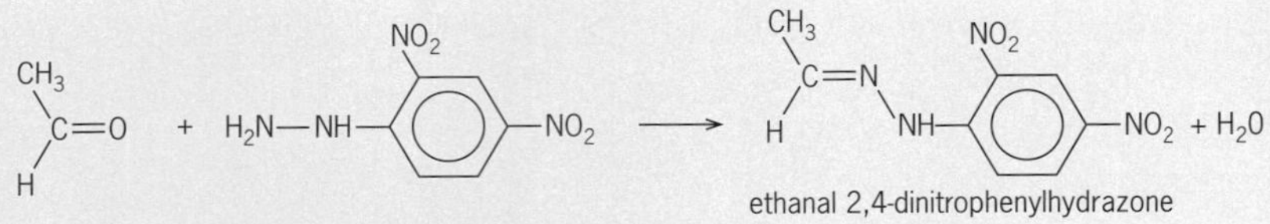

$CH_3(H)C{=}O + H_2N{-}NH{-}C_6H_3(NO_2)_2 \longrightarrow CH_3(H)C{=}N{-}NH{-}C_6H_3(NO_2)_2 + H_2O$

ethanal 2,4-dinitrophenylhydrazone

Fig 16.17 Production of ethanal 2,4-dinitrophenylhydrazone

For information on X-ray crystallography, see page 78.

For information on thin-layer chromatography, see page 233.

The production of 2,4-dinitrophenylhydrazone derivatives is still important for X-ray crystallography, which requires solid compounds.

A mixture of carbonyl compounds can be separated by reacting the mixture with 2,4-dinitrophenylhydrazine and then using thin-layer chromatography. The various 2,4-dinitrophenylhydrazones separate when a suitable solvent is used.

STRETCH AND CHALLENGE

Addition–elimination mechanism of the 2,4-dinitrophenylhydrazine reaction

The two nucleophilic addition reactions in the previous section produce stable products. The reaction between the carbonyl group and 2,4-dinitrophenylhydrazine gives an unstable product, which spontaneously reacts with the elimination of a water molecule.

The mechanism of the reaction (Fig 16.18) shows that the lone pair of electrons on the first nitrogen atom provides the basis of the nucleophilic addition, and the intermediate compound is then formed by an internal rearrangement. The final product, a 2,4-dinitrophenylhydrazone (Figs 16.17 and 16.18), is formed from the unstable intermediate by the elimination of a water molecule. So this reaction involves both an addition reaction and an elimination reaction – hence addition–elimination.

2,4-dinitrophenylhydrazine; addition; proton transfer; unstable intermediate; elimination; a 2,4-dinitrophenylhydrazone $+ H_2O$

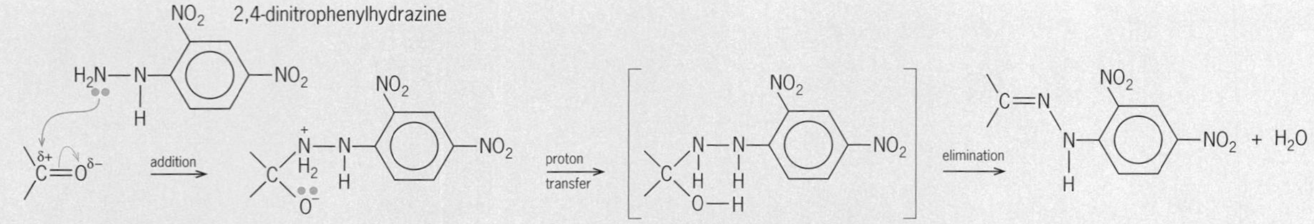

Fig 16.18 Addition–elimination reaction of 2,4-dinitrophenylhydrazine

SCIENCE IN CONTEXT

Sun tan from a bottle

Over-exposure to sunlight is known to cause skin cancer, yet millions of people still work hard to give themselves a suntan because they think it makes them look healthier and more attractive.

Fig 16.19 Despite grim warnings about the consequences, sunbathing still has millions of devotees

However, suntans have not always been fashionable. Until about 200 years ago, upper-class women who went on leisurely walks would protect their pale complexions with bonnets and parasols. This distinguished them from the bronzed working men and women who laboured outdoors in the sunshine.

Attitudes gradually began to change with the industrial revolution. While factory and office workers toiled in buildings that shut out the sunshine all day, better off people began to think of a suntan as a sign of wealth and position in society. This trend grew when it was the fashion in the early 1900s to go on touring holidays of continental Europe – the forerunners of today's package holidays.

A natural tan is caused by the dark pigment melanin, which is produced in the skin to absorb harmful ultraviolet rays. A similar tan-like effect can be produced after a few hours using DHA, a colourless ketone that reacts with the protein in the outer skin to produce a brown pigment. DHA, traditionally called dihydroxyacetone, is 1,3-dihydroxypropanone:

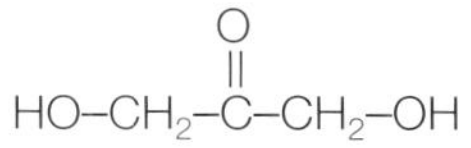

$$HO{-}CH_2{-}\overset{\overset{\large O}{\|}}{C}{-}CH_2{-}OH$$

There are problems with a 'chemical' tan. Cells of the outer layer of skin are dead and so within a few weeks these have rubbed off – and so has the tan! Also, because only dead cells react with DHA, where these are in a thicker layer, such as at the elbows and knees, the tan is darker than elsewhere. It is also worth noting that a chemical tan does nothing to protect the skin from harmful ultraviolet rays.

STRETCH AND CHALLENGE

Grignard reagents and aldehydes and ketones

If magnesium turnings are added to a solution of halogenoalkane (RX) in ethoxyethane, an exothermic reaction produces a solution of Grignard reagent:

$$R{-}X + Mg \rightarrow R{-}Mg{-}X$$

R can be either an alkyl or an aryl group and X is a halogen.
The reagents are named after their discoverer Victor Grignard, a French chemist. One of the most important uses of Grignard reagents is to synthesise alcohols from aldehydes and ketones. The mechanism for this reaction is nucleophilic addition. The carbon–magnesium bond is highly polar (Fig 16.20) and attacks the carbonyl group (Fig 16.21) to form an intermediate that is decomposed by adding dilute acid (Fig 16.22).

Fig 16.20 Highly polar carbon–magnesium bond

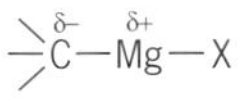

$$\rangle\overset{\delta-}{C}{-}\overset{\delta+}{Mg}{-}X$$

Fig 16.21 Nucleophilic attack on the carbonyl group

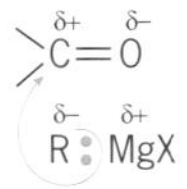

$$\rangle\overset{\delta+}{C}{=}\overset{\delta-}{O} \qquad \overset{\delta-}{R}:\overset{\delta+}{MgX}$$

$$-\underset{R}{\overset{|}{\underset{|}{C}}}-OMgX \xrightarrow{\text{dilute acid}} -\underset{R}{\overset{|}{\underset{|}{C}}}-OH$$

Fig 16.22 Intermediate is decomposed by dilute acid

Primary, secondary and tertiary alcohols can be made using this reaction (Figs 16.23 to 16.25).

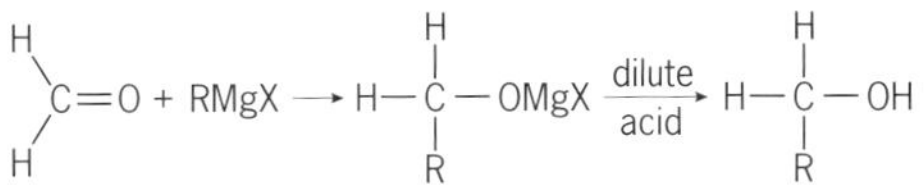

$$(H)(H)C{=}O + RMgX \rightarrow H{-}\underset{R}{\overset{H}{C}}{-}OMgX \xrightarrow[\text{acid}]{\text{dilute}} H{-}\underset{R}{\overset{H}{C}}{-}OH$$

Fig 16.23 Primary alcohols are produced when methanal reacts with Grignard agents

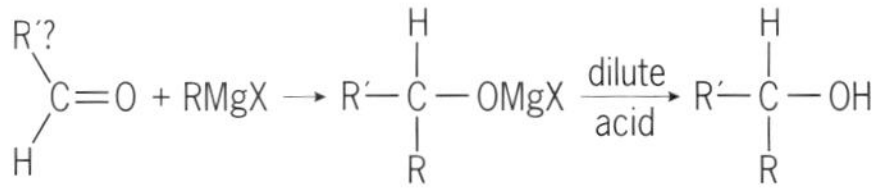

$$(R')(H)C{=}O + RMgX \rightarrow R'{-}\underset{R}{\overset{H}{C}}{-}OMgX \xrightarrow[\text{acid}]{\text{dilute}} R'{-}\underset{R}{\overset{H}{C}}{-}OH$$

Fig 16.24 Secondary alcohols are produced using other aldehydes other than methanal

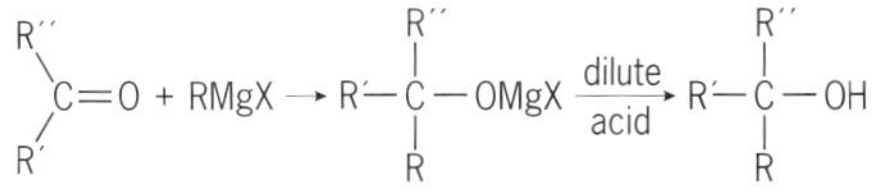

$$(R'')(R')C{=}O + RMgX \rightarrow R'{-}\underset{R}{\overset{R''}{C}}{-}OMgX \xrightarrow[\text{acid}]{\text{dilute}} R'{-}\underset{R}{\overset{R''}{C}}{-}OH$$

Fig 16.25 Tertiary alcohols are produced using ketones

? QUESTION 7

7 What alcohols would be produced from the reaction of the following reagents?

a) $CH_3CH_2MgBr + HCHO$

b) $CH_3CH_2CH_2CH_2MgBr + CH_3COCH_3$

c) $CH_3CH_2MgBr + CH_3CHO$

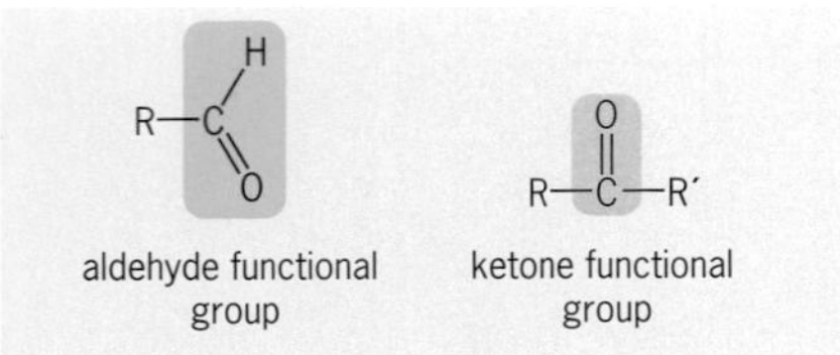

Fig 16.26 The aldehyde and ketone functional groups

> **? QUESTION 8**
>
> 8 In Fig 16.26, what do R and R′ stand for? (Hint: if you are not sure, see Figs 16.2 and 16.3.)

> **REMEMBER THIS**
>
> Primary alcohols have an R–CH_2OH group. Secondary alcohols have an R–CHOHR′ group. For further information about alcohols, see Chapter 13.

> **? QUESTION 9**
>
> 9 **a)** What species does the reducing when $NaBH_4$ is used? (Hint: see page 319.)
>
> **b)** Using [H], write balanced equations for the reduction of CH_3CHO and CH_3COCH_3

4 DIFFERENCES BETWEEN ALDEHYDES AND KETONES

So far, we have treated carbonyl compounds together and looked at the reactions of the C=O functional group in both aldehydes and ketones. They do, however, differ in their reactions with oxidising and reducing agents. Therefore, we now look at reactions of the aldehyde functional group separately from those of the ketone functional group.

REDUCTION OF ALDEHYDES AND KETONES

With a reducing agent, such as $NaBH_4$ or $LiAlH_4$, aldehydes give **primary alcohols** and ketones give **secondary alcohols**, as shown in Fig 16.27. Notice that the reducing H in the equation is in brackets (to balance the equation simply – it does not represent atomic hydrogen). The reaction is not that simple, as you can see from its mechanism on page 319.

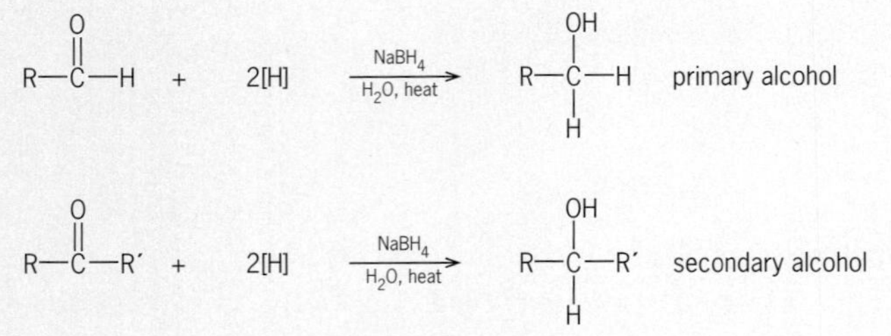

$$R-\overset{\overset{O}{\|}}{C}-H \;+\; 2[H] \xrightarrow[H_2O,\ heat]{NaBH_4} R-\underset{\underset{H}{|}}{\overset{\overset{OH}{|}}{C}}-H \quad \text{primary alcohol}$$

$$R-\overset{\overset{O}{\|}}{C}-R' \;+\; 2[H] \xrightarrow[H_2O,\ heat]{NaBH_4} R-\underset{\underset{H}{|}}{\overset{\overset{OH}{|}}{C}}-R' \quad \text{secondary alcohol}$$

Fig 16.27 Reduction of aldehydes and ketones

As an example, benzaldehyde gives the balanced equation shown in Fig 16.28.

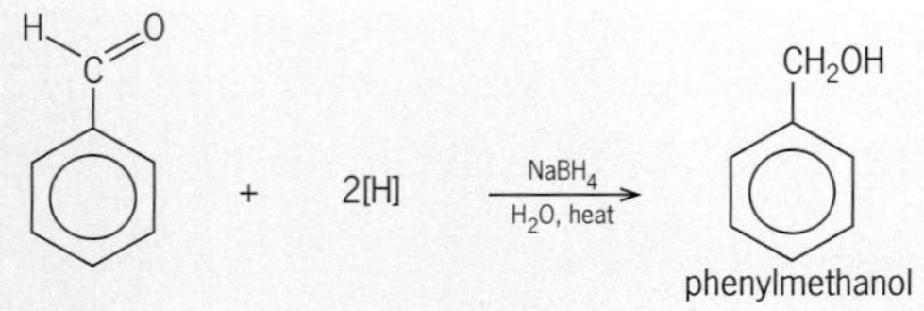

$$C_6H_5CHO \;+\; 2[H] \xrightarrow[H_2O,\ heat]{NaBH_4} C_6H_5CH_2OH$$

phenylmethanol

Fig 16.28 Reduction of benzaldehyde

The reduction reactions of aldehydes and ketones are essentially the reverse of the oxidation reactions of primary and secondary alcohols (see page 267).

OXIDATION OF ALDEHYDES

> **REMEMBER THIS**
>
> Primary alcohols are usually mixed with acidified potassium dichromate and the aldehyde distilled off as soon as it is formed to prevent further oxidation. If the primary alcohol is refluxed with $K_2Cr_2O_7(aq)/(H^+(aq))$, then a carboxylic acid results.

Aldehydes are prepared by oxidising primary alcohols, and ketones by oxidising secondary alcohols. In the case of primary alcohols, if the aldehyde is not removed from the oxidising agent immediately it is formed, it is further oxidised to give the carboxylic acid (Fig 16.29). The aldehyde is removed from the reacting mixture by distillation.

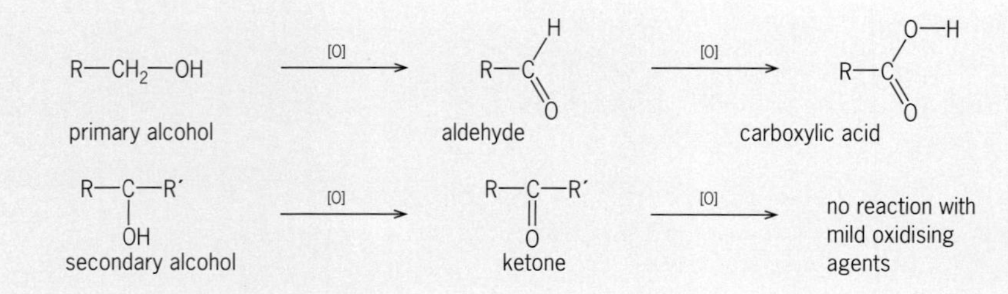

Fig 16.29 Oxidation of primary and secondary alcohols to form aldehydes and ketones, respectively

Ketones resist further oxidation, because they have no oxidisable hydrogen bonded to the carbonyl group. However, with prolonged heating they can be oxidised by strong oxidising agents.

By contrast, the oxidation of aldehydes occurs even in air, so that bottles of opened aldehydes soon contain amounts of carboxylic acids. This difference in the reactivities of aldehydes and ketones to oxidation is one of the main reasons they are considered as separate classes of compounds. It also provides an ideal way to distinguish between them.

Oxidation of aldehydes by acidified potassium dichromate(VI)

When an acidified solution of the orange dichromate(VI) ion is warmed with an aldehyde, it is reduced to the green chromium(III) ion. The effect of this reaction is shown in Fig 16.30, in which the aldehyde is ethanal.

The equation for the reaction with ethanal is given in Fig 16.31.

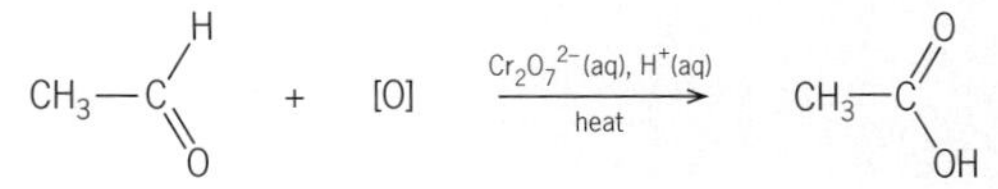

$$CH_3CHO + [O] \xrightarrow[\text{heat}]{Cr_2O_7^{2-}(aq),\ H^+(aq)} CH_3COOH$$

Fig 16.31 Oxidation of ethanal

Fig 16.30 Left: acidified potassium dichromate(VI) before ethanal is added and the mixture heated. Right: after the oxidation of ethanal, the green Cr^{3+} is clearly visible

STRETCH AND CHALLENGE

Using half equations

The half equations for this reaction are:

$$Cr_2O_7^{2-}(aq) + 14H^+(aq) + 6e^- \rightarrow 2Cr^{3+}(aq) + 7H_2O(l)$$
$$CH_3CHO(aq) + H_2O(l) \rightarrow CH_3COOH(aq) + 2H^+(aq) + 2e^-$$

Combining these to cancel out the electrons gives the redox equation:

$$3CH_3CHO(aq) + Cr_2O_7^{2-}(aq) + 8H^+(aq) \rightarrow 3CH_3COOH(aq) + 2Cr^{3+}(aq) + 4H_2O(l)$$

The use of half equations in redox reactions is covered in Chapters 5 and 25.

Oxidation of aldehydes by Fehling's and Benedict's solutions

Either or both of these reagents may be used in your chemistry course. They are alkaline solutions that contain copper(II) as **complex ions** (see Chapter 24), so the colour of both is blue (Fig 16.32). The copper(II) ions act as a mild oxidising agent. When the reagent is mixed with an aldehyde and heated, the aldehyde is oxidised to give a carboxylic acid, while the copper(II) ions are reduced to a brick-red precipitate of copper(I) oxide, Cu_2O.

A balanced equation for this reaction is:

$$RCHO + 2Cu^{2+}(aq) + 2H_2O \rightarrow RCOOH + Cu_2O(s) + 4H^+(aq)$$

The reaction for ethanal is shown in Fig 16.33.

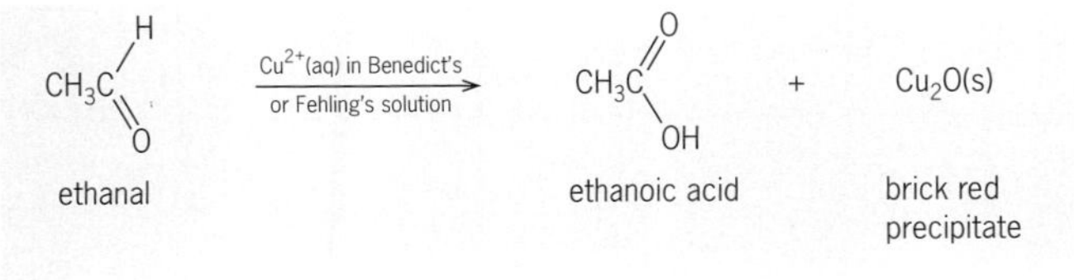

$$\underset{\text{ethanal}}{CH_3CHO} \xrightarrow[\text{or Fehling's solution}]{Cu^{2+}(aq)\text{ in Benedict's}} \underset{\text{ethanoic acid}}{CH_3COOH} + \underset{\text{brick red precipitate}}{Cu_2O(s)}$$

Fig 16.33 The oxidation of ethanal by Fehling's and Benedict's solutions

Fig 16.32 Left: The copper(II) complex ion is responsible for the blue colour of Fehling's solution. Right: After oxidation of the aldehyde, the copper(II) ion is reduced to Cu_2O, in the form of a brick-red precipitate

This reaction can be used to distinguish between an aldehyde and a ketone. While this reaction is less used now that we have spectrometers, it is still an important test in some situations, such as diagnosing diabetes (see next page).

Tollens' reagent and the oxidation of aldehydes

Tollens' reagent provides yet another way to test for aldehydes. This time, the oxidising agent is a complex of silver(I) ions, which are reduced to silver in the test. Fig 16.34 shows silver coating the inside of a test tube, making a 'silver mirror'.

The complex, which has the formula $[Ag(NH_3)_2]^+$, is made by mixing together aqueous solutions of ammonia and silver nitrate. The resulting solution is called Tollens' reagent, or ammoniacal silver nitrate. When warmed with an aldehyde, it produces the 'silver mirror'.

Fig 16.34 Left: ammoniacal silver nitrate before the addition of the aldehyde. Right: the aldehyde reduces the $[Ag(NH_3)_2]^+$ to Ag to give the silver mirror effect

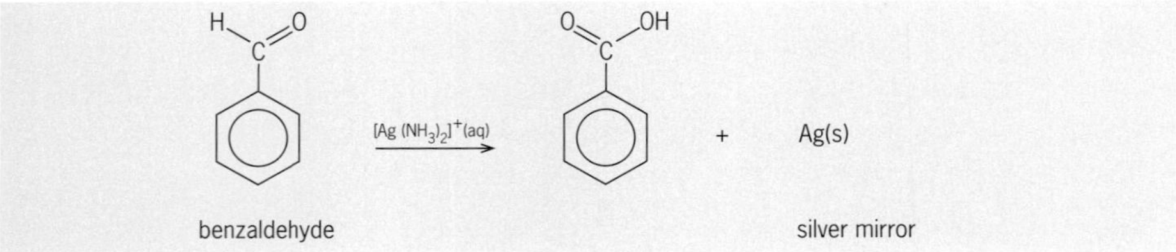

benzaldehyde $\xrightarrow{[Ag(NH_3)_2]^+(aq)}$ benzoic acid + Ag(s) (silver mirror)

Fig 16.35 Producing a silver mirror by reacting benzaldehyde with Tollens' reagent

This is a simplified balanced equation for the reaction:

$$RCHO + 2Ag^+(aq) + H_2O \rightarrow RCOOH + 2Ag(s) + 2H^+$$

The reaction provides one of the ways in which mirrors are silvered.

> **? QUESTION 10**
>
> 10 What is the name of the aldehyde formed in the bodies of meths drinkers?

SUGARS

Glucose gives a positive test with both Fehling's and Benedict's solutions. As it has this reducing property, glucose is known as a **reducing sugar**. In one of its forms, glucose contains an aldehyde group, which accounts for this property.

Diabetes is easily diagnosed using Benedict's or Fehling's reagent, which detects glucose in urine samples; and Benedict's reagent in tablet form, or impregnated into a strip of paper, is used by diabetics to monitor their blood sugar levels.

HOW SCIENCE WORKS

Alcoholic drinks and the formation of aldehydes

When you have an alcoholic drink, ethanol passes into your bloodstream. The liver has to break down the ethanol. In the first stage, it is oxidised to ethanal:

$$CH_3CH_2OH + [O] \xrightarrow{\text{enzyme in liver}} CH_3CHO + H_2O$$

Ethanal is then oxidised into other products. However, if you drink a lot of ethanol in a short space of time, the ethanal that enters the bloodstream is distributed throughout the body. Your face goes red, you may feel an unpleasant tingling in the limbs and nausea, and your blood pressure may drop.

Methanol is added to ethanol that is intended for other uses, to make it unfit for drinking. This mixture is commonly known as methylated spirits (meths). Once inside the body, methanol is oxidised into an aldehyde, which rapidly causes liver damage and blindness. So, unfortunately, if people are determined drink meths they are risking their lives.

TRI-IODOMETHANE TEST

This is a useful test for the CH_3CO group in carbonyl compounds. An alkaline solution of aqueous iodine is warmed with the substance suspected to contain this group (Fig 16.36). (Iodine has a very low solubility in water. Therefore, the aqueous solution is made up by dissolving it in KI solution.) If a yellow precipitate of tri-iodomethane is produced, it is highly likely that CH_3CO is present.

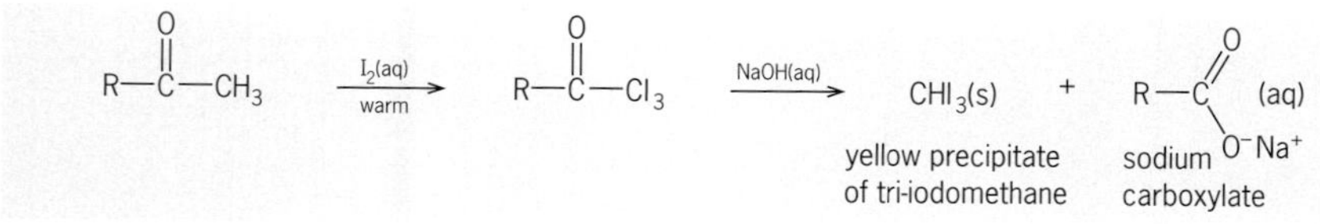

$$R{-}CO{-}CH_3 \xrightarrow[\text{warm}]{I_2(aq)} R{-}CO{-}CI_3 \xrightarrow{NaOH(aq)} CHI_3(s) + R{-}COO^-Na^+(aq)$$

yellow precipitate of tri-iodomethane; sodium carboxylate

Fig 16.36 Tri-iodomethane test. Note: R could also be hydrogen in this case

As $I_2(aq)/NaOH(aq)$ is an oxidising agent, the tri-iodomethane test also gives a positive result with alcohols that contain the $CH_3CH(OH)$ group (Fig 16.37).

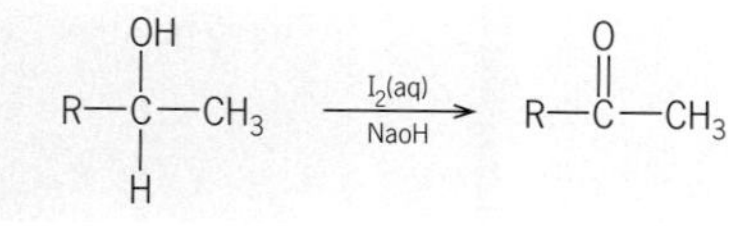

$$R{-}CH(OH){-}CH_3 \xrightarrow[\text{NaoH}]{I_2(aq)} R{-}CO{-}CH_3$$

Fig 16.37 Note that R could also be hydrogen in this case

? QUESTION 11

State which of the following compounds will give a positive result with the tri-iodomethane test:

a) CH_3CHO

b) the compound in Fig 16.38

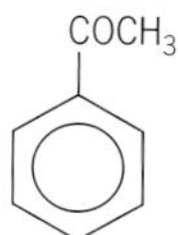

Fig 16.38

c) $C_2H_5COCH_3$

d) C_2H_5CHO

e) C_2H_5OH

(Hint: draw the structures out more fully before you decide. Only one of the five compounds does not give a positive test.)

SUMMARY

After studying this chapter, you should know the following:

- The carbonyl group C=O is found in both aldehydes (RCHO) and ketones (RCOR'), so they are called carbonyl compounds.
- The polar nature of the carbon–oxygen double bond in the carbonyl group makes it susceptible to nucleophilic addition reactions. This contrasts with the electrophilic addition reactions of the alkene carbon–carbon double bond.
- 2,4-dinitrophenylhydrazine gives characteristic orange precipitates with carbonyl compounds. The resulting 2,4-dinitrophenylhydrazone derivatives can be used to identify specific aldehydes and ketones from their precise melting points.
- Both aldehydes and ketones undergo nucleophilic addition reactions with HCN and with H^- ions (from, for example, $NaBH_4$).
- Aldehydes (but not ketones) are oxidised to give carboxylic acids by mild oxidising agents, such as $Cr_2O_7^{2-}/H^+$, alkaline solutions of Cu^{2+} (Fehling's solution) and Ag^+ (Tollens' reagent). These reactions can be used to distinguish between aldehydes and ketones.
- Alkaline aqueous iodine gives tri-iodomethane (CHI_3) with CH_3CO compounds, and also with alcohols that contain $CH_3CH(OH)$.

Practice questions for this chapter are available at www.collinseducation.co.uk/CAS

17

Carboxylic acids and pH

17 CARBOXYLIC ACIDS AND pH

Blood chemistry It is essential for the chemistry of blood that the pH value of the plasma remains fairly constant. Blood contains complex proteins and simple inorganic ions that act as buffers and together control the plasma's acidity

pH Balance in blood, foods and industry

Blood is very intimately connected with the biochemistry of all our tissues, which function correctly only while the blood is maintained at pH 7.4. In fact, death is likely to follow if the blood pH departs from this value for any length of time.

Chemical reactions in the blood itself continually produce acids and alkalis, which would seriously upset the pH level but for the fact that blood also contains buffers. These chemicals resist changes in blood pH and keep its fluctuations within safe limits.

Just as our bodies need a steady pH, so we take steps to control the pH of a wide range of foods and drinks we consume, for their stability as well as for our taste preferences and health. Wines, beers and lagers contain acids, such as ethanoic acid and tartaric acid. On the one hand, acidity helps to preserve them, but on the other too much acid makes them sour and vinegary, so food chemists ensure that the pH strikes the right balance.

When designing new chemical processes, too, industrial chemists often need to find ways to stabilize the pH values of solutions so that they will not significantly change when small amounts of acid or alkali are added. Again, for this buffers are required: often these are carboxylic acids and their soluble salts, the carboxylates, the subjects of this chapter.

Fig 17.1 Vinegar is a 5 per cent solution of ethanoic acid in water. It is acidic because ethanoic acid donates a proton – a hydrogen ion – to a water molecule. The sour taste is due to the presence of aqueous hydrogen ions

1 WHAT IS AN ACID?

Carboxylic acids are organic chemicals that contain a group of atoms known as the carboxyl group, COOH. Many common chemicals, such as citric acid in lemon juice and ethanoic acid in vinegar, are carboxylic acids and are described as weak acids. So before we go into the chemistry of carboxylic acids, we need to be clear what is meant by the term 'acid' and why chemists talk of 'weak acids' and 'strong acids'.

Chemists knew about the properties of acids long before they understood what makes an acid. In the late 1700s, they believed that oxygen was the essential element in acids. Then in 1810, Humphry Davy disproved this hypothesis when he discovered that hydrochloric acid consists of only hydrogen and chlorine. So, hydrogen was established as the element common to all acids.

One early definition says that an acid is a compound whose molecule has at least one hydrogen atom that can be replaced by a metal atom. For example, hydrochloric acid, HCl(aq), forms sodium chloride, NaCl, when hydrogen is replaced by sodium. In the same way, ethanoic acid, CH_3COOH(aq), is an acid because it forms sodium ethanoate, CH_3COONa. However, as theoretical chemistry developed, more sophisticated and generalized definitions of an acid – and a base – were introduced.

REMEMBER THIS

CH_3COONa is an ionic salt and is sometimes written as $CH_3COO^-Na^+$.

BRØNSTED–LOWRY THEORY

In 1923, the Swedish chemist Johannes Brønsted and the English chemist Thomas Lowry defined acids and bases thus:

An acid is a proton donor and a base is a proton acceptor.

According to this definition, when an acid reacts with a base, a proton is transferred from the acid to the base.

REMEMBER THIS

A proton remains when a hydrogen atom loses an electron. So a proton is the same as a hydrogen ion, H^+.

2 ACID–BASE REACTIONS

Put simply, an acid–base reaction is one in which an acid is neutralised by a base, or vice versa. During an acid–base reaction, a proton is transferred from the acid to the base. The Brønsted–Lowry theory of acids and bases really focuses on the behaviour of acids and bases in water. The water molecule is **amphoteric**, which means that it can behave as either an acid or a base (see page 330).

REMEMBER THIS

An alkali is just a base that dissolves in water.

REMEMBER THIS

When an acid dissolves in water it produces hydrogen ions in solution.

BEHAVIOUR OF BRØNSTED–LOWRY ACIDS IN WATER

When an acid, HA(aq), is added to water, a proton is transferred to a water molecule:

$$HA(aq) + H_2O(l) \rightleftharpoons H_3O^+(aq) + A^-(aq)$$

In $H_3O^+(aq)$, a single water molecule forms a dative covalent bond with a hydrogen ion (see page 76). It is called an **aqueous hydrogen ion** or an **oxonium ion**. (The term hydronium ion is also sometimes used.)

So when, for example, pure nitric acid is added to water, there is an immediate reaction in which a proton is donated from the acid to a water molecule:

$$HNO_3(aq) + H_2O(l) \rightarrow H_3O^+(aq) + NO_3^-(aq)$$

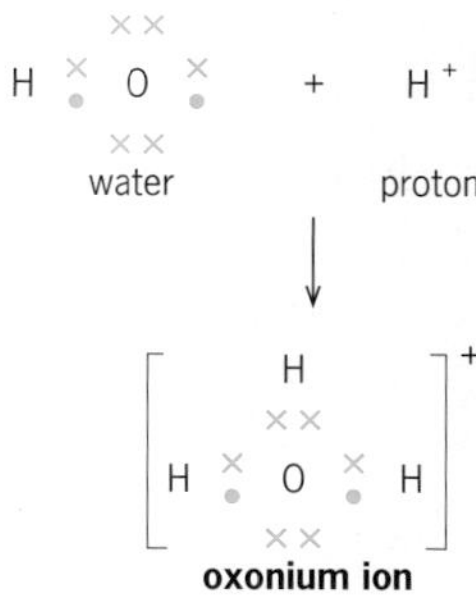

Fig 17.2 Dot-and-cross diagram of an oxonium ion, also known as an aqueous hydrogen ion or a hydronium ion

Be careful with the use of the formula HCl. The formula HCl(g) is hydrogen chloride, a colourless gas, and HCl(aq) is hydrochloric acid. HCl without (g) or (aq) does not, strictly speaking, refer to either of the two substances.

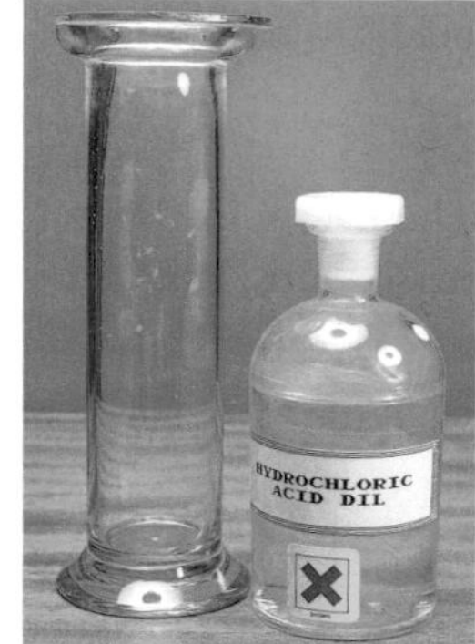

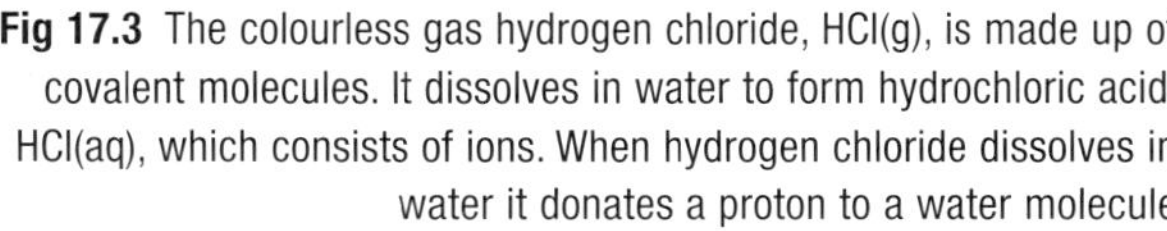

Fig 17.3 The colourless gas hydrogen chloride, HCl(g), is made up of covalent molecules. It dissolves in water to form hydrochloric acid, HCl(aq), which consists of ions. When hydrogen chloride dissolves in water it donates a proton to a water molecule

QUESTION 1

1 Write equations to show what happens when:

a) hydrogen chloride dissolves in water;

b) hydrogen iodide dissolves in water.

BEHAVIOUR OF BRØNSTED–LOWRY BASES IN WATER

When a base is added to water, it accepts a proton from the water to form a hydroxide ion. For example, when sodium oxide, Na_2O, is added to water, it is ionised and the oxide ion O^{2-} accepts a proton from a water molecule to form a hydroxide ion:

$$O^{2-}(s) + H_2O(l) \rightarrow OH^-(aq) + OH^-(aq)$$

The equation has been written with two separate hydroxide ions to emphasise that the oxide ion becomes part of a hydroxide ion when it accepts a proton from a water molecule.

As another example, ammonia, a colourless gas, dissolves in water to form an alkaline solution:

$$NH_3(g) + H_2O(l) \rightleftharpoons NH_4^+(aq) + OH^-(aq)$$

This time, the whole ammonia molecule accepts a proton from a water molecule to form the ammonium ion.

REMEMBER THIS

Notice that the equation with NH_3 does not have an arrow, but has the symbol for a reversible reaction. There is more about the significance of this symbol in relation to bases in Chapter 28.

REMEMBER THIS

For the rest of the chapter, the symbol $H^+(aq)$ is used to represent an aqueous hydrogen ion, rather than $H_3O^+(aq)$.

AMPHOTERIC NATURE OF WATER

In water, there is always a very small proportion of $OH^-(aq)$ and of $H^+(aq)$. This is the result of an acid–base reaction between two water molecules:

$$H_2O(l) + H_2O(l) \rightleftharpoons H_3O^+(aq) + OH^-(aq)$$

One water molecule donates a proton and the other accepts a proton. This demonstrates that water can behave both as an acid and as a base, so it is an amphoteric compound.

This acid–base reaction can be written in a simpler form as the dissociation of water:

$$H_2O(l) \rightleftharpoons H^+(aq) + OH^-(aq)$$

DISSOCIATION IN BRØNSTED–LOWRY ACIDS

$$A{-}H + H_2O \longrightarrow A^- + H_3O^+$$

Fig 17.4 Dissociation of an acid

When a Brønsted–Lowry acid is put into water, a chemical reaction called **dissociation** or **ionisation** takes place. A covalent bond between an electronegative atom and a hydrogen atom is broken by heterolytic fission, to leave a proton and a negative ion. So, when the gas hydrogen chloride dissolves in water, it dissociates to form a proton or hydrogen ion, and a chloride ion:

$$HCl(g) \rightarrow H^+(aq) + Cl^-(aq)$$

The arrow in the equation indicates that, in water, every hydrogen chloride molecule dissociates.

In water, sulfuric acid dissociates to form two hydrogen ions per sulfuric acid molecule:

$$H_2SO_4(aq) \rightarrow 2H^+(aq) + SO_4^{2-}(aq)$$

The arrow in the equation indicates that, in water, every sulfuric acid molecule dissociates, but in reality many of the sulfuric acid molecules only dissociate to form the hydrogensulfate ion, HSO_4^-.

QUESTION 2

2 Write down the equations that show the dissociation of hydrogen bromide and of hydrogen iodide in water.

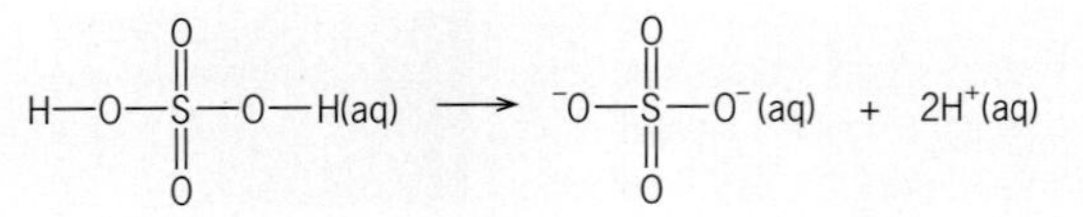

$$H{-}O{-}S(=O)_2{-}O{-}H(aq) \longrightarrow {}^-O{-}S(=O)_2{-}O^-(aq) + 2H^+(aq)$$

Fig 17.5 Dissociation of sulfuric acid represented by structural formulas

Conjugate bases

During the dissociation of hydrogen chloride, all the molecules dissociate to form aqueous hydrogen ions and chloride ions. The chloride ion is known as the **conjugate base** of hydrochloric acid. This does not mean that the chloride ion is a base, but that in theory it could accept a proton. That is, it could behave as a base. The conjugate base of any acid is the anion formed after dissociation.

Note that the conjugate base of sulfuric acid is the hydrogensulfate ion, HSO_4^-, rather than the sulfate ion, SO_4^{2-}(see Table 17.1).

Table 17.1 Five acids and their dissociations

Acid				Conjugate base
$HCl(aq)$	→	$H^+(aq)$	+	$Cl^-(aq)$
$HNO_3(aq)$	→	$H^+(aq)$	+	$NO_3^-(aq)$
$H_2SO_4(aq)$	→	$H^+(aq)$	+	$HSO_4^-(aq)$
$HBr(aq)$	→	$H^+(aq)$	+	$Br^-(aq)$
$HI(aq)$	→	$H^+(aq)$	+	$I^-(aq)$

Conjugate acids

Just as a Brønsted–Lowry acid has a conjugate base, so a Brønsted–Lowry base has a conjugate acid, which is the particle formed once the base has accepted a proton. For example, the conjugate acid of ammonia, NH_3, is the ammonium ion, NH_4^+, and the conjugate acid of the hydroxide ion, OH^-, is the water molecule, H_2O.

Strong acids

Table 17.1 shows how five common covalently bonded acids dissociate when in water. In each case, almost all the molecules of the acid dissociate to form ions. In fact, we assume that all of the molecules dissociate, and represent the dissociation by an arrow in the equation. Such acids are called **strong acids**, a term meaning that almost all of the acid's molecules dissociate when it is dissolved in water. Do not confuse a strong acid with a concentrated acid. A strong acid is fully dissociated in water, but a concentrated acid has a high number of moles of acid in 1 dm^3 of solution.

Weak acids

Other acids, such as ethanoic acid, dissolve in water, but do not fully dissociate. Only a very small percentage of their molecules may be dissociated at any one time. They are known as **weak acids**. For example, in a solution containing 0.1 mol dm^{-3} of ethanoic acid, only 1.3 per cent of the molecules dissociate. Ethanoic acid is therefore a weak acid. Do not confuse a weak acid with a dilute acid. The term weak acid means that only a small proportion of molecules dissociate when the acid is dissolved in water:

$$CH_3COOH(aq) \rightleftharpoons CH_3COO^-(aq) + H^+(aq)$$

Note that the dissociation equation includes the symbol for a reversible reaction instead of an arrow. This indicates that the dissociation of ethanoic acid and the association of the ethanoate ion and the hydrogen ion occur at the same time.

As in Figs 17.6 and 17.7, once ethanoic acid has dissolved in water, the rate of the dissociation reaction soon reaches that of the association reaction, so that there is no net change in the concentrations of the undissociated ethanoic acid, the ethanoate ion and the hydrogen ion. We say that the reaction reaches a state of **dynamic equilibrium**, in which the concentration of each species remains the same.

QUESTIONS 3–4

3 **a)** What is the conjugate base of hydrofluoric acid, HF?
b) What is the conjugate base of water?
c) What is the conjugate base of the hydrogensulfate ion?

4 **a)** What is the conjugate acid of water?
b) What is the conjugate acid of the oxide ion, O^{2-}?
c) What is the conjugate acid of the hydrogensulfate ion?

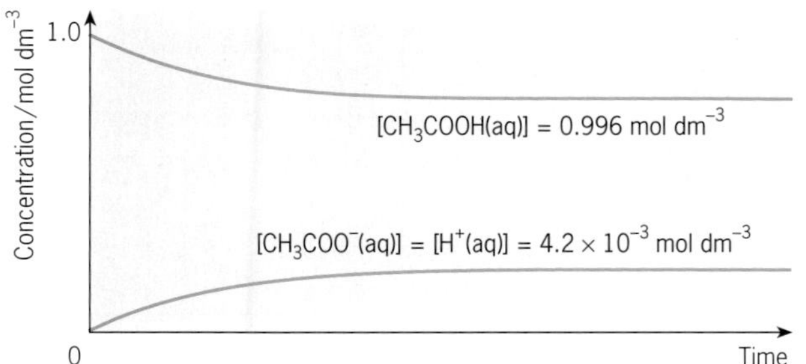

Fig 17.7 A short time after the ethanoic acid is mixed with water at 25 °C, the concentration of each species remains constant. The concentrations are called equilibrium concentrations. The reaction is in equilibrium because the concentration of each species is constant.

This is not the whole story, however. Chemical reactions are still taking place and particles are colliding and reacting, but the rate of the forward reaction is matched by the rate of the backward reaction. Note the *y* (concentration) axis has no scale between 0 and 1.0. This is because on a uniform y-axis scale, the (equal) concentrations of the aqueous hydrogen ion and the ethanoate ion are so small compared with the concentration of the undissociated ethanoic acid that they would not show as a curve above the *x*-axis

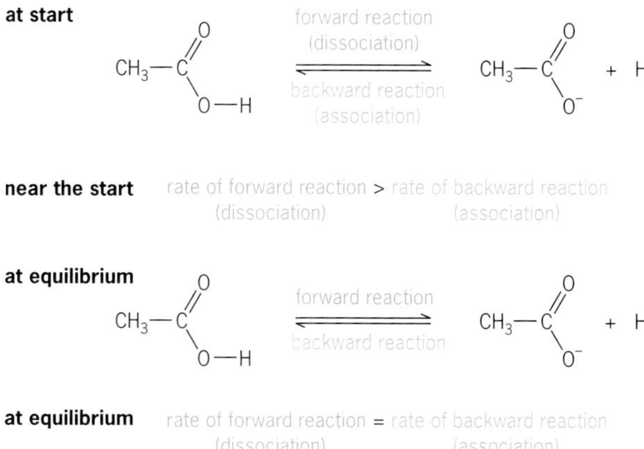

Fig 17.6 Dissociation and association reactions of ethanoic acid in water

ACID DISSOCIATION CONSTANT

Once a weak acid has dissolved in water, it quickly reaches the state of dynamic equilibrium. It is possible to gauge just how weak an acid is by determining the **acid dissociation constant, K_a**. The $_a$ in K_a stands for 'acid'.

A weak acid HA dissociates in water and reaches a state of dynamic equilibrium:

$$HA(aq) \rightleftharpoons H^+(aq) + A^-(aq)$$

The acid dissociation constant is given by:

$$K_a = \frac{[H^+(aq)][A^-(aq)]}{[HA(aq)]}$$

REMEMBER THIS

The acid dissociation constant is a modified equilibrium constant. Equilibrium constants are covered on page 352.

The square brackets represent the concentrations of the various particles measured in mol dm^{-3}. For example, $[H^+(aq)]$ is the concentration of aqueous hydrogen ions in mol dm^{-3}. Note that the concentrations used are those at the equilibrium position.

The acid dissociation constant for ethanoic acid is:

$$K_a = \frac{[H^+(aq)][CH_3COO^-(aq)]}{[CH_3COOH(aq)]} \qquad \text{Units} = \frac{\cancel{(mol\ dm^{-3})}\ (mol\ dm^{-3})}{\cancel{(mol\ dm^{-3})}}$$

The smaller the value of the acid dissociation constant, the weaker the acid. Ethanoic acid, CH_3COOH, has a K_a of 1.8×10^{-5} mol dm^{-3}, whereas nitrous acid, HNO_2, has a K_a of 4.5×10^{-4} mol dm^{-3}. Notice that the acid dissociation constant includes a unit. It is not just a numerical value.

QUESTION 5

5 **a)** Write down the equation to show the dissociation of nitrous acid in water.
b) Write down an expression for the acid dissociation constant for nitrous acid.
c) Which is the weaker acid, ethanoic or nitrous?

pK_a

The numerical values of K_a for strong acids normally lie in the range from 10^{-1} to 10^2, but for weak acids they are below 10^{-4}. To avoid dealing with the small numbers associated with weak acids, we use pK_a. This is its definition:

pK_a is the negative logarithm to base 10 of the acid dissociation constant:

$$pK_a = -\log_{10} K_a$$

Note that as the acid strength decreases, pK_a increases. So a high pK_a value means a very weak acid.

REMEMBER THIS

pK_a values are easier to compare than numerical K_a values, since they are not very small numbers expressed in standard form (log to base *e*).

REMEMBER THIS

All scientific calculators have a key labelled lg or log. Do not use the key labelled ln, which is the 'natural logarithm'.

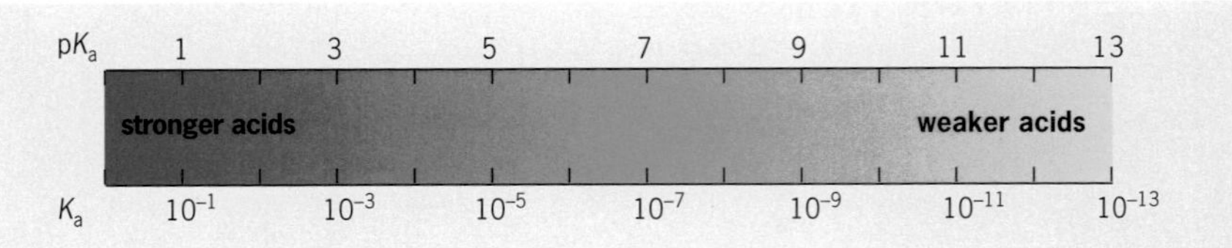

Fig 17.8 The connection between pK_a and K_a

QUESTION 6

6 **a)** What is the pK_a value for each of these acids?
i) Phenol, $K_a = 1.3 \times 10^{-10}$ mol dm^{-3}.
ii) Butanoic acid, $K_a = 1.5 \times 10^{-5}$ mol dm^{-3}.
b) Suggest why ethanedioic acid, $(COOH)_2$, has two pK_a values.

EXAMPLE

Q The acid dissociation constant for nitrous acid is 4.5×10^{-4} mol dm^{-3}. What is the pK_a value for nitrous acid?

A $pK_a = -\log_{10} K_a$
Substituting the value for K_a gives:

$$pK_a = -\log_{10} (4.5 \times 10^{-4})$$
$$pK_a = 3.3$$

Propanedioic acid has two pK_a values, because one molecule can donate two hydrogen ions. The first is donated more easily than the second:

$$HOOCCH_2COOH(aq) \rightleftharpoons HOOCCH_2COO^-(aq) + H^+(aq) \qquad pK_a = 2.77$$
$$HOOCCH_2COO^-(aq) \rightleftharpoons {}^-OOCCH_2COO^-(aq) + H^+(aq) \qquad pK_a = 5.66$$

Why is sulfuric acid strong and ethanoic acid weak?

The ability of a carboxylic acid, such as ethanoic acid, to donate a hydrogen ion depends on the strength of the O–H bond in the carboxyl group. Likewise, the ability of sulfuric acid to donate a hydrogen ion depends on the strength of its O–H bonds. In simple terms, the stronger the O–H bond(s), that is, the greater the bond energy, the weaker the acid. In sulfuric acid the presence of many highly electronegative oxygen atoms withdraws electrons away from the O–H bond and makes it weaker than the O–H bond in ethanoic acid. This is not the whole story, since the stability of the conjugate bases formed is also important. So the sulfate ion and the hydrogensulfate ion are more stable conjugate bases than the ethanoate ion.

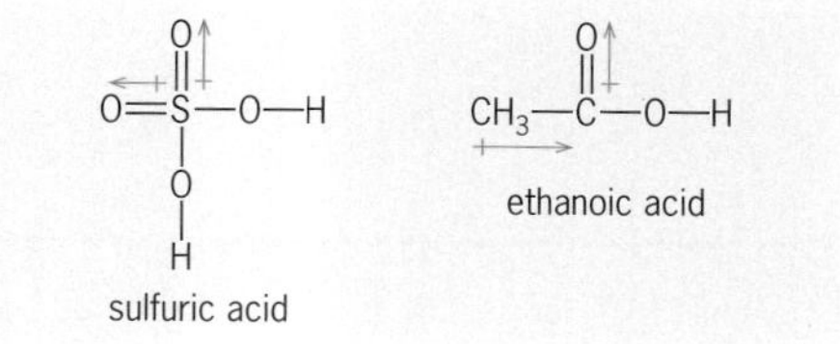

Fig 17.9 The displayed formulas of sulfuric acid and ethanoic acid. The O—H bond in sulfuric acid is very weak compared with the O—H bond in ethanoic acid. The electronegative oxygen atoms in sulfuric acid draw the electron cloud away from the O—H bonds, but this effect is less marked in ethanoic acid. So sulfuric acid is a strong acid and ethanoic acid is a weak acid

Table 17.2 The relative strengths of some acids. Chloric(VII) acid is the strongest acid of all. It fully dissociates in water. All acids stronger than H_3O^+ fully dissociate in water. Notice that they include most of the common laboratory acids. Acids that are weaker than water are not usually considered to be acids. Ammonia, for example, is normally classified as a base

increasing acid strength (upwards) | increasing base strength (downwards)

Acid			Conjugate base
chloric(VII) acid	$HClO_4$	ClO_4^-	chlorate(VII) ion
sulfuric acid	H_2SO_4	HSO_4^-	hydrogensulfate ion
hydrogen iodide	HI	I^-	iodide ion
hydrogen bromide	HBr	Br^-	bromide ion
hydrogen chloride	HCl	Cl^-	chloride ion
nitric acid	HNO_3	NO_3^-	nitrate ion
oxonium ion	H_3O^+	H_2O	water
hydrogensulfate ion	HSO_4^-	SO_4^{2-}	sulfate ion
phosphoric(V) acid	H_3PO_4	$H_2PO_4^-$	dihydrogenphosphate(V) ion
hydrogen fluoride	HF	F^-	fluoride ion
nitrous acid	HNO_2	NO_2^-	nitrite ion
ethanoic acid	CH_3COOH	CH_3COO^-	ethanoate ion
carbonic acid	H_2CO_3	HCO_3^-	hydrogencarbonate ion
ammonium ion	NH_4^+	NH_3	ammonia
water	H_2O	OH^-	hydroxide ion
ethanol	C_2H_5OH	$C_2H_5O^-$	ethoxide ion
ammonia	NH_3	NH_2^-	amide ion

QUESTION 7

a) What is the connection between the strength of an acid and the strength of its conjugate base?

b) Explain the difference in acid strength of the hydrogen halides HBr, HCl, HF and HI. (Hint: think about the bond energy of the bond that must break for dissociation to take place.)

3 pH

An acid is a proton donor – when it dissolves in water the $H^+(aq)$ concentration increases. That is, the higher the $H^+(aq)$ concentration, the more acidic a solution becomes.

However, the problem in measuring hydrogen ion concentration is that for many solutions it is extremely small numerically. For example, the hydrogen ion concentration in wine is only about 5×10^{-4} mol dm^{-3}. In blood, it is even smaller, being about 4×10^{-8} mol dm^{-3}.

In 1909, the Danish chemist Sören Sörensen, working for the brewers Carlsberg, suggested an alternative way to measure the acidity of a solution – the **pH scale**.

Table 17.3 The relationship between [H+(aq)] and pH

$[H^+(aq)]$/ mol dm^{-3}	pH	Sample solution
10^1	–1	
10^0 or 1	0	1 mol dm^{-3} H_2SO_4
10^{-1}	1	
10^{-2}	2	
10^{-3}	3	vinegar
10^{-4}	4	stomach acid, wine
10^{-5}	5	
10^{-6}	6	black coffee
10^{-7}	7	
10^{-8}	8	pure water, blood
10^{-9}	9	
10^{-10}	10	
10^{-11}	11	milk of magnesia
10^{-12}	12	household ammonia
10^{-13}	13	
10^{-14}	14	1 mol dm^{-3} NaOH

REMEMBER THIS

Each change of one pH unit represents a 10-fold increase in the aqueous hydrogen ion concentration.

The pH scale is constructed from the negative of the logarithm to base 10 of the $[H^+(aq)]$. Thus, aqueous hydrogen-ion concentrations that may differ by, say, a hundred million million, are converted into numbers on a scale which runs from just below 0 to just above 14. Accordingly, the pH of a solution is described by the equation:

$$\mathbf{pH = -log_{10}\,[H^+(aq)]}$$

where the aqueous hydrogen ion concentration is in mol dm^{-3}.

Measuring the pH of a solution

Our sense of taste is very sensitive to pH, and it provided the earliest way to distinguish between acids and alkalis. We can detect a sour acidic taste in solutions whose pH values are as high as 5, which corresponds to H+(aq) concentrations as low as 10^{-5} mol dm^{-3}. The familiar acidic taste of most fruit juices and soft drinks is caused by the presence of weak acids, which give these drinks a pH value of about 3.

When the pH is below 2, we find the taste distinctly unpleasant and the liquid is sufficiently acid to burn the lining of the mouth. In the same way, we can detect alkalis with a pH value of 9 to 10 as they taste bitter. Liquids with higher pH values than this are dangerous to taste and also burn the lining of the mouth.

The pH of a solution is measured accurately using a pH meter (Fig 17.10), which converts the difference in electrical potential between two electrodes into a pH reading. (Electrode potentials and electrochemistry are covered in Chapter 25.)

An indicator, such as universal indicator, which gives a range of colour in solutions of different pH values (Fig 17.11), can be used to estimate the pH value of a solution.

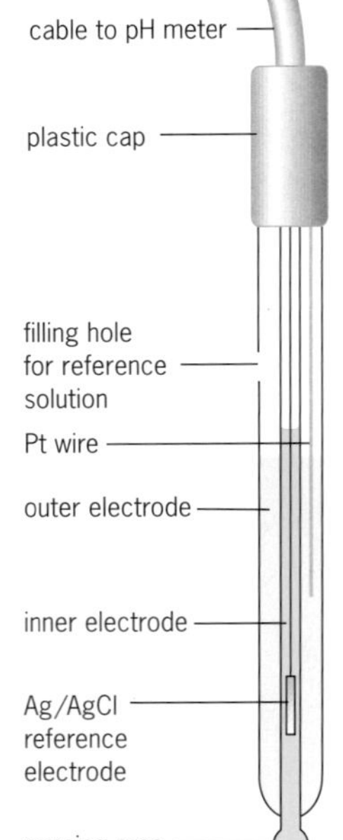

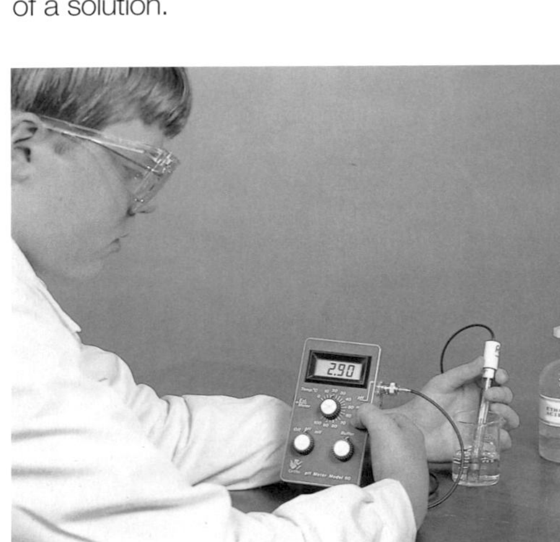

Fig 17.10 The pH meter. When the probe is dipped into the solution, an electrical potential difference is established between the two electrodes.
This potential difference is amplified and converted into a digital pH reading

Fig 17.11 Universal indicator gives a colour that depends on pH

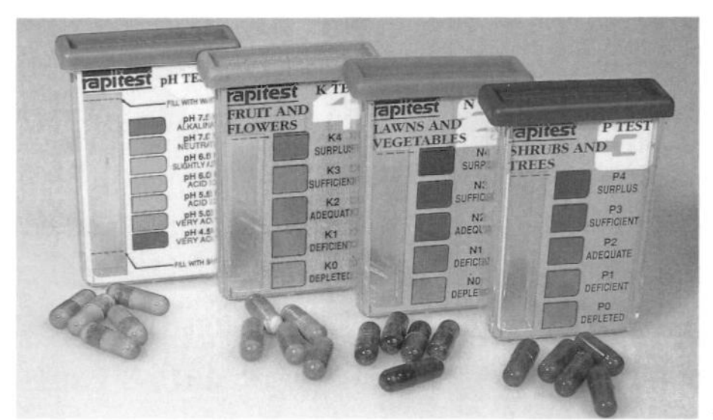

Fig 17.12 A commercial kit that gardeners use to test the pH of soil

CALCULATING THE pH OF STRONG ACIDS

To calculate the pH of a strong acid, substitute the appropriate $H^+(aq)$ concentration in the equation for pH. This can be deduced from the concentration of the acid itself, since it is assumed that every acid molecule has dissociated. It is also possible to calculate the H^+concentration given the pH.

REMEMBER THIS

A negative pH value means that the concentration of aqueous hydrogen ions, $[H^+(aq)]$, in mol dm^{-3}, is very high.

EXAMPLE

Q What is the pH of a solution of 1.00 mol dm^{-3} sulfuric acid?

A In water, sulfuric acid is assumed to be fully dissociated because it is a strong acid:

$$H_2SO_4(aq) \rightarrow 2H^+(aq) + SO_4^{2-}(aq)$$

From the equation, every mole of H_2SO_4 gives 2 moles of $H^+(aq)$. The concentration of 1.00 mol dm^{-3} refers to moles of the undissociated H_2SO_4, so the concentration of $H^+(aq)$ must be twice this. Therefore,

$$[H^+(aq)] = 2 \times 1.00 \text{ mol dm}^{-3} = 2.00 \text{ mol dm}^{-3}$$

Substituting this value in the equation for pH gives:

$$pH = -\log_{10}[2.00] = -0.301$$

EXAMPLE

Q A sample of acid rain has a pH of 3.41. What is the molar concentration of hydrogen ions

A The relationship between pH and $[H^+(aq)]$ is:

$pH = -\log_{10} [H^+(aq)]$

Substituting the pH value gives

$3.41 = -\log_{10} [H^+(aq)]$

So $-3.41 = \log_{10} [H^+(aq)]$

So $\log{-3.41} = [H^+(aq)]$

Using a calculator $[H^+(aq)] = 3.39 \times 10^{-4}$ mol dm^{-3}

STRETCH AND CHALLENGE

The pH of boiling water

At 298 K, the pH of pure water is 7.00, which is designated neutral on the pH scale. So the concentration of $H^+(aq)$ in pure water is 1.00×10^{-7} mol dm^{-3}. An acid has a pH below 7. That is, the concentration of $H^+(aq)$ is greater than 1.00×10^{-7} mol dm^{-3}.

When a pH meter is placed in boiling pure water, the reading is pH 6.12. So, apparently, boiling water is acidic. But how can it be? In boiling water, the concentration of $H^+(aq)$ has increased, but so has the concentration of the aqueous hydroxide ion. So, there should be no surplus of either type of ion:

$$H_2O(l) \rightleftharpoons H^+(aq) + OH^-(aq)$$

Now, this is the same situation as that in water at 298 K, which is considered to be neutral. The paradox is easy to explain. In boiling water, a greater proportion of water molecules dissociates at any one time, but there is no surplus of hydrogen ions.

QUESTIONS 8–9

8 **a)** Calculate the pH of
i) 1.00 mol dm^{-3} hydrochloric acid
ii) 1.00×10^{-4} mol dm^{-3} hydrochloric acid;
iii) 0.0520 mol dm^{-3} sulfuric acid.
b) Assuming that in 0.100 mol dm^{-3} nitrous acid, 6.5 per cent of the nitrous acid molecules have dissociated, calculate the pH of 0.100 mol dm^{-3} nitrous acid.

9 **a)** What is the $[H^+(aq)]$ and the $[OH^-(aq)]$ in boiling water?
b) The dissociation of water is an endothermic process. Use this to help explain why the pH of water is below 7.

CALCULATING THE pH OF WEAK ACIDS

Ethanoic acid has already been described as a weak acid that reaches a state of dynamic equilibrium when dissolved in water. When this state is reached, the concentration of aqueous hydrogen ions is very low compared with the concentration of the undissociated ethanoic acid. We can calculate the pH of aqueous ethanoic acid provided that we know the numerical value of the acid dissociation constant, or the percentage of ethanoic acid molecules that have dissociated.

EXAMPLES

Q What is the pH of 0.10 mol dm^{-3} ethanoic acid, given that 1.3 per cent of the ethanoic acid molecules have dissociated?

A In water, ethanoic acid forms an equilibrium mixture very quickly:

$$CH_3COOH(aq) \rightleftharpoons H^+(aq) + CH_3COO^-(aq)$$

The 0.10 mol dm^{-3} is the concentration of CH_3COOH before any has dissociated.

Given that only 1.3 per cent of ethanoic acid molecules dissociate, then the concentration of ethanoic acid that has dissociated must be 1.3 per cent of 0.100 mol dm^{-3}, which is 1.3×10^{-3} mol dm^{-3}.

From the equation, every mole of $CH_3COOH(aq)$ that dissociates gives 1 mole of $H^+(aq)$. So, $[H^+(aq)]$ must be 1.3×10^{-3} mol dm^{-3}.

Substituting this value into the equation for pH gives:

$$pH = -\log_{10}(1.3 \times 10^{-3}) = 2.9$$

Q What is the pH of 0.10 mol dm^{-3} ethanoic acid given that the acid dissociation constant, K_a, for ethanoic acid is 1.8×10^{-5} mol dm^{-3}?

A Since ethanoic acid is a weak acid, it forms an equilibrium mixture in water:

$$CH_3COOH(aq) \rightleftharpoons CH_3COO^-(aq) + H^+(aq)$$

Therefore, the expression for the acid dissociation constant is:

$$K_a = \frac{[CH_3COO^-(aq)][H^+(aq)]}{[CH_3COOH(aq)]}$$

From the equation, every mole of CH_3COOH that dissociates gives 1 mole of CH_3COO^- and 1 mole of $H^+(aq)$.

So, if $[H^+(aq)]$ at equilibrium is x, then it follows that $[CH_3COO^-(aq)]$ must also be x.

It also follows that $[CH_3COOH(aq)]$ must be $0.10 - x$, since some of the ethanoic acid has dissociated.

This can be summarised as follows:

	$CH_3COOH(aq)$ undissociated ethanoic acid	$\rightleftharpoons$	$CH_3COO^-(aq)$	$+ H^+(aq)$
At start, ie before equilibrium/mol dm^{-3}	0.10		0	0
At equilibrium/mol dm^{-3}	$0.10 - x$		x	x

These values can be substituted in the equation for the acid dissociation constant. Therefore,

$$1.8 \times 10^{-5} = \frac{x^2}{0.10 - x}$$

(The dissociation of water to form $H^+(aq)$ is ignored since its concentration is so small.)

The equation may be simplified by reference to Fig 17.6 on page 331. The concentration of the $H^+(aq)$ is very small compared with that of the undissociated acid, and therefore the concentration of the undissociated acid is very nearly 0.10. So we can assume that $(0.10 - x)$ is approximately 0.10. Using this assumption, we can write the equation as:

$$1.8 \times 10^{-5} = \frac{x^2}{0.10}$$

which can be rearranged to give:

$$x^2 = 1.8 \times 10^{-6}$$

Therefore:

$$x = \sqrt{(1.8 \times 10^{-6})} = 1.34 \times 10^{-3} \text{ (ignoring sig. figs)}$$

Since x is the $H^+(aq)$ concentration, we can substitute this value of x into the equation for pH:

$$\begin{aligned} pH &= -\log[H^+(aq)] \\ &= -\log(1.34 \times 10^{-3}) \\ &= 2.9 \end{aligned}$$

Notice that even with the assumption made in the second calculation, the value obtained for the pH is the same as that obtained in the Example before this.

? QUESTION 10

10 What is the pH of 0.10 mol dm^{-3} methanoic acid, HCOOH, whose K_a is 1.8×10^{-4} mol dm^{-3}?

4 CARBOXYLIC ACIDS

Carboxylic acids are weak acids. Therefore, they do not fully dissociate when dissolved in water. Many of them will not dissolve in water at all, although they can be shown to be acids since they react with alkalis.

CARBOXYLIC ACIDS IN NATURE

Many naturally occurring organic compounds are carboxylic acids. Familiar examples already mentioned are citric acid in citrus fruits and ethanoic acid in vinegar. Three more examples are given in Figs 17.13 to 17.15.

Methanoic acid is the simplest carboxylic acid. Discovered in 1670, it was originally called formic acid, after the Latin word *formicus* for ant, since it is one of the substances responsible for the sting of an ant bite (Fig 17.14).

Fig 17.13 Two naturally occurring carboxylic acids are ethanedioic acid and lactic acid. Ethanedioic acid is a poisonous white crystalline solid. As a metal salt, ethanedioic acid is found in rhubarb (above) and sorrel, and gives them a sour taste. Lactic acid is formed when lactose, found in milk, ferments. It gives the sour taste of soured milk products such as yoghurt. It is a preservative, deterring the growth of micro-organisms, and is added to many foods, such as confectionery and salad dressings

Fig 17.14 Methanoic acid, the simplest carboxylic acid, is one of the substances responsible for the irritation caused by an ant bite. It was once prepared by distilling red ants

Fig 17.15 Benzoic acid is an aromatic carboxylic acid found in raspberries and in some tree barks. It is used as a preservative and an antioxidant in numerous foods, including soft drinks, pickles, salad dressings and fruit products

THE CARBOXYL GROUP

A carboxylic acid contains the carboxyl group, COOH, which you may see written as CO_2H. The carboxyl group contains a carbonyl bond C=O, and a hydroxyl bond O–H. Each bond is attached to the same carbon atom. This arrangement modifies the characteristic behaviour of both the carbonyl bond and the hydroxyl bond, which is why the carboxyl group is considered a functional group in its own right.

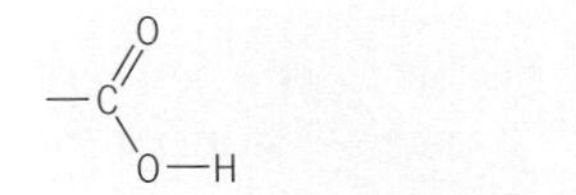

Fig 17.16 The carboxyl group

NAMING CARBOXYLIC ACIDS

Carboxylic acids are named by adding the suffix 'oic acid' to the name of the carbon chain (Fig 17.17). Unlike many functional groups, the carboxyl group contains one of the carbon atoms of the chain. The carboxyl group is almost always on the number 1 carbon atom, so this carbon atom rarely has to be given any other position number.

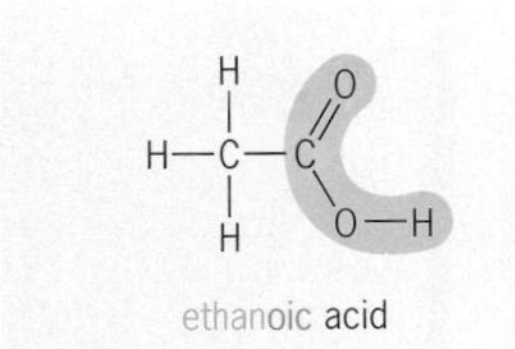

Fig 17.17 The derivation of the name ethanoic acid

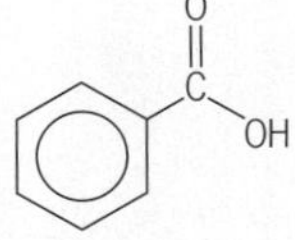

benzoic acid

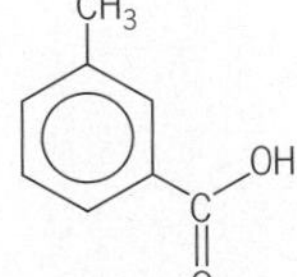

3–methylbenzoic acid

CH_3CH_2COOH
propanoic acid

$CH_3CHOHCOOH$
2-hydroxypropanoic acid

$HOOCCH_2COOH$
propane-1,3-dioic acid

$ClCH_2COOH$
chloroethanoic acid

Fig 17.18 Six carboxylic acids and their systematic names

QUESTION 11

a) What is the systematic name of each of the following carboxylic acids?

i) $CH_3CH_2CH_2CH_2COOH$
ii) $CH_3CH_2CH_2COOH$
iii) $HOOCCH_2CH_2CH_2COOH$

b) Draw the structural formula of each of the following carboxylic acids.

i) 3-chloropropanoic acid
ii) 3-chlorobenzoic acid
iii) 4-methylbenzoic acid
iv) benzene-1,4-dicarboxylic acid

QUESTION 12

12 **a)** Methanoic acid, ethanoic acid and propanoic acid are members of a homologous series. What is the general formula for this homologous series?

b) i) Predict the boiling point of pentanoic acid.

ii) Explain why it is difficult to predict the melting point of pentanoic acid.

REMEMBER THIS

The volatility of carboxylic acids decreases as the relative formula mass increases. Remember that liquids with low boiling point are very volatile, which means they have a high vapour pressure at room temperature.

REMEMBER THIS

The solubility of carboxylic acids in water decreases as the relative formula mass increases. This is because the ability to form hydrogen bonds with molecules decreases as the non-polar alkyl group of the carboxylic acid gets longer.

The effect of concentration on reaction rate is described in Chapter 26.

PHYSICAL PROPERTIES OF CARBOXYLIC ACIDS

Carboxylic acids are either solids or liquids at room temperature (Table 17.4).

The melting points of methanoic acid and ethanoic acid are quite high, and so it is not unusual in cold weather for a bottle of concentrated ethanoic acid to be solid. In methanoic acid and ethanoic acid, the intermolecular forces are predominantly hydrogen bonds. However, as the carbon chain increases in length, there is also a considerable contribution from induced dipole–induced dipole attractions.

Table 17.4 Melting and boiling points of six carboxylic acids

Acid	Formula	Melting point/°C	Boiling point/°C
methanoic acid	HCOOH	8.4	110.5
ethanoic acid	CH_3COOH	16.6	118
propanoic acid	CH_3CH_2COOH	−22	141
butanoic acid	$CH_3CH_2CH_2COOH$	−5	163
ethanedioic acid	$(COOH)_2$	187	decomposes
benzoic acid	C_6H_5COOH	122	249

5 ACIDIC PROPERTIES OF THE CARBOXYL FUNCTIONAL GROUP

As mentioned already, carboxylic acids are weak acids, a property caused by the heterolytic fission of the hydroxyl bond in the COOH group (Fig 17.19).

A carboxylic acid that can dissolve in water forms a weakly acidic solution in water. If a carboxylic acid cannot dissolve in water, it is almost impossible for any of its molecules to donate a hydrogen ion to a water molecule, and so the water remains about neutral. Nevertheless, insoluble carboxylic acids are acids, because they will react with bases, such as aqueous sodium hydroxide.

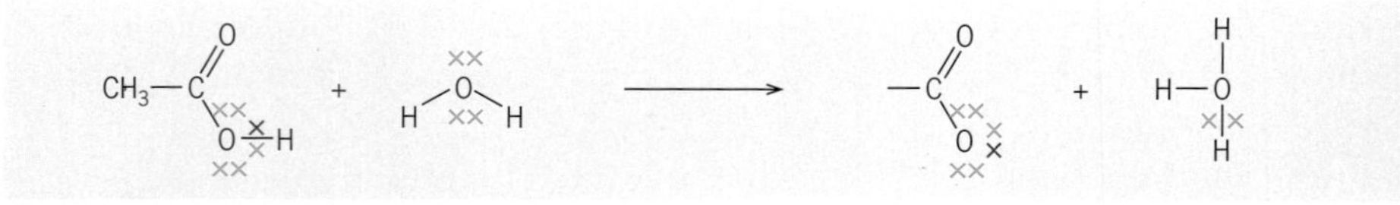

Fig 17.19 Dissociation of the carboxyl group involves heterolytic fission of the O–H bond and the subsequent transfer of a hydrogen ion (a proton) to a water molecule. For clarify, the charges on COO^- and H_3O^+ have been omitted

ACID–BASE REACTIONS OF CARBOXYLIC ACIDS

The typical reactions of aqueous acids arise from the presence of the aqueous hydrogen ion. Since carboxylic acids are weak acids, such reactions are much slower than they are with strong acids, such as sulfuric acid.

As Fig 17.20 shows, the acidic reactions of ethanoic acid always form the ethanoate ion.

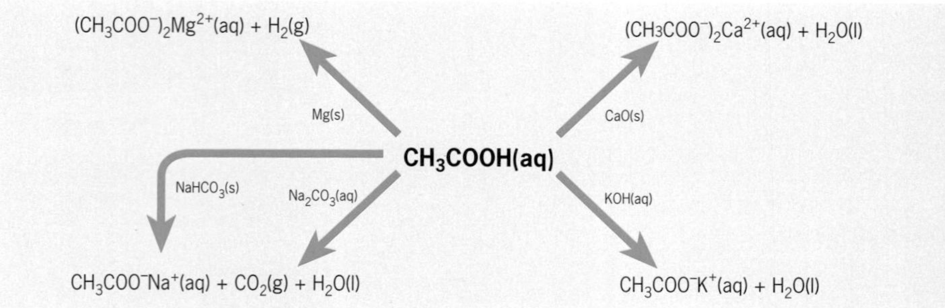

Fig 17.20 Acidic reactions of ethanoic acid

REACTION OF CARBOXYLIC ACIDS WITH ALKALIS AND BASES

Carboxylic acids, such as ethanoic acid, can be neutralised by an alkali, such as aqueous sodium hydroxide:

$$CH_3COOH(aq) + NaOH(aq) \rightarrow CH_3COO^-Na^+(aq) + H_2O(l)$$

This reaction is often carried out in the laboratory as a titration. An aqueous solution of the acid is added from a burette to a known volume of the alkali. At neutralisation, an indicator changes colour.

Bases such as magnesium oxide react slowly with aqueous carboxylic acids to form carboxylate salts. For example, dilute ethanoic acid reacts to form aqueous magnesium ethanoate:

$$MgO(s) + 2CH_3COOH(aq) \rightarrow Mg^{2+}(CH_3COO^-)_2(aq) + H_2O(l)$$

This reaction is much slower than the reaction of magnesium oxide with dilute hydrochloric acid because ethanoic acid is a weak acid and so has a much lower concentration of $H^+(aq)$ than does hydrochloric acid, which is a strong acid.

Reaction with sodium carbonate and sodium hydrogencarbonate

Sodium hydrogencarbonate and sodium carbonate will react with aqueous solutions of carboxylic acids to form carbon dioxide:

$$NaHCO_3(s) + CH_3COOH(aq) \rightarrow CH_3COO^-Na^+(aq) + CO_2(g) + H_2O(l)$$

$$Na_2CO_3(s) + 2CH_3COOH(aq) \rightarrow 2CH_3COO^-Na^+(aq) + CO_2(g) + H_2O(l)$$

TITRATION CURVES

The changes in pH can be monitored at regular intervals as an alkali is added to an acid. The pH can then be plotted against the volume of alkali added, to obtain a graph known as a **titration curve**.

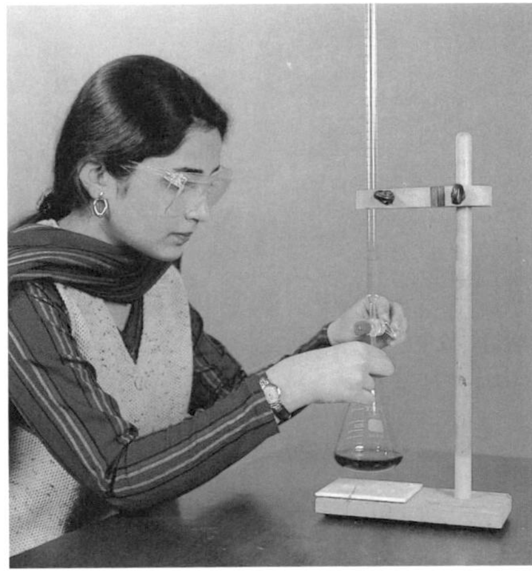

Fig 17.21 In titration, an alkali is added dropwise from a graduated burette to an acid of known volume and concentration, which contains an indicator. The addition continues until one drop changes the colour of the indicator, which signals that the alkali has just neutralised the acid. Often, it is preferable to reverse the procedure, and add the acid to the alkali

Fig 17.22 (overleaf) is such a curve for the addition of aqueous sodium hydroxide to ethanoic acid.

Note that, as the alkali is added, the pH at first changes only slightly. Then, near the point of neutralisation, it shoots up. With the addition of more alkali, it quickly levels off. The exact neutralisation point, or **equivalence point**, can be estimated as the midpoint of the near-vertical portion of the graph, that is, where there is a large change of pH for a very small addition of alkali. This is why you have to add the alkali one drop at a time when you are close to the end-point in a titration.

QUESTIONS 13–14

Write down the name of the products of the following reactions:

a) calcium oxide and ethanoic acid;
b) calcium and propanoic acid;
c) magnesium carbonate and ethanoic acid;
d) calcium carbonate and ethanoic acid;
e) sodium hydrogencarbonate and benzoic acid.

Calcium reacts with 50 cm^3 of 1.0 mol dm^{-3} ethanoic acid. The reaction is much slower than the reaction between the same mass of calcium and 50 cm^3 of 1.0 mol dm^{-3} hydrochloric acid.

a) Write down the balanced symbol equation for the reaction between calcium and ethanoic acid.
b) Explain the difference in the rate of reaction with ethanoic acid and hydrochloric acid.

There are four types of titration curve (Figs 17.22 to 17.25), depending on whether strong or weak acids and bases are used.

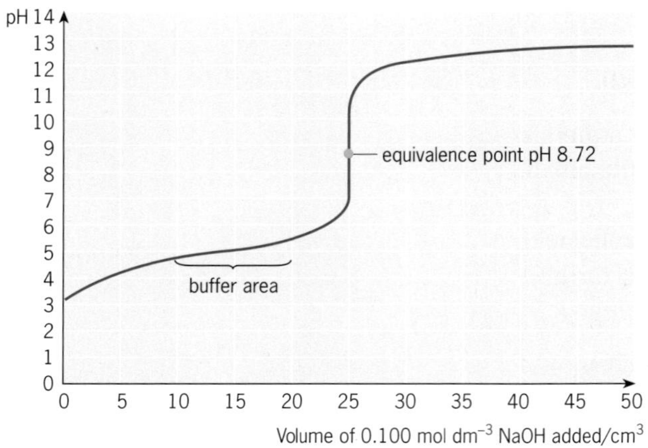

Fig 17.22 The titration curve of a strong base, NaOH(aq), against a weak acid, CH_3COOH(aq). This titration used 0.100 mol dm^{-3} NaOH with 25.0 cm^3 of 0.100 mol dm^{-3} CH_3COOH. In the buffer area there is a mixture of unreacted aqeuous ethanoic acid and aqeuous sodium ethanoate. The pH changes only slightly as more alkali is added. For more information about buffer solutions see page 343.

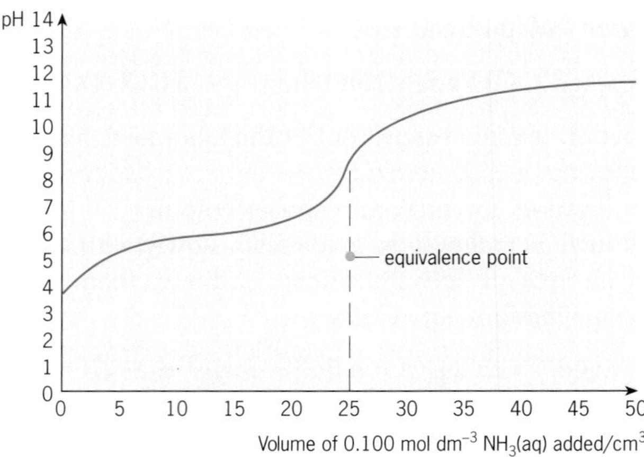

Fig 17.23 The titration curve of a weak base, NH_3(aq), against a weak acid, CH_3COOH(aq). This titration used 0.100 mol dm^{-3} NH_3(aq) with 25.0 cm^3 of 0.100 mol dm^{-3} CH_3COOH. Note that the pH changes gradually over almost the entire titration with no very sudden change in the pH at neutralisation. It is impossible to determine the exact end point in this type of titration

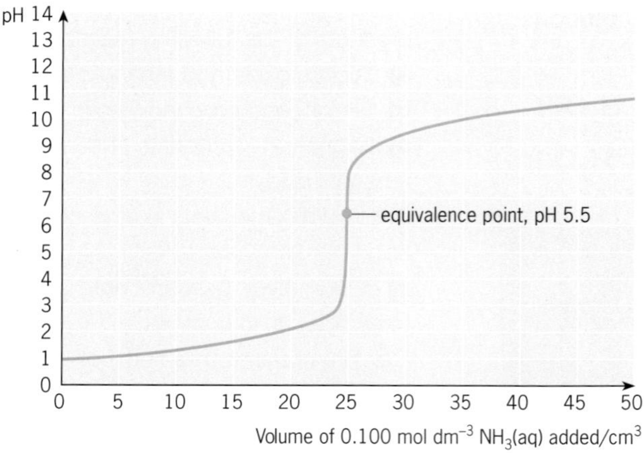

Fig 17.24 The titration curve of a weak base, NH_3(aq), against a strong acid, HCl(aq). This titration used 0.100 mol dm^{-3} NH_3(aq) with 25.0 cm^3 of 0.100 mol dm^{-3} HCl(aq). Note that as the alkali is added, the pH hardly changes until neutralisation is close, at which point it shoots up from pH 2 to pH 7.5. Thereafter, the pH gradually levels off

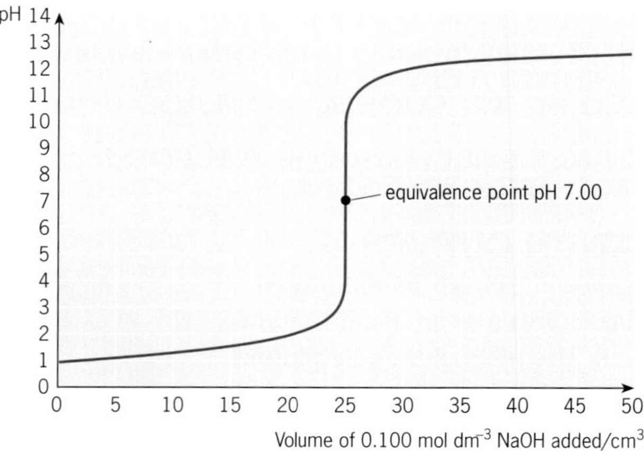

Fig 17.25 The titration curve of a strong base, NaOH(aq), against a strong acid, HCl(aq). This titration used 0.100 mol dm^{-3} NaOH(aq) with 25.0 cm^3 of 0.100 mol dm^{-3} HCl(aq). Note again that as the alkali is added, the pH hardly changes until neutralisation is close, when it shoots up

An indicator is chosen that changes colour within the pH range of the vertical portion of the titration curve. However, as the graph in Fig 17.23 shows, when a weak acid is titrated with a weak base, there is no sudden change in pH. This makes it very difficult to estimate the equivalence point, since any indicator changes colour within a range of pH values.

? QUESTIONS 15–16

Read the Science in Context box about indicators (opposite page) before answering these questions.

15 Write down an expression for the K_a of an indicator.

16 **a)** Name an indicator suitable for the titration of a strong acid with a weak base.

b) Estimate the pK_a of phenolphthalein and of thymolphthalein (see Table 17.5).

SCIENCE IN CONTEXT

Indicators

The approximate pH value of a solution can be determined using an indicator solution, or indicator paper, and comparing the colour obtained against a colour chart.

An **indicator** is a substance that changes colour when placed in an alkaline or acidic solution. The change of colour is because of changes in the complex organic structure of the indicator molecule, which affect the absorption of visible light by the molecule (see Chapter 27).

Litmus, a common indicator, is a natural product extracted from a lichen, while methyl orange, another common indicator, is a synthetic organic molecule known as an azo dye.

Indicators are often weak acids, and so in aqueous solutions they dissociate to a hydrogen ion and the conjugate base. It is convenient to use the simplified formula HIn for an indicator, where In^- stands for the conjugate base. In aqueous solution, the indicator dissociates:

$$HIn(aq) \rightleftharpoons H^+(aq) + In^-(aq)$$

Normally, the indicator, HIn, has one colour and the conjugate base, In^-, has another. In the case of litmus, HIn is red and In^- is blue.

In an acidic solution (one with an excess of aqueous hydrogen ions), the position of the equilibrium shifts to the left to minimise the effect of the increase of hydrogen ions. So, as the concentration of HIn increases, the solution takes on the colour of the HIn molecule, which for litmus is red. This is an example of Le Chatelier's principle (see chapters 1, 18 and 28).

When an indicator is added to an alkaline solution, the position of equilibrium moves to the right, because $H^+(aq)$ reacts with the alkali. Hence, there is a large concentration of In^- and the solution takes on the colour of this species.

The pH values for these colour changes vary with different indicators, as shown in Table 17.5.

To be useful in titration, an indicator must change colour within the almost vertical part of the titration curve. Almost all the indicators in Table 17.5 show the end point of a titration between a strong acid and a strong alkali. However, with a weak acid and a strong base, an indicator such as phenolphthalein should be used.

Fig 17.26 The colour of some flowers depends on the pH of the soil, as shown by these hydrangeas which are pink in acid soil and blue in alkaline soil

Table 17.5 The main properties of eight widely used indicators

Indicator	pK_a	Effective pH range	Colour of acid form	Colour of base form
methyl violet	1.6	0.0–3.0	yellow	violet
methyl orange	3.2	2.1–4.4	red	yellow
bromocresol green	4.8	4.0–5.6	yellow	blue
methyl red	5.2	4.2–6.2	red	yellow
bromothymol blue	6.9	6.0–7.8	yellow	blue
thymol blue	8.7	7.9–9.4	yellow	blue
phenolphthalein	(see q.16)	8.3–10.0	colourless	magenta
thymolphthalein	(see q.16)	9.3–10.5	colourless	blue

STRETCH AND CHALLENGE

Determining pK_a from a titration curve

To determine the K_a of a weak acid can be difficult. However, its pK_a can be easily estimated by analysis of its titration curve. The pK_a is the pH value at the half-equivalence point, that is, the point at which only half of the volume of alkali needed to reach the equivalence point has been added.

The reason for this is that, at the half-equivalence point, the concentration of the conjugate base, A^-, and that of the undissociated acid, HA, are almost equal. Therefore, they cancel out in the expression for K_a:

$$K_a = \frac{[H^+(aq)][A^-(aq)]}{[HA(aq)]}$$

So: $K_a = [H^+(aq)]$ and $pK_a = pH$.

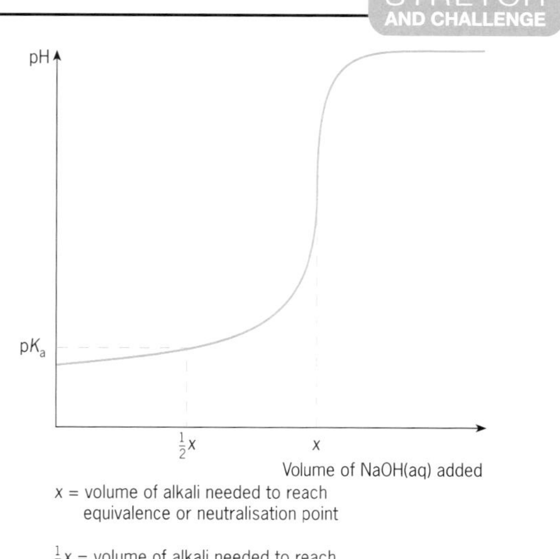

Fig 17.27 Measuring the pK_a of a weak acid from its titration curve

STRETCH AND CHALLENGE

Titration curves for diprotic acids

In a **monoprotic acid**, such as ethanoic acid, an acid molecule can donate only one proton to water. In a **diprotic acid**, such as ethanedioic acid, an acid molecule can donate two protons to water. Thus, ethanedioic acid has two K_a values, designated K_{a1} and K_{a2}. Their significance is shown in Fig 17.28, in which K_{a1} refers to equilibrium 1 and K_{a2} to equilibrium 2. The titration curve for a diprotic weak acid has two equivalence points (Fig 17.29).

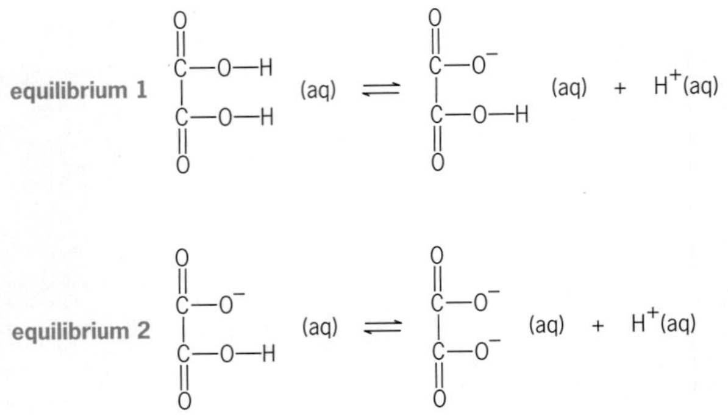

QUESTION 17

17 Write expressions for K_{a1} and K_{a2} for ethanedioic acid.

equilibrium 1 $HO{-}C(=O){-}C(=O){-}O{-}H(aq) \rightleftharpoons {}^-O{-}C(=O){-}C(=O){-}O{-}H(aq) + H^+(aq)$

equilibrium 2 ${}^-O{-}C(=O){-}C(=O){-}O{-}H(aq) \rightleftharpoons {}^-O{-}C(=O){-}C(=O){-}O^-(aq) + H^+(aq)$

Fig 17.28 The dissociation of ethanedioic acid

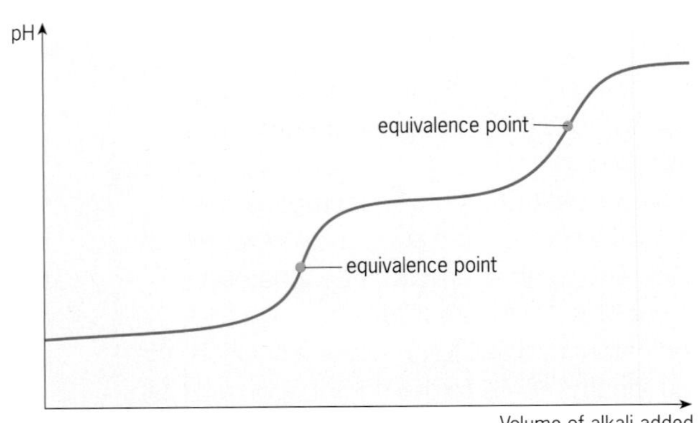

Fig 17.29 The titration curve of a diprotic weak acid against a strong alkali. Note that there are two equivalence points

STRETCH AND CHALLENGE

Enthalpy of neutralisation

The **enthalpy change of neutralisation** is defined as the enthalpy change when one mole of water is produced during the neutralisation of an aqueous acid with an aqueous alkali. The neutralisation of an acid with an alkali is exothermic and represents the reaction between an aqueous hydrogen ion and an aqueous hydroxide ion:

$$H^+(aq) + OH^-(aq) \rightarrow H_2O(l)$$

This reaction is common to all acid–alkali neutralisation reactions, so by Hess's law (see pages 167–168) the enthalpy of neutralisation should be independent of the acid and the alkali used. When a strong acid is neutralised by a strong alkali the enthalpy of neutralisation is -57 kJ mol^{-1}, which does not change even if the strong acid or strong base is changed. This value is considerable smaller, however, when a weak acid or a weak base is neutralised. The reason is quite simple – a weak acid is not fully dissociated, so that before the neutralisation takes place the acid must dissociate and dissociation is an endothermic process. So the enthalpy of neutralisation in this case has the exothermic contribution of the neutralisation and an endothermic contribution needed to dissociate the acid or the base.

BUFFER SOLUTIONS

A buffer solution resists changes in its pH when a small amount of either acid or alkali is added to it. A buffer solution is therefore used to control the pH of another solution to which a small amount of either acid or alkali is to be added. A typical buffer solution is a mixture of either a weak acid and its conjugate base, or a weak base and its conjugate acid. For example, aqueous solutions of a carboxylic acid and its sodium or potassium salt will act as a buffer.

The chemical processes that take place in a solution containing a weak acid, HA, and its sodium salt, NaA, will produce large concentrations of A^- and HA:

$$HA(aq) \rightleftharpoons A^-(aq) + H^+(aq)$$

This equilibrium lies almost entirely on the left. So, we can assume that the concentration of HA at equilibrium is that of undissociated HA. The sodium salt completely ionises in water, to form $A^-(aq)$. Notice that this process is shown by a right-pointing arrow:

$$NaA(aq) \rightarrow A^-(aq) + Na^+(aq)$$

We can therefore assume that a solution consisting of, for example, 0.100 mol dm^{-3} ethanoic acid and 0.500 mol dm^{-3} sodium ethanoate has a concentration of 0.100 mol dm^{-3} of undissociated acid, CH_3COOH, and 0.500 mol dm^{-3} of ethanoate ion, CH_3COO^-.

What happens when an acid is added to a buffer solution?

When acid is added to a buffer solution, the extra H^+ ions react with the conjugate base present, A^-, to produce undissociated acid, HA. This is an example of Le Chatelier's principle, since the equilibrium of the weak acid moves to the left to minimise the effect of the increase in the $H^+(aq)$ concentration. Since most of the extra $H^+(aq)$ added is converted into undissociated HA, the pH hardly decreases.

What happens when hydroxide ions are added to a buffer solution?

When extra alkali is added to a buffer solution, the extra OH^- ions react with the H^+ present to form water:

$$H^+(aq) + OH^-(aq) \rightarrow H_2O(l)$$

Extra H^+ ions are provided by more of the weak acid as it dissociates to maintain the equilibrium between HA(aq), $A^-(aq)$ and $H^+(aq)$. The overall buffering action can be represented by:

$$HA(aq) + OH^-(aq) \rightarrow H_2O(l) + A^-(aq)$$

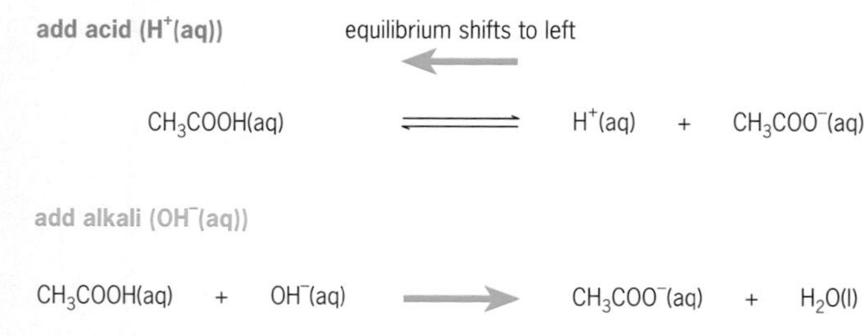

Fig 17.30 A summary of the buffering action of a mixture of sodium ethanoate and ethanoic acid

QUESTION 18

A buffer solution is 0.750 mol dm^{-3} in propanoic acid and 0.350 mol dm^{-3} in sodium propanoate. What are the approximate concentrations in the buffer solution of undissociated propanoic acid and of propanoate ions?

For more on the way that a weak base and its conjugate acid together act as a buffer, see Chapter 28.

There is more information on Le Chatelier's principle on pages 19 and 354, and in Chapter 28.

Calculating the pH of a buffer solution

Buffer solutions with different pH values can be prepared by changing the relative proportions of HA(aq) and A^-(aq). It is easy to calculate the pH of a buffer solution provided the K_a of the weak acid and the concentrations of the weak acid and its conjugate base are known.

EXAMPLE

Q A buffer solution is 0.10 mol dm^{-3} in ethanoic acid and 0.20 mol dm^{-3} in sodium ethanoate. Calculate its pH.

A Let x be the concentration of the H^+(aq). We then have:

	$CH_3COOH(aq)$ $\rightleftharpoons$	$CH_3COO^-(aq)$ +	$H^+(aq)$
At start/mol dm^{-3}	0.10	0.20	0
At equilibrium/ mol dm^{-3}	$(0.10 - x)$	$(0.20 + x)$	x

We can assume that x is very small compared with 0.10 and 0.20, since ethanoic acid is a weak acid. Therefore, at equilibrium:

$$[CH_3COOH(aq)] = 0.10 \text{ mol dm}^{-3}$$

$$[CH_3COO^-(aq)] = 0.20 \text{ mol dm}^{-3}$$

Substituting these values into the expression for the K_a for ethanoic acid:

$$K_a = \frac{[CH_3COO^-][H^+]}{[CH_3COOH]} = \frac{0.20 \times x}{0.10} = 2x$$

Now: $K_a = 1.8 \times 10^{-5}$

So: $x = 9.0 \times 10^{-6}$ mol dm^{-3}

$$\text{pH} = -\log_{10} [H^+(aq)]$$
$$= -\log_{10} (9.0 \times 10^{-6})$$
$$= 5.0$$

The pH of the buffer solution is therefore 5.0 (to 2 sig. figs.).

SCIENCE IN CONTEXT

Buffering in blood

To remain healthy, the acid–base balance of our blood has to be maintained at a constant pH of 7.4. If, for instance, the blood became acidic and this value dropped (as it does in a medical condition called acidosis), we would have to breathe rapidly to expel more of the acidic gas carbon dioxide. The main mechanism that maintains pH at 7.4 is the buffering action of several conjugate acid–base pairs – including H_2CO_3(aq) and HCO_3^-(aq), and $H_2PO_4^-$(aq) and HPO_4^{2-}(aq) – together with the buffering action of plasma proteins and haemoglobin. The action of the proteins is because of the carboxyl and amino groups in the side-chains of some of the amino acids that make up proteins. The concentrations of $H_2PO_4^-$(aq) and HPO_4^{2-}(aq) are too low for this conjugate acid–base pair to have a large buffering effect.

It is the equilibrium between carbon dioxide, carbonic acid, hydrogencarbonate ions and carbonate ions that acts as the most important buffering system in blood plasma:

$$H_2CO_3(aq) \rightleftharpoons HCO_3^-(aq) + H^+(aq)$$

When the pH of blood decreases, the concentration of H^+(aq) increases and the position of equilibrium shifts to the left to minimize this increase. The carbonic acid concentration does not rise indefinitely, since this too is in equilibrium with carbon dioxide and water, and so the concentration of dissolved carbon dioxide increases:

$$H_2CO_3(aq) \rightleftharpoons CO_2(aq) + H_2O(l)$$

Dissolved carbon dioxide is removed from blood by gas exchange in the lungs:

$$CO_2(aq) \rightleftharpoons CO_2(g)$$

When the pH of blood begins to rise, the $H_2CO_3(aq)/HCO_3^-(aq)$ equilibrium shifts to the right to produce more H^+(aq).

Strenuous exercise increases the metabolic rate, which produces more carbon dioxide in tissue respiration. Representing the overall oxidation of glucose:

$$C_6H_{12}O_6(aq) + 6O_2(aq) \rightarrow 6CO_2(aq) + 6H_2O(l)$$

Buffering in blood (Cont.)

Therefore, the concentration of dissolved carbon dioxide in the blood starts to increase and the net effect tends to lower the blood pH. However, this effect is counteracted by an increase in breathing rate, to speed up gas exchange and thereby reduce the concentration of dissolved carbon dioxide. The blood pH therefore does not decrease.

Fig 17.31 Strenuous sport generates acidic blood, which the body's buffer system counteracts

QUESTION 19

What happens to the concentration of undissolved carbon dioxide if the pH of blood rises?

6 CARBOXYLATES

Carboxylates are the salts of a carboxylic acid. They contain the carboxylate anion (Fig 17.32).

Measurements have established that both the carbon–oxygen bonds in the carboxyl group have the same length, which is longer than the typical C=O bond, but shorter than the typical C–O bond. The reason they are not different lengths is that the negative charge on the anion is delocalised and does not reside on either of the two oxygen atoms. Fig 17.33 shows an alternative way to describe the carboxylate ion.

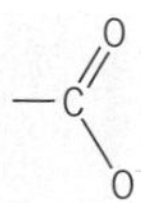

Fig 17.32 The structure of the carboxylate ion

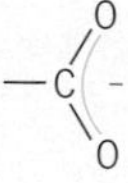

Fig 17.33 An alternative way to represent the carboxylate ion, which shows the delocalised nature of the ion. The curved line indicates that the electron is somewhere between the two oxygen atoms

QUESTION 20

a) Suggest why sodium benzoate has a greater solubility than benzoic acid in cold water.

b) Suggest why it is important for food preservatives, such as sodium benzoate, to be soluble in water.

Delocalised electrons are covered in more detail on page 298.

SCIENCE IN CONTEXT

Properties and uses of carboxylates

Most sodium and potassium carboxylates are soluble in water. How soluble they are depends on the length of the carbon chain of the carboxylate ion, and the solubility decreases as the chain becomes longer.

As they are soluble, carboxylates are widely used in the food industry, which adds them to foods as preservatives and as acid regulators (buffers). As well as carboxylates, the food industry uses carboxylic acids; Table 17.6 summarises these applications. You can probably identify some of them in processed foods by their E numbers.

Table 17.6 Carboxylic acids and carboxylates as food additives

Name	E number	Use
sorbic acid	E200	Occurs naturally in some fruits, but is also made synthetically. Used as a preservative in such foods as soft drinks, cakes and frozen pizzas
sodium sorbate	E201	Preservatives
potassium sorbate	E202	Preservatives
calcium sorbate	E203	Preservatives
benzoic acid	E210	Occurs naturally in cherry bark, raspberries and tea, but normally made synthetically. Used as a preservative and as an antioxidant in fruit products, soft drinks, pickles and salad dressings
sodium benzoate	E212	Preservatives
potassium benzoate	E213	Preservatives
calcium benzoate	E214	Preservatives
ethanoic acid	E260	Vinegar
sodium ethanoate	E281	Preservatives
calcium ethanoate	E282	Preservatives
potassium ethanoate	E283	Preservatives
lactic acid	E270	Found naturally in soured milk and yoghurt, it acts as a preservative and a flavouring. Used in biscuits, confectionery and cakes
propanoic acid	E280	Preservative in baked foods
potassium lactate	E326	Acid regulator

7 PREPARATION OF CARBOXYLIC ACIDS

Dilute sulfuric acid is normally used to acidify aqueous potassium dichromate(VI) or potassium manganate(VII).

The oxidation of primary alcohols is covered in more detail on page 267.

Phenylmethanol is sometimes called benzyl alcohol.

QUESTION 21

21 Write balanced symbol equations for the reactions between:
a) propanal and warm acidified potassium dichromate(VI);
b) butan-1-ol and hot acidified potassium dichromate(VI).

More information on the hydrolysis of esters is given on pages 357–359.

QUESTION 22

22 Write the names of the products of the following hydrolysis reactions:
a) heating aqueous sodium hydroxide with propylethanoate;
b) refluxing hydrochloric acid with ethylbenzoate;
c) heating aqueous sodium hydroxide with $CH_3CH_2CONH_2$;
d) refluxing dilute hydrochloric acid with $C_6H_5CONH_2$.

The hydrolysis of proteins and polypeptides is covered in more detail on page 380.

Aromatic acids can be prepared by side chain oxidation of alkyl benzenes, see page 313.

Carboxylic acids are either prepared by an oxidation reaction or by the hydrolysis of a derivative of a carboxylic acid.

Oxidation of primary alcohols and aldehydes

Carboxylic acids can be prepared by the oxidation of primary alcohols. For example, ethanol can be oxidised to give ethanoic acid by heating it under reflux with either acidified potassium dichromate(VI) or acidified potassium manganate(VII):

$$CH_3CH_2OH + 2[O] \rightarrow CH_3COOH + H_2O$$

In a similar way, aldehydes can be oxidised using the same reagents to form carboxylic acids. So phenylmethanal is oxidised to benzoic acid when refluxed with acidified potassium dichromate(VI):

$$C_6H_5CHO + [O] \rightarrow C_6H_5COOH$$

HYDROLYSIS OF ESTERS

When an ester is refluxed with either dilute acid or dilute alkali, the ester is hydrolysed to give the carboxylic acid or the carboxylate ion. For example, ethyl ethanoate can be hydrolysed to give ethanoic acid (with acid hydrolysis) or sodium ethanoate (with aqueous sodium hydroxide):

$$CH_3COOCH_2CH_3(l) + H_2O(l) \xrightleftharpoons{\text{acid hydrolysis}} CH_3COOH(aq) + CH_3CH_2OH(aq)$$

$$CH_3COOCH_2CH_3(l) + NaOH(aq) \xrightarrow[\text{hydrolysis}]{\text{alkaline}} CH_3COO^-Na^+(aq) + CH_3CH_2OH(aq)$$

The sodium ethanoate can be converted into ethanoic acid by reaction with dilute hydrochloric acid.

HYDROLYSIS OF AMIDES

Amides can be hydrolysed in the same way as esters. Again, there is the opportunity to use either acid- or base-catalysed hydrolysis. For example, ethanamide, CH_3CONH_2, can be hydrolysed to give ethanoic acid and ammonium ion with acid hydrolysis, and ethanoate ion and ammonia with alkaline hydrolysis:

$$CH_3CONH_2(s) + H_2O(l) + HCl(aq) \xrightarrow{\text{acid hydrolysis}} CH_3COOH(aq) + NH_4Cl(aq)$$

$$CH_3CONH_2(s) + NaOH(aq) \xrightarrow[\text{hydrolysis}]{\text{alkaline}} CH_3COO^-Na^+(aq) + NH_3(g)$$

Proteins and polypeptides are polymers that contain the amide linkage. They can be hydrolysed either by refluxing with concentrated hydrochloric acid or by enzymatic action to give amino acids.

HYDROLYSIS OF NITRILES

Nitriles are hydrolysed by refluxing in either dilute acid or aqueous alkali. Acid hydrolysis gives the ammonium ion and a carboxylic acid. Alkaline hydrolysis gives ammonia and the carboxylate ion. For example, ethanenitrile can be hydrolysed by refluxing with hydrochloric acid to give ethanoic acid, but with aqueous sodium hydroxide it forms sodium ethanoate:

$$CH_3CN(l) + 2H_2O(l) + HCl(aq) \rightarrow CH_3COOH(aq) + NH_4Cl(aq)$$

$$CH_3CN(l) + H_2O(l) + NaOH(aq) \rightarrow CH_3COO^-Na^+(aq) + NH_3(g)$$

8 REACTIONS OF CARBOXYLIC ACIDS

The two most important synthetic reactions of carboxylic acids are the production of esters and the production of acyl chlorides.

ESTERIFICATION

When an alcohol and a carboxylic acid are refluxed together in the presence of an acid catalyst, such as concentrated sulfuric acid, an ester is produced. The reaction does not go to completion, but results in an equilibrium mixture that must be separated to isolate the ester. For example, ethanol, ethanoic acid and a trace of concentrated sulfuric acid give the ester ethyl ethanoate:

$$CH_3CH_2OH(l) + CH_3COOH(l) \rightleftharpoons CH_3COOCH_2CH_3(l) + H_2O(l)$$

FORMATION OF ACYL CHLORIDES

In acyl chlorides (also known as acid chlorides), the carboxyl's hydroxyl group is replaced by a chlorine atom (Fig 17.34). An acyl chloride is always named after the corresponding carboxylic acid. For example, the acyl chloride of ethanoic acid is called ethanoyl chloride and that of benzoic acid is called benzoyl chloride.

The conversion of carboxylic acids into acyl chlorides is very useful for the synthesis of esters and amides, because acyl chlorides are much more susceptible to nucleophilic attack than are carboxylic acids. Acyl chlorides are therefore used as synthetic intermediates.

Carboxylic acids are converted into acyl chlorides by reaction with sulfuryl(IV) chloride, $SOCl_2$, or phosphorus(V) chloride, PCl_5 (Fig 17.35). Normally, the carboxylic acid is refluxed with the $SOCl_2$ or PCl_5 and the acyl chloride product is isolated by fractional distillation. For example, ethanoic acid can easily be converted into ethanoyl chloride, and benzoic acid into benzoyl chloride:

$$CH_3COOH(l) + SOCl_2(l) \rightarrow CH_3COCl(l) + SO_2(g) + HCl(g)$$

$$C_6H_5COOH(l) + PCl_5(s) \rightarrow C_6H_5COCl(l) + POCl_3(l) + HCl(g)$$

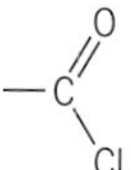

Fig 17.34 The acyl chloride functional group, in which the highly electronegative chlorine atom is attached to the carbonyl carbon

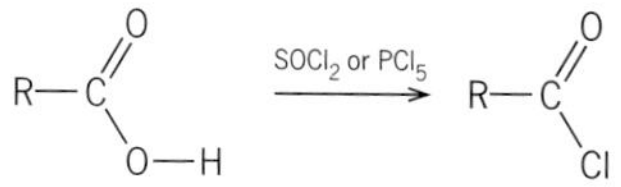

$$R{-}COOH \xrightarrow{SOCl_2 \text{ or } PCl_5} R{-}COCl$$

Fig 17.35 The formation of acyl chlorides

QUESTION 23

Write down the symbol equation for the reaction between:

a) ethanoic acid and PCl_5

b) benzoic acid and $SOCl_2$

REDUCTION OF CARBOXYLIC ACIDS

Powerful reducing agents, such as lithium tetrahydridoaluminate, $LiAlH_4$, reduce carboxylic acids to the corresponding alcohols (this is opposite to the preparation of carboxylic acids by oxidising primary alcohols):

$$RCOOH + 4[H] \xrightarrow{\text{(i) } LiAlH_4\text{/ether (ii) } HCl(aq)} RCH_2OH + H_2O$$

DECARBOXYLATION

Decarboxylation involves a reaction in which the carboxyl group is removed from a carboxylic acid, essentially as carbon dioxide, leaving behind a hydrocarbon. The reaction is typically carried out by strongly heating a mixture of powdered acid and sodium hydroxide.

$$RCOOH + 2NaOH \rightarrow RH + Na_2CO_3 + H_2O$$

Alternatively the solid sodium carboxylate can be heated with solid sodium hydroxide. Typically the hydrocarbon formed, RH, will burn. If it burns with a smoky flame the R group probably contains a benzene ring.

STRETCH AND CHALLENGE

Salt hydrolysis

A solution of sodium chloride is neutral and at 298 K has a pH of 7. This is typical behaviour of salts made from the reaction of strong acids with strong bases.

A solution of sodium ethanoate is not neutral. It has a pH above 7 and is therefore an alkaline solution. This is typical behaviour of salts made from the reaction of weak acids with strong bases.

The explanation is due to a process called salt hydrolysis. In essence the following equilibrium is set up, which although it lies very much on the left-hand side does contribute to increasing the concentration of hydroxide ions.

$$CH_3COO^-(aq) + H_2O(l) \rightleftharpoons CH_3COOH(aq) + OH^-(aq)$$

In water, sodium ethanoate is fully dissociated into Na^+ (aq) and CH_3COO^-(aq). The CH_3COO^-(aq) ions react reversibly with the small concentration of H^+ ions naturally present in all water samples to form undissociated ethanoic acid, a weak acid.

$$CH_3COO^-(aq) + H^+(aq) \rightleftharpoons CH_3COOH(aq)$$

This in turn will affect the equilibrium involving the dissociation of water.

$$H_2O(l) \rightleftharpoons H^+(aq) + OH^-(aq)$$

Since H^+(aq) are used up the dissociation of water will move to the right to try and replace those lost. As a result there is an increase concentration of OH^-(aq) and the solution becomes alkaline.

SUMMARY

After studying this chapter, you should know the following.

- Brønsted–Lowry acids are proton donors and Brønsted–Lowry bases are proton acceptors.
- Strong acids are assumed to dissociate fully when dissolved in water. The dissociation of weak acids is an equilibrium process, in which the position of equilibrium lies very much to the side of the undissociated weak acid.
- K_a is the acid dissociation constant, which for the acid HA is given by the formula:

$$K_a = \frac{[H^+][A^-]}{[HA]}$$

- pH is the negative logarithm to base 10 of the aqueous hydrogen ion concentration.
- A buffer solution resists changes in pH when small amounts of acid or alkali are added to it.
- Many buffer solutions are a mixture of an aqueous weak acid and its conjugate base.
- An indicator is a compound that changes colour in a solution at a certain pH value. Many indicators are weak acids.
- Carboxylic acids are typical weak acids; they react slowly with reactive metals such as magnesium to give hydrogen, react with bases to give carboxylate salts and release carbon dioxide from carbonates or hydrogencarbonates.
- Carboxylic acids can be prepared by the oxidation of primary alcohols or aldehydes by heating under reflux with acidified potassium dichromate(VI), or by the acid- or base-catalysed hydrolysis of esters, amides or nitriles.
- Carboxylic acids react with alcohols in the presence of an acid catalyst, such as concentrated sulfuric acid, to give esters.

Practice questions and a How Science Works assignment for this chapter are available at www.collinseducation.co.uk/CAS

18

Esters, acyl chlorides, amides and equilibria

18 ESTERS, ACYL CHLORIDES, AMIDES AND EQUILIBRIA

SCIENCE IN CONTEXT

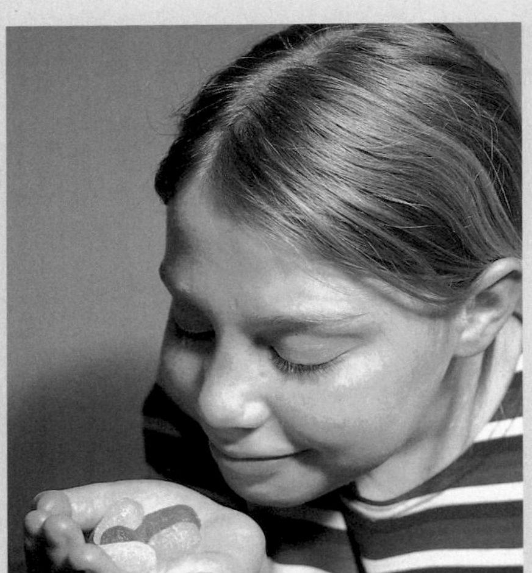

Odours The characteristic flavour of pear drops is caused by a cocktail of organic compounds, which include carboxylic acid derivatives called esters. The organic compounds have molecules that trigger responses in sensory cells in the nose chamber and the tongue

Odours

Our noses can detect hundreds of different odours at unbelievably low concentrations, perhaps just a few molecules per sniff. Some noses are so sensitive that their owners – humans and dogs included – can learn to identify scores of different substances by their smell alone! How does the nose do it?

The chamber of the nose is lined with thousands of different types of receptor cells that respond to the different shapes of molecules. It is likely that a single substance may trigger several types of receptor cell, which in some way together produce the sensation of the odour of that substance. Many natural odours are not just single substances but complex cocktails of substances, and we do not understand the complicated interaction of the receptor cells involved. Also, we don't know why many substances whose structures differ widely give the same odour, as we have yet to discover the connection between molecular structure and odour.

What we do know, however, is that to have an odour a substance must be volatile enough for its vapour to diffuse into the nostrils; it must also be water soluble to ensure that it makes its way into the mucus lining the nostrils; and it must be fat soluble so that it can penetrate the receptor cells. Many esters – derivatives of carboxylic acids – possess these properties, which is why they are used a lot to synthesise substances we like to smell, particularly those in food products. The subtle odours of fruits, for instance, are reproduced by blending the appropriate esters. There is an expanding market in attractive smells, such as baked bread and coffee, to distribute in supermarkets and encourage shoppers to buy more food.

1 DERIVATIVES OF CARBOXYLIC ACIDS

As explained on page 337, carboxylic acids have a carbonyl group that is directly attached to a hydroxyl group. Derivatives of carboxylic acids still have the carbonyl group, but the hydroxyl group is replaced by the atom of a highly electronegative element. The main derivatives of carboxylic acids are esters, amides (sometimes called acid amides), acid anhydrides and acyl chlorides.

Fig 18.1 A generalised carboxylic acid and its derivatives. R' is an alkyl group such as C_2H_5, an aryl group such as C_6H_5, or a cycloalkyl group

POLAR CARBONYL GROUPS

Look at Fig 18.1, which shows the general structure of a carboxylic acid and its derivatives. They all possess a highly polar carbonyl (C=O) group whose normal polarity is enhanced because the extra atom attached to it is highly electronegative. In carboxylic acids and esters, this atom is an oxygen atom; in amides, it is a nitrogen atom; and in acyl chlorides, it is a chlorine atom.

? QUESTION 1

1 What is electronegativity?

USE OF CARBOXYLIC ACID DERIVATIVES

Esters and amides have numerous commercial applications, including perfumes, polymers and pharmaceuticals (Figs 18.3 and 18.4). Acyl chlorides are important synthetic intermediates, but do not have many everyday uses in their own right.

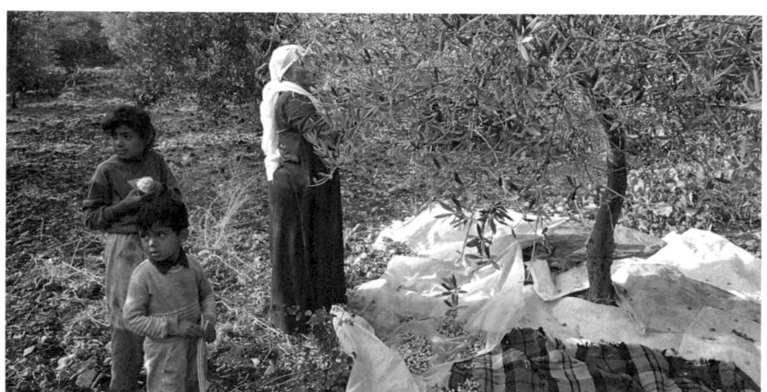

Fig 18.2 Oils from plants, such as olive oil, are esters of long-chain carboxylic acids and the alcohol propane-1,2,3-triol

CH_2OH
HO
H
H
ester linkage
C=O
OH OH

Fig 18.3 Among the functional groups in vitamin C (ascorbic acid) is a lactone, which is a cyclic ester

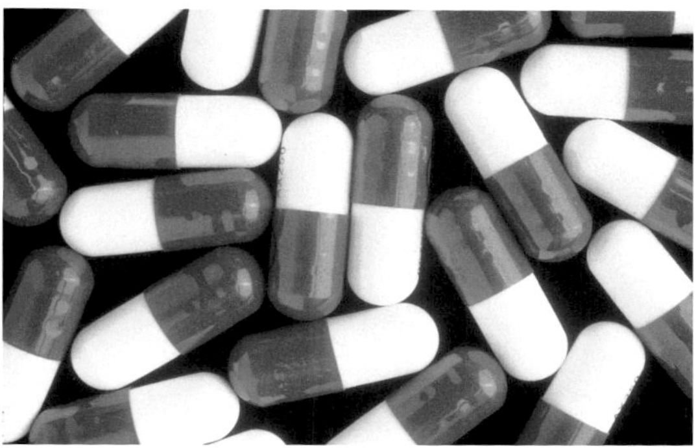

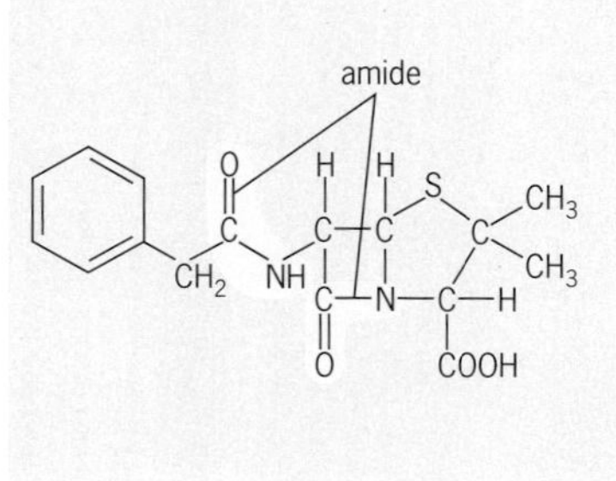

Fig 18.4 Penicillins are powerful bactericides whose molecules contain an amide functional group

2 ESTERS

An ester is a carboxylic acid derivative in which the hydroxyl group of the acid is replaced by an oxygen atom attached to a carbon chain. Fig 18.5 shows how esters are related to carboxylic acids and to alcohols. An ester has two carbon chains, one related to a carboxylic acid and the other related to an alcohol or a phenol.

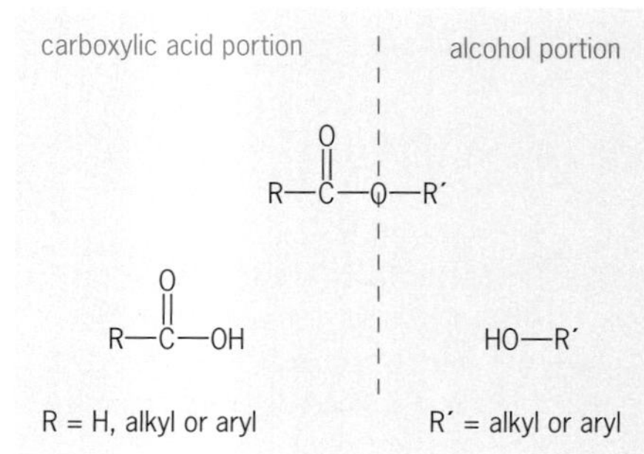

Fig 18.5 Esters have a structure that consists of a carboxylic acid portion and an alcohol portion

NAMING ESTERS

As esters have two carbon chains, they are more difficult to name than other classes of compounds. For this reason, the main carbon chain of an ester is taken to be the one related to the carboxylic acid, so the main (second) part of the name of an ester is based on the name of this carboxylic acid (like a surname).

R′—O—C(=O)—R
alkyl alkanoate

$CH_3—O—C(=O)—CH_2—CH_2—CH_3$
For example, methyl butanoate

Fig 18.6 The name of an ester is derived from the name of its alcohol (alkyl) group and the name of its carboxylic acid (alkanoate) group. Note that the ester linkage is reversed in relation to that shown in Fig 18.5

Take, for example, methyl propanoate (Fig 18.7). The carbon chain that is related to the carboxylic acid (the one that contains the carbonyl group) has

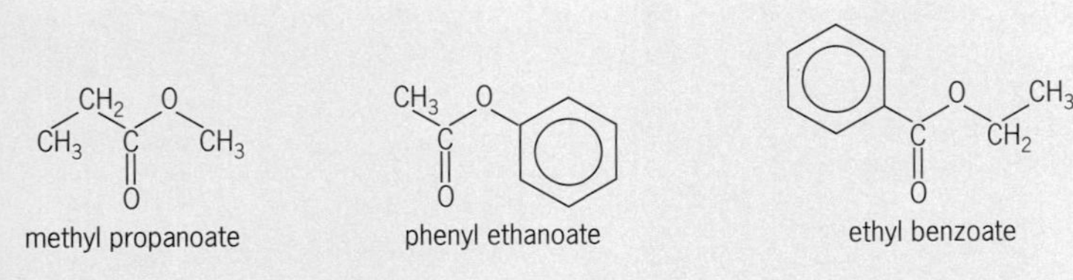

Fig 18.7 Structural formulae of methyl propanoate, phenyl ethanoate and ethyl benzoate

three carbon atoms, and the chain related to the alcohol has one carbon atom. Its formula is therefore written as:

$$CH_3CH_2COOCH_3 \quad \text{or} \quad CH_3CH_2CO_2CH_3$$

Note that in the structures of Fig 18.7, the main (carboxylic acid) part is on the left, rather than on the right as in their names.

In the same way, phenyl ethanoate (Fig 18.7) consists of two portions: a two-carbon unit that contains the carbonyl carbon, and the phenyl group (the benzene ring). It has the formula:

$$CH_3COOC_6H_5 \quad \text{or} \quad CH_3CO_2C_6H_5$$

Ethyl benzoate is related to benzoic acid and has the formula:

$$C_6H_5COOC_2H_5 \quad \text{or} \quad C_6H_5CO_2C_2H_5.$$

? QUESTION 2

2 Write down the displayed formulae for each of these esters: methyl ethanoate, propyl propanoate, pentyl butanoate and methyl benzoate.

3 ESTERIFICATION AND EQUILIBRIUM CONSTANTS

To make an ester should be simple enough, since when heated in the presence of an acid catalyst, a carboxylic acid and an alcohol react to give an ester and water. This is known as esterification.

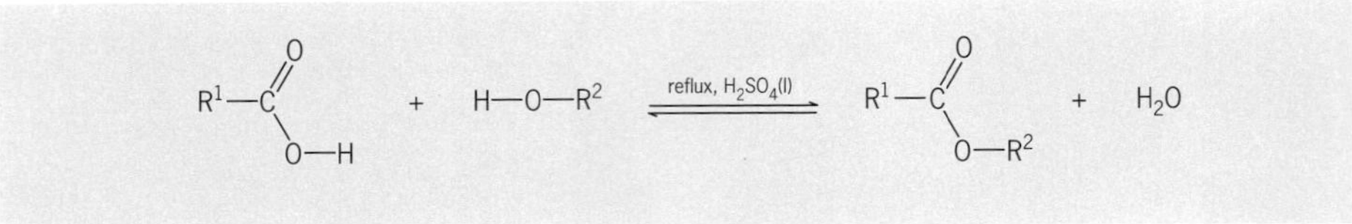

$$R^1\text{—}C(=O)\text{—}O\text{—}H \; + \; H\text{—}O\text{—}R^2 \xrightarrow{\text{reflux, } H_2SO_4(l)} R^1\text{—}C(=O)\text{—}O\text{—}R^2 \; + \; H_2O$$

Fig 18.8 The esterification reaction. R^1 is an alkyl, an aryl or hydrogen. R^2 is an alkyl group

However, the equation in Fig 18.8 does not really indicate the problems with the reaction. To achieve a good yield of ester, the reaction has to be managed carefully since it is reversible. Fig 18.9 shows that when a carboxylic acid is reacted with an alcohol, the reaction does not go to completion. It reaches a balance point at which the concentrations of all four substances (carboxylic acid, alcohol, ester and water) remain constant. The reaction has not stopped, but the concentrations do not change, which means that the rate of the forward reaction then equals the rate of the backward reaction. A situation known as a **dynamic equilibrium** has been reached.

We have already described dynamic equilibria and Le Chatelier's principle in qualitative terms in Chapter 1. In this chapter equilibria are explored further using a quantitative approach.

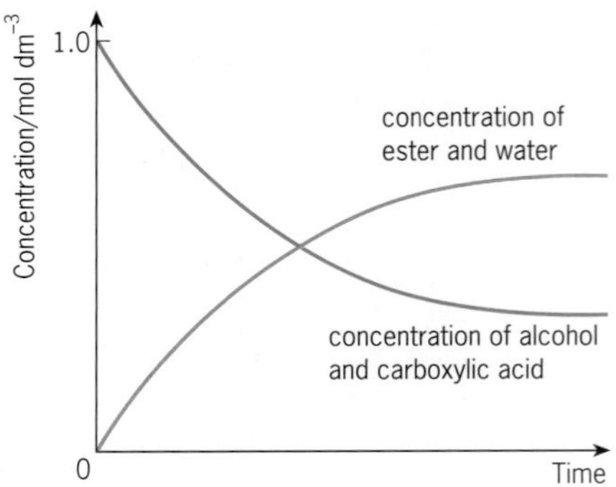

Fig 18.9 The concentration changes that occur during esterification, when 1 mole of alcohol is refluxed with 1 mole of carboxylic acid. Note that each concentration eventually reaches a constant value known as the equilibrium concentration

Another equilibrium constant – the acid dissociation constant, K_a – is discussed on page 332.

EQUILIBRIUM CONSTANTS

Any dynamic equilibrium can be described in terms of its **equilibrium constant, K_c**.

For the following reaction at equilibrium

$$wA + xB \rightleftharpoons yC + zD$$

the equilibrium constant is given by:

$$K_c = \frac{[C]^y[D]^z}{[A]^w[B]^x}$$

where [ ] is the concentration of the species in mol dm^{-3}.

REMEMBER THIS

The term 'stoichiometric equation' is equivalent to 'balanced equation'. That is, the mole ratios of the reactants and the products are stated.

The equilibrium constant relates the molar concentrations at equilibrium of each species in the reactions, thus:

The equilibrium constant is the (mathematical) product of the molar concentrations of the products each raised to the power of its coefficient in the stoichiometric equation, divided by the product of the molar concentrations of the reactants, each raised to the power of its coefficient in the stoichiometric equation.

An equilibrium constant is independent of the concentrations of the reactants and the products, but will change its numerical value as the temperature of a reaction changes.

A reaction whose equilibrium constant has a large numerical value is one in which, at equilibrium, the concentrations of the products are much higher than the concentrations of the reactants.

For esterification, the general reaction is shown by the equation:

$$RCOOH + R'OH \rightleftharpoons RCOOR' + H_2O$$

where R and R' represent carbon skeletons, such as alkyl or aryl groups.

The equilibrium constant is thus given by:

$$K_c = \frac{[RCOOR']^1[H_2O]^1}{[RCOOH]^1[R'OH]^1}$$

This equilibrium constant does not have a unit, since the units for concentration cancel out. The units for any equilibrium constant depend on the equation it is based on.

Notice that, although esterification normally requires an acid catalyst, the catalyst is not shown in the expression for the equilibrium constant. This is because it is not shown in the stoichiometric equation. The catalyst decreases the time it takes to reach equilibrium, but *it does not change the concentrations* obtained at equilibrium.

Calculations that involve the equilibrium constant

Despite the problem of having to extract the required product from an equilibrium mixture, esters are made via the acid-catalysed reaction of an alcohol and a carboxylic acid. The reaction is carried out in a batch process, so the ester is not made continuously but only as needed. When using this equilibrium process on an industrial scale, the chemists need to estimate the yield of ester, and the amounts of starting materials left unreacted, and for this they use the equilibrium constant.

EXAMPLE

Q The K_c for the esterification reaction between ethanol and ethanoic acid is 4.0. Calculate the yield of ester when 1.0 mole of ethanol and 1.0 mole of ethanoic acid are left to reach equilibrium.

A First, write down the stoichiometric equation. At the start of the reaction there is no product. However, once equilibrium has been reached, the reaction vessel contains ethanoic acid, ethanol, ethyl ethanoate and water.

$$CH_3COOH + C_2H_5OH \rightleftharpoons CH_3COOC_2H_5 + H_2O$$

From the equation, every mole of ethanoic acid requires 1 mole of ethanol and makes 1 mole of ethyl ethanoate and 1 mole of water. This means that if the yield at equilibrium is x moles of ethyl ethanoate, then x moles of each of ethanol and ethanoic acid must have reacted. This leaves behind $(1.0 - x)$ moles of each of ethanoic acid and ethanol.

REMEMBER THIS

a) The concentrations used to calculate K_c must be those after equilibrium has been reached.

b) To deduce the units for a K_c value you must substitute the units for concentration ($mol\ dm^{-3}$) into the K_c expression instead of numerical concentrations, then simplify the expression.

? QUESTION 3

3 **a)** Write down expressions for the equilibrium constant for each of the following reactions:

i) $(COOH)_2 + 2C_2H_5OH \rightleftharpoons (COOC_2H_5)_2 + 2H_2O$;

ii) $2NO_2 \rightleftharpoons N_2O_4$.

Show that in i) K_c has no unit and in ii) the unit is $dm^3\ mol^{-1}$.

b) Ethanoic acid and pentan-1-ol are refluxed together to obtain an equilibrium mixture. The concentration of each substance at equilibrium is shown in Table 18.1.

Table 18.1

Substance	Molar concentration at equilibrium/mol dm^{-3}
ethanoic acid	2.12
pentan-1-ol	2.12
pentyl ethanoate	4.32
water	4.32

i) Write down a balanced equation for the esterification reaction.

ii) Write down an expression for the equilibrium constant.

iii) Calculate the numerical value for the equilibrium constant.

QUESTION 4

4 The equilibrium constant, K_c, for the reaction between propanoic acid and ethanol is 4.3 at a particular temperature. 1.0 mole of water and 1.0 mole of ethyl propanoate are refluxed together with an acid catalyst. Calculate the number of moles at equilibrium of:

a) water;
b) ethyl propanoate;
c) propanoic acid;
d) ethanol.

EXAMPLE (Cont.)

This part of the calculation is summarised thus:

	CH_3COOH	+	C_2H_5OH	$\rightleftharpoons$	$CH_3COOC_2H_5$	+	H_3O
At start of reaction/moles	1.0		1.0		0		0
At equilibrium/moles	$1.0 - x$		$1.0 - x$		x		x

The equilibrium constant is: $K_c = \dfrac{[CH_3COOC_2H_5][H_2O]}{[CH_3COOH][C_2H_5OH]}$

The concentration of each substance must be substituted into the above expression, so the moles at equilibrium must be converted into concentrations.

Remember (page 130) that
concentration is measured in mol $dm^{-3} = \dfrac{\text{moles}}{\text{volume}\ (dm^3)}$

So be careful to substitute concentrations into the expression for K_c, *not* the number of moles.

Assuming that the volume of the esterification mixture is V dm^3, this gives:

$$K_c = 4.0 = \frac{\left(\frac{x}{V}\right)\left(\frac{x}{V}\right)}{\left(\frac{1.0 - x}{V}\right)\left(\frac{1.0 - x}{V}\right)}$$

Cancelling V gives: $4.0 = \dfrac{x^2}{(1.0 - x)^2}$

Taking the square root of both sides gives: $2.0 = \dfrac{x}{1.0 - x}$

So: $2.0 - 2.0x = x$

That is, $x = 0.67$ moles

This means that the exact composition of the equilibrium mixture can be estimated.

The Example above shows that when an equimolar mixture of ethanol and ethanoic acid are refluxed together, only about 67 per cent of the starting ethanol and ethanoic acid are converted into ethyl ethanoate. In the laboratory, the ester can be separated from the starting materials, but industrially this represents a large loss of valuable reactants, unless they can be recycled. Therefore, an industrial chemist needs to find a way for this reaction to give a maximum yield, especially when the ester is derived from an expensive starting material.

LE CHATELIER'S PRINCIPLE

Le Chatelier's principle is described in Chapter 1, and states:

The position of the equilibrium of a system changes to minimise the effect of any imposed change in conditions.

Le Chatelier's principle describes the effect that a change in the pressure, temperature or concentration of a substance in an equilibrium mixture will have on the position of equilibrium. Le Chatelier's principle applies to any reaction that is in equilibrium and allows chemists to manage and manipulate equilibrium reactions. We describe here the consequences of Le Chatelier's principle on the esterification reaction.

Effect of concentration changes on the esterification equilibrium

Changing the **concentration** of a reactant or a product **does not change the numerical value of the equilibrium constant,** K_c. However, it does change the position of equilibrium. As an example, let's look at what happens to the equilibrium mixture formed by the reaction of an alcohol with a carboxylic acid when extra alcohol is added. Temporarily, the reaction is not in equilibrium, but equilibrium is quickly restored by reaction of the alcohol to form the ester. A new equilibrium mixture is produced that has the same equilibrium constant. Fig 18.10 shows what happens to the concentrations of each substance during this process.

The effect of pressure change on equilibrium is discussed in Chapter 28.

If you want to read more about energy changes that occur during chemical reactions, see pages 10 and 152.

? QUESTION 5

5 Suggest why the esterification of fatty acids by methanol is carried out in the presence of an excess of methanol.

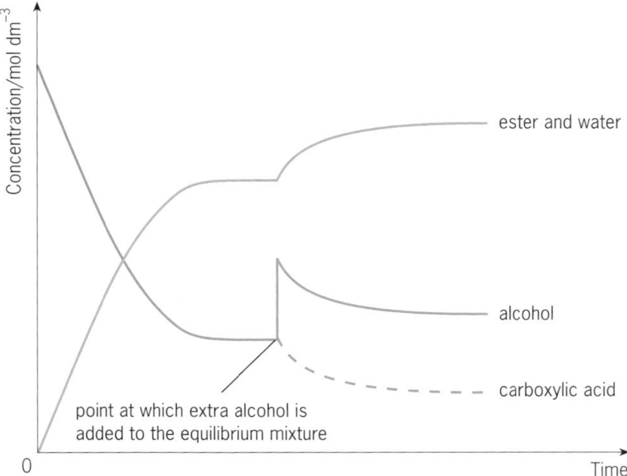

Fig 18.10 The effect of increasing the concentration of alcohol on the esterification equilibrium. Note that even higher equilibrium concentrations can be achieved if even more alcohol is added to the mixture

This effect is an example of how Le Chatelier's principle operates in practice. The change is the increase of alcohol concentration, which disturbs the equilibrium of the mixture. However, once a new state of equilibrium is established, some of the extra alcohol is changed into ester.

Industrially, this means that if an ester is to be made starting from an expensive alcohol, it is more economical to use an excess of the cheaper carboxylic acid. This ensures that the equilibrium is shifted to the right side and that as much as possible of the expensive alcohol is converted into the ester.

Another way to manage the reversible reaction is to *remove* one of the products the moment it has formed. This prevents the reaction from achieving equilibrium, but it will nevertheless 'keep trying' to, as long as the product continues to be removed from the reaction mixture. In esterification, water is the easier product to remove.

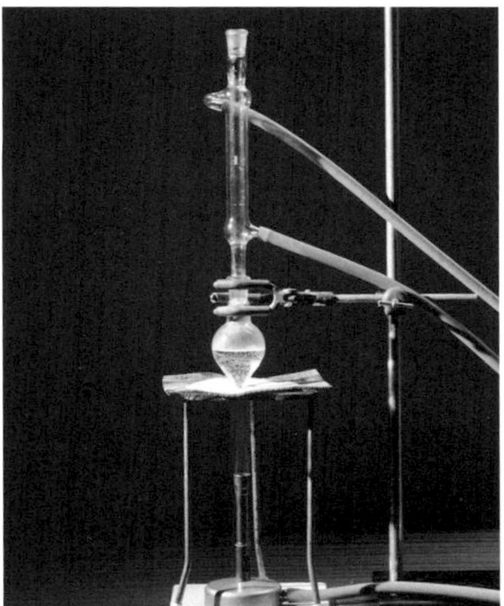

Fig 18.11 In the laboratory, esterification is usually carried out by refluxing an alcohol and a carboxylic acid in the presence of an acid catalyst, such as concentrated sulfuric acid

Effect of temperature changes on the esterification equilibrium and K_c

Although the equilibrium constant does not change when concentrations of the reactants and products change, it will change as the temperature of the equilibrium is changed. For an **exothermic** reaction the K_c becomes **numerically smaller** as the temperature increases, and for an **endothermic** process the K_c becomes **numerically larger** as the temperature increases. Esterification is an exothermic process: when a carboxylic acid and an alcohol react, energy is transferred to the

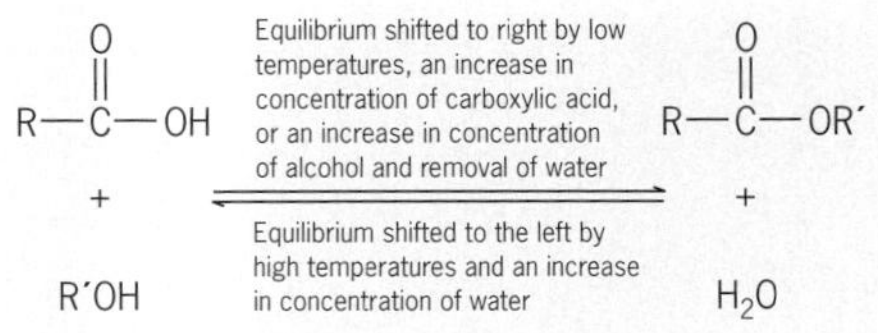

Fig 18.12 A summary of the effect of Le Chatelier's principle on esterification

? QUESTIONS 6–7

6 You want to make an ester starting from ethanol and an expensive carboxylic acid. Suggest conditions that will maximise the conversion of the carboxylic acid into the corresponding ester.

7 Predict the effect of increasing separately the temperature, the catalyst concentration and the concentration of alcohol on the equilibrium constant for an esterification reaction.

surroundings. So, if the temperature of the esterification is raised, the equilibrium shifts to absorb energy, that is, in the direction of the endothermic reaction (the left side). As the temperature is increased, the yield of ester decreases, though the rate of production of ester increases. For esterification in industry, a compromise temperature is used so that the rate and the yield are both acceptable, and it is normal for the alcohol and the carboxylic acid to be refluxed together in the presence of concentrated sulfuric acid as a catalyst. The sulfuric acid catalyst has no effect on the position of equilibrium, but reduces the time to reach equilibrium.

APPLICATIONS OF THE ESTERIFICATION REACTION

The importance of esters as a source of pleasant-smelling substances and food flavourings has been mentioned earlier (see page 351). Almost all these esters are manufactured using the esterification reaction just described with the formation of the ester linkage.

There are several examples of the esterification reaction used to link two carbon chains to form a molecule that has a particular shape associated with biological activity. Two such molecules, benzocaine and procaine, are anaesthetics based on 4-aminobenzoic acid. Among other uses, there are ester solvents (including ethyl ethanoate), ester plasticisers in PVC and ester synthetic fibres (polyesters).

? QUESTION 8

8 Write down the names of the carboxylic acid and alcohol used to make each of the following esters:

a) ethyl ethanoate;

b) ethyl butanoate;

c) propenyl ethanoate.

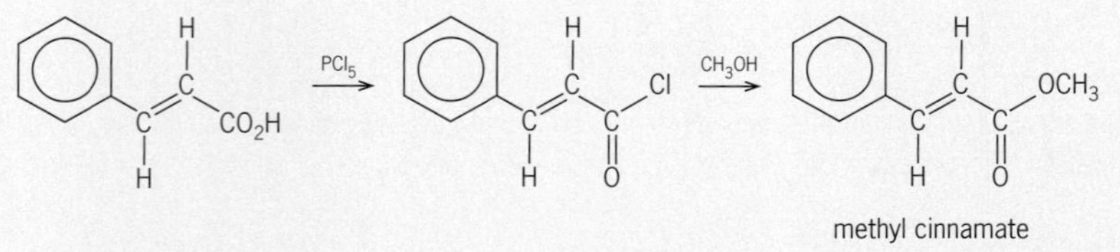

H H H C C C C CO_2H H PCl_5 C C Cl C H O CH_3OH C C OCH_3 C H O methyl cinnamate

Fig 18.13 The preparation of methyl cinnamate, an ester responsible for the spicy aroma of the matsutake mushroom. It can be prepared from cinnamic acid (3-phenylpropenoic acid), the corresponding carboxylic acid. Note that the esterification occurs via the acyl chloride

SCIENCE IN CONTEXT

Esters and the food flavourings industry

The simple esters (that is, those with only a few carbon atoms per molecule) tend to have pleasant odours. The characteristic flavours of fruits and the fragrances of flowers often result from esters – commonly a subtle blend of esters and other odoriferous compounds.

When we talk about 'flavour' we mean a combination of taste and odour by receptors on the tongue and in the nose. Almost always, it is a combination of the substances that our various receptors detect that allows us to recognise a particular flavour.

Table 18.2 shows the 11 chemicals, and the amount of each, that chemists have put together to imitate the flavour of pineapple. Notice that many of the esters listed are still known by their traditional (pre-systematic) names – the result of long-established use in the food industry. Almost all the esters in this formulation are made by the reaction of the appropriate alcohol and carboxylic acid.

Table 18.2 Components of imitation pineapple flavouring

Compound	% in formulation	Compound	% in formulation
allyl caproate	5	ethyl ethanoate	15
butanoic acid	12	isoamyl acetate	3
caproic acid	8	isoamyl isovalerate	3
ethanoic acid	5	terpinyl propanoate	3
ethyl butanoate	22	other essential oils	19
ethyl crotonoate	5		

SCIENCE IN CONTEXT

Local anaesthetics

Benzocaine and procaine are local anaesthetics (Fig 18.14), which means that they make only small areas of the body insensitive to touch and pain. They were used a lot until recently – benzocaine as an ointment, drug or aerosol to relieve painful conditions of the skin, mouth and respiratory tract, and procaine for dental injections. Both compounds are usually prepared by esterification reactions.

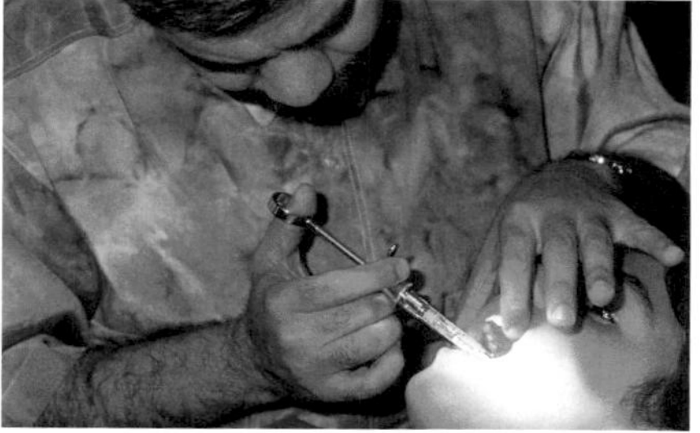

Fig 18.14 A local anaesthetic is used to numb just the area being treated

Benzocaine has the systematic name of ethyl 4-aminobenzoate. It is synthesised from 4-aminobenzoic acid (Fig 18.15). The acid is esterified with ethanol in the presence of a little concentrated sulfuric acid to speed up the attainment of equilibrium. An excess of ethanol is used to drive the equilibrium to the right, according to Le Chatelier's principle.

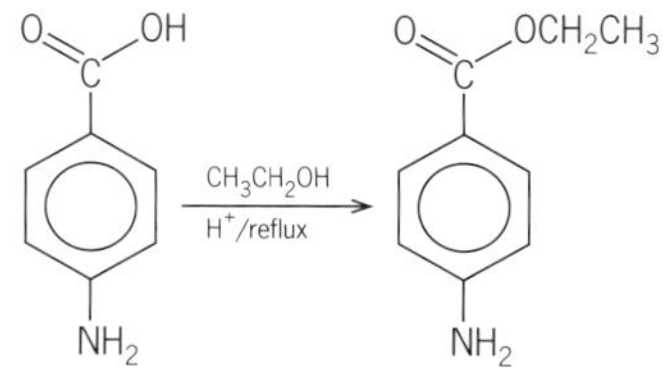

Fig 18.15 The preparation of benzocaine by the esterification of 4-aminobenzoic acid with ethanol

In a very similar process, procaine is also synthesised from 4-aminobenzoic acid (Fig 18.16). This time the alcohol has a much more complicated structure, but the chemistry behind the reaction is the same.

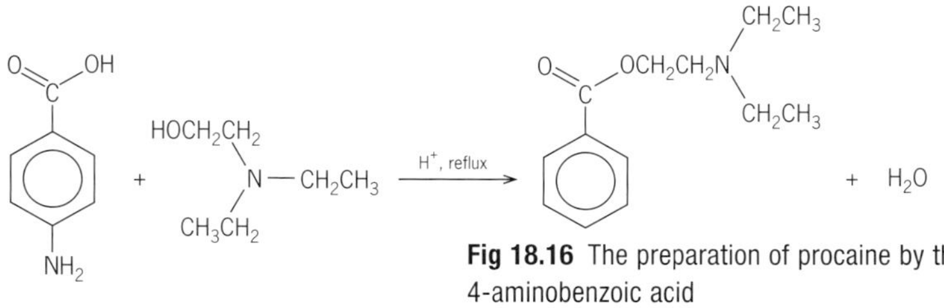

Fig 18.16 The preparation of procaine by the esterification 4-aminobenzoic acid

? QUESTION 9

9 Explain how an excess of ethanol drives the equilibrium reaction between ethanol and 4-aminobenzoic acid to the right-hand side.

4 REACTIONS OF ESTERS

The carbonyl carbon in the ester group is susceptible to nucleophilic attack. The principal nucleophiles that react with it are the water molecule, the hydroxide ion and the hydride ion.

HYDROLYSIS OF ESTERS

In the hydrolysis of esters, water reacts with the carbonyl carbon and an alcohol is eliminated. The reaction is very slow because the water molecule is a poor nucleophile. It is also a reversible reaction, because it is essentially the reverse of the esterification reaction:

$$RCOOR' + H_2O \rightleftharpoons RCOOH + R'OH$$

It is this reaction with water that prevents esters from being used in some perfumes and deodorants. Pleasant-smelling esters are liable to be hydrolysed by chemicals in perspiration to form carboxylic acids, many of which have an unpleasant smell. They include butanoic acid, which gives the odour we detect in rancid butter. These carboxylic acids are among the components of body odour and are in the scent that dogs pick up when tracking humans (Fig 18.17).

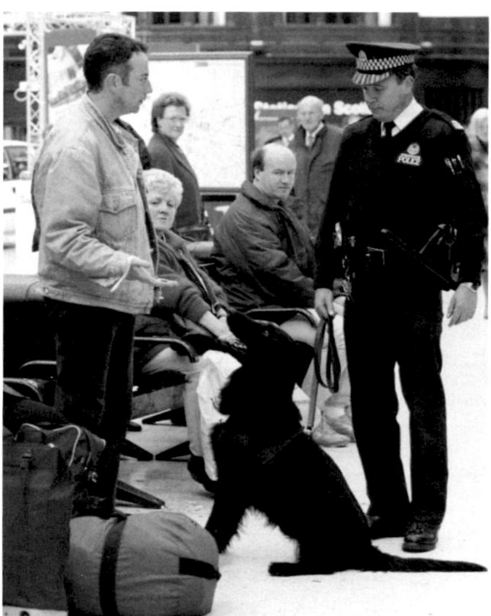

Fig 18.17 Esters in perspiration are hydrolysed to form unpleasant-smelling carboxylic acids. These are in the scent that a sniffer dog picks up when tracing someone

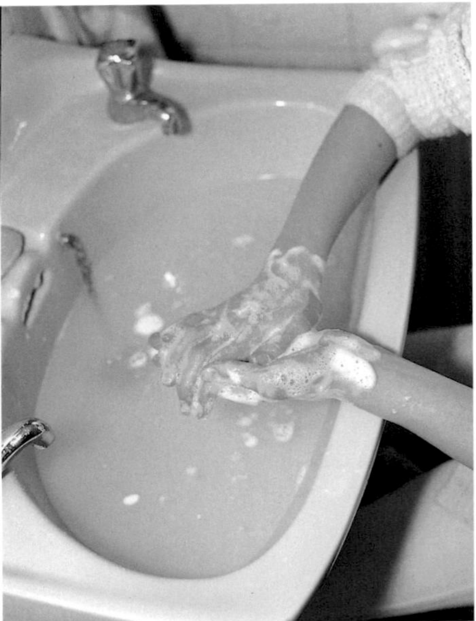

Fig 18.18 Soaps are mainly the sodium salts of carboxylic acids known as fatty acids

BASE-CATALYSED HYDROLYSIS OF ESTERS

The base-catalysed hydrolysis reaction involves boiling an ester with aqueous sodium hydroxide to form the sodium salt of the acid and the corresponding alcohol. Fats and oils are natural esters and their alkaline hydrolysis is the basis of making soap (saponification).

During saponification of a natural fat or oil, propane-1,2,3-triol (glycerol) and the sodium salt of a long-chain fatty acid are formed (Fig 18.20). The sodium salt is a major constituent of soaps. Its carboxylate ion is negatively charged at one (hydrophilic) end and non-polar at the other (hydrophobic) end.

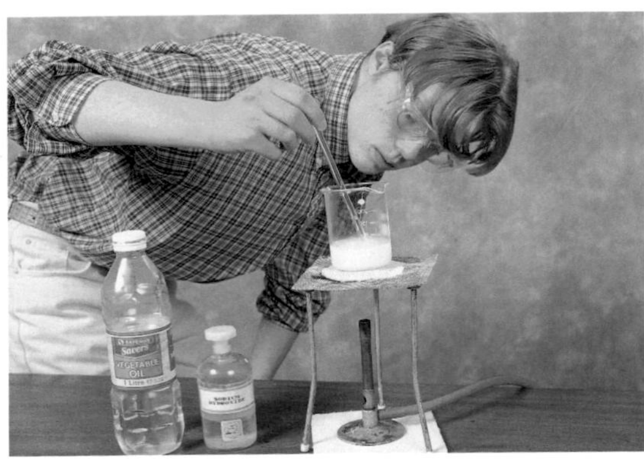

Fig 18.19 Soaps are made by the hydrolysis of fats and oils during boiling with aqueous sodium hydroxide. The reaction is called saponification

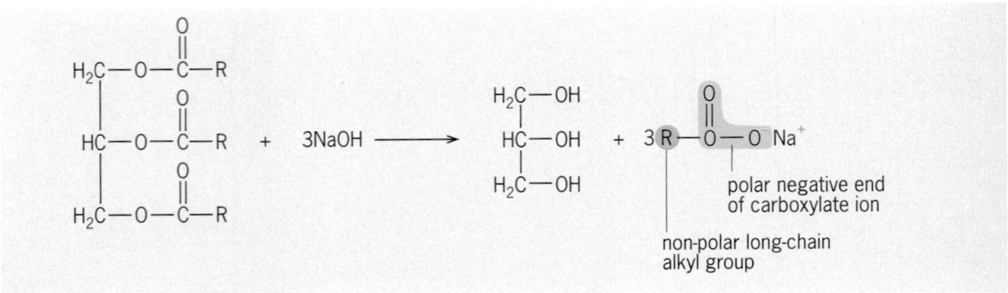

$$\text{(RCOOCH}_2)_2\text{CH(OCOR)} + 3\text{NaOH} \longrightarrow \text{H}_2\text{C(OH)–HC(OH)–H}_2\text{C(OH)} + 3\,\text{R–CO–O}^-\text{Na}^+$$

polar negative end of carboxylate ion

non-polar long-chain alkyl group

Fig 18.20 Saponification. The fats and oils used are esters in which the alcohol part is propane-1,2, 3-triol and the carboxylic acid part is a fatty acid. Hydrolysis of such an ester forms a sodium carboxylate, the main constituent of soap

An ester such as an ethyl benzoate is hydrolysed to form sodium benzoate and ethanol (Fig 18.21).

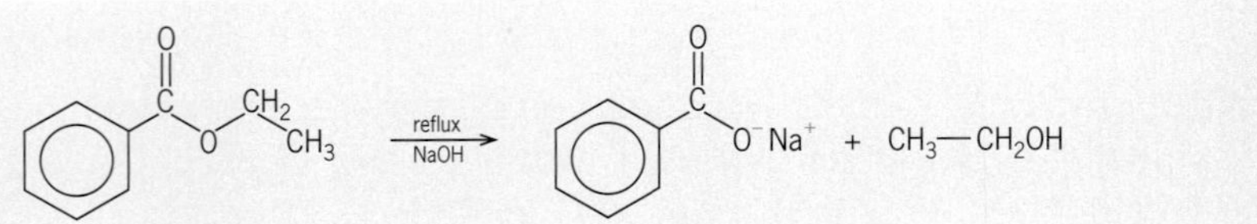

Fig 18.21 The hydrolysis of ethyl benzoate to form sodium benzoate and ethanol

Phenyl benzoate is hydrolysed to give two ionic products, since the phenol formed is sufficiently acidic to react with sodium hydroxide to give sodium phenoxide:

$$C_6H_5COOC_6H_5(l) + 2NaOH(aq) \rightarrow C_6H_5COO^-Na^+(aq) + C_6H_5O^-Na^+(aq) + H_2O(l)$$

Acid-catalysed hydrolysis of esters

It is also possible to hydrolyse an ester by refluxing with a dilute acid. Normally, dilute sulfuric acid or concentrated hydrochloric acid is used as the acid catalyst. This is really the reverse reaction of esterification, and in theory should lead to an equilibrium mixture that contains carboxylic acid, alcohol, ester and water. However, the presence of excess water in the dilute acid

? QUESTION 10

10 Write down the names of the products of the reactions between the following esters and hot aqueous sodium hydroxide:

a) ethyl ethanoate;
b) propyl propanoate;
c) ethyl benzoate.

You can read more about the acidic properties of phenol on page 311.

? QUESTION 11

11 Write down the structural formulae for the products of the reactions of the following esters with hot concentrated hydrochloric acid:

a) butyl ethanoate;
b) phenyl benzoate;
c) diethyl ethanedioate, $(COOC_2H_5)_2$

drives the reaction to completion and so the ester is virtually all hydrolysed. Notice that in acidic hydrolysis, the carboxylic acid is obtained rather than the carboxylate salt.

Ethyl methanoate is hydrolysed to give methanoic acid and ethanol:

$$HCOOC_2H_5(l) + H_2O(l) \underset{}{\overset{H^+(aq)}{\rightleftharpoons}} HCOOH(l) + C_2H_5OH(l)$$

SCIENCE IN CONTEXT

Fuels from oils and fats

Both fats and vegetable oils are used by living systems to transfer energy. They are converted into the corresponding carboxylic acids, which are then oxidised in a complicated series of reactions to form carbon dioxide and water.

Fats and vegetable oils will combust in excess oxygen to give carbon dioxide and water. This reaction is highly exothermic and releases lots of heat energy. This reaction can be harnessed when vegetable oils are used as alternative biofuels.

Vegetable oils and animal fats are made up of molecules called triglycerides (Fig 18.22). These are esters of the alcohol propane-1,2,3-triol and long-chain carboxylic acids. Vegetable oils are easily extracted from the seeds of plants, and several of them are used in cooking, the most popular ones being olive oil, sunflower oil, soya oil and palm oil.

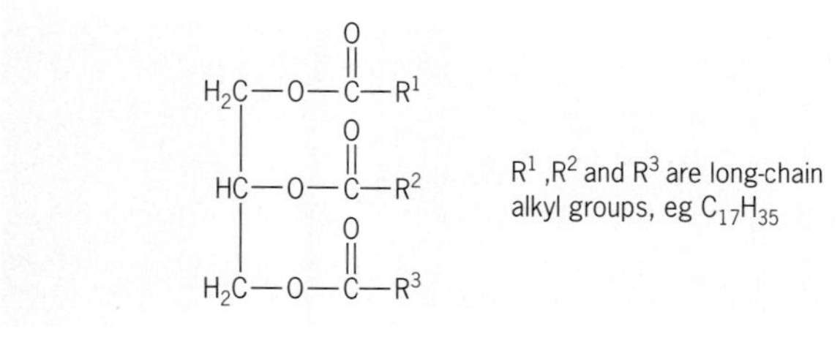

$H_2C—O—C(=O)—R^1$

$HC—O—C(=O)—R^2$

$H_2C—O—C(=O)—R^3$

R^1, R^2 and R^3 are long-chain alkyl groups, eg $C_{17}H_{35}$

Fig 18.22 The structure of triglycerides, the constituents of vegetable oils. When the carbon chain contains a double bond, it is an unsaturated oil or fat

The long carbon chain of the carboxylic acid portion of the triglyceride resembles a long-chain alkane, and hence it can be used to make a renewable source of fuel. The vegetable oil as extracted can be used as biodiesel in specially modified diesel engines. However, if used in normal diesel engines, any partially combusted residues clog up the engine and severely reduce efficiency. So the vegetable oil is hydrolysed with an alkali and then acidified, and the carboxylic acid products are isolated. They are then reacted with methanol to form methyl esters (Fig 18.23). The methyl esters are much more volatile than the original oil and can be used in unmodified diesel engines. In the United Kingdom, over one million tonnes of rapeseed is produced each year (Fig 18.24). The oil from rapeseed is easily converted into

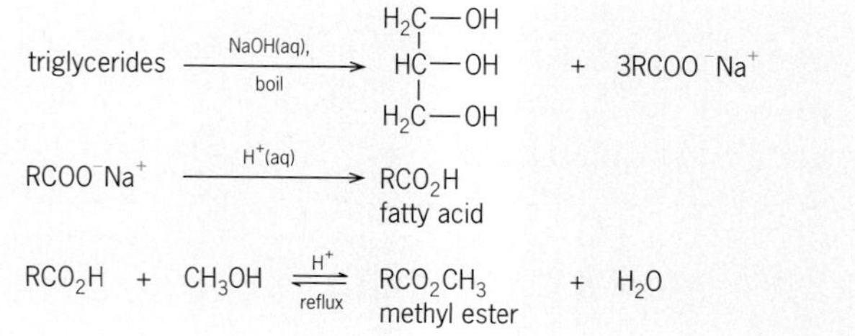

Fig 18.23 The synthesis of methyl esters of fatty acids from vegetable oils. The triglycerides in vegetable oils can be hydrolysed to form sodium salts of fatty acids. These salts can be converted into the fatty acids, and finally esterification occurs to form a methyl ester. Methyl esters of fatty acids burn easily and therefore look promising as substitutes for diesel fuel

methyl esters known as rape methyl ester (RME). Rape methyl ester can be used as a diesel substitute and offers several environmental advantages over conventional diesel fuel. It does not form sulfur dioxide and emits fewer sooty particles during combustion. It may even be possible to utilise used cooking oils as diesel substitute, because they can be treated in the same way as rape oil to make methyl esters that can be used in normal diesel engines.

Methyl esters of long-chain fatty acids are also being manufactured to make low-fat spreads. This avoids the use of hydrogenated unsaturated fats and the risks of producing *E* or *trans* double bonds described on page 287.

Fig 18.24 Rape is easily grown in the United Kingdom. Its seed, commonly called rapeseed, contains the oil that is converted into the diesel fuel known commercially as RME (rape methyl ester)

Fig 18.25 A bus in Reading, UK that runs on RME biodiesel

5 ACYL CHLORIDES

Fig 18.26 The structure of an acyl chloride. As a result of the highly electronegative chlorine atom, its carbonyl group is highly polar, too

An acyl chloride is a carboxylic acid in which the hydroxyl group has been replaced with a chlorine atom (Fig 18.26).

Acyl chlorides are highly reactive compounds that make very useful synthetic intermediates. The preparation of acyl chlorides by reacting sulfuryl(IV) chloride, $SOCl_2$, or phosphorus(V) chloride, PCl_5, with a carboxylic acid is described on page 347.

Acyl chlorides are named after the corresponding carboxylic acid, using the suffix *-oyl* followed by chloride. For example, CH_3COCl is ethanoyl chloride, and C_6H_5COCl is benzoyl chloride.

? QUESTION 12

12 Draw the structures of:

a) benzoyl chloride;

b) propanoyl chloride;

c) hexanedioyl dichloride.

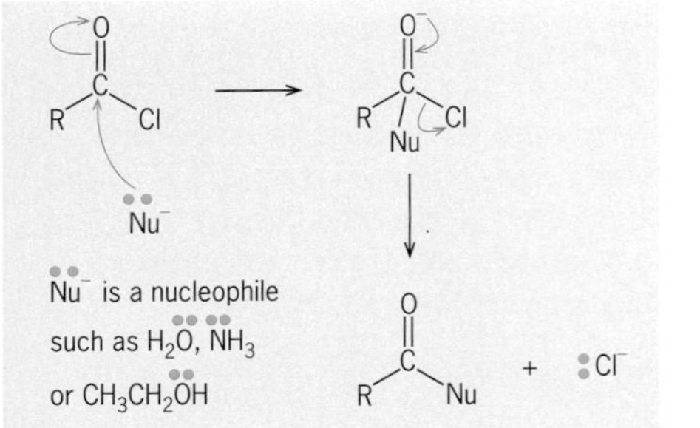

ADDITION–ELIMINATION REACTIONS OF ACYL CHLORIDES

An acyl chloride has an extremely electron-deficient carbonyl carbon because of the electron-withdrawing effect of the highly electronegative chlorine atom. This enhances the reaction of a nucleophile with the carbonyl carbon. Nucleophilic addition occurs and the carbonyl double bond is broken. The addition product then eliminates the chloride as an ion to regenerate the carbonyl double bond. This mechanism is shown in Fig 18.27. The curly arrows show the movement of an electron pair.

The addition–elimination reaction is an example of a **condensation reaction**, in which two molecules react together with the elimination of a *simple* molecule, such as water, hydrogen chloride or an alcohol. With an acyl chloride, hydrogen chloride is eliminated.

Nu^- is a nucleophile such as H_2O, NH_3 or CH_3CH_2OH

Fig 18.27 The mechanism of the addition–elimination reaction of an acyl chloride

HYDROLYSIS OF ACYL CHLORIDES

An acyl chloride reacts readily with water to form the corresponding carboxylic acid. Most acyl chlorides fume in air because of this reaction with water vapour: the hydrogen chloride gas formed dissolves in atmospheric moisture to form tiny droplets of hydrochloric acid.

For instance, ethanoyl chloride fumes in moist air to form ethanoic acid and hydrogen chloride:

$$CH_3COCl(l) + H_2O(g) \rightarrow CH_3COOH(l) + HCl(g)$$

In water, hydrochloric acid would be formed instead.

This reaction of the carbon–chlorine bond is much more rapid than reactions involving other carbon–chlorine bonds. The highly electron-deficient carbonyl carbon encourages nucleophilic addition, to be followed later by elimination.

Compare the hydrolysis of acyl chlorides with the reaction of halogenoalkanes with water described on page 250.

When an acyl chloride is hydrolysed in alkaline solution, the carboxylate ion is formed instead of the acid:

$$RCOCl + 2NaOH \rightarrow RCOO^-Na^+ + H_2O + NaCl$$

? QUESTIONS 13–14

13 Write equations for the hydrolysis of:

a) benzoyl chloride;

b) propanoyl chloride.

14 Benzoyl chloride reacts with cold aqueous sodium hydroxide to form sodium benzoate.

a) Write an equation to show the hydrolysis of benzoyl chloride in aqueous sodium hydroxide.

b) Explain why benzoyl chloride is hydrolysed rapidly by aqueous sodium hydroxide, but the hydrolysis of chloroethane is much slower. (Hint: read pages 309–310.)

Preparation of esters from acyl chlorides

As already noted on page 352, the reaction of a carboxylic acid with an alcohol is an equilibrium process, which makes it difficult to obtain good yields of esters. An alternative way to make an ester from a carboxylic acid involves the formation of an acyl chloride and the subsequent reaction of the acyl chloride with an alcohol (Fig 18.28). Acyl chlorides react completely with

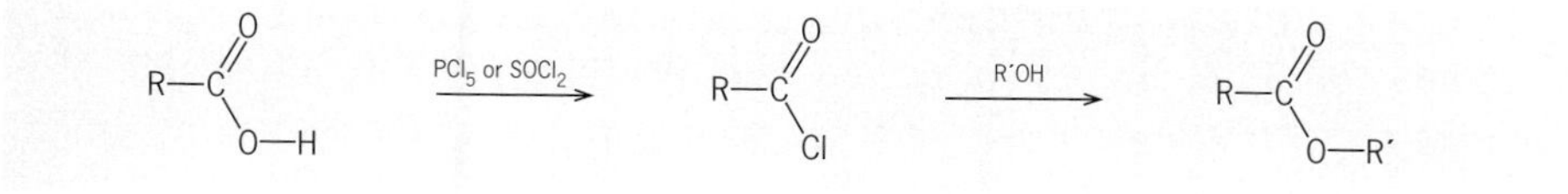

Fig 18.28 Synthesis of an ester via an acyl chloride. An ester can be prepared from a carboxylic acid in a two-stage reaction. First, the acid is converted into an acyl chloride by reaction with either phosphorus(V) chloride or sulfuryl(IV) chloride. Next, the acyl chloride is used in a condensation reaction with an alcohol to give an ester. The advantage of this method is that there are no equilibrium reactions

alcohols and do not form an equilibrium mixture. This provides, in the laboratory, a powerful synthetic route to make esters. Both steps of the process can be achieved with yields over 90 per cent.

HOW SCIENCE WORKS

The addition–elimination reaction between acyl chlorides and alcohols

Chemists use models to explain chemical reactions. The 'curly arrow' model, the movement of electron pairs, is a powerful way of explaining reaction mechanisms. Acyl chlorides react with alcohols and phenols to form esters. These reactions have a mechanism known as **nucleophilic addition–elimination**.

In the reaction between an acyl chloride and an alcohol the first stage is nucleophilic addition of the alcohol. An alcohol molecule can act as a nucleophile by donating a lone pair of electrons from the oxygen atom to the electron-deficient carbonyl carbon atom, thereby forming a covalent bond (Fig 18.29).

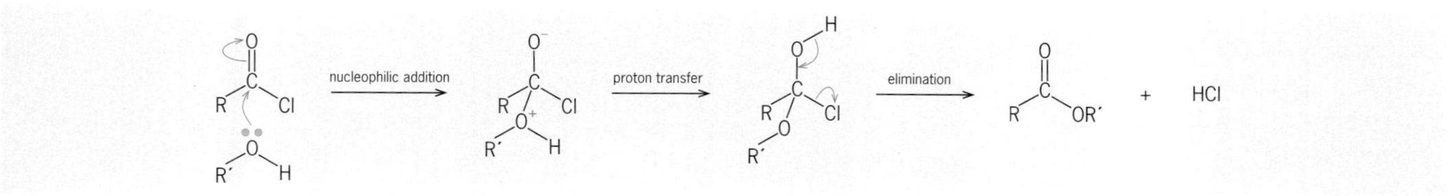

Fig 18.29 The mechanism of the reaction of an alcohol with an acyl chloride

The mechanism's next stage involves the transfer of a proton from one highly electronegative atom to another highly electronegative atom. Following proton transfer, hydrogen chloride is eliminated. When the reaction is carried out in the presence of an alkali, the hydrogen chloride reacts to form a chloride ion and it becomes impossible for the reaction to be reversed.

Phenol reacts with acyl chlorides in the same way, and the presence of an alkali can assist the reaction in a second way (Fig 18.30). Phenol is sufficiently acidic for a reaction with an alkali, such as aqueous sodium hydroxide, to form the phenoxide ion, $C_6H_5O^-$. This ion is a much better nucleophile than the neutral phenol molecule, and so the first stage of the reaction, nucleophilic addition, occurs easily.

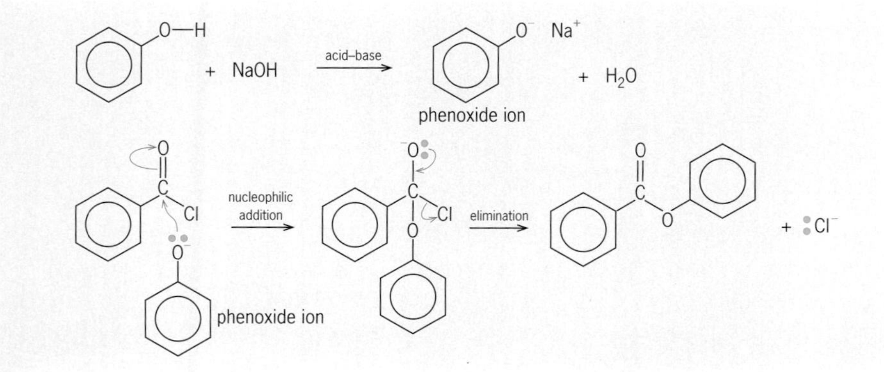

Fig 18.30 The mechanism of the reaction of phenol with benzoyl chloride

Reactions of acyl chlorides with ammonia and amines

Ammonia and other amines are nucleophiles because they have a lone pair of electrons on the nitrogen atom. They can react with acyl chlorides in the same way as alcohols. This time, the products formed are amides. Amides are difficult to make from carboxylic acids, since carboxylic acids react with ammonia in an acid–base reaction to form an ammonium salt instead of an amide:

$$RCOOH + NH_3 \rightarrow RCOO^-NH_4^+$$

So the use of an acyl chloride as a synthetic intermediate solves this problem. And it is easy to produce amides by the reaction of an amine or ammonia with an acyl chloride (Fig 18.31). It is normal to use an excess of the amine

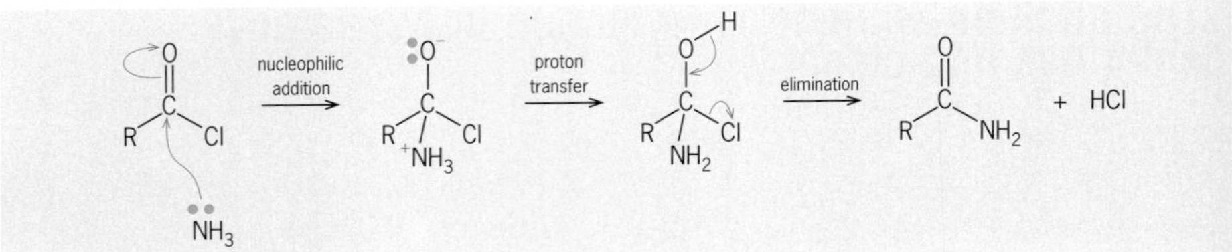

Fig 18.31 The mechanism of the reaction of ammonia with acyl chlorides

? QUESTION 15

15 **a)** Draw the mechanism of the reaction of phenylamine, $C_6H_5NH_2$, with benzoyl chloride to give N-phenyl benzamide, $C_6H_5CONHC_6H_5$.

b) Predict the products of each of the following reactions:

i) ethanoyl chloride and aqueous sodium hydroxide;

ii) ethanoyl chloride and excess ammonia;

iii) benzoyl chloride and ammonia;

iv) benzoyl chloride and methanol;

v) benzoyl chloride and sodium benzoate.

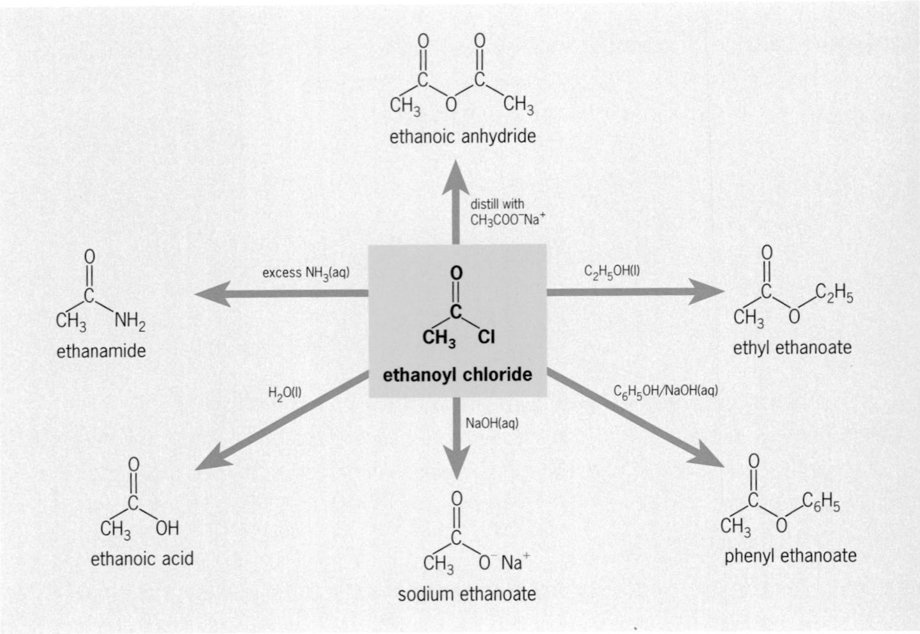

Fig 18.32 Six reactions of ethanoyl chloride

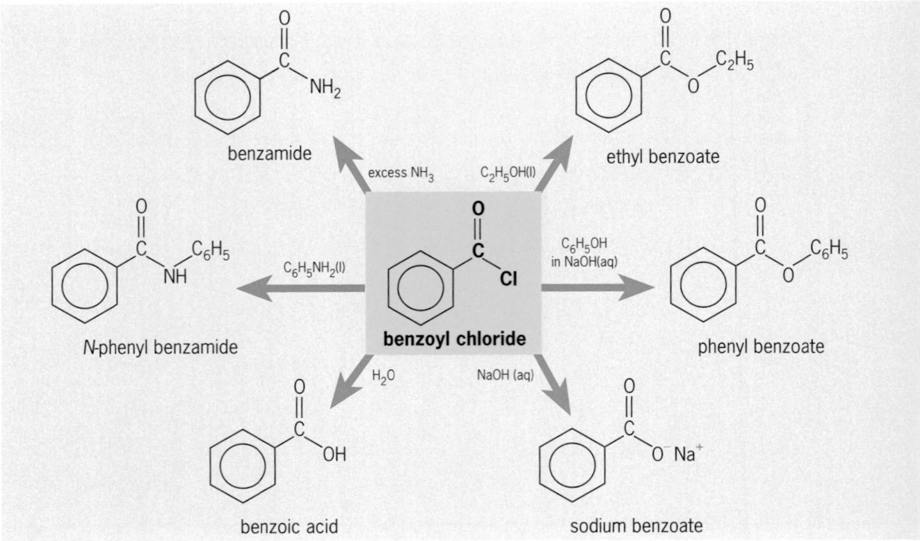

Fig 18.33 Six reactions of benzoyl chloride

or ammonia, so that the excess is available to remove the hydrogen chloride eliminated at the end in an acid–base reaction.

Ethanoyl chloride reacts with excess ammonia to make ethanamide. Notice from this equation that the excess of ammonia leads to the formation of ammonium chloride:

$$CH_3COCl + 2NH_3 \rightarrow CH_3CONH_2 + NH_4Cl$$

Ethanoyl chloride reacts with methylamine to make N-methylethanamide:

$$CH_3COCl + CH_3NH_2 \rightarrow CH_3CONHCH_3 + HCl$$

6 ACID ANHYDRIDES

Acid anhydrides are very similar to acyl chlorides and are also very useful synthetic intermediates. Acid anhydrides can be prepared by distilling a sodium carboxylate with an acid chloride. So sodium ethanoate and ethanoyl chloride will form ethanoic anhydride and leave behind sodium chloride as an involatile salt.

$$CH_3COO^-Na^+ + CH_3COCl \rightarrow (CH_3CO)_2O + NaCl$$

Acid anhydrides are normally named after the corresponding acid, so that $(CH_3CO)_2CO$ is ethanoic anhydride.

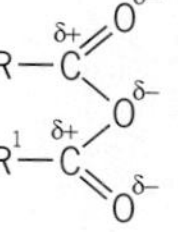

Fig 18.34 Acid anhydrides can either have R = R′ or be a mixed anhydride where R is not R′. Note the polar carbonyl bonds attached to oxygen atoms

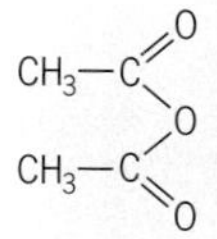

Fig 18.35 Structure of ethanoic anhydride

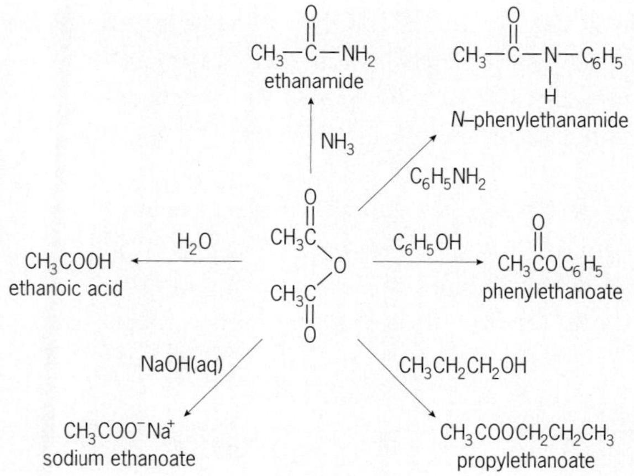

Fig 18.36 Some reactions of ethanoic anhydride

QUESTION 16

16 Draw the structure for propanoic anhydride.

QUESTION 17

17 Write equations for each of the following reactions:

a) ethanoic anhydride with ammonia;

b) ethanoic anhydride with ethanol;

c) ethanoic anhydride with phenylamine;

d) ethanoic anhydride with water.

SCIENCE IN CONTEXT

Manufacture of aspirin

Aspirin is a well-known painkiller. It is made from phenol in a two-step process, shown in Fig 18.37. The formation of the ester linkage in stage two of the synthesis uses ethanoic anhydride. 2-hydroxybenzoic acid and ethanoic anhydride are refluxed together and the aspirin is precipitated by placing the reaction mixture in water. The aspirin can then be filtered and purified before it is ready to be sold commercially. Ethanoic anhydride is used to synthesise the ester rather than ethanoyl chloride, since ethanoic acid (the co-product formed) is less volatile than hydrogen chloride (the co-product with ethanoyl chloride). Another advantage with ethanoic anhydride is that it is much cheaper to manufacture than ethanoyl chloride.

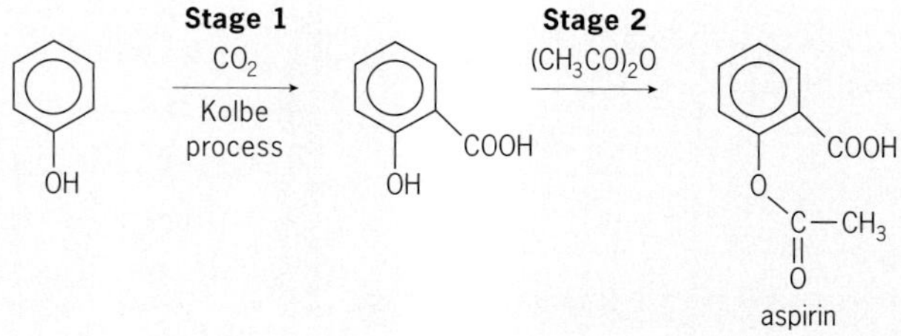

Fig 18.37 The synthesis of aspirin

REACTIONS OF ACID ANHYDRIDES

Nucleophiles such as water, hydroxide ions, ammonia, alcohols and amines react with acid anhydrides by an addition–elimination mechanism with the elimination of a carboxylic acid.

A typical reaction is that of ethanoic anhydride with phenol to make phenyl ethanoate:

$$(CH_3CO)_2O + C_6H_5OH \rightarrow CH_3COOC_6H_5 + CH_3COOH$$

Figure 18.36 shows the range of compounds that can be made from ethanoic anhydride.

7 AMIDES

Amides have an amino group directly attached to the carbonyl carbon (Fig 18.39). Proteins, polypeptides and synthetic polymers such as nylon contain the amide linkage.

Fig 18.38 This mountaineer's rope is made from nylon, which is a polyamide. This is a polymer molecule that contains repeating units linked by amide bonds. It is made by the reaction of an acyl chloride (hexanedioyl dichloride) and an amine (1,6-diaminohexane)

You can read more about secondary amides in proteins and polymers on pages 401 and 402.

NAMING OF AMIDES

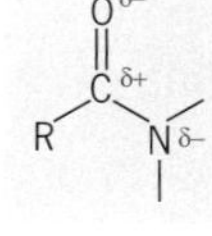

Fig 18.39 The displayed formula of amides. Note that they contain a polar carbonyl bond attached to a nitrogen atom

Amides are named by adding the suffix *amide* to the name of the carbon skeleton. Take, for example, ethanamide, pentanamide and benzamide. Ethanamide is a two-carbon chain in which the number 1 carbon is part of the amide functional group. Its formula can be written as CH_3CONH_2. Pentanamide is a five-carbon chain in which the number 1 carbon is part of the amide group. Benzamide consists of a benzene ring attached to an amide functional group (Fig 18.40). It has the formula $C_6H_5CONH_2$.

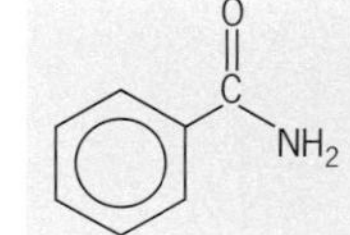

Fig 18.40 The structure of benzamide

These amides are known as primary amides, because they have only one carbon atom attached to the nitrogen atom of the amide. Secondary amides have two carbon atoms attached, and tertiary amides have three carbon atoms attached, as Fig 18.41 shows.

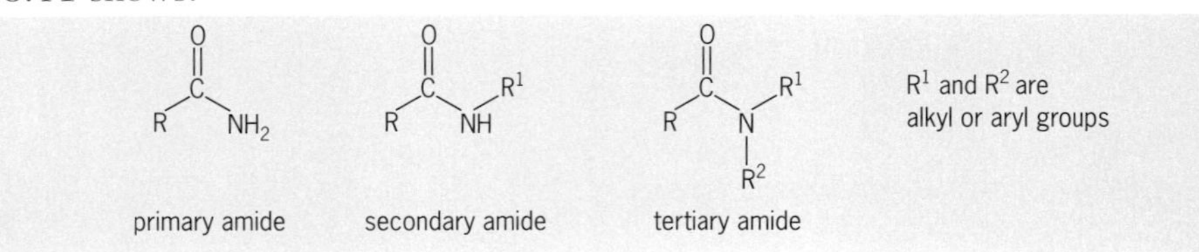

Fig 18.41 The structure of the three types of amide

Note that the extra carbon atoms are not part of the carbon chain that contains the carbonyl group, so they are considered as substituents. Their position is indicated by *N*-, which signifies they are bonded to the nitrogen atom. For example, *N*-methyl ethanamide has the formula $CH_3CONHCH_3$. It has a two-carbon chain, which includes the carbonyl carbon, and a methyl group, which is attached to the nitrogen atom.

? QUESTIONS 18–19

18 Write down the formula for pentanamide.

19 What is the structure of *N*-ethyl ethanamide?

REACTIONS OF AMIDES

Amides show the same types of reaction as the other derivatives of carboxylic acids, but they are rather less reactive.

The hydrolysis of secondary amides in peptides is discussed on page 380.

Hydrolysis of amides

Just as with esters, the reaction of an amide with water is extremely slow, but the reaction can be catalysed by an acid or a base (Fig 18.42). For example,

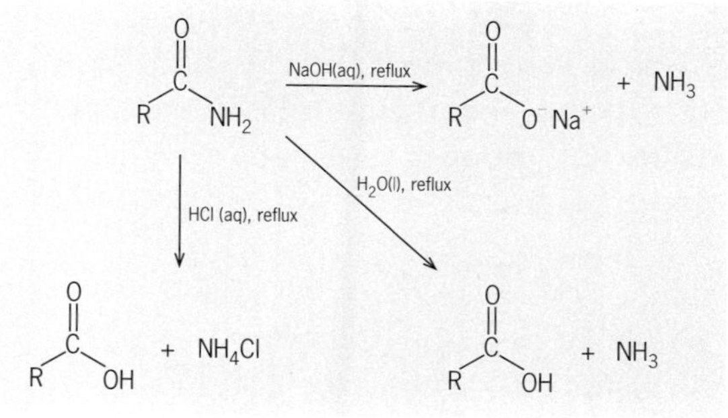

Fig 18.42 The hydrolysis of amides. The reactions catalysed by acids and bases are much faster than with water alone

? QUESTION 20

20 **a)** Draw the structure of the major organic product of the reaction of ethanamide with boiling hydrochloric acid.
b) Draw the structure of the major organic product of the reaction of propanamide with boiling aqueous sodium hydroxide.

when hydrolysed using boiling hydrochloric acid, ethanamide forms ethanoic acid and ammonium chloride. When refluxed with aqueous sodium hydroxide, it forms sodium ethanoate and ammonia:

$$CH_3CONH_2(s) + H_2O(l) + HCl(aq) \rightarrow CH_3COOH(aq) + NH_4Cl(aq)$$

$$CH_3CONH_2(s) + NaOH(aq) \rightarrow CH_3COO^-Na^+(aq) + NH_3(g)$$

8 NITRILES

Nitriles are not acid derivatives, but they do behave similarly to acid derivatives because of the presence of a polar carbon–nitrogen triple bond (Fig 18.43). They are important synthetic intermediates and are often involved in synthetic routes that need to extend the carbon skeleton.

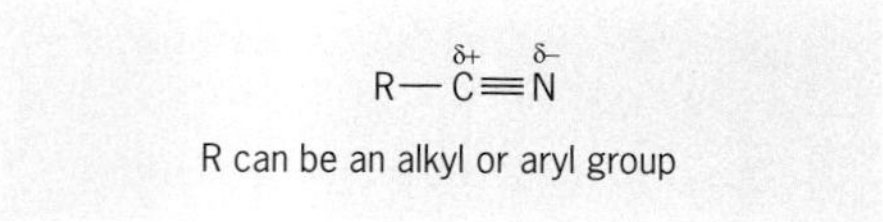

$R{-}\overset{\delta+}{C}{\equiv}\overset{\delta-}{N}$

R can be an alkyl or aryl group

Fig 18.43 Nitriles contain a polar triple bond which is susceptible to nucleophilic attack at the electron deficient carbon atom

NAMING OF NITRILES

Nitriles are named by adding the suffix 'nitrile' to the parent alkane. So CH_3CN would be called ethanenitrile and CH_3CH_2CN would be propanenitrile.

PREPARATION OF NITRILES

Nitriles are normally prepared using hydrogen cyanide, HCN, or the cyanide ion, CN^-. Either nucleophilic substitution of a halogenoalkane using CN^-, or nucleophilic addition of HCN to an aldehyde or ketone will make a nitrile.

Nucleophilic substitution of halogenoalkanes

On page 252 we described how halogenoalkanes can react by nucleophilic substitution with potassium cyanide using ethanol as a solvent – so propanenitrile can be prepared by reacting bromoethane with ethanolic potassium cyanide.

$$CH_3CH_2Br + KCN \rightarrow CH_3CH_2CN + KBr$$

Notice that in this reaction the carbon skeleton has been extended by one carbon.

Nucleophilic addition of aldehydes and ketones

On page 318 we described the reaction of HCN, in the presence of CN^-, with aldehydes and ketones. This reaction made a cyanohydrin, which is a hydroxynitrile; so propanal will react with HCN to make 2-hydroxybutanenitrile.

$$CH_3CH_2CHO + HCN \rightarrow CH_3CH_2CHOHCN$$

Notice that in this reaction the carbon skeleton has again been extended by one carbon.

You can read more about nucleophilic substitution to make nitriles on page 252 and nucleophilic addition to make cyanohydrins on page 318

You can read more about the use of $LiAlH_4$ to reduce compounds on page 319.

? QUESTION 21

21 Benzenenitrile has the formula C_6H_5CN. What are the products of the reaction of benzonitrile with the following reagents?

a) concentrated hydrochloric acid under reflux;

b) aqueous potassium hydroxide under reflux;

c) $LiAlH_4$.

? QUESTION 22

22 Describe, using a flow chart showing reaction intermediates and reagents, how you would convert:

a) propanal into 2-hydroxybutanoic acid;

b) 1-iodopropane into butylamine;

c) 2-bromopropane into 2-methylpropanoic acid.

REACTIONS OF NITRILES

Nitriles have similar properties to carbonyl compounds, because both have polar multiple bonds: the nitriles have a triple bond and carbonyl compounds a double bond. The carbon atom of the CN group is susceptible to attack by nucleophiles.

Hydrolysis of nitriles

Nitriles are hydrolysed extremely slowly by water, but the hydrolysis occurs much faster in either acidic or basic conditions. With acidic hydrolysis a carboxylic acid and an ammonium salt are formed.

$$RCN + 2H_2O + HCl \rightarrow RCOOH + NH_4Cl$$

With basic hydrolysis a carboxylate and ammonia are formed.

$$RCN + H_2O + NaOH \rightarrow RCOO^-Na^+ + NH_3$$

Reduction of nitriles

Nitriles can be reduced by reaction with the powerful reducing agent $LiAlH_4$ to form a primary amine. In the equation the reducing agent is represented by [H] to show that it provides hydrogen atoms for the reduction.

$$RCN + 4[H] \rightarrow RCH_2NH_2$$

Alternatively a nitrile can be hydrogenated by using hydrogen under pressure in the presence of a nickel catalyst.

$$RCN + 2H_2 \rightarrow RCH_2NH_2$$

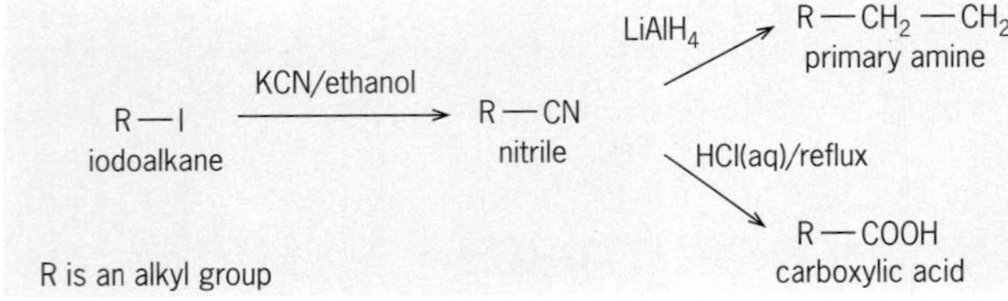

Fig 18.44 Some important reaction involving nitriles

SYNTHESIS INVOLVING NITRILES

Nitriles are often used as synthetic intermediates. Ethanal can be converted into 2-hydroxypropanoic acid, which most people would more easily remember by its trivial name of lactic acid. The synthetic intermediate is 2-hydroxypropanenitrile, as shown in Fig 18.45.

HCN/KCN

etanal

2-hydroxypropanenitrile

HCl (aq)/reflux

2-hydroxypropanoic acid

Fig 18.45 Nitriles are useful synthetic intermediates. In this synthesis ethanal is converted into lactic acid (2-hydroxgypropanoic acid)

SUMMARY

After studying this chapter, you should know that:

- Esters, acyl chlorides, acid anhydrides and amides are all derivatives of carboxylic acids.
- Esters can be prepared by refluxing a carboxylic acid and an alcohol in the presence of a small amount of concentrated sulfuric acid as a catalyst.
- Esterification is a reversible reaction that reaches a dynamic equilibrium.
- At equilibrium, the concentrations of all the substances involved remain constant, and the rate of the forward reaction is equal to the rate of the backward reaction.
- Le Chatelier's principle states that the position of equilibrium changes to minimise any change imposed on the conditions at equilibrium.
- The equilibrium constant for a reversible reaction is the product of the molar concentrations of the products, each raised to the power of its coefficient in the stoichiometric equation, divided by the product of the molar concentrations of the reactants, each raised to the power of its coefficient in the stoichiometric equation.
- The numerical value of the equilibrium constant only changes if the temperature of the equilibrium process changes.
- Acyl chlorides and acid anhydrides react with alcohols to form esters in reactions that go to completion.
- All derivatives of carboxylic acids react by nucleophilic addition–elimination (i.e. by condensation reactions).
- Acyl chlorides and acid anhydrides can be hydrolysed by water to give the corresponding carboxylic acid; whereas amides and esters need to be hydrolysed using an acid or a base catalyst.
- Amides can be made by the reaction between acid anhydrides or acyl chlorides and ammonia or amines.
- Nitriles are good synthetic intermediates because their formation involves the extension of the carbon skeleton.
- Nitriles can be hydrolysed to make carboxylic acids and reduced to make primary amines.

Practice questions and a How Science Works assignment for this chapter are available at www.collinseducation.co.uk/CAS

19

Amino acids and proteins

19 AMINO ACIDS AND PROTEINS

Extremoenzymes Organisms such as the algae that colour this volcanic spring in Yellowstone National Park, USA, survive because their enzymes can tolerate high temperatures

Enzymes

Without enzymes – biomolecule catalysts – the chemical processes of life that take place in each cell of our bodies would be impossibly slow. When in action, each of these special proteins catalyses its specific chemical reaction with a turnover of thousands – even millions – of molecules per minute.

The suggestion is that enzymes could possibly revolutionise large-scale industrial processes. They are thousands of times faster than inorganic catalysts and most work best at body temperature. Industry uses vast amounts of energy to speed up many reactions, so enzymes could save precious energy. Moreover, they are environmentally friendly because it is easy to dispose of them safely.

But there is a major snag. Enzymes lack stability over the wide range of reaction conditions encountered in industry. So the race is on to find 'extremozymes' – enzymes that can tolerate higher temperatures and a wider range of other conditions. As their enzymes must be able to withstand high temperatures, organisms that survive in volcanic springs are a promising starting point. By understanding the structure of these enzymes, biochemists can work out ways to synthesise extremosymes that will make industrial reactions more efficient – and greener, too.

1 WHY PROTEINS ARE IMPORTANT

Proteins make up about 15 per cent of our body weight. They are the major components of skin, muscle, nails and hair, giving structural support and holding cells together. While these may be the obvious proteins, others are just as essential. These include:

- enzymes that catalyse the chemical reactions going on inside our bodies;
- many protein hormones that act as chemical messengers within and between cells;
- protein antibodies that protect us from disease;
- the protein haemoglobin, which transports oxygen through our arteries.

So, proteins are crucial biological molecules (biomolecules). Yet, despite being so diverse, they all are made up from just 20 different small molecules called amino acids.

REMEMBER THIS
Not all hormones are proteins. Some, including the sex hormones, are steroids (see page 316).

2 NATURALLY OCCURRING AMINO ACIDS

As their name implies, amino acid molecules have two functional groups, an amine group and a carboxylic acid group (Fig 19.1). All naturally occurring amino acids have the amine group on the second carbon atom of the molecule (next to the carboxylic acid group). This is often called the **α-carbon**, see Fig 19.2.

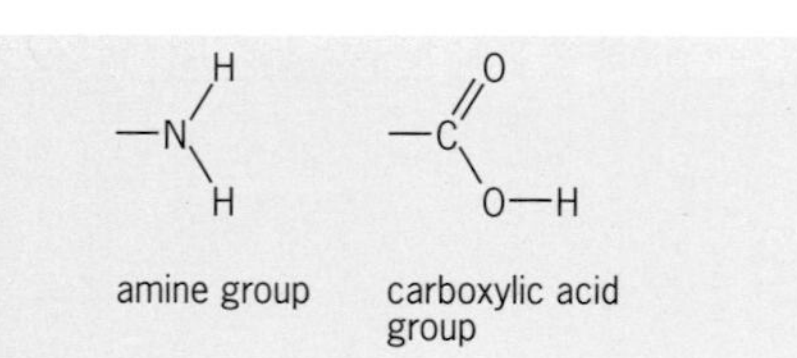

Fig 19.1 Functional groups in an amino acid

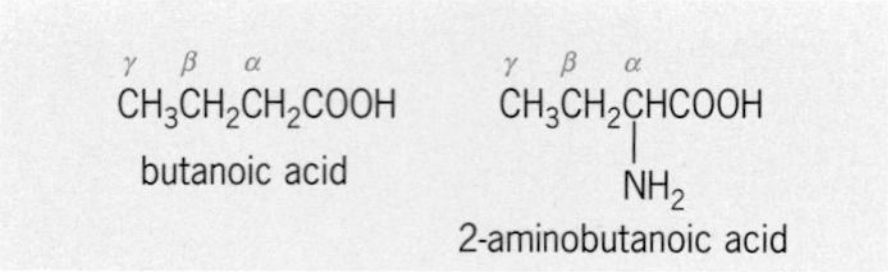

γ β α $CH_3CH_2CH_2COOH$ butanoic acid	γ β α $CH_3CH_2CHCOOH$ (with NH_2 on the α-carbon) 2-aminobutanoic acid

Fig 19.2 The Greek lettering of the carbon atoms. 2-aminobutanoic acid is an α-amino acid

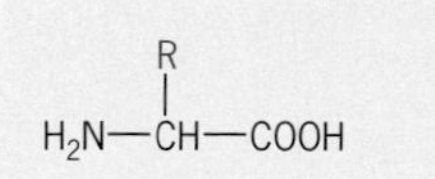

Fig 19.3 General formula for α-amino acids

2-aminobutanoic acid is an **α-amino acid**, or a **2-aminocarboxylic acid**. The more commonly used term is α-amino acid. In this type of acid, the amine and the carboxylic acid functional groups are both bonded to the same carbon atom. α-amino acids are represented by the *general formula* given in Fig 19.3.

The side chain R varies considerably, as Fig 19.4 shows. The composition of the R group confers an individual set of properties to each amino acid, and this, of course, affects the properties of the proteins in which they are found.

Of the 20 amino acids needed to make up our proteins, eight cannot be synthesised in our bodies. These eight are called essential amino acids and must be part of our diet.

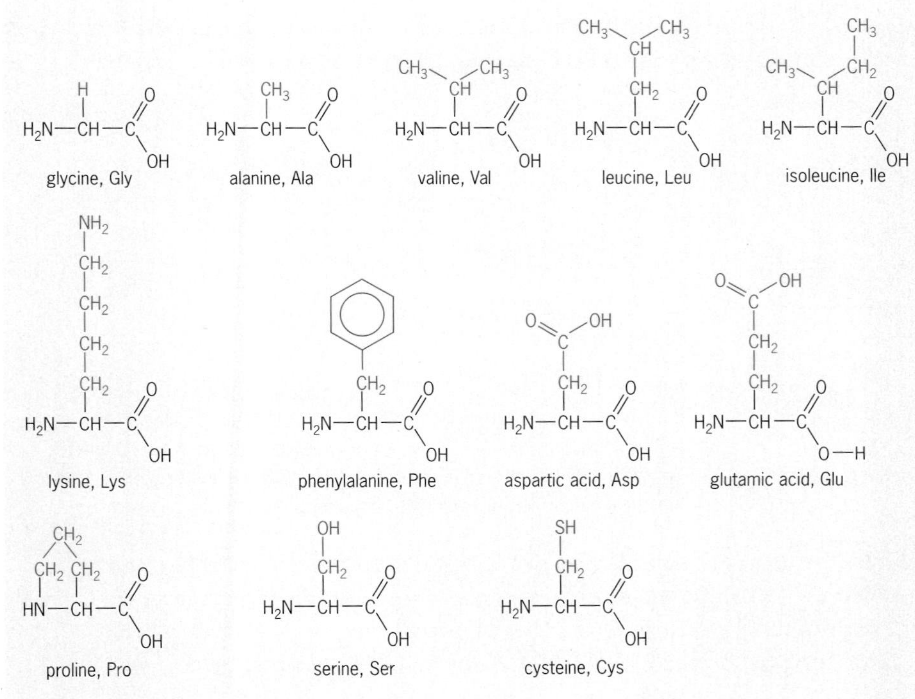
glycine, Gly
alanine, Ala
valine, Val
leucine, Leu
isoleucine, Ile
lysine, Lys
phenylalanine, Phe
aspartic acid, Asp
glutamic acid, Glu
proline, Pro
serine, Ser
cysteine, Cys

Fig 19.4 Shown are 12 of the 20 naturally occurring amino acids that make up most proteins. In each case, the R group is in colour

REMEMBER THIS

The reactions of the amine functional group can be found on pages 251 and 590.

The reactions of carboxylic acids can be found on pages 347 and 350.

? QUESTIONS 1–2

1 **a)** Draw the α-amino acid based on butanoic acid.

b) 4-aminobutanoic acid is also known as gamma-aminobutanoic acid (GABA for short) and is involved in transmitting nerve impulses. Draw this amino acid.

2 **a)** Which amino acids in Fig 19.4 have polar side chains?

b) Look at leucine and isoleucine. Why is isoleucine so named?

c) i) The R group of threonine (Thr), another amino acid found in proteins, is $CH_3CH(OH)$–. Draw the displayed formula of this amino acid. (Hint: remember that the displayed formula shows all the atoms and bonds (including those in OH groups).)

ii) What type of alcohol side chain is this?

iii) How does the R group in Thr differ from that in serine?

3 OPTICAL ISOMERISM AND CHIRAL CARBONS

As discussed on page 182, there are two types of isomerism: **structural isomerism** and **stereoisomerism** (Fig 19.5). The amino acids leucine and isoleucine (see Fig 19.4) are structural isomers because they both have the same molecular formula, but different structural formulas (which answers part b) of Question 2). However, with the exception of glycine, all the α-amino acids can exhibit stereoisomerism, which means that their atoms are bonded in the same order but arranged differently in space. One form of stereoisomerism, known as E–Z (*cis–trans*) isomerism, is described on page 276–277. The other form of isomerism that is shown by these α-amino acids is **optical isomerism**.

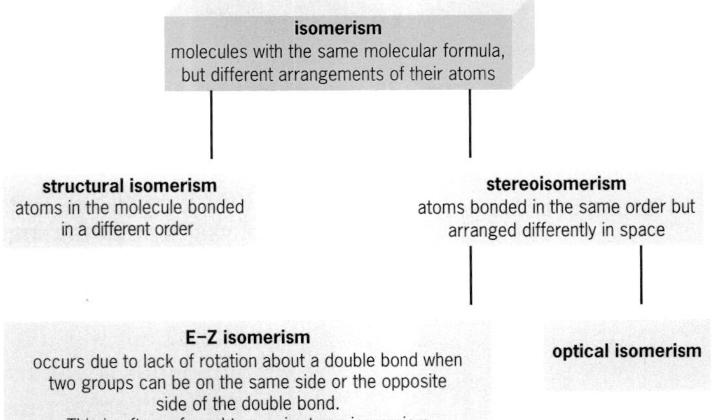
isomerism
molecules with the same molecular formula, but different arrangements of their atoms
structural isomerism
atoms in the molecule bonded in a different order
stereoisomerism
atoms bonded in the same order but arranged differently in space
E–Z isomerism
occurs due to lack of rotation about a double bond when two groups can be on the same side or the opposite side of the double bond.
This is often referred to as cis–trans isomerism
optical isomerism

Fig 19.5 Different types of isomerism

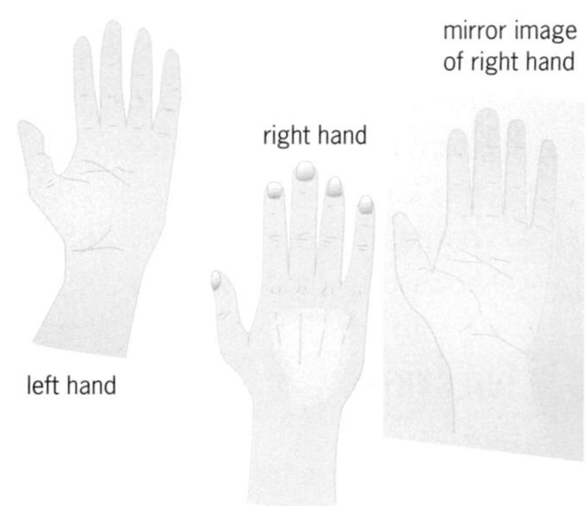

Fig 19.6 Since your hand is not symmetrical, it cannot be superimposed on its mirror image

CHIRALITY

Look at both your hands together with their palms facing you. One is a mirror image of the other. Now put one hand palm up on this page and the other on top of it, also palm up. You will see that you cannot superimpose them – the thumbs stick out in opposite directions. This property is known as **chirality** and exists because hands are not symmetrical. Chiral objects cannot be superimposed on their mirror images – they are **non-superimposable**. The term chirality is derived from *kheir*, the Greek word for hand.

Molecules, too, can be chiral and so have non-superimposable mirror images called **enantiomers**. In organic compounds, the usual reason for a molecule being chiral is that it has a carbon atom bonded to *four different groups*. When this occurs, the molecule cannot be symmetrical and the carbon atom is called an **asymmetric carbon atom** or a **chiral carbon**.

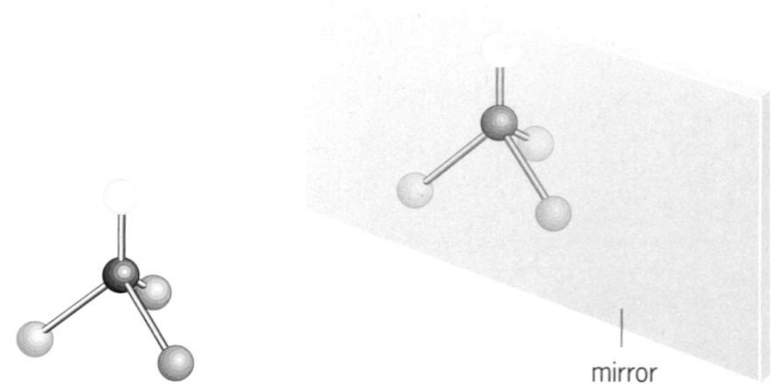

Fig 19.7 A common way for a molecule to be chiral is when it contains a carbon atom bonded to four different groups. The mirror image of this molecule is non-superimposable and is called an enantiomer

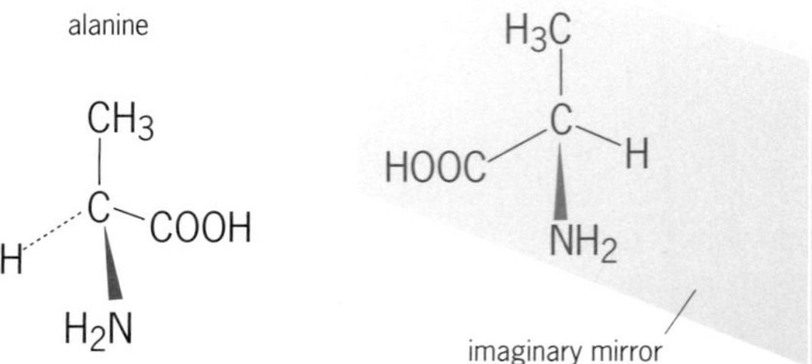

Fig 19.8 The enantiomers of alanine are shown using an imaginary mirror. Note that, although the central carbon atom is bonded to two carbon atoms, these atoms are part of two different groups

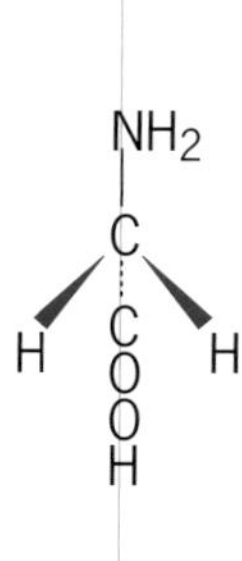

Fig 19.9 Glycine

Take, for example, alanine. It has a chiral carbon because this atom is bonded to four different groups. It is therefore a chiral molecule, having a pair of enantiomers that cannot be superimposed.

Now look at glycine in Fig 19.9. Note that its carbon is no longer asymmetric because it has two identical groups attached (the hydrogen atoms). The glycine molecule and its mirror image can be superimposed, so they cannot be enantiomers of each other.

The enantiomers of chiral molecules have the same chemical properties in ordinary test-tube reactions, which might be expected given that their atoms are bonded in the same order. However, because their atoms are arranged differently in space, we might expect differences in their physical properties such as melting point, boiling point and solubility. However, these too, are identical, except in one unusual way which gives rise to optical isomerism.

REMEMBER THIS

Although enantiomers have identical chemical properties, they do interact differently with the enantiomers of other chiral molecules. Hence, some enantiomers do not smell and taste the same, because they interact differently with the chiral taste and smell receptors.

WHAT IS OPTICAL ISOMERISM?

With any pair of enantiomers, one enantiomer rotates plane-polarised light in one direction, and the other enantiomer rotates it in the opposite direction. But the angles of rotation are equal. In 1815, Jean-Baptiste Biot, a French physicist, was the first to discover that the crystals of certain substances could rotate a beam of plane-polarised light either to the right (clockwise) or to the left (anticlockwise), looking at the incoming light. At the time, the reason for this rotation was a mystery. It was later discovered that solutions of certain compounds could also exhibit this **optical activity**.

Molecules that rotate plane-polarised light to the left (anticlockwise) are said to be **laevorotatory**. Those that rotate it to the right (clockwise) are called **dextrorotatory**.

HOW SCIENCE WORKS

Light and optical activity

Normal light consists of electromagnetic waves that vibrate in all directions perpendicular to the direction of travel, as shown in Fig 19.10. (The electromagnetic nature of light is covered on page 50.) Certain crystals allow light with vibrations in one plane only to pass through them. Such crystals are known as polarisers. The light that emerges from a polariser is called plane-polarised light, or just polarised light. The lenses of Polaroid sunglasses provide a good example of polarisers.

When certain substances are placed between two aligned polarisers through which plane-polarised light is passing, the second polariser stops transmitting light. This is because the substance has rotated the plane of polarisation of the light. The substance is said to be optically active. Rotating the second polariser through a certain angle restores its transmission of light. The angle of rotation caused by a substance is measured in this way, which is the principle of the polarimeter.

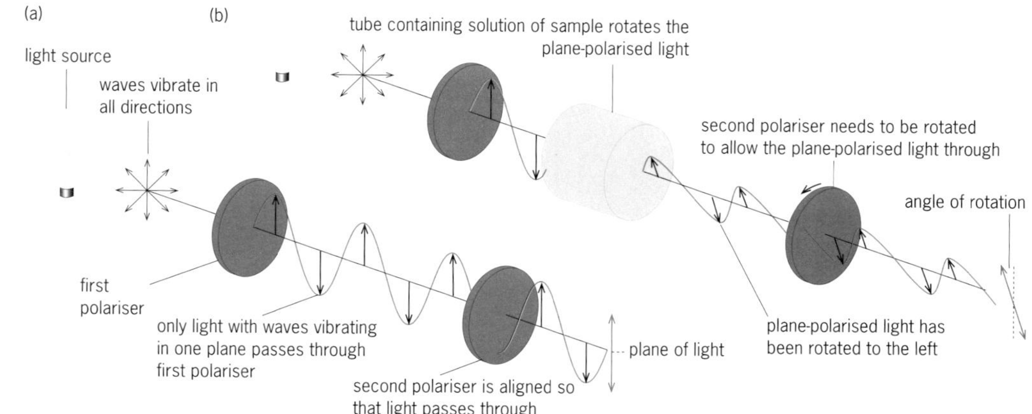

Fig 19.10(a) Unpolarised light from a source passes through the first polariser and the light emerges from it vibrating in one plane only. This plane-polarised light passes through the second polariser (analyser) only when both polarisers are aligned. **(b)** A solution of an optically active sample placed in the beam of plane-polarised light rotates the beam. For light to pass through the second polariser, this too must be rotated until it aligns with the plane of polarisation of the light

In 1848, the French scientist Louis Pasteur (see next page) discovered the important connection between the structure of crystals and their optical activity. He was examining salts of tartaric acid when he noticed that samples of the sodium salt included two sorts of crystal, which were mirror images of each other. Pasteur separated the two sorts of crystal and made a solution of each. He then examined both solutions in a polarimeter, an instrument used to measure optical activity. He found that one sort of crystal rotated plane-polarised light clockwise, while the other rotated it anticlockwise. He also found that, when the concentrations were the same, the two angles of rotation were equal, but in opposite directions.

Pasteur proposed that the molecules making up the crystals must also be mirror images of one another. This proved to be an advance of fundamental importance. Less than 20 years later, the tetrahedral model was proposed for those carbon-containing molecules in which each carbon atom has four single bonds (see the description of methane on page 80). It was then recognised that when a carbon atom is bonded to four different groups, it is asymmetric or chiral.

Many molecules have more than one chiral carbon. For example, the chain form of the glucose molecule has four chiral carbon atoms (Figs 19.11 and 19.12).

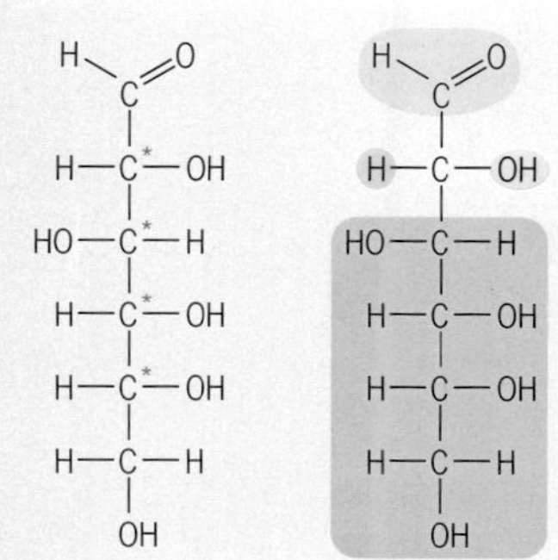

Fig 19.11 (Far left) The structural formula of the chain form of glucose showing the four chiral carbon atoms, each identified by an asterisk

Fig 19.12 (Left) Although each chiral carbon is bonded to two other carbon atoms, to decide if a carbon atom is chiral you must consider the whole of each group attached to it. This has been done for one carbon atom, showing clearly that there are four different groups around it. Can you see why the other three carbon atoms, indicated in Fig 19.11, are chiral?

? QUESTIONS 3–4

3 Optical isomers of the molecule carvone produce different taste sensations. One enantiomer gives the taste of spearmint and is used in spearmint chewing gum. The other enantiomer gives the taste of caraway seeds, which are used to flavour seed cakes.

a) Explain what is meant by the term enantiomer.

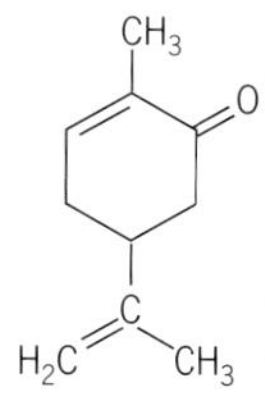

b) Copy the displayed formula of carvone and mark the chiral carbon. If you are in doubt, look at Fig 19.12 and its caption.

Fig 19.13 Carvone

4 Copy the fructose molecule shown in Fig 19.14 and identify the chiral carbons using asterisks.

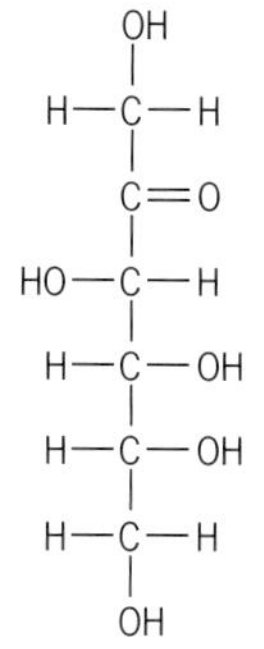

Fig 19.14 Fructose

EXAMPLE

Q Lactic acid, $CH_3CH(OH)COOH$, is isolated from milk. It is an optical isomer of the lactic acid present in muscles.

a) Explain what is meant by the term optical isomerism.

b) Draw the full structural formulas of the two enantiomers, showing clearly the feature that causes optical isomerism.

A a) Optical isomerism occurs when a molecule has a non-superimposable mirror image (Fig 19.15). The molecules of such a pair are called enantiomers. The two enantiomers rotate plane-polarised light in opposite directions.

b) It is often helpful to include an imaginary mirror when drawing optical isomers. In Fig 19.15 the groups around the chiral carbon atom are shown as mirror images, but this is not essential.

The feature that causes optical isomerism in lactic acid is a chiral carbon, which is often indicated by an asterisk (C*).

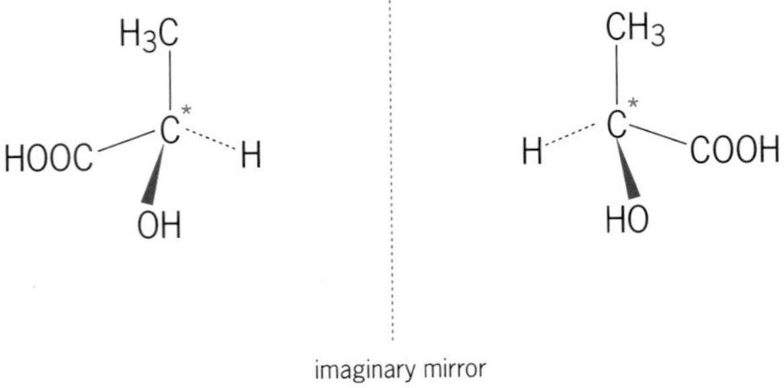

Fig 19.15 Enantiomers of the same compound, lactic acid

? QUESTIONS 5

5 What is the systematic name of lactic acid?

HOW SCIENCE WORKS

Louis Pasteur

Most people have heard of Louis Pasteur as the inventor of the so-called pasteurisation of milk, heat treatment to destroy the bacteria that are naturally in milk and that would otherwise make it 'go off'. This process was just one aspect of his revolutionary work on microorganisms and how they cause food spoilage and fermentation.

Fig 19.16 Louis Pasteur (1822–1895)

Among Pasteur's other achievements was his intense research on silkworm disease, which led to his pioneering germ theory. Putting his theory into practice, he produced vaccines for anthrax and rabies.

Pasteur's achievements outside scientific medicine were just as significant. For instance, he solved the puzzle of why certain crystals and solutions are able to rotate plane-polarised light. He observed that the sodium ammonium salt of tartaric acid (crystals of tartaric acid are found on wine casks) rotated plane-polarised light to the right, while the sodium ammonium salt of racemic acid (another acid found on wine casks during fermentation) had no effect. Apart from this observation, the salts of racemic acid and tartaric acid appeared to be the same, having identical chemical composition and properties. The shapes of the crystals apparently looked the same. However, under the microscope he noticed right- and left-handed crystals in racemic acid.

Pasteur pressed on, painstakingly sorting the two sorts of crystals with the aid of a microscope and tweezers. He made a solution of each sort and examined both solutions in his polarimeter.

Pasteur's discovery of right- and left-handed crystals in racemic acid owed much to luck. He was doing his experiments in winter. Above 26 °C, racemic acid would not have crystallised out into two different forms. Also, the salt of racemic acid he had chosen to investigate is the only salt whose mirror-image crystals have faces clear enough to allow separation with the aid of a microscope (which would not have been as powerful as today's instruments). However, as Pasteur himself remarked, 'Chance favours the prepared mind.'

Racemic acid contains a 50:50 mixture of dextrorotatory and laevorotatory crystals of tartaric acid. Until Pasteur's separation of racemic acid, the laevorotatory form of tartaric acid was unknown. Any 50:50 mixture of left- and right-handed molecules is called a **racemic mixture** after the salt of racemic acid that Pasteur separated. In a racemic mixture, the optical activity of one isomer cancels out the optical activity of the other isomer.

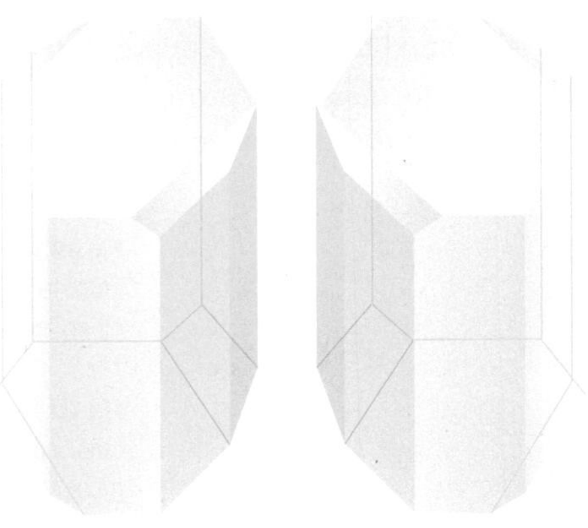

Fig 19.17 Mirror-image crystals of the sodium ammonium salt of tartaric acid

STRETCH AND CHALLENGE

L- and D-enantiomers of α-amino acids

Nearly all the amino acids in every organism are made up from one type of enantiomer, known as the L-enantiomer. In the L-enantiomer, the arrangement of the four groups spell CORN (**CO**OH, **R**, **N**H_2) when you look directly down on the H atom and go clockwise round the molecule (Fig 19.18).

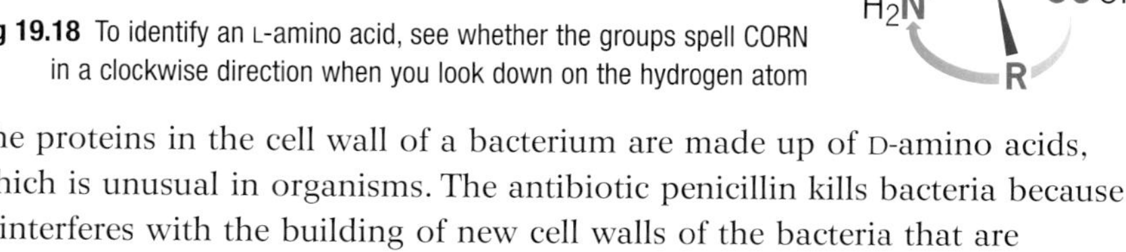

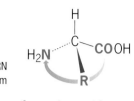

Fig 19.18 To identify an L-amino acid, see whether the groups spell CORN in a clockwise direction when you look down on the hydrogen atom

The proteins in the cell wall of a bacterium are made up of D-amino acids, which is unusual in organisms. The antibiotic penicillin kills bacteria because it interferes with the building of new cell walls of the bacteria that are dividing. Penicillin does not destroy cell walls in our bodies because our protein is made of L-amino acids.

REMEMBER THIS

To depict a pair of optical isomers, draw the 3-D structure of one molecule and then swap the positions of two of the groups in the other molecule.

SCIENCE IN CONTEXT

Ageing a skull by its smile

When we are born, the dentine inside our teeth contains L-aspartic acid (see Fig 19.4). As we grow older, this starts to change to the D-form. When we die, this process continues at the same rate for hundreds of years. This means that we can accurately date a skull that has been removed from the earth just by knowing the proportions of D- to L-aspartic acid in its teeth – provided that the skull is not too old. This technique is reliable back to the tenth century AD.

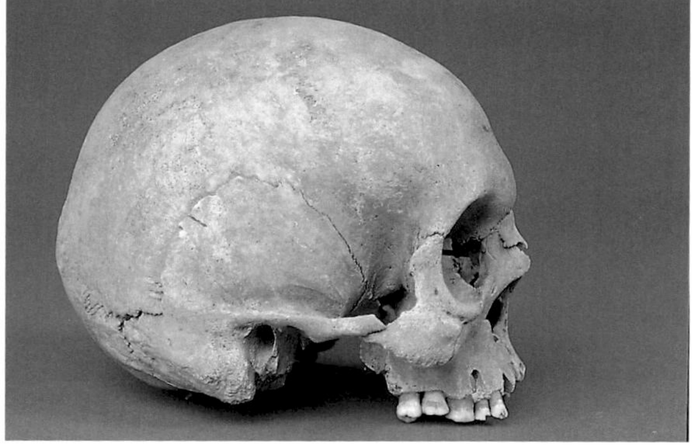

Fig 19.19 As a skull ages, there is an increase in the ratio of the L- to D-isomers of aspartic acid in its teeth. Exhumed skulls have been dated by determining this ratio. This is the skull of a medieval English man

CHIRAL DRUGS

Pharmaceutical drugs work on complex chemical sites around the body because they have the correct shape to fit the receptor sites. L-dopa is just such a drug (Fig 19.20). It is used to treat Parkinson's disease, which is characterised by uncontrollable shaking in the hands and loss of balance. The L-dopa is absorbed by the brain, where it is converted into dopamine. In the brain, high concentrations of dopamine improve conduction by nerves and thus help to control the shaking. The D-form of the chiral dopa molecule causes very unpleasant side-effects.

Molecules prepared synthetically in the laboratory often contain mixtures of optical isomers. An example is the drug Thalidomide. Between 1956 and

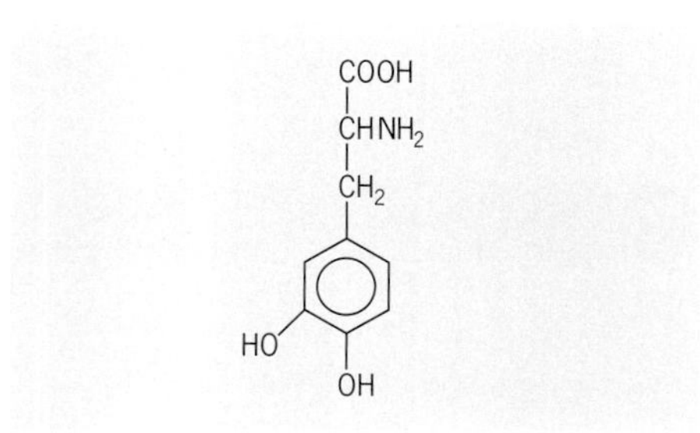

Fig 19.20 The dopa molecule

? QUESTIONS 6

6 Identify the chiral centre on the dopa molecule (Fig 19.20).

1961 it was prescribed to women who suffered badly from sickness during early pregnancy. While one of the optical isomers was an effective treatment, the other form caused severe deformities. Pharmaceutical companies want to ensure this never happens again.

If a drug is prepared naturally by enzymes in living systems, then often only one optical isomer is produced. Bacteria or enzymes are often used to promote stereoselectivity – that is, the production of one optical isomer. Another way of producing the desired optical isomer is to use a chemical synthesis route using a reagent that promotes the formation of this isomer. Chiral catalysts can also be used. These catalyse the formation of one optical isomer but not the other. Lastly, a single optical isomer may be made in a series of reactions starting with a naturally occurring chiral molecule, such as an amino acid or a sugar.

A positive aspect of understanding the chiral nature of some drugs is that smaller doses can be given, since only half as much is required (the other half being the other optical isomer). Also, side effects caused by the wrong optical isomer are eliminated and the pharmacological activity is much improved.

✓ REMEMBER THIS

Bases are proton acceptors and acids are proton donors (see page 329).

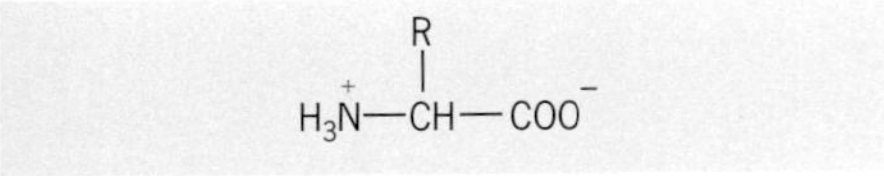

ZWITTERIONS

Amino acids have two functional groups. The chemical properties of compounds with two functional groups are often the sum of the chemical properties of each individual group. However, with amino acids, the basic amine group and the acidic carboxylic acid group can react with each other. A proton from COOH can be donated to the NH_2 group of the same molecule to give a **zwitterion** (Fig 19.21), a molecule that carries both a positive and a negative charge. (*Zwitter* is the German term for hybrid.)

$$H_3\overset{+}{N}-\underset{}{CH}(R)-COO^-$$

Fig 19.21 An amino acid zwitterion

Their relatively high melting point and high solubility in water indicate that amino acids exist as zwitterions, both in the solid state and in solution. When a dilute acid is added to an aqueous solution of an amino acid (Fig 19.22), the COO^- group accepts a proton to form COOH. This leaves a positive ion (a cation). When the pH of the aqueous solution is raised by adding OH^- (Fig 19.23), a proton from NH_3^+ is removed to form NH_2 and a negative ion (an anion).

The reaction of COOH with alkalis is covered on page 339, and the reaction of NH_2 with acids on page 583.

$$H_3\overset{+}{N}-CH(R)-COO^-(aq) + H_3O^+(aq) \longrightarrow \underset{\text{cation}}{H_3\overset{+}{N}-CH(R)-COOH(aq)} + H_2O(l)$$

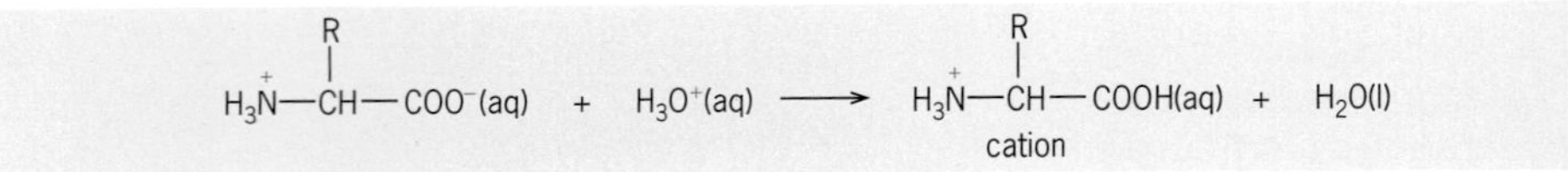

Fig 19.22 Dilute acid added to an aqueous solution of an amino acid results in a positive ion and COOH

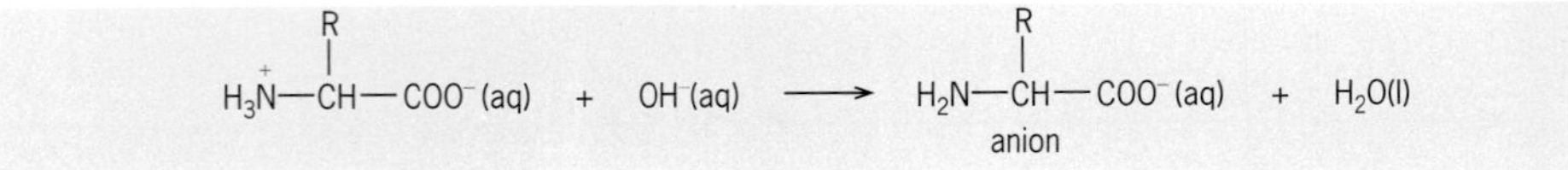

$$H_3\overset{+}{N}-CH(R)-COO^-(aq) + OH^-(aq) \longrightarrow \underset{\text{anion}}{H_2N-CH(R)-COO^-(aq)} + H_2O(l)$$

Fig 19.23 Adding OH^- raises the pH of the aqueous solution and results in a proton being removed from NH_3^+ to form NH_2 and a negative ion (an anion)

? QUESTION 7

7 Write balanced equations for the reactions of:

a) glycine with sodium hydroxide solution;

b) leucine with dilute hydrochloric acid.

Salts are produced in both cases, so remember to show the charges on the ions. (Hint: See Fig 19.4 on page 371 for the formulae of amino acids.)

The pH at which zwitterions have the highest concentration in solution in equal amounts is known as the **isoelectric point**. For those amino acids with a neutral R group, the isoelectric point is about pH 6. By contrast, lysine has a basic side chain that contains an extra NH_2 group (Fig 19.24). This means that a solution of lysine is alkaline and its isoelectric point is about pH 9.5. Those amino acids with an additional COOH in the side chain are acidic and have isoelectric points with low pH.

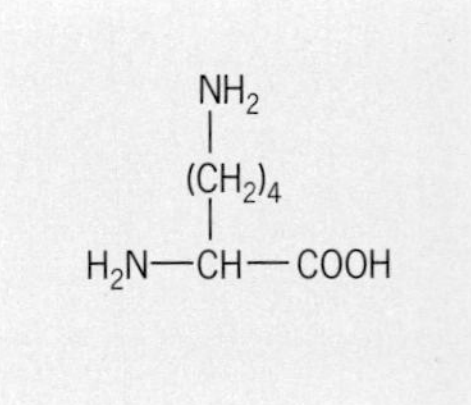

$$H_2N-CH((CH_2)_4NH_2)-COOH$$

Fig 19.24 Lysine has two NH_2 groups, making it a basic amino acid

? QUESTION 8

8 Look at Fig 19.4, page 371. Which amino acids have an acidic side chain?

HOW SCIENCE WORKS

Electrophoresis

Electrophoresis can be used to separate a mixture of amino acids according to the ionic charge at a particular pH. The mixture is placed in a gel or on filter paper (Fig 19.25) in an electric field.

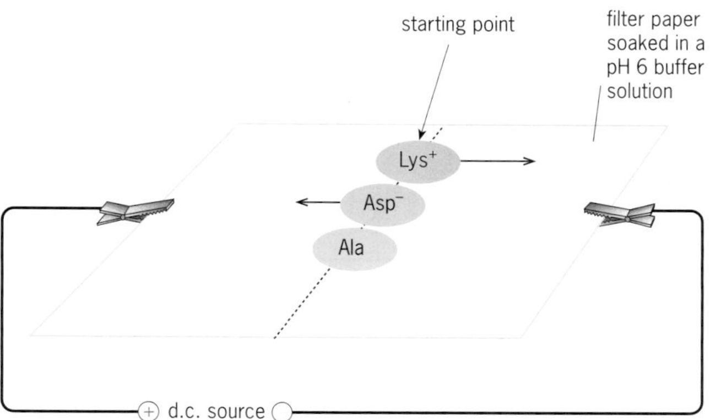

Fig 19.25 Electrophoresis of a mixture of three amino acids. The pH used in this example is 6, so alanine exists as a zwitterion. At this pH, aspartic acid is negatively charged and lysine is positively charged

Positively charged ions migrate to the cathode, and negatively charged ions to the anode. When the pH is that at which zwitterions are formed with equal positive and negative charges, the zwitterions cannot move in either direction. So, the pH value must be carefully chosen before the mixture of amino acids can be separated and analysed successfully.

Electrophoresis can also be used to separate and identify different proteins in a gel. The large number of different amino acids that make up a protein give it an overall charge at a particular pH. This, combined with the protein's mass and shape, enables it to be separated from other proteins by electrophoresis. (The higher the mass and more irregular their shape, the slower they move.)

Genetic fingerprinting

Genes, too, can be analysed by electrophoresis. DNA, the molecule that carries genetic information, is broken up into small fragments by enzymes. These fragments are negatively charged and can be separated into bands using gel electrophoresis. The bands are made visible by tagging them with molecules that contain radioactive phosphorus, ^{32}P, and allow them to fog a photographic film. This forms the basis of genetic fingerprinting used by forensic scientists to identify criminals from blood, saliva or other samples that contain cells. The chances that two samples of DNA from different people will actually match is estimated to be of the order of one in 25 million. In this way, innocent people can be eliminated from criminal investigations and compelling evidence can be provided against the guilty.

The technique is not only used to catch criminals. It can be used to match parents to their children (see Fig 19.27), and also to identify cancerous cells in bone marrow.

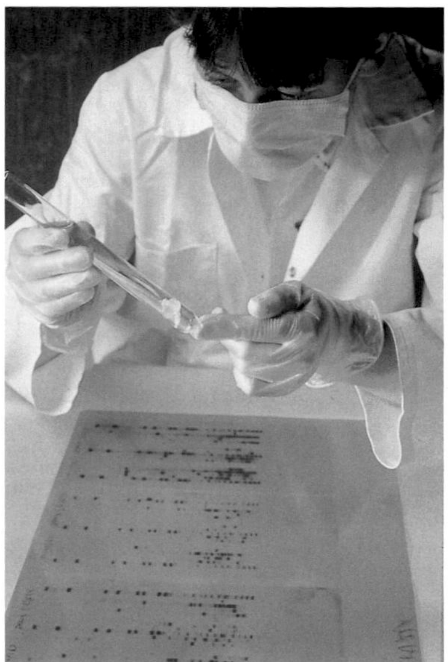

Fig 19.26 A researcher with autoradiograms of DNA fragments separated by gel electrophoresis

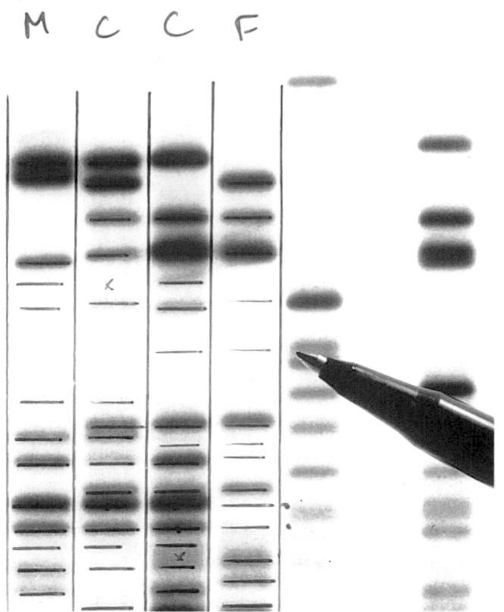

Fig 19.27 Genetic fingerprints to confirm parenthood

SCIENCE IN CONTEXT

Piecing together the Dead Sea Scrolls

Another fascinating use of electrophoresis is by archaeologists. The first seven of the Dead Sea Scrolls were discovered in a limestone cave by a Bedouin shepherd close to the Dead Sea in 1947 (Fig.19.28). They were stored in a clay jar and were remarkably well preserved; radiocarbon dating has found them to be more than 2000 years old. Following this discovery a total of 800–900 manuscripts have been found in 11 caves. The manuscripts are in 15 000 fragments, and they are still being pieced together to reveal more about the Old Testament of the Bible and the history of the Jews. Each manuscript is thought to be written on the skin of one animal, either that of a sheep or of a goat. By using genetic fingerprinting the fragments can be identified as belonging to one particular document, which speeds up the re-assembling of the scrolls.

Fig 19.28 These fragments were found in one of the eleven caves close to the Dead Sea. Piecing them together is made much easier by the use of genetic fingerprinting

4 FORMING PEPTIDES

When two amino acids react together, the compound formed is known as a **dipeptide**. The reaction is between the COOH of one amino acid and the NH_2 of the other (Fig 19.29), to give a dipetide. By convention, the amino acid that contains the free NH_2 group is always placed on the left. A molecule of water is eliminated and so the reaction is called a **condensation reaction**. can keep reacting to form longer and longer amino acid chains, known as **polypeptides**. The amino acids in peptides and polypeptides are called **amino acid residues**. **Proteins** are made up of one or more long polypeptide chains. Thus, polypeptides and proteins are condensation polymers of amino acids. In our bodies, the condensation reaction is catalysed by enzymes.

REMEMBER THIS

A condensation reaction is also called an addition–elimination reaction. The small molecule eliminated can be H_2O, NH_3 or HCl. (See page 360.)

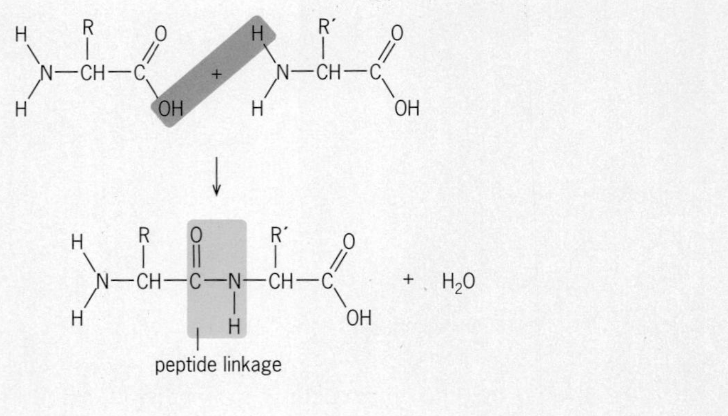

Fig 19.29 The formation of a peptide linkage in a condensation reaction between two amino acids

QUESTION 9

9 When alanine reacts with glycine, two dipeptides can be formed: Ala–Gly and Gly–Ala. Draw the structural formulae of both dipeptides.

The shorthand notation for a dipeptide made from alanine and glycine is Ala–Gly. The peptide link is shown by the dash. Three-letter codes or abbreviations are used for the names of the amino acids, as in Fig 19.4.

The CO–NH group is a **secondary amide** functional group, often referred to as just an **amide** group. The **peptide group** or **peptide link** is the amide functional group that joins two amino acids. Either end of the dipeptide can react with another amino acid to form a **tripeptide**. Amino acids can keep reacting to form longer and longer amino acid chains, known as **polypeptides**. The amino acids in peptides and polypeptides are called **amino acid residues**. **Proteins** are made up of one or more long polypeptide chains. Thus, polypeptides and proteins are condensation polymers of amino acids. In our bodies, the condensation reaction is catalysed by enzymes.

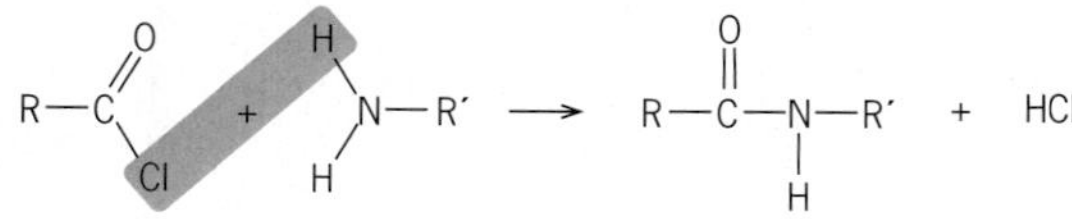

Fig 19.30 A primary amide reacts with an acyl chloride to form a secondary amide

REMEMBER THIS

Secondary amides are usually made in the laboratory by reacting a primary amine (RNH_2) with an acyl chloride (R′COCl), as in Fig 19.30. This is also called a condensation reaction because a small molecule is eliminated. (See page 360).

HOW SCIENCE WORKS

Peptides don't have to be large

Many biologically important peptides contain just a few amino acid residues. For example, the human brain produces a peptide called leucine enkephalin that contains just five amino acid residues. This peptide was first discovered when the pain-killing actions of morphine and codeine were being investigated in the 1970s. Both morphine and codeine fit into a brain receptor site, which was found to be the start of their pain-killing action.

The mystery scientists set out to unravel was how these two drugs, both obtained from the dried sap of poppies (called opium), could fit into brain receptor sites. They looked for compounds made in the body that might fit into these sites and discovered that the brain produces its own pain-killers, called enkephalins, of which leucine enkephalin is one (Fig 19.31).

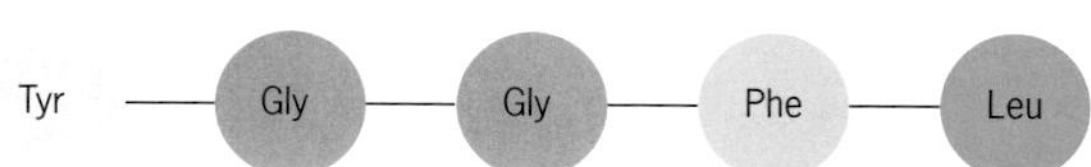

Fig 19.31 Leucine enkephalin. By convention, the amino acid with the free NH_2 group is shown at the left end of the molecule, while the amino acid with the COOH group is shown at the right end

Oxytocin, which contains nine amino acid residues, is another small peptide. It is a hormone secreted by the human pituitary gland and is responsible for inducing the uterus to contract at the end of pregnancy. It was the first natural peptide to be synthesised in a laboratory. Vincent du Vigneaud, the biochemist who made this important breakthrough in 1954, was awarded the Nobel Prize for Chemistry for his achievement.

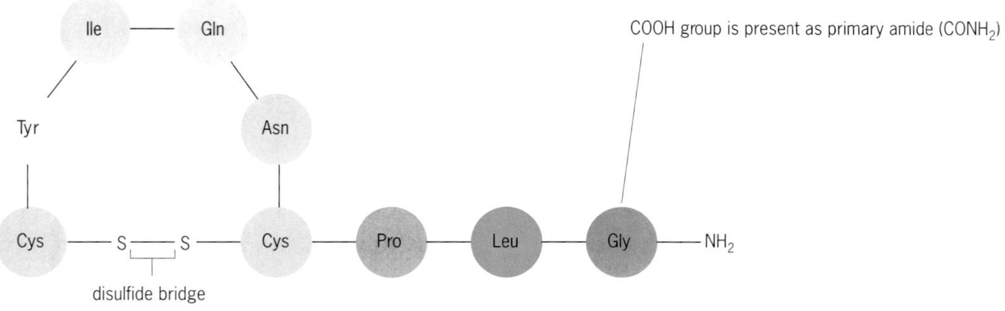

Fig 19.32 Oxytocin: the two sulfur atoms form a bridge holding six of the amino acids in a ring. (Disulfide bridges are covered on page 380)

5 PROTEIN STRUCTURE

Proteins are very large molecules, with values of M_r that range from 5000 to several million. When proteins are formed inside a cell, they are too large to pass through the cell membrane and so stay trapped inside. That is, unless tissues are damaged. For example, certain enzymes (which we have seen are proteins) can be found outside cells after a heart attack. Normally, these enzymes would be confined inside the cells of the heart, but during a heart attack some of the cells rupture, which allows these enzymes to escape. The

REMEMBER THIS

Proteins are condensation polymers and the amino acids that make them up are the monomers. See Chapter 20 for more information on condensation polymerisation.

? QUESTION 10

10 Human insulin, which consists of 51 amino acid residues, is a relatively small protein. It has a molecular formula of $C_{254}H_{377}N_{65}S_6$. What is its relative molecular mass?

more major the heart attack, the more cells split open and the higher the concentration of the enzymes found in the blood. The presence of these enzymes in the blood is used by doctors to diagnose the seriousness of a heart attack. In the same way, certain proteins in urine indicate that a kidney has been damaged.

PRIMARY STRUCTURE OF PROTEINS

Although only 20 amino acids, usually, make up our proteins, the number of ways in which they are combined is vast. Take, for example, lysozyme, an enzyme found in tears, saliva, nasal mucous and milk, which destroys the cell walls of bacteria by breaking certain bonds through hydrolysis. It has 129 amino acid residues and uses all 20 amino acids (Fig 19.33). The number of different arrangements of the 20 amino acids in a protein chain of this length is 20^{129} (7.0×10^{167}), which is estimated to be more than twice the number of atoms in our galaxy.

The sequence of the amino acids in a protein is called its **primary structure** and this forms the backbone of the protein.

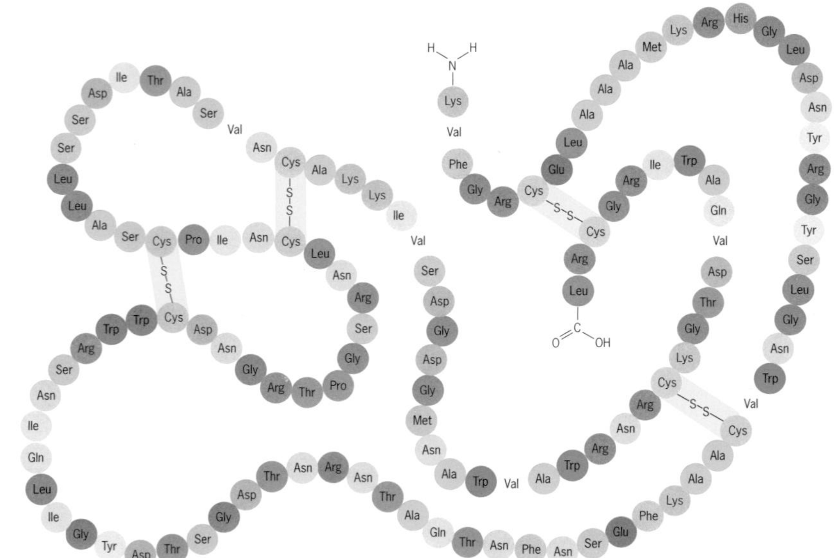

Fig 19.33 The primary structure of the enzyme lysozyme. Notice the disulfide bridges

REMEMBER THIS

Disulfide bridges are formed between two cysteines by the oxidation of two S–H groups to S–S.

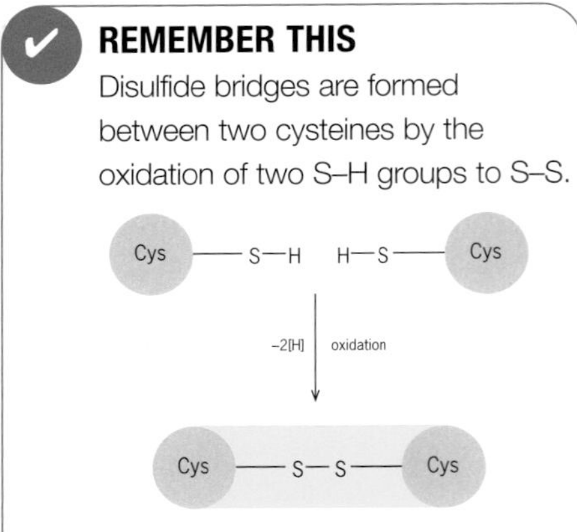

Fig 19.34 Formation of disulfide bridges

HYDROLYSIS OF PEPTIDE BONDS

To find the sequence of amino acids in a protein, the peptide links are **hydrolysed** (split by the action of water). This is done by boiling the protein with an acidic or alkaline solution (Fig 19.35), or by using certain enzymes. The protein chain may be hydrolysed completely, to leave a mixture of individual amino acids that can be separated using chromatography. This gives information as to how many different amino acids are present.

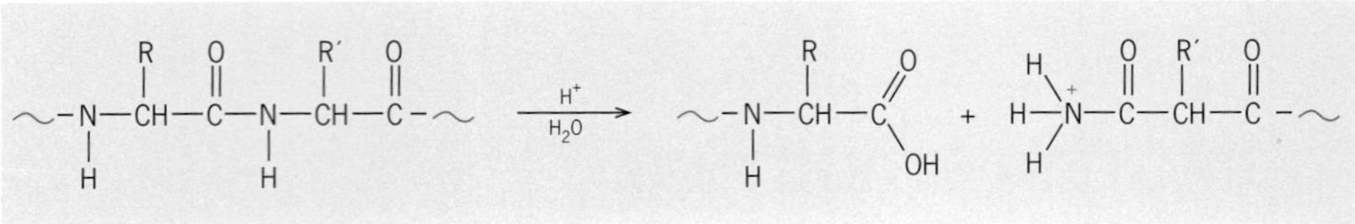

$$\sim\text{NH}-\text{CH(R)}-\text{C(=O)}-\text{NH}-\text{CH(R')}-\text{C(=O)}-\sim \xrightarrow[\text{H}_2\text{O}]{\text{H}^+} \sim\text{NH}-\text{CH(R)}-\text{C(=O)OH} + \text{H}_2\overset{+}{\text{N}}\text{H}-\text{C(=O)}-\text{CH(R')}-\text{C(=O)}-\sim$$

Fig 19.35(a) The acid hydrolysis of one peptide link in a protein chain is often carried out with hot HCl(aq). This gives a carboxylic acid

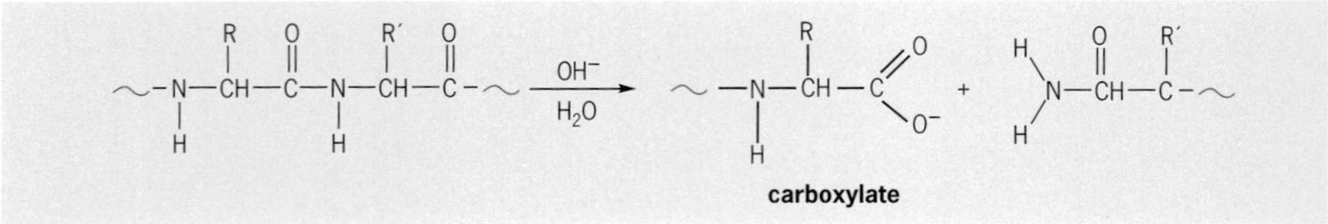

$$\sim\text{NH}-\text{CH(R)}-\text{C(=O)}-\text{NH}-\text{CH(R')}-\text{C(=O)}-\sim \xrightarrow[\text{H}_2\text{O}]{\text{OH}^-} \sim\text{NH}-\text{CH(R)}-\text{C(=O)O}^- + \text{H}_2\text{N}-\text{CH}-\text{C(=O)}-\sim$$

carboxylate

Fig 19.35(b) The alkaline hydrolysis of a peptide link using hot NaoH(aq). This forms a carboxylate

If different hydrolysing agents are used, the protein chain is split in different places. Some of the amino acid sequences overlap, which enables the overall sequence of amino acids to be worked out.

Primary structure of insulin

Insulin was the first protein to have its amino acid sequence (primary structure) worked out. This was done by the English biochemist Frederick Sanger, who began in 1944 and spent ten years on the task. One of the techniques he used was to completely hydrolyse the insulin into its component amino acids by heating a mixture of 6 mol dm^{-3} hydrochloric acid with insulin in a sealed tube for 24 hours.

Sanger then separated the amino acids, using paper chromatography (Fig 19.36). Instead of using just one solvent to separate the amino acids, he used two, one after the other, in a technique called two-way chromatography. A chromatogram is made using the first solvent and is left to dry. Then it is turned on its side and the second solvent is applied.

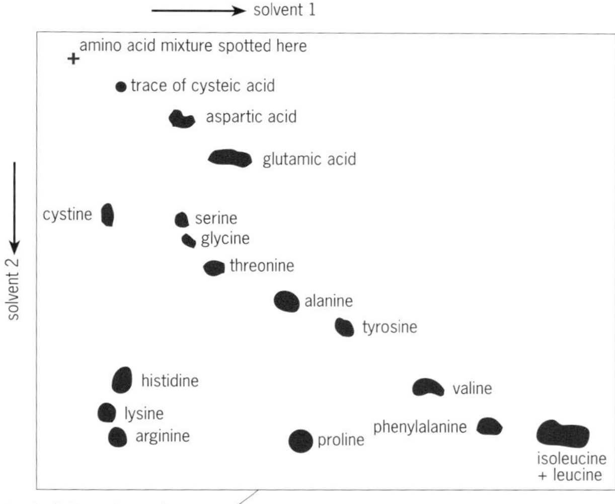

Fig 19.36 Insulin contains 17 amino acids. A spot of the amino acid mixture obtained by hydrolysing insulin was placed at the cross. The first solvent was allowed to run almost to the top of the paper, causing some of the amino acids to separate out. To complete the separation, after drying, the chromatography paper was turned through 90° and a second solvent was run up the paper. The amino acids are made visible as purple spots using ninhydrin solution

For his painstaking research, Sanger received the Nobel Prize for Chemistry in 1958. In 1980, he was awarded a second Nobel Prize jointly with two American scientists for work on DNA, and so he became only the third person in history to win two Nobel prizes in scientific disciplines.

SECONDARY STRUCTURE OF PROTEINS AND HYDROGEN BONDING

The shape of a protein molecule is what makes it able to perform its enzyme function.

Linus Pauling, the eminent American scientist, first turned his attention to protein structure in the 1930s (see also page 85). After 15 years, he and another American, Robert Corey, proposed that the primary structure of proteins could have one of two orderly arrangements, both held together by hydrogen bonds. The first arrangement is a regular coiling of part of the polypeptide chain, called an **α-helix**. The second arrangement is a folding of the polypeptide chain to make sheets, called a **β-pleated sheet**. These are known as the **secondary**

? QUESTION 11

11 Work out the sequence of this simple peptide, which contains five amino acid residues. It is an enkephalin, similar to the one on page 379. The chain has been partially hydrolysed by two different reagents:
Reagent A produces Gly–Phe–Met and Tyr–Gly
Reagent B produces Tyr–Gly–Gly and Phe–Met
(Hint: remember the convention that the amino acid with a free NH_2 group is always placed on the left of the molecule, while the amino acid with the COOH group is at the right end.)

For more information on two-way chromatography see page 232.

REMEMBER THIS

Ninhydrin reacts with amino acids to give a characteristic purple colour. This is how amino acids that have been separated by chromatography can be made to show up (see Fig 19.36).

REMEMBER THIS

The hydrogen bond is a weak electrostatic attraction between an H atom bonded to a highly electronegative atom (N, O or F), which gives it a partial positive charge, and a lone pair of electrons on a neighbouring highly electronegative atom (N, O or F).

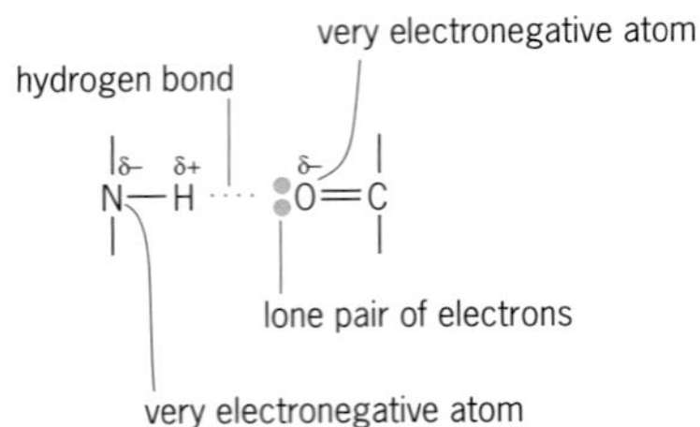

Fig 19.37 Hydrogen bonding

For more information on hydrogen bonds, see page 129.

structures of a protein. Both were subsequently discovered through X-ray crystallography.

The hydrogen bonding that holds a secondary structure together is between the N–H of one peptide link and the C=O of another. Since this is within the molecule, it is known as **intramolecular** hydrogen bonding. The α-helix is a right-handed spiral (Fig 19.38) and hydrogen bonds form between every fourth amino acid residue. It is found extensively in wool fibres and allows wool to stretch. When it is pulled, the α-helix elongates, which breaks the hydrogen bonds. When the α-helix is released, the hydrogen bonds re-form as the α-helix returns to its usual shape.

The β-pleated sheet occurs when the amino-acid backbone of the primary structure folds to give sections of parallel chains of amino acids, which again form hydrogen bonds through their peptide links (Fig 19.39). Although the β-pleated sheet is fairly flexible, it cannot be stretched, as the chains of amino acids are already extended. Silk, for example, is composed of the protein silk fibroin. This is almost entirely a β-pleated sheet, which is the reason why silk cannot be stretched like wool.

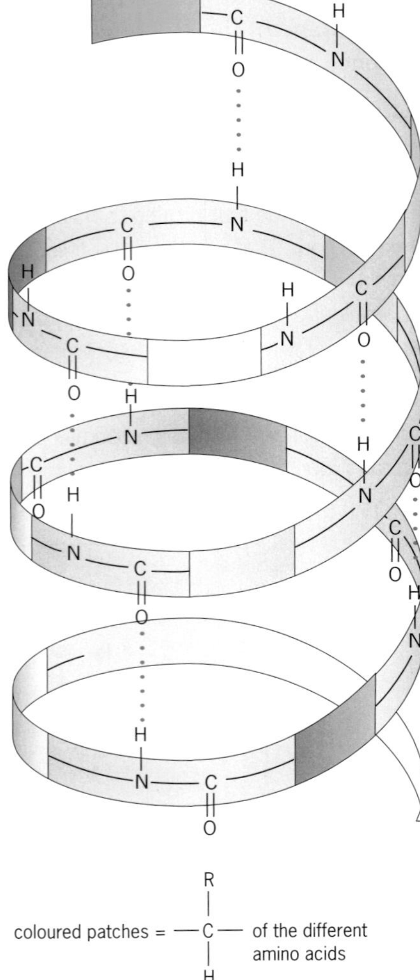

Fig 19.38 The shape of the α-helix is maintained by hydrogen bonds

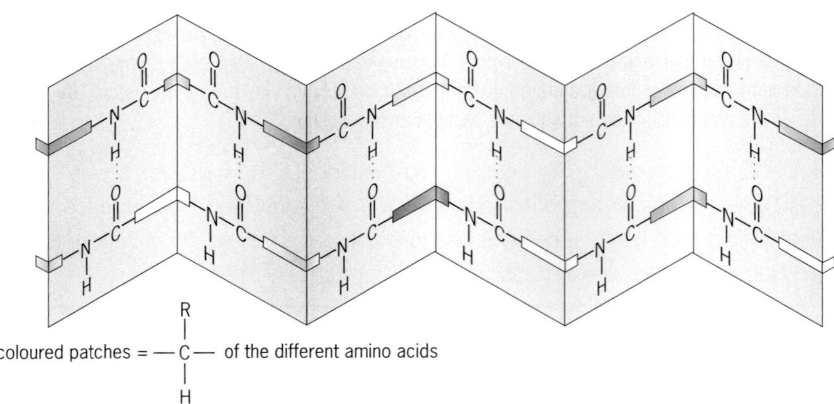

Fig 19.39 The β-pleated sheet results from hydrogen bonding between parallel chains of amino acids

STRETCH AND CHALLENGE

Tertiary structure of proteins

The overall three-dimensional shape of a protein is the result of yet more folding and bending to give what looks like a random structure, but most definitely is not. This is the **tertiary structure** and is specific to a particular polypeptide chain. It is this shape that determines how each protein will function.

Proteins can be divided into two sorts: **globular proteins** and **fibrous proteins**.

Globular proteins, as their name implies, are roughly spherical (e.g. myoglobin, Fig 19.40). Our enzymes and protein hormones are globular proteins, and their polypeptide chains are folded extensively to give a very compact structure.

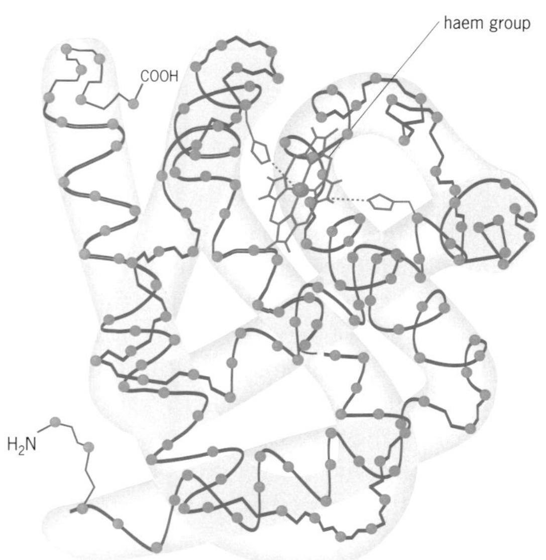

Fig 19.40 The tertiary structure of the globular protein myoglobin. Note that sections of the secondary α-helix structure have been folded to give the overall three-dimensional shape. In the middle of the protein is a haem group that contains an iron atom. Myoglobin is found in muscles and is responsible for binding oxygen and releasing it as needed. It was the first protein for which the three-dimensional structure was worked out

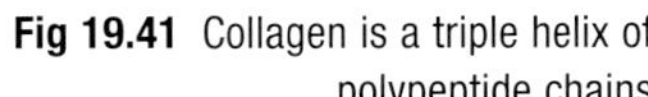

Fig 19.41 Collagen is a triple helix of polypeptide chains

These proteins are usually soluble because they are arranged in a way that leaves the hydrophilic R-groups of individual amino acid residues on the outside of their structures.

Fibrous proteins have their polypeptide chains arranged in bundles to form fibres. They have a structural function. Collagen is a fibrous protein in tendons and muscles that consists of intertwining polypeptide chains (Fig 19.41). The α-keratin of hair, wool and claws is a fibrous protein in which the α-helix is twisted so that several strands can intertwine. The silk fibroin protein is also fibrous, only this time β-pleated sheets become bound together.

The tertiary structure of a protein is held together by four different types of interaction between the various R side chains of the different amino acids. These are shown in Fig 19.42.

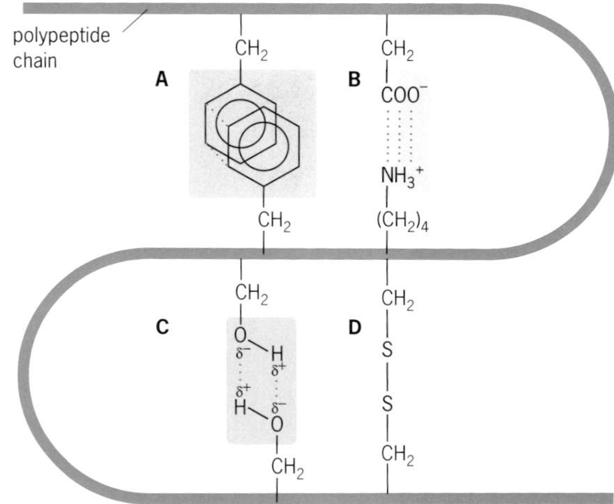

Fig 19.42 The four types of interaction that hold the tertiary structure of a protein together:
A Van der Waals forces (induced dipole–induced dipole bonding) exist when non-polar R groups come close together. They are usually found on the inside of globular proteins where, because they are hydrophobic, they do not interfere with solubility
B Ionic linkages occur between COO^- and NH_3^+
C Hydrogen bonding occurs between polar R groups, such as those that contain OH and NH_2
D Disulfide bridges result from two cysteine residues coming close together and the two S–H groups oxidizing to S–S

? QUESTION 12

12 Hydrophilic means water-loving, whereas hydrophobic means water-hating. Hydrophilic groups can interact with water through, for example, in hydrogen bonding.

a) Look at Fig 19.4 and work out which amino acids have R groups that could form hydrogen bonds with water.

b) Non-polar R groups tend to be hydrophobic. Which amino acids in Fig 19.4 have hydrophobic R groups?

SCIENCE IN CONTEXT

It makes your hair curl!

Our hair is made up of strands of the protein α-keratin. Its shape is maintained by hydrogen bonds, ionic linkages and disulfide bridges between the keratin strands. The very act of washing our hair disrupts these forces because water molecules can get in between the strands and affect all three interactions – in particular, the hydrogen bonds.

As hair dries, the water molecules leave the keratin and the original forces reassert themselves. So, if we do not immediately comb and set our hair in the shape we want, hydrogen bonds may re-form to curl our hair in a way we don't want.

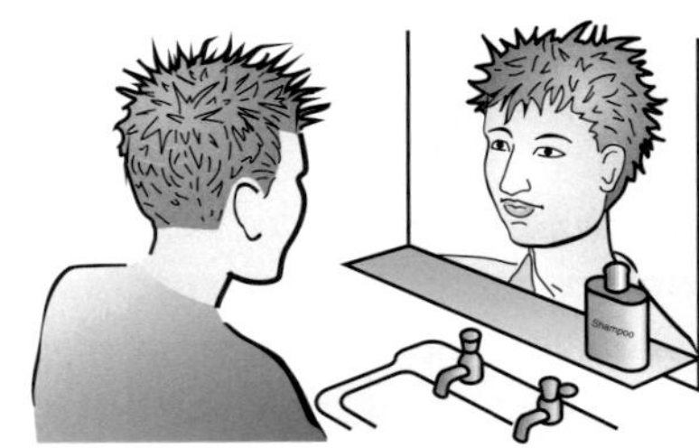

Fig 19.43 The very act of washing hair disrupts some of the weaker forces holding the protein molecules of the hair together

6 ENZYMES AS CATALYSTS

An enzyme is a biomolecule catalyst. Almost all enzymes are proteins, and most are found inside cells. Like all catalysts, enzymes alter the energy of activation of a reaction by providing an alternative route for reactants to interact without becoming permanently involved in the reaction. Enzymes work by having a three-dimensional active site that is part of the tertiary structure of the enzyme. The molecule that an enzyme catalyses is called the **substrate**. The substrate molecule forms intermolecular bonds, such as hydrogen bonds, that bind it to the active site. This reduces the activation energy required for the particular reaction. When the product molecules are formed they have weaker intermolecular forces binding them to the active site, and are able to leave easily to be replaced by more substrate molecules.

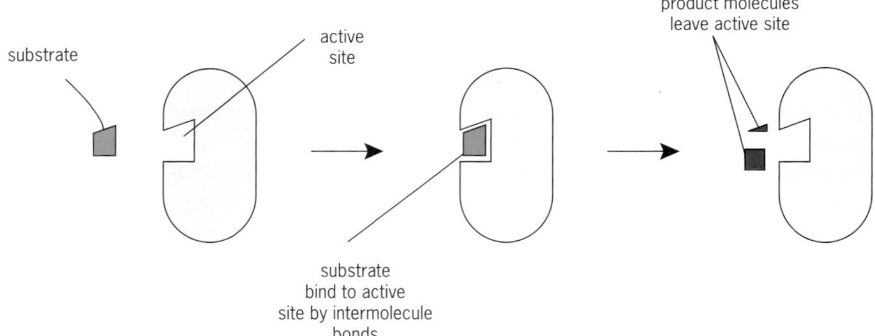

Fig 19.44 The substrate fits into the active site like a key fits a lock

Enzymes differ from other catalysts in their amazingly high activity, being between a million and 20 million times more efficient than inorganic catalysts. Sometimes, each enzyme molecule can catalyse the reaction of thousands of molecules per second. An example is the enzyme catalase, which catalyses the decomposition of hydrogen peroxide:

$$2H_2O_2(l) \xrightarrow{\text{catalase}} 2H_2O(l) + O_2(g)$$

One molecule of catalase catalyses about 50 000 molecules of H_2O_2 per second at 0 °C, whereas manganese(IV) oxide, an inorganic catalyst for this reaction, works very much less efficiently at 0 °C.

Enzymes can also be very specific in the reactions they catalyse. For example, out of the vast number of chemicals in the body, an enzyme may catalyse just one particular reaction of one chemical.

The three-dimensional shape of an enzyme is critical to its catalytic ability. This shape can be denatured (disrupted) by changes of temperature, by altering the pH of the environment of the enzyme or by exposing it to heavy metal ions, such as mercury(II) ions (Hg^{2+}) and lead(II) ions (Pb^{2+}). Enzymes have an optimum temperature and pH at which they work best. Their catalytic activity decreases markedly outside a narrow range, which is why it is dangerous to have a very high temperature for too long during an illness. Most enzymes work efficiently in the range of 25–40 °C, and their shape is irreversibly changed at about 50–60 °C. The optimum pH for most of our body enzymes is about pH 7, but pepsin, a digestive enzyme, has an optimum pH of 2, with its maximum activity in the acid conditions of the stomach.

It is also possible for molecules, which also fit into the active site and form intermolecular bonds to it, to inhibit the enzyme's catalytic ability. As they do not react, the inhibitor molecules stay in the active sites and block them to other substrate molecules. (You can read more about catalysis on pages 166 and 496.)

SCIENCE IN CONTEXT

Immobilising enzymes

At the start of this chapter, you read about the advantages of using enzymes in industry. Unfortunately, if an enzyme is simply mixed with the reactants to be catalysed, it is difficult to separate at the end of the reaction and is usually destroyed.

However, if the enzyme is attached to an inert surface across which the reactants pass, the enzyme functions as a heterogeneous catalyst (see page 497) and can be used several times over.

Various surfaces can hold the enzyme – from polystyrene to glass beads – and immobilised enzymes are used increasingly in food and chemical manufacture. They are also important in medical diagnosis: clinical tests for glucose and cholesterol rely on immobilised enzymes that react with these compounds.

7 DNA AND ITS ROLE IN PROTEIN SYNTHESIS

Deoxyribonucleic acid (DNA) is found in almost all living things. It carries genetic information that can be passed from one generation to the next. But what exactly does DNA do? It carries information to synthesise our body's proteins from our enzymes to our skin and hair.

THE STRUCTURE OF DNA

DNA is a **polymer** made up from monomers called **nucleotides**. The nucleotides have three components: **deoxyribose** (a sugar), **phosphate** and one of four **bases**.

We can see how a strand of DNA is formed in Fig 19.45. Because water molecules are lost when the sugar, phosphate and base react to form DNA this is known as a **condensation polymerisation**.

Fig 19.45 shows a single strand of DNA, but in 1953 two scientists, Francis Crick and James Watson, working at Cambridge University realised that DNA forms a double strand, held together by hydrogen bonding between A and T and between C and G.

You can read more about condensation polymerisation in Chapter 20 page 399.

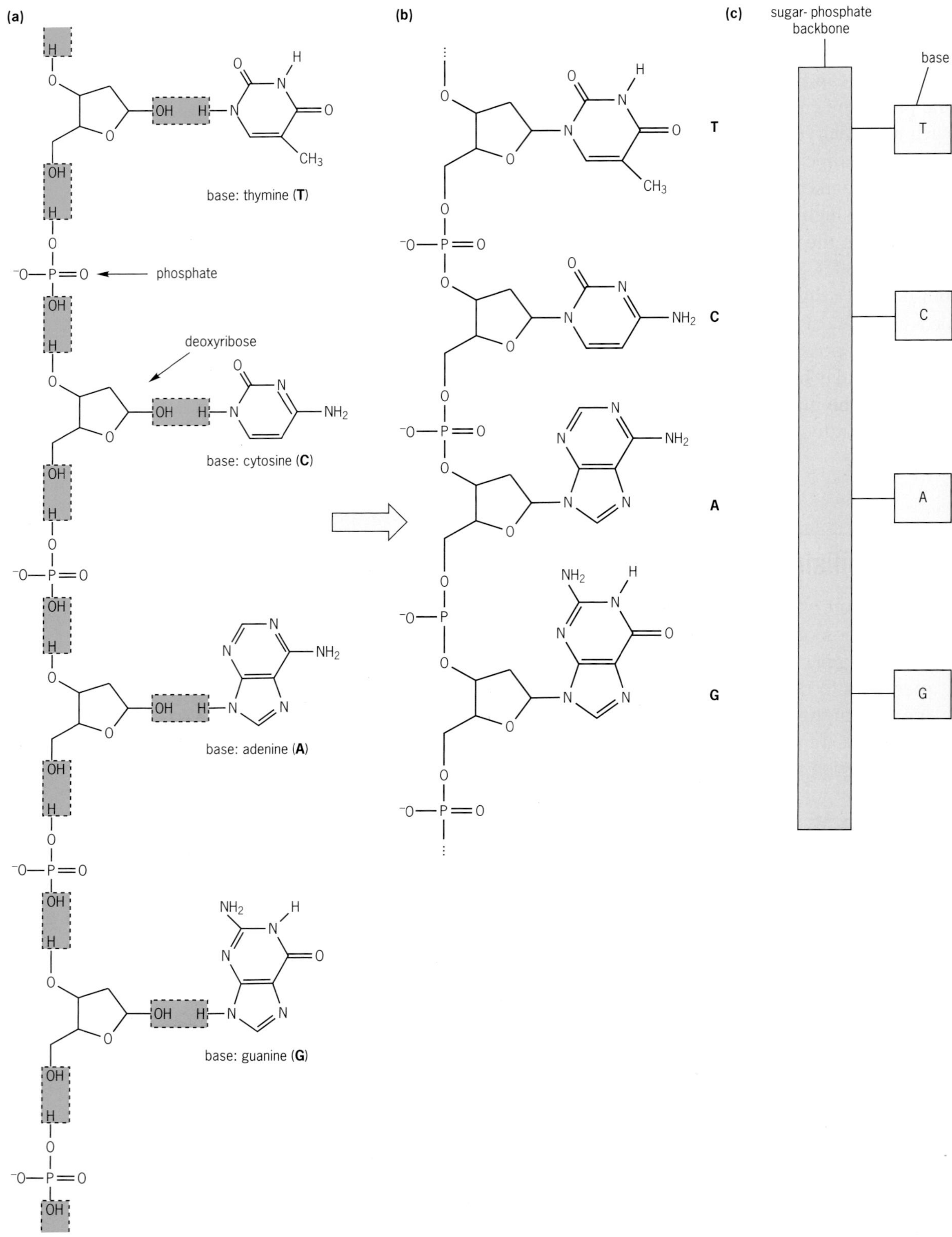

Fig 19.45 The structure of DNA and its bases
(a) Deoxyribose, phosphate and four bases combine together
(b) The skeletal formula of a strand of DNA
(c) A simplified diagram of a strand of DNA showing the four bases attached to the sugar-phosphate backbone

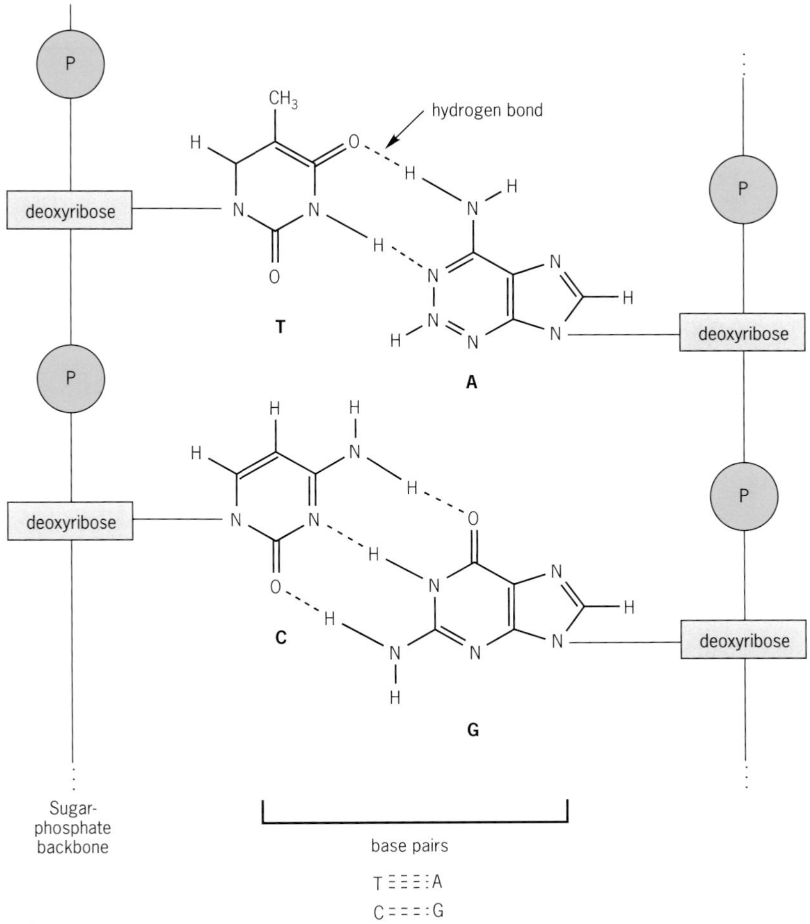

Fig 19.46 The DNA double helix

Modelling DNA

HOW SCIENCE WORKS

A fascination for discovering the structure of DNA started, for James Watson (an American biologist), after he saw an X-ray diffraction photograph of the molecule in 1951, while researching in a Copenhagen laboratory. (You can read more about the technique of X-ray diffraction on page 78.) James Watson realised that if DNA had a diffraction pattern then perhaps the structure of genes could be worked out. He switched his research to the Cavendish Laboratory in Cambridge, where much work was being carried out on determining the structure of large molecules. Here he met Francis Crick, a British physicist. They hit it off immediately and became great friends, sharing a passion for discovering DNA's structure; yet it is interesting to note that neither had a brief to do any research related to DNA, in fact neither actually did any direct research on the DNA molecule. They made models of what they thought DNA's structure might be, based on observations and information from other scientists.

They already knew that:

- DNA was made up of monomers called nucleotides.
- Nucleotides are made up from a sugar, a phosphate and a base.
- There are four different bases in a DNA molecule: adenine, thymine, guanine and cytosine.
- There are always the same number of adenine bases as there are thymine bases.
- There are always the same number of guanine bases as there are cytosine bases.
- The basic structure was probably based on a helix.

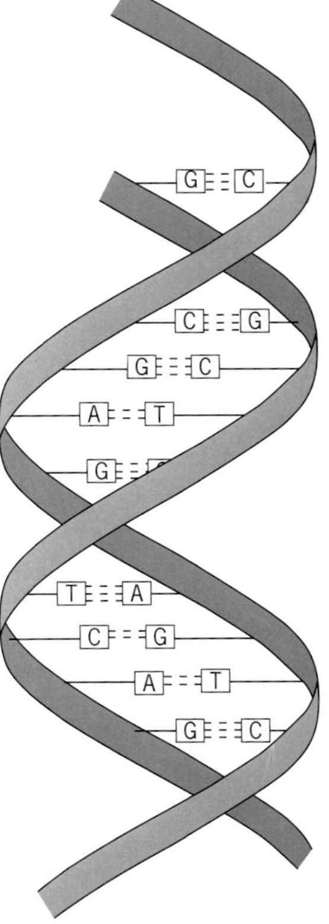

Fig 19.47 A simplified model of DNA showing the hydrogen bonds between the bases, making the double helix possible

Modelling DNA (Cont.)

Fig 19.48 Photos of Crick (right) and Watson (left), at the time of the discovery

Crucial to the discovery of the now accepted model of DNA was the work of Maurice Wilkins (British chemist) and Rosalind Franklin (British chemist) at Kings College, London. They took an experimental approach, using X-ray diffraction images, largely taken by Franklin. In 1951, Watson attended a lecture given by Franklin, where she showed some of her superb X-ray photographs. Franklin had already used her diffraction patterns to work out the basic dimensions of the DNA strands and had realised that the phosphate groups must be on the outside of the strands in what was probably a helical structure. Because Watson took no notes of the lecture his recollection was hazy; however, Crick was friendly with Wilkins, who showed him some of Franklin's photographs. Watson and Crick made a model based on Franklin's results and Watson's recollections of her lecture. But when Franklin saw it she told them that it must be wrong. She pointed out that the phosphate-sugar backbone must be on the outside of any DNA model, because only in that way could DNA interact with water, since the backbone was hydrophilic (water-loving).

Linus Pauling, the great American scientist (see page 85) proposed a triple-helix model for DNA early in 1953 but when Watson saw his research paper, which was never published, he knew it did not fit the data. This stimulated them to go back to making yet another model of DNA. Without Franklin's knowledge, another X-ray photograph was shared with Crick and Watson. It was then that they realised that DNA must be a double helix, which resembled a spiral staircase – the steps being the paired bases. They started to make their ball-and-stick model, which eventually reached two metres tall. They were uncertain as to how the bases paired up in their staircase structure, until a colleague pointed out that A must go with T, and C with G.

They finished their model and published their paper in Nature in April 1953. Because it so readily fitted the facts, the scientific community immediately accepted the structure. Watson and Crick received a Nobel Prize together with Wilkins, in 1962. Rosalind Franklin died in 1958 from cancer. Nobel prizes are awarded only to living recipients so she did not receive one. There are two tragedies here: the first is that Rosalind Franklin's work with X-rays almost certainly caused her cancer; and second, her role in the discovery of DNA, though crucial, was not rewarded with a Nobel Prize. However, her acheivements as a brilliant experimental scientist are now fully recognised.

QUESTION 13

13 The skeletal formula of ribose is

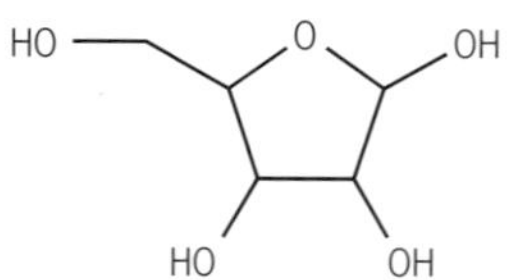

Fig 19.49 Skeletal formula of ribose

a) Draw out the displayed formula for this molecule.

b) What is the key difference between a molecule of uracil and a molecule of thymine?

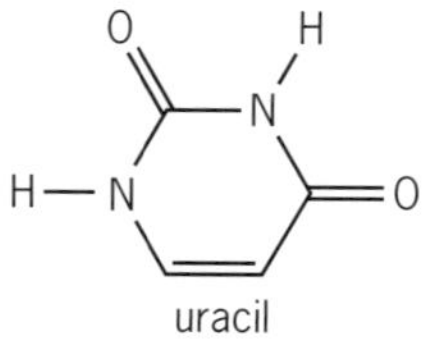

uracil

Fig 19.50 Skeletal formula of uracil

(Hint: to remind yourself about skeletal formulae see Chapter 9 page 181.)

PROTEIN SYNTHESIS AND RNA

The code for synthesising proteins is carried by the sequence of base pairs on the inside of the double helix. Although DNA may carry this genetic coding, DNA itself is not directly involved in protein synthesis. That is the function of another set of molecules called ribonucleic acid, or RNA.

RNA has a structure very similar to DNA, except that the sugar in the sugar-phosphate backbone is ribose (instead of deoxyribose) and thymine is replaced by uracil.

Messenger RNA

The molecule that takes the code for the amino acid sequence of a protein to the site where the amino acids are assembled to make a protein is called **messenger RNA (mRNA)**. Each protein has a different mRNA molecule. These molecules are made by the sequence of bases on the DNA called a gene. The DNA unzips, revealing its bases, and this sequence is transcribed onto the RNA in a process called **transcription**. The base A on a strand of DNA will produce a U on the messenger RNA and T will produce an A.

Codons

Each amino acid in a protein is determined by a sequence of three bases. This is called the **triplet code**. Each triplet is called a **codon**. The codons tell the

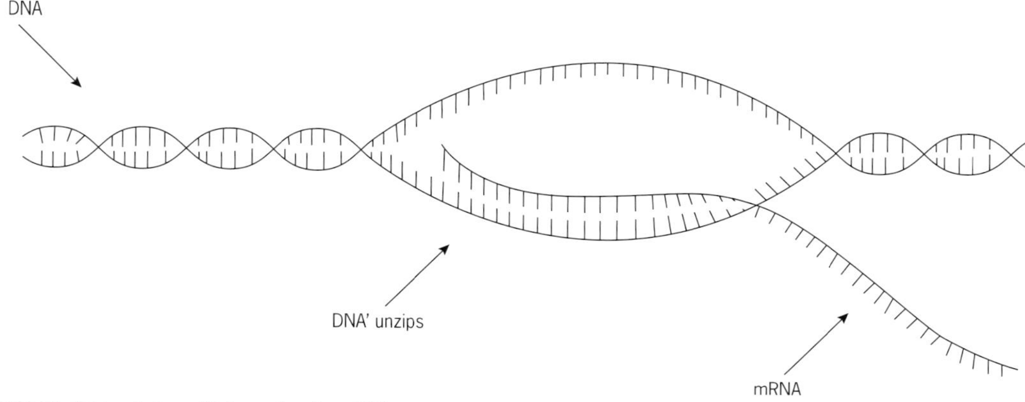

Fig 19.51 Transcription of information to mRNA

cell which amino acids to assemble into a protein chain. Sometimes more than one codon can code for an amino acid. For example the amino acid serine has six possible codons: UCU, UCC, UCA, UCG, AGU and AGC, but the amino acid tryptophan has only one codon: UGG.

Some codons will tell a cell to stop making the protein chain any longer. An example is UAA.

Ribosomes and transfer RNA (tRNA)

The messenger RNA carries the code to make a protein from the DNA in the nucleus of the cell to a structure in the cell called a **ribosome**. The ribosome provides the site where amino acids are assembled into proteins. **Transfer RNA (tRNA)** is the molecule that selects the correct amino acids to make up a protein chain.

Just as the bases on mRNA recognise and pair with bases on DNA, so the same thing happens with tRNA. The three bases that recognise the bases in a codon on messenger RNA are called an **anti-codon.**

This tRNA molecule bonds to serine, the amino acid coded for by AGC in the codon on mRNA. Serine is then assembled into the protein chain in the ribosome. The ribosome travels down the mRNA chain like a bead on a necklace. Each tRNA codes for a particular amino acid. Sometimes more than one tRNA will code for the same amino acid. The tRNA brings the appropriate amino acid to the front of the ribosome, and the protein chain grows out of the back. The process is shown in Fig 19.53.

? QUESTION 14

14 A sequence of bases on a DNA strand is GATC.

a) What will be the sequence of bases transcribed onto the mRNA molecule?

b) What sequence of bases pair with this DNA sequence on the other strand of DNA in the double helix?

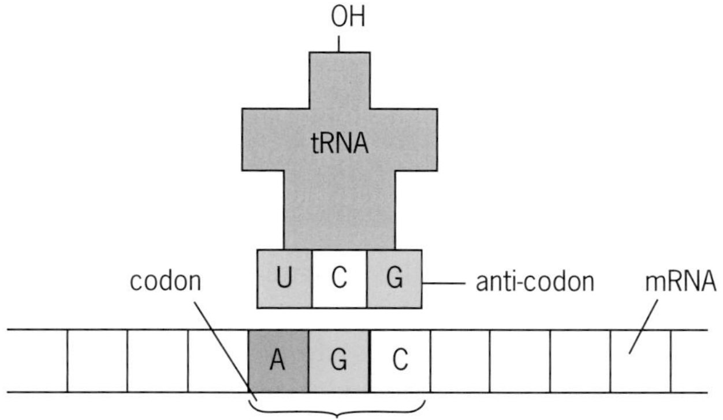

Fig 19.52 tRNA, showing the anti-codon that lines up with the codon

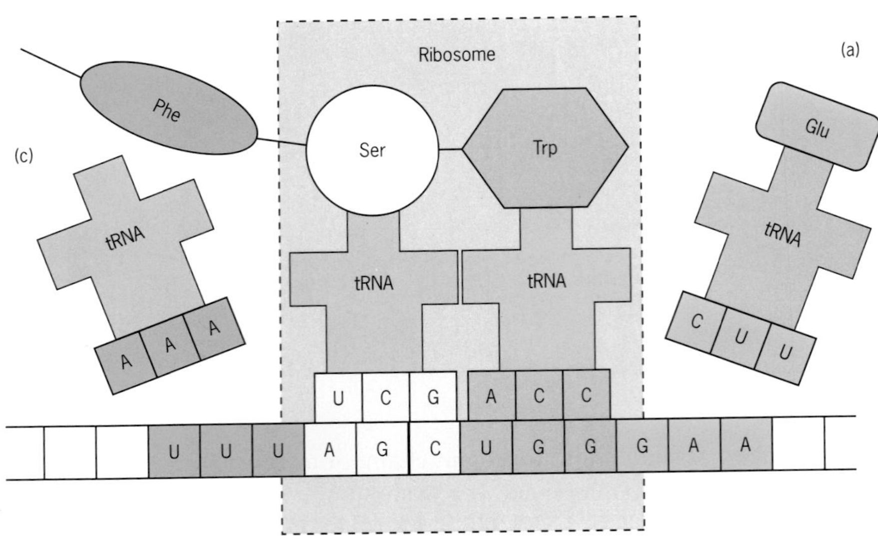

Fig 19.53 Protein synthesis in the ribosome.
At (a) the tRNA arrives at the front of the ribosome carrying its amino acid.
At (b) the amino acids are assembled into the protein chain in the ribosome.
At (c) the tRNA leaves the ribosome and the growing protein chain emerges.

When production of a particular protein is sufficient for a cell's needs the mRNA is destroyed. The permanent record is still stored in the DNA ready to make more mRNA when the need arises. The tRNA remains in the cytoplasm of the cell bringing up amino acids to make the different proteins as required.

GENETIC MUTATIONS

Sometimes the DNA sequence can become slightly muddled. This causes a mutation. An incurable disease caused by a genetic mutation is sickle cell anaemia. It affects millions of people in tropical Africa and people of African descent around the world. A tiny defect in the protein haemoglobin causes it to be shaped differently, misshaping red blood cells. These red blood cells become stuck in narrow blood vessels called capillaries, causing blockages, severe pain, and damage to the kidneys, lungs and liver. The red blood cells also have a shorter life span causing anaemia. The fault is a very simple one; a T replaces an A in the gene for haemoglobin, causing it to have a characteristic sickle shape. Now that the cause of this mutation is known there is the very real possibility of repairing this gene sequence and promising research is underway.

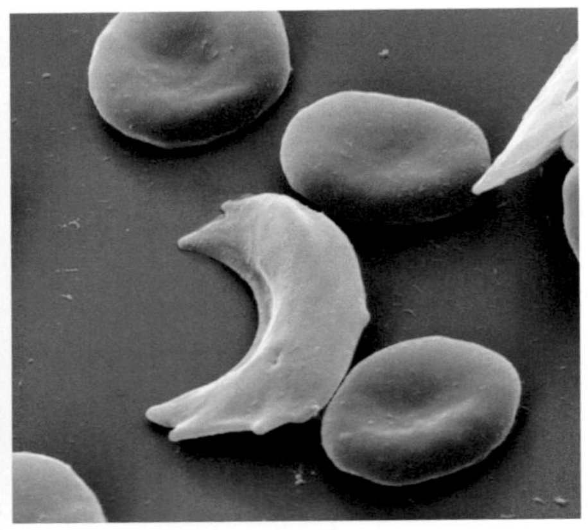

Fig 19.54 Photo of a sickle cell

Sometimes mutations in the DNA structure can lead to beneficial modification to the primary structure of proteins, causing them to have a different structure and perhaps a different function. It may be that the protein that confers colour to a butterfly's wings changes, making it more difficult for the butterfly to be seen by predators so it survives and passes on this mutation to the next generation. Perhaps haemoglobin structure is altered in such a way that allows the molecule to transport more oxygen around the body making an individual better at endurance activities such as running long distances. The primary structure of an enzyme may be altered so that the active site becomes differently shaped, giving the enzyme a different function.

We know that in about 5-14 per cent of the European population there is a base pair missing on a particular gene. This has an effect on the shape of the protein coating of particular cells that are susceptible to entry by the HIV virus. Through this change of shape a receptor site that the HIV virus uses to gain entry is missing, making the individual more resistant to the onset of AIDS.

SUMMARY

After studying this chapter, you should know the following:

- Amino acids have two functional groups: COOH and NH_2.
- Naturally occurring α-amino acids have the general formula in Fig 19.55.

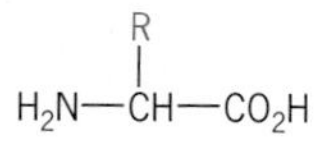

$$H_2N\text{—}\underset{}{\overset{\displaystyle R \atop |}{CH}}\text{—}CO_2H$$

Fig 19.55 General formula for amino acids

- Optical isomerism occurs when two molecules, called enantiomers, exist as non-superimposable mirror images of each other. These molecules are said to be chiral.
- A carbon atom bonded to four different groups is called a chiral carbon. Such carbon atoms give rise to optical isomers.
- A pair of enantiomers rotates plane-polarised light in equal but opposite directions.
- Pharmaceutical companies often require drugs to be made up of one optical isomer (enantiomer) to avoid unpleasant side-effects. Chemicals synthesised in the laboratory usually contain a mixture of optical isomers, whereas chemicals synthesised using enzymes or bacteria often contain a single optical isomer.
- A zwitterion is formed when a COOH group donates a proton to an NH_2 group on the same molecule, to give an ion that is both positively and negatively charged.
- CO–NH is called a peptide group or peptide link when it joins two amino acids together.
- Polypeptides are long-chain molecules that contain many amino acids joined by peptide links. Proteins are very large molecules that have M_r values of from 5000 to several million.
- The primary structure of a protein (either an α-helix or a β-pleated sheet) is the sequence of its constituent amino acids. It forms the protein backbone.
- The secondary structure of a protein (either an α-helix or a β-pleated sheet) is held together by hydrogen bonding between C=O and N–H in different peptide groups.
- The tertiary structure is the overall three-dimensional shape of a polypeptide in a protein. It is the result of the secondary structure folding and bending.
- Enzymes are biomolecule catalysts and nearly all are proteins. They have a very high activity and usually a high specificity. They are also very sensitive to changes in pH and temperature.
- Enzymes have a three-dimensional active site to which substrate molecules can bind. This lowers the activation energy for a particular reaction. Some molecules can act as enzyme inhibitors occupying the enzyme's active site.
- DNA is a condensation polymer made up of monomers called nucleotides. These nucleotides are themselves made up of deoxyribose, phosphate and one of four bases: A, T, C or G.
- DNA exists as a double helix and other models were put forward before the currently accepted version produced by Crick and Watson. Bases pair in the double helix A to T and C to G. These pairings are held together by hydrogen bonds.
- Messenger RNA (mRNA) lines up with bases on DNA to transcribe the information required for protein synthesis. A sequence of three bases is required to code for a particular amino acid. This triplet code is called a codon.
- Transfer RNA (tRNA) contains three bases that line up with the codon; these three bases are called an anti-codon. Each tRNA is also bonded to a specific amino acid.
- The ribosome travels down mRNA and is the site of protein synthsis. tRNA molecules bring their amino acids to be bonded into a growing protein chain.
- Changes to the base pairs in DNA, called mutations, can give rise to genetic diseases such as sickle cell anaemia. However, modification to base pairs may alter the primary structure of proteins in a beneficial way, altering structure and/or function.

Practice questions and a How Science Works assignment for this chapter are available at www.collinseducation.co.uk/CAS

20 Polymers

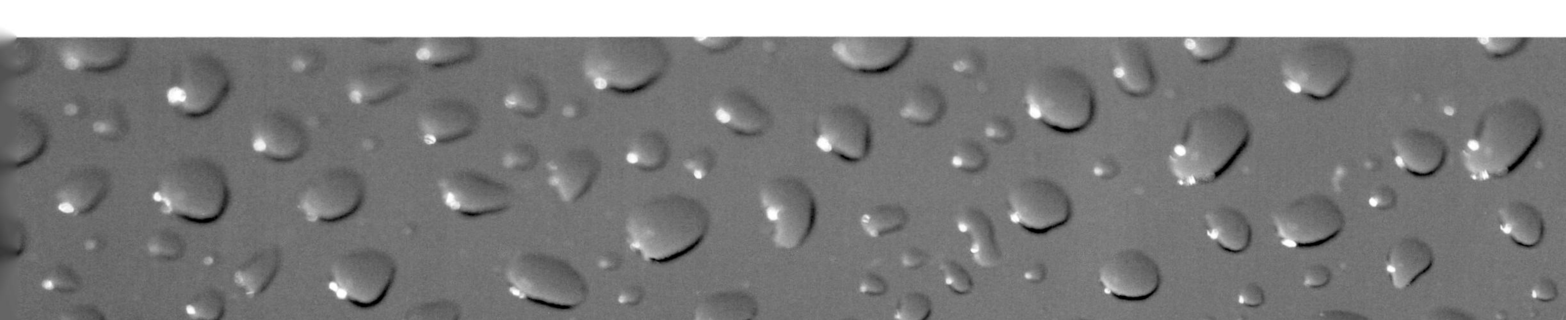

20 POLYMERS

Uses for Biopol Biopol bottles at stages of increasing biodegradation

Polymers from bacteria

Most polymers that have revolutionised our everyday living are manufactured from chemicals that come from non-renewable crude oil. As the reserves of this precious resource are used up, the hunt is on for other sources of polymers, and growing crops that provide them is one promising option. We are all familiar with the natural polymers starch, cellulose protein and DNA; certain bacteria produce other polymers that are the subject of intense research throughout the world.

Biopol, a polyester, was the world's first bacterial polymer to be produced commercially. The polymer makes up 80 per cent of the dry weight of the bacterium *Alcaligenes eutrophus*, which uses it as an energy store in the same way that we use fat. The first articles made of Biopol, bottles for hair-care products, appeared in 1990 and it found uses in coating paper bags and in producing disposable cutlery. However, bacterial polymers are much more costly to produce than synthetic polymers from crude oil. This meant that it did not take off in the way the manufacturer had hoped and production virtually ceased. But another way to make this plastic was discovered.

We waste millions of tonnes of food each year from canteens and restaurants. Most of it ends up in landfill sites and produces the potent greenhouse gas, methane. Researchers in Hawaii have used bacteria to convert this waste food into Biopol, also known as BHA, and other plastics, depending on the conditions used. For every 100 kg of waste food slurry, 25 kg of polymer can be produced. This has created renewed interest in polymers made by bacteria.

The outstanding advantage of bacterial polymers is that they are totally biodegradable. In a world in which the disposal of synthetic polymers is reaching crisis levels, as landfill sites become scarce and the incineration of plastic waste adds to atmospheric pollution, this is a very important virtue. Even if it is incinerated the only combustion products are carbon dioxide and water, and it releases more energy in the process than its synthetic counterparts.

1 WHAT IS A POLYMER?

For more information on proteins and DNA, see pages 378–390.

More chemists work with polymers than with any other type of material. Polymers enter every aspect of our lives. The DNA that carries the genetic code is a natural polymer, as are proteins, starch and cellulose. Chemists started to make synthetic polymers in the early 1900s and now we are dependent on them. They are used in clothing, packaging, furniture, adhesives, inks, coatings and electrical equipment, to mention just a few.

Polymers are very large molecules with very high molecular masses. It took chemists some time to realise that they were dealing with such large molecules. The German chemist Hermann Staudinger, working in the early 1900s, was the first to realise that compounds such as rubber were not just collections of tiny molecules held together by intermolecular forces, but were made up of huge molecules that contained many thousands of atoms covalently bonded together. It was thought at the time that a molecular mass of 5000 was the limit for any molecule, so it took some time for Staudinger's model of a polymer to become accepted. In 1953, Hermann Staudinger (1881–1963) was awarded the Nobel Prize in Chemistry for his pioneering work with polymers.

The word polymer is derived from two Greek words: *polys*, which means many, and *meros*, which means a part. Polymers, then, are large molecules made up of many small molecules called **monomers**. The process of forming polymers from monomers is called **polymerisation**. **Plastics** – derived from the Greek word *plastikos*, which means shaped – is the general name for materials that, when heated under pressure, can be shaped or moulded or extruded (extruded means forced through a hole, for example to make rods or tubes). Not all polymer materials possess this property, so to avoid confusion the word 'plastics' is not used in this chapter until environmental issues are considered on page 406.

2 ETHENE AND ADDITION POLYMERISATION

Poly(ethene) – or polythene as it is popularly known – was discovered by accident in 1933. Two British chemists working at ICI, Eric Fawcett and Reginald Gibson, were trying to produce a ketone using benzaldehyde and ethene under very high pressure (Fig 20.1). They decided to leave the reaction mixture at high pressure over a weekend, even adding more ethene when the reaction vessel sprang a leak.

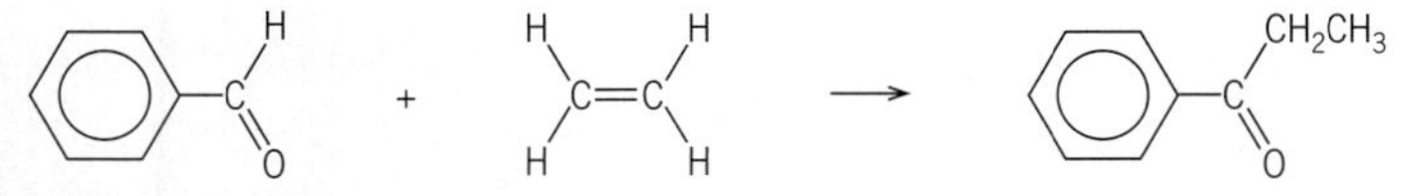

Fig 20.1 The reaction Fawcett and Gibson were expecting

By Monday morning, the hoped-for ketone had not been produced. Instead, there was a minute amount of a white waxy solid. They tried to repeat the experiment, but the reaction mixture exploded. The unpredictability of the reaction meant that making this unusual substance was very hazardous, so they stopped.

However, it was eventually realised that this was a polymer. Interest grew, and in 1935 Fawcett and Gibson made another attempt to produce it, this time with reactor vessels that could withstand very high pressures. By December, they had made 8 grams. They continued their experiments to find the optimum conditions. When they added cold ethene at a controlled rate, the reaction stopped going out of control. Soon they had enough solid polymer to establish that it could be melted and moulded.

Another British chemist, Michael Perrin, who took charge of the project in 1935, realised that oxygen was essential to produce this white waxy solid, polythene. (It is probable that oxygen had entered the original reaction vessel when it sprang its leak.) Perrin also showed that polythene was produced without the benzaldehyde of the 1933 experiment.

? QUESTION 1

1 If the reaction between benzaldehyde and ethene had worked, it would have been an addition reaction. Explain what is meant by addition reaction. (Hint: to check your answer, see page 282.)

SCIENCE IN CONTEXT

Polythene and the Second World War

The claim is often made that penicillin, the first antibiotic, was one of the chemicals that won the Second World War because it prevented terrible suffering and needless death from wound infections for thousands of soldiers. Polythene, too, had a valued role in the war. It was commercially produced in 1939, three years before penicillin. Unlike other insulators, it was impervious to rainwater and sea water, and one of its first uses was to insulate the electrical wiring of aircraft radar sets – radar being crucial to Britain's survival in the war's early years.

QUESTION 2

2 **a)** Draw the displayed formula of part of a poly(ethene) chain that shows three ethene units.

b) Explain why poly(ethene) may be classed as an alkane. (Hint: see page 192 to remind yourself about alkanes.)

There is more about addition reactions and addition polymerisations on pages 282 and 288.

Poly(ethene) is called an **addition polymer** because, when the ethene monomers join together in the reaction, no small molecules are eliminated: the polymer is the only product, as Fig 20.2 shows.

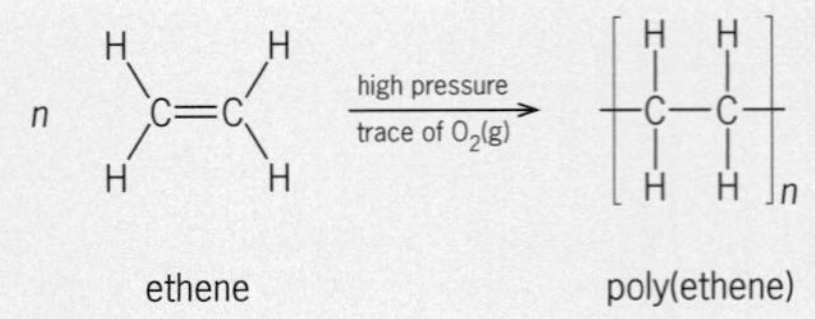

$$n\,H_2C{=}CH_2 \xrightarrow[\text{trace of } O_2(g)]{\text{high pressure}} \left[\!-CH_2-CH_2-\!\right]_n$$

ethene → poly(ethene)

Fig 20.2 Polymerising ethene

In the equation of Fig 20.2, n represents a very large number. Poly(ethene) chains can be made up of thousands of ethene monomers, hence its systematic name. In the formula for the poly(ethene), the ethene unit is enclosed in brackets which cut the covalent bonds to show that we are representing the **repeating unit** of a very long chain.

Types of poly(ethene) and poly(propene)

POLY(ETHENE)

Most people refer to poly(ethene) by its ICI brand name of polythene. It is the world's leading polymer in terms of annual production, with more than 50 million tonnes of worldwide production and 650 thousand tonnes in the UK alone. However there is more than one type of poly(ethene) that makes up this annual production.

LOW-DENSITY POLY(ETHENE)

The process invented at ICI to manufacture polythene produces branched chains. This means that the polymer chains do not all lie in parallel, as unbranched molecules could. Instead, they form a tangled mass (Fig 20.3). As the chains cannot lie close together, the density of this polythene is low. It is therefore called **low-density poly(ethene)** or **LDPE**.

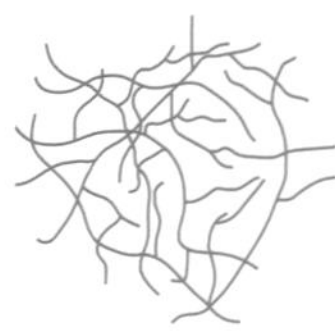

Fig 20.3 Chains of low-density poly(ethene)

The intermolecular forces that hold the poly(ethene) chains together are induced dipole–induced dipole forces (see page 195). As the chains get longer, these weak forces act over a larger surface area. And as they become tangled, the tensile strength increases because when one chain is pulled it drags other chains along with it. However, because the branching keeps the chains further apart, LDPE has a lower tensile strength than polymers with no branched chains. Branching also keeps the melting point fairly low, at about 130 °C.

The process requires very high pressures of 2000 atmospheres, with a trace of oxygen or an organic peroxide.

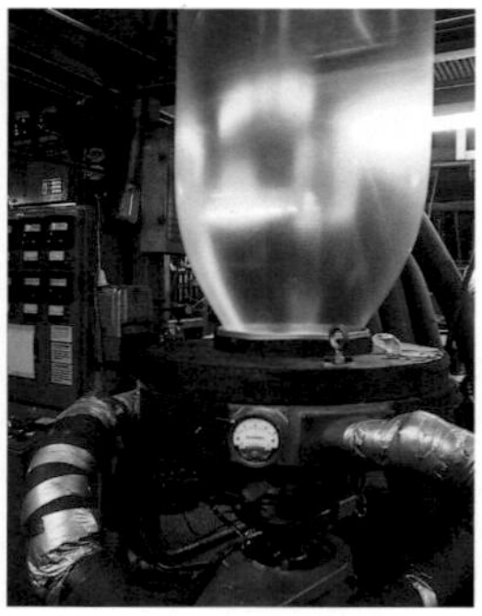

Fig 20.4 Hot low-density poly(ethene) being blown to form a bag

Among the uses of LDPE are electrical insulation, tough transparent film in packaging, dustbin liners and carrier bags (Fig 20.4). However, the high pressure and a temperature of 200 °C are very costly in terms of energy. A new process is producing a new form of poly(ethene) called LLDPE (linear low-density poly(ethene)) with similar properties to LDPE but at a pressure as low as 1 atmosphere and a reduced temperature of 100 °C. This makes the process greener (and safer too, as it avoids dangerously high pressures, with the potential to burst pipes in the plant). You can read more about LLDPE later in this section.

HIGH-DENSITY POLY(ETHENE)

Linear chains of poly(ethene) – chains with no branching – can lie closer together than branched chains. This increases the density. **High-density poly(ethene)**, **HDPE**, was first produced in 1953 in Germany by Karl Ziegler while he was experimenting with an organometallic catalyst. This catalyst consisted of ethyl groups covalently bonded to aluminium atoms (Fig 20.5). Ziegler found that a little titanium(IV) chloride, $TiCl_4$, mixed with triethylaluminium in an alkane solvent caused ethene to polymerise at room temperature and atmospheric pressure. Long linear chains of about 100 000 ethene units were formed. Today the conditions for HDPE production have changed little.

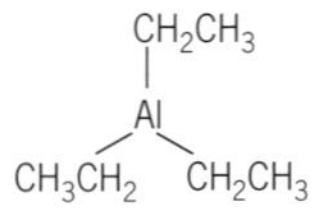

CH_2CH_3 / Al / CH_3CH_2 CH_2CH_3

Fig 20.5 Triethylaluminium is an organometallic catalyst

$$nC_2H_4 \xrightarrow[50\,^\circ C,\ 1.5\ atm]{(C_2H_5)_3Al/TiCl_4} \left[\!-CH_2-CH_2-\!\right]_n$$

Being linear, HDPE chains can form ordered structures by aligning themselves. This leads to large regions of **crystallinity** within HDPE (Fig 20.6), which give it a higher density than LDPE.

Types of poly(ethene) and poly(propene) (Cont.)

The melting point and tensile strength of HDPE are also higher, and HDPE is harder than LDPE. The molecular regions that are not ordered are known as **amorphous** regions.

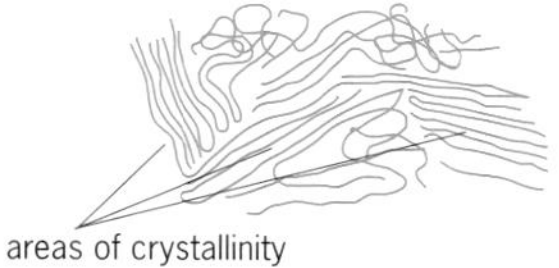

Fig 20.6 The crystalline regions in HDPE result from linear chains that become aligned

The uses of HDPE relate to its different properties. HDPE is more chemically inert and retains its shape at higher temperatures better than LDPE. This makes it ideal for containers for industrial and household chemicals and for milk. Almost half of the HDPE produced is blow-moulded to make bottles and other containers for a range of chemicals from shampoos to bleaches (Fig 20.7). Another quarter is injected into moulds (injection moulding) to make food storage containers, car petrol tanks, buckets and crates. It is also used for pipes that carry drinking water. Medical appliances, such as bedpans, are made from HDPE because its higher melting point allows high-temperature sterilisation without loss of shape.

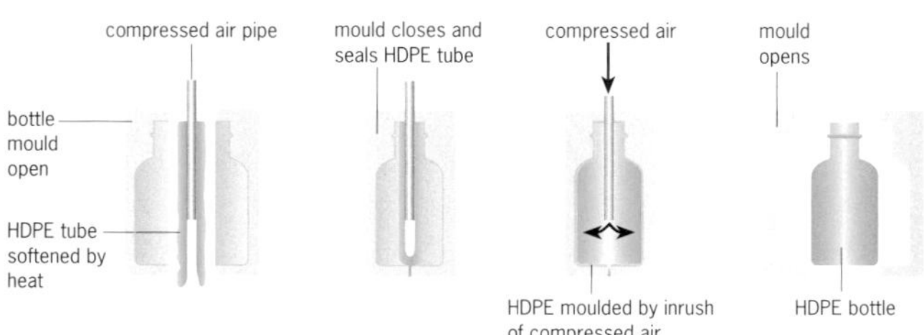

Fig 20.7 Blow-moulding a bottle

Fig 20.8 Two bottles made from poly(ethene). The one on the right was made from LDPE, the other from HDPE

LINEAR LOW-DENSITY POLY(ETHENE) (LLDPE)

The method of production of **LLDPE** is the same as for HDPE, using the same catalyst. This catalyst is now given the name **Ziegler–Natta catalyst** in honour of Ziegler and an Italian chemist, Giulo Natta, who used it to produce a particular type of another addition polymer, poly(propene). The polymer chains are linear, so to produce short branches to keep the chains apart and make a lower-density poly(ethene), a second alkene monomer is mixed with the ethene. This monomer is called a **co-monomer** and the polymer produced is a **co-polymer**. Co-polymerisation is often used to modify the properties of polymers. Two examples of alkene co-monomers are shown in Fig 20.9.

$CH_3{-}CH_2{-}CH{=}CH_2$ but-1-ene

$CH_3{-}(CH_2)_3{-}CH{=}CH_2$ hex-1-ene

Fig 20.9 The co-monomers inserted into the poly(ethene) chain

These co-monomers give short branches at intervals along the polymer chains, giving the polymer a low density because the chains cannot lie close to one another. Ziegler–Natta catalysts are of great economic and environmental significance, because they allow the use of much lower presures temperatures to produce polymers. For their contribution to polymer chemistry Ziegler and Natta jointly received the Nobel Prize for Chemistry in 1963.

POLY(PROPENE): ANOTHER ADDITION POLYMER

The equation for the addition polymerisation of propene is given in Fig 20.10.

$$n\,\mathrm{CH_2{=}CH(CH_3)} \longrightarrow \left[-\mathrm{CH_2}-\mathrm{CH(CH_3)}- \right]_n$$

propene → poly(propene)

Fig 20.10 The polymerisation of propene

If all the methyl groups are oriented in the same direction, the polymer chains can get very close together and be crystalline (Fig 20.11). This form of poly(propene) is called **isotactic** to signify that all the methyl groups point in the same direction. It was produced by Giulio Natta using a Ziegler–Natta catalyst.

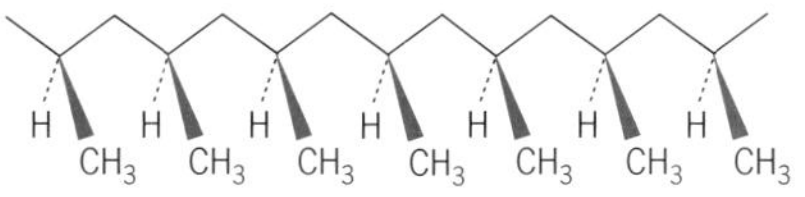

Fig 20.11 Isotactic poly(propene). Notice that the methyl groups are all pointing in the same direction

In atactic poly(propene), which has no ordered orientation of its CH_3 groups, the polymer chains cannot lie very close together. As would be expected, this form of poly(propene) has a lower melting point and is softer than the isotactic form. Isotactic poly(propene) is moulded into objects such as car bumpers and battery cases, and drawn into fibres for carpets and clothing. It is also made in sheet form for packaging. Isotactic poly(propene) is particularly useful for athletics wear because it does not absorb perspiration, but allows it to evaporate, unlike cotton clothing which absorbs and holds moisture. The atactic form is used in weather-proofing materials and sealants.

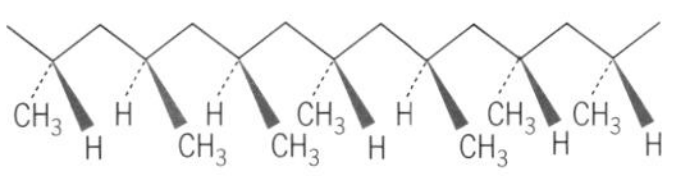

Fig 20.12 Atactic poly(propene): the random orientation of the methyl groups prevents the polymer chains lying close together

QUESTIONS 3–4

3 After reading the How Science Works box on types of poly(ethene), explain why linear polymer chains have higher melting points than branched chains that contain similar numbers of carbon atoms.

4 Draw a section of LLDPE polymer chain produced from two ethene monomers and one but-one monomer.

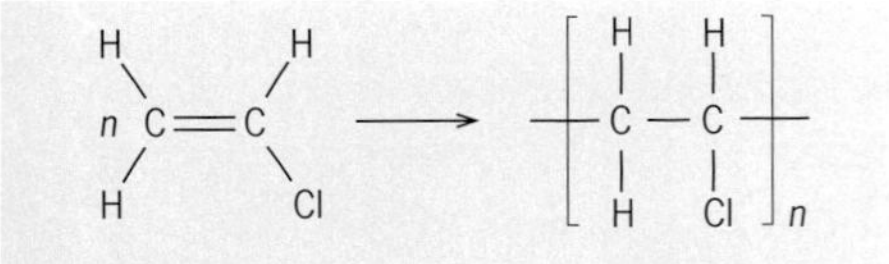

$$n\,\mathrm{CHH{=}CHCl} \longrightarrow \left[\mathrm{-CH_2-CHCl-}\right]_n$$

chloroethene — poly(chloroethene)

Fig 20.13 The polymerisation of chloroethene

QUESTION 5

5 Poly(phenylethene) is the systematic name of polystyrene. Part of the polymer chain is shown in Fig 20.15. Draw the structural formula of the monomer.
(Hint: check your answer by looking at page 288. There is also more information about this important polymer on page 305.)

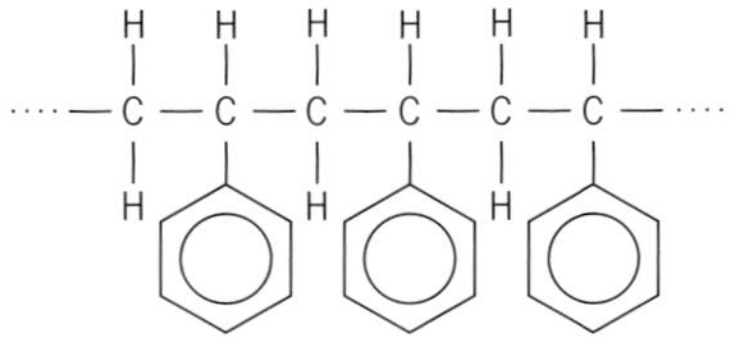

Fig 20.15 Poly(phenylethene)

PVC AND ADDITIVES

Look at Table 14.3 (page 288) in which examples are given of other addition polymers whose monomers are based on the ethene structure and add together in a similar way. PVC or poly(chloroethene) is made from chloroethene monomers, as shown in Fig 20.13.

The non-systematic name for chloroethene is vinyl chloride, hence the name by which the polymer was first known: polyvinyl chloride or PVC. When first made in 1912 in Germany, PVC was hard and brittle. These properties come from the chlorine atom.

Being highly electronegative, the chlorine atom forms a dipole with the carbon atom and the partial negative charge on the chlorine atom attracts a slightly positive hydrogen atom, which also forms a dipole with the carbon atom. This gives rise to permanent dipole–permanent dipole forces which are stronger than induced dipole–induced dipole forces, making PVC stronger than poly(ethene).

Fig 20.14 The non-glass surfaces of these flats are made of unplasticised PVC (uPVC)

The brittleness arises because the larger chlorine atoms tend to catch on each other when the polymer chains are pulled apart. Add to this that PVC often decomposed before it could be moulded, it is small wonder that the German company let its patent lapse in 1926.

However, in the same year it was discovered that certain additives change the properties of PVC, making it more flexible and easier to mould. As a result, PVC is now the world's most versatile polymer and second only to poly(ethene) in the amount produced.

The additives that make PVC more flexible and softer are called plasticisers. Plasticisers are molecules that get in between the chains, and enable them to slide over one another more easily. There is, however, concern that plasticisers in PVC food wrapping film may migrate into fatty foods and be harmful.

The PVC used in door and window frames is called PVC where the 'u' stands for unplasticised, since flexible frames would not usually be a good idea. Another additive prevents the decomposition of PVC by ultraviolet light.

The use of additives to alter its properties means that uPVC is found everywhere. Credit cards are made from it, as are precision engineering items, flooring and footwear. If your house has polymer gutters, down-pipes and plumbing, the material is almost certainly PVC.

STRETCH AND CHALLENGE

Glass transition temperature

Most polymers contain some crystalline regions and some amorphous regions. Below a certain temperature (called the glass transition temperature, T_g) the long polymer chains are fixed in position and cannot move over one another. As the polymer warms up, the chains in the amorphous regions are able to slide over one another and so the polymer softens and becomes more flexible. But the crystalline regions stay rigid and do not start to move until the polymer melts.

Poly(propene) bumpers on cars become brittle in very cold winters when the temperature falls below its T_g of −10 °C. Even a slight knock can cause a bumper to shatter. Another example is poly(propene) food containers kept in a refrigerator. They, too, can become brittle and so crack when someone tries to open them.

HOW SCIENCE WORKS

Teflon: the non-stick polymer

Teflon was discovered by accident in 1938 when Roy Plunkett, a chemist working for the Du Pont company on tetrafluoroethene gas, could not get any of the gas from one of the steel storage cylinders he wanted to use. Instead of jumping to the conclusion that the gas must have escaped, he weighed the cylinder and found that it weighed the same as if it were full. He sawed it open and found a white powder, which turned out to be poly(tetrafluoroethene) or PTFE – better known as Teflon. Ten years later it was in commercial production.

Teflon has a virtually friction-free surface, is very resistant to heat and chemicals, and is an excellent electrical insulator. Apart from its use as a coating on non-stick frying pans (Fig 20.16), Teflon is used to coat bearings (even those that support loads as heavy as bridges) and in human joint replacements. Unlike other polymers, its properties remain constant over a wide temperature range (–70 to 350 °C).

Fig 20.16 Teflon makes this frying pan non-stick

QUESTION 6

6 The displayed formula of tetrafluoroethene is shown in Fig 20.17. Write an equation for the production of poly(tetrafluoroethene).

F F
C=C
F F

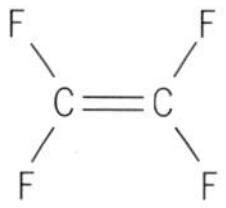

Fig 20.17 Tetrafluoroethene

DISSOLVING POLYMERS

The addition polymers we have considered so far do not dissolve in water, but if groups along the polymer chain can hydrogen bond to water then the polymer may be soluble. Poly(ethenol) is one such example. It is used to make dissolving laundry bags, particularly in hospitals, where soiled laundry can go straight into the wash without having to be handled. It is also used in liquid detergent capsules that can be put directly into washing machines and dishwashers. Poly(ethanamide) can also hydrogen bond with water. One of the many uses of this polymer is in soft contact lenses, where its ability to absorb water makes the lens soft.

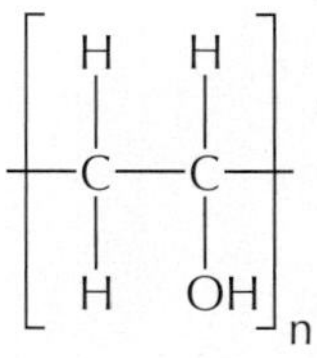

H H
—C—C—
H OH
n

Fig 20.18 Poly(ethenol)

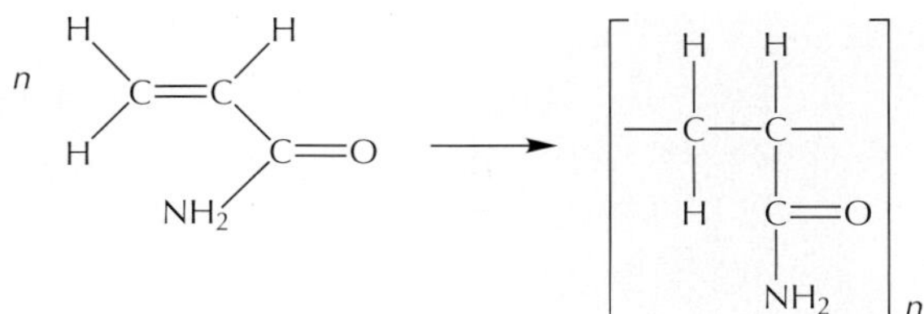

H H
—C—C—
H C=O
NH2
n

Fig 20.19 The polymerisation of poly(ethanamide)

QUESTION 7

7 **a)** Draw the displayed formula structure of the monomer of poly(ethenol).

b) The number of OH groups along the chain can be varied to change the solubility of the polymer. If there are a very large number of OH groups in the polymer chain, then poly(ethenol) is insoluble in water. Explain why this is so. (Hint: think about hydrogen bonds between the polymer chains.)

c) If there are too few OH groups, then the polymer also becomes insoluble. Explain why.

To remind yourself about hydrogen bonding see page 129 of Chapter 6 and page 261 of Chapter 13.

3 CONDENSATION POLYMERISATION

In 1928 a brilliant young American chemist, Wallace Carothers, was invited to join the Du Pont company to head a team researching into polymers. Here he made an immense contribution to our understanding of polymer science, which was probably his most valuable work. However, this is not what posterity remembers him for, because in 1935 he produced the first nylon (see page 401) – a wholly synthetic fibre that mimics the protein, silk.

Within three years of Carothers joining Du Pont, his group produced the first commercial synthetic rubber, Neoprene (*neo* means new), an addition polymer. Then they switched their attention to **condensation polymerisation**. (When monomers form condensation polymers, small molecules, such as water, are eliminated.) Carothers' group investigated two different types of polymer: **polyesters** and **polyamides**. Their first success was the production of a polyester fibre – the world's first wholly synthetic fibre.

Before reading about polyesters you will find it helpful to turn to pages 352–356 and revise how esters are made.

POLYESTERS

Carothers realised that for condensation polymers to be formed, the monomers that make them up need two reactive ends. So monomer A (Fig 20.21) could contain two COOH groups, while monomer B could contain two OH groups, to give polyesters with monomer units joined by ester functional groups.

Fig 20.20 Forming ester links between monomers in a polyester

H—O—C(=O)—A—C(=O)—O—H H—O—B—O—H H—O—C(=O)—A—C(=O)—O—H H—O—B—O—H

↓

····—C(=O)—A—C(=O)—O—B—O—C(=O)—A—C(=O)—O—B—O—····

H_2O H_2O H_2O

QUESTION 8

8 Write an equation for the formation of ethyl ethanoate from ethanoic acid and ethanol.

REMEMBER THIS

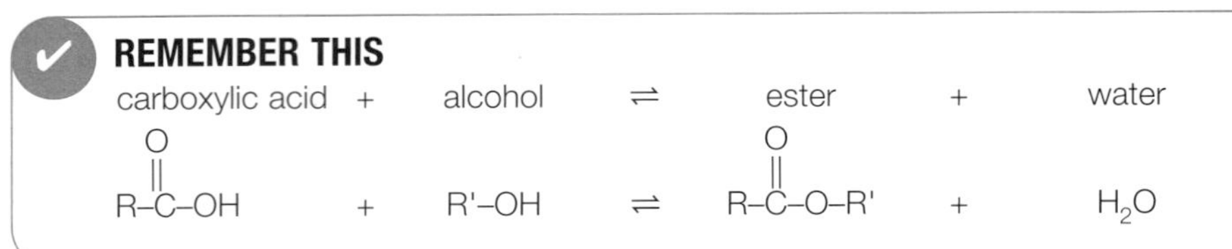

Fig 20.21 Julian Hill demonstrates the birth of the first completely synthetic fibre

The polyesters that the Du Pont team produced were thought to be of theoretical interest, but of little practical use. The water eliminated in the condensation reaction appeared to prevent the formation of very long chains.

This problem was solved by Carothers, who invented a 'molecular still' that evaporated the water molecules as they were produced. This enabled very long polymer chains to be produced with M_r values of about 10 000. However, the polyesters made of these were just sticky masses when hot, and tough, opaque solids when cold. Then, in a happy accident, one of the team, Julian Hill, pulled a stirring rod out of a hot, sticky ball of polyester. The result was a long, thin fibre of the polymer that, when it cooled, could be stretched considerably and was very strong.

X-ray analysis revealed what had occurred. Pulling the polymer into a long filament had aligned the polymer chains, which increased the tensile strength of the material (Fig 20.22). The pulling process is known as cold drawing.

REMEMBER THIS

The M_r of polymers is an average value, since not all the polymer chains have the same M_r.

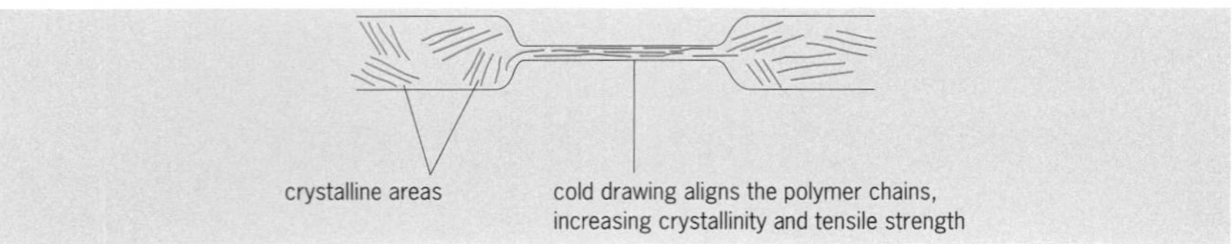

Fig 20.22 Cold drawing of a polymer aligns its chains in the fibre produced

REMEMBER THIS

Examples of happy accidents occur throughout the development of science. Another name for this is **serendipity**. However, it is recognising the implications of serendipity that marks out great scientists. It was Louis Pasteur who said, 'Chance favours the prepared mind.' Pasteur's serendipitous discovery is featured on page 374.

The polyester fibres that the Du Pont group led by Carothers produced were not destined for commercial success. Although the fibres were strong and pliable, they melted at too low a temperature to be of practical use in clothing, which would have to be ironed. They were also slightly water soluble, another considerable disadvantage!

The polyester story now switches to the laboratories of the Calico Printers' Association in England, where in 1941 J Whinfield and J Dickson invented Terylene, building on the foundations laid by Wallace Carothers. This polyester is produced from benzene-1,4-dicarboxylic acid and the alcohol ethane-1,2-diol (Fig 20.23). The systematic name for this polyester need not concern us. In industry, it is usually referred to simply as polyester or PET.

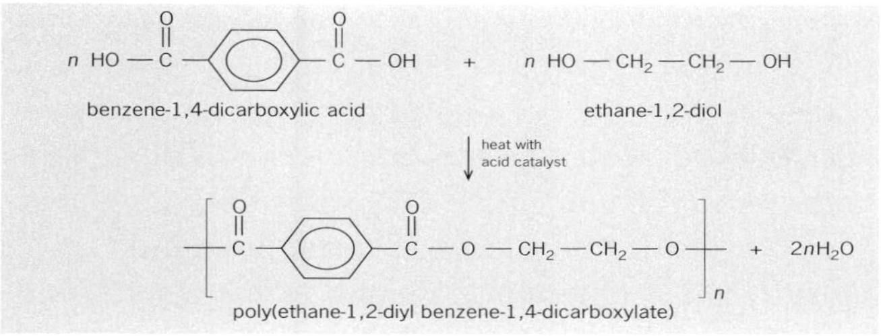

$$n\ \mathrm{HO{-}\overset{\displaystyle O}{\overset{\|}{C}}{-}C_6H_4{-}\overset{\displaystyle O}{\overset{\|}{C}}{-}OH} \quad + \quad n\ \mathrm{HO{-}CH_2{-}CH_2{-}OH}$$

benzene-1,4-dicarboxylic acid + ethane-1,2-diol

$\downarrow$ heat with acid catalyst

$$\mathrm{\left[\overset{\displaystyle O}{\overset{\|}{C}}{-}C_6H_4{-}\overset{\displaystyle O}{\overset{\|}{C}}{-}O{-}CH_2{-}CH_2{-}O \right]_{\mathit{n}}} \quad + \quad 2n\mathrm{H_2O}$$

poly(ethane-1,2-diyl benzene-1,4-dicarboxylate)

Fig 20.23 Condensation reaction for the polyester Terylene

The names Terylene and PET are derived from the traditional names for the two monomers and the polymer, respectively. Benzene-1,4-dicarboxylic acid is also called **tere**phthalic acid and ethane-1,2-diol is also called ethy**lene** glycol – hence Terylene. The traditional name for the polymer is **p**oly(**e**thylene **t**erephthalate), which gives the acronym PET.

Polyester is the leading synthetic fibre with worldwide production of about 24 million tonnes. Its principal use is in clothing, because it is crease-resistant. Often, it is mixed with other synthetic or natural fibres. For example, a mixture of cotton and polyester is very popular because the cotton absorbs moisture. Since polyester is a good thermal insulator, duvets and anoraks are filled with its fibres. It is also becoming the dominant material in packaging and is used in some carbonated drinks bottles.

? QUESTION 9

9 **a)** The dimethyl ester of benzene-1,4-dicarboxylic acid also produces Terylene.

i) Draw the structural formula of this molecule.

ii) Predict the small molecule that is lost during the condensation polymerisation reaction with ethane-1,2-diol.

iii) Predict the products of the acid and alkaline hydrolysis of this type of Terylene by referring to pages 357–359.

b) Another polyester can be produced using benzene-1,4-dicarboxylic acid and propane-1,3-diol. Draw the structural formula of the repeating unit of this polymer.

Fig 20.24 PET is an ideal material for carbonated drinks bottles because it is light, tough and won't shatter when dropped

? QUESTIONS 10–11

10 Terylene and similar polyesters are water repellent. Why is it an advantage that, when blended with cotton fibres, the resulting fabric should absorb moisture?

11 Polyesters also find many uses in medicine. For example, as mesh tubes they can replace blood vessels, with human tissue growing into and around the mesh. While sometimes a permanent mesh is required, it is an advantage at other times to insert a mesh that can be degraded by enzymes in the body. A polymer made from 2-hydroxypropanoic acid (lactic acid, Fig 20.25) forms a degradable mesh.

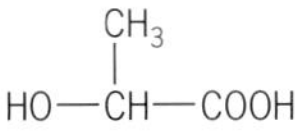

$$\mathrm{HO{-}\underset{}{\overset{\displaystyle CH_3}{\overset{|}{C}}}H{-}COOH}$$

Fig 20.25 2-Hydroxypropanoic acid lactic acid

Write an equation for the production of this polyester.

NYLON: A POLYAMIDE

SCIENCE IN CONTEXT

Wallace Carothers

Working at the Du Pont company in Delaware, Wallace Carothers led the research team that produced the world's first nylon. He was an internationally renowned expert on polymers, and his theoretical ideas are the foundation of today's polymer science. He was the first industrial chemist to become a member of the prestigious National Academy of Sciences in the USA.

However, his success masked the depression that had plagued him since childhood and, in 1937, feeling that his life's work had been a failure, he committed suicide by drinking a solution of cyanide. Yet, only three years later, nylon was proclaimed an outstanding commercial success.

Fig 20.26 Wallace Carothers (1896–1937)

After working on polyesters, Carothers and his team focused their efforts on another type of polymer, in which the monomer units are joined through secondary amide linkages (Fig 20.27), instead of ester linkages. These secondary amide linkages are the same linkages as those found in proteins such as silk.

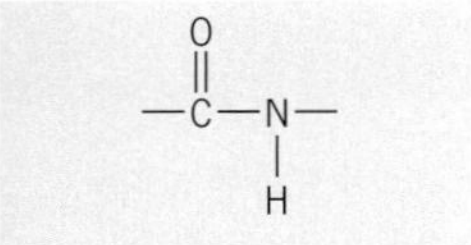

$$-\overset{\overset{\displaystyle O}{\|}}{C}-\underset{\underset{\displaystyle H}{|}}{N}-$$

Fig 20.27 The secondary amide linkage

For secondary amide links – called peptide linkages in proteins – see page 378.

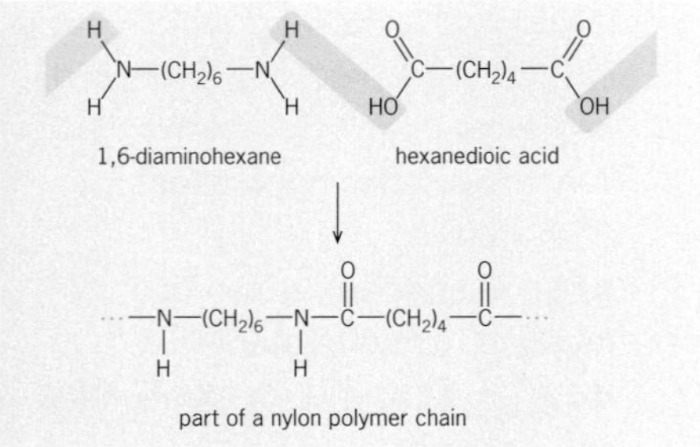

$H_2N-(CH_2)_6-NH_2$ 1,6-diaminohexane + $HOOC-(CH_2)_4-COOH$ hexanedioic acid

↓

$\cdots-NH-(CH_2)_6-NH-CO-(CH_2)_4-CO-\cdots$ part of a nylon polymer chain

Fig 20.28 The reaction of Carothers and his team that gave rise to nylon 6,6

In 1935, the team produced a polymer, now called nylon 6,6, from a diamine and a dicarboxylic acid (Fig 20.28). Notice the similarity of this condensation reaction to that which produces a polyester. Both use a dicarboxylic acid monomer, but the alcohol groups at each end of the other monomer are replaced with primary amine functional groups (NH_2).

Nylon 6,6 is not the only polyamide that Carothers' team produced, and so each is distinguished by a pair of numbers unique to that nylon. The first number after the word nylon is the number of carbon atoms in the diamine. The second number is the number of carbon atoms in the dicarboxylic acid.

When nylon 6,6 is cold drawn, it is stretched to four times its original length, which gives the fibres so produced a high tensile strength and elasticity. As a fibre is drawn, the polymer chains align parallel to one another and the tensile strength comes from the hydrogen bonding between the CO and NH groups of adjacent chains (Fig 20.29). The cold drawing also increases the lustre (shininess) of nylon. In a commercial nylon plant, molten nylon is forced through tiny holes to produce the fibres, which are then cold drawn (Fig 20.30).

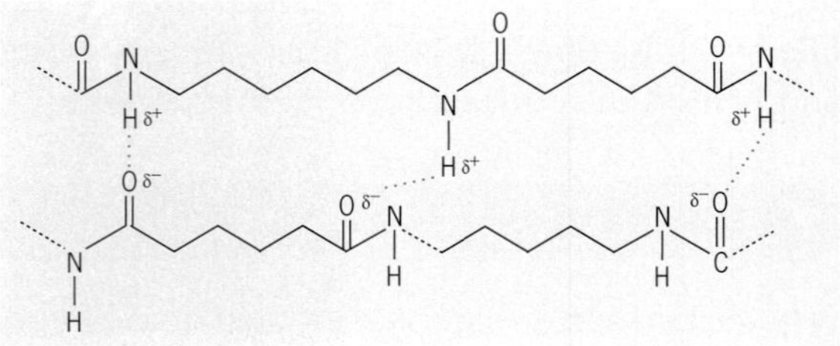

Fig 20.29 Part of two linear chains of nylon, showing intermolecular hydrogen bonding

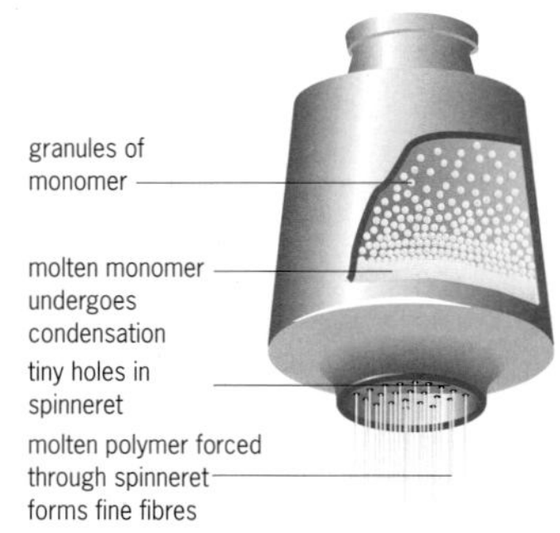
granules of monomer
molten monomer undergoes condensation
tiny holes in spinneret
molten polymer forced through spinneret forms fine fibres

Fig 20.30 The spinneret through which molten nylon is forced before being cold drawn

Du Pont went from the laboratory preparation of nylon 6,6 to its manufacture in the remarkably short time of less than five years. By 1937 a pilot plant was operational, and by the end of 1939 a full-scale plant had been built and was in production – all this when the techniques of large-scale condensation polymerisation were unknown and there were no bulk supplies of either monomer. Add to this the different technologies required to spin nylon fibres, and we have a remarkable achievement.

The first article to contain nylon was Dr West's Miracle Toothbrush, in which nylon bristles replaced animal bristles. Meanwhile, Du Pont had test-marketed women's nylon stockings and recognised their enormous sales potential. Skirt lengths had become shorter and silk stockings were highly fashionable, but were very expensive. When nylon stockings became widely available in 1940, the commercial success of nylon was assured. However, there was not enough production capacity to supply both the consumer market and the United States war requirements, because nylon was in great demand for ropes and, in particular, for parachutes, which had previously been made from silk.

Nylon stockings were rationed in the US until the end of the Second World War, and it took until the early 1950s before production capacity in Western Europe matched the high demand for nylon stockings and other nylon goods. Today, nylon goods account for 95 per cent of the women's hosiery market.

A variety of machine parts are now made from nylon instead of metals. For these, its properties of toughness, strength and abrasion resistance are much in demand, and specific properties are enhanced by the use of **fillers**, such as glass fibre. Fillers are widely used to tailor the properties of polymers to specific functions. Fillers are also added to provide bulk and make a cheaper product.

? QUESTION 12

12 Nylon 5,10 is another polyamide produced by Carothers and his team. He favoured this as the nylon to be developed commercially, but he was overruled and nylon 6,6 became the first nylon to be manufactured. Write the structural formulae of the monomers used for nylon 5,10.

? QUESTION 13

13 With reference to page 380, what are the products of the complete hydrolysis of nylon 6,6? What reagents and conditions could be used to carry out this reaction in the laboratory?

Thermoplastic and thermosetting polymers

All the polymers discussed so far have one property in common – they are **thermoplastic**. Thermoplastic polymers soften when heated and can be moulded. The process of heating and moulding can be repeated many times. At a molecular level, the forces between the polymer chains are weak, such as induced dipole–induced dipole forces or hydrogen bonds (Fig 20.31). These are sometimes referred to as **secondary bonds**. When a thermoplastic polymer is warmed, these forces are broken and the polymer chains can slide over one another, which allows the polymer to be shaped. When the polymer cools, the forces reform with the chains in their new positions.

This ability to be remoulded makes these polymers very versatile and therefore in much demand. Consequently, the annual world production of thermoplastic polymers is seven times that of thermosetting polymers.

Thermosetting polymers can be moulded only when they are first produced. With strong heating, they decompose and cannot be softened and remoulded. A thermosetting polymer has covalent cross-links between its chains, so the polymer is a three-dimensional network of strong covalent bonds (Fig 20.32). On heating, these break, which causes the polymer to decompose before melting can occur.

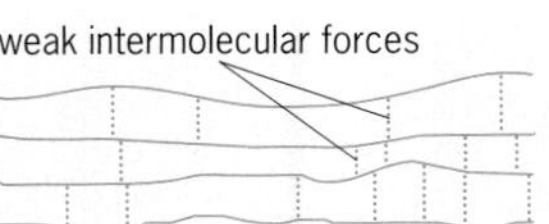

Fig 20.31 Diagrammatic representation of a thermoplastic showing the weak intermolecular forces between the polymer chains, which can be overcome on heating to allow the chains to slide over one another. On cooling, the intermolecular forces reform

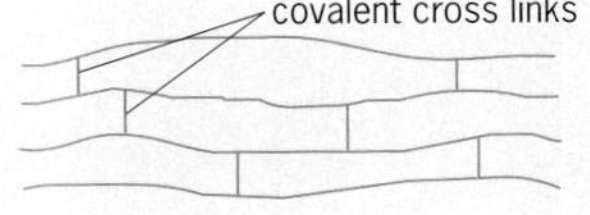

Fig 20.32 Diagrammatic representation of a thermosetting polymer. The covalent bonds form a three-dimensional network that, once formed, cannot be melted again

Fig 20.33 78 r.p.m. records were made from Bakelite

BAKELITE: THE FIRST SYNTHETIC POLYMER

Leo Baekeland, a Belgian who had emigrated to the United States, invented the first wholly synthetic polymer, which he called Bakelite. He made it from the reaction of phenol with methanal. Bakelite was first manufactured in 1910.

The initial reaction involves a substitution in the benzene ring of phenol, which can be at either position 2 or position 4 (Fig 20.34(a)). Once formed, the product can react with another molecule of phenol with the elimination of a water molecule (Fig 20.34(b)). The three-dimensional network is then built up through a series of similar reactions (Fig 20.34(c)).

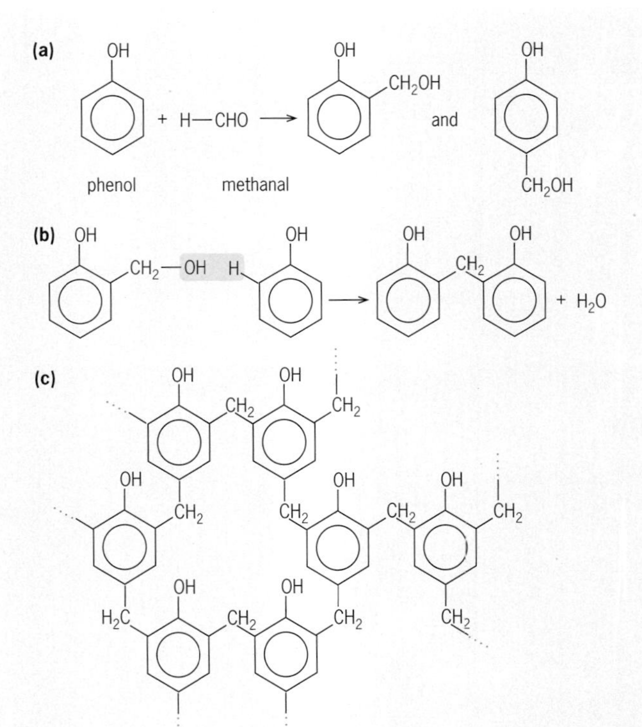

Fig 20.34 Reactions in the formation of Bakelite

As it has a high number of cross-links, Bakelite forms a hard, rigid structure. Like most polymers, it is a good electrical and heat insulator. It was widely used in all types of articles, among them electric sockets and plugs, pan handles, and even music records, before PVC superseded it. But phenol–methanal resins are still important.

? QUESTION 14

14 **a)** Is Bakelite:

i) a condensation polymer or an addition polymer?

ii) a thermoplastic or a thermosetting polymer?

b) What properties of Bakelite would make it a suitable material for pan handles?

? QUESTION 15

15 **a)** Draw the repeating unit for Kevlar and predict its monomers.

b) What is the intermolecular force that holds Kevlar fibres together?

c) Why is Kevlar less dense than steel? (Hint: think about what density means.)

KEVLAR

Du Pont have been at the forefront of research on aromatic polyamides, now called aramids, for a number of years. Kelvar is one these aramids and is developed from monomers with functional groups in the 1,4 position on the benzene ring (Fig 20.35).

Kevlar polymer chains can be aligned to make very strong fibres. A 7 cm diameter steel cable has a breaking strain of about 40 tonnes. An identical Kevlar cable has the same breaking strain, but is five times lighter than the steel cable. Kevlar is embedded in tyres in place of steel reinforcement, which has reduced the weight of a typical truck tyre by 9 kg. However, the market for Kevlar-reinforced tyres has been slow to take off because tyre manufacturers have invested in equipment to produce steel reinforcement, believing that customers would prefer the strength that the word steel implies.

Kevlar is renowned as the protective padding in bullet-resistant vests (Fig 20.37). It is also incorporated into aircraft wings, in which strength combined with lightness is essential.

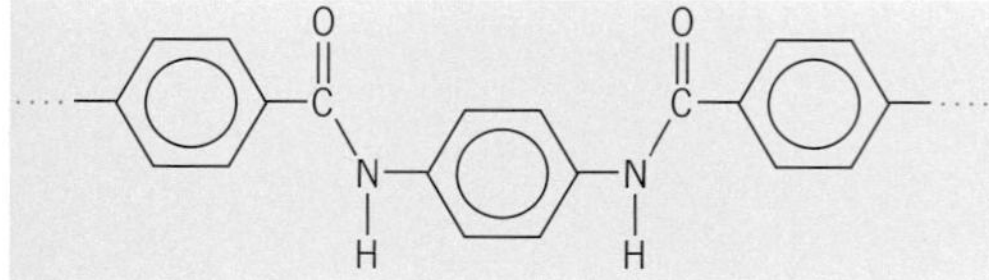

Fig 20.35 Kevlar

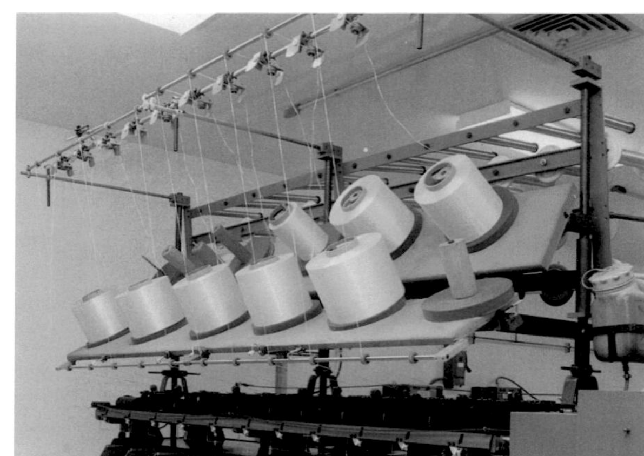

Fig 20.36 Kevlar fibre being twisted into cord for radial tyres at a Goodyear tyre plant

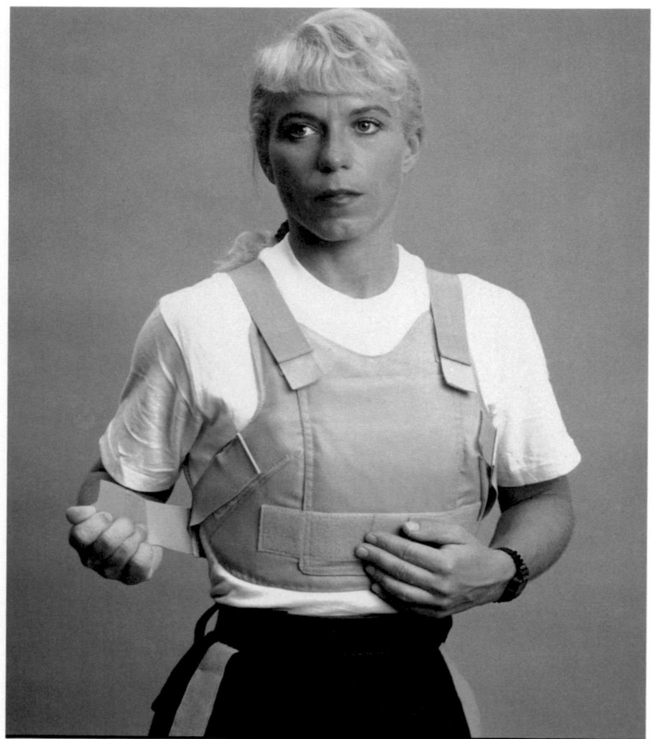

Fig 20.37 A bullet-resistant vest made from Kevlar

Fig 20.38 The yacht sails are made from Kevlar

4 TWO NATURAL POLYMERS: STARCH AND CELLULOSE

Natural polymers are mentioned in the How Science Works box on page 394. Proteins, which are polymers of amino acids, are discussed in some detail on pages 378–385 and DNA on pages 385–390. Two other important polymers are the polysaccharides, starch and cellulose, both made from glucose monomers.

Glucose can exist in two distinct forms, as Fig 20.39 shows. Note the difference between these two glucose molecules by looking at carbon 1. This seemingly minor difference has a marked effect on the properties of starch and cellulose.

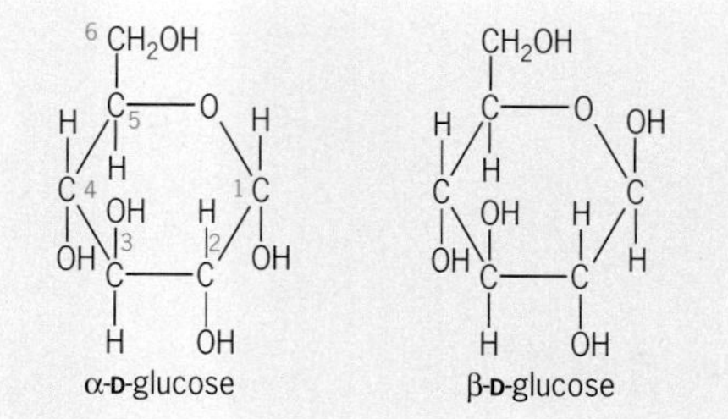

Fig 20.39 The two ring forms of glucose

Starch is a polymer of α-glucose monomers. It is linked by α-1,4 linkages (Fig 20.40), also called glycosidic linkages. This form of starch is called amylose and is a straight-chain polymer.

We digest starch by hydrolysing it with enzymes to give glucose. Starch is therefore an energy-storage compound, since once we have broken it down to glucose it can be oxidised in the body to give energy during respitation. Starch can also be hydrolysed by acids in the laboratory to give α-glucose:

$$(C_6H_{10}O_5)_n + nH_2O \xrightarrow{H^+(aq)} nC_6H_{12}O_6$$

Cellulose, a linear polymer formed from β–glucose (Fig 20.41), is probably the world's most abundant organic chemical. It has a structural role in plants and is the source of cotton fibre.

In common with most animals, we cannot digest cellulose because we do not have the necessary enzymes to hydrolyse the ß-1,4 linkage. However, cellulose forms the fibre we require in our diet.

? QUESTIONS 16–17

16 Another form of starch, called amylopectin, has branches formed by reaction between the OH groups of carbon 1 and carbon 6. Draw part of this polymer, showing the carbon 1,6 linkage.

17 Cellulose has considerable strength because its linear chains are able to pack closely together. What intermolecular forces give cellulose its strength?

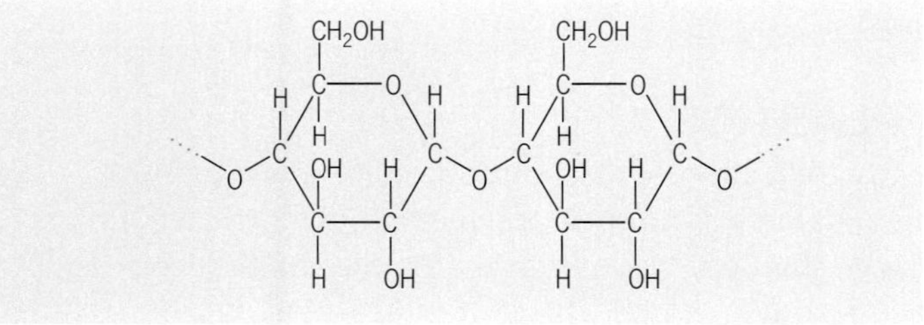

Fig 20.40 Starch

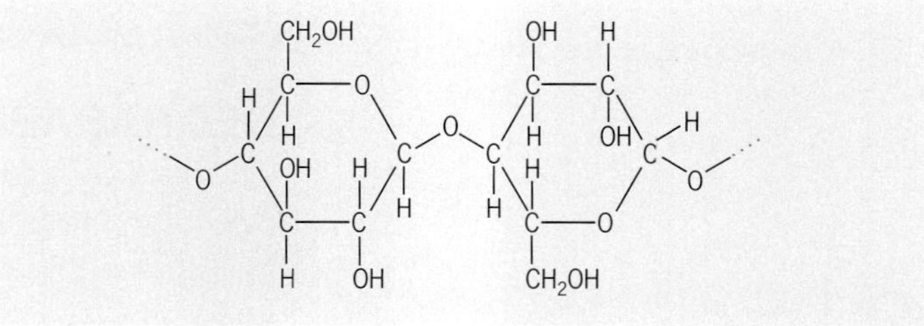

Fig 20.41 Cellulose

5 WHAT DO WE DO ABOUT PLASTIC WASTE?

This problem will not go away, mainly because most of the polymers we throw away do not degrade (break down). In the UK, most plastic waste goes into landfill (Fig 20.42) and that is where it will stay. So, what are the options?

Most polymers come from non-renewable crude oil, and it takes energy to produce them. In fact, of the energy that goes into making a plastic article, most is used in producing the polymer. So, why not recycle the polymer? The majority of polymers thrown out are thermoplastics, which could be melted down and reused. However, with some exceptions, all kinds of rubbish are thrown away together. It would take much energy and be fairly costly to sort it. Even if plastic waste were kept separate from other household rubbish,

Fig 20.42 Most of the plastic waste in the UK goes into landfill sites

the several different types of polymer involved would have to be further separated before they could be reprocessed.

However, research is under way into cracking polymers, in much the same way as fractions of oil are cracked, to form a chemical feedstock from which new polymers can be produced. Plastic waste is also being cracked to produce petrol and diesel. But to be economically viable and environmentally friendly, any recycling scheme must not use up more energy than it takes to make the polymer in the first place.

Some articles themselves can be reused, which is achieved successfully for PET (polyester) bottles in some countries in Europe. Again, though, the energy consumption of recycling must also be taken into account.

Another way of dealing with plastic waste is to burn it and use the energy it provides. This is very controversial. Those who support this, point to the fact that most plastics do not biodegrade in landfill and it is better to at least extract the energy from them through incineration. However, opponents point to the environmental disadvantages of doing this. This is a non-renewable energy resource and it prevents recycling of plastic products so that they have to be made all over again from crude oil, expending considerable energy in the process and probably releasing carbon dioxide into the atmosphere. The very act of burning the plastic waste will produce more carbon dioxide, contributing to global warming and climate change. There are also real concerns about the toxic nature of some of the other emissions that occur during burning. Supporters of incineration point to the advances chemists have made in removing some of the toxic combustion products, such as hydrogen chloride from burning PVC.

? QUESTIONS 18–20

18 Explain why the molecular structure of thermoplastics allows them to be melted and remoulded.

19 Epoxy resin is a common component of household waste, but it is a major obstacle to plastics recycling because it is a thermosetting polymer. Why is this a problem?

20 What processes cost energy in recycling a PET coke bottle?

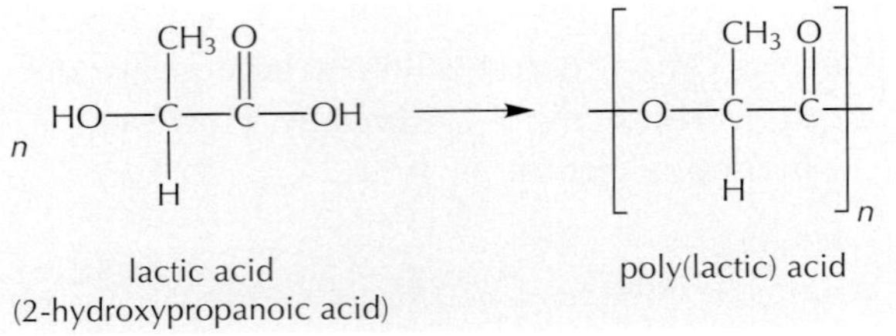

$$n\ \mathrm{HO{-}C(CH_3)(H){-}C({=}O){-}OH} \longrightarrow \left[\mathrm{{-}O{-}C(CH_3)(H){-}C({=}O){-}}\right]_n$$

lactic acid (2-hydroxypropanoic acid) → poly(lactic) acid

Fig 20.43 Polymerisation of poly(lactic) acid

DEGRADABLE PLASTICS

Chemists are playing a significant role in minimising plastic waste by developing degradable polymers. One such polymer is poly(lactic acid) or PLA. It has the advantage that it is made from renewable sources: corn or

? QUESTION 21

21 If plastics biodegrade, the energy of their bonds is not reused. An alternative is to burn plastic waste to give steam, which is used to generate electricity. What are the environmental and energy issues associated with incinerating the plastics in household rubbish?

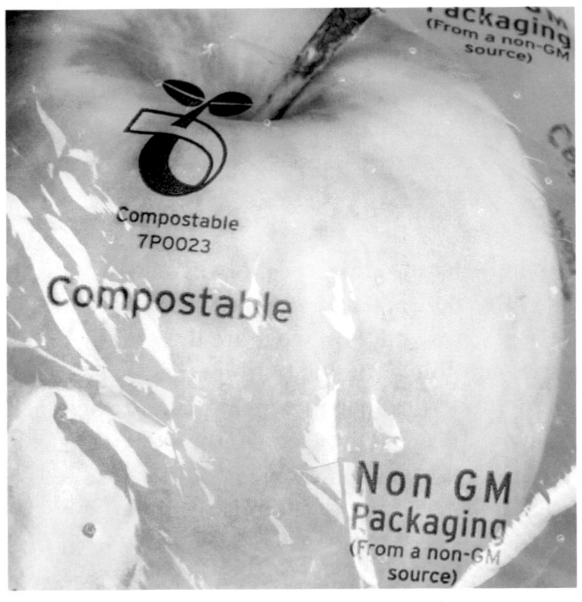

Fig 20.44 This food packaging is made from a plastic inpregnated with starch, so it is biodegradable

sugar cane. Bacteria ferment the sugar to produce lactic acid, which is polymerised to make the degradable polymer. PLA has found many uses from disposable cutlery and waste sacks to internal stitches in the body.

There are two ways in which PLA degrades. The first is by hydrolysis at the ester linkage (see page 357). This can be done by enzymes from microorganisms in the soil; this process is called biodegradation. The second is the absorption of radiation from light by the C=O bond, which splits PLA apart. This is called photodegradation.

Apart from biopolymers, there are other ways of making synthetic polymers degradable. For example, cellulose, starch or protein can be impregnated into plastic articles such as poly(ethene) bags. When such an article is buried in a landfill site, microorganisms break down the cellulose, starch or protein. The article then disintegrates. The synthetic polymer chains that remain have an increased surface area, which speeds up their decomposition.

Poly(alkenes) do not biodegrade in landfill sites if they are not impregnated with a substance that makes them biodegrade. However, polyesters and polyamides are hydrolysable at their ester and amide linkages. Enzymes from bacteria or acid catalysts will break the long polymer chains into shorter and shorter fragments, and so eventually they will biodegrade in landfill sites.

Another method with some polymers is to impregnate them with chemical activators so that on exposure to ultraviolet light the polymer chains break down into much shorter chains, which can then be biodegraded. Poly(ethene) can be made so that it will photodegrade. This is done by inserting carbonyl groups (C=O) into the polymer chain.

There are environmental problems even with degradable plastics. If polymers degrade, then the articles made of them may not be recyclable. Also, there is some concern over the unknown nature of the decomposition products, which may cause more long-term damage to the environment than the original plastics themselves.

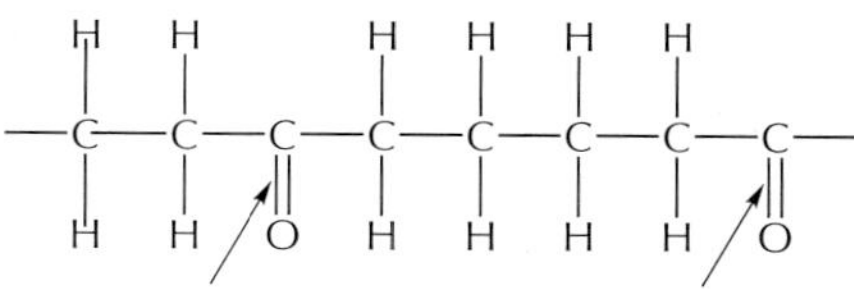
These bonds absorb energy from UV radiation

Fig 20.45 C=O bonds inserted into poly(ethene) absorb UV radiation and split.

SUMMARY

After studying this chapter, you should know:

- Polymers are very large molecules made up of many small molecules called monomers.
- Addition polymers form when monomers react together without the elimination of small molecules.
- Poly(ethene), PTFE, PVC (poly(chloroethene)) and poly(phenylethene) are all synthetic addition polymers.
- Polymers such as poly(ethenol) are water solutle provide the number of hydrogen bonds is not too great or too small.
- Polymers such as poly(propere) can be manufactured using Ziegler–Natta catalysts which use less energy and so are of great economic and environmental significance.
- Condensation polymers are formed from monomers with the elimination of small molecules such as water.
- Polyesters such as Terylene (PET) and polyamides such as nylon 6,6 and kelvar are synthetic condensation polymers.
- Starch, cellulose and proteins are natural polymers.
- Many synthetic polymers are not biodegradable and so environmental issues surround the disposal or recycling of plastic waste.
- Condensation polymers can photodegrade by absorbing radiation at the C=O bond.
- Condensation polymers can degrade through hydrolysis at the ester or amide group.

Practice questions and a How Science Works assignment for this chapter are available at www.collinseducation.co.uk/CAS

21

Patterns across the Periodic Table

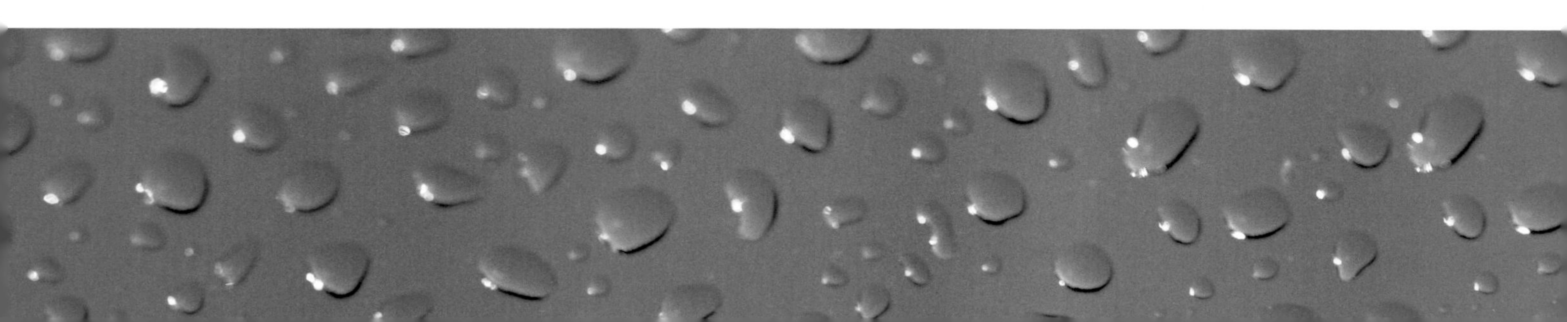

21 PATTERNS ACROSS THE PERIODIC TABLE

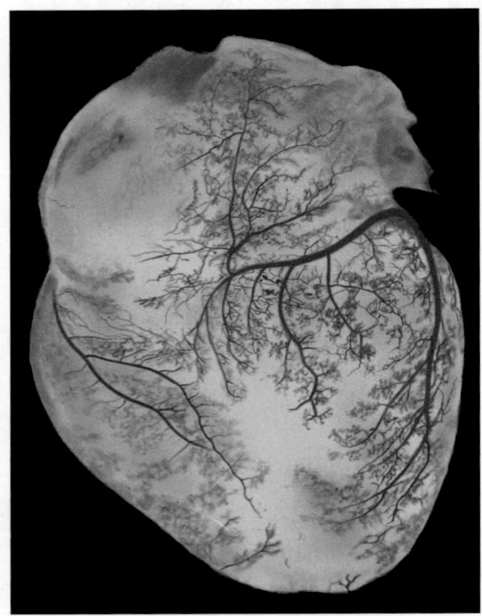

Angina When the coronary arteries become narrowed, glyceryl trinitrate can relax them, avoiding the pain of angina

A solution for heart pain

Glyceryl trinitrate tablets have been used for many years to relieve the gripping pain of angina, a heart condition caused by the narrowing of the coronary arteries in the heart. How glyceryl trinitrate works had been unclear. But now biochemists have the answer.

The muscles in the walls of arteries are partly responsible for controlling blood pressure. When the muscles are relaxed, the arteries enlarge and blood pressure drops. Instructions to the artery muscles to relax are carried from the brain by a number of messenger molecules.

One messenger is a gaseous oxide of nitrogen called nitrogen monoxide. This is a very small molecule and so is able to diffuse quickly into muscle cells, where it binds with an iron atom in the enzyme responsible for muscle relaxation. The enzyme is activated by the binding, and so the muscle relaxes.

Research evidence suggests that when someone takes glyceryl trinitrate, it is converted into nitrogen monoxide, which is carried to the muscles in the coronary artery and relaxes them. Blood pressure is reduced, thereby relieving the pain of angina.

1 PERIODICITY OF PROPERTIES OF THE ELEMENTS

On pages 97 to 99, the periodic classification of the elements is described in relation to some of the properties of the atoms of elements. The similarities between the elements within a group are explained in terms of the number of electrons in the outer electron shell. Similarities in the structure of the outer electron shell lead to a periodicity in the properties of elements, so that the pattern of properties shown by the elements in one period is repeated in the next period. The idea of periodicity is extended in this chapter to include properties of the *compounds* of elements, rather than the elements themselves.

Oxides and chlorides are two of the most common classes of inorganic compounds. Within a single period in the Periodic Table, a bewildering variety of oxides and chlorides occurs – including crystalline solids, materials that look solid but are really supercooled liquids, and colourless gases. Behind this variety, however, is a pattern that we can explain in terms of the structure and bonding of the oxides and chlorides themselves.

This chapter describes and explains the change in the properties of oxides and chlorides of the elements in Period 3 as the atomic (proton) number of the element increases. It relates these properties to both structure and bonding, and to some technological and environmental aspects of chemistry.

REACTIONS OF ELEMENTS

Almost all the known elements, even some of the noble gases, react to form compounds. When an element reacts, it forms one of the following types of bond: ionic, covalent or dative covalent. This means that there must be a redistribution of electrons around the atom of the element. This redistribution often leads to a particle that is isoelectronic with a noble gas. Some atoms lose electrons to form cations, others gain electrons to form anions, and others share electrons.

When an element reacts, a type of reaction called **redox** occurs since the oxidation number of an element will necessarily change from zero to either a positive number (with metals and some non-metals) or a negative number (with non-metals). We describe redox reactions in Chapter 5 as a reaction that involves electron transfer and a change in the oxidation number of an element.

Oxidation and reduction

Redox reactions necessarily involve both an oxidation and a reduction. The oxidation number is a most useful way to explain whether a reaction is an example of a redox reaction. During oxidation, the oxidation number of an element becomes more positive and electrons are lost. During reduction, the oxidation number becomes more negative and electrons are gained. Do not confuse this with what happens to the oxidising agent and to the reducing agent. The oxidising agent can be identified because it contains an element whose oxidation number becomes more negative during the reaction. The reducing agent can be identified because it contains an element whose oxidation number becomes more positive during the reaction.

Metals are reducing agents and non-metals are often oxidising agents. Since metals are positioned on the left-hand side of the Periodic Table and non-metals on the right-hand side as the atomic number increases across a period, the elements change from reducing agents to oxidising agents.

Reaction of elements with oxygen

Many elements react with oxygen in a reaction often called combustion since it involves elements burning. Oxygen is the oxidising agent and the element will be oxidised, and at the same time oxygen is reduced.

For example, magnesium will burn to form magnesium oxide.

$$2Mg(s) + O_2(g) \rightarrow 2MgO(s)$$

During this reaction the oxidation state of magnesium increases as magnesium atoms lose electrons to form Mg^{2+}. These electrons are gained by oxygen to form the oxide ion, O^{2-}.

Reaction of elements with chlorine

Chlorine is also an oxidising agent and it will react with many elements to form chlorides. For example, phosphorus will react with excess chlorine to form phosphorus(V) chloride:

$$P_4(s) + 10Cl_2(g) \rightarrow 4PCl_5(s)$$

Phosphorus is oxidised, as its oxidation number increases from 0 to +5; at the same time the oxidation number of chlorine changes from 0 to –1.

QUESTION 4

Phosphorus(III) chloride reacts with excess chlorine to form phosphorus(V) chloride:

$$PCl_3(l) + Cl_2(g) \rightarrow PCl_5(s)$$

a) What are the changes in the oxidation number for each element?
b) Which substance is the oxidising agent?

QUESTION 1

Identify the oxidation number of each atom in the following compounds:

a) NaCl **b)** $MgCl_2$ **c)** $AlCl_3$ **d)** $SiCl_4$
e) PCl_3 **f)** PCl_5 **g)** S_2Cl_2

REMEMBER THIS

Within a period, the best reducing agent is always in Group 1 and the best oxidising agent is in Group 7.

QUESTION 2

Carbon combusts in air to make carbon dioxide:

$$C(s) + O_2(g) \rightarrow CO_2(g)$$

a) Which substance is the oxidising agent in the combustion of carbon?
b) Which substance is the reducing agent in the combustion of carbon?

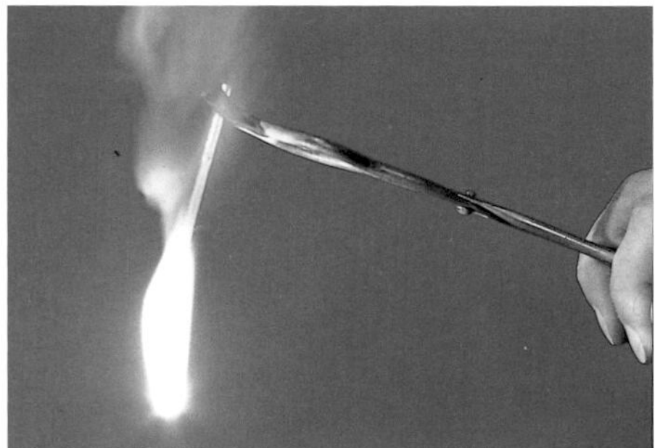

Fig 21.1 Magnesium ribbon burning in air

QUESTION 3

Silicon reacts with chlorine to form silicon(IV) chloride:

$$Si(s) + 2Cl_2(g) \rightarrow SiCl_4(l)$$

a) What is the change of oxidation number of silicon during the reaction?
b) What is the change in oxidation number of chlorine during the reaction?
c) Identify, with a reason, which element is the oxidising agent.
d) Identify, with a reason, which element is the reducing agent.

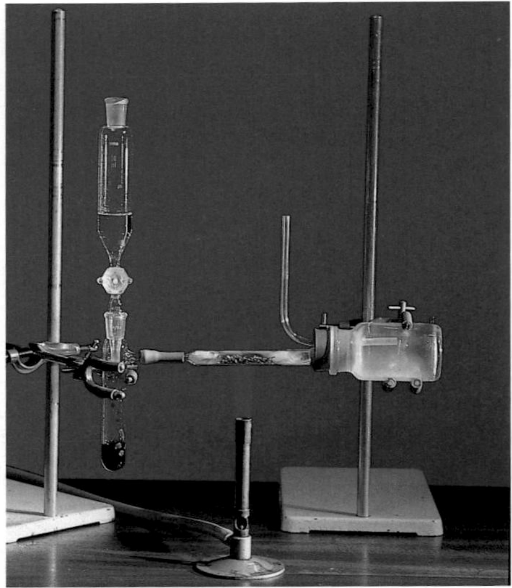

Fig 21.2 Chlorine reacting with aluminium to form white aluminium chloride

2 PERIODICITY IN CHLORIDES

On page 99, the first ionisation energy is described as a periodic function of the atomic (proton) number, with each noble gas having the highest ionisation energy within a period. Periodicity is not restricted to the properties of the elements themselves. It is also reflected in the properties of compounds of the elements. By comparing the properties of a range of similar compounds, we can see that the properties are a periodic function of the atomic number of the element in the compound. The two classes of compounds most studied are the oxides and the chlorides of elements. We concentrate here on the chlorides formed by the elements in Period 3, but a similar pattern of properties is shown by the chlorides of elements in other periods.

CHLORIDES OF ELEMENTS IN PERIOD 3

Chlorine reacts with almost all the elements of this period. The formulae of the chlorides and the oxidation numbers of the elements in the chlorides are given in Table 21.1.

Table 21.1

Element	Formula of chloride	Oxidation number of element in chloride	Appearance of chloride
sodium	$NaCl$	+1	white solid
magnesium	$MgCl_2$	+2	white solid
aluminium	Al_2Cl_6	+3	white solid
silicon	$SiCl_4$	+4	colourless liquid
phosphorus	PCl_3	+3	colourless liquid
	PCl_5	+5	white solid
sulfur	S_2Cl_2	+1	pale yellow liquid

Table 21.1 clearly shows that there is a predictable change in the oxidation number of each element as the atomic number increases. It is also evident that the highest oxidation number corresponds to the *group number* of the element involved (except for sulfur). This is easy to explain, because each successive element in a period has one more electron available in its outer shell to form a bond with chlorine. Notice that the oxidation state of sulfur in S_2Cl_2 does not fit the pattern, but sulfur does form the halide SF_6, in which the oxidation number of sulfur is +6.

REMEMBER THIS

The terms 'oxidation number' and 'oxidation state' are interchangeable. Both terms are used in this book.

QUESTION 5

5 The equation for the reaction between sodium and chlorine is:

$$2Na(s) + Cl_2(g) \rightarrow 2NaCl(s)$$

Write down the equations for the formation of all the other chlorides shown in Table 21.1.

MELTING POINTS OF CHLORIDES

Fig 21.3 shows the wide range of melting points of chlorides. There is only one interpretation of this: namely, the internal structures of the solids are notably different, and the magnitude of the force of attraction between the particles in the different crystal lattices varies widely. We describe here the different crystal lattices in some detail, as well as how each lattice can explain the melting point of the chloride.

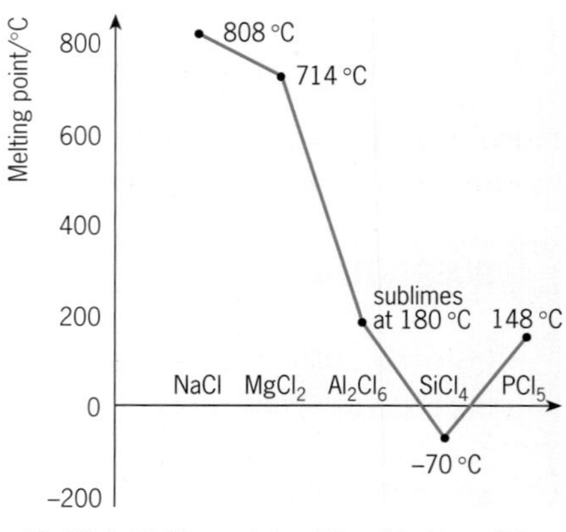

Fig 21.3 Melting points of the chlorides of five elements in Period 3

The different types of solid and how the size of the force of attraction between their particles changes the value of the melting point are covered on page 137.

GIANT IONIC LATTICES

Metals react with chlorine to give ionic chlorides. Metal atoms lose electrons, and chlorine molecules accept them to become chloride ions. There is a strong electrostatic force of attraction between positive ions and negative ions. This force is not in any particular direction, as in the case of a covalent bond. The result is that positive ions attract negative ions to produce a regular pattern of ions in which each positive ion has several negative ions as its nearest neighbours, and each negative ion has several positive ions as its nearest neighbours. Eventually, a giant structure is produced that consists of the regular repetition of a unit cell in three dimensions.

The exact nature of the unit cell, and the coordination number of the positive and negative ions, depend on several factors: certain geometric considerations, the ionic radii, the formula of the ionic compound (the ratio of the number of positive to negative ions) and the magnitude of the forces of attraction, which is related to the charge on the ions.

In Period 3 both magnesium chloride and sodium chloride form a giant ionic lattice. The properties of these two compounds are typical of compounds that have a giant ionic structure.

These properties include:

- high melting and boiling points;
- solubility in water;
- electrical non-conductors as solids;
- electrical conductors when an aqueous solution or a molten liquid.

The properties of ionic compounds can be explained by their giant ionic lattice structure. We use the giant ionic lattice of sodium chloride to illustrate how these properties may be explained.

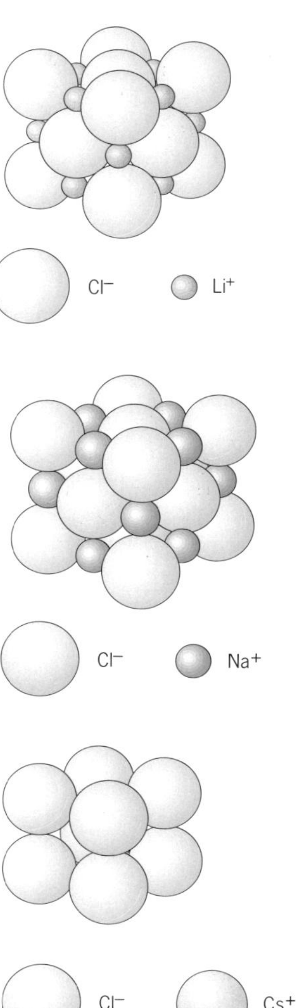

Fig 21.4 Lithium chloride and sodium chloride are face-centred cubic structures, whereas caesium chloride is a body-centred cubic structure. The caesium ion is too large to fit in a face-centred cubic structure

Ionic lattice of sodium chloride

Composed of sodium ions and chloride ions, sodium chloride is a typical ionic chloride. Each sodium ion has a coordination number of six. That is, it has six chloride ions as its nearest neighbours. Each chloride ion also has a coordination number of six. Since the force of attraction between the sodium ions and the chloride ions is very strong, separating them is difficult and therefore needs lots of energy. Hence, sodium chloride has a high melting point.

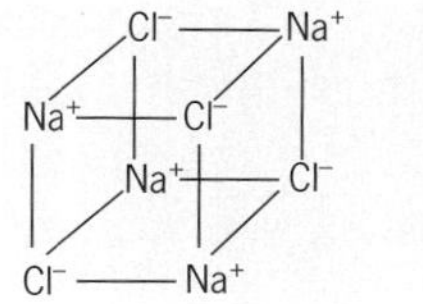

Fig 21.5 The sodium chloride unit cell

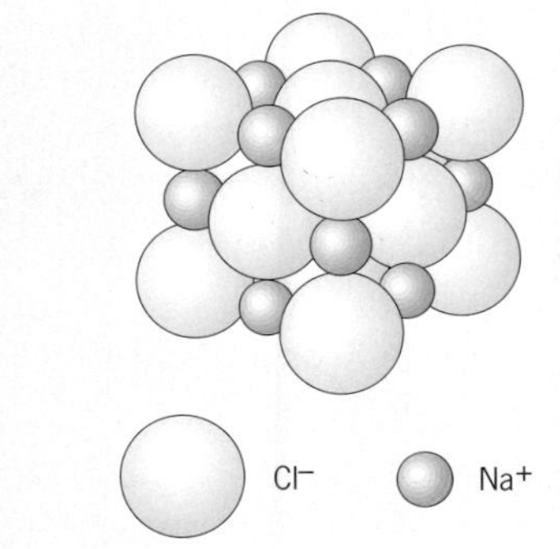

Fig 21.6 Sodium chloride crystal lattice

QUESTION 6

What are the differences in the lattice structure of lithium chloride, sodium chloride and caesium chloride? Suggest why you think there are differences.

Solubility in water

Many ionic compounds dissolve in water to form solutions. The regular pattern of the cations and anions in the lattice is broken down as aqueous or hydrated ions are formed. The energy required to break the strong electrostatic attraction between the cation and the anion is compensated by the energy released when polar water molecules are attracted to the ions. We discuss this energy of dissolution in Chapter 22.

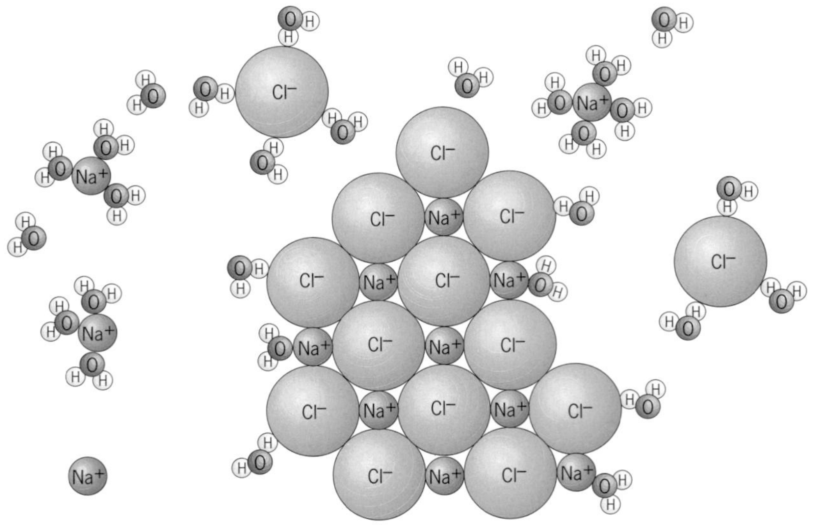

Fig 21.7 The dissolution of a sodium chloride lattice. Water is a polar molecule and the slightly negative oxygen end is attracted to the positive ions, while the slightly positive hydrogen part is attracted to the negative ions

How and why ionic substances dissolve in water are explained on page 443.

The pH of aqueous sodium chloride is 7.0. Magnesium chloride also dissolves in water, but it forms a solution with a pH slightly less than 7.0. This is because of the polarisation of water molecules by the magnesium ion (which is highly charged and has a small ionic radius) forms a small concentration of aqueous hydrogen ions. In solution, aluminium chloride behaves as an ionic chloride, but its aqueous solution is very acidic since an aluminium ion, Al^{3+}, has a higher positive charge and a smaller ionic radius than Mg^{2+}. The chemistry of aqueous aluminium compounds is described in more detail later in this chapter and in Chapter 22.

Electrical conductivity

Solid ionic substances such as sodium chloride cannot conduct electricity because their ions are not free to move. When an ionic substance is melted the ions are free to move, so the molten ionic compound is able to conduct electricity. In the same way, in an aqueous solution of sodium chloride the ions are free to move so the solution conducts electricity.

Solutions of ionic chlorides, or molten ionic chlorides, can be electrolysed because their ions are free to move and act as charge carriers.

SIMPLE MOLECULAR CHLORIDES

The melting points of all the other chlorides of the elements of Period 3 are quite low. This indicates that the forces of attraction between the particles in their crystal lattices are not particularly strong. The lattices are composed of molecules that are bonded to one another by intermolecular forces. These chlorides have a simple molecular structure.

The properties of simple molecular chlorides such as silicon(IV) chloride is typical of all compounds that have a simple molecular structure. These properties are:

- low melting point, because the intermolecular forces are weak and so easily broken;
- non-conduction of electricity, since there are no free electrons to carry the charge.

We now describe the structure of several of the chlorides to illustrate the formation of simple molecular lattices and weak intermolecular forces.

Aluminium chloride

As a metal, aluminium might be expected to form an ionic chloride. Indeed, a bottle of aluminium chloride picked from the chemical store cupboard will probably be ionic. This is because it is hydrated aluminium chloride rather than anhydrous aluminium chloride.

An anhydrous crystalline solid does not have any molecules of water as part of the structure and bonding of the crystal. A hydrated crystal, however, has a lattice that incorporates water molecules. It is important to realise that hydrated crystals are not wet crystals. They are completely dry, since the water is *chemically* bound into the crystal lattice.

Anhydrous aluminium chloride is a simple molecular solid. The formula of the molecule is Al_2Cl_6 and it contains two bridging chlorine atoms (Fig 21.8). The intermolecular forces between each Al_2Cl_6 unit are fairly weak and, once broken, individual Al_2Cl_6 molecules can escape from the lattice. Anhydrous aluminium chloride **sublimes**. That is, when heated, it changes directly from a solid to a gas without having a liquid phase.

The structure of aluminium chloride is discussed on page 77.

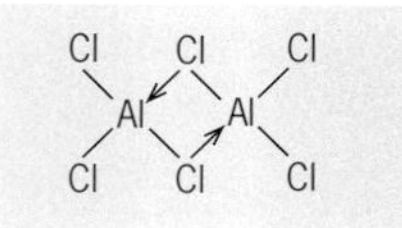

Fig 21.8 The displayed formula of aluminium chloride

QUESTION 7

Draw the dot and cross diagram for silicon(IV) chloride.

Silicon(IV) chloride

Often referred to as silicon tetrachloride, silicon(IV) chloride is a colourless liquid with a low boiling point. Its molecules have the formula $SiCl_4$ (Fig 21.9). Each molecule has a tetrahedral shape. The silicon–chlorine bond is polar because the two elements have different electronegativities. This means that the chlorine end of the bond is slightly negative and the carbon atom is slightly positive. Since $SiCl_4$ is a symmetrical molecule, the individual bond dipoles cancel out and the molecule is non-polar. As a result, $SiCl_4$ molecules are attracted to each other by weak intermolecular forces. These are van der Waals forces of attraction or induced dipole–induced dipole attractions. To melt or boil $SiCl_4$ only the weak intermolecular forces need to be broken, so that the melting point and the boiling point will be low. The atoms within the molecule remain strongly covalently bonded to each other.

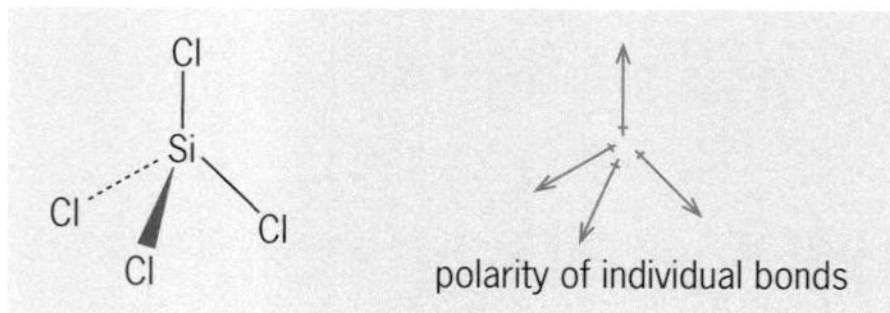

Fig 21.9 The displayed formula of silicon(IV) chloride. This molecule is symmetrical so, even though it has four polar covalent bonds, it has no overall dipole moment

Phosphorus(III) chloride

Also known as phosphorus trichloride, phosphorus(III) chloride (Fig 21.10) is a liquid at room temperature with a low boiling point. Its molecule is polar with an overall dipole moment. That is, one end of the molecule is slightly positive and the other end is slightly negative. This happens because the molecule has polar covalent bonds and is not symmetrical, so the individual bond dipoles do not cancel out.

Since one end of the molecule is slightly positive, it can form a weak electrostatic attraction to the negative end of another phosphorus(III) chloride molecule. The electrostatic force of attraction in the phosphorus(III) chloride lattice is not nearly as strong as that in an ionic lattice, since the magnitude of the positive and negative charges is quite small. The molecule has a permanent dipole and it is this that causes the intermolecular attraction. This type of weak intermolecular attraction is known as **permanent dipole–permanent dipole attraction**. Permanent dipole–permanent dipole attractions are stronger than induced dipole–induced dipole interactions.

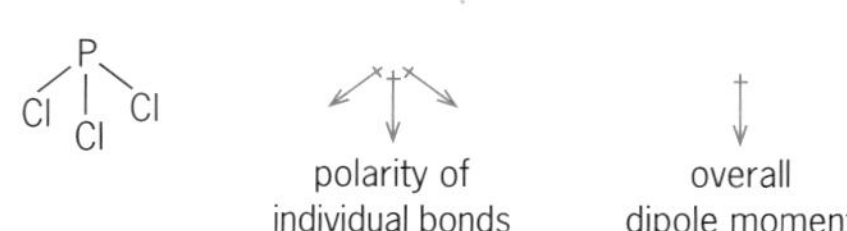

Fig 21.10 The displayed formula of phosphorus(III) chloride

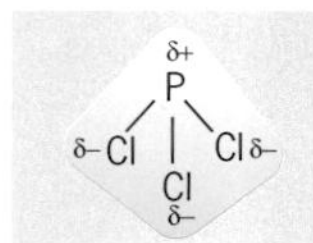

Fig 21.11 The polar nature of the phosphorus(III) chloride molecule

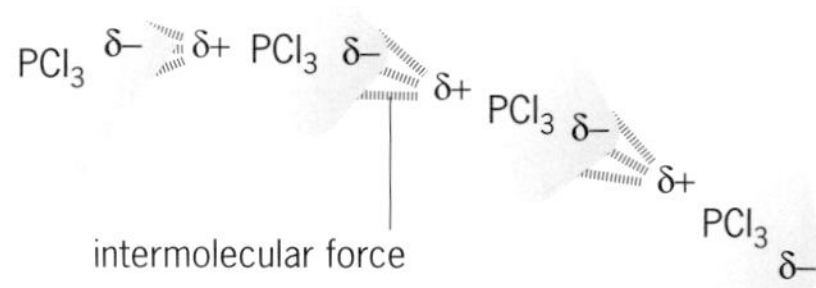

Fig 21.12 Intermolecular forces in phosphorus(III) chloride

Phosphorus(V) chloride

Also known as phosphorus pentachloride, phosphorus(V) chloride is a white crystalline solid at room temperature. This means that the forces of attraction between particles within the crystal lattice must be stronger than those in the liquid phosphorus(III) chloride.

Phosphorus(V) chloride is unusual in that the particles present in the solid and gaseous phase are different. In the gaseous phase, it is molecular phosphorus(V) chloride, PCl_5, whereas in the crystalline solid there is an interesting ionic type of interaction involving PCl_4^+ and PCl_6^- (Fig 21.14).

QUESTION 8

Boron trichloride, BCl_3, does not have a dipole moment. Why?

Cl
|
B
Cl Cl

Fig 21.13 The structure of the BCl_3 molecule

The dot-and-cross diagram for molecular phosphorus(V) chloride is given on page 77.

For further details on ionic bonding, read pages 71–73.

QUESTION 9

9 Why does phosphorus(V) chloride have a higher melting point than phosphorus(III) chloride?

These positive and negative ions are arranged in an ordered pattern. As the positive and negative charges are spread over a large particle, the attraction between the PCl_4^+ and PCl_6^- ions is much less than that in an ionic solid like sodium chloride, with ions that have small radii.

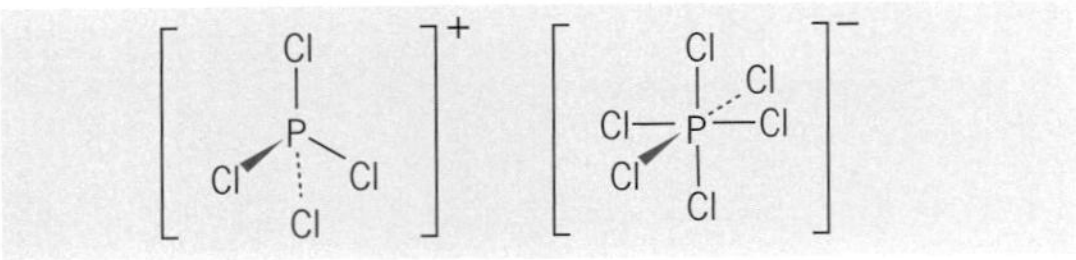

Fig 21.14 The structure of the particles present in solid phosphorus(V) chloride

Chlorides of sulfur

We have already noted in Table 5.6 (page 109) that the oxidation state of the most stable chloride of sulfur is +1 rather than the oxidation state that would be predicted from looking at the other chlorides of the elements in Period 3. Sulfur forms S_2Cl_2 and the much less stable chlorides SCl_2 and SCl_4. S_2Cl_2 is a pale yellow, fuming liquid formed when chlorine reacts with molten sulfur:

$$2S + Cl_2 \rightarrow S_2Cl_2$$

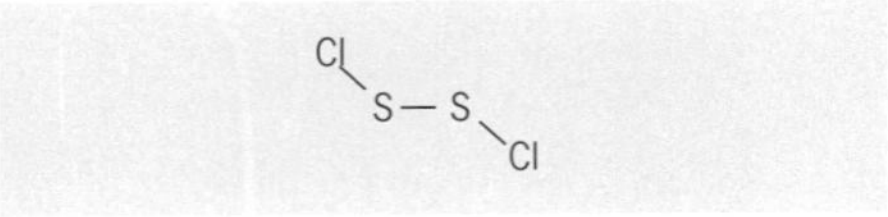

Fig 21.15 S_2Cl_2 has a similar shape to that of H_2O_2

S_2Cl_2 is a simple molecule with weak intermolecular attractions, and so it has a low melting point and does not conduct electricity.

Hydrolysis of molecular chlorides

Covalently bonded chlorides behave in a completely different way from ionic chlorides when they are added to water. Ionic chlorides dissolve in water, but covalently bonded chlorides are hydrolysed. Hydrolysis involves the reaction of a compound with water, which results in the decomposition or splitting up of the water.

Often an acidic solution is formed after hydrolysis of a molecular chloride. For example, when silicon(IV) chloride is dropped into water, an immediate reaction takes place with the formation of a strongly acidic solution and a white precipitate of silicon dioxide (silica):

$$SiCl_4(l) + 2H_2O(l) \rightarrow SiO_2(s) + 4HCl(aq)$$

REMEMBER THIS

Carbon has no 2d subshells.

Since carbon is the same group as silicon, you might expect tetrachloromethane (carbon tetrachloride) to behave in the same way and be hydrolysed to give carbon dioxide and hydrochloric acid. This is not the case. In fact, tetrachloromethane is inert towards water or steam. What causes this difference in reactivity?

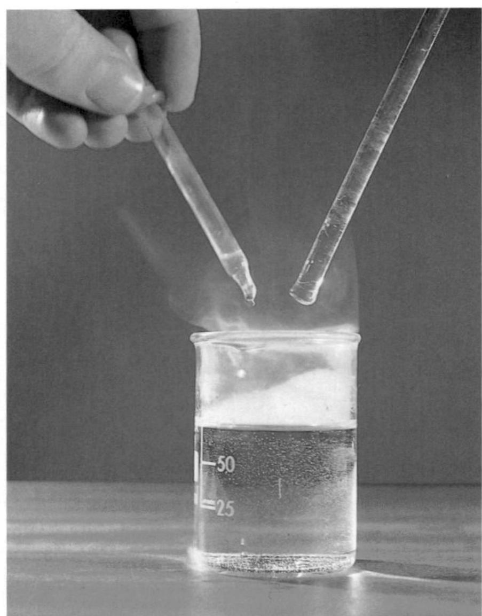
50
25

Fig 21.16 The reaction of silicon(IV) chloride with water

The mechanism of the hydrolysis of silicon(IV) chloride is believed to involve water molecules forming temporary bonds with the central silicon atom. Electrons from the lone pair on oxygen can be donated into vacant 3d subshells, which are sufficiently low in energy to be available. In the case of carbon, there are no energy levels of sufficiently low energy available. This means that the water molecule cannot temporarily form a bond with the carbon atom, and so no reaction takes place.

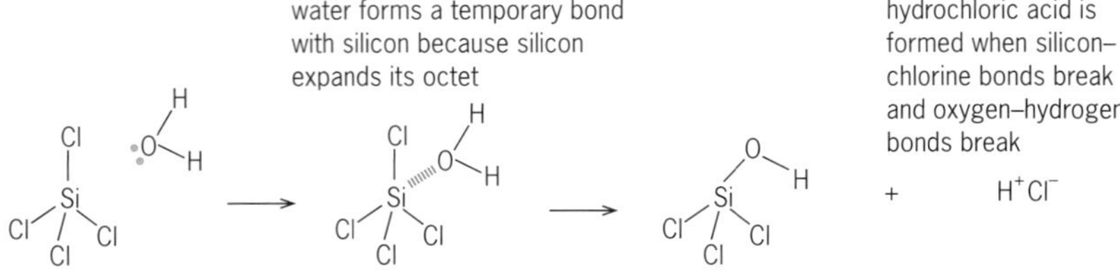
water forms a temporary bond with silicon because silicon expands its octet
hydrochloric acid is formed when silicon–chlorine bonds break and oxygen–hydrogen bonds break
+ H⁺Cl⁻

Fig 21.17 Possible mechanism of hydrolysis

You may want to read about expanding the octet on page 77.

Both phosphorus chlorides are hydrolysed by cold water to form highly acidic solutions, since hydrochloric acid and either phosphoric(III) acid or phosphoric(V) acid are formed:

$$PCl_3(l) + 3H_2O(l) \rightarrow H_3PO_3(aq) + 3HCl(aq)$$

$$PCl_5(s) + 4H_2O(l) \rightarrow H_3PO_4(aq) + 5HCl(aq)$$

Tetrachloromethane

Many household 'spot' cleaners used to be composed of chlorinated hydrocarbons such as tetrachloromethane. Its use has been banned as part of the London revision of the Montreal Protocol of 1987. This international agreement was one of the first pieces of global environmental legislation to result in the banning of chemicals believed to be responsible for environmental damage, in this case ozone depletion. In fact, in the United Kingdom, even the manufacture of tetrachloromethane has been banned.

Tetrachloromethane is a tetrahedral molecule with the same shape as silicon(IV) chloride and methane. It is non-polar, although it does contain polar covalent bonds because of the presence of the very electronegative chlorine atoms. Tetrachloromethane is able to dissolve other non-polar materials, such as fats and greases (Fig 21.18). The tetrachloromethane molecule forms intermolecular bonds with the fat molecule of the induced dipole–induced dipole type. This helps the dissolving process.

Water-based solvents often cannot dissolve such stains. Water is a polar molecule and so forms strong intermolecular attractions, often hydrogen bonds, with other molecules that are polar. But fat molecules are non-polar, so cannot form strong intermolecular forces of attraction with water and are therefore not dispersed throughout water.

The use of tetrachloromethane or similar non-aqueous solvents to clean fat and grease off clothes is called dry cleaning, since it does not involve water. The use of tetrachloromethane was widespread because it is not hydrolysed by water. However, with the realisation of its implication in ozone depletion, coupled with the fact that it is a carcinogen and also a toxic substance associated with liver damage, there has been much research into finding a safer but just as effective solvent.

Liquid carbon dioxide is one of the new dry-cleaning fluids that have been developed.

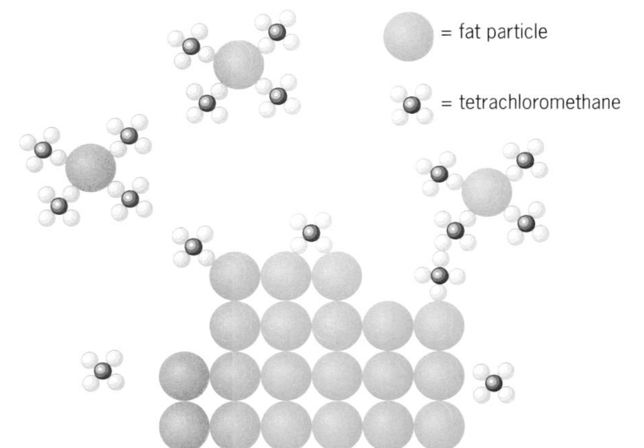

Fig 21.18 The dissolution of fat with tetrachloromethane

Hydrolysis of aluminium chloride

Anhydrous aluminium chloride is described on pages 414–415 as a simple molecular substance, so it would be expected to hydrolyse when added to water to give an acidic solution. When aluminium chloride is added to water, the molecular lattice is broken and the bonding changes to ionic, which produces aqueous aluminium ions $Al^{3+}(aq)$. These ions are surrounded by water molecules, with the polar negative end of the water molecule being attracted to the positive aluminium ion. So, it is more accurate to use the formula $[Al(H_2O)_6]^{3+}(aq)$ for the ions.

The aluminium ion has a high charge density, which distorts the electron clouds in a water molecule and so weakens one of the O–H bonds. This distortion is called polarisation (Fig 21.20 overleaf). The polarisation is so great that the O–H bond breaks, to form an aqueous hydrogen ion:

$$[Al(H_2O)_6]^{3+}(aq) \rightleftharpoons [Al(H_2O)_5(OH)]^{2+}(aq) + H^+(aq)$$

So the solution becomes acidic. In fact, it is sufficiently acidic to fizz immediately and give off carbon dioxide when sodium carbonate is added to it.

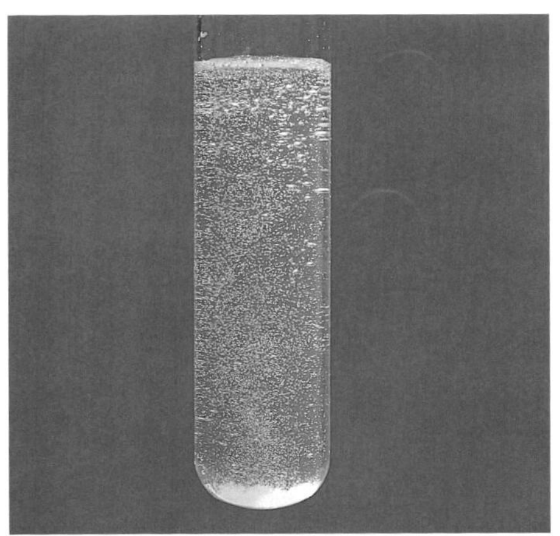

Fig 21.19 The reaction when sodium carbonate is put into aluminium chloride solution: bubbles of carbon dioxide appear

REMEMBER THIS

An ion with a high charge density has a small ionic radius and a large positive or negative charge. You can read more about charge density on page 440.

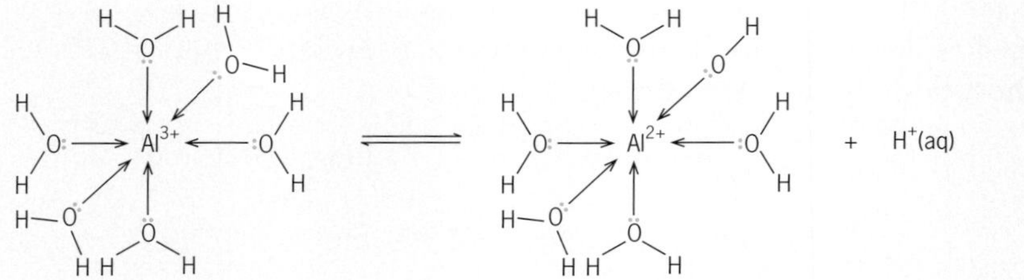

Fig 21.20 The polarisation of water molecules by the aluminium ion. Aqueous aluminium ions are surrounded by six water molecules. The high charge density breaks one of the O–H bonds, releasing a proton

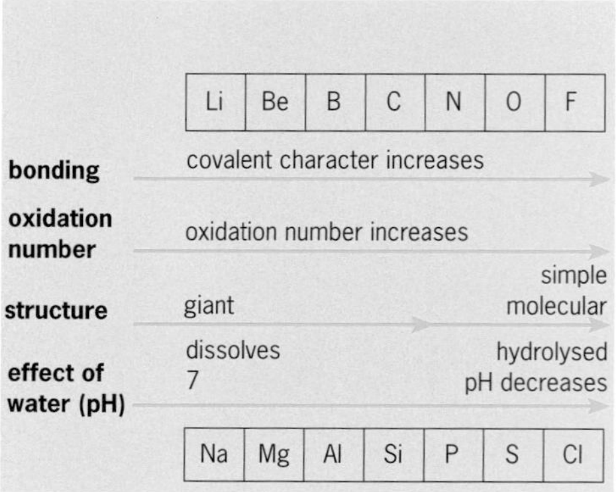

	Li	Be	B	C	N	O	F
bonding	covalent character increases →						
oxidation number	oxidation number increases →						
structure	giant → simple molecular						
effect of water (pH)	dissolves 7 → hydrolysed pH decreases						
	Na	Mg	Al	Si	P	S	Cl

Fig 21.21 A summary of the behaviour of chlorides across periods 2 and 3

The pH of aqueous magnesium chloride is slightly less than 7. This is because magnesium ions, Mg^{2+}, polarise water molecules in the same way as Al^{3+}. Mg^{2+} has a lower charge density than Al^{3+} because Mg^{2+} has a larger ionic radius and a smaller charge than Al^{3+}. So Mg^{2+} only weakly polarises water molecules and a much lower concentration of hydrogen ions are produced than with Al^{3+}.

3 PERIODICITY IN OXIDES

Fig 21.22 Sodium burning in oxygen

Fig 21.24 Aluminium burns in oxygen to form aluminium oxide, a white solid powder

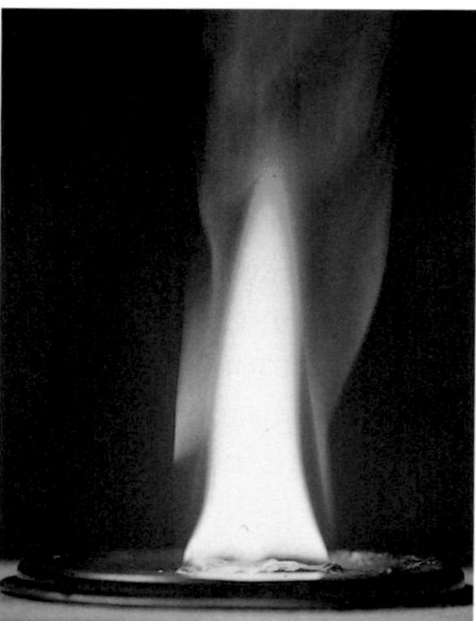

Fig 21.23 Phosphorus burns in air to give phosphorus(V) oxide, P_4O_{10}, a white solid powder

Fig 21.25 Sulfur burns in oxygen with a blue flame to give a colourless gas

Table 21.2

Element	Formula of oxide	Oxidation number of element in oxide	Appearance of oxide
sodium	Na_2O	+1	white solid
magnesium	MgO	+2	white solid
aluminium	Al_2O_3	+3	white solid
silicon	SiO_2	+4	white solid
phosphorus	P_4O_6	+3	white solid
	P_4O_{10}	+5	white solid
sulfur	SO_2	+4	colourless gas
	SO_3	+6	colourless gas
chlorine	Cl_2O	+1	brown-yellow gas
	Cl_2O_7	+7	colourless liquid

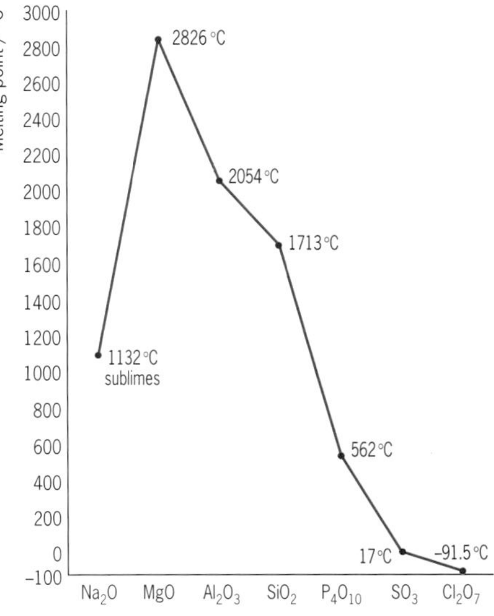

Fig 21.26 Graph of melting points of the oxides with the highest oxidation numbers

Oxygen reacts with almost all the elements of Period 3. Normally, oxides have oxygen with an oxidation number of −2, but there are oxides in which oxygen has an oxidation number of −1. Such oxides are called peroxides.

Over the next few pages, the oxides are discussed in terms of the change in their properties as one goes from one element to another across the period.

As in the case of the chlorides, there is a clear pattern in the maximum oxidation number of each element. Remember that elements in Period 3 can expand their octets, so that the highest oxidation state occurs when all of the outer electrons are used in bonding. Once the oxidation state of the element reaches 4 or above, the bonding between the atoms in the oxide becomes covalent. It is impossible to supply the energy during a chemical reaction for an atom to lose four or more electrons, so that sharing of electrons is the only option left.

The wide range of melting points (Fig 21.26) indicates that there are several types of structure and bonding within these oxides. In fact, they can have giant ionic structures, giant molecular structures and simple molecular structures. The structure with the highest melting point clearly has the strongest attraction between its particles.

QUESTION 10

a) Sodium burns in air to form sodium oxide:

$4Na(s) + O_2(g) \rightarrow 2Na_2O(s)$

Write down the equations for the formation of the following oxides by burning the element in oxygen:

MgO, Al_2O_3, SiO_2, P_4O_{10} and SO_2.

b) Which of the oxides in Table 21.2 involve bonding in which an atom has had to expand its octet?

c) Give the systematic names for each of the oxides in Table 21.2.

IONIC OXIDES

The metals sodium, magnesium and aluminium form ionic oxides. The metal atoms lose electrons, which are accepted to form oxide ions. As in the case of sodium chloride, the ions are packed in a giant lattice that consists of positive ions surrounded by negative ions, and vice versa.

Magnesium oxide has the same structure as sodium chloride. However, the magnesium ion has a smaller ionic radius than a sodium ion, because of its increased nuclear charge. Also, it has a +2 charge rather than a +1 charge. This means that the positive charge is much more concentrated around a magnesium ion than around a sodium ion. It is said to have a higher charge density. In exactly the same way, the oxide ion has a −2 charge rather than a −1 charge, and its radius is smaller than that of the chloride ion. So, the oxide ion has a greater charge density than the chloride ion.

Ionic radii are dealt with on page 98.

Fig 21.27 Relative sizes and charges of the sodium ion and the magnesium ion

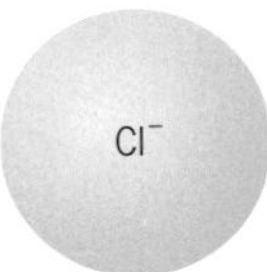

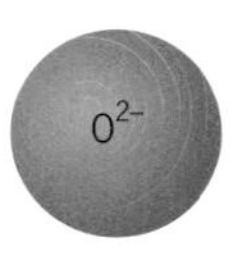

Fig 21.28 Relative sizes of the chloride ion and the oxide ion

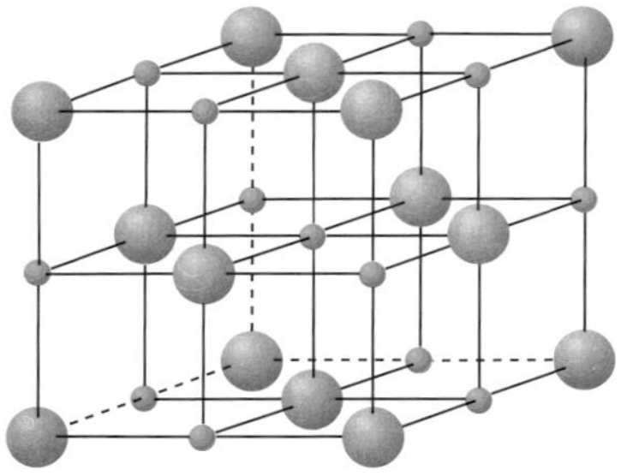

Mg^{2+}

O^{2-}

Fig 21.29 A section of the ionic lattice of magnesium oxide

With the magnesium ions and the oxide ions both having high charge densities, they have an extremely strong attraction for each other. Hence, it is difficult to separate the magnesium ions from the oxide ions in the crystal lattice of magnesium oxide. So this oxide has a very high melting point. Magnesium oxide is therefore widely used as a refractory material, such as for the linings of high-temperature furnaces.

Aluminium oxide

Aluminium oxide is another ionic lattice, although there is some degree of covalent character to the bonds. This happens because the aluminium ion has a large positive charge and a very small radius. It can therefore polarise oxide ions. Nevertheless, aluminium oxide has a very high melting point.

REMEMBER THIS

Aluminium oxide is said to have intermediate bonding (ionic with some degree of covalent character).

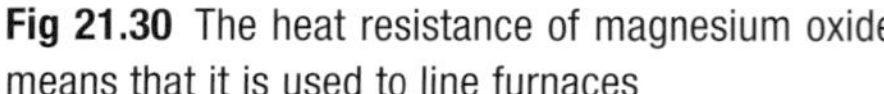

Fig 21.30 The heat resistance of magnesium oxide means that it is used to line furnaces

Aluminium oxide and anodising

Over 3000 different articles in daily use are made from aluminium. Among these are cooking utensils, kitchen appliances, kitchen foil, electrical conductors, and engineering and building components. The metal has found so many applications because of its low density coupled with a remarkable resistance to corrosion. The plaque carried on board Pioneer 10 – the first constructed object to escape from the Solar System – was made from gold and anodised aluminium because of these two properties.

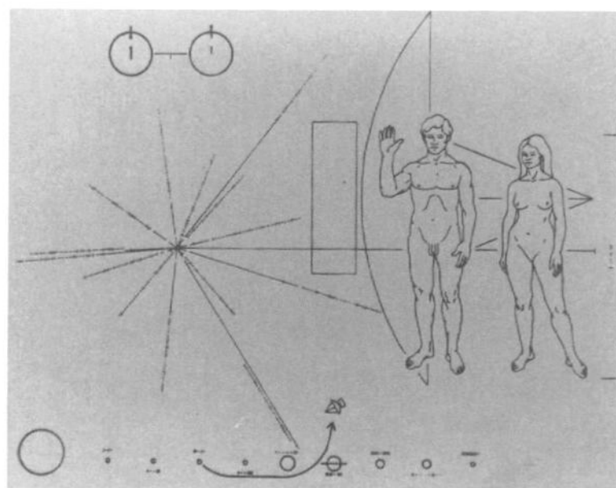

Fig 21.31 The plaque carried on board Pioneer 10

Aluminium normally has a dull lustre because it is coated in a thin, transparent surface film of aluminium oxide that forms on exposure to air. This layer of oxide protects the underlying metal from further reaction and oxidation. It is in complete contrast with the formation of rust on iron, which leads to further rusting. Aluminium can be made even more corrosion-resistant by increasing the thickness of the layer of aluminium oxide. This process is called anodising, in which the aluminium is the anode during the electrolytic decomposition of sulfuric acid.

Another advantage of anodising is that the layer of aluminium oxide formed can act as a 'mordant' and adsorb (bond to) coloured dyes. In this way, the aluminium surface can be given a more attractive coloured finish.

Fig 21.32 Eros, at Piccadilly Circus, London, is made of aluminium

Basic character of ionic oxides

Ionic oxides behave as bases. That is, they are able to accept a proton. This is because they contain the oxide ion, which reacts with two protons to give water:

$$O^{2-} + 2H^+ \rightarrow H_2O$$

Metals tend to form ionic oxides, which explains why metal oxides are basic.

If a metal oxide dissolves in water, it forms an alkaline solution. Most metal oxides are insoluble in water, or at the most are sparingly soluble. Sodium oxide, however, not only dissolves in water, but reacts with it to form sodium hydroxide:

$$Na_2O(s) + H_2O(l) \rightarrow 2NaOH(aq)$$

The ionic equation shows the formation of hydroxide ions:

$$O^{2-}(s) + H_2O(l) \rightarrow 2OH^-(aq)$$

QUESTION 11

a) Given that magnesium oxide and sodium oxides are basic, what type of substance do they react with?

b) What is the name of the salt formed when sodium oxide reacts with nitric acid?

c) What is the name of the salt formed when sodium oxide reacts with hydrochloric acid?

Amphoteric nature of aluminium oxide

Aluminium oxide does not dissolve in or react with water, but it does show basic properties in its reactions with acids. Aluminium oxide also shows acidic properties, since it reacts with alkalis.

An oxide that shows both basic and acidic properties is called an **amphoteric oxide**. Often, there is some degree of covalent character in the bonding in amphoteric oxides. When aluminium oxide is heated with aqueous sodium hydroxide, it forms a salt called sodium aluminate:

$$2NaOH(aq) + Al_2O_3(s) \rightarrow 2NaAlO_2(aq) + H_2O(l)$$

The formula of sodium aluminate is open to speculation. A common formula used is $Na_3Al(OH)_6$, as is $NaAlO_2$. Note that in this compound, aluminium is found in the *anion*.

Covalent character in ionic compounds is covered on page 87.

SILICON DIOXIDE: A GIANT MOLECULAR OXIDE

Fig 21.33 The geometric shape of quartz crystals indicates the regular pattern of their particles

Fig 21.34 A quartz crystal in electronic watches ensures accurate time

Silicon dioxide (silicon(IV) oxide) is also known as silica or quartz, and is the main constituent of sand. Quartz is a hard, brittle, clear, colourless solid. Among its many applications are architectural decorations, semi-precious jewels, optical components and frequency controllers in radio transmitters. It melts to form a viscous liquid. When this liquid is cooled, the particles have difficulty in taking up a regular pattern, and it supercools to form a glass called silica glass. Silica glass is also a useful substance, being inert towards most acids. A mixture of boron oxide and silicon dioxide, heated into a liquid and then cooled, forms borosilicate glass, which is heat resistant.

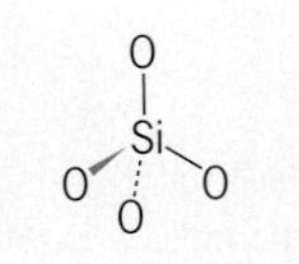

Fig 21.35 The displayed formula of the unit cell of silicon dioxide

The high melting point of quartz is attributed to its giant molecular structure. Each silicon atom is covalently bonded to four oxygen atoms in a structure like that of diamond (Fig 21.35). Like diamond, silicon dioxide is hard, insoluble in water and a non-conductor of electricity.

The structure of silicon dioxide is in complete contrast to that of carbon dioxide, which is a simple molecular lattice held together by weak intermolecular forces. Theoretically, carbon could form four C–O bonds or two C=O bonds. More energy is released in making two C=O bonds than four C–O bonds, so carbon dioxide has the displayed formula shown in Fig 21.36. But more energy is released in making four Si–O bonds than two Si=O bonds, so silicon(IV) oxide has the structure shown in Fig 21.37.

You can read more about the giant molecular structures of graphite and diamond on pages 143–144.

QUESTION 12

12 What are the differences between the structure of diamond and the structure of silicon dioxide?

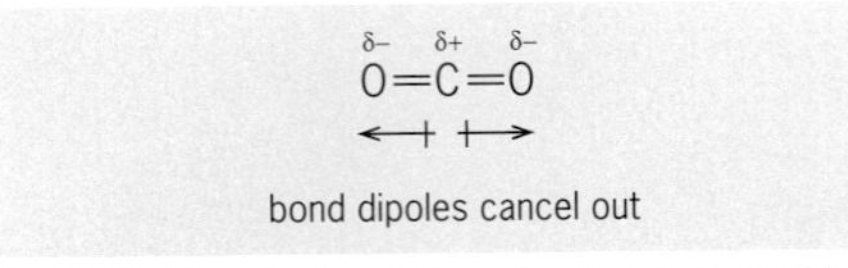

Fig 21.36 The displayed formula of carbon dioxide showing the bond dipoles

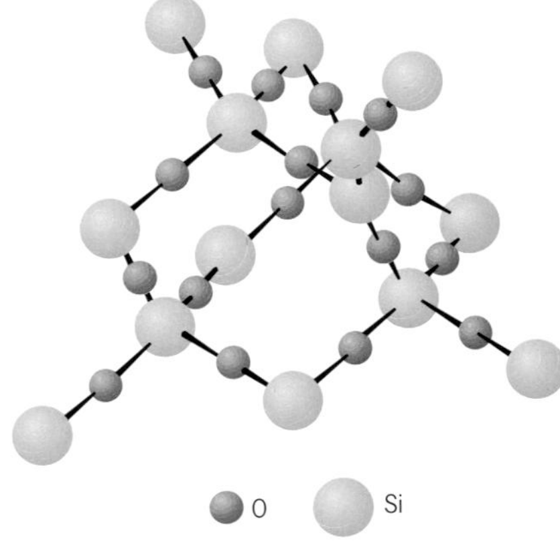

Fig 21.37 The structure of silicon(IV) oxide

SIMPLE MOLECULAR OXIDES

A simple molecular oxide has simple molecules that are held together in a solid by weak intermolecular forces. Just as for the simple molecular chlorides in the preceding section, the physical properties of these oxides will be similar; they have relatively low melting and boiling points and are non-conductors of electricity.

Oxides of sulfur, phosphorus and chlorine

Sulfur dioxide, SO_2 (Fig 21.38), sulfur trioxide, SO_3, phosphorus(III) oxide, P_4O_6, and phosphorus(V) oxide, P_4O_{10} (Fig 21.39), and the oxides of chlorine are all simple molecules and so form simple molecular lattices in the solid state. All these molecules have a permanent dipole. The negative end of one molecule can attract the positive end of another, which results in the formation of a weak intermolecular force. This force is a permanent dipole–permanent dipole attraction.

Since only weak intermolecular forces exist in all of these oxides, they almost all have relatively low melting points. P_4O_{10} has quite a high melting point because it is a large molecule. Therefore, in addition to the permanent dipole–permanent dipole attraction, there are significant forces of attraction through the induced dipole–induced dipole interaction.

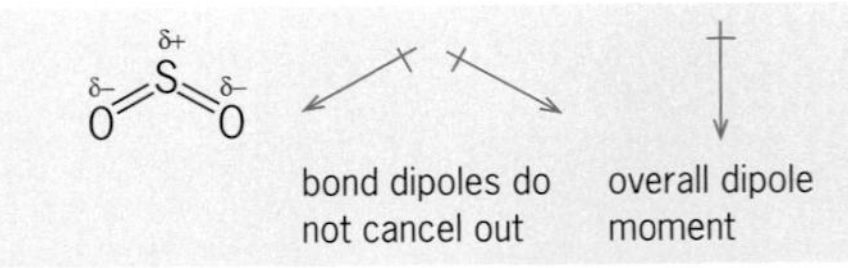

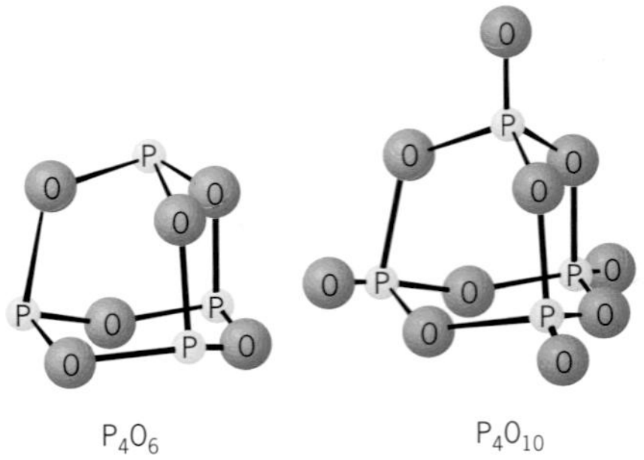

Fig 21.38 The structure of sulfur dioxide and its dipole moments

P_4O_6 P_4O_{10}

Fig 21.39 The structures of phosphorus oxides

Acidic behaviour of covalent oxides

In Period 3 the non-metals form covalent oxides and these oxides are acidic. If an acidic oxide dissolves in water, it will form an acidic solution.

For example, sulfur dioxide reacts reversibly with water to make sulfurous acid whereas sulfur trioxide reacts to give sulfuric acid.

$$SO_2(g) + H_2O(l) \rightleftharpoons H_2SO_3(aq)$$

$$SO_3(g) + H_2O(l) \rightarrow H_2SO_4(aq)$$

Notice that because these reactions are not redox, the oxidation number of the sulfur remains unchanged during the reactions.

The two phosphorus oxides, P_4O_6, and P_4O_{10}, will react in a similar way to form phosphoric(III) acid and phosphoric(V) acid respectively.

$$P_4O_6(s) + 6H_2O(l) \rightarrow 4H_3PO_3(aq)$$

$$P_4O_{10} + 6H_2O(l) \rightarrow 4H_3PO_4$$

Chlorine(VII) oxide reacts with water to form chloric(VII) acid:

$$Cl_2O_7 + H_2O \rightarrow 2HClO_4$$

This is perhaps the strongest acid that has so far been discovered.

When an oxide does not dissolve in or react with water, its acidic properties are less obvious. For example, silicon dioxide does not react with water – if it did, there would be no sandy beaches. But it does react when heated with a basic oxide. So, when sodium oxide and silicon dioxide are heated together, sodium silicate is formed:

$$Na_2O(s) + SiO_2(s) \rightarrow Na_2SiO_3(s)$$

Strong alkalis should not be left for a long time in glass burettes and in bottles with glass stoppers, because the stoppers and stopcocks become fused. This happens because of the slow reaction of the alkali with the silica in the glass, which makes a silicate.

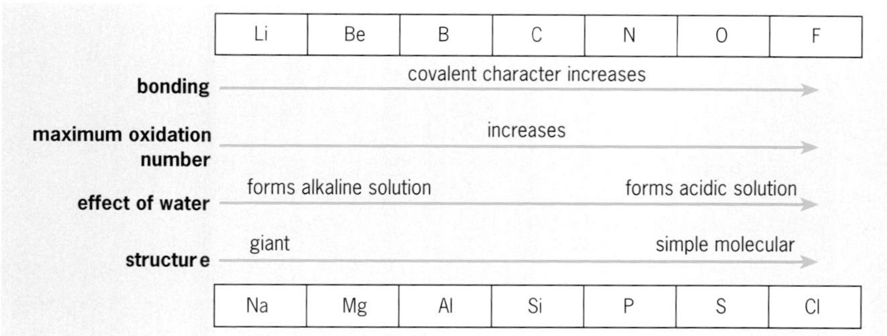

Fig 21.40 A summary of the properties of the oxides of the elements of periods 2 and 3

On page 13, there is a description of the reaction of calcium oxide with sand (silicon dioxide) to make calcium silicate or slag.

REMEMBER THIS

Some non-metals form neutral oxides, for example carbon monoxide. Metal oxides with the metal in a high oxidation state, e.g. chromium(VI) oxide, are also acidic because the bonding in the oxide is covalent.

QUESTION 13

What acid do you get when the following oxides are reacted with water?

a) dichlorine(I) oxide, Cl_2O

b) carbon dioxide

Write equations for the reactions.

QUESTION 14

Nitrogen dioxide reacts with water to form two acids, nitric(III) acid (nitrous acid) and nitric(V) acid (nitric acid).

a) Write an equation for this reaction.

b) Use changes in oxidation state to explain why this reaction is an example of a redox reaction.

SUMMARY

After studying this chapter, you should know the following:

- Non-metals often react by electron gain and are oxidising agents, whereas metals react by electron loss and are reducing agents.
- When any element reacts in a chemical reaction or is formed in a chemical reaction, a redox or electron transfer reaction has taken place.
- The maximum oxidation number attained by elements when combined with chlorine increases across Period 3. To obtain an oxidation number above +4, the atom has to expand its stable set of outer electrons beyond an octet.
- When a metal and a non-metal react together, the compound formed is normally ionic. When two non-metals react the compound formed is normally covalent.
- Ionic compounds are composed of a giant lattice of positive and negative ions held in place by strong electrostatic attraction. This results in ionic crystals that have high melting points.
- Ionic solids do not conduct electricity, since they have no mobile charge carriers. However, when molten or in solution, they can be electrolysed and conduct electricity because the ions are free to move.
- Giant molecular compounds are composed of atoms held together by strong covalent bonds. Giant molecular compounds have high melting and boiling points, do not dissolve in water and do not conduct electricity.
- Simple molecular compounds are composed of simple molecules held together by weak intermolecular forces.
- Simple molecular compounds have low melting and boiling points, and are non-conductors of electricity.
- The structures of the chlorides of the elements in Period 3 vary from giant ionic to simple molecular.
- Ionic chlorides usually dissolve in water to form neutral solutions, whereas molecular chlorides are hydrolysed by water to form hydrogen chloride or hydrochloric acid.
- The structures of the oxides of the elements in Period 3 vary from giant ionic to giant molecular to simple molecular.
- Metal oxides are basic or amphoteric in nature, non-metal oxides are acidic or neutral.
- Ionic oxides are basic and covalent oxides are acidic or neutral. An ionic oxide with a high degree of covalent character is amphoteric.
- Many non-metal oxides react with water to form aqueous acids.

Practice questions and a How Science Works assignment for this chapter are available at www.collinseducation.co.uk/CAS

22

The reactive metals

22 THE REACTIVE METALS

Calcium carbonate Coral reefs are composed mainly of the insoluble compound calcium carbonate

Calcium carbonate Chalk cliffs are composed of calcium carbonate

A carbon dioxide sink

Some people think of carbon dioxide as just the greenhouse gas that causes global warming problems, although of course carbon dioxide is essential for life as well. From earliest times on Earth and continuously since then, vast amounts of carbon dioxide from the atmosphere have been locked up in rocks of the Earth's crust as carbonates.

Several different carbonates occur in rocks, but by far the most abundant is that of the highly reactive metal calcium, calcium carbonate, which we see in the enormous chalk cliffs and widespread chalklands in the UK, and also in limestone deposits and coral reefs around the world.

Chalk, limestone and coral reefs exist only because calcium carbonate is an insoluble material, while other calcium salts are soluble. The calcium carbonate comes from the activity of marine animals and microscopic organisms that synthesise the carbonate as part of their life processes, but they can only do this by using calcium salts that are soluble in the waters around them. The calcium carbonate forms their shells, bones or coral protective structures, and as the organisms die and reach the ocean floor, these deposits feed into the rock-forming cycle.

Calcium carbonate formed at the bottom of seas and oceans is therefore a 'sink' for carbon dioxide. It has been suggested that, as atmospheric carbon dioxide increases, this will to some extent be offset by an increase in the carbon dioxide content of the oceans and, in turn, by an increase in the conversion of it into calcium carbonate by marine creatures.

1 REACTIVE METALS

The classification of elements into metals and non-metals is based upon the physical and chemical properties of the elements and their compounds. To most people, metals are hard, strong and shiny and are good thermal and electrical conductors. These properties describe the metals we are familiar with outside the chemistry laboratory, such as iron, lead, silver, gold, zinc, copper and nickel.

Other elements are also classified as metals, but they are less recognisable as such, in that they are soft and have a low melting point. It is only when their chemical properties are described that they are clearly seen to be metals. These include lithium, sodium, potassium, calcium and barium, which are found in the s block of the Periodic Table and are often collectively called the reactive metals.

Aluminium, in Group 3, is sometimes also referred to as a reactive metal. So aluminium is included in this chapter.

CHEMICAL PROPERTIES OF METALS

The main chemical property of a metal atom is its ability to lose one, two or sometimes three electrons to form a positive ion. When an atom loses electrons, it often attains a stable, noble-gas electron configuration. It is this single property that can be used to explain the reactivity of metals. Atoms of reactive metals easily lose electrons; atoms of unreactive metals lose them with difficulty.

? QUESTION 1

1 **a)** Francium, atomic (proton) number 87, is an alkali metal. Predict values for francium for its: i) melting point ii) boiling point; iii) atomic radius; iv) ionic radius.
b) In alkali metals, what does the change in the melting point suggest about the change in the strength of the metallic bonding? (Hint: read page 427.)
c) Water has a density of 1000 kg m^{-3}. Which of the alkali metals sink when placed in water? (See Table 22.1.)

The ability of a metal atom to lose electrons to form a positive ion explains why:

- metals are reducing agents;
- metals form ionic compounds with non-metals;
- metals do not normally form compounds with other metals.

The most reactive of all metals are those in the s block of the Periodic Table, since they have an electron configuration that contains only 1 or 2 electrons more than the nearest noble gas. This means that atoms of these elements lose electrons easily to form cations.

REMEMBER THIS

OIL RIG: Oxidation is Loss of electrons and Reduction is Gain of electrons. So metals are reducing agents because they give away electrons easily to an oxidising agent.

Table 22.1 Some properties of the first five alkali metals

Metal	Atomic (proton) number	Common oxidation state	Atomic radius/pm	Ionic radius (M^+)/pm	Density/kg m^{-3}	Melting point/°C	Boiling point/°C	Electron configuration
lithium	3	+1	152	60	534	180	889	$1s^2\ 2s^1$
sodium	11	+1	186	95	968	98	757	$1s^2\ 2s^2\ 2p^6\ 3s^1$
potassium	19	+1	231	133	856	63	679	$1s^2\ 2s^2\ 2p^6\ 3s^2\ 3p^6\ 4s^1$
rubidium	37	+1	244	148	1532	39	1326	$[Kr]5s^1$
caesium	55	+1	262	169	1870	29	690	$[Xe]6s^1$

2 GROUP 1: THE ALKALI METALS

The metals in Group 1 are collectively known as the alkali metals. There are six of them: lithium, sodium, potassium, rubidium, caesium and francium. They are the most reactive metals in their appropriate period of the Periodic Table.

These six elements are all remarkably similar in terms of both their chemical and physical properties. Being in the same group, we would expect the elements to have similar chemical properties (see page 93). However, that many of their physical properties are also similar and with, in some cases, an observable and predictable trend is unusual for most groups within the Periodic Table. For example, the melting and boiling points decrease with increasing atomic (proton) number.

It is interesting that the melting points of these elements are very low for metals, which makes them easy to melt. Advantage is taken of this property in some nuclear reactors, in which liquid sodium is used as the primary coolant because it also has a relatively high specific heat capacity and can be pumped easily through the pipes of the cooling system.

REMEMBER THIS

Lithium has the strongest metallic bonding and caesium the weakest metallic bonding of the alkali metals, because the lithium ions have the largest charge density and so the strongest attraction to the sea of delocalised electrons. This is why lithium has the highest boiling point and melting point and caesium the lowest.

OCCURRENCE AND EXTRACTION

Table 22.2 The natural abundances of the alkali metals in the Earth's crust

Element	Abundance (% by mass)	Common mineral
lithium	1.8×10^{-3}	lepidolite
sodium	2.63	rock salt and Chile saltpetre
potassium	2.40	sylvite and carnallite
rubidium	7.8×10^{-3}	lepidolite
caesium	3×10^{-4}	pollucite

The physical properties of the alkali metals, coupled with their high reactivity, limit the number of large-scale applications. Nevertheless, there is significant demand for lithium, sodium and potassium.

REMEMBER THIS

Specific heat capacity measures the energy needed to raise the temperature of 1 kg of a substance by 1 °C.

Fig 22.1 Sodium chloride is the principal source of the reactive metal sodium. Most sodium chloride is mined

The alkali metals are so reactive that they occur naturally only as compounds. It is impossible to reduce alkali metal compounds using a chemical process such as heating with carbon. The only successful method of reduction involves electrolytic decomposition. Francium, the last of the alkali metals to be discovered, is highly radioactive and, although its chemistry is easily predicted, it has not been fully determined experimentally. This is hardly surprising, since it is estimated that only 15 grams of francium occur in the whole of the Earth's crust.

Manufacture of sodium

Sodium is manufactured by the electrolysis of molten sodium chloride. The melting point of sodium chloride is 801 °C and so calcium chloride is added to lower the melting point to about 600 °C. The overall reaction is represented by the equation:

$$2NaCl(l) \rightarrow 2Na(l) + Cl_2(g)$$

Electrolysis is the only practical way to achieve this decomposition. Sodium ions are reduced at the cathode and chloride ions are oxidized at the anode:

$$\text{Cathode: } Na^+ + e^- \rightarrow Na$$
$$\text{Anode: } 2Cl^- \rightarrow Cl_2 + 2e^-$$

Since sodium is a highly reactive metal, it is important to ensure that the sodium and chlorine formed are not allowed to recombine. They must be produced in different parts of the electrolytic cell. Fig 22.2 shows the Down's cell used to electrolyse molten sodium chloride. A fine screen prevents the chlorine from diffusing to the cathode.

All the alkali metals are costly to make because of the amount of electrical energy needed.

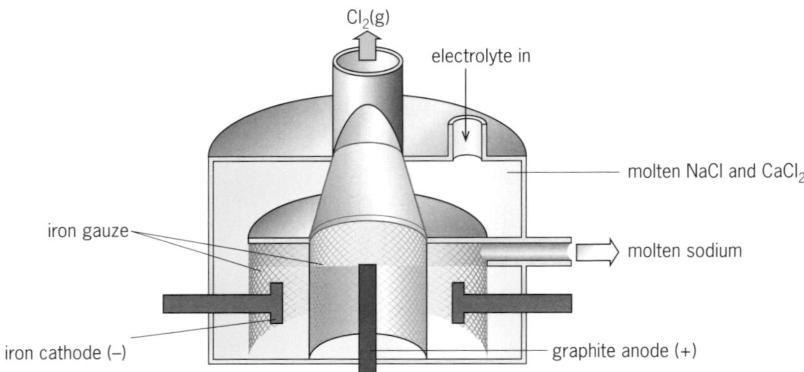

Fig 22.2 A Down's cell is used to manufacture sodium. An electrolyte of molten sodium chloride and calcium chloride is electrolysed using a graphite anode and an iron cathode. Molten sodium is collected as it floats on top of the molten electrolyte. A fine screen or gauze prevents the chlorine and sodium from recombining. The chlorine produced is collected and stored

? QUESTIONS 2–3

2 **a)** What would happen in the Down's cell if sodium and chlorine are allowed to come into contact with each other? Write an equation for the reaction.

b) Chlorine does not react with the graphite anode in the Downs cell. Suggest why the anode cannot be made from a metal such as iron.

3 Suggest how lithium can be extracted from lithium chloride.

3 REACTIONS OF THE ALKALI METALS

The alkali metals are good reducing agents because their atoms can easily transfer the outer electrons to non-metal atoms. Once electrons are lost, the resulting positive ions have noble gas electron configurations.

For example, the sodium atom ($1s^2\ 2s^2\ 2p^6\ 3s^1$) loses an electron to form the sodium ion ($1s^2\ 2s^2\ 2p^6$) with the same electron configuration as argon. This means that in all compounds the elements of Group 1 have an oxidation state of +1.

FIRST IONISATION ENERGY AND ELECTRODE POTENTIALS

Lithium has the highest of the first ionisation energy values, so it is expected to be the least reactive of the alkali metals.

Table 22.3 First ionisation energies of the Group 1 elements

Element	First ionisation energy/kJ mol^{-1}
lithium	519
sodium	494
potassium	418
rubidium	400
caesium	380

The first ionisation energy refers to the loss of an electron from a gaseous atom, which does not represent the situation for most reactions. A more appropriate way to describe the ease of electron loss from an alkali metal atom is to refer to its oxidation potential, E_{oxid}, since this refers to the reaction of a metal atom to form an aqueous metal ion. For example, the oxidation potential for lithium refers to the following half-equation:

$$Li(s) \rightarrow Li^{+}(aq) + e^{-}$$

This is precisely the reaction that occurs when lithium metal reacts with water or dilute acid.

The values of the oxidation potentials show that all the elements in Group 1 are highly reactive in that they can lose electrons easily.

Table 22.4 Oxidation potentials

Half-equation	Oxidation potential, E_{oxid}/volts
$Li(s) \rightarrow Li^{+}(aq) + e^{-}$	+3.05
$Na(s) \rightarrow Na^{+}(aq) + e^{-}$	+2.71
$K(s) \rightarrow K^{+}(aq) + e^{-}$	+2.93
$Rb(s) \rightarrow Rb^{+}(aq) + e^{-}$	+2.92
$Cs(s) \rightarrow Cs^{+}(aq) + e^{-}$	+2.92

The more positive an oxidation potential, the more feasible the reaction. (More information on oxidation potentials is given on page 507.)

Reaction with water

The alkali metals are so-called because of their reaction with water. All the metals react vigorously with water to form hydrogen and an alkaline solution.

Take lithium as an example (Fig 22.3). During the reaction, the metal reduces the water by losing electrons to form lithium cations and hydrogen. Lithium reacts with water to form aqueous lithium hydroxide and hydrogen:

$$2Li(s) + 2H_2O(l) \rightarrow 2LiOH(aq) + H_2(g)$$

The ionic equation more clearly describes why an alkali is formed, since aqueous hydroxide ions are formed:

$$2Li(s) + 2H_2O(l) \rightarrow 2Li^{+}(aq) + 2OH^{-}(aq) + H_2(g)$$

Fig 22.3 Lithium reacts with water to form aqueous lithium hydroxide and hydrogen. The water has turned pink because it contains phenolphthalein indicator

QUESTION 4

a) Write down the electron configuration for each of the following: Li^{+}, K^{+} and Rb^{+}.
b) Why is sodium more reactive than lithium? Use Table 22.3 to help you.

QUESTION 5

Predict the first ionisation energy for francium.

REMEMBER THIS

As the atomic (proton) number increases down the group, the number of inner-shell shielding electrons increases and the atomic radius increases. The outer electron is therefore attracted less strongly to the nucleus and so less energy is needed to remove it from the atom. This means that the first ionisation energy decreases within a group as the atomic (proton) number increases.

You can read about first ionisation energy and electron configuration on page 59 and page 95.

? QUESTION 6

6 **a)** Write down the balanced equations for the reaction of sodium and of potassium with water.
b) Predict what would happen if a piece of francium were added to cold water.

The reactivity of the alkali metals increases with increasing atomic number, so that the reaction of potassium with water is often accompanied by a lilac flame as the hydrogen produced burns (Fig 22.5).

Fig 22.4 Sodium reacts with water to form aqueous sodium hydroxide and hydrogen. A yellow flame occurs when the hydrogen produced burns. Notice that sufficient energy is transferred to the surroundings to melt the sodium into a sphere

Fig 22.5 Potassium is more reactive than sodium or lithium. It reacts violently with water, to produce hydrogen and aqueous potassium hydroxide. The reaction is exothermic and the energy transferred is sufficient to melt the potassium and to ignite the hydrogen formed

Reactions with acids

The alkali metals react explosively with dilute acids, such as hydrochloric acid and sulfuric acid:

$$2Li(s) + 2HCl(aq) \rightarrow 2LiCl(aq) + H_2(g)$$

$$2Na(s) + 2H^+(aq) \rightarrow H_2(g) + 2Na^+(aq)$$

They will even displace hydrogen from very weak acids such as alcohols, to form compounds called alkoxides. (The reactions of alkali metals with alcohols are covered on page 270.)

Reaction with air

Fig 22.6 Sodium, potassium and lithium can be easily cut by a knife to reveal a shiny, silver metal. This surface tarnishes quickly, with the formation of the oxide, hydroxide and eventually carbonate

Unlike most metals, the alkali metals are very soft and easy to cut with a knife. The surface obtained after cutting is shiny, but almost immediately tarnishes by reaction with moisture and/or oxygen from the air (Fig 22.6).

All the alkali metals react with air to form a complex mixture of compounds, including the corresponding carbonate. The following reaction scheme shows one way in which sodium may react with air:

$$4Na(s) + O_2(g) \rightarrow 2Na_2O(s)$$

$$Na_2O(s) + H_2O(g) \rightarrow 2NaOH(s)$$

$$2NaOH(s) + CO_2(s) \rightarrow Na_2CO_3.H_2O(s)$$

Reaction of alkali metals with oxygen

All alkali metals burn when heated in oxygen, to produce oxides. The combustion is always accompanied by a coloured flame characteristic of the element. Lithium burns with a red flame, sodium with a yellow flame and potassium with a lilac flame.

The oxides formed are not always the predicted M_2O, containing the M^+ and O^{2-} ions. Peroxides that contain the O_2^{2-} ion and superoxides that contain the ion O_2^- are formed with the more reactive alkali metals. The stability of the peroxides and the superoxides increases with the size of the cation. So, caesium oxide forms CsO_2 when burnt in excess oxygen, whereas lithium forms Li_2O:

$$4Li(s) + O_2(g) \rightarrow 2Li_2O(s)$$

$$Cs(s) + O_2(g) \rightarrow CsO_2(s)$$

REMEMBER THIS

The colour of a flame is a result of electron excitation and the consequent release of energy as the electron falls back to a lower energy level. You can read more about electron excitation on pages 53 and 557.

SCIENCE IN CONTEXT

Potassium superoxide

Superoxides are solids. They are very powerful oxidising agents, and can oxidise water to form oxygen. This is the reaction in some types of breathing mask used in mine rescue, where potassium superoxide is the source of the emergency oxygen supply to the wearer. Moisture reacts with the superoxide to provide this oxygen:

$$4KO_2(s) + 2H_2O(l) \rightarrow 4KOH(s) + 3O_2(g)$$

Note that 4 moles of potassium superoxide provide 3 moles of oxygen. This means that one gram of potassium superoxide can provide approximately 250 cm^3 of oxygen at room temperature and atmospheric pressure.

The attraction of this reaction is that, as the superoxide is used up, it makes potassium hydroxide, which removes carbon dioxide. This prevents the user from breathing in large quantities of carbon dioxide:

$$2KOH(s) + CO_2(g) \rightarrow K_2CO_3(s) + H_2O(l)$$

The overall effect is the removal of carbon dioxide and formation of oxygen. Potassium superoxide can also be used in submarines to provide oxygen, while at the same time it prevents the build-up of dangerous levels of carbon dioxide.

Fig 22.7 Gas masks protect miners and potholers from excess carbon dioxide in the air

QUESTION 7

7 Write down equations to show the reaction of:

a) sodium to form sodium peroxide, Na_2O_2;

b) potassium to form potassium peroxide;

c) rubidium to form rubidium superoxide.

4 COMPOUNDS OF THE ALKALI METALS

The common feature of almost all compounds of the alkali metals is their high solubility in water. All the compounds of alkali metals are ionic and so they have a giant ionic lattice. As a result, they have high melting points and can be electrolysed both in aqueous solution and as molten liquids.

REMEMBER THIS

A few ionic compounds of lithium with anions that have a large ionic radius have a small degree of covalent character.

You can read about the conductivity of ionic compounds on page 414 in Chapter 21.

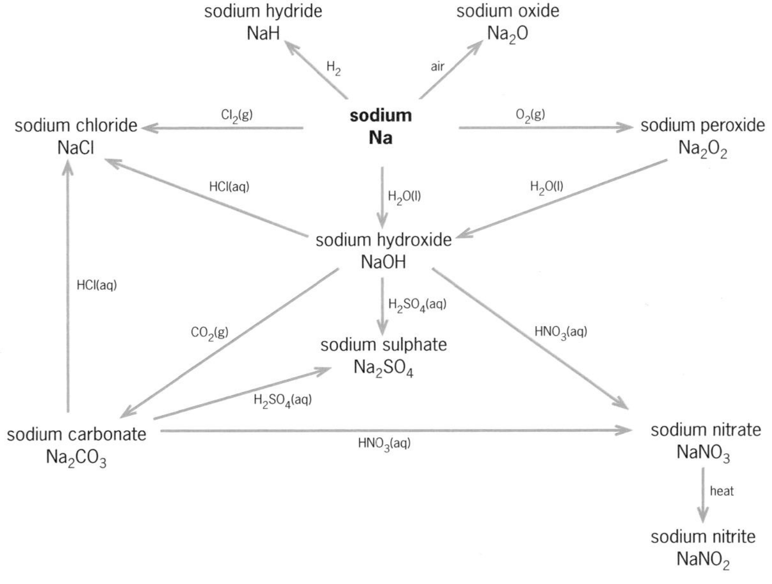

REMEMBER THIS

Sodium nitrate is sometimes called sodium nitrate (V) and $NaNO_2$ sodium nitrate (III)

Fig 22.8 Some important reactions of sodium and its compounds

? QUESTION 8

8 Write an equation to show the reaction of rubidium oxide with cold water. Include state symbols in your answer.

REMEMBER THIS

A soluble base forms an alkaline solution.

You can read about strong and weak acids on page 331 in Chapter 17 and about strong and weak bases on page 584 in Chapter 28.

The reaction of sodium hydroxide with chlorine is described on page 462 in Chapter 23.

ALKALI METAL OXIDES

The normal oxides of the alkali metals contain the O^{2-} ion, and so are basic and react with water to give aqueous hydroxides – alkalis. Bases are proton acceptors, so the oxide ion accepts a proton from water to form the hydroxide ion:

$$O^{2-}(aq) + H_2O(l) \rightarrow 2OH^-(aq)$$

Take sodium oxide, Na_2O, as an example:

$$Na_2O(s) + H_2O(l) \rightarrow 2NaOH(aq)$$

HYDROXIDES

Sodium hydroxide – commonly called caustic soda – is a strong alkali. In water, it is fully dissociated into aqueous sodium ions and aqueous hydroxide ions and is referred to as a strong base. All the alkali metal hydroxides are strong bases and react with both weak and strong acids to give a salt and water:

$$OH^-(aq) + H^+(aq) \rightarrow H_2O(l)$$

$$KOH(aq) + HNO_3(aq) \rightarrow KNO_3(aq) + H_2O(l)$$

Sodium hydroxide in qualitative analysis

In the laboratory, aqueous sodium hydroxide is used as a reagent to test for metal ions in solutions. Aqueous sodium hydroxide precipitates metal ions as their insoluble hydroxides. Very often, the hydroxides have characteristic colours that aid identification of the metal ion. Table 22.5 shows the reactions of aqueous hydroxide ions with various aqueous metal ions.

Table 22.5 Reactions of aqueous sodium hydroxide with metal ions in solutions

Metal ion in solution	Reaction with aqueous sodium hydroxide	Equation
calcium	white precipitate	$Ca^{2+}(aq) + 2OH^-(aq) \rightarrow Ca(OH)_2(s)$
magnesium	white precipitate	$Mg^{2+}(aq) + 2OH^-(aq) \rightarrow Mg(OH)_2(s)$
copper(II)	pale blue precipitate	$Cu^{2+}(aq) + 2OH^-(aq) \rightarrow Cu(OH)_2(s)$
iron(II)	green precipitate that slowly darkens	$Fe^{2+}(aq) + 2OH^-(aq) \rightarrow Fe(OH)_2(s)$
iron(III)	rust-red precipitate	$Fe^{3+}(aq) + 3OH^-(aq) \rightarrow Fe(OH)_3(s)$
cobalt(II)	blue precipitate that turns pink in excess aqueous sodium hydroxide with warming	$Co^{2+}(aq) + 2OH^-(aq) \rightarrow Co(OH)_2(s)$
nickel(II)	apple-green precipitate	$Ni^{2+}(aq) + 2OH^-(aq) \rightarrow Ni(OH)_2(s)$
manganese(II)	white precipitate that rapidly darkens in air	$Mn^{2+}(aq) + 2OH^-(aq) \rightarrow Mn(OH)_2(s)$ followed by $2Mn(OH)_2(s) + O_2(g) \rightarrow 2MnO_2.H_2O(s)$
chromium(III)	green precipitate that redissolves in excess to give a green solution of aqueous chromate(III) ion	$Cr^{3+}(aq) + 3OH^-(aq) \rightarrow Cr(OH)_3(s)$ followed by $Cr(OH)_3(aq) + 3OH^-(aq) \rightarrow [Cr(OH)_6]^{3-}(aq)$
silver(I)	dark-brown precipitate	$2Ag^+(aq) + 2OH^-(aq) \rightarrow Ag_2O(s) + H_2O(l)$
zinc	white precipitate that redissolves into a colourless solution in excess aqueous sodium hydroxide because of the formation of a soluble zincate ion	$Zn^{2+}(aq) + 2OH^-(aq) \rightarrow Zn(OH)_2(s)$ followed by $Zn(OH)_2(s) + 2OH^-(aq) \rightarrow [Zn(OH)_4]^{2-}(aq)$
lead(II)	white precipitate that redissolves into a colourless solution in excess sodium hydroxide because of the formation of soluble plumbate(II) ion	$Pb^{2+}(aq) + 2OH^-(aq) \rightarrow PbO(s) + H_2O(l)$ followed by $PbO(s) + 2OH^-(aq) \rightarrow PbO_2^{2-}(aq) + H_2O(l)$
aluminium	white precipitate that redissolves into a colourless solution in excess sodium hydroxide because of the formation of soluble aluminate ion	$Al^{3+}(aq) + 3OH^-(aq) \rightarrow Al(OH)_3(s)$ followed by $Al(OH)_3(s) + OH^-(aq) \rightarrow [Al(OH)_4]^-(aq)$

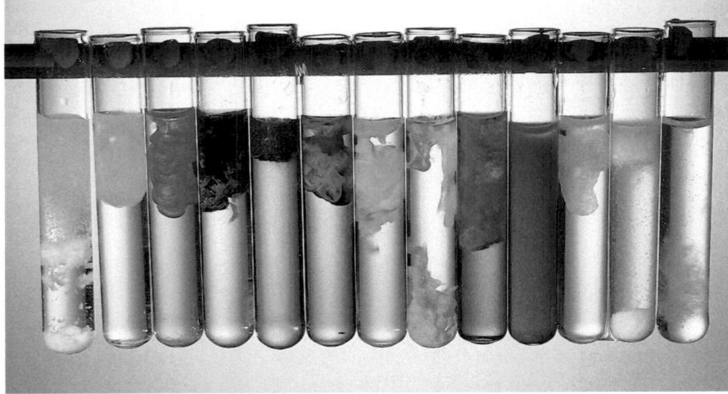

Fig 22.9 Aqueous sodium hydroxide can be used in qualitative analysis, because it gives characteristic coloured solutions and precipitates (lower row) with aqueous metal ions

OXY-SALTS

The nitrates, sulfates and carbonates are much more thermally stable than those of other metals. The sulfates and carbonates are thermally stable at Bunsen-burner temperatures, with the exception of lithium carbonate. Even the nitrates decompose to form nitrites and oxygen rather than the oxide and oxygen. This contrasts with most other metal nitrates, which thermally decompose to give the oxide or the metal itself. Again, the exception is provided by a lithium compound, since lithium nitrate forms lithium oxide:

$$4LiNO_3(s) \rightarrow 2Li_2O(s) + 4NO_2(g) + O_2(g)$$

Potassium nitrate thermally decomposes to form potassium nitrite:

$$2KNO_3(s) \rightarrow 2KNO_2(s) + O_2(g)$$

The alkali metals are the only group of metals that form stable, solid hydrogencarbonates. Sodium hydrogencarbonate is an important constituent in baking powder. It decomposes when heated to form sodium carbonate, carbon dioxide and water:

$$2NaHCO_3(s) \rightarrow Na_2CO_3(s) + H_2O(l) + CO_2(g)$$

The thermal stabilities of some of the carbonates and nitrates are described in more detail later in this chapter on pages 442 and 443.

REMEMBER THIS

Nitrates are sometimes called nitrate(V) and nitrites are called nitrate(III).

Fig 22.10 Baking powder contains sodium hydrogencarbonate. When heated in an oven, sodium hydrogencarbonate decomposes to form carbon dioxide, which helps the sponge cake to rise

Table 22.6 Occurrence of four Group 2 elements in the Earth's crust

Element	Abundance (% by mass)	Mineral
magnesium	1.93	magnesite ($MgCO_3$) dolomite ($CaCO_3.MgCO_3$) sea-water and brines
calcium	3.39	dolomite ($CaCO_3.MgCO_3$) marble and limestone ($CaCO_3$)
strontium	0.02	celestite ($SrSO_4$)
barium	0.04	barite ($BaSO_4$)

? QUESTION 9

9 **a)** Use the trends in the physical properties of the Group 2 elements to predict the density, melting point and boiling point of radium.

b) Explain the trend in the melting points in terms of structure and bonding.

5 GROUP 2: ALKALINE EARTH METALS

The elements in Group 2 are collectively called the alkaline earth metals. With the exception of beryllium, these elements are all closely similar. The metallic character (ease of loss of electrons) of the elements increases with increasing atomic (proton) number and, with the exception of beryllium, their compounds are almost all ionic.

OCCURRENCE AND EXTRACTION

Their compounds, particularly those of calcium and magnesium, are found extensively in the Earth's crust.

Electrolysis of the molten chloride of the element is the normal method of manufacture. The process is similar to that used to manufacture sodium. Magnesium is produced from molten magnesium chloride, much of it obtained from sea-water. It is estimated that 800 tonnes of sea-water are processed to obtain 1 tonne of magnesium.

Table 22.7 Properties of Group 2 elements

Metal	Atomic number	Oxidation number	Atomic radius/pm	Ionic radius (M^{2+})/pm	Density/ kg m^{-3}	Melting point/°C	Boiling point/°C	Electron configuration
beryllium	4	+2	111	31	1850	1278	2970	$1s^2 2s^2$
magnesium	12	+2	160	65	1740	651	1107	$1s^2 2s^2 2p^6 3s^2$
calcium	20	+2	197	99	1550	850	1490	$1s^2 2s^2 2p^6 3s^2 3p^6 4s^2$
strontium	38	+2	215	113	2540	770	1384	$[Kr]5s^2$
barium	56	+2	217	135	3500	704	1638	$[Xe]6s^2$
radium	88	+2	220	152	see Q9	see Q9	see Q9	$[Rn]7s^2$

SCIENCE IN CONTEXT

Teeth and bones

'Calcium is good for your bones and teeth.' Is this popular saying correct?

Teeth and bone are composed of two main constituents. One is a protein called collagen, and the other is a complex calcium phosphate compound with the approximate formula $Ca_{10}(PO_4)_6.(OH)_2$. This complex calcium phosphate is called hydroxyapatite. Bone acts as the body's calcium store, but calcium can be removed from it. Unless bones are kept under load, they begin to lose their calcium, which happens to astronauts who stay in space stations for a long time.

In teeth, the complex calcium phosphate undergoes changes in the presence of fluoride ions, since there is a partial replacement of the hydroxide ion in hydroxyapatite with fluoride ions to form fluoroapatite. The fluoroapatite is much more resistant to acids in the mouth and so its presence in teeth reduces decay. This is the reason why fluoride ions are added to toothpastes and some water supplies. It is also known that fluoride ions promote bone growth, but the mechanism is not yet fully understood.

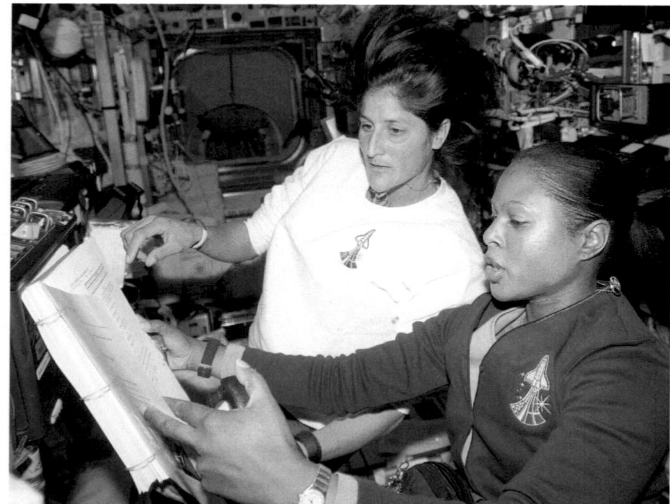

Fig 22.11 Prolonged space travel can lead to a drastic reduction in bone mass, because the calcium ions within the bone are reabsorbed. These astronauts are working in the International Space Station

6 REACTIONS OF THE GROUP 2 ELEMENTS

REMEMBER THIS

Be^{2+} is a very small, highly charged ion. It will strongly polarise negative ions, see page 87.

Atoms of Group 2 elements lose two electrons to form a stable electron configuration. Their reactivity increases with increasing atomic number. This is explained by the increasing ease with which electrons can be lost from the atoms. Beryllium is unusual in that it does not easily lose two electrons to form Be^{2+}, and many compounds of beryllium are covalent. So, with the exception of beryllium, Group 2 elements are reactive metals because they can very easily lose two electrons per atom. This means that the oxidation number of Group 2 elements in their compounds is always +2.

Table 22.8 Ionisation energy and oxidation potential of five alkaline earth metals. The sum of the first and second ionisation energies represents the energy transferred when two electrons are lost from each atom in a mole of gaseous atoms:

$$M(g) \rightarrow M^{2+}(g) + 2e^-$$

The oxidation potential is a measure of the ease of electron loss in the reaction:

$$M(s) \rightarrow M^{2+}(aq) + 2e^-$$

Metal	First ionisation energy/kJ mol^{-1}	Second ionisation energy/kJ mol^{-1}	Sum of first and second ionisation energies/kJ mol^{-1}	Oxidation potential, E_{oxid}/V
beryllium	900	1760	2660	+1.85
magnesium	736	1450	2186	+2.37
calcium	590	1150	1740	+2.87
strontium	548	1060	1608	+2.89
barium	502	966	1468	+2.91

The data in Table 22.8 demonstrate the increasing ease with which electrons can be lost. The reactivity of each Group 2 element is less than that of the Group 1 element in the same period because two electrons are lost per atom rather than one, which requires more energy. For example, potassium is considerably more reactive than calcium.

Since Group 2 elements lose two electrons per atom in reactions, it is the sum of the first and second ionisation energies that must be considered when comparing reactivities. The ionisation energy data and the electrode potential data both indicate that the reactivity of the Group 2 elements increases with atomic number.

You can read about ionisation energy and electron loss on page 59, and about oxidation potentials on page 507.

REACTION WITH WATER

As already noted, the reactivity of each Group 2 metal is lower than that of the Group 1 metal in the same period. This is exemplified by the reaction of the metals with water. There is virtually no reaction between magnesium and cold water, but with hot water it slowly forms magnesium hydroxide and hydrogen whereas sodium reacts violently with cold water. When hot steam is passed over heated magnesium (Fig 22.12), an exothermic reaction occurs with the formation of hydrogen:

$$Mg(s) + H_2O(g) \rightarrow MgO(s) + H_2(g)$$

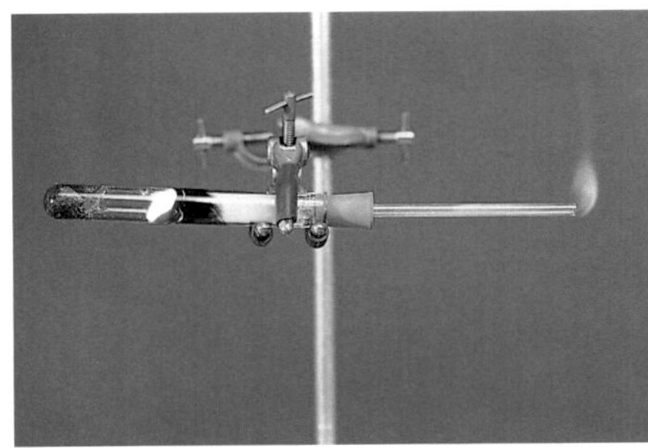

Fig 22.12 The reaction of magnesium with steam

Note that magnesium oxide is produced rather than magnesium hydroxide.

Calcium, the next metal in Group 2, is more reactive than magnesium. It reduces cold water to form an alkaline solution, aqueous calcium hydroxide, commonly known as lime-water:

$$Ca(s) + 2H_2O(l) \rightarrow Ca(OH)_2(aq) + H_2(g)$$

> **? QUESTION 10**
>
> 10 Write down equations to show the reaction of barium and strontium with cold water.

The ionic equation shows the formation of an alkaline solution with calcium, strontium, barium or radium.

$$M(s) + 2H_2O(l) \rightarrow M^{2+}(aq) + 2OH^-(aq) + H_2(g)$$

This means that the pH of water increases as the reaction proceeds, and since the basicity and solubility of the hydroxides increases down the group, the reaction with barium and water will give a higher pH value than that with calcium (approximately pH 9–11 for calcium).

REACTION WITH AIR AND OXYGEN

All the Group 2 elements tarnish in air to form a coating of the oxide. This reaction is rapid with those elements having a high atomic number. When these metals are heated in air or oxygen, they burn vigorously to produce the white ionic oxide. Magnesium burns in air and produces a brilliant white light (Fig 22.13):

$$2Mg(s) + O_2(g) \rightarrow 2MgO(s)$$

> **? QUESTIONS 11–13**
>
> 11 Explain why the reactivity of the Group 2 elements increases down the group. Use ideas about electron loss.
>
> 12 Explain why the reaction between magnesium and sulfuric acid is a redox reaction in terms of:
> **a)** electron transfer;
> **b)** changes in oxidation number.
>
> 13 Write balanced equations for the reactions between:
> **a)** calcium and hydrochloric acid;
> **b)** barium and hydrochloric acid;
> **c)** magnesium and dilute ethanoic acid, CH_3COOH.

Fig 22.13 Magnesium burns with a brilliant white flame

Fig 22.14 Calcium oxide emits white light when it is heated strongly. Blocks of calcium oxide heated by gas burners were used as stage lights in theatres before electricity was available. It was from this that the phrase 'being in the limelight' originated

REACTIONS WITH ACIDS

The Group 2 elements react with dilute acids to release hydrogen. The reactivity of the metals increases down the group, as it does with water. The reactions are an example of redox reactions, since the metal atom loses electrons (oxidation) and hydrogen ions gain electrons (reduction). This is represented by the overall ionic equation:

$$M(s) + 2H^+(aq) \rightarrow M^{2+}(aq) + H_2(g)$$

where M = Mg, Ca, Sr, Ba or Ra.

The redox reaction can also be explained in terms of the change in oxidation number. So when magnesium reacts with dilute hydrochloric acid, the oxidation number of magnesium changes from 0 to +2 (oxidation) and that of hydrogen from +1 to 0 (reduction):

$$Mg(s) + 2HCl(aq) \rightarrow MgCl_2(aq) + H_2(g)$$

7 COMPOUNDS OF THE GROUP 2 METALS

With the exception of beryllium, most of the compounds of the Group 2 metals are ionic. Although it is a property of ionic compounds that they are soluble in water, some compounds are insoluble and there are distinct trends in solubility with similar compounds. For example, the solubilities of the carbonates and sulfates decrease, but the solubilities of the hydroxides increase as the atomic number of the Group 2 element increases.

OXIDES AND HYDROXIDES

The oxides and hydroxides of Group 2 elements are basic and neutralise acids. The oxide ion or the hydroxide ion reacts with aqueous protons to form water:

$$O^{2-} + 2H^+ \rightarrow H_2O$$

$$OH^- + H^+ \rightarrow H_2O$$

The solubility of the hydroxides in water increases with increasing atomic number. Magnesium hydroxide is insoluble in water, whereas barium hydroxide is soluble. The reason for this change in solubility is explained on page 443. When water is added to the oxides, a comparable difference in solubility is observed, but this is because a chemical reaction takes place. For example, calcium oxide (quicklime) reacts with water to form aqueous calcium hydroxide or lime-water, which is sparingly soluble in water:

$$CaO(s) + H_2O(l) \rightarrow Ca(OH)_2(aq)$$

When water is dropped slowly onto the oxide, slaked lime (solid calcium hydroxide) is formed:

$$CaO(s) + H_2O(l) \rightarrow Ca(OH)_2(s)$$

Notice the importance of the state symbols in these two equations – without the symbols the equations would be identical, but they describe different processes.

REMEMBER THIS

When the oxides of Group 2 elements are added to water an alkaline solution results. The pH increases from pH 8–9 with magnesium oxide to pH 13–14 with barium oxide.

Uses of the oxides and hydroxides

The basic properties of the oxides and hydroxides are used extensively to neutralise acids in a variety of applications. Slaked lime is used to neutralize acid soils and lakes. A suspension of magnesium hydroxide in water, called 'milk of magnesia', is used to cure indigestion.

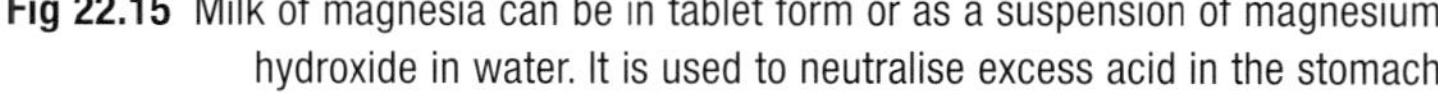

Fig 22.15 Milk of magnesia can be in tablet form or as a suspension of magnesium hydroxide in water. It is used to neutralise excess acid in the stomach

Fig 22.16 Farmers use slaked lime to neutralise acid soils

Fig 22.17 Most lakes in Scandinavia are highly acidic due to acid rain. The acid can be neutralised by spreading slaked lime

? QUESTION 14

14 **a)** Stomach acid is dilute hydrochloric acid. Write an equation to show the neutralization of stomach acid by milk of magnesia.
b) Acid rain contains acids, including dilute sulfuric acid and nitric acid. Write equations to show the neutralisation of acid rain with calcium carbonate.

REMEMBER THIS

The solubility of a substance gives the maximum amount of the substance that will dissolve in a known volume. The solubility of a substance varies with temperature, so it is important to compare solubilities at the same temperature.

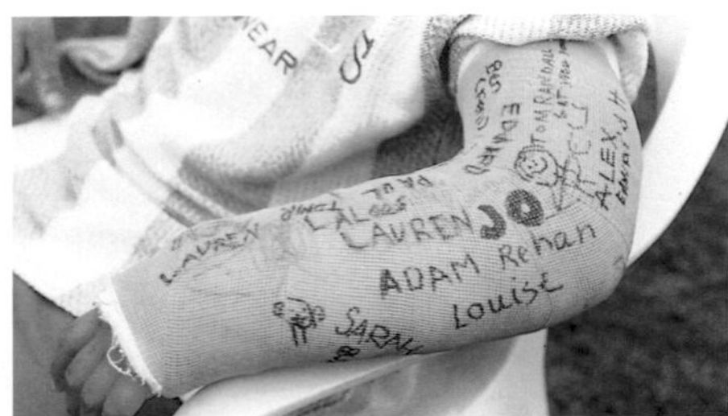

Fig 22.18 Plaster of Paris is a hydrate of calcium sulfate $(CaSO_4)_2.H_2O$. When water is added to this hydrate, it forms gypsum – a different hydrate of calcium sulfate, $CaSO_4.2H_2O$ – which sets solid. Plaster of Paris is used to make moulds and casts

Lime-water test for carbon dioxide

Aqueous calcium hydroxide (lime-water) is used to test for carbon dioxide. It gives a characteristic white precipitate often described as 'milky'. This is a further example of the basic behaviour of the hydroxides, since carbon dioxide is an acidic gas and reacts to form insoluble calcium carbonate:

$$Ca(OH)_2(aq) + CO_2(g) \rightarrow CaCO_3(s) + H_2O(l)$$

When more carbon dioxide is bubbled into the white precipitate, it redissolves to form a colourless solution of aqueous calcium hydrogencarbonate:

$$CaCO_3(s) + CO_2(g) + H_2O(l) \rightarrow Ca(HCO_3)_2(aq)$$

OXY-SALTS

The oxy-salts (such as the nitrates and sulfates) of the Group 2 elements are white ionic substances. They show distinct trends in their solubility and thermal decomposition, which are explained on pages 442 to 443 in terms of the bonding between ions.

Sulfates

The solubility of the sulfates of the Group 2 elements decreases as the atomic number of the element increases. The reason for this trend is described on pages 442 to 446.

The aqueous barium chloride or barium nitrate test for sulfate ions rests on the insolubility of barium sulfate. A solution of the test chemical is mixed with aqueous barium ions and dilute nitric acid. The sulfate ion is present in the test solution if a white precipitate is formed and no sulfur dioxide is formed:

$$Ba^{2+}(aq) + SO_4^{2-}(aq) \rightarrow BaSO_4(s)$$

Carbonates

The carbonates of the Group 2 elements are all considered to be insoluble in water. Therefore, each can be prepared through precipitation by reacting a soluble carbonate, such as aqueous sodium carbonate, with a soluble salt of the element:

$$CO_3^{2-}(aq) + M^{2+}(aq) \rightarrow MCO_3(s)$$

The thermal stability of the carbonates increases with increasing atomic number of the Group 2 elements. Beryllium carbonate decomposes so easily that it cannot be isolated at room temperature. Magnesium carbonate decomposes at 540 °C and barium carbonate at about 1360 °C. The reason for this variation in thermal stability is described on page 439.

$$MCO_3(s) \rightarrow MO(s) + CO_2(g)$$

where M = Be, Mg, Ca, Sr, Ba or Ra.
Calcium oxide (quicklime) is made in large quantities by the thermal decomposition of calcium carbonate (limestone). A temperature of between 900 and 1200 °C is needed, and it is important to remove the carbon dioxide produced, since the decomposition is reversible:

$$CaCO_3(s) \rightleftharpoons CaO(s) + CO_2(g)$$

Fig 22.19 Cement is made from calcium carbonate. Calcium carbonate and clay are heated strongly together to form a mixture of calcium silicate and aluminium silicate. When cement is mixed, first with a ballast of sand and gravel and then with water, it makes concrete

SCIENCE IN CONTEXT

Calcium carbonate, acid rain and FGD plants

Acid rain results from the presence in the air of sulfur dioxide and oxides of nitrogen. These gases react with the water and oxygen in the air to produce a mixture of acids, including dilute sulfuric acid and dilute nitric acid. A major source of these gases is the power station that uses fossil fuels contaminated with sulfur.

Coal and orimulsion (a mixture of a tar and water imported from South America) are particularly rich in sulfur and cause the most problems at power stations. Oxides of nitrogen are formed in the very high temperatures reached when the fossil fuels burn. They are sufficiently high to allow nitrogen and oxygen from the air to react directly to form nitrogen monoxide. When cool, the nitrogen monoxide reacts with more oxygen to give acidic nitrogen dioxide.

It has now become a high priority to remove these acidic gases from the waste or flue gases before they are emitted. A process called flue-gas desulfurisation (FGD) is being introduced, in which cold waste gases are treated with powdered calcium carbonate. The acidic gases react with the calcium carbonate to give calcium nitrate, calcium nitrite ($Ca(NO_2)_2$), and, mostly, calcium sulfite ($CaSO_3$):

$$SO_2(g) + CaCO_3(s) \rightarrow CaSO_3(s) + CO_2(g)$$

$$4NO_2(g) + 2CaCO_3(s) \rightarrow Ca(NO_2)_2(s) + Ca(NO_3)_2(s) + 2CO_2(g)$$

Almost 90 per cent of the sulfur dioxide and nitrogen dioxide produced by a power station can be removed by this process. The calcium sulfite produced is converted by reaction with oxygen into calcium sulfate, which is sold as gypsum and used in the building industry. The downside of the process is that large quantities of limestone have to be quarried to provide the calcium carbonate.

Fig 22.20 Power stations remove acidic gases, such as sulfur dioxide, by reacting them with powdered calcium carbonate

Nitrates

The nitrates of the elements of Group 2 are all white soluble solids that thermally decompose to produce the metal oxide, oxygen and nitrogen dioxide:

$$2M(NO_3)_2(s) \rightarrow 2MO(s) + 4NO_2(g) + O_2(g)$$

where M = Mg, Ca, Sr, Ba and Ra.

The thermal stability of the nitrates increases as the atomic number of the Group 2 elements increases. This trend is explained later on page 443.

REMEMBER THIS

The more thermally stable a compound, the higher the temperature required to decompose it.

STRETCH AND CHALLENGE

The atypical behaviour of beryllium compounds

Beryllium is not a typical Group 2 metal; in fact, it resembles aluminium in Group 3 more than magnesium in Group 2. The reason for this is the very small ionic radius of Be^{2+}, which gives the ion a similar charge density to that of Al^{3+}.

The beryllium ion, Be^{2+}, is very polarising and distorts the electron cloud of any anion to which it is bonded. This results in either decomposition of the compound, as in beryllium carbonate (which decomposes even at room temperature) or the compound being covalent rather than ionic. Beryllium chloride is a covalent chloride with a layer lattice as shown in Fig 22.21. This structure has each beryllium atom surrounded by four chlorine atoms, two bonded by dative bonds and two by covalent bonds. As a result the beryllium atom has a stable octet of electrons.

Beryllium hydroxide is also atypical in that it is amphoteric, whereas all the other hydroxides of Group 2 metals are basic, so it reacts with both acids and alkalis:

$$Be(OH)_2(s) + 2H_2O(l) + H_2SO_4(aq) \rightarrow [Be(H_2O)_4]^{2+}SO_4^{2-}(aq)$$

$$Be(OH)_2(s) + 2NaOH(aq) \rightarrow Na_2Be(OH)_4(aq)$$

Notice that in both the reactions a product is obtained in which beryllium has a coordination number of four, namely four groups around it that donate a pair of electrons to give a stable octet of electrons.

Cl Cl Cl Cl Cl
Be Be Be Be Be
Cl Cl Cl Cl Cl

Fig 22.21 The structure of beryllium chloride

REMEMBER THIS

The enthalpy change for the breakdown of an ionic lattice into gaseous ions is numerically the same as the lattice enthalpy, but positive. It is endothermic.

? QUESTION 15

15 Write down the equation for the reaction that corresponds to the lattice enthalpy for each of the following ionic solids:

a) MgO
b) $MgSO_4$
c) $Fe_2(SO_4)_3$
d) Mg_3N_2

REMEMBER THIS

A third factor that affects the magnitude of the lattice enthalpy is the type of ionic lattice formed. A discussion of this factor lies beyond the scope of this book.

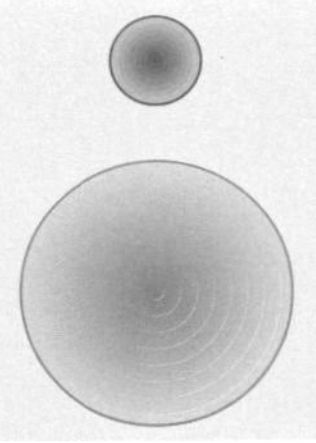
small ionic radius and high charge on ion produce a strong attraction for ion of opposite charge
large ionic radius and low charge on ion produce a weak attraction for ion of opposite charge

Fig 22.22 The magnitude of the attractive force between ions in a crystal lattice depends on both the ionic radius and the charge on the ion

Table 22.9 Charge density of 15 ions

Ion	Charge	Ionic radius/nm	Charge density
lithium	+1	0.060	277
sodium	+1	0.095	111
potassium	+1	0.133	57
rubidium	+1	0.148	46
beryllium	+2	0.031	2081
magnesium	+2	0.065	473
calcium	+2	0.099	204
strontium	+2	0.113	78
aluminium	+3	0.050	1200
oxide	−2	0.140	102
sulphide	−2	0.184	59
fluoride	−1	0.136	54
chloride	−1	0.181	31
bromide	−1	0.195	26
iodide	−1	0.216	21

8 LATTICE ENTHALPY

To explain the thermal decomposition and solubility of an ionic substance, the forces that exist between positive and negative ions in its ionic lattice must be understood. Chemists estimate the strength of attraction between a positive ion and a negative ion in an ionic lattice from the enthalpy change called **lattice enthalpy** or **lattice energy**.

Lattice enthalpy is the energy released into the surroundings when one mole of an ionic lattice is made from its constituent gaseous ions.

Therefore, the lattice energy is always negative. The more exothermic the lattice enthalpy, the greater the attraction between the positive and the negative ions.

For sodium chloride, the lattice enthalpy corresponds to the following reaction:

$$Na^+(g) + Cl^-(g) \rightarrow NaCl(s) \qquad \Delta H^{\ominus}_{le} = \text{lattice enthalpy}$$

FACTORS THAT AFFECT THE MAGNITUDE OF THE LATTICE ENTHALPY

The lattice enthalpy of each ionic compound is different and depends on the degree of attraction between the negative and the positive ions. This attraction depends on two factors: the ionic radius and the charge on the ion.

The smaller the ionic radius, the greater the density of the positive or negative charge. This gives a more exothermic value for the lattice enthalpy.

The greater the charge on the ion, the greater the attraction for an ion of the opposite charge. This also gives a more exothermic value for the lattice enthalpy.

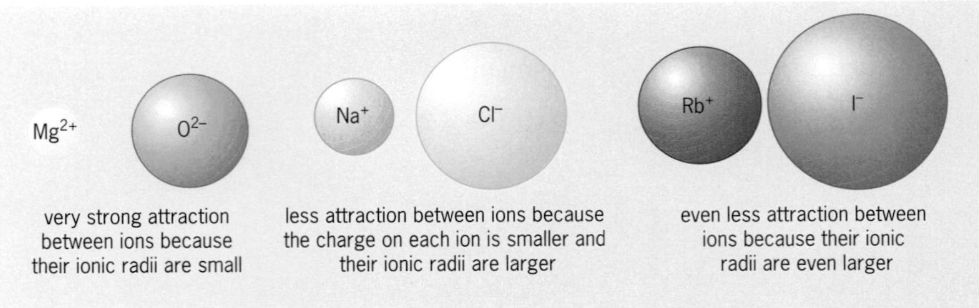
Mg2+
O2–
Na+
Cl–
Rb+
I–
very strong attraction between ions because their ionic radii are small
less attraction between ions because the charge on each ion is smaller and their ionic radii are larger
even less attraction between ions because their ionic radii are even larger

Fig 22.23 Relative sizes of, and attractions between, three pairs of ions. MgO has a greater lattice enthalpy than NaCl, which has a greater lattice enthalpy than RbI (scale: 1 mm to 20 nm)

Charge density

The ionic radius and the charge on the ion can be brought together and treated as a single property of the ion called the **charge density**. The charge density is based on the assumption that the charge is spread over the outer surface of the ion, and defined as the charge per unit surface area. Although ions are not solid, we can still think of ions as having a boundary surface.

The surface area is proportional to the square of the ionic radius, and so the charge density of an ion is given by:

$$\text{charge density} = \frac{\text{charge on ion}}{(\text{ionic radius})^2}$$

BORN–HABER CYCLES

Experimentally, it is impossible to use a direct method to determine the lattice enthalpy of an ionic solid. So an indirect method that involves Hess's law has to be used. Hess's law states that if a change can be brought about by more than one route, then the enthalpy change for each route must be the same, provided that the starting and finishing conditions are the same for each route. (There is more about Hess's law on page 167.)

Applying Hess's law to the determination of lattice enthalpy, there is more than one route from the gaseous ions to the ionic lattice. The direct route and the indirect route shown in Fig 22.24 is known as a Born–Haber cycle. All the enthalpy changes in the indirect route can be measured experimentally.

There are many ways to draw Born–Haber cycles, of which two are shown in Figs 22.24 and 22.25. Others appear elsewhere in this book.

REMEMBER THIS

Lattice enthalpies from Born–Haber cycles are experimental values. It is possible to calculate lattice enthalpies theoretically, but these are often quite different from experimentally determined values. Theoretical lattice enthalpies assume that the ionic lattice is totally ionic when, in fact, many ionic substances have a degree of covalent character.

QUESTION 16

16 Calculate the lattice enthalpy for calcium oxide given the following data:

- Standard enthalpy change of formation of calcium oxide = –635 kJ mol^{-1}
- Standard enthalpy change of atomisation of calcium = +178 kJ mol^{-1}
- First ionisation energy plus second ionisation energy of calcium = +1735 kJ mol^{-1}
- Standard enthalpy change of atomisation of oxygen = +249 kJ mol^{-1} of oxygen atoms
- First electron affinity plus second electron affinity of oxygen = +657 kJ mol^{-1}

One way to draw a Born–Haber cycle for sodium chloride is first to put down the equation for the formation of the lattice from its gaseous ions, and then to construct an alternative pathway. The pathway will always involve the energy changes shown for sodium chloride.

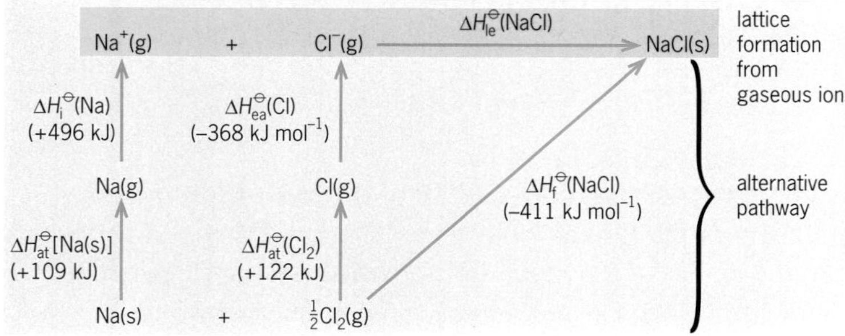

Key

$\Delta H^\ominus_{le}$ (NaCl) = lattice enthalpy of sodium chloride

$\Delta H^\ominus_i$(Na) = first ionisation energy for sodium

$\Delta H^\ominus_{at}$(Na) = enthalpy change of atomisation for sodium. This is the energy needed to make 1 mole of gaseous atoms from the element in its standard state

$\Delta H^\ominus_{at}(Cl_2)$ = enthalpy change of atomisation of chlorine

$\Delta H^\ominus_{ea}$(Cl) = first electron affinity for chlorine. This is the enthalpy change to make 1 mole of gaseous negative ions $X^-(g)$ from 1 mole of gaseous atoms X(g)

$\Delta H^\ominus_f$(NaCl) = enthalpy change of formation of sodium chloride

By Hess's Law:

$$\Delta H^\ominus_{le}(NaCl) = -\Delta H^\ominus_i(Na) - \Delta H^\ominus_{at}(Na) - \Delta H^\ominus_{ea}(Cl) - \Delta H^\ominus_{at}(Cl_2) + \Delta H^\ominus_f(NaCl)$$
$$= -496 - 109 - 122 + 368 - 411$$
$$= -770 \text{ kJ mol}^{-1}$$

An alternative way to draw a Born–Haber cycle for sodium chloride is to use a vertical energy axis.

The elements sodium and chlorine in their standard states have an enthalpy of formation of 0 kJ mol^{-1}. All endothermic reactions are shown by an arrow pointing upwards, and all exothermic reactions by an arrow pointing downwards. Notice that the lattice energy is exothermic.

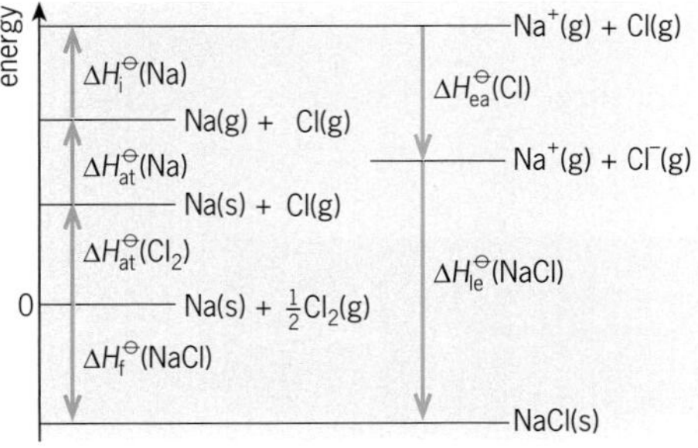

Fig 22.24 Born–Haber cycles for working out the lattice enthalpy of the ionic solid NaCl

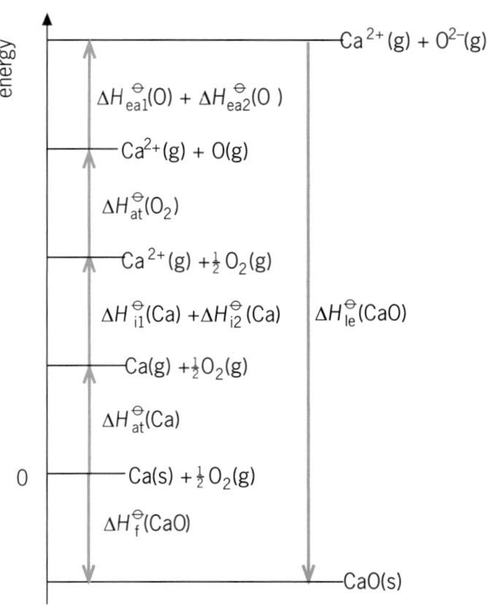

Fig 22.25 Born–Haber cycle for the formation of calcium oxide. Notice that the ionisation energy term for calcium involves both the first and second ionisation energies. In the same way, the electron affinity term involves both the first and second electron affinities: $O(g) + e^- \longrightarrow O^-(g)$ $\Delta H^\ominus_{ea1}(O)$
$O^-(g) + e^- \longrightarrow O^{2-}(g)$ $\Delta H^\ominus_{ea2}(O)$

9 THERMAL DECOMPOSITION OF IONIC SALTS

The thermal decomposition of an ionic salt must necessarily involve the breakdown of one ionic lattice and the formation of another. For example, when magnesium carbonate is decomposed, it forms magnesium oxide and carbon dioxide. This requires that the lattice of magnesium carbonate is completely broken down followed by the formation of the lattice of magnesium oxide. It therefore follows that the enthalpy change that occurs during this decomposition must be related to the lattice energies of the two ionic substances involved.

DECOMPOSITION OF THE CARBONATES

The carbonates of Group 2 elements decompose as follows:

$$MCO_3(s) \rightarrow MO(s) + CO_2(g)$$

where M = Be, Mg, Ca, Sr, Ba or Ra.

As already stated, the thermal stability of the carbonates increases as the atomic numbers of the Group 2 elements increase. This trend is best explained in terms of the ability of the metal ion to **polarise** the carbonate ion and of the exothermicity of the lattice energy for the oxide formed.

Polarisation

There is more information on polarisation on page 85.

Small, highly charged, positive ions can polarise negative ions. That is, they are able to distort the electron cloud of the negative ion. Negative ions that have a large ionic radius, such as the carbonate ion, are much more easily polarised than ions that have a small ionic radius, such as an oxide ion. Since the cations M^{2+} (where M is one of the Group 2 elements) all have the same charge, it is the ion with the smallest ionic radius that polarises the carbonate ion the most. Be^{2+} has the smallest ionic radius and so it causes most polarisation of the carbonate ion. Once the carbonate ion is highly polarised, it forms an oxide ion and carbon dioxide. This means that $BeCO_3$ is the least thermally stable and $RaCO_3$ the most thermally stable, since Ra^{2+} has the largest ionic radius.

Lattice enthalpy

An additional factor is the exothermicity of the lattice enthalpy of the oxide produced. This lattice enthalpy is more exothermic with a small cation than with a large cation. So, in the decomposition of magnesium carbonate, the formation of magnesium oxide with a highly exothermic lattice enthalpy can be considered as the driving force for the thermal decomposition.

Fig 22.26 Magnesium oxide is often used to line furnaces. The exceptionally large exothermic value for the lattice energy of magnesium oxide explains why it has a very high melting point

Decomposition of the carbonates of Group 1

The carbonates of Group 1 metals are much more thermally stable than those of Group 2 metals, because the cation has only a single rather than a double positive charge. Lithium carbonate is the least thermally stable, since Li^+ has a very small ionic radius, and hence a large charge density, and so can strongly polarise the carbonate ion:

$$Li_2CO_3(s) \rightarrow Li_2O(s) + CO_2(g)$$

THERMAL DECOMPOSITION OF NITRATES OF GROUPS 1 AND 2

The thermal stabilities of the nitrates of Group 2 elements show the same trend as the carbonates: namely, the stability increases with increasing atomic (proton) number. The reason for this trend in thermal stability is again that the polarisation power of the M^{2+} decreases as the ionic radius increases:

$$2M(NO_3)_2(s) \rightarrow 2MO(s) + 4NO_2(g) + O_2(g)$$

where M = Mg, Ca, Sr, Ba and Ra.

Since the Group 1 nitrates have ions that normally are much less polarising, the decomposition products are the nitrites rather than the oxides:

$$2MNO_3(s) \rightarrow 2MNO_2(s) + O_2(g)$$

where M = Na, K, Rb and Cs.

REMEMBER THIS

Lithium nitrate thermally decomposes to form lithium oxide, nitrogen dioxide and oxygen.

? QUESTION 17

Write balanced equations for the thermal decomposition of:

a) lithium carbonate, Li_2CO_3, to form lithium oxide, Li_2O;

b) lithium nitrate, $LiNO_3$, to form lithium oxide;

c) francium nitrate, $FrNO_3$, to form francium nitrite, $FrNO_2$.

10 SOLUBILITY OF IONIC SALTS

Most ionic salts dissolve in polar solvents such as water. A **solvent** is a liquid that can dissolve other substances. The most common solvent is water. A **solute** is the substance that is dissolved in the solvent. A **solution** is the name given to the mixture of the solute and the solvent. For example, a sodium chloride solution is simply a mixture of water (the solvent) and sodium chloride (the solute).

? QUESTION 18

What is the solvent and the solute in:

a) a mixture of alcohol and excess water;

b) a solution of excess alcohol and water?

SOLUBILITY

The solubility of a substance is defined as the maximum concentration that can be obtained when the substance (the solute) is dissolved in a solvent. This concentration is usually measured either in grams of solute per 100 grams of solvent or in mol dm^{-3}. Throughout this section of the book, the chosen unit of solubility is mol dm^{-3}.

Solubility changes with temperature, so it should always be quoted at a particular temperature, usually 298 K.

THE DISSOLVING PROCESS

This section considers what happens when an ionic salt dissolves in water.

In the regular arrangement of an ionic lattice, the cations are electrostatically attracted to the anions (see page 413). When an ionic salt dissolves, these electrostatic forces of attraction are broken and the anions and cations are free to move throughout the solution. In a polar solvent, such as water, the ions are surrounded by water molecules. Since the electronegativity of oxygen is much higher than that of hydrogen, the oxygen atom of a water molecule carries a small negative charge and so is weakly attracted to the cations. The hydrogen atoms of a water molecule carry a small positive charge and so they are weakly attracted to the anions. As a result, both cations and anions are surrounded by many water molecules.

When sodium chloride dissolves, aqueous sodium ions and aqueous chloride ions are formed:

$$NaCl(s) \rightarrow Na^+(aq) + Cl^-(aq)$$

The symbols $Na^+(aq)$ and $Cl^-(aq)$ represent the respective ion surrounded by many water molecules.

Overall, the dissolving process can be considered to be two processes:

- the breaking of the electrostatic attraction between ions in the ionic lattice;
- the formation of aqueous ions with attractive forces being established between the water molecules and these ions.

Fig 22.27 The dissolving process for an ionic solid dissolving in water

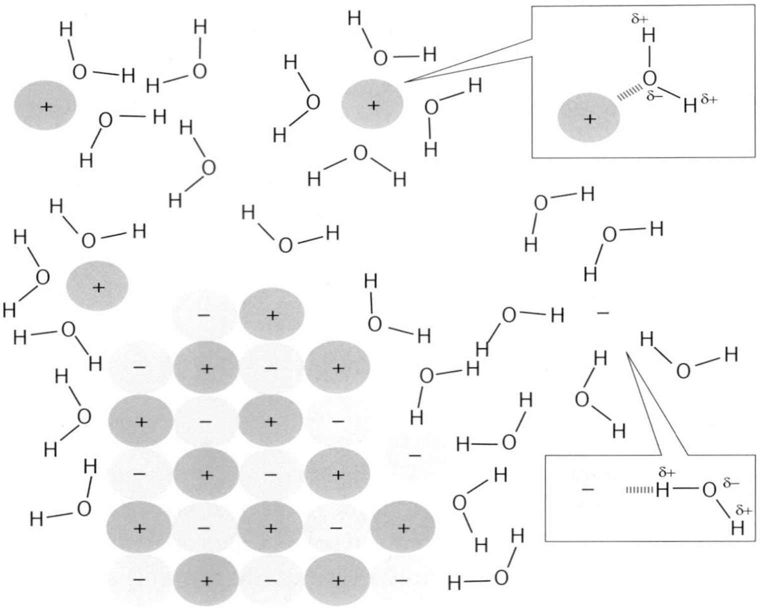

ENERGY TRANSFERS DURING DISSOLVING

When an ionic compound dissolves in water, there is always an energy transfer. Normally, the energy transfer process is modest and therefore there is only a small change in the temperature of the water. Energy is required to break the electrostatic attraction between ions, and energy is released when the attractive forces between water molecules and the ions are established. The numerical value for these two energy transfers (one endothermic and the other exothermic) is often very similar so that the overall energy change when making a solution is small. Occasionally, a dramatic change occurs in water temperature, but this is nearly always because a chemical reaction is taking place during the formation of the solution.

The **enthalpy change of solution,** $\Delta H_{\mathbf{soln}}$, is defined as the enthalpy change when one mole of solute is dissolved in a solvent and extra dilution causes no further change in enthalpy.

Entropy and dissolving

On page 165, entropy is described as a measure of the disorder of a system. When an ionic solid dissolves to make a solution, the system becomes more disordered (increases in entropy), because all the particles in the solution are free to move, whereas in the solid the particles were in fixed positions (lower entropy). Processes in which the entropy – the amount of disorder – increases are favoured over those in which the entropy does not increase. This means that all substances should dissolve in a solvent. But this is not the case because entropy changes in the surroundings must also be taken into account. The energy transfer processes offer a simpler explanation.

BORN–HABER CYCLES FOR DISSOLVING

Fig 22.28 shows that the enthalpy change of solution is equal to the sum of three different energy transfer processes. The first of these processes is the lattice enthalpy, the second is the enthalpy change of hydration of the cation and the third the enthalpy change of hydration of the anion. The **enthalpy change of hydration** ΔH_{hyd} is the enthalpy change that occurs when 1 mole of gaseous ion is dissolved in excess water to make one mole of aqueous ion.

Although the enthalpy change of hydration is very much a theoretical energy change, it does give a clear indication of the strength of the interaction of the water molecules with the ions. The enthalpy change of hydration is always exothermic, since it necessarily involves bond-making between water molecules and the ions. The lattice enthalpy contribution to the enthalpy change of solution is always endothermic, since it involves the breaking of the strong electrostatic interactions between the positive and the negative ions. The lattice enthalpy and the sum of two enthalpy changes of hydration have roughly the same magnitude (but are of opposite sign), so that the enthalpy change of solution is usually very small.

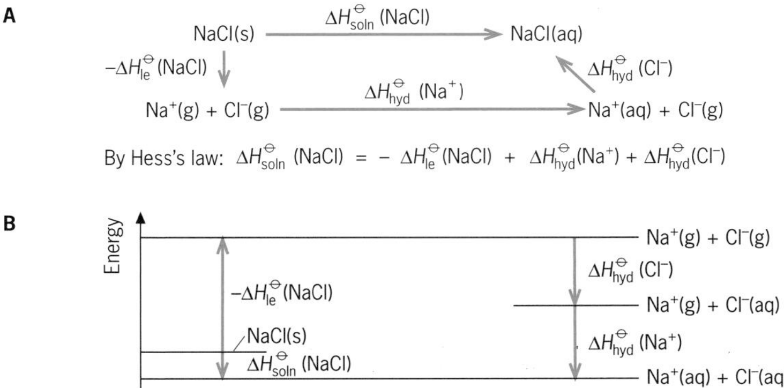

Fig 22.28 Two ways (A and B) of showing energy cycles for dissolving. The overall enthalpy change of solution is often small, since the exothermic contributions of the enthalpy changes of hydration are often cancelled out by the endothermic contribution of the lattice enthalpy

Enthalpy change of hydration and charge density

The magnitude of the enthalpy change of hydration depends on the ionic radius and the charge on the ion. The smaller the ionic radius and the higher the charge on the ion, the more exothermic the enthalpy change of hydration. This is because an ion with a small radius and a high charge has a high charge density, and therefore water molecules are strongly attracted to it. Conversely, an ion with a large radius and a small charge has a low charge density, and therefore water molecules are weakly attracted to it.

Within Group 2, the enthalpy change of hydration of M^{2+} becomes less exothermic as the atomic number increases:

$$M^{2+}(g) \xrightarrow{H_2O(l)} M^{2+}(aq) \quad \Delta H^{\ominus}_{hyd} = \text{enthalpy change of hydration}$$

SOLUBILITY TRENDS OF THE GROUP 2 SULFATES

On page 437, it is stated that the solubility of the sulfates decreases as the atomic number of the Group 2 elements increases. This trend can be explained in terms of the energy transfer processes that take place during dissolving.

As Fig 22.29 shows, the enthalpy change of hydration of the $SO_4^{2-}(g)$ is common to all Born–Haber cycles for all of the Group 2 sulfates, so it cannot be responsible for the solubility trend. The lattice energy of the Group 2 sulfates, $MSO_4(s)$, becomes less exothermic as the atomic number of M increases, but the change in lattice energy from $MgSO_4$ to $BaSO_4$ is quite small because of the large size of the sulfate ion.

? QUESTION 19

19 Write down the equation for the reaction that corresponds to the enthalpy change of hydration of:

a) a magnesium ion;

b) a sulfate ion.

Remember to include the state symbols in your equation.

? QUESTION 20

20 Calculate the enthalpy change of solution for LiCl and for NaCl, given the following enthalpy changes in kJ mol^{-1}:

ΔH_{le}(LiCl) = −848

$\Delta H_{hyd}(Li^+)$ = −499

$\Delta H_{hyd}(Cl^-)$ = −381

ΔH_{le}(NaCl) = −776

$\Delta H_{hyd}(Na^+)$ = −390

Table 22.10 Solubility of the Group 2 sulphates

Compound	Solubility at 298 K/mol dm^{-3}
magnesium sulfate	2.8
calcium sulfate	1.4×10^{-3}
strontium sulfate	7.6×10^{-5}
barium sulfate	1.1×10^{-6}

? QUESTION 21

21 Calculate the solubility of each of the Group 2 sulfates in grams per dm^3.

However, the difference in the enthalpy change of hydration of $M^{2+}(g)$ from Mg^{2+} to Ba^{2+} is large because of the big difference in ionic radii between the very small ion Mg^{2+} and the large ion Ba^{2+}. Referring to Fig 22.29, it is the enthalpy change of hydration of $M^{2+}(g)$ that determines the solubility trend. The more exothermic this process, the more likely $MSO_4(s)$ is to dissolve.

To sum up: the enthalpy change of solution of $MSO_4(s)$ becomes less exothermic as the atomic number increases, and hence the solubility decreases.

For the same reason, all salts of Group 2 elements that have an anion with a large ionic radius show the same solubility trend as the sulfates.

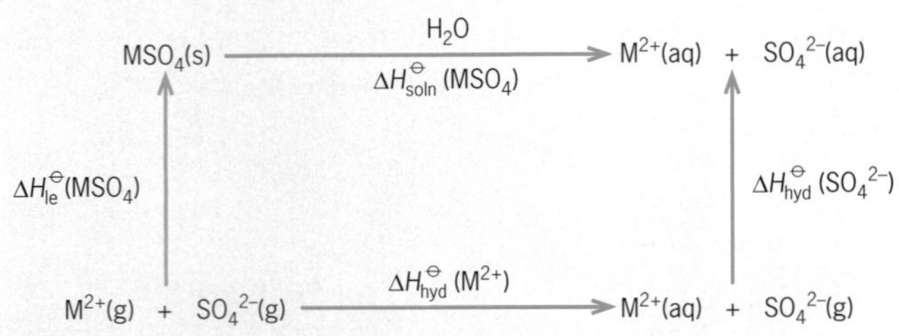

Fig 22.29 Born–Haber cycle for the determination of the enthalpy of solution, ΔH_{soln}, for the sulfates of Group 2. Note that the enthalpy change of hydration of $SO_4^{2-}(g)$ is common to all the sulfates of Group 2

SOLUBILITY TREND OF THE GROUP 2 HYDROXIDES

The solubility of hydroxides, $M(OH)_2(s)$, shows the opposite trend to that of the sulfates. Namely, as the atomic number increases, the solubility increases. This time, the key factor that determines solubility is the lattice enthalpy of the $M(OH)_2(s)$ rather than the enthalpy change of hydration. Since the hydroxide ion is small, there is a considerable difference between the very highly exothermic lattice enthalpy of magnesium hydroxide and the relatively small exothermic lattice enthalpy of barium hydroxide.

As Fig 22.30 shows, a large exothermic lattice enthalpy makes a large endothermic contribution to the enthalpy change of solution and so reduces solubility. Other salts of Group 2 elements with small anions, such as the fluorides, display a solubility trend similar to that of the hydroxides.

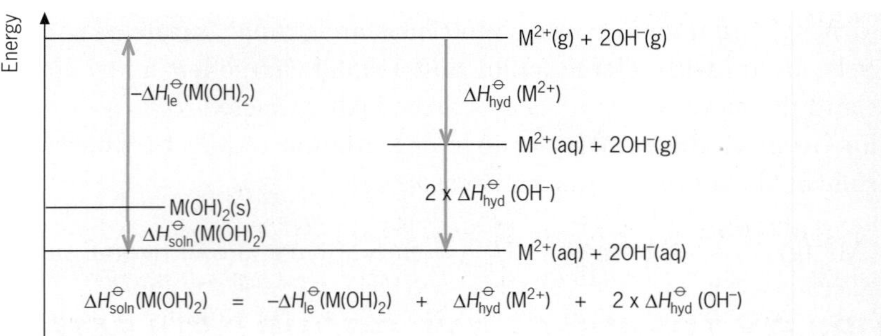

Fig 22.30 The key factor in determining the solubility of the hydroxides of Group 2 is the change in the lattice enthalpy from $Mg(OH)_2$ to $Ba(OH)_2$. $Mg(OH)_2$ has a much larger lattice enthalpy than $Ba(OH)_2$, so the enthalpy change of solution of $Mg(OH)_2$ is less exothermic than that of $Ba(OH)_2$

Table 22.11 Summary of solubility trends

Anion	Magnesium salt	Barium salt
large, e.g. carbonate and sulfate	high solubility	very low solubility
small, e.g. hydroxide and fluoride	low solubility	high solubility

11 SOLUBILITY PRODUCT

The term 'insoluble' is not very suitable, since no ionic substance is completely insoluble in water. The dissociation process takes place even with substances referred to as insoluble, such as barium sulfate and calcium carbonate. Sometimes, the term 'sparingly soluble' is used to express the fact that the solubility of a substance is low. The term 'insoluble' should be taken to mean that the solubility of a substance is extremely low.

SOLUBILITY PRODUCT

When calcium carbonate is shaken up with distilled water and left to settle, the water contains aqueous calcium ions and aqueous carbonate ions in very small concentrations. A dynamic equilibrium is set up that is best represented by the following equation:

$$CaCO_3(s) \rightleftharpoons Ca^{2+}(aq) + CO_3^{2-}(aq)$$

An expression can be written for the equilibrium constant for this process:

$$K_c = \frac{[Ca^{2+}(aq)][CO_3^{2-}]}{[CaCO_3(s)]}$$

The concentration of a solid is a constant and therefore can be included in a modified equilibrium constant known as the **solubility product, K_{sp}:**

$$K_{sp} = [Ca^{2+}(aq)][CO_3^{2-}(aq)]$$

The solubility product is the product of the concentrations of the aqueous ions formed when an insoluble or sparingly soluble ionic substance dissolves in water. Each concentration is raised to the power shown in the dissociation equation. For example, the dissociation equation for barium chromate(VI) is:

$$BaCrO_4(s) \rightleftharpoons Ba^{2+}(aq) + CrO_4^{2-}(aq)$$

Therefore: $K_{sp}(BaCrO_4) = [Ba^{2+}(aq)][CrO_4^{-}(aq)]$

For lead(II) iodide, the dissociation equation is:

$$PbI_2(s) \rightleftharpoons Pb^{2+}(aq) + 2I^-(aq)$$

Therefore: $K_{sp}(PbI_2) = [Pb^{2+}(aq)][I^-(aq)]^2$

CALCULATING SOLUBILITY PRODUCTS

The solubility product of a salt can be calculated by substituting the appropriate concentrations into the expression for the solubility product.

Notice in the first example on page 488 that the solubility product has a unit that depends on the expression for the solubility product. In this case, it is concentration squared. The unit can be worked out by substituting the initial units into the expression for the solubility product:

$$K_{sp}(CaCO_3) = [Ca^{2+}(aq)][CO_3^{2-}(aq)]$$
$$= (mol\ dm^{-3})(mol\ dm^{-3})$$
$$= mol^2\ dm^{-6}$$

See page 352 for an explanation of the equilibrium constant.

? QUESTION 22

22 Write down expressions for the solubility product of the following sparingly soluble salts:
a) calcium sulfate, $CaSO_4$;
b) barium sulfate, $BaSO_4$;
c) silver chromate(VI), Ag_2CrO_4;
d) lead(II) chloride, $PbCl_2$.

? QUESTION 23

23 **a)** The solubility of barium chromate(VI), $BaCrO_4$, at 298 K is 1.4×10^{-5} mol dm^{-3}. Calculate the solubility product for barium chromate(VI) at this temperature.
b) The solubility of calcium hydroxide at 298 K is 1.25×10^{-2} mol dm^{-3}. Show that the solubility product for calcium hydroxide at this temperature is 7.81×10^{-6} mol^3 dm^{-9}

EXAMPLE

Q What is the solubility product for calcium carbonate, given that the solubility of calcium carbonate at 298 K is 6.9×10^{-5} mol dm^{-3}?

A Assume that excess calcium carbonate is shaken with distilled water until equilibrium is attained. The solubility of calcium carbonate corresponds to the concentration of the aqueous calcium ions (or the aqueous carbonate ions) in the equilibrium solution:

	$CaCO_3(s) \rightleftharpoons$	$Ca^{2+}(aq)$ +	$CO_3^{2-}(aq)$
At start (before shaking)/mol dm^{-3}		0	0
At equilibrium/mol dm^{-3}		6.9×10^{-5}	6.9×10^{-5}

$$K_{sp}(CaCO_3) = [Ca^{2+}(aq)][CO_3^{2-}(aq)]$$
$$= (6.9 \times 10^{-5})(6.9 \times 10^{-5})$$
$$= 4.8 \times 10^{-9} \text{ mol}^2 \text{ dm}^{-6}$$

Calculating solubility

When the solubility product is known, it is possible to work out the solubility of the solute.

? QUESTION 24

24 Calculate the solubility at 298 K of each of the following:
a) barium fluoride, BaF_2, $K_{sp} = 1.7 \times 10^{-6}$ mol^3 dm^{-9}
b) calcium fluoride, CaF_2, $K_{sp} = 3.9 \times 10^{-11}$ mol^3 dm^{-9}
c) strontium sulfate, $SrSO_4$, $K_{sp} = 2.8 \times 10^{-7}$ mol^2 dm^{-6}

EXAMPLE

Q The solubility product of magnesium fluoride at 298 K is 6.4×10^{-9} mol^3 dm^{-9}. What is the solubility of magnesium fluoride at 298 K?

A Let s be the solubility of magnesium fluoride (in moldm^{-3}):

$$MgF_2(s) \rightleftharpoons Mg^{2+}(aq) + 2F^-(aq)$$

At equilibrium: s $2s$

$$K_{sp}(MgF_2) = [Mg^{2+}(aq)][F^-(aq)]^2$$

Therefore: $6.4 \times 10^{-9} = s(2s)^2 = 4s^3$

Therefore: $s = 1.2 \times 10^{-3}$ mol dm^{-3} (to 2 sig. fig)

SCIENCE IN CONTEXT

The barium meal

Doctors often need to use X-rays to look inside the intestine and the stomach for cancers, ulcers and blockages. However, using X-rays to diagnose such conditions does pose a problem, because the stomach and the intestine are transparent to X-rays (they pass through), so no image is detected – unlike bone, which is opaque to X-rays.

To overcome this, doctors give their patients a barium meal that contains insoluble barium sulfate and water. The stomach and intestine become coated with the barium sulfate, which is opaque to X-rays, and so a shadow of these organs is cast on an X-ray photographic film.

Using barium sulfate has one drawback: aqueous barium ions are highly toxic. Even though barium sulfate is said to be insoluble, there is still a small concentration of aqueous barium ions in the stomach, which could lead to poisoning. The equilibrium equation shows the production of barium ions:

$$BaSO_4(s) \rightleftharpoons Ba^{2+}(aq) + SO_4^{2-}(aq)$$

The concentration of the aqueous barium ions is reduced by the common-ion effect (see box opposite), brought about by magnesium sulfate added to the barium meal. The aqueous sulfate ions from the soluble magnesium sulfate shift the equilibrium to the left to lower the concentration of aqueous barium ions.

Fig 22.31 The image of someone who has had a barium meal shows the stomach on the right and intestine on the left

STRETCH AND CHALLENGE

Common-ion effect

Since the process involved in the solubility product is a dynamic equilibrium, it is possible to change the position of equilibrium according to Le Chatelier's principle (see pages 19 and 354).

Consider the solubility of barium carbonate. Aqueous barium ions and aqueous carbonate ions are in equilibrium with solid barium carbonate:

$$BaCO_3(s) \rightleftharpoons Ba^{2+}(aq) + CO_3^{2-}(aq)$$

When extra carbonate ions are added to the equilibrium mixture, the equilibrium shifts to the left to remove the extra carbonate ions. This causes a reduction in the solubility of barium carbonate, the concentration of aqueous barium ions having decreased. This is an example of the common-ion effect. The solubility of barium carbonate is also reduced by the addition of extra aqueous barium ions.

The common-ion effect can be demonstrated quantitatively by means of the solubility product.

Consider the solubility of barium chromate(VI) in a) water and b) 1.0 mol dm^{-3} potassium chromate(VI). The solubility product of barium chromate is 2×10^{-10} mol^2 dm^{-6} at 298 K.
It is possible to calculate the solubility in a) and b).

a) Let s be the solubility of barium chromate(VI).

$$BaCrO_4(s) \rightleftharpoons Ba^{2+}(aq) + CrO_4^{2-}(aq)$$

At equilibrium: s s

Therefore: $K_{sp} = [Ba^{2+}(aq)][CrO_4^{2-}(aq)]$
$= s^2$

This gives: $s = 1.4 \times 10^{-5}$ mol dm^{-3}.

b) In this calculation, it is necessary to consider dissolving barium chromate(VI) in aqueous potassium chromate(VI):

$$BaCrO_4(s) \rightleftharpoons Ba^{2+}(aq) + CrO_4^{2-}(aq)$$

At start/mol dm^{-3}: 0 1.0
At equilibrium/mol dm^{-3}: s $(1.0 + s)$

Barium chromate(VI) is sparingly soluble in water. Therefore s is small compared with 1.0, and so $(s + 1.0)$ is approximately equal to 1.0.

Now: $K_{sp} = [Ba^{2+}(aq)][CrO_4^{2-}(aq)]$
$2 \times 10^{-10} = s(1.0)$

This gives: $s = 2 \times 10^{-10}$ mol dm^{-3}.

Note that the solubility of barium chromate(VI) is very much smaller in potassium chromate(VI) than in water alone. This is the common-ion effect in operation.

PRECIPITATION

Precipitation is the formation of an insoluble solid when two solutions react together. For example, when a solution that contains aqueous barium ions is mixed with a solution that contains aqueous carbonate ions, a white precipitate of insoluble barium carbonate is formed:

$$Ba^{2+}(aq) + CO_3^{2-}(aq) \rightarrow BaCO_3(s)$$

Note that this is the opposite process to the dissolving of sparingly soluble barium carbonate in water. This means that the solubility product predicts the conditions for a precipitation reaction. In fact, precipitation occurs when the **ionic product** for the compound exceeds its solubility product. The ionic product of barium carbonate is the product of the concentration of the aqueous barium ion and of the aqueous carbonate ion.

QUESTION 25

25 Calculate the solubility of barium chromate in 0.1 mol dm^{-3} barium chloride.

REMEMBER THIS

Ionic product is the same expression as for K_{sp}, but uses the actual concentrations, not those at equilibrium.

EXAMPLE

Q Will a precipitate of barium carbonate ($K_{sp}(BaCO_3) = 8.1 \times 10^{-9}$ mol^2 dm^{-6}) be formed from a solution that is both 0.1 mol dm^{-3} in aqueous barium ion and 1.0 mol dm^{-3} in aqueous carbonate ion?

A Ionic product $(BaCO_3) = [Ba^{2+}(aq)][CO_3^{2-}(aq)]$
$= 0.1 \times 1.0$
$= 0.1$ mol^2 dm^{-6}
$K_{sp}(BaCO_3) = 8.1 \times 10^{-9}$ mol^2 dm^{-6}

The ionic product is larger than the solubility product, so a precipitate will be formed.

QUESTION 26

26 Will a precipitate of magnesium fluoride, $K_{sp}(MgF_2) = 6.4 \times 10^{-9}$ mol^3 dm^{-9}, be formed from a solution initially containing 1.0×10^{-3} mol dm^{-3} aqueous magnesium ions and 1.0×10^{-3} mol dm^{-3} aqueous fluoride ions? Explain your answer fully.

12 ALUMINIUM AND BORON

So far, this chapter has concentrated on the behaviour of the Group 1 and 2 elements and their compounds. But there is one other metal that can reasonably be called reactive and that is aluminium. The surface of aluminium is covered by an impermeable layer of aluminium oxide that masks the true reactivity of the metal. Once this layer has been removed, the metal resembles magnesium in its reactivity.

Boron is also in Group 3 but it is very different to aluminium, because it is a non-metal rather than a reactive metal.

OCCURRENCE AND MANUFACTURE OF ALUMINIUM

Aluminium is estimated to make up 7.5 per cent of the Earth's crust, most of it as complex aluminates and aluminosilicates in clays, and also as bauxite, which is a form of aluminium oxide and hydroxide. It is not possible commercially to extract aluminium from clays and so bauxite is the major source of all aluminium metal used today.

Electrolytic manufacture of aluminium

The process dates back to 1886, when Charles Hall and Paul Heroult independently discovered a way to extract aluminium from bauxite using electrolysis. The bauxite mineral is mined and then purified to form pure aluminium hydroxide, which is in turn converted into pure aluminium oxide. This is the substance that is electrolysed to form aluminium.

The aluminium oxide is first dissolved in molten sodium aluminium fluoride (cryolite) and some calcium fluoride, and is then electrolysed. Electrolysis of pure aluminium oxide would be too expensive since very high temperatures are needed to melt it. Introducing the two fluorides allows a much lower temperature to be used. As a result of shortages of sodium aluminium fluoride, some plants use sodium fluoride and aluminium fluoride instead.

A carbon (graphite) anode is used. Oxygen is produced by the discharge of oxide ions:

$$2O^{2-} \rightarrow O_2 + 4e^-$$

The high temperature of production is a nuisance, because the oxygen reacts with the carbon anode to form carbon dioxide. So the carbon anode is continually being replaced.

The carbon lining of the cell acts as the cathode, and aluminium is formed by the discharge of aluminium ions. The temperature in the cell is above the melting point of aluminium, so molten aluminium is formed. This sinks to the bottom of the cell, where it is run off and allowed to cool. The aluminium obtained is exceptionally pure, being at least 99.9 per cent aluminium.

QUESTION 27

27 **a)** Write down the equation to show the discharge of aluminium ions, Al^{3+}, during the electrolysis of aluminium oxide.

b) Explain why the gaseous emissions from an aluminium smelter contain traces of hydrogen fluoride and fluorine.

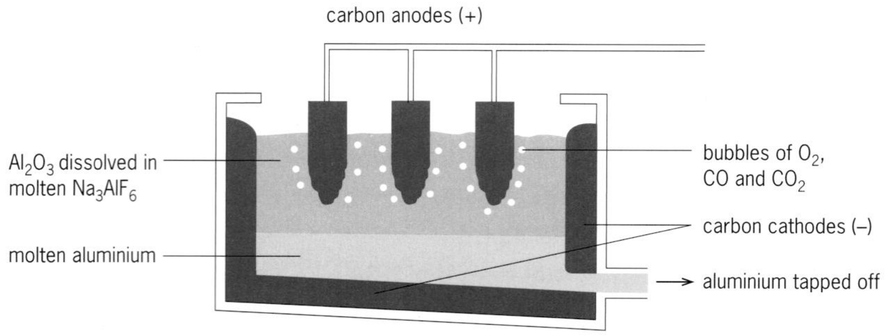

carbon anodes (+)
Al_2O_3 dissolved in molten Na_3AlF_6
molten aluminium
bubbles of O_2, CO and CO_2
carbon cathodes (–)
aluminium tapped off

Fig 22.32 The cell used for the manufacture of aluminium by electrolysis

Applications of aluminium

As soon as it was discovered, aluminium found many applications associated with its low density and its resistance to corrosion. More recent applications include foil (often incorrectly called tin foil) for use in the home and in packaging, double-glazed window frames and now car engines and bodies.

Fig 22.33 Aluminium and its alloys are used in the aviation industry because of their high strength and low density

CHEMICAL PROPERTIES OF ALUMINIUM

The electronic configuration of aluminium (see Table 22.12) suggests that an aluminium atom should lose three electrons from an atom to form an aluminium ion. The loss of three electrons is always energetically more difficult than the loss of one or two electrons, and the cation formed will always be small and have a large charge density. This means that the small and highly charged cation is able to attract back the electrons it has lost in forming an ionic bond. An aluminium ion is said to polarise the anions, and so many aluminium compounds have a covalent character.

In aqueous solution, six water molecules surround an aluminium ion to form a hydrated ion, $[Al(H_2O)_6]^{3+}(aq)$. For simplicity, this is often written as $Al^{3+}(aq)$.

Table 22.12 Some properties of aluminium

electron configuration	$1s^2\ 2s^2\ 2p^6\ 3s^2\ 3p^1$
melting point	660 °C
boiling point	2300 °C
oxidation potential, E_{oxid}	+1.67 V
density	2.70 g cm^{-3}

Reaction with air and oxygen

As already noted, aluminium forms an oxide layer on its surface which protects the rest of the metal from further corrosion. When heated in air or oxygen, aluminium powder burns to form aluminium oxide, a white ionic solid:

$$4Al(s) + 3O_2(g) \rightarrow 2Al_2O_3(s)$$

If the oxide layer is removed, for example by wiping the surface with mercury, then aluminium will react with water to form hydrogen.

Reaction with acids

Aluminium powder reacts with dilute sulfuric acid and dilute hydrochloric acid to give the corresponding aqueous aluminium salts. These reactions are initially very slow and need heat to accelerate the reaction rate. But, once started, they become quite violent. The initial slow rate of reaction arises from the time it takes for the acid to react with the protective oxide layer before it exposes the active metal.

$$2Al(s) + 6H^+(aq) \rightarrow 2Al^{3+}(aq) + 3H_2(g)$$

Surprisingly, aluminium does not react with pure nitric acid. It is therefore used as the lining of vessels to transport this acid.

Fig 22.34 The thermite process involves the reaction of aluminium powder and iron(III) oxide. Aluminium is sufficiently reactive to displace iron from its oxide

Reaction with alkalis

Aluminium is unusual in that it reacts with both acids and alkalis. This property makes aluminium an amphoteric metal. Aluminium powder reacts vigorously – after a slow start – with warm aqueous sodium hydroxide to form hydrogen and aqueous sodium aluminate:

$$OH^-(aq) + 3H_2O(l) + Al(s) \rightarrow [Al(OH)_4]^-(aq) + H_2(g)$$

Reaction with halogens

When one of the halogens is passed over heated aluminium, the corresponding halide is formed by direct combination. The halides formed have a large degree of covalent character:

$$2Al(s) + 3X_2(g) \rightarrow Al_2X_6(s)$$

where X = a halogen atom.

REMEMBER THIS

The formula for the aluminate ion is sometimes written as $[Al(OH)_6]^{3-}$ or as $[Al(H_2O)_2(OH)_4]^-$.

COMPOUNDS OF ALUMINIUM

Most aluminium compounds have some covalent character, but when dissolved in water or as hydrated crystals, the compounds are ionic, and contain the hydrated aluminium ion.

Aluminium oxide and aluminium hydroxide

Both these are white ionic amphoteric solids. With acids, they form the hydrated aluminium ion $[Al(H_2O)_6]^{3+}$; with alkalis they form the aluminate ion $[Al(OH)_4]^-$ in which the aluminium atom forms part of the anion.

Aluminium oxide is used as an abrasive and as a dehydrating agent, particularly in the formation of alkenes from alcohols. Aluminium hydroxide is used in the dyeing industry, since it can adsorb coloured materials onto its surface. This property is also used to remove particulate matter from water during its purification. In this application, a solution of aluminium sulfate is added to the water and forms aluminium hydroxide, which, as it settles, collects particulate matter and removes it from the water. The amount of aluminium sulfate used in this process has to be carefully controlled to prevent excess aluminium ions reaching domestic water supplies, since aluminium ions may be linked to Alzheimer's disease.

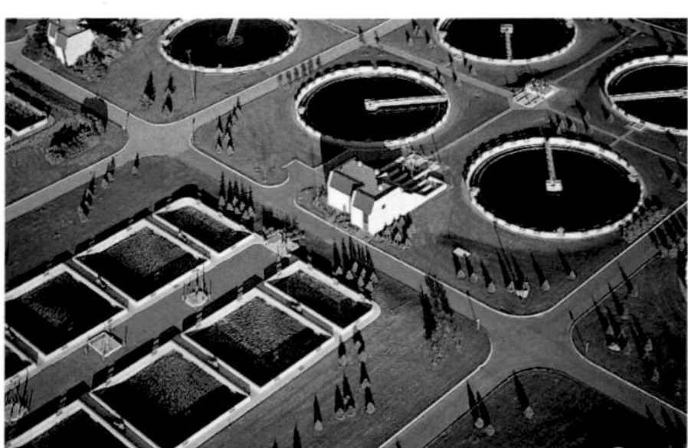

Fig 22.35 Aluminium sulfate is used in water purification plants to remove solid particles, and its level has to be carefully monitored so as not to reach toxic levels

Solutions of aluminium salts

Aqueous solutions of aluminium salts are highly acidic, so much so that they can release carbon dioxide from sodium hydrogencarbonate. This acidity is caused by hydrolysis and involves one of the water molecules in the aqueous aluminium cation being polarised to form an aqueous proton:

$$[Al(H_2O)_6]^{3+}(aq) \rightleftharpoons [Al(H_2O)_5OH]^{2+}(aq) + H^+(aq)$$

All aqueous solutions of aluminium salts react with aqueous hydroxide ions to give first a white precipitate of hydrated aluminium hydroxide and then, with excess hydroxide ions, a colourless solution of the aluminate ion. This reaction involves **deprotonation**, in which each reaction step removes one more proton from the aluminium species:

$$[Al(H_2O)_6]^{3+}(aq) + OH^-(aq) \rightleftharpoons [Al(H_2O)_5OH]^{2+}(aq) + H_2O(l)$$

$$[Al(H_2O)_5OH]^{2+}(aq) + OH^-(aq) \rightleftharpoons [Al(H_2O)_4(OH)_2]^+(aq) + H_2O(l)$$

$$[Al(H_2O)_4(OH)_2]^+(aq) + OH^-(aq) \rightleftharpoons [Al(H_2O)_3(OH)_3](s) + H_2O(l)$$

$$[Al(H_2O)_3(OH)_3](s) + OH^-(aq) \rightleftharpoons [Al(H_2O)_2(OH)_4]^-(aq) + H_2O(l)$$

? QUESTION 28

28 Would you expect aluminium carbonate to exist? Explain your answer.

REMEMBER THIS

Remember that the H^+ ion is really hydrated and could be written as $H_3O^+(aq)$. It is this ion that makes an aqueous solution acidic.

HOW SCIENCE WORKS

Ionic solvents

One of the typical properties of an ionic salt is that it has a high melting point. Often an ionic salt has to be heated to above 700°C before it will melt and produce an ionic liquid. However recent research has developed ionic liquids at room temperature. All that needs to be done is to mix together and warm gently certain powdered organic salts with anhydrous aluminium chloride; the result is a clear, colourless, 'ionic liquid'. The ionic liquids contain an organic cation and a tetrachloroaluminate anion, $AlCl_4^-$.

These ionic liquids may well be a possible alternative to using standard organic solvents to dissolve substances. Many industrial processes use volatile organic compounds, known as VOCs, as solvents. These solvents are often toxic and flammable; they vaporise easily and are greenhouse gases and/or ozone-depleting pollutants.

The new ionic liquids are non-volatile and are being used to develop new processes that use clean-technology solvents. Uses in chemical syntheses, catalysis and metal finishing are all being developed.

Ionic liquids may reduce atmospheric pollution, but chemists have to research any possible environmental concern about the disposal of large quantities of these ionic liquids.

BORON

Even though boron is in Group 3, it is a non-metal and it forms compounds by sharing electrons rather than by losing electrons. Boron trichloride, BCl_3, is a typical simple molecular chloride that hydrolyses in water; this is the typical behaviour of a non-metal chloride.

HOW SCIENCE WORKS

Forms of boron nitride

When chemists discovered the two forms of boron nitride they noticed how similar the structures were to graphite and diamond. As a result, the chemists realised that they would have very similar physical properties as well.

Boron nitride can be used as a lubricant in the same way as graphite. The layers of boron–nitrogen hexagons can slide over one another very easily, since they are held in place by weak van der Waals forces. Boron nitride remains a lubricant even at high temperatures, unlike graphite. Another advantage over graphite is that it is much less likely to react with oxygen, even at high temperature. It is used in areas of metal extrusion, plastic extrusion and the hot pressing of metals.

Boron nitride can also be used as an abrasive and in glass cutting, in the same way as diamond. One disadvantage of boron trinitride is that is more brittle than diamond.

Knowing that carbon forms fullerenes and nanotubes, as described on page 145, chemists developed boron trinitride nanotubes as shown in Figure 22.37.

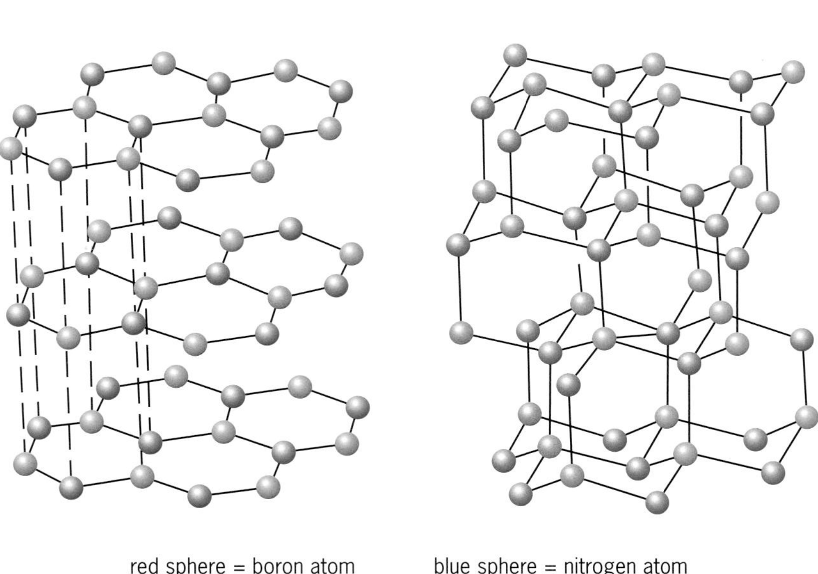

Fig 22.36 The structure of boron nitride. The structure on the left resembles graphite and that on the right, diamond

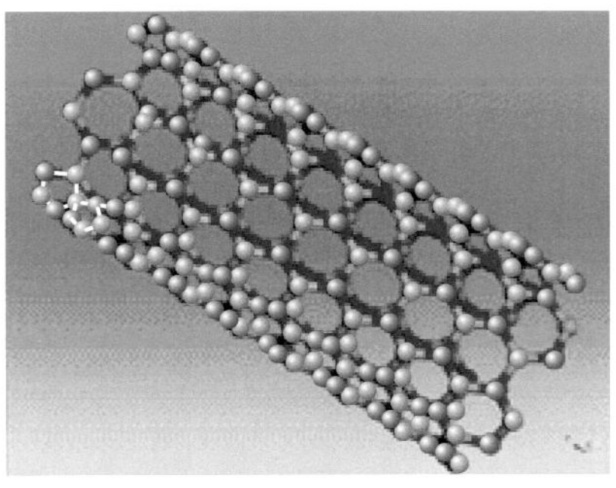

Fig 22.37 A boron nitride nanotube

SUMMARY

- Group 1 metals are highly reactive metals. Each atom loses one electron when it reacts. In compounds, the oxidation state of Group 1 metals is always +1.
- Group 2 elements are less reactive than the Group 1 element in the same period. In compounds, the Group 2 elements always have an oxidation number of +2.
- The ease with which an atom loses electrons increases with increasing atomic (proton) number in both Group 1 and Group 2.
- The oxides and hydroxides of Group 1 and Group 2 elements are basic and when they dissolve in water, they form strongly alkaline solutions.
- The nitrates of Group 1 (except lithium nitrate) thermally decompose to form the corresponding nitrites and oxygen, whereas the nitrates of Group 2 thermally decompose to give nitrogen dioxide, oxygen and the corresponding metal oxides. The thermal stability of the nitrates increases with increasing atomic (proton) number of the metal.
- The carbonates of Group 1 (other than lithium carbonate) do not decompose at Bunsen burner temperatures, but most carbonates of Group 2 do to give carbon dioxide and the metal oxides. The thermal stability of the carbonates increases with increasing atomic (proton) number of the metal.
- Lattice enthalpy is the enthalpy change when 1 mole of an ionic lattice is formed from its constituent ions in the gas phase. Lattice enthalpy depends on the charge densities of the ions involved.
- The solubility of the Group 2 sulfates, MSO_4, and the Group 2 carbonates, MCO_3, decreases with increasing atomic (proton) number of M. This trend is explained by the decrease in the magnitude of the enthalpy change of hydration of M^{2+}.
- The solubility of the Group 2 hydroxides, $M(OH)_2$, increases with increasing atomic (proton) number of M. This trend is explained by the decrease in magnitude of the lattice enthalpy of $M(OH)_2$.
- The thermal decomposition of nitrates and carbonates is determined by the ability of the cation present to polarise the large carbonate and nitrate ions.
- Aluminium is manufactured by the electrolytic decomposition of molten aluminium oxide dissolved in sodium aluminium fluoride.
- Aqueous aluminium salts are acidic because of the polarisation of water molecules by the very small and highly charged aluminium ion.
- The solubility product can be used to explain the common-ion effect and precipitation.

Practice questions and a How Science Works assignment for this chapter are available at www.collinseducation.co.uk/CAS

23

The halogens and electrolysis

23 THE HALOGENS AND ELECTROLYSIS

SCIENCE IN CONTEXT

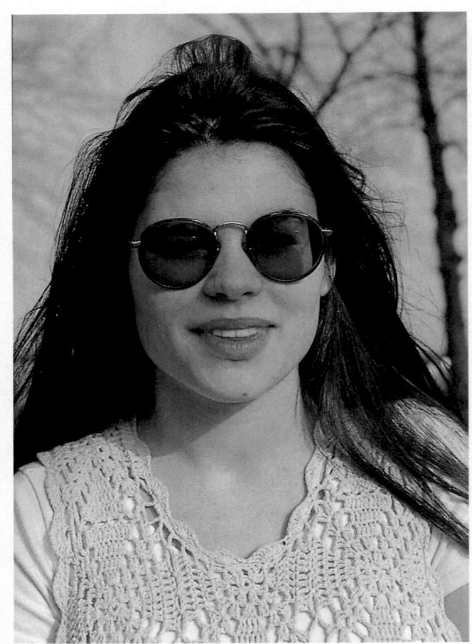

Sunglasses Photochromic spectacles are easy on the eyes

Photochromic sunglasses

We find it quite uncomfortable on the eyes to move rapidly in and out of brilliant sunlight and shadow, even if we are wearing sunglasses. Our eyes cannot adjust quickly enough to sudden changes of light intensity – that is, unless the sunglasses are made from photochromic glass, which compensates for changes in brightness.

Lens glass is usually made photochromic by adding tiny amounts of silver chloride and copper(I) chloride to the molten glass as it cools, which traps the crystals within the structure of the glass. When bright sunlight strikes the glass, the silver chloride decomposes to form metallic silver, which darkens the glass. This reduces the intensity of the light that reaches the eyes. At the same time, chlorine atoms are formed and they react with copper(I) ions to form copper(II) ions and chloride ions. As soon as the exposure to bright light ends, the copper(II) ions are reduced by silver atoms to reform silver chloride and copper(I) chloride. This lightens the glass and allows all the available light to reach the eyes.

In this way, spectacle lenses can be made to darken or lighten according to the intensity of the light, and so make it comfortable for people to keep their sunglasses on, whether they are in bright light or shadow.

1 WHAT ARE THE HALOGENS?

Fig 23.1 Bromomethane is used as a soil fumigator for high-value crops such as strawberries

Halogens is the collective name given to the non-metal elements in Group 7 of the Periodic Table. The name is derived from *hals* (the Greek for salt) and the suffix *gen* (meaning producer). It was first used by the Swedish scientist Jöns Berzelius (1779–1884) to indicate that chlorine, bromine and iodine occurred in sea water. Chlorine was first prepared in 1774 by the Swedish chemist Carl Wilhelm Scheele, but he did not recognise it as an element. Humphry Davy in 1810 recognised chlorine as an element and named it after the Greek word for green, *chloros*. In the same way, iodine was derived from *ioides* (the Greek word for violet-like) and bromine from *bromos* (the Greek word for stench). It was not until 1940 that astatine, a radioactive element and the halogen with the highest atomic number, was prepared artificially.

SOME USES OF THE HALOGENS AND THEIR COMPOUNDS

The halogens are commercially of great importance, although the use of many chlorine-containing compounds is controversial for environmental reasons. Chlorine and fluorine are both used in the production of polymers, such as polyvinyl chloride (PVC) and polytetrafluoroethene (PTFE), and fluorine is also used in toothpaste. Many insecticides contain chlorine, but there is concern over the use of these compounds, despite their obvious usefulness. Bromine has a variety of applications, including the manufacture of fuel additives and of soil fumigators such as 1,2-dibromoethane and bromomethane, which kill pests found in the soil. Silver bromide and silver iodide both have applications in traditional photographic film. You can read more about the uses of halogens and their compounds later in this chapter.

Fig 23.2 We take it for granted that the water we drink is sterile. Chlorine is used to remove bacteria from our water supply

2 PHYSICAL PROPERTIES OF THE HALOGENS

The halogens have physical properties that are typical of non-metals with a simple molecular structure. The halogens have relatively low melting and boiling points and are very poor conductors of heat and electricity.

Table 23.1 Properties of four halogens

Halogen	Melting point/°C	Boiling point/°C	Atomic (covalent) radius/pm	Electronegativity	Electron configuration
fluorine	−223	−188	64	4	$1s^2\ 2s^2\ 2p^5$
chlorine	−101	−34	99	3	$1s^2\ 2s^2\ 2p^6\ 3s^2\ 3p^5$
bromine	−7	59	114	2.8	$[Ar]3d^{10}\ 4s^2\ 4p^5$
iodine	114	187	133	2.4	$[Kr]4d^{10}\ 5s^2\ 5p^5$

You can read more about the physical properties of substances with a simple molecular structure in chapters 7 and 21.

Fig 23.3 The volatility of the halogens decreases moving down Group 7. At room temperature and pressure, chlorine is a greenish gas, bromine is an orange liquid and iodine is a dark purple-grey solid

SIMPLE MOLECULAR STRUCTURE OF THE HALOGENS

All the halogens have diatomic molecules, which in the solid state are arranged in a simple molecular lattice. Halogen molecules are held in place by weak intermolecular forces known as **induced dipole–induced dipole attraction**. These forces result from an asymmetric distribution of electrons within each halogen molecule. This produces an instantaneous dipole, which induces dipoles in neighbouring molecules. Such a weak attractive force between molecules is easily overcome, so the elements have relatively low melting points and boiling points. The simple molecular structure has no free electrons, so a halogen, either as a solid or as a liquid, cannot conduct electricity.

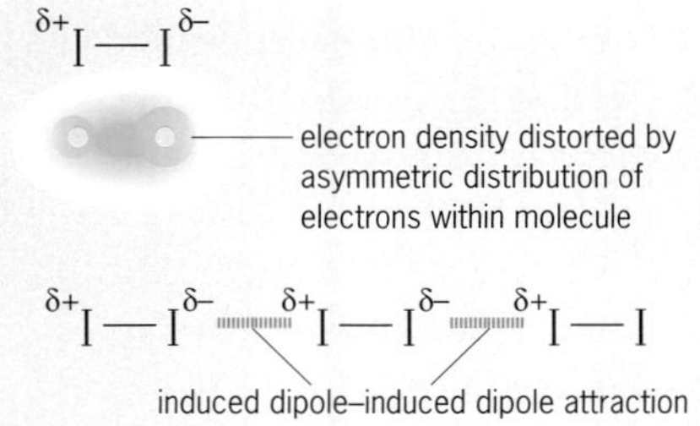

Fig 23.4 A temporary dipole is set up in an iodine molecule when its electrons move such that they become distributed asymmetrically, with the electron density at one end of the molecule greater than that at the other end. The top diagram shows how the electron density is temporarily distorted. The temporary dipole induces dipoles in neighbouring molecules and an intermolecular force is set up (bottom diagram). This force is quite weak

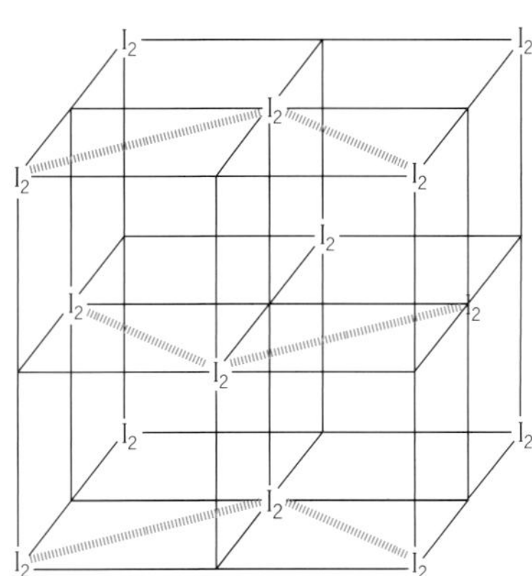

Fig 23.5 Solid iodine has a cubic structure. The lines of short strokes represent the weak intermolecular forces that hold the iodine molecules in place within the lattice. The other solid halogens have a similar simple molecular lattice

All the halogens form solids with this type of intermolecular force. The larger the halogen molecule (and the more electrons), the easier it is to distort the electron cloud and increase the intermolecular forces. Hence, the melting point of the halogens increases with increasing atomic number.

QUESTIONS 1–2

1. The appearance of the halogens at room temperature and atmospheric pressure has an observable trend. Predict the colour and physical state of fluorine and of astatine under these conditions.
2. Astatine is the last member of the halogens.
   a) i) What is the atomic (proton) number for astatine?
   ii) How many electrons does astatine have in the outer shell of its atom?
   iii) Predict the melting point, boiling point, covalent radius and electronegativity of astatine.
   b) What is meant by the term electronegativity? Why does the electronegativity of the halogens decrease with increasing atomic (proton) number?

Check your answers on page 85.

Weak intermolecular forces in elements are discussed on page 195.

REMEMBER THIS

Volatility measures the ease of evaporation of a substance. Liquids with weak intermolecular forces that are easy to overcome are very volatile, whereas liquids with strong intermolecular forces are not very volatile.

REMEMBER THIS

Some anhydrous metal halides, such as aluminium chloride and iron(III) chloride, have a simple molecular structure.

? QUESTION 3

3 Write down the electron configurations for
a) chloride ion, Cl^-;
b) bromide ion, Br^-.

Table 23.2 Reduction potentials ($E^{\ominus}_{red}$) for the reaction $X_2 + 2e^- \rightarrow 2X^-$

Half reaction	Reduction potential $E^{\ominus}_{red}$/V
$F_2 + 2e^- \rightarrow 2F^-$	+2.87
$Cl_2 + 2e^- \rightarrow 2Cl^-$	+1.36
$Br_2 + 2e^- \rightarrow 2Br^-$	+1.07
$I_2 + 2e^- \rightarrow 2I^-$	+0.54

? QUESTION 4

4 State whether a reaction takes place when the following substances are mixed. If a reaction takes place, write a balanced ionic equation for the reaction:
a) aqueous iodine and aqueous potassium bromide;
b) aqueous chlorine and aqueous sodium iodide;
c) aqueous chlorine and aqueous calcium fluoride;
d) aqueous iodine and aqueous sodium astatide.

3 CHEMICAL PROPERTIES OF THE HALOGENS

Halogens react by gaining electrons to form an anion in ionic halides, or by sharing electrons to form a covalent bond in molecular halides. In both cases the halogen atom attains a noble gas electron configuration. Most metals form ionic halides and most non-metals form molecular halides. The ability to gain an electron is typical of a non-metal atom.

THE HALOGENS AS OXIDISING AGENTS

Since their atoms accept electrons, the halogens are oxidising agents and in a reaction they are reduced. Of the halogens, fluorine is the most powerful oxidising agent and astatine the least. This can be explained by the relative size of their atoms and their ability to capture an electron. The fluorine atom is the smallest, with fewer inner-shell shielding electrons, so its nucleus can have a greater attraction or ability to attract an electron. The reduction potentials, $E^{\ominus}_{red}$, of the halogens also illustrate this trend. The reaction of a fluorine molecule to give a fluoride ion has a very positive reduction potential, which indicates that this is a highly feasible process. (The more positive an electrode potential the more feasible the reaction. More details on electrode potentials can be found in Chapter 25.)

Fluorine always has an oxidation number of –1 in its compounds, since a fluorine atom either gains one electron to form an ion or contributes one electron to a shared pair of electrons in a covalent bond. Fluorine oxidises other substances by gaining electrons from them.

Displacement reactions

Since fluorine is the most reactive halogen, in theory it can react with the halide ion of any of the other halogens. Fluorine becomes the fluoride ion and the free halogen (chlorine, bromine or iodine) is formed from the halide ion. This is called a **displacement reaction**. Experiments are not normally carried out with fluorine because its extraordinary reactivity makes it extremely dangerous. For example, if inhaled, fluorine can seriously damage the respiratory tract.

In aqueous solution, chlorine displaces bromide ions to form bromine:

$$Cl_2(aq) + 2KBr(aq) \rightarrow 2KCl(aq) + Br_2(aq)$$

During this reaction, the colour of the aqueous mixture becomes orange, which indicates that the element bromine has been produced.

Chlorine will also displace iodide ions to form iodine, and again an orange–brown solution is formed.

$$Cl_2(aq) + 2I^-(aq) \rightarrow 2Cl^-(aq) + I_2(aq)$$

In the same way, aqueous bromine displaces aqueous iodide ion to form iodine. (The presence of iodine in a solution can be confirmed by the addition of starch solution. A dark blue coloration is produced.)

In general, the presence of the displaced halogen can be demonstrated easily if hexane is added after the displacement reaction, since bromine will preferentially dissolve in the hexane to give an orange–red solution, while iodine will give a purple solution.

The displacement reaction is an example of a redox reaction. This is because the halogen that reacts is reduced because it gains electrons to form halide ions, and halide ions are oxidised by the loss of electrons to form the halogen.

SCIENCE IN CONTEXT

Manufacture of bromine

The concentration of bromide ion in normal sea water is between 65 and 70 ppm. Therefore, a large quantity of sea water has to be processed to make significant quantities of bromine. Inland seas, such as the Dead Sea, have considerably higher concentrations of bromide ion and provide a much better feedstock (Fig 23.6).

Chlorine is used to liberate bromine by a displacement reaction. It is bubbled through the acidified sea water and the bromine formed is removed as a gas from the water by blowing air through it:

$$Cl_2(g) + 2Br^-(aq) \rightarrow Br_2(aq) + 2Cl^-(aq)$$

The bromine vapour formed is difficult to handle and so is converted back into bromide ion by reaction with sulfur dioxide and water:

$$Br_2(g) + SO_2(g) + H_2O(l) \rightarrow H_2SO_4(aq) + 2HBr(aq)$$

This makes a much more concentrated solution of bromide ion, from which it is easier to produce high-purity bromine. A second displacement reaction with chlorine yields bromine vapour, which can be condensed and then purified by distillation.

Fig 23.6 The Dead Sea is rich in dissolved minerals. It is about 60 times more concentrated in bromide ions than normal sea water, and has a higher density

REACTIONS OF HALOGENS WITH ELEMENTS

We have already described some of the reactions of metals and non-metals with chlorine in Chapter 21. In every case the halogen oxidises the element. Fluorine will react the most violently, and astatine the least.

Reaction of halogens with non-metals

Almost all non-metals react with fluorine and chlorine to form covalent halides. The halogen atom shares electrons to make a single covalent bond (Fig 23.7). Very often, the halide contains the element in its highest possible oxidation state. In the reaction between phosphorus and chlorine, the reaction can be controlled to make either phosphorus(III) chloride or phosphorus(V) chloride:

$$P_4(s) + 6Cl_2(s) \rightarrow 4PCl_3(l)$$

$$P_4(s) + 10Cl_2(g) \rightarrow 4PCl_5(s)$$

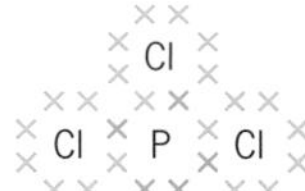

Fig 23.7 Phosphorus(III) chloride is a covalent compound that forms a simple molecule, PCl_3. Electrons are shared so that all the atoms attain a stable octet of electrons. All non-metal halides are covalent compounds, and the halogen atom attains a stable octet. Sometimes, the other atom involved expands its octet. Note: only the outer electrons are shown in the dot-and-cross diagram

Reactions of the halogens with metals

Most metals react with the halogens to form ionic halides. This reaction is necessarily a redox reaction, in which the metal is oxidised and the non-metal is reduced. Electrons are transferred to halogen atoms from metal atoms. The reaction of a metal with the halogens varies with the halogen. For instance, the reaction between a metal and fluorine is faster, more exothermic and more violent than that between the same metal and iodine.

Since halogens are good oxidising agents, if the metal has more than one oxidation state then in the halide formed the metal is often in one of its higher oxidation states. So, when chlorine is passed over hot iron wire it forms iron(III) chloride, $FeCl_3$, rather than iron(II) chloride. Similarly, copper and chlorine form copper(II) chloride, $CuCl_2$, rather than copper(I) chloride.

QUESTION 5

5 **a)** Write equations for the reactions between:

i) iron and chlorine;

ii) copper and chlorine;

iii) iron and fluorine.

b) What is the name of the product formed in the reactions between:

i) chromium and chlorine;

ii) zinc and chlorine;

iii) barium and iodine?

The reaction of the halogens with hydrocarbons to form halogenoalkanes and hydrogen halide is described on page 240.

REACTIONS OF CHLORINE WITH COMPOUNDS

Chlorine and the other halogens have an affinity for hydrogen. Chlorine reacts with compounds that contain hydrogen to form hydrogen chloride, or hydrochloric acid when the reaction is carried out in aqueous solution. Chlorine behaves as an oxidising agent.

Reaction of chlorine with water

Chlorine reacts with water to form an acidic solution that is a mixture of two acids, hydrochloric acid and chloric(I) acid. This is an example of a **disproportionation** reaction, in which chlorine is both oxidised and reduced during the reaction:

$$Cl_2(aq) + H_2O(l) \rightarrow HCl(aq) + HOCl(aq)$$

This reaction is the basis of one of the chemical tests for chlorine: namely, it turns moist blue litmus paper first red and then white. Chlorine reacts with the moisture to form the acidic solution which turns the litmus paper red. The chloric(I) acid acts as a bleach and turns the paper white.

Reaction of chlorine with cold dilute alkali

Cold aqueous sodium hydroxide can react with chlorine to produce chlorate(I) ions, ClO^-, and chloride ions, Cl^-:

$$Cl_2(g) + 2OH^-(aq) \rightarrow ClO^-(aq) + Cl^-(aq) + H_2O(l)$$
$$Cl_2(g) + 2NaOH(aq) \rightarrow NaClO(aq) + NaCl(aq) + H_2O(l)$$

A solution of aqueous sodium chlorate(I) is the liquid bleach found in most homes.

Reaction of chlorine with hot concentrated alkali

When chlorine is added to hot concentrated aqueous sodium hydroxide, the overall reaction is:

$$3Cl_2(g) + 6OH^-(aq) \rightarrow ClO_3^-(aq) + 5Cl^-(aq) + 3H_2O(l)$$

This reaction occurs because of the **disproportionation** of chlorate(I) ions. Chlorate(I) ions lose electrons to give chlorate(V) ions, and gain electrons to form chloride ions:

$$3ClO^-(aq) \rightarrow ClO_3^-(aq) + 2Cl^-(aq)$$

Sodium chlorate(V), $NaClO_3$, is used as a weedkiller. It is also reduced to form ClO_2, chlorine dioxide, which is a safer bleach than elemental chlorine for paper and textiles, though it is less efficient.

Oxidation of iron(II) ions

Chlorine oxidises aqueous iron(II) chloride to form aqueous iron(III) chloride:

$$Cl_2(g) + 2Fe^{2+}(aq) \rightarrow 2Fe^{3+}(aq) + 2Cl^-(aq)$$

Reaction of halogens with aqueous sodium thiosulfate

Aqueous chlorine and aqueous bromine oxidise aqueous thiosulfate ions into sulfate ions:

$$4Cl_2(aq) + S_2O_3^{2-}(aq) + 5H_2O(l) \rightarrow 10H^+(aq) + 8Cl^-(aq) + 2SO_4^{2-}(aq)$$

Iodine is a less powerful oxidising agent than either chlorine or bromine, and so the reaction does not give sulfate ions. Instead, it gives $S_4O_6^{2-}(aq)$:

$$I_2(aq) + 2S_2O_3^{2-}(aq) \rightarrow 2I^-(aq) + S_4O_6^{2-}(aq)$$

? QUESTION 6

6 **a)** Determine the oxidation number of each of the chlorine species for the following reaction:
$Cl_2(aq) + H_2O(l) \rightarrow HCl(aq) + HOCl(aq)$
b) Use the oxidation numbers to show that the reaction is an example of disproportionation

For some more examples of disproportionation, read page 111.

? QUESTION 7

7 **a)** Construct the ionic equation for the reaction between aqueous chlorine and aqueous sulfur dioxide and water to form a chloride ion and a sulfate ion.
b) Suggest why astatine does not oxidise aqueous iron(II) ions into iron(III) ions.

? QUESTION 8

8 In a determination to find the amount of iodine liberated by an oxidising agent, it was found that the number of moles of sodium thiosulfate needed to react with the iodine in a titration was 2.1×10^{-3}. What was the number of moles of iodine liberated by the oxidising agent?

Chemical test for an oxidising agent

One chemical test for an oxidising agent uses aqueous potassium iodide. An oxidising agent liberates iodine from aqueous potassium iodide, and the amount of iodine can be determined using volumetric analysis with a titration against aqueous sodium thiosulfate. The presence of iodine is shown by a blue–black coloration with starch solution.

REMEMBER THIS

The chemical test for iodine involves aqueous starch turning a blue–black coloration

QUESTION 9

9 Show, by using change of oxidation numbers, that the reaction between iodine and sodium thiosulfate is an example of a redox reaction.

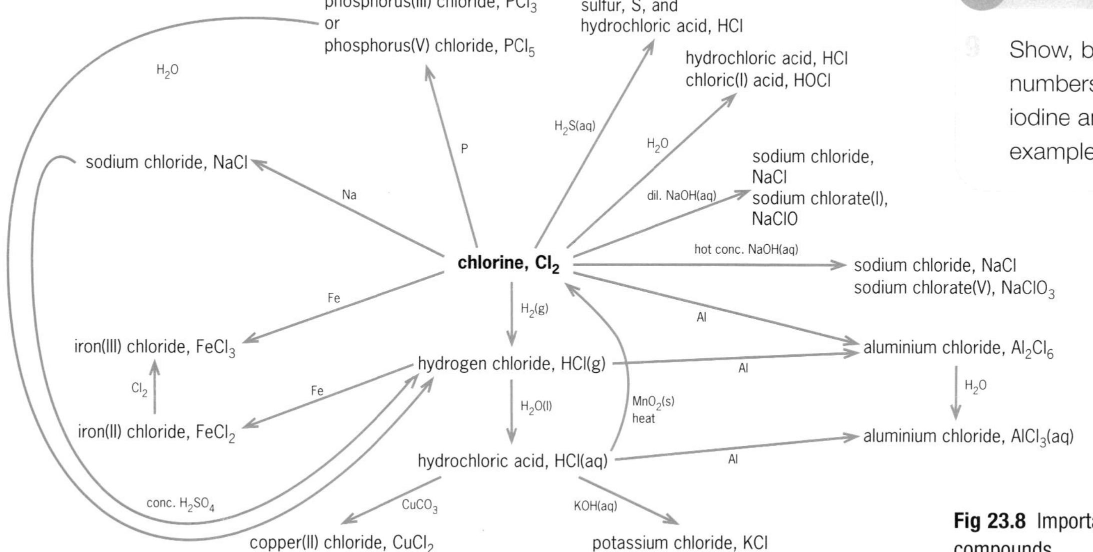

Fig 23.8 Important reactions of chlorine and its compounds

QUESTION 10

10 **a)** What is the oxidation number of xenon in xenon tetrafluoride and in xenon hexafluoride? (See Stretch and Challenge box below).

b) Write equations to show the reaction of fluorine with xenon to make xenon tetrafluoride and to make xenon hexafluoride.

4 COVALENT HALIDES

All non-metal halides are covalent and have a simple molecular structure. Typically, non-metal halides are liquids at room temperature with a low boiling point. Some metals will also form covalent halides, which is the case when the oxidation state of the metal is +3 or above. In such cases the metal ion that could be formed, e.g. Al^{3+}, has a very small ionic radius and is highly charged, and consequently distorts the electron cloud around the anion so much that the resulting bond has sufficient covalent character to be considered as being covalent.

STRETCH AND CHALLENGE

Noble gas compounds

Until 1962, chemists believed that the noble gases, such as xenon, did not react. Since then, several compounds that contain xenon and fluorine have been prepared. Xenon difluoride, for example, is formed when excess xenon is heated with fluorine gas at 400 °C:

$$Xe(g) + F_2(g) \rightarrow XeF_2(g)$$

Xenon difluoride forms colourless crystals, which are stable at room temperature in a dry atmosphere. It is quite surprising that this compound should be so stable. Xenon has a stable octet of electrons, so by reacting with fluorine and attaining an oxidation number of +2 in the difluoride, it should lose its stability.

By changing the mole ratios of xenon to fluorine, it is also possible to prepare xenon tetrafluoride, XeF_4, and xenon hexafluoride, XeF_6. XeF_4 crystals are shown on page 77.

All the xenon fluorides are very powerful oxidising agents, since xenon in a positive oxidation state is less stable than elemental xenon with an oxidation number of 0. Xenon difluoride oxidises water to form xenon, hydrofluoric acid and oxygen:

$$2XeF_2(s) + 2H_2O(l) \rightarrow 2Xe(g) + 4HF(aq) + O_2(g)$$

Read more about compounds of noble gases on page 70.

You can read about the hydrolysis of covalent chlorides of Period 2 and Period 3 elements and the resistance to hydrolysis of tetrachloromethane in Chapter 21.

? QUESTION 11

11 Construct the equation to show the hydrolysis of phosphorus(III) bromide, PBr_3, to make phosphoric(III) acid, H_3PO_3, and hydrogen bromide.

? QUESTION 12

12 Explain why all the hydrogen halides are polar molecules. (Hint: look at the electronegativities of the atoms involved on page 86).

Table 23.3 Boiling points of hydrogen halides

hydrogen halide	boiling point / °C
HF	17
HCl	−81
HBr	−63
HI	−38

Hydrogen fluoride has a relatively high boiling point because of the strong intermolecular hydrogen bonds between HF molecules. The other hydrogen halides have much weaker permanent dipole–permanent dipole intermolecular forces. These increase in strength from HCl to HI, because the molecular size and number of electrons increase.

HYDROLYSIS OF COVALENT HALIDES

Almost all covalent chlorides can be hydrolysed to give an acidic solution. It is a typical property of covalently bonded chlorides that they can be hydrolysed to form hydrochloric acid or hydrogen chloride if only a limited supply of water is available. One exception to this rule is tetrachloromethane, which does not react with water at all.

The hydrolysis of covalent halides, such as phosphorus(III) bromide and phosphorus(III) iodide, provides a suitable way to prepare hydrogen bromide and hydrogen iodide.

5 HYDROGEN HALIDES

All the hydrogen halides have the formula HX, where X is F, Cl, Br, I or At. They all are colourless acidic gases that are highly soluble in water, in which they form an acidic solution. The bonding in a hydrogen halide is covalent, but the molecule is polar with its halogen end being slightly negative and its hydrogen end slightly positive.

REACTION OF HALOGENS WITH HYDROGEN

The halogens react with hydrogen to give hydrogen halides, but the rate of reaction decreases with increasing atomic (proton) number of the halogen. The reaction between hydrogen and fluorine takes place very rapidly, even at low temperatures, whereas the reaction between hydrogen and iodine is reversible and needs elevated temperatures:

$$H_2(g) + F_2(g) \rightarrow 2HF(g)$$

When heated, hydrogen reacts with chlorine, and when a mixture of the two gases is subjected to ultraviolet light the reaction can be explosive.

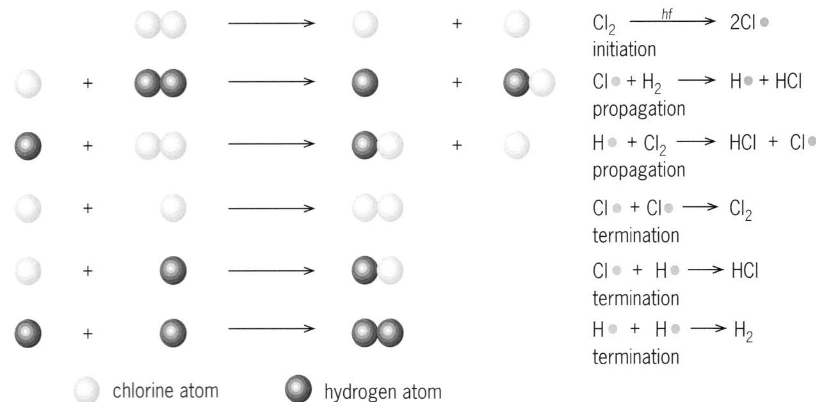

Fig 23.9 The reaction between hydrogen and chlorine in the presence of sunlight has a free-radical mechanism. The initiation reaction involves the homolytic fission of a chlorine–chlorine single bond to form highly reactive chlorine atoms. The chlorine atoms then collide with hydrogen molecules to form hydrogen chloride and hydrogen atoms. The hydrogen atoms collide with chlorine molecules to form hydrogen chloride molecules and regenerate chlorine atoms. These two stages are propagation steps, since there is no net loss of chlorine atoms. In the termination steps, pairs of atoms collide to form molecules. (Initiation, propagation and termination are covered on pages 242–243.)

Thermal decomposition of hydrogen halides

The thermal stability of the hydrogen halides decreases with increasing relative molecular mass, in a similar manner to the decrease in the bond energy from H–F to H–I. In the decomposition of hydrogen iodide, it is better to describe the reaction as an equilibrium process:

$$2HI(g) \rightleftharpoons H_2(g) + I_2(g) \quad \Delta H \text{ is positive}$$

Le Chatelier's principle (see pages 19 and 354) states that the position of equilibrium of a system shifts to minimise the effect of any change in the external conditions, such as increasing the temperature or the pressure. In the equilibrium process above, pressure has no effect whatsoever, because there is no volume change during the reaction – the number of moles of gas is the same on both sides of the equation. Since the decomposition is endothermic, an increase in temperature favours the formation of the two elements, because the reaction from left to right absorbs energy.

QUESTION 13

a) Write down an expression for the equilibrium constant for the thermal decomposition of hydrogen iodide.
b) At 600 K, hydrogen iodide is 19.1 per cent dissociated. Calculate the mole ratios of hydrogen, iodine and hydrogen iodide when a sample of hydrogen iodide is allowed to reach equilibrium at 600 K.

CHEMICAL TEST FOR HYDROGEN HALIDES

Hydrogen halides are colourless acidic gases that are able to react with bases such as ammonia gas. This reaction is interesting in that two gases react together to make a white solid dispersed in a gas, correctly described as a smoke (Fig 23.10). This can be used as a chemical test for the hydrogen halides:

$$NH_3(g) + HCl(g) \rightarrow NH_4Cl(s)$$

Fig 23.10 Ammonia gas and hydrogen chloride gas react to form a white smoke of ammonium chloride (a test either for ammonia or for hydrogen chloride)

AQUEOUS SOLUTIONS OF THE HYDROGEN HALIDES

All the hydrogen halides dissolve in water to form acidic solutions. Hydrogen chloride forms hydrochloric acid, hydrogen bromide forms hydrobromic acid and hydrogen iodide forms hydroiodic acid:

$$HX(g) + H_2O(l) \rightarrow H_3O^+(aq) + X^-(aq) \quad \text{where X is Cl, Br or I.}$$

The acid strength increases from hydrochloric acid to hydroiodic acid. This is because the H–I bond is weaker than the H–Cl bond (see Table 23.4). The bonding changes from covalent in the hydrogen halide to ionic in the corresponding acid.

All three acids are strong and display the typical reactions of strong acids (see page 331):

- They form hydrogen with metals above hydrogen in the electrochemical series.
- They form carbon dioxide with carbonates and hydrogencarbonates.
- They form salts with bases, such as metal oxides and hydroxides.
- They fully dissociate to form aqueous hydrogen ions.

Concentrated hydrochloric, hydrobromic and hydroiodic acids contain a large percentage of water, since if an attempt is made to concentrate them further, the gaseous hydrogen halide is evolved.

Table 23.4 Bond energies of hydrogen halides

Hydrogen halide	Bond energy (H–X)/kJ mol^{-1}
HF	562
HCl	431
HBr	366
HI	299

Oxidation of hydrohalic acids

Hydrochloric acid, hydrobromic acid and hydroiodic acid can all be oxidised to form the elemental halogen. The reaction involves the removal of hydrogen and an increase in the oxidation number of the halogen atom. For example:

$$PbO_2(s) + 4HCl(aq) \rightarrow PbCl_2(aq) + 2H_2O(l) + Cl_2(g)$$

The ease of oxidation increases as the atomic (proton) number of the halogen increases. So it is quite easy to reduce hydroiodic acid, but much more difficult to reduce hydrochloric acid.

Hydrochloric acid can be oxidised by lead(IV) oxide, manganese(IV) oxide or potassium manganate(VII) to make chlorine, but less powerful oxidising agents can be used to convert hydrobromic acid and hydroiodic acid into bromine and iodine, respectively. In other words, the reducing power decreases from hydroiodic to hydrochloric acid.

QUESTION 14

Predict the products of the reaction of:
a) hydrochloric acid with sodium carbonate;
b) hydroiodic acid with magnesium oxide;
c) hydrobromic acid with sodium hydroxide.

SCIENCE IN CONTEXT

Hydrofluoric acid and hydrogen fluoride

Hydrofluoric acid is unique among acids in that it reacts with glass. Fortunately, it does not react with plastics; if it did, it would be difficult to find a suitable container in which to store it. During the reaction with glass, a volatile silicon compound, SiF_4, is produced:

$$SiO_2 + 4HF \rightarrow SiF_4 + 2H_2O$$
$$CaSiO_3 + 6HF \rightarrow CaF_2 + SiF_4 + 3H_2O$$

Hydrofluoric acid is extremely dangerous to handle, because it acts as a local anaesthetic while burning into the skin and flesh. So if it comes into contact with your skin, you are not aware of the severity of the damage until it is too late. You are left with acutely painful and slow-healing burns.

Dilute hydrofluoric acid is a weak acid. Only about 10 per cent of the HF molecules are dissociated in a 0.1 mol dm^{-3} solution. This is because the H–F bond is very strong and the presence of strong intermolecular hydrogen bonds hinders dissociation.

Hydrogen fluoride is used mainly to produce fluorocarbons and the sodium aluminium fluoride required to manufacture aluminium. It is also used to prepare a variety of important synthetic chemicals and catalysts.

Hydrogen fluoride also attacks glass, and so it is used to etch glassware.

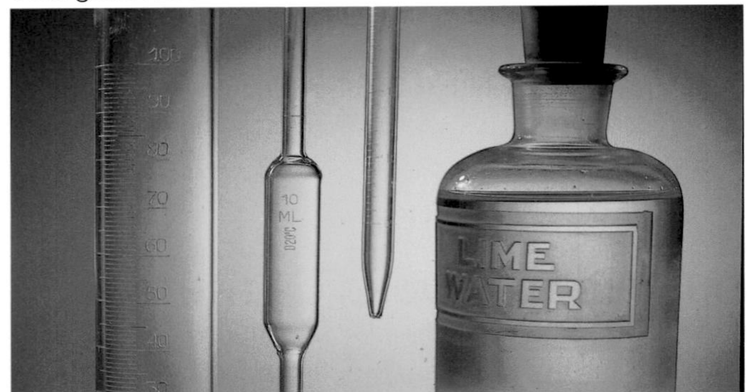

Fig 23.11 Hydrogen fluoride is used to etch the markings on thermometers and burettes

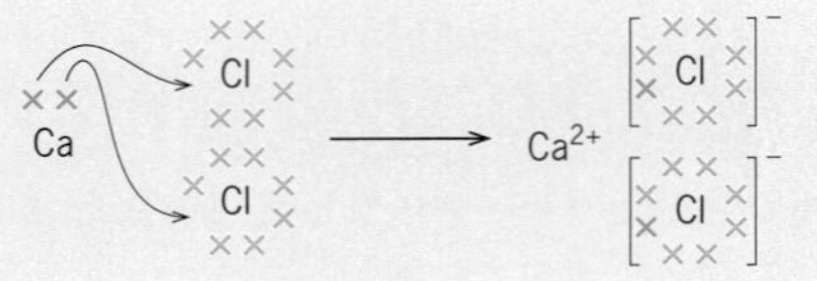

Fig 23.12 Calcium chloride is an ionic compound. Two electrons are lost by the calcium atom, one to each chlorine atom. This transfer of electrons means that both chlorine atoms and the calcium atom have a stable octet of electrons. The Ca^{2+} and two Cl^- ions are held together by strong electrostatic attractions, called ionic bonds. Most metal halides are ionic compounds and have metal atoms that transfer electrons to halogen atoms. Note: only the outer electrons are shown in the dot-and-cross diagram

6 IONIC HALIDES

Ionic halides are made when halogens react with metals. Most metals react when the hot or burning metal is plunged into an atmosphere of the halogen. Ionic halides show the properties typical of compounds with a giant ionic structure. They have high melting and boiling points, do not conduct electricity as solids, often dissolve in water and the solution will conduct electricity. However, as mentioned on page 461, some metals can form ionic halides with covalent character. These have much lower melting points than those of the other ionic halides, and often sublime instead of melting.

ACTION OF WATER ON IONIC CHLORIDES

Ionic chlorides, such as sodium chloride, dissolve in water to form a neutral solution. However, ionic chlorides with a covalent character dissolve to form an acidic solution because of hydrolysis. Metal halides in which the metal has an oxidation state of +3 have a high degree of covalent character. These halides have a metal cation with a charge of 3+, which is highly polarising because it has a small ionic radius and a high charge. That is, it has a high charge density (see pages 440).

When metal chlorides with covalent character are added to water, the ionic lattice breaks up and the chloride ion and the metal ion are surrounded by water molecules. The highly charged metal ion polarises one of the water molecules to form a proton, so the solution becomes acidic. Chromium(III) chloride, iron(III) chloride and aluminium chloride all form acidic solutions when they are added to water, because of the polarisation of a water molecule by an M^{3+} ion:

$$[Cr(H_2O)_6]^{3+}(aq) \rightleftharpoons [Cr(H_2O)_5OH]^{2+}(aq) + H^+(aq)$$

QUESTION 15

15 Construct the equation to show the reaction of potassium manganate(VII) with concentrated hydrochloric acid. The products are manganese(II) chloride, potassium chloride, water and chlorine.

REACTION OF SOLID HALIDES WITH CONCENTRATED SULFURIC ACID

Concentrated sulfuric acid reacts with solid halides to form the corresponding hydrogen halide. Since concentrated sulfuric acid is an oxidising agent, the reaction is often complicated by oxidation of the hydrogen halide.

Solid sodium chloride reacts to form hydrogen chloride. This reaction is often used to prepare hydrogen chloride in the laboratory:

$$NaCl(s) + H_2SO_4(l) \rightarrow NaHSO_4(s) + HCl(g)$$

The same reaction with sodium bromide yields a collection of products, which include sulfur dioxide, hydrogen bromide and bromine. This is a result of the oxidation of hydrogen bromide formed initially in the reaction by concentrated sulfuric acid:

$$NaBr(s) + H_2SO_4(l) \rightarrow NaHSO_4(s) + HBr(g)$$

$$2HBr(g) + H_2SO_4(l) \rightarrow 2H_2O(l) + SO_2(g) + Br_2(g)$$

When the reaction with concentrated sulfuric acid is repeated with sodium iodide, an even more complicated set of reactions takes place, with the formation of hydrogen iodide, sulfur dioxide, iodine and hydrogen sulfide. This time, violet fumes of iodine are produced (Fig 23.14):

$$8HI(g) + H_2SO_4(l) \rightarrow H_2S(g) + 4H_2O(l) + 4I_2(g)$$

These three reactions provide a way to distinguish between solid ionic chlorides, bromides and iodides.

The reactions of solid ionic halides with concentrated sulfuric acid illustrate the relative ease with which hydrogen halides and halide ions can be oxidised. Hydrogen iodide and iodide ions are the easiest to oxidise, hydrogen fluoride and fluoride ions the most difficult. The reason for this trend in reducing power of the halide ions lies in the change in the ionic radius of the halide. It is easier to remove an electron from the large iodide ion than from the much smaller fluoride ion.

Fig 23.14 Fumes of iodine are produced when concentrated sulfuric acid is added to sodium iodide

The action of water on aluminium chloride is discussed on page 417.

Fig 23.13 Fumes of bromine are produced when concentrated sulfuric acid is added to sodium bromide

QUESTION 16

a) Write an equation to show the reaction between concentrated sulfuric acid and sodium iodide to form hydrogen iodide.

b) Write an equation to show the reaction between concentrated sulfuric acid and hydrogen iodide to form iodine and sulfur dioxide.

c) Predict the reaction products for the reaction between sodium astatide and concentrated sulfuric acid.

d) Predict the reaction products for the reaction between sodium fluoride and concentrated sulfuric acid.

REACTIONS OF THE AQUEOUS HALIDE IONS

In addition to the displacement reactions described on page 500, the aqueous halide ions also take part in precipitation reactions. Most halides are soluble, but silver halides are insoluble. Therefore, silver halides can be precipitated by mixing together aqueous solutions of silver nitrate and the appropriate halides:

$$Ag^+(aq) + X^-(aq) \rightarrow AgX(s)$$

where X is Cl, Br or I. These precipitation reactions are very useful in qualitative analysis, and are summarised in Table 23.5.

Fig 23.15 Aqueous silver nitrate forms characteristic coloured precipitates with aqueous chlorides, bromides and iodides. Silver chloride is white, silver bromide is cream and silver iodide is pale yellow

Table 23.5 Action of dilute nitric acid followed by aqueous silver nitrate and aqueous ammonia on aqueous halide ions

Fluoride	Chloride	Bromide	Iodide
No precipitate	White precipitate of silver chloride that turns violet in sunlight, redissolves in dilute aqueous ammonia to form a colourless solution through the formation of diamminesilver(I) ion $[Ag(NH_3)_2]^+(aq)$	Cream precipitate of silver bromide redissolves in concentrated aqueous ammonia to form a colourless solution through the formation of $[Ag(NH_3)_2]^+(aq)$	Yellow precipitate of silver iodide that does not redissolve in concentrated aqueous ammonia, but turns white

In figure 23.15 you will notice that the white precipitate of silver chloride has a grey tinge. This is because silver halides decompose slowly when left in sunlight. The grey tinge is metallic silver. It is this decomposition that is used in black and white photography. Silver iodide is also used to seed rain clouds to produce rain.

Fig 23.16 In order to avoid drought, rain is made to fall in Australia by seeding clouds with silver iodide crystals

Black and white photography

Silver halides are sensitive to light. For example, in the precipitation reaction given in Table 23.5, the silver chloride formed turns grey-violet in the presence of light. This is because of the decomposition of silver chloride to give silver and chlorine:

$$2AgCl(s) \rightarrow 2Ag(s) + Cl_2(g)$$

The longer the exposure to light, the greater is the amount of silver produced.

Traditional black and white photography works on a similar principle. The film consists of a very thin layer of gelatine and silver bromide crystals deposited on clear plastic. As light shines on the film, an image is captured by the silver bromide crystals. This image is developed by immersing the film in a developer that reacts only with those silver bromide crystals that have been exposed to the light. The developer reduces the silver bromide to metallic silver:

$$2AgBr(s) + C_6H_4(OH)_2(aq) \rightarrow 2Ag(s) + C_6H_4O_2(aq) + 2HBr(aq)$$

Next, the film is placed in a fixer, where the silver bromide crystals not exposed to light, and hence unaffected by the developer, are removed from the film. This leaves the film as a negative with black areas of silver where the film was exposed to light.

The negative is then converted into a positive picture on photographic paper. Light is shone through the negative onto the photographic paper, which is coated with silver bromide. The action of light on the silver bromide produces an image that again has to be developed and fixed to give a permanent photograph.

Fig 23.17 The negative image

Fig 23.18 The positive image

7 MANUFACTURE AND USES OF CHLORINE

Chlorine is made by the chlor-alkali industry. The chlor-alkali industry is so called because it produces chlorine and the alkali sodium hydroxide from the electrolysis of concentrated, aqueous sodium chloride (brine). The industry is more than 100 years old and is amongst the largest in the world, consuming vast quantities of electricity and producing more than 25 000 tonnes of chlorine every day. The UK alone produces 800 000 tonnes a year.

There are three types of electrolysis cell: the mercury cell, the diaphragm cell and, more recently, the membrane cell. All three are in use today. This section concentrates on the diaphragm cell.

Diaphragm cell

In concentrated aqueous sodium chloride, four ions are present: Na^+, Cl^-, $H^+(aq)$ and OH^-.

At the cathode: $2H^+(aq) + 2e^- \rightarrow H_2(g)$
At the anode: $2Cl^-(aq) \rightarrow Cl_2(g) + 2e^-$

REMEMBER THIS

The electrolysis of *molten* sodium chloride makes sodium and chlorine, whereas the diaphragm cell involves the electrolysis of *aqueous* sodium chloride, so hydrogen and chlorine are formed.

$H^+(aq)$ ions are discharged at the cathode, which causes more water molecules to dissociate to produce more $H^+(aq)$ ions. At the same time, more OH^- ions are produced:

$$H_2O(1) \rightleftharpoons H^+(aq) + OH^-(aq)$$

Combining the two reactions gives:

$$2H^+(aq) + 2e^- \rightarrow H_2(g)$$

$$[H_2O(l) \rightarrow H^+(aq) + OH^-(aq)] \times 2$$

$$2H_2O(l) + 2e^- \rightarrow H_2(g) + 2OH^-(aq)$$

REMEMBER THIS

This is an example of Le Chatelier's principle. As $H^+(aq)$ ions are discharged, the equilibrium is shifted to the right. See pages 19 and 354.

The ions remaining in solution are Na^+ and OH^-, which tend to concentrate near the cathode to produce sodium hydroxide. In the commercial electrolysis of brine, the hydroxide ions forming must be kept away from the chlorine produced at the anode. This is because chlorine reacts with OH^- to produce chlorate(I) ions, ClO^-:

$$Cl_2(g) + 2OH^-(aq) \rightarrow Cl^-(aq) + ClO^-(aq) + H_2O(l)$$

In the diaphragm cell (Fig 23.19), a porous asbestos partition (the diaphragm), placed between the electrodes, keeps the sodium hydroxide forming at the cathode away from the chlorine at the anode. Purified fresh brine (sodium chloride) solution is fed continuously into the anode compartment and the level is kept above that of the cathode compartment. This allows the brine solution to seep into the cathode compartment and also prevents OH^- ions migrating to the anode. The brine used is purified to remove Ca^{2+} and Mg^{2+} ions. They would react with OH^- ions to form insoluble hydroxides which would block the pores of the diaphragm.

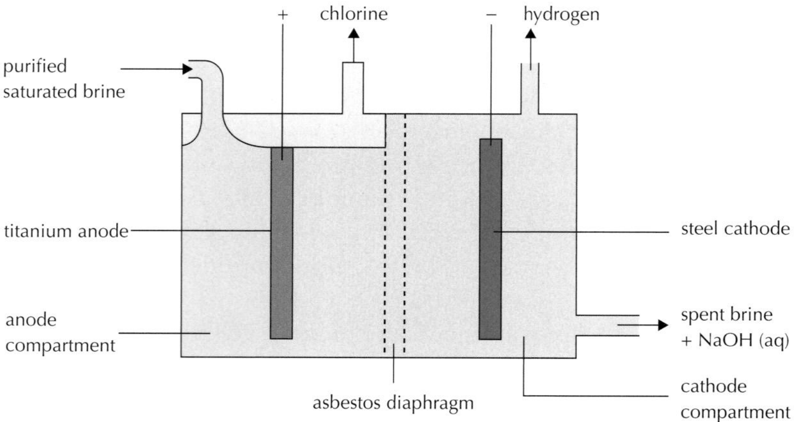

Fig 23.19 The diaphragm cell

The solution produced in the cathode compartment contains about 10 per cent sodium hydroxide and 15 per cent sodium chloride. The sodium chloride is separated from the sodium hydroxide by evaporating the solution to one fifth of its original volume. This causes sodium chloride to crystallise out, leaving a 50 per cent sodium hydroxide solution contaminated by 1 per cent sodium chloride.

HOW SCIENCE WORKS

The mercury cell and the minamata tragedy

Much of the chlorine and sodium hydroxide manufactured in the UK used to be produced using a flowing mercury cathode (Fig 23.20).

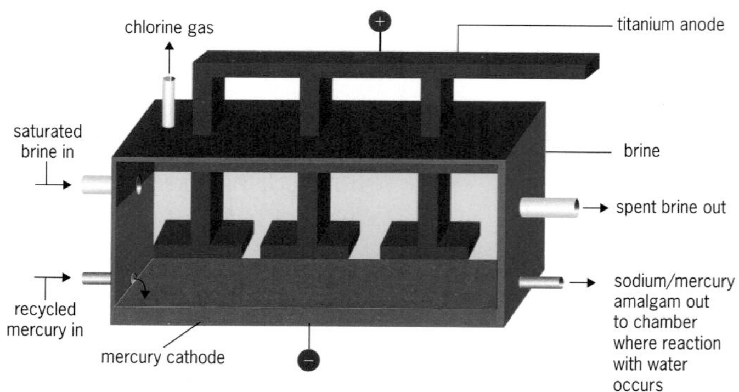

Fig 23.20 The mercury cell

A saturated brine solution is used as the electrolyte, but in this process sodium ions are discharged at the cathode. The sodium produced dissolves in the mercury as it passes through the cell, forming an amalgam:

$$Na^+(aq) + e^- \xrightarrow{Hg} \underset{\text{sodium amalgam}}{Na/Hg(l)}$$

Therefore, hydroxide ions do not build up in the cell and react with the chlorine produced. The sodium amalgam flows out of the cell before the sodium has a chance to react with the water in the brine solution. It is piped to a chamber where it is allowed to react with pure water to form sodium hydroxide and hydrogen:

$$2Na/Hg(l) + 2H_2O(l) \rightarrow 2NaOH(aq) + H_2(g) + 2Hg(l)$$

Once all the sodium has reacted the mercury is recirculated into the electrolysis cell. Fifty per cent sodium hydroxide solution of high purity is produced.

It was because of the purity of the products that the mercury cell once dominated world-wide production of sodium hydroxide. However, mercury and its compounds are highly toxic. In theory, no mercury should escape during this process, but it does through waste solutions, which pollute the environment. One of the most tragic incidents occurred in the 1950s at Minamata Bay in Japan. Mercury entered the food chain after reacting with organic compounds which were ingested by fish and shellfish. The first to suffer from what came to be known as Minamata disease were fishermen's cats, which became paralysed and eventually died. A month or so later, the first human case was diagnosed and over the next decade 43 people died and another 60 were brain damaged.

Much work has been done to reduce mercury leakage, and it may be that as little as 0.25 g now escapes for every tonne of chlorine produced. However, such is the concern that this process is in sharp decline. For example, manufacturers around the North Sea have pledged to phase out its use by the year 2010. In the United Kingdom there are no large factories producing chlorine using the mercury cell.

USING THE PRODUCTS OF THE CHLOR-ALKALI INDUSTRY

For every tonne of chlorine manufactured, 1.1 tonnes of sodium hydroxide are produced. There is no viable alternative to electrolysis for the manufacture of these chemicals. If the market were to decline for either chemical, there would be a surplus of that chemical. This would be more of a problem if it were chlorine, since its disposal presents significant problems.

In 1993, demand for sodium hydroxide slumped in the USA and some manufacturers were almost paying customers to take it away. A year later, as Fig 23.21 shows, there was an upsurge in demand and an increase in price. Forecasting demand for these two products is a difficult balancing act, requiring a knowledge of how and where they are used.

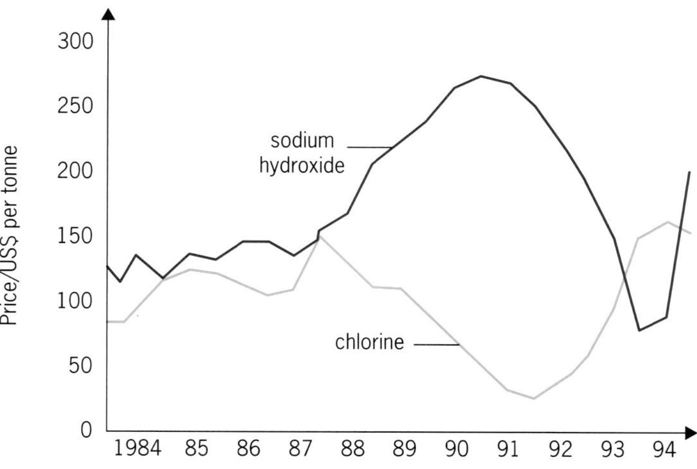

Fig 23.21 Fluctuating prices of chlorine and sodium hydroxide

Uses of chlorine

One of the applications of chlorine as an element is to bleach the wood pulp used to make paper products and textiles. Highly toxic dioxins are by-products of this process, most of which are carried away in the waste-water. So, there is growing concern that this process is helping to increase dioxin levels in the environment. There is also evidence that dioxins do enter the paper products themselves – another cause for concern. This application of chlorine is being gradually reduced and could eventually be phased out altogether. There is also a move to find an alternative to chlorine for destroying harmful microbes in the water supply. But any alternative must be proved to be as effective as chlorine before the switch can be made. Ozone is a possibility, but it breaks down rapidly and loses its disinfectant properties. Therefore, it offers no protection against reinfection of the water.

PVC manufacture and its uses are covered on page 398.

Almost 30 per cent of all chlorine produced is used to make the monomer chloroethene (CH_2=CHCl), from which PVC – the world's most versatile plastic – is manufactured. Demand for PVC is likely to continue to grow. There are some fears over the safety of the plasticisers in this polymer, particularly in its use as food wrapping. Another issue is how it should be disposed of. Incineration can produce minute quantities of dioxins. However, there would seem to be no case for banning this chlorine compound, since virtually all of its uses are safe and its disposal could be made as safe.

REMEMBER THIS

The many uses of halogens mean that large quantities of each halogen must be manufactured, stored and transported. Any leakage will be extremely dangerous because halogens are corrosive, highly reactive and can cause extensive damage to the respiratory tract and lungs.

Solvents are among the organic products made using chlorine. The production is being phased out of those known to cause depletion of the ozone layer in the upper atmosphere. All chlorinated organic solvents are bound to come under increasing scrutiny.

While the dangers of chlorine and its compounds must never be denied, blanket condemnation of them is not realistic. Each of the 15 000 chlorine-based compounds now produced deserves to be assessed individually according to its advantageous properties, its toxicity and its environmental effects. Many of the most effective medicines and pesticides are chlorine-based compounds, and without them, disease would be more widespread and crop production would fall.

HOW SCIENCE WORKS

Dioxins

Dioxins are present naturally in minute quantities and form whenever wood and certain other substances burn. However, the chemical industry has added a great deal more to the environment, much of it before the danger posed by this group of chemicals was recognised.

For seven years during the Vietnam war in the 1960s, US aircraft sprayed the jungle with a mixture of herbicides known as Agent Orange. The plan was to destroy the foliage under which the Viet Cong could hide. In all, some 50 000 tonnes of Agent Orange were used. However, Agent Orange was contaminated with dioxins, in concentrations of about 2 parts per million. So, about 100 kg of dioxins entered the Vietnamese jungles. The subsequent births of babies with abnormalities provided the very first indication that dioxins cause genetic defects. Also much in evidence was a terrible skin complaint called chloracne, which was caused by exposure to dioxins (Fig 23.22).

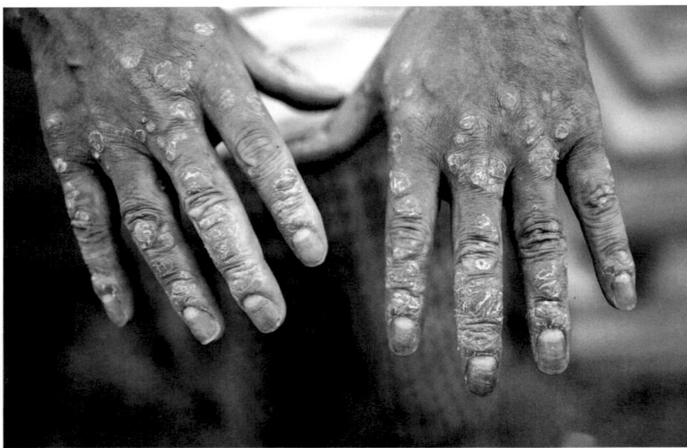

In 1976 an accident at a chemical plant in Seveso, Italy, released dioxins into the air. There was an immediate outbreak of chloracne and 600 people were evacuated from the area. This brought dioxins to public attention.

Some dioxins are harmless, but many are not. One of the most deadly is TCDD (Fig 23.23). Bleaching wood pulp with chlorine to produce paper products, such as newsprint and disposable babies' nappies, is now known to produces minute quantities of dioxins, including TCDD. This has led to a phasing out of chlorine as a bleaching agent. Chlorine dioxide provides a safer alternative.

Should we continue to be alarmed about dioxins? Now everyone is aware of the dangers of TCDD and the other deadly dioxins, the means of their production are being reduced, which is bringing levels down significantly. For example, dioxins are produced by leaded petrol, so the switch to unleaded petrol has led to a notable reduction in air pollution by dioxins. Attention is now focusing on improving the incineration of waste – another source of dioxins in the environment.

Fig 23.22 The skin condition caused by exposure to Agent Orange

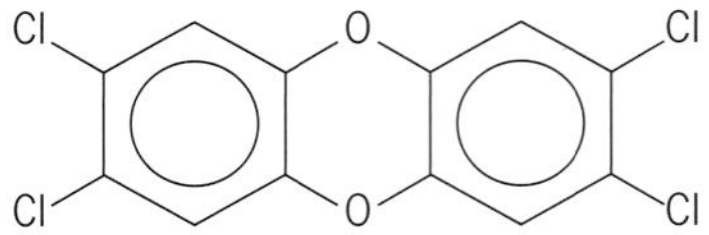

Fig 23.23 TCDD (2,3,7,8-tetrachlorodibenzodioxin)

SCIENCE IN CONTEXT

Chlorinating water supplies and swimming pools

Chlorine is pumped into water as the final stage in its treatment before it enters the domestic supply. Chlorine reacts with water to form chloric(I) acid, which is the chemical which kills bacteria:

$$(1)\ Cl_2(g) + H_2O(l) \rightleftharpoons HOCl(aq) + H^+(aq) + Cl^-(aq)$$

HOCl is a weak acid and dissociates slightly:

$$(2)\ HOCl(aq) \rightleftharpoons H^+(aq) + ClO^-(aq) \text{ chlorate(I) ion}$$

HOCl(aq) is the bactericide rather than $ClO^-(aq)$, as it is 80 times more effective. It is thought that the negative charge on ClO^- hinders its penetration into the bacterial cell wall.

A decrease in $H^+(aq)$ ions causes equilibrium (2) to shift to the right, according to Le Chatelier's principle. So, more

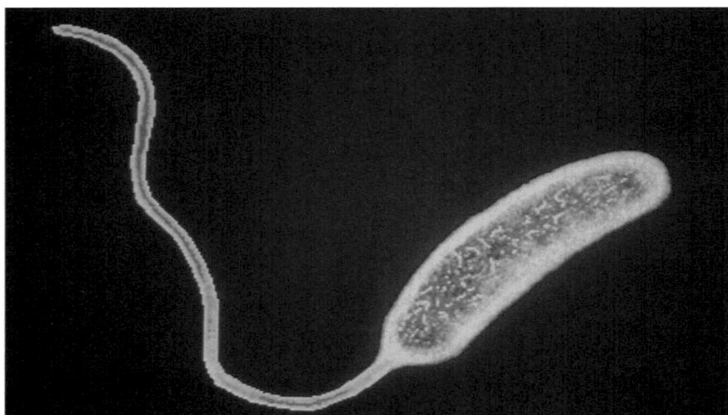

Fig 23.24 *Vibrio cholerae*, the bacterium which causes cholera. It is found in many untreated water supplies. Cholera is spread by infected food and water

SCIENCE IN CONTEXT

HOCl(aq) ionises, thereby reducing the HOCl(aq) concentration. In a water treatment works, the pH is adjusted to about 7.3. This produces a fairly high concentration of HOCl(aq). A lower, more acidic pH would start to dissolve harmful substances from the water pipes.

Unfortunately chlorination of water also produces some chloroalkanes as a result of the reaction of hydrocarbon residues with Cl_2. These chloroalkanes may be harmful to humans.

A public swimming pool makes an ideal breeding ground for microorganisms, so it must be sterilised (Fig 23.25). The water used to be sterilised using chlorine gas from cylinders. However slight, there is a risk that chlorine will escape and be a health hazard. So health and safety regulations ban the use of cylinders in public pools. Instead, solid compounds of chlorine, such as Trichlor, are used to produce the chloric(I) acid.

$+ \ 3H_2O \rightleftharpoons \quad + \ 3HOCl$

Fig 23.26 The reaction of Trichlor with water to produce chloric(I) acid

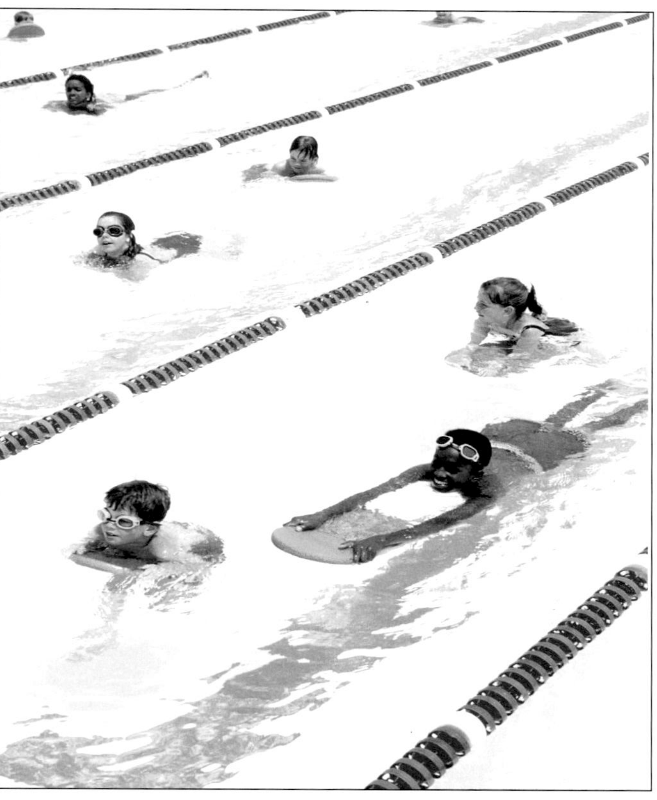

Fig 23.25 With many people using this swimming pool, sterilisation of the water with chlorine keeps it bacteria free

8 ELECTROLYSIS

Electrolysis is the decomposition of a liquid by passing an electric current through it. It is a special type of redox reaction where the oxidation and the reduction take place in different regions of an electrolytic cell. Electrolysis is extensively used to manufacture substances, although the use of electricity invariably makes these substances expensive to produce. Metals such as sodium and aluminium are manufactured by the electrolysis of an appropriate molten liquid. Copper is purified by electrolysis, and earlier on in the chapter we described how aqueous sodium hydroxide and chlorine can be manufactured by the electrolysis of aqueous brine.

ELECTROLYSIS CELL

Much of the early work on electrolysis was done by Michael Faraday in the 1830s. He coined the word **electrolytes** to describe those compounds which, when molten or in solution, conduct electricity and are decomposed by it. The conducting rods dipping into the electrolyte he called **electrodes**. Electrolysis works only with **direct current (d.c.)**. This means that the charge flows in one direction only, whereas the current in the mains in your house is **alternating current (a.c.),** where the flow of electrons (charge) reverses, or alternates, fifty times every second. Direct current gives a positive charge to one electrode, which Faraday called the **anode,** and a negative

REMEMBER THIS

Oxidation always occurs at the anode and reduction always occurs at the cathode.

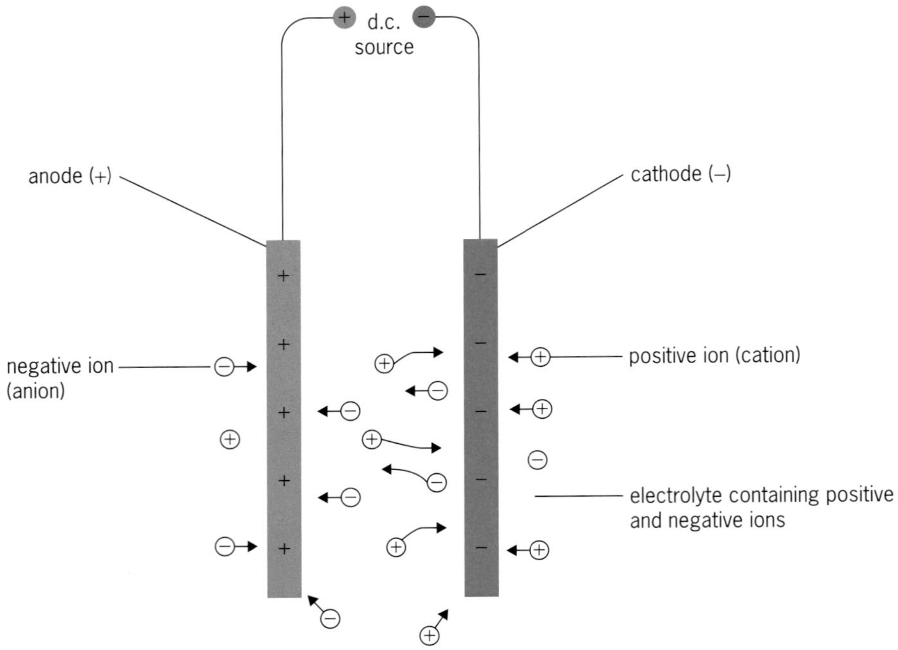

Fig 23.27 An electrolysis cell. The positive ions (cations) are attracted to the cathode and the negative ions (anions) are attracted to the anode

charge to the other electrode, which he called the **cathode**. These terms are still in use today.

Note from Fig 23.27 that it is the anions (negative ions) and the cations (positive ions) which move in the electrolyte. They carry the charge between the electrodes and cause the electrolyte to conduct. **Anions** are so called because during electrolysis they are attracted to the anode. Electrons are the charge carriers bringing the current to and from the battery.

HALF EQUATIONS AND REDOX IN ELECTROLYSIS

The electrolysis of molten sodium chloride (page 428) provides a suitable starting point to explaining what happens during electrolysis. Sodium ions are attracted to the negatively charged cathode, where they each pick up an electron to become sodium atoms.

$$\text{At the cathode: } Na^+(l) + e^- \rightarrow Na(l)$$

Because each sodium ion is gaining an electron, **reduction** is occurring at the cathode. Simultaneously, electrons are drawn off from the anode by the d.c. source (a battery or power pack), giving the anode a positive charge. Chloride ions are attracted to the positively charged anode, where they give up electrons.

$$\text{At the anode: } 2Cl^-(l) \rightarrow Cl_2(g) + 2e^-$$

So **oxidation** of the chloride ions occurs at the anode. The electrons released by the chloride ions travel back to the d.c. source. The equations for the reactions at the electrodes are called **half equations**, because they represent **half reactions** (oxidation and reduction). These combine to give the complete reaction occurring during electrolysis. (We use 2 × cathode reaction to cancel out electrons.)

? QUESTION 17

17 **a)** Why are postive ions called cations?

b) Why is solid sodium chloride not called an electrolyte?

REMEMBER THIS

OIL RIG – **O**xidation **I**s **L**oss (of electrons). **R**eduction **I**s **G**ain (of electrons). This and redox reactions are covered on page 102 and in Chapter 25.

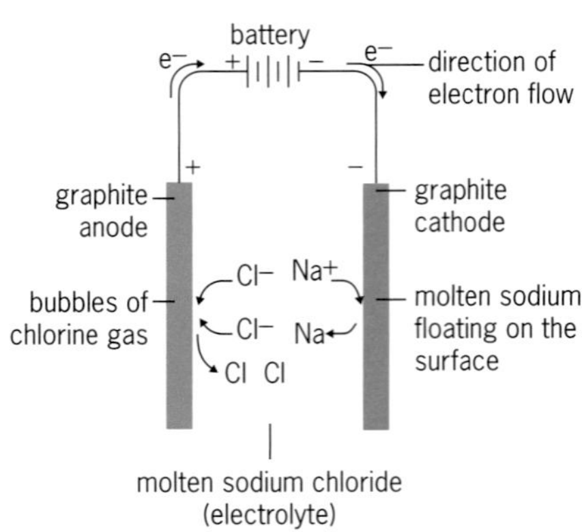

Fig 23.28 The overall process of electrolysis using molten sodium chloride as an example. Note the direction of electron flow

$$2Na^+(l) + 2e^- \rightarrow 2Na(l)$$

$$2Cl^-(l) \rightarrow Cl_2(g) + 2e^-$$

$$2Na^+(l) + 2Cl^-(l) \rightarrow 2Na(l) + Cl_2(g)$$

So, electrolysis involves **red**uction and **ox**idation reactions, usually abbreviated to **redox** reactions.

QUESTION 18

Molten lead(II) bromide consists of Pb^{2+} ions and Br^- ions. Write the half equations for the reactions occurring at the anode and cathode during electrolysis and state which is oxidation or reduction.

PREDICTING THE PRODUCTS OF ELECTROLYSIS OF AQUEOUS SOLUTIONS

In the case of molten sodium chloride, there is one cation (Na^+) and one anion (Cl^-), so the products of electrolysis are easy to predict. But what happens when sodium chloride is dissolved in water? As well as Na^+ and Cl^- ions, there are also $H^+(aq)$ and OH^- ions from the water:

$$H_2O(l) \rightleftharpoons H^+(aq) + OH^-(aq)$$

Even though the concentration of $H^+(aq)$ is extremely low at 10^{-7} mol dm^{-3}, it is enough to cause hydrogen to be discharged (released) at the cathode in preference to sodium. Why is this?

To answer this question, we must look at which of Na^+ or $H^+(aq)$ is most easily reduced to form atoms. Sodium is a very reactive metal, because it readily forms Na^+ ions by losing electrons. Thus, Na^+ ions do not accept electrons easily. Hydrogen ions are much more readily reduced than sodium ions, so hydrogen ions are discharged in preference to sodium ions:

$$2H^+(aq) + 2e^- \rightarrow H_2(g)$$

REMEMBER THIS

$H^+(aq)$ is more correctly written as $H_3O^+(aq)$. However, the form $H^+(aq)$ is used in this chapter, where the state symbol (aq) is consistently included to remind you that we are dealing with the $H_3O^+(aq)$ ion.

A list of cations can be drawn up in order of increasing ease of discharge at the cathode. A shortened version is given in Table 23.6. This table is sometimes referred to as the **electrochemical series**, **redox series** or **reactivity series**. It is discussed further on pages 511, in connection with electrode potentials.

As a rule, the ion nearer the bottom of the electrochemical series is the more easily discharged. However, like all rules, there are exceptions and if the concentration of a particular ion is very high, it may affect which ion is discharged at the cathode.

Table 23.6 Increasing ease of discharge of cations

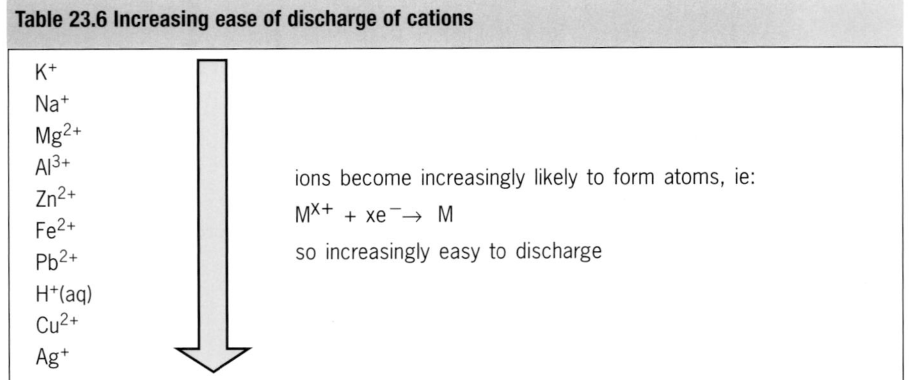

Cations	
K^+	
Na^+	
Mg^{2+}	
Al^{3+}	
Zn^{2+}	ions become increasingly likely to form atoms, ie:
Fe^{2+}	$M^{x+} + xe^- \rightarrow M$
Pb^{2+}	so increasingly easy to discharge
$H^+(aq)$	
Cu^{2+}	
Ag^+	

A similar list can be drawn up for the anions discharging at the anode, but this time the interest is in the ease of oxidation of ions. As with cations, when two anions are present in equal amounts, usually the lower one in the table is discharged. But, once again, the relative concentrations of different anions can have an effect. This is particularly true of halide solutions, such as aqueous

QUESTION 19

In the electrolysis of aqueous copper(II) sulfate, use Table 23.6 to predict which cation will be discharged.

Table 23.7 Increasing ease of discharge of anions. In practice , SO_4^{2-} and NO_3^- are never discharged from aqueous solutions

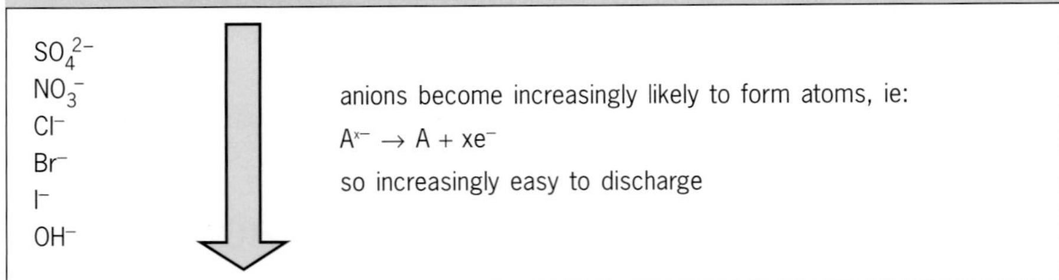
SO_4^{2-}
NO_3^-
Cl^-
Br^-
I^-
OH^-

anions become increasingly likely to form atoms, ie:

$A^{x-} \rightarrow A + xe^-$

so increasingly easy to discharge

sodium chloride, where the two anions present are OH^- and Cl^-. If a very dilute aqueous solution of sodium chloride is electrolysed, OH^- ions are discharged as predicted by Table 23.7 and oxygen gas bubbles off at the anode.

$$\text{At the anode: } 4OH^-(aq) \rightarrow 2H_2O(l) + O_2(g) + 4e^-$$

However, when the sodium chloride solution is concentrated, the Cl^- concentration is much greater than the OH– concentration. So, Cl^-ions are discharged, producing chlorine instead of oxygen at the anode:

$$\text{At the anode: } 2Cl^-(aq) \rightarrow Cl_2(g) + 2e^-$$

In the case of concentrated nitrate and sulfate solutions, even though the concentration of the hydroxide ion is extremely low, it is still discharged preferentially as predicted by Table 23.7.

EXAMPLE

Q What products are formed at the anode and cathode from the electrolysis of aqueous sodium sulfate? Include in your answer the relevant half equations.

A Ions present in aqueous sodium sulfate:

from Na_2SO_4: $Na^+(aq)$ and $SO_4^{2-}(aq)$
from water: $H^+(aq)$ and $OH^-(aq)$

Ions attracted to the cathode (cations):

$Na^+(aq)$ and $H^+(aq)$

Ions discharged at the cathode:

$H^+(aq)$ is discharged because it is lower in the electrochemical series than Na^+

Half equation: $2H^+(aq) + 2e^- \rightarrow H_2(g)$

Ions attracted to the anode (anions):

$SO_4^{2-}(aq)$ and $OH^-(aq)$

Ions discharged at the anode:

$OH^-(aq)$ is discharged in preference to $SO_4^{2-}(aq)$

Half equation: $4OH^-(aq) \rightarrow 2H_2O(l) + O_2(g) + 4e^-$

So, the products are hydrogen at the cathode and oxygen at the anode.

? QUESTION 20

20 What products are formed at the anode and cathode during the electrolysis of dilute sulfuric acid? Include in your answer the reasons and relevant half equations.

REMEMBER THIS

The number of particles (atoms, molecules, ions or electrons) in 1 mole is called the **Avogadro constant** (see page 6).

WORKING OUT AMOUNTS PRODUCED DURING ELECTROLYSIS

It was Faraday who again led the way in calculating how much of a substance is produced from a given current in a given time during electrolysis. His work was done in 1832, before the electron had been identified. Faraday deduced that:

The quantity of electricity passed is proportional to the amount of substance discharged at an electrode.

This relationship is sometimes referred to as Faraday's first law.

The quantity of electricity is measured in **coulombs (C)**. One amp of current passes one coulomb of charge every second. This means that:

quantity of electricity (charge) = current × time
coulombs = amps × seconds

One mole of electrons has a charge of 96 500 C, and this quantity of charge is called the **Faraday constant**, ***F***, in honour of Michael Faraday's pioneering work. Thus, the Faraday constant is related to the Avogadro constant, *L*, and the charge on an electron, *e*:

$$F = L \times e$$

Consider the amount of sodium produced by 1 mole of electrons in the electrolysis of molten sodium chloride. From the half equation:

$$\begin{array}{ccccc} Na^+(l) & + & e^- & \rightarrow & Na(l) \\ 1\ mol & & 1\ mol & & 1\ mol \end{array}$$

it follows that 1 mole of electrons produces 1 mole of sodium atoms. At the same time, 1 mole of electrons is released from ½ mole of chlorine molecules, since:

$$\begin{array}{ccccc} Cl^-(l) & \rightarrow & \tfrac{1}{2}Cl_2(g) & + & e^- \\ 1\ mol & & \tfrac{1}{2}mol & + & 1\ mol \end{array}$$

Now consider ions with a double charge, such as Cu^{2+}. It follows that it takes 2 moles of electrons to deposit 1 mole of copper:

$$\begin{array}{ccccc} Cu^{2+} & + & 2e^- & \rightarrow & Cu \\ 1\ mol & & 2\ mol & & 1\ mol \end{array}$$

Thus:

The number of moles of electrons required to discharge 1 mole of ions is equal to the charge on the ion.

This is sometimes called Faraday's second law. We can use the relationships first discovered by Faraday to calculate amounts of substances produced during electrolysis.

QUESTION 21

a) How many electrons will there be in one mole of electrons? (See page 6 if you are not sure.)

b) What is the charge, in coulombs, on 1 electron if 1 mole of electrons has a charge of 96 500 C?

QUESTION 22

a) How many moles of electrons and how many Faradays are required to produce:

i) 1 mole of aluminium (the aluminium ion is Al^{3+}),

ii) 1 mole of potassium?

b) Write down the half equation for the production of oxygen from OH^- ions. How many Faradays are required to produce 1 mole of oxygen molecules?

QUESTION 23

Aqueous copper(II) sulfate is electrolysed using graphite electrodes under room conditions. What mass of copper is produced at the cathode and what volume of oxygen at the anode if a constant current of 2.68A flows for 2 hours?

(A_r of Cu = 64, A_r of O = 16)

Note: graphite electrodes are regarded as inert because they do not take part in cell electrolysis reactions.

EXAMPLE

Q An aqueous solution of sulfuric acid is electrolysed in a laboratory using platinum electrodes. The current is kept constant at 2 A for 1 hour. Write the electrode reactions and calculate the mass and volume of the products formed at the electrodes, assuming room conditions.

A The mention of platinum electrodes in the question simply tells you that the electrodes are inert (unreactive).

Ions present: from H_2SO_4:	$H^+(aq)$ and $SO_4^{2-}(aq)$
from water:	$H^+(aq)$ and $OH^-(aq)$
Ions attracted to the cathode:	$H^+(aq)$
Ions discharged:	$H^+(aq)$
Half equation:	$2H^+(aq) + 2e^- \rightarrow H_2(g)$
Ions attracted to the anode:	$SO_4^{2-}(aq)$ and $OH^-(aq)$
Ions discharged at the anode:	$OH^-(aq)$ is discharged in preference to $SO_4^{2-}(aq)$ (See Table 23.7)
Half equation:	$4OH^-(aq) \rightarrow 2H_2O(1) + O_2(g) + 4e^-$

EXAMPLE (Cont.)

The amount of hydrogen and oxygen formed can now be calculated by working out how many moles of electrons have passed through the electrolysis cell.

Step 1 Calculate the quantity of charge passed:

Current = 2A Time = 1 hour = 60 min = $1 \times 60 \times 60$ s = 3600 s

Quantity of charge (C) = current (A) × time (seconds)
= 2.00×3600
= 7200 C

Step 2 Work out moles of electrons passing through the electrolysis cell:

1 mole of electrons carries 1 Faraday of charge = 96 500 C

Moles of electrons carrying 7200 C $= 1 \times \frac{7200}{96\,500} = 0.0746$ mol

For hydrogen production at the cathode, follow steps 3–5.

Step 3 Convert the half equation to amounts:

$$2H^+(aq) + 2e^- \rightarrow H_2(g)$$
$$2\text{mol} + 2\text{mol} \rightarrow 1 \text{ mol}$$

Step 4 Scale the amounts in the half equation:

2mol e^- produce 1 mol $H_2(g)$

Therefore: 0.0746 mol e^- produces $\frac{0.0746}{2} = 0.0373 \text{ mol } H_2(g)$

Step 5 Convert amounts (moles) to masses and volumes:

mass in grams = amount in moles × mass of 1 mole
= $0.0373 \times 2 = 0.0746$ g H_2

Remember: 1 mole of any gas occupies 24 dm^3 under room conditions.

volume of H_2 = $0.0373 \times 24 = 0.895$ dm^3 = 895 cm^3

For oxygen production at the anode, repeat Steps 3–5:

Step 3 Convert the half equation to amounts:

$$4OH^-(aq) \rightarrow 2H_2O(l) + O_2(g) + 4e^-$$
$$4 \text{ mol} \quad\quad 2 \text{ mol} + 1 \text{ mol} + 4 \text{ mol}$$

Step 4 Scale the amounts in the half equation. Since 4mol e^- is produced when 1 mol $O_2(g)$ is formed, then:

0.0746 mol e^- is produced when $\frac{0.0746}{4} = 0.0187 \text{ mol } O_2(g)$ is formed

Step 5 Convert amounts (moles) to masses and volumes:

mass in grams = amount in moles × mass of 1 mole
= $0.0187 \times 32 = 0.0598$ g O_2

Since 1 mol of any gas occupies 24 dm^3 under room conditions:

volume of O_2 = $0.0187 \times 24 = 0.449$ dm^3 = 449 cm^3

SUMMARY

After studying this chapter, you should know that:

- The reactivity of the halogens decreases with increasing atomic (proton) number.
- The oxidising power of the halogens decreases with increasing atomic (proton) number and that this can be explained in terms of the ability to capture electrons.
- The oxidation state of a halogen in a compound is normally −1.
- A halogen atom can gain an electron through ionic bonding or share electrons through covalent bonding to attain a noble-gas electron configuration.
- Metals react with halogens to form ionic halides and non-metals react with them to form covalent halides.
- The hydrogen halides are colourless acidic gases that dissolve in water to form acid solutions. Hydrogen chloride forms hydrochloric acid, hydrogen fluoride forms hydrofluoric acid, hydrogen bromide forms hydrobromic acid and hydrogen iodide forms hydroiodic acid.
- Covalent halides or ionic halides with covalent character are hydrolysed to form acidic solutions.
- The thermal stability of the hydrogen halides decreases with increasing atomic (proton) number of the halogen, which can be explained in terms of the bond energy of the hydrogen–halogen bond.
- The ease of oxidation of the hydrogen halides increases with increasing atomic (proton) number of the halogen.
- In aqueous solution, the halogen with the lower atomic (proton) number can displace the halide ion with the higher atomic number.
- Acidified silver nitrate can be used to distinguish between aqueous halide ions. The silver halides formed can be distinguished by their differing solubilities in aqueous ammonia.
- Electrolysis is the process of decomposing a liquid using electricity.
- The products of electrolysis depend on whether the electrolyte is molten or in aqueous solution.
- The products of electrolysis depend on the position of ions present in the electrochemical series and on the concentration of the ions.
- The quantity of electricity (charge) in coulombs = amps × seconds.
- The amount of substance discharged at an electrode is directly proportional to the charge passed through the electrolyte.
- 96 500 C = the Faraday constant, F.
- $F = Le$, where L is the Avogadro constant and e is the charge on an electron.
- The number of moles of electrons required to discharge 1 mole is equal to the charge on the ion.
- The chlor-alkali industry is based on the electrolysis of brine (concentrated aqueous sodium chloride).
- Chlorine and its products have tremendous industrial importance, but the use of some chlorine compounds is becoming controversial owing to their environmental significance.
- Chlorine is used to make organic solvents, PVC, bleaches and hydrochloric acid.

Practice questions and a How Science Works assignment for this chapter are available at www.collinseducation.co.uk/CAS

24

The transition elements

24 THE TRANSITION ELEMENTS

HOW SCIENCE WORKS

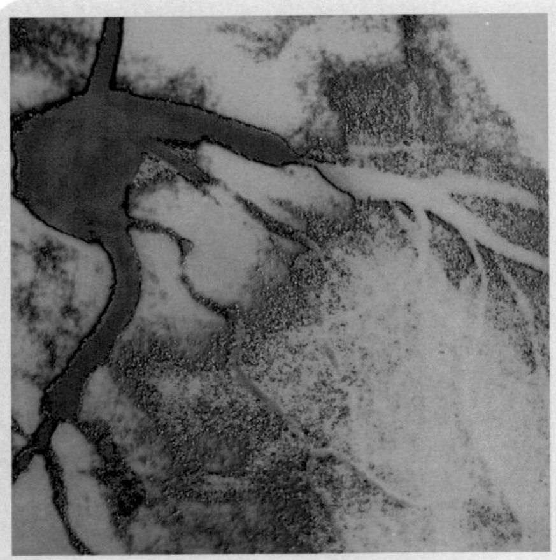

Artery repair Arteries can collapse or become constricted. This image shows a severe constriction in a coronary artery, which restricts blood flow from the red to the orange part of the artery. To overcome this, a coil of Nitinol wire can be inserted to hold it open and allow the blood to flow

Nitinol – a 'shape memory' alloy

Nitinol is a remarkable alloy that can 'remember' shapes. Deform a piece of the alloy and it will return to its original shape after being heated.

Nitinol is an alloy of nickel and titanium, both transition elements. It was discovered in the USA in 1962 quite by chance, when researchers in the defence industry were seeking ways to make titanium less brittle by adding nickel to it. They found that the temperature at which Nitinol returns to its original shape can be set anywhere between –100 °C and 100 °C by altering the amount of nickel in the alloy.

Nitinol wire can be used to repair damaged or blocked arteries. The wire is wound round a tube of the same diameter as the inside of an artery. It is heated so that this coiled shape is remembered. The wire is then cooled and straightened. It is passed into the artery to be repaired. The temperature of the blood warms up the wire, which returns to its coiled shape. So an artery with a weak wall can be reinforced and a constriction in an artery can be unblocked. Nitinol does not react with body chemicals, and the arterial wall soon grows round it.

1 WHAT IS A TRANSITION ELEMENT?

More information on iron and steel can be found on pages 3 and 12.

REMEMBER THIS

The terms transition element and transition metal are used interchangeably.

REMEMBER THIS

Many transition elements are required in tiny amounts to keep the body functioning. These **trace elements** include iron and cobalt.

Ask someone to name the first metal that comes to mind, and the chances are that a transition element will be named. It may be iron, essential to the construction of many buildings, most vehicles and almost all machinery. It may be gold, a precious metal used in jewellery and other forms of ornamentation, and as a protective coating. It may be copper, used in electric wiring and for indoor water pipes, or it may be silver or platinum, nickel or chromium. Everyone is familiar with the transition elements through their use, not in compounds, but on their own or in alloys, such as brass, bronze and steel.

The transition metal compound with the world's largest annual production (4.5 million tonnes) is titanium(IV) oxide, TiO_2. This compound is bright white when pure, and so is used as the pigment in white paint, white paper and white plastics, as well as in sunscreens.

Transition metals are essential to life. Cobalt is found in vitamin B1, where it acts as a catalyst. The iron in haemoglobin is involved in transporting oxygen around the body. And nickel and copper are essential components of several enzymes.

Fig 24.1 The pigment in white paint is TiO_2

So what is a transition element? The central block of the Periodic Table is the **d block**. This chapter deals with the ten d block elements in Period 4, sometimes called the 'first row transition elements' as they are the top row of the d block. These elements have a closely similar set of properties, so it is tempting to call them all transition elements. However, on further examination, scandium at one end of the first row and zinc at the other are obviously different from the rest. This has led chemists to define transition elements as **elements that form one or more ions with a partially filled d subshell**. This definition excludes zinc and scandium.

? QUESTION 1

1 Why is the central block of elements called the d block?
(Hint: if you are not sure, look at page 66.)

ELECTRON CONFIGURATION

Table 24.1 shows the electron configurations of the first-row transition elements. Notice that the 4s subshell has already been filled because this is at a lower energy than the 3d subshell. The full electron configuration of titanium (atomic number 22) is $1s^2\,2s^2\,2p^6\,3s^2\,3p^6\,3d^2\,4s^2$.

To review electron configurations, see page 63.

Table 24.1 Arrangement of electrons in the outer 4s and inner 3d subshells of the first row of the d block. Each atomic orbital is represented by a box, and each electron by an arrow

Element	Symbol	Atomic (proton) number	Electron configuration 3d	4s
scandium	Sc	21	[Ar] [↑] [ ] [ ] [ ] [ ]	[↑↓]
titanium	Ti	22	[Ar] [↑] [↑] [ ] [ ] [ ]	[↑↓]
vanadium	V	23	[Ar] [↑] [↑] [↑] [ ] [ ]	[↑↓]
chromium	Cr	24	[Ar] [↑] [↑] [↑] [↑] [↑]	[↑]
manganese	Mn	25	[Ar] [↑] [↑] [↑] [↑] [↑]	[↑↓]
iron	Fe	26	[Ar] [↑↓] [↑] [↑] [↑] [↑]	[↑↓]
cobalt	Co	27	[Ar] [↑↓] [↑↓] [↑] [↑] [↑]	[↑↓]
nickel	Ni	28	[Ar] [↑↓] [↑↓] [↑↓] [↑] [↑]	[↑↓]
copper	Cu	29	[Ar] [↑↓] [↑↓] [↑↓] [↑↓] [↑↓]	[↑]
zinc	Zn	30	[Ar] [↑↓] [↑↓] [↑↓] [↑↓] [↑↓]	[↑↓]

? QUESTION 2

2 **a)** Write down the full electron configuration for manganese and nickel.
b) Look at Table 24.1. Why are atomic orbitals in the d subshell occupied by single, unpaired electrons first? (Hint: see page 63 if you are not sure.)

Since the noble gas core is argon ($1s^2\,2s^2\,2p^6\,3s^2\,3p^6$), titanium can be abbreviated to $[Ar]3d^2 4s^2$.

For each element from scandium to zinc, the number of protons increases by one. This increases the positive charge on the nucleus. However, because the electrons are added to the inner 3d subshell, the outer 4s electrons tend to be shielded from the increasing charge. This partly explains the similarity in physical and chemical properties. Also, because the 3d and 4s subshells have similar energies, the electrons from both can take part in bonding, which gives rise to some characteristic transition metal properties.

Electrons occupy first the orbitals of lowest energy. The orbitals in a subshell are first occupied singly by electrons that spin in the same direction, which helps to minimize the interelectron repulsion. Only when all the d orbitals in the d subshell are singly occupied do electrons start to pair up. For the build-up of electron configurations, see page 63.

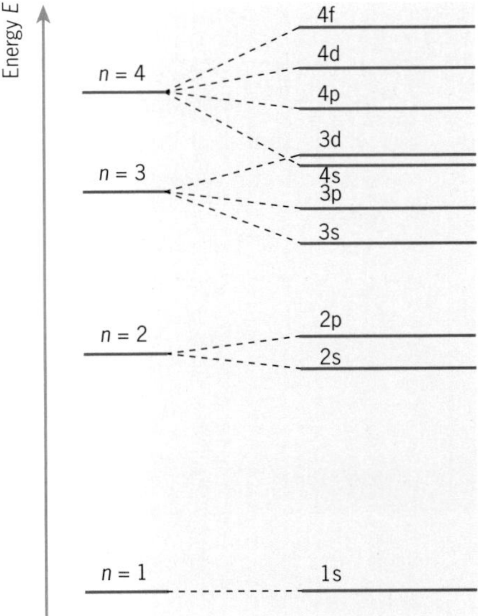

Fig 24.2 Energy levels of the various subshells in a many-electron atom

Fig 24.3 Shapes of d orbitals. Four of the five d orbitals have the shape shown on the left and one has the shape shown on the right. Remember that an orbital's shape represents the boundary surface and that there is a 90 per cent probability of finding an electron within it

Electron configurations of chromium and copper

Look at Table 24.1. The chromium atom and the copper atom each have only one electron in the 4s subshell. As the number of protons increases, the number of electrons in the d subshell also increases by one each time, until chromium is reached. At chromium, there is a jump of two electrons, one of which is from the 4s subshell. Why is this? The half-filled subshell has five

REMEMBER THIS

The more stable an arrangement, the lower its energy.

singly occupied orbitals. So, it is a lower energy arrangement to have $3d^5$ and only one 4s electron, as this removes the paired electron in the 4s orbital and thereby reduces repulsion between electrons. At copper, the most stable arrangement is to have $3d^{10}4s^1$, since a full 3d subshell is a stable arrangement.

? QUESTION 3

3 What is the electron configuration of:
a) Co^{2+} and Co^{3+}
b) Cu^+ and Cu^{2+}
c) Zn^{2+}?
(Hint: see Table 24.1.)

FORMATION OF TRANSITION METAL IONS

Metal ions are always formed by the loss of electrons, which requires energy. In the case of transition metal ions, the first electrons to be lost are in the 4s subshell. This is because, once the 3d subshell starts to fill, the 4s electrons are repelled further from the positive nucleus. This can be seen with Mn^{2+}:

	3d	4s			3d	4s
Mn [Ar]	[↑][↑][↑][↑][↑]	[↑↓]	$\xrightarrow{-2e}$	Mn^{2+} [Ar]	[↑][↑][↑][↑][↑]	[ ]

Thus, the electron configuration of Mn^{2+} is $[Ar]3d^5$.

Fe loses its 4s electrons to form Fe^{2+}:

	3d	4s			3d	4s
Fe [Ar]	[↑↓][↑][↑][↑][↑]	[↑↓]	$\xrightarrow{-2e}$	Fe^{2+} [Ar]	[↑↓][↑][↑][↑][↑]	[ ]

which has an electron configuration of $[Ar]3d^6$.

As the energies of the 4s and 3d subshells are still fairly close together, 3d electrons may also be lost to form ions with charges of 3+. In forming Fe^{3+} a 3d electron is lost together with the two 4s electrons to give an electron configuration of $[Ar]3d^5$:

	3d	4s			3d	4s
Fe [Ar]	[↑↓][↑][↑][↑][↑]	[↑↓]	$\xrightarrow{-3e}$	Fe^{3+} [Ar]	[↑][↑][↑][↑][↑]	[ ]

? QUESTION 4

4 Scandium forms only 3+ ions. What is the electron configuration of Sc^{3+} and why is scandium not considered to be a transition element?

As already noted, scandium and zinc do not show typical transition metal properties. Zinc has an electron configuration $[Ar]3d^{10}4s^2$. The stable full d subshell means that zinc forms only Zn^{2+} ions, $[Ar]3d^{10}$. Recalling the definition of transition metals as those that form one or more ions with a partially filled d subshell, zinc does not qualify as a transition element.

2 PHYSICAL PROPERTIES OF TRANSITION ELEMENTS

? QUESTION 5

5 What other typical properties of metals are there?
To check your answer, look at page 101.

In terms of their physical properties, transition elements are typical metals. They are hard and dense, have high melting points and are good conductors of heat and electricity. They tend to be strong and durable, and have high tensile strengths, as well as other useful mechanical properties. So they find many applications, from bridges (see the Humber Bridge, page 2) to cooking utensils. Contrast this with the s block metals considered on page 427, some of which are soft and have low melting points.

A more detailed examination of the physical properties reveals their close similarity. This is again explained by the electron configurations.

REMEMBER THIS

Electrons in metallic bonding are delocalised. They are free to move throughout the metal and so do not 'belong' to any particular atom.

MELTING POINTS AND HARDNESS

The high melting points and the hardness of transition metals are caused by their strong metallic bonds. As described on pages 140–142 metallic bonding is caused by the delocalisation of outer electrons, leaving positive metal ions surrounded by a sea of electrons. The more electrons that are delocalised, the stronger the metallic bond. In the case of transition metals, some 3d electrons can be delocalised along with the 4s. This explains why they have higher melting points compared with s block elements, such as potassium and calcium.

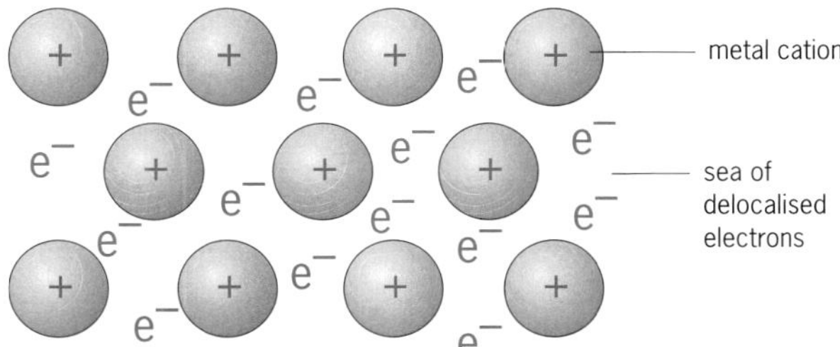

Fig 24.4 Metallic bonding in transition metals involves 3d as well as 4s electrons

HOW SCIENCE WORKS

Ionisation energies provide evidence for electronic configuration

Ionisation energy is discussed in several chapters of this book. Although it is a physical property, it is fundamental in determining the chemical properties of elements.

Fig 24.5 shows the first four successive ionisation energies of the elements of the first transition series Sc to Zn. Ca and K are also included for comparison. Note how similar all the first ionisation energies are across this series. The same similarity is seen with the second ionisation energies. This is because the 4s electrons are the first to be removed, even though it is the 3d subshell that is being filled along the series. This shows that the inner 3d electrons shield the 4s electrons from the increasing nuclear charge. In the case of the third and fourth ionisation energies, the 3d electrons are being removed from the same subshell. So, the effective nuclear charge is increasing significantly along the period.

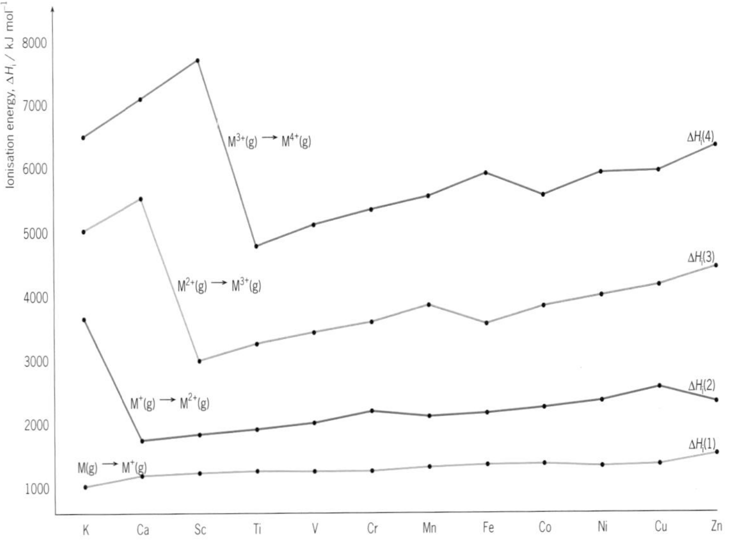

Fig 24.5 First four successive ionisation energies of the elements of the first transition series, with K and Ca included for comparison

REMEMBER THIS

The first ionisation energy is the energy required to remove 1 mole of electrons from 1 mole of gaseous atoms to form 1 mole of gaseous ions:

$M(g) \rightarrow M^{+}(g) + e^{-} \quad \Delta H_i(1)$

The second ionisation energy is the energy required to remove the second mole of electrons, the first mole having already been removed:

$M^{+}(g) \rightarrow M^{2+}(g) + e^{-} \quad \Delta H_i(2)$

CONDUCTION OF ELECTRICITY

As explained on page 141, the conductivity of metals results from delocalisation of the outer electrons in the metallic bond, which makes the electrons free to move under the influence of a potential difference. Transition metal atoms also have the 3d electrons as part of their metallic bond, which makes them better conductors than calcium, which can only use its 4s electrons. In fact, copper and silver are the best metallic conductors at room temperature.

? QUESTION 6

6 **a)** Why is the second ionisation energy of chromium higher than expected? (Hint: think about the electron configuration of Cr^{+}.)
b) The third ionisation energy of manganese is also higher than expected. Why is this? (Hint: consider the electron configuration of Mn^{2+}.)
c) Why is there a big jump between the second and third ionisation energies of calcium?

3 CHEMICAL PROPERTIES

There is a close similarity of the physical properties of the transition elements, so what about their chemical properties? There are four that characterise transition elements:

- variable oxidation states in their compounds;
- formation of complexes;
- formation of coloured ions;
- catalysis of reactions both as elements and compounds.

Each of these is now examined.

VARIABLE OXIDATION STATES

The terms oxidation state and oxidation number tend to be used interchangeably (see page 106). However, when considering the transition elements, the term oxidation state is usually preferred.

Lower oxidation states in ionic compounds

There is a wide range of oxidation states in the transition metals. Contrast this with Group 1 and Group 2 elements, each of which has only one oxidation state (+1 and +2 for Groups 1 and 2, respectively). The difference is explained by the closeness of the 4s and 3d energy levels. In calcium, the two 4s electrons are easily removed, but to remove another electron means breaking into the 3p subshell, which requires much more energy. This is not the case with the transition metals. Remember that the first electrons to be lost in forming a transition metal ion are the 4s electrons, which makes the +2 oxidation state common, as in Fe^{2+} and Co^{2+}. However, the 3d electrons may also be lost, which means that ions such as V^{3+}, Fe^{3+} and Cr^{3+} are also common. Table 24.2 shows this.

Table 24.2 Comparison of the first four ionisation energies of calcium and iron

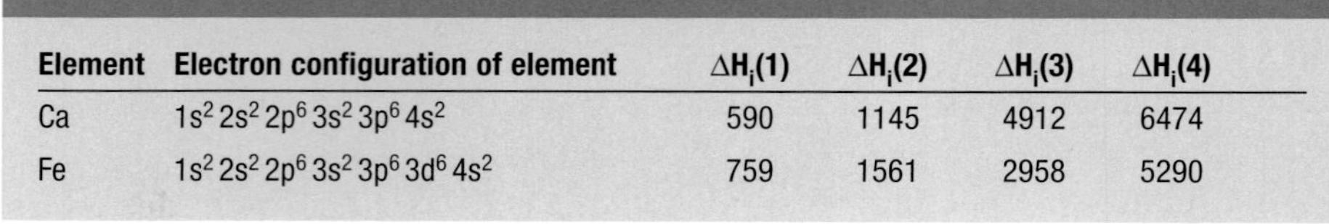

Element	Electron configuration of element	$\Delta H_i(1)$	$\Delta H_i(2)$	$\Delta H_i(3)$	$\Delta H_i(4)$
Ca	$1s^2 2s^2 2p^6 3s^2 3p^6 4s^2$	590	1145	4912	6474
Fe	$1s^2 2s^2 2p^6 3s^2 3p^6 3d^6 4s^2$	759	1561	2958	5290

Notice the big jump in the third ionisation energy of calcium as the 3p electron is removed. In the case of iron, the increase is more gradual and the simple ions Fe^{2+} and Fe^{3+} are both formed. The electron configurations of Fe^{2+} and Fe^{3+} are:

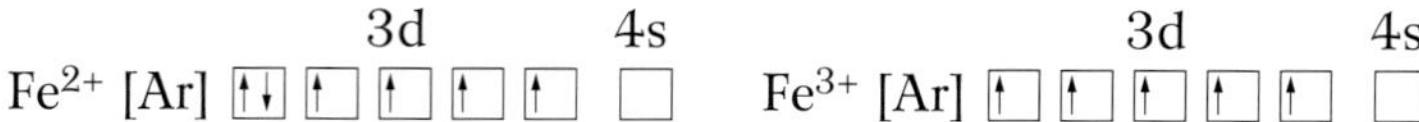

The electron configurations of manganese in the +2 and +3 oxidation states are:

Mn^{2+} [Ar] 3d: ↑ ↑ ↑ ↑ ↑ 4s: □ Mn^{3+} [Ar] 3d: ↑ ↑ ↑ ↑ □ 4s: □

Fig 24.6 shows all the common oxidation states of the first transition series. Note that both scandium and zinc have only one oxidation state, which is yet another reason why they should not be classified as transition metals. From titanium to copper, both the +2 and +3 oxidation states exist. At first,

? QUESTION 7

7 **a)** Solutions of Fe^{3+}(aq) are more stable than Fe^{2+}(aq). Suggest why.
b) Why is Mn^{2+} more stable than Mn^{3+}?
c) Using box diagrams, write down the electrons for Cu^{+} and Cu^{2+}. Although you might predict Cu^{+} to be the more stable, in fact the +2 oxidation state of copper is the more stable. The reasons for the stability of copper(II) are explained on page 528.

REMEMBER THIS

Half filled d subshells have singly occupied orbitals. This avoids electron–electron repulsion within an orbital. See page 481.

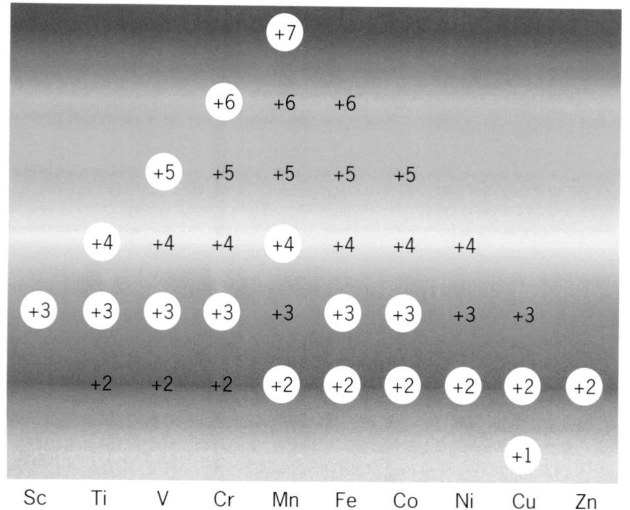

Fig 24.6 Oxidation states of the first transition series. The most common oxidation states are ringed

the +3 oxidation state is common, which shows the greater stability of the M^{3+} ion. But from manganese onwards the +2 oxidation state becomes more common. The removal of a third electron from the d subshell is energetically more difficult after manganese, because the effective nuclear charge becomes so much greater than it is up to manganese. The reason for the relative stabilities of the +2 and +3 oxidation states is explored further when electrode potentials are considered (see Chapter 25).

Higher oxidation states

From Fig 24.6, the maximum oxidation state increases across the first transition series, reaching +7 at manganese. Note that this corresponds to the total number of electrons in the 4s and 3d subshells. However, a Mn^{7+} ion or a Cr^{6+} ion would be too polarising to exist. Also, too much energy would be needed to remove so many electrons. The higher oxidation states are found covalently bonded either in simple compounds, such as TiO_2, V_2O_5, CrO_3 and Mn_2O_7, or as ions, such as VO_2^+, $Cr_2O_7^{2-}$ and MnO_4^-. The availability of partially filled or empty orbitals, particularly in the 4s and 3d subshells, allows the higher oxidation states to be reached.

After manganese, the maximum oxidation state of each element decreases because of the increasing energy needed to involve 3d electrons in bonding, and because singly filled orbitals are required in order to form a covalent bond.

Naming the transition metal compounds and ions

Ionic NaCl is easy to name. Sodium has only one oxidation state, so the compound can be called sodium chloride, and Na^+ can be called the sodium ion. However, in the case of, for example, $FeCl_3$ and $FeCl_2$ the name iron chloride does not distinguish the two compounds, so they have to be given systematic names that include the oxidation state of the transition element:

$FeCl_3$, iron(III) chloride which contains the ion Fe^{3+} iron(III) ion

$FeCl_2$, iron(II) chloride which contains the ion Fe^{2+} iron(II) ion

Note that the oxidation state is shown by a roman numeral in brackets and that there is no space between the metal name and the oxidation number.

? QUESTION 8

8 All the oxidation states in Fig 24.6 are positive. Why is this? If you are not sure, see page 108.

? QUESTION 9

9 Many of the higher oxidation states are reached by covalent bonding to oxygen, which is highly electronegative.
a) What does the term electronegative mean?
b) What other element is most likely to produce these higher oxidation states?
Hint: to remind yourself about electronegativity see pages 85 and 86.

? QUESTIONS 10–11

10 **a)** Use oxidation states to name the following transition metal compounds: i) TiO_2 ii) Fe_2O_3 iii) $Mn(OH)_3$ iv) CrO_3 v) V_2O_5
b) Write down the formulae of the following compounds:
i) copper(II) hydroxide;
ii) manganese(II) carbonate;
iii) titanium(IV) chloride;
iv) copper(II) nitrate;
v) iron(II) bromide.

11 **a)** What is the formula of sodium ferrate(VI)?
b) What is the name of the oxyanion CrO_4^{2-}?

A table of typical oxidation states and oxidation numbers is given on page 109.

REMEMBER THIS

In MnO_4^-, since O has an oxidation number of –2, the total oxidation number due to O is –2 × 4 = –8. Therefore, the oxidation state of manganese is +7, because this leaves –1 as the charge on the anion.

REMEMBER THIS

Redox means oxidation (oxidation number increases) and reduction (oxidation number decreases). To find out more about redox reactions and half equations, read Chapter 25.

Here are two more examples of compounds with systematic names:

CuO, copper(II) oxide MnO_2, manganese(IV) oxide

Ions such as MnO_4^- and $Cr_2O_7^{2-}$ are called **oxyanions**, because they are negative ions that contain oxygen. MnO_4^- is called the manganate(VII) ion. The (VII) signifies that the oxidation state of manganese in this oxyanion is +7. $Cr_2O_7^{2-}$ is the dichromate(VI) ion, so the oxidation state of chromium in this oxyanion is +6. Thus, $KMnO_4$ is called potassium manganate(VII) and $Na_2Cr_2O_7$ is called sodium dichromate(VI).

REDOX REACTIONS AND THE STABILITY OF OXIDATION STATES

Since the transition elements have variable oxidation states in their compounds, they can undergo **redox** reactions. Compounds that contain transition elements in high oxidation states, such as potassium manganate(VII), tend to be oxidising agents. This is because they are usually less stable than compounds with lower oxidation states. In acid solution, the MnO_4^- ion is reduced to Mn^{2+} by gaining electrons, as can be seen in this **half equation**:

$$MnO_4^-(aq) + 8H^+(aq) + 5e^- \rightarrow Mn^{2+}(aq) + 4H_2O(l)$$

Look again at Fig 24.6 on page 485 which shows that, across the first transition series, the +3 oxidation state is more stable than the +2 state until manganese is reached. This means that Ti^{2+}, V^{2+} and Cr^{2+} are highly reducing, because they are easily oxidised to the +3 state. For example:

$$Cr^{2+}(aq) \rightarrow Cr^{3+}(aq) + e^-$$

Traces of oxygen can be removed from other gases by bubbling the gases through a solution of $Cr^{2+}(aq)$.

From manganese to copper, the +2 state becomes more stable. So, compounds with oxidation states of +3 are highly oxidising. For example:

$$Co^{3+}(aq) + e^- \rightarrow Co^{2+}(aq)$$

The relative stabilities of the different oxidation states are predicted from **standard electrode potentials, $E^\ominus$** (see Chapter 25).

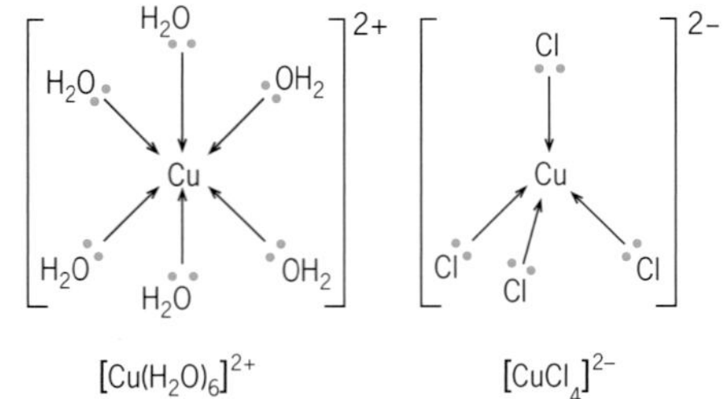

Fig 24.7 Two complex ions

COMPLEXES

A **complex** is formed when a central metal atom or ion is surrounded by species that donate lone pairs of electrons. The actual bonding in complexes is complicated and beyond the scope of this book. However, it is assumed that the lone pairs from the ligands form dative covalent (coordinate) bonds.

A species that donates a lone pair of electrons is called a ligand.

Ligands are usually molecules such as water or negative ions such as Cl^-. Fig 24.7 shows two complex ions.

Many complexes are positively or negatively charged, but some are neutral. In the formula, the central metal atom or ion is written first followed by the ligands. The charge on a complex is the charge of the central metal ion and the charges on the surrounding ligands added together. The overall charge of $[Cu(H_2O)_6]^{2+}$ results from the 2+ charge of Cu^{2+}, since the water molecules carry no charge. In the case of $[CuCl_4]^{2-}$, there are four Cl^- ions, to give a charge of 4–, which when added to the 2+ charge of Cu^{2+} gives an overall charge of 2– [i.e., 4(1–) + (2+) = 2–]. The charge on a complex ion is delocalised over the whole ion and is usually shown outside square brackets. However, complex ions are sometimes drawn with a charged central metal ion.

? QUESTION 12

12 What is the charge on the central metal ion in the following complexes?

a) $[Ag(NH_3)_2]^+$ **b)** $[NiCl_4]^{2-}$
c) $[Fe(H_2O)_6]^{3+}$ **d)** $[Ni(CN)_4]^{2-}$
e) $[Fe(CN)_6]^{3-}$
(Hint: the cyanide ligand is CN^-.)

Coordination number and shape

The number of dative covalent (coordinate) bonds from the ligands to the central metal ion is called the **coordination number**. The coordination number determines the shape of the complex. The most common coordination numbers are 4 and 6. Table 24.3 shows the shapes that are often associated with these two numbers, together with the shape of coordination number 2. Note that a coordination number of 4 gives two possible shapes. The more common is the tetrahedral structure. On page 79, the electron pair repulsion theory is discussed, which predicts the tetrahedral, octahedral and linear shapes in Table 24.3. However, d orbital electrons affect the shape in a different way to s and p orbital electrons, and many shapes are not as predicted by this theory. For example, the shape of $[Cu(H_2O)_6]^{2+}$ is not a regular octahedron because four of the copper–oxygen bonds are short and the other two are longer, which gives a distorted octahedron.

REMEMBER THIS

A dative covalent bond is a covalent bond in which the shared pair of electrons come from the same atom. It is also called a coordinate bond. See page 76.

Table 24.3 Shapes of four complexes

Coordination number	Shape	Example	Structure
2	linear	$[Ag(NH_3)_2]^+$	$[H_3N \rightarrow Ag \leftarrow NH_3]^+$
4	tetrahedral	$[CuCl_4]^{2-}$	Cu with four Cl ligands; overall charge 2–
4	square planar	$[Ni(CN)_4]^{2-}$	Ni with four CN ligands (NC, CN); overall charge 2–
6	octahedral	$[Co(NH_3)_6]^{2+}$	Co with six NH_3 ligands (H_3N, NH_3); overall charge 2+

REMEMBER THIS

Transition metal ions commonly form octahedral complexes with small ligands (eg. H_2O and NH_3) and tetrahedral complexes with larger ligands (eg. Cl^-).

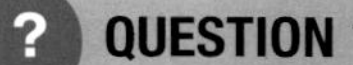

QUESTION 13

13 For each of the complex ions in the How Science Works box below work out:

a) the charge on the central silver ion;

b) the shapes of the complex ions.

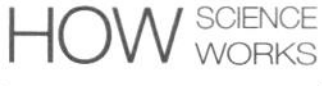

The usefulness of two silver complexes

$[Ag(CN)_2]^-$ in electroplating

Silver is an unreactive precious metal. If aqueous potassium cyanide is added to solutions of $Ag^+(aq)$ ions then the soluble $[Ag(CN)_2]^-$ complex is formed. This complex is used to electroplate a layer of silver onto objects (Fig 24.8). The object to be plated is made the cathode. During electrolysis $[Ag(CN)_2]^-$ decomposes to release Ag^+ ions, which are attracted to the negatively charged cathode where they accept electrons to form the layer of silver required.

$[Ag(NH_3)_2]^+$ in testing for aldehydes

We first met this reaction in Chapter 16 (see page 324). Tollens reagent contains the complex ion $[Ag(NH_3)_2]^+$, made by adding an excess of aqueous ammonia to aqueous silver nitrate. It is this complex that is reduced to silver in the silver mirror test for the aldehyde functional group.

Fig 24.8 This decorative box has been electroplated with silver

Fig 24.9 The silver mirror produced when $[Ag(NH_3)_2]^+$ is reduced to Ag is used as a test for aldehydes

An example of this reaction is shown in the equation and in Fig 24.9.

$$CH_3CHO + 2Ag^+(aq) + H_2O \rightarrow CH_3CO_2H + 2Ag(s) + 2H^+$$

POLYDENTATE LIGANDS

All four ligands discussed so far (H_2O, NH_3, Cl^- and CN^-) donate one lone pair of electrons to the central metal atom, and so are called **monodentate**. This means 'single toothed' and is derived from *monos*, the Greek for one, and *dens*, the Latin for tooth. **Bidentate** ligands can form two dative covalent bonds to the metal ion, so they are 'two toothed', as shown in Fig 24.10. The lone pairs come from the nitrogen and oxygen, respectively. When these ligands attach to the central metal ion, they form five-membered rings and the complexes are called **chelates** (from *chele*, the Greek for claw).

REMEMBER THIS

When a ligand forms more than one dative covalent (co-ordinate) bond with a central metal atom it is called **polydentate**. So 1, 2-diaminoethane is a polydentate ligand.

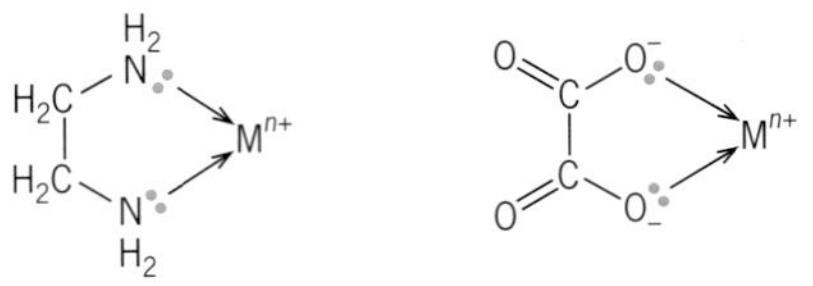

1,2-diaminoethane called en for short

ethanedioate traditional name: oxalate

Fig 24.10 Two examples of bidentate ligands

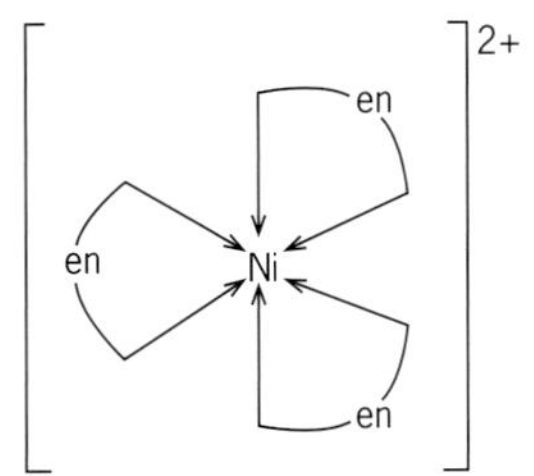

Fig 24.11 An example of chelation with 1,2-diaminoethane. Shown on the far right is the simplified version of this bidentate ligand, 'en'. The complex is written using 'en': $[Ni(en)_3]^{2+}$

EDTA – a most important ligand

Ethylene**d**iamine**t**etra**a**cetate ion (EDTA) is the traditional name for a very important chelating agent. As Fig 24.12 shows, six lone pairs of electrons are available to bond with a metal ion, so it is called a hexadentate ligand. This ligand surrounds and encloses metal ions to form a very stable complex ion, which explains why it is so useful.

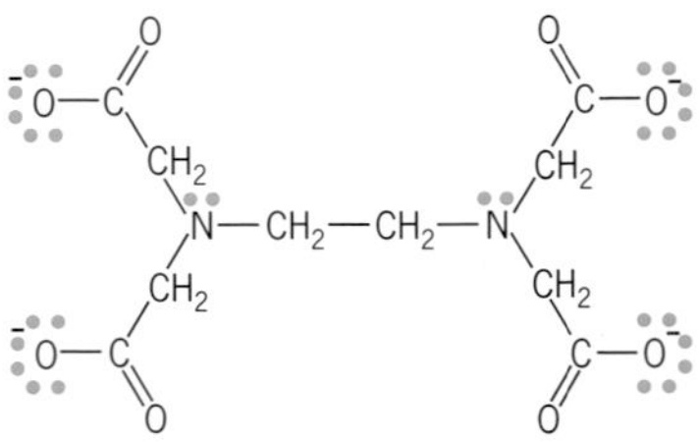

Fig 24.12 $EDTA^{4-}$ ion

Lead(II) ions (Pb^{2+}) are toxic in the human body; if they are not removed quickly they can affect the enzymes that catalyse the formation of haemoglobin. Lead(II) ions can cause permanent brain damage in children and young people. EDTA is administered to remove lead(II) ions from the blood and tissue as a stable lead–EDTA complex ion (Fig 24.13), which is excreted in urine. It used equally effectively to treat mercury poisoning. EDTA is so effective at trapping and removing metal ions that it also removes calcium and important trace metal ions that are essential to normal body functioning. These must be replaced during the treatment.

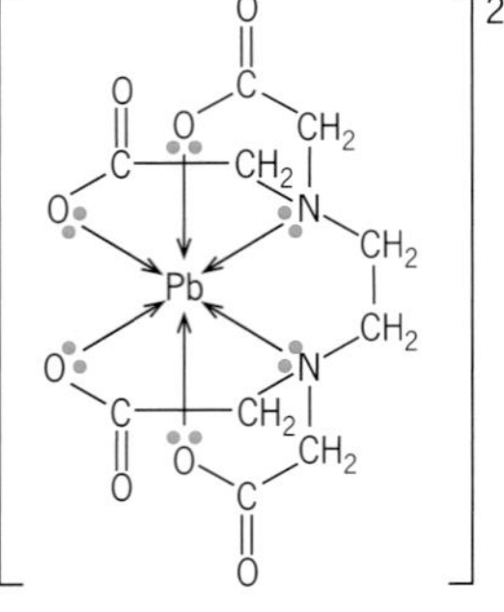

Fig 24.13 Lead–EDTA complex ion. Since the charge on the lead ion is 2+ and EDTA has a charge of 4–, the overall charge of the complex ion is 2–

EDTA is added to human blood when it is stored for transfusions because it keeps the calcium ions trapped in the EDTA complex and so prevents the blood from clotting. It is also used during operations to prevent clotting.

However, EDTA's applications extend far beyond the medical. It is added to some foods to remove traces of metal ions that catalyse air oxidation. In salad dressings, for example, EDTA prevents rancidity by removing the ions that catalyse the oxidation of oil. EDTA is even used in liquid fertilizers because it forms a chelate complex with the Fe^{3+} ions. Thus, OH^- ions are prevented from reacting with the Fe^{3+} to form insoluble $Fe(OH)_3$. This allows the soluble iron complex to reach and enter the plant. EDTA is also added to bathroom products – particularly shampoos – to remove calcium ions from the water, and thereby prevent the formation of scum.

HOW SCIENCE WORKS

Haemoglobin

Haemoglobin, a protein molecule, is a vital component of blood. It transports molecular oxygen from the lungs around the body via the arteries. This giant of a molecule is made up of four protein groups (called globulin groups). In the centre of each globulin group is the ion Fe^{2+} bound in a complex. Four sites of attachment around the Fe^{2+} are occupied by a planar ring structure, called a porphyrin (Fig 24.14).

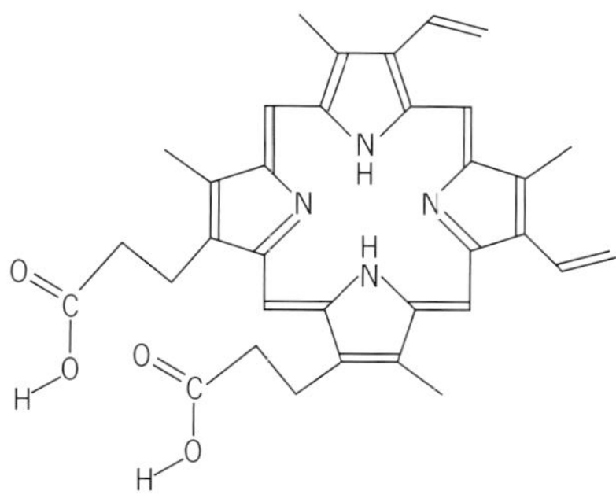

Fig 24.14 Porphyrin ring

The iron–porphyrin part of the complex is called the haem group (Fig 24.15). The fifth ligand site is occupied by a dative covalent bond to the protein globin, while molecular oxygen loosely attaches to the sixth site using a lone pair, so it too is a ligand.

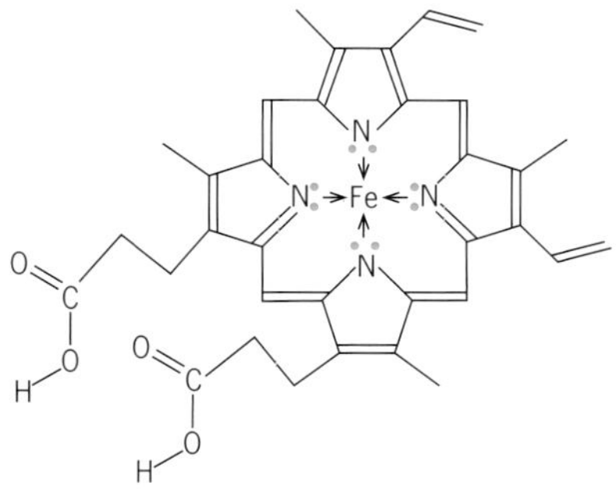

Fig 24.15 Haem group

When oxygen is bonded to the Fe^{2+}, the whole complex, called oxyhaemoglobin, takes on a bright red colour. Once the oxygen is removed, it is replaced by a water ligand, which changes the colour of the complex to blue. The complex is now called deoxyhaemoglobin. Deoxyhaemoglobin gives the blood in our veins its dark red colour as it returns to the lungs to pick up more oxygen.

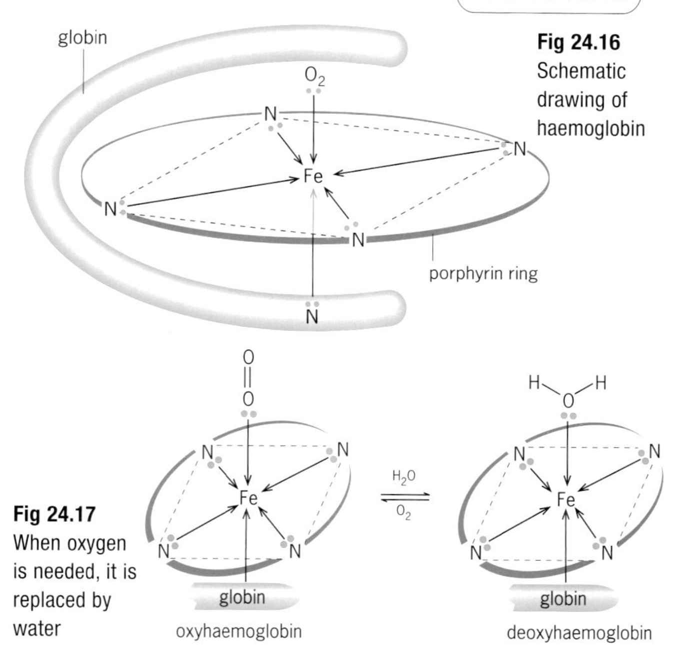

Fig 24.16 Schematic drawing of haemoglobin

Fig 24.17 When oxygen is needed, it is replaced by water

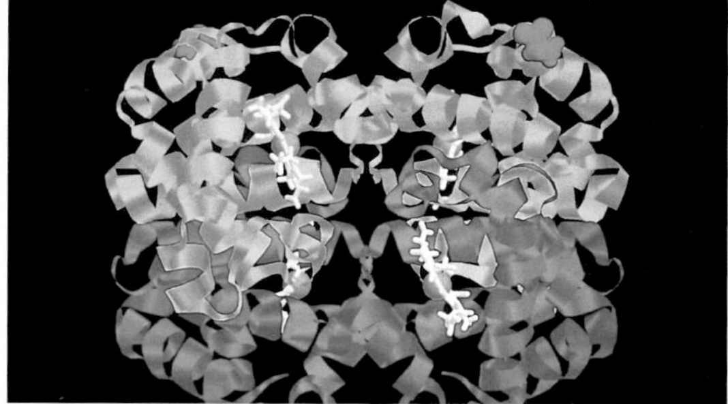

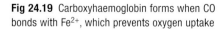

Fig 24.18 Computer graphic of the haemoglobin molecule. The four globulin groups (coloured yellow and blue) each contain a haem group (white)

Carbon monoxide gas is sometimes referred to as the 'silent killer'. The CO molecule has a lone pair and can form dative covalent bonds with Fe^{2+} in the same place as oxygen bonds with haemoglobin, to form carboxyhaemoglobin. This is called ligand substitution or ligand exchange (see page 492). As the bonds formed by CO are stronger than those formed by oxygen the uptake of oxygen is prevented, which is why carbon monoxide is so deadly.

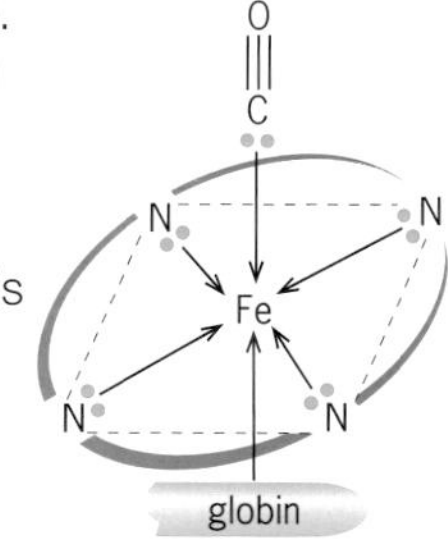

Fig 24.19 Carboxyhaemoglobin forms when CO bonds with Fe^{2+}, which prevents oxygen uptake

QUESTION 14

14 **a)** What is the coordination number of iron in the complex shown in Fig 24.16?

b) What type of ligand is the porphyrin ring? (Hint: how many points of attachment does it have?)

c) Explain why blood in most arteries is bright red.

ISOMERISM IN COMPLEXES

Isomerism occurs when molecules have the same molecular formula, but different ways of arranging their atoms. As explained on pages 182, 277 and 371, there are two principal types of isomerism: **structural isomerism** and **stereoisomerism**.

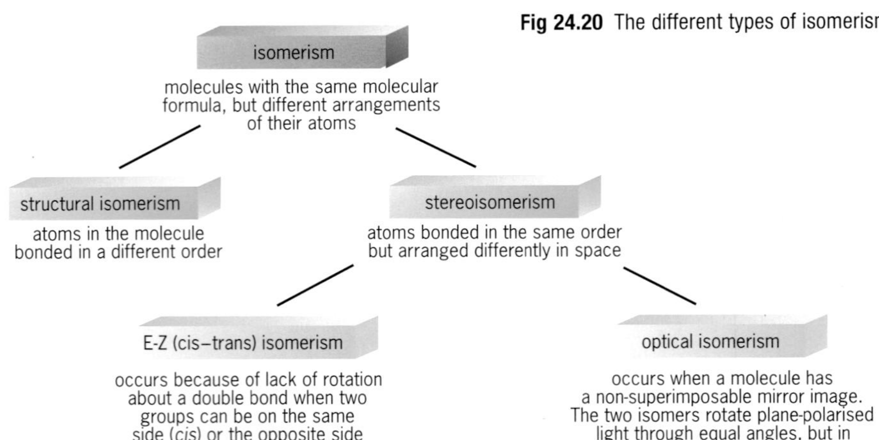

isomerism
molecules with the same molecular formula, but different arrangements of their atoms
structural isomerism
atoms in the molecule bonded in a different order
stereoisomerism
atoms bonded in the same order but arranged differently in space
E-Z (cis–trans) isomerism
occurs because of lack of rotation about a double bond when two groups can be on the same side (cis) or the opposite side (trans) of the double bond
optical isomerism
occurs when a molecule has a non-superimposable mirror image. The two isomers rotate plane-polarised light through equal angles, but in opposite directions

Fig 24.20 The different types of isomerism

REMEMBER THIS

In organic chemistry, the double carbon–carbon bond prevents rotation and gives rise to *cis-trans* isomers (see page 277).

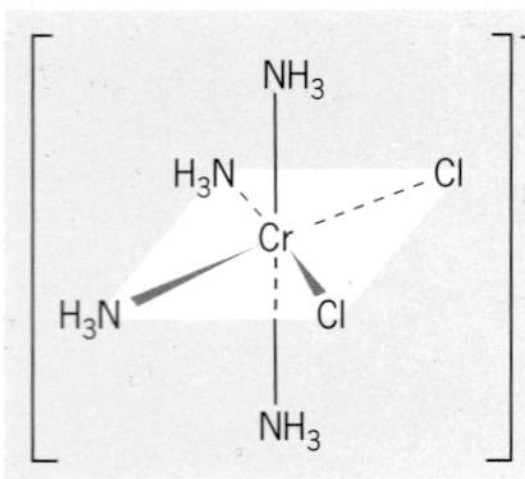

NH3
H3N
Cl
Cr
H3N
Cl
NH3
+

Fig 24.21(a) The *cis* isomer of $[Cr(NH_3)_4Cl_2]^+$ is violet

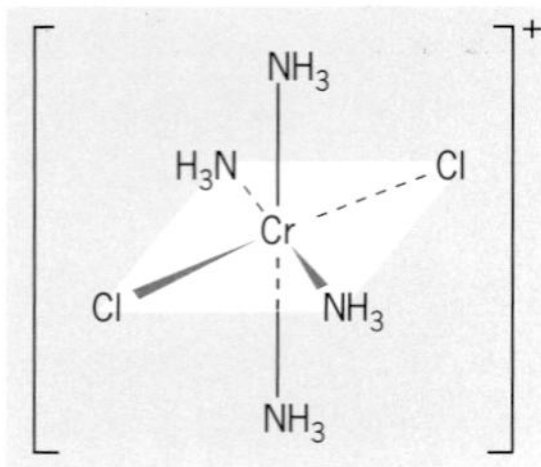

NH3
H3N
Cl
Cr
Cl
NH3
NH3
+

Fig 24.21(b) The *trans* isomer of $[Cr(NH_3)_4Cl_2]^+$ is green

The discussions on pages 182, 277 and 371 are confined to organic molecules, but transition metal complexes can also exhibit isomerism. Both types of stereoisomerism – *cis-trans* and **optical isomerism** – occur in transition metal complexes, as the next two sections show.

Cis–trans isomerism

Stereoisomers have their atoms bonded in the same order, but they are arranged differently in space. *Cis–trans* isomers occur when two different kinds of ligands are bonded in different positions around the central metal cation. The octahedral complex ion $[Cr(NH_3)_4Cl_2]^+$ shows *cis–trans* isomerism. In the *cis* form, the chloride ligands are on the same side, while in the *trans* form they lie on opposite sides. (*Trans* is the Latin word for 'across'.) The *cis* and *trans* isomers shown in Fig 24.21 have different colours.

Optical isomerism

When a stereoisomer has a non-superimposable mirror image, it exhibits optical isomerism. Optical isomers have identical physical and chemical properties, but they rotate plane-polarised light in opposite directions (see page 371). Bidentate ligands, such as 1,2-diaminoethane (called en for short), can give rise to optical isomerism in octahedral complexes, such as tri-1,2-diaminoethanecobalt(III) shown in Fig 24.22.

QUESTION 15

15 The complex $Ni(NH_3)_2Cl_2$ is square planar in shape. Draw diagrams of the *cis* and *trans* isomers.

QUESTION 16

16 Draw the optical isomers for $[Ni(NH_2CH_2CH_2NH_2)_3]^{2+}$.

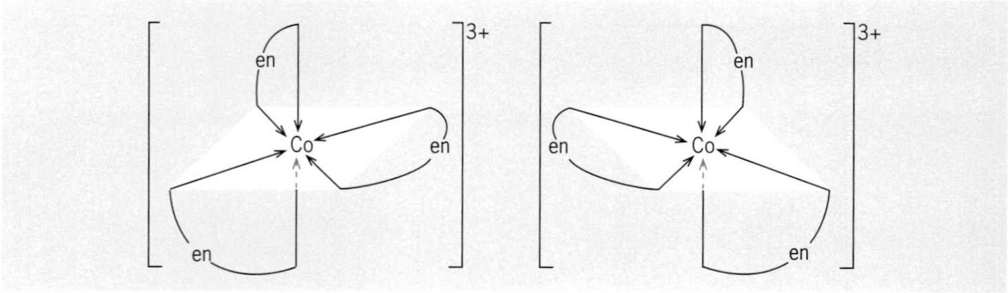

en
Co
en
en
3+
en
Co
en
en
3+

Fig 24.22 $[Co(en)_3]^{3+}$ exists as two mirror image forms or enantiomers

HOW SCIENCE WORKS

Platinum complexes fight cancer

A momentous discovery in chemistry was made in 1964, in a biophysics department at Michigan State university. Like many discoveries in science, it was made by accident.

Physicist Barnett Rosenberg, who had set up the department, was investigating the effect of an alternating current on mammalian cell division. To test the apparatus, he first passed a current through a culture of bacterial cells growing and dividing in a nutrient medium. To his amazement, the cells stopped dividing and started growing into long filaments.

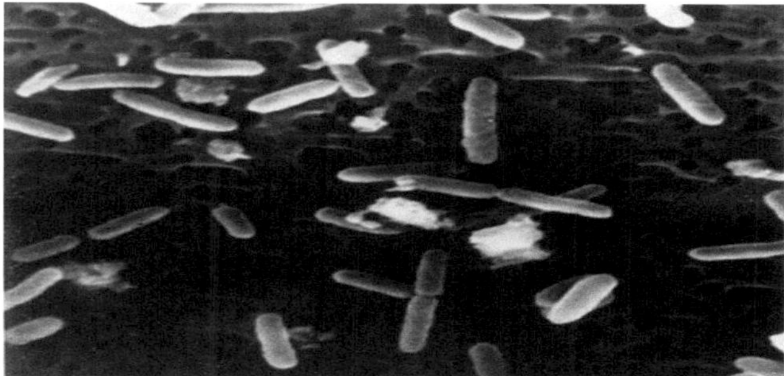

Fig 24.23 Bacterial cells before the passage of an electric current

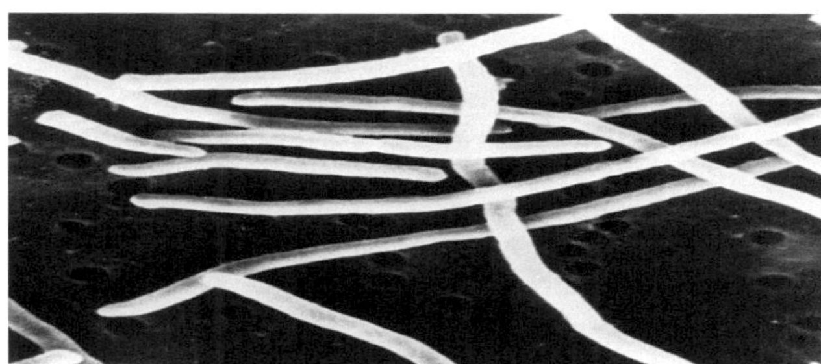

Fig 24.24 Long filaments seen after the electric current had passed

On further investigation, Rosenberg and his team discovered that the platinum electrodes inserted in the nutrient medium – chosen because they were believed to be unreactive – were forming a product by electrolysis. It was this product that was inhibiting cell division, but not cell growth. After much research, it appeared that platinum was forming complex ions with ammonia and chloride ions. Two likely candidates were $[PtCl_2(NH_3)_2]$ and $[PtCl_4(NH_3)_2]$.

English chemist Andy Thompson, working with the Rosenberg team, synthesised trans-$[Pt(NH_3)_2Cl_2]$ and dissolved some into the bacterial solution. To everyone's disappointment, the bacterial cells continued to divide. He subsequently made the *cis*-isomer and this did inhibit cell division. The active chemical had been identified.

? QUESTION 17

17 The platinum complexes formed in the nutrient medium are both neutral.

a) What is the oxidation state of platinum in:
i) $[PtCl_2(NH_3)_2]$ and ii) $[PtCl_4(NH_3)_2]$?

b) Predict the shape of $[PtCl_4(NH_3)_2]$.

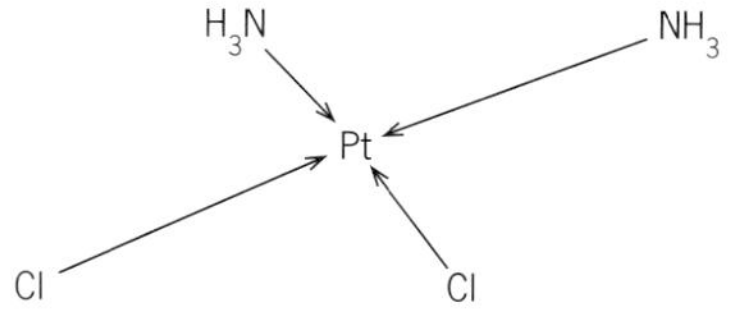

Fig 24.25 cis-$[Pt(NH_3)_2Cl_2]$, now commonly known as cisplatin

Cancer cells divide in an uncontrolled way, and often give rise to tumours. Rosenberg approached the USA National Cancer Institute and suggested that *cis*-$[Pt(NH_3)_2Cl_2]$, now called cisplatin, might have the same effect on cancer cells as it had on bacterial cell division. At first the NCI were sceptical. However, Rosenberg demonstrated that cisplatin could destroy tumours in mice. One of the major problems to overcome was cisplatin's predicted toxicity to the kidneys. By giving the patient large amounts of water before and after the injection of cisplatin, the toxic effects were reduced.

? QUESTIONS 18–19

18 a) *Cis–trans* isomerism is a form of stereoisomerism. What is meant by stereoisomerism?

b) Draw the *trans* isomer of $[Pt(NH_3)_2Cl_2]$.

c) $[PtCl_4(NH_3)_2]$ can also exist as *cis* and *trans* isomers. Draw these two isomers and label them.

19 a) Why is 1,1-cyclobutanedicarboxylic acid called a bidentate ligand?

b) The complex formed is called a chelate. What is meant by this term?

Cisplatin is now one of the most widely used cancer drugs, but the toxicity and its side effects remain. This has stimulated the search for other complexes that may be equally effective but safer. The way cisplatin works has recently been determined by X-ray crystallography. The chloride ligands are exchanged for dative covalent bonds with nitrogen atoms of DNA. This bends the DNA, which causes a loss of function, so cell division is prevented. However, the ease with which chloride ligands can be exchanged causes the toxicity of cisplatin. Carboplatin, a new drug, replaces the chloride ligands with a bidentate ligand called 1,1-cyclobutanedicarboxylic acid.

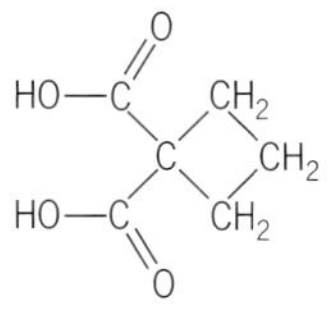

Fig 24.26 1,1-Cyclobutanedicarboxylic acid

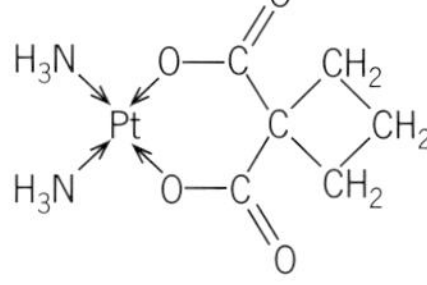

Fig 24.27 Carboplatin

REMEMBER THIS

NH_3 and H_2O are similar in size and are uncharged so ligand exchange occurs without change of co-ordination number. In the case of Cl^- which is larger than H_2O and NH_3, ligand exchange can involve a change in co-ordination number.

Table 24.4 The colours and stability of some copper complexes

Complex Ion	Colour	Stability
$[Cu(H_2O)_6]^{2+}$	blue	increases ↓
$[Cu(Cl_4)]^{2-}$	yellow	
$[Cu(NH_3)_4(H_2O)_2]^{2+}$	deep blue	
$[Cu(EDTA)(H_2O)_2]^{2-}$	pale blue	

For an explanation of why ligand exchange often results in a change in colour, see page 566 of Chapter 27.

LIGAND EXCHANGE

In aqueous solutions, transition metal ions exist as aqua-complexes. The water around the central metal ion can be substituted by other ligands, which is called **ligand exchange** or **ligand substitution**. Fig 24.28(a) shows copper(II) sulfate dissolved in water, so the complex ion present is $[Cu(H_2O)_6]^{2+}$. When concentrated hydrochloric acid is added to this solution, it turns green (Fig 24.28(b)). This occurs because the water ligands of the copper(II) complex have been exchanged for chloride ions. $[CuCl_4]^{2-}$ is a more stable complex because chloride ligands have replaced the water ligands:

$$\underset{\text{blue}}{[Cu(H_2O)_6]^{2+}(aq)} + 4Cl^-(aq) \rightleftharpoons \underset{\text{yellow}}{[CuCl_4]^{2-}(aq)} + 6H_2O(l)$$

The reason for the green colour that is observed is explained in the Stretch and Challenge box on page 493.

$[Cu(NH_3)_4(H_2O)_2]^{2+}$ is more stable than the chloro-complex (see Table 24.5 in the box on stability constants). So, when concentrated ammonia solution is added, ammonia ligands replace chloride ligands to produce a deep blue solution. This can be seen in Fig 24.28(c). Fig 24.28(d) shows the lightening of the solution that occurs when EDTA is added to the solution of ammine complex, which is again as predicted from Table 24.4.

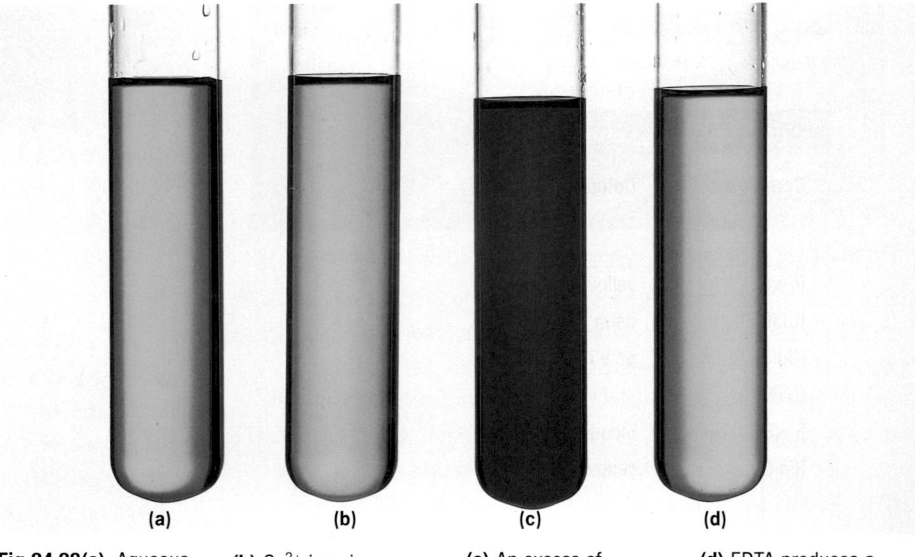

Fig 24.28(a) Aqueous solution of copper(II) sulfate containing $[Cu(H_2O)_6]^{2+}(aq)$ ions. **(b)** Cu^{2+} ions in concentrated HCl, giving the complex ion $[CuCl_4]^{2-}(aq)$. **(c)** An excess of ammonia has now been added, producing $[Cu(NH_3)_4(H_2O)_2]^{2+}$. **(d)** EDTA produces a much more stable complex, $[Cu(EDTA)(H_2O)_2]^{2-}$

Another example of ligand exchange is the reaction that occurs when thiocyanate ions, SCN^-, are added to an aqueous solution of iron(III) ions. A blood red solution of $[Fe(SCN)(H_2O)_5]^{2+}$ replaces the yellow $[Fe(H_2O)_6]^{3+}$, as is seen in Fig 24.36 on page 496. This is a very sensitive test for the presence of $Fe^{3+}(aq)$ ions.

$$\underset{\text{yellow}}{[Fe(H_2O)_6]^{3+}(aq)} + SCN^-(aq) \rightleftharpoons \underset{\text{blood-red}}{[Fe(SCN)(H_2O)_5]^{2+}(aq)} + H_2O(l)$$

STRETCH AND CHALLENGE

Stability constants

If we examine the ligand substitution reaction that occurs when concentrated hydrochloric acid is added to aqueous copper(II) sulfate, we do not obtain the expected yellow colour.

$$\underset{\text{blue}}{[Cu(H_2O)_6]^{2+}(aq)} + 4Cl^-(aq) \rightleftharpoons \underset{\text{yellow}}{[CuCl_4]^{2-}(aq)} + 6H_2O(l)$$

This is because the reaction has not gone to completion, since the solution is green – a combination of the blue and yellow complexes. As this is an equilibrium reaction, the equilibrium constant, K_c, can be calculated:

$$K_c = \frac{[CuCl_4{}^{2-}(aq)]}{[Cu(H_2O)_6{}^{2+}(aq)]\,[Cl^-(aq)]^4}$$

The equilibrium constant is a measure of how far the position of the equilibrium is over to the right. The further the position of the equilibrium is to the right, the higher the value of K_c and the more stable the complex ion compared with the aqua-complex. K_c is called the **stability constant K_{stab}**. As the numbers for K_{stab} are often very large, they are made more manageable by using a logarithm scale. Five values of log K_{stab} are shown in Table 24.5.

Table 24.5 Stability constants

Complex ion	Colour	Stability constant, log K_{stab}
$[Cu(Cl_4)]^{2-}$	yellow	5.6
$[Cu(NH_3)_4(H_2O)_2]^{2+}$	deep blue	13.1
$[Cu(EDTA)(H_2O)_2]^{2-}$	pale blue	18.8
$[Ni(NH_3)_6]^{2+}$	purple	8.6
$[Fe(SCN)(H_2O)_5]^{2+}$	blood-red	3.0
$[Fe(CN)_6]^{3-}$	yellow	31.0

The larger K_{stab}, the more stable the complex ion.

ENTROPY CONSIDERATIONS AND K_{STAB}

When a monodentate ligand is replaced by a bidentate ligand there will be an increase in entropy.

$$[Ni(H_2O)_6]^{2+}(aq) + 3\underset{\text{1,2-diaminoethane or en}}{NH_2CH_2CH_2NH_2}(aq) \rightarrow [Ni(NH_2CH_2CH_2NH_2)_3]^{2+}(aq) + 6H_2O(l)$$

If you look at the reactants, there are four particles, whereas on the right (products) there are seven particles. This means that the number of possible arrangements of the particles has increased, so there is an increase in entropy. This means the stability constant is greater for $[Ni(en)_3]^{2+}$ than it is for the complex in which water is the ligand. This increase in stability is sometimes known as the **chelate effect**. It is even more pronounced when EDTA replaces the water ligands.

$$[Ni(H_2O)_6]^{2+}(aq) + EDTA^{4-} \rightarrow [Ni(EDTA)]^{2-}(aq) + 6H_2O(l)$$

Table 24.6 A Stability constants for nickel complex ions

Complex ion	log K_{stab}
$[Ni(H_2O)_6]^{2+}$	0*
$Ni(NH_3)_6]^{2+}$	8.6
$[Ni(en)_3]^{2+}$	18.3
$[Ni(eda)]^{2-}$	19.3

*Notice that the stability constant of the complex with water ligands is given a value of 1 and the log of 1 is zero. The other stability constants are then quoted relative to this value.

Notice in the expression for K_c that the complex ions are shown without square brackets to avoid confusion with the square brackets that indicate concentration. Also, notice that water is not included in the K_c equation because, as the solvent, its concentration is considered to be constant.

For more information on K_c, see pages 352–353.

AQUEOUS COMPLEXES AND ACIDITY

As already stated, in aqueous solution, transition metal ions form aqua-complexes. When the pH values of various aqueous solutions of transition metal salts are measured, a pattern emerges. There is an increase in acidity as the charge of the transition metal ion increases. So, for example, $[Fe(H_2O)_6]^{3+}(aq)$ is more acidic than $[Fe(H_2O)_6]^{2+}(aq)$. Why is this?

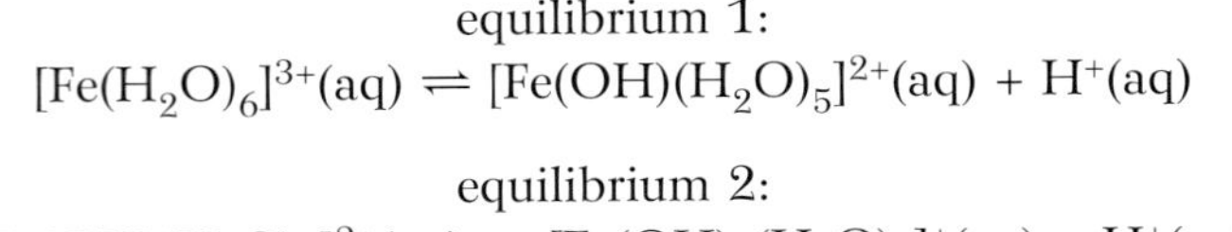

Fig 24.29 The highly charged cation pulls electrons from the water molecule, which weakens the O–H bond. A proton is thereby released, to produce H_3O^+ with water. The H_3O^+ causes acidity

The more highly charged the central ion, the more highly polarising it is (see page 87). This attracts the electrons of the surrounding water molecules, which allows the release of protons and thus makes the solution acidic. This process is called **deprotonation** of the cation. A number of equilibria exist in solution, two of which are:

equilibrium 1:

$$[Fe(H_2O)_6]^{3+}(aq) \rightleftharpoons [Fe(OH)(H_2O)_5]^{2+}(aq) + H^+(aq)$$

equilibrium 2:

$$[Fe(OH)(H_2O)_5]^{2+}(aq) \rightleftharpoons [Fe(OH)_2(H_2O)_4]^{+}(aq) + H^+(aq)$$

? QUESTION 20

20 Non-transition metal ions in solution can also be acidic. For example, $[Al(H_2O)_6]^{3+}$(aq) releases protons into an aqueous solution. Write an equation to show the equilibrium that exists when one proton is released. You can check your answer on page 452.

The polarising power of ions with charges greater than 3+ explains why oxyanions are formed. For example, $[Cr(H_2O)_6]^{6+}$ and $[Mn(H_2O)_6]^{7+}$ are theoretically possible, but the central ions are too polarising. This causes the loss of both protons from some of the surrounding ligand water molecules to produce O^{2-} ions, which form dative covalent bonds to give CrO_4^{2-} and MnO_4^-.

ADDITION OF AQUEOUS SOLUTIONS OF SODIUM HYDROXIDE OR AMMONIA

Aqueous sodium hydroxide contains hydroxide ions, which remove H^+(aq) ions by forming water:

$$H^+(aq) + OH^-(aq) \rightarrow H_2O(l)$$

This shifts equilibria 1 and 2 (above) to the right and causes the formation of the neutral complex $[Fe(OH)_3(H_2O)_3]$ from a third equilibrium. This is an example of Le Chatelier's principle (see pages 19 and 354). The complex forms as a rusty brown, gelatinous precipitate:

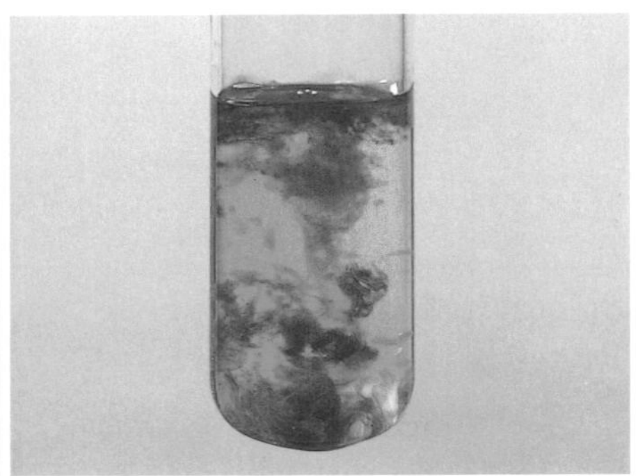

Fig 24.30 The rusty brown precipitate produced when aqueous sodium hydroxide is added to Fe^{3+}(aq) ions.
$Fe^{3+} + 3OH^-(aq) \rightarrow Fe(OH)_3(s)$

equilibrium 3:

$$[Fe(OH)_2(H_2O)_4]^{+}(aq) \rightleftharpoons \underset{\text{rusty brown precipitate}}{[Fe(OH)_3(H_2O)_3](s)} + H^+(aq)$$

? QUESTION 21

21 Co^{2+}(aq), when it reacts with sodium hydroxide, forms a pink precipitate. Write the ionic equation for this reaction.

Similar reactions occur between hydroxide ions and other aqueous transition metal ions, to produce coloured, gelatinous precipitates (Figs 24.31 to 24.33). In the equations below, the aqua-complexes are represented by (aq) and the hydroxoaqua-complexes by the simple metal hydroxide formulas:

$$Cr^{3+}(aq) + 3OH^-(aq) \rightarrow \underset{\text{grey–green}}{Cr(OH)_3(s)}$$

$$Mn^{2+}(aq) + 2OH^-(aq) \rightarrow \underset{\text{cream}}{Mn(OH)_2(s)}$$

$$Fe^{2+}(aq) + 2OH^-(aq) \rightarrow \underset{\text{green}}{Fe(OH)_2(s)}$$

For the colour of other metal hydroxides produced when aqueous sodium hydroxide is added to solutions of metal ions, see pages 432 and 589.

Fig 24.31 $Cr(OH)_3(s)$ is produced when $Cr^{3+}(aq)$ reacts with $NaOH(aq)$

Fig 24.32 $Mn(OH)_2(s)$ is precipitated when $OH^-(aq)$ ions react with $Mn^{2+}(aq)$ ions

Fig 24.33 The addition of $OH^-(aq)$ to $Fe^{2+}(aq)$ precipitates $Fe(OH)_2(s)$

When excess $OH^-(aq)$ is added to chromium hydroxide, the precipitate redissolves because further deprotonation takes place, as shown in equilibrium 4, to form a soluble complex:

equilibrium 4:

$$[Cr(OH)_3(H_2O)_3](s) \rightleftharpoons [Cr(OH)_4(H_2O)_2]^-(aq) + H^+(aq)$$

Aqueous ammonia also produces coloured gelatinous precipitates because hydroxide ions are present:

$$NH_3(aq) + H_2O(l) \rightleftharpoons NH_4^+(aq) + OH^-(aq)$$

However, ammonia is also a ligand, and in some cases the precipitates dissolve because a soluble ammine complex is formed (an ammine complex contains ammonia ligands). This occurs when aqueous ammonia is added to $Cu^{2+}(aq)$:

$$\underset{\text{blue solution}}{[Cu(H_2O)_6]^{2+}(aq)} + 2OH^-(aq) \rightarrow \underset{\text{pale blue precipitate}}{[Cu(OH)_2(H_2O)_4](s)}$$

On addition of an excess of ammonia solution, the precipitate dissolves to form the deep blue ammine complex $[Cu(NH_3)_4(H_2O)_2]^{2+}(aq)$:

$$[Cu(OH)_2(H_2O)_4]^{2+}(s) + 4NH_3(aq) \rightarrow [Cu(NH_3)_4(H_2O)_2]^{2+}(aq) + 2H_2O(l) + 2OH^-(aq)$$

REMEMBER THIS

When the chromium hydroxide dissolves in an excess of alkali we call this **amphoteric behaviour** since it can both accept and donate $H^+(aq)$ i.e act as a base and an acid. This is not the same reaction as ligand exchange (see page 492)

For more examples of reactions with aqueous ammonia, see page 589.

Fig 24.34(a) When concentrated ammonia solution is added dropwise to aqueous copper(II) ions, a precipitate of copper(II) hydroxide forms

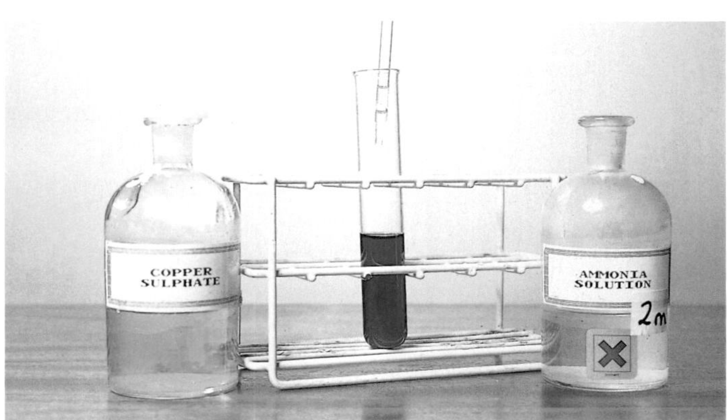

Fig 24.34(b) On the further addition of excess ammonia solution, the precipitate dissolves as the soluble ammine complex $[Cu(NH_3)_4(H_2O)_2]^{2+}(aq)$ is formed

Fig 24.35 Cobalt chloride paper before and after the addition of water

? QUESTION 22

22 **a)** Write the equation for the formation of $[Co(H_2O)_6]^{2+}(aq)$ from $[CoCl_4]^{2-}(aq)$. These are in equilibrium, so remember to include the equilibrium sign.

b) Heating moist cobalt chloride paper turns it blue again. Why does this happen? (Hint: you will need to have understood Le Chatelier's principle on page 392.)

We discuss the origin of colour in more detail in Chapter 27.

Even the way the ligands are arranged around the central cation can affect the colour. Turn to page 490 to see the different colours of the cis and trans isomers of $[Cr(NH_3)_4Cl_2]^+$.

Fig 24.36 The blood-red colour of $[Fe(H_2O)_5SCN]^{2+}(aq)$ is a very sensitive test for the presence of $Fe^{3+}(aq)$ ions

Fig 24.37 Turnbull's blue precipitate

4 THE FORMATION OF COLOURED IONS

The formation of coloured ions is another characteristic property of transition metals. Most of their compounds are coloured and they form coloured solutions. The colours of nine transition metal ions in aqueous solution are given in Table 24.7. Scandium and zinc ions are also included.

Being colourless is another reason why $Sc^{3+}(aq)$ and $Zn^{2+}(aq)$ are not usually regarded as transition metal ions. A coloured ion results from an incomplete d subshell, which neither of these ions has.

The type of ligand also affects the colour, as already noted in the case of $[Cu(H_2O)_6]^{2+}(aq)$, which is a paler blue than the ammine complex $[Cu(NH_3)_4(H_2O)_2]^{2+}(aq)$, which is deep blue. The cobalt chloride test for water is probably a test you have done yourself in the laboratory. Cobalt chloride paper is blue because it contains the complex ion $[CoCl_4]^{2-}$. When water is added, $[Co(H_2O)_6]^{2+}$ is formed, which is pink.

Table 24.7 Colours of some common aqueous ions in the first transition series

Ion	Colour	Outer 3d electrons
$Sc^{3+}(aq)$	colourless	$3d^0$
$Ti^{3+}(aq)$	purple	$3d^1$
$V^{3+}(aq)$	green	$3d^2$
$Cr^{3+}(aq)$	violet	$3d^3$
$Mn^{2+}(aq)$	pink	$3d^5$
$Fe^{3+}(aq)$	yellow*	$3d^5$
$Fe^{2+}(aq)$	green	$3d^6$
$Co^{2+}(aq)$	pink	$3d^7$
$Ni^{2+}(aq)$	green	$3d^8$
$Cu^{2+}(aq)$	blue	$3d^9$
$Zn^{2+}(aq)$	colourless	$3d^{10}$

*$[Fe(H_2O)_6]^{3+}(aq)$ is violet, but because of the deprotonation (see page 528) the yellow ion $[Fe(OH)(H_2O)_5]^{2+}(aq)$ predominates.

The colours of ions are affected too by the number of different types of ligand, as shown by these chromium(III) complexes:

$$\underset{\text{violet}}{[Cr(H_2O)_6]^{3+}(aq)} \quad \underset{\text{green}}{[CrCl(H_2O)_5]^{2+}(aq)} \quad \underset{\text{dark green}}{[CrCl_2(H_2O)_4]^{+}(aq)}$$

Many transition metal complexes have characteristic colours that are so distinctive they can be used to test for these particular ions. For example, $Fe^{3+}(aq)$ forms a blood-red complex with thiocyanate ions (SCN^-). The addition of aqueous potassium thiocyanate, KSCN, to a solution of $Fe^{3+}(aq)$ is a very sensitive test, since tiny amounts give the typical blood-red coloration:

$$[Fe(H_2O)_6]^{3+}(aq) + SCN^-(aq) \rightarrow \underset{\text{blood-red}}{[Fe(H_2O)_5SCN]^{2+}(aq)}$$

$Fe^{2+}(aq)$ ions, however, do not give a coloured complex with thiocyanate ions.

Another way to distinguish between $Fe^{2+}(aq)$ and $Fe^{3+}(aq)$ is to use the complex ion $[Fe(CN)_6]^{3-}(aq)$, which produces a deep blue precipitate, called Turnbull's blue, with $Fe^{2+}(aq)$.

5 HETEROGENEOUS AND HOMOGENEOUS CATALYSIS

Catalysis is the final characteristic property of transition elements to be considered here. A catalyst is a substance that alters the rate of a chemical reaction without becoming permanently involved in the reaction. A catalyst works by providing an alternative reaction pathway with a lower energy of activation. (See pages 164 and 206 for more information.)

There are two types of catalyst. **Homogenous catalysts** are in the same phase as the reactants they catalyse, while **heterogeneous catalysts** are in a different phase from the reactants they catalyse. Transition metals and their compounds can act as either of these two types.

REMEMBER THIS

A phase is defined as a homogeneous (uniform) portion of matter separated from other portions of matter by a boundary surface.

HETEROGENEOUS CATALYSIS

Platinum, for example, plays a crucial role in reducing emissions from car exhausts through its use in catalytic converters. Here it acts as a heterogeneous catalyst because it is in a different phase from the gaseous reactants. One reaction that is catalysed by platinum is the oxidation of carbon monoxide:

$$2CO(g) + O_2(g) \xrightarrow{\text{Pt catalyst}} 2CO_2(g)$$

Nickel catalyses the hydrogenation of carbon–carbon double bonds, an important reaction in the production of fats from oils (see page 286).

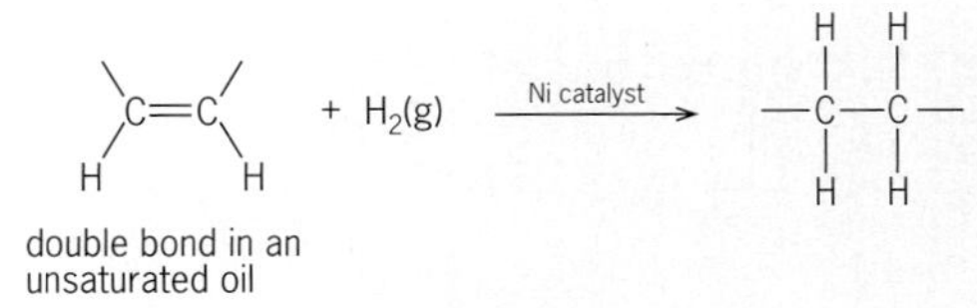

$$\text{>C=C<} + H_2(g) \xrightarrow{\text{Ni catalyst}} \text{–CH–CH–}$$

double bond in an unsaturated oil

Fig 24.38 Hydrogenation of carbon–carbon double bonds

The Haber process for the production of ammonia (featured on page 20) uses finely divided iron to catalyse the reaction between nitrogen and ammonia:

$$N_2(g) + 3H_2(g) \overset{\text{Fe catalyst}}{\rightleftharpoons} 2NH_3(g)$$

This ranks among the most important industrial reactions because ammonia is used to produce nitrogenous fertilisers. Fig 24.39 shows a proposed mechanism for its heterogeneous catalysis.

The Contact process, which leads to the production of sulfuric acid (see pages 611–613), is another important catalysed reaction:

$$2SO_2(g) + O_2(g) \overset{V_2O_5 \text{ catalyst}}{\rightleftharpoons} 2SO_3(g)$$

A heterogeneous catalyst works by providing a surface on which the reactants can form weak bond. These are called **active sites** (Fig 24.40). Transition metals use their 3d and 4s electons to make these bonds.

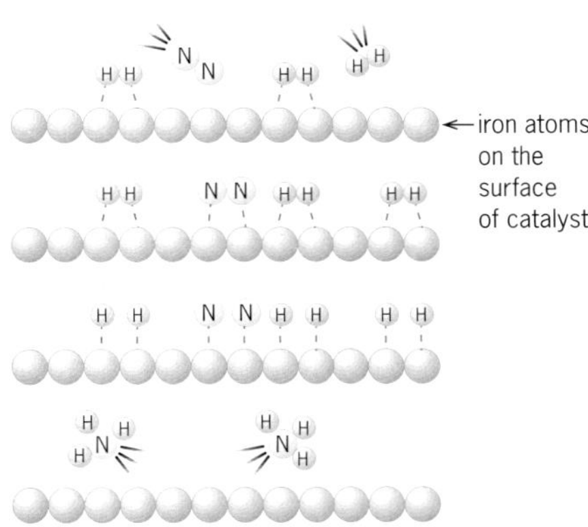

Fig 24.39 Proposed mechanism for the functioning of the iron catalyst in the Haber process. The nitrogen and hydrogen molecules are adsorbed on the surface of the catalyst. The bonds within the molecules break, to leave separate nitrogen and hydrogen atoms. The atoms then rearrange and form new bonds. The ammonia molecules fly off. This mechanism offers an alternative reaction pathway of lower activation energy

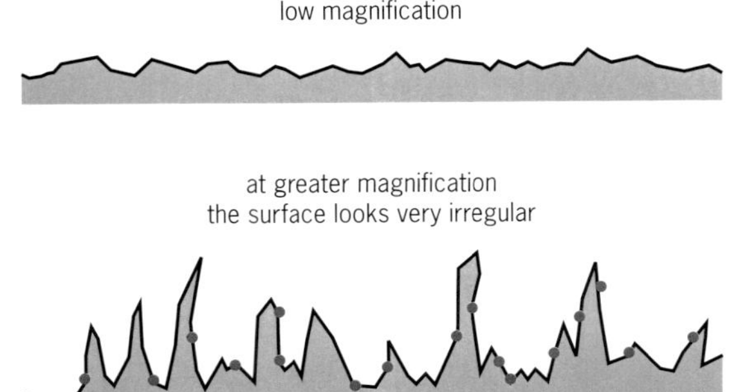

Fig 24.40 The active sites (shown by the red dots) occur on the surface of the catalyst

You can find out more about heterogeneous catalysts in Chapter 10, page 206 (catalytic converters) and in Chapter 14, page 286 (nickel in the hydrogenation of ethene)

REMEMBER THIS

Adsorb is the word used when reactants are weakly bonded to a surface. The bonds are formed by a similar mechanism to that when ligands bond to a central metal cation. Desorbtion is the opposite of adsorption.

? QUESTION 23

23 Give reasons why:

a) Catalysts should have rough surfaces rather than smooth ones.

b) Platinum is applied as a powder on a cheap solid support in a catalytic converter.

c) In some catalysed reactions gases are blown through a *fluidised bed* of finely powdered catalyst so that the catalyst particles float on the gases being blown in.

? QUESTION 24

24 In the catalysis of sulfur trioxide from sulfur dioxide with vanadium (see text), what are the oxidation states of sulfur in SO_3 and SO_2 and why is this an oxidation of sulfur? (Hint: if you need to remind yourself about oxidation states, look back at Chapter 5, 106–109.)

REMEMBER THIS

Homogeneous catalysts are completely exposed to the reactants but heterogeneous catalysts are only effective at the surface. However, homogeneous catalysts can be difficult to remove from the products for re-use, which is not usually the case with heterogeneous catalysts, which are often removed by filtration.

However not all transition metals make good heterogeneous catalysts. For example, tungsten bonds strongly to molecules adsorbed on its surface and will not release product molecules, while silver's bonds with adsorbed molecules are too weak to hold them on the surface for long enough for a reaction to occur.

It is the availability of the 3d and 4s electrons coupled with the ability to use variable oxidation states that make transition metals and their compounds such good catalysts.

Fig 24.41 Gauze of platinum and rhodium provides a huge surface area on which to catalyse the oxidation of ammonia in the production of nitric acid

Let's examine how variable oxidation states allow vanadium(V) oxide to catalyse the reaction that produces sulfur trioxide from sulfur dioxide.

The first reaction that occurs is the oxidation of SO_2 into SO_3:

$$V_2O_5(s) + SO_2(g) \rightarrow V_2O_4(s) + SO_3(g)$$

Oxidation state of vanadium +5 +4

Vanadium is reduced from oxidation state +5 to +4 and at the same time the sulfur is oxidised.

A catalyst must remain chemically unchanged at the end of the reaction, which occurs through a second reaction in which oxygen oxidises the vanadium back to its +5 oxidation state.

$$V_2O_4(s) + \tfrac{1}{2}O_2(g) \rightarrow V_2O_5(s)$$

Oxidation state of vanadium +4 +5

HOMOGENEOUS CATALYSIS

Transition metal compounds also make excellent **homogeneous catalysts**. Homogeneous catalysts are in the same physical state as the reactants, which usually means in solution. While these reactions are not as commercially significant as heterogeneous catalysis, they are fundamental to many biological reactions. It is the ability of transition metals to exist in more than one oxidation state that is the key to these reactions, since transition metal ions can take part in electron-transfer reactions that provide an alternative pathway of lower activation energy.

The reaction between iodide ions, $I^-(aq)$, and peroxodisulfate(VI) ions, $S_2O_8^{2-}(aq)$, is catalysed by $Fe^{2+}(aq)$ ions:

$$S_2O_8^{2-}(aq) + 2I^-(aq) \rightarrow 2SO_4^{2-}(aq) + I_2(aq)$$

$Fe^{2+}(aq)$ is first oxidised into $Fe^{3+}(aq)$:

$$S_2O_8^{2-}(aq) + 2Fe^{2+}(aq) \rightarrow 2SO_4^{2-}(aq) + 2Fe^{3+}(aq)$$

$Fe^{3+}(aq)$ is then reduced back to $Fe^{2+}(aq)$ by $I^-(aq)$:

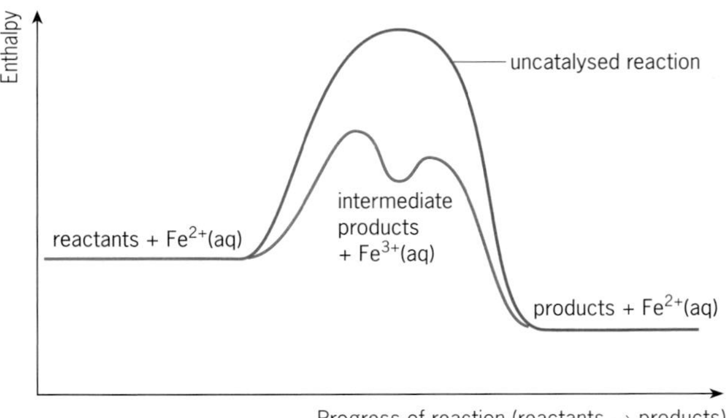

Fig 24.42 The reaction between $S_2O_8^{2-}(aq)$ and $2I^-(aq)$ is slow without the addition of an $Fe^{2+}(aq)$ catalyst, which provides an alternative route of lower activation energy

$$2I^-(aq) + 2Fe^{3+}(aq) \rightarrow 2Fe^{2+}(aq) + I_2(aq)$$

Exciting research is continuing to discover new homogeneous catalysts. One promising area involves organometallic catalysts, which could build up complex organic molecules from carbon monoxide or carbon dioxide. Compared with heterogeneous catalysts, which often involve the use of precious metals such as platinum, homogeneous catalysts generally cost

HOW SCIENCE WORKS

The search for new catalysts

The development of new catalysts is a priority area of research today. If a new catalyst means that a process can take place at a lower temperature or pressure (or both), then energy is saved and this energy usually comes from burning fossil fuels. Also the **carbon footprint** (the amount of carbon dioxide released into the atmosphere) is reduced.

Ethanoic acid, CH_3COOH, is a very important industrial chemical with many uses, from the manufacture of polymers to the making of perfumes, flavourings and pharmaceuticals. Until the 1970s it was manufactured by oxidising naphtha or butane. This process took place at 200 °C and 50 atmospheres pressure using a catalyst of cobalt(II) ethanoate. There were a number of by-products, however, which meant that the atom economy (see page 9 in Chapter 1) was only about 35 per cent.

Another process was developed in 1960, which had a 100 per cent atom economy, at least theoretically. This used cobalt and iodine catalysts.

$$CH_3OH + CO \rightarrow CH_3COOH$$

However, the temperature required was 300 °C with 700 atmospheres pressure. Then in 1966 Monsanto used the same reaction, but with a new rhodium/iodine-based catalyst. This reduced the temperature required to between 150 °C and 200 °C, and pressures to between 30 and 60 atmospheres, making altogether a greener process. The atom economy was high, but not as high as the equation suggests, as some by-products are still produced. The majority of ethanoic acid was produced in this way until another catalyst system produced an even greener route. This was developed by BP in 1986 and replaced the transition metal rhodium with another transition metal, iridium. The advantage is that iridium is much cheaper than rhodium and it works very much faster, so much less is required in the first place. Also there are very few by-products produced, meaning that the atom economy is even higher. Using this process, called the Cativa process, 75 per cent more ethanoic acid is actually produced each day than with the process devised by Monsanto.

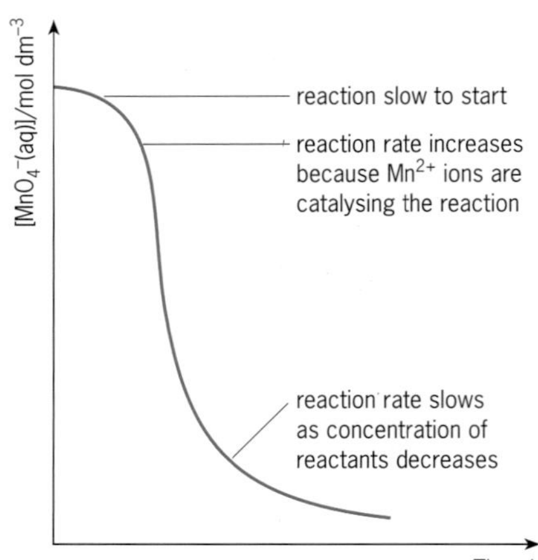

$[MnO_4^-(aq)]$/mol dm^{-3}
reaction slow to start
reaction rate increases because Mn^{2+} ions are catalysing the reaction
reaction rate slows as concentration of reactants decreases
Time/s

Fig 24.43 The change of concentration of $MnO_4^-(aq)$ when it reacts with $C_2O_4^{2-}(aq)$. This graph is typical of autocatalysed reactions

less and their catalytic action often takes place at lower temperatures and pressures.

AUTOCATALYSIS

Sometimes, catalysts are produced by the reaction itself. This is known as **autocatalysis**. One example is on page 582 of Chapter 26. Another is the reaction of $MnO_4^-(aq)$ with $C_2O_4^{2-}(aq)$:

$$2MnO_4^-(aq) + 5C_2O_4^{2-}(aq) + 16H^+(aq) \rightarrow 2Mn^{2+}(aq) + 10CO_2(g) + 8H_2O(l)$$

The reaction is slow to begin with, but speeds up once $Mn^{2+}(aq)$ ions are formed (see Fig 24.43). The $Mn^{2+}(aq)$ is acting as a homogeneous catalyst.

? QUESTION 25

25 Sketch a graph of the volume of CO_2 produced against time for the reaction of $MnO_4^-(aq)$ with $C_2O_4^{2-}(aq)$.

SUMMARY

After studying this chapter, you should know that:

- A transition element is an element that forms one or more ions with a partially filled d subshell.
- When a transition metal atom forms an ion, the first electrons to be lost are in the 4s subshell.
- The 3d and 4s subshells have closely similar energies, which results in ions with 3+ charges.
- Transition elements are typical metals. They are hard and dense, have high melting points and are good conductors of heat and electricity.
- The chemical properties that characterise transition elements are: variable oxidation states in their compounds, the formation of complexes, the formation of coloured ions and the catalysis of reactions, both as elements and as compounds.
- A complex is formed when a central metal atom or ion is surrounded by species that donate lone pairs of electrons.
- A ligand is a species that donates a lone pair of electrons to a central metal atom or ion.
- The coordination number is the number of dative covalent (coordinate) bonds from ligands to the central metal ion. It determines the shape of the complex.
- A monodentate ligand forms a single dative covalent bond with the central metal cation.
- A bidentate ligand can form two dative covalent bonds with the central metal ion. When the ligands form five-membered rings with the metal ion, they are called chelates.
- Some complex ions show isomerism.
- Ligand exchange (ligand substitution) may occur if the complex ion produced is more stable.
- The stability of complex ions is measured using the stability constant, K_{stab}. If, during ligand exchange the entropy increases, then the stability constant is greater.
- Hydrated cations (aqua-complexes) may become deprotonated, which causes acidity.
- The addition of aqueous solutions of sodium hydroxide or ammonia may result in the production of coloured hydroxide precipitates. Some of these precipitates dissolve in excess alkali, showing amphoteric behaviour, while others dissolve in excess ammonia.
- Many transition metal complexes have characteristic colours that are so distinctive they can be used to test for these particular ions.
- There are two types of catalyst: homogeneous, which is in the same phase as the reactants, and heterogeneous, which is in a different phase.
- Variable oxidation states are important in both heterogeneous and homogeneous catalysis.
- Some reactions are autocatalysed.

Practice questions and a How Science Works assignment for this chapter are available at www.collinseducation.co.uk/CAS

25

Electrode potentials, transition elements and feasibility of reactions

25 ELECTRODE POTENTIALS, TRANSITION ELEMENTS AND FEASIBILITY OF REACTIONS

Electric vehicles An electrically powered bus in Sweden

Electric vehicles

Electric cars and vans would dramatically reduce atmospheric pollution. So, why haven't they yet supplanted petrol and diesel-driven vehicles?

Their widespread adoption has been stunted by the absence of a battery able to provide sufficient motive power for a sufficient length of time. Much is being done to rectify this, with the development of lightweight batteries that can run a vehicle for at least 500 km between charges. It is hoped that these batteries can be recharged many times using cheap-rate electricity.

There has also been progress with the development of fuel cells that use hydrogen or petrol in a reaction with oxygen. No burning occurs and electric current is generated directly. Appropriate technology does exist, but there are still problems with the availability, storage and transport of hydrogen.

1 REDOX REACTIONS

Redox is a type of chemical reaction that involves electron transfer. At its simplest, one atom loses electrons and another atom gains electrons. Atoms of metals lose electrons, usually to attain a stable electron configuration, and atoms of non-metals gain electrons, usually to attain a stable electron configuration.

HALF-EQUATIONS

A typical redox reaction is the reaction of the metal magnesium with dilute hydrochloric acid to form magnesium chloride (Fig 25.1):

$$Mg(s) + 2HCl(aq) \rightarrow MgCl_2(aq) + H_2(g)$$

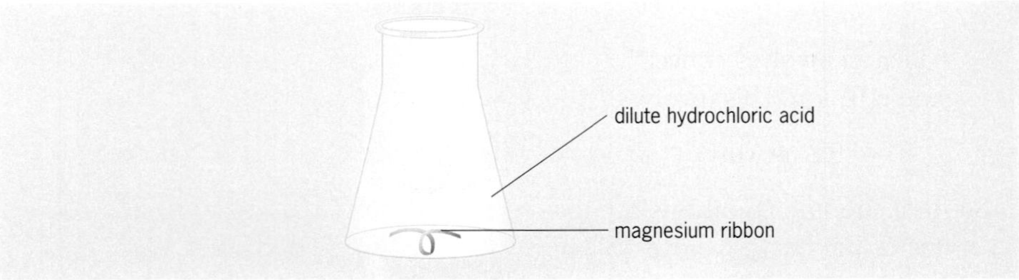

Fig 25.1 The redox reaction between magnesium and hydrochloric acid forms hydrogen and aqueous magnesium chloride

Strong acids and dissociation are covered on pages 330 to 333.

This reaction is a redox reaction because electron transfer occurs. An atom of magnesium loses two electrons during the reaction, but it is not immediately obvious where these electrons go. Hydrochloric acid is a strong acid and it is completely dissociated into aqueous chloride ions and aqueous hydrogen ions. It is the hydrogen ions that gain electrons: two hydrogen ions gain one electron each, eventually to form a hydrogen molecule. The hydrogen ions are said to be **reduced** to hydrogen. The chloride ions do not take part in the reaction at all. They are sometimes called **spectator ions**, to indicate that they are present in the solution during the reaction, but take no part in it.

The overall redox reaction can be represented by two half-equations, one to show the reduction and the other to show the oxidation:

Oxidation half-equation: $Mg(s) \rightarrow Mg^{2+}(aq) + 2e^-$
Reduction half-equation: $2H^+(aq) + 2e^- \rightarrow H_2(g)$

By combining these two half-equations, an ionic equation can be written, which does not show the spectator ions:

$$Mg(s) + 2H^+(aq) \rightarrow Mg^{2+}(aq) + H_2(g)$$

For many redox reactions, it is possible to write two half-equations, one for the oxidation and the other for the reduction.

Working out the half-equations

Before writing down the two half-equations, the species being oxidised and the species being reduced must be established. This is best achieved using oxidation numbers (see pages 106–111). During oxidation, the oxidation number of an atom increases. During reduction, it decreases. The change in oxidation number also gives the number of electrons lost (oxidation) or gained (reduction).

Consider the reaction of aqueous iron(II) ions with acidified manganate(VII) ions. The overall ionic equation is:

$$5Fe^{2+}(aq) + MnO_4^-(aq) + 8H^+(aq) \rightarrow 5Fe^{3+}(aq) + Mn^{2+}(aq) + 4H_2O(l)$$

The oxidation state of Fe changes from +2 to +3 (Fig 25.2). This is oxidation and involves the loss of one electron. The oxidation state of Mn changes from +7 to +2. This is reduction and involves the gain of five electrons. We are now in a position to write the half-equations.

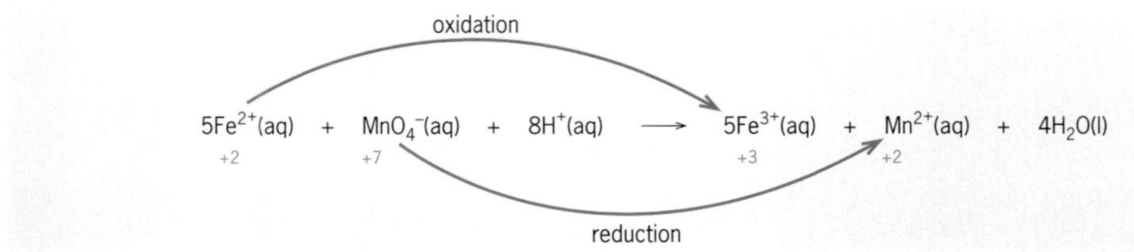

Fig 25.2 The reaction of acidified manganate(VII) ions, $MnO_4^-(aq)$, with aqueous iron(II) ions, $Fe^{2-}(aq)$. The oxidation numbers that change during the reaction are shown in red

The oxidation involves conversion of Fe^{2+} into Fe^{3+}, and is written as:

$$Fe^{2+}(aq) \rightarrow Fe^{3+}(aq) + e^-$$

The reduction involves conversion of the manganate(VII) ion into the manganese(II) ion. This involves the gain of five electrons per manganate(VII) ion:

$$5e^- + MnO_4^-(aq) + 8H^+(aq) \rightarrow Mn^{2+}(aq) + 4H_2O(l)$$

Note that the half-equations are balanced, both in terms of their symbols and in terms of the charge. Note also that the aqueous hydrogen ions are not spectator ions, but are needed for the reduction to occur, and that the hydrogen ion is not itself reduced.

Electron transfer

The two half-equations clearly demonstrate the electron-transfer processes that occur during a redox reaction. The processes shown by the half-equations cannot take place independently, because reduction cannot occur without the simultaneous occurrence of oxidation. It is best to think of the half-equations as a 'book-keeping' exercise to show where electrons are lost and where they are gained.

? QUESTION 1

1 Write down the half-equations that correspond to the following overall reactions:

a) $Cu(s) + 2Ag^+(aq) \rightarrow 2Ag(s) + Cu^{2+}(aq)$

b) $6Fe^{2+}(aq) + 14H^+(aq) + Cr_2O_7^{2-}(aq) \rightarrow 6Fe^{3+}(aq) + 2Cr^{3+}(aq) + 7H_2O(l)$

? QUESTION 2

2 Write down the two half-equations (electrode reactions) that occur during the electrolysis of:
a) molten sodium chloride;
b) molten lead(II) bromide.

Electrolysis reactions

Electrolysis can be defined thus:

Electrolysis is the decomposition of a liquid by the passage of an electric current, which enters and leaves the compound through electrodes.

Electrolysis is necessarily a redox reaction, since it involves electron transfer. The reactions represented by the two half-equations occur in different parts of an electrolytic cell. Reduction always occurs at the cathode (negative electrode) and oxidation always occurs at the anode (positive electrode).

Consider the electrolysis of molten lead(II) iodide. The two half-equations are:

Oxidation half-equation: $2I^- \rightarrow I_2 + 2e^-$ occurs at the anode
Reduction half-equation: $Pb^{2+} + 2e^- \rightarrow Pb$ occurs at the cathode

In electrolysis, these half-equations are referred to as **electrode reactions.**

OXIDISING AGENTS

An **oxidising agent,** or **oxidant,** is the substance or species that gains electrons during a redox reaction. It oxidises another substance by gaining electrons from it. So an oxidising agent is reduced during a redox reaction.

Typically, oxidising agents are non-metal elements or compounds that have an element in a high, positive oxidation state. For example, fluorine is an oxidising agent, as is potassium dichromate(VI) in which chromium has an oxidation state of +6. In every reaction, the oxidising agent gains electrons. Fluorine molecules gain electrons to form fluoride ions, and dichromate(VI) ions gains electrons to form chromium(III) ions:

$$F_2 + 2e^- \rightarrow 2F^-$$
$$Cr_2O_7^{2-} + 14H^+ + 6e^- \rightarrow 2Cr^{3+} + 7H_2O$$

Note that these two half-equations are written to show that the oxidising agent gains electrons.

Sometimes, we want to compare the oxidising power of substances. That is, we want to compare the ability of oxidising agents to gain electrons. For example, fluorine is the better oxidising agent of the two shown above, but it is impossible to tell just from the half-equation. This is dealt with on page 510.

REDUCING AGENTS

A **reducing agent,** or **reductant,** is a substance or species that loses electrons during a redox reaction. It reduces another substance by losing electrons to it. So a reducing agent is oxidised during a redox reaction.

Typically, reducing agents are metal elements or compounds that have an element in a low, positive oxidation state or a negative oxidation state. For example, potassium is a reducing agent, as is the sulfide ion in which sulfur has an oxidation state of −2. In every reaction, the reducing agent loses electrons. A potassium atom loses an electron to form a potassium ion, and a sulfide ion loses two electrons to form a sulfur atom.

$$K \rightarrow K^+ + e^-$$
$$S^{2-} \rightarrow S + 2e^-$$

Note that the two half-equations are written to show that the reducing agent loses electrons.

Sometimes, we want to compare the reducing power of substances. That is, we want to compare the ability of reducing agents to lose electrons. For example, potassium is the better reducing agent of the two shown above, but it is impossible to tell this just from the half-equations. This is also discussed on page 511.

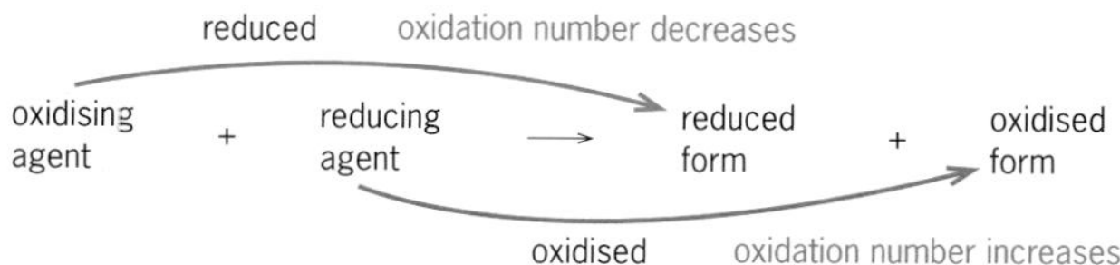

Fig 25.3 A summary of the action of oxidising agents and reducing agents

2 SIMPLE CELLS

It has been known for more than two centuries that a potential difference is generated when two dissimilar metal electrodes are put into an electrolyte. What takes place in this so-called electrochemical cell is a redox reaction, in which oxidation occurs at one electrode and reduction at the other.

This poses a question, since in a redox reaction an electron transfer must occur. Given that the two half-reactions (oxidation and reduction) are at different places, the transfer of electrons cannot take place during collision between the oxidising and the reducing agents. So where does it take place? Instead, the electrons are transferred via an external circuit (a metal wire). This may be regarded as electrolysis in reverse. Instead of a reaction occurring because of the passage of an electric current, an electric current (electron transfer) is produced because a redox reaction is occurring.

DANIELL CELL

In 1836, the English chemist John Fredrick Daniell constructed a simple battery that employed this idea. He used zinc and copper plates as the electrodes and two different electrolytes – aqueous copper(II) sulfate and aqueous zinc sulfate (Fig 25.4). (It is now known that in the Daniell cell zinc is in equilibrium with aqueous zinc ions and copper is in equilibrium with copper(II) ions.) When the two metal plates are connected by a metal wire, a current flows through the wire. The current is a result of the gain and loss of electrons that occur in the cell.

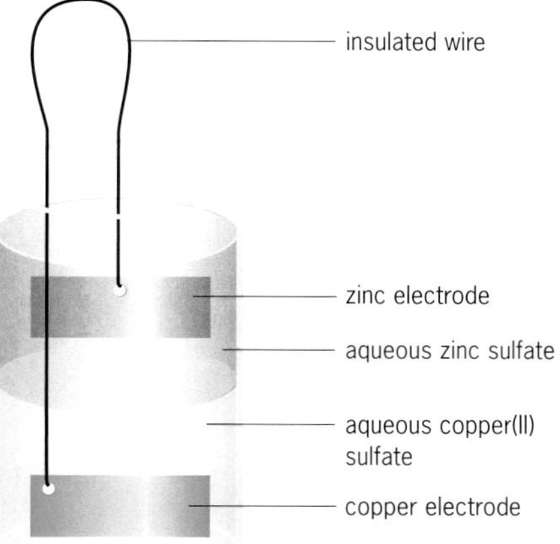

Fig 25.4 A Daniell cell consists of a copper plate surrounded by saturated aqueous copper(II) sulfate solution and a zinc plate surrounded by saturated aqueous zinc sulfate. The aqueous zinc sulfate must be added carefully so that it floats on top of the more dense aqueous copper(II) sulfate solution, but does not mix with it. When the two different metals are connected by a wire, a current flows. Eventually, the cell stops producing a potential difference as the solutions mix through diffusion

During the operation of the cell, the zinc plate forms zinc ions and loses electrons. This corresponds to the oxidation half-equation:

$$Zn(s) \rightarrow Zn^{2+}(aq) + 2e^-$$

Meanwhile, at the copper plate, copper(II) ions gain electrons to form copper. This corresponds to the reduction half-equation:

$$Cu^{2+}(aq) + 2e^- \rightarrow Cu(s)$$

Now, a redox reaction can occur only if electrons can be transferred. Only when the two plates are connected by a metal wire can this happen. The zinc plate, at which oxidation takes place, is the anode; the copper, at which reduction takes place, is the cathode.

The overall reaction that occurs in the cell is a combination of the two half-equations:

$$Zn(s) + Cu^{2+}(aq) \rightarrow Zn^{2+}(aq) + Cu(s)$$

When zinc is placed directly into aqueous copper(II) sulfate, the same reaction takes place, but the electron transfer occurs during a collision between a zinc atom and a copper(II) ion (Fig 25.5).

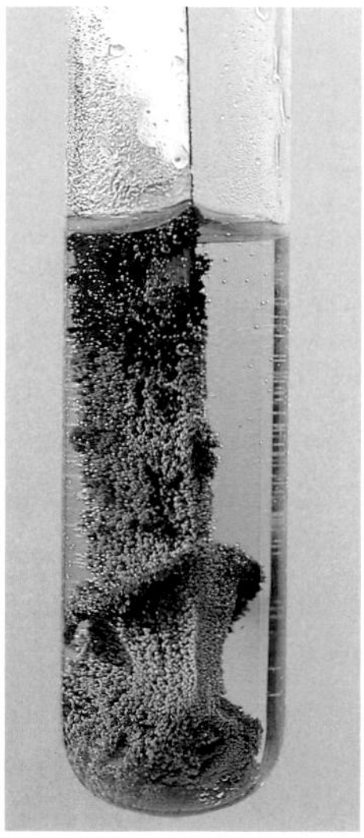

Fig 25.5 When zinc is dipped into aqueous copper(II) sulfate, a spontaneous redox reaction occurs to form copper and aqueous zinc sulfate

ANOTHER SIMPLE CELL

It is possible to arrange the reaction between magnesium and hydrochloric acid as two half-reactions that occur in separate places, with the electrons being transferred via an external metal wire.

The arrangement is shown in Fig 25.6. Note the salt bridge. This allows ions in the two electrolytes to migrate from one half-cell to the other without the electrolytes themselves mixing. Without a salt bridge, there would not be a complete circuit for the charge to flow round, and so no current would be produced. A salt bridge can be as simple as a piece of filter paper dipped in saturated aqueous potassium nitrate. Potassium nitrate is usually used because it does not react with most of the electrolytes used in electrochemical cells.

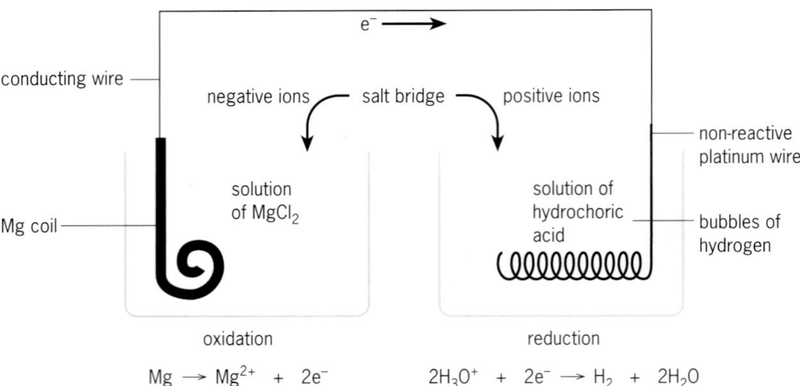

Fig 25.6 The oxidation of magnesium ions and reduction of hydrogen ions occur in separate parts of the cell. There is no reaction unless both the salt bridge and the external wire are in place

CELL POTENTIAL

For a current to flow, there must be a potential difference between the two halves of the cell. This is known as the **cell potential** or the electromotive force (e.m.f.). It is given the symbol $E^{\ominus}_{\mathbf{cell}}$ and is measured in volts. The cell potential is independent of the amount of each substance present in the cell, but is dependent on each of their concentrations. The effect of concentration on cell potentials and electrode potentials is described on page 513.

Measuring the cell potential with a voltmeter

The potential of a cell can be measured using a high-resistance voltmeter, as shown in Fig 25.7. This simple set-up does not quite measure the true maximum cell potential. As the current flows from one electrode to the other, the circuit heats up because of the resistance of the wire. So, the cell transfers some of its energy as heat rather than as electric current.

To measure the true maximum cell potential possible, no energy should be transferred as heat. That is, the potential difference should be measured when no current is flowing. This is why a high-resistance voltmeter must be used, so that almost no current is drawn from the cell.

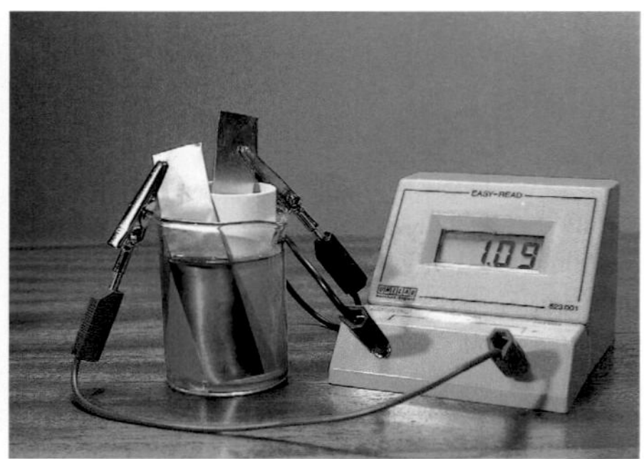

Fig 25.7 A cell that consists of a zinc electrode dipped into aqueous zinc ions, and a copper electrode dipped into aqueous copper(II) ions, should have a cell potential of 1.10 V (measured on a high-resistance voltmeter) when solutions of 1.0 mol dm^{-3} are used at a temperature of 298 K

ELECTRODE POTENTIALS

Generally, the term 'electrode' is used to describe the conductor that allows the passage of electric current in and out of the cell. In some applications, the electrode is inert, as in the use of graphite or platinum for the electrolysis of acidified water. In other applications, it actually takes part in a reaction. In work on electrochemical cells, however, the term 'electrode' is extended to include what is known as a half-cell. So it refers not only to the conductor, but also to the conducting solution in which it is placed.

As explained earlier, a cell consists of an oxidation and a reduction reaction. Therefore, the overall cell potential should be the sum of the potential of the oxidation process and the potential of the reduction process. The potential of the oxidation process is called the **oxidation potential** of the anode, which, for convenience, is shortened to $E^{\ominus}_{\mathbf{oxid}}$ or oxidation potential. Likewise, the potential of the reduction process, which is called the reduction potential of the cathode, is shortened to $E^{\ominus}_{\mathbf{red}}$ or the reduction potential. So, in an electrochemical cell:

$$E^{\ominus}_{\mathbf{cell}} = E^{\ominus}_{\mathbf{oxid}} + E^{\ominus}_{\mathbf{red}}$$

Standard electrode potentials

The word 'standard' in the term **standard electrode potential** signifies that the electrode potential has been measured (or calculated) under standard conditions, which are defined as 298 K, 101 kPa (1 atmosphere) and all aqueous solutions at a concentration of 1.0 mol dm^{-3}. (Standard conditions in thermochemistry are discussed on page 153.)

Standard reduction potential

This is the potential difference between a cathode and the solution into which it is dipped, measured under standard conditions. Its symbol is $E^{\ominus}_{\text{red}}$. It is impossible to measure an absolute value for $E^{\ominus}_{\text{red}}$, so data books give a value of the potential of the cathode with reference to the standard hydrogen electrode, which is assigned a cell potential value of 0.00 V. (The standard hydrogen electrode is discussed in more detail on page 508.) In a Daniell cell, the copper(II) ions are in equilibrium with the copper electrode:

$$Cu^{2+}(aq) + 2e^- \rightleftharpoons Cu(s)$$

The value of $E^{\ominus}_{\mathbf{red}}$ gives an indication of the position of this equilibrium. The greater its positive value, the more the position of equilibrium lies to the right (i.e. to the side of the reduced form).

Standard oxidation potential

This is the potential difference between an anode and the solution into which it is dipped, measured under standard conditions. Again, it is impossible to measure absolute values for $E^{\ominus}_{\mathbf{oxid}}$, so a value is used that is the potential of the anode with reference to the standard hydrogen electrode, which is assigned a cell potential value of 0.00 V. In a Daniell cell, the zinc ions are in equilibrium with the zinc anode:

$$Zn(s) \rightleftharpoons Zn^{2+}(aq) + 2e^-$$

The value $E^{\ominus}_{\text{oxid}}$ gives an indication of the position of this equilibrium. The greater its positive value, the more the position of equilibrium lies to the right (i.e. to the side of the oxidised form).

MEASURING ELECTRODE POTENTIALS

As already stated, it is not possible to measure absolute values for the oxidation and reduction potentials. However, if the potential for one of the processes is assigned a value, it is possible to determine a value for any other electrode potential. The chosen process is the redox reaction involving aqueous hydrogen ions and gaseous hydrogen under standard conditions (temperature 298 K, pressure of the hydrogen gas 101 kPa, concentration of the aqueous hydrogen ion 1.00 mol dm^{-3}). Both $E^{\ominus}_{\text{oxid}}$ and the $E^{\ominus}_{\text{red}}$ are arbitrarily given a value of zero. Therefore:

Oxidation:	$H_2(g)$	$\rightarrow 2H^+(aq) + 2e^-$	$E^{\ominus}_{\text{oxid}} = 0.00V$
Reduction:	$2H^+(aq) + 2e^-$	$\rightarrow H_2(g)$	$E^{\ominus}_{\text{red}} = 0.00V$

? QUESTION 3

3 You want to measure the standard reduction potential for the following reaction:

$$Cl_2(g) + 2e^- \rightarrow 2Cl^-(aq)$$

Write down the conditions that you must have to measure the standard reduction potential.

REMEMBER THIS

All electrochemical cells consist of two half-cells (one where reduction takes place and the other where oxidation takes place), connected by a salt bridge and an external wire.

Hence, all other electrode potentials are compared with either the oxidation potential of gaseous hydrogen or the reduction potential of aqueous hydrogen ions. Remember that the oxidation and reduction reactions are part of the reference equilibrium reaction:

$$2H^+(aq) + 2e^- \rightleftharpoons H_2(g)$$

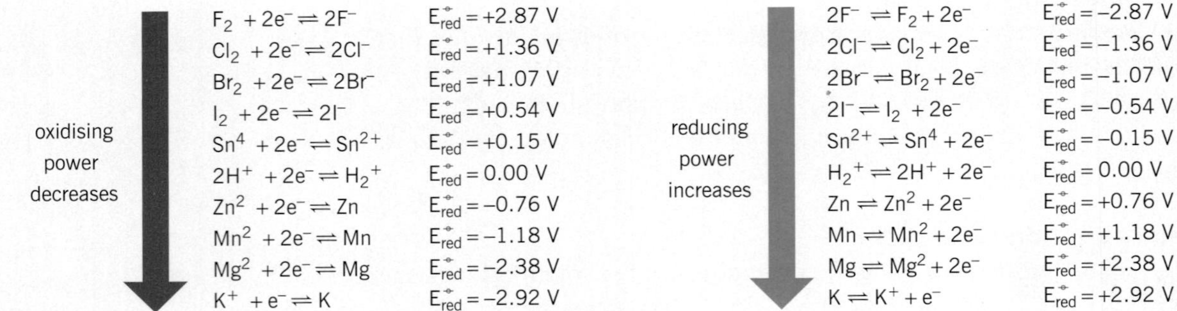

Fig 25.8 A summary of the values of electrode potentials

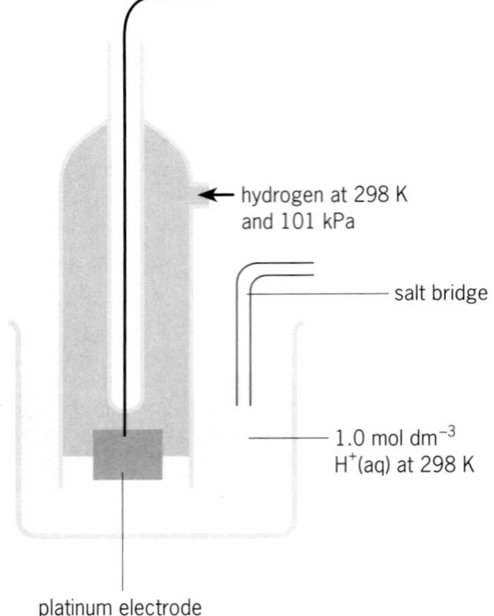

Fig 25.9 In the standard hydrogen electrode, hydrogen gas under a pressure of 101 kPa is in contact with aqueous hydrogen ions at a concentration of 1.00 mol dm^{-3}. The platinum electrode allows electrical current to enter and leave the hydrogen electrode. The platinum does not react with the hydrogen or the aqueous hydrogen ions. The aqueous hydrogen ions are normally provided by dilute sulfuric acid of the appropriate concentration

Standard hydrogen electrode

The standard hydrogen electrode is a half-cell that allows $H_2(g)$ to be in equilibrium with $H^+(aq)$. As an electrode the half-cell must also allow electric current to flow in and out. This poses a problem, because gaseous hydrogen is not an electrical conductor. The problem is overcome by using platinum foil (which is inert) as the conducting part of the electrode (Fig 25.9).

Using the standard hydrogen potential to measure other standard electrode potentials

Since the reference equilibrium reaction involves $H_2(g)$ and $H^+(aq)$, it is possible to determine the standard electrode potential of other half-cells just by combining them with a standard hydrogen electrode, as shown in Fig 25.10.

In Fig 25.10, the two half-equations are:

At the anode: $H_2(g) \rightarrow 2H^+(aq) + 2e^-$

At the cathode: $Cu^{2+}(aq) + 2e^- \rightarrow Cu(s)$

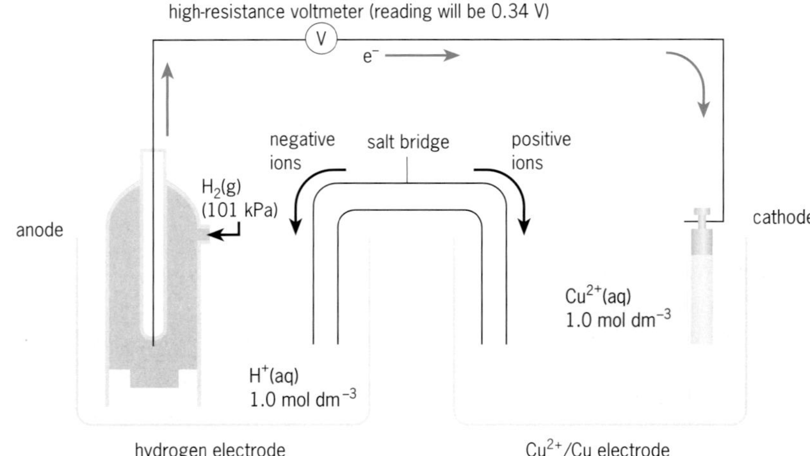

Fig 25.10 A hydrogen electrode is connected to a copper/copper(II) ion electrode using a salt bridge. A high-resistance voltmeter measures the electrode potential of the cell. The arrows show the electron flow from the hydrogen electrode to the copper electrode via the external wire (if the voltmeter were replaced by a lower-resistance component, e.g. a light bulb)

Remember: oxidation occurs at the anode and reduction at the cathode. The value of $E^{\ominus}_{cell}$ is +0.34 V, and:

$$E^{\ominus}_{cell} = E^{\ominus}_{oxid} + E^{\ominus}_{red}$$

Therefore:

$$+0.34 = 0.00 + E^{\ominus}_{red}$$

The $E^{\ominus}_{red}$ for the copper electrode is +0.34 V. So $E^{\ominus}_{oxid} = -0.34$ V. It is a general rule that, for the same electrode, the values of $E^{\ominus}_{red}$ and $E^{\ominus}_{oxid}$ have the same magnitude but are of opposite sign.

It is also worth stressing that the electrode potentials do not depend on the amounts of reactants and products involved in redox equations. So, for instance, the reduction potential for:

$$Cu^{2+}(aq) + 2e^- \rightarrow Cu(s)$$

is the same as that for:

$$2Cu^{2+}(aq) + 4e^- \rightarrow 2Cu(s)$$

CELL CONVENTION

It is cumbersome to keep drawing the electrodes used in electrochemical cells. Therefore, a system of notation has been devised whereby a whole cell can be described on a single line. The anode is written on the left and the cathode on the right. A single vertical bar distinguishes components that are in different phases: for example, a solid electrode and the aqueous ions with which it is in contact. The salt bridge is shown by two dashed vertical bars. Thus, the cell of Fig 25.10 can be represented by:

$$Pt(s) \mid H_2(g) \mid H^+(aq) \,\|\, Cu^{2+}(aq) \mid Cu(s)$$

Note that the half-cell with the greater positive value of $E^{\ominus}_{red}$ is put on the right.

For the Daniell cell, the notation is:

$$Zn(s) \mid Zn^{2+}(aq) \,\|\, Cu^{2+}(aq) \mid Cu(s)$$

OTHER WAYS TO MEASURE STANDARD ELECTRODE POTENTIALS

The standard hydrogen electrode does not always have to be used to measure the standard electrode potential of another half-cell. Because it is known for certain that a copper electrode has a standard reduction potential of 0.34 V, this can be used to determine the standard electrode potential of another half-cell.

An example of this method is illustrated in Fig 25.11. Magnesium is more reactive than copper. That is, it loses electrons more easily than copper. This makes the magnesium half-cell the anode, since this is where oxidation occurs. So, the notation for the cell in Fig 25.11 is:

$$Mg(s) \mid Mg^{2+}(aq) \,\|\, Cu^{2+}(aq) \mid Cu(s)$$

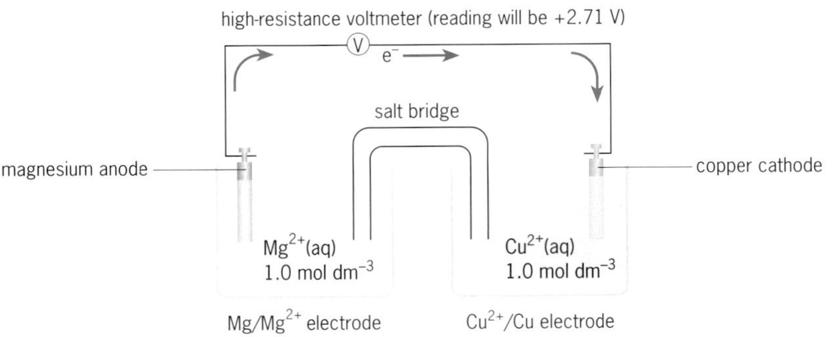

Fig 25.11 To determine the standard electrode potential for magnesium, two half-cells are required. One cell contains a copper strip and 1.0 mol dm^{-3} copper(II) ions, and the other a magnesium strip and 1.0 mol dm^{-3} magnesium ions. A high-resistance voltmeter records the cell potential, from which the electrode potential of magnesium can be determined:

$$E^{\ominus}_{cell} = E^{\ominus}_{oxid} + E^{\ominus}_{red}$$
$$+2.71 = E^{\ominus}_{oxid} + 0.34$$

So:

$$E^{\ominus}_{oxid} = +2.37V$$

It is possible to experimentally determine any electrode potential, provided a suitable cell is set up. The blue arrows show the direction of electron flow if the voltmeter were replaced by a lamp

? QUESTION 4

4 What concentration of dilute sulfuric acid would be appropriate to use in a standard hydrogen electrode?

? QUESTION 5

5 Write the following electrochemical cells using the cell convention. In each case predict the direction of the flow of electrons in the external circuit.

a) A cell made up of a copper electrode and a silver electrode.

b) A cell made up of a hydrogen electrode and a zinc electrode.

REMEMBER THIS

Sometimes a comma is used to separate components that are in the same phase.

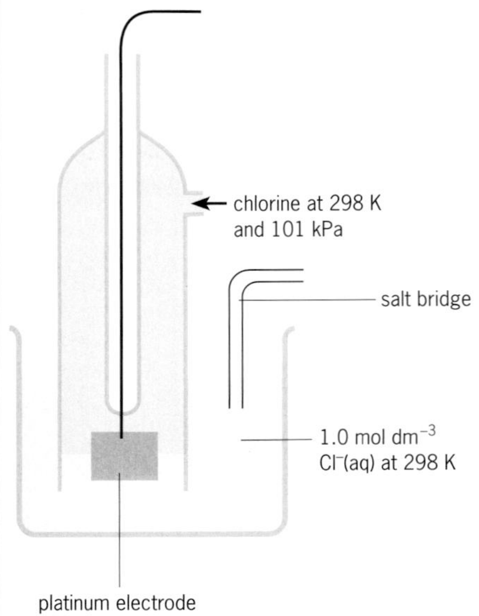

Fig 25.12 A chlorine electrode consists of chlorine gas at 101 kPa, aqueous chloride ions at a concentration of 1.0 mol dm^{-3} and an inert platinum electrode to conduct current into or out of the solution

ELECTRODES THAT INVOLVE GASES OR SOLUTIONS

Many redox half-reactions do not involve a metal, but it is still possible to construct a half-cell using the reagents shown in the half-equation. For example, with the aid of gas electrodes, gases can be involved in redox half-equations. Since gases do not conduct electricity, an inert metal (usually platinum) has to be used as the conducting part of the gas electrode. Platinum does not react with dilute acids or most aqueous solutions. For example, the chlorine electrode shown in Fig 25.12 allows $Cl_2(g)$ and $Cl^-(aq)$ to be in equilibrium:

$$Cl_2(g) + 2e^- \rightleftharpoons 2Cl^-(aq)$$

Some half-cells involve only solutions. Take, for example, a half-cell that involves the acidified manganate(VII) ions. It must allow the following equilibrium to be set up:

$$MnO_4^-(aq) + 8H^+(aq) + 5e^- \rightleftharpoons Mn^{2+}(aq) + 4H_2O(l)$$

None of the species given in the equation can act as the electrode. Therefore, a platinum electrode is again needed. Despite the stoichiometry of the equation, all the concentrations of the aqueous species should be 1.0 mol dm^{-3}, the temperature 298 K and the pressure 101 kPa, for the half-cell to be at standard conditions.

3 USING STANDARD ELECTRODE POTENTIALS

Knowledge of electrode potentials allows predictions to be made about oxidising and reducing agents, and about electrolysis products. It also allows speculation on the feasibility of redox reactions.

When the external circuit in an electrochemical cell is closed, electrons flow from the anode to the cathode. This is always the direction of flow of electrons. The reason is that oxidation, accompanied by the release of electrons, always takes place at the anode. The electrons then travel along the external circuit until, at the cathode, they are used in the reduction process.

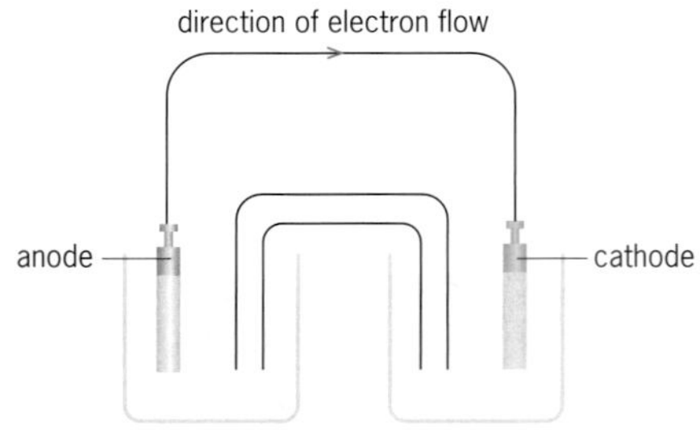

Fig 25.13 A summary of electron flow in an electrochemical cell

COMPARING OXIDISING AGENTS

To compare oxidising agents, it is necessary to look at the $E^{\ominus}_{red}$ of the species involved. This is because an oxidising agent must accept electrons so as to be reduced. Since the figures are all comparative, the most powerful oxidising agent is the species with the highest positive (or lowest negative) $E^{\ominus}_{red}$. (The significance of the sign and size of electrode potentials is summarised in Fig 25.8, page 508.)

EXAMPLE

Q Which is the more powerful oxidising agent, F_2 or Cl_2?

A The relevant reduction half-equations and electrode potentials (given in Appendix 4) are:

$2F^- \rightleftharpoons F_2 + 2e^-$ $\quad E^{\ominus}_{red} = +2.87$ V

$2Cl^- \rightleftharpoons Cl_2 + 2e^-$ $\quad E^{\ominus}_{red} = +1.36$ V

So fluorine is the more powerful oxidising agent.

COMPARING REDUCING AGENTS

To compare reducing agents, it is necessary to look at the $E^{\ominus}_{oxid}$ of the species involved. This is because a reducing agent must give away electrons so as to be oxidised. Since the figures are all comparative, the most powerful reducing agent is the species with the highest positive (or lowest negative) value of $E^{\ominus}_{oxid}$.

EXAMPLE

Q Which is the better reducing agent, I^- or Cl^-?

A The relevant oxidation half-equations and electrode potentials (given in Appendix 4) are:

$2I \rightleftharpoons I_2 + 2e \qquad E^{\ominus}_{oxid} = -0.54\ V$

$2CI \rightleftharpoons CI_2 + 2e \qquad E^{\ominus}_{oxid} = -1.36\ V$

So the iodide ion is the better reducing agent, since its $E^{\ominus}_{oxid}$ has a lower negative value than that of the chloride ion.

? QUESTION 6

6 In each of the following pairs, which is the better oxidising agent:

a) Cu^{2+} or Cr^{2+}

b) Acidified MnO_4^- or acidified $Cr_2O_7{}^{2-}(aq)$?

? QUESTION 7

7 In each of the following pairs, decide which is the better reducing agent:

a) Ni or Sn^{2+}

b) F^- or Au.

ELECTROCHEMICAL SERIES AND ELECTRODE POTENTIALS

Metals are reducing agents because they lose electrons. The most reactive metals lose electrons easily, whereas the least reactive metals lose electrons with difficulty. The electrochemical series ranks metals in order of their reactivities, with the most reactive metal at the top. Since the most reactive metals are also the best reducing agents, electrode potentials can be used to deduce the electrochemical series, as Table 25.1 shows.

Note that there is a good correlation between the position of the metal in the electrochemical series and its $E^{\ominus}_{oxid}$.

Table 25.1 Electrochemical series

Metal	Half-equation	Reduction potential $E^{\ominus}_{red}$/V	Oxidation potential $E^{\ominus}_{oxid}$/V
potassium	$K^+ + e^- \rightleftharpoons K$	−2.92	+2.92
calcium	$Ca^{2+} + 2e^- \rightleftharpoons Ca$	−2.87	+2.87
magnesium	$Mg^{2+} + 2e^- \rightleftharpoons Mg$	−2.38	+2.38
aluminium	$Al^{3+} + 3e^- \rightleftharpoons Al$	−1.66	+1.66
zinc	$Zn^{2+} + 2e^- \rightleftharpoons Zn$	−0.76	+0.76
iron	$Fe^{2+} + 2e^- \rightleftharpoons Fe$	−0.44	+0.44
lead	$Pb^{2+} + 2e^- \rightleftharpoons Pb$	−0.13	+0.13
hydrogen	$2H^+ + 2e^- \rightleftharpoons H_2$	0.00	0.00
copper	$Cu^{2+} + 2e^- \rightleftharpoons Cu$	+0.34	−0.34
silver	$Ag^+ + e^- \rightleftharpoons Ag$	+0.80	−0.80

REMEMBER THIS

For the same redox system, $E_{red} = -E_{oxid}$

PREDICTING THE PRODUCTS OF ELECTROLYSIS

As mentioned on page 504, electrolysis involves two half-reactions, which are known as electrode reactions. In solutions that contain several ions, it can be difficult to predict which ion will react at the anode and which at the cathode.

? QUESTION 8

8 **a)** Predict the products of the electrolysis of aqueous zinc nitrate solution using inert electrodes of graphite.
b) The electrolysis of concentrated aqueous sodium chloride with inert electrodes gives chlorine and hydrogen. Is this what you would expect from inspection of the relevant electrode potentials? If it is not, suggest possible reasons for the discrepancy.

Since oxidation always occurs at the anode, it is possible to compare the relevant $E^{\ominus}_{oxid}$ to predict which ion is most likely to react there. In the same way, it is possible to predict which ion will react at the cathode by comparing the relevant $E^{\ominus}_{red}$. There is, however, one major drawback with this approach, which is that the electrode potentials are quoted for standard conditions, which involve concentrations of 1.0 mol dm^{-3} of aqueous ionic species. Therefore, unless these are the conditions of electrolysis, there will always be some doubt about accuracy of the prediction. On page 513 the effect of concentration on electrode potential is considered, and the same approach can be used to make better predictions about the electrolysis products of an aqueous solution.

EXAMPLE

Q Predict the products of the electrolysis of aqueous copper(II) nitrate using carbon electrodes.

A Ions present: $Cu^{2+}(aq)$ and $NO_3^-(aq)$, and from water $H^+(aq)$ and $OH^-(aq)$.

Ions attracted to cathode: $Cu^{2+}(aq)$ and $H^+(aq)$

Possible cathode reactions: $Cu^{2+}(aq) + 2e^- \rightarrow Cu(s)$ $E^{\ominus}_{red} = +0.34V$
$2H^+(aq) + 2e^- \rightarrow H_2(g)$ $E^{\ominus}_{red} = 0.00$ V

By inspection, the more favourable reduction involves copper(II) ions, since its electrode potential has the higher positive value.

Ions attracted to anode: $OH^-(aq)$ and $NO_3^-(aq)$

Anode reaction: $4OH^-(aq) \rightarrow O_2(g) + 2H_2O(l) + 4e^-$ $E^{\ominus}_{oxid} = 0.40$ V

The nitrate ion is attracted to the anode, but is not oxidised.

CALCULATING CELL POTENTIALS

When any simple cell is set up, its cell potential, $E^{\ominus}_{cell}$ can be calculated from the electrode potentials of the two half-reactions taking place. $E^{\ominus}_{cell}$ is given by:

$$E^{\ominus}_{cell} = E^{\ominus}_{oxid} + E^{\ominus}_{red}.$$

EXAMPLE

Q Work out the cell potential for the following cell:

$$Co(s) \mid Co^{2+}(aq) \parallel Cr^{3+}(aq) \mid Cr(s)$$

A By the cell convention, the anode is on the left side of the cell, so this is where the oxidation occurs:

$$Co(s) \rightarrow Co^{2+}(aq) + 2e^-$$

From the table of electrode potentials, $E^{\ominus}_{oxid} = +0.28V$ (since $E^{\ominus}_{red} = -0.28V$ for $Co^{2+}(aq) + 2e^- \rightarrow Co(s)$).

At the cathode, reduction takes place, so the half-reaction must involve the gain of electrons:

$$Cr^{3+}(aq) + 3e^- \rightarrow Cr(s)$$

From the table of electrode potentials, $E^{\ominus}_{red} = -0.74$ V.

Since $E^{\ominus}_{cell} = E^{\ominus}_{oxid} + E^{\ominus}_{red}$.

$E^{\ominus}_{cell} = +0.28 + (-0.78) = -0.50$ V

Q Calculate the cell potential of the cell that has the following overall reaction:

$$Zn + I_2 \rightarrow Zn^{2+} + 2I^-$$

A The oxidation half-equation is:

$$Zn \rightarrow Zn^{2+} + 2e^-$$

for which $E^{\ominus}_{oxid} = +0.76$ V

The reduction half-equation is:

$$I_2 + 2e^- \rightarrow 2I^-$$

for which $E^{\ominus}_{red} = +0.54$ V

Using $E^{\ominus}_{cell} = E^{\ominus}_{oxid} + E^{\ominus}_{red}$ gives:

$E^{\ominus}_{cell} = +0.76 + 0.54 = +1.30$ V

The values of some reduction potentials are given in Appendix 4. The values and those quoted in most data books are standard electrode potentials. Therefore, any cell potential calculated from them will be a standard cell potential. Most data books list only the reduction half-equations together with their $E^{\ominus}_{red}$ values. The oxidation half-equation is the reverse of the reduction half-equation and the corresponding $E^{\ominus}_{oxid}$ values are obtained by changing the sign of the $E^{\ominus}_{red}$ values.

If you wish to read more about the quantitative effects of changing temperature, pressure and concentration on electrode potentials, you should look up the Nernst equation in a more advanced textbook.

ELECTRODE POTENTIALS UNDER NON-STANDARD CONDITIONS

So far, the assumption has been that the reactions occur under standard conditions. That is, all solutions are 1.0 mol dm^{-3} in concentration, the temperature is 298 K and the pressure is 101 kPa. This is not the case for most reactions. Even if standard temperature and pressure could be maintained, as soon as the cell reaction starts, the concentrations of the reactants and products would necessarily change. The effect of changing conditions can be predicted qualitatively by assuming that the reaction taking place is an equilibrium and applying Le Chatelier's principle.

Temperature and cell potential

If the cell reaction as written is exothermic, then as the temperature increases the position of equilibrium shifts to the left. This means that the cell potential assumes a lower positive (or larger negative) value.

Concentration and cell potential

If the concentrations of the reactants are increased, then the position of equilibrium shifts to the right and the cell potential assumes a higher positive (or lower negative) value. If the concentration of one of the products increases, then the opposite effect is observed and the cell potential assumes a lower positive (or larger negative) value.

Concentration and electrode potential

Le Chatelier's principle can also be applied to half-cells, to see how $E^{\ominus}_{red}$ or $E^{\ominus}_{oxid}$ values are affected. Consider the equilibrium between hydrogen ions and hydrogen:

$2H^+(aq) + 2e^- \rightleftharpoons H_2(g)$ Under standard conditions $E^{\ominus}_{red} = E^{\ominus}_{oxid} = 0.00$ V

If the concentration of the aqueous hydrogen ion is increased, the position of the equilibrium shifts to the right and the reduction potential becomes positive. At the same time, the oxidation potential for the reaction becomes negative.

Pressure and cell potential

A change in pressure affects only those cell reactions that involve a gas. Essentially, increasing the pressure of a gas increases its concentration. Therefore, the effect of increasing the pressure of a gaseous reactant or a product on cell potential is the same as that with increasing the appropriate concentration.

Consider the hydrogen half-cell equilibrium again:

$2H^+(aq) + 2e^- \rightleftharpoons H_2(g)$ Under standard conditions $E^{\ominus}_{oxid} = E^{\ominus}_{red} = 0.00$ V

If the pressure of the hydrogen is increased, the position of the equilibrium shifts to the left. Therefore, the oxidation potential becomes positive and the reduction potential negative.

? QUESTION 9

9 Consider the following cell reaction:
$Zn(s) + 2H^+(aq)$ $Zn^{2+}(aq) + H_2(g)$
a) Calculate the standard cell potential for this reaction.
b) Predict qualitatively the effect on $E^{\ominus}_{cell}$ of a cell with:
i) $[Zn^{2+}(aq)]$ = 2.0 mol dm^{-3} $[H^+(aq)]$ = 1.0 mol dm^{-3} and pressure of H_2 = 101 kPa
ii) $[Zn^{2+}(aq)]$ = 1.0 mol dm^{-3} $[H^+(aq)]$ = 2.0 mol dm^{-3} and pressure of H_2 = 101 kPa
iii) $[Zn^{2+}(aq)]$ = 1.0 mol dm^{-3} $[H^+(aq)]$ = 1.0 mol dm^{-3} and pressure of H_2 = 202 kPa

There is more information on spontaneous reactions on pages 164 and 536.

? QUESTION 10

10 **a)** Explain whether each one of the following reactions will occur spontaneously:

i) $Zn^{2+} + Mg \rightarrow Mg^{2+} + Zn$

ii) $2Fe^{3+} + 2F^{-} \rightarrow 2Fe^{2+} + F_2$

b) Work out the cell potential for the following cell:

$Mn(s) \mid Mn^{2+}(aq) \parallel Fe^{2+}(aq) \mid Fe(s)$

Hence write down the overall cell reaction for the spontaneous reaction that occurs.

? QUESTION 11

11 Predict the direction of the electron flow in each of the following electrochemical cells, given the overall cell reaction:

a) $Cu + Br_2 \rightarrow Cu^{2+} + Br^{-}$

b) $Cu^{2+} + Cr \rightarrow Cu + Cr^{2+}$

REMEMBER THIS

Remember: electrons have a negative charge.

4 FEASIBILITY OF REACTIONS AND ELECTRODE POTENTIALS

The absolute value of the cell potential determines the feasibility of the reaction. When the cell potential is positive, the reaction is spontaneous. When it is negative, the reaction is not spontaneous. The description 'spontaneous' refers to the tendency of a reaction to occur. It says nothing about the rate of reaction. So it is possible for a spontaneous reaction to be so slow as not to occur. In an electrochemical cell, if $E^{\ominus}_{cell}$ is positive, the reaction is spontaneous; if $E^{\ominus}_{cell}$ is negative, the reaction is not spontaneous.

ELECTRON FLOW IN THE EXTERNAL CIRCUIT

As already stated, the external flow of electrons is from the anode to the cathode. The cell potential can be used to identify the anode. Provided the cell potential is positive, the cell works spontaneously and the anode is the electrode at which the oxidation occurs.

One battery being developed uses liquid sulfur and sodium. The overall cell reaction is:

$$2Na + S \rightarrow 2Na^{+} + S^{2-}$$

The two half-equations are:

$Na \rightarrow Na^{+} + 2e^{-}$ $\quad E^{\ominus}_{oxid} = +2.71\ V$

$S + 2e^{-} \rightarrow S^{2-}$ $\quad E^{\ominus}_{red} = +0.14\ V$

So, the cell potential is +2.85 V. This suggests that the cell reaction is proceeding in the correct direction. Therefore, electrons flow from the sodium electrode to the sulfur electrode.

Now, consider a cell that has a theoretical overall cell reaction of:

$$2Ag + Cu^{2+} \rightarrow 2Ag^{+} + Cu$$

The half-equations are:

$Ag \rightarrow Ag^{+} + e^{-}$ $\quad E^{\ominus}_{oxid} = -0.80\ V$

$Cu^{2+} + 2e^{-} \rightarrow Cu$ $\quad E^{\ominus}_{red} = +0.34\ V$

So $E^{\ominus}_{cell} = -0.46\ V$ which indicates that, as written, the reaction will not proceed spontaneously. If the equation is reversed, then the cell potential becomes positive, which indicates that electrons will flow from the copper electrode to the silver electrode:

$$Cu(s) + 2Ag^{+}(aq) \rightarrow Cu^{2+}(aq) + 2Ag(s) \qquad E^{\ominus}_{cell} = +0.46\ V$$

CELL POTENTIAL AND FREE ENERGY

Cells are a source of electrical energy that can be used to do work. A potential difference of 1 volt imparts 1 joule of energy to a charge of 1 coulomb. If the energy is used to do work, such as running an electric motor, then:

work = charge × potential difference

In this case, the potential difference is the cell potential. Since the cell potential is necessarily positive in order to produce a current, the work done is negative (because charge is negative). This means that work is done on the surroundings.

Consider now the amount of work done when the molar quantities as shown in the overall cell reaction are used:

work = (charge on number of electrons transferred during cell reaction) × $E^{\ominus}_{cell}$

The charge on the number of electrons transferred is the charge on n moles of electrons, where n is the number of moles transferred in the cell reaction. Therefore:

$$\text{work} = -nFE^{\ominus}_{cell}$$

where n is the number of electrons transferred in the cell reaction, and F is the Faraday constant, which is the charge on 1 mole of electrons in coulombs.

The work done according to this equation is the maximum amount of work that can be done. (In reality, some electrical energy is transferred by the heating of the wires.) The maximum amount of work available from a process carried out under standard conditions is known as the free-energy change, ΔG (G was chosen to celebrate the American Willard Gibbs, who developed the concept of free energy). So, in an electrochemical cell:

$$\Delta G = -nFE^{\ominus}_{cell}$$

The unit of ΔG is J mol^{-1}.

REMEMBER THIS

The Faraday constant F is 96 500 coulombs mol^{-1}.

EXAMPLE

Q Calculate the free-energy change for an electrochemical cell based on the following cell reaction:

$$Mn(s) + Cd^{2+}(aq) \rightarrow Mn^{2+}(aq) + Cd(s)$$

A From the overall reaction, the two half-equations are:

Oxidation half-equation: $Mn(s) \rightarrow Mn^{2+}(aq) + 2e^-$ $E^{\ominus}_{oxid} = +1.18$ V
Reduction half-equation: $Cd^{2+}(aq) + 2e^- \rightarrow Cd(s)$ $E^{\ominus}_{red} = -0.40$ V

Since $E^{\ominus}_{cell} = E^{\ominus}_{oxid} + E^{\ominus}_{red}$, then:

$E^{\ominus}_{cell} = +1.18 - 0.40 = +0.78$ V

From the two half-equations, it can be seen that 2 moles of electrons are transferred during the reaction.
So: $\Delta G = -nFE^{\ominus}_{cell}$

$\Delta G = -2 \times 96\,500 \times 0.78$ J mol^{-1}
$= -150.5$ kJ mol^{-1}

? QUESTION 12

12 Calculate the free-energy changes, ΔG, for each of the following reactions:

a) $Cd(s) + Pb^{2+}(aq) \rightarrow Cd^{2+}(aq) + Pb(s)$

b) $2Fe^{3+}(aq) + Cu(s) \rightarrow 2Fe^{2+}(aq) + Cu^{2+}(aq)$

c) $Cl_2(g) + 2Br^-(aq) \rightarrow Br_2(l) + 2Cl^-(aq)$

FREE ENERGY AND FEASIBILITY OF REACTIONS

Just as the cell potential must be positive for a reaction to occur spontaneously, so the value of ΔG must be negative. The free-energy change of a reaction can be used in non-electrochemical cells to predict the feasibility of reactions. (On page 533 is a description of the application of free-energy changes in predicting the feasibility of the reduction of some metal oxides.) Any discussion of the feasibility of reactions must be taken with caution, since the $E^{\ominus}_{cell}$ or ΔG values do not contain any information about the rate of reaction. Therefore, a reaction may have a large negative ΔG, but still not take place because its activation energy is too large.

Activation energy is covered on pages 163–166.

? QUESTION 13

13 Predict whether the following reactions will take place spontaneously. In each case, work out the cell potential and the free-energy change for the reaction.

a) $Cl_2(g) + 2F^-(aq) \rightarrow F_2(g) + 2Cl^-(aq)$

b) $MnO_4^- + 5Cr^{2+}(aq) + 8H^+(aq) \rightarrow Mn^{2+}(aq) + 5Cr^{3+}(aq) + 4H_2O(l)$

EXAMPLE

Q Predict whether the following cell reaction should proceed spontaneously:

$$Cr^{3+}(aq) + Fe^{2+}(aq) \rightarrow Cr^{2+}(aq) + Fe^{3+}(aq)$$

A From the overall reaction, the two half-equations are:

Oxidation half-equation:
$Fe^{2+}(aq) \rightarrow Fe^{3+}(aq) + e^-$ $E^\ominus_{oxid} = -0.77$ V

Reduction half-equation:
$Cr^{3+}(aq) + e^- \rightarrow Cr^{2+}(aq)$ $E^\ominus_{red} = -0.41$ V

Since the cell potential is given by $E^\ominus_{cell} = E^\ominus_{oxid} + E^\ominus_{red}$, then

$$E^\ominus_{cell} = -0.77 - 0.41 = -1.18 \text{ V}$$

A negative cell potential predicts that the reaction should not take place spontaneously.

The free-energy change for this reaction can also be calculated to show that it is positive:

$$\Delta G = -nFE^\ominus_{cell}$$
$$= -1 \times 96\,500 \times (-1.18) \text{ J mol}^{-1}$$
$$= +114 \text{ kJ mol}^{-1}$$

The positive value of the free energy change is another indicator that the reaction will not take place spontaneously.

Fig 25.14 Batteries containing lithium chloride are often used in torches

5 BATTERIES AND FUEL CELLS

A battery is an electrochemical cell that is used as an energy source. Some kinds of battery have several electrochemical cells in series, and thereby combine the individual cell potentials to give a large potential at the output terminals. Batteries are normally designed to deliver a reasonably large current and to be able to withstand fairly rough handling.

Batteries are of two types: primary cells and secondary cells.

PRIMARY CELLS

Primary cells cannot be re-charged. They are the most familiar type of battery on sale today.

LECLANCHÉ OR DRY CELL

The dry cell (Fig 25.15) consists of a zinc anode (also the cell's casing) and a graphite cathode. A paste of ammonium chloride, manganese(IV) oxide and zinc chloride surrounds the graphite electrode. The porous separator acts as a salt bridge.

When the cell is connected into an external circuit, electrons flow from the zinc anode to the graphite cathode:

Oxidation half-equation:

$Zn \rightarrow Zn^{2+} + 2e^-$

Reduction half-equation:

$MnO_2 + NH_4^+ + e^- \rightarrow MnO(OH) + NH_3$

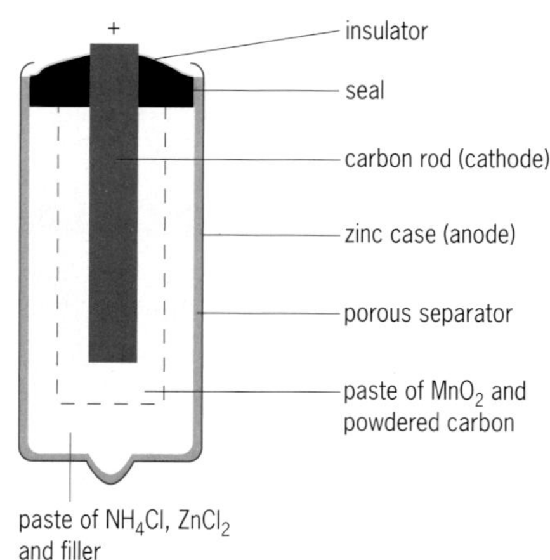

Fig 25.15 A cross-section through a dry-cell battery

The cell potential is about 1.5 V, which is impossible to verify using standard electrode potentials because the cell reaction occurs under non-standard conditions. The zinc ions present in the paste prevent ammonia being given off by forming an ammine complex, $[Zn(NH_3)_4]^{2+}$. In addition to the nuisance of having to replace this type of battery at frequent intervals, there is the disadvantage that the acidic ammonium chloride corrodes the zinc casing, even when the battery is not being used.

SECONDARY CELLS

These are batteries that can be recharged and therefore have a much longer life than primary cells. A typical rechargeable battery is the lead–acid battery used in most motor vehicles (Figs 25.16 and 25.17).

The lead-acid battery consists of an aqueous electrolyte (30 per cent by volume sulfuric acid) and two sets of plates which form the electrodes. The positive plates are lead grids filled with a paste of spongy lead. The negative plates are lead grids filled with a paste of lead(IV) oxide. At first, the lead reacts with the dilute sulfuric acid to form an insoluble coating of lead(II) sulfate, on both sets of plates:

Fig 25.16 Topping up a lead–acid car battery with distilled water

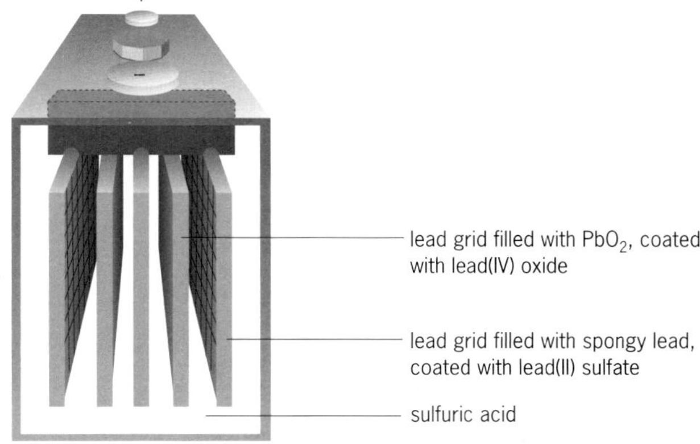

Fig 25.17 A lead–acid storage battery

$$Pb(s) + H_2SO_4(aq) \rightarrow PbSO_4(s) + H_2(g)$$

The battery is initially charged by direct current. During the charging process, the plates become chemically different. A layer of lead(IV) oxide is deposited on each negative plate, but the positive plates retain their layer of lead(II) sulfate. The charging half-equations are:

Oxidation half-equation:

$$PbSO_4(s) + 2e^- \rightarrow Pb(s) + SO_4^{2-}(aq)$$

Reduction half-equation:

$$PbSO_4(s) + 2H_2O(l) \rightarrow PbO_2(s) + 4H^+(aq) + SO_4^{2-}(aq) + 2e^-$$

The overall reaction during charging is:

$$2PbSO_4(s) + 2H_2O(l) \rightarrow Pb(s) + PbO_2(s) + 4H^+(aq) + 2SO_4^{2-}(aq)$$

Electrolysis takes place during charging.

Once the battery is charged, the half-equations can be reversed to provide an external current or flow of electrons. The discharging half-equations are:

Oxidation half-equation:

$$Pb(s) + SO_4^{2-}(aq) \rightarrow PbSO_4(s) + 2e^-$$

Reduction half-equation:

$$PbO_2(s) + 4H^+(aq) + SO_4^{2-}(aq) + 2e^- \rightarrow PbSO_4(s) + 2H_2O(l)$$

? QUESTION 14

14 Car batteries are normally described as 12 V batteries. They contain six individual cells, each connected in series.

a) Estimate the cell potential of one lead–acid electrochemical cell.

b) Use the electrode potentials in Appendix 4 to estimate the cell potential of a lead–acid electrochemical cell.

c) Comment on your answers to parts a) and b).

The overall discharging reaction is:

$$Pb(s) + PbO_2(s) + 4H^+(aq) + 2SO_4^{2-}(aq) \rightarrow 2PbSO_4(s) + 2H_2O(l)$$

Note that sulfuric acid is used up during the discharging process and so its concentration decreases. This provides a convenient test for a lead–acid battery. Being proportional to concentration, the density of the sulfuric acid in the battery is simply measured with a hydrometer.

FUEL CELLS

A fuel cell is an electrochemical cell designed so that the reactants, often gases, are replenished continuously. This enables the cell to supply electric current continuously without recharging.

A fuel cell is a much more efficient way of using a fuel to generate electrical energy. An electrical power station that burns a fuel is a very inefficient way of converting chemical energy into electrical energy. There have to be several energy transfer processes, all of which will reduce the efficiency of the overall energy transfer. With a fuel cell chemical energy is transferred directly into electrical energy. The hydrogen–oxygen fuel cell is one of the most promising fuel cells being developed. The only chemical product from the fuel cell is non-polluting water; however, disposal of damaged or old fuel cells could cause heavy-metal pollution. One potential drawback at present with the hydrogen–oxygen fuel cell concerns the availability, transport and storage of liquid hydrogen. People think of the Hindenburg airship disaster in 1936 and they remain sceptical that the necessary safety measures can be put in place to

SCIENCE IN CONTEXT

Hydrogen–oxygen fuel cell

A spacecraft needs a continuous supply of electrical energy. Much of it is provided by a hydrogen–oxygen fuel cell. Essentially, the overall cell reaction is that of oxygen and hydrogen to form water:

$$2H_2(g) + O_2(g) \rightarrow 2H_2O(l)$$

This is a redox reaction, in which electrons are transferred from hydrogen to oxygen. In a fuel cell, this transfer takes place via an external circuit, rather than directly during a collision between particles.

The electrolyte is aqueous sodium hydroxide. This is contained within the cell using porous electrodes, which allow the passage of water, hydrogen and oxygen. During their passage through the cathode, oxygen molecules are reduced to hydroxide ions, which enter the electrolyte.

Reduction half-equation:

$$O_2(g) + 2H_2O(l) + 4e^- \rightarrow 4OH^-(aq)$$

During their passage through the anode, hydrogen molecules are oxidised to give water molecules, and thereby hydroxide ions are removed from the electrolyte.

Oxidation half-equation:

$$H_2(g) + 2OH^-(aq) \rightarrow 2H_2O(l) + 2e^-$$

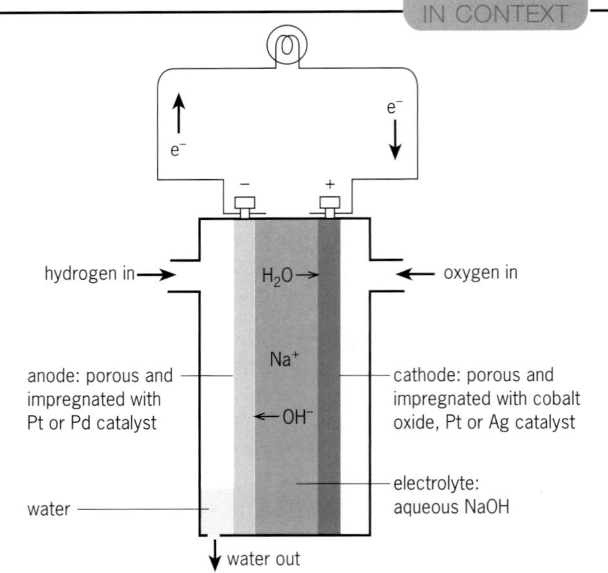

Fig 25.18 A hydrogen–oxygen fuel cell

? QUESTION 15

15 Construct the overall cell reaction that takes place within the hydrogen–oxygen fuel cell to convince yourself that all that is happening is the reaction of hydrogen and oxygen to make water.

ensure the safe use of hydrogen. No doubt in the future the use of hydrogen will become acceptable and fuel cells will be widely used.

Fuel cells do not have to rely just on hydrogen and oxygen and research is proceeding to use methanol–oxygen fuel cells and even hydrocarbon–oxygen ones. Again the fuel cell comprises two half reactions producing a current without the need for combustion.

6 ELECTRODE POTENTIALS AND TRANSITION ELEMENTS

The first-row transition elements are a series of metals that share several key characteristics. One of these is that they can have more than one oxidation state in their compounds. This is related to the electron configurations of these elements. The existence of several oxidation states of the transition elements is described fully on pages 484–486.

REDOX REACTIONS AND TRANSITION ELEMENTS

Since transition elements can have several different oxidation states, it is possible for their compounds, and the elements themselves, to take part in redox reactions. That is, the lower oxidation states of a transition element can be converted into one of the higher states by oxidation, and its higher oxidation states can be converted into one of the lower states by reduction. These changes necessarily involve the loss and gain of electrons, so they must be redox reactions.

It is shown on page 507 how $E^{\ominus}_{red}$ and $E^{\ominus}_{oxid}$ can be used to explain redox reactions. So, it follows that the chemistry of transition elements and their compounds involves electrode potentials.

7 CHEMISTRY OF SIX FIRST-ROW TRANSITION ELEMENTS

The first-row transition elements start with titanium and finish with copper. (See page 480 for the definition of a transition element.) The important redox reactions of six of these elements are the subject of this section, particular attention being paid to the electrode potentials involved.

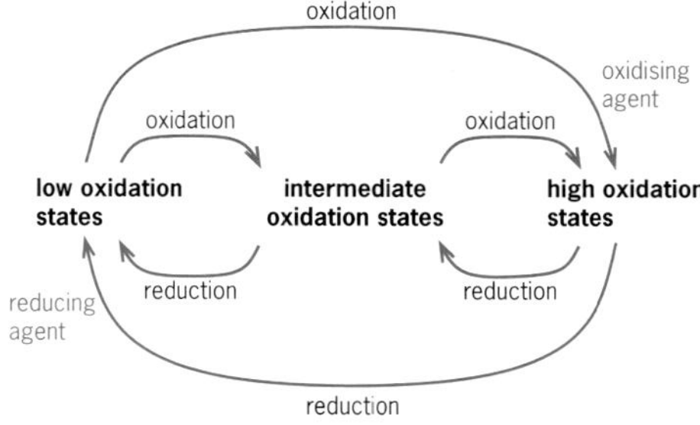

Fig 25.19 A summary of redox reactions of the transition elements

VANADIUM AND ITS COMPOUNDS

Table 25.2 Electrode potentials of some vanadium species

Reduction half-equation	$E^{\ominus}_{red}$/V
$V^{2+}(aq) + 2e^- \rightleftharpoons V(s)$	−1.18
$V^{3+}(aq) + e^- \rightleftharpoons V^{2+}(aq)$	−0.26
$VO^{2+}(aq) + 2H^+(aq) + e^- \rightleftharpoons V^{3+}(aq) + H_2O(l)$	+0.34
$VO_2^+(aq) + 2H^+(aq) + e^- \rightleftharpoons VO^{2+}(aq) + H_2O(l)$	+1.00

Vanadium exhibits four common oxidation states, +2, +3, +4 and +5. Under ordinary conditions, vanadium(IV) is considered to be the most stable oxidation state. Vanadium in its lower oxidation states is a reducing agent and in the highest oxidation state it is an oxidising agent. Therefore the chemistry of vanadium compounds typically involves a conversion from one oxidation state into another.

Vanadium in the +5 oxidation state

In the +5 oxidation state, the bonding to the vanadium atom is covalent in character, since it is impossible for a vanadium atom to lose five electrons to form V^{5+}. Probably the most important compound in this oxidation state is vanadium(V) oxide, which is used as a catalyst in the manufacture of sulfuric acid in the contact process.

The aqueous chemistry of vanadium(V) is centred around the vanadyl(V) ion, VO_2^+ (aq), and the vanadate(V) ion, VO_3^- (aq).

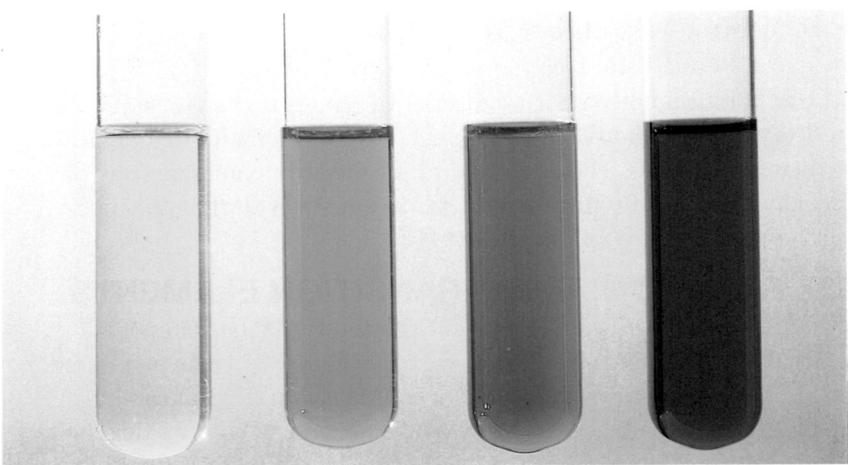

Fig 25.20 When the yellow vanadyl(V) ion, VO_2^+(aq), is reduced by zinc, the vanadium is reduced through the +4 and +3 oxidation states, and finally forms the +2 oxidation state. The reduction can be followed by the changes in colour of the solution: +5 yellow, +4 blue, +3 green and +2 purple

Since vanadium(V) is the highest oxidation state of vanadium, one would expect it to be highly oxidising, and to be reduced to one of the lower oxidation states. For example, zinc can reduce VO_2^+(aq) to V^{2+}(aq) in a series of steps that involves the formation of different coloured solutions as the oxidation state of vanadium changes. Each of these steps can be explained by use of the appropriate electrode potential.

The first step is the reduction of vanadyl(V) ion, VO_2^+(aq):

$$Zn(s) \rightarrow Zn^{2+}(aq) + 2e^- \qquad E^\ominus_{oxid} = +0.76\ V$$

$$VO_2^+(aq) + 2H^+(aq) + e^- \rightarrow VO^{2+}(aq) + H_2O(l) \qquad E^\ominus_{red} = +1.00\ V$$

Therefore, $E^\ominus_{cell} = +1.76\ V$

The overall equation can be obtained by combining the two half-equations:

$$Zn(s) + 2VO_2^+(aq) + 4H^+(aq) \rightarrow 2VO^{2+}(aq) + 2H_2O(l) + Zn^{2+}(aq)$$

The second step is the reduction of vanadyl(IV) ion, VO^{2+}(aq):

$$VO^{2+}(aq) + 2H^+(aq) + e^- \rightarrow V^{3+}(aq) + H_2O(l)\ 0.34 \qquad E^\ominus_{red} = +0.34V$$

Therefore, $E^\ominus_{cell} = +1.10\ V$

> **? QUESTION 16**
>
> 16 **a)** Construct the overall equation for the reaction between acidified VO^{2+}(aq) and zinc to give V^{3+}(aq).
>
> **b)** By use of the appropriate electrode potentials, show that zinc can reduce acidified V^{3+}(aq) to form V^{2+}(aq).

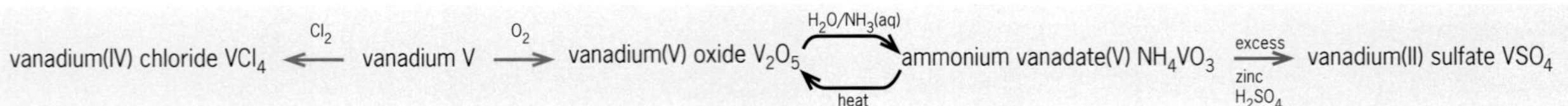

Fig 25.21 A summary of the reactions of vanadium and its compounds. Reactions that involve oxidation are represented by red arrows, and the reaction that involves reduction by a blue arrow

CHROMIUM AND ITS COMPOUNDS

Table 25.3 Electrode potentials of some chrodium species

Reduction half-equation	$E^\ominus_{red}$/V
$Cr^{2+}(aq) + 2e^- \rightleftharpoons Cr(s)$	−0.91
$Cr^{3+}(aq) + 3e^- \rightleftharpoons Cr(s)$	−0.74
$Cr^{3+}(aq) + e^- \rightleftharpoons Cr^{2+}(aq)$	−0.41
$CrO_4^{2-}(aq) + 4H_2O(l) + 3e^- \rightleftharpoons Cr(OH)_3(s) + 5OH^-(aq)$	−0.13
$Cr_2O_7^{2-}(aq) + 14H^+(aq) + 6e^- \rightleftharpoons 2Cr^{3+}(aq) + 7H_2O\ (l)$	+1.33

Chromium exhibits three common oxidation states, +2, +3 and +6, of which chromium(III) is the most stable. Chromium is manufactured by the reduction of chromium(III) oxide by a reactive metal (which must be above chromium in the reactivity series). Chromium reacts with dilute acids, such as hydrochloric or sulfuric acid, to give a sky-blue solution that contains the aqueous chromium(II) ion, as indicated by the electrode potential data. The reaction must be carried out in an inert atmosphere to avoid atmospheric oxidation, which would give chromium(III).

QUESTION 17

17 **a)** Write down an equation to show the reaction between chromium metal and dilute sulfuric acid in an inert atmosphere.

b) Use the electrode potential to work out $E^{\ominus}_{cell}$ for the following electrochemical cell:

$Cr(s) \mid Cr^{2+}(aq) \parallel H^+(aq) \mid H_2(g) \mid Pt$

Oxidation of aqueous chromium(iii) ions

Aqueous chromium(III) ions can be oxidised to form chromate(VI). First, they are reacted with aqueous sodium hydroxide to form chromium(III) hydroxide, and then the mixture obtained is boiled with hydrogen peroxide:

$$Cr^{3+}(aq) + 3OH^-(aq) \rightarrow Cr(OH)_3(s)$$
$$2Cr(OH)_3(s) + 3H_2O_2(aq) + 4OH^-(aq) \rightarrow 2CrO_4^{2-}(aq) + 8H_2O(l)$$

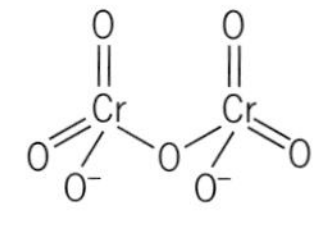

Fig 25.22 The dichromate(VI) ion has one bridging oxygen atom

Chromium in the +6 oxidation state

The highest oxidation number of a transition element often involves covalent bonding to the transition metal atom. Such is the case with compounds of chromium(VI). These compounds are also very powerful oxidising agents.

As Table 25.3 shows, $E^{\ominus}_{red}$ for dichromate(VI) has a fairly large positive value, which indicates the ability of dichromate(VI) to oxidise other substances. The equation shows that the product of this reduction is normally the aqueous chromium(III) ion. The use of aqueous dichromate(VI) ion as an oxidising agent is always associated with the presence of aqueous hydrogen ions (dilute sulfuric acid).

In an alkaline solution, the oxidising ability is greatly reduced, since dichromate(VI) is converted into chromate(VI). As Table 25.3 shows, $E^{\ominus}_{red}$ for the reduction of the CrO_4^{2-} ion has a low negative value:

$$2OH^-(aq) + Cr_2O_7^{2-}(aq) \rightleftharpoons 2CrO_4^{2-}(aq) + H_2O(l)$$

Information about acidified dichromate(VI) in the oxidation of alcohols and aldehydes is given on pages 267 and 323.

The conversion of dichromate(VI) into chromate(VI) is pH dependent. In acid, the equilibrium shifts to the left to form orange dichromate(VI). In alkali, the equilibrium shifts to the right to form yellow chromate(VI) ion.

Iron(II) ions are oxidised by acidified $Cr_2O_7^{2-}(aq)$ to form $Fe^{3+}(aq)$, and $Cr_2O_7^{2-}(aq)$ is reduced to $Cr^{3+}(aq)$. The colour changes in this reaction are complicated, since green $Fe^{2+}(aq)$ reacts with orange $Cr_2O_7^{2-}(aq)$ to form orange $Fe^{3+}(aq)$ and blue–green $Cr^{3+}(aq)$:

$$6Fe^{2+}(aq) + Cr_2O_7^{2-}(aq) + 14H^+(aq) \rightarrow 6Fe^{3+}(aq) + 2Cr^{3+}(aq) + 7H_2O(l)$$

Acidified potassium dichromate(VI) can be reduced by zinc in an inert atmosphere to give chromium(II). The reaction must be carried out in an inert atmosphere to prevent oxidation of $Cr^{2+}(aq)$ to $Cr^{3+}(aq)$:

$$4Zn(s) + Cr_2O_7^{2-}(aq) + 14H^+(aq) \rightarrow 4Zn^{2+}(aq) + 2Cr^{2+}(aq) + 7H_2O(l)$$

Fig 25.23 A summary of the reactions of chromium and its compounds. Reactions that involve oxidation are represented by red arrows, and those that involve reduction by blue arrows

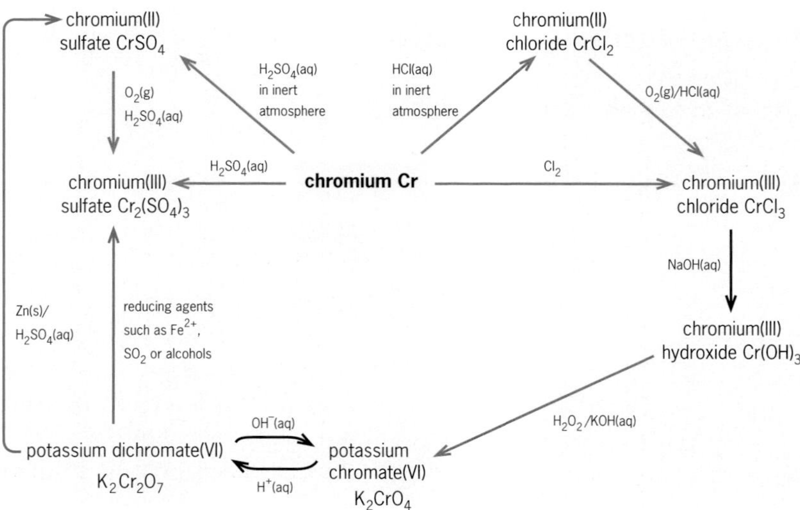

? QUESTION 18

18 **a)** Use the electrode potentials to suggest why manganese(III) is considered to be a very unstable oxidation state with respect to the +2 oxidation state.

b) Predict whether you would expect manganese to react with dilute hydrochloric acid. Explain your answer using electrode potentials.

MANGANESE AND ITS COMPOUNDS

Table 25.4 Electrode potentials of some manganese species

Reduction half-equation	$E^{\ominus}_{red}$/V
CMn^{2+}(aq) + $2e^-$ ⇌ Mn(s)	−1.18
Mn^{3+}(aq) + e^- ⇌ Mn^{2+}(aq)	+1.49
MnO_2(s) + $2H_2O$(l) + $2e^-$ ⇌ $Mn(OH)_2$(s) + $2OH^-$(aq)	+0.05
MnO_2(s) + $4H^+$(aq) + $2e^-$ ⇌ Mn^{2+}(aq) + $2H_2O$(l)	+1.23
MnO_4^-(aq) + $2H_2O$(l) + $3e^-$ ⇌ MnO_2(s) + $4OH^-$(aq)	+0.59
MnO_4^-(aq) + $8H^+$(aq)+$5e^-$ ⇌ Mn^{2+}(aq) + $4H_2O$(l)	+1.52

Manganese exhibits several common oxidation states, including +2, +4, +6 and +7. The +2 oxidation state is generally considered to be the most stable under normal conditions.

Manganese(IV) oxide as an oxidising and a reducing agent

Manganese(IV) oxide has an intermediate oxidation number. Therefore, it can act both as a reducing agent and as an oxidising agent. Manganese(IV) oxide can oxidise hydrochloric acid to form chlorine:

$$MnO_2(s) + 4HCl(aq) \rightarrow MnCl_2(aq) + 2H_2O(l) + Cl_2(g)$$

Manganese(IV) oxide can also be oxidised to give manganate(VI). This is usually achieved by heating the oxide with an oxidising agent, such as potassium nitrate or potassium chlorate(VI), with potassium hydroxide.

SCIENCE IN CONTEXT

Lithium–manganese(IV) oxide cell

The lithium–manganese(IV) oxide cell – popularly referred to as the lithium cell – cannot have a water-based electrolyte because the lithium would react with the water to form hydrogen. The great advantage of using lithium as one of the electrodes is that its oxidation potential is very large, which gives the battery a high voltage. The battery can therefore deliver a low current for a long time, which makes it an ideal power supply for such items as watches and heart pacemakers.

Oxidation half-equation at the anode:

Li → Li^+ + e^-

Reduction half-equation at the cathode:

Li^+ + MnO_2 + e^- → $LiMnO_2$

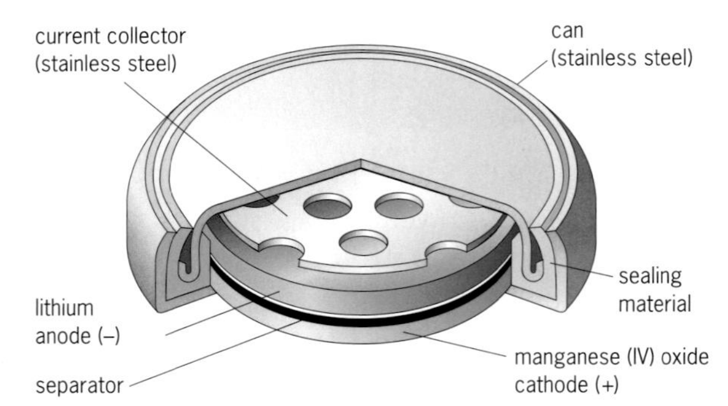

Fig 25.24 A lithium-manganese(IV) oxide battery

Higher oxidation states of manganese

Both the +6 and the +7 oxidation states are highly oxidising, since the manganese can be reduced to manganese(IV) or manganese(II). Aqueous manganate(VI) is a green solution, which is easily oxidised to give the familiar purple colour of the manganate(VII) ion. The manganate(VII) ion can be prepared directly by reacting manganese(II) ions with oxidising agents such as sodium bismuthate, $NaBiO_3$, or lead(IV) oxide. This reaction is used in the estimation of the percentage of manganese in a sample of steel. The steel is reacted with dilute acid to form a solution that contains aqueous iron(II) ions and aqueous manganese(II) ions. The aqueous manganese(II) ions are oxidised to give purple manganate(VII). The concentration of the manganate(VII) can be determined using ultraviolet–visible spectroscopy (see pages 564–565).

Volumetric analysis using potassium manganate(VII)

Aqueous potassium manganate(VII) is used in laboratories as an oxidising agent for organic preparations, volumetric analysis and qualitative analysis.

Acidified potassium manganate(VII) is used to test for reducing agents: it reacts to form manganese(II) ions and the distinct colour of the manganate(VII) ions disappears. For example, iodide ions reduce acidified manganate(VII) ions to form manganese(II) ions and iodine.

In volumetric analysis, acidified potassium manganate(VII) is used to determine the concentration of reducing agents. Since it changes colour during the titration, there is no need to have an indicator. If potassium manganate(VII) is used in the burette, the end-point is the first appearance of a purple-pink colour. A typical example (Fig 25.25) would be the reaction of aqueous iron(II) ions with acidified potassium manganate(VII):

$$5Fe^{2+}(aq) + MnO_4^-(aq) + 8H^+(aq) \rightarrow 5Fe^{3+}(aq) + Mn^{2+}(aq) + 4H_2O(l)$$

EXAMPLE

Q A solution of a moss-killer contained aqueous iron(II) ions. A 25.0 cm^3 sample of this solution was acidified with dilute sulfuric acid and titrated with 0.0200 mol dm^{-3} potassium manganate(VII). It was found that 21.0 cm^3 of the $KMnO_4$(aq) was needed to react fully with the Fe^{2+}(aq) in the solution of the moss-killer. Calculate the concentration of Fe^{2+}(aq) in the solution of the moss-killer.

A $5Fe^{2+}(aq) + MnO_4^-(aq) + 8H^+(aq) \rightarrow 5Fe^{3+}(aq) + Mn^{2+}(aq) + 4H_2O(l)$

moles of MnO_4^- = volume in dm^3 × concentration

= 0.0210 × 0.0200 = 4.20×10^{-4}

From the equation:

moles of Fe^{2+} = 5 × moles of MnO_4^- = 2.10×10^{-3}

Therefore:

$$[MnO_4^-(aq)] = \frac{\text{mole of } MnO_4^-}{\text{volume of } MnO_4^- \text{ (in dm}^3)}$$

$$= \frac{2.10 \times 10^{-3}}{0.0250} = 0.0840 \text{ mol dm}^{-3}$$

In this description of the oxidising abilities of manganate(VII), references are made to acidified conditions. This is important, because the pH at which the reaction is carried out affects the nature of the reduction product. If the pH is increased so that the reaction is carried out in neutral or alkaline conditions, then manganese(IV) oxide is produced instead of manganese(II) ions. As Table 25.4 shows, in alkaline conditions there is a different half-equation.

? QUESTION 19

19 **a)** Write down:

i) the half-equation to show the reduction of acidified manganate(VII) ions to give manganese(II) ions;

ii) the oxidation of aqueous iodine ions to form aqueous iodine.

b) Hence construct the ionic equation for the reaction between acidified manganate(VII) ions and aqueous iodide ions.

Fig 25.25 A titration involving potassium manganate(VII) and iron(II) ions

For more information about the concentration of aqueous solutions, read page 130.

? QUESTION 20

20 Potassium manganate(VII) was titrated with acidified iron(II) sulfate solution. 25.0 cm^3 of $KMnO_4$(aq) required 35.9 cm^3 of 0.100 mol dm^{-3} $FeSO_4$(aq) to react fully. Calculate the concentration of MnO_4^-(aq).

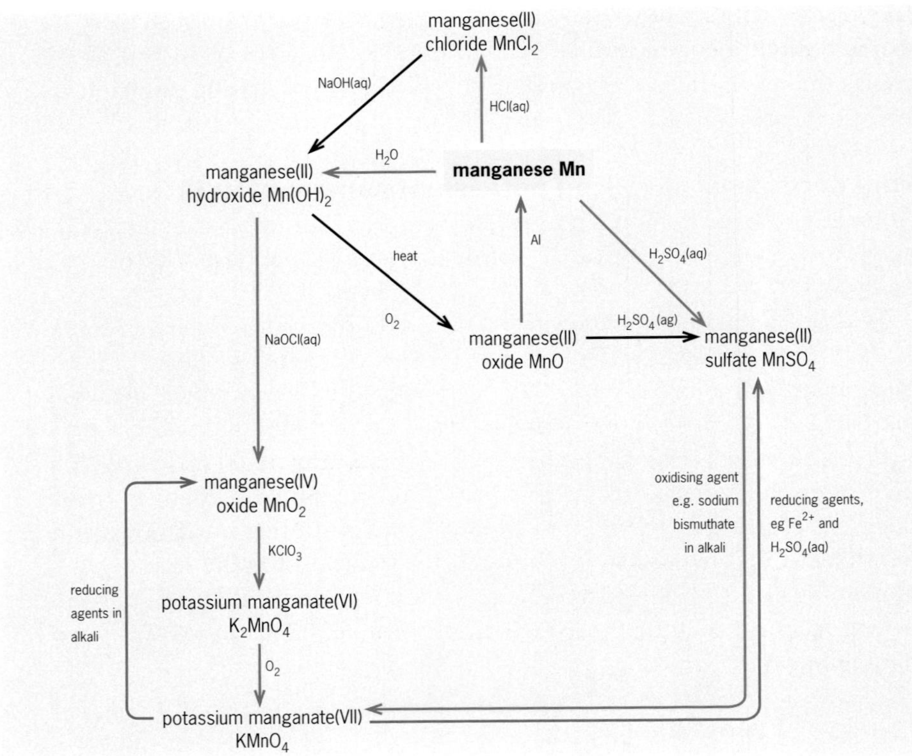

Fig 25.26 A summary of the reactions of manganese and its compounds. Reactions that involve oxidation are represented by red arrows, and those that involve reduction by blue arrows

IRON AND ITS COMPOUNDS

Table 25.5 Electrode potentials of some iron species

Reduction half-equation	$E^\ominus_{red}$/V
$Fe^{2+}(aq) + 2e^- \rightleftharpoons Fe(s)$	−0.44
$Fe^{3+}(aq) + 3e^- \rightleftharpoons Fe(s)$	−0.04
$Fe^{3+}(aq) + e^- \rightleftharpoons Fe^{2+}(aq)$	+ 0.77
$Fe(OH)_3(s) + e^- \rightleftharpoons Fe(OH)_2(s) + OH^-(aq)$	−0.56

Iron exhibits two common oxidation states, +2 and +3, both of which are ionic. Iron does form another oxidation state, +6, but it occurs in only a few compounds.

Relative stability of the +2 and +3 oxidation states

As the electrode potentials in Table 25.5 suggest, iron reacts with dilute acids, such as sulfuric acid, to form iron(II) salts:

$$Fe(s) + 2H^+(aq) \rightarrow Fe^{2+}(aq) + H_2(g) \qquad E^\ominus_{cell} = +0.44\text{ V}$$

The iron(II) ions are very susceptible to aerial oxidation to form iron(III) ions. The stability of these two oxidation states is very much dependent on the pH. In acidic conditions, aqueous iron(II) ions are oxidised to give iron(III) ions. The appropriate half-equations are:

$$Fe^{2+}(aq) \rightarrow Fe^{3+}(aq) + e^- \qquad E^\ominus_{oxid} = -0.77\text{ V}$$
$$O_2(g) + 2H^+(aq) + e^- \rightarrow 2H_2O(l) \qquad E^\ominus_{red} = +1.23\text{ V}$$

In alkaline conditions, the oxidation is even more favourable:

$$Fe(OH)_2(s) + OH^-(aq) \rightarrow Fe(OH)_3(s) + e^- \qquad E^\ominus_{oxid} = +0.56\text{ V}$$
$$O_2(g) + 2H_2O(1) + 2e^- \rightarrow 4OH^-(aq) \qquad E^\ominus_{red} = +0.40\text{ V}$$

The relative stability of the +2 and +3 oxidation states can also be affected by the formation of complexes.

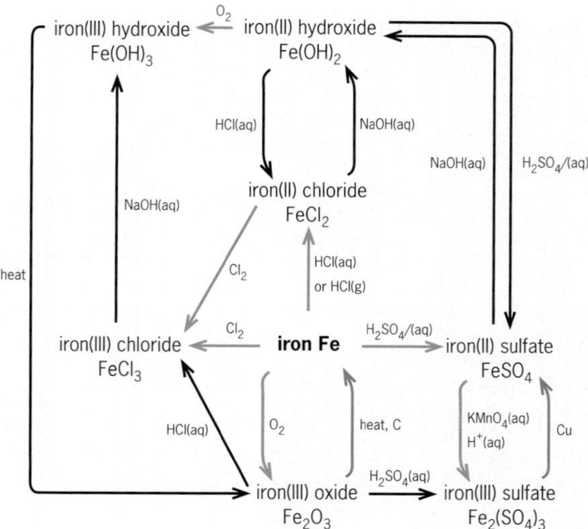

Fig 25.27 A summary of the reactions of iron and its compounds. Reactions that involve oxidation are represented by red arrows, and those that involve reduction by blue arrows

REMEMBER THIS

For convenience, aqueous metal ions such as $[M(H_2O)_6]^{2+}$(aq) are written as M^{2+}(aq).

QUESTION 21

21 **a)** i) Write down the overall equation for the aerial oxidation of acidified iron(II) ions.

ii) What is the $E^\ominus_{cell}$ value for this reaction?

b) What is the $E^\ominus_{cell}$ value for the oxidation of iron(II) hydroxide into iron(III) hydroxide in alkaline conditions?

Rusting of iron

Any unpainted iron object shows evidence of rusting. The surface of iron forms a flaky orange solid, the composition of which is best described as hydrated iron(III) oxide. Rusting is, in fact, a complicated electrochemical process that occurs on the surface of iron.

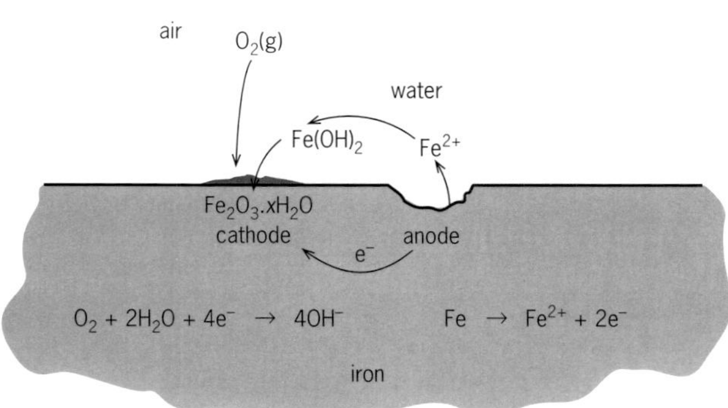

Fig 25.28 How rusting occurs

Fig 25.29 The railway bridge over the River Tyne at Newcastle is an iron structure which has to be painted to prevent rust damage

Fig 25.28 summarises some of the complex electrochemical processes that take place when iron rusts. When iron is in contact with a drop of water, a redox reaction occurs:

Oxidation half-equation: $Fe(s) \rightarrow Fe^{2+}(aq) + 2e^-$

Reduction half-equation: $O_2(g) + 2H_2O(l) + 4e^- \rightarrow 4OH^-(aq)$

These two reactions take place in different areas of the iron, which results in the formation of an anodic region and a cathodic region. The electrons move through the metal from the anodic region to the cathodic region. The circuit is completed by ions that move through the water. Without the water, the circuit is not complete and rusting cannot take place. Essentially, the water is acting as a salt bridge. If the water contains electrolytes, such as sodium chloride, then the concentration of ions in the droplet is higher and so the rate of rusting increases. The iron(II) ions and the hydroxide in the water droplet are precipitated as iron(II) hydroxide, which is further oxidised to form hydrated iron(III) oxide or rust:

$$Fe^{2+}(aq) + 2OH^-(aq) \rightarrow Fe(OH)_2(s)$$

$$4Fe(OH)_2(s) + O_2(g) + 2H_2O(l) \rightarrow 4Fe(OH)_3(s)$$

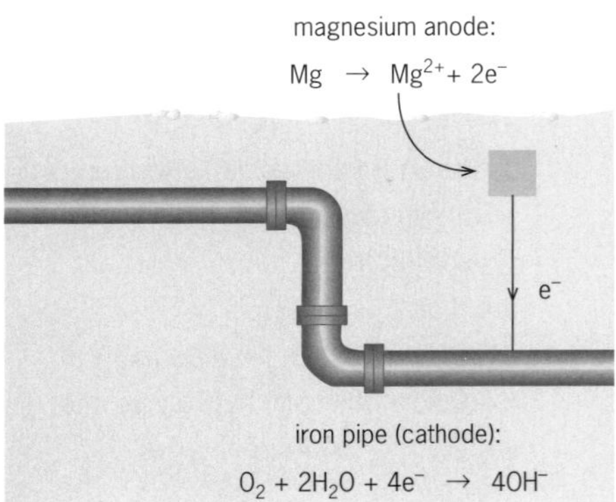

Fig 25.30 Sacrificial protection

Rust protection

Any layer that is impervious to water protects iron from rusting. However, the protection provided by paints and other widely used materials is of limited duration, because of such factors as physical damage, chemical deterioration and weathering. So, alternative methods of rust protection use electrochemical principles instead. One of the best known involves giving the iron a protective layer of zinc by a process called galvanisation, which is described below.

Iron pipes and tanks in the ground are often protected by a block of magnesium or zinc (Fig 25.30). The block of magnesium (or zinc) is attached to the iron object and, since magnesium (or zinc) has a higher positive oxidation potential than iron, it oxidises in preference (sacrificially) to the iron. The iron object acts as the cathode in this method.

Tin-plated cans used in the food industry provide another example of electrochemical protection. The oxidation potential for tin has a lower positive value than that for iron. Therefore, tin is less likely to react with moist oxygen and so protects the surface of the iron. There is one drawback with tin plating. It is easily scratched to reveal the iron, in which case the can rusts very rapidly. What happens is that once the iron is in contact with moist air, it gives sacrificial protection to the less reactive tin.

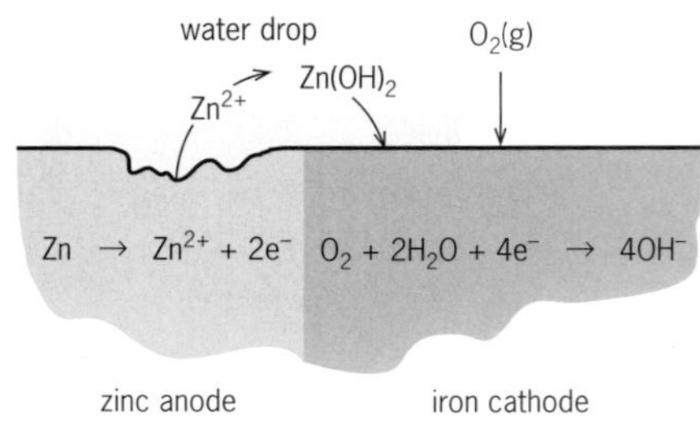

Fig 25.31 Galvanised iron. The protective layer of zinc forms an electrochemical cell with the overall cell reaction:

$$2Zn + O_2 + 2H_2O \rightarrow 2Zn(OH)_2$$

Galvanising iron

As already stated, a coating of zinc protects iron from rusting. Zinc-coated iron is known as galvanised iron.

Galvanised iron is protected mainly because, in the electrochemical cell formed by galvanisation, the zinc is preferentially oxidised. (The oxidation potential for zinc has a larger positive value than that for iron.) However, were it not for another reaction taking place at the same time, all the zinc would eventually be oxidised and rusting could start. The zinc hydroxide produced reacts with carbon dioxide in the air to form a layer of a zinc hydroxide-zinc carbonate compound that adheres firmly to the iron to give further protection.

Fig 25.32 Galvanised iron does not rust as fast as iron on its own. The coating of zinc corrodes in preference to the iron, in the course of which it forms a further protective layer over the iron

COBALT

Table 25.6 Electrode potentials of some cobalt species

Reduction half-equation	$E^{\ominus}_{red}$/V
$Co^{2+}(aq) + 2e^- \rightleftharpoons Co(s)$	−0.28
$Co^{3+}(aq) + e^- \rightleftharpoons Co^{2+}(aq)$	+1.82
$[Co(NH_3)_6]^{3+}(aq) + e^- \rightleftharpoons [Co(NH_3)_6]^{2+}(aq)$	+0.10
$Co(OH)_3(s) + e^- \rightleftharpoons Co(OH)_2(s) + OH^-(aq)$	+0.17

Cobalt exhibits only two common oxidation states, +2 and +3, both of which are ionic in nature. As expected from the reduction potential of Co^{2+}(aq), cobalt reacts slowly with dilute acids to form aqueous cobalt(II) ion:

$$Co(s) + 2H^+(aq) \rightarrow Co^{2+}(aq) + H_2(g) \qquad E^{\ominus}_{cell} = +0.28\ V$$

Redox reactions involving cobalt(III)

The absence of a higher oxidation state suggests that redox reactions are not particularly significant, but this would be some way from the truth. Just a glance at the electrode potential for Co^{3+}(aq) in Table 25.6 indicates that it should be easy to reduce aqueous cobalt(III) ions to give aqueous cobalt(II) ions. In aqueous solution, water reduces cobalt(III) ions to give cobalt(II) ions and oxygen:

$$Co^{3+}(aq) + e^- \rightarrow Co^{2+}(aq) \qquad E^{\ominus}_{red} = +1.82\ V$$

$$2H_2O(L) \rightarrow 4H^+(aq) + O_2(g) + 4e^- \qquad E^{\ominus}_{oxid} = -1.23\ V$$

The overall equation is:

$$4Co^{3+}(aq) + 2H_2O(l) \rightarrow 4Co^{2+}(aq) + 4H^+(aq) + O_2(g) \qquad E^{\ominus}_{cell} = +0.59\ V$$

As already mentioned, pH can affect the relative stabilities of oxidation states, an example of which is provided by cobalt. Co^{3+}(aq) is much more stable in basic conditions than in acidic conditions. This means that the oxidation of Co^{2+}(aq) to form Co^{3+}(aq) must be carried out in the presence of aqueous sodium hydroxide. The oxidising agent of choice is hydrogen peroxide. The oxidation can also be achieved by air oxidising an ammoniacal solution of Co^{2+}

The formation of complex ions can also affect the stability of an oxidation state. This is exemplified by the stability of aqueous hexaamminecobalt(III), $[Co(NH_3)_6]^{3+}$(aq), as compared with the aqueous hexaaquacobalt(III) ion, $[Co(H_2O)_6]^{3+}$(aq). As Table 25.6 shows, the redox potential for the hexaammine complex is only just above zero, and therefore it is much more difficult to reduce than Co^{3+}(aq).

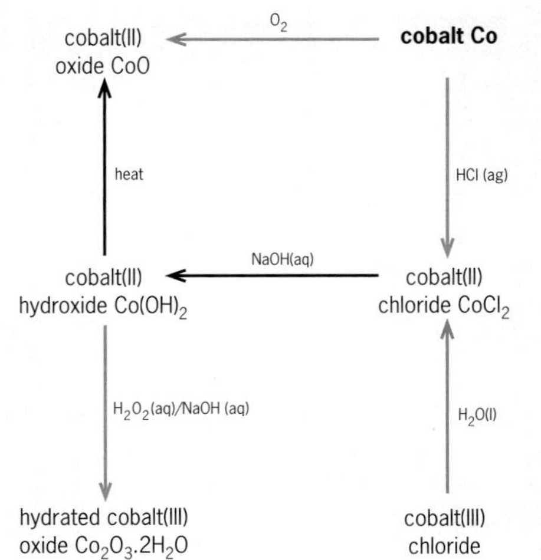

Fig 25.33 A summary of the reactions of cobalt and its compounds. Reactions involving oxidation are represented by red arrows, and those involving reduction by blue arrows

COPPER

The main oxidation states of copper are +1 and +2. Both are ionic oxidation states, and involve the loss of either one electron or two electrons per copper atom.

Table 25.7 Electrode potentials of some copper species

Reduction half-equation	$E^{\ominus}_{red}$/V
$Cu^{2+}(aq) + 2e^- \rightleftharpoons Cu(s)$	+ 0.34
$Cu^{+}(aq) + e^- \rightleftharpoons Cu(s)$	+ 0.52
$Cu^{2+}(aq) + e^- \rightleftharpoons Cu^{+}(aq)$	+ 0.15

Reactions of copper

The reduction potentials for both $Cu^+(aq)$ and $Cu^{2+}(aq)$ indicate that copper does not react with dilute acids, such as sulfuric or hydrochloric, to form hydrogen. The presence of dissolved oxygen in the acid does, however, allow dilute acids to oxidise copper to give the copper(II) ion:

$$Cu(s) + 2H^+(aq) + \tfrac{1}{2}O_2(g) \rightarrow Cu^{2+}(aq) + H_2O(l)$$

$$E^{\ominus}_{cell} = E^{\ominus}_{oxid} + E^{\ominus}_{red}$$

$$= -0.34 + (+1.23)$$

$$= +0.89\ V$$

Stability of the +1 oxidation state

Read more about Benedict's or Fehling's test on page 323.

Aqueous copper(I) compounds often disproportionate to give copper and copper(II) compounds, but insoluble copper(I) compounds, such as the halides, are much more stable:

$$Cu_2SO_4(aq) \rightarrow Cu(s) + CuSO_4(aq)$$

Stability of copper(II) compounds

Copper(II) compounds are not really considered to be oxidising agents, but there a few examples of their reduction to copper(I) compounds. Aldehydes reduce complexed copper(II) ions into copper(I) oxide, which forms a red-brown precipitate. This is the basis of Benedict's or Fehling's test for aldehydes and reducing sugars.

Fig 25.34 A summary of the reactions of copper and its compounds. Reactions that involve oxidation are represented by red arrows, and those that involve reduction by blue arrows

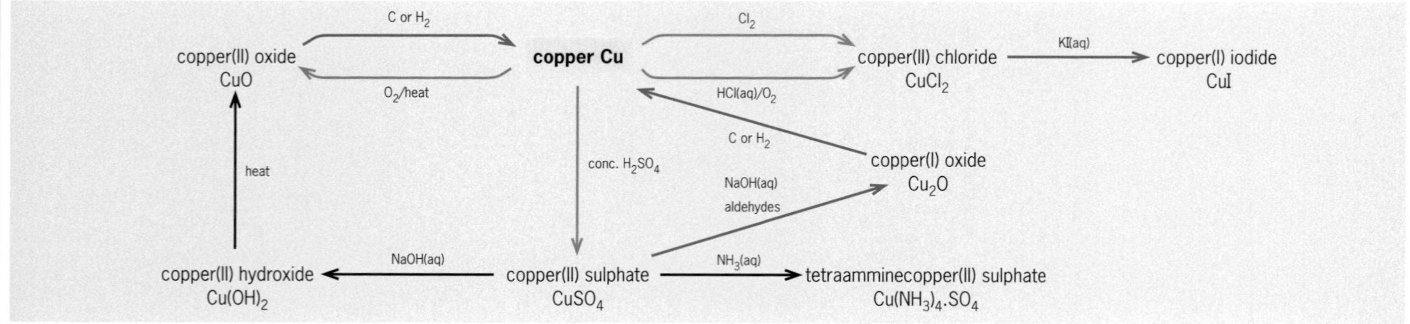

Iodide ion reduces aqueous copper(II) ions to give a white precipitate of copper(I) iodide, and so copper(II) iodide does not exist:

$$2Cu^{2+}(aq) + 4I^-(aq) \rightarrow 2CuI(s) + I_2(s)$$

This reaction is the basis of the estimation of the concentration of copper(II) ions in solution, since the iodine produced can be titrated against aqueous sodium thiosulfate.

8 ENTROPY AND FEASIBILITY OF REACTIONS

On page 133 entropy is described as a measure of the disorder of a system. The second law of thermodynamics states:

Any spontaneous change that occurs in the universe must be accompanied by an increase in the entropy of the universe.

This means that a reaction is spontaneous if the total entropy change is positive. The total entropy change includes the entropy of the reaction (the entropy of the system) and the entropy of the rest of the surroundings.

$$\Delta S_{total} = \Delta S_{system} + \Delta S_{surroundings}$$

ENTROPY OF SUBSTANCES

The third law of thermodynamics states:

At absolute zero the entropy of a pure substance is zero.

As the temperature increases, so does the entropy of the substance. The more the particles move, the greater the entropy becomes. At the same temperature, one mole of a gas has a much greater entropy than one mole of a liquid, and one mole of liquid has a much greater entropy than one mole of solid. This allows information about the entropy change of a reaction to be predicted just by looking at the number of moles of gas shown in the equation. The decomposition of calcium carbonate makes carbon dioxide:

$$CaCO_3(S) \rightarrow CaO(S) + CO_2(g)$$

This means that the entropy change for the reaction is positive, because there are no moles of gas on the left-hand side of the equation but one mole of gas on the right-hand side. So even though the reaction is endothermic, the reaction will be spontaneous above a certain temperature.

ENTROPY CHANGE OF SURROUNDINGS

In a chemical reaction only the reactants and products themselves are part of the system; everything else – the surrounding air, the reaction container, even water if it is used as the solvent – is part of the surroundings. In an exothermic reaction energy is transferred into the surroundings and the surroundings

EXAMPLE

Q Predict whether the decomposition of copper carbonate is spontaneous at 298 K.

$CuCO_3(s) \rightarrow CuO(s) + CO_2(g)$

$\Delta H = +357.5 \text{ kJ mol}^{-1}$

$\Delta S_{system} = +169 \text{JK}^{-1} \text{ mol}^{-1}$

A $\Delta S_{surroundings} = \dfrac{-\Delta H}{T}$

$\Delta S_{surroundings} = \dfrac{-357500}{298}$ (note that Δ H must be converted in J mol^{-1})

$\Delta S_{surroundings} = -1120$ to 3 sig. fig.

So as $\Delta S_{total} = \Delta S_{system} + \Delta S_{surroundings}$

$\Delta S_{total} = +169 + (-1120)$

$\Delta S_{total} = -951 \text{ JK}^{-1} \text{ mol}^{-1}$

So at 298 K the reaction is not spontaneous (from the second law of thermodynamics).

? **QUESTION 22**

22 **a)** Predict whether the decomposition of copper(II) carbonate is spontaneous at 1000 K. Assume that the values for ΔH and ΔS_{system} do not change with temperature. Use data from the Example box on this page.
b) Calculate the equilibrium constant for this decomposition at 1000 K.

You can read more about position of equilibrium on page 354.

REMEMBER THIS

The description of a reaction as 'spontaneous' does not mean that the rate of reaction has to be fast. In fact, the rate of reaction may be so slow that the reaction does not actually take place. In this case, one of the reactants is said to be kinetically stable, but energetically unstable.

become warmer. As a result, the entropy of the surroundings will increase. We can calculate this increase in entropy using the following equation:

$$\Delta S_{surroundings} = -\Delta H$$

You can see from this equation that if ΔH is negative (the reaction is exothermic), then the entropy change of the surroundings must be positive. It also follows that for an endothermic reaction the entropy change of the surroundings is negative.

Total entropy change and endothermic reactions

With an endothermic reaction ΔH is positive, so $\Delta S_{surroundings}$ will be negative. The reaction cannot be spontaneous unless ΔS_{system} is positive. Even then it may well take a very high temperature before the ΔS_{system} contribution outweighs the $\Delta S_{surroundings}$ component of the ΔS_{total}. An endothermic reaction in which ΔS_{system} is negative can never be spontaneous.

Total entropy change and exothermic reactions

With an exothermic reaction ΔH is negative, so $\Delta S_{surroundings}$ will be positive. The reaction must be spontaneous at all temperatures if ΔS_{system} is positive. Even if the ΔS_{system} is negative, then the positive $\Delta S_{surroundings}$ will outweigh it, but only below a certain temperature.

Table 25.8 A summary of whether reactions are spontaneous or not

	Exothermic reaction	Endothermic reaction
positive entropy change of reaction	spontaneous at all temperatures	spontaneous below a certain temperature
negative entropy change of reaction	spontaneous above a certain temperature	never spontaneous

STRETCH AND CHALLENGE

Equilibrium constants and the total entropy change

The total entropy change is also related to the equilibrium constant for the reaction. The equilibrium constant gives a measure of the position of equilibrium. In qualitative terms the greater the numerical value of the equilibrium constant, the more positive the total entropy change and the more spontaneous the reaction will be mathematically:

$$\Delta S_{total} = R\ln K_c$$

EXAMPLE

Q Calculate the numerical value of the equilibrium constant for the reaction between hydrogen and bromine at 600 K.

$H_2(g) + Br_2(g) \rightleftharpoons 2HBr(g)$

$\Delta H = -72\ kJ\ mol^{-1}$

$\Delta S_{system} = +22 JK^{-1}\ mol^{-1}$

A $\Delta S_{surroundings} = \dfrac{-\Delta H}{T}$

$\Delta S_{surroundings} = \dfrac{-72000}{600}$ note that ΔH must be converted into J mol^{-1}

$\Delta S_{surroundings} = -120$

So as $\Delta S_{total} = \Delta S_{system} + \Delta S_{surroundings}$

$S_{total} = +22 + (-120)$

$S_{total} = -98\ JK^{-1}\ mol^{-1}$

$S_{total} = R\ln K_c = 8.314\ln K_c$ where ln is the natural log (base e)

$1nK_c = \dfrac{-98}{-8.314}$

Antilogging this gives

$K_c = 7.57\times10^{-6}$

9 EXTRACTION OF METALS

Several different methods of extracting metals have already been covered. On page 12 is a description of the reduction of iron(III) oxide with carbon in the blast furnace, and Chapter 22 contains a description of the electrolytic extraction of the alkali metals and of aluminium. In this section, some of the free-energy changes associated with the extraction of metals, such as aluminium and iron, are reviewed.

Although this section is not really connected with electrode potentials, it does use the concept of free energy, ΔG, introduced on page 515.

FREE-ENERGY CHANGE

Reactions proceed spontaneously if they are energetically downhill (ΔH is negative) and if the process leads to more disorder (ΔS is positive). These two ideas of entropy change and enthalpy change can be combined to give the free-energy change of a reaction, ΔG, expressed as:

$$\Delta G = \Delta H - T\Delta S$$

For this equation ΔS is the entropy change of the system.
As already stated, reactions in which ΔG is negative are spontaneous. Therefore, provided the enthalpy change, the entropy change and the temperature of a reaction are known, it can be predicted whether a reaction will be spontaneous.

REMEMBER THIS

Free energy is the maximum available work that could be carried out on the surroundings, and a reaction should occur spontaneously when the free-energy change is negative.

EXAMPLE

Q Predict whether the reaction between hydrogen and chlorine is spontaneous at 298 K:

$H_2(g) + Cl_2(g) \rightarrow 2HCl(g)$ $\quad \Delta H^{\ominus} = -185 KJ^{-1}\ mol^{-1}$, $\Delta S^{\ominus} = +141 JK^{-1}\ mol^{-1}$

A For these figures, there is no need to do a calculation. Since $\Delta H^{\ominus}$ is negative and $-T\Delta S^{\ominus}$ is negative (absolute temperature cannot be negative),
then $\Delta G^{\ominus} = \Delta H^{\ominus} - T\Delta S^{\ominus}$ must also be negative. Therefore, the reaction is spontaneous.

Q Predict whether the thermal dissociation of ammonium chloride will proceed spontaneously at 298 K:

$NH_4Cl(s) \rightarrow NH_3(g) + HCl(g)$ $\quad \Delta H^{\ominus} = -176\ KJ^{-1}\ mol^{-1}$, $\Delta S^{\ominus} = +284\ JK^{-1}\ mol^{-1}$

A $\Delta G^{\ominus} = \Delta H^{\ominus} - T\Delta S^{\ominus}$
$= +\ 176 - 298\ (+\ 284 \times 10^{-3})$
$= +\ 91\ kJ\ mol^{-1}$

[Remember: kJ must be used throughout.]

Since the free-energy change is positive, the reaction cannot be spontaneous at 298 K.

FREE-ENERGY CHANGE AND TEMPERATURE

Fig 25.35 shows that the free-energy change for a reaction, ΔG, alters with the temperature. Assuming that the enthalpy change is a constant, then the change in the value of ΔG results from the effect of entropy. This means that a reaction involving a positive entropy change always has a range of temperatures in which the reaction is spontaneous.

In the Example on the next page, in reality, both the enthalpy change and the entropy change are not constants. So the actual temperature calculated is not reliable. However, the calculation does demonstrate that this reaction only becomes spontaneous at high temperatures.

Read more about entropy and spontaneous reactions on pages 529–530.

Fig 25.35 The four graphs show how the free-energy change, ΔG, depends on temperature. They refer to every combination of ΔH and ΔS. The blue sections indicate that the reaction should be spontaneous, since ΔG is negative. Note that when ΔS is negative and ΔH is positive, the reaction cannot be spontaneous

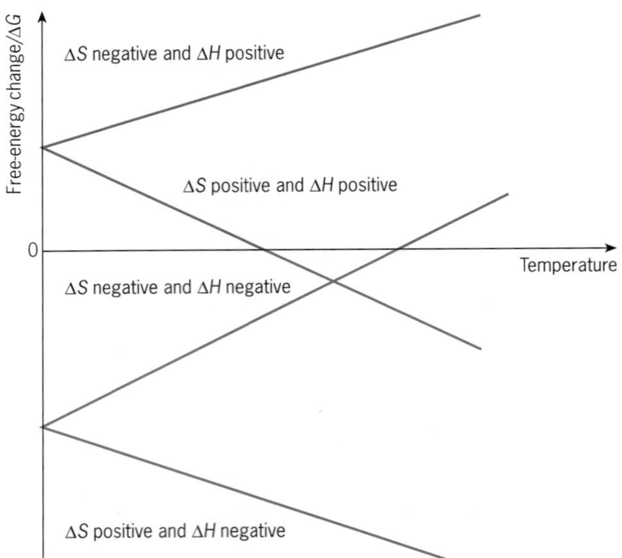

EXAMPLE

Q Predict the temperature range over which the reaction between nitrogen and oxygen becomes spontaneous:

$N_2(g) + O_2(g) \rightarrow 2NO(g)$ $\Delta H = +180\ KJ^{-1}\ mol^{-1}$

$\Delta S = +25\ J\ K^{-1}\ mol^{-1}$

A Assuming that both ΔH and ΔS are constants, the reaction becomes spontaneous when $\Delta G = 0$.

That is:

$$0 = \Delta H - T\Delta S$$

$$0 = +184 - (T \times 25 \times 10^{-3})$$

(Note: energy is measured in kJ throughout.)

Therefore:

$$T = \frac{184}{25 \times 10^{-3}} = 7360\ K$$

The temperature range is therefore any temperature above 7360 K.

QUESTION 23

23 Predict whether the following reactions are spontaneous at 298 K:

a) $3O_2(g) \rightarrow 2O_3(g)$

$\Delta H = +286$ kJ mol^{-1}

$\Delta S = -137$ JK^{-1} mol^{-1}

b) $2SO_2(g) + O_2(g) \rightarrow 2SO_3(g)$

$\Delta H = -198$ kJ mol^{-1}

$\Delta S = -187$ J K^{-1} mol^{-1}

ELLINGHAM DIAGRAMS

Since the free-energy change is a function of temperature, one way to describe the change is to draw an Ellingham diagram (Fig 25.36). An Ellingham diagram shows ow the free-energy change for the oxidation of an element depends on temperature. Note that in each case the free-energy change refers to the equation with 1 mole of oxygen. Consider, for example, the oxidation of carbon to carbon monoxide:

$$2C(s) + O_2(g) \rightarrow 2CO(g)$$

The entropy change for this reaction is positive, since the 2 moles of gas are made from 1 mole of gaseous reactant. It follows that, as the temperature increases, the free-energy change for this reaction has increasing negative values, as shown in Fig 25.36.

The oxidation of carbon to carbon dioxide is much more difficult to assess, since there is 1 mole of gaseous reactant and 1 mole of gaseous product:

$$C(s) + O_2(g) \rightarrow CO_2(g)$$

The entropy change for this reaction has a very small negative value. Therefore, the free-energy change for the oxidation of carbon dioxide remains fairly constant.

For the oxidation of metals by gaseous oxygen to form metal oxides, it is much easier to assess the entropy change because there are no gaseous products. So, the entropy change must be negative. It follows from this that the free-energy change for such reactions assumes smaller and smaller negative values as the temperature increases.

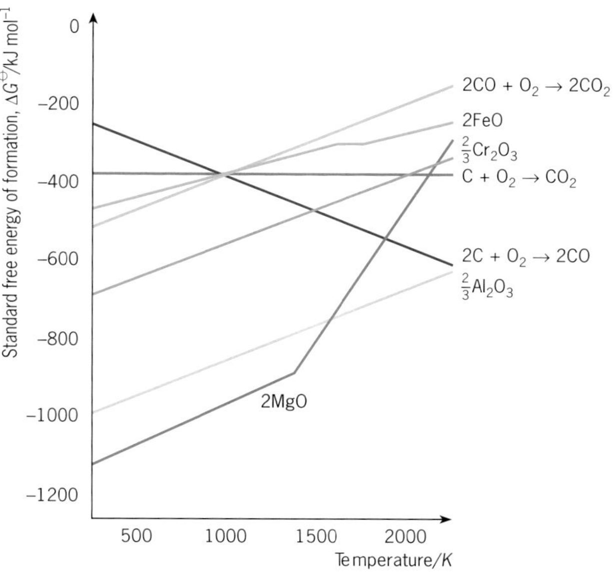

Fig 25.36 An Ellingham diagram for the formation of some oxides

Free-energy changes during the extraction of metals

The extraction of a metal by reduction of its oxide with carbon can be considered to be the sum of two reactions. The first reaction is the oxidation of carbon, and the second is the decomposition of the oxide to give the metal and oxygen.

For example, the reduction of iron(II) oxide to give iron has an overall reaction of:

$$2FeO + 2C \rightarrow 2Fe + 2CO$$

This can be written as the combination of two reactions:

$2C + O_2 \rightarrow 2CO$ $\quad \Delta G$ is negative

$2FeO \rightarrow 2Fe + O_2$ $\quad \Delta G$ is positive because it is the opposite of that shown in the Ellingham diagram.

Note that the overall equation is written so that it involves 1 mole of oxygen in each of the reactions.

A reaction is spontaneous when the free-energy change is negative. The minimum temperature at which this happens is given by the point of intersection of the two appropriate lines on the Ellingham diagram. Above that temperature, the magnitude of the negative free energy of oxidation of carbon is greater than the magnitude of the positive free-energy change of decomposition.

Table 25.9 $Ti(s)+O_2(g) \rightarrow TiO_2(s)$

Temperature/K	Free-energy change, ΔG/kJ mol^{-1}
300	−856
600	−801
1200	−690
2000	−542

Table 25.10 $2C(s)+O_2(g) \rightarrow 2CO(g)$

Temperature/K	Free-energy change, ΔG/kJ mol^{-1}
300	−276
600	−330
1200	−440
2000	−582

? QUESTION 24

24 **a)** Which metals in the Ellingham diagram in Fig 25.36 could be obtained by reduction with carbon at 1500°C?

b) The free-energy changes at various temperatures for the following reaction are shown in Table 25.9:

$Ti(s) + O_2(g) \rightarrow TiO_2(s)$

The free energy changes at various temperatures for the following reaction are shown in Table 25.10:

$2C(s) + O_2(g) \rightarrow 2CO(g)$

i) Plot both sets of data on graph paper.

ii) Use your results to predict the likelihood of obtaining titanium by the reduction of titanium(IV) oxide with carbon.

SUMMARY

After studying this chapter, you should know that:

- A redox reaction involves the transfer of electrons from a reducing agent to an oxidising agent.
- All redox reactions can be written as a sum of two half-equations, one for oxidation and the other for reduction.
- An electrochemical cell is a way to carry out a redox reaction such that the two half reactions take place separately and the electrons are transferred via an external circuit.
- In an electrochemical cell, oxidation always takes place at the anode and reduction always takes place at the cathode.
- An electrode potential is the potential difference between an electrode and the ion-containing solution into which the electrode dips. All electrode potentials are compared to the standard hydrogen electrode, which is assigned a standard electrode potential of 0.00 V.
- The standard hydrogen electrode involves the reversible half-reaction of aqueous hydrogen ions and electrons to form hydrogen gas.
- The chemistry of transition elements can be explained by using electrode potentials.
- The cell potential of an electrochemical cell is the sum of the oxidation potential and the reduction potential. A process that has a positive cell potential takes place spontaneously.
- A primary cell is a battery that cannot be recharged. A secondary cell can be recharged.
- The higher oxidation states of the transition elements are oxidising agents, and the lower oxidation states are reducing agents.
- Electrode potentials and cell potentials are affected by changes in concentration of the aqueous species, temperature, pressure of gases, pH and the complexing of ions.
- Free-energy change is a measure of the maximum available work that can be done on the surroundings. When a process has a negative free-energy change, it should take place spontaneously.
- The standard free-energy change, ΔG, is related to the standard cell potential, $E^{\ominus}_{cell}$ by the equation $\Delta G = -nFE^{\ominus}_{cell}$
- The free-energy change, ΔG, is related to the enthalpy change, ΔH, entropy change, ΔS, and the absolute temperature, T, by the following equation $\Delta G = \Delta H - T\Delta S$.
- For a reaction to be spontaneous the total entropy change ($\Delta S_{system} + \Delta S_{surroundings}$) must be positive.
- The entropy change of the surroundings is related to the enthalpy change of a reaction by the following relationship:
  $$\Delta S_{surroundings} = \frac{-\Delta H}{T}$$
- The total entropy change is related to the equilibrium constant of a reaction by the following relationship:
  $\Delta S_{total} = R\ln Kc$

Practice questions and a How Science Works assignment for this chapter are available at www.collinseducation.co.uk/CAS

26

Reaction kinetics – a more detailed consideration

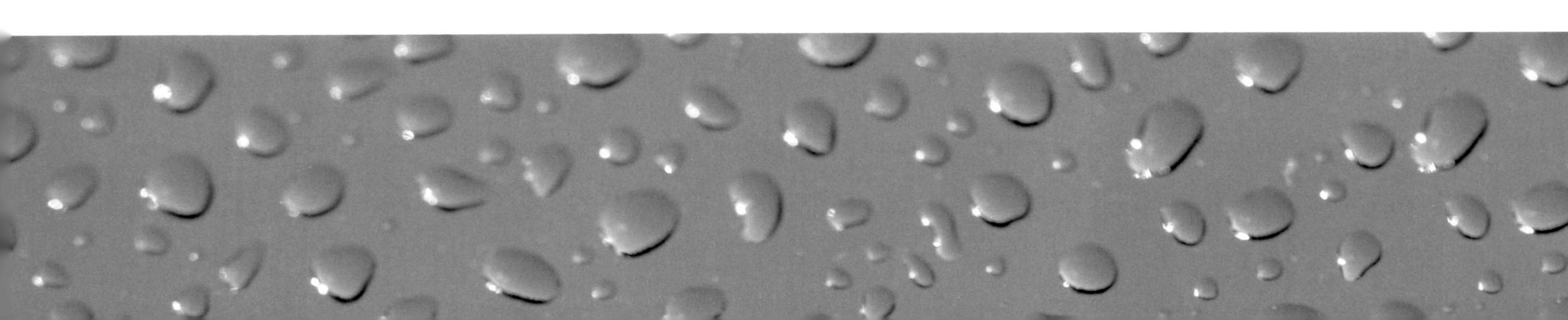

26 REACTION KINETICS – A MORE DETAILED CONSIDERATION

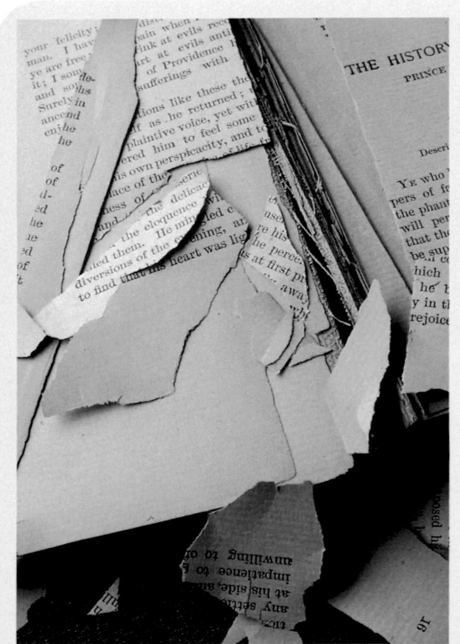

Paper deterioration The long-term effect of hydrogen ions on paper

Paper preservation

Unlike industrial chemists, who mostly want to accelerate chemical reactions, curators of the world's great libraries are desperate to stop reactions. If they fail, they will be the keepers of a growing collection of crumbling paper. Millions of books, documents, drawings and photographs are already falling to pieces. Many rare and valuable publications are close to total disintegration as their paper becomes brittle and starts to crumble. This applies not just to publications of many centuries ago, but to those as recent as the mid-19th century. What is the cause, and is it preventable?

Untreated paper is too porous to print on. Ink would seep into the pores to produce fuzzy print. So it has to be 'sized', a process that fills in the pores with a compound that leaves the paper with a smooth surface. Since 1850, the bulk of book paper has been sized with aluminium sulfate – and this is the culprit.

Aluminium ions in the paper react with moisture to produce hydrogen ions. These hydrogen ions break down the cellulose molecules that the paper is composed of, and so it becomes brittle and eventually crumbles. Neutralising the acidity caused by hydrogen ions will halt this slow decomposition, and there are ways of doing this that do not damage the books. However, the cost can be prohibitive – sometimes over £1000 per book.

Paper manufacturers are helping to prevent this problem in future by producing more 'acid-free' paper treated with sizing agents that do not produce hydrogen ions. Papers treated with these should have a life expectancy of well over 200 years – nearly ten times that of the crumbling paper.

The rate of the reaction that destroys these millions of publications is certainly slow, but already too many important and irreplaceable books have been damaged. Understanding reaction kinetics has helped to save some of the world's most treasured volumes, but for others it is too late.

1 WHY BOTHER WITH REACTION KINETICS?

REMEMBER THIS

For a reaction to occur spontaneously, there must be an overall increase in entropy. This is the Second Law of Thermodynamics. This information can be used to predict whether a reaction is spontaneous, but it cannot predict how fast a reaction will be.

Reaction kinetics is the study of rates of reaction. In Chapter 1 we first discussed the factors that affect the rate of a chemical reaction (page 9). We studied reaction rates in more detail in Chapter 8 (page 170) and considered how **collision theory** could explain why some factors, such as concentration and temperature, could accelerate or reduce rates of reaction.

We describe a reaction as **spontaneous** if it tends to occur. You met this concept first on pages 164–165. A very important spontaneous reaction is that between petrol and air in a car engine (see page 197). Petrol and air do not react in a petrol tank at 25 °C because the reaction rate is so slow that it cannot be measured. But petrol explodes in air when the energy from a spark is added, as in the internal combustion engine. Being spontaneous does not mean that a reaction is necessarily fast.

The reaction between nitrogen and hydrogen is another spontaneous one, yet nothing seems to happen at room temperature. However, under the conditions of the Haber process, ammonia is produced – one of the world's most important industrial chemical reactions. Industrial chemists need to know just how fast they can make a reaction proceed. A reaction that is too slow is unlikely to be a commercial proposition.

The Haber process is covered on pages 19–23.

With a knowledge of the kinetics of a reaction, you can also understand how the reaction takes place and which species are involved in each step. The series of steps involved in a reaction is called the **reaction mechanism**. When the reaction mechanism is known, then ways to alter the rate of reaction by changing the conditions can be determined. This information is, of course, essential to industrial chemists seeking cheaper ways to bring about higher yields. Also important is that information about the reaction mechanism of a drug in the body can tell doctors how long the drug will remain effective before another dose is required. This often helps pharmaceutical companies to produce drugs with fewer side effects from unwanted reactions.

2 RATES OF REACTION AND THEIR MEASUREMENT

This is a definition for the rate of a reaction:

The rate of a reaction is the change in the concentration of product formed per unit of time, or the change in the concentration of reactant used per unit of time.

The reaction rate for a reactant that is consumed is negative.

Fig 26.1 shows how the concentration of a reactant changes during the course of a typical chemical reaction. Note that concentration is denoted by the use of square brackets. So, the concentration of a reactant is shown as [reactant]. The similar graph in Fig 26.2 shows the increase in concentration of a product of a chemical reaction. However, in any given reaction, the decrease in the concentration of the reactant is not necessarily the same as the increase in concentration of the product. For example, if 2 moles of reactant produce 1 mole of product, then the initial concentration of the reactant in Fig 26.1 would be twice that of the final concentration of the product in Fig 26.2.

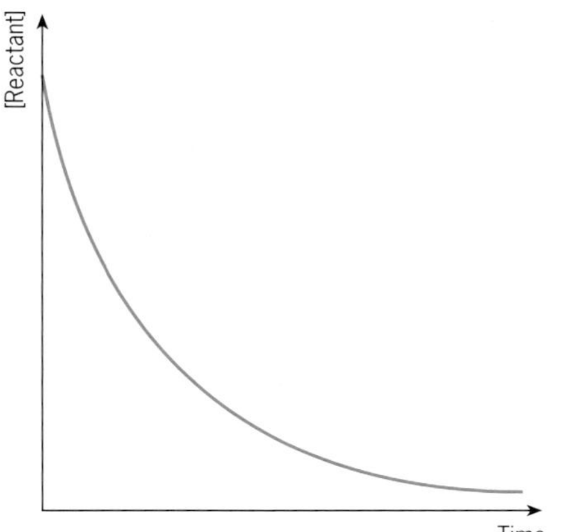

Fig 26.1 The decrease in the concentration of a reactant, or [reactant], during the course of a typical chemical reaction

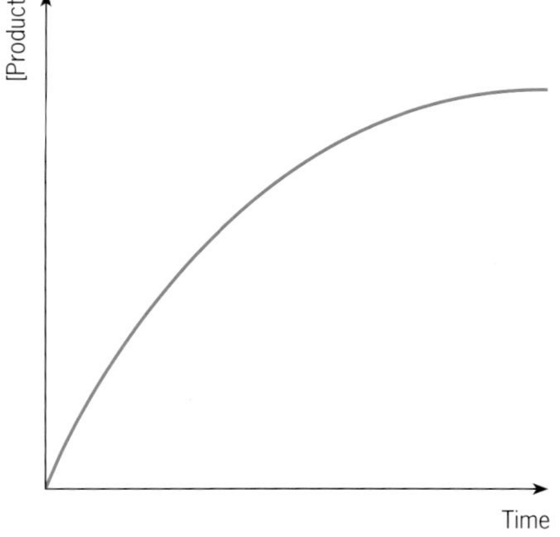

Fig 26.2 The increase in the concentration of a product, or [product], during the course of a typical chemical reaction

QUESTION 1

1 **a)** What do Figs 26.1 and 26.2 tell you about how the rate of reaction changes during the course of a reaction?

b) Why should the temperature be kept constant throughout the experiment?

? QUESTION 2

2 What is the initial rate of the reaction shown in Fig 26.3? (Hint: the tangent at time zero has been drawn for you.)

FINDING THE RATE OF REACTION AT A PARTICULAR TIME

Look again at Fig 26.2. If the rate of reaction were constant, the graph would be a straight line. Clearly, it is changing with time and will eventually fall to zero when the reaction is finished. The graph shows the change of concentration during the course of the reaction and is therefore known as a rate curve. The rate of reaction at any instant in time is given by the gradient of the curve at that instant.

The gradient at any point on a curve is found by drawing the tangent to the curve at that point and taking its gradient. Fig 26.3 shows two tangents. The tangent at time zero is called the initial rate, which occurs when the reactants are first mixed. The gradient of this tangent is the steepest of any taken along the rate curve, which means that the reaction is fastest at the start. The reaction rate at 50 seconds is calculated as follows:

$$\text{rate at 50 s} = \text{gradient of tangent at 50 s} = \frac{3.0 \text{ mol}}{40 \text{ s}} = 0.075 \text{ mol s}^{-1}$$

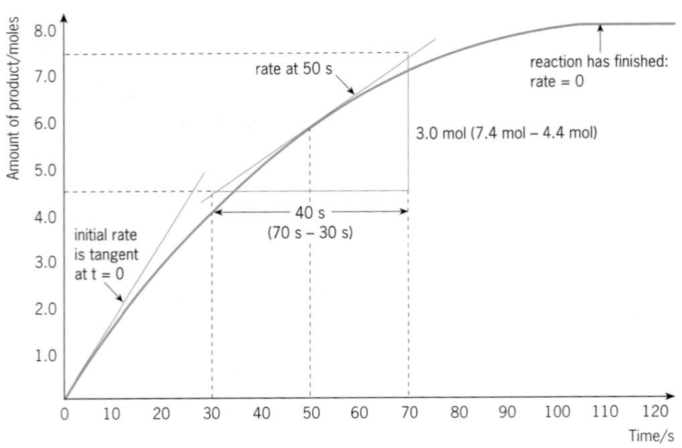

Fig 26.3 The rate curve for a reaction. The method for calculating the rate at 50 s is shown

MEASURING REACTION RATES

When the rate of a reaction is being studied, it is crucial to know what the reactants are, what the products are and what state these substances are in. This is conveniently found in the **stoichiometric equation**. This is a balanced chemical equation which states the amount of each reactant that reacts and the amount of each product formed (see page 7). For example, the stoichiometric equation for the formation of nitrogen dioxide, an atmospheric pollutant, from nitrogen monoxide and oxygen in a car exhaust is:

$$2NO(g) + O_2(g) \rightarrow 2NO_2(g)$$

Although this does not detail the steps by which the reaction occurs, it clearly states that only one product is formed and that only two reactants are involved. It also states that three volumes of reactants produce two volumes of product.

Provided we know the stoichiometric equation, we can decide on how to measure the concentration changes that occur as the reaction proceeds. In the above case there is a reduction in volume, which we could use to measure changes in the rate of reaction with time. If the reaction is carried out at a constant volume in a pressure vessel, then change in pressure could be measured.

Anything that changes during a reaction and can be measured may be used to determine a reaction rate, provided that it is proportional in some way to the concentration of a particular reactant or product.

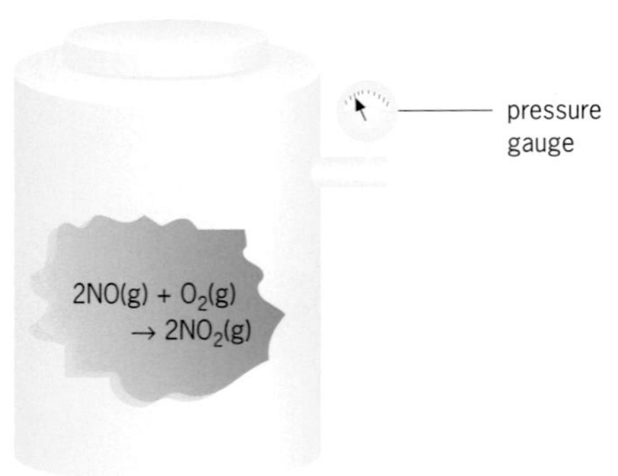

Fig 26.4 The reaction between nitrogen monoxide and oxygen could be followed by measuring the variation of pressure with time

? QUESTION 3

3 What is the name given to the steps by which a chemical reaction takes place? (If you want to check your answer, see page 537.)

Measuring change in volume of gas produced

The reaction between dilute hydrochloric acid and magnesium ribbon produces hydrogen according to this stoichiometric equation:

$$Mg(s) + 2HCl(aq) \rightarrow MgCl_2(aq) + H_2(g)$$

The volume of hydrogen produced could be used to follow (monitor) changes in the rate of this reaction. Apparatus for doing this is shown in Fig 26.5.

> **REMEMBER THIS**
> The rate of a reaction may be defined as the change in concentration of product formed or the change in concentration of reactant used per unit of time.

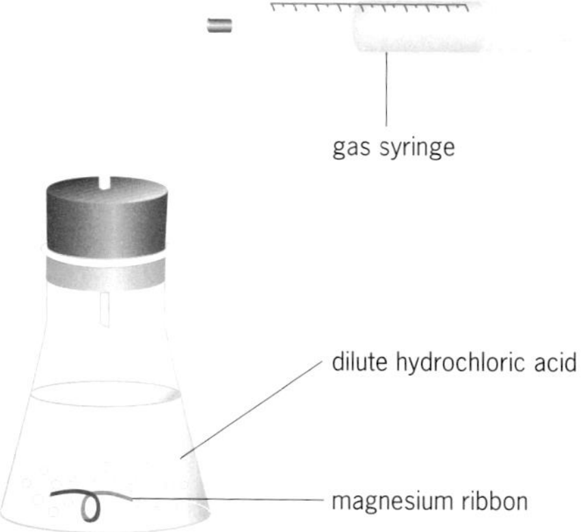

Fig 26.5 The change in volume of this reaction due to the evolution of hydrogen gas can be used to measure the reaction rate

Changes in colour

A colorimeter can be used to measure the change in colour of a reaction. This instrument (see pages 568–569) measures the amount of electromagnetic radiation absorbed by substances in the visible part of the spectrum. In the reaction between zinc and aqueous copper(II) sulfate, for example, the blue coloration of copper(II) sulfate disappears as the colourless zinc sulfate solution is formed:

$$\underset{\text{blue solution}}{Zn(s) + CuSO_4(aq)} \rightarrow \underset{\text{colourless solution}}{ZnSO_4(aq)} + Cu(s)$$

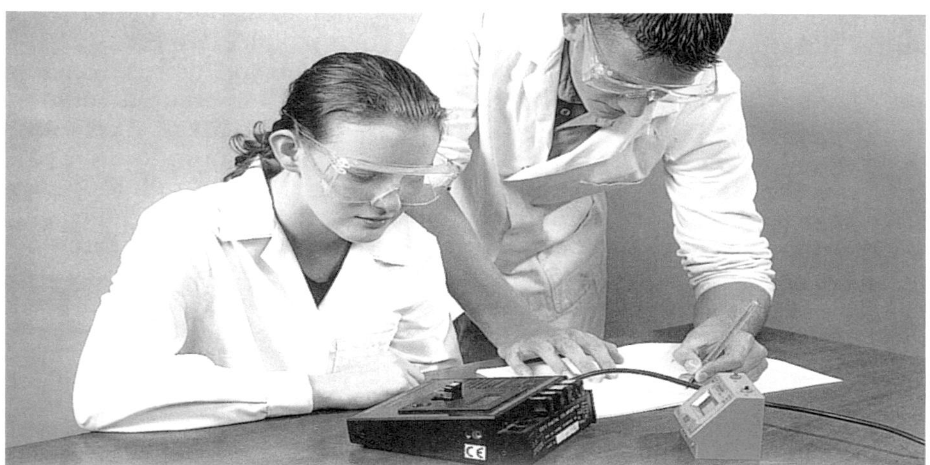

Fig 26.6 When one of the reactants or products is coloured, the reaction can be followed by measuring the change in absorbance with a colorimeter

> **QUESTION 4**
> 4 As hydrogen is being lost in the Mg/HCl(aq) reaction, what other properties change during the course of the reaction and how can they be measured?

Changes in electrical conductivity

Many reactions involve a change of conductivity because the number of ions in the reaction mixture changes during the reaction. In the following reaction, the number of ions decreases as the products are formed and the conductivity of the reaction can be followed using a conductivity cell (Fig 26.7):

$$BrO_3^-(aq) + 6I^-(aq) + 6H^+(aq) \rightarrow 3I_2(s) + 3H_2O(l) + Br^-(aq)$$

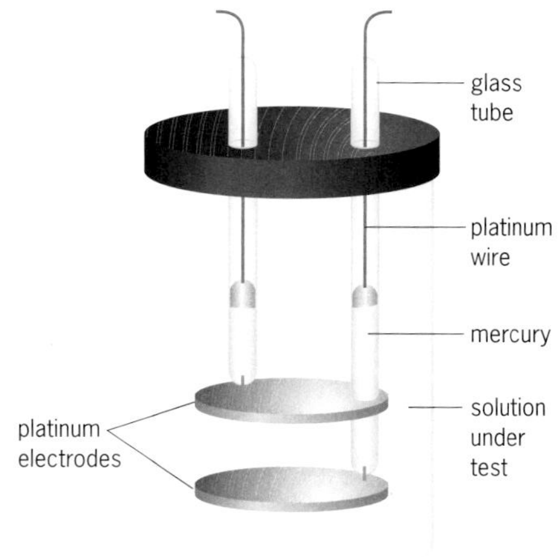

Fig 26.7 A conductivity cell

Chemical analysis

All the techniques so far described follow a reaction continuously and do not interfere with the progress of the reaction. Chemical analysis, however, involves taking samples of the reaction mixture at regular intervals of time. The reaction in the sample is stopped as soon as it is withdrawn. This can be accomplished by rapid cooling, by removing one of the reactants or the catalyst, or by diluting the reaction mixture. The process of stopping a reaction (or slowing it down to a rate of almost zero) is called **quenching**.

Methyl ethanoate, CH_3COOCH_3, an ester, is an important industrial solvent. It can be hydrolysed using aqueous sodium hydroxide:

$$CH_3COOCH_3(l) + NaOH(aq) \rightarrow CH_3COO^-Na^+(aq) + CH_3OH(aq)$$

Hydrolysis of esters is covered on page 357.

Fig 26.8 The change in mass as carbon dioxide is evolved can be used to measure the reaction rate

The rate of reaction can be followed by monitoring the concentration of sodium hydroxide as it is used up during the course of the reaction. Several identical reaction solutions are set to react. The temperature is kept constant, since changes in temperature affect the reaction rate. They are quenched by dilution with ice-cold water, at different time intervals from the start of the reaction. The concentration of sodium hydroxide that remains in each reaction solution is determined by titrating it with an acid, such as dilute hydrochloric acid.

Fig 26.9 These students are titrating a reaction mixture. The reaction has been quenched by adding ice while the titration is performed

QUESTIONS 5–6

5 A 10.0 cm^3 sample of the reaction solution of methyl methanoate and aqueous sodium hydroxide was withdrawn after 2.00 minutes and titrated with 1.0×10^{-3} mol dm^{-3} dilute hydrochloric acid. 15.0 cm^3 of HCl(aq) was required to neutralise the OH^- ions remaining.

a) How many moles of H^+(aq) ions would be present in 15 cm^3 HCl(aq)?

b) Write down the ionic equation for the neutralisation of H^+(aq) ions by OH^-(aq).

c) How many moles of OH^-(aq) ions were present in the 10.0 cm^3 sample?

d) What was the concentration of OH^-(aq) ions after 2.00 minutes? (Hint: if you are not sure how to do these calculations, look at page 130.)

6 Suggest a method by which you could follow the rates of the following reactions:

a) $SO_2Cl_2(g) \rightarrow SO_2(g) + Cl_2(g)$

b) $2H_2O_2(aq) \rightarrow 2H_2O(l) + O_2(g)$

c) $BrO_3^-(aq) + 5Br^-(aq) + 6H^+(aq) \rightarrow 3Br_2(aq) + 3H_2O(l)$

d) $CH_3Br + OH^- \rightarrow CH_3OH + Br^-$

e) $CH_3COCH_3 + I_2(aq) \rightarrow CH_3COCH_2I(aq) + H^+(aq) + I^-(aq)$

Use a different method for each reaction, if you can.

STRETCH AND CHALLENGE

Sonochemistry

In the 1920s it was first discovered that ultrasound – sound with a frequency above 18 kHz – produces chemical effects. The study of sonochemistry, as it is called, did not really take off until the 1980s, when reliable and inexpensive ultrasound generators became readily available. Now, there is a host of interesting and important applications of ultrasound in chemistry.

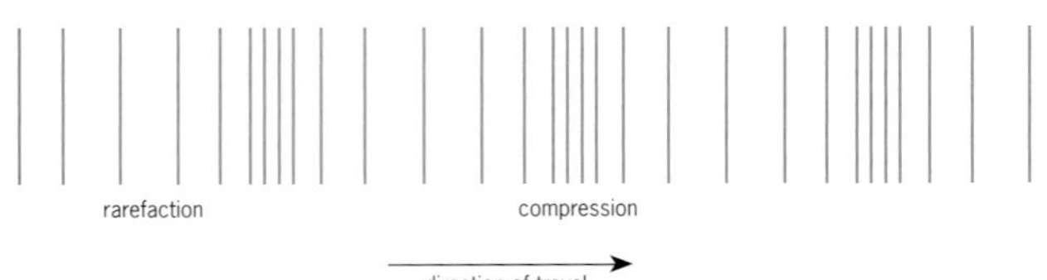

Fig 26.10 The propagation of a sound wave through a liquid. As the wave travels through the liquid, it creates alternate regions of high pressure (compression) and low pressure (rarefaction). These travelling pressure changes are exceptionally rapid when the liquid is excited by ultrasound

When a liquid is excited by ultrasound, the rapid changes in pressure (see Fig 26.10) produce an effect known as cavitation. As the rarefactions travel through the liquid, they pull its molecules apart to produce tiny cavities or bubbles. The compressions cause the bubbles to collapse, which releases tremendous amounts of energy. It has been estimated that the temperature near the collapse may be about 7000 K, which is the temperature at the surface of the Sun. Even higher temperatures – up to 2×10^6 K – may be generated as cavitation bubbles implode. The pressure created could make a gas as dense as a metal. However, the rate of cooling is astonishing at 10^{10} K per second, so overall the liquid does not become hot.

These localised energy hot spots are used to increase the rates of chemical reactions. They also produce highly

STRETCH AND CHALLENGE

Sonochemistry (Cont.)

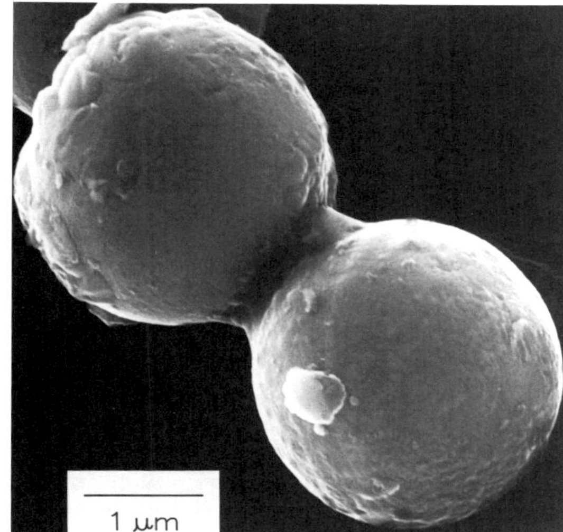

Fig 26.11 Cavitation bubbles produced by ultrasonic waves

reactive radicals. For example, the water molecule can be torn apart to produce H• and OH• radicals. Radicals have unpaired electrons, which makes them highly reactive. So these radicals can combine to produce hydrogen gas and hydrogen peroxide. The OH• radical is also a potent oxidising agent, which can react with other chemicals placed in the water. (For more about radicals – often called free radicals – see page 240.)

Early on, sonochemistry led to the production of catalysts that have particles so minute that they are called nanostructured catalysts. (A nanometre is 10^{-9} metres.) The surface area of these catalyst particles is huge. Nowadays, special combinations of metals in catalysts produced with the aid of ultrasound are making chemical processes more efficient, and it is likely that alternatives to platinum-based catalysts will be found.

Tailoring polymer molecules with ultrasound to enhance particular properties is another exciting prospect. The polymer chains are dissolved in a solvent, in which they are subjected to the awesome energy of cavitation bubbles. The chains break into smaller structures that, under the action of the bubbles, recombine to form different monomers in blocks along the chains.

There is so much unexplored potential in sonochemistry. Another possibility is that some organochlorine pollutants in water supplies could be broken down by ultrasound into harmless products. In a contrasting application, tiny haemoglobin spheres have been synthesised and may make it possible to produce artificial blood.

3 RATE EQUATIONS

Figs 26.1 and 26.2 show that the rate of reaction is fastest at the beginning of a reaction. This is because the concentrations of the reactants are at their highest value at the start. As the concentrations of the reactants fall, so does the rate of reaction. The **rate equation** states the relationship between the rate of reaction and the concentration of each reactant. It is also known as the **rate law**. As shown later, the rate equation helps chemists to work out how a reaction takes place. The dependence of the rate of reaction on the concentration of a reactant can be expressed mathematically in the form:

$$\text{rate} \propto [\text{reactant}]^n$$

which gives the rate equation as:

$$\text{rate} = k \times [\text{reactant}]^n \qquad \text{or} \qquad \text{rate} = k[\text{reactant}]^n$$

where:

- k is a constant called the **rate constant**: the larger the rate constant, the faster the reaction,
- n is called the **order of the reaction** with respect to the given reactant:

The order of the reaction is the power to which the concentration of the reactant is raised in the experimentally determined rate equation.

? QUESTION 7

7 For a reaction A + B → C + D the rate equation is:

$$\text{rate} = k[A]^m[B]^n$$

What are the orders of reaction with respect to the reactants A and B?

Table 26.1 The effect of changing the concentration on the rate of reaction between calcium carbonate and hydrochloric acid

Mixture	[HCl(aq)]/ mol dm^{-3}	Relative rate
A	2.0	8
B	1.0	4
C	0.5	2
D	0.25	1

? QUESTION 8

8 Explain how the calcium carbonate/hydrochloric acid reaction can be followed.

The definition of order of reaction states that the rate equation must be determined by experiment. It has nothing to do with the stoichiometric equation. For example, the equation for the reaction of dilute hydrochloric acid with marble chips is:

$$CaCO_3(s) + 2HCl(aq) \rightarrow CaCl_2(aq) + CO_2(g) + H_2O(l)$$

Yet, when experiments are done to find how the rate of reaction varies with the concentration of each reactant, it is discovered that the rate is directly proportional to the concentration of hydrochloric acid (Table 26.1). So, when the concentration doubles, so does the reaction rate. In this case:

$$\text{rate} \propto [HCl(aq)]^1$$

Therefore, the rate equation is:

$$\text{rate} = k[HCl(aq)]^1 = k[HCl(aq)]$$

(since power 1 is conventionally omitted)

This reaction is **first order** with respect to hydrochloric acid, because the concentration is raised to the power 1 in the experimentally determined rate equation.

EXAMPLE

Q An experiment is carried out to determine the reaction kinetics of this reaction:

$$BrO_3^-(aq) + 6I^-(aq) + 6H^+(aq) \rightarrow 3I_2(s) + 3H_2O(l) + Br^-(aq)$$

The results obtained are given in Table 26.2.

Table 26.2 Results of a study of the kinetics of the reaction between bromate(V) ions and iodide ions in acid solution

Mixture	[BrO_3^-(aq)]/mol dm^{-3}	[I^-(aq)]/mol dm^{-3}	[H^+(aq)]/mol dm^{-3}	Relative initial rate
A	0.10	0.60	0.60	1
B	0.20	0.60	0.60	2
C	0.20	1.20	0.60	4
D	0.20	0.60	1.20	8

Work out the orders of reaction with respect to each reactant, and hence write the overall rate equation.

A Comparing the relative rates for A and B shows that the rate in B doubles when [BrO_3^-(aq)] doubles. Note that the other concentrations remain the same. Therefore:

$$\text{rate} \propto [BrO_3^-(aq)]$$

Comparing the relative rates for B and C shows that [BrO_3^-(aq)] and [H^+(aq)] are the same in both experiments, while [I^-(aq)] doubles in C. This doubles the rate in C. Therefore:

$$\text{rate} \propto [I^-(aq)]$$

Comparing the relative rates of B and D shows that, while the other two concentrations stay the same, doubling [H^+(aq)] quadruples the rate in D. Therefore:

$$\text{rate} \propto [H^+(aq)]^2$$

Combining the three reaction rates gives:

$$\text{rate} \propto [BrO_3^-(aq)] \times [I^-(aq)] \times [H^+(aq)]^2$$

So, the rate equation is:

$$\text{rate} = k[BrO_3^-(aq)][I^-(aq)][H^+(aq)]^2$$

The reaction is **first order** with respect to [BrO_3^-(aq)], because in the rate equation the power to which this concentration is raised is 1.

It is also **first order** with respect to [I^-(aq)], because the power to which this concentration is raised is also 1.

The reaction is **second order** with respect to [H^+(aq)], because in the rate equation the power to which this concentration is raised is 2.

The overall order of the reaction is the sum of the individual orders, which gives 1+1+2 = 4.

Hence, the reaction is **fourth order**.

DETERMINING THE RATE EQUATION BY THE INITIAL-RATE METHOD

In the previous worked example, the reaction rates quoted are initial reaction rates. These are the rates at the start of the reaction. They were found by doing four separate experiments and plotting rate curves for each experiment. The tangent at time zero is drawn for each experiment and the gradient of each tangent is determined. This was the procedure followed to find the initial rate in Fig 26.3, page 538.

In the next Example, the initial-rate method is used to find the rate equation and the rate constant of a reaction that is believed to occur in the exhaust gases of car engines to produce the atmospheric pollutant nitrogen dioxide. In each experiment, the concentration of only one reactant is varied.

? QUESTION 9

9 When using the initial-rate method, chemists often stop the experiment long before the reaction is complete. Why is this?

EXAMPLE

Q The reaction between nitrogen monoxide and oxygen was investigated using the initial rate method. The stoichiometric equation is:

$$2NO(g) + O_2(g) \rightarrow 2NO_2(g)$$

The results obtained at a particular temperature are given in Table 26.3 below.

a) Deduce the orders of the reaction with respect to NO(g) and $O_2(g)$.
b) Write the expression for the rate equation.
c) Determine the overall order of the reaction.
d) Work out the value of the rate constant, showing its units.

Table 26.3

Experiment	[NO(g)]/mol dm^{-3}	[O_2(g)]/mol dm^{-3}	Initial rate of production of NO_2(g)/mol dm^{-3} s^{-1}
1	1.00×10^{-3}	3.00×10^{-3}	4.00×10^{-4}
2	1.00×10^{-3}	6.00×10^{-3}	8.00×10^{-4}
3	2.00×10^{-3}	3.00×10^{-3}	1.60×10^{-3}

A a) Comparing experiments 1 and 3, the concentration of NO(g) doubles, which quadruples the rate of reaction.

The concentration of $O_2(g)$ is constant. Therefore:

$$\text{rate} \propto [NO(g)]^2$$

So, the reaction is **second order** with respect to NO(g).

Comparing experiments 1 and 2, the concentration of $O_2(g)$ is doubled, which doubles the rate of reaction.

The concentration of NO(g) is the same in both experiments. Therefore:

$$\text{rate} \propto [O_2(g)]$$

So, the reaction is **first order** with respect to $O_2(g)$.

b) Since rate $\propto [NO(g)]^2[O_2(g)]$, the rate equation is:

$$\text{rate} = k[NO(g)]^2[O_2(g)]$$

Note that on this occasion the stoichiometric coefficients (the numbers of moles of each substance in the balanced equation) match those of the orders of the reaction. But do remember that the rate equation can only be determined by experiment.

c) The overall order of the reaction is the sum of the powers in the experimentally determined rate equation. This is 2+1 = 3. ([O_2(g)] is to the power 1.) So, the overall order is third.

d) The rate equation is: rate = $k[NO(g)]^2[O_2(g)]$ which gives:

$k = \dfrac{\text{rate}}{[NO(g)]^2[O_2(g)]}$ and units of $k = \dfrac{\text{mol dm}^{-3}\text{ s}^{-1}}{(\text{mol dm}^{-3})^2(\text{mol dm}^{-3})}$

Now substitute one of the sets of values into the equation to determine k. For example, take those for experiment 1:

$k = \dfrac{4.00 \times 10^{-4}}{(1.00 \times 10^{-3})^2 \times 3.00 \times 10^{-3}}$ and units of k = dm^6 mol^{-2} s^{-1}

Therefore:

$$k = 1.33 \times 10^5 \text{ dm}^6 \text{ mol}^{-2} \text{ s}^{-1}$$

Note that dm^6 is placed first when quoting the units. This follows the convention whereby a unit raised to a positive power is placed first.

? QUESTION 10

10 The antiseptic chlorophenol is made using $SO_2Cl_2(g)$. The latter breaks down according to the equation:

$$SO_2Cl_2(g) \rightarrow SO_2(g) + Cl_2(g)$$

Three experiments were performed to determine the rate equation at a particular temperature. The results obtained are given in Table 26.4.

Table 26.4 Kinetic data used to determine the rate equation for the decomposition of $SO_2Cl_2(g)$

$[SO_2Cl_2(g)]$/mol dm^{-3}	Initial rate of formation of $Cl_2(g)$/mol dm^{-3} s^{-1}
0.02	4.4×10^{-7}
0.04	8.8×10^{-7}
0.06	1.32×10^{-6}

Deduce the rate equation for this reaction and calculate the rate constant, stating the units of k.

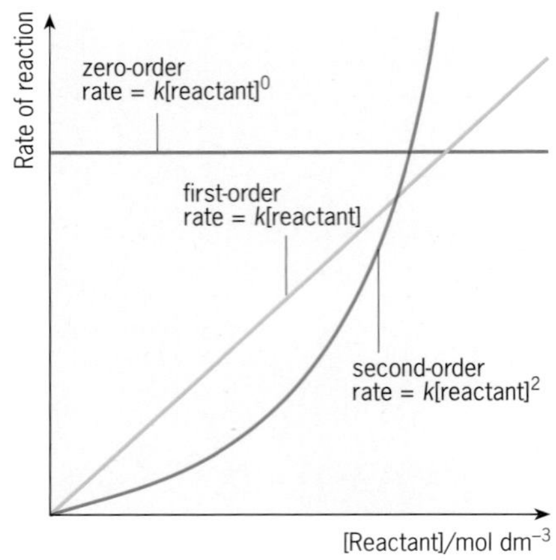

ZERO ORDER AND OTHER ORDERS

In the Examples so far, reactions have been first or second order with respect to each reactant. Another common order of reaction is **zero order** (sometimes called **zeroth order**):

A zero-order reaction is one in which the concentration of a reactant has no effect on the rate of reaction.

For example, the decomposition of ammonia on a tungsten wire takes place according to the equation:

$$2NH_3(g) \xrightarrow{W} N_2(g) + 3H_2(g)$$

The rate equation for this reaction is:

$$\text{rate} = k[NH_3(g)]^0 = k \qquad (\text{since } [\text{reactant}]^0 = 1)$$

It does not matter what the concentration of ammonia is; the rate is always the same at a particular temperature.

Fig 26.12 illustrates the effect of reactant concentration on the rate of reaction for zero, first and second orders. For a zero-order reaction, the concentration has no effect on the rate, and hence the graph is a horizontal line. For a first-order reaction, the rate is directly proportional to the reactant concentration, which gives a straight-line graph. Since rate = k[reactant], the gradient is equal to the rate constant, k. For a second-order reaction, where:

$$\text{rate} \propto [\text{reactant}]^2$$

the graph is a curve.

Fig 26.12 The effect of reactant concentration on the rate of zero, first- and second-order reactions

The reaction of propanone (CH_3COCH_3) with iodine is interesting. Although in the stoichiometric equation hydrogen ions are not involved as reactants, they catalyse the reaction. As the reaction itself produces these ions, the reaction is said to be **auto-catalysed** (self-catalysed):

$$CH_3COCH_3(aq) + I_2(aq) \rightarrow CH_3COCH_2I(aq) + H^+(aq) + I^-(aq)$$

The rate equation for this reaction can be written:

$$\text{rate} = k[CH_3COCH_3(aq)]^1[H^+(aq)]^1[I_2(aq)]^0$$
$$= k[CH_3COCH_3(aq)]\,[H^+(aq)] \qquad (\text{since } [\ ]^0 = 1)$$

Since the concentration of iodine has no effect on the rate, the reaction is zero order with respect to iodine. But it is first order with respect to both propanone and hydrogen ions.

An order of reaction need not be zero, first or second order. An order of reaction may be a fraction, but this aspect lies outside the scope of this book.

? QUESTIONS 11–12

11 What is the overall order for the reaction of iodine with propanone?

12 The rate equations are given for the following reactions. Give the order with respect to each reactant and the overall order for the reaction.

a) $BrO_3^-(aq) + 3SO_3^{2-}(aq) \rightarrow Br^-(aq) + 3SO_4^{2-}(aq)$
rate = $k[BrO_3^-(aq)][SO_3^{2-}(aq)]$

b) $2H_2(g) + 2NO(g) \rightarrow 2H_2O(g) + N_2(g)$
rate = $k[NO(g)]^2[H_2(g)]$

c) $NO_2(g) + CO(g) \rightarrow NO(g) + CO_2(g)$
rate = $k[NO_2(g)]^2$

Work out the units of the rate constant for each reaction. In all cases, concentration is measured in mol dm^{-3}, and rate in mol dm^{-3} s^{-1}.

HALF-LIVES AND FIRST-ORDER REACTIONS

The half-life for a chemical reaction is defined as follows:

The half-life of a chemical reaction is the time taken for the concentration of a reactant to decrease to half of its initial value.

In the case of first-order reactions, the half-life is constant. This means that, whatever the starting concentration of a reactant, it will always take the same time for this concentration to be halved.

The decomposition of hydrogen peroxide using manganese(IV) oxide catalyst is a first-order reaction (Fig 26.13):

$$H_2O_2(aq) \rightarrow H_2O(l) + \tfrac{1}{2}O_2(g)$$

$$\text{rate} = k[H_2O_2(aq)]$$

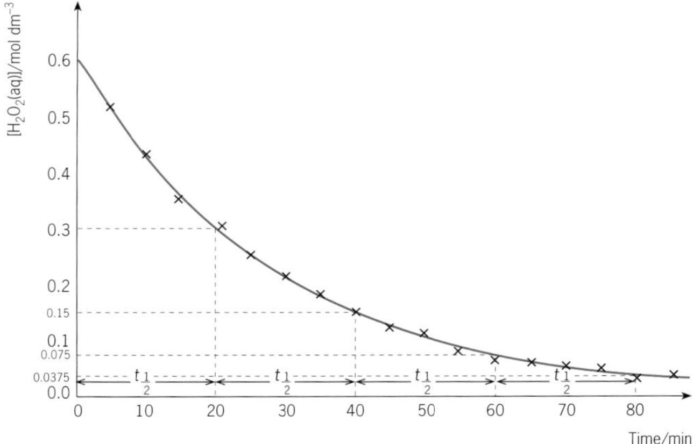

Fig 26.13 The change in concentration of hydrogen peroxide with time in the catalytic decomposition of hydrogen peroxide, using MnO_2 as the catalyst at 20 °C

? QUESTION 13

13 How can the decomposition of hydrogen peroxide reaction be followed? (Hint: one of the products is a gas.)

Table 26.5 Four successive half-lives from the decomposition of hydrogen peroxide

Initial concentration H_2O_2(aq)/mol dm^{-3}	Half-life $t_{\frac{1}{2}}$/min
0.60	20
0.30	20
0.15	20
0.075	20

Four successive half-lives from the graph in Fig 26.13 are given in Table 26.5. Note that the four half-lives are all constant. This is a characteristic of first-order reactions.

Using successive half-lives may involve very small concentrations, which can lead to inaccuracies. In Fig 26.13, the time taken for other concentrations of H_2O_2(aq) to decrease by half (e.g. 0.5 mol dm^{-3} falling to 0.25 mol dm^{-3}) could have been used.

Therefore, half-lives can be used to identify a first-order reaction from a single kinetics experiment. If the half-life of a reactant is known, the rate constant of a first-order reaction can be calculated using the following relationship between the half-life and the rate constant:

$$k = \frac{0.693}{t_{\frac{1}{2}}}$$

You do not have to know how to derive this equation, but you should note its usefulness in calculating rate constants.

REMEMBER THIS

You do not have to measure successive half-lives to verify that the half-lives are constant. Instead you can measure at least three different half-lives, starting from three different reactant concentrations.

? QUESTION 15

15 Radioactive decay is another first-order reaction, because successive half-lives are constant. The half-life of sodium-24 is 15 hours.

a) Starting with 1 kg of a radioactive sample of sodium-24, how much of it will be left after 30 hours?

b) What is the rate constant for this decay?

? QUESTION 14

14 The determination of the catalytic decomposition of hydrogen peroxide as a first-order reaction can be made with a single experiment. Another method used to study the kinetics of this reaction involves initial rates. Explain how you would study this reaction using the initial-rates method. How would you show that the reaction is first order?

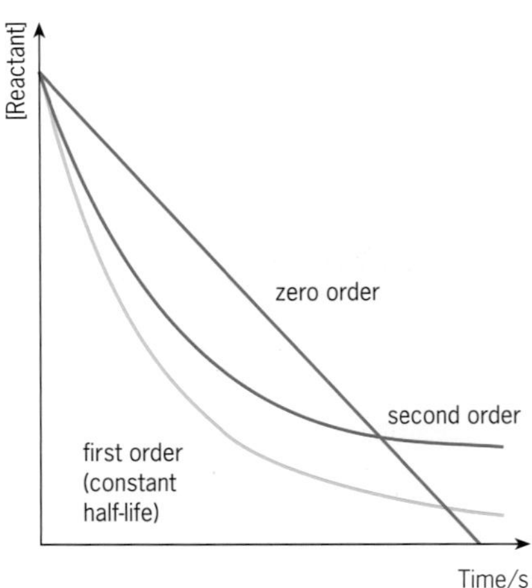

Fig 26.14 Concentration–time graphs for zero, first- and second-order reactions

CONCENTRATION–TIME GRAPHS

When a reaction is zero order with respect to a reactant, the reactant concentration has no effect on the rate of reaction:

$$\text{rate} = k[\text{reactant}]^0 = k \qquad (\text{since } [\text{reactant}]^0 \text{ is } 1)$$

In Fig 26.14, the concentration–time graph for a zero-order reaction shows a constant decline in concentration as the reaction proceeds.

For a first-order reaction:

$$\text{rate} = k[\text{reactant}]^1 = k[\text{reactant}] \qquad (\text{since } [\text{reactant}]^1 \equiv [\text{reactant}])$$

As already stated, irrespective of the starting concentration, the half-life has a fixed value for a first-order reaction, so the concentration–time graph has the characteristic shape shown in Fig 26.14.

For a second-order reaction:

$$\text{rate} = k[\text{reactant}]^2$$

While the starting rate is the same as that of a first-order reaction, the concentration–time graph approaches zero much more slowly, as Fig 26.14 shows. This is because in a second-order reaction the half-life increases as the concentration diminishes, unlike in a first-order reaction for which the half-life is independent of the concentration.

4 RATE EQUATIONS IN THE DETERMINATION OF REACTION MECHANISMS

There is more about the reaction mechanism of ozone depletion by CFCs in the How Science Works assignment at www.collinseducation.co.uk

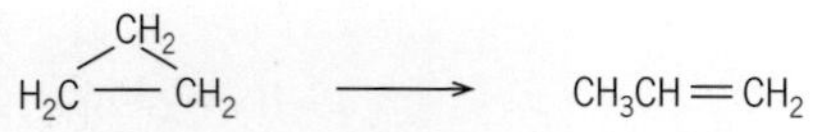

As stated earlier, one of the reasons for studying the kinetics of a reaction is to obtain the information that enables chemists to decide how the reaction occurs – the **reaction mechanism**. Although some reactions do take place in one step, most proceed in a series of steps that lead to the formation of products. In each step, bonds are broken or made.

Rate equations provide evidence of what the reaction mechanism might be. For example, knowledge of the likely reaction mechanism of ozone depletion in the upper atmosphere by chlorofluorocarbons (CFCs) proved to be crucial in helping governments to agree to ban their manufacture.

For a sequence of bond-breaking and bond-making steps to occur, the reactants must collide with sufficient energy. This is the **collision theory** of chemical reactions, which is discussed in more detail in Chapter 8 (pages 169 and 170). Each step in a reaction mechanism usually involves one or two molecules.

CH_2 / H_2C — CH_2 ⟶ $CH_3CH{=}CH_2$

Fig 26.15 The conversion of cyclopropane into propene

Consider a proposed single-step reaction mechanism: the high-temperature conversion of cyclopropane into propene (Fig 26.15). The experimentally derived rate equation for this reaction is:

$$\text{rate} = k[\text{cyclopropane}]$$

The reaction is first order. So, given the single-step mechanism, only one molecule of cyclopropane can be involved. Hence, the energy given to this molecule must cause it to shake itself apart. This is called a **unimolecular step**. The **molecularity** of a step is the number of species (atoms, molecules or ions) involved.

? QUESTION 16

16 The conversion of cyclopropane into propene is sometimes called an isomerisation reaction. Can you explain why? (Hint: if you are not sure, refer to page 182.)

It is unusual to find reactions that have just one step. Most involve at least two, even when they are first-order reactions. For example, the stoichiometric equation for the hydrolysis of 2-bromo-2-methylpropane is given on the right, and its rate equation is:

$$\text{rate} = k[(CH_3)_3CBr]$$

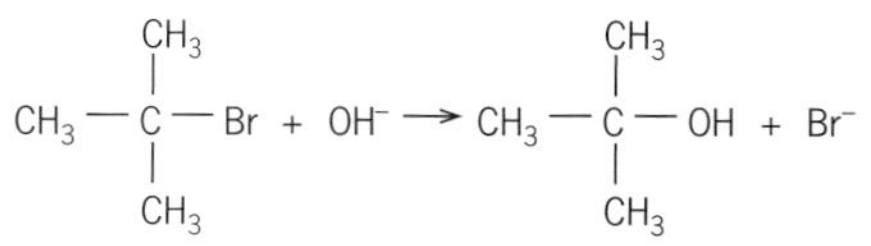

The stoichiometric equation gives no clue as to the rate equation, which always has to be determined by experiment. A proposed mechanism that fits this rate equation is:

$$(CH_3)_3CBr \xrightarrow{\text{slow}} (CH_3)_3C^+ + Br^-$$

$$(CH_3)_3C^+ + OH^- \xrightarrow{\text{fast}} (CH_3)_3COH$$

It is the slower step that governs the rate of reaction. It is called the **rate-determining step**. Suppose you are throwing a party and send out hand-written invitations. You have three people helping you. You write the invitations, which is the slow step. One friend folds them, which does not take long, while another friend puts them in envelopes. The last person addresses the envelopes and sticks on the stamps. It is your step that takes the longest. It does not matter how fast your friends work, the rate is determined by the slowest step. Writing the invitations is the rate-determining step. It is the same with chemical reactions: the slowest step determines the rate of the entire reaction.

The rate-determining step for this hydrolysis reaction is unimolecular because it involves only one molecule. The fast step may be referred to as **bimolecular** because it involves two reacting species. Although the stoichiometry of the overall reaction does not give the rate equation, the stoichiometry of the rate-determining step of the reaction can.

In the rate-determining step:

$$(CH_3)_3CBr \xrightarrow{\text{slow}} (CH_3)_3C^+ + Br^-$$

the rate of reaction is governed by the breaking of a chemical bond in the reactant molecule. As only this molecule is involved, the rate is proportional to the concentration. The greater the concentration, the faster the reaction. So, the rate equation for this step is:

$$\text{rate} = k[(CH_3)_3CBr]$$

which is also the rate of the overall reaction. The overall order is zero for OH^- ions, since these react after the rate-determining step and their concentration therefore has no effect on the reaction rate. Reacting species that occur before the slow step may feature in the overall rate equation.

This reaction is sometimes called an S_N1 reaction. S signifies that it is a substitution reaction, N that it is nucleophilic, and 1 that it is first order with one molecule involved in the rate-determining step. For more information on this reaction, turn to page 249.

Whatever reaction mechanism is proposed, the sum of its steps must equal the overall stoichiometric equation. In the case of the hydrolysis of 2-bromo-2-methylpropane, it does (Fig 26.17).

$$CH_3-C(CH_3)_2-Br + OH^- \longrightarrow CH_3-C(CH_3)_2-OH + Br^-$$

Fig 26.16 Hydrolysis of 2-bromo-2-methylpropane

QUESTION 17

17 NO_2F is an explosive compound. The mechanism for its production is thought to be:

$$NO_2 + F_2 \xrightarrow{\text{slow}} NO_2F + F$$

$$NO_2 + F \xrightarrow{\text{fast}} NO_2F$$

What is the overall stoichiometric equation for this reaction?

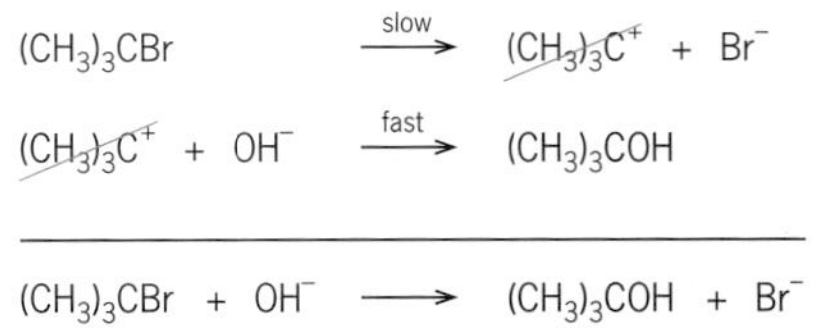

$$(CH_3)_3CBr \xrightarrow{\text{slow}} (CH_3)_3C^+ + Br^-$$

$$(CH_3)_3C^+ + OH^- \xrightarrow{\text{fast}} (CH_3)_3COH$$

$$(CH_3)_3CBr + OH^- \longrightarrow (CH_3)_3COH + Br^-$$

Fig 26.17 Mechanism of the hydrolysis of 2-bromo-2-methylpropane

QUESTION 18

18 2-bromo-2-methylpropane is called a tertiary halogenoalkane. Primary halogenoalkanes such as 1-bromobutane, $CH_3CH_2CH_2CH_2Br$, are believed to react by the mechanism shown in Fig 26.18.

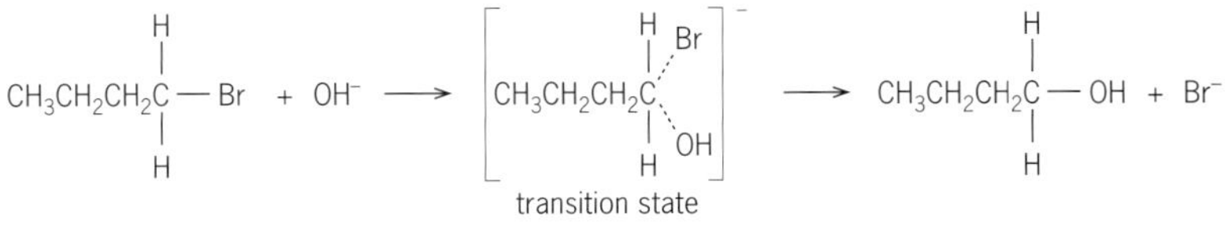

$$CH_3CH_2CH_2CH_2-Br + OH^- \longrightarrow [HO\cdots CH_2(CH_2CH_2CH_3)\cdots Br]^- \longrightarrow CH_3CH_2CH_2CH_2-OH + Br^-$$

transition state

Fig 26.18 The S_N2 reaction mechanism for the hydrolysis of 1-bromobutane

The rate equation is: $\text{rate} = k[CH_3CH_2CH_2CH_2Br][OH^-]$

Why do you think this is called an S_N2 reaction?

HOW SCIENCE WORKS

Working out another mechanism

Reaction mechanisms are developed as a result of analysis of experimental observations and scientific guesses. To illustrate this, consider the reaction of carbon monoxide and nitrogen dioxide. This reaction is believed to take place in the exhaust gases of car engines. To determine the mechanism, start with the stoichiometric equation:

$$NO_2(g) + CO(g) \rightarrow NO(g) + CO_2(g)$$

Note that this equation gives no information about how the reaction occurs, so the next stage is to carry out experiments to determine the overall rate equation. This has been found to be:

$$\text{rate} = k[NO_2]^2$$

The reaction is zero order with respect to carbon monoxide, which means it does not take part in the rate-determining step.

Now a mechanism has to be proposed that is both consistent with the rate equation and, when the steps are added together, gives the overall stoichiometric equation. This is really an educated guess:

$$NO_2 + NO_2 \xrightarrow{\text{slow}} N_2O_4$$
$$N_2O_4 + CO \xrightarrow{\text{fast}} NO + CO_2 + NO_2$$
$$\overline{\overline{\quad NO_2(g) + CO(g) \rightarrow NO(g) + CO_2(g) \quad}}$$

The slower step is the rate-determining step. This involves a collision between two nitrogen dioxide molecules, so the rate equation for this step is rate = $k[NO_2]^2$, and the overall rate equation is also rate = $k[NO_2]^2$. Carbon monoxide appears in the fast step, so its concentration does not affect the rate. Therefore, the proposed mechanism is consistent with the rate equation and the steps when added together give the overall equation.

However, when this reaction is analysed, a short-lived intermediate, NO_3, is discovered. Also, no N_2O_4 is found. This means that the educated guess was incorrect and another mechanism must be proposed, such as:

$$NO_2 + NO_2 \xrightarrow{\text{slow}} NO_3 + NO$$
$$NO_3 + CO \xrightarrow{\text{fast}} NO_2 + CO_2$$
$$\overline{\overline{\quad NO_2(g) + CO(g) \rightarrow NO(g) + CO_2(g) \quad}}$$

This still gives a bimolecular rate-determining step and the rate equation is consistent with this. Also, the overall stoichiometric equation is obtained when the two steps are added together. This is a good example of a mechanism being proposed that is consistent with the rate equation, but is not consistent with other experimental evidence and therefore has to be amended accordingly. No-one can prove that this mechanism *is* correct. What can be said is that, based on current evidence, it is thought to be likely.

Sometimes mechanisms are proposed that have a fast step followed by a slow step. In such cases, a reaction intermediate builds up and could be sufficiently long-lived to be detected. This would give another clue to the mechanism.

? QUESTIONS 19–20

19 The decomposition of hydrogen peroxide can be catalysed by bromide ions:

$$2H_2O_2(aq) \xrightarrow{Br^-(aq)} 2H_2O(l) + O_2(g)$$

The rate equation is:

$$\text{rate} = k[H_2O_2][Br^-(aq)]$$

Which of the following mechanisms is consistent with this rate equation? Give reasons for your choice.

Mechanism 1

$$H_2O_2 \xrightarrow{\text{slow}} 2HO$$
$$2HO + Br^- \xrightarrow{\text{fast}} BrO^- + H_2O$$
$$H_2O_2 + BrO^- \xrightarrow{\text{fast}} H_2O + O_2 + Br^-$$

Mechanism 2

$$H_2O_2 + Br^- \xrightarrow{\text{slow}} H_2O + BrO^-$$
$$BrO^- + H_2O_2 \xrightarrow{\text{fast}} H_2O + O_2 + Br^-$$

20 The stoichiometric equation for a reaction is:

$$NO(g) + N_2O_5(g) \rightarrow 3NO_2(g)$$

A proposed mechanism is:

$$N_2O_5 \xrightarrow{\text{slow}} NO_2 + NO_3$$
$$NO + NO_3 \xrightarrow{\text{fast}} 2NO_2$$

a) What is the rate equation for each step?
b) Which is the rate-determining step?
c) Predict the overall rate equation.
d) Why is the overall reaction zero order with respect to NO?

PREDICTING RATE EQUATIONS FROM STEPS IN THE REACTION MECHANISM

Table 26.6 shows three possible steps in reaction mechanisms, together with the rate equations that can be derived from them. The molecularity (number of molecules) of each step and the order are the same. So when the step is unimolecular, it is a first-order reaction. Similarly, a bimolecular reaction gives a second-order reaction. On very rare occasions, a step that contains three molecules is thought to be involved, but this requires three particles to collide simultaneously, which is improbable. When these are the slow steps, they govern the overall rate of the reaction. Therefore, the overall rate equation can be predicted to be the same as this rate-determining step.

REMEMBER THIS

The rate equation can be predicted from the rate-determining step, but not from the overall stoichiometric equation.

Table 26.6 Possible steps in reaction mechanisms

Step	Molecularity	Rate equation
$A \rightarrow$ products	unimolecular	rate = $k[A]$
$A + A \rightarrow$ products	bimolecular	rate = $k[A]^2$
$A + B \rightarrow$ products	bimolecular	rate = $k[A][B]$

Explosive reactions and Nobel Prizes

For centuries, gunpowder was the only high explosive. Then, in 1846, the compounds nitrocellulose (gun cotton) and nitroglycerine were made. Both compounds have very low activation energies for their decomposition and so are highly unstable. (For information about activation energies see Chapter 8, pages 163–165.)

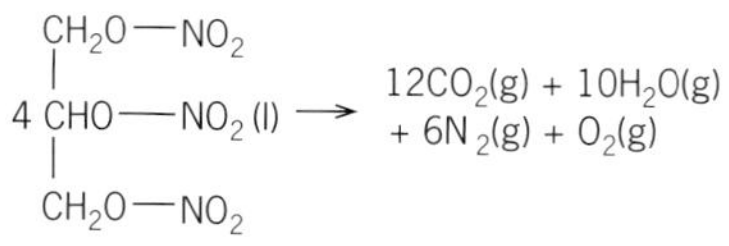

$$4\ \begin{matrix} CH_2O—NO_2 \\ | \\ CHO—NO_2 \\ | \\ CH_2O—NO_2 \end{matrix}(l) \rightarrow 12CO_2(g) + 10H_2O(g) + 6N_2(g) + O_2(g)$$

nitroglycerine

Fig 26.19 Four moles of liquid give 29 moles of gases. So it is the expansion of the hot gases from a small volume of liquid that produces the explosion

Fig 26.20 Alfred Nobel (1833–1896) in his laboratory

The Swedish Nobel family were manufacturers of explosives. In 1863, Alfred Nobel invented a detonator that would set off liquid nitroglycerine. However, in 1864 a nitroglycerine explosion at the Nobel factory killed Alfred's younger brother, Emil, and four other people. Alfred set about making nitroglycerine safe, and within four years he had produced dynamite. (The name is from *dynamis*, the Greek for 'power'.)

In dynamite, the nitroglycerine is absorbed by an inert solid called kieselguhr, which renders it stable until it is set off by a detonator. In 1875, Nobel produced gelignite – an even more powerful explosive. This, too, contained nitroglycerine. This time it was mixed with nitrocellulose in a gel. These explosives were soon put to use in civil engineering projects, such as blasting routes for new roads and excavating the ground for canals and the foundations of buildings.

Fig 26.21 Dynamite is sometimes used to demolish unwanted buildings and other structures, such as these cooling towers. The explosive has to be correctly positioned to ensure that the building will collapse in on itself and not be scattered over the surrounding area

Alfred Nobel was convinced that the only way to stop wars was to produce an explosive so powerful that no-one would dare use it. In this he did not succeed, but in his will he left most of the vast fortune he had amassed from selling explosives to the establishment of five prizes to be awarded annually for outstanding achievements in chemistry, physics, physiology or medicine, literature and peace. In 1969, a sixth prize was added for economics. These Nobel prizes are now acknowledged to be the world's most prestigious accolade. They cannot be awarded posthumously, and no more than three people can share each prize.

STRETCH AND CHALLENGE

The iodination of propanone

On page 544 we described the reaction of iodine with propanone. This reaction is catalysed by the presence of hydrogen ions:.

$CH_3COCH_3(aq) + I_2(aq) \quad CH_3COCH_2I(aq) + HI(aq)$

The reaction can be investigated in two ways.

- The first way involves using colorimetry (see page 608), since as the reaction proceeds the brown colour of the aqueous iodine fades away. It is possible to continuously monitor the change in iodine concentration by this method.
- The second way involves chemical analysis, by taking samples of the reaction mixture at different times. The reaction is quenched by adding excess aqueous sodium hydrogencarbonate, which neutralises the acid catalyst. The iodine in the mixture can then be titrated against aqueous sodium thiosulphate to determine the iodine concentration.

When investigating the reaction it is important that only one variable is changed at a time. So when investigating the effect of the $[CH_3COCH_3]$ all other concentrations must remain constant. But how is this possible, when the acid concentration will increase during the reaction? The answer is simple: use a much higher concentration of the acid, so that even if some acid is made it is very small compared to the acid already there. In the same way, to study the effect of the acid concentration on the rate of reaction use a much higher concentration of propanone.

Rate studies have shown that the rate equation is:

$\text{Rate} = k[CH_3COCH_3(aq)][H^+(aq)]$

This means that the rate-determining step must involve a collision between a hydrogen ion and a propanone molecule. Chemists guess that this is the first step, and then later on there is a slower step that involves an iodine molecule.

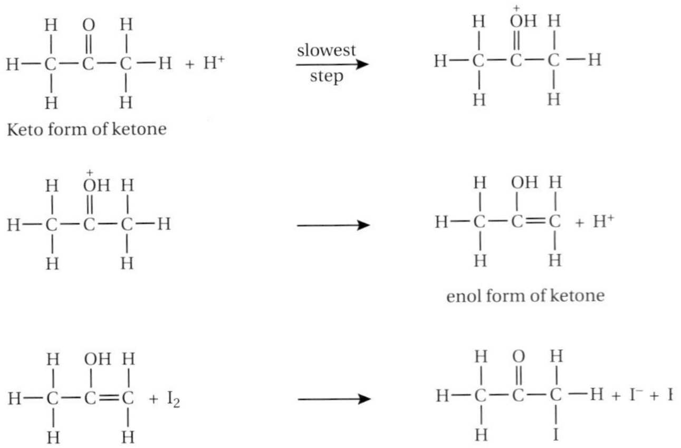

```
   H  O  H                                    H  +OH H
   |  ‖  |                   slowest          |  ‖   |
H—C—C—C—H  + H+            ————————→       H—C—C—C—H
   |     |                    step            |      |
   H     H                                    H      H
Keto form of ketone

   H  +OH H                                   H  OH  H
   |  ‖   |                                   |  |   |
H—C—C—C—H                  ————————→       H—C—C=C  + H+
   |      |                                   |      |
   H      H                                   H      H
                                             enol form of ketone

   H  OH  H                                   H  O   H
   |  |   |                                   |  ‖   |
H—C—C=C  + I2              ————————→       H—C—C—C—H + I− + I
   |      |                                   |      |
   H      H                                   H      I
```

Fig 26.22 A possible mechanism for the reaction between propanone and iodine. The reaction is believed to take place via an intermediate called an enol

5 EFFECT OF TEMPERATURE ON THE RATE CONSTANT

To remind yourself about energy distributions in molecules of a gas, see page 172 in chapter 8.

REMEMBER THIS

The rate equation can only be determined by doing experiments.

? QUESTION 21

21 If a 10 °C rise in temperature doubles the rate of a chemical reaction at constant concentrations, what is the effect on the rate constant, k?

As Fig 26.23 shows, increasing the temperature increases the proportion of molecules that have the minimum activation energy needed for them to react when they collide. This means that the rate of reaction increases with temperature. For every 10 °C rise, the rate of most reactions approximately

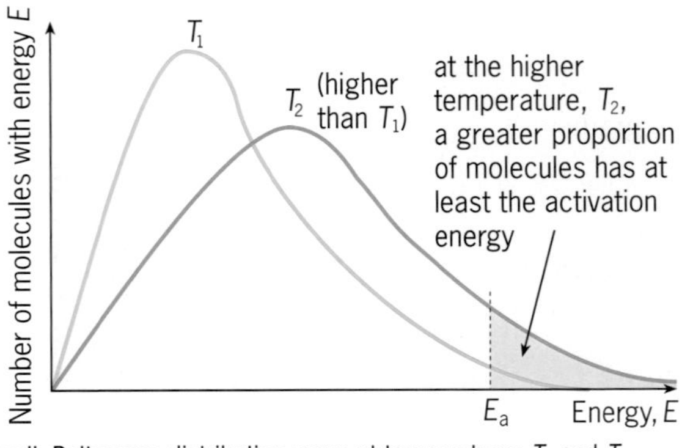

Fig 26.23 Maxwell–Boltzmann distribution curve at temperatures T_1 and T_2

doubles.

Consider the reaction:

$$A + B \rightarrow C + D$$

for which the rate equation is:

$$\text{rate} = k[A]^m[B]^n$$

If the concentrations of A and B are kept constant, then the rate constant must increase with increasing temperature.

This can be clearly seen if we consider how the rate constant varies for the decomposition of hydrogen iodide (Table 26.7). The stoichiometric equation is:

$$2HI(g) \rightarrow H_2(g) + I_2(g)$$

with a rate equation:

$$\text{rate} = k[HI(g)]^2$$

From Table 26.7, the rate constant increases with increasing temperature.

Table 26.7 Variation of the rate constant over a range of temperatures for the decomposition of hydrogen iodide

Temperature/K	Rate constant, $k/dm^3\ mol^{-1}\ s^{-1}$
550	2.9×10^{-7}
600	7.2×10^{-6}
650	1.1×10^{-4}
700	1.2×10^{-3}
750	1.1×10^{-2}
800	7.1×10^{-2}

STRETCH AND CHALLENGE

The Arrhenius equation

The Swedish chemist Svante Arrhenius was the first to discover a mathematical relationship between temperature and the rate constant. The **Arrhenius equation** is:

$$k = Ae^{-E_a/RT}$$

where k is the rate constant.

A, also a constant, is the collision frequency. When molecules collide, they must have the correct orientation to react, a factor that is included in the collision frequency. A is sometimes called the Arrhenius constant.

The term $e^{E_a/RT}$ gives the fraction of collisions that have the minimum activation energy at a particular temperature, where E_a is the activation energy in joules, R is the gas constant 8.31 J K^{-1} mol^{-1} (see page 121), T is the absolute temperature in K and e means it is an exponential relationship. It expresses the condition that an increase in the temperature increases the reaction rate exponentially.

Another way to express the Arrhenius equation is to take natural logarithms (base e) of both sides:

$$\ln k = \ln A - \frac{E_a}{RT}$$

A straight-line graph of $\ln k$ against $1/T$ can be plotted, as in Fig 26.24. Its gradient is $-E_a/R$, so E_a can be determined.

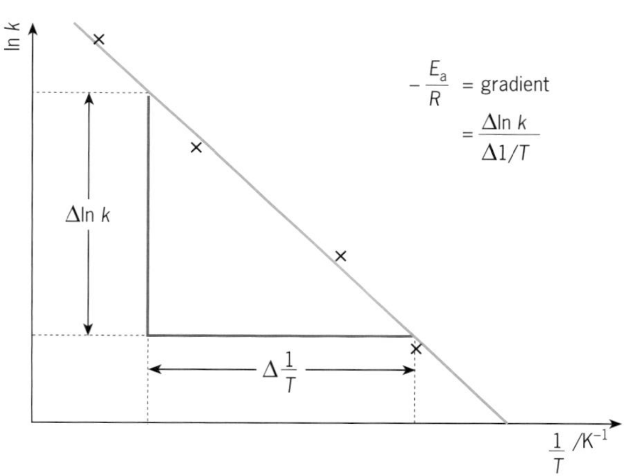

Fig 26.24 The activation energy can be determined from the gradient of this graph

Fig 26.25 Svante Arrhenius (1859–1927). As well as for his equation, he is famous for formulating the idea that electrolytes dissociate in solution to give charged particles

SUMMARY

After studying this chapter, you should know that:

- Reaction kinetics is the study of rates of reactions.
- The rate of a reaction is the change in concentration of a product, or the change in concentration of a reactant, per unit of time.
- Concentration is indicated by square brackets, as in [reactant].
- The gradient of a graph of reactant (or product) concentration against time gives the rate of the reaction at any particular instant in time. It is usually worked out using tangents. The initial rate is the gradient at time zero.
- Anything that changes during a reaction, which can be measured, may be used to determine a reaction rate provided that it is proportional in some way to the concentration of a particular reactant or product.
- Rate of reaction = $k[A]^m[B]^n$ is the rate equation for the reaction:

  $$A + B \rightarrow C + D$$

  where m and n are the orders of the reaction with respect to A and B, and k is the rate constant.
- The overall rate equation must be experimentally determined. It cannot be worked out from the stoichiometric equation.
- When a reaction is zero order with respect to a reactant, the reactant's concentration has no effect on the rate of the reaction.
- The half-life of a chemical reaction is the time taken for the concentration of a reactant to fall to half of its initial value. First-order reactions have constant half-lives.
- The series of steps involved in a reaction is called the reaction mechanism.
- Rate equations provide evidence for proposed reaction mechanisms.
- The rate-determining step is the slowest step in a reaction.
- A unimolecular rate-determining step produces a first-order reaction, while a bimolecular rate-determining step produces a second-order reaction.
- Increasing the temperature increases the rate constant. The relationship between the rate constant and temperature is given by the Arrhenius equation, $k = Ae^{-E_a/RT}$.

Practice questions and a How Science Works assignment for this chapter are available at www.collinseducation.co.uk/CAS

27

Colour in organic and inorganic compounds

27 COLOUR IN ORGANIC AND INORGANIC COMPOUNDS

Denim jeans These seem to be here to stay

Indigo – the dye for denim jeans

The blue of denim jeans has a long and colourful history. Levi Strauss emigrated to America in 1850 during the gold rush, taking with him some heavy cotton canvas called *serge de Nîmes*. Instead of making tents with the canvas, he used the hard-wearing cloth to make trousers for the gold miners. This was the start of a multi-million pound industry.

The blue dye Levi Strauss used was indigo, which has been used as a dye for at least 3000 years. In 1850, all dyes were extracted from plants or animals, and indigo came almost exclusively from indigo-bearing plants grown in huge plantations in India. The indigo market brought in £4 000 000 of revenue per annum, a fortune at that time.

Six years after Strauss dyed his first pairs of denim jeans, a chance discovery in England by William Perkin led to the world's first synthetic dye. In 1880, Adolph von Baeyer synthesized indigo, and by 1897 synthetic indigo was being manufactured. Within a few years the cultivation of natural indigo had ceased, wiping out one of India's major export industries.

Indigo is a dye that fades, and during the first half of the 20th century its popularity waned as chemists synthesised new dyes that did not fade. However, in the 1960s blue denim jeans came back into fashion. The fact that synthetic indigo fades with wear became part of its attraction.

The synthesis of indigo is a chemical success story, with chemists first determining indigo's structure and then finding ways to manufacture it much more cheaply than it could be extracted from plants.

1 WHY ARE THINGS COLOURED?

Colours play an enormous part in your life. Your eyes detect the colours of objects and send messages to your brain, providing you with a constant stream of information. The colour of traffic lights helps to control traffic flow and the blue flashing light of the emergency services alerts you to their presence. The colour of a food determines how appetising it appears and may indicate how fresh it is. The colours of your clothes make a statement about your personality.

Advertisers are well aware of how we are influenced by colours. They use warm colours such as orange and red to make us feel at home with a product, whereas blue is a colder colour used to give the hint of sophistication. Whether we realise it or not we are very influenced every waking moment by colour. So just what is colour?

We know from Chapter 3 that visible light is electromagnetic radiation with wavelengths between approximately 400 nm and 700 nm. Visible light forms a very small part of the electromagnetic spectrum. Isaac Newton, in 1666, was the first to realise that what we perceive as white light (such as sunlight) is actually made up of different colours – including red, orange, yellow, green, blue and violet. These, and the range of colours in between, form the visible spectrum shown in Fig 27.1.

Fig 27.1 The visible part of the electromagnetic spectrum is formed by passing light through a prism

Newton passed sunlight through a prism to obtain the visible spectrum. He also demonstrated that when this spectrum of colours passed through a second prism it produced white light once more. Thomas Young and George Palmer, 150 years later, independently suggested that receptors in our eyes are sensitive to blue, green or red light and that different stimulations of these three colour receptors (now called cones) enable us to perceive all the different colours. In 1861, James Clerk Maxwell, a Scottish physicist, combined beams of these three coloured lights to produce white light.

The three colours – blue, green and red – are called **additive primary colours** because they cannot be produced by the combination of other coloured lights. However, yellow can be made from the addition of green and red lights, cyan (a bluish green) from green and blue lights and magenta from red and blue (Fig 27.2). Yellow, cyan and magenta are called **secondary colours**. Colour televisions (both flat screen and older models) make use of the three additive primary colours. The screen contains millions of phosphor dots that glow either blue, red or green when activated. These dots combine to form coloured images.

The reason why the world appears so colourful is that the multitude of chemical compounds around us absorb and reflect different wavelengths from the light that falls onto them. For example, a leaf appears green because it absorbs red light and reflects light of other wavelengths. A substance that is white reflects all visible wavelengths of electromagnetic radiation. However, a substance that appears black absorbs all visible wavelengths. A substance appears coloured if it absorbs some of the electromagnetic radiation from white light, but not all of it.

When a compound absorbs wavelengths of one particular colour, a **complementary colour** appears. Pairs of complementary colours are represented on the colour wheel in Fig 27.3. The coloured wavelengths that are absorbed lie opposite their complementary colours. For example, crystals of hydrated copper(II) sulfate appear blue because they absorb light in the orange region of the spectrum.

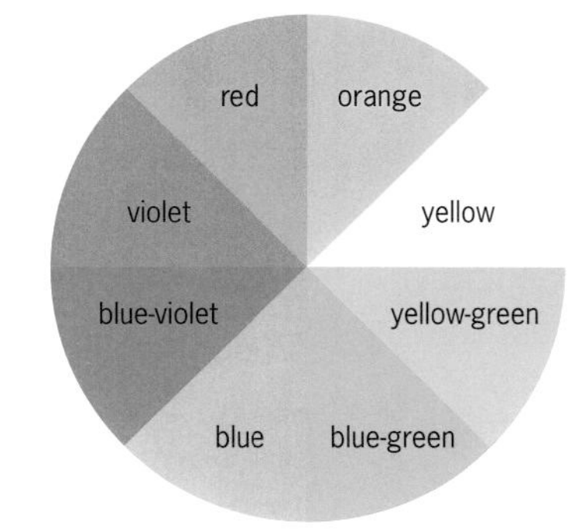

Fig 27.3 A colour wheel. Complementary colours are opposite one another on this colour wheel

You can read about receptors in the eye on page 279.

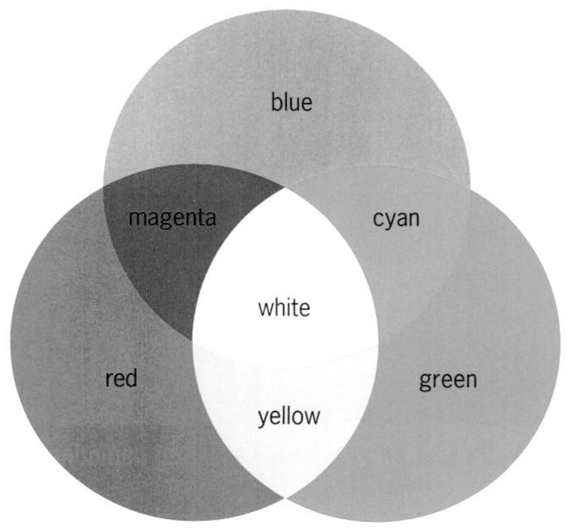

Fig 27.2 Mixing coloured light produces secondary colours and white light

QUESTION 1

A solution of potassium dichromate(VI), $K_2Cr_2O_7$, appears orange. What colour does the dichromate(VI) ion absorb?

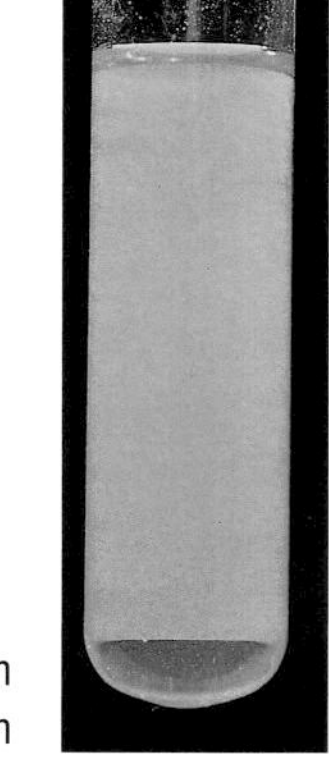

Fig 27.4 Potassium dichromate(VI) solution

SCIENCE IN CONTEXT

Dyes and pigments

Dyes and pigments differ in one important way: dyes are usually soluble in the medium in which they are applied and pigments are insoluble. Humans have used dyes and pigments since early times. Some cave paintings discovered in southern France and northern Spain are up to 30 000 years old. The artists of these paintings used mineral pigments to colour them:

- iron(III) oxide provided the red colour;
- iron(II) carbonate provided the yellow colour;
- either soot or manganese(IV) oxide were used for black.

The pigments were applied to surfaces by first mixing them into a paste using mud or oil. Neanderthal tribes painted the bodies of their dead using iron(III) oxide, also known as red ochre. The Ancient Egyptians used some chemical reactions to extend the range of pigments available. For example, red lead, Pb_3O_4, was produced by heating together lead and white lead, $PbCO_3.Pb(OH)_2$.

Fig 27.5 A primitive cave painting depicting animals

Pigments are spread as a surface layer in paints, and in coloured plastic articles they are dispersed throughout the plastic. Until the 19th century, nearly all pigments were inorganic. Artists' paints used to contain lead and chromium compounds, but these are now known to pose a health hazard. They have been replaced by safer red and yellow organic pigments that can be synthesised.

In contrast to pigments, dyes are soluble. Dye molecules attach themselves to the molecules of the substance they are colouring. In some cases, dye molecules use ionic or covalent bonds, but more usually they attach themselves by hydrogen bonds or induced dipole–induced dipole forces (van der Waals forces or non-polar forces). Sometimes a fabric is first treated with a mordant, an intermediate substance that bonds to the fabric and to the dye. The mordant most commonly used in ancient times was potassium aluminium sulfate (also called potash alum), but other metal salts were also used. The metal ions bonded to the fabric and – through the formation of complexes – to the dye molecules. Potassium aluminium sulfate may have been the first chemical to be purified to avoid contaminants, such as iron(III) salts, which introduced other colours into the dyeing process.

The dyeing of fabrics has a very long history and until the end of the 19th century dyes came from animals or plants. This meant that colouring clothes could be an expensive business. One of the most highly prized dyes in the days of the Roman Empire was Tyrian purple. A quarter of a million Mediterranean molluscs were required to produce just 30 g of the dye. Only members of the Emperor's family could wear togas dyed with Tyrian purple, so it became known as royal purple.

Fig 27.6(a) The structural formula of Tyrian purple

Early Britons used the woad plant, which contains the chemical indigo, to dye their clothes blue. About 3000 years ago, Mexicans started to use a red dye, cochineal, produced from crushing the Coccus insect. It was not until the 16th century that the Spanish brought cochineal to Europe. The production of this dye was a painstaking process that involved collecting about 150 000 insects by hand for every kilogram of dye produced.

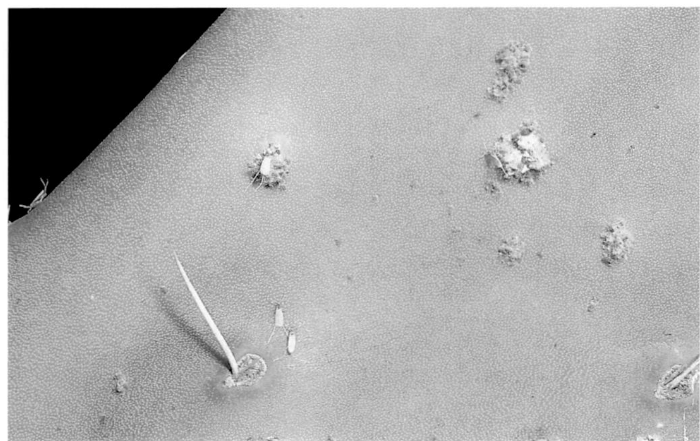

Fig 27.6(b) The insects that provide cochineal dye are tiny

Another red dye, alizarin, was used by the Ancient Egyptians. It came from the roots of the madder plant, which was very widely cultivated. One of the reasons why many military uniforms were red was the availability of this dye and the fact that it did not fade very quickly in sunlight.

2 FLAME TESTS AND ATOMIC EMISSION SPECTROSCOPY

You have probably carried out some flame tests on inorganic compounds yourself. The presence of certain metal ions gives rise to characteristic colours. Fig 27.7 shows the characteristic flame colour caused by the potassium cation, which shows that the substance being tested must contain a potassium compound. Table 27.1 gives the characteristic flame colours of some other cations.

Table 27.1 Characteristic flame colours of metal ions in Groups 1 and 2

Group 1:		Group 2:	
Metal ion	**Flame colour**	**Metal ion**	**Flame colour**
Li^+	deep red	Be^{2+}	no colour
Na^+	yellow	Mg^{2+}	no colour
K^+	lilac	Ca^{2+}	brick red
Rb^+	red	Sr^{2+}	deep red
Cs^+	blue	Ba^{2+}	apple green

Fig 27.7 The lilac colour indicates that the compound in the flame contains potassium

In Chapter 3, we said that flame colours resulted from electron transitions. The energy from the Bunsen flame causes electrons in the cation to jump up to higher energy levels by absorbing amounts of energy. As they fall back down to their original energy levels they each release a certain amount of energy. If the energy corresponds to a frequency (and wavelength) in the visible spectrum, a colour is observed. You will notice from Table 27.1 that some Group 2 cations do not produce a coloured flame. Many ions do not emit light with frequencies in the visible part of the spectrum, so these do not colour the flame.

REMEMBER THIS

Remember that $E = h \times f$, where E is energy, h is Planck's constant and f is the frequency. Light is electromagnetic radiation; you can read more about the nature of electromagnetic radiation in Chapter 3.

When viewed through a spectroscope, the flame colour produced by a particular cation gives rise to a series of coloured lines. This series of lines is called a **line emission spectrum**. Other emission lines are found in other parts of the spectrum, such as the ultraviolet and infrared. A line emission spectrum is characteristic of a particular cation and can be used to identify elements in compounds, rather like fingerprints can be used to identify individuals. It is possible to determine the amount of an element present in a sample by measuring the intensities of the different emission lines in its spectrum.

You can see some line emission spectra on page 49.

If bright white light is shone onto a coloured flame and observed through a spectroscope, a series of black lines result, forming a **line absorption spectrum**. The metal ions in the flame remove the frequencies from the white light, which they would normally emit when excited.

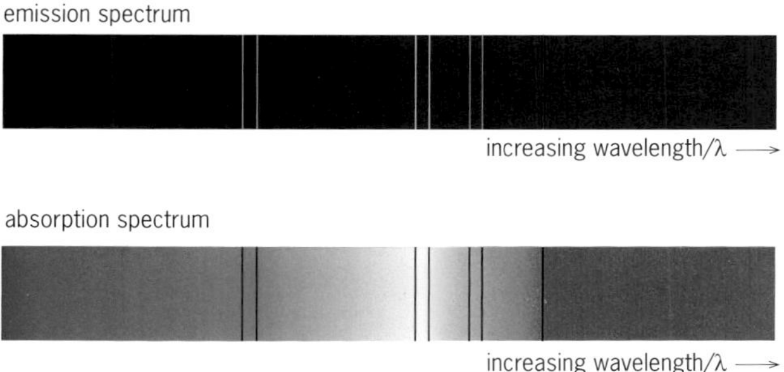

Fig 27.8 Emission and absorption spectra for sodium. Notice that the frequencies are the same in the emission and absorption spectra because electrons absorb photons of a particular frequency when they become excited and release photons of the same frequency when they return to their ground state

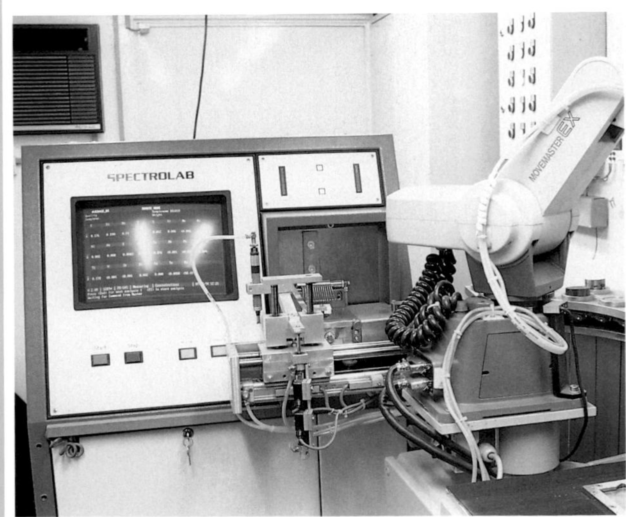

Fig 27.9 Atomic emission spectrometers are used in the steel industry to determine the presence and percentage content of certain elements in steel

You can read more about steel manufacture on pages 13–16.

Atomic emission spectroscopy is used in the steel manufacturing process to determine the composition of different steels. Up to 20 elements are used in steels; the amount and presence of these elements has a profound effect on the properties of a particular steel.

Atomic emission spectroscopy is also used to monitor the levels of potassium and sodium in blood. To do this for sodium the wavelength chosen is 589 nm. This is the wavelength of one of the bright yellow emission lines, which is best seen in Fig. 3.4 on page 49. The atomic emission spectrometer is then calibrated using solutions of known sodium ion concentrations. The higher the concentration, the more intense the emission line. The blood serum is then analysed and the intensity of the emission reading at 589 nm is compared with the intensities from known concentrations of sodium ions used to calibrate the instrument. This allows the concentration of sodium ions in the blood serum to be calculated.

QUESTION 2

2 Work out the following frequency and the energy of the emission line chosen to analyse sodium in blood. (Hint: $c = f\lambda$ and $E = hf$. To remind yourself about the use of these equations, see Chapter 3.)

3 DIAZONIUM COMPOUNDS AND AZO DYES

The first synthetic dye, **mauve**, was discovered in 1856 by an 18-year-old Englishman, William Perkin. Perkin's discovery led to a search for new dyes and prompted the start of the organic chemical industry. One of the chemicals Perkin used in the production of mauve was phenylamine (Fig 27.10).

You can read more about Perkin's discovery of mauve in the How Science Works box on page 560.

Phenylamine is an **aromatic amine** (**arylamine**). It has a benzene ring with an NH_2 (amino) functional group directly attached to it.

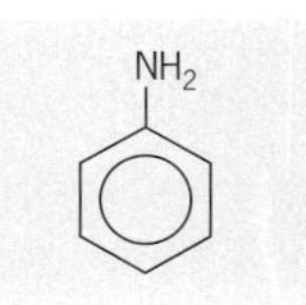

Fig 27.10 The structural formula of phenylamine

DIAZOTISATION

Six years after the discovery of mauve, Johann Peter Griess used aromatic amines to produce **diazonium compounds** (also known as **diazonium salts**). Diazonium compounds are produced when nitrous acid, HNO_2, reacts with an aromatic amine. The reaction mixture must be kept below 5 °C because the diazonium salt is unstable. Nitrous acid is also unstable and it has to be made *in situ* by reacting sodium nitrite with concentrated hydrochloric acid:

QUESTION 3

3 What is meant by the term functional group?

$$NaNO_2(aq) + HCl(aq) \rightarrow HNO_2(aq) + NaCl(aq)$$

For more information on functional groups, see page 190.

The nitrous acid reacts with the aryl amine and more of the concentrated hydrochloric acid to give the diazonium salt (Fig 27.11):

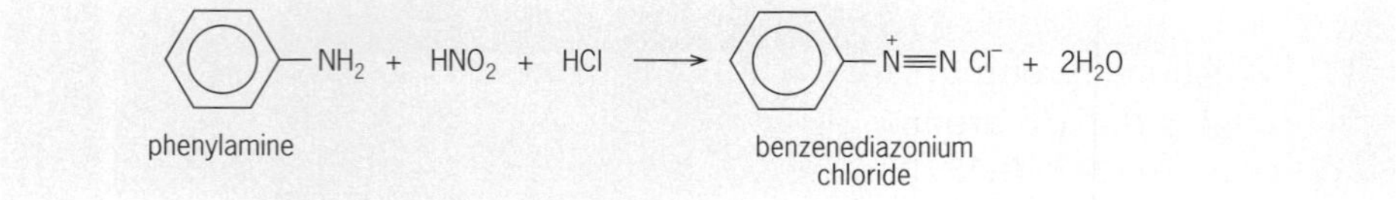

$$C_6H_5NH_2 + HNO_2 + HCl \longrightarrow C_6H_5\overset{+}{N}{\equiv}N\ Cl^- + 2H_2O$$

phenylamine → benzenediazonium chloride

Fig 27.11

QUESTION 4

4 Write the equations for the production of nitrous acid and the diazonium salt formed from phenylamine using sulfuric acid.

This reaction is called a **diazotisation**. The usual way to carry this out in the laboratory is to add a cold aqueous solution of sodium nitrite to a solution of phenylamine dissolved in concentrated hydrochloric acid. The reaction mixture is kept below 5 °C by using ice.

The benzene ring stabilises the diazonium ion by **delocalisation** of its electrons. Alkylamines, such as ethylamine ($C_2H_5NH_2$), produce diazonium salts that decompose immediately, even below 5 °C.

Above 5 °C, the solution of benzenediazonium compound rapidly decomposes, giving off nitrogen gas according to the equation:

$\overset{+}{N}{\equiv}N$

Fig 27.12 The benzenediazonium ion

$$C_6H_5N_2{}^+Cl^- + H_2O \rightarrow C_6H_5OH + N_2 + HCl$$

You can find out more about delocalisation on pages 298.

QUESTION 5

Why are solid benzenediazonium compounds explosive?
(Hint: think about the relative volumes of solid and gas produced on decomposition.)

THE PRODUCTION OF AZO DYES USING COUPLING REACTIONS

A diazonium ion may act as an **electrophile** and attack the benzene ring of another compound. This reaction is called a **coupling reaction**. The first coupling reaction was observed by Griess: an aromatic amine coupled with a diazonium salt to produce a stable **azo compound**.

$\overset{+}{N}{\equiv}N$ + phenylamine (NH_2) → yellow azo dye (–N=N–, NH_2) + H^+

Fig 27.13

REMEMBER THIS

An electrophile is a species that accepts a lone pair of electrons to form a covalent bond. It is attracted to an electron-rich centre. See pages 311 and 334.

In the 1870s, coupling reactions involving alkaline solutions of phenols were first observed (Fig 27.14a).

$\overset{+}{N}{\equiv}N$ + phenol (OH) → orange azo dye (–N=N–, OH) + H^+

Fig 27.14(a) The coupling reaction of a diazonium salt with an alkaline solution of phenol

QUESTION 6

a) Why must a diazo coupling reaction be carried out below 5 °C?
b) What happens to phenol molecules when they react with aqueous sodium hydroxide? Draw the structure of the species produced. (Hint: look back to page 343 to help you answer this question.)
c) Why is the species you have drawn in **b)** more reactive with electrophiles than with benzene?

Fig 27.14(b) The orange azo dye produced when benzenediazonium chloride couples with phenol in the test tube. This can be used as a fabric dye

The functional group of an azo compound is the azo group, –N=N–. In a compound R–N=N–R′, if the R groups are aromatic (aryl) groups, the azo group becomes part of the delocalised systems of the adjoining benzene rings. This gives extra stability to the molecule and is also responsible for the bright colours of azo dyes.

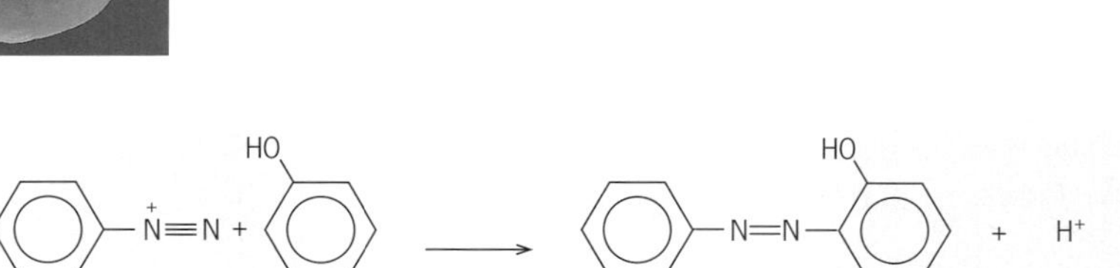

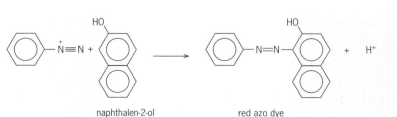

Fig 27.14(c) The coupling reaction of a diazonium salt with alkaline naphthalen-2-ol

Perkin and the start of the dyestuffs industry

William Perkin was the son of a carpenter in East London. As William had some artistic talent, his father hoped his son would train as an architect. However, a fascination with chemistry led him to become a student at the Royal College of Chemistry. At the age of 18 years he made an accidental discovery that was to prove to be the start of not only the dyestuffs industry but also the whole organic chemicals industry.

Fig 27.15 William Henry Perkin (1838–1907) and the original bottle of dye that he produced

At the Royal College of Chemistry, Perkin became an assistant of Professor August von Hofmann, a German chemist. Hofmann suggested that he might like to try synthesise the important, naturally occurring drug quinine, which was used to treat malaria. The starting material was coal tar, because it was known to contain arylamines.

At this time, the structures of organic chemicals had not been deduced. However, the molecular formula of quinine was known, as was the empirical formula of the arylamine he intended to use. Perkin thought that if he oxidised the arylamine he would produce the reaction:

$$2C_{10}H_{13}N + 3[O] \rightarrow \underset{\text{quinine}}{C_{20}H_{24}N_2O_2} + H_2O$$

All Perkin managed to produce was a dirty brown precipitate. Undeterred, he set about oxidising a simpler compound, phenylamine. This time he produced a black precipitate that, when dried and dissolved in ethanol, produced a brilliant purple solution. The product's structure was nothing like that of quinine, but Perkin was quick to see its potential as a dye. He dyed some silk with the purple dye and sent it to a firm of dyers.

The reply came back, 'If your discovery does not make the goods too expensive, it is decidedly one of the most valuable that has come out for a very long time.'

One of the most important properties of the dye was that it was 'fast' – it didn't fade or change colour, even when exposed to light and air. Perkin named the dye mauve after a French flower.

He left the Royal College of Chemistry and, with the help of his father and brother, set about building a factory for the large-scale production of mauve. Perkin used coal tar as the starting material because it was a cheap, plentiful by-product of the coal gas industry. Fig 27.16 shows the steps involved in his synthesis of mauve.

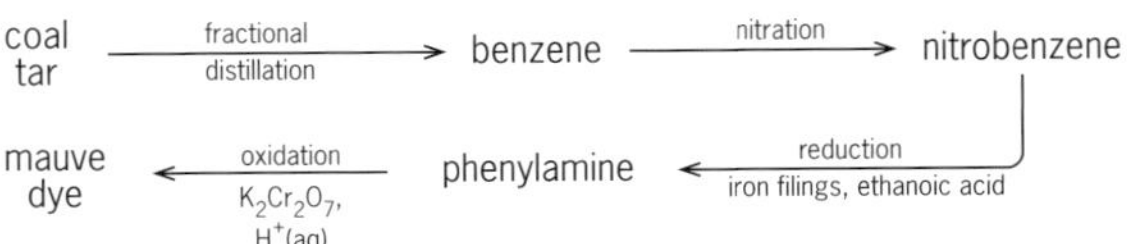

Fig 27.16 A flow chart showing Perkin's synthesis of mauve

He produced benzene by fractionally distilling coal tar. He then nitrated the benzene to produce nitrobenzene. The nitrobenzene was reduced using a mixture of iron filings and ethanoic acid to form phenylamine. This phenylamine he then oxidized using acidified potassium dichromate(VI).

Queen Victoria wore a dress dyed with mauve to the International Exhibition of 1862. *Punch,* the satirical magazine, proclaimed that policeman could be heard telling people to 'get a mauve on!'

It was not long before other dyes of different colours were made using phenylamine. Although the synthesis of dyes began in England, Germany soon became the centre of the dyestuffs industry. Perkin's mauve was superseded because it cost too much to produce. One of its last applications was in the production of the stamp shown in Fig 27.17(a).

Fig 27.17(a) The 1d stamp, printed in 1881, was dyed with Perkin's mauve

At 36 years of age, Perkin sold his factory and returned to his first love, chemistry research. He made many important contributions to organic chemistry, including the production of a perfume from coal tar. However, the synthesis of the quinine molecule was not finally achieved until 1944, almost 90 years after Perkin's attempt.

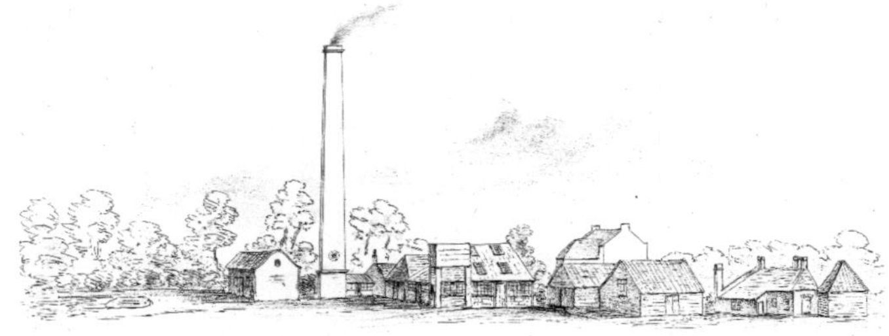

Fig 27.17(b) A sketch by William Perkin of his first synthetic dye factory which made Perkin's mauve

QUESTIONS 7–8

7 **a)** What reagents are used to nitrate benzene?
b) Perkin used ethanoic acid and iron filings to produce phenylamine from nitrobenzene. What reagents are used to produce phenylamine from nitrobenzene in the laboratory?
(Hint: if you have trouble answering these questions, see Fig 15.46 on page 306.)

8 **a)** If a substance reflects all light with frequencies in the visible range of the spectrum, what colour will it be?
b) If a substance absorbs all the radiation in the visible spectrum, what colour will it be?

4 COLOUR IN ORGANIC MOLECULES

What gives rise to the bright colours of azo dyes? We have already seen on page 555 that colour results from electron transitions. If a substance absorbs electromagnetic radiation with frequencies in the visible spectrum, then it will be coloured. This is because the substance reflects or transmits only part of the visible spectrum.

The absorption of a photon (a quantum of electromagnetic radiation) causes an electron to jump from its **ground state** (the lowest possible energy level) to an **excited state** (a higher energy level). If the difference in energy between the ground state and the excited state is equivalent to a photon with a frequency (or wavelength) in the visible part of the spectrum, the substance will absorb at that frequency when white light falls on it. Only outer-shell electrons are excited by visible and ultraviolet radiation. Electrons from inner shells are held much more firmly by the nucleus, so they require much more energy to become excited.

REMEMBER THIS

$E = h \times f$, where E is the energy of the photon, h is Planck's constant and f is the frequency of radiation. A photon with a particular energy has a corresponding frequency. You can read more about photons and the excitation of electrons in Chapter 3.

STRETCH AND CHALLENGE

Electron transitions in organic molecules and ultraviolet–visible absorption

On page 83, we discussed how atoms bond together by the overlap of atomic orbitals to form covalent bonds. A **σ bond** (sigma bond) forms when two atomic orbitals overlap at one point. There is always a σ bond between two covalently bonded atoms.

Double and triple bonds consist of not only a σ bond, but also **π bonds** (pi bonds). π bonds form by the sideways overlap of p orbitals. A π bond is usually at a higher energy level than a σ bond. Electron transitions between these bonded orbitals in organic molecules gives rise to absorptions, either in the visible part of the spectrum, which leads to colour, or in the ultraviolet part of the spectrum, which causes the compound to appear colourless. **Lone pairs** (non-bonded pairs) of electrons are also found in orbitals and these, too, contribute electron transitions that occur in the ultraviolet–visible part of the spectrum.

Ethane (Fig 27.18) is composed of σ bonds and electron transitions in these orbitals occur in the ultraviolet part of the spectrum, because more energy is required to excite the electrons in these bonding orbitals.

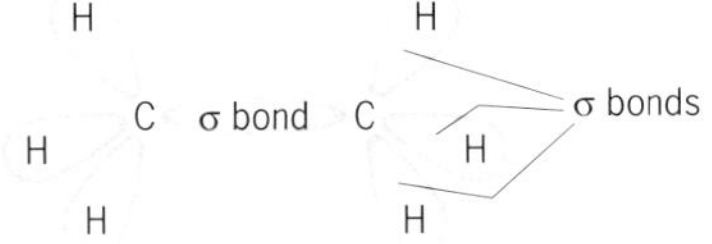

Fig 27.18 Bonding in ethane comprises only σ bonds

CONJUGATED SYSTEMS

In ethene, we have a π bond as well as a σ bond. Ethene still appears colourless, because the double bond is restricted to the two carbon atoms (Fig 27.19).

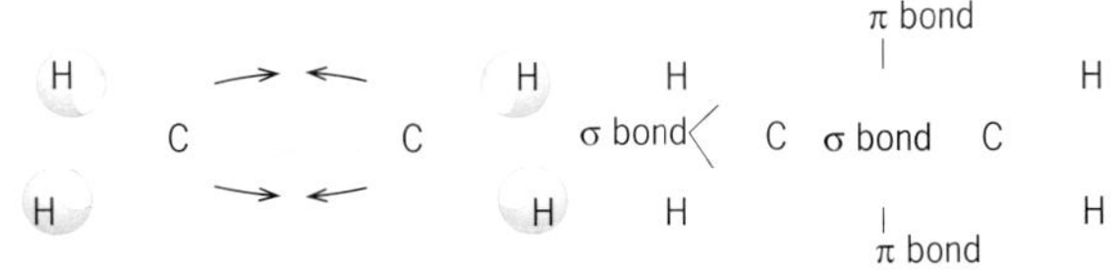

Fig 27.19 Bonding in ethene. A π bond is formed by the sideways overlap of the p orbitals

However, in buta-1,3-diene, $CH_2{=}CH{-}CH{=}CH_2$, there are two double bonds. The p orbitals can overlap so that the two π bonds interact with one another to form a **delocalised system**. The system is **conjugated**. Such

Electron transitions in organic molecules and ultraviolet–visible absorption (Cont.)

systems have alternating double and single bonds and the delocalisation that occurs lowers the energy of some of the electron transitions.

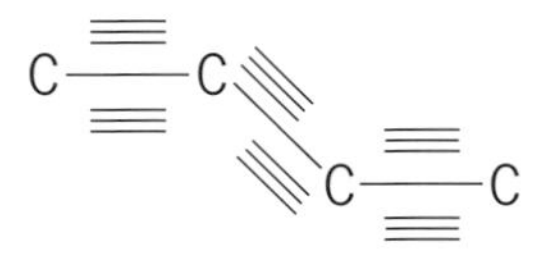

Fig 27.20 Conjugation in buta-1,3-diene

Although buta-1,3-diene absorbs in the ultraviolet region, its maximum absorption wavelength is 220 nm (compared with 185 nm in ethene) and the **intensity** of absorption also increases. The intensity depends on how many photons are absorbed from the radiation that falls on the substance. Carotene (Fig 27.21) is a vitamin found in carrots that is used as a food colorant. Its conjugated system involves 11 double bonds, which shifts its maximum absorption into the blue part of the visible spectrum. As blue light is removed from the white light that falls on carotene, it appears orange. The part of a molecule responsible for absorbing coloured radiation is called the **chromophore**. It is usually an extended delocalised electron system.

Fig 27.21 The skeletal formula of carotene showing its extended conjugated system, which forms the chromophore

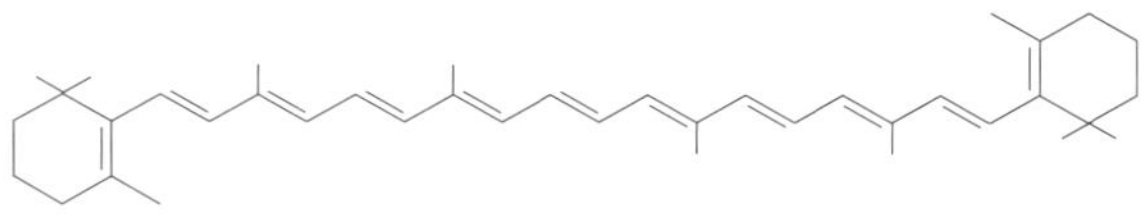

REMEMBER THIS

$E = h \times f$ and $c = f \times \lambda$. So the higher the wavelength, λ, the *lower* the frequency, f, and the lower the energy, E. (c is the speed of light and h is Planck's constant).

Orange is the complementary colour to blue. Refer back to the colour wheel on page 555 to remind yourself about complementary colours.

The more conjugated a system the smaller the gap between energy levels, which means the absorptions that excite electrons here occur at longer wavelengths.

COLOUR IN AZO COMPOUNDS

The first commercially successful azo dye was Chrysoidine (Fig 27.22). The azo group, –N=N–, acts as a 'delocalisation bridge' between the two benzene rings to form an extended delocalised system, which is the chromophore.

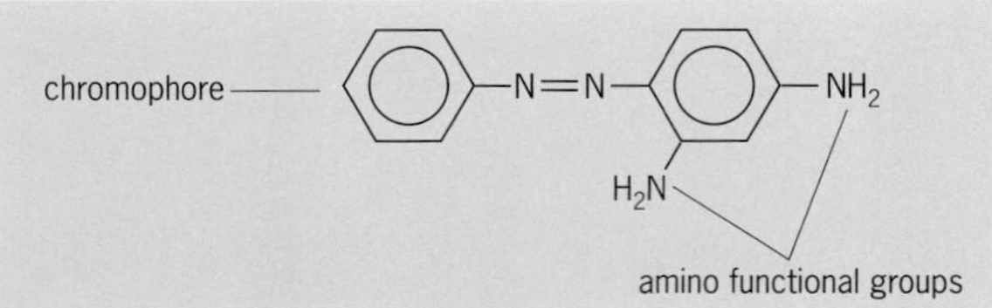

Fig 27.22 Chrysoidine, an orange azo dye

The two amine functional groups of Chrysoidine have lone pairs of electrons on the nitrogen atoms. These lone pairs interact with the delocalised system.

The nature of the functional groups that interact with the chromophore can dramatically alter the colour of the azo dye molecule by causing a shift in the electrons of the chromophore. This electron shift alters the energy required to promote them into an excited state, and so shifts the wavelength of light absorbed.

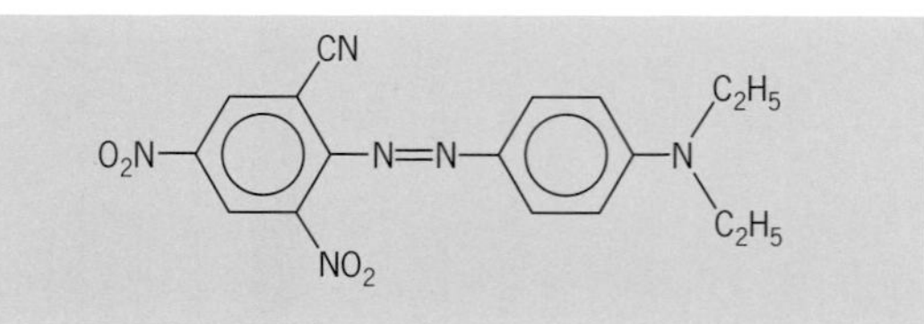

Fig 27.23 Blue azo dye. The substituent groups on the left of the molecule tend to accept electrons, while those on the right are electron-donating groups. This shifts the electrons of the chromophore considerably, making the molecule appear blue

QUESTION 9

9 What wavelength range is light in the visible spectrum? (Hint: if you are not sure look back to page 50.)

QUESTION 10

10 In which region of the electromagnetic spectrum does Chrysoidine absorb?

HOW SCIENCE WORKS

How are dyes stuck onto textiles?

The dye of blue denim jeans is indigo and it is called a vat dye. A vat dye is usually soluble in its reduced form, but when oxidised becomes insoluble and precipitates in the pores of the denim cotton fibres. This property means that vat dyes do not wash out of clothes. (To learn a little more about the history of blue denim and indigo, read the Science in Context box at the start of this chapter.)

Vat dyes are particularly effective for cotton fibres and other fabrics that contain cellulose. The large number of hydroxyl groups on the cellulose molecules mean that the fabric readily absorbs water and hence the water-soluble dye. This is due to the formation of hydrogen bonds between water molecules and the hydroxyl groups.

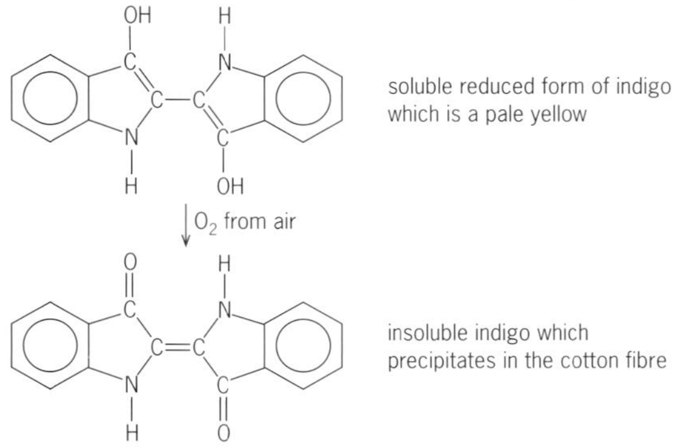

Fig 27.24 The soluble and insoluble forms of indigo

Another type of cotton dye is called a **direct dye**. Direct dyes are long planar molecules that can lie alongside the cellulose polymer chains, and form intermolecular hydrogen bonds and induced dipole–induced dipole forces. Such large dye molecules are made water-soluble using $SO_3^-Na^+$ groups. Direct dyes are usually diazo or triazo dyes (Fig 27.25).

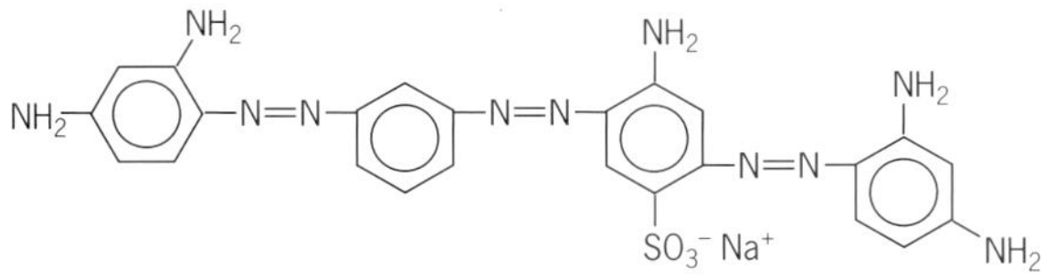

Fig 27.25 C.I. Direct Brown 138, a direct dye. C.I. are the initials for the Colour Index, an internationally recognised publication that lists and categorises almost 40 000 commercial pigments and dyes

The intermolecular forces between direct dyes and cotton are relatively weak, so the water-fastness of the dye is poor and some of the dye comes out when the material is washed. However, in 1956 chemists at ICI produced a dye that would form covalent bonds with the cellulose fibres of cotton. The covalent bonds are formed by reactive groups that first bond to the dye molecules (Fig 27.26). These groups then react with the hydroxyl groups on the cellulose fibres (Fig 27.27). The dye is water-fast and the colour does not run when the textile is washed because covalent bonds have formed with the cellulose fibres. Such a dye is called **fibre-reactive**.

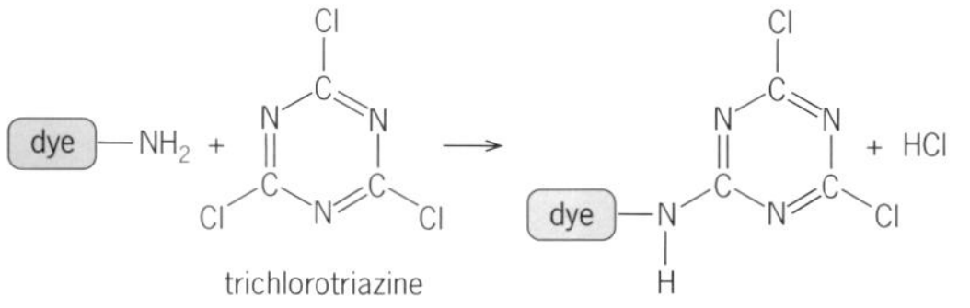

Fig 27.26 Trichlorotriazine reacts with a dye molecule

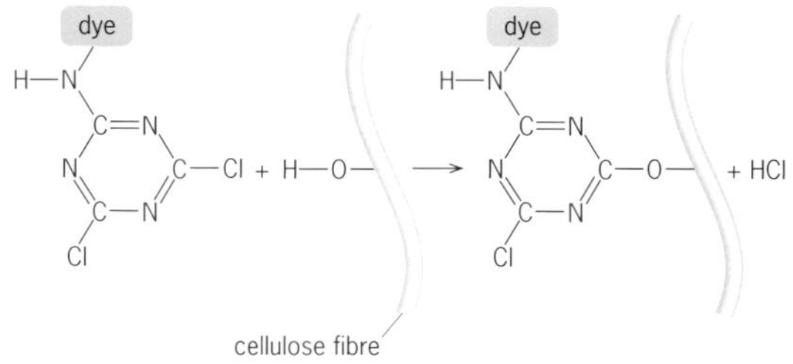

Fig 27.27 The reactive chlorotriazine group forms a covalent bond with cotton

Wool and silk contain many amino functional groups (NH_2). These groups are basic and react with acid dyes that contain sulfonic acid groups (SO_3H) to form ionic bonds (Fig 27.28).

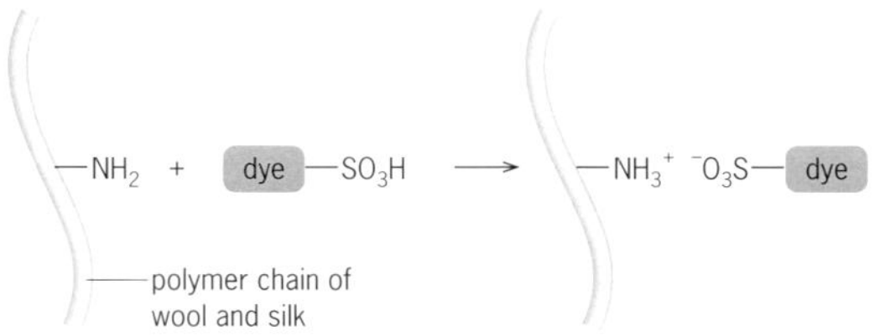

Fig 27.28 Acidic dyes bond to fibres that contain basic NH_2 groups, so this type of dye is used for wool and silk.

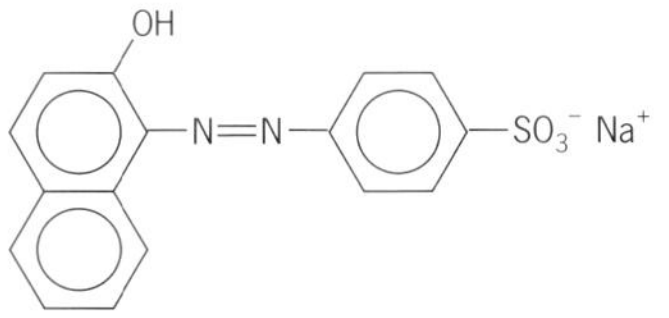

Fig 27.29 C.I. Acid Orange 7 is a salt of sulfonic acid, which makes it more soluble.

QUESTIONS 11–13

11 **a)** The reduced form of indigo is pale yellow, while indigo is blue. Both molecules have conjugated systems. What is meant by the term conjugated system?
b) Which form of indigo absorbs the longest wavelength of light?
c) Which form of indigo absorbs the highest energy photons in the visible spectrum?

12 **a)** Why is Direct Brown called a triazo dye?
b) What is meant by the term chromophore?

13 Why does the covalent bond between the dye and cotton make it water-fast?

HOW SCIENCE WORKS

How are dyes stuck onto textiles? (Cont.)

Poly(propenenitrile) fibres (acrylics) contain CO_2H and SO_3H groups, which are acidic. These can form ionic bonds with basic dyes (Fig 27.30).

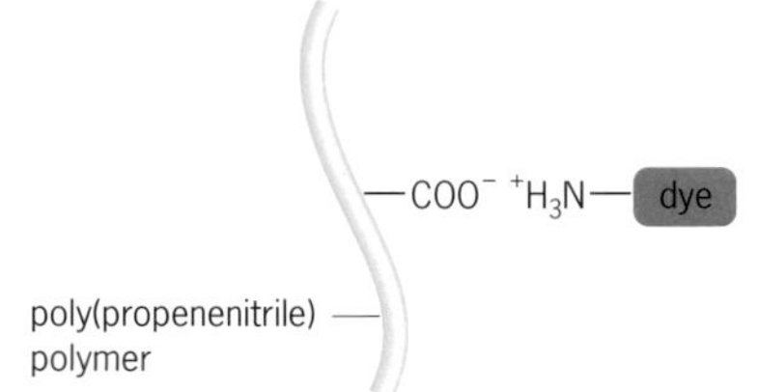

Fig 27.30 Basic dyes form ionic bonds with acid functional groups on poly(propenenitrile) fibres

The last group of dyes we consider are **disperse dyes**. All the other dye groups are water-soluble when applied, but polyester fibres do not form hydrogen bonds with water and are known as **hydrophobic** (water-hating). They do not allow water molecules to penetrate them. Disperse dyes are a fine suspension of dye particles that are absorbed by the fibres and held there by induced dipole–induced dipole forces and some hydrogen bonding. As the dye is only sparingly soluble it tends to stay in the fibres, and so it is water-fast.

5 PH INDICATORS

Methyl orange is an azo compound that has different colours depending on the pH of the solution it is in. This property means that it is used as an indicator in acid–base reactions. Adding or removing an H^+ ion causes an electron shift in the molecule, and so alters the wavelength at which methyl orange absorbs.

yellow above pH 4.4 ⇌ red below pH 3.2

Fig 27.31 Methyl orange is used as an acid–base indicator. Its colour depends on the pH of the solution it is in.

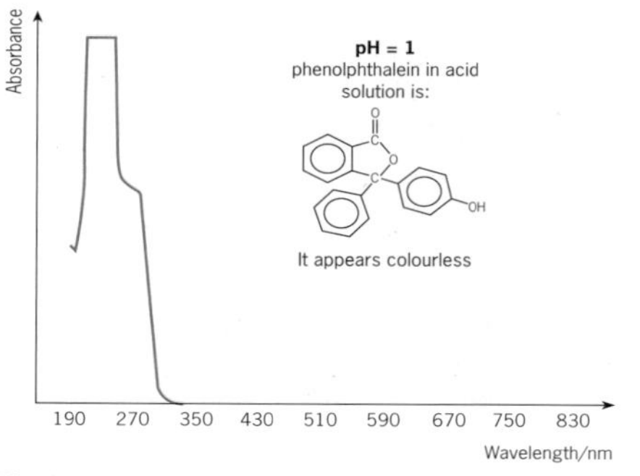

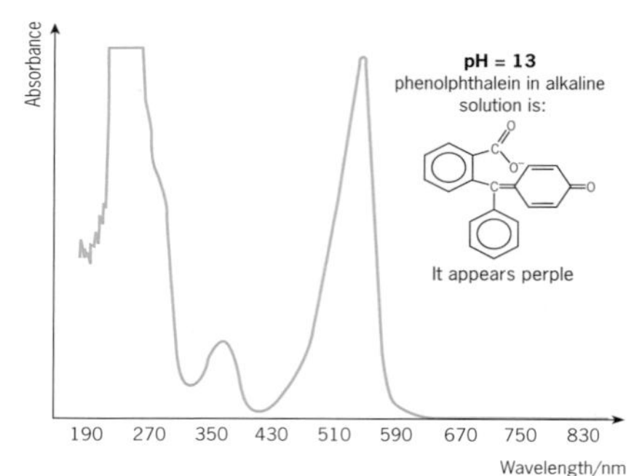

Fig 27.32 The ultraviolet–visible spectra of phenolphthalein at pH 1 and pH 13.

On page 334, we saw that indicators change colour depending on the concentration of $H^+(aq)$ ions in the solution. Each indicator changes colour at a specific pH; using the correct indicator, any neutralisation reaction can be monitored to its end point. Methyl orange changes colour between pH 3.2 and pH 4.4. It can be used to determine the neutralisation point of a strong acid and a weak base.

Phenolphthalein is colourless below pH 8.2, but above this it changes to purple. This indicator can be used in titrations with strong alkalis and weak acids.

QUESTION 14

14 What feature of the structure of phenolphthalein in alkaline solution makes it coloured?

HOW SCIENCE WORKS

Ultraviolet and visible spectroscopy

In Chapter 11, we discussed two types of spectroscopy that involve absorption of electromagnetic radiation by a substance under investigation. In infrared spectroscopy, light is absorbed in the infrared part of the spectrum because of the increased vibration of different bonds within a molecule. In nuclear magnetic resonance spectroscopy, radio waves are absorbed through the excitation of nuclei within molecules.

Ultraviolet and visible spectroscopy are possible because outer electrons of atoms or ions in compounds absorb in the ultraviolet or visible part of the spectrum when they are excited. Compounds that absorb *only* in the ultraviolet part of the spectrum are colourless. In the spectrometer, a beam of electromagnetic radiation passes through a monochromator, which selects varying wavelengths. The beam is then split and one beam passes through a solution of the substance under investigation, while the other passes through the pure solvent.

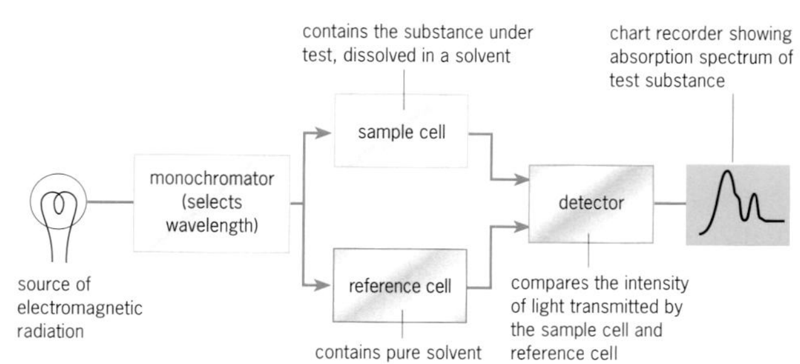

Fig 27.33 A simplified schematic diagram of how an ultraviolet–visible spectrometer works

The spectra produced usually have broad absorption bands (see Figs 27.34, 27.35, 27.36), in contrast to atomic absorption spectra in which gaseous atoms and ions have definite sharp lines. The broad absorption bands occur because in solution a number of vibrational and rotational energy levels are possible for each energy level of the electrons. You can read more about vibrational energy levels on page 221.

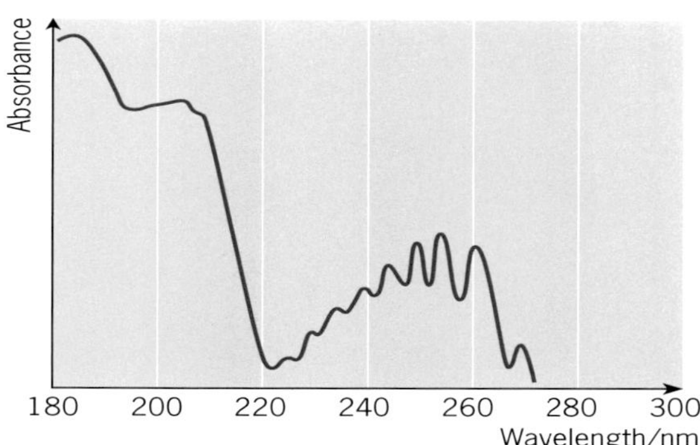

Fig 27.34 The ultraviolet–visible absorption spectrum of benzene. Notice that all the absorption appears in the ultraviolet part of the spectrum, so the compound appears colourless

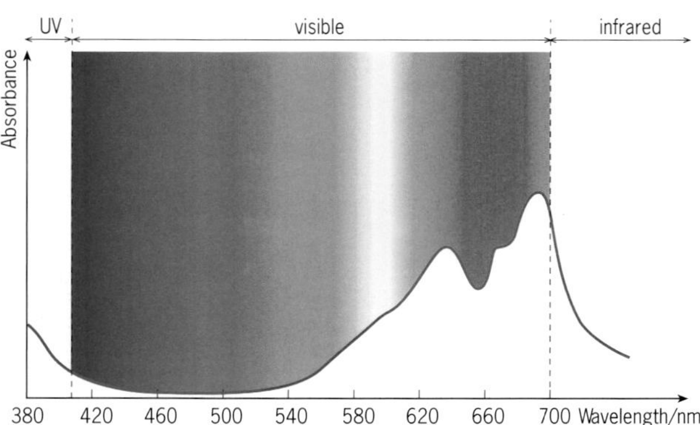

Fig 27.35 The absorption spectrum of a blue dye. Red, orange and yellow light is absorbed by the dye, so the dye appears blue. (See the colour wheel on page 555)

INTERPRETING ULTRAVIOLET-VISIBLE ABSORPTION SPECTRA

The horizontal axis of an ultraviolet–visible absorption spectrum gives the wavelength in nanometres (nm). The shape of the absorption peak is usually characteristic of a particular compound and so the spectrum can be used to help identify the compound.

However, a more common use of this type of spectroscopy is to measure concentrations accurately from the intensity of absorption on the *y*-axis. For example, the uptake of a drug at different sites around the body can be monitored using ultraviolet–visible spectroscopy by taking samples from these sites and analysing their solutions.

In the steel industry, the amount of trace metal, such as manganese, can be analysed by reacting the metal to form an identifiable coloured ion and measuring the absorption of its solution. In this case, the manganese may be oxidised to form the purple MnO_4^- ion.

The food industry also uses ultraviolet–visible spectroscopy to determine how much nitrite has been added to meat.

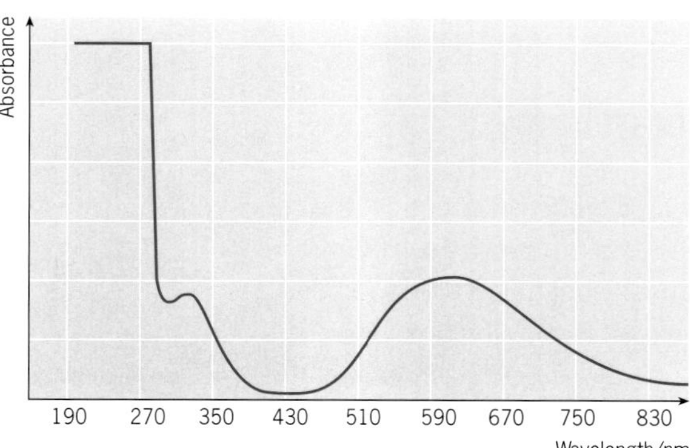

Fig 27.36 The ultraviolet–visible spectrum of $[Cu(NH_3)_4(H_2O)_2]^{2+}$. The solution of this complex appears blue–violet

QUESTIONS 15–17

15 **a)** Why is the ideal solvent in an ultraviolet–visible spectrometer one which does not absorb in the ultraviolet–visible range?
b) Why is it rare to find a solvent which is ideal?

16 In Fig 27.36, which colour(s) is/are being absorbed by a solution of $[Cu(NH_3)_4(H_2O)_2]^{2+}$?

17 Sketch the absorption spectrum of β-carotene, the orange pigment found in carrots.

6 TRANSITION METAL IONS AND COLOUR

A characteristic of transition metals is that many of their compounds are coloured (see Chapter 24). Transition metal compounds are responsible for the colours in gemstones, stained glass windows and pottery glazes. From reading this chapter, you will realize that transition metal ions appear coloured because they absorb some wavelengths in the visible spectrum, but transmit or reflect the rest. The electrons absorb photons of a specific wavelength, become excited and jump to higher energy levels. In compounds of transition elements, colour results from a difference in energies between d orbitals.

d–d TRANSITIONS

You are probably wondering how d orbitals can be at different energy levels. When we discussed d orbitals in Chapter 24, they were drawn at the same energy level (degenerate) on energy level diagrams. (See Fig 24.2, page 481.)

However, this is only true for gaseous transition metal ions. When ligands bond to transition metal ions they cause a splitting of the energy level of the d orbitals (see Fig 27.37). The energy difference between the two sets of d orbitals is often such that the wavelength of photons absorbed is in the coloured part of the spectrum.

QUESTION 18

18 Which photons have the lowest energy: photons of red light or photons of blue light? (Remember the equations $E = h \times f$ and $c = f \times \lambda$.)

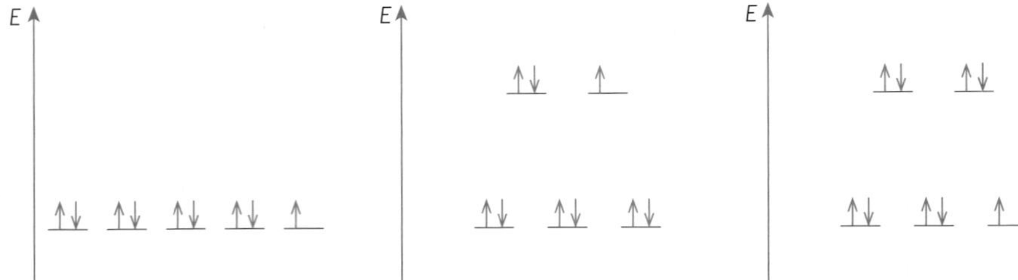

Fig 27.37(a) d orbitals are all at the same energy level in a gaseous Cu^{2+} ion. We say they are degenerate

Fig 27.37(b) In $[Cu(H_2O)_6]^{2+}$, the water ligands cause the d orbitals to split, so they are now non-degenerate

Fig 27.37(c) An electron is excited from a lower energy level d orbital to a higher energy level d orbital

QUESTION 19

19 The difference in energy of the two sets of d orbitals in $[Cu(H_2O)_6]^{2+}$ causes absorption at the red end of the spectrum. What colour is the $[Cu(H_2O)_6]^{2+}$ ion? (Hint: look at the colour wheel on page 555.)

Copper(I) compounds are white because the Cu^+ ion has a $3d^{10}$ outer electron configuration. Since the 3d subshell is full, no transition of electrons between d orbitals can occur. The same is true of scandium(III) compounds, which have no electrons in the 3d subshell ($3d^0$). But not all colour in transition metals results from d–d transitions: sometimes, electrons can jump from the ligand to the metal. This is called **charge transfer** or **electron transfer**, and it is responsible for the bright colours of Prussian blue and chrome yellow.

REMEMBER THIS

A **complex** is formed when a central metal atom or ion is surrounded by species that donate lone pairs of electrons. A species that donates a lone pair of electrons is called a **ligand**. You can read more about this on page 486.

FACTORS THAT AFFECT d–d SPLITTING AND COLOUR

The colour of a transition metal complex depends chiefly on the central metal cation. Each transition metal is different and has a different nuclear charge. The larger the nuclear charge, the more firmly electrons are held in their d orbitals. This affects the energy levels of the split d orbitals and thus the amount of energy required to excite an electron from a lower energy d orbital to a higher energy d orbital. This in turn affects the colour.

The oxidation state of the metal also affects the splitting of the d orbitals and, as a result, the colour. This is well illustrated by vanadium complexes in aqueous solution.

QUESTION 20

20 **a)** What is the full electron configuration of Zn^{2+}?

b) Explain why zinc compounds are white.

Table 27.2 The relationship between the oxidation state of vanadium and the colour of its complexes

Ion	Oxidation state	Colour of aqueous solution
VO_2^+	+5	yellow
VO^{2+}	+4	blue
V^{3+}	+3	green
V^{2+}	+2	violet

The nature of the ligand also has an effect on d–d splitting. Different ligands cause different separations of energy between d orbitals. The **spectrochemical series** is a list of ligands arranged in order of their ability to cause d–d splitting:

$$I^- < Br^- < Cl^- < F^- < OH^- < H_2O < (CO_2)^{2-} < NH_3 < en < CN^-$$

smallest splitting ⟶ greatest splitting
smallest energy gap ⟶ greatest energy gap
longest wavelength ⟶ shortest wavelength

$[Cu(H_2O)_6]^{2+}$ absorbs at the red end of the spectrum, which makes an aqueous solution of copper(II) ions appear pale blue (Fig 27.38).

However, by substituting H_2O with NH_3 ligands to form the ion $[Cu(NH_3)_4(H_2O)_2]^{2+}$, the difference in energy of the d orbital split increases. This shifts the absorption into the yellow part of the spectrum, which makes the solution deep blue.

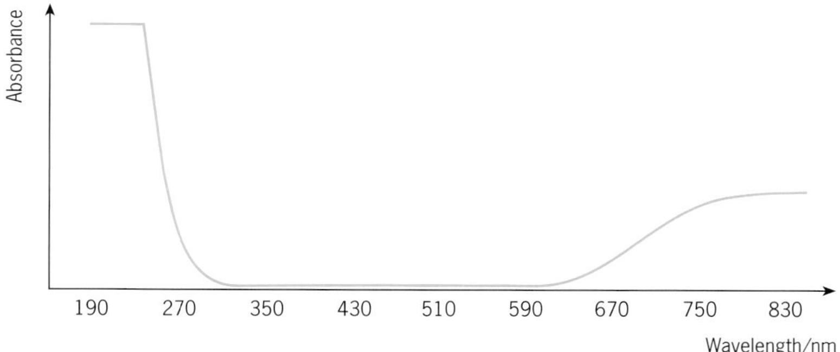

Fig 27.39 The ultraviolet–visible spectrum of $[Cu(H_2O)_6]^{2+}$. Compare this with the ultraviolet–visible spectrum of $[Cu(NH_3)_4(H_2O)_2]^{2+}$ in Fig 27.36

Another factor that has an effect on d–d splitting is the number of each type of ligand in the complex:

$[Cr(H_2O)_6]^{3+}$	$[Cr(H_2O)_5Cl]^{2+}$	$[Cr(H_2O)_4Cl_2]^+$
violet	green	dark green

The arrangement of ligands around the central metal cation can also have an effect on the colour of a complex (Fig 27.41).

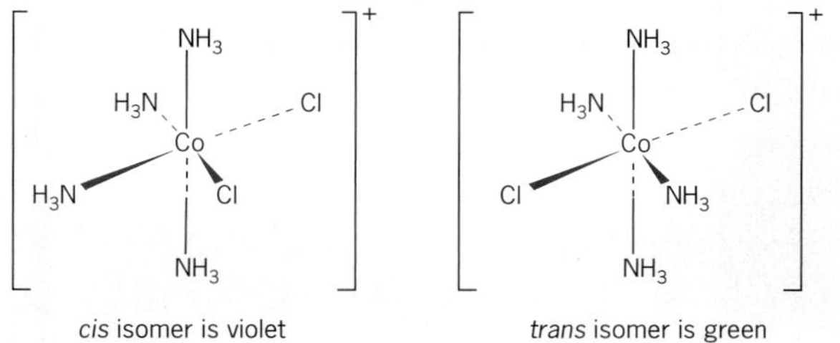

Fig 27.41 The *cis* and *trans* isomers of $[Co(NH_3)_4Cl_2]^+$

QUESTION 21

Using the colour wheel on page 555, work out which colour is being absorbed by each of the vanadium ions in Table 27.2.

You can see these colours of vanadium ions on page 520.

QUESTION 22

$[CuCl_4]^{2-}$(aq) is a yellow colour. Sketch its ultraviolet–visible spectrum.

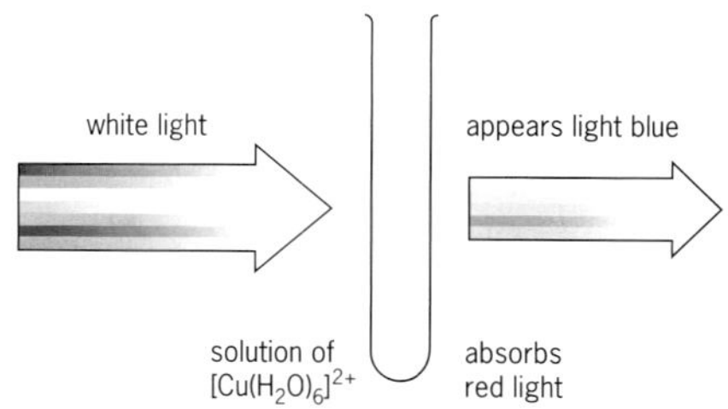

Fig 27.38 Absorbtion of wavelength of visible light by $[Cu(H_2O)_6]^{2+}$

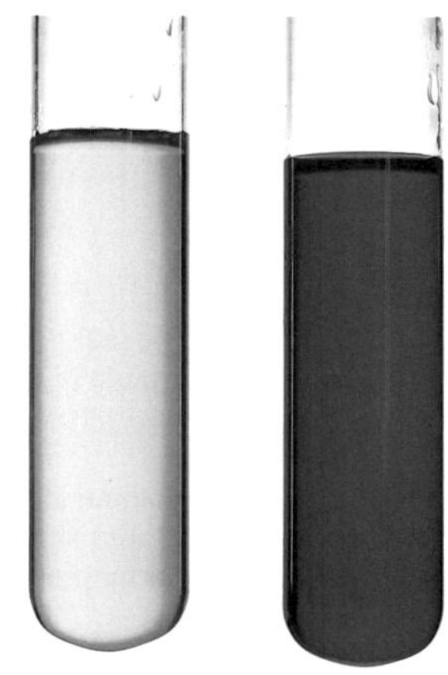

Fig 27.40 Left: An aqueous solution of copper(II) sulfate contains the $[Cu(H_2O)_6]^{2+}$ ion. Right: In this solution, concentrated ammonia has been added, which replaces some water ligands to form the amine complex $[Cu(NH_3)_4(H_2O)_2]^{2+}$. The ammonia ligand increases the energy gap between the split d orbitals. This affects the energy of the electron transition and therefore the colour of the complex solution

Fig 24.21 on page 490 is another example of *cis–trans* isomerism that affects the colour of a complex.

USING A COLORIMETER TO DETERMINE THE FORMULA OF A COMPLEX ION

A colorimeter is a simple form of visible spectrometer. A fixed band of wavelengths of visible light is selected using a filter. This is then passed through the solution under test. The light that has not been absorbed by the solution is transmitted to a photocell. The light generates an electric current, which is measured by a meter. The more light that is transmitted by the solution, the greater the electric current produced. However, most colorimeter meters are calibrated so that they record the light absorbed by the solution rather than the light transmitted. The absorbance of a solution is proportional to the concentration of the coloured compound in the solution.

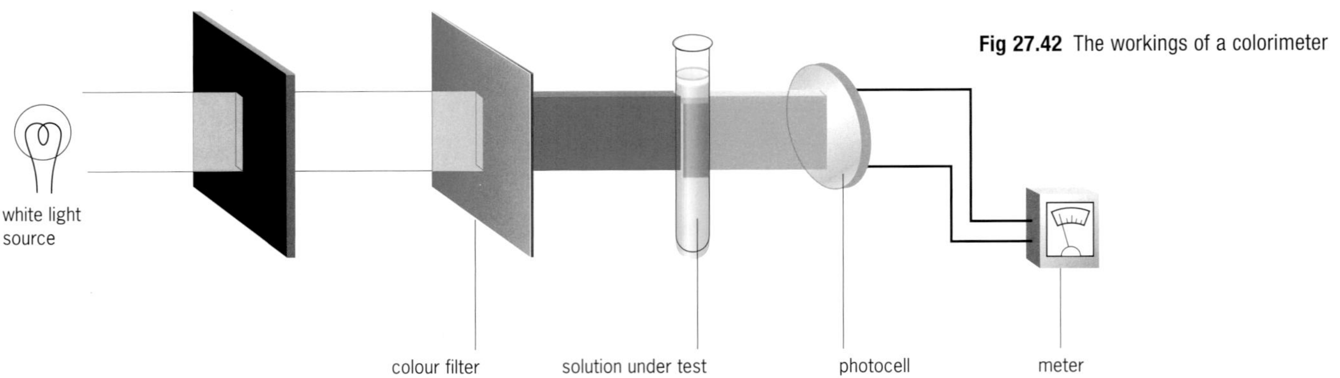

Fig 27.42 The workings of a colorimeter

QUESTION 23

23 **a)** What is the colour of the $[Fe(H_2O)_6]^{3+}$ ion? (To remind yourself turn to page 492.)

b) Which colour is it absorbing?

c) What is its shape?

Table 27.3 Mixtures of 0.1 mol dm^{-3} of solutions of Fe^{3+}(aq) and SCN^-(aq) that could be used to determine the formula of the iron thiocyanate complex using a colorimeter

Volume of 0.1 mol dm^{-3} Fe^{3+}(aq)/cm^3	Volume of 0.1 mol dm^3 SCN^-(aq)/cm^3
10	0
9	1
8	2
7	3
6	4
5	5
4	6
3	7
2	8
1	9
0	10

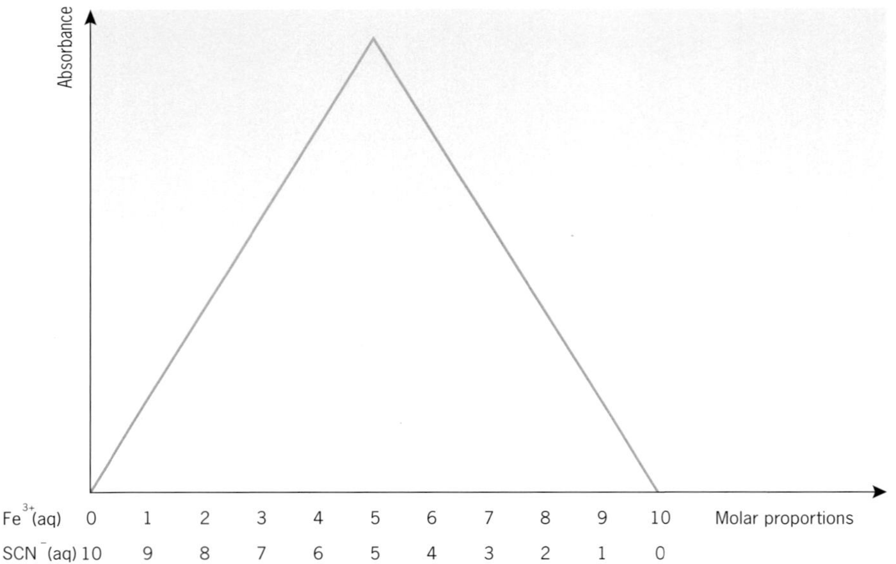

Fig 27.43 Absorbance versus molar proportions of Fe^{3+}(aq) and $SCN^{-(aq)}$

The colorimeter is set to zero by inserting a tube of water and pressing a zeroing button. Standard solutions of known concentrations of a species whose concentration you wish to determine are then placed in the colorimeter and their absorbances measured. A graph of absorbance against concentration is then plotted. This is called a calibration curve. The absorbance of a solution of unknown concentration can now be measured and the reading checked against the **calibration curve** to determine the concentration of the particular species under investigation.

In Chapter 24 (page 496) we noted that the addition of aqueous potassium thiocyanate to a solution of aqueous iron(III) ions $[Fe(H_2O)_6]^{3+}$ causes a distinctive blood-red complex to form. We can use the colorimeter to determine the formula of this complex.

The first task is to select a filter for the colorimeter that will allow the solution to absorb most strongly. Since the iron thiocyanate complex is red it will absorb most in the blue wavelengths of the spectrum, so a blue filter is chosen. Solutions to be analysed are next made up to give different molar proportions of $Fe^{3+}(aq)$ and $SCN^-(aq)$: 10:0, 9:1, 8:2, 7:3, 6:4, 5:5, etc.

If the formula of the complex is $[Fe(H_2O)_5(SCN)]^{2+}$, then 1 mole of $[Fe(H_2O)_6]^{3+}$ would need to be added to 1 mole of SCN^- ions, that is the 5:5 mixture. This is, in fact, the correct formula. If these different molar proportions were placed in turn into the colorimeter, a graph similar in shape to that in Fig 27.43 would be obtained. You will notice that the maximum absorbance corresponds to the molar proportions:

$$5Fe^{3+}(aq):5SCN^-(aq)$$

which is the 1 mole:1 mole ratio. Table 27.3 shows how the different solutions may be prepared.

QUESTION 24

a) Write the method you would use to determine the formula of the nickel(II)–EDTA (**e**thylene**d**iamine**t**etra**a**cetic ion) complex. You are provided with 0.05 mol dm^{-3} solutions of $Ni^{2+}(aq)$ and EDTA.

b) The formula of the complex under investigation is $[Ni(EDTA)]^{2-}$. Sketch the graph of absorbance against molar proportions you would expect to obtain from the colorimeter readings.

SCIENCE IN CONTEXT

The colours of gemstones

Some of the most beautiful and highly prized examples of the colours that transition metal ions can impart are found in gemstones. In many cases, the transition metal ions are actually impurities. For example, the deep red of rubies comes from traces of Cr^{3+} ions embedded in a lattice of aluminium oxide. Cr^{3+} ions have the same charge as Al^{3+} ions and are about the same size, and they occupy about 5 per cent of the Al^{3+} positions.

Emeralds are coloured green also because Cr^{3+} ions are present. This time the Cr^{3+} ions are embedded in a lattice that contains large silicate anions $Si_6O_{18}{}^{12-}$ together with the cations Be^{2+} and Al^{3+}.

So how can Cr^{3+} impart different colours? The answer lies in the splitting of the d orbitals. There are three 3d electrons in Cr^{3+} (Fig 27.46). One of these d electrons becomes excited by absorbing a photon of light with a wavelength in the visible spectrum. The wavelengths that are transmitted through the gemstone are predominantly red, which is why a ruby is red. In the case of emerald, the different environment of the Cr^{3+} ion leads to a different energy gap between the two sets of d orbitals, one that allows the transmission of blue–green light while absorbing violet, yellow and red wavelengths.

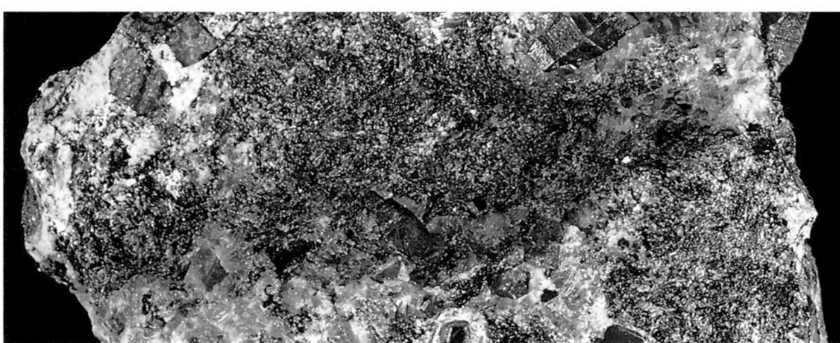

Fig 27.44 The deep red of these rubies results from the presence of Cr^{3+} ions

Fig 27.45 The green colour of the emeralds in this specimen is caused by Cr^{3+} ions present as an impurity in $Be_3Al_2Si_6O_{18}$

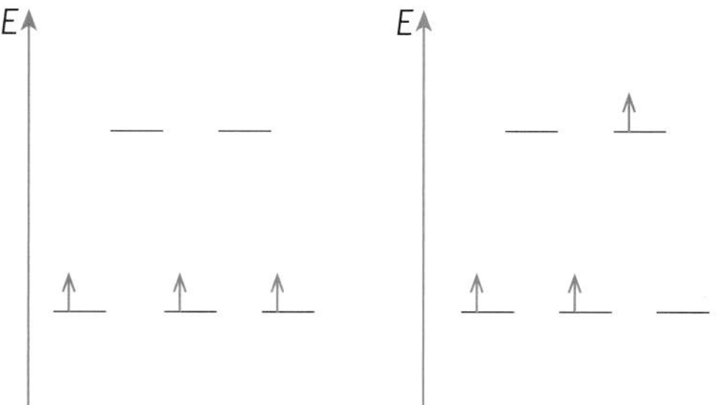

Fig 27.46 Excitation of electrons in Cr^{3+}. Cr^{3+} has three d electrons. The energy difference between the two sets of d orbitals in emerald or ruby causes the absorption of some of the visible spectrum wavelengths as photons excite the electrons to the higher energy d orbitals.

Other precious gemstones are coloured by different transition metal ions. The blue of sapphire results from Ti^{4+}, V^{3+} or Co^{3+} ions. Aquamarine, named after the colour of a tropical sea, owes its pale blue colour to Fe^{3+} ions. The Fe^{3+} ion also gives a yellow colour to topaz and a red colour to garnet. The purple of amethyst is caused by Mn^{3+} ions.

HOW SCIENCE WORKS

The amazing Monastral Blue

Through your studies in science you may have realised that many important discoveries occur by accident. An example in this chapter is the discovery of the first synthetic dye by William Perkin, which he called mauve (see the Science in context box on page 560). The English writer Horace Walpole invented a term for this in 1754; he called accidental discoveries that lead to happy outcomes *serendipity*. The word has its origin in a fairy tale about the three princes of Serendip who were constantly making chance but fortunate discoveries. The story of the blue pigment Monastral Blue is an example of serendipity.

The story begins in 1928 when a researcher at Scottish Dyes Ltd, A. G. Dandridge, noticed blue crystals on the lid of a glass-lined iron vat designed to make the compound phthalimide from molten phthalic anhydride and ammonia. Phthalimide is a white compound and was a necessary compound in the manufacture of some dyestuffs. Sometimes, when the glass lining became damaged, similar blue crystals would form on the sides of the vat and contaminate the white molten phthalimide. It is fortunate that Dandridge was curious enough to examine and analyse these crystals. He realised that they were made of a compound of iron and he named it iron phthalocyanine (Fig 27.47). He also understood that it had potential as a pigment.

Fig 27.48 Steam locomotives were often painted with Monastral Blue

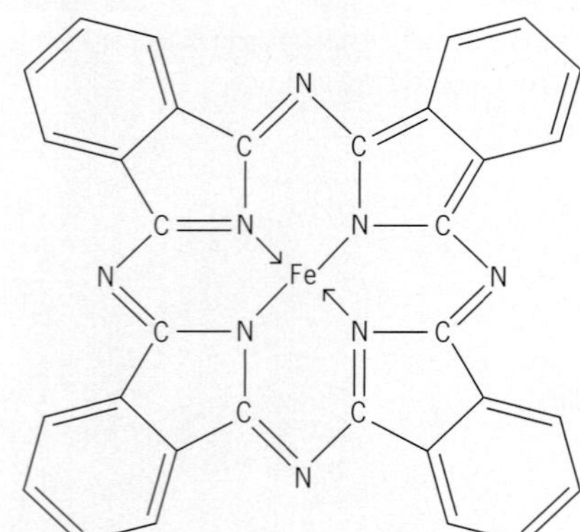

Fig 27.47 Iron phthalocyanine

The ring structure that surrounds the iron atom is very similar to those found in nature, called **porphyrins**. Iron porphyrin occurs in haemoglobin (see the How Science Works box on page 489) and the plant pigment chlorophyll also contains a porphyrin ring. The ring structures contribute to the colour of the molecules. If you examine the iron phthalocyanine's structure closely, you will see it has a conjugated structure that forms a large delocalised system.

Once the structure of iron phthalocyanine was known researchers at Imperial College, London, set about making other phthalocyanines, the best of which was copper phthalocyanine, first made in 1934. This intense blue pigment goes under the trade name Monastral Blue. It is used extensively by printers, paint manufacturers and to colour plastics. It is non-toxic and is even used to stain cells in the sclera membrane of the eye to facilitate laser surgery. The reason it is so useful in laser treatments is that Monastral Blue strongly absorbs in the infrared region of the spectrum.

QUESTION 25

25 **a)** Porphyrins and phthalocyanines are examples of *chromophores*, which have extensive *conjugated systems*. Explain the meanings of the terms in italics.

b) Iron phthalocyanine is a blue colour. What wavelengths of light is it absorbing?

c) Draw a possible visible spectrum for iron phthalocyanine.

SUMMARY

After studying this chapter, you should know that:

- Electron energy levels exist in atoms and molecules.
- Electron transitions between these levels emit or absorb radiation.
- Colour arises when a substance absorbs or emits electromagnetic radiation partly in the visible spectrum.
- Absorption spectra record the wavelengths of photons that excite electrons and so cause them to jump to higher energy levels. Emission spectra record the wavelengths of photons emitted when excited electrons return to lower energy levels.
- Atomic emission spectroscopy can be used to determine the concentrations of metal ions in biological fluids, such as blood serum.
- Dyes are soluble in the medium in which they are applied; pigments are insoluble.
- Azo dyes are produced by coupling reactions, such as that between benzenediazonium chloride and phenol.
- The functional group of an azo compound is the azo group, –N=N–.
- Electron transitions in organic molecules give rise to ultraviolet or visible absorptions, because the molecules contain double or triple bonds, delocalised systems or lone pairs of electrons
- Ultraviolet–visible spectroscopy is possible because when the outer electrons of atoms or ions in compounds are excited they absorb in the ultraviolet or visible part of the spectrum.
- Conjugated systems have alternating double and single bonds; the delocalisation that occurs lowers the energy of electron transitions, so conjugated compounds may appear coloured.
- The part of a molecule responsible for absorbing coloured radiation is called the chromophore. It is usually an extended delocalised electron system.
- The colour changes of acid–base indicators results from a change in the delocalisation of the chromophore when H^+ ions are added or removed.
- The colour of a transition element's ions results from a difference in energies between d orbitals caused by the ligands around the metal cation. This splitting allows d–d electron transitions.
- Factors that affect d–d splitting include the oxidation state of the metal, the nature and number of the ligands, and the arrangement of the ligands around the metal cation.
- Complexes of Zn^{2+} and Cu^+ are white because the 3d subshell is full.
- The formula of a complex can be determined using colorimetry.

Practice questions and a How Science Works assignment for this chapter are available at www.collinseducation.co.uk/CAS

28

Ammonia, bases and food production

28 AMMONIA, BASES AND FOOD PRODUCTION

Population and food production

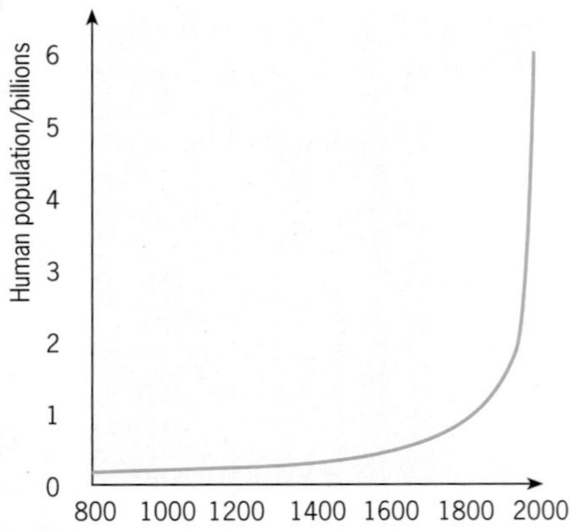

Population explosion The graph shows the population explosion that has taken place since the early 19th century

With the world's population increasing at more than 250 000 people per day, the question arises: how will enough food for these extra mouths be produced? The overall output of staple foods, such as grain, has risen significantly since 1950 through better farming methods and the use of fertilisers. But in many African and Latin American countries food production is actually falling.

As crops grow, they take essential elements such as nitrogen and phosphorus from the soil. Farmers in poorer countries cannot afford fertilisers to restore these elements to the soil. As a result, crop yields fall and in time the soil becomes unable to sustain useful growth.

In richer countries, the pressure to build homes, factories and roads on good agricultural land has meant farmers have sought to obtain higher crop yields from smaller amounts of land by using pesticides and fertilisers. But the use of these agrochemicals has its problems: the over-use of fertilisers has polluted groundwater, and hence rivers, which feed into reservoirs.

There are moves to develop fertilisers that could be used in smaller amounts to maintain maximum crop yields. These would also reduce the harmful effects on the environment. If the manufacturers do manage to find the right formula, and can convince farmers everywhere that these new fertilisers are effective, then there could be a better chance of meeting the needs of the world's swelling population.

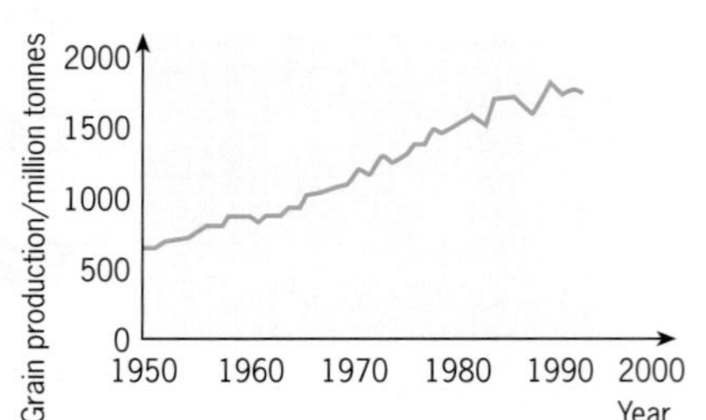

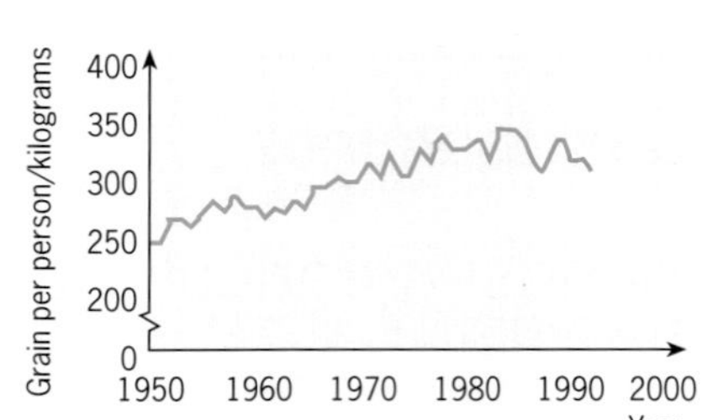

Food source Grains are the world's leading food crops, but it is not clear whether the increase in production will continue to keep up with the exponential increase in demand

1 THE NITROGEN CYCLE

For nearly all manufactured fertilizers, the starting substance is ammonia – a compound of hydrogen and nitrogen. Nitrogen is readily available since it forms about 80 per cent by volume of the lower atmosphere. However, it is a very unreactive gas because of the strong triple covalent bond in each nitrogen molecule (Fig 28.1). The bond energy for this triple bond is 994 kJ mol^{-1}.

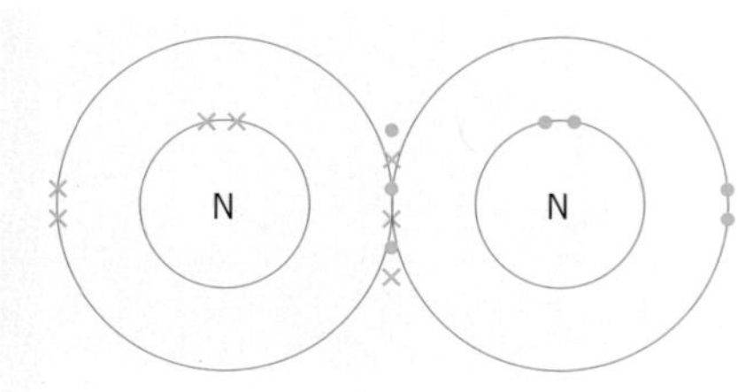

Fig 28.1 Dot-and-cross diagram for a nitrogen molecule

All plants and animals contain nitrogen in amino acids, polypeptides, proteins and nucleic acids. Plants cannot obtain nitrogen directly from the air. Instead, they take up nitrogen-containing compounds from the soil. The **nitrogen cycle** (Fig 28.2) describes the way that nitrogen and its compounds circulate within the environment and living organisms.

You can read about amino acids, polypeptides and proteins in Chapter 19.

Fig 28.2 shows the processes that remove nitrogen-containing compounds from the soil. There must be a constant supply of nitrates to the soil for plants to grow healthily. Much of the nitrogen cycle has taken place naturally for many millions of years. In recent times, however, human activity has added some extra stages to the cycle. The advent of intensive farming has affected the delicate balance of the nitrogen cycle. Intensive farming removes more nitrogen-containing compounds than are naturally put back into the soil, so it has become necessary to supplement the supply of nitrogen by adding synthetic fertilisers such as ammonium nitrate.

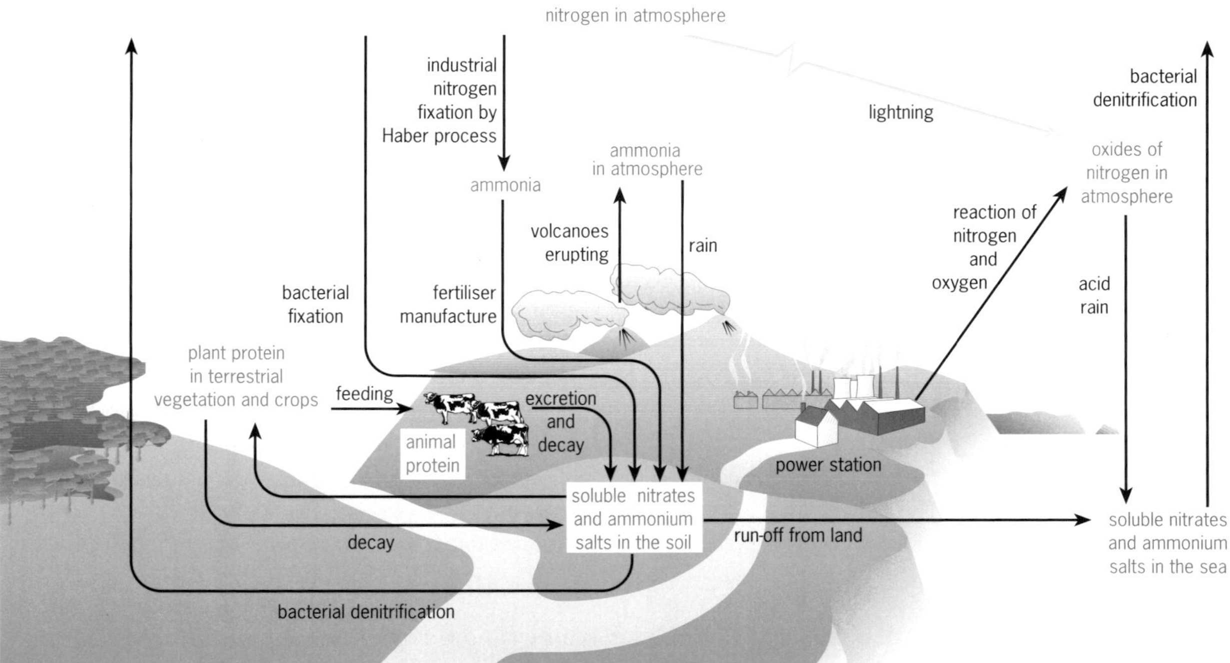

Fig 28.2 The nitrogen cycle

NITROGEN FIXATION

Processes that convert atmospheric nitrogen into nitrogen-containing substances are a very important part of the nitrogen cycle. They are known as **nitrogen-fixation** processes. Nitrogen fixation occurs by natural processes and as a result of human activity. The most important synthetic nitrogen-fixation process is the manufacture of ammonia. This chapter shows how the synthesis of such a simple molecule has had a profound effect on agriculture and the chemical industry.

Leguminous plants and nitrogen fixation

Some plants, such as peas and clover, are able to convert atmospheric nitrogen into ammonium ions, NH_4^+, by the action of bacteria found in root nodules. The ammonium ions may be directly absorbed into the roots and metabolised by plants to form plant protein. Alternatively, the ammonium ions may be oxidised by nitrifying bacteria to give nitrate ions:

$$NH_4^+(aq) + 1\tfrac{1}{2}O_2(g) \rightarrow NO_2^-(aq) + 2H^+(aq) + H_2O(l)$$

$$NO_2^-(aq) + \tfrac{1}{2}O_2(g) \rightarrow NO_3^-(aq)$$

The aqueous nitrate ions can then be absorbed through the roots by osmosis and later assimilated into plant protein.

Lightning and nitrogen fixation

Molecular nitrogen only reacts if enough energy is available to break the very strong triple covalent bond. Lightning provides sufficient energy to break the bond, and allows the direct combination of nitrogen and oxygen to form nitrogen monoxide:

$$N_2(g) + O_2(g) \rightarrow 2NO(g)$$

QUESTION 1

The oxidation state of nitrogen in the ammonium ion is –3. Is atmospheric nitrogen oxidised or reduced within root nodules to make ammonium ions? Explain your answer.

Fig 28.3 The bacteria in root nodules of peas, clover and beans convert atmospheric nitrogen into ammonium ions and nitrates

QUESTION 2

2 The compounds or ions listed below form part of the nitrogen cycle. Write down the oxidation state of nitrogen in each case:
a) NO_2^- **b)** NO_3^- **c)** KNO_3
d) NO **e)** NO_2

For how to determine oxidation numbers, see page 109.

REMEMBER THIS

Nitrogen monoxide, NO, is also known as nitric oxide or nitrogen(II) oxide.

Nitrogen monoxide oxidises into nitrogen dioxide even at low temperatures. Nitrogen dioxide may be further oxidised in the presence of water and oxygen to give dilute nitric acid, a solution that contains nitrate ions. These processes are represented by the equations below.

$$NO(g) + \frac{1}{2}O_2(g) \rightarrow NO_2(g)$$

$$2NO_2(g) + H_2O(l) + \frac{1}{2}O_2(g) \rightarrow 2HNO_3(aq)$$

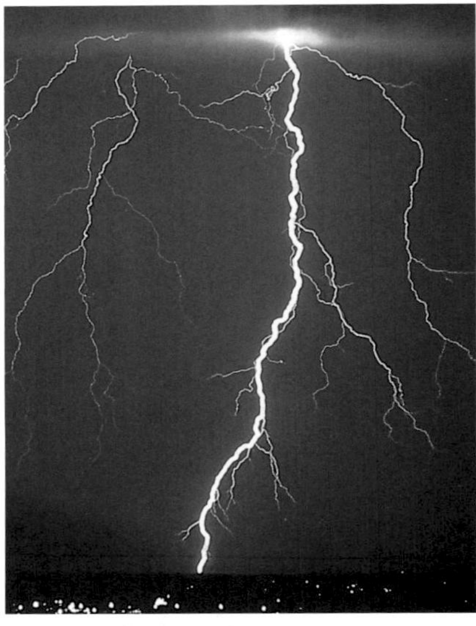

Fig 28.4 The enormous quantity of electrical energy available within a lightning bolt is sufficient to allow nitrogen to react with oxygen. This is one of the most important natural nitrogen-fixation processes

2 THE HABER PROCESS AND NITROGEN FIXATION

In Chapter 1, we describe the manufacture of ammonia by the Haber process.

$$N_2(g) + 3H_2(g) \rightleftharpoons 2NH_3(g)$$

Ammonia is a starting material in the production of many fertilisers. The presence of the triple bond in each nitrogen molecule means that nitrogen and hydrogen react very slowly under normal conditions. You will also notice that the reaction is reversible. The choice of conditions used in the manufacture of ammonia determines the rate of reaction and the equilibrium position of the reaction mixture.

The position of equilibrium for the reaction between nitrogen and hydrogen can be estimated by looking at the numerical value of the equilibrium constant. The equilibrium constant, K_c, for this reaction is represented by the equation:

$$K_c = \frac{[NH_3(g)]^2}{[H_2(g)]^3[N_2(g)]}$$

The concentrations refer to the molar concentrations at equilibrium.

We have already described some equilibria in chapters 17 and 18. You may want to read pages 331 and 352 before continuing.

Table 28.1 K_c at different temperatures for the equilibrium $N_2(g) + 3H_2(g) \rightleftharpoons 2NH_3(g)$

Temperature/K	K_c/dm^6 mol^{-2}
298	4.05×10^8
400	4.39×10^3
500	5.98×10
600	4.03
700	2.56×10^{-1}
800	2.98×10^{-2}
900	5.45×10^{-3}

Table 28.1 illustrates that at low temperatures the numerical value of K_c is higher – the equilibrium position lies well to the right. At higher temperatures the equilibrium position lies to the left, the side of the reactants nitrogen and hydrogen. In contrast, the rate of reaction is very slow at low temperatures, but accelerates at higher temperatures. This conflict between rate of reaction and position of equilibrium is explored in some detail in Chapter 1.

In a gas-phase reaction it is not always appropriate to measure the concentrations of the reactants and the products. It is often easier to measure the total pressure of the equilibrium mixture and the partial pressures of the reactants and of the products.

PARTIAL PRESSURE

For a mixture of gases it is convenient to express the composition in terms of **partial pressures** rather than concentrations.

The partial pressure of a gas in a container of several gases is the pressure that particular gas would have if it were the only gas present in the container.

Consider a mixture of neon and helium gases. The partial pressure of neon, pNe, is proportional to the **mole fraction** of neon, xNe, in the mixture, and:

$$p\text{Ne} = x\text{Ne} \times P$$

where P is the total pressure of the mixture of gases.

The mole fraction of neon is the number of moles of neon gas in the mixture divided by the sum of the number of moles of each gas present:

$$x\text{Ne} = \frac{n\text{Ne}}{n\text{He} + n\text{Ne}}$$

where nNe = number of moles of neon and nHe = number of moles of helium.

EXAMPLE

Q A mixture of gases at a total pressure of 1.0×10^5 Pa contains 3.4 mol of carbon dioxide, 4.5 mol of oxygen and 1.1 mol of nitrogen. What is the partial pressure of each gas in the mixture?

A The mole fraction of $CO_2 = \frac{3.4}{3.4 + 4.5 + 1.1} = 0.38$

So: $pCO_2 = 0.38 \times 1.0 \times 10^5$ Pa $= 3.8 \times 10^4$ Pa

The mole fraction of $O_2 = \frac{4.5}{3.4 + 4.5 + 1.1} = 0.50$

So: $pO_2 = 0.50 \times 1.0 \times 10^5$ Pa $= 5.0 \times 10^4$ Pa

The mole fraction of $N_2 = \frac{1.1}{3.5 + 4.5 + 1.1} = 0.12$

So: $pN_2 = 0.12 \times 1.0 \times 10^5$ Pa $= 1.2 \times 10^4$ Pa

THE EQUILIBRIUM CONSTANT, K_p

In reactions that involve gaseous reactants or products, the equilibrium constant is often stated in terms of the partial pressures of the gases present. The expression for this equilibrium constant resembles that for K_c, except that the partial pressures of the gases in the equilibrium mixture are used instead of the molar concentrations of the components in the mixture. The expression for the equilibrium constant, K_p, for the Haber process is:

$$K_p = \frac{(pNH_3)^2}{(pH_2)^3(pN_2)}$$

where pNH_3 is the partial pressure of ammonia gas.

Notice that the partial pressures are raised to the same powers as the concentrations in the expression for K_c. The equilibrium constant K_p has both a numerical value and a unit. The higher the numerical value, the greater the proportion of ammonia in the equilibrium mixture and the more the position of equilibrium lies on the right-hand side. If the partial pressures are measured in Pascals, the units of K_p for this equilibrium will be Pa^{-2}. The numerical value of this equilibrium constant will not be affected by changes in concentration and pressure.

If an equilibrium involves solids, liquids and solutions, then only the gases in the process are used in the K_p expression. The unit for the equilibrium constant in the reaction:

$$NH_4Cl(s) \rightleftharpoons NH_3(g) + HCl(g) \quad K_p = (pNH_3)(pHCl) \quad \text{is} \quad Pa \times Pa = Pa^2.$$

REMEMBER THIS

In a gaseous mixture, the mole ratio of each gas equals the ratio of the partial pressures of the gases.

QUESTION 3

A cylinder contains 2.0 mol of oxygen gas, and 5.6 mol of gaseous dinitrogen oxide. The pressure inside the cylinder is 450 kPa. What is the partial pressure of oxygen in the cylinder?

Later in this chapter, we discuss the factors that affect the position the equilibrium reaches and the numerical value of the equilibrium constant.

QUESTIONS 4–5

Write down expressions for the equilibrium constant, K_p, for each of the following equilibria:

a) $H_2(g) + I_2(g) \rightleftharpoons 2HI(g)$;

b) $2NO(g) + O_2(g) \rightleftharpoons 2NO_2(g)$;

c) $CaCO_3(s) \rightleftharpoons CaO(s) + CO_2(g)$.

For each of the equilibria in Question 4 state the units of the equilibrium constant K_p. Assume that the partial pressures are all measured in Pa.

QUESTION 6

6 A mixture of nitrogen and hydrogen were heated at a constant temperature in a reaction vessel until equilibrium was reached. The partial pressures of each gaseous component in the reaction mixture at equilibrium were:
$pNH_3 = 8.3 \times 10^5$ Pa
$pN_2 = 3.5 \times 10^4$ Pa
$pH_2 = 1.1 \times 10^5$ Pa
Calculate the numerical value for the equilibrium constant, K_p.

QUESTION 7

7 Sulfur dioxide and oxygen were mixed in a mole ratio of 2:1 at a total initial pressure of 300 kPa. The mixture was held at a temperature of 430 °C in the presence of a suitable catalyst until it reached equilibrium:

$$2SO_2(g) + O_2(g) \rightleftharpoons 2SO_3(g)$$

At equilibrium, the partial pressure of sulfur trioxide was 190 kPa.
a) Calculate the initial partial pressures of SO_2 and O_2.
b) Calculate the partial pressures of SO_2 and O_2 at equilibrium.
c) Calculate the value of K_p for this equilibrium.

REMEMBER THIS

In economic terms, the best conditions for the Haber process are a compromise between achieving a high rate of reaction and a high percentage of ammonia in the equilibrium mixture.

Not all ammonia produced by the Haber process is used to make fertilisers; nitric acid and polyamides, such as nylon, are other important derivatives.

EXAMPLE

Q Fritz Haber studied the reaction between hydrogen and ammonia. He mixed nitrogen and hydrogen in a mole ratio of 1:3, at 300 °C and in the presence of a catalyst. The reaction was allowed to reach equilibrium. At equilibrium, the total pressure of the equilibrium mixture was 800 kPa and the partial pressure of the ammonia was 75 kPa. Calculate the numerical value for K_p under these conditions.

A As in other equilibrium questions, we need to start with a balanced equation. Under each term we write the partial pressure at equilibrium.

	$N_2(g)$	+	$3H_2(g)$	$\rightleftharpoons$	$2NH_3(g)$
Partial pressures at equilibrium:	x		$3x$		75 kPa

If the partial pressure of nitrogen is x kPa, then the partial pressure of hydrogen must be $3x$. This follows because the initial mole ratio was 1:3, which remains unchanged throughout the reaction, since the stoichiometry of the reaction indicates that for every mole of nitrogen that reacts, three moles of hydrogen react.

The total pressure is the sum of the partial pressures:

$$x + 3x + 75 = 800 \text{ kPa}$$

So: $x = 181$ kPa

Therefore: $pN_2 = 181$ kPa

and $pH_2 = 543$ kPa.

Substituting these values into the expression for the equilibrium constant gives:

$$K_p = 1.93 \times 10^{-7} \text{ kPa}^{-2}.$$

MANUFACTURING AMMONIA

Most manufacturers of ammonia use:

- a temperature of about 450 °C;
- a pressure of between 200 and 400 atmospheres (2×10^4 kPa and 4×10^4 kPa);
- an iron catalyst with potassium hydroxide promoter.

The conditions are chosen for economic reasons. A typical factory produces up to 1500 tonnes of ammonia per day. Some manufacturers use pressures as high as 1000 atmospheres, but this presents technological difficulties in the construction of the plant. The use of better catalysts has allowed some modern plants to reduce the operating pressures to as low as 80 atmospheres.

LE CHATELIER'S PRINCIPLE AND THE HABER PROCESS

Le Chatelier's principle predicts how changes in temperature, pressure and concentration will affect a system at equilibrium. Fig 28.5 summarises the effects on the equilibrium position when the conditions are changed in the Haber process.

The position of equilibrium

Le Chatelier's principle suggests that low-temperature and high-pressure conditions would maximise ammonia production. In terms of rate of reaction, if the temperature is too low the reaction proceeds too slowly. Therefore, manufacturers use a combination of a moderate temperature and a catalyst to ensure that the rate of ammonia production is high.

High pressure gives a fast rate of reaction for a gas-phase reaction, but using higher pressures drastically increases the cost of building and running the plant.

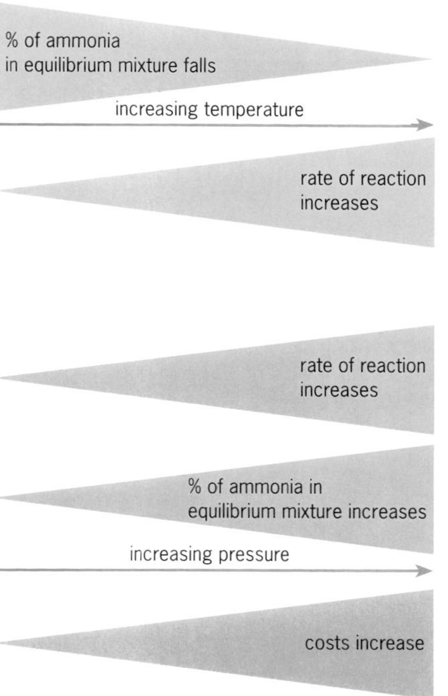

Fig 28.5 The effects of changing conditions on the Haber process

Equilibrium constants

The equilibrium constants K_c and K_p do not change when the concentration of any species in the equilibrium changes. However, the numerical values will change with temperature. For an endothermic reaction, as the temperature increases the numerical value of the equilibrium constant increases. For an exothermic reaction, as the temperature rises the numerical value of the equilibrium constant decreases. Catalysts have no effect whatsoever on the numerical value of the equilibrium constant.

3 NITRIC ACID

Nitric acid is manufactured by the oxidation of ammonia. It is a strong acid that is also a very powerful oxidising agent.

MANUFACTURING NITRIC ACID

Some of the ammonia that is produced in the Haber process is used to manufacture nitric acid. The **Ostwald** process uses cheap materials – ammonia, air and water – to manufacture nitric acid (Fig 28.6).

The flow diagram illustrates the three key steps in the manufacture of nitric acid:

- Step one involves the catalytic oxidation of ammonia.

$$4NH_3(g) + 5O_2 \rightleftharpoons 4NO(g) + 6H_2O(g) \quad \Delta H = -909 \text{ kJ mol}^{-1}$$

The conditions used are a high temperature of 900 °C, a low pressure of between 4 to 10 atmospheres and a platinum–rhodium catalyst.

These conditions give a 96 per cent yield but, unlike the Haber process, it has a poor atom economy.

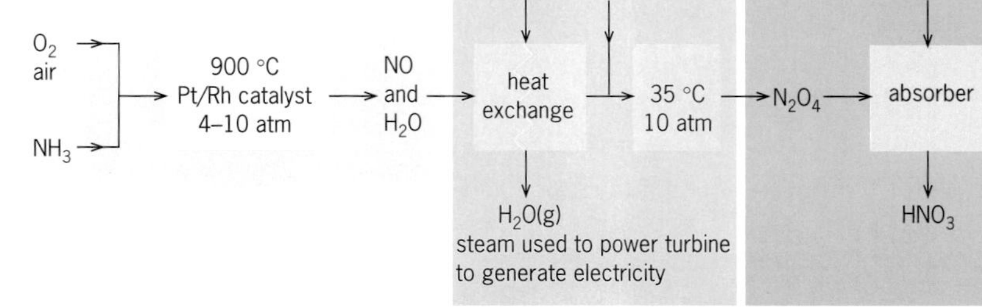

Fig 28.6 A flow diagram showing the Ostwald process for manufacturing nitric acid

- Step two involves the oxidation of nitrogen monoxide into nitrogen dioxide and the subsequent dimerisation to make dinitrogen tetroxide.

$$2NO(g) + O_2(g) \rightleftharpoons 2NO_2(g) \quad \Delta H = -115 \text{ kJ mol}^{-1}$$

$$2NO_2(g) \rightleftharpoons N_2O_4(g) \quad \Delta H = -58 \text{ kJ mol}^{-1}$$

The conditions used are a low temperature of 25 °C and a low pressure, again in the range of 4 to 10 atmospheres.

- Step three involves the reaction of dinitrogen tetroxide with water. A disproportionation reaction occurs, with the formation of nitric acid.

$$3N_2O_4(g) + 2H_2O(l) \rightarrow 4HNO_3(aq) + 2NO(g) \quad \Delta H = -103 \text{ kJ mol}^{-1}$$

The conditions used are similar to those in step two.

QUESTIONS 8–12

8 **a)** Using Le Chatelier's principle, predict the conditions that would give the maximum conversion of ammonia to nitrogen monoxide.
b) Predict conditions that would give the maximum rate of reaction.
c) Comment on the conditions actually used for the catalytic oxidation.
d) Write down an expression for the equilibrium constant, K_p, for the oxidation of ammonia.

9 At 25 °C and 100 kPa, the partial pressures in an equilibrium mixture of N_2O_4 and NO_2 are pN_2O_4 = 70 kPa and pNO_2 = 30 kPa.
a) Calculate the numerical value for the equilibrium constant K_p and state the units for the constant.
b) Describe the effect on the numerical value of K_p if the temperature of the equilibrium mixture increases, assuming the total pressure remains at 100 kPa.
c) Describe the effect on the numerical value of K_p if the total pressure of the equilibrium mixture increases, assuming the temperature remains at 25 °C.

10 Le Chatelier's principle suggests that the conversion of NO_2 into N_2O_4 should be carried out at high pressure. Suggest advantages of carrying out this reaction at low pressures.

11 Write down the oxidation state of nitrogen in: **a)** N_2O_4 **b)** HNO_3 **c)** NO

12 **a)** By using the change in oxidation state, confirm that the reaction of dinitrogen tetroxide, N_2O_4, with water is a disproportionation.
b) Explain why the reaction of N_2O_4 with water is carried out at low temperature.

REACTIONS OF NITRIC ACID

Dilute nitric acid is a typical strong acid. It will react with bases and carbonates to give nitrate salts, as the following equations exemplify.

$$CaO(s) + 2HNO_3(aq) \rightarrow Ca(NO_3)_2(aq) + H_2O(l)$$

$$NaOH(aq) + HNO_3(aq) \rightarrow NaNO_3(aq) + H_2O(l)$$

$$ZnCO_3(s) + 2HNO_3(aq) \rightarrow Zn(NO_3)_2(aq) + CO_2(g) + H_2O(l)$$

As we described in Chapter 15, concentrated nitric acid reacts with arenes to make nitroarenes. This reaction can be used to make explosives such as TNT.

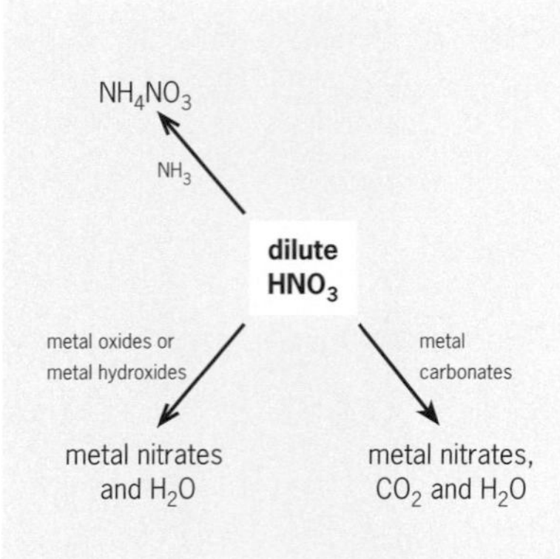

Fig 28.7 Some reactions of dilute nitric acid. Dilute nitric acid acts as a strong acid

Fig 28.8 Nitric acid is used to make explosives such as TNT and nitroglycerine

4 FERTILISERS AND THE NITROGEN CYCLE

All plants require essential elements for healthy growth. The three most important elements required are nitrogen, phosphorus and potassium (NPK). When fertilisers are sold they often quote an NPK value, which gives the percentage by mass of each of these three elements in the fertiliser.

As you can see from Table 28.2, no single compound gives all three of the essential elements. To provide all three essential elements, commercial fertilisers are normally a mixture of at least two chemicals.

Table 28.2 Some chemical fertilisers and their NPK values (percentages of the elements by mass)

Fertiliser	Formula	%N	%P	%K
ammonium nitrate	NH_4NO_3	35	0	0
ammonium phosphate	$(NH_4)_3PO_4$	28	21	0
potassium nitrate	KNO_3	39	0	14
ammonium sulfate	$(NH_4)_2SO_4$	21	0	0
urea	$(NH_2)_2CO$	47	0	0

Fertilisers need to be soluble in water so that they can be absorbed through the roots of plants. It is also preferable if they can be manufactured in pellet form to enable easy application to farm land.

MANUFACTURE OF FERTILISERS

Most fertilisers are manufactured by reacting either ammonia or nitric acid with another compound (Fig 28.9). Ammonia is a base. It reacts with acids to give ammonium salts, some of which are listed in Table 28.2. For example, ammonia and phosphoric acid react to form ammonium phosphate, which contains two of the essential elements needed by plants.

Problems arise when fertilisers such as ammonium nitrate are applied to alkaline soils. Under these conditions, ammonia can be formed either as a gas or as aqueous ammonia:

$$(NH_4)_2SO_4(s) + 2OH^-(aq) \rightarrow 2NH_3(g) + SO_4^{2-}(aq) + 2H_2O(l)$$

Addition of an excess of calcium oxide or calcium hydroxide to neutralise acidic soils can have the same effect on fertilisers that contain the ammonium ion; again, ammonia gas is liberated and the soil loses available nitrogen.

OVER-USE OF FERTILISERS

Although fertilisers can increase crop yields, it is important to use the correct amount of fertiliser to ensure that most of it is absorbed by the plants. If too much fertiliser is applied, it is leached from the ground by water and eventually it can reach rivers and even domestic water supplies. This is a particular problem with nitrate fertilisers, which are more likely to dissolve into the soil solution. In East Anglia there is some concern because the nitrate level in domestic water supplies exceeds the acceptable concentration of 50 mg dm^{-3} recommended by the European Union.

Much research into the level of fertiliser run-off and leaching is still required. It is not only the amount and type of fertiliser used that affect nitrate levels – the time of year, type of crop, soil type and underlying bedrock are all factors that must be considered.

Eutrophication

Leaching of fertilisers from the ground by water causes aqueous nitrate and phosphate ions to enter rivers. Nitrate ions and, to a greater extent, phosphate ions help the growth of all plants in a river. Green algae reproduce in huge numbers using the nitrate and phosphate nutrients and form an algal bloom that covers the surface and clogs up the river.

QUESTION 13

13 A 1 kg bag of fertiliser is made up of 500 g of potassium nitrate and 500 g of ammonium phosphate. Calculate the NPK values for this bag of fertiliser.

QUESTION 14

14 Write down the equation to show the formation of the fertiliser ammonium nitrate from nitric acid and aqueous ammonia.

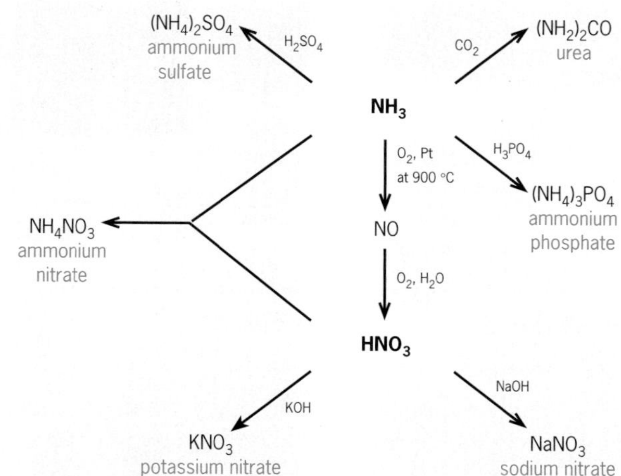

Fig 28.9 Some reactions of ammonia that give rise to many fertilizers

The algal bloom prevents sunlight from reaching plants that grow beneath the surface of the water. Consequently, these underwater plants cannot photosynthesise and they die. Colonies of bacteria feed on the decaying plant material. The bacteria's respiration uses up most of the dissolved oxygen. As a result, other aerobic forms of life, such as fish, die. The whole process – which results in the death of a river – is known as **eutrophication**.

Fig 28.10 Ammonium nitrate fertiliser being sprayed onto a field. In some parts of the world, liquid ammonia is pumped directly into the ground as a fertiliser. This can be quite dangerous – liquid ammonia can cause chemical burns and irritate the respiratory tract

Fig 28.11 Excess phosphate and nitrate nutrients in river water cause the accelerated growth of green algae. The algae form a thick surface layer, which causes underwater plants and fish to die. Increased algal growth in the sea near river estuaries is also of concern

SCIENCE IN CONTEXT

The nitrate–nitrite debate

Excessive levels of nitrates in tap water can cause methaemoglobinaemia (blue-baby syndrome) in bottle-fed babies. In this syndrome, fetal haemoglobin stays in the baby's blood after birth instead of being replaced at the usual rate by normal haemoglobin.

Fetal haemoglobin has a greater affinity for nitrogen monoxide (NO) than normal haemoglobin; as a result the oxygen-carrying efficiency of the blood is reduced. Breast-fed babies are at a much lower risk of this happening. The NO that enters a bottle-fed baby's blood is thought to come from nitrite and nitrate ions in the water that the milk powder is mixed with.

It is believed that the nitrate and nitrite ions in tap water are probably from fertiliser run-off that enters the water supply, but this is by no means the whole story. There are nitrates in many vegetables, particularly leaf and root crops, and other products such as beer. Nitrite ions are also in the body as a result of the biochemical reduction of nitrate ions.

Nitrates are added to cured meats to prevent bacterial spoilage and food poisoning, as the labels on the packaging indicate. An average-sized ham may contain several grams of potassium nitrate or sodium nitrate. Most cured meats also contain sodium nitrite or potassium nitrite to inhibit the growth of the toxin-producing bacterium *Clostridium botulinum*.

Some scientists also think that nitrates in our diet are responsible for some stomach cancers. When nitrite ions reach the stomach, the acidic conditions are sufficient to produce nitrous acid, which in turn can undergo protonation to form $[H_2ONO]^+$.

$$NO_2^-(aq) + H^+(aq) \rightleftharpoons HNO_2(aq)$$

$$HNO_2(aq) + H^+(aq) \rightleftharpoons [H_2ONO]^+(aq)$$

The protonated form of nitrous acid reacts with nitrogen-containing compounds in food to produce suspected carcinogens (chemicals that cause cancer).

Scrutiny of nitrate and nitrite ions in food preservation and tap water will continue until we fully understand the implications for our health.

Fig 28.12 Sodium nitrite is responsible for the pink colour of ham. It inhibits the growth of dangerous bacteria

Organic fertilisers

Many farmers now use organic fertilisers, such as manure, fish blood and bone. These fertilisers derive from living organisms rather than the synthetic fixation of nitrogen via the Haber process. Bacterial decay of animal and plant material produces ammonium ions and nitrate ions. Organic farming also uses neither pesticides nor herbicides, so it produces crops that contain fewer toxins. However, the use of organic fertilisers, such as manure or silage, can still cause water pollution and eutrophication if they are allowed to enter a river.

5 AMMONIA AS A BASE

Ammonia is one of the most important bases. In aqueous conditions, an ammonia molecule can accept a proton from a water molecule to form an ammonium ion and a hydroxide ion:

$$NH_3(aq) + H_2O(l) \rightleftharpoons NH_4^+(aq) + OH^-(aq)$$

The presence of OH^- ions means that the solution is alkaline and that ammonia shows basic properties.

The basic properties of ammonia allow it to be identified by:

- its reaction with hydrogen chloride, with which it forms a white smoke of ammonium chloride;
- its effect on moist red litmus, which turns blue.

We discuss the basic characteristics of ammonia in more detail later in the chapter.

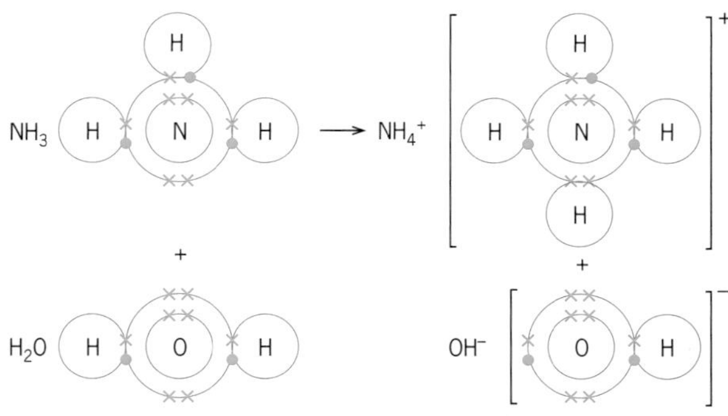

Fig 28.13 A dot-and-cross diagram for ammonia, water, the ammonium ion and the hydroxide ion

BASE DISSOCIATION CONSTANT

In Chapter 17, we defined Brønsted–Lowry acids as proton donors and Brønsted–Lowry bases as proton acceptors. Fig 28.13 shows that when ammonia dissolves in water, ammonia behaves as a Brønsted–Lowry base. Look at the equation:

$$NH_3(aq) + H_2O(l) \rightleftharpoons NH_4^+(aq) + OH^-(aq)$$

NH_3 accepts a proton from H_2O to form NH_4^+, which explains why NH_3 is considered a base. In this case, H_2O is the conjugate acid and donates a proton to the NH_3 base.

You can read more about conjugate acids on page 331.

Although one of the four bonds in NH_4^+ is a dative covalent bond (see Fig 28.13), the four bonds are indistinguishable.

The reaction of ammonia with water is an equilibrium process, so it is possible to write an expression for the equilibrium constant:

$$K_c = \frac{[NH_4^+(aq)][OH^-(aq)]}{[NH_3(aq)][H_2O(l)]}$$

We have included water in the expression because it is a reactant, not just a solvent. However, since water is a solvent to all components of the equilibrium, its concentration is almost constant and can be ignored. By removing the water term from the expression, we can write down a new equilibrium constant called the **base dissociation constant, K_b**.

$$K_b = \frac{[NH_4^+(aq)][OH^-(aq)]}{[NH_3(aq)]}$$

REMEMBER THIS

Do not confuse the idea of strong and weak bases with concentrated and dilute bases. Strong and weak in this context refer to the extent of dissociation of the base in water.

REMEMBER THIS

Amines are called primary, secondary or tertiary, depending on how many alkyl or aryl groups are bonded to the nitrogen atom; for example, diethylamine is a secondary amine.

The numerical value for K_b is often very small, so pK_b, the negative logarithm to base 10 of the base dissociation constant, is sometimes used instead:

$$pK_b = -\log_{10}K_b$$

Strong and weak bases

We have already described the reaction of ammonia with water to form aqueous hydroxide ions. In aqueous solutions, a **base** is a substance that increases the concentration of $OH^-(aq)$ either by reacting with water or by adding extra OH^- ions. Sodium hydroxide is a base, since it fully dissociates in water to form aqueous sodium ions and aqueous hydroxide ions:

$$NaOH(s) + \xrightarrow{H_2O(l)} Na^+(aq) + OH^-(aq)$$

Bases such as sodium hydroxide that dissolve in water are called **alkalis**.

Sodium hydroxide and ammonia show an important difference in their behaviour with water. Sodium hydroxide fully dissociates in water, so it is known as a **strong base**; ammonia only partially dissociates and forms a dynamic equilibrium, so it is known as a **weak base**:

$$NH_3(aq) + H_2O(l) \rightleftharpoons NH_4^+(aq) + OH^-(aq)$$

The position of equilibrium lies almost completely on the side of the undissociated NH_3, so a solution of a weak base contains only a small proportion of hydroxide ions.

Base strength of amines

Amines are a class of organic molecules closely related to ammonia. Amines are derived from ammonia by the replacement of one, two or three hydrogen atoms by alkyl and or aryl groups (Fig 28.14).

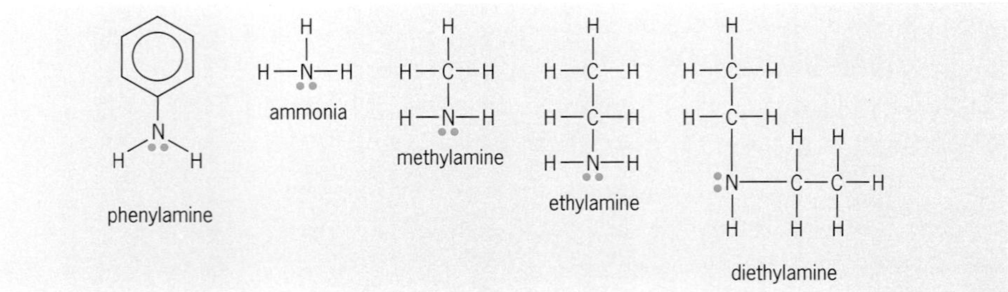

phenylamine ammonia methylamine ethylamine diethylamine

Fig 28.14 The structure of some amines

Amines are bases because they have a lone pair of electrons on the nitrogen atom that can be donated to $H^+(aq)$. This means they are proton acceptors:

$$R^1R^2R^3N(l) + H_2O(l) \rightleftharpoons R^1R^2R^3NH^+(aq) + OH^-(aq)$$

R^1, R^2 and R^3 can be H, an aryl or an alkyl group.

The strength of amines can be compared through their base dissociation constants. More dissociation occurs in a strong base, so it has a larger base dissociation constant.

Table 28.3 The base dissociation constant of some amines

Amine	Formula	Base dissociation constant, K_b/mol dm^{-3}	pK_b
ammonia	NH_3	1.8×10^{-5}	4.7
methylamine	CH_3NH_2	4.38×10^{-4}	3.4
ethylamine	$C_2H_5NH_2$	5.6×10^{-4}	3.3
diethylamine	$(C_2H_5)_2NH$	1.3×10^{-3}	2.9
phenylamine	$C_6H_5NH_2$	3.8×10^{-10}	9.4

? QUESTION 15

15 **a)** Write an equation to show the dissociation of methylamine in water.

b) Write an expression for the base dissociation constant for methylamine.

The base strengths of the amines in Table 28.3 can be explained in terms of the availability of the lone pair on nitrogen; the more available the lone pair, the stronger the base. In methylamine and ethylamine, the alkyl groups are electron-releasing groups, which make the lone pair on the nitrogen atom more available for donation, and therefore for accepting protons. Substituting another hydrogen atom with an alkyl group further increases the availability of the lone pair. Therefore, diethylamine is a stronger base than ethylamine.

Phenylamine is a much weaker base than ammonia because the lone pair on the nitrogen atom is delocalised into the π system of the benzene ring. This makes the lone pair less available for donation to a $H^+(aq)$.

All amines react with dilute acids, such as hydrochloric acid, to give a salt that is soluble in water. Phenylamine reacts with dilute hydrochloric acid to give a solution of phenylammonium chloride:

$$C_6H_5NH_2(l) + HCl(aq) \rightleftharpoons C_6H_5NH_3^+(aq) + Cl^-(aq)$$

QUESTION 16

a) Write an equation for the reaction of aqueous ammonia with sulfuric acid.
b) Write an equation to show the reaction between ethylamine and dilute hydrochloric acid.

pH OF AN ALKALINE SOLUTION

In Chapter 17, we defined the pH of an aqueous solution as the negative logarithm to base 10 of the hydrogen-ion concentration:

$$pH = -\log_{10}[H^+(aq)]$$

We can determine the pH of a solution by knowing the hydrogen ion concentration. When bases dissolve in water they increase the aqueous hydroxide ion concentration. To find out what effect this has on the aqueous hydrogen ion concentration, we must study the acid–base behaviour of water.

REMEMBER THIS

Remember: $H^+(aq)$ is used to represent $H_3O^+(aq)$.

Self-ionisation of water

Water is considered to be a covalent substance, but even in the most pure sample of water there is a very small concentration of $H^+(aq)$ and $OH^-(aq)$. These ions are a result of **self-ionisation**:

$$\underset{\text{acid}}{H_2O(l)} + \underset{\text{base}}{H_2O(l)} \rightleftharpoons \underset{\text{conjugate acid}}{H_3O^+(aq)} + \underset{\text{conjugate base}}{OH^-(aq)}$$

The equation shows that two molecules of water react with one another; one behaves as a base and the other as an acid. In this way, a proton is transferred from one water molecule to the other one. The self-ionisation is an equilibrium process and is often written in a simpler form:

$$H_2O(l) \rightleftharpoons H^+(aq) + OH^-(aq)$$

The position of this equilibrium lies very much to the left. The concentration of $H^+(aq)$ and of $OH^-(aq)$ is extremely small.

QUESTION 17

The ionisation of water is an endothermic process.
a) Using Le Chatelier's principle, predict what will happen to the position of equilibrium as the temperature of the water increases.
b) What will happen to the numerical value of the ionic product of water as the temperature of the water increases?
c) The electrical conductivity of water increases with increasing temperature. Suggest a reason why.

Ionic product of water

The equilibrium constant for the ionisation of water is:

$$K_c = \frac{[H^+(aq)][OH^-(aq)]}{[H_2O(l)]}$$

Since the concentration of water is so large, it is considered to be constant. The expression simplifies to:

$$K_w = [H^+(aq)][OH^-(aq)]$$

where K_w **is the ionic product of water**. The ionic product is defined as the product of the concentrations, in mol dm^{-3}, of aqueous hydrogen ions and of aqueous hydroxide ions in water. Its unit is $mol^2\ dm^{-6}$.

K_w is a constant at a particular temperature, so it will not change even if the concentration of $H^+(aq)$ changes. At 298 K, the ionic product of water is

QUESTIONS 18–20

18 A sample of stomach acid has an aqueous hydrogen concentration of 5.78×10^{-3} mol dm^{-3}. Calculate the OH^-(aq) concentration in the sample.

19 The pH of an aqueous solution at 298 K is 7.6.

a) Calculate the value of $[H^+(aq)]$.

b) Hence calculate the value of $[OH^-(aq)]$.

20 The pH of pure water at 298 K is 7.0. Calculate the concentrations of the H^+(aq) and of the OH^-(aq) in pure water.

REMEMBER THIS

For an aqueous solution

$pH + pOH = pK_w$

pK_w is $-\log_{10}K_w$, which is 14 at room temperature.

Also for an acid–base pair such as ammonia and the ammonium ion:

$pK_a + pK_b = pK_w$

1.0×10^{-14} mol^2 dm^{-6}. This means that the concentrations of aqueous hydroxide ions and aqueous hydrogen ions are mathematically linked. In any aqueous solution, if one of the two concentrations is known then the other one can be calculated.

EXAMPLE

Q The concentration of hydrogen ions in a sample of tap water at a temperature of 298 K is 2.5×10^{-8} mol dm^{-3}. Calculate the concentration of aqueous hydroxide ions in the water sample.

A Use the ionic product of water:

$$K_w = [H^+(aq)][OH^-(aq)]$$

Substitute into this equation the hydrogen ion concentration and the numerical value for the ionic product:

$$1.0 \times 10^{-14} = 2.5 \times 10^{-8} \times [OH^-(aq)]$$

Rearranging this equation gives:

$$[OH^-(aq)] = \frac{1.0 \times 10^{-14}}{2.5 \times 10^{-8}}$$
$$= 4.0 \times 10^{-7} \text{ mol dm}^{-3}$$

EXAMPLE

Q What is the pH of 0.150 mol dm^{-3} aqueous potassium hydroxide?

A Potassium hydroxide is a strong base that fully dissociates into its constituent aqueous ions:

$$KOH(aq) \rightarrow K^+(aq) + OH^-(aq)$$

The stoichiometry of the equation shows that in 0.150 mol dm^{-3} aqueous potassium hydroxide, $[OH^-(aq)] = 0.150$ mol dm^{-3}. Using the ionic product of water gives:

$$[H^+(aq)] = \frac{K_w}{[OH^-(aq)]} = \frac{1.00 \times 10^{-14}}{0.150} = 6.67 \times 10^{-14} \text{ mol dm}^{-3}$$

So:

$$pH = -\log_{10}(6.67 \times 10^{-14})$$
$$= 13.2$$

pOH

The concentration of OH^-(aq) in pure water is 1.0×10^{-7} mol dm^{-3}. In most aqueous solutions, $[OH^-(aq)]$ has a small value, so it is convenient to use a logarithmic scale like the one used to measure the aqueous hydrogen ion concentration. The scale that is used is called the **pOH scale**. pOH is the negative logarithm to base 10 of the hydroxide ion concentration in mol dm^{-3}:

$$\mathbf{pOH = -\log_{10}[OH^-(aq)]}$$

There is a simple connection between the pH and pOH of an aqueous solution. The sum of the pH and the pOH values is 14:

$$pOH + pH = 14$$

This equation is very useful in calculating the pH of alkaline solutions.

Determining the pH of aqueous solutions of strong bases

The pH of a strong base, such as aqueous sodium hydroxide, can be determined simply if the overall concentration of the base is known. We have described two ways that link the $H^+(aq)$ and the $OH^-(aq)$ concentrations, and either can be used to calculate the pH of a basic solution (see, for example, the second Example on page 586).

The next Example shows another way to work out the pH of an alkaline solution. This time it uses the idea of the pOH value of a solution. You must decide for yourself which of the two methods for calculating the pH of an alkaline solution you find easier to use.

EXAMPLE

Q What is the pH of 0.200 mol dm^{-3} aqueous barium hydroxide?

A Aqueous barium hydroxide is a strong base. It fully dissociates in water to give $Ba^{2+}(aq)$ and $OH^-(aq)$:

$$Ba(OH)_2(aq) \rightarrow Ba^{2+}(aq) + 2OH^-(aq)$$

1 mol of barium hydroxide contains 2 mol of hydroxide ions, so:

$$[OH^-(aq)] = 2 \times [Ba(OH)_2(aq)] = 0.400 \text{ mol dm}^{-3}$$

$$pOH = -\log_{10}(0.400) = 0.40$$

Using the relationship between pOH and pH gives:

$$pH = 14 - pOH = 13.60$$

Table 28.4 pH and pOH values for aqueous solutions

[H⁺(aq)]	[OH⁻(aq)]	pH	pOH
10	1×10^{-15}	–1	15
1	1×10^{-14}	0	14
1×10^{-1}	1×10^{-13}	1	13
1×10^{-2}	1×10^{-12}	2	12
1×10^{-3}	1×10^{-11}	3	11
1×10^{-4}	1×10^{-10}	4	10
1×10^{-5}	1×10^{-9}	5	9
1×10^{-6}	1×10^{-8}	6	8
1×10^{-7}	1×10^{-7}	7	7
1×10^{-8}	1×10^{-6}	8	6
1×10^{-9}	1×10^{-5}	9	5
1×10^{-10}	1×10^{-4}	10	4
1×10^{-11}	1×10^{-3}	11	3
1×10^{-12}	1×10^{-2}	12	2
1×10^{-13}	1×10^{-1}	13	1
1×10^{-14}	1	14	0

STRETCH AND CHALLENGE

Determining the pH of the weak bases

A similar method to that used in Chapter 17 for determining the hydrogen ion concentration, and hence the pH of a weak acid, can be used to calculate the pH of a weak acid. This time the hydroxide ion concentration and pOH can be determined. From this we have already shown how to calculate the pH of the solution.

The following example shows how this calculation can be done for 0.10 mol dm^{-3} aqueous ammonia (K_b = 1.8×10^{-5} mol dm^{-3}).

First write down the equation.

	$NH_3(aq)$ + $H_2O(l)$	⇌	$NH_4^+(aq)$	+	$OH^-(aq)$
At start:	0.10 mol dm^{-3}		0 mol dm^{-3}		0 mol dm^{-3}
At equilibrium:	$0.10 - x$ mol dm^{-3}		x mol dm^{-3}		x mol dm^{-3}

Using the expression for K_b: $1.8 \times 10^{-5} = \dfrac{x \times x}{0.10 - x}$

Ammonia is a weak base, so the position of equilibrium lies to the left. Therefore, x is small compared to 0.10, so:

$$1.8 \times 10^{-5} = \frac{x \times x}{0.10}$$

Solving this equation gives:

$$x = 1.34 \times 10^{-3} \text{ mol dm}^{-3}$$

Since x is $[OH^-(aq)]$, then pOH = 2.88 and using the expression pOH + pH = 14 gives the pH = 11.12.

QUESTIONS 21–22

Calculate the pH of each of the following aqueous solutions at 298 K:

a) a solution with a pOH of –0.34,

b) a solution with $[OH^-(aq)]$ = 3.67×10^{-4} mol dm^{-3};

c) 0.100 mol dm^{-3} aqueous caesium hydroxide, CsOH;

d) 2.00×10^{-3} mol dm^{-3} aqueous calcium hydroxide, $Ca(OH)_2$.

The pH of an aqueous solution of calcium hydroxide is 12.1. Calculate the concentrations in mol dm^{-3} of each of the following ions in the solution:

a) $H^+(aq)$;

b) $OH^-(aq)$;

c) $Ca^{2+}(aq)$.

QUESTION 23

Use the K_b value in Table 28.3 to calculate the pH of 0.020 mol dm^{-3} methylamine.

STRETCH AND CHALLENGE

The titration of carbonate ion with dilute hydrochloric acid

In Chapter 17, various titration curves that involve strong and weak acids are described. Titrating sodium carbonate with aqueous hydrochloric acid shows two rapid changes in pH in the titration curve (Fig 28.15).

When sodium carbonate reacts with hydrochloric acid, first hydrogencarbonate ion is formed and then carbonic acid, which decomposes to give carbon dioxide.

$$Na_2CO_3(aq)+HCl(aq) \rightarrow NaCl(aq)+NaHCO_3(aq)$$
$$NaHCO_3(aq)+HCl(aq) \rightarrow NaCl(aq)+H_2CO_3(aq)$$
$$H_2CO_3(aq) \rightarrow CO_2(g) + H_2O(l)$$

In the first reaction, $CO_3^{2-}(aq)$ acts as a base and accepts an aqueous proton, $H^+(aq)$. In the second reaction, $HCO_3^-(aq)$ acts a base. The two-step reaction means that the carbonate ion has two K_b values. With the use of appropriate indicators, it is possible to titrate sodium carbonate to form sodium hydrogencarbonate (using phenolphthalein) or sodium carbonate (methyl orange).

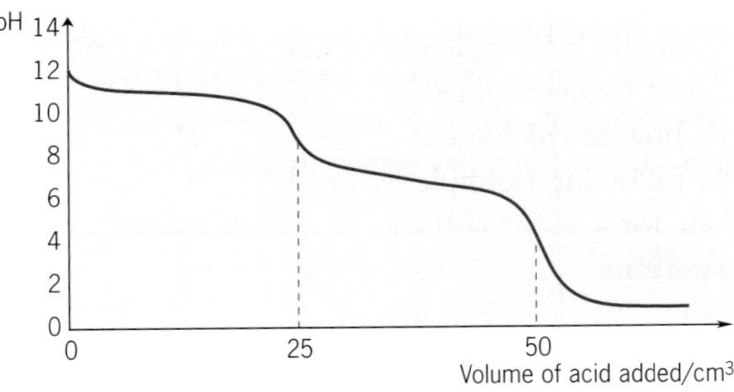

Fig 28.15 Titration curve for the titration of 25 cm^3 of 0.100 mol dm^{-3} Na_2CO_3 with 0.100 mol dm^{-3} HCl

BUFFERS

On page 343 a description is given of how buffer solutions can resist a change of pH when small volumes of acid or alkali are added to them. Many buffers are made from a mixture of a weak acid and its conjugate base. Another form of buffer can be made from a mixture of a weak base and its conjugate acid. Ammonia is a weak base and its conjugate acid is the ammonium ion, so a mixture of $NH_3(aq)$ and $NH_4Cl(aq)$ is a buffer solution.

When $H^+(aq)$ is added to this buffer solution it reacts with $NH_3(aq)$, so the pH of the solution hardly changes:

$$NH_3(aq) + H^+(aq) \rightarrow NH_4^+(aq)$$

If hydroxide ions are added, the equilibrium below shifts to the right so as to minimise the increase:

$$OH^-(aq) + NH_4^+(aq) \rightleftharpoons NH_3(aq) + H_2O(l)$$

There is a vast excess of $NH_4^+(aq)$ to react with any extra $OH^-(aq)$ added to the buffer. The $NH_4^+(aq)$ is provided by the complete dissociation of the ionic salt NH_4Cl.

? QUESTION 24

24 Using the value for K_b in Table 28.3 calculate the pH of a buffer solution that is 0.200 mol dm^{-3} methylamine and 0.100 mol dm^{-3} methylammonium chloride.

STRETCH AND CHALLENGE

Salt hydrolysis

Aqueous ammonium chloride is a salt of a weak base and a strong conjugate acid; however, even though it is a salt, the pH of the aqueous solution is well below 7.

This can be explained by looking at the ions that are present in the solution. Ammonium chloride completely dissociates in water, forming aqueous ammonium ions and aqueous chloride ions:

$$NH_4Cl(aq) \rightarrow NH_4^+(aq) + Cl^-(aq)$$

The ammonium ion will react with hydroxide ions present from the partial dissociation of water. This makes ammonia molecules in a reversible reaction, in which the position of equilibrium lies on the side of the ammonia molecules:

$$NH_4^+(aq) + OH^-(aq) \rightarrow NH_3(aq) + H_2O(l)$$

As a result, more water molecules dissociate to produce more hydroxide and hydrogen ions. As the hydroxide ions react with the ammonium ions the hydrogen ions are left in solution – hence the pH value will fall.

STRETCH AND CHALLENGE

Calculating the pH of a buffer solution

A similar method to that used in Chapter 17 for determining the pH of a buffer made from a weak acid and its conjugate base can be used to calculate the pH of a buffer made from a weak base and its conjugate acid. This time the hydroxide ion concentration, and so the pOH, can be determined. From this we have already shown how to calculate the pH of the solution.

The following example shows how this calculation can be done for a buffer solution that contains 0.10 mol dm^{-3} aqueous ammonia and 0.20 mol dm^{-3} ammonium chloride.

The $NH_3(aq)$ will be in equilibrium with its conjugate acid.

At start: $NH_3(aq) + H_2O(l) \rightleftharpoons NH_4^+(aq) + OH^-(aq)$
0.10 mol dm^{-3} 0.20 mol dm^{-3} 0 mol dm^{-3} (from the dissociation of NH_4Cl)

At equilibrium:

$$NH_3(aq) + H_2O(l) \rightleftharpoons NH_4^+(aq) + OH^-(aq)$$

0.10 − x mol dm^{-3} 0.20 + x mol dm^{-3} x mol dm^{-3}

Since $NH_3(aq)$ is a weak base, the equilibrium lies almost completely on the left. Therefore, x is very small and can be ignored compared to 0.10 and 0.20.

At equilibrium: $[NH_3(aq)] = 0.10$ mol dm^{-3}
$[NH_4^+(aq)] = 0.20$ mol dm^{-3}
$[OH^-(aq)] = x$ mol dm^{-3}

Substituting these values into the expression for the base dissociation constant gives:

$$K_b = 1.8 \times 10^{-5} = \frac{0.20 \times x}{0.10}$$

Solving for x gives: $x = 9 \times 10^{-6}$ mol dm^{-3}
So pOH = 5.0 and pH = 9.0.

THE USE OF AQUEOUS AMMONIA TO IDENTIFY METAL IONS

Aqueous ammonia is in equilibrium with aqueous ammonium ions and aqueous hydroxide ions. The aqueous hydroxide ions react with any aqueous metal ions present to form insoluble precipitates of the metal hydroxides.

These hydroxides have characteristic colours, as shown in Table 28.5, and they help chemists to identify metal ions in solution. The hydroxides formed sometimes redissolve in excess ammonia through the formation of ammine complexes with the central metal ions. You can read more about complex ions on page 486.

Table 28.5 Reaction of metal ions with aqueous ammonia (ppt = precipitate)

Metal ion	Effect of adding aqueous ammonia	Effect of adding excess ammonia	Chemical equations
aluminium	white ppt	white ppt	$Al^{3+}(aq) + 3OH^-(aq) \rightarrow Al(OH)_3(s)$
calcium	none	none	
chromium(III)	green ppt	green ppt	$Cr^{3+}(aq) + 3OH^-(aq) \rightarrow Cr(OH)_3(s)$
cobalt(II)	blue ppt	redissolves giving a brownish–yellow solution	$Co^{2+}(aq) + 2OH^-(aq) \rightarrow Co(OH)_2(s)$ $Co(OH)_2(s) + 6NH_3(aq) \rightarrow [Co(NH_3)_6]^{2+}(aq) + 2OH^-(aq)$
copper(II)	light-blue ppt	redissolves giving a dark blue solution	$Cu^{2+}(aq) + 2OH^-(aq) \rightarrow Cu(OH)_2(s)$ $Cu(OH)_2(s) + 4NH_3(aq) \rightarrow [Cu(NH_3)_4]^{2+}(aq) + 2OH^-(aq)$
iron(II)	green ppt	green ppt	$Fe^{2+}(aq) + 2OH^-(aq) \rightarrow Fe(OH)_2(s)$
iron(III)	red–brown ppt	red–brown ppt	$Fe^{3+}(aq) + 3OH^-(aq) \rightarrow Fe(OH)_3(s)$
lead(II)	white ppt	white ppt	$Pb^{2+}(aq) + 2OH^-(aq) \rightarrow Pb(OH)_2(s)$
lithium	none	none	
magnesium	white ppt	white ppt	$Mg^{2+}(aq) + 2OH^-(aq) \rightarrow Mg(OH)_2(s)$
manganese(II)	white ppt	white ppt	$Mn^{2+}(aq) + 2OH^-(aq) \rightarrow Mn(OH)_2(s)$
nickel(II)	green ppt	redissolves giving a blue solution	$Ni^{2+}(aq) + 2OH^-(aq) \rightarrow Ni(OH)_2(s)$ $Ni(OH)_2(s) + 6NH_3(aq) \rightarrow [Ni(NH_3)_6]^{2+}(aq) + 2OH^-(aq)$
potassium	none	none	
silver	brown ppt	redissolves to give a colourless solution	$2Ag^+(aq) + 2OH^-(aq) \rightarrow Ag_2O(s) + H_2O(l)$ $Ag_2O(s) + H_2O(l) + 4NH_3(aq) \rightarrow 2[Ag(NH_3)_2]^+(aq) + 2OH^-(aq)$
sodium	none	none	
zinc	white ppt	redissolves giving a colourless solution	$Zn^{2+}(aq) + 2OH^-(aq) \rightarrow Zn(OH)_2(s)$ $Zn(OH)_2(s) + 6NH_3(aq) \rightarrow [Zn(NH_3)_6]^{2+}(aq) + 2OH^-(aq)$

You can read more about nucleophilic substitution in Chapter 12.

QUESTION 25

25 Predict the products of the following reactions:
a) iodomethane and excess ammonia;
b) 1-bromobutane and ethylamine;
c) 2-chloropropane and methylamine.

QUESTION 26

26 Predict the products of the reactions between:
a) CH_3COCl and $CH_3CH_2NH_2$
b) C_6H_5COCl and CH_3NH_2
c) $(CH_3CO)_2O$ and NH_3

6 THE CHEMISTRY OF AMMONIA AND AMINES

AMMONIA AND AMINES AS NUCLEOPHILES AND BASES

We have already described ammonia and amines as bases, since they can donate the lone pair of electrons on nitrogen to accept a proton. The same lone pair on nitrogen also allows ammonia and amines to behave as nucleophiles. The lone pair is donated to make a covalent bond with an electron-deficient centre, normally a carbon atom.

Nucleophilic substitution

Ammonia reacts with halogenoalkanes to form amines. An excess of ammonia is used to ensure that a primary amine is formed. Normally a halogenoalkane is heated with excess ammonia in ethanol solution in a sealed tube. Ammonia will react with iodoethane to form ethylamine:

$$CH_3CH_2I + NH_3 \rightarrow CH_3CH_2NH_2 + HI$$

The excess ammonia reacts with the hydrogen iodide produced to form ammonium iodide, which helps to drive the reaction to completion.

Ethylamine will react with iodoethane to form diethylamine, which is another reason why an excess of ammonia is used to try and ensure that the product of the first nucleophilic substitution does not react with the iodoethane reactant:

$$CH_3CH_2NH_2 + CH_3CH_2I \rightarrow (CH_3CH_2)_2NH + HI$$

If an excess of the halogenoalkane is used, then a tertiary amine and finally a quaternary ammonium salt will be formed:

$$3CH_3CH_2I + NH_3 \rightarrow (CH_3CH_2)_3N + 3HI$$
$$4CH_3CH_2I + NH_3 \rightarrow (CH_3CH_2)_4N^+I^- + 3HI$$

Amines are a class of organic molecule containing a nitrogen atom attached to hydrogen, alkyl or aryl groups. The nitrogen atom possesses a lone pair, and earlier on in this chapter we have described the importance of this lone pair in terms of the basic properties of amines.

Amines are classified as primary, secondary or tertiary. This classification depends on the number of alkyl and aryl groups attached to the nitrogen atom. If one alkyl or aryl group is attached, it is a primary amine; if there are two attached, it is a secondary amine; and if three groups are attached, it is a tertiary amine.

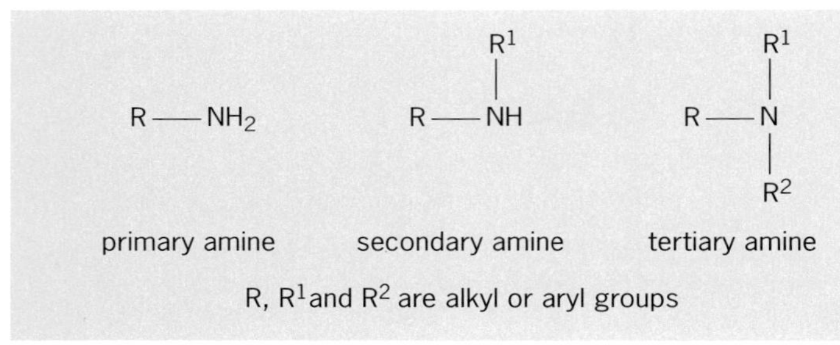

Fig 28.16 Primary, secondary and tertiary amines

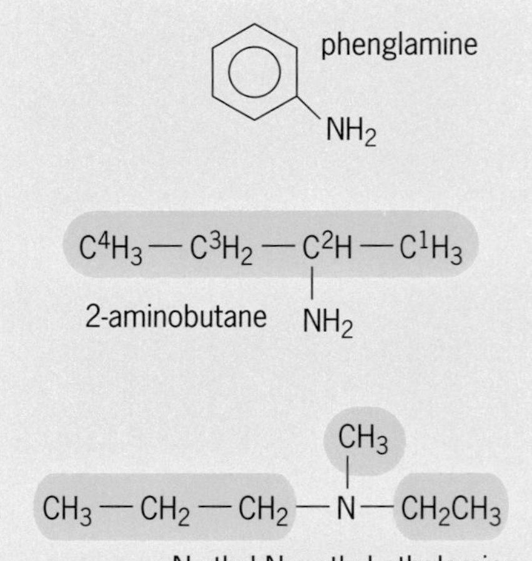

Fig 28.17 Naming amines

NOMENCLATURE

Amines are named by using either a prefix of amino or a suffix of amine. So ethylamine, $CH_3CH_2NH_2$, could also be called aminoethane. With secondary and tertiary amines all of the carbon skeletons need to be named. So diethylamine is $CH_3CH_2NHCH_2CH_3$ and trimethylamine is $(CH_3)_3N$.

PREPARATION OF AMINES

Amines can be prepared by reduction of nitriles or nitro compounds or by nucleophilic substitution of halogenated alkanes.

Reduction of nitriles

The nitrile functional group can be reduced either by $LiAlH_4$ or high pressure hydrogen in the presence of a nickel catalyst. Unfortunately this method can only make primary amines.

$$CH_3CH_2CH_2CN + 4[H] \rightarrow CH_3CH_2CH_2CH_2NH_2$$

Remember that [H] represents hydrogen atoms supplied by the $LiAlH_4$ reducing agent. With hydrogen as the reducing agent the equation should be written with H_2 instead of [H].

Reduction of nitro compounds

This method is only used to make aromatic amines. So nitrobenzene can be reduced to make phenylamine. The reagents and conditions are reflux with tin and concentrated hydrochloric acid.

$$C_6H_5NO_2 + 6[H] \rightarrow C_6H_5NH_2 + 2H_2O$$

Again only primary amines can be made by this method.

Nucleophilic addition and elimination with ammonia and amines

We described the chemistry of acyl chlorides and acid anhydrides in Chapter 18. Both have very polar carbonyl groups and react by condensation with ammonia or amines to form amides. The reaction is often carried out in the presence of excess ammonia or amine to react with the hydrogen chloride or hydrochloric acid that is formed as a co-product. The mechanism of the reaction is known as addition–elimination, as shown in Fig 28.18.

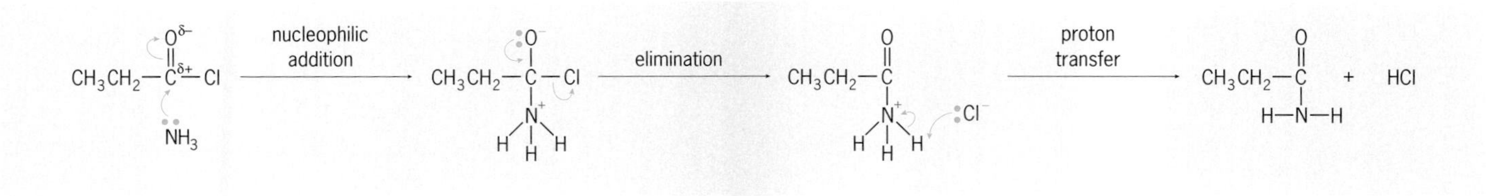

Fig 28.18 The addition–elimination mechanism for the condensation reaction between propanoyl chloride and ammonia

AMINES AS BASES

Amines will react with dilute acids to make salts similar to ammonium salts. Ammonia reacts with dilute hydrochloric acid to make aqueous ammonium chloride. In the same way, phenylamine reacts with dilute hydrochloric acid to make aqueous phenylammonium chloride:

$$C_6H_5NH_2(l) + HCl(aq) \rightarrow C_6H_5NH_3Cl(aq)$$

Notice in the equation that insoluble phenylamine is changed into soluble phenylammonium chloride. The phenylammonium chloride is soluble because it is ionically bonded and it is more correct to write its formula as $C_6H_5NH_3^+Cl^-$. Careful evaporation of the aqueous solution reveals a white, crystalline solid – another typical property of an ionic salt.

REACTION OF AMINES WITH NITROUS ACID

Nitrous acid, HNO_2, is always made in situ. This means that two other chemicals are mixed together to make nitrous acid within the actual reaction mixture. Nitrous acid is made by mixing sodium nitrite and dilute hydrochloric acid at low temperatures typically –5 to 5°C.

We describe the reaction of nitrous acid with aromatic amines on page 558, where a diazonium salt is made.

$$C_6H_5NH_2 + HCl + HNO_2 \rightarrow C_6H_5N_2^+Cl^- + 2H_2O$$

If this reaction is tried with alkyl amines then an alcohol is often formed because the diazonium salt formed immediately decomposes even at 0°C to give nitrogen gas.

$$CH_3CH_2NH_2 + HNO_2 \rightarrow CH_3CH_2OH + N_2 + H_2O$$

REACTIONS OF AMMONIA

As noted earlier in the chapter, ammonia is a weak base and a nucleophile. It will react as a base with acids to make ammonium salts.

$$NH_3(aq) + HNO_3(aq) \rightarrow NH_4NO_3(aq)$$

As a nucleophile it will react with halogenoalkanes in a substitution reaction, and with acid chlorides and acid anhydrides in an addition-elimination reaction (condensation). Fig 28.19 shows some of these reactions of ammonia.

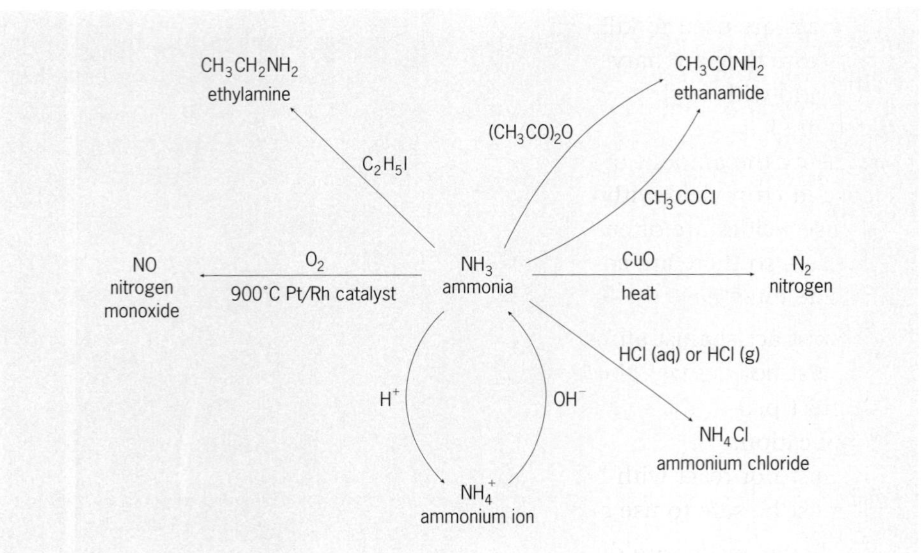

Fig 28.19 Some important reactions of ammonia

7 USE OF PHOSPHORUS COMPOUNDS IN AGRICULTURE

In this chapter we have concentrated on the chemistry of nitrogen, but phosphorus compounds also have an important role to play in increasing food production. Unlike nitrogen, which is available from the unreactive reservoir of atmospheric nitrogen, phosphorus occurs naturally in a combined form as phosphate rocks. Elemental phosphorus does not appear in the phosphorus cycle. Nitrogen exists in protein, DNA and RNA, whereas phosphorus is principally found in cell membranes as phospholipids.

PHOSPHATE FERTILISERS

Some fertilisers contain phosphorus(V) oxide as a source of phosphorus. When it is applied to soil, phosphorus(V) oxide reacts with water to form phosphoric acid. This, in turn, reacts with basic components of the soil, such as metals, to form phosphates.

Rocks that contain phosphates are plentiful. However, many are insoluble or only sparingly soluble, so they release aqueous phosphate ions to the soil very slowly. Therefore, phosphate rocks are treated with sulfuric acid to make soluble calcium hydrogenphosphate. When controlled amounts of sulfuric acid are added, a solid called superphosphate forms. Superphosphate is a mixture of calcium sulfate and calcium dihydrogenphosphate. This fertiliser is rich in soluble phosphate:

$$Ca_3(PO_4)_2(s) + 2H_2SO_4(aq) + 5H_2O(l) \rightarrow Ca(H_2PO_4)_2.H_2O(s) + 2CaSO_4.2H_2O(s)$$

QUESTION 27

27 Calculate the percentage by mass of phosphorus in superphosphate, $Ca(H_2PO_4)_2.2CaSO_4.5H_2O(s)$.

HERBICIDES

Agrochemicals not only increase crop yields by supplying the essential elements, but they can also kill off unwanted plants or weeds, and so reduce the competition for the essential nutrients. Such agrochemicals are known as herbicides. Herbicides are often polychlorinated compounds, but there are examples of herbicides that contain phosphorus.

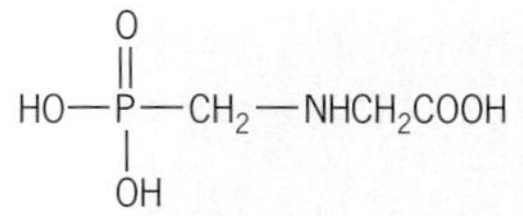

HO—P(=O)(OH)—CH_2—$NHCH_2COOH$

glyphosate

NH_2CH_2COOH

glycine

Fig 28.20 The herbicide glyphosate is a derivative of the amino acid glycine

INSECTICIDES

Insecticides are used to kill insects that reduce crop yields by feeding on the crops before they are harvested. It is estimated that over £2500 million is spent world-wide each year on insecticides; more than 15 per cent of crops planted are lost to feeding insects.

Reducing the amount of crop lost to insect feeding causes a dramatic increase in crop yield without the need for extra fertilisers or extra agricultural land. Insecticides are often toxic – not only to insects, but also to mammals and birds – so there are concerns about their continued use. A good insecticide must have the following characteristics:

- it must act against all the insects that infest the particular crop;
- it must not damage the crop;
- it must persist for a sufficient time to avoid the need for repeat spraying or application;
- it must not react with herbicides that may be used on the crop;
- it must be safe to use and must leave no harmful residues.

These characteristics are difficult to achieve, so a compromise is often reached. Future development of new insecticides will try to achieve all five characteristics.

Fig 28.21 Insecticides help farmers to reduce the number of insects that feed on crops

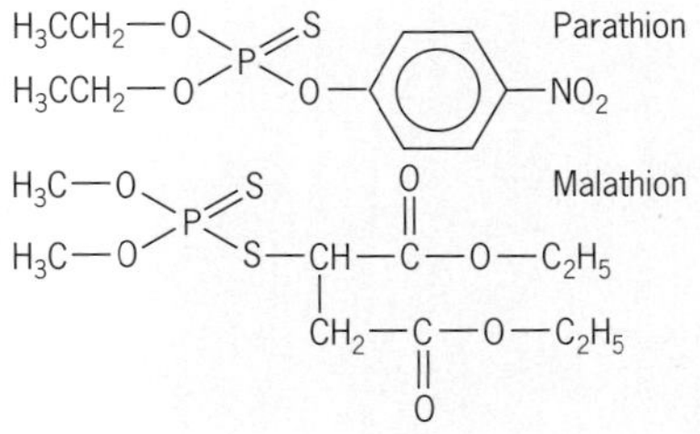

Organophosphorus compounds are often used as insecticides (Fig 28.22). These organophosphorus insecticides are less persistent in nature than polychlorinated compounds, since they can be broken down by bacterial enzymes in the soil.

Parathion: $(H_3CCH_2—O)_2P(=S)—O—C_6H_4—NO_2$

Malathion: $(H_3C—O)_2P(=S)—S—CH(—C(=O)—O—C_2H_5)—CH_2—C(=O)—O—C_2H_5$

Fig 28.22 Two organophosphorus insecticides

Future developments in insecticides include the development of insect antifeedants, which do not kill the insect directly, but inhibit feeding so that the insect dies through starvation. Another possibility is to develop an insecticide that uses an aggregation pheromone to bring all the insects to a particular location at which they are treated with another insecticide. Using this method, most of the crop remains unaffected by the application of the insecticide.

SUMMARY

After studying this chapter, you should know the following:

- Nitrogen is a very unreactive element because of the presence of a triple covalent bond in its molecule.
- Nitrogen, phosphorus and potassium are essential elements for plant growth.
- The nitrogen cycle describes the processes involved in the passage of nitrogen and its compounds through both the living and the non-living environment.
- The Haber process is the only viable synthetic way of converting atmospheric nitrogen to nitrogen-containing compounds. In the Haber process, nitrogen and hydrogen react to form ammonia. The synthesis typically takes place at 450 °C and 200 atmospheres, in the presence of an iron catalyst.
- In an industrial process, a compromise is always made between the position of equilibrium and the rate of reaction to ensure that the chemical is produced using the most economic conditions.
- An increase of temperature shifts the position of equilibrium to the left-hand side and decreases the numerical value of the equilibrium constant in an exothermic reaction.
- An increase in pressure shifts the position of equilibrium to the side of the stoichiometric equation that has the smaller number of moles of gaseous substances, but has no effect on the numerical value of the equilibrium constant.
- For an equilibrium that involves at least one gaseous component, you can write an equilibrium constant, K_p, based upon partial pressures.
- Changes in pressure and concentration have no effect on the numerical value of the equilibrium constant.
- The partial pressure of a gas in a mixture of gases is equal to the product of its mole fraction and the total pressure.
- Ammonia is a weak base and reacts with acids, such as sulfuric acid, phosphoric acid and nitric acid, to form salts that are used as fertilisers.
- Ammonia is converted into nitric acid by its catalytic oxidation to nitrogen monoxide and the subsequent reaction of nitrogen monoxide with oxygen and water.
- The ionic product of water, K_w, is the product of the molar aqueous hydrogen ion concentration and the molar aqueous hydroxide ion concentration in water.
- Strong bases fully dissociate when dissolved in water, whereas weak bases form an equilibrium mixture. The equilibrium constant for a reaction between a weak base and water is called the base dissociation constant, K_b.
- The larger the value of K_b, the stronger the base. Phenylamine is a weaker base than ammonia, which is a weaker base than ethylamine.
- Herbicides and insecticides are used to increase crop yields; they are often organophosphorus compounds.
- Amines react as nucleophiles with halogenoalkanes, acyl chlorides and acid anhydrides.
- Amines are made by the reduction of nitriles and nitroarenes.

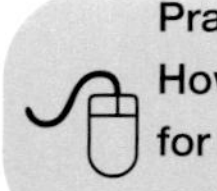

Practice questions and a How Science Works assignment for this chapter are available at www.collinseducation.co.uk/CAS

29

Group 4 and Group 6 elements

29 GROUP 4 AND GROUP 6 ELEMENTS

Silicon Fundamental to computers

Silicon and material science

Silicon compounds are at the centre of materials science and vital to the glass, cement and quartz industries. The element silicon is now being used on a smaller scale in the manufacture of semiconductors, solar cells and microchips. The impact of silicon in computing has been enormous. The introduction of micromachines that involve silicon technology means that the applications of silicon-based materials will continue to expand.

Specialised silicon compounds are being developed to withstand harsh conditions. For example, materials in spacecraft have to perform over a wide temperature range and while bombarded by high levels of cosmic radiation. Among the materials that involve silicon that chemists have been developing are:

- ceramics, such as silicon carbide, that can withstand extremely high temperatures;
- glasses, some of which are used to store radioactive waste;
- silicones that can be used as oils or water-resistant sealants;
- aerogels that are solids with a density only four times that of air.

Almost all of these materials contain silicon–oxygen single bonds in their structure. This bond is fundamental to large areas of materials science.

1 GROUP 4 AND GROUP 6 ELEMENTS

The chemistry of oxygen and oxides has been covered in detail in earlier chapters, see pages 418–423.

This chapter describes the chemistry of some non-metals, a semi-metal and some metals. We consider the elements in Group 4, one of the most diverse groups of elements in the Periodic Table. At the top of Group 4 is carbon, a non-metal; at the bottom of the group is lead, a metal. We identify some of the trends and patterns within the group, and concentrate on the chemistry of silicon and its manufacturing applications.

In addition, we describe some aspects of the chemistry of sulfur, one of the non-metals in Group 6.

2 GROUP 4 ELEMENTS

Table 29.1 Group 4 Elements

Element	Symbol
carbon	C
silicon	Si
germanium	Ge
tin	Sn
lead	Pb

When we described the properties of the halogens in Chapter 23 and the elements of Groups 1 and 2 in Chapter 22, we were able to emphasise the similarity in chemical properties and the predictable variation of physical properties with increasing atomic number. This is not really possible with the elements of Group 4, because the elements vary from a typical non-metal at the top of the group to typical metals at the bottom of the group.

OCCURRENCE OF THE ELEMENTS

Carbon occurs naturally in the Earth's crust as carbonate minerals, crude oil, natural gases and coals. Furthermore, all living organisms contain compounds of carbon. Elemental carbon occurs in small quantities as graphite or diamond deposits.

Silicon comprises almost a quarter of the Earth's crust and is almost always found combined with oxygen. Clays, sands and most of the components of soils contain compounds of silicon and oxygen. Despite the abundance of silicon in the Earth's crust it is still a difficult element to isolate. The extraction process for most of the Group 4 elements is quite similar. The favoured method is to reduce the oxide of the Group 4 element using carbon or a reducing agent of similar reactivity:

$$2PbO(s) + C(s) \rightarrow 2Pb(s) + CO_2(g)$$

Some of the minerals that contain silicon and oxygen are described later in the chapter.

Fig 29.1 Clays are complex silicates and aluminosilicates. Silicates contain strong silicon–oxygen bonds

STRETCH AND CHALLENGE

The manufacture of silicon

Silicon is manufactured by the high temperature reduction of silica (silicon dioxide) with either magnesium or carbon:

$$SiO_2 + 2C \rightarrow 2CO + Si$$

The silicon obtained by the reduction of silica is not pure enough to use in the manufacture of semiconductors, microchips and solar cells. It is purified by converting silicon into silicon(IV) chloride, and then reducing it back to silicon:

$$Si + 2Cl_2 \rightarrow SiCl_4$$

$$SiCl_4 + 2Mg \rightarrow Si + 2MgCl_2$$

It is easy to remove the soluble magnesium chloride from the silicon by washing with hot water. A further process, called zone refining, yields very pure silicon (Fig 29.2).

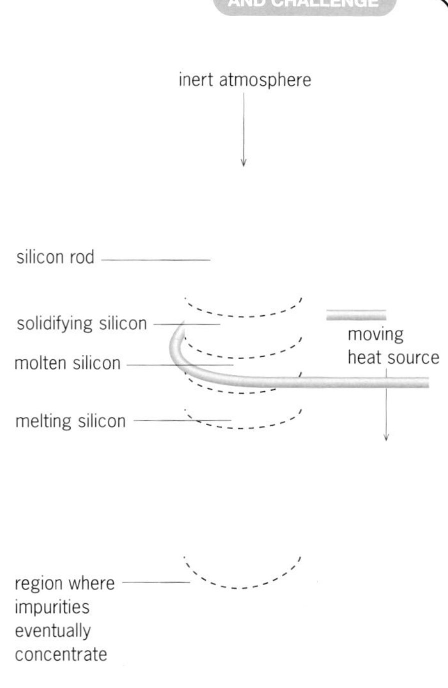

Fig 29.2 The zone-refining process. Zone refining is used to make ultra-pure silicon. A rod of silicon is heated at one end by a moving heat source. The heat source produces a thin cross-section of molten silicon. As the heat source moves downwards so does the cross section of molten silicon, taking dissolved impurities with it. Eventually, the impurities are concentrated at the bottom of the rod while the rest of the silicon is ultra-pure

QUESTIONS 1–3

1 Write down the equation for the reaction of magnesium with silicon dioxide.

2 Suggest reasons why silicon is very expensive to isolate, even though the raw material silicon dioxide is both cheap and plentiful.

3 **a)** Why must zone refining take place in an inert atmosphere rather than in air?

b) Suggest a gas that could be used to provide the inert atmosphere.

QUESTIONS 4–5

4 Write balanced equations for the reduction by carbon of:

a) lead(II) oxide, PbO;

b) tin(IV) oxide, SnO_2.

5 Lead(II) oxide is obtained from galena, PbS. Galena is roasted in air and forms lead(II) oxide and sulfur dioxide.

a) Suggest one environmental problem with this roasting process.

b) Write down the balanced chemical equation for the roasting process.

Extraction of tin and lead

Both tin and lead are extracted by reduction of an oxide (lead(II) oxide or tin(IV) oxide) with carbon. In both cases, the chemistry of the reduction is similar to that in the blast furnace during the extraction of iron.

Read about the blast furnace on page 12.

PHYSICAL PROPERTIES OF GROUP 4 ELEMENTS

Table 29.2 Physical properties of Group 4 elements

	Carbon	Silicon	Germanium	Tin	Lead
atomic number	6	14	32	50	82
electron configuration	$1s^2\ 2s^2\ 2p^2$	$1s^2\ 2s^2\ 2p^6\ 3s^2\ 3p^2$	$[Ar]3d^{10}\ 4s^2\ 4p^2$	$[Kr]4d^{10}\ 5s^2\ 5p^2$	$[Xe]4f^{14}\ 5d^{10}\ 6s^2\ 6p^2$
covalent atomic radius/pm	77	117	122	140	146
electronegativity	2.5	1.7	2.0	1.7	1.6
first ionisation energy/kJ mol^{-1}	1090	786	762	707	716
melting point/°C	3570 (diamond)	1414	959	232	328
boiling point/°C	sublimes above 3700	3309	2837	2687	1755
density/kg m^{-3}	3510 (diamond); 2220 (graphite)	2330	5360	7310 (grey tin); 5750 (white tin)	11350
electrical conductivity	good (graphite); poor (diamond)	semiconductor	semiconductor	good	good
structure of solid	giant molecular	giant molecular	giant molecular	giant metallic	giant metallic

Read about the structure and properties of diamond, graphite and silicon on pages 143–144.

Read more about metallic bonding and the properties of metals on pages 140–142.

? QUESTION 6

6 What properties of tin and lead make them suitable for producing an alloy to be used in soldering?

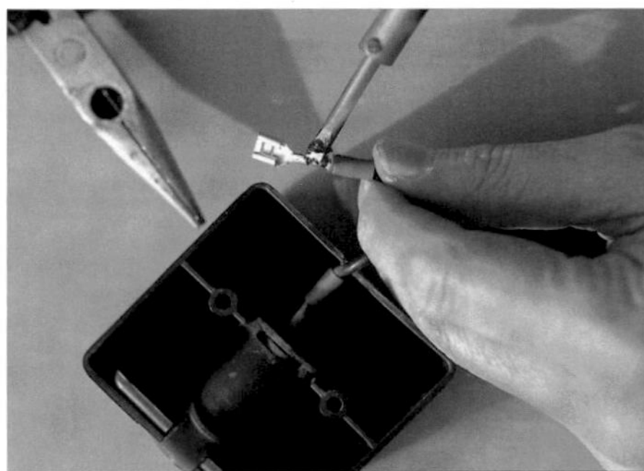

Fig 29.3 An alloy of tin and lead can be used to mend broken electrical connections

We described in Chapter 7 how the physical properties of an element are related to its structure. Giant molecular structures have high melting points and simple molecular structures have low melting points. Giant metallic structures conduct heat and electricity very well, but molecular structures in general do not.

The metallic character of the elements increases as the atomic (proton) number increases. With a larger atomic radius, the outer electrons are further from the nucleus and are better shielded by inner shells of electrons. As a result, tin and lead have structures that involve positive lead and tin ions in a sea of delocalised electrons.

Tin and lead have the typical physical properties of metals. They are:

- dense;
- lustrous;
- good conductors of heat and electricity;
- malleable;
- ductile.

Tin and lead have relatively low melting points, which indicates that the metallic bonding is not as strong as in the transition elements. Transition elements are able to use d electrons in the sea of delocalized electrons, but tin and lead use only p electrons from the 5p and 6p subshells, respectively.

HOW SCIENCE WORKS

Silicon carbide

As soon as chemists discovered the structure and bonding of silicon carbide they realised it resembled that of diamond. They predicted that silicon carbide would share many physical properties with diamond.

Silicon carbide indeed does have a very high melting point and is very hard. These two properties explain its use as an abrasive. Chemists also predicted that silicon carbide could be used as a replacement for silicon in electronic devices, providing that flawless crystals could be obtained. Research carried out in Japan has now developed procedures to make near-perfect silicon carbide crystals. Silicon carbide can carry more efficiently than silicon, can

HOW SCIENCE WORKS

Silicon carbide (Cont.)

withstand higher voltages and can work at very high temperatures. Now chemists are speculating that electronic devices such as fuel regulators in jet engines will soon be made from silicon carbide.

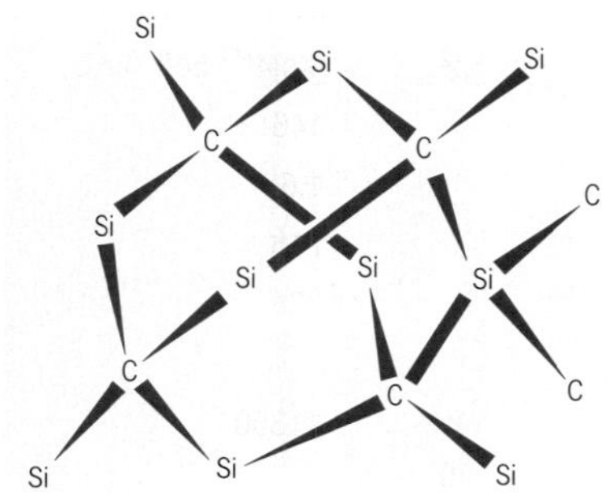

Fig 29.4 The structure of silicon carbide. Silicon carbide has a diamond-like structure; each silicon atom is surrounded by four carbon atoms and each carbon atom is surrounded by four silicon atoms

3 CHEMICAL PROPERTIES OF THE ELEMENTS OF GROUP 4

You can see from Table 29.2 that all the elements in Group 4 have four electrons in the outer shell. So one atom of the element could either gain or lose four electrons to attain a stable electron configuration. It is virtually impossible for any atom to lose four electrons to form an X^{4+} ion. Firstly, the ion's high charge density would polarise any anion in its vicinity and so would form a covalent bond. Secondly, the energy required to remove four electrons is extremely large.

This means that an atom of a Group 4 element must gain four electrons to give the X^{4-} ion. But the ion is not very stable. Four electrons are more likely to be gained by covalent bonding, giving the elements an oxidation state of +4 in some of their compounds. In fact, almost all the compounds of silicon and carbon are covalent with an oxidation state of +4.

THE INERT PAIR EFFECT

Germanium, tin and lead form both ionic compounds and covalent compounds. In ionic compounds, an atom of these elements loses two electrons to form Ge^{2+}, Sn^{2+} or Pb^{2+} ions. For example, if a lead atom loses two electrons it attains an electron configuration that involves a filled outer sub-shell, $6s^2$. The two electrons in the 6s orbital are known as the **inert pair**, because they are in the outer shell of electrons but are not used in bonding. As a result lead, germanium and tin can show two oxidation states:

- the ionic, +2 oxidation state;
- the covalent, +4 oxidation state.

Table 29.3 The ionisation energies of Group 4 elements

Element	First ionisation energy/kJ mol^{-1}, $X(g) \rightarrow X^+(g) + e^-$	Second ionisation energy/kJ mol^{-1}, $X^+(g) \rightarrow X^{2+}(g) + e^-$	Energy needed to make 1 mole of X^{2+}(g) ions from X atoms/kJ mol^{-1} $X(g) \rightarrow X^{2+}(g) + 2e^-$
carbon	799	2420	
silicon	786	1580	
germanium	762	1540	
tin	707	1410	
lead	716	1450	

QUESTION 7

List the similarities and the differences between the structures of diamond and silicon carbide.

Read more about charge density and polarisation of anions on pages 440 and 87.

QUESTION 8

Write down the electronic configuration for:

a) Sn and Sn^{2+}
b) Ge and Ge^{2+}
c) Si and Si^{2+}
d) Use electron configurations to show the existence of an inert pair of electrons in tin and germanium, but not in silicon.

QUESTION 9

a) Copy out and complete the third column in Table 29.3.
b) Explain why the second ionisation energy is always larger than the corresponding first ionisation energy.
c) Explain why the first ionisation energy of carbon is larger than the first ionisation energy of germanium.
d) Comment on any unexpected pattern in the first and second ionisation energies.

It is impossible for carbon and silicon atoms to behave in the same way, because the loss of two electrons does not give a sufficiently stable electron configuration.

Metallic character in Group 4

We described in Chapter 7 how the loss of electrons from an atom to form a cation is characteristic of a metal atom and the gain of electrons to form either an anion or a covalent bond is characteristic of a non-metal atom. The loss of electrons results in a +2 oxidation state and shows the metallic character of germanium, tin and lead. The metallic character of the elements increases with increasing atomic number, so that lead has the most metallic character of all the Group 4 elements.

Elements with higher atomic (proton) numbers have more shielding electron shells between the outer electrons and the nucleus. This means that the outer electrons experience a reduced effective nuclear charge. In addition, the outer electrons are further away from the nucleus, so an atom of lead can easily lose the two 6p electrons.

Carbon and silicon show no evidence whatsoever of metallic character chemically; in fact, the +2 oxidation state is virtually absent from the chemistry of both elements.

QUESTION 10

10 Draw a dot-and-cross diagram to show the covalent bonding in lead(IV) chloride. You need only draw the electrons in the outermost shell.

Non-metallic character in Group 4

The non-metallic character of the elements decreases as atomic number increases. However, there is evidence of some non-metallic character as far down the group as lead. Lead(IV) compounds and, in particular, the lead(IV) halides are covalent in nature; they involve a lead atom that gains four electrons through covalent bonding to achieve a stable electron configuration.

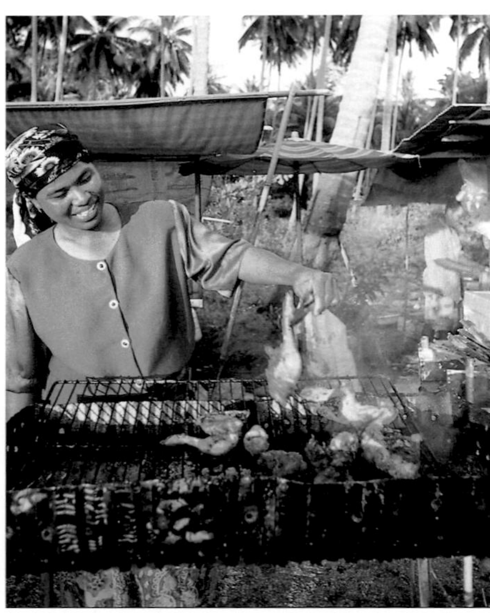

Fig 29.5 Charcoal is mainly carbon. It burns to give carbon dioxide in a highly exothermic reaction

REACTIONS OF GROUP 4 ELEMENTS

Whether metallic or non-metallic character is considered, the elements in Group 4 are members of a relatively unreactive group. The elements will react with non-metals, such as oxygen and the halogens, but such reactions normally require heating.

Reactions with oxygen

Most Group 4 elements react with air or oxygen when heated to form a dioxide in which the element has a +4 oxidation state. Carbon forms a gaseous dioxide, whereas the other elements form solid dioxides:

$$C(s) + O_2(g) \rightarrow CO_2(g)$$
$$X(s) + O_2(g) \rightarrow XO_2(s)$$

where X is Si, Ge and Sn.

Lead forms lead(II) oxide when it is heated in air:

$$2Pb(s) + O_2(g) \rightarrow 2PbO(s)$$

Even if lead(IV) oxide were to form, it would thermally decompose at Bunsen flame temperatures to give lead(II) oxide.

QUESTION 11

11 Write down equations for the reactions between each pair of elements:

a) lead and iodine;

b) tin and fluorine;

c) germanium and bromine.

Reactions with halogens

All the elements except carbon react when heated with halogens to give a halide:

$$X(s) + 2Cl_2(g) \rightarrow XCl_4(l)$$

where X = Si, Ge and Sn. Lead forms the chloride $PbCl_2$, rather than the tetrachloride, because of the thermal instability of lead(IV) chloride.

$$Pb(s) + Cl_2(g) \rightarrow PbCl_2(s)$$

4 OXIDES OF GROUP 4 ELEMENTS

Most of the Group 4 elements form two oxides. These are an oxide of formula XO with the element in the +2 oxidation state and an oxide of formula XO_2 with the element in the +4 oxidation state.

Read about why carbon dioxide forms a simple molecule and silicon dioxide forms a giant molecule on page 422.

Read more about quartz and silicon dioxide on pages 421–422.

XO_2 OXIDES (THE DIOXIDES)

All Group 4 elements form a dioxide with the empirical formula XO_2, where X is C, Si, Ge, Sn or Pb. These dioxides are often referred to as the element(IV) oxide, such as silicon(IV) oxide. There is a distinct change in the chemical properties of these oxides as the atomic number of the Group 4 element increases. This reflects a change in the stability of the +4 oxidation state.

Table 29.4 The dioxides of Group 4 elements

Oxide	Formula	Structure and bonding	Acid–base behaviour	Thermal stability	Oxidising power
carbon dioxide	CO_2	simple molecular	acidic	does not decompose on heating	very weak oxidising agent
silicon(IV) oxide	SiO_2	giant molecular	acidic	does not decompose on heating	↓ increasing oxidising power
germanium(IV) oxide	GeO_2	giant molecular	amphoteric	does not decompose on heating	
tin(IV) oxide	SnO_2	giant molecular with ionic character	amphoteric	does not decompose at Bunsen burner temperatures	
lead(IV) oxide	PbO_2	giant molecular with ionic character	amphoteric	decomposes on heating	powerful oxidising agent

Thermal stability of the dioxides

The thermal stability of the dioxides decreases as the atomic number of the Group 4 element increases. Strong heating of lead(IV) oxide gives oxygen and lead(II) oxide:

$$2PbO_2(s) \rightarrow 2PbO(s) + O_2(g)$$

Silicon(IV) oxide is very thermally stable; it is often a component of ceramics. One crystalline form of silicon dioxide is quartz. Quartz is a hard, brittle, clear and colourless solid. It is used for architectural decorations, semi-precious jewels, optical instruments and as a frequency controller in radio transmitters.

Quartz melts at 1600 °C to form a viscous liquid. When this liquid cools, it does not crystallize and the internal structure remains random. A substance with this type of structure is known as a glass.

Fig 29.6 Radioactive material is easily stored in glass enclosed in steel. As long as it remains dry, the radioactive substance is confined by this process (called vitrification)

Oxidising power of the dioxides

Since Group 4 elements form two sets of oxides, XO_2 (oxidation state of X +4) and XO (oxidation state of X +2); it is possible to convert XO_2 into XO, which is a reduction. This means that the dioxides XO_2 are oxidising agents. Lead(IV) oxide and tin(IV) oxide are the most powerful of these oxidising agents and both can oxidise hydrochloric acid to form chlorine:

$$PbO_2(s) + 4HCl(aq) \rightarrow PbCl_2(aq) + 2H_2O(l) + Cl_2(g)$$

QUESTION 13

11 Explain why lead(IV) oxide is a better oxidising agent than silicon dioxide. (Hint: think about the relative stabilities of the +2 oxidation states.)

QUESTION 12

Write down the equation for the reaction of aqueous sodium silicate, Na_2SiO_3, with dilute hydrochloric acid. (Hint: in this reaction the sodium silicate behaves in a similar way to sodium carbonate.)

SCIENCE IN CONTEXT

Aerogels

Aerogels are an exciting new class of synthetic materials that have an extremely low density. They consist of 10 per cent silica or silicon(IV) oxide, the same material found in sand, and 90 per cent gas. They have such a low density that they can float on whisked egg white. If filled with helium they can be less dense than air. In addition, they have unusual optical, thermal, electrical and acoustic properties.

It has been known for some time that an aqueous solution of sodium silicate and hydrochloric acid react to form a gelatinous solid. Heating this mixture eventually gives a solid called silica gel, which has a large surface area and readily adsorbs water. Aerogels are similar in nature, but are solids with gas, rather than liquid, trapped within a cage of atoms.

Fig 29.7 Silica gel absorbs water from the air to make a dry atmosphere

When aerogels were first discovered, to prepare them was a long and difficult process that involved using potentially dangerous methanol to make the gel. Later on, a method was developed to make an aerogel filled with carbon dioxide. Such aerogels could be made into dry pellets that contain strings of silica tetrahedra (see Fig 29.24) encircling the gas spaces.

Aerogels are excellent insulators because there are fewer solid atoms to conduct heat than in conventional insulators and the gas pores are too small for the gas inside to convect heat effectively. Some aerogels have one-fifth of the thermal conductivity of poly(phenylethene) (expanded polystyrene)

One application of aerogels is in windows. A layer of aerogel can be sandwiched between two layers of glass to improve the insulation properties of the glass. The aerogel also scatters light and produces a frosted glass effect.

Research is underway to improve the thermal insulation properties of aerogels still further by removing all the gas from the pores, to leave a vacuum. There is also scope to make aerogel catalysts – the reaction between gases takes place within the gel itself.

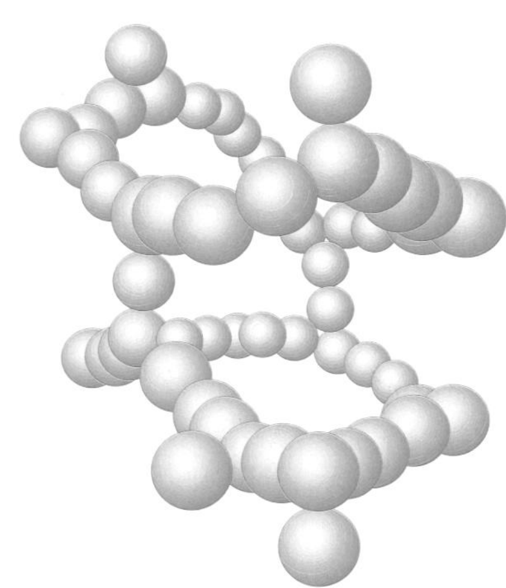

Fig 29.8 An aerogel is made of silica chains that encircle air-filled pores. The circles in the diagram represent SiO_4 tetrahedra

REMEMBER THIS

If carbon dioxide is bubbled through lime water for several minutes, a white precipitate, $CaCO_3(s)$, is formed first. This reacts with further carbon dioxide to form a colourless solution of calcium hydrogencarbonate.

QUESTION 14

14 Write down the equation for the reaction of carbon dioxide with lime water to form insoluble calcium carbonate.

Acid–base reactions of the dioxides

In Chapter 21 we describe the acid–base reactions of oxides of the elements in Period 3. Covalent oxides are often acidic and ionic oxides are basic. Most metal oxides are ionic and most non-metal oxides are covalent, so we can say that metal oxides are normally basic and non-metal oxides are often acidic.

On this basis, we would expect most Group 4 dioxides to show some acidic behaviour, because bonding in carbon dioxide and silicon dioxide is covalent, and even in lead(IV) oxide there is a high degree of covalent bonding. This acidic behaviour is shown by the reactions of the dioxides. The test for carbon dioxide, in which lime water (aqueous calcium hydroxide) turns milky, is an example of an acidic reaction of carbon dioxide. The precipitate formed is calcium carbonate. In the same way, sodium hydroxide pellets react with carbon dioxide to form sodium carbonate:

$$2NaOH(s) + CO_2(g) \rightarrow Na_2CO_3(s) + H_2O(l)$$

Limestone (mostly calcium carbonate) is added to blast furnaces. In these, it is decomposed into calcium oxide, which reacts with SiO_2 (sand) impurity to remove it as calcium silicate, $CaSiO_3$. Here, silicon dioxide behaves as an acidic oxide.

Even lead(IV) oxide shows some acidic character when it reacts with concentrated aqueous sodium hydroxide to form the plumbate(VI) ion.

$$PbO_2(s) + 2NaOH(aq) \rightarrow Na_2PbO_3(aq) + H_2O(l)$$

As the dioxides of lead and tin also have a high degree of ionic character, they also show basic properties, but their reactions with acids are often complicated by subsequent redox reactions. Lead(IV) oxide and tin(IV) oxide are known as **amphoteric** oxides, since they have both acidic and basic properties.

XO OXIDES (THE MONOXIDES)

Table 29.5 The oxides of Group 4 elements (SiO is omitted from the table because it does not exist under standard conditions)

Oxide	Formula	Structure and bonding		Acid–base behaviour	
carbon monoxide	CO	simple molecular	increasing ionic character ↓	neutral	increasing basic character ↓
germanium(II) oxide	GeO	giant ionic		amphoteric	
tin(II) oxide	SnO	giant ionic		amphoteric	
lead(II) oxide	PbO	giant ionic		amphoteric	

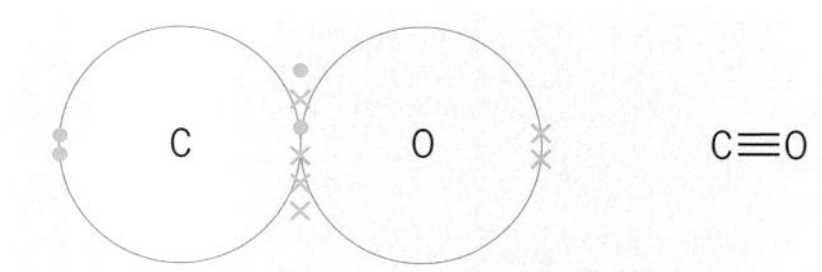

Fig 29.9 Carbon monoxide is a simple molecule and is iso-electronic with a nitrogen molecule. It has two covalent bonds and one dative covalent bond

Carbon monoxide has completely different properties to those of the other monoxides formed by Group 4 elements. It is a covalently bonded gaseous compound that is a neutral oxide. Germanium(II) oxide, tin(II) oxide and lead(II) oxide are ionically bonded solid oxides that show increasing basic properties as the ionic character of the oxide increases.

SCIENCE IN CONTEXT

Metal carbonyls

Carbon monoxide forms a range of compounds with transition metals. They are known as carbonyls. Carbon monoxide molecules act as ligands and donate pairs of electrons to an empty orbital of an atom of the transition element. An interesting feature of many of these carbonyls is that the oxidation state of the transition element is often 0.

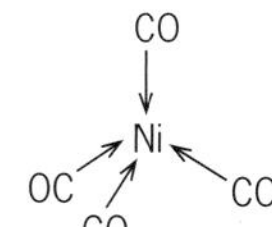

Fig 29.10 The structure of tetracarbonylnickel(0). Tetracarbonylnickel(0) is a tetrahedral neutral complex. Carbon monoxide ligands form dative bonds with the central nickel atom

The first metal carbonyl was discovered by Ludwig Mond in 1889. It was tetracarbonylnickel(0), $Ni(CO)_4$, a colourless, volatile liquid. The discovery of $Ni(CO)_4$ opened up the way for large-scale production of pure nickel.

Carbon monoxide reacts with nickel at 50 °C to give volatile $Ni(CO)_4$, which evaporates to leave behind impurities:

$$Ni(s) + 4CO(g) \rightarrow Ni(CO)_4(l)$$

The $Ni(CO)_4$ decomposes at 200 °C to yield pure nickel and CO. Fortunately for Mond, the discovery of stainless steel at that time created a demand for large quantities of nickel, so he was able to develop a large manufacturing plant that produced pure nickel for a ready-made market.

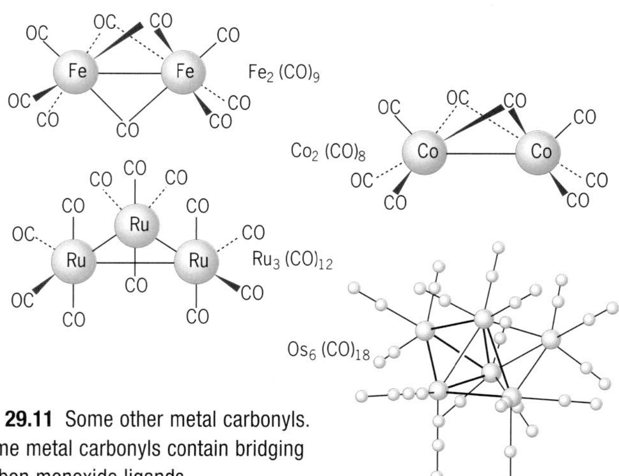

Fig 29.11 Some other metal carbonyls. Some metal carbonyls contain bridging carbon monoxide ligands

Fig 29.12 Nickel prepared via $Ni(CO)_4$ is used as an alloy with tungsten and other metals to make the engines for this airliner

QUESTIONS 15–16

15 Suggest the likely products of the following reactions:
a) Lead(II) oxide and concentrated ethanoic acid, CH_3COOH.
b) Tin(II) oxide and aqueous sodium hydroxide.

16 Red lead has the formula Pb_3O_4. Its name is dilead(II)lead(IV) oxide. It acts as though it has 2 moles of lead(II) oxide and 1 mole of lead(IV) oxide per mole of red lead. Suggest the names of the products formed when:
a) red lead is heated strongly in air;
b) concentrated hydrochloric acid is gently heated with red lead;
c) red lead is gently heated with aqueous sodium hydroxide.
In each case, write an equation to show the reactions taking place.

Oxidation of the monoxides

Since lead(IV) oxide is the least thermally stable of the dioxides, lead forms the most stable monoxide.

The monoxides CO, GeO and SnO all readily react with oxygen to give the respective dioxides. For example, carbon monoxide combusts to give carbon dioxide, and anhydrous tin(II) oxide reacts with oxygen in the air to give SnO_2:

$$2SnO(s) + O_2(g) \rightarrow 2SnO_2(s)$$

Lead(II) oxide will not form lead(IV) oxide when heated in air, but at 400 °C with prolonged heating it will form red lead (see Question 16).

Acid–base reactions of the monoxides

The oxides become more basic as the atomic number of the Group 4 element increases. This is because bonding in the oxide becomes more ionic in character, and ionic oxides show basic properties. Even though the basic character increases, lead(II) oxide is not a basic oxide; it is amphoteric because it reacts with acids and bases. In acids, lead(II) oxide forms lead(II) salts and with aqueous alkalis it forms plumbate(II), PbO_2^{2-}:

$$PbO(s) + 2HNO_3(aq) \rightarrow Pb(NO_3)_2(aq) + H_2O(l)$$
$$PbO(s) + 2NaOH(aq) \rightarrow Na_2PbO_2(aq) + H_2O(l)$$

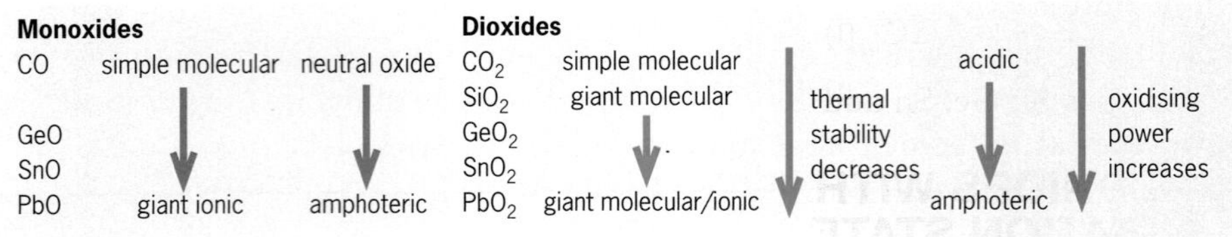

Fig 29.13 A summary of the properties of Group 4 oxides

5 CHLORIDES OF GROUP 4 ELEMENTS

Earlier in the chapter we described how the stability of Group 4 dioxides (with the element in the +4 oxidation state) decreases as the atomic number of the element increases, whereas the stability of the monoxide (element in the +2 oxidation state) increases. Exactly the same trend in stability is observed with the two sets of chlorides, XCl_4 (with the element in the +4 oxidation state) and XCl_2 (with the element in the +2 oxidation state). The +4 oxidation state involves covalent bonding and the +2 oxidation state involves ionic bonding.

THE TETRACHLORIDES

Table 29.6 Group 4 tetrachlorides

Formula	Structure and bonding	Bond length of X–Cl/nm	Thermal stability	Action of cold water
CCl_4	simple molecular	0.177	thermally stable	none
$SiCl_4$	simple molecular	↓ increases	↓ thermal stability decreases	rapidly hydrolysed to form SiO_2 and HCl
$GeCl_4$	simple molecular			rapidly hydrolysed to form GeO_2 and HCl
$SnCl_4$	simple molecular			rapidly hydrolysed to form SnO_2 and HCl
$PbCl_4$	simple molecular		thermally decomposes at room temperature to form $PbCl_2$ and Cl_2	rapidly hydrolysed to form PbO_2 and HCl

With the exception of carbon, the tetrachlorides (XCl_4) are named element(IV) chloride. In the case of carbon, the name tetrachloromethane is preferred to show its relationship to the hydrocarbon methane. All of the tetrachlorides have a simple molecular structure and so have low boiling points and melting points.

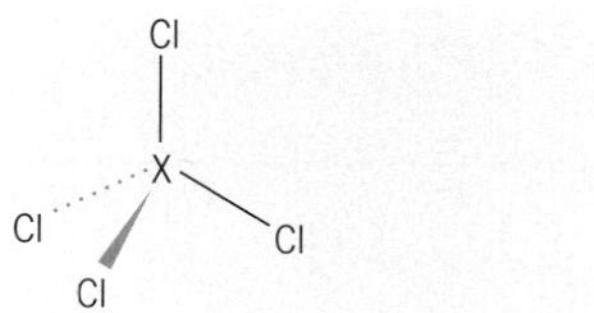

Fig 29.14 The shape of the covalent tetrachlorides. The tetrachlorides of Group 4 elements are all simple, tetrahedral molecules

Thermal stability of the tetrachlorides

The thermal stability of the tetrachlorides decreases with increasing atomic number of the Group 4 element. This trend in thermal stability is explained in terms of the bond lengths and bond strengths of the X–Cl bond. We have described that a covalent bond is formed by the overlap of atomic orbitals; in the case of the Pb–Cl bond, there is very little overlap because of the difference in energy between the orbitals involved. With very little overlap, the covalent bond is extremely weak and is therefore easy to break. The Pb–Cl bond is much longer than the other X–Cl bonds, which contributes to the weakness of the Pb–Cl bond.

? QUESTION 17

17 Write an equation to show the thermal decomposition of lead(IV) chloride.

Hydrolysis of the tetrachlorides

In Chapters 21 and 23 we describe the hydrolysis of covalent chlorides. With the exception of tetrachloromethane, all the tetrachlorides are readily hydrolysed to form the dioxide as a precipitate and either hydrogen chloride or hydrochloric acid, depending on the amount of water available during the hydrolysis:

$$XCl_4(l) + 2H_2O(l) \rightarrow XO_2(s) + 4HCl(aq)$$

where X is Si, Ge, Sn and Pb.

Read why tetrachloromethane cannot be hydrolysed on page 416.

CHLORIDES WITH THE ELEMENT IN THE +2 OXIDATION STATE

We have already described the inert-pair effect to explain why germanium, tin and lead can have an oxidation state of +2. The three elements form chlorides of the general formula XCl_2 in this oxidation state. Germanium(II) chloride is not really ionic and anhydrous tin(II) chloride has a great deal of covalent character. The only solid chloride for which the term ionic is really appropriate is lead(II) chloride.

REMEMBER THIS

Solutions that contain aqueous iodide ions and lead(II) ions react to form yellow lead(II) iodide precipitate:

$$Pb^{2+}(aq) + 2I^-(aq) \rightarrow PbI_2(s)$$

This could be used as a test for $Pb^{2+}(aq)$.

Reducing properties

Germanium(II) chloride and tin(II) chloride are both reducing agents. Aqueous solutions of tin(II) chloride are best stored with a little tin metal to prevent aerial oxidation. Aqueous acidified tin(II) chloride reduces iron(III) ions and dichromate(VI) ions:

$$3Sn^{2+}(aq) + 14H^+(aq) + Cr_2O_7^{2-}(aq) \rightarrow 2Cr^{3+}(aq) + 3Sn^{4+}(aq) + 7H_2O(l)$$

The aqueous Sn^{4+} ions have too high a charge density to exist on their own and so they form complex hydrated ions.

Lead(II) chloride is not a reducing agent, as shown by the following electrode potentials:

$$Pb^{4+}(aq) + 2e^- \rightleftharpoons Pb^{2+}(aq) \qquad E_{red} = +1.69\text{ V}$$

$$Sn^{4+}(aq) + 2e^- \rightleftharpoons Sn^{2+}(aq) \qquad E_{red} = +0.15\text{ V}$$

The highly positive E_{red} for the reduction of $Pb^{4+}(aq)$ indicates that lead(IV) is very easily reduced to Pb^{2+} and so the reverse reaction, Pb^{2+} to Pb^{4+}, will be much more difficult than the corresponding reaction of Sn^{2+} to Sn^{4+}.

? QUESTIONS 18–19

18 Construct an ionic equation to show the reaction of $Fe^{3+}(aq)$ with $Sn^{2+}(aq)$ to form $Fe^{2+}(aq)$.

19 Write an ionic equation for the reaction that occurs when chlorine bubbles through a solution of acidified aqueous tin(II) chloride. Suggest the name(s) of the product(s). (Hint: you can read about the reactions of chlorine in Chapter 23.)

6 SILICONES AND SILANES

STRETCH AND CHALLENGE

Silicones

Silicones are polymers that have a backbone of alternating silicon and oxygen atoms with carbon-based side chains. They are characteristically non-toxic and are stable over a wide range of temperatures.

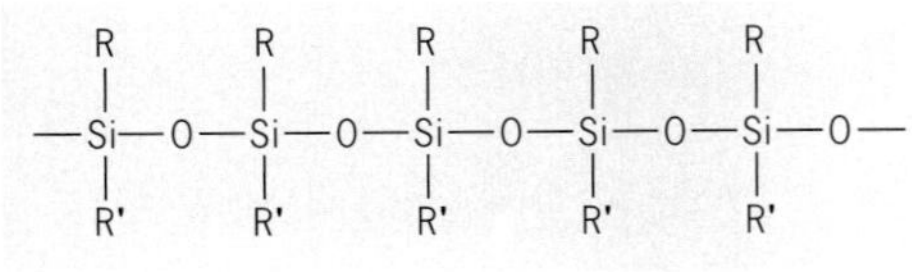

$$-\overset{R}{\underset{R'}{Si}}-O-\overset{R}{\underset{R'}{Si}}-O-\overset{R}{\underset{R'}{Si}}-O-\overset{R}{\underset{R'}{Si}}-O-\overset{R}{\underset{R'}{Si}}-O-$$

Fig 29.15 Silicones have a polymer chain of alternate oxygen and silicon atoms joined by a single bond. The side chains R and R[1] are alkyl groups or aryl groups

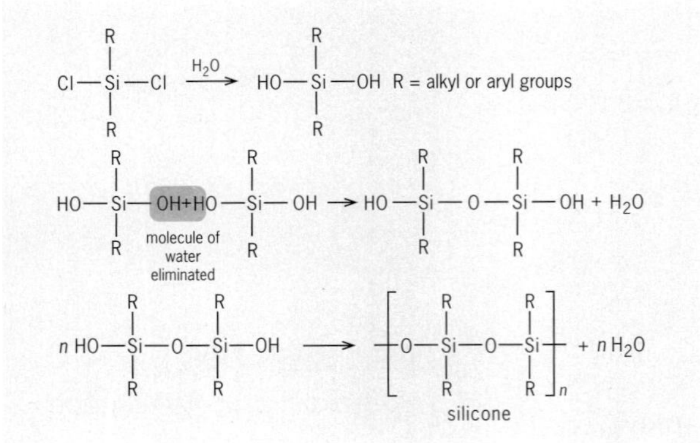

$$Cl-\overset{R}{\underset{R}{Si}}-Cl \xrightarrow{H_2O} HO-\overset{R}{\underset{R}{Si}}-OH \quad R = \text{alkyl or aryl groups}$$

$$HO-\overset{R}{\underset{R}{Si}}-OH + HO-\overset{R}{\underset{R}{Si}}-OH \rightarrow HO-\overset{R}{\underset{R}{Si}}-O-\overset{R}{\underset{R}{Si}}-OH + H_2O$$

molecule of water eliminated

$$n\,HO-\overset{R}{\underset{R}{Si}}-O-\overset{R}{\underset{R}{Si}}-OH \rightarrow \left[-O-\overset{R}{\underset{R}{Si}}-O-\overset{R}{\underset{R}{Si}}-\right]_n + n\,H_2O$$

silicone

Fig 29.16 Silicones are manufactured by the hydrolysis of dichlorosilanes followed by condensation of the resulting dihydroxysilanes. A molecule of water is eliminated during each condensation step

The side chains shown in Fig 29.15 can be varied to develop a wide range of silicones with different properties. If the side chain of the silicone is a methyl group, then the resulting polymer is a liquid. These polymers are linear with no cross-linking between polymer chains. This means that the polymer chains only have weak intermolecular forces between them, so the silicone has a relatively low melting point.

Fig 29.17 shows that there are no double bonds in the polymer chain, so all the bonds in the polymer molecule can freely rotate. Such a feature gives this type of silicone the physical properties required to make the non-stick coating on pans.

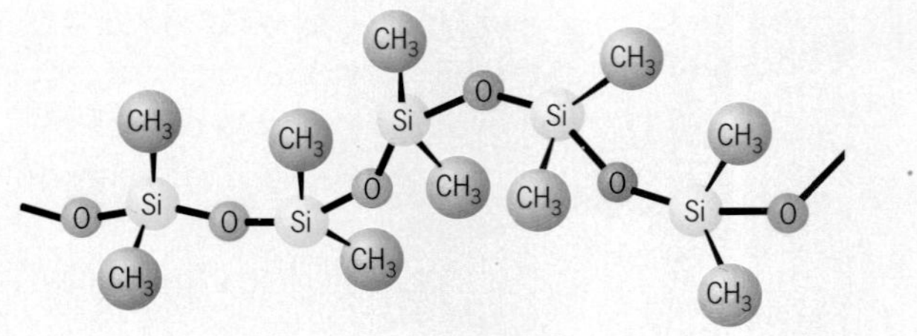

Fig 29.17 The structure of silicone fluid. A silicone fluid has an R group such as a methyl or ethyl group. There are no cross-links with other polymer chains and only weak intermolecular forces exist between the polymer chains

Another use of silicone fluids is to protect masonry from rain damage. Silicone bonds to masonry and the hydrophobic (water-hating) methyl groups repel water molecules from the surface of the building.

Fig 29.18 Many bathroom sealants used to seal the sides of baths, showers and handbasins are made from silicone rubber

An advantage of silicones is the way that the side chain can be manipulated to change the property of the material. If the side chains contain more reactive groups – such as alkenes or hydroxyl groups – then some degree of cross-linking between the polymer chains can occur. Providing the cross-linking is not extensive, the silicone will be solid at room temperature, but still retain a great deal of flexibility and resistance to water.

Silicones are electrical insulators and they can be used to insulate electric cables, as long as an inert filler is added to increase the rigidity of the material.

Fig 29.19 An extensive network of cross-links are present in more rigid and less flexible silicone resins. Some racing helmets are made of silicone resin mixed with glass fibre

QUESTION 20

20 The relative molecular mass of a silicone fluid with the structure shown in Fig 29.17 and an R group of CH_3 is approximately 1500. Estimate the number of silicon atoms there are in the polymer chain.

STRETCH AND CHALLENGE

The silanes

Carbon forms an almost infinite number of compounds that contain carbon–hydrogen single bonds. Among these compounds are the alkanes, which have the general formula C_nH_{2n+2}. Silicon forms a range of compounds similar in terms of general formula to the alkanes, which are collectively called the silanes. The chain length of carbon–carbon bonds is not limited. However, the longest chain silane that has been isolated at present has six silicon atoms.

All silanes are simple molecular compounds and the single bonds are arranged tetrahedrally about the central silicon atoms.

The vacant 3d orbitals in silicon atoms make the properties of the silanes different from those of the corresponding alkanes. The enthalpy changes of formation of silane (SiH_4) and methane indicate that silane is considerably less stable with respect to the elements silicon and hydrogen than methane is to carbon and hydrogen. In fact, the enthalpy change of formation of silane is positive.

$$Si(s) + 2H_2(g) \rightarrow SiH_4(g) \qquad \Delta H_f = +34 \text{ kJ mol}^{-1}$$
$$C(s) + 2H_2(g) \rightarrow CH_4(g) \qquad \Delta H_f = -75 \text{ kJ mol}^{-1}$$

Silane spontaneously ignites in air to form silicon dioxide and water:

$$SiH_4(g) + 2O_2(g) \rightarrow SiO_2(s) + 2H_2O(g) \qquad \Delta H_f = -1429 \text{ kJ mol}^{-1}$$

The enthalpy change of combustion of silane is much higher than that of methane (–802 kJ mol^{-1}), because of the weakness of the Si–H single bonds broken and the strength of the Si–O bonds formed in silicon dioxide. Silane reacts with aqueous sodium hydroxide to form hydrogen and the silicate ion.

$$SiH_4(g) + 2OH^-(aq) + H_2O(l) \rightarrow SiO_3^{2-}(aq) + 4H_2(g)$$

Again, the formation of the very strong silicon–oxygen single bond acts as a driving force for this reaction. In contrast, methane is inert to aqueous sodium hydroxide.

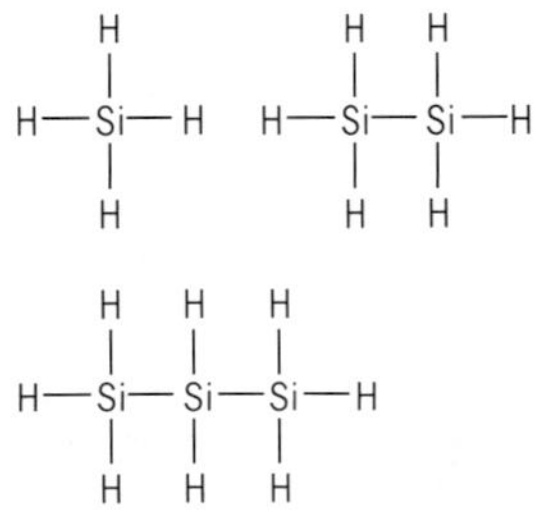

Fig 29.20 The structures of some silanes. Notice the similarity to the homologous series called the alkanes

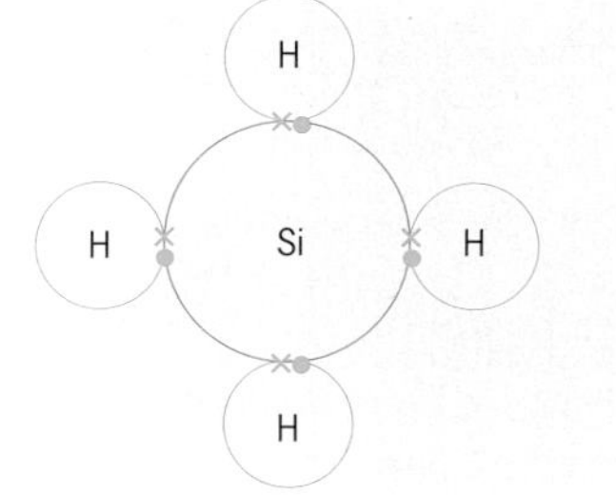

Fig 29.21 The dot-and-cross diagram for silane

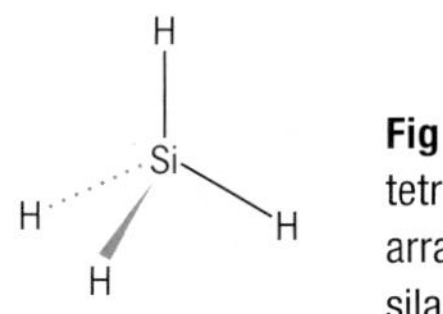

Fig 29.22 The tetrahedral arrangement of the silane molecule

? QUESTION 21

a) What is the molecular formula of the silane that contains six silicon atoms per molecule?

b) Draw a possible structural formula for this silane.

7 OXY-SALTS

The metal elements of Group 4 form oxy-salts such as lead(II) nitrate and lead(II) carbonate, which contain the cation Pb^{2+}. The non-metal elements also form oxy-salts, but they are present as the anion rather than as the cation. So carbon forms carbonates and silicon forms silicates. There are examples of the other elements in Group 4 that form anions similar to carbonates, such as plumbate(II), which illustrates the amphoteric nature of lead and its compounds.

Fig 29.23 Lead(II) chromate(VI) is used to make the yellow lines on roads

OXY-SALTS OF LEAD

Most lead(II) salts are insoluble, but two of the oxy-salts – lead(II) ethanoate, $Pb(CH_3COO)_2$, and lead(II) nitrate, $Pb(NO_3)_2$ – are soluble in water. They are often used in qualitative analysis to precipitate anions as insoluble lead(II) compounds.

When aqueous lead(II) nitrate is added to any soluble sulfate, white lead(II) sulfate forms as a precipitate:

$$Pb^{2+}(aq) + SO_4^{2-}(aq) \rightarrow PbSO_4(s)$$

Lead(II) chromate(VI) forms as a yellow precipitate when aqueous lead(II) nitrate is added to aqueous sodium chromate(VI):

$$Pb^{2+}(aq) + CrO_4^{2-}(aq) \rightarrow PbCrO_4(s)$$

CARBONATES

Carbonates are derived from the weak acid carbonic acid, formed when carbon dioxide dissolves in water.

$$CO_2(aq) + H_2O(l) \rightleftharpoons H_2CO_3(aq)$$

Carbonates are salts that contain the CO_3^{2-} ion and hydrogencarbonates contain the HCO_3^- ion. Only carbonates of the Group 1 elements and ammonium carbonate are soluble in water. Therefore, most carbonates are prepared by precipitation:

$$M^{2+}(aq) + CO_3^{2-}(aq) \rightarrow MCO_3(s)$$

where M is a metal in the +2 oxidation state.

Metals that form a positive ion with a high charge density, such as Fe^{3+}, Al^{3+} and Cr^{3+}, do not form carbonates. In solution, these ions polarise water molecules to form sufficient aqueous hydrogen ions to react with the carbonate ion to form carbon dioxide:

$$CO_3^{2-}(s) + 2H^+(aq) \rightarrow CO_2(g) + H_2O(l)$$

QUESTION 22

22 Write an equation to show the decomposition of lead(II) carbonate.

Thermal decomposition

With the exception of the Group 1 carbonates, all carbonates thermally decompose to give carbon dioxide. The energy changes that accompany this decomposition are discussed in detail in Chapter 22.

Reaction with acid

Treating any carbonate with acid evolves carbon dioxide and produces a salt. Lead(II) carbonate reacts with dilute nitric acid to form lead(II) nitrate, water and carbon dioxide:

$$PbCO_3(s) + 2HNO_3(aq) \rightarrow Pb(NO_3)_2(aq) + CO_2(g) + H_2O(l)$$

SILICATES

Carbonates contain the anion CO_3^{2-}; the corresponding silicon compounds, called the silicates, have a very different structure and contain more complex negative ions. The simplest silicate structure contains the tetrahedral SiO_4^{4-} ion.

The structure of the anion in silicates is based on an SiO_4 tetrahedron that contains covalent silicon–oxygen single bonds. If two SiO_4 tetrahedral units share an oxygen atom, the silicate contains the $Si_2O_7^{6-}$ ion.

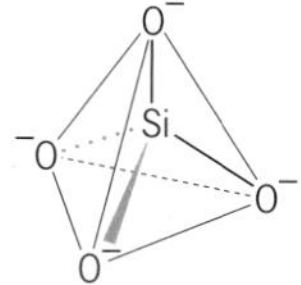

Fig 29.24 Minerals such as olivine, Mg_2SiO_4, and zircon, $ZrSiO_4$, contain the tetrahedral silicate ion, SiO_4^{4-}

The asbestos problem

Asbestos is a silicate mineral that is fibrous. It has long been used for fireproofing and insulating buildings. Since it became known that exposure to asbestos fibres can cause serious respiratory disorders such as asbestosis, asbestos has been removed from many buildings.

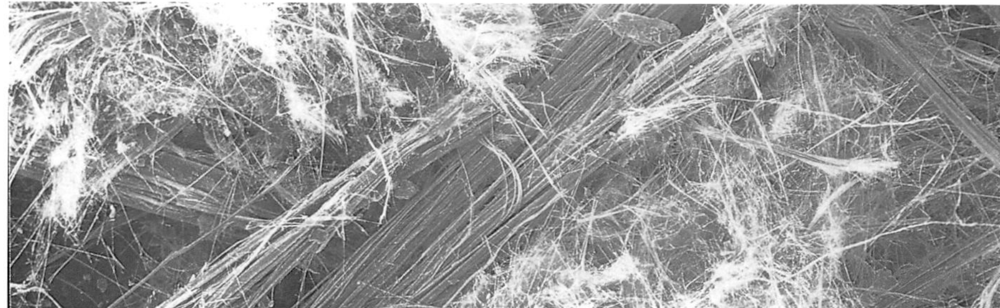

Fig 29.25 A scanning electron micrograph of the fibres of blue asbestos. The lung disease asbestosis is caused by breathing in these fibres

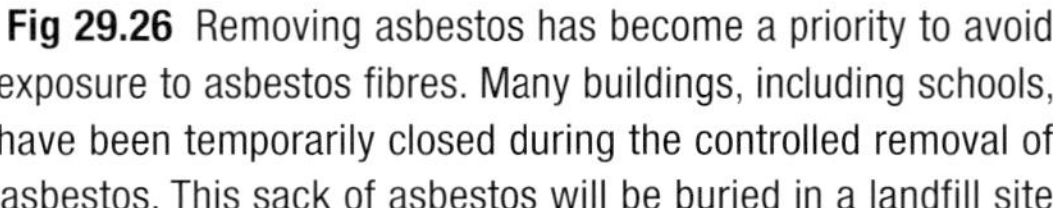

Fig 29.26 Removing asbestos has become a priority to avoid exposure to asbestos fibres. Many buildings, including schools, have been temporarily closed during the controlled removal of asbestos. This sack of asbestos will be buried in a landfill site

8 GROUP 6 – OXYGEN AND SULFUR

Group 6 contains two well-known elements – oxygen and sulfur – together with three less well-known elements – selenium, tellurium and polonium. (We do not discuss selenium, tellurium and polonium, because their chemistry is well beyond the scope of this book.)

The chemistry of oxygen and oxides is described in earlier chapters, but Figs 29.27 and 29.28 give some general information about oxygen and oxides. The discussion in this chapter concentrates on sulfur.

Table 29.7 The elements of Group 6

Element	Symbol	Electron configuration
oxygen	O	$1s^2\ 2s^2\ 2p^4$
sulfur	S	$1s^2\ 2s^2\ 2p^6\ 3s^2\ 3p^4$
selenium	Se	$[Ar]3d^{10}\ 4s^2\ 4p^4$
tellurium	Te	$[Kr]4d^{10}\ 5s^2\ 5p^4$
polonium	Po	$[Xe]4f^{14}\ 5d^{10}\ 6s^2\ 6p^4$

basic oxides	Ionic oxides, contain O^{2-}. React with acids to give salt and water e.g. Na_2O, CuO
amphoteric oxides	Contain O^{2-}. React with both acids and alkalis to give salts, e.g. Al_2O_3, PbO, ZnO
acidic oxides	Covalent oxides. Often react with bases to give salts e.g. CO_2, SO_2, SO_3
neutral oxides	Covalent oxides, often having a non-metal in a low oxidation state e.g. CO, NO
peroxides and superoxides	Ionic, containing O_2^{2-} (peroxide) and O_2^- (superoxide)

Fig 29.27 A summary of the oxides

Details on the chemistry of oxygen can be found on the following pages:
reactions of oxygen with elements on page 418;
basic oxides on page 421;
acidic oxides on page 422.

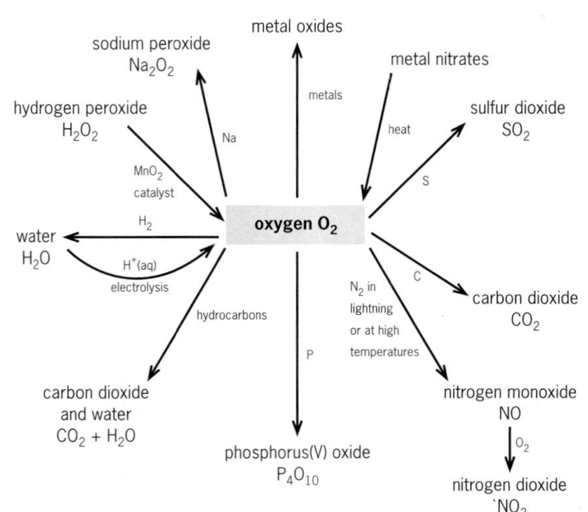

Fig 29.28 Some of the reactions involving oxygen

Fig 29.29 About two-thirds of the sulfur found in the atmosphere comes from sulfurous gases emitted by volcanoes

OCCURRENCE OF SULFUR

Sulfur is an element that has been known to humans since the earliest times. It was once known as brimstone. Sulfur deposits are widespread, as are sulfur's compounds with metals and non-metals. Many of the ores of metals are sulfides, a combination of a metal and sulfur.

CHEMICAL PROPERTIES OF SULFUR

Sulfur has the electron configuration $1s^2\ 2s^2\ 2p^6\ 3s^2\ 3p^4$; it has six electrons in its outermost shell. To obtain a stable octet of electrons, the sulfur atom gains two electrons either by sharing electrons with the atoms of a non-metal to form covalent bonds or by gaining electrons lost by metal atoms to form ionic bonds. The ionic compounds contain S^{2-} ions and are sulfur's equivalent to oxides. The oxidation number of sulfur in a sulfide is –2.

A sulfur atom can also expand its octet by accepting electrons into its vacant 3d orbitals to form compounds in which sulfur has an oxidation state of +4 or +6.

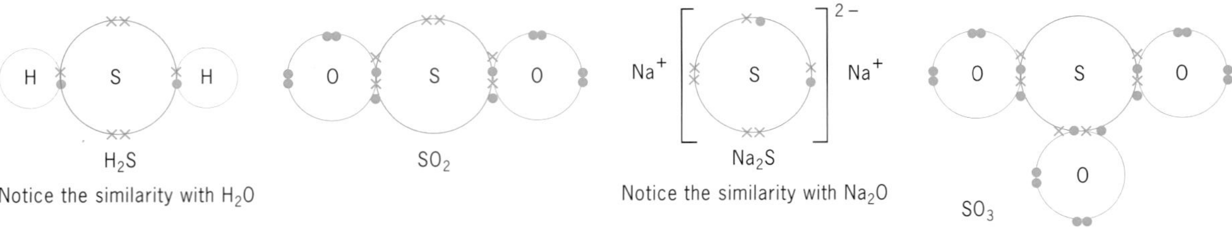

Fig 29.30 Some dot-and-cross diagrams of sulfur compounds. In H_2S and Na_2S, sulfur has a stable octet and an oxidation number of –2. In SO_2 and SO_3, the sulfur atom has expanded its octet and as a result the oxidation numbers of sulfur are +4 in SO_2 and +6 in SO_3

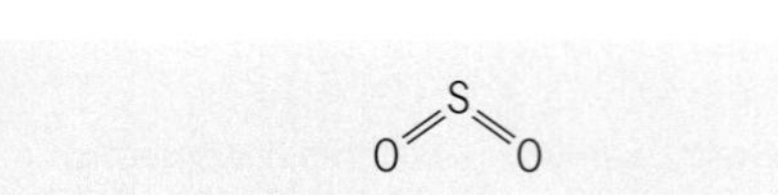

Fig 29.31 Sulfur dioxide is a bent or V-shaped molecule

SULFUR DIOXIDE AND SULFUROUS ACID

Sulfur dioxide is produced when almost any compound that contains sulfur burns in air. In Chapter 10 we discussed the formation of sulfur dioxide in power stations and indicated how it is implicated in the formation of acid rain.

Acidic properties of sulfur dioxide

Sulfur dioxide dissolves in water to form sulfurous acid, $H_2SO_3(aq)$, a moderately strong acid. An aqueous solution of sulfurous acid normally has about 25 per cent of the acid molecules dissociated to form HSO_3^- ions. Only a small proportion of these ions dissociate further to form SO_3^{2-} ions:

$$SO_2(aq) + H_2O(l) \rightleftharpoons H_2SO_3(aq)$$

$$H_2SO_3(aq) \rightleftharpoons HSO_3^-(aq) + H^+(aq)$$

$$HSO_3^-(aq) \rightleftharpoons SO_3^{2-}(aq) + H^+(aq)$$

Alkalis, such as aqueous potassium hydroxide, neutralise sulfurous acid to give the metal sulfite:

$$2KOH(aq) + H_2SO_3(aq) \rightarrow K_2SO_3(aq) + 2H_2O(l)$$

It is also possible to obtain potassium hydrogensulfite by controlling the mole ratios of acid and alkali that react.

? QUESTION 23

23 What is the mole ratio of potassium hydroxide to sulfurous acid needed to make:

a) potassium sulfite;

b) potassium hydrogensulfite.

Reducing properties of sulfur dioxide

Sulfurous acid and sulfur dioxide are both reducing agents. Sulfur dioxide is used as an antioxidant to preserve food: it is oxidised easily, removing oxygen in the air that could have oxidised the food. Aqueous chlorine, aqueous dichromate(VI) and aqueous iron(III) are all reduced by sulfurous acid. The reaction of moist sulfur dioxide and acidified potassium dichromate(VI) is a **test for sulfur dioxide**:

$$Cr_2O_7^{2-}(aq) + 3SO_2(aq) + 2H^+(aq) \rightarrow 2Cr^{3+}(aq) + 3SO_4^{2-}(aq) + H_2O(l)$$

Filter paper previously dipped into acidified potassium dichromate(VI) changes colour from orange to a blue–green in the presence of sulfur dioxide.

QUESTION 24

24 Write down the equations for the reduction by sulfur dioxide of:

a) aqueous chlorine;

b) aqueous iron(III) ions. (Hint: use the half equations in Appendix 4 to help.)

Sulfites

Salts of sulfurous acid are known as sulfites. These salts react with dilute acid to release sulfur dioxide:

$$K_2SO_3(s) + 2HNO_3(aq) \rightarrow 2KNO_3(aq) + SO_2(g) + H_2O(l)$$

Many solid sulfites thermally decompose when heated:

$$ZnSO_3(s) \rightarrow ZnO(s) + SO_2(g)$$

In this way, sulfite salts behave in a similar way to carbonates.

SULFUR TRIOXIDE

Sulfur trioxide is a typical acidic oxide and it dissolves in water to form the strong acid sulfuric acid:

$$H_2O(l) + SO_3(g) \rightarrow H_2SO_4(aq)$$

It also reacts directly with solid bases to form sulfates; barium oxide reacts to form barium sulfate:

$$BaO(s) + SO_3(g) \rightarrow BaSO_4(s)$$

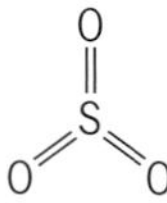

Fig 29.32 Gaseous sulfur trioxide is a trigonal planar molecule

9 SULFURIC ACID

Sulfuric acid is one of the world's key industrial chemicals. Justus von Liebig wrote in 1843 that the consumption of sulfuric acid is a barometer of a nation's commercial prosperity. Sulfuric acid's wide range of uses means that Liebig's statement still holds today. Sulfuric acid is widely used in the oil and petroleum industry. It is also involved in making:

- fertilisers, such as ammonium phosphate and superphosphate;
- paints, pigments and dyes;
- soaps and detergents;
- plastics and fibres;
- general chemicals, such as salts and battery acid.

THE CONTACT PROCESS

Sulfuric acid is manufactured using the Contact process. It was patented in 1831 by Peregrine Phillips and later improved by Rudolf Messel. The contact process is an efficient and economical way of converting relatively cheap raw materials into sulfuric acid.

Raw materials

The raw materials chosen in the United Kingdom are sulfur, air and water. Water and air are very cheap raw materials, but most sulfur needs to be imported.

REMEMBER THIS

The contact process is really just the catalytic oxidation of sulfur dioxide, to form sulfur trioxide, rather than all the processes that take place to make sulfuric acid.

About 10 per cent of the sulfur comes from the petroleum industry; sulfur is an impurity that must be removed from crude oil before the oil can be processed. Selling this impurity adds to the economic efficiency of an oil refinery.

The three stages of the manufacture of sulfuric acid are:

- stage one, the oxidation of molten sulfur to give sulfur dioxide:

$$S + O_2 \rightarrow SO_2 \quad \Delta H = -297 \text{ kJ mol}^{-1}$$

- stage two, the actual Contact process, the reaction of sulfur dioxide with oxygen to give sulfur trioxide:

$$SO_2(g) + \tfrac{1}{2}O_2(g) \rightleftharpoons SO_3(g) \quad \Delta H = -98 \text{ kJ mol}^{-1}$$

- stage three, the reaction of sulfur trioxide with water:

$$SO_3(g) + H_2O(l) \rightarrow H_2SO_4(l) \quad \Delta H = -130 \text{ kJ mol}^{-1}$$

In theory, it is a simple process. However, there are a number of problems involved with these three stages that must be overcome before the yield of sulfuric acid is economically viable.

Oxidation of sulfur

In the combustion of molten sulfur the enthalpy change of reaction is negative. The temperature used ensures that the rate of reaction is sufficient to provide an economic rate of production of sulfur dioxide. The rate of reaction must be controlled to ensure that the reaction does not go too fast because of the release of heat energy during the reaction.

The Contact process – catalytic conversion of sulfur dioxide into sulfur trioxide

The second stage is the oxidation of sulfur dioxide to give sulfur trioxide. The reaction is very slow and needs a catalyst.

Originally, the precious metal platinum was used, but now manufacturers use vanadium(V) oxide. The reaction is exothermic, so Le Chatelier's principle suggests that the forward reaction is favoured by low temperatures. The volume of products is smaller than that of the reactants, so high pressure also favours the forward reaction. To maintain an excess of reactants and continually remove the product from the reaction mixture is another way to shift the equilibrium position to the right.

A low temperature slows down the rate of reaction and high pressures are expensive to maintain, so a compromise set of conditions is chosen that gives an economic yield of sulfur trioxide. Manufacturers use a temperature of between 450 °C and 600 °C. A pressure of around 10 atmospheres is sufficient to push the gases around the plant. There is no need to use a high pressure in this reaction, despite the Le Chatelier prediction, because even at low pressure the position of equilibrium lies well to the right.

Vanadium(V)oxide, V_2O_5, is a heterogeneous catalyst. It is the ability of vanadium to have more than one oxidation state that enables it to act as a catalyst. One suggested mechanism for its action is that V_2O_5 oxidises SO_2 to make V_2O_4. Later on the V_2O_4 is oxidised by oxygen to remake V_2O_5.

$$V_2O_5 + SO_2 \rightarrow V_2O_4 + SO_3$$

$$V_2O_4 + \tfrac{1}{2} O_2 \rightarrow V_2O_5$$

Adding these two equations together gives the oxidation of SO_2 by O_2 to make SO_3.

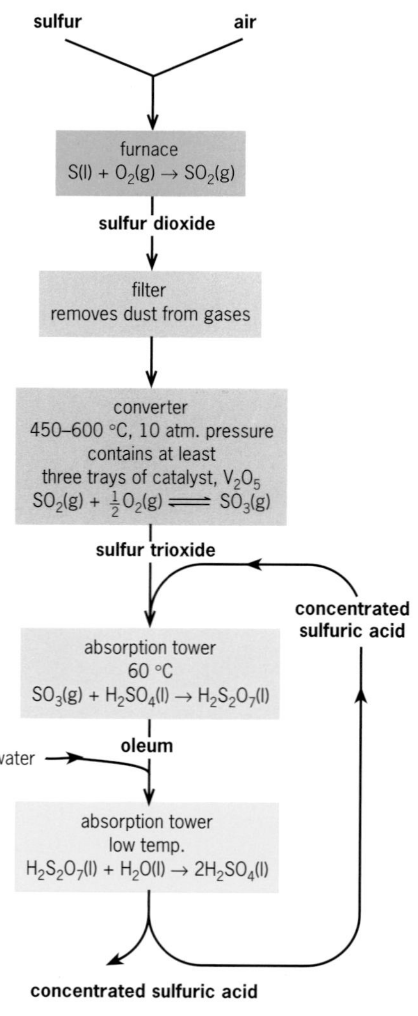

Fig 29.33 Flow chart to show the manufacture of sulfuric acid

At least three catalyst chambers or beds are used to ensure maximum conversion of sulfur dioxide (Fig 29.33). The conversion yield can exceed 98 per cent. As well as being economically viable, this high yield has environmental benefits, because very little sulfur dioxide waste enters the atmosphere.

The hydration of sulfur trioxide

The reaction of sulfur trioxide with water does not work with large quantities of sulfur trioxide. The normal way to dissolve a gas industrially is to use a counter-flow system in which water is sprayed downwards and gas flows upwards. In the case of sulfur trioxide, this technique causes clouds of sulfuric acid droplets to form. These droplets are difficult to condense, so a different method is used.

The sulfur trioxide is dissolved in concentrated sulfuric acid to form oleum, $H_2S_2O_7$. The oleum is diluted with measured amounts of water to give concentrated sulfuric acid:

$$SO_3(g) + H_2SO_4(l) \rightarrow H_2S_2O_7(l)$$

$$H_2S_2O_7(l) + H_2O(l) \rightarrow 2H_2SO_4(l)$$

Both of the processes are exothermic and are carried out at low temperature.

Throughout the plant, the flow of energy is vital. The reactions are exothermic and the conditions needed vary from high to low temperatures. Complex heat exchangers ensure that little heat is wasted.

You can read about the use of sulfuric acid in the dehydration of alcohols on page 280.

REACTIONS OF SULFURIC ACID

Sulfuric acid has a wide range of reactions in addition to those expected of a strong acid. They include:

- redox reactions with both dilute and concentrated sulfuric acid;
- dehydration reactions;
- reactions as a non-volatile acid.

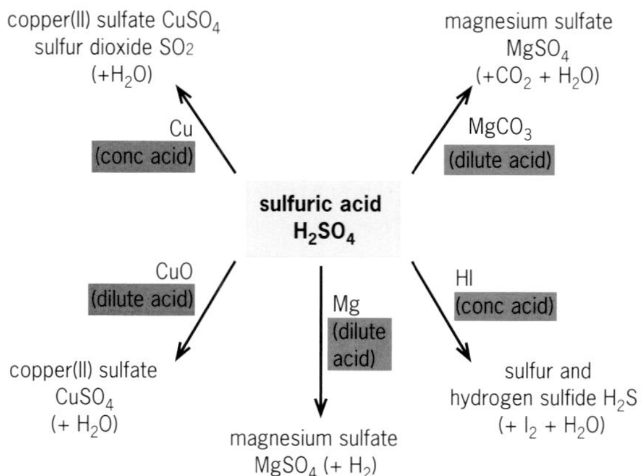

Fig 29.34 Some of the reactions of sulfuric acid

QUESTION 25

25 Look at Fig 29.34. In which reactions does sulfuric acid act as:

a) an acid;

b) an oxidising agent?

QUESTION 26

26 What is the name of the acid formed when concentrated sulfuric acid reacts with:

a) potassium benzoate;

b) magnesium ethanoate?

Reactions as a non-volatile acid

Sulfuric acid reacts with the salts of many other acids to liberate the acid. We have already described in Chapter 23 the reaction of concentrated sulfuric acid with sodium chloride, in which it forms hydrogen chloride. In the same way, concentrated sulfuric acid reacts when heated with sodium nitrate to form nitric acid:

$$NaNO_3(s) + H_2SO_4(l) \rightarrow NaHSO_4(s) + HNO_3(l)$$

Sulfates

Sulfuric acid forms a class of oxy-salts called sulfates. Sulfates are generally soluble in water and are relatively resistant to thermal decomposition. Many sulfates do not change when heated at Bunsen burner temperatures, but at higher temperatures they decompose to form the corresponding oxide.

We describe the trend of solubility of the sulfates of the Group 2 elements in Chapter 22. The lack of solubility of barium sulfate in water is the basis of the **test for aqueous sulfate** ions.

Table 29.8 The tests for sulfate, carbonate, sulfite and thiosulfate ions

	Add aqueous barium nitrate followed by excess dilute nitric acid
sulfate	white precipitate of barium sulfate that does not redissolve in the acid
carbonate	white precipitate of barium carbonate that redissolves in acid with the formation of carbon dioxide
sulfite	white precipitate of barium sulfite that redissolves slowly and forms sulfur dioxide
thiosulfate	white precipitate of barium thiosulfate that changes to a yellow precipitate on addition of acid, with the formation of sulfur dioxide

For more information on thiosulphates, see page 110.

Thiosulfates

Thiosulfates contain the $S_2O_3^{2-}$ ion. This ion is really a sulfate ion in which one oxygen atom has been replaced by a sulfur atom. On reaction with dilute acids, thiosulfates form yellow sulfur as a precipitate:

$$S_2O_3^{2-}(aq) + 2H^+(aq) \rightarrow S(s) + H_2O(l) + SO_2(g)$$

SUMMARY

After studying this chapter, you should know that:

- The metallic character of the elements in Group 4 increases with atomic number, and the non-metallic character decreases with atomic number.
- The stability of the +4 oxidation state in Group 4 decreases with increasing atomic number, and the stability of the +2 oxidation state increases with increasing atomic number.
- All the tetrachlorides of Group 4 elements except carbon are readily hydrolysed by cold water to form the corresponding dioxide and HCl.
- The thermal stability of the Group 4 tetrachlorides decreases (as the stability of the +4 oxidation state decreases and the strength of the covalent bond with chlorine weakens) with increasing atomic number of the Group 4 element.
- The tetrachlorides of Group 4 are all simple molecular covalent molecules.
- The thermal stability of the Group 4 dioxides decreases (as the stability of the +4 oxidation state decreases) with increasing atomic number of the Group 4 element.
- The acidic character of Group 4 oxides decreases (as the ionic character increases) with increasing atomic number of the Group 4 element.
- The oxidising power of the dioxides of Group 4 elements increases (as the stability of the +4 oxidation state decreases) with increasing atomic number of the element.
- The ionic character and the basic character of the monoxides of Group 4 elements increase (as the stability of the +2 oxidation state increases) with increasing atomic number of the element.
- Sulfuric acid is manufactured by the catalysed oxidation of sulfur dioxide by oxygen, followed by a controlled two-step hydration with water.
- The presence of sulfates, sulfites, thiosulfates and carbonates in solution can be tested by using aqueous barium chloride followed by dilute hydrochloric acid or by using aqueous barium nitrate solution followed by dilute nitric acid.

Practice questions and a How Science Works assignment for this chapter are available at www.collinseducation.co.uk/CAS

APPENDIX 1: SIGNIFICANT FIGURES AND DECIMAL PLACES

Students and teachers alike have difficulty with the use of significant figures and decimal places in calculations. This appendix contains simple rules that will enable you to give answers to calculations to the appropriate number of decimal places or significant figures.

Significant figures

Whenever you make a measurement of a physical quantity in the laboratory, such as the temperature of a liquid or the mass of a test tube, there will be an uncertainty in the measurement. When you use an electronic balance to measure the mass of a test tube, the last figure on the balance reading almost always fluctuates. One moment it may be 16.41 g and the next it is 16.40 g. This shows that the last figure in the mass must be uncertain. We say that 16.40 g has four significant figures and the last figure, in this case the zero, is uncertain.

Whenever you measure a physical quantity, always quote the figure to include the first one which is uncertain. Sometimes this will be a zero, but it must be quoted.

Determining the number of significant figures in a measurement

In this book you must assume that the measurements have been determined correctly and that they have the correct number of significant figures. To determine the number of significant figures, use the following rules.

- Find the first non-zero digit on the left and count the total number of digits,
  e.g. 0.00234 has three significant figures and 234.12 has five significant figures.
- If the number has a decimal point, count all the digits to the right, even if they are zero,
  e.g. 0.120 has three significant figures.
- If the number is written in standard form, do not count the exponential part of the number,
  e.g. 1.23×10^{-3} has three significant figures and 9.560×10^{6} has four significant figures.

These rules do not apply to quantities such as 100 m. You cannot tell if this has one, two or three significant figures. However if this was written in standard form then the number of significant figures can be specified. So 1.0×10^{2} m has 2 significant figures, but 1.0×10^{2} m has 3 significant figures.

Other numbers are exact, for example there are exactly 1000 cm^3 in 1 dm^3. There is no uncertainty with the 1000; it cannot be anything else.

Table A1.1

Measurement	Number of significant figures
24.0 °C	3
5.67×10^{-5} m^3	3
0.00560 moles	3
2500 cm^3	uncertain

Decimal places

To determine the number of decimal places in a number, simply count the number of digits to the right of the decimal place. Even zeros will be counted if they are significant. So if the mass of a sample of copper (II) oxide is measured as 0.300 g (we assume that the mass was determined to the nearest milligram), then this number has three decimal places. Remember, the number of significant figures is normally different from the number of decimal places.

Calculations

You must be very careful during a calculation not to quote too many or too few decimal places or significant figures. Do not just copy the answer from your calculator. Think about how many significant figures or decimal places should be used. These are some simple rules.

- When adding or subtracting numbers, it is the number of decimal places that is important. Decide which number in the calculation has the least number of decimal places. This tells you the number of decimal places in the answer.
  For example, the M_r of NO is 14.0 + 16.0 = 30.0.
- When multiplying or dividing numbers, it is the number of significant figures that is important. Decide which number in the calculation has the least number of significant figures. This tells you how many significant figures are in the answer.
  For example, the number of moles of carbon in 14.20 g is: $\frac{14.2}{12.0} = 1.183333$
  but the answer should be quoted only to three significant figures: 1.18.
- Remember that exact numbers are just that, and so do not affect the number of significant figures.

APPENDIX 2: THE PERIODIC TABLE

Relative atomic masses given in brackets refer to the isotopic mass of the most abundant isotope of the elements concerned.

Key: atomic no / **symbol** / name / relative atomic mass — metal (unshaded); non–metal (shaded)

Period \ Group	1	2											3	4	5	6	7	0
1	1 **H** hydrogen 1.0																	2 **He** helium 4.0
2	3 **Li** lithium 6.9	4 **Be** beryllium 9.0											5 **B** boron 10.8	6 **C** carbon 12.0	7 **N** nitrogen 14.0	8 **O** oxygen 16.0	9 **F** fluorine 19.0	10 **Ne** neon 20.2
3	11 **Na** sodium 23.0	12 **Mg** magnesium 24.3											13 **Al** aluminium 26.9	14 **Si** silicon 28.1	15 **P** phosphorus 31.0	16 **S** sulphur 32.1	17 **Cl** chlorine 35.5	18 **Ar** argon 39.9
4	19 **K** potassium 39.1	20 **Ca** calcium 40.1	21 **Sc** scandium 45.0	22 **Ti** titanium 47.8	23 **V** vanadium 50.9	24 **Cr** chromium 52.0	25 **Mn** manganese 54.9	26 **Fe** iron 55.9	27 **Co** cobalt 58.9	28 **Ni** nickel 58.7	29 **Cu** copper 63.5	30 **Zn** zinc 65.4	31 **Ga** gallium 69.7	32 **Ge** germanium 72.6	33 **As** arsenic 74.9	34 **Se** selenium 79.0	35 **Br** bromine 79.9	36 **Kr** krypton 83.8
5	37 **Rb** rubidium 85.5	38 **Sr** strontium 87.6	39 **Y** yttrium 88.9	40 **Zr** zirconium 91.2	41 **Nb** niobium 92.9	42 **Mo** molybdenum 95.9	43 **Tc** technetium (98)	44 **Ru** ruthenium 101.1	45 **Rh** rhodium 102.9	46 **Pd** palladium 106.4	47 **Ag** silver 107.9	48 **Cd** cadmium 112.4	49 **In** indium 114.8	50 **Sn** tin 118.7	51 **Sb** antimony 121.8	52 **Te** tellurium 127.6	53 **I** iodine 126.9	54 **Xe** xenon 131.3
6	55 **Cs** caesium 132.9	56 **Ba** barium 137.3	57 **La** lanthanum 138.9	72 **Hf** hafnium 178.5	73 **Ta** tantalum 181.0	74 **W** tungsten 183.9	75 **Re** rhenium 186.2	76 **Os** osmium 190.2	77 **Ir** iridium 192.2	78 **Pt** platinum 195.1	79 **Au** gold 197.0	80 **Hg** mercury 200.6	81 **Tl** thallium 204.4	82 **Pb** lead 207.2	83 **Bi** bismuth 209.0	84 **Po** polonium (209)	85 **At** astatine (210)	86 **Rn** radon (222)
7	87 **Fr** francium (223)	88 **Ra** radium (226)	89 **Ac** actinium (227)	104 **Rf** Rutherfordium (267)	105 **Db** Dubnium (268)	106 **Sg** Seaborgium (271)	107 **Bh** Bohrium (272)	108 **Hs** Hassium (270)	109 **Mt** Meitnerium (276)	110 **Ds** Darmstadtium (281)	111 **Rg** Roentgenium (280)	112 **Uub** Ununbium (285)	113 **Uut** Ununtrium (284)	114 **Uuq** Ununquadium (289)	115 **Uup** Ununpentium (288)	116 **Uuh** Ununhexium (293)	118 **Uuo** Ununoctium (294)	

Lanthanides	58 **Ce** cerium 140.1	59 **Pr** praseodymium 140.9	60 **Nd** neodymium 144.2	61 **Pm** promethium (145)	62 **Sm** samarium 150.4	63 **Eu** europium 152.0	64 **Gd** gadolinium 157.3	65 **Tb** terbium 158.9	66 **Dy** dysprosium 162.5	67 **Ho** holmium 164.9	68 **Er** erbium 167.3	69 **Tm** thulium 168.9	70 **Yb** ytterbium 173.0	71 **Lu** lutetium 175.5
Actinides	90 **Th** thorium 232.0	91 **Pa** protactinium (231)	92 **U** uranium 238.1	93 **Np** neptunium (237)	94 **Pu** plutonium (244)	95 **Am** americium (243)	96 **Cm** curium (247)	97 **Bk** berkelium (247)	98 **Cf** californium (251)	99 **Es** einsteinium (254)	100 **Fm** fermium (253)	101 **Md** mendelevium (258)	102 **No** nobelium (259)	103 **Lr** lawrencium (262)

APPENDIX 3: IONISATION ENERGIES

First ionisation energies of the elements

These values are in kJ mol^{-1}.

H 1310																	He 2370
Li 519	Be 900											B 799	C 1090	N 1400	O 1310	F 1680	Ne 2080
Na 494	Mg 736											Al 577	Si 786	P 1060	S 1000	Cl 1260	Ar 1520
K 418	Ca 590	Sc 632	Ti 661	V 648	Cr 653	Mn 716	Fe 762	Co 757	Ni 736	Cu 745	Zn 908	Ga 577	Ge 762	As 966	Se 941	Br 1140	Kr 1350
Rb 402	Sr 548	Y 636	Zr 669	Nb 653	Mo 694	Tc 699	Ru 724	Rh 745	Pd 803	Ag 732	Cd 866	In 556	Sn 707	Sb 833	Te 870	I 1010	Xe 1170
Cs 376	Ba 502	La 540	Hf 531	Ta 577	W 770	Re 762	Os 841	Ir 887	Pt 866	Au 891	Hg 1010	Tl 590	Pb 716	Bi 774	Po 812	At	Rn 1040
Fr 381	Ra 510	Ac 669															

Ce 665	Pr 556	Nd 607	Pm 556	Sm 540	Eu 548	Gd 594	Tb 648	Dy 657	Ho	Er	Tm	Yb 598	Lu 481
Th 674	Pa	U 385	Np	Pu	Am	Cm	Bk	Cf	Es	Fm	Md	No	Lr

Ionisation energies of selected elements

Table A1.2

Element	Atomic (proton) number	First ionisation energy/kJ mol^{-1}	Second ionisation energy/kJ mol^{-1}	Third ionisation energy/kJ mol^{-1}	Fourth ionisation energy/kJ mol^{-1}
K	19	420	3070	4600	5860
Ca	20	590	1150	4940	6480
Sc	21	630	1240	2390	7110
Ti	22	660	1310	2720	4170
V	23	650	1370	2870	4600
Cr	24	650	1590	2990	4770
Mn	25	720	1510	3250	5190
Fe	26	760	1560	2960	540
Co	27	760	1640	3230	5100
Ni	28	740	1750	3390	5400
Cu	29	750	1960	3350	5690
Zn	30	910	1730	3830	6190

APPENDIX 4: STANDARD ELECTRODE POTENTIALS

Electrode reaction	$E^{\ominus}$/V
$Ag^{+} + e^{-} \rightleftharpoons Ag$	+0.80
$Ag^{2+} + e^{-} \rightleftharpoons Ag^{+}$	+1.98
$AgBr + e^{-} \rightleftharpoons Ag + Br^{-}$	+0.07
$AgCN + e^{-} \rightleftharpoons Ag + CN^{-}$	−0.04
$Ag(CN)_2^{-} + e^{-} \rightleftharpoons Ag + 2CN^{-}$	−0.38
$AgCl + e^{-} \rightleftharpoons Ag + Cl^{-}$	+0.22
$AgI + e^{-} \rightleftharpoons Ag + I^{-}$	−0.15
$Ag(NH_3)_2^{+} + e^{-} \rightleftharpoons Ag + 2NH_3$	+0.37
$Ag_2O + H_2O + 2e^{-} \rightleftharpoons 2Ag + 2OH^{-}$	+0.34
$Al^{3+} + 3e^{-} \rightleftharpoons Al$	−1.66
$Al(OH)_4^{-} + 3e^{-} \rightleftharpoons Al + 4OH^{-}$	−2.35
$As + 3H^{+} + 3e^{-} \rightleftharpoons AsH_3$	−0.38
$H_3AsO_4 + 2H^{+} + 2e^{-} \rightleftharpoons H_3AsO_3 + H_2O$	+0.56
$Au^{+} + e^{-} \rightleftharpoons Au$	+1.68
$Au^{3+} + 3e^{-} \rightleftharpoons Au$	+1.50
$H_3BO_3 + 3H^{+} + 3e^{-} \rightleftharpoons B + 3H_2O$	−0.73
$Ba^{2+} + 2e^{-} \rightleftharpoons Ba$	−2.90
$Ba^{2+} + 2e^{-} \rightleftharpoons Be$	−1.85
$BiO^{+} + 2H^{+} + 3e^{-} \rightleftharpoons Bi + H_2O$	+0.28
$Br_2 + 2e^{-} \rightleftharpoons 2Br^{-}$	+1.07
$2HOBr + 2H^{+} + 2e^{-} \rightleftharpoons Br_2 + 2H_2O$	+1.59
$2BrO_3^{-} + 12H^{+} + 10e^{-} \rightleftharpoons Br_2 + 6H_2O$	+1.52
$CO_2 + H^{+} + e^{-} \rightleftharpoons \frac{1}{2}H_2C_2O_4$	−0.49
$Ca^{2+} + 2e^{-} \rightleftharpoons Ca$	−2.87
$Cd^{2+} + 2e^{-} \rightleftharpoons Cd$	−0.40
$Ce^{4+} + e^{-} \rightleftharpoons Ce^{3+}$	+1.45
$Cl_2 + 2e^{-} \rightleftharpoons 2Cl^{-}$	+1.36
$2HOCl + 2H^{+} + 2e^{-} \rightleftharpoons Cl_2 + 2H_2O$	+1.64
$2ClO_3^{-} + 12H + 10e^{-} \rightleftharpoons Cl_2 + 6H_2O$	+1.47
$CO^{2+} + 2e^{-} \rightleftharpoons CO$	−0.28
$CO^{3+} + e^{-} \rightleftharpoons Co^{2+}$	+1.82
$Co(NH_3)_6^{2+} + 2e^{-} \rightleftharpoons Co + 6NH_3$	−0.43
$Cr^{2+} + 2e^{-} \rightleftharpoons Cr$	−0.91
$Cr^{3+} + 3e^{-} \rightleftharpoons Cr$	−0.74
$Cr^{3+} + e^{-} \rightleftharpoons Cr^{2+}$	−0.41
$Cr_2O_7^{2-} + 14H^{+} + 6e^{-} \rightleftharpoons 2Cr^{3+} + 7H_2O$	+1.33
$Cs^{+} + e^{-} \rightleftharpoons Cs$	−2.92
$Cu^{+} + e^{-} \rightleftharpoons Cu$	+0.52
$Cu^{2+} + 2e^{-} \rightleftharpoons Cu$	+0.34
$Cu^{2+} + e^{-} \rightleftharpoons Cu^{+}$	+0.15
$Cu^{2+} + I^{-} + e^{-} \rightleftharpoons CuI$	+0.86
$Cu(NH_3)_4^{2+} + 2e^{-} \rightleftharpoons Cu + 4NH_3$	−0.05
$2D^{+} + 2e^{-} \rightleftharpoons D_2$	−0.003
$F_2 + 2e^{-} \rightleftharpoons 2F^{-}$	+2.87
$Fe^{2+} + 2e^{-} \rightleftharpoons Fe$	−0.44
$Fe^{3+} + 3e^{-} \rightleftharpoons Fe$	−0.04
$Fe^{3+} + e^{-} \rightleftharpoons Fe^{2+}$	+0.77
$Fe(CN)_6^{3-} + e^{-} \rightleftharpoons Fe(CN)_6^{4-}$	+0.36
$FeO_4^{2-} + 8H^{+} + 3e^{-} \rightleftharpoons Fe^{3+} + 4H_2O$	+2.20
$2H^{+} + 2e^{-} \rightleftharpoons H_2$	0.00
$H_2 + 2e^{-} \rightleftharpoons 2H^{-}$	−2.25
$Hg_2^{2+} + 2e^{-} \rightleftharpoons 2Hg$	+0.79

Electrode reaction	$E^{\ominus}$/V
$Hg^{2+} + 2e^{-} \rightleftharpoons Hg$	+0.85
$2Hg^{2+} + 2e^{-} \rightleftharpoons Hg_2^{2+}$	+0.91
$Hg_2Cl_2 + 2e^{-} \rightleftharpoons 2Hg + 2Cl^{-}$	+0.27
$I_2 + 2e^{-} \rightleftharpoons 2I^{-}$	+0.54
$2HOI + 2H^{+} + 2e^{-} \rightleftharpoons I_2 + 2H_2O$	+1.45
$2IO_3^{-} + 12H^{+} + 10e^{-} \rightleftharpoons I_2 + 6H_2O$	+1.19
$K^{+} + e^{-} \rightleftharpoons K$	−2.92
$Li^{+} + e^{-} \rightleftharpoons Li$	−3.04
$Mg^{2+} + 2e^{-} \rightleftharpoons Mg$	−2.38
$Mn^{2+} + 2e^{-} \rightleftharpoons Mn$	−1.18
$MnO_2 + 4H^{+} + 2e^{-} \rightleftharpoons Mn^{2+} + 2H_2O$	+1.23
$MnO_4^{-} + e^{-} \rightleftharpoons MnO_4^{2-}$	+0.56
$MnO_4^{-} + 4H^{+} + 3e^{-} \rightleftharpoons MnO_2 + 2H_2O$	+1.67
$MnO_4^{-} + 8H^{+} + 5e^{-} \rightleftharpoons Mn^{2+} + 4H_2O$	+1.52
$N_2 + 8H^{+} + 6e^{-} \rightleftharpoons 2NH_4^{+}$	+0.27
$HNO_2 + H^{+} + e^{-} \rightleftharpoons NO + H_2O$	+0.99
$NO_3^{-} + 2H^{+} + e^{-} \rightleftharpoons NO_2 + H_2O$	+0.81
$NO_3^{-} + 3H^{+} + 2e^{-} \rightleftharpoons HNO_2 + H_2O$	+0.94
$NO_3^{-} + 4H^{+} + 3e^{-} \rightleftharpoons NO + 2H_2O$	+0.96
$2NO_3^{-} + 10H^{+} + 8e^{-} \rightleftharpoons N_2O + 5H_2O$	+1.11
$2NO_3^{-} + 12H^{+} + 10e^{-} \rightleftharpoons N_2 + 6H_2O$	+1.24
$NO_3^{-} + 10H^{+} + 8e^{-} \rightleftharpoons NH^{+}_{4} + 3H_2O$	+0.87
$Na^{+} + e^{-} \rightleftharpoons Na$	−2.71
$Ni^{2+} + 2e^{-} \rightleftharpoons Ni$	−0.25
$Ni(NH_3)_6^{2+} + 2e^{-} \rightleftharpoons Ni + 6NH_3$	−0.51
$H_2O_2 + 2H^{+} + 2e^{-} \rightleftharpoons 2H_2O$	+1.77
$O_2 + 4H^{+} + 4e^{-} \rightleftharpoons 2H_2O$	+1.23
$O_2 + 2H^{+} + 2e^{-} \rightleftharpoons H_2O_2$	+0.68
$O_2 + 2H_2O + 4e^{-} \rightleftharpoons 4OH^{-}$	+0.40
$O_3 + 2H^{+} + 2e^{-} \rightleftharpoons O_2 + H_2O$	+2.07
$P + 3H^{+} + 3e^{-} \rightleftharpoons PH_3$	−0.04
$H_3PO_4 + 2H^{+} + 2e^{-} \rightleftharpoons H_3PO_3 + H_2O$	−0.28
$Pb^{2+} + 2e^{-} \rightleftharpoons Pb$	−0.13
$PB^{4+} + 2e^{-} \rightleftharpoons PB^{2+}$	+1.69
$PbO_2 + 4H^{+} + 2e^{-} \rightleftharpoons Pb^{2+} + 2H_2O$	+1.47
$PbO_2 + H_2O + 2e^{-} \rightleftharpoons PBO + 2OH^{-}$	+0.28
$Ra^{2+} + 2e^{-} \rightleftharpoons Ra$	−2.92
$Rb^{+} + e^{-} \rightleftharpoons Rb$	−2.92
$S + 2e^{-} \rightleftharpoons S^{2-}$	−0.51
$S + 2H^{+} + 2e^{-} \rightleftharpoons H_2S$	+0.14
$SO_4^{2-} + 4H^{+} + 2e^{-} \rightleftharpoons H_2SO_3 + H_2O$	+0.17
$S_2O_8^{2-} + 2e^{-} \rightleftharpoons 2SO_4^{2-}$	+2.01
$S_4O_6^{2-} + 2e^{-} \rightleftharpoons 2S_2O_3^{2-}$	+0.09
$Sb + 3H^{+} + 3e^{-} \rightleftharpoons SbH_3$	−0.51
$SbO^{+} + 2H^{+} + 3e^{-} \rightleftharpoons Sb + H_2O$	+0.21
$Sn^{2+} + 2e^{-} \rightleftharpoons Sn$	−0.14
$Sn^{4+} + 2e^{-} \rightleftharpoons Sn^{2+}$	+0.15
$V^{2+} + 2e^{-} \rightleftharpoons V$	−1.2
$V^{3+} + e^{-} \rightleftharpoons V^{2+}$	−0.26
$VO^{2+} + 2H^{+} + e^{-} \rightleftharpoons V^{3+} + H_2O$	+0.34
$VO^{2+} + 2H^{+} + e^{-} \rightleftharpoons VO^{2+} + H_2O$	+1.00
$VO^{3-} + 4H^{+} + e^{-} \rightleftharpoons VO^{2+} + 2H_2O$	+1.00
$Zn^{2+} + 2e^{-} \rightleftharpoons Zn$	−0.76

PHOTOGRAPH CREDITS

Every effort has been made to contact the holders of copyright material, but if any have been inadvertently overlooked the publishers will be pleased to make the necessary arrangements at the first opportunity.

The publishers would like to thank the following for permission to reproduce photographs (T = Top, B = Bottom, C = Centre, L= Left, R = Right):

AIP Emilio Segre Visual Archives/ William G Myers Collection, 2.7; *Allsport/*M Hewitt, 17.31, M Cooper, 29.19; *Ann Ronan Pictures/Heritage-Images*, 2.1a, 2.13, 3.2b, 3.3, 5.3, 5.5; *Argonne National Laboratory, managed and operated by the University of Chicago for the US Department of Energy*, 4.19a; *BASF Group*, 1.21; *Barnaby's Picture Library*, 16.19; *Doris J Beck, Dept of Biological Sciences, Bowling Green State University*, 24.23, 24.24; *Biopol: Trademark and Property of Monsanto plc*, p.394; *John Birdsall Social Issues Photo Library*, 17.1, p.350, p.456T; *Blackpool Pleasure Beach Ltd*, 1.14d; *BMW*, p.152R; *Chris Bonington Picture Library*, 18.38; *British Aerospace*, 22.33; *British Steel*, p.2B, 1.3, 1.4, 1.16, 27.9; *Broadleaf Design and Marketing*, 10.32; *J A Cash Ltd*, 14.47; *Martyn Chillmaid*, 6.27a, 10.23; *Bruce Coleman Ltd/*K Rushby, p.212, N Myers, 11.2, H Lange, 13.4, K Taylor, 17.14; *Collections/*B Boswell, 29.1; *Complete Weed Control, Ltd*, 28.10; *P Duckers, Assoc. Member of the Guild of Railway Artists*, 27.48; *Du pont*, 20.21, 20.26, 20.36, 20.37; *Vivien Fifield*, 2.2, 2.11a, 3.7, 3.11a; *flpa-images.co.uk/* N Cattlin, 1.22, 17.12, 17.15, 23.1, 28.3; *Ford Motor Company Ltd.*, 10.22; *Fundamental Photos, NYC/* K Brochmann, p.536; *GSF Photo Library*, 4.2b; *Leslie Garland Photo Library*, 22.19, 24.1, p.502, 25.29, 25.32, 29.3; *Geophotos/ Tony Waltham*, p.2T, 1.8, 1.9, 1.12L, 1.12C, 1.12R, 1.13R, 10.26, 15.54, p.370, 21.33, p.426B, 22.1; *Getty Images*, 1.13L, 7.16, 21.32, 26.21, Hulton Archive, 2.8, p.152TL, S Westmorland, 8.21, p.426T, D Woodfall, 10.9, 18.24, A Husmo, 14.2, P & K Smith, 14.39R, P Harris, 15.26, J Edwards, 22.35, A Puzey, 23.6, P Chesley, 23.16, C Somodevilla, 23.22, M Williams, 29.5; *Stanley Gibbons, Ltd*, 27.17a; *Peter Gould*, 6.20a; *Sally & Richard Greenhill*, 20.14; *IBM Almaden Research Center*, 15.8; *IBM UK*, 3.16b; *ICCE/Boulton*, 11.25; *iStockphoto*, 3.1, 3.12, 24.8, 25.14, L Ha, 3.8, B Doty, 6.20e, T Sullivan, 7.15, M Kolbe, 12.17, D Stein, 12.41, J Pitcher, 14.42, N Loran, 14.46, T Bercic, 14.49, S Bolt, 15.39, Slobo Mitic, p.554, A Green AGMIT, 27.5 ; © *2008 Jupiterimages Corporation*, 12.1; *Andrew Lambert Photographs*, 4.26c, 4.27c, 4.28a, 4.30c, 4.31c, 4.32b, 4.33b, 4.34b, 4.37b, 4.38b, 5.16, 5.17, 5.18, 5.19, 5.21, 5.22, 6.20b, 6.20d, p.136CR, 7.8, 7.9, 7.21R, p.176CR, 11.4, 11.19, 11.20, 11.48, 12.29, 13.30, 13.33, 14.20, 14.39L, 15.13, 15.40, 16.6, 16.16, 16.30, 16.32,16.34, 17.3, 17.10, 17.11, 17.21, 18.3, 18.18, 18.19, 20.8, 20.16, 20.24, 21.1, 21.2, 21.16, 21.19, 21.22, 21.23, 21.24, 21.25, 21.34, 22.3, 22.4, 22.5, 22.6, 22.9, 22.10, 22.12, 22.13, 22.14, 22.15, 22.20, 22.34, 23.2, 23.3, 23.10, 23.11, 23.13, 23.14, 23.15, 24.9, 24.28a, 24.28b, 24.28c, 24.28d, 24.30, 24.31, 24.32, 24.33, 24.34a, 24.34b, 24.35, 24.36, 24.37, 25.5, 25.7, 25.16, 25.20, 25.25, 26.6, 26.9, 27.4, 27.7, 27.14bR, 27.40, 28.12, 29.7, 29.18, 29.23; *NASA*, 22.11; © *National Gallery, London, Sassoferratto 'The Virgin and Child Embracing'*, 11.46; *Natural History Museum, London*, 7.3, 7.21L; *Natural Visions/Heather Angel*, 27.6b; *The Nobel Foundation*, 26.20; *PA Photos*, p.192, PA Archive/D Kendall, 8.20, AP, 10.19, PA Archive/C Bacon, 18.17, 28.8; *Panos Pictures/*P Fryer, 18.2; *Photoshot/*Woodfall Wild Images/M Hamblin, 8.2, Woodfall Wild Images/D Woodfall, p.294, 20.42, 22.16, 28.11, 28.21, NHPA/ S Dalton, 17.13, NHPA/A Ackerley, 17.26; *Phototake NYC/* B Masini, 20.4; *Premaphotos/*K G Preston-Mafham, 16.14; *Princeton University*, 2.16a, 2.16b; *Reading Buses*, 18.25; *Rex Features Ltd*, p.176BR, 27.14bL; *Charlotte Roberts, Calvin Wells Laboratory, Department of Archaeological Sciences, University of Bradford*, 19.19; *Rolls-Royce plc*, 29.12; *Gavin Rowe*, 20.33; *Science Photo Library (SPL)*, 1.14a, 4.25, 5.1, 6.3, 6.9, p.410, 26.25, 27.15L, p.596, SPL/J King-Holmes, 1.1, 2.21, 18.4, SPL/A Bartel, 1.14b, SPL/T Kinsbergen, 1.14c, SPL/NASA, p.28, 2.15, 7.28, p.152CL, 8.18, 10.14, 21.31, SPL/NOAA, 2.17, 12.40, SPL/G Tortoli, 2.22, SPL/W & D McIntyre, 2.23, SPL/P Plailly, 2.24, SPL/Royal Observatory, Edinburgh/AAO, p.48, SPL/ R Megna/Fundamental Photos, 3.9a, SPL/NOAO, 3.14, SPL/S Camazine, 3.16a, SPL/G Tompkinson, p.70, p.136BL, 11.40, SPL/P Plailly/Eurelios, 4.24c, p.90C, SPL/C Priest, 4.36, SPL/T Hollyman, 4.47, SPL/D Parker, p.90T, 7.14, 19.27, SPL/J L Charmet, 5.2, SPL/ V Habbick Visions, 5.20, SPL/Kairos, Latin Stock, p.114, SPL/Science Source, 6.2, SPL/C D Winters, 6.22, 6.27b, SPL/BSIP LECA, 6.23, SPL/G Muller, Struers GMBH, 7.7, SPL/Laguna Design, 7.25, SPL/Rosenfeld Images LTD, p.176C, 21.30, 22.26, SPL/ESA, 10.15, SPL/ US Geological Survey, 10.16, SPL/A & H Frieder Michler, 10.21, SPL/J Mclean, 11.7, SPL/S Fraser Mauna Loa Observatory, 11.26, SPL/S Fraser/Royal Victoria Infirmary, Newcastle Upon-Tyne, 11.39, SPL/E Young, 11.42, SPL/C Molloy, 12.8, SPL/GJLP, p.258, SPL/ D Lovegrove, 14.18, SPL/J Durham, 15.3, SPL/Saturn Stills, p.316, SPL/J C Revy, p.328, 27.44, SPL/Andrew Lambert Photography, 18.11, SPL/H Pincis, 18.14, SPL/Custom Medical Stock Photo, 19.16, SPL/BSIP Laurent, 19.26, SPL/A Barrington Brown, 19.48, SPL/Eye of Science, 19.54, SPL/P Rapson, 20.44, SPL/ J Selby, 22.18, SPL/A Pasieka, 22.31, SPL/Moredun

Animal Health Ltd, 23.24, SPL/B Bachmann, 23.25, SPL/S Fraser, p.480, SPL/K Eward/Biografx, 24.18, SPL/M Fielding, Johnson Matthey PLC, 24.41, SPL/R de Gugliemo, 27.45, SPL/K Wood, 28.4, SPL/Dr J Burgess, 29.25, SPL/A Hart-Davis, 29.26, SPL/S Summerhays, 29.29 ; *Science and Society Picture Library,* 27.17b, Science Museum Library, 2.1b, 2.3, 27.15R; *Science Vu/Visuals Unlimited,* 8.26; *Roger Scruton,* 13.19; *Shell Photographic Services,* 9.1, 9.2, 9.3; *Shout Pictures,* 9.28, 11.24, p.236, 22.7; *South American Pictures,* 8.3; *Southeastern Technology Center, Augusta, Georgia,* 8.12; *Still Pictures/*M Edwards, 1.2, 10.31, 15.59, 22.17, M Schroder, 7.29, D Garcia, p.176CL, D Drain, p.176BL, 14.1, Mitchell-UNEP, 10.18; *SuperStock,* 8.10; *Kenneth Suslick, Department of Chemistry, University of Illinois,* 26.11; *C & S Thompson,* 23.17, 23.18; *UKAEA Photographic Library,* 29.6; *Robin Whymer,* P.136TL; *Yachtshots/*K Lehman, 20.38

Chapter contents and unit photo, © 2008 Jupiterimages Corporation

Cover photo, Science Photo Library/D Roberts

ANSWERS TO QUESTIONS

CHAPTER 1

1 (a) $CaCO_3 \rightarrow CaO + CO_2$; **(b)** $2Mn + O_2 \rightarrow 2MnO$; **(c)** $4P + 5O_2 \rightarrow P_4O_{10}$
2 (a) 56; **(b)** 2; **(c)** 3.5
3 (a) 1.0 mol of iron; 0.50 mol of calcium; 2.0 mol of copper; 0.125 (0.1 to 1 sig. fig.) mol of sulfur; **(b)** 120 g of calcium; 6 g of neon
4 (a) P_4O_{10}, 284; **(b)** O_2, 32; **(c)** $CaSiO_3$, 116
5 (a) 90; **(b)** 342; **(c)** 102
6 (a) $Fe_2O_3 + 3C \rightarrow 3CO + 2Fe$; **(b)** 0.525 tonnes
7 (a) $ZnO + C \rightarrow Zn + CO$; **(b)** Moles ZnO = 1×10^5 and moles of C = 6.75×10^5 so zinc oxide is the limiting reagent
8 96%
9 (a) $Fe_3O_4 + 4CO \rightarrow 3Fe + 4CO_2$; **(b)** 7.24 tonnes; **(c)** 89.8%; **(d)** 49%
10 (a) 66.7%; **(b)** 100%; **(c)** Acid rain or an effect of acid rain such as corrosion of metal work, erosion of marble buildings etc
11 (a) Exothermic

$S(s) + O_2(g)$ → $\Delta H = -297$ kJ mol^{-1} → $SO_2(g)$

(b) Endothermic

$CaCO_3(s)$ → $\Delta H = +178$ kJ mol^{-1} → $CaO(s) + CO_2(g)$

(c)

$Fe_2O_3(s) + 3CO(g)$ → $\Delta H = -27$ kJ mol^{-1} → $2Fe(s) + 3CO_2(g)$

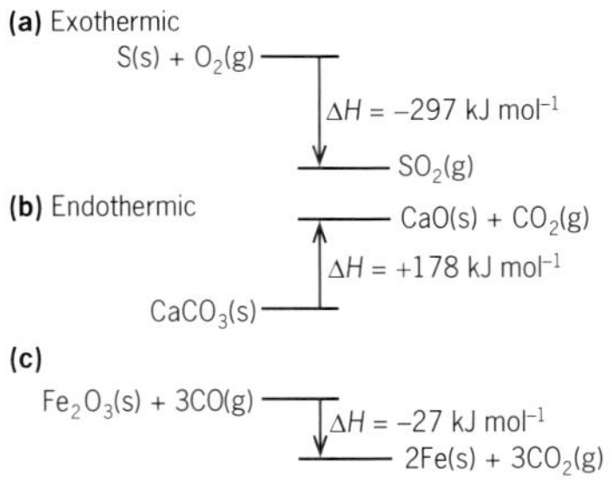

12 Does not decompose. Melting point in excess of 2000°C; does not react with substances present in the blast furnace
13 (a) To ensure thorough mixing of reactants; larger surface area; faster rate of reaction; **(b)** Allows easy passage of gases through the furnace from bottom to top
14 Continuous processes need less labour, no need to close and reset reactor vessel, the product formed has a more consistent quality
15 (a) $Mg + S \rightarrow MgS$; **(b)** 0.24 kg; **(c)** Incomplete reaction; may react with any oxygen or oxide still present in molten iron; **(d)** Larger surface area; can be dispersed throughout the molten iron; so faster rate of reaction; **(e)** Iron can be tapped off from beneath the symolten magnesium sulfide; molten magnesium sulfide can be skimmed off the top
16 (a) Rate of reaction becomes so low that the reaction appears to stop; **(b)** 2.00 mol; **(c)** 1.56 mol
17 (a) Decreases, since concentration of HI decreases; **(b)** Shifts to the right to minimise the decrease in concentration of HI
18 (a) and **(c)**
19 100%
20 Rate of reaction too slow at low temperatures
21 (a) Position of equilibrium shifts to the left (the endothermic direction) to minimise the temperature increase; **(b)** Position of equilibrium shifts to the left (the endothermic direction) to minimise the temperature increase; **(c)** Position of equilibrium shifts to the right (the endothermic direction) to minimise the temperature increase
22 (a) Position of equilibrium shifts to the left since there are fewer moles of gas on the left and so this will minimise the increase in pressure; **(b)** No effect on position of equilibrium since the number of moles of gas on both sides of the equation is the same; **(c)** Position of equilibrium shifts to the right since there are fewer moles of gas on the right and so this will minimise the increase in pressure
23 (a) Reaction is endothermic so that an increase in temperature will shift the position of equilibrium to the right and absorb heat and so minimise the increase in temperature; **(b)** Position of equilibrium shifts to the left because there are fewer moles of gas on the left and so this will minimise the increase in pressure; **(c)** The higher the pressure the faster the rate of reaction because there are more collisions per second; easier to move gases around a plant under pressure
24 (a) Increases surface area and so a faster rate of reaction; heterogeneous catalyst with surface action so needs the greatest amount of surface to be available; **(b)** Shifts the equilibrium to the right the next time the gas goes through the synthesis reaction vessel

CHAPTER 2

1 2000
2 (a) α, positive; β, negative; γ, neutral; **(b)** β has a smaller mass than α particles
3 3 km
4 (a) atomic number = number of protons = number of electrons = 11 **(b)** atomic number = number of protons = 11 and number of electrons = 10
5 (a) Iron, A = 56, Z = 26, 26 protons, 26 electrons and 30 neutrons; **(b)** Mercury, A = 200, Z = 80, 80 protons, 80 electrons and 120 neutrons
6 $^{12}_{6}C$, A = 12, Z = 6, 6 protons, 6 electrons and 6 neutrons; $^{13}_{6}C$, A = 13, Z = 6, 6 protons, 6 electrons and 7 neutrons; $^{14}_{6}C$, A = 14, Z = 6, 6 protons, 6 electrons and 8 neutrons
7 A_r, Br = 80.0; A_r, Mg = 24.3
8 $^{4}_{2}He + ^{4}_{2}He \rightarrow ^{8}_{4}Be$ + energy; the element is beryllium
9 $^{226}_{88}Ra \rightarrow ^{4}_{2}He + ^{222}_{86}Rn$; the element is radon
10 (a) Since α particles can only penetrate a few centimetres in air they will not escape from the detector; **(b)** The americium oxide in the detector will still be radioactive
11 $^{90}_{38}Sr \rightarrow ^{0}_{-1}e + ^{90}_{39}Y$; the element is yttrium
12 9.96×10^{-16} g
13 (a) So that the body is not subjected to ionising radiation over a prolonged length of time. **(b)** 10 half-lives so mass is 0.0977 g
14 About $6\frac{1}{2}$ half lives so about 37000 years ago

CHAPTER 3

1 5.50×10^{-7} m; green
2 (a) 3.0×10^{-2} m; **(b)** 2.0×10^{-35} J; **(c)** 1.2×10^{-11} J

3 Since frequency of gamma radiation is greater than that of ultraviolet, a photon of gamma radiation will have more energy than that of ultraviolet

4 Higher energy end of spectrum

5 (a)

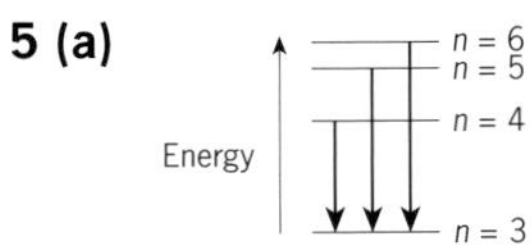

(b) $n = 4 \rightarrow n = 3$

6 Yellow

7 Formula, $2n^2$; number of electrons, 50

8

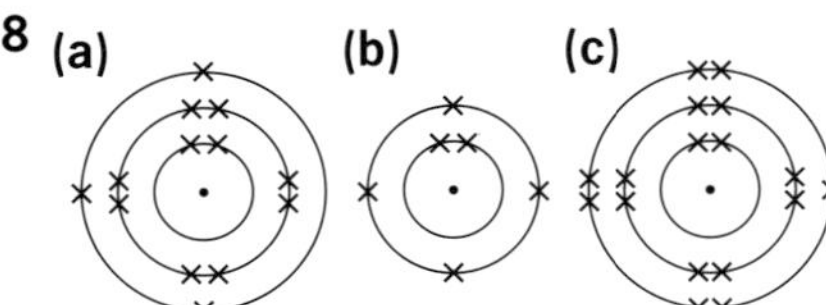

9 (a) $Mg^{10+}(g) \rightarrow Mg^{11+}(g) + e^-$;
(b) Mg^{10+} has a noble gas configuration $1s^2$, which does not have any inner-shell shielding electrons so the two electrons are very strongly attracted to the nucleus; in Mg^{9+} the electron configuration is $1s^22s^1$, which has an inner shell of shielding electrons, so the electron in the outer shell is much less firmly attracted to the nucleus than the outer electrons in Mg^{10+}

10

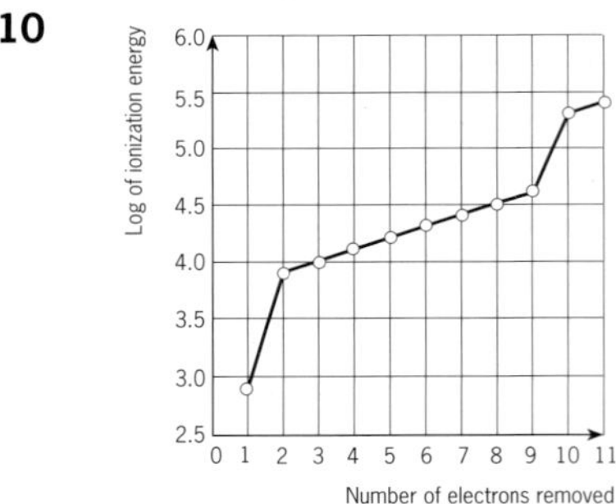

First electron lost is from the outer shell and has the lowest ionisation energy, because it is shielded by two sets of inner shells. It is also the furthest electron away from the nucleus. The next eight electrons lost all have the same number of inner shielding electrons, but the ionic radius decreases as the charge on the ion increases, so there is more attraction towards the nucleus. The number of electrons is decreasing, but the nuclear charge is the same, so there is a greater attraction for the remaining electrons. The last two electrons lost are the inner-shell electrons that are not shielded and are very close to the nucleus. The large increases in ionisation energy occur when an electron is removed from a new shell, so there are fewer shielding shells of electrons

11 (a) The enthalpy change when one electron is lost from each atom in a mole of gaseous atoms, M, to form gaseous positive ions, $M^+(g)$;
(b) Lithium has a core charge of +1 and an inner shielding shell of two electrons, whereas neon has a core charge of +8 and an inner shielding shell of two electrons; this means that the outer electrons in a neon atom are attracted much more strongly to the nucleus than those in lithium. The atomic radius of lithium is also greater than that of neon; **(c)** As you go the group atomic radius and number of shielding shells increases so less attraction between outer electrons and nucleus despite increase in nuclear charge. This provides evidence for different electron shells.

12 (a) Same number of inner-shell shielding electrons, but the core charge of magnesium is +2 whereas in sodium it is +1, so outer electrons in magnesium are much more firmly attracted and closer to the nucleus; **(b)** (i) In magnesium the electron is removed from the 3s orbital whereas in aluminium the electron is removed from the 3p energy level, which is higher in energy (and further away from the nucleus) than the 3s energy level and so requires less energy; (ii) In phosphorus each 3p orbital contains just one electron so that there is no electron–electron repulsion in these orbitals, but in sulfur one 3p orbital contains two electrons so that there is some electron–electron repulsion and therefore it is easier to remove one electron from a sulfur atom than from a phosphorus atom

13 Be ($1s^22s^2$) has a full 2s subshell, whereas B ($1s^22s^22p^1$) has an extra electron in the 2s subshell. The 2p subshell has a higher energy level and the electron in this subshell is further away from the nucleus so that it requires less energy to be removed from the atom

14 Answer is fig 3.24 but does not have the 3s, 3p and 3d showing. It will have the 1s, 2s and 2p together with the arrows

15 (a) Na $1s^22s^22p^63s^1$; Cl $1s^22s^22p^63s^23p^5$; Al $1s^22s^22p^63s^23p^1$
(b) Na^+ $1s^22s^22p^6$; Cl^- $1s^22s^22p^63s^23p^6$; Al^{3+} $1s^22s^22p^6$

16 As, $1s^22s^22p^63s^23p^63d^{10}4s^24p^3$; Se, $1s^22s^22p^63s^23p^63d^{10}4s^24p^4$; Br, $1s^22s^22p^63s^23p^63d^{10}4s^24p^5$

17 (a) Rb $1s^22s^22p^63s^23p^63d^{10}4s^2 4p^65s^1$;
(b) Y $1s^22s^22p^63s^23p^63d^{10}4s^24p^64d^15s^2$; **(c)** Cd $1s^22s^22p^63s^23p^63d^{10}4s^24p^64d^{10}5s^2$;
(d) In $1s^22s^22p^63s^23p^63d^{10}4s^24p^64d^{10}5s^25p^1$; **(e)** I $1s^22s^22p^63s^23p^63d^{10}4s^24p^64d^{10}5s^25p^5$;
(f) Xe $1s^22s^22p^63s^23p^63d^{10}4s^24p^64d^{10}5s^25p^6$

18 (a) Po $6s^26p^4$; **(b)** Ra $7s^2$; **(c)** As $6s^26p^5$; **(d)** Rn $6s^26p^6$

CHAPTER 4

1 ns^2np^6

2 Na^+ has an electron configuration of Ne; Cl^- has an electron configuration of Ar

3

(a) KF K^+ $[F]^-$

(b) MgO Mg^{2+} $[O]^{2-}$

(c) $CaCl_2$ Ca^{2+} $[Cl]^-$ $[Cl]^-$

(b) Na_2O Na^+ Na^+ $[O]^{2-}$

4 (a) 7; **(b)** (i) 1; (ii) 2

5 Na^+ is isoelectronic with Ne; Cl^- is isoelectronic with Ar

6 (a) K, $1s^22s^22p^63s^23p^64s^1 \rightarrow K^+$, $1s^22s^22p^63s^23p^6$ or K, $[Ar]4s^1 \rightarrow K^+$, [Ar]
F, $1s^22s^22p^5 \rightarrow F^-$, $1s^22s^22p^6$ or F, $[He]2s^22p^5 \rightarrow F^-$, [Ne];
(b) Mg, $1s^22s^22p^63s^2 \rightarrow Mg^{2+}$ $1s^22s^22p^6$; or Mg, $[Ne]3s^2 \rightarrow Mg^{2+}$, [Ne]

$2 \times Cl$, $1s^2 2s^2 2p^6 3s^2 3p^5 \rightarrow$
$2 \times Cl^-$, $1s^2 2s^2 2p^6 3s^2 3p^6$; or
$2 \times Cl$, $[Ne]3s^2 3p^5 \rightarrow 2 \times Cl^-$, [Ar];
(c) $2 \times Al$, $1s^2 2s^2 2p^6 3s^2 3p^1 \rightarrow$
$2 \times Al^{3+}$, $1s^2 2s^2 2p^6$; or
$2 \times Al$, $[Ne]3s^2 3p^1 \rightarrow 2 \times Al^{3+}$, [Ne]
$3 \times O$, $1s^2 2s^2 2p^4 \rightarrow 3 \times O^{2-}$,
$1s^2 2s^2 2p^6$; or
$3 \times O$, $[He]2s^2 2p^4 \rightarrow 3 \times O^{2-}$, [Ne]

7 It is exothermic: energy is transferred into the surroundings when a gaseous fluorine atom gains an electron

8

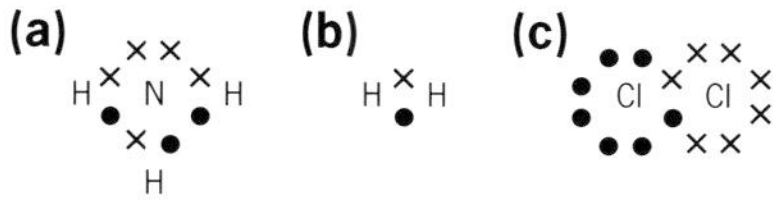

9

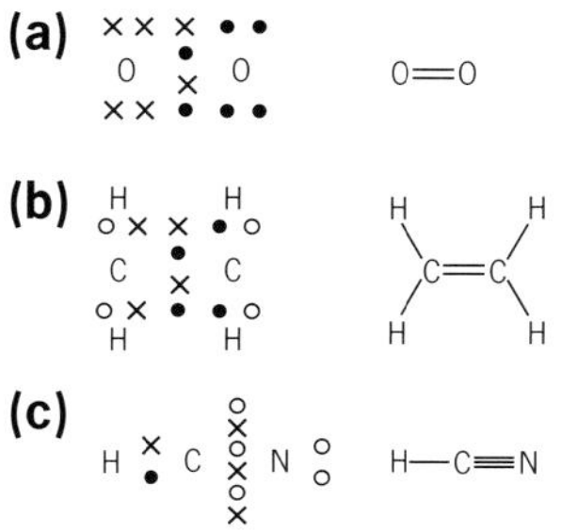

10

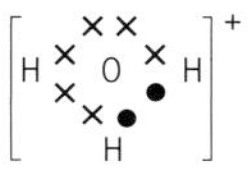

11

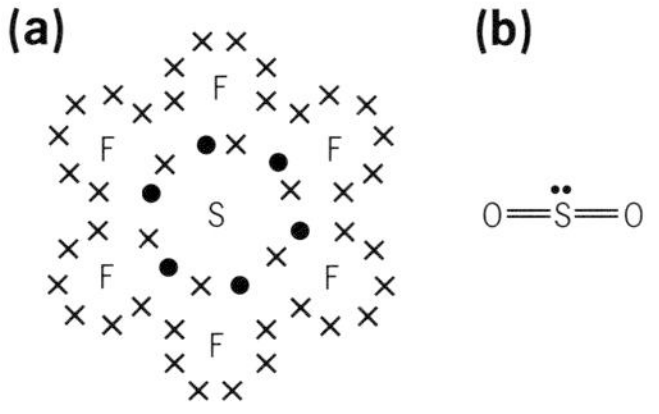

12

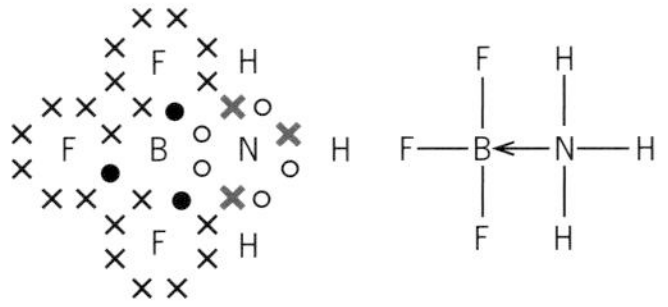

13 Outer shell of the carbon atom is surrounded by one single bond (a bonding pair of electrons) and a triple bond (which acts as one bonding pair of electrons) and no lone pairs. To minimise repulsion of the bonding pairs of electrons the electron pairs are arranged as far away from each other as possible, i.e. at an angle of 180°

14 Trigonal planar because the central carbon atom is surrounded by two bonding pairs of electrons from single bonds and a double bond pair which is counted as one region of electron density. Three 'bonding pairs' repel each other to minimise repulsion to give a trigonal planar shape

15

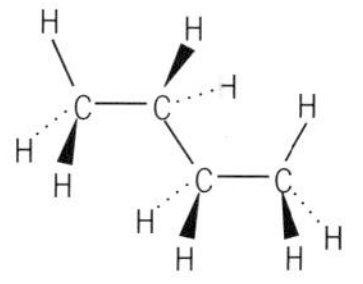

16 Tetrahedral (outer shell of aluminium surrounded by four bonding pairs which repel each other equally to minimise repulsion)

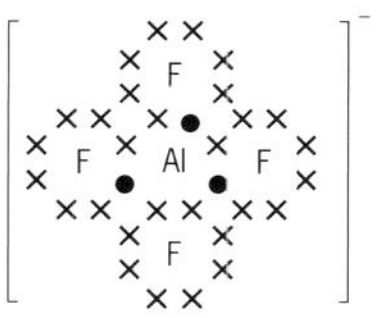

17 Beryllium ion surrounded by four bond pairs that equally repel each other to minimise repulsion and so the shape is tetrahedral

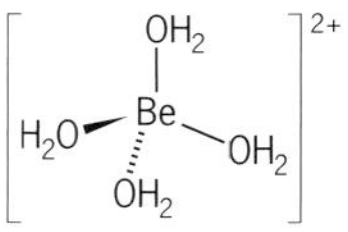

18 (a) Four bond pairs and no lone pairs

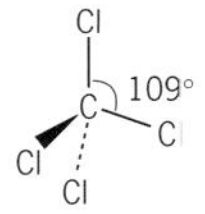

(b) Four bond pairs and no lone pairs

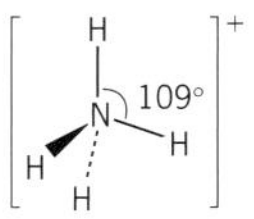

(c) Three bond pairs and one lone pair

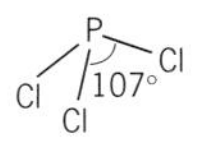

(d) V-shaped with a bond angle of about 120°. Two double bond pairs and one lone pair

19 sp^2 – because hybrid orbitals are formed from one s orbital and 2 p orbitals sp – because hybrid orbitals are formed from one s orbital and one p orbitals

20 F–H, O–H, N–H, Cl–C, H–P, and C–I, F–F; the first atom is slightly negative and the second atom is slightly positive – the last three bonds are not polar

21 +6; core charge remains constant but as the atomic number decreases so does the atomic radius and the number of inner shielding electron shells, so the nucleus has a greater attraction for the electron pair within a covalent bond

22 Aluminium chloride

CHAPTER 5

1 Either unreactive so found as elements in the Earth's crust rather than compounds; or found as ores easy to reduce using carbon

2 Discovery of radioactivity

3 Difference in A_r between the first two elements is the same or nearly the same as the difference between A_r of the second two elements; difference in atomic number between first two elements is the same as that between the second two elements

4 (a) They are in atomic number order;
(b) Argon (A_r = 40.0) and potassium (A_r = 39.1);
tellurium (A_r = 127.6) and iodine (A_r = 126.9)

5 Selenium

6 Rf after the scientist named Rutherford; Db after the place in Russia, Dubna, where research was carried out; Sg after the scientist named Seaborg; Bh after the scientist named Bohr; Hs after the scientist named Hass; Mt after the scientist named Meitner; Ds after the place in Germany called Darmstadt where research was carried out; Rg named after the scientist called Roentgen

7 N, $1s^2 2s^2 2p^3$;
Cl, $1s^2 2s^2 2p^6 3s^2 3p^5$;
Ca, $1s^2 2s^2 2p^6 3s^2 3p^6 4s^2$;
Ti, $1s^2 2s^2 2p^6 3s^2 3p^6 3d^2 4s^2$

8 (a) 7; **(b)** 3; **(c)** 4

9 (a) p; **(b)** p; **(c)** d; **(d)** d

10 (a) Increase; **(b)** Decrease

11 (a) (i) OH = 96 pm; (ii) NH = 100 pm;
(b) Atomic radii are determined from single bonds;
(c) Does not form He_2 so cannot measure a He–He bond

12 As the atomic number increases, even though the nuclear charge increases there are more shielding shells of electrons, so that the electrons in the outer shell are less firmly attracted towards the nucleus and are further away from the nucleus

13 **(a)** Greater nuclear charge; **(b)** (i) S has equal numbers of electrons and protons, S^{2-} has two more electron than protons so the outer electrons are less firmly attracted to nucleus; (ii) As the atomic number increases, atoms have same number of inner shell shielding electrons, but increased nuclear charge

14 Both have the same number of shielding shells of electrons, but sodium has a larger atomic radius and less protons in the nucleus, i.e. a smaller nuclear charge

15 No shielding shells of electrons, the smallest atomic radius and one more proton than the only other atom with no shielding shells

16 **(a)** The np sub-shell is higher in energy than the ns sub-shell, so that less energy is needed to remove an electron from an element in Group 3, which has one electron in an np orbital whereas a Group 2 element does not; from Group 3 to Group 4 there is no increase in shielding, but an increase in the nuclear charge so the outer electron is more firmly attracted to the nucleus with a Group 4 element; **(b)** Elements in Group 5 have one electron in each of the three outer p orbitals, so minimising electron–electron repulsion within an orbital, whereas in an element of Group 6 one outer p orbital has to have two electrons and so there is electron–electron repulsion in that orbital. A Group 7 element has an atom with the same number of shielding shells of electrons as the element in Group 6, but a greater nuclear charge, so the outer electrons are held more strongly

17 Has a full 3d sub-shell

18 Francium has the largest atomic radius and the most number of shielding shells of all the Group 1 elements, so it has the lowest first-ionisation energy and thus a francium atom can lose electrons very easily

19 **(a)** Oxidation; **(b)** Oxidation; **(c)** Reduction; **(d)** Neither; **(e)** Oxidation; **(f)** Oxidation; **(g)** Reduction

20 **(a)** $Ca \rightarrow Ca^{2+} + 2e^-$, $Cl_2 + 2e^- \rightarrow 2Cl^-$; **(b)** $Mg \rightarrow Mg^{2+} + 2e^-$, $F_2 + 2e^- \rightarrow 2F^-$; **(c)** $Mg \rightarrow Mg^{2+} + 2e^-$, $O_2 + 4e^- \rightarrow 2O^{2-}$; **(d)** $Na \rightarrow Na^+ + e^-$, $N_2 + 6e^- \rightarrow 2N^{3-}$

21 **(a)** $Al \rightarrow Al^{3+} + 3e^-$, $O_2 + 4e^- \rightarrow 2O^{2-}$, $4Al + 3O_2 \rightarrow 4Al^{3+} + 6O^{2-}$; **(b)** (i) $Zn + 2Fe^{3+} \rightarrow Zn^{2+} + 2Fe^{2+}$; (ii) $MnO_4^- + 8H^+ + 5Fe^{2+} \rightarrow Mn^{2+} + 4H_2O + 5Fe^{3+}$; (iii) $Cr_2O_7^{2-} + 14H^+ + 6Fe^{2+} \rightarrow 2Cr^{3+} + 6Fe^{3+} + 7H_2O$

22 **(a)** Al is +3 and O is –2; **(b)** Ca is +2 and Cl is –1; **(c)** Mg is +2 and N is –3; **(d)** Cu is +2 and Cl is –1; **(e)** K is +1 and S is –2; **(f)** Ba is +2 and F is –1

23 **(a)**

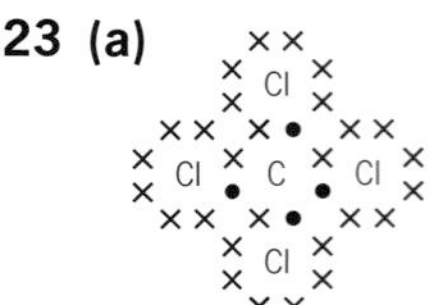

(b) Chlorine; **(c)** C is +4 and Cl is –1

24 **(a)** H is +1, O is –2 and N is +5; **(b)** (i) H is +1, O is –2 and Cl is +5; (ii) It is the oxidation number of the chlorine

25 **(a)** K is +1, S is +6 and O is –2; **(b)** K is +1, S is +4 and O is –2; **(c)** +2; **(d)** –1; **(e)** +2; **(f)** +2.5

26 **(a)** +6; **(b)** Ferrate(VI); **(c)** Mn is +7 in MnO_4^- and Mn is +6 in MnO_4^{2-}; **(d)** (i) +2; (ii) –3; (iii) +2

27 **(a)** Oxygen; **(b)** Carbon

28 Sodium atoms lose electrons (oxidation) to form sodium ions; the electrons lost are gained by hydrogen molecules (reduction) to form hydride ions

29 **(a)** (i) Cl in Cl_2 changes from 0 to –1, Cl in PCl_3 has no change, P changes from +3 to +5; (ii) Cl_2; **(b)** (i) Cu from +2 to +1, I from –1 to 0, oxidising agent is Cu^{2+}, reducing agent is I^-; (ii) Mn from +4 to +2, Cl from –1 to 0, oxidising agent is MnO_2, reducing agent is HCl; (iii) Cu from 0 to +2, O from 0 to –2, oxidising agent is O_2, reducing agent is Cu; (iv) Cl from 0 to +1 and to –1, oxidising agent is Cl_2, reducing agent is Cl_2

30 Oxidation state of Cl in Cl_2 is 0, in NaCl is –1 and in $NaClO_3$ is +5; chlorine, Cl_2, is oxidized to give $NaClO_3$ and reduced to give NaCl; simultaneous oxidation and reduction of the same element is disproportionation

31 Selenium is a non-metal, forms Se^{2-}, forms an acidic oxide, will be a solid at room temperature and pressure, reacts with metals to form ionic compounds, is less reactive than sulfur, is a poor oxidising agent, is a poor conductor of heat and electricity, and has a hydride with the formula H_2Se

CHAPTER 6

1 $10Na(s) + 2KNO_3(s) \rightarrow 5Na_2O(s) + K_2O(s) + N_2(g)$

2 2:1:2 ($H_2:O_2:H_2O(g)$)

3 Using Avogadro's hypothesis, 1 molecule of hydrogen reacts with $\frac{1}{2}$ molecule of oxygen to make one molecule of water; assuming both water and hydrogen are diatomic, the formula of water must be H_2O

4 25 °C

5 0.25 mole

6 1.7×10^3 dm^3 (to 2 sig. fig.)

7 29 dm^3

8 25 °C (298 K) and 101 kPa

9 2 dm^3

10 100 cm^3

11 **(a)** 298 K; **(b)** 373 K; **(c)** 223 K

12 63.7 cm^3

13 0.0224 m^3

14 **(a)** 4.9×10^4 cm^3; **(b)** 3.1×10^4 dm^3

15 146

16 So that it can be subtracted from the volume of the gaseous compound under test, otherwise this volume will include the small volume of air

17 46

18 **(a)** There is no change in the mean kinetic energy of the particles. Particles will be more spread out, so that fewer particles collide per unit area of container wall per second, i.e. pressure decreases; **(b)** 2.41×10^{22}; **(c)** If the amount of gas increases, the number of particles increases, and so to keep the number of particles that collide with the container wall per second constant, the volume of gas will have to increase to lengthen the distance between particles. If the amount doubles then the volume will need to double to keep the particles the same distance apart and thus keep the number of collisions per second the same

19 **(a)** Steam is a gas, so particles are very spread out with lots of space between them, and when it condenses the particles are close together with only very little space between them; **(b)** Lots of space between particles in a gas, but little space between particles in a liquid; **(c)** More particles per unit volume in a liquid, since the particles are much closer together than in a gas; **(d)** Particles in a liquid are free to move and are not in fixed positions; **(e)** Particles in a solid are not free to move, but particles in gases and liquids are free to move

20 ΔH_m is much smaller than ΔH_b

21 **(a)** Ethanol; **(b)** Force between water molecules stronger than between ethanol molecules

22 **(a)** 190 to 210 K; **(b)** –83 to –63°C

23

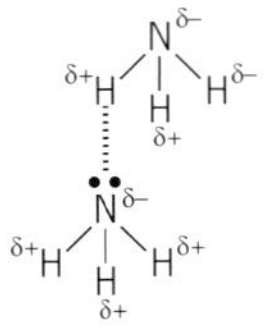

$^{\delta+}H$–$N^{\delta-}$(–$H^{\delta+}$)–$H^{\delta-}$ ······ $N^{\delta-}$ (with $^{\delta+}H$, $H^{\delta+}$, $H^{\delta+}$)

24 Suggestions could include: much less liquid water present in the oceans, so fewer ocean currents such as the Gulf Stream; weather may be drier since less water is available to be evaporated; there may not be a polar ice-cap at the North Pole, since it will sink under the surface of the water once formed; possibly warmer weather as a result since ice reflects the Sun's energy back into space; because ice floats on water it insulates the water underneath. This has a surprising consequence since in winter the very cold polar air would be warmed significantly by the energy from the warmer ocean. This warming could be as much as 20-40 °C. This would have an effect on the surface temperatures at the poles melting more ice; sea-water may become saltier and so it would be less likely that marine life could survive; if the oceans froze from the bottom up then life would only be possible only in the top few metres of seawater that were unfrozen by the Sun's rays; in freshwater lakes the densest water is at 4 °C, when surface water reaches this temperature it sinks taking dissolved oxygen from the surface down to the bottom of the lakes enabling life to continue

25 **(a)** (i) 0.2 mole; (ii) 0.125 mole; (iii) 1 mole; (iv) 5×10^{-3} mole; (v) 2.5×10^{-3} mole; **(b)** (i) 1.6 mol dm^{-3}; (ii) 1.0 mol dm^{-3}

26 **(a)** 159.6 g; **(b)** 26.5 g; **(c)** 3.5 g

27 0.951 mol dm^{-3} and 57.06 g/cm^3

28 **(a)** Increase; **(b)** Decrease; **(c)** Decrease; **(d)** Increase; **(e)** Increase; **(f)** Decrease

CHAPTER 7

1 Particles of a liquid are in random motion, collide with one another and with the container wall, are not in an ordered pattern, are attracted to one another, but less so than in a solid, are moving faster than in a solid and slower than in a gas and the distance between particles is very small; particles of a gas are in random motion, collide with one another and with the container wall, are not in an ordered pattern, are only weakly attracted towards each other, are moving faster than in a liquid and the distance between particles is large

2 Particles of a solid gain energy and vibrate more energetically until eventually the inter-particle force is broken and the particles become free to move around; as soon as the substance has melted the particles will move in any direction (random motion), but they will still be close together

3 **(a)** $Zn(s) + 2Ag^+(aq) \rightarrow Zn^{2+}(aq) + 2Ag(s)$; **(b)** $Zn(s) + Pb^{2+}(aq) \rightarrow Zn^{2+}(aq) + Pb(s)$

4 **(a)** 12; **(b)** 12; **(c)** 8

5 Stronger metallic bonding because of electrostatic attraction between two moles of delocalised electrons for each mole of Mg^{2+}, i.e. Mg^{2+} higher charge of cation, smaller ionic radius and more delocalised electrons

6 Aluminium has 3 moles of electrons per mole of metal compared to 2 moles for Mg and 1 mole for Na

7 Carbon dioxide

8 **(a)** The layers of carbon atoms can easily slide over each other because of the weak van der Waals forces between each layer; **(b)** Has a very high melting point; **(c)** Diamond, since each carbon atom is bonded covalently to four other carbon atoms, but in graphite a carbon atom is only covalently bonded to three others

9 **(a)** Silicon is a giant molecule with each silicon atom bonded covalently to four other silicon atoms; covalent bonds are much stronger than intermolecular attractions so that a large amount of energy is needed to break the many covalent bonds; **(b)** (i) Giant molecular structure;

(ii) Si bonded to four C atoms (C, C, C, C)

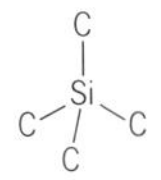

10 As the temperature is increased, more thermal energy is available so that a greater fraction of the electrons can be promoted to an excited energy level

11 **(a)** Si has four electrons in its outer shell and As has five electrons; **(b)** Each As atom will provide one extra electron, since only four electrons are needed for bonding; these extra electrons are easier to excite

12 **(a)** $1s^22s^22p^1$; **(b)** Each B atom will provide one electron hole since it has one less electron in its outer shell than Si

13 Astatine, since it has a larger molecule with more electrons than the other halogens, so it is more likely to have an asymmetric distribution of electrons and hence there are stronger induced dipole–induced dipole attractions between molecules

14 All have weak intermolecular forces that are induced dipole–induced dipole attractions, but P_4 and S_8 have larger molecules and more electrons than N_2 and O_2, so an asymmetric distribution of electrons is easier in P_4 and S_8 and thus the intermolecular forces in P_4 and S_8 are stronger and they have much higher melting points and boiling points than N_2 and O_2

15 Radon

CHAPTER 8

1 The energy released cannot be used to do useful work

2 The enthalpy change of combustion will be different if gaseous water is produced, and also under standard conditions water must be a liquid

3

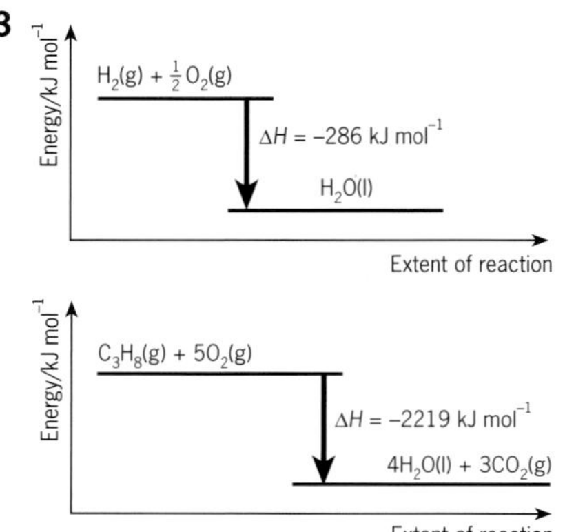

4 **(a)** 4.2 kJ/4200 J; **(b)** 84 kJ/84 000 J

5 **(a)** Metal is a much better conductor of heat than glass; **(b)** Temperature rise is a difference in temperature and, since absolute temperature is temperature in °C + 273, then the difference in temperature on both scales is identical; **(c)** Use a lid, calculate the energy absorbed by the metal calorimeter using the specific heat capacity of the metal, have the calorimeter at the optimum height above the burner

6 Incomplete combustion; loss of energy to the surroundings; no account taken of the energy absorbed by the calorimeter itself; experiment not carried out under standard conditions; possible loss of liquid fuel through evaporation

7 **(a)** (i) Lower value; (ii) Weigh inside the bomb calorimeter so that no vapour can escape; **(b)** (i) Insulated container which prevents energy loss by conduction from the water into the surroundings, water in the calorimeter completely surrounds the chamber where the fuel is burnt so almost all the energy will be transferred into the water; (ii) To ensure that the temperature of the water is constant all through the calorimeter

8 51.7 kJ mol^{-1}

9 N_2 has an extremely high bond energy since it has a triple covalent bond

10 **(a)** H–H and O=O since they are the bonds in the molecules H_2 and O_2, which are necessarily identical, C=O figure refers to the C=O bond in carbon dioxide and these must also be identical; **(b)** +2061 kJ mol^{-1}; **(c)** Bond energies are quoted for one mole of gaseous bonds so that all the reactants and products must be gases; if methanol is not a gas in the calculation then the enthalpy change for $CH_3OH(l)$ to $CH_3OH(g)$ must be accounted for

11 **(a)** By experiment $\Delta H = -286$ kJ mol^{-1} and by calculation using bond energies it is $\Delta H = -243$ kJ mol^{-1} (the calculation uses average O–H bond energy, not the bond energy for O–H in H_2O); **(b)** The energy or enthalpy of the products is higher than the energy or enthalpy of the reactants

12 –658 kJ mol^{-1}

13 **(a)** The fire will not be put out, and often it will become worse because more gaseous petrol is produced; **(b)** Methanol has a low energy-density value

14 **(a)** $C_8H_{18}(l) + 12\frac{1}{2}O_2(g) \rightarrow 8CO_2(g) + 9H_2O(l)$; **(b)** Per mole of C_8H_{18} 16 moles of C=O and 18 moles of O–H

15 **(a)** 50 432 kJ kg^{-1}; **(b)** (i) 12 000 dm^3; (ii) 1.35 litres

16 **(a)**

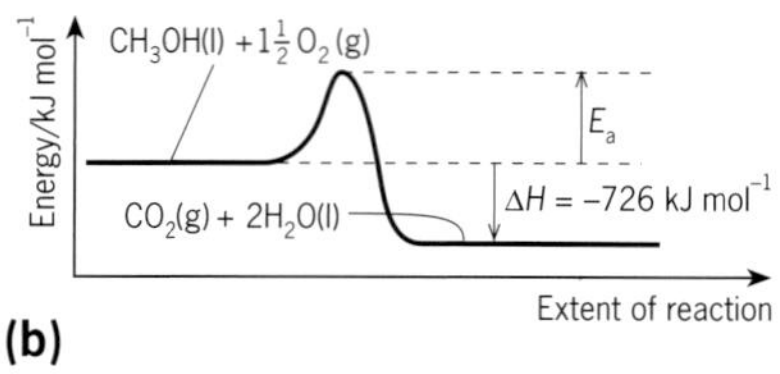

Energy/kJ mol^{-1}; $CH_3OH(l) + 1\frac{1}{2}O_2(g)$; E_a; $CO_2(g) + 2H_2O(l)$; $\Delta H = -726$ kJ mol^{-1}; Extent of reaction

(b)

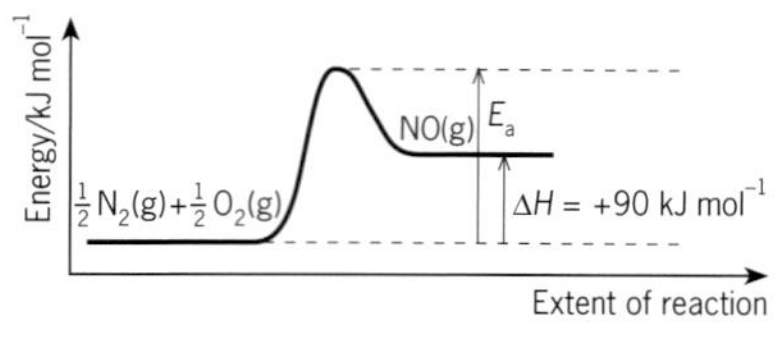

Energy/kJ mol^{-1}; $\frac{1}{2}N_2(g) + \frac{1}{2}O_2(g)$; $NO(g)$; E_a; $\Delta H = +90$ kJ mol^{-1}; Extent of reaction

(c) It is not exothermic and it does not involve the reaction of one mole of N_2

17 **(a)** $C(s) + O_2(g) \rightarrow CO_2(g)$; **(b)** $C(s) + 2H_2(g) \rightarrow CH_4(g)$; **(c)** Both transfer energy to the surroundings and so are more stable than the elements that make them up

18 The standard state of carbon is graphite since it is energetically more stable than diamond

19 **(a)** –104 kJ mol^{-1}; **(b)** –110 kJ mol^{-1}; **(c)** It is impossible to react carbon with oxygen without at least making some carbon dioxide

20 **(a)** Enthalpy change of formation of N_2 is 0 since it is an element; **(b)** –129 kJ mol^{-1}

CHAPTER 9

1 Methane, CH_4

2 Loss of four electrons from a carbon atom to form C^{4+} requires too much energy because as more electrons are lost the carbon ion gets smaller so the remaining outer electrons are more strongly attracted to the nucleus. Gain of four electrons to form C^{4-} also requires too much energy because of the repulsion an electron will experience from the negative carbon ions

3 Methane Water

4 Dative covalent bond/coordinate bond

5 C_8H_{18}

6 Ethane

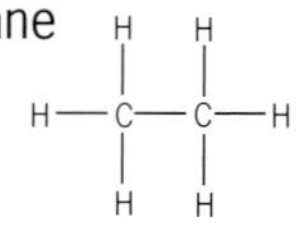

Pentane

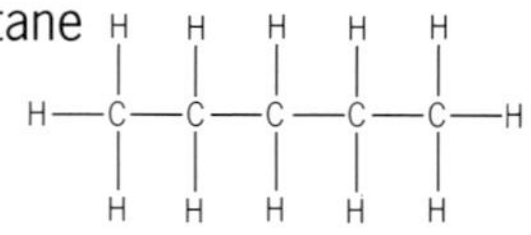

7 **(a)** C_7H_{14} and C_9H_{18} are unsaturated

(b)

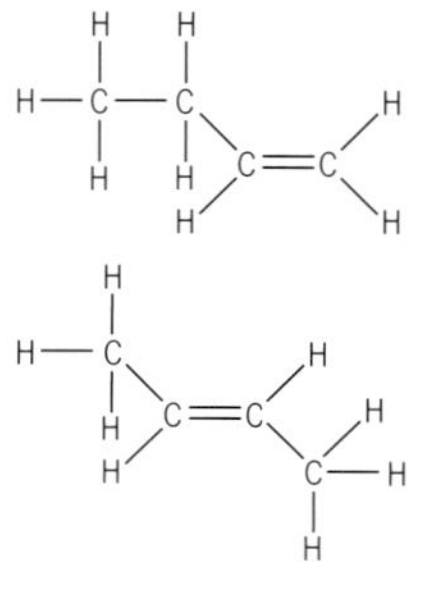

Both have the same molecular formula

8 (a)

```
    H   H
    |   |
H—C—C—H
    |   |
H—C—C—H
    |   |
    H   H
```

(b) Same molecular formula

9 (a) (zigzag skeletal formula)

(b) (skeletal formula)

(c) (skeletal formula) or (skeletal formula)

10 (i)

```
    H   H   H   H   H
    |   |   |   |   |
H—C—C—C—C—C—H
    |   |   |   |   |
    H   H   H   H   H
```

```
    CH2    CH2
   /   \  /   \
CH3    CH2    CH3
```

(ii)

```
    H   H   H
    |   |   |
H—C—C—C—H        CH3—CH—CH3
    |   |   |               |
    H   |   H              CH3
     H—C—H
        |
        H
```

(iii)

```
    H   H
    |   |              H
H—C—C               /
    |   |  \ C=C
    H   H  /        \
        H             H
```

CH_3—CH_2—CH=CH_2

(iv)

```
    H   H
    |   |
H—C—C—H     CH2—CH2
    |   |        |      |
H—C—C—H     CH2—CH2
    |   |
    H   H
```

11 (a) The free rotation about C–C bonds means this is an alternative way to draw butane; **(b)** This has a molecular formula of C_5H_{12} rather than C_4H_{10}

12 $CH_3CH_2CH_2CH_2CH_2CH_3$; $(CH_3)_2CHCH_2CH_2CH_3$; $CH_3CH_2CH(CH_3)CH_2CH_3$; $CH_3CH(CH_3)CH(CH_3)CH_3$; $CH_3C(CH_3)_2CH_2CH_3$

13 (a) C_4H_{10}; **(b)** It has molecular formula C_4H_6 not C_4H_{10}

14 (a)

```
    H       H            H   H
    |       |            |   |
H—C—O—C—H     H—C—C—O—H
    |       |            |   |
    H       H            H   H
```

(b) Alcohol

15 Isomers

16 C_9H_{20}

17 (a) Kerosene and/or diesel oil; **(b)** (i) $C_{12}H_{26} \rightarrow CH_3(CH_2)_8CH_3 + C_2H_4$; (ii) $C_{12}H_{26} \rightarrow CH_3CH(CH_3)(CH_2)_5CH_3 + C_3H_6$

18 On breaking a C–C bond both carbon atoms need to form an extra covalent bond, but there are no extra hydrogen atoms available, so a carbon=carbon double bond is formed to provide the hydrogen atoms

19 (a) Hexane, cyclohexane, benzene. $CH_3(CH_2)_4CH_3 \rightarrow C_6H_{12} + H_2$; $C_6H_{12} \rightarrow C_6H_6 + 3H_2$; **(b)** Hydrogen

20 (a) Because the molecular formula of the product is the same as the reactant, but they have different structures so they are isomers; **(b)** Unbranched hydrocarbons fit into a zeolite pore, but branched hydrocarbons do not

CHAPTER 10

1 $C_{11}H_{24}$, $CH_3CH_2CH_2CH_2CH_2CH_2CH_2CH_2CH_2CH_2CH_3$; $C_{12}H_{26}$, $CH_3CH_2CH_2CH_2CH_2CH_2CH_2CH_2CH_2CH_2CH_2CH_3$

2 (a) $CH_3CH_2CH_2CH_2CH_2CH_2CH_2CH_2CH_3$

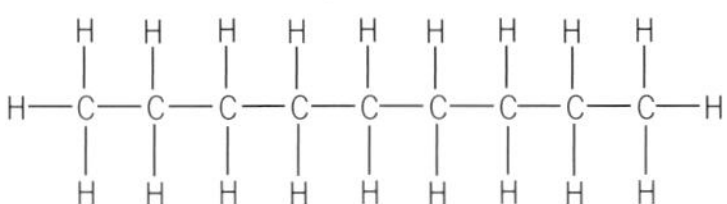

(b) Decane

3 (a) Pentyl; **(b)** C_nH_{2n+1}

4 $CH_3CH_2CH_2CH_3$

5 (a) It is another way of drawing methylbutane; **(b)** The methyl group can only be on carbon 2; if it were on carbon 1 then the molecule would be pentane

6 $CH_3C(CH_3)_2CH(C_2H_5)CH_2CH_2CH_3$; $CH_3CH_2CH_2CH(C_3H_7)CH_2CH_2CH_3$

7 Two

8 (a) Methane, ethane, propane and butane; **(b)** (i) Boiling point is about 216 °C and density is about 0.748 g cm^{-3}; (ii) 18

9 The δ + end of the molecule and the δ – end are quite large, so there is a fairly strong electrostatic intermolecular attraction

10 As the atomic number increases a noble gas atom has more electrons and a greater atomic radius, so that the formation of an instantaneous dipole is easier and thus the strength of the induced dipole–induced dipole intermolecular attractive force increases

11 (a) It is hotter in Spain so the petrol is more volatile, and as temperature increases vapour pressure increases; **(b)** Use branched-chain alkanes

12 (a) Exothermic; **(b)** Spontaneous reactions are ones that have a tendency to occur (and the entropy change is positive), activation energy is the minimum energy per mole of particles needed for a collision between particles to lead to a reaction; **(c)** The reaction of alkanes and oxygen is exothermic so alkanes and oxygen have more enthalpy than carbon dioxide and water (energetically unstable), but the activation energy is sufficiently high that at room temperature the rate of reaction between petrol and oxygen is very slow (kinetically stable); **(d)** Reacts with oxygen to release lots of energy (large value of energy released per gram of fuel), easily stored, easily transported, non-toxic; readily available

13 (a) $C_nH_{2n+1}OH$; **(b)** Same chemical properties, and physical properties vary with increasing M_r (boiling point, melting point and density all increase with increasing M_r)

14 $C_{10}H_{22} + 15\frac{1}{2}O_2 \rightarrow 10CO_2 + 11H_2O$

15 (a) Visible light; **(b)** Absorbed by coloured substances and by gases in the atmosphere

16 Plants that can photosynthesise using up carbon dioxide are not present on Mars and Venus

17 (a) In summer months, during which green plants grow very fast, there is a high rate of photosynthesis that uses up carbon dioxide; **(b)** In winter months there is an increase in the mass of fuel burnt and the rate of growth by green plants is low, so not much photosynthesis occurs

18 (a) EU nations agreed in Kyoto in December 1997 to reduce emissions of greenhouse gases to that of 1990 by the year 2012. This is about an 8%

reduction per annum; **(b)** (i) 700 g; (ii) $C_8H_{18} + 12\frac{1}{2}O_2 \rightarrow 8CO_2 + 9H_2O$; (iii) 2160 g; (iv) 8.6×10^{11} g

19 **(a)** The catalyst is a solid but the reactants are gases, so a large surface area is required to maximize the number of collisions between reactant and catalyst, and also the catalyst works by adsorption (surface action) so it needs a large surface area; **(b)** This gives an even more larger surface area

20 At the start of a journey when the engine is cold or when the car is stationary

21 **(a)** $C_6H_6 + 7\frac{1}{2}O_2 \rightarrow 6CO_2 + 3H_2O$; **(b)** Greenhouse effect

22 $N_2 + O_2 \rightarrow 2NO$

23 Greenhouse effect

24 NO_2 is regenerated at the end of the reaction steps, which speeds up the rate of reaction and the mechanism of the oxidation of SO_2 is changed

25 $1s^22s^22p^4$, when occupying the three 2p atomic orbitals the electrons are arranged so as to minimise the electron–electron repulsion that occurs if two electrons are in the same orbital, so only one orbital, the $2p_x$, is occupied by two electrons while the other two, $2p_y$ and $2p_z$, have one electron in each, and so it has two unpaired electrons

CHAPTER 11

1 Organic compounds that contain mostly carbon and hydrogen are more likely to be completely combusted, and the amounts of products of the combustion, H_2O and CO_2,can be easily determined

2 For ethane, CH_3, and for ethanoic acid, CH_2O

3 **(a)** C_5H_{11}; **(b)** $C_2H_5O_2N$

4 CH_2O

5 **(a)** C_6H_6; **(b)** CH_2O_2; **(c)** $C_8H_{10}N_4O_2$

6 Prevents collisions between positive ions and other particles, and avoids undue contamination of the sample with air, since the molecules in air will also be ionised by electron bombardment

7 72.8

8 CH_2O_2

9 **(a)** Since all the particles have a charge of +1 then $m/e = m$ since $e = 1$; **(b)** $C_2H_4^+$; **(c)** $m/e = 43$, $C_3H_7^+$

10 **(a)** If a molecule with all paired electrons loses one electron then it must end up with one unpaired electron, indicated by a dot for the a free radical and a '+' for the positive ion, but simplified it is important to show the '+' and so $C_4H_{10}^+$ is also used; **(b)** Only positive particles are detected by the mass spectrometer and $CH_3\bullet$ is neutral

11 **(a)** CO or C_2H_4; **(b)** $CH_3CH_2O\bullet$; **(c)** $C_6H_5\bullet$

12 **(a)** $OH\bullet$; **(b)** $m/e = 45$ is CH_2O+, $m/e = 29$ is CHO^+, $m/e = 28$ is CO^+
(c)

13 **(a)** 10; **(b)** (i) $m/e = 49$ is $CH_2{}^{35}Cl^+$ and $m/e = 51$ is $CH_2{}^{37}Cl^+$; (ii) $m/e = 64$ is $CH_3CH_2{}^{35}Cl^+$ and $m/e = 66$ is $CH_3CH_2{}^{37}Cl^+$; (iii) $CH_3\bullet$; **(c)** m/e values are shown next to the peaks:

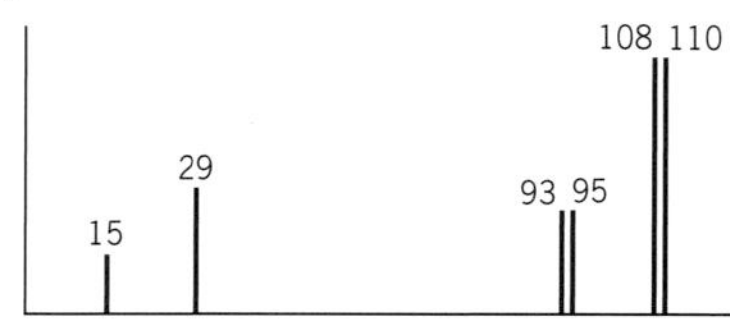

(d) (i) $m/e = 74$ is $(^{37}Cl_2)^+$, $m/e = 72$ is $(^{35}Cl^{37}Cl)^+$, $m/e = 70$ is $(^{35}Cl_2)^+$; (ii) ratio of $^{35}Cl:^{37}Cl$ is 3:1 so the ratio of $(^{35}Cl_2)^+:(^{35}Cl^{37}Cl)^+:(^{37}Cl_2)^+$ is 9:6:1 so the peak at $m/e = 74$ is the smallest

14 **(a)** H_2 does not have a dipole moment even during vibrations; **(b)** Only one vibration is possible with a diatomic molecule a stretch; **(c)** Bond energy of HCl is greater than that of HI

15 A is the hydrogen bond in O–H stretch, and B is the C–O stretch

16 **(a)** O–H; **(b)** Confusion between the stretching frequency for O–H in water and in ethanol; **(c)** Propanone also contains C–H bonds so that the C–H stretching frequency at 3000 cm^{-1} can result from ethanol or propanone; **(d)** 0.02%

17 **(a)** Mass numbers are all odd; **(b)** Fewer ^{13}C nuclei in molecule

18 Proton magnetic resonance

19 **(a)** 1; **(b)** 1; **(c)** 2

20 Signal at $\delta = 2.1$ (will be a doublet) integrated trace three protons, and signal at $\delta = 9.7$ (will be a quartet) integrated trace one proton

21 CHO protons have four arrangements and CH_3 protons have two peaks

22 Compound Y

23 **(a)** Pencil will not dissolve in any solvent used; **(b)** So that as the solvent moves up the chromatography paper it will not evaporate; **(c)** 0.15 and 0.50; **(d)** It is not soluble in the mobile phase

24 Adsorption

25 **(a)** The liquid must not boil when the GLC column is in a hot oven; **(b)** GLC works by the difference in solubility between the liquid and the carrier gas.

26 Since retention time is dependent on the flow rate, temperature, packing of the column, and length of the column, any data-base used must have the same GLC characteristics as used in the chromatographic separation

27 Adsorption

CHAPTER 12

1 Five atoms (four Cl and one I)

2 **(a)** C is 1,2-dibromo-1,2-dichloroethane and D is 1,1-dibromo-1,2-dichloroethane; **(b)** (i)

(ii)

(iii)

(iv)

(v)

3 **(a)** Perfluorobutane or decafluorobutane; **(b)** 2-Chlorobutane; **(c)** Perchlorohexane

4 (a) (i) CF_3Cl; (ii) $C_2F_2Cl_4$;
(b) Tetrachlorodifluoroethane

5 This property means that they can smother and prevent oxygen from reaching a flame in a fire

6 (a) 47–60 °C; **(b)** Stronger intermolecular attraction in 1-chlorobutane; **(c)** Isomers that have branched chains are less able to pack efficiently with one another and have less surface contact. So overall there is a weaker intermolecular attraction than in an unbranched isomer.

7

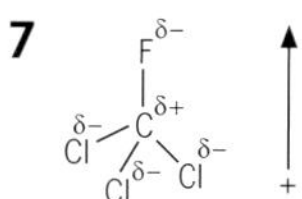

8 CH_3● is the only free radical

Cl^- Cl^+

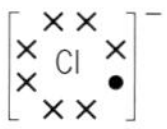

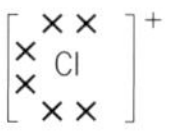

BF_3 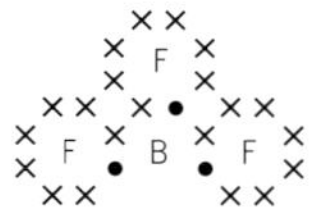 CH_3

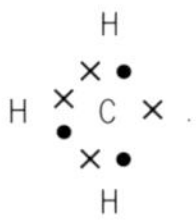

9 (a) 79 kJ mol^{-1}; **(b)** Fluorine

10 There is no net change when a methyl free radical collides with a methane molecule, CH_3● + CH_4 → CH_4 + CH_3●

11 (a) Have a small mole ratio of Cl_2:CH_4 and carry out chlorination over a short time period; **(b)** Have a high mole ratio Cl_2:CH_4 and carry out reaction over a long time period

12 Any two reactions that involve two free radicals present in the reaction mixture, for example,
Cl● + CH_3● → CH_3Cl and
Cl● + $CHCl_2$● → $CHCl_3$

13 Initiation step, Cl_2 → 2Cl●
Propagation steps:
Cl● + C_2H_6 → C_2H_5● + HCl
C_2H_5● + Cl_2 → C_2H_5Cl + Cl●
Cl● + C_2H_5Cl → C_2H_4Cl● + HCl
C_2H_4Cl● + Cl_2 → $C_2H_4Cl_2$ + Cl●
Cl● + $C_2H_4Cl_2$ → $C_2H_3Cl_2$● + HCl
$C_2H_3Cl_2$● + Cl_2 → $C_2H_3Cl_3$ + Cl●
Cl● + $C_2H_3Cl_3$ → $C_2H_2Cl_3$● + HCl
$C_2H_2Cl_3$● + Cl_2 → $C_2H_2Cl_4$ + Cl●
Cl● + $C_2H_2Cl_4$ → C_2HCl_4● + HCl
C_2HCl_4● + Cl_2 → C_2HCl_5 + Cl●
Cl● + C_2HCl_5 → C_2Cl_5● + HCl
C_2Cl_5● + Cl_2 → C_2Cl_6 + Cl●
Termination steps:
Cl● + C_2H_5● → C_2H_5Cl
C_2H_5● + C_2H_5● → C_4H_{10}

14 Initiation step: F_2 → 2F●
Propagation steps:
F● + CH_4 → CH_3● + HF
CH_3● + F_2 → CH_3F + F●
F● + CH_3F → CH_2F● + HF
CH_2F● + F_2 → CH_2F_2 + F●
F● + CH_2F_2 → CHF_2● + HF
CHF_2● + F_2 → CHF_3 + F●
F● + CHF_3 → CF_3● + HF
CF_3● + F_2 → CF_4 + F●
Termination steps:
F● + CH_3● → CH_3F
CH_3● + CH_3● → C_2H_6

15 1-bromohexane

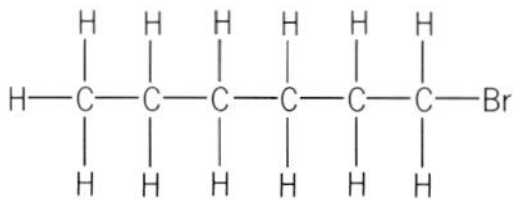

2-bromohexane

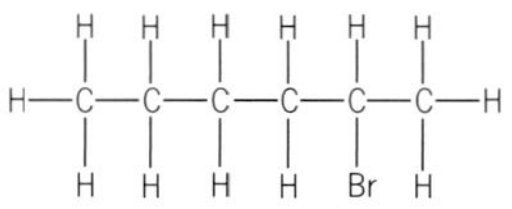

3-bromohexane

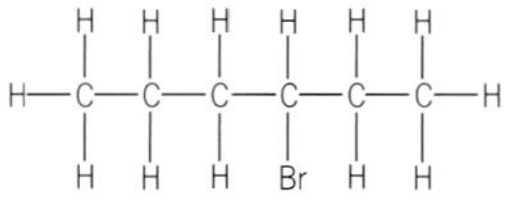

16 Bubble chlorine into refluxing methylbenzene in the presence of ultraviolet light over a long period of time or excess chlorine over a long time,
$C_6H_5CH_3$ + $3Cl_2$ → $C_6H_5CCl_3$ + 3HCl

17 (a) 2-iodo-3-methylhexane;
(b) (i) Ethanol, red phosphorus and iodine; (ii) Hexan-3-ol, sodium bromide and concentrated sulfuric acid, or hexan-3-ol, concentrated hydrobromic acid and concentrated sulfuric acid, or hexan-3-ol and phosphorus(III) bromide

18 Fluoromethane

19 (a) NH_2^- nucleophile

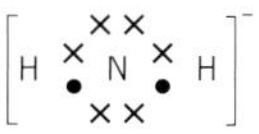

(b) CN^- nucleophile

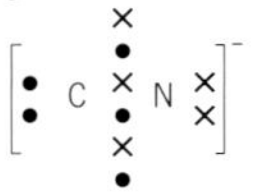

(c) BCl_3 not a nucleophile

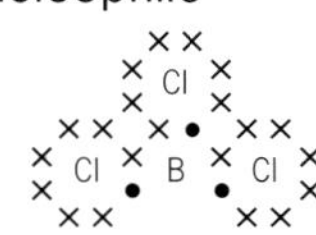

20 (a) +90 kJ mol^{-1}; **(b)** –80 kJ mol^{-1}

21 Pentan-2-ol

22 (a)

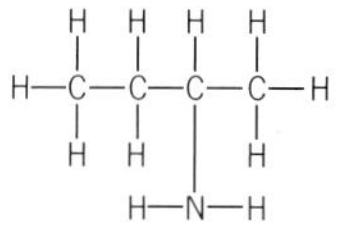

(b)

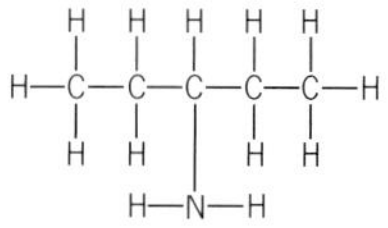

(c)

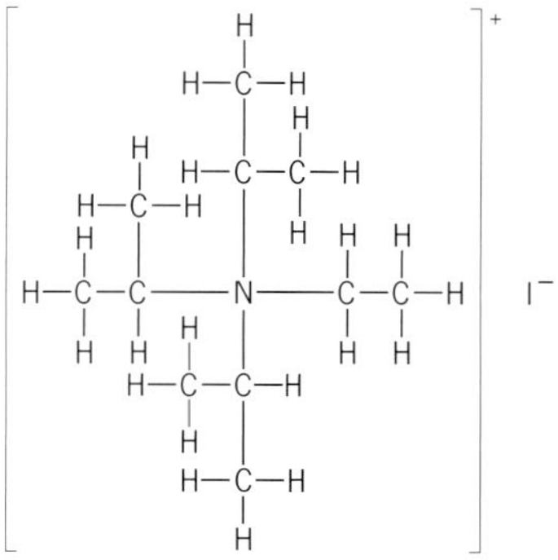

23

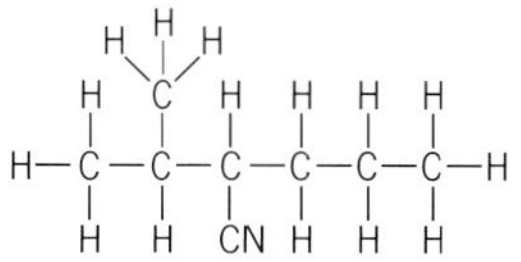

24 (a) Substitution:
$CH_3CH_2CH_2I + NaOH \rightarrow CH_3CH_2CH_2OH + NaI$
Elimination:
$CH_3CH_2CH_2I + NaOH \rightarrow CH_3CH{=}CH_2 + NaI + H_2O$

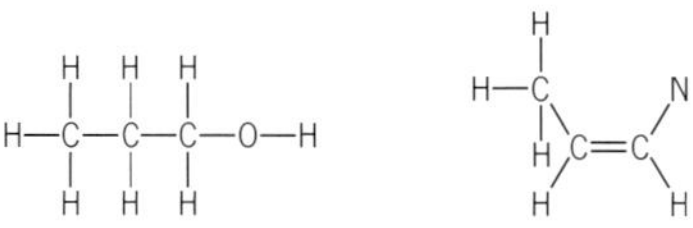

(b) Pent-1-ene

25

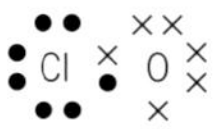

(outer electrons shown only)

26 A homogeneous catalyst

CHAPTER 13

1

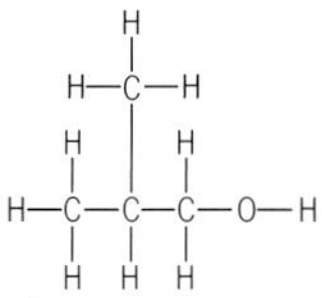

(a) 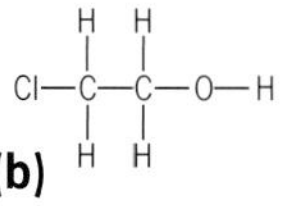 **(b)** 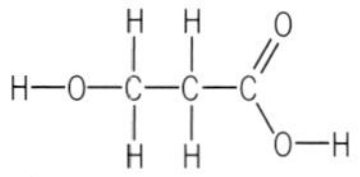 **(c)**

2 **(a)**

primary (i) secondary (ii)

primary and secondary (iii) secondary (iv)

(b) Geraniol is primary, nerol is primary, linalool is tertiary

3 **(a)** Dihydric

(b) and

and

(c) Geraniol is monohydric, nerol is monohydric, linalool is monohydric, maltose is polyhydric

4 97 °C

5 **(a)** Both have strong intermolecular hydrogen bonding, but butan-1-ol is a larger molecule with more electrons so it has stronger van der Waals forces as well; **(b)** Decane and decan-1-ol are structurally similar since they both have a long-chain alkyl group that is non-polar. As a result both have van der Waals forces of approximately the same strength. The extra hydrogen bonding in decan-1-ol is not significant, since the molecules are large. In methane the intermolecular force is a weak van der Waals force, since it is a small molecule, but in methanol it forms strong hydrogen bonding

6 Nonan-1-ol is more soluble in an alkane solvent than water because it has a very long non-polar alkyl group that can form van der Waals intermolecular bonds with the solvent molecules; methanol is a polar molecule and cannot easily form van der Waals forces – it forms hydrogen bonds with water molecules when dissolved in water

7 The chain form has a free aldehyde group but the ring form does not; the ring form has a C–O–C bond, which is not present in the chain form

8 **(a)** The glucose molecule has an aldehyde at carbon number 1, but sorbitol has a hydroxyl group; **(b)** sorbitol can form many intermolecular hydrogen bonds between the hydroxyl groups and water molecules

9 **(a)** $C_2H_4 + H_2SO_4 \rightarrow CH_3CH_2OSO_3H$; **(b)** $CH_3CH_2OSO_3H + H_2O \rightarrow CH_3CH_2OH + H_2O$; **(c)** Cyclopentanol

10 **(a)** $CO_2 + 3H_2 \rightarrow CH_3OH + H_2O$; **(b)** Carbon dioxide – fermentation/combustion of fuels, hydrogen – cracking/water; **(c)** It will depend on the source of carbon dioxide if carbon dioxide comes from fermentation then the fuel will be renewable.

11 **(a)** 2-Bromopentane; **(b)** Iodocyclohexane; **(c)** 1-Chloropropane

12 **(a)** Reflux, under heating, ethanol with PI_3 (or phosphorus and iodine); **(b)** Reflux, under heating, pentan-2-ol with PCl_5 or $SOCl_2$, or reflux, under heating, pentan-2-ol, NaCl(s) and H_2SO_4(l); **(c)** Reflux under heating 3-methylhexan-2-ol with PI_3 (or phosphorus and iodine)

13 CH_3OCH_3 (methoxymethane)

14 **(a)**

(b) $CH_2{=}CHCH_2CH_3$

(c) But-2-ene

15 $CH_3CH_2CH_2CH_2CH_2OH + [O] \rightarrow CH_3CH_2CH_2CH_2CHO + H_2O$

16 Geraniol is oxidised to compound **A** and nerol is oxidised to compound **B**

A **B**

17

18 **(a)** $CH_3CH_2CH_2CH_2OH + 2[O] \rightarrow CH_3CH_2CH_2COOH + H_2O$

(b)

19 **(a)** Cyclohexanone; **(b)** $CH_3CHOHCH_2CH_3 + [O] \rightarrow CH_3COCH_2CH_3 + H_2O$

20 Possible oxidation products include

and

21 **(a)** Butanal; **(b)** Butanone; **(c)** No reaction

22 **(a)** So fuel will not contribute to increasing concentration of atmospheric carbon dioxide and possible climate change; **(b)** (i) $CH_3OH_2 + 1\frac{1}{2}O_2 \rightarrow CO_2 + 2H_2O$ (ii) If methanol is made from carbon dioxide by reaction with hydrogen then it for every mole of carbon dioxide used to make methanol one mole of carbon dioxide is made. So it will then be carbon-neutral

23 $CH_3CHOHCH_2CH_2CH_3$

24 The highly electronegative chlorine atom withdraws electrons away from the O–H bond, which makes it weaker than the O–H bond in ethanol; the weaker the O–H bond the stronger the acid

25 $2Na + 2CH_3OH \rightarrow 2CH_3O^-Na^+ + H_2$; The alkoxide produced is sodium methoxide

CHAPTER 14

1 **(a)** C_7H_{16}; **(b)** $C_{10}H_{22} \rightarrow C_2H_4 + C_8H_{18}$

2 $C_2H_4 + 3O_2 \rightarrow 2CO_2 + 2H_2O$; $C_3H_6 + 4\frac{1}{2}O_2 \rightarrow 3CO_2 + 3H_2O$

3 **(a)** $CH_4 + 2O_2 \rightarrow CO_2 + 2H_2O$; **(b)** Keep the mole ratio of oxygen to methane in the ratio of 1:2, since combustion requires a ratio of 2:1

4 Each carbon atom is surrounded by one double bond (which counts as one bond pair) and two single bonds (two bond pairs) and no lone pairs. The three 'bond pairs' repel each other to minimise repulsion and so take up a trigonal planar arrangement, which has an angle of 120° between the bond pairs

5 **(a)**

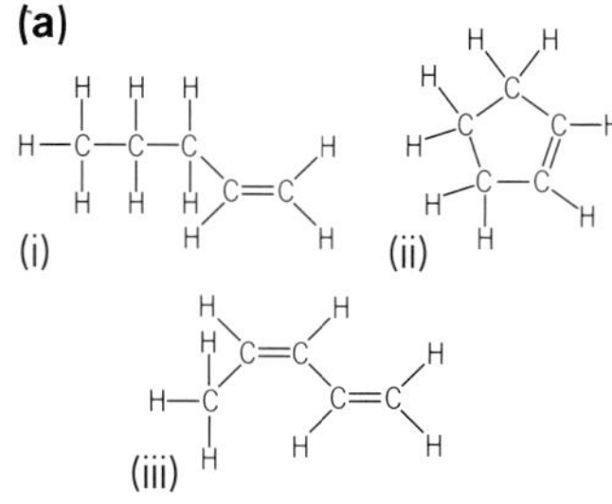

(iii)

(b) (i) Oct-1-ene;
(ii) 3-ethyl-2-methylpent-2-ene;
(iii) Cyclohexene

6 **(a)**

(i) (ii)

(b) (i) E-(or *trans*)-Hept-3-ene;
(ii) Z-(or *cis*)-2,3-Dichloropropent-2-ene;
(c) Hex-2-ene and hex-3-ene

7 265 kJ mol^{-1}

8 **(a)** **(b)**

(c) (i) Z-3-choloro-2-fluorobut-2-ene;
(ii) E-2-chloro-3methylpent-2-ene;
(iii) Z-1-bromo-1-fluoro-2-methylbut-1-ene; (iv) Z-1-bromo-1,2–dichloroethene

9 **(a)** Molecules are larger and have more electrons so that the strength of the induced dipole–induced dipole (van der Waals) intermolecular force increases;
(b) about 60 °C

10 **(a)** (i) Ethanol; (ii) Propan-1-ol or propan-2-ol; (iii) Cycloheptanol; **(b)** (i) Pent-1-ene, *cis*-pent-2-ene and *trans*-pent-2-ene; (ii) Pent-2-ene

11 Hex-1-ene, *cis*-hex-2-ene and *trans*-hex-2-ene

12 Lower bond energy than σ bond

13 Br^+ and CH_3^+

14

15

16 **(a)** 2-bromobutane;
(b) bromocyclohexane;
(c) 2-bromopentane;
(d) 2-bromohexane;
(e) 1-bromo-1,2-dichloroethane

17 To make a primary alcohol the carbocation formed during the addition must be a primary carbocation, but these are less stable than secondary carbocations and so the addition of a proton to an alkene normally occurs via a secondary or tertiary carbocation rather than a primary one

18 **(a)** **(b)** **(c)**

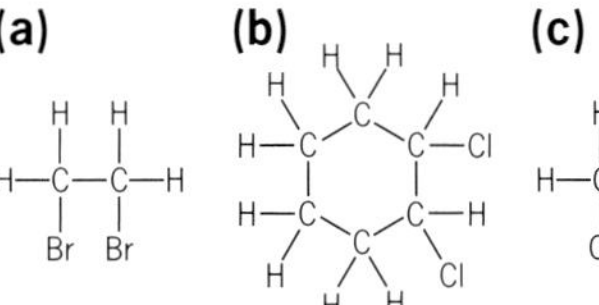

19

20 The carbocation formed can react with either a bromide ion or water to give either $CH_3CHBrCHBrCH_3$ or $CH_3CHOHCHBrCH_3$

21 5

22 **(a)** Propane-1,2-diol; **(b)** Butane-2, 3-diol; Cyclopentane-1,2-diol

23 **(a)** Propene with Cl_2, H_2O, C_3H_6 and C_6H_6;
(b) Propene → poly(propene)

CHAPTER 15

1 **(a)** **(b)**

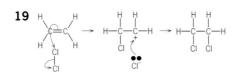

(c) and and or

2 **(a)** (i) $C_6H_{10} + H_2 \rightarrow C_6H_{12}$;
(ii) $C_6H_8 + 2H_2 \rightarrow C_6H_{12}$;
(iii) $C_6H_6 + 3H_2 \rightarrow C_6H_{12}$;
(b) (i) –120 kJ mol^{-1};
(ii) –240 kJ mol^{-1}; (iii) –360 kJ mol^{-1}

3 **(a)** $C_6H_6 + 7\frac{1}{2}O_2 \rightarrow 6CO_2 + 3H_2O$;
(b) It would be less than that calculated for Kekule's structure since it requires more energy to break the bonds in a benzene ring than three C=C and three C–C bonds

4 **(a)** $C_{10}H_8$; **(b)** $C_{12}H_9$

5 **(a)** **(b)** **(c)**

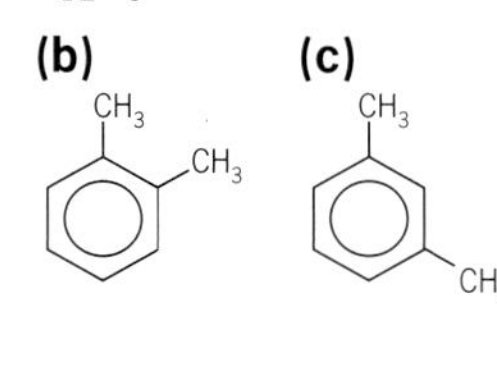

(b) Because it will be 1,3-dimethylbenzene since the position numbers are as low as possible

6 **(a)** **(b)** **(c)**

7 **(a)**

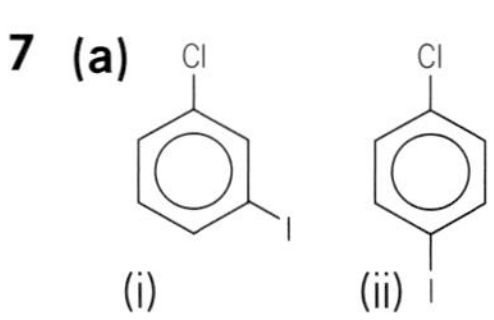

(b) Position numbers must be as low as possible so it must be 1,2-dichlorobenzene

8 **(a)**

(i) (ii)

(b) 1-Phenylpropan-1-one, 1-phenylpropan-2-one, 3-phenylpropan-1-ol, 3-phenylpropanal

9 **(a)** **(b)**

10 C_6H_6, $C_{10}H_8$, $C_{12}H_9$ and $C_{14}H_{10}$

11 **(a)** Methylbenzene
(b)

12 **(a)** $C_6H_6 + Cl^+ \rightarrow C_6H_5Cl + H^+$
(b)

13 –1

14 $H^+ + AlCl_4^- \rightarrow HCl + AlCl_3$

15 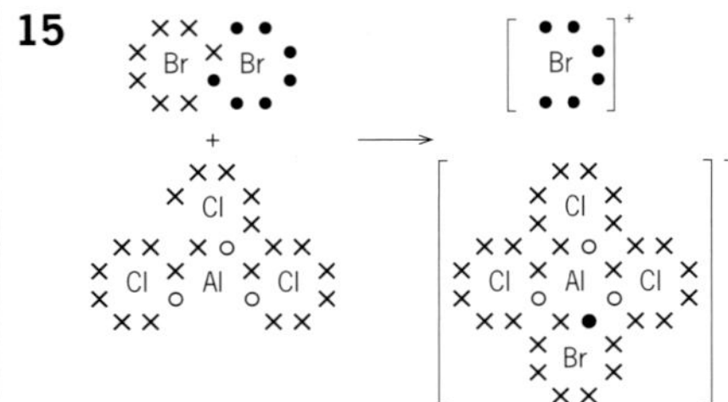

16 CH_3

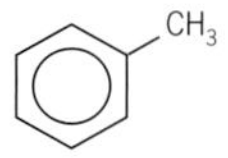

17 **(a)** $C_6H_5COCH_2CH_3$; **(b)** $CH_3CH_2CH_2COCl$

18 $C_6H_5SO_3H + 3NaOH \rightarrow C_6H_5O^-Na^+ + Na_2SO_3 + 2H_2O$

19 $C_{10}H_{18}$

20 $C_6H_6 + 3Cl_2 \rightarrow C_6H_6Cl_6$

20 **(a)**

21

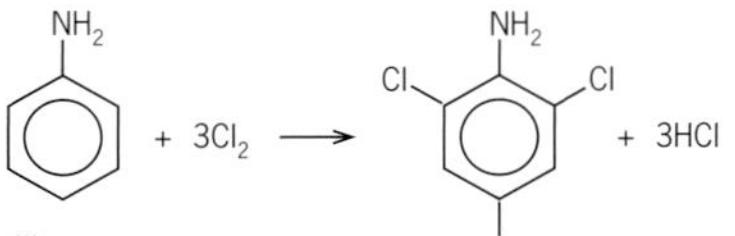

(i) NH_2 + $3Cl_2$ → NH_2, Cl, Cl, Cl + 3HCl

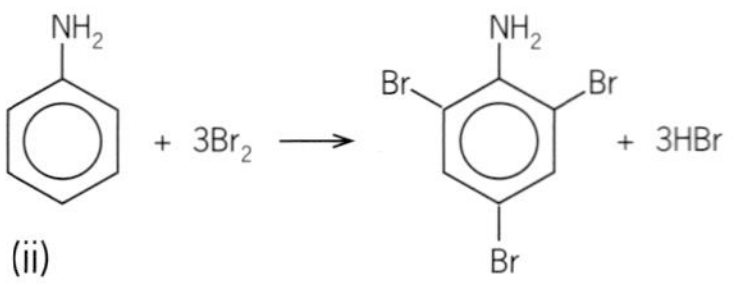

(ii) NH_2 + $3Br_2$ → NH_2, Br, Br, Br + 3HBr

(b)

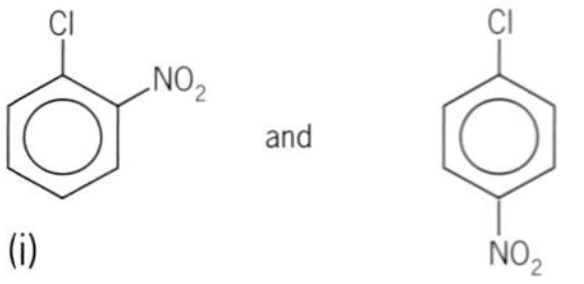

(i) Cl, NO_2 and Cl, NO_2

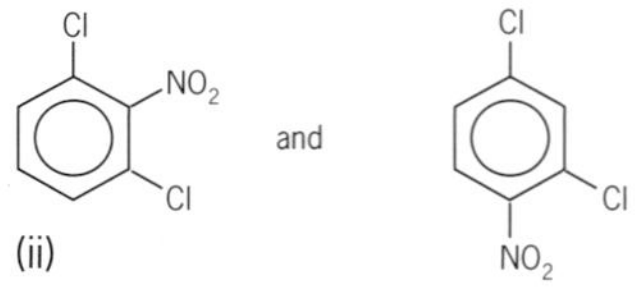

(ii) Cl, NO_2, Cl and Cl, Cl, NO_2

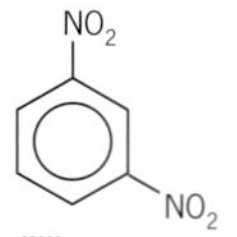

(iii) NO_2, NO_2

22 **(a)** 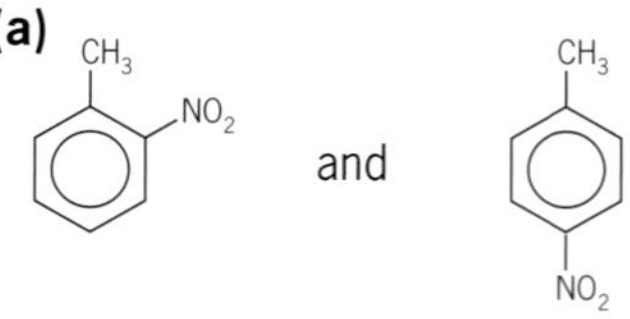 CH_3, NO_2 and CH_3, NO_2

(b) 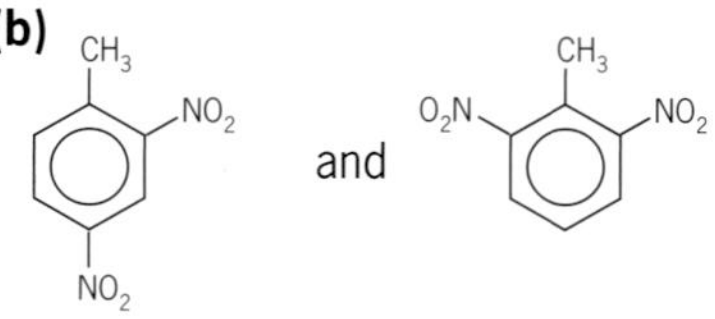 CH_3, NO_2, NO_2 and CH_3, O_2N, NO_2

23 **(a)** $CH_3CH_2COOC_6H_5$;
(b) $C_6H_5COOC_6H_5$

24 **(a)** 4-bromophenol, 4-chlorophenol and 4-fluorophenol because there is an increase in electronegativity from Br to F. So the fluorine atom will withdraw electrons from the benzene ring much more effectively than the bromine atom hence the O–H bond in 4-fluorophenol is a stronger acid. **(b)** 2,4,6-trimethylphenol, 2,4-dimethylphenol and 4-methylphenol. Methyl groups are electron releasing so that they increase the density of the benzene ring and strengthen the O–H bond. The more methyl groups the greater the electron releasing effect. **(c)** 3-nitrophenol, 2-nitrophenol and 2,4-dinitrophenol because nitro groups at positions 2 and 4 can withdraw electron density from O–H bond much better than a nitro group at position 3.
As more electron density is removed the O–H bond becomes weaker and so it can break easier to make a proton

25 **(a)** 1,4-Dimethylbenzene;
(b) (i) C_6H_5COOH; (ii) C_6H_5COOH;
(iii) 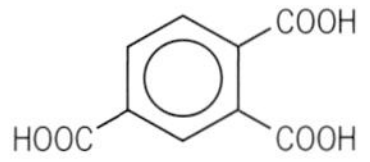 COOH, HOOC, COOH

26 **(a)** 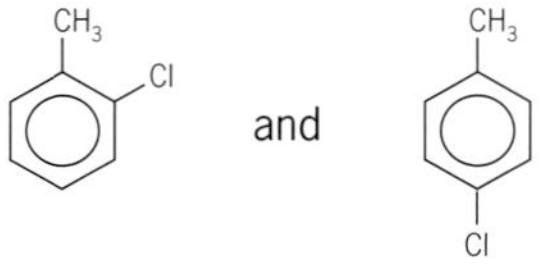 CH_3, Cl and CH_3, Cl

(b) $C_6H_5CH_2Cl$, $C_6H_5CHCl_2$ and $C_6H_5CCl_3$

CHAPTER 16

1 **(a)** Aldehydes are citral and $CH_3(CH_2)_9CHO$, and ketones are menthone and oestrone; **(b)** Phenol

2 **(a)** (i) Both have the same molecular formula, C_3H_6O, and the name for such compounds is structural isomers;
(b) $CH_3CH_2CH_2CHO$;
(c) $CH_3CH_2COCH_2CH_3$

3 **(a)** Hydrogen bonding;
(b) 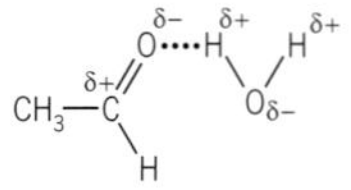 CH_3–C(δ+)(H)=O(δ–)····H(δ+)–O(δ–)–H(δ+)

(c) The benzene ring is non-polar and cannot form hydrogen bonds with water molecules; **(d)** Alcohols have stronger hydrogen bonds between molecules but weaker permanent dipole-permanent dipole intermolecular attraction between aldehyde or ketone molecules

4 $C_6H_5CHO + HCN \rightarrow C_6H_5CH(CN)OH$

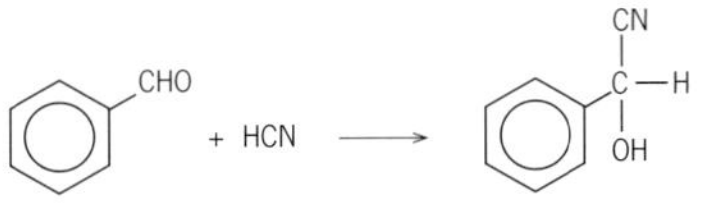

CHO + HCN → CN, C–H, OH

5 **(a)** Ethanal gives ethanol, CH_3CH_2OH, and propanone gives propan-2-ol, $CH_3CHOHCH_3$; **(b)** Ethanol is a primary alcohol and propan-2-ol is a secondary alcohol; **(c)** $CH_3COCH_3 + 2[H] \rightarrow CH_3CHOHCH_3$

6

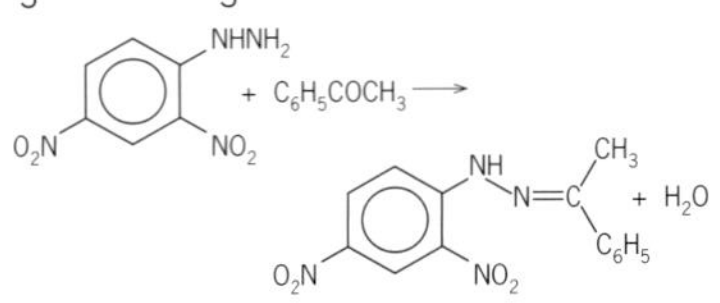

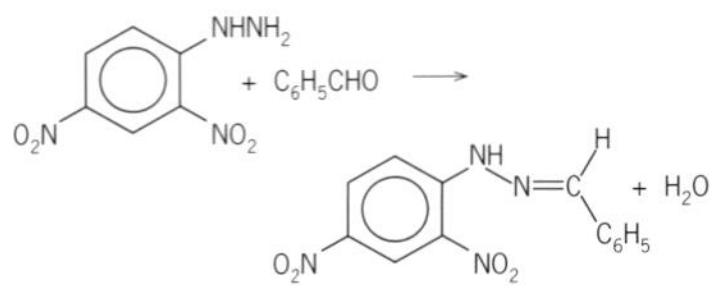

$NHNH_2$, O_2N, NO_2 + $C_6H_5COCH_3$ → NH–N=C(CH_3)(C_6H_5), O_2N, NO_2 + H_2O

$NHNH_2$, O_2N, NO_2 + C_6H_5CHO → NH–N=C(H)(C_6H_5), O_2N, NO_2 + H_2O

7 **(a)** Propan-1-ol; **(b)** 2-Methylhexan-2-ol;
(c) Butan-2-ol

8 Alkyl, cycloalkyl or aryl

9 **(a)** H^-; **(b)** $CH_3CHO + 2[H] \rightarrow CH_3CH_2OH$, $CH_3COCH_3 + 2[H] \rightarrow CH_3CHOHCH_3$

10 Methanal (and ethanal)

11 **(a)** CH_3CHO; **(b)** $C_6H_5COCH_3$;
(c) $C_2H_5COCH_3$; **(d)** C_2H_5OH

CHAPTER 17

1 **(a)** $HCl(g) + H_2O(l) \rightarrow H_3O^+(aq) + Cl^-(aq)$;
(b) $HI(g) + H_2O(l) \rightarrow H_3O^+(aq) + I^-(aq)$

2 $HBr(g) + aq \rightarrow H^+(aq) + Br^-(aq)$, $HI(g) + aq \rightarrow H^+(aq) + I^-(aq)$

3 **(a)** F^-; **(b)** OH^-; **(c)** SO_4^{2-}

4 **(a)** H_3O^+; **(b)** OH^-; **(c)** H_2SO_4

5 **(a)** $HNO_2(aq) = H^+(aq) + NO_2^-(aq)$;
(b) $K_a = \dfrac{[H^+(aq)][NO_2^-(aq)]}{[HNO_2(aq)]}$;
(c) Ethanoic acid

6 **(a)** (i) 9.9; (ii) 4.8; **(b)** Acid is diprotic and the protons are lost sequentially, the first proton being transferred to a water molecule before the second proton

7 **(a)** The stronger the acid the weaker the conjugate base; **(b)** Acid strength HI > HCl > HF, since the H–I bond is the weakest bond and easiest to break and the H–F bond is the strongest

8 **(a)** (i) 0.0; (ii) 4.0; (iii) 0.98; **(b)** 2.2

9 (a) Both are 6.6×10^{-7} mol dm^{-3};
(b) pH = $-\log10(6.6 \times 10^{-7}) = 6.2$

10 2.37

11 (a) (i) Pentanoic acid; (ii) Butanoic acid; (iii) Pentanedioic acid
(b) (i) $ClCH_2CH_2COOH$
(ii) (iii) (iv)

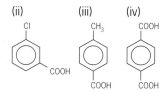

Cl, COOH | CH_3, COOH | COOH, COOH

12 (a) $C_nH_{2n+1}COOH$; **(b)** (i) 185 °C; (ii) No observable trend

13 (a) Calcium ethanoate and water; **(b)** Calcium propanoate and hydrogen; **(c)** Magnesium ethanoate, carbon dioxide and water; **(d)** Calcium ethanaote, carbon dioxide and water; **(e)** Sodium benzoate, carbon dioxide and water

14 (a) $Ca(s) + 2CH_3COO(aq) \rightarrow Ca^{2+}(CH_3COO^-)_2(aq) + H_2(g)$;
(b) Ethanoic acid is a weak acid and hydrochloric acid is a strong acid, so there is a greater concentration of hydrogen ions with dilute hydrochloric acid than with ethanoic acid and, since the rate of reaction is proportional to the concentration of hydrogen ions, reaction with hydrochloric acid is faster

15 $K_a = \dfrac{[H^+(aq)][In^-(aq)]}{[HIn(aq)]}$

16 (a) Methyl orange, bromocresol green or methyl red; **(b)** Phenolphthalein = 9.15 and thymolphthalein = 9.9

17 $K_{a1} = \dfrac{[H^+(aq)][COOHCOO^-(aq)]}{[(COOH)_2]}$

$K_{a2} = \dfrac{[H^+(aq)][(COO^-)_2(aq)]}{[COOHCOO^-(aq)]}$

18 $[C_2H_5COOH] = 0.750$ mol dm^{-3} and $[C_2H_5COO^-] = 0.350$ mol dm^{-3}

19 The concentration decreases

20 (a) Sodium benzoate is ionic and the negative carboxylate groups can form strong electrostatic attractions with polar water molecules, whereas benzoic acid is not ionic and as a result forms much weaker electrostatic interactions with water molecules; **(b)** Many foods contain water or are stored in water, so to control acidity the additive must dissolve in water

21 (a) $CH_3CH_2CHO + [O] \rightarrow CH_3CH_2COOH$;
(b) $CH_3CH_2CH_2CH_2OH + 2[O] \rightarrow CH_3CH_2CH_2COOH + H_2O$

22 (a) Propan-1-ol and sodium ethanoate; **(b)** Ethanol and benzoic acid; **(c)** Ammonia and sodium propanoate; **(d)** Benzoic acid and ammonium chloride

23 (a) $CH_3COOH + PCl_5 \rightarrow CH_3COCl + HCl + POCl_3$; **(b)** $C_6H_5COOH + SOCl_2 \rightarrow C_6H_5COCl + SO_2 + HCl$

24

CHAPTER 18

1 Electronegativity is the ability of an atom in a covalent bond to attract the bonding pair of electrons towards itself

2 Methyl ethanoate

```
   H  O     H
   |  ||    |
H—C—C—O—C—H
   |        |
   H        H
```

Propyl propanoate

```
   H  H  O     H  H  H
   |  |  ||    |  |  |
H—C—C—C—O—C—C—C—H
   |  |        |  |  |
   H  H        H  H  H
```

Pentyl butanoate

```
   H  H  H  O     H  H  H  H  H
   |  |  |  ||    |  |  |  |  |
H—C—C—C—C—O—C—C—C—C—C—H
   |  |  |        |  |  |  |  |
   H  H  H        H  H  H  H  H
```

Methyl benzoate

3 (a)
(i) $K_c = \dfrac{[ester][water]^2}{[acid][alcohol]^2} = \dfrac{[conc][conc]^2}{[conc][conc]^2}$

so all the units cancel out, as $(conc)^0$ means no units;

(ii) $K_c = \dfrac{[N_2O_4]}{[NO_2]^2} = \dfrac{(conc)}{(conc)^2} = (conc)^{-1}$

= $(mol\ dm^{-3})^{-1}$, so the units are dm^3 mol^{-1}; **(b)** (i) $CH_3COOH + C_5H_{11}OH = CH_3COOC_5H_{11} + H_2O$;

(ii) $K_c = \dfrac{[CH_3COOC_5H_{11}][H_2O]}{[CH_3COOH][C_5H_{11}OH]}$; (iii) 4.15

4 (a) 0.67; **(b)** 0.67; **(c)** 0.33; **(d)** 0.33

5 To shift the position of equilibrium onto the product side to try and minimise the excess methanol.

6 Reflux under heating excess ethanol with carboxylic acid in the presence of a little concentrated sulfuric acid as a catalyst and remove the water as it is formed

7 Increasing the temperature shifts the position of equilibrium towards the left; increasing catalyst concentration has no effect on the position of equilibrium; increasing alcohol concentration shifts the position of equilibrium towards the right

8 (a) Ethanol and ethanoic acid; **(b)** Ethanol and butanoic acid; **(c)** Prop-2-en-1-ol and ethanoic acid

9 To minimise the increase of ethanol the position of the equilibrium shifts towards the right to try to use up ethanol

10 (a) Ethanol and sodium ethanoate; **(b)** Propan-1-ol and sodium propanoate; **(c)** Ethanol and sodium benzoate

11 (a) $CH_3CH_2CH_2CH_2OH$ and CH_3COOH; **(b)** C_6H_5OH and C_6H_5COOH; **(c)** CH_3CH_2OH and $(COOH)_2$

12 (a) C_6H_5COCl **(b)** CH_3CH_2COCl **(c)** $ClOCCH_2CH_2CH_2CH_2COCl$

13 (a) $C_6H_5COCl + H_2O \rightarrow C_6H_5COOH + HCl$; **(b)** $CH_3CH_2COCl + H_2O \rightarrow CH_3CH_2COOH + HCl$

14 (a) $C_6H_5COCl + 2NaOH \rightarrow C_6H_5COO^-Na^+ + NaCl + H_2O$; **(b)** C–Cl in chlorobenzene is much stronger than the C–Cl in benzoyl chloride because the delocalised π system of the benzene ring is extended over the C–Cl bond and so is much more difficult to break

15 (a)

nucleophilic addition — elimination — proton transfer

$C_6H_5COCl + H_2N{-}C_6H_5 \rightarrow C_6H_5C(O^-)(Cl)N^+H_2C_6H_5 \rightarrow C_6H_5CON^+H_2C_6H_5 + Cl^- \rightarrow C_6H_5CONHC_6H_5 + H^+ + Cl^-$

(b) (i) Sodium ethanoate, water and sodium chloride; (ii) Ethanamide and ammonium chloride; (iii) benzamide and hydrogen chloride (or ammonium chloride); (iv) Methyl benzoate and hydrogen chloride; (v) Benzoic anhydride and sodium chloride

16 $CH_3-CH_2-C(=O)-O-C(=O)-CH_2-CH_3$

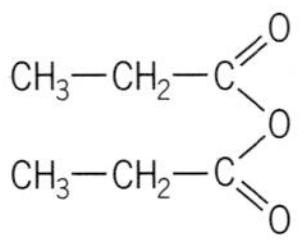

17 (a) $(CH_3CO)_2O + 2NH_3 \rightarrow CH_3CONH_2 + CH_3COO^-NH_4^+$;
(b) $(CH_3CO)_2O + CH_3CH_2OH \rightarrow CH_3COOCH_2CH_3 + CH_3COOH$;
(c) $(CH_3CO)_2O + C_6H_5NH_2 \rightarrow CH_3CONHC_6H_5 + CH_3COOH$;
(d) $(CH_3CO)_2O + H_2O \rightarrow 2CH_3COOH$

18 $CH_3CH_2CH_2CH_2CONH_2$ or $C_5H_{11}ON$

19 $CH_3CONHCH_2CH_3$

20 (a) $CH_3-C(=O)-OH$

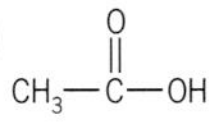

(b) $CH_3-CH_2-C(=O)-O^-\ Na^+$

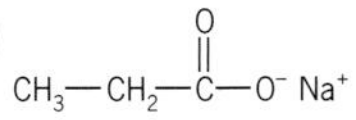

21 (a) Benzoic acid and ammonium chloride; **(b)** Potassium benzoate and ammonia; **(c)** $C_6H_5CH_2NH_2$

22

CHAPTER 19

1 (a) $H-CH_2-CH_2-CH(NH_2)-C(=O)OH$

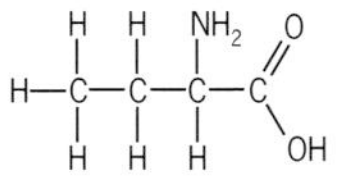

(b) $H-CH(NH_2)-CH_2-CH_2-C(=O)OH$

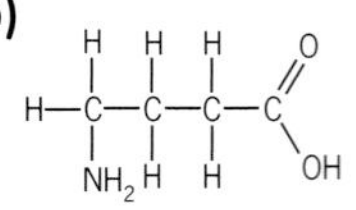

2 (a) Lysine, serine, aspartic acid and glutamic acid; **(b)** Leucine and isoleucine are structural isomers;
(c) (i) $H-CH_2-CH(O-H)-C(H)(N(H)-H)-C(=O)O-H$

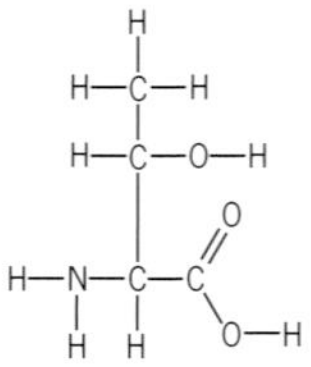

(ii) Secondary alcohol; (iii) The R group in serine has one less carbon atom and it is a primary alcohol

3 (a) An enantiomer is one of a pair of optical isomers; its mirror image is non-superimposable;
(b) [structure: ring with CH_3, =O, and *C(=CH_2)CH_3]

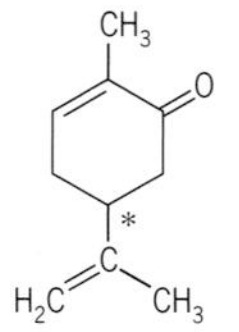

4 [structure: $CH_2OH-C(=O)-C^*H(OH)-C^*H(OH)-C^*H(OH)-CH_2OH$, drawn as H–C(OH)–H, C=O, HO–C*–H, H–C*–OH, H–C*–OH, H–C–OH, H]

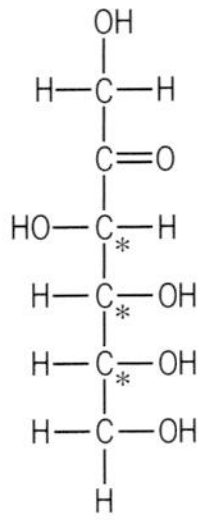

5 2-hydroxypropanoic acid

6 [structure: $COOH-{}^*CHNH_2-CH_2$–ring with HO and OH]

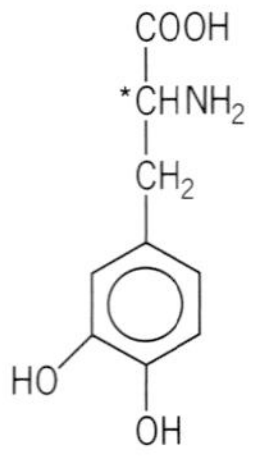

7 (a) $NH_2CH_2COOH + NaOH \rightarrow NH_2CH_2COO^-Na^+ + H_2O$, or $^+NH_3CH_2COO^- + NaOH \rightarrow NH_2CH_2COO^-Na^+ + H_2O$;
(b) $HOOCCHRNH_2 + HCl \rightarrow HOOCCHRNH_3^+Cl^-$, or $^-OOCCHRNH_3^+ + HCl \rightarrow HOOCCHRNH_3^+Cl^-$
where $R = (CH_3)_2CHCH_2-$

8 Aspartic acid and glutamic acid

9 $NH_2-CH_2-C(=O)-N(H)-CH(CH_3)-COOH$

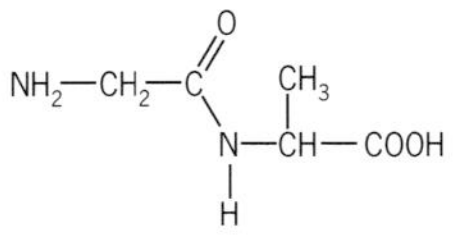

and $NH_2-CH(CH_3)-C(=O)-N(H)-CH_2-COOH$

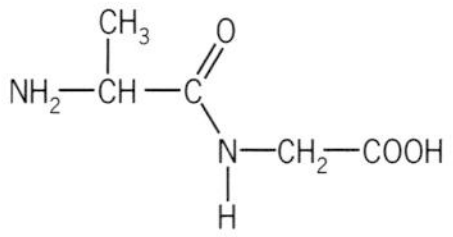

10 4527

11 Tyr–Gly–Gly–Phe–Met

12 (a) Lysine, serine, aspartic acid and glutamic acid; **(b)** Alanine, valine, leucine, isoleucine, proline and phenylalanine

13 (a) (b) Thyamine has an extra methyl group

14 (a) C U A G; **(b)** C T A G

CHAPTER 20

1 Two substances, one with a double bond (or triple bond), react together to form one product in which the double bond has been changed into a single bond (or the triple bond is changed into a double bond or a single bond)

2 (a) $-CH_2-CH_2-CH_2-CH_2-CH_2-CH_2-$

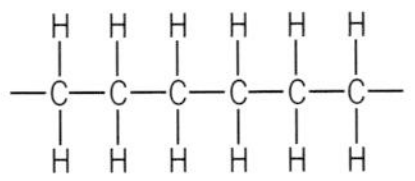

(b) Polymer is a hydrocarbon that is fully saturated with no carbon–carbon double bonds

3 Linear polymers can lie closer to one another, which allows more contact between molecules so that the induced dipole–induced dipole forces of attraction can operate over a larger surface, and hence intermolecular force is stronger than in a branched chain, in which the molecules cannot lie as close to one another

4

5 $C_6H_5CH=CH_2$

6 $nCF_2{=}{=}CF_2 \rightarrow -(CF_2CF_2)_n-$

7 (a) CH2==CHOH / CH2==CHOOCCH3; **(b)** Forms intermolecular hydrogen bonds with other polymer chains rather than with water molecules; **(c)** Not enough hydrogen bonds with water molecules

8 $CH_3COOH + CH_3CH_2OH = CH_3COOCH_2CH_3 + H_2O$

9 (a) (i) $CH_3OOC-C_6H_4-COOCH_3$

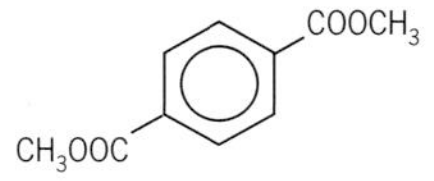

(ii) CH_3OH

(iii) Base hydrolysis

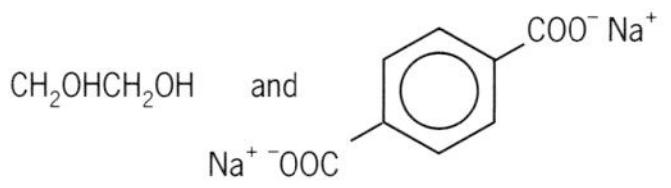

CH_2OHCH_2OH and $Na^+\ {}^-OOC-C_6H_4-COO^-\ Na^+$

Acid hydrolysis

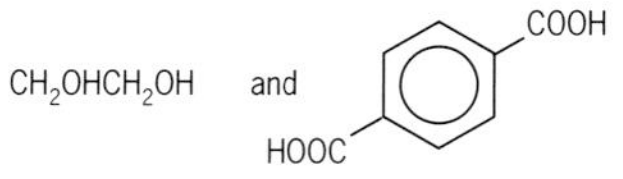

CH_2OHCH_2OH and $HOOC-C_6H_4-COOH$

(b) $-C(=O)-C_6H_4-C(=O)-O-CH_2-CH_2-O-$

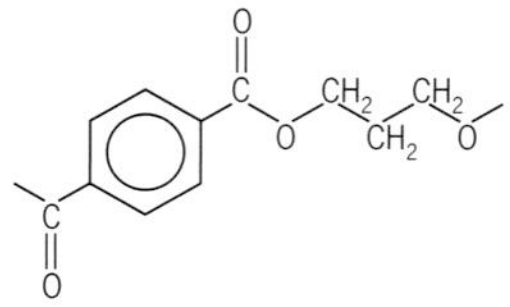

10 When blended with cotton fibres perspiration will be absorbed, which will make the fabric more comfortable to wear

11 $nCH_3CHOHCOOH \rightarrow -(OCH(CH_3)CO)_n- + nH_2O$

12 $H_2NCH_2CH_2CH_2CH_2CH_2NH_2$ and $HOOCCH_2CH_2CH_2CH_2CH_2CH_2CH_2CH_2COOH$

13 NaOH(aq)/reflux under heating, $H_2NCH_2CH_2CH_2CH_2CH_2NH_2$ and $Na^{+-}OOCCH_2CH_2CH_2CH_2COO^-Na^+$, 6 mol dm^{-3} HCl(aq)/reflux under heating; $Cl^-H_3N^+CH_2CH_2CH_2CH_2CH_2NH_3^+Cl^-$ and $HOOCCH_2CH_2CH_2CH_2COOH$

14 **(a)** (i) Condensation; (ii) Thermosetting; **(b)** High melting point, strong and rigid, relatively low density and a good thermal conductor

15 **(a)**

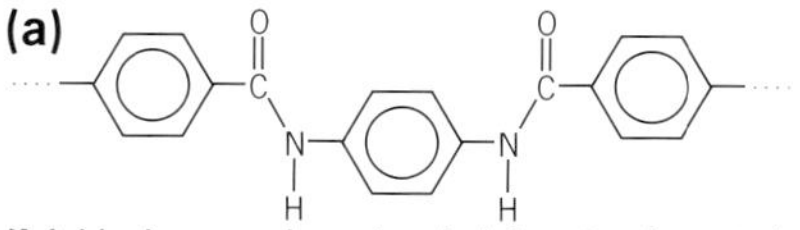

(b) Hydrogen bonds; **(c)** In steel metal ions are closely packed. In Kevlar polymer chains cannot be closely packed.

16

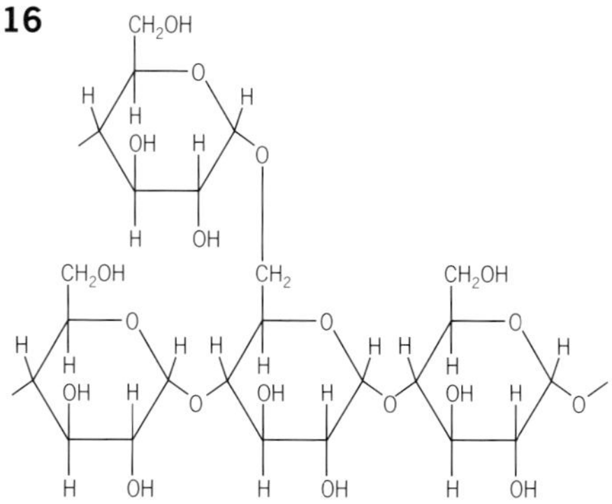

17 Intermolecular hydrogen bonds

18 Thermoplastics have weak intermolecular forces, such as induced dipole–induced dipole interactions, which are easily broken and overcome so that the melting point is low; once melted the polymer chains can take up new positions, and on cooling the intermolecular forces are easily reformed

19 It cannot be remoulded because it decomposes when heated strongly, so the plastic cannot be recycled

20 Collection of the bottles, e.g. fuel; cleaning of bottles; relabelling of bottles, melting and remoulding

21 Issues include: toxic gases, such as hydrogen chloride and hydrogen cyanide, will be produced during incineration; high temperatures are often required for the incineration process to prevent the formation of toxic materials; incineration means that fewer landfill sites are needed for plastic waste; need to sort out household waste; less crude oil needed for fuels; energy locked in the bonds of plastics is not wasted

CHAPTER 21

1 **(a)** Na is +1 and Cl is −1; **(b)** Mg is +2 and Cl is −1; **(c)** Al is +3 and Cl is −1; **(d)** Si is +4 and Cl is −1; **(e)** P is +3 and Cl is −1; **(f)** P is +5 and Cl is −1; **(g)** S is +1 and Cl is −1

2 **(a)** Oxygen; **(b)** Carbon

3 **(a)** 0 to +4; **(b)** 0 to −1; **(c)** Chlorine, as it gains electrons and its oxidation number decreases; **(d)** Silicon, since it loses electrons and its oxidation number increases

4 **(a)** P from +3 to +5 and Cl from 0 to −1; **(b)** Chlorine

5 $Mg(s) + Cl_2(g) \rightarrow MgCl_2(s)$; $2Al(s) + 3Cl_2(g) \rightarrow Al_2Cl_6(s)$; $Si(s) + 2Cl_2(g) \rightarrow SiCl_4(l)$; $P_4(s) + 6Cl_2(g) \rightarrow 4PCl_3(l)$; $P_4(s) + 10Cl_2(g) \rightarrow 4PCl_5(s)$; $2S(l) + Cl_2(g) \rightarrow S_2Cl_2(l)$

6 LiCl and NaCl both have a face-centred cubic structure, but CsCl has a body-centred cubic structure; coordination number of Li^+ and Na^+ = 6, but that of Cs^+ = 8; change in structure because of the increase in the ratio of the cation radius to the anion radius from LiCl to CsCl

7

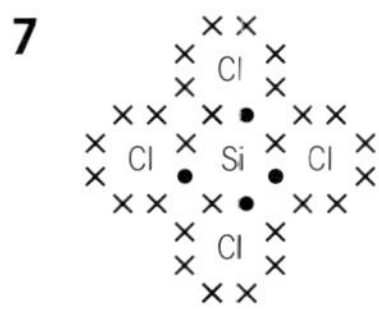

8 BCl_3 has a trigonal planar shape so that all the individual bond dipoles cancel out

9 Ionic attraction between PCl_4^+ and PCl_6^- in PCl_5 is stronger than the permanent dipole–permanent dipole intermolecular attraction in PCl_3

10 **(a)** $2Mg(s) + O_2(g) \rightarrow 2MgO(s)$; $4Al(s) + 3O_2(g) \rightarrow 2Al_2O_3(s)$; $Si(s) + O_2(g) \rightarrow SiO_2(s)$; $P_4(s) + 5O_2(g) \rightarrow P_4O_{10}(s)$; $S(s) + O_2(g) \rightarrow SO_2(g)$; **(b)** P_4O_{10}, SO_2, SO_3 and Cl_2O_7; **(c)** Na_2O is sodium oxide, MgO is magnesium oxide, Al_2O_3 is aluminium oxide, SiO_2 is silicon(IV) oxide or silicon dioxide; SO_2 is sulfur dioxide or sulfur(IV) oxide; SO_3 is sulfur trioxide or sulfur(VI) oxide; P_4O_6 is phosphorus(III) oxide; P_4O_{10} is phosphorus(V) oxide; Cl_2O_7 is chlorine(VII) oxide

11 **(a)** Acids; **(b)** Sodium nitrate; **(c)** Sodium chloride

12 Diamond contains all carbon atoms and SiO_2 contains silicon and oxygen; coordination number of all atoms is four in diamond, but in SiO_2 it is four around Si and two around O; the C–C bond length is shorter than the Si–O bond length in SiO_2

13 **(a)** Chloric(I) acid, $CO_2 + H_2O \rightarrow 2HOCl$; **(b)** Carbonic acid, $CO_2 + H_2O == H_2CO_3$

14 **(a)** $2NO_2 + H_2O \rightarrow HNO_2 + HNO_3$; **(b)** NO_2 (N = +4) is oxidised to HNO_3 (N = +5) and reduced to HNO_2 (N = +3)

CHAPTER 22

1 **(a)** (i) ~23 °C; (ii) ~680 °C; (iii) ~275 nm; (iv) ~187 nm; **(b)** Strength of metallic bonding decreases as atomic (proton) number increases; **(c)** Rubidium, caesium and francium

2 **(a)** Reacts (recombines) to give sodium chloride, $2Na + Cl_2 \rightarrow 2NaCl$; **(b)** Chlorine will react with iron to give iron(III) chloride

3 Electrolysis of molten lithium chloride

4 **(a)** Li^+ is $1s^2$, K^+ is $1s^22s^22p^63s^23p^6$, Rb^+ is $1s^22s^22p^63s^23p^63d^{10}4s^24p^6$; **(b)** Na atoms require less energy to lose an electron than Li atoms, as the outer electrons are less strongly attracted to the nucleus because of the increased atomic radius and the increased number of shielding inner shells of electrons

5 ~360 kJ mol^{-1}

6 **(a)** $2Na(s) + 2H_2O(l) \rightarrow 2NaOH(aq) + H_2(g)$, $2K(s) + 2H_2O(l) \rightarrow 2KOH(aq) + H_2(g)$; **(b)** Francium sinks in water, and there is an extremely rapid (possibly explosively) exothermic reaction to form hydrogen and an alkaline solution of francium hydroxide, $2Fr(s) + 2H_2O(l) \rightarrow 2FrOH(aq) + H_2(g)$

7 (a) $2Na(s) + O_2(g) \rightarrow Na_2O_2(s)$; **(b)** $2K(s) + O_2(g) \rightarrow K_2O_2(s)$; **(c)** $Rb(s) + O_2(g) \rightarrow RbO_2(s)$

8 $Rb_2O(s) + H_2O(l) \rightarrow 2RbOH(aq)$

9 (a) Density is ~4500 kg m^{-3}, melting point is ~640 °C, boiling point is ~1500 °C (difficult to estimate since there is no obvious trend); **(b)** As ionic radius of M^{2+} ion increases (charge density decreases) the metallic bonding generally gets weaker because the electrostatic attraction between the delocalised electrons and the positive ions becomes weaker

10 $Ba(s) + 2H_2O(l) \rightarrow Ba(OH)_2(aq) + H_2(g)$; $Sr(s) + 2H_2O(l) \rightarrow Sr(OH)_2(aq) + H_2(g)$

11 As you go down the group, the atomic radius increases and the number of shielding shells increases, so the attraction between the positive nucleus and the outer shell electrons decreases; the numerical addition of the first- and second-ionisation energies is smaller down the group, so electrons are lost more easily from the atoms

12 (a) Magnesium atoms lose electrons so are oxidised, hydrogen ions gain electrons so are reduced; **(b)** Magnesium changes from 0 to +2, which is oxidation, hydrogen changes from +1 to 0, which is reduction

13 (a) $Ca(s) + 2HCl(aq) \rightarrow CaCl_2(aq) + H_2(g)$; **(b)** $Ba(s) + 2HCl(aq) \rightarrow BaCl_2(aq) + H_2(g)$; **(c)** $Mg(s) + 2CH_3COOH(aq) \rightarrow Mg(CH_3COO)_2(aq) + H_2(g)$

14 (a) $Mg(OH)_2(s) + 2HCl(aq) \rightarrow MgCl_2(aq) + 2H_2O(l)$; **(b)** $CaCO_3 + H_2SO_4 \rightarrow CaSO_4 + H_2O$, $CaCO_3 + 2HNO_3 \rightarrow Ca(NO_3)_2 + CO_2 + H_2O$

15 (a) $Mg^{2+}(g) + O^{2-}(g) \rightarrow MgO(s)$; **(b)** $Mg^{2+}(g) + SO_4^{2-}(g) \rightarrow MgSO_4(s)$; **(c)** $2Fe^{3+}(g) + 3SO_4^{2-}(g) \rightarrow Fe_2(SO_4)_3(s)$; **(d)** $3Mg^{2+}(g) + 2N^{3-}(g) \rightarrow Mg_3N_2(s)$

16 –3454 kJ mol^{-1}

17 (a) $Li_2CO_3(s) \rightarrow Li_2O(s) + CO_2(g)$; **(b)** $2LiNO_3(s) \rightarrow Li_2O(s) + 2NO_2(g) + O_2(g)$; **(c)** $2FrNO_3(s) \rightarrow 2FrNO_2(s) + O_2(g)$

18 (a) Solvent is water and solute is alcohol; **(b)** Solvent is alcohol and solute is water

19 (a) $Mg^{2+}(g) + aq \rightarrow Mg^{2+}(aq)$; **(b)** $SO_4^{2-}(g) + aq \rightarrow SO_4^{2-}(aq)$

20 NaCl is +5 kJ mol^{-1}, LiCl is –32 kJ mol^{-1}

21 $MgSO_4$ is 337 g dm^{-3}, $CaSO_4$ is 1.9×10^{-1} g dm^{-3}, $SrSO_4$ is 1.4×10^{-3} g dm^{-3}, $BaSO_4$ is 2.6×10^{-4} g dm^{-3}

22 (a) $K_{sp}(CaSO_4) = [Ca^{2+}(aq)]\,[SO_4^{2-}(aq)]$; **(b)** $K_{sp}(BaSO_4) = [Ba^{2+}(aq)][SO_4^{2-}(aq)]$; **(c)** $K_{sp}(Ag_2CrO_4) = [Ag^+(aq)]^2[CrO_4^{2-}(aq)]$; **(d)** $K_{sp}(PbCl_2) = [Pb^{2+}(aq)][Cl^-(aq)]^2$

23 (a) 2.0×10^{-10} mol^2 dm^{-6}; **(b)** $[Ca^{2+}(aq)] = 1.25 \times 10^{-2}$ mol dm^{-3}, $[OH^-(aq)] = 2 \times 1.25 \times 10^{-2} = 2.50$ mol dm^{-3}, $K_{sp}(Ca(OH)_2) = [Ca^{2+}(aq)][OH^-(aq)]^2 = (1.25 \times 10^{-2})(2.50 \times 10^{-2})^2 = 7.81 \times 10^{-6}$ mol^3 dm^{-9}

24 (a) 7.5×10^{-3} mol dm^{-3}; **(b)** 2.1×10^{-4} mol dm^{-3}; **(c)** 5.3×10^{-4} mol dm^{-3}

25 2×10^{-9} mol dm^{-3}

26 No: the ionic product 1.0×10^{-9} is less than the K_{sp}

27 (a) $Al^{3+} + 3e^- \rightarrow Al$; **(b)** Some F^- will be oxidised at the anode to form F_2 and some of the fluorine will react with moisture in the atmosphere to give HF

28 No, since aluminium carbonate would have to be formed in aqueous conditions and in aqueous solution $Al^{3+}(aq)$ is highly acidic because Al^{3+} polarizes water molecules to form $H^+(aq)$ and so the solution will react with carbonate ions to give carbon dioxide and water; also Al^{3+} has a very high charge density so it will polarize the carbonate ion, weakening the covalent bonds in the anion so it will decompose

CHAPTER 23

1 Fluorine will be a pale yellow gas, and astatine will be a black solid

2 (a) (i) 85; (ii) 7; (iii) Melting point ~210 °C, boiling point ~280 °C, covalent radius ~150 pm, electronegativity ~2.2; **(b)** Electronegativity is the ability of an atom in a covalent bond to attract the bonding pair of electrons; in the halogens, the increased number of inner shielding shells and same core charge of +7 means the effective nuclear charge decreases as atomic number increases, so the nucleus has a less powerful attraction for electrons in the shared bonding pair

3 (a) $1s^22s^22p^63s^23p^6$; **(b)** $1s^22s^22p^63s^23p^63d^{10}4s^24p^6$

4 (a) No reaction; **(b)** $Cl_2(aq) + 2I^-(aq) \rightarrow I_2(aq) + 2Cl^-(aq)$; **(c)** No reaction; **(d)** $I_2(aq) + 2At^-(aq) \rightarrow At_2(s) + 2I^-(aq)$

5 (a) (i) $2Fe(s) + 3Cl_2(g) \rightarrow 2FeCl_3(s)$; (ii) $Cu(s) + Cl_2(g) \rightarrow CuCl_2(s)$; (iii) $2Fe(s) + 3F_2(g) \rightarrow 2FeF_3(s)$; **(b)** (i) Chromium(III) chloride; (ii) Zinc chloride; (iii) Barium iodide

6 (a) Oxidation states of Cl_2, HCl and HOCl are 0, –1 and +1, respectively; **(b)** Oxidation number of chlorine has increased and decreased, showing that one atom in Cl_2 has been oxidised and the other reduced

7 (a) $Cl_2(aq) + SO_2(aq) + 2H_2O(l) \rightarrow 2Cl^-(aq) + SO_4^{2-}(aq) + 4H^+(aq)$; **(b)** Astatine is the least powerful oxidising agent of the halogens, and iodine also cannot oxidise $Fe^{2+}(aq)$

8 1.05×10^{-3} mol dm^{-3}

9 Iodine changes oxidation state from 0 to –1, whereas the oxidation state of sulfur changes from +2 to +2.5. So I_2 is reduced and $S_2O_3^{2-}$ is oxidised

10 (a) +4 in XeF_4 and +6 in XeF_6; **(b)** $Xe + 2F_2 \rightarrow XeF_4$, $Xe + 3F_2 \rightarrow XeF_6$

11 $PBr_3(l) + 3H_2O(l) \rightarrow H_3PO_3(aq) + 3HBr(aq)$, NB: HBr(g) also acceptable

12 Halogens have high electronegativities so that the halogen end of the H–X (where X is F, Cl, Br and I) will be δ – and the hydrogen δ +

13 (a) $K_c = \frac{[H_2][I_2]}{[HI]^2}$; **(b)** Mole ratio $HI:H_2:I_2$ is 8.47:1:1

14 (a) Sodium chloride, water and carbon dioxide; **(b)** Magnesium iodide and water; **(c)** Sodium bromide and water

15 $KMnO_4(s) + 8HCl(aq) \rightarrow KCl(aq) + MnCl_2(aq) + 4H_2O(l) + 2\frac{1}{2}Cl_2(g)$

16 (a) $NaI(s) + H_2SO_4(l) \rightarrow NaHSO_4(s) + HI(g)$; **(b)** $2HI(g) + H_2SO_4(l) \rightarrow SO_2(g) + I_2(s) + 2H_2O(l)$; **(c)** Initially, hydrogen astatide and sodium hydrogensulfate, then hydrogen astatide reduces sulfuric acid to form astatine, hydrogen sulfide, sulfur dioxide, sulfur and water; **(d)** Hydrogen fluoride and sodium hydrogensulfate

17 **(a)** Cations are attracted towards the cathode; **(b)** Ions cannot move in a solid so the solid cannot conduct electricity.

18 Anode $2Br^- \rightarrow Br_2 + 2e^-$ oxidation
Cathode $Pb^{2+} + 2e^- \rightarrow Pb$ reduction

19 Cu^{2+} will be discharged since it accepts electrons more easily than H^+

20 Oxygen is made at the anode.
$4OH^- \rightarrow O_2 + 2H_2O + 4e^-$
OH^- are discharged in preference to SO_4^{2-}.
Hydrogen is made at the cathode.
$2H^+ + 2e^- \rightarrow H_2$
H^+ is the only positive ion present

21 **(a)** 6.023×10^{23}; **(b)** 1.60×10^{-19} C

22 **(a)** (i) 3 moles of electrons (3 F)
(ii) 1 mole of electrons (1 F); **(b)** $4OH^- \rightarrow O_2 + 2H_2O + 4e^-$ so needs 4 F

23 Charge supplied = 19296 C, Moles of electrons = 0.200
Mass of copper = 6.4 g and volume of oxygen (at room temperature and pressure) = 1.2 dm^3

CHAPTER 24

1 As the atomic number of each element in the period increases, electrons fill the 3d sub-shell rather than the outer sub-shells

2 **(a)** Mn is $1s^22s^22p^63s^23p^63d^54s^2$ and Ni is $1s^22s^22p^63s^23p^63d^84s^2$;
(b) It minimises electron–electron repulsion within the 3d orbitals

3 **(a)** Co^{2+} is [Ar]$3d^7$ and Co^{3+} is [Ar]$3d^6$;
(b) Cu^+ is [Ar]$3d^{10}$ and Cu^{2+} is [Ar]$3d^9$;
(c) Zn^{2+} is [Ar]$3d^{10}$

4 Sc^{3+} is $1s^22s^22p^63s^23p^6$; Sc^{3+} does not have any 3d electrons, but a transition element needs a partially filled d sub-shell in the electron configuration of one of its oxidation states

5 Metals form cations; metals normally form basic oxides; metals are reducing agents; metals normally react with non-metals to form ionic compounds; metals are malleable and ductile

6 **(a)** Cr^+, [Ar]$3d^5$, has a stable half-filled sub-shell with no electron–electron repulsion within a 3d orbital; **(b)** Mn^{2+}, [Ar]$3d^5$, also has a stable half-filled sub-shell with no electron–electron repulsion within a 3d orbital; **(c)** The third ionisation energy refers to an electron lost from the third shell rather than the fourth shell, so there are fewer shielding shells and so a much greater attraction between the outer electrons and the nucleus

7 **(a)** Fe^{3+} is [Ar]$3d^5$, which has a stable half-filled sub-shell, whereas Fe^{2+} is [Ar]$3d^6$ and has one 3d orbital with electron–electron repulsion, and so Fe^{2+} will lose electrons to become Fe^{3+}; **(b)** Mn^{2+} is [Ar]$3d^5$ which has a stable half-filled sub-shell, whereas Mn^{3+} is [Ar]$3d^4$ and so the loss of electrons is not compensated by the formation of a much more stable electron configuration;
c)

		3d	4s
Cu^+	[Ar]	↑↓ ↑↓ ↑↓ ↑↓ ↑↓	☐
Cu^{2+}	[Ar]	↑↓ ↑↓ ↑↓ ↑↓ ↑	☐

8 Transition-metal atoms always lose electrons in reaction since they often have only two electrons in their outer shell, and metals have a low electronegativity

9 **(a)** The ability of an atom in a covalent bond to attract the bonding pair of electrons; **(b)** Fluorine

10 **(a)** (i) Titanium(IV) oxide; (ii) Iron(III) oxide; (iii) Manganese(III) hydroxide; (iv) Chromium(VI) oxide; (v) Vanadium(V) oxide;
(b) (i) $Cu(OH)_2$; (ii) $MnCO_3$; (iii) $TiCl_4$; (iv) $Cu(NO_3)_2$; (v) $FeBr_2$

11 **(a)** Na_2FeO_4; **(b)** Chromate(VI)

12 **(a)** +1; **(b)** +2; **(c)** +3; **(d)** +2; **(e)** +3

13 **(a)** $[Ag(CN)_2]^-$ contains a central Ag^+, $[Ag(NH_3)_2]^+$ contains a central Ag^+ and $[Ag(S_2O_3)_2]^{3-}$ contains a central Ag^+;
(b) $[Ag(CN)_2]^-$ is linear, $[Ag(NH_3)_2]^+$ is linear and $[Ag(S_2O_3)_2]^{3-}$ is linear

14 **(a)** 6; **(b)** Tetradentate;
(c) Most arteries contain a high concentration of oxyhaemoglobin, in which an oxygen molecule is complexed with the Fe^{2+}, which is red in colour

15
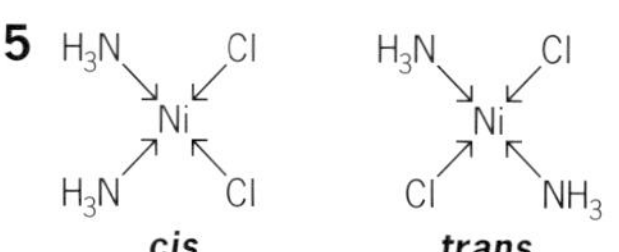

16
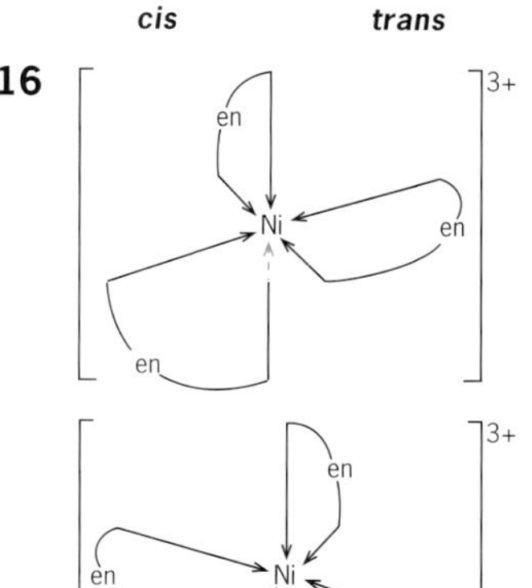

17 **(a)** (i) +2; (ii) +4; **(b)** Octahedral

18 **(a)** Same structure but different three dimensional arrangement of atoms
(b)
(c)
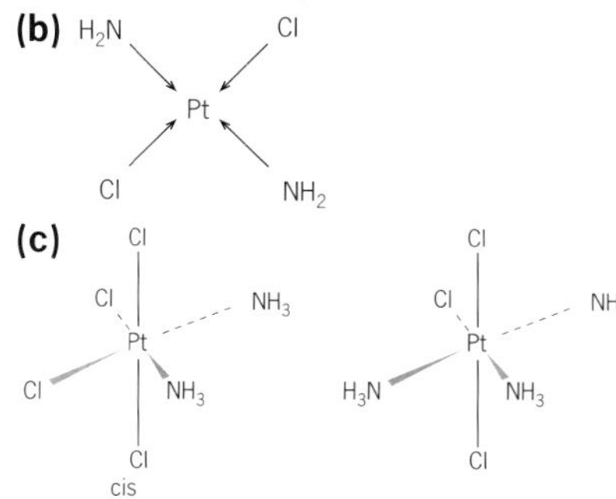

19 **(a)** Ligand has donated two lone pairs to make two dative bonds to central metal ion; **(b)** Chelates have the ligand and the central metal ion in a ring

20 $[Al(H_2O)_6]^{3+}(aq) = [Al(H_2O)_5(OH)]^{2+}(aq) + H^+(aq)$

21 $[Al(H_2O)_6]^{3+}(aq) = [Al(H_2O)_5(OH)]^{2+}(aq) + H^+(aq)$

22 **(a)** $[CoCl_4]^{2-}(aq) + 6H_2O(l) = [Co(H_2O)_6]^{2+}(aq) + 4Cl^-(aq)$;
(b) On heating, water evaporates and so the position of equilibrium shifts to the left side to try to minimise this loss, which results in an increase in $[CoCl_4]^{2-}(aq)$ and so the solution takes on the colour of this complex, which is blue

23 **(a)** Increases the surface area, which increases rate of reaction; **(b)** Increases surface area and minimises the mass of platinum, which is very expensive, that needs to be used; **(c)** Although a powder has a large surface area, unless gas can permeate all the way through the powder most of the powder will not be in contact with the gas. In a fluidised bed all the surface area will be available to the gas

24 In SO_2 sulfur is +4 and in SO_3 it is +6, as sulfur loses electrons the oxidation state of sulfur increases

25 The graph is a mirror image of Fig 24.40, starting at the origin with a very small gradient, then the gradient increases and finally, at the end of the reaction, the gradient decreases to zero

CHAPTER 25

1 **(a)** $Cu(s) \rightarrow Cu(aq) + 2e^-$, and $Ag^+(aq) + e^- \rightarrow Ag(s)$;
(b) $Fe^{2+}(aq) \rightarrow Fe^{3+}(aq) + e^-$, and $Cr_2O_7{}^{2-}(aq) + 14H^+(aq) + 6e^- \rightarrow 2Cr^{3+}(aq) + 7H_2O(l)$

2 **(a)** $Na^+ + e^- \rightarrow Na$, and $2Cl^- \rightarrow Cl_2 + 2e^-$; **(b)** $Pb^{2+} + 2e^- \rightarrow Pb$, and $2Br^- \rightarrow Br_2 + 2e^-$

3 298 K, $[Cl^-(aq)]$ = 1.0 mol dm^{-3} and pressure of $Cl_2(g)$ = 101 kPa

4 0.5 mol dm^{-3}

5 **(a)** Cu(s) I Cu^{2+}(aq) II Ag^+(aq) I Ag(s), electrons flow from copper to silver;
(b) Zn(s) I Zn^{2+}(aq) II H^+(aq) I H_2(g) I Pt(s), electrons flow from zinc to platinum (hydrogen)

6 **(a)** Cu^{2+}; **(b)** Acidified $MnO_4{}^-$

7 **(a)** Ni; **(b)** Au

8 **(a)** Hydrogen at cathode and oxygen at anode; **(b)** Expect hydrogen at cathode and oxygen at anode, a prediction based on $[OH^-(aq)] = [Cl^-(aq)]$ = 1.0 mol dm^{-3}, but in concentrated NaCl(aq) the $[Cl^-(aq)]$ and $[OH^-(aq)]$ << 1.0 mol dm^{-3}

9 **(a)** +0.76 V; **(b)** (i) Less positive; (ii) More positive; (iii) Less positive

10 **(a)** (i) E = +1.62 V, so spontaneous; (ii) E = –2.10 V, so not spontaneous;
(b) +0.74 V, $Mn(s) + Fe^{2+}(aq) \rightarrow Mn^{2+}(aq) + Fe(s)$

11 **(a)** From Cu to Br_2;
(b) From Cr to Cu^{2+}

12 **(a)** –52 kJ mol^{-1}; **(b)** –83 kJ mol^{-1};
(c) 56 kJ mol^{-1}

13 **(a)** E = –1.51 V, ΔG = +291 kJ mol^{-1}, not spontaneous;
(b) E = +1.93 V, ΔG = –931 kJ mol^{-1}, spontaneous

14 **(a)** 2 V; **(b)** +1.60 V; **(c)** The lead–acid battery does not operate under standard conditions, e.g. $[H_2SO_4(aq)]$ > 1.0 mol dm^{-3} and the temperature will not be 298 K

15 $O_2(g) + 2H_2O(l) + 4e^- \rightarrow 4OH^-(aq)$ and $2H_2(g) + 4OH^-(aq) \rightarrow 4H_2O(l) + 4e^-$ gives $O_2(g) + 2H_2(g) + 2H_2O(l) + 4e^- + 4OH^-(aq) \rightarrow 4OH^-(aq) + 4e^- + 4H_2O(l)$, cancelling gives $O_2(g) + 2H_2(g) \rightarrow 2H_2O(l)$

16 **(a)** $2VO^{2+}(aq) + 4H^+(aq) + Zn(s) \rightarrow Zn^{2+}(aq) + 2V^{3+}(aq) + 2H_2O(l)$;
(b) $2V^{3+}(aq) + Zn(s) \rightarrow Zn^{2+}(aq) + 2V^{2+}(aq)$, and E = +0.50 V, i.e. the process should be spontaneous

17 **(a)** $Cr(s) + H_2SO_4(aq) \rightarrow CrSO_4(aq) + H_2(g)$; **(b)** +0.91 V

18 **(a)** The reduction potential for $Mn^{3+}(aq) + e^- = Mn^{2+}(aq)$ is +1.49 V, and this highly positive nature of the reduction potential indicates that manganese(III) is a powerful oxidising agent and so is easily reduced to manganese(II), i.e. manganese(II) is more stable than manganese(III);
(b) Manganese should react with HCl(aq) to give hydrogen since the standard electrode potential for $Mn(s) + 2H^+(aq) \rightarrow Mn^{2+}(aq) + H_2(g)$ is positive (+1.18 V)

19 **(a)** (i) $MnO_4{}^-(aq) + 8H^+(aq) + 5e^- \rightarrow Mn^{2+}(aq) + 4H_2O(l)$;
(ii) $2I^-(aq) \rightarrow I_2(aq) + 2e^-$;
(b) $2MnO_4{}^-(aq) + 16H^+(aq) + 10I^-(aq) \rightarrow 5I_2(aq) + 2Mn^{2+}(aq) + 8H_2O(l)$

20 0.0287 mol dm^{-3}

21 **(a)** (i) $Fe^{2+}(aq) + O_2(g) + 4H^+(aq) \rightarrow 2Fe^{3+}(aq) + 2H_2O(l)$; (ii) +0.46 V;
(b) +0.96 V

22 **(a)** $\Delta S_{surroundings}$ = –357.5 and ΔS_{total} = –188.5 J K^{-1} mol^{-1};so entropy decreasing not spontaneous; **(b)** K = 1.43×10^{-10}

23 **(a)** Not spontaneous, since ΔH is positive and ΔS is negative;
(b) Spontaneous, since ΔG = –142 kJ mol^{-1}

24 **(a)** Iron and possibly chromium;
(b) (ii) Should be feasible above approximately 1770 K

CHAPTER 26

1 **(a)** Rate of reaction decreases during the course of a reaction;
(b) Rate of reaction changes as temperature changes

2 0.17 mol s^{-1}

3 Reaction mechanism

4 Mass of reaction mixture, measured by placing the reaction container on a balance, and concentration of H^+(aq), measured with a pH meter, since pH will change during reaction

5 **(a)** 1.50×10^{-5}; **(b)** $H^+(aq) + OH^-(aq) \rightarrow H_2O(l)$; **(c)** 1.50×10^{-5};
(d) 1.50×10^{-3} mol dm^{-3}

6 **(a)** Monitor the increase in pressure if the reaction is carried out in a sealed container; **(b)** Measure the volume of oxygen produced using a gas syringe;
(c) Monitor the change in conductivity of the reaction mixture and use colorimetry to monitor the appearance of bromine;
(d) Take aliquots out of the reaction mixture and titrate against an acid of known concentration, and also take aliquots out of the reaction mixture and acidify with HNO_3(aq), then add $AgNO_3$(aq) to measure the mass of AgBr precipitate formed; **(e)** Monitor the change in conductivity of the solution, and monitor the change in I_2(aq) concentration using colorimetry or by taking out aliquots and titrating with $Na_2S_2O_3$(aq)

7 m with respect to A, and n with respect to B

8 By the change in total mass of the reaction mixture and by the change in total volume of carbon dioxide collected

9 Since the rate at time zero is all that is required; once sufficient results are obtained to be able to construct the gradient at time zero then no more are needed

10 Rate = $k[SO_2Cl_2]$, and $k = 4.5 \times 10^4$ s^{-1}

11 Second

12 **(a)** First order with respect to $BrO_3{}^-$ and first order with respect to $SO_3{}^{2-}$, to give a total order of 2, with rate constant units of dm^3 mol^{-1} s^{-1};
(b) Second order with respect to NO and first order with respect to H_2, to give a total order of 3, with rate constant units dm^6 mol^{-2} s^{-1};
(c) Second order with respect to NO_2 and zero order with respect to CO, giving a total order of 2, with rate constant units dm^3 mol^{-1} s^{-1}

13 Loss of mass from the reaction mixture, and volume of oxygen formed

14 Record the change in either the mass lost by the reaction mixture or volume of oxygen formed at regular time intervals over a set time period, using the same mass of catalyst, same volume of H_2O_2(aq) and same temperature; do experiment five times using different concentrations of H_2O_2(aq); plot graph of time against either the mass loss or volume of oxygen formed for each concentration; determine the initial rate of reaction by drawing a tangent to the curve at time zero; plot the initial rates obtained against the concentration of H_2O_2(aq), and first-order kinetics will give a straight line through the origin

15 **(a)** 0.25 kg; **(b)** $k = 4.62 \times 10^{-2}\ h^{-1}$

16 Cyclopropane and propene are structural isomers, since they have the same molecular formula but different structural formulae

17 $2NO_2 + F_2 \rightarrow 2NO_2F$

18 It is a substitution reaction that involves a nucleophile, and it has second-order kinetics, i.e. two particles are involved in the rate-determining step

19 Mechanism 2, since the slowest step involves a collision between H_2O_2 and Br^-, i.e. it will be first order with respect to both H_2O_2 and Br^-

20 **(a)** Slow step rate = $k[N_2O_5]$, fast step rate = $k[NO][NO_3]$; **(b)** $N_2O_5 \rightarrow NO_2 + NO_3$; **(c)** Rate = $k[N_2O_5]$; **(d)** Because NO is not involved in the slowest step of the reaction

21 It doubles

CHAPTER 27

1 Blue

2 $f = 5.09 \times 10^{14}$ Hz, and $E = 3.37 \times 10^{-19}$ J

3 A group of atoms in a molecule that gives the molecule a characteristic set of reactions

4 $2NaNO_2(aq) + H_2SO_4(aq) \rightarrow Na_2SO_4(aq) + 2HNO_2(aq)$; $C_6H_5NH_2(l) + H_2SO_4(aq) + HNO_2(aq) \rightarrow C_6H_5N_2^+(aq) + HSO_4^-(aq) + 2H_2O(l)$; or $2C_6H_5NH_2(l) + H_2SO_4(aq) + 2HNO_2(aq) \rightarrow 2C_6H_5N_2^+(aq) + SO_4^{2-}(aq) + 4H_2O(l)$

5 They decompose very easily to produce nitrogen gas, so that a small volume of solid will produce a large volume of gas

6 **(a)** At a higher temperature the diazonium salt will react with water to give phenol and nitrogen; **(b)** Phenol reacts in an acid–base reaction to give sodium phenoxide, $C_6H_5OH + OH^- \rightarrow C_6H_5O^- + H_2O$

Sodium phenoxide: 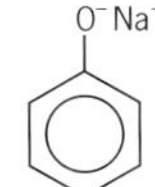

(c) A lone pair on the oxygen atom is donated into the system of the benzene ring to extend the system over seven atoms, which increases the electron density in the benzene ring so that the species is more reactive towards electrophiles than benzene is

7 **(a)** Concentrated sulfuric acid and concentrated nitric acid; **(b)** Tin and concentrated hydrochloric acid

8 **(a)** White; **(b)** Black

9 400–700 nm

10 Blue

11 **(a)** An extended π system over many atoms which has alternating carbon–carbon double bonds and carbon–carbon single bonds; **(b)** Blue; **(c)** Yellow

12 **(a)** It has three diazo groups per molecule; **(b)** A group of atoms responsible for the absorption of ultra-violet and/or visible light.

13 Covalent bonds are strong and so are difficult to break

14 In alkali the is a bigger conjugated system with all carbon atoms being part of the π system

15 **(a)** So that the only absorbance is due to the solution under test; **(b)** Many solvent molecules will contain chromophores that absorb ultraviolet and/or visible light, for example carbonyl groups, carbon–carbon double bonds or benzene rings

16 Yellow

17

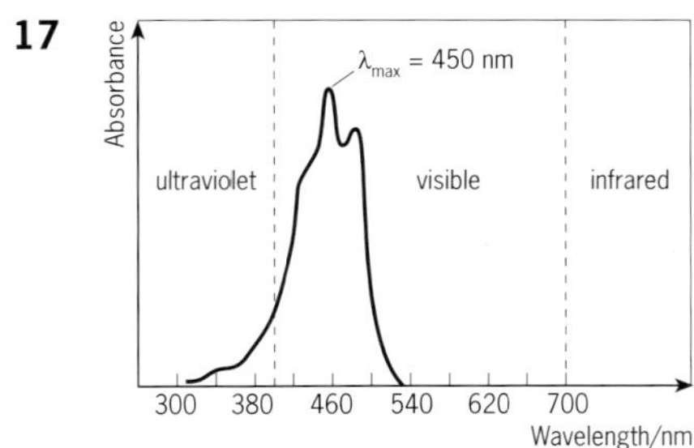

18 Red light

19 Blue

20 **(a)** $1s^2 2s^2 2p^6 3s^2 3p^6 3d^{10}$; **(b)** Even though the energy of the 3d orbitals can be split in Zn^{2+} there is no possibility of a d $\rightarrow$ d transition since all the 3d orbitals are full

21 VO_2^+ absorbs blue–violet. VO^{2+} absorbs orange, V^{3+} absorbs red–violet, V^{2+} absorbs yellow–green

22

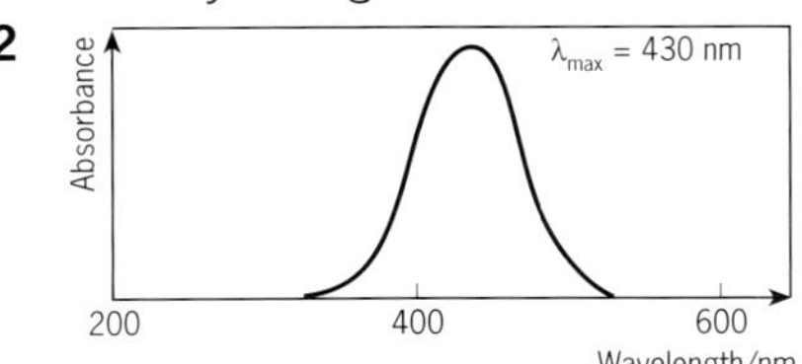

23 **(a)** Orange; **(b)** Blue; **(c)** Octahedral

24 **(a)** Use colorimetry by making up solutions that contain different molar ratios of nickel(II) ion and EDTA, analyse each solution with a colorimeter, and find the solution with the maximum absorbance, which should have the stoichiometric molar ratio of the complex; **(b)** Either an inverted V-shape graph as in Fig 27.43, or the absorbance increases until it reaches a maximum and then remains at the maximum absorbance (this is obtained if the number of moles of one ion is kept constant and the number of moles of the other ion is changed)

25 **(a)** A chromophore is a group of atoms responsible for the absorption of ultraviolet and/or visible light, and a conjugated system is an arrangement of atoms with an extended π system over many atoms in which there are alternating single and double bonds; **(b)** λ_{max} = ~620 nm; **(c)** Drawing of a spectrum with a λ_{max} = ~620 nm

CHAPTER 28

1 Reduced, since the oxidation number of nitrogen changes from 0 to –3, i.e. nitrogen gains electrons

2 **(a)** +3; **(b)** +5; **(c)** +5; **(d)** +2; **(e)** +4

3 118 kPa

4 **(a)** $K_p = \dfrac{(pHI)^2}{(pH_2)(pI_2)}$; **(b)** $K_p = \dfrac{(pNO_2)^2}{(pNO)^2(pO_2)}$; **(c)** $K_p = pCO_2$

5 **(a)** None; **(b)** Pa^{-1}; **(c)** Pa

6 1.5×10^{-8}

7 **(a)** pSO_2 = 200 kPa and pO_2 = 100 kPa; **(b)** pSO_2 = 10 kPa and pO_2 = 5 kPa; **(c)** 72.2 kPa^{-1}

8 **(a)** Low temperature since it is an exothermic reaction, low pressure since the volume of the products is more than the volume of the reactants (more moles of gas on product side), and increase the concentration of oxygen; **(b)** High pressure, high temperature and the addition of a catalyst; **(c)** Very high temperature is used to ensure a rapid rate of reaction, a low pressure is used as this also reduces the operating costs (high-pressure apparatus is expensive), a catalyst is used to increase the rate of reaction, and excess air is used to ensure that the concentration of oxygen is higher than the mole ratio predicted from the equation; **(d)** $K_p = \frac{(pNO)^4(pH_2O)^6}{(pNH_3)^4(pO_2)^5}$

9 (a) For $2NO_2(g) = N_2O_4(g)$, $K_p = 7.7 \times 10^{-2}\ kPa^{-1}$; **(b)** K_p decreases; **(c)** K_p stays the same

10 Lower maintenance costs as no need to have the expensive apparatus required to withstand high pressure

11 (a) +4; **(b)** +5; **(c)** +2

12 (a) N_2O_4 (N has +4 oxidation state) is oxidised to HNO_3 (N has +5 oxidation state) and reduced to NO (N has +2 oxidation state); **(b)** Shifts the position of equilibrium to the right, since the reaction is exothermic, and to ensure that the water remains liquid

13 N:P:K ratio is 33.5:10.5:7

14 $HNO_3(aq) + NH_3(aq) \rightarrow NH_4NO_3(aq)$

15 (a) $CH_3NH_2(aq) + H_2O(l) = CH_3NH_3^+(aq) + OH^-(aq)$; **(b)** $K_b = \frac{[CH_3NH_3^+(aq)][OH^-(aq)]}{[CH_3NH_2(aq)]}$

16 (a) $NH_3(aq) + H_2SO_4(aq) \rightarrow NH_4HSO_4(aq)$, or $2NH_3(aq) + H_2SO_4(aq) \rightarrow (NH_4)_2SO_4(aq)$; **(b)** $C_2H_5NH_2(aq) + HCl(aq) \rightarrow C_2H_5NH_3^+(aq) + Cl^-(aq)$

17 (a) Position of equilibrium shifts towards the right; **(b)** Increases; **(c)** The position of equilibrium moves to the right so $[H^+(aq)]$ and $[OH^-(aq)]$ increase, which are the mobile charge carriers and so the electrical conductivity increases

18 $1.73 \times 10^{-12}\ mol\ dm^{-3}$

19 (a) $2.51 \times 10^{-8}\ mol\ dm^{-3}$; **(b)** $3.98 \times 10^{-7}\ mol\ dm^{-3}$

20 $[H^+(aq)] = [OH^-(aq)] = 1.0 \times 10^{-7}\ mol\ dm^{-3}$

21 (a) 14.34; **(b)** 10.56; **(c)** 13.00; **(d)** 11.60

22 (a) $7.94 \times 10^{-13}\ mol\ dm^{-3}$; **(b)** $1.26 \times 10^{-2}\ mol\ dm^{-3}$; **(c)** $6.3 \times 10^{-3}\ mol\ dm^{-3}$

23 11.5

24 10.9

25 (a) Methylamine, CH_3NH_2; **(b)** *N* - Ethylbutylamine, $CH_3CH_2CH_2CH_2NHCH_2CH_3$; **(c)** *N* - Methyl-2-propylamine, $(CH_3)_2CHNHCH_3$

26 (a) $CH_3CONHCH_2CH_3$; **(b)** $C_6H_5CONHCH_3$; **(c)** CH_3CONH_2

27 10.4%

CHAPTER 29

1 $SiO_2 + 2Mg \rightarrow Si + 2MgO$

2 High temperatures are needed (high energy costs), magnesium is a very expensive metal to buy, and the silicon produced must be ultra-pure, which requires further expensive processing.

3 (a) To stop silicon oxidizing to form SiO_2; **(b)** Neon, argon or nitrogen

4 (a) $PbO + C \rightarrow Pb + CO$, or $2PbO + C \rightarrow 2Pb + CO_2$; **(b)** $SnO_2 + 2C \rightarrow Sn + 2CO$, or $SnO_2 + C \rightarrow Sn + CO_2$

5 (a) Acid rain; **(b)** $2PbS + 3O_2 \rightarrow 2PbO + 2SO_2$

6 Low melting points and high electrical conductivity.

7 Both have a giant molecular structure held together by strong covalent bonds, atoms with a coordination number of 4, a tetrahedral arrangement about each atom in the lattice, and the bond angles in both are 109°; Differences are that the C–C bond length is shorter than Si–C, and in diamond each carbon atom is surrounded by four carbon atoms, whereas in SiC a carbon atom is surrounded by four silicon atoms

8 (a) Sn is $[Kr]4d^{10}5s^25p^2$ and Sn^{2+} is $[Kr]4d^{10}5s^2$; **(b)** Ge is $[Ar]3d^{10}4s^24p^2$ and Ge^{2+} is $[Kr]3d^{10}4s^2$; **(c)** Si is $1s^22s^22p^63s^23p^2$ and Si^{2+} is $1s^22s^22p^63s^2$; **(d)** The inert pairs of electrons are the $5s^2$ and $4s^2$ electrons in Sn^{2+} and Ge^{2+}, respectively, which are in the outer shell of electrons but are not used in bonding, whereas Si^{2+} is not formed because silicon uses all four outer electrons in bonding

9 (a)

Element	Energy for $X(g) \rightarrow X^{2+}(g)/kJ\ mol^{-1}$
Carbon	3219
Silicon	2366
Germanium	2302
Tin	2117
Lead	2166

(b) $X^+(g)$ has more protons than electrons so the electrons are held more strongly to the nucleus and so are more difficult to remove than in X(g), in which the number of electrons and protons are equal; **(c)** Both have the same core charge of +4 but in Ge there are more inner shielding shells of electrons than in C and since the atomic radius of Ge is greater than that of C the outer electrons are at a greater distance from the nucleus; **(d)** 1st and 2nd ionisation energies of tin are lower than expected, or those for Pb are greater than expected, which is accounted for by presence of inner shell 4f electrons in Pb

10

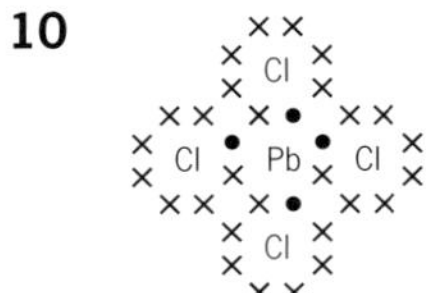

11 (a) $Pb + I_2 \rightarrow PbI_2$; **(b)** $Sn + 2F_2 \rightarrow SnF_4$; **(c)** $Ge + 2Br_2 \rightarrow GeBr_4$

12 $Na_2SiO_3(aq) + 2HCl(aq) \rightarrow 2NaCl(aq) + SiO_2(s) + H_2O(l)$

13 PbO_2 is less stable than SiO_2 (stability of +4 oxidation state decreases down the group), but PbO is much more stable than SiO (stability of +2 oxidation state increases down the group), so that PbO_2 can accept electrons (act as an oxidising agent) to form Pb^{2+}, whereas SiO_2 will not accept electrons to form Si^{2+}

14 $CO_2(g) + Ca(OH)_2(aq) \rightarrow CaCO_3(s) + H_2O(l)$

15 (a) $Pb(CH_3COO)_2(aq)$ and $H_2O(l)$; **(b)** $Na_2SnO_2(aq)$ and $H_2O(l)$

16 (a) $Pb_3O_4(s) \rightarrow 3PbO(s)$ (lead(II) oxide) + $\frac{1}{2}O_2(g)$ (oxygen); **(b)** $Pb_3O_4(s) + 8HCl(aq) \rightarrow 3PbCl_2(aq)$ (lead(II) chloride) + $Cl_2(g)$ (chlorine) + $4H_2O(l)$ (water); **(c)** $Pb_3O_4(s) + 6NaOH(aq) \rightarrow 2Na_2PbO_2(aq)$ (sodium plumbate(IV)) + $Na_2PbO_3(aq)$ (sodium plumbate(VI)) + $3H_2O(l)$ (water)

17 $PbCl_4 \rightarrow PbCl_2 + Cl_2$

18 $2Fe^{3+}(aq) + Sn^{2+}(aq) \rightarrow 2Fe^{2+}(aq) + Sn^{4+}(aq)$

19 $Cl_2(g) + Sn^{2+}(aq) \rightarrow Sn^{4+}(aq) + 2Cl^-(aq)$, tin(IV) chloride

20 About 20

21 (a) Si_6H_{14}; **(b)** $SiH_3SiH_2SiH_2SiH_2SiH_2SiH_3$ or any other structural isomer

22 $PbCO_3(s) \rightarrow PbO(s) + CO_2(g)$

23 (a) 2 mole KOH to 1 mole H_2SO_3; **(b)** 1 mole KOH to 1 mole H_2SO_3

24 (a) $Cl_2(aq) + SO_2(g) + 2H_2O(l) \rightarrow 2HCl(aq) + H_2SO_4(aq)$; **(b)** $2Fe^{3+}(aq) + SO_2(g) + 2H_2O(l) \rightarrow 2Fe^{2+}(aq) + SO_4^{2-}(aq) + 4H^+(aq)$

25 (a) With magnesium carbonate, magnesium, and copper(II) oxide; **(b)** With copper, hydrogen iodide, magnesium and benzene

26 (a) Benzoic acid; **(b)** Ethanoic acid

INDEX